SMITHSONIAN

MISCELLANEOUS COLLECTIONS.

METEOROLOGICAL

AND

PHYSICAL TABLES.

Smithsonian Miscellaneous Collections.

TABLES,

METEOROLOGICAL AND PHYSICAL,

PREPARED FOR

THE SMITHSONIAN INSTITUTION.

BY

ARNOLD GUYOT, P. D., LL. D.,
PROFESSOR OF GEOLOGY AND PHYSICAL GEOGRAPHY, COLLEGE OF NEW JERSEY.

SECOND EDITION,
REVISED AND ENLARGED.

WASHINGTON:
SMITHSONIAN INSTITUTION.
1858.

CAMBRIDGE:
METCALF AND COMPANY, ELECTROTYPERS AND PRINTERS.

PREFACE

TO THE FIRST EDITION.

To PROF. JOSEPH HENRY,

Secretary of the Smithsonian Institution.

SIR, —

IN compliance with your instructions, I have prepared the collection of Meteorological Tables contained in the following pages. I have endeavored to render it useful, not only to the observers engaged in the system of Meteorological Observations now in operation under the direction of the Smithsonian Institution, for whom it was immediately designed, but also to any Meteorologist who may desire to compare and to work out portions of the vast amount of Meteorological Observations already accumulated in the stores of science.

The reduction of the observations and the extensive comparisons, without which Meteorology can do but little, require an amount of mechanical labor which renders it impossible for most observers to deduce for themselves the results of their own observations. The difficulty is still further increased by the diversity of the thermometrical and barometrical scales which Meteorologists, faithful to old habits rather than to science and to reason, choose to retain, notwithstanding the additional labor they thus gratuitously assume to themselves. To relieve the Meteorologist of a great portion of this labor, by means of tables sufficiently extensive to render calculations and even interpolations unnecessary, is to save his time and his forces in favor of science itself, and thus materially contribute to its advancement. But most of the tables useful in Meteorology being scattered through many volumes, which are often not of easy access, this collection will be, it is hoped, acceptable to the friends of Meteorology, and will supply a want very much felt in this department of the physical sciences.

In the selection of the matter, I have been guided by the idea that the tables which I sought for my own use might also be those most likely to be wanted by others. But I wish the following to be considered as a first collection, containing only the tables most appropriate to the present purpose. They are, therefore, arranged in different and independent series, with distinct paging, but constituting together a frame-work into which any tables may be readily inserted when wanted, either to make the collection more complete, or to present a choice of tables calculated from somewhat different elements, or adapted to various methods of calculation.

The measurement of heights by means of the barometer being intimately connected with Meteorology, it was thought not inappropriate to admit into this collection Hypsometrical Tables, destined to render this kind of calculations more easy and more rapid, and thus to increase the taste for a method so useful in physical geography. I have preferred the tables of Delcros, as uniting in the greatest degree simplicity and accuracy. Those of Gauss, Bessel, and Baily may be given afterwards.

Every table contains directions for its use, when necessary; moreover, the indication of the elements used in its calculation, and of the source from which it has been taken. When no remark is made as to this last point, the table has been expressly calculated for this volume.

Very respectfully,

Your obedient servant,

A. GUYOT.

CAMBRIDGE, MASS., *December 15th,* 1851.

PREFACE

TO THE SECOND EDITION.

To PROF. JOSEPH HENRY,

Secretary of the Smithsonian Institution.

SIR, —

IN sending to you the Meteorological Tables composing the first edition of this volume, published in 1852, I expressed the desire that they be considered as a first collection, containing the tables most needed at the time by the meteorological observers engaged in the system carried on under the supervision of the Smithsonian Institution, but destined to be increased. It was in that expectation, I remarked, that the tables had been arranged in independent series, as a kind of framework, into which a larger number could readily be inserted. It seemed, indeed, highly desirable to offer to the Meteorologist and Physical Geographer, not only the tables they daily need for working out the results of their observations, but also such a variety of tables, computed from different elements, or by different methods, or adapted to different measures, as to enable every one to choose among them those that he most approves, and at the same time properly to compare and to appreciate the results obtained by others.

Thanks to the congenial spirit with which the elevated views of the founder of the Smithsonian Institution are carried out, that character of general usefulness is not wanting in the present volume. With your agreement, the present edition contains more than three times as much matter as the first; and a rapid indication of the additions will suffice to justify them, and to show that, in selecting or calculating the new tables, the object just mentioned was constantly kept in view

As to the tables in the first edition, I must remark that, several of them having been printed in my absence, the copy prepared for the printer, in which decimals had to be left out, failed to give always the nearest value. Though these errors are too small to have any importance whatsoever in Meteorology, a careful revision of all the tables on the original computations was made, and they were corrected in the present edition The few actual misprints which were discovered are indicated in a table of *errata* to the first edition.

In the Thermometrical series six small tables have been added; they were prepared for converting into each other differential results given in degrees of any one of the three thermometrical scales, irrespective of their zero point.

The Hygrometrical series has been entirely reorganized. It only contained five tables, all in French measures, and the Appendix. It is now composed of twenty-seven, arranged in three divisions. In the first are found ten tables, based on Regnault's hygrometrical constants, both in French and in English measures, in two corresponding sets, for the use of the psychrometer, the dew-point instruments, and for computing the weight of vapor in the air. The whole set in English measures, and Table V. in French measures, have been prepared for this edition. Being based on the best elements we now possess, they are given here for ordinary use. The second division contains the seven most important tables published in the *Greenwich Observations*, and Glaisher's extensive Psychrometrical Table. These tables being much used in England, and the results obtained by them exhibiting no inconsiderable differences from those derived from the preceding ones, they are indispensable for comparing these results. The third division, composed of ten miscellaneous tables, furnishes the means of comparing the different values of the force and the weight of vapor, especially those which have frequently been used in Germany, and also of reducing the indications of Saussure's Hair-Hygrometer to the ordinary scale of moisture. The Appendix has remained as in the first edition, but all the tables have been revised and corrected.

The Barometrical series, now in four divisions, has been increased from twelve to twenty-eight tables. Excepting three small tables for capillary action, all the new ones have been computed for this edition. The comparison, now so much needed, of the Russian barometer with the other scales, appears here for the first time.

The Hypsometrical series is almost entirely new. It contained only Delcros's table for barometric and Regnault's table for thermometric measurements, besides two auxiliary tables and the thirteen small tables of the Appendix. It now offers twenty-three tables for barometrical measurement of heights, in which all the principal formulæ and scales are represented; three for the measurement of heights by the thermometer, in French and in English measures; and a rich Appendix of forty-four tables, more extensive and convenient than those in the old set, which afford the means of readily converting into each other all the measures usually employed for indicating altitudes.

The series of Meteorological Corrections for periodic and non-periodic variations, for all parts of the world, mostly due to the untiring industry of Professor Dove, is an addition which will surely be appreciated by those who know how difficult access to the original tables is for most Meteorologists. A few tables have been added to Dove's collection, computed by Glaisher, Captain Lefroy, and by myself. Most of the tables refer to temperature, only two to moisture. Two tables of Barometrical Corrections have been placed in the Hypsometrical series, where they were needed, until they can be joined by others to make a set in this series, which still awaits new contributions, especially for these last two departments.

The Miscellaneous series is but begun. I have prepared a list of useful tables, which were no doubt welcome to the lovers of Terrestrial Physics, and which may be published at some future occasion, if you should then find it expedient.

The present collection being designed, not for the scientific only, but for the observers at large, the propriety of the explicit and popular form of the explanations which accompany the tables, and of the directions for using them, will readily be understood.

I close by the remark, that, in every instance, the works from which the tables were taken have been carefully noted, and due credit given to their authors. For all the tables without author's names, I am personally responsible.

I remain, Sir,

Very respectfully, yours,

A. GUYOT.

PRINCETON, N. J., *December*, 1857.

CONTENTS.

The Tables contained in this collection are divided into six series, as follows: —

I.	Thermometrical Tables,	marked	A.
II.	Hygrometrical Tables,	"	B.
III.	Barometrical Tables,	"	C.
IV.	Hypsometrical Tables,	"	D.
V.	Meteorological Corrections,	"	E.
VI.	Miscellaneous Tables,	"	F.

Each series has an independent paging running through all the tables that it contains.

The letters A, B, C, D, E, F, at the bottom of each page, indicate the series, and the figure the folio of the series to which the page belongs.

The figure at the top of the page indicates the folio of the particular table of which the page is a part.

At the head of each series is found a detailed table of its contents.

ERRATA IN THE FIRST EDITION.

A, page 7, line below the title, instead of $(32 + \frac{4}{5}x^\circ)$ read $(32 + \frac{9}{5}x^\circ)$.

A, " 21, on the line beginning with 30, in the last four columns,

instead of	100.75	100.97	101.20	101.42
read	100.85	101.07	101.30*	101.52.

B, " 7, on the line beginning with 23°, second column, instead of 20.410, read 20.888.

B, " 23, on the line beginning with 12°, third column, instead of 5.87, read 5.37.

B, " 30-32, at the head of each first column, Temperature of the Air, add "in Centigrade degrees."

B, " 40, Table II., first part, on the line beginning with 60, column headed 7, instead of 24.11,01, read 24.9,01.

B, " " Table II., first part, on the line beginning with 70, column headed 5, instead of 27.4,47, read 27.8,47 ; and column headed 8, instead of 28.11,77, read 28.9,77.

B, " " Table II., second part, on the line beginning with 70, column headed 5, instead of 328.47, read 332.47.

B, " 41, Table III., line beginning with 20, the five last columns,

instead of	63.54	66.08	68.62	71.16	73.70,
read	63.50	66.04	68.58	71.12	73.66.

B, " " Table III, line beginning with 200, column headed 3, instead of 515.11, read 515.61.

B, " 42, Table V., line beginning with 180, column headed 7, instead of 516.21, read 506.21.

B, " 43, Table VI., first part, on line beginning with 70, column headed 2, instead of 76.635, read 76.735.

B, " " Table VI., second part, on line beginning with 7, the last eight columns,

instead of	0.6483	0.6572	0.6661	0.6750	0.6839	0.6927	0.7016	0.7105,
read	0.6395	0.6483	0.6572	0.6661	0.6750	0.6839	0.6927	0.7016.

B, " 43, Table VI., second part, on line beginning with 12, column headed 5, instead of 1.1018, read 1.1102.

C, " 11, on line beginning with 26.5 inches, column headed 6, instead of 674.41, read 674.61.

C, " " on line of 27.1 inches, column headed 1, instead of 688.38, read 688.58.

C, " 12, on line of 30.5 inches, column headed 2, instead of 778.20, read 775.19.

C, " 39 and 41, at the head of table, instead of "Tenths of Degrees," read "English Inches."

D, " 28, 29, and 30, head of page, instead of "Tenths of a Degree," read "Hundredths of a Degree."

D, " 35, note at the bottom, instead of "Geology," read "Geodesy."

D, " 36, on line beginning with 160, columns headed 8 and 9, instead of 550.19 and 553.47, read 551.19 and 554.47.

D, " 36, on line beginning with 260, columns headed 2, 3, 5, and 6,

instead of	860.59	863.87	879.43	882.72,
read	859.60	862.88	869.44	872.72.

D,	"	37, on line beginning with	620,	column headed	4,	instead of	2048.28,	read	2047.28.
D,	"	" " " " "	770,	"	0,	"	2526.39,	"	2526.29.
D,	"	38, " " " "	880,	"	5,	"	2903.69,	"	2903.60.
D,	"	" " " " "	890,	"	6,	"	2939.79,	"	2939.69.
D,	"	" " " " "	930,	"	5,	"	3069.64,	"	3067.64.
D,	"	" " " " "	990,	"	4,	"	3261.71,	"	3261.21.
D,	"	" " " " "	990,	"	5,	"	3264.59,	"	3264.49.

D, " 39, on line beginning with 1380, columns headed 3, 4, 5, 6, 7, 8,

instead of	4537.28	4540.56	4543.85	4547.13	4550.41	4553.69,
read	4537.48	4540.76	4544.05	4547.33	4550.61	4553.89.

D, " 40, on line beginning with 1610, column headed 5, instead of 5292.65, read 5298.65.

D, " 44, Table X., on line beginning with 3, column headed 6, instead of 21.0205, read 23.0205.

D, " 45, Table XII., on line beginning with column 0, column headed 5, instead of 0.83333, read 0.08333.

ERRATA IN THIS EDITION.

B, page 25, on line beginning with 12, third column, for 5.87, read 5.37.
D, " 9, line 10 from bottom, for "coefficiency," read "coefficient."
D, " 30, line 4 from bottom, for "calculating," read "computing by."
D, " 73, line 7, for *l*, read L.
D, " " line 9, for "dilatation," read "expansion."
D, " 122, line beginning with 930, column headed 5, for 3069.64, read 3067.64.
E, " 65, line 4, for "Degrees of Fahrenheit," read "Degrees of Reaumur."

METEOROLOGICAL TABLES.

I.

THERMOMETRICAL TABLES.

CONTENTS.

COMPARISON OF THE THERMOMETRICAL SCALES.

(The figures refer to the folio at the bottom of the page.)

I. – III.

GENERAL COMPARISON

OF

THE THERMOMETRICAL SCALES,

OR

TABLES

SHOWING THE CORRESPONDING VALUES OF EACH FULL DEGREE OF FAHRENHEIT'S, CENTIGRADE, AND REAUMUR'S THERMOMETERS, FROM +212° TO —39° FAHRENHEIT.

COMPARISON OF THE THERMOMETRICAL SCALES.

The first three tables of this set give a simultaneous comparison of the three scales mostly used at present in Meteorology, and especially of the portion of the scales not comprised in the more extensive tables which follow them. They form thus a complement to these last tables; but as most of the temperatures contained in them do not occur in Meteorology, the comparison of the full degrees was found sufficient.

These three tables have been taken from E. L. *Schubarth's Collection of Physical Tables.* Berlin, 1836.

Tables IV. to IX. being more useful to the Meteorologist, the calculation has been carried out for every tenth of a degree. Tables VII. and IX. are from the *Annuaire Météorologique de France;* the others have been calculated.

A comparison of the Centigrade and Fahrenheit degrees near the boiling point, for every tenth of a degree, for the sake of the comparison of standard thermometers, will be found at the end of Table VI.

Tables X. to XV. will be found useful for comparing differential results, such as ranges of temperature, and any relative amount expressed in degrees of different scales, without reference to their respective zeros.

I. COMPARISON OF FAHRENHEIT'S THERMOMETRICAL SCALE WITH THE CENTIGRADE AND REAUMUR'S.

x° Fahr. $= (x^\circ - 32^\circ)\,\frac{5}{9}$ Centig. $= (x^\circ - 32^\circ)\,\frac{4}{9}$ Reaum.

Fahren.	Centigrade.	Reaumur.	Fahren.	Centigrade.	Reaumur.	Fahren.	Centigrade.	Reaumur.
+212	+100.00	+80.00	+172	+77.78	+62.22	+132	+55.56	+44.44
211	99.44	79.56	171	77.22	61.78	131	55.00	44.00
210	98.89	79.11	170	76.67	61.33	130	54.44	43.56
209	98.33	78.67	169	76.11	60.89	129	53.89	43.11
208	97.78	78.22	168	75.56	60.44	128	53.33	42.67
207	97.22	77.78	167	75.00	60.00	127	52.78	42.22
206	96.67	77.33	166	74.44	59.56	126	52.22	41.78
205	96.11	76.89	165	73.89	59.11	125	51.67	41.33
204	95.56	76.44	164	73.33	58.67	124	51.11	40.89
203	95.00	76.00	163	72.78	58.22	123	50.56	40.44
202	94.44	75.56	162	72.22	57.78	122	50.00	40.00
201	93.89	75.11	161	71.67	57.33	121	49.44	39.56
200	93.33	74.67	160	71.11	56.89	120	48.89	39.11
199	92.78	74.22	159	70.56	56.44	119	48.33	38.67
198	92.22	73.78	158	70.00	56.00	118	47.78	38.22
197	91.67	73.33	157	69.44	55.56	117	47.22	37.78
196	91.11	72.89	156	68.89	55.11	116	46.67	37.33
195	90.56	72.44	155	68.33	54.67	115	46.11	36.89
194	90.00	72.00	154	67.78	54.22	114	45.56	36.44
193	89.44	71.56	153	67.22	53.78	113	45.00	36.00
192	88.89	71.11	152	66.67	53.33	112	44.44	35.56
191	88.33	70.67	151	66.11	52.89	111	43.89	35.11
190	87.78	70.22	150	65.56	52.44	110	43.33	34.67
189	87.22	69.78	149	65.00	52.00	109	42.78	34.22
188	86.67	69.33	148	64.44	51.56	108	42.22	33.78
187	86.11	68.89	147	63.89	51.11	107	41.67	33.33
186	85.56	68.44	146	63.33	50.67	106	41.11	32.89
185	85.00	68.00	145	62.78	50.22	105	40.56	32.44
184	84.44	67.56	144	62.22	49.78	104	40.00	32.00
183	83.89	67.11	143	61.67	49.33	103	39.44	31.56
182	83.33	66.67	142	61.11	48.89	102	38.89	31.11
181	82.78	66.22	141	60.56	48.44	101	38.33	30.67
180	82.22	65.78	140	60.00	48.00	100	37.78	30.22
179	81.67	65.33	139	59.44	47.56	99	37.22	29.78
178	81.11	64.89	138	58.89	47.11	98	36.67	29.33
177	80.56	64.44	137	58.33	46.67	97	36.11	28.89
176	80.00	64.00	136	57.78	46.22	96	35.56	28.44
175	79.44	63.56	135	57.22	45.78	95	35.00	28.00
174	78.89	63.11	134	56.67	45.33	94	34.44	27.56
173	78.33	62.67	133	56.11	44.89	93	33.89	27.11

x° Fahr. $= (x^\circ - 32^\circ)\frac{5}{9}$ Centig. $= (x^\circ - 32^\circ)\frac{4}{9}$ Reaum.

Fahren.	Centigrade.	Reaumur.	Fahren.	Centigrade.	Reaumur.	Fahren.	Centigrade.	Reaumur.
+92	+33.33	+26.67	+48	+ 8.89	+ 7.11	+ 4	−15.56	−12.44
91	32.78	26.22	47	8.33	6.67	3	−16.11	−12.89
90	32.22	25.78	46	7.78	6.22	2	−16.67	−13.33
89	31.67	25.33	45	7.22	5.78	1	−17.22	−13.78
88	31.11	24.89	44	6.67	5.33	0	−17.78	−14.22
87	30.56	24.44	43	6.11	4.89	− 1	−18.33	−14.67
86	30.00	24.00	42	5.56	4.44	− 2	−18.89	−15.11
85	29.44	23.56	41	5.00	4.00	− 3	−19.44	−15.56
84	28.89	23.11	40	4.44	3.56	− 4	−20.00	−16.00
83	28.33	22.67	39	3.89	3.11	− 5	−20.56	−16.44
82	27.78	22.22	38	3.33	2.67	− 6	−21.11	−16.89
81	27.22	21.78	37	2.78	2.22	− 7	−21.67	−17.33
80	26.67	21.33	36	2.22	1.78	− 8	−22.22	−17.78
79	26.11	20.89	35	1.67	1.33	− 9	−22.78	−18.22
78	25.56	20.44	34	1.11	0.89	−10	−23.33	−18.67
77	25.00	20.00	33	0.56	0.44	−11	−23.89	−19.11
76	24.44	19.56	32	0.00	0.00	−12	−24.44	−19.56
75	23.89	19.11	31	− 0.56	− 0.44	−13	−25.00	−20.00
74	23.33	18.67	30	− 1.11	− 0.89	−14	−25.56	−20.44
73	22.78	18.22	29	− 1.67	− 1.33	−15	−26.11	−20.89
72	22.22	17.78	28	− 2.22	− 1.78	−16	−26.67	−21.33
71	21.67	17.33	27	− 2.78	− 2.22	−17	−27.22	−21.78
70	21.11	16.89	26	− 3.33	− 2.67	−18	−27.78	−22.22
69	20.56	16.44	25	− 3.89	− 3.11	−19	−28.33	−22.67
68	20.00	16.00	24	− 4.44	− 3.56	−20	−28.89	−23.11
67	19.44	15.56	23	− 5.00	− 4.00	−21	−29.44	−23.56
66	18.89	15.11	22	− 5.56	− 4.44	−22	−30.00	−24.00
65	18.33	14.67	21	− 6.11	− 4.89	−23	−30.56	−24.44
64	17.78	14.22	20	− 6.67	− 5.33	−24	−31.11	−24.89
63	17.22	13.78	19	− 7.22	− 5.78	−25	−31.67	−25.33
62	16.67	13.33	18	− 7.78	− 6.22	−26	−32.22	−25.78
61	16.11	12.89	17	− 8.33	− 6.67	−27	−32.78	−26.22
60	15.56	12.44	16	− 8.89	− 7.11	−28	−33.33	−26.67
59	15.00	12.00	15	− 9.44	− 7.56	−29	−33.89	−27.11
58	14.44	11.56	14	−10.00	− 8.00	−30	−34.44	−27.56
57	13.89	11.11	13	−10.56	− 8.44	−31	−35.00	−28.00
56	13.33	10.67	12	−11.11	− 8.89	−32	−35.56	−28.44
55	12.78	10.22	11	−11.67	− 9.33	−33	−36.11	−28.89
54	12.22	9.78	10	−12.22	− 9.78	−34	−36.67	−29.33
53	11.67	9.33	9	−12.78	−10.22	−35	−37.22	−29.78
52	11.11	8.89	8	−13.33	−10.67	−36	−37.78	−30.22
51	10.56	8.44	7	−13.89	−11.11	−37	−38.33	−30.67
50	10.00	8.00	6	−14.44	−11.56	−38	−38.89	−31.11
49	9.44	7.56	5	−15.00	−12.00	−39	−39.44	−31.56

For the Continuation see Table IV. and V.

II. COMPARISON OF THE CENTIGRADE THERMOMETER WITH REAUMUR'S AND FAHRENHEIT'S.

x° Centig. $= (32 + \frac{9}{5} x^\circ)$ Fahr. $= \frac{4}{5} x^\circ$ Reaum.

Centig.	Reaumur.	Fahrenheit.	Centig.	Reaumur.	Fahrenheit.	Centig.	Reaumur.	Fahrenheit.
+100	+80.0	+212.0	+83	+66.4	+181.4	+66	+52.8	+150.8
99	79.2	210.2	82	65.6	179.6	65	52.0	149.0
98	78.4	208.4	81	64.8	177.8	64	51.2	147.2
97	77.6	206.6	80	64.0	176.0	63	50.4	145.4
96	76.8	204.8	79	63.2	174.2	62	49.6	143.6
95	76.0	203.0	78	62.4	172.4	61	48.8	141.8
94	75.2	201.2	77	61.6	170.6	60	48.0	140.0
93	74.4	199.4	76	60.8	168.8	59	47.2	138.2
92	73.6	197.6	75	60.0	167.0	58	46.4	136.4
91	72.8	195.8	74	59.2	165.2	57	45.6	134.6
90	72.0	194.0	73	58.4	163.4	56	44.8	132.8
89	71.2	192.2	72	57.6	161.6	55	44.0	131.0
88	70.4	190.4	71	56.8	159.8	54	43.2	129.2
87	69.6	188.6	70	56.0	158.0	53	42.4	127.4
86	68.8	186.8	69	55.2	156.2	52	41.6	125.6
85	68.0	185.0	68	54.4	154.4	51	40.8	123.8
84	67.2	183.2	67	53.6	152.6	50	40.0	122.0

For the Continuation see Tables V. and VI.

III. COMPARISON OF REAUMUR'S THERMOMETER WITH FAHRENHEIT'S AND THE CENTIGRADE.

x° Reaum. $= (32^\circ + \frac{9}{4} x^\circ)$ Fahr. $= \frac{5}{4} x^\circ$ Centig.

Reaumur.	Fahrenheit.	Centigrade.	Reaumur.	Fahrenheit.	Centigrade.	Reaumur.	Fahrenheit.	Centigrade.
+80	+212.00	+100.00	+66	+180.50	+82.50	+52	+149.00	+65.00
79	209.75	98.75	65	178.25	81.25	51	146.75	63.75
78	207.50	97.50	64	176.00	80.00	50	144.50	62.50
77	205.25	96.25	63	173.75	78.75	49	142.25	61.25
76	203.00	95.00	62	171.50	77.50	48	140.00	60.00
75	200.75	93.75	61	169.25	76.25	47	137.75	58.75
74	198.50	92.50	60	167.00	75.00	46	135.50	57.50
73	196.25	91.25	59	164.75	73.75	45	133.25	56.25
72	194.00	90.00	58	162.50	72.50	44	131.00	55.00
71	191.75	88.75	57	160.25	71.25	43	128.75	53.75
70	189.50	87.50	56	158.00	70.00	42	126.50	52.50
69	187.25	86.25	55	155.75	68.75	41	124.25	51.25
68	185.00	85.00	54	153.50	67.50	40	122.00	50.00
67	182.75	83.75	53	151.25	66.25	39	119.75	48.75

For the Continuation see Tables VIII. and IX.

IV. - V.

COMPARISON

OF

FAHRENHEIT'S THERMOMETER

WITH

THE CENTIGRADE AND WITH THAT OF REAUMUR,

OR

TABLES

FOR CONVERTING THE DEGREES OF FAHRENHEIT INTO CENTIGRADE DEGREES AND INTO DEGREES OF REAUMUR;

GIVING THE CORRESPONDING VALUES FOR EACH TENTH OF A DEGREE, FROM +122° TO —76° FAHRENHEIT.

Degrees of Fahrenheit.	Tenths of Degrees.									
	0.	**1.**	**2.**	**3.**	**4.**	**5.**	**6.**	**7.**	**8.**	**9.**
	Centig.	Centig.	Centig.	Centig.	Centig.	Centig.	Centig.	Centig.	Centig.	Centig.
+122	+50.00	+50.06	+50.11	+50.17	+50.22	+50.28	+50.33	+50.39	+50.44	+50.50
121	49.44	49.50	49.56	49.61	49.67	49.72	49.78	49.83	49.89	49.94
120	48.89	49.94	49.00	49.06	49.11	49.17	49.22	49.28	49.33	49.39
119	48.33	48.39	48.44	48.50	48.56	48.61	48.67	48.72	48.78	48.83
118	47.78	47.83	47.89	47.94	48.00	48.06	48.11	48.17	48.22	48.28
117	47.22	47.28	47.33	47.39	47.44	47.50	47.56	47.61	47.67	47.72
116	46.67	46.72	46.78	46.83	46.89	46.94	47.00	47.06	47.11	47.17
115	46.11	46.17	46.22	46.28	46.33	46.39	46.44	46.50	46.56	46.61
114	45.56	45.61	45.67	45.72	45.78	45.83	45.89	45.94	46.00	46.06
113	45.00	45.06	45.11	45.17	45.22	45.28	45.33	45.39	45.44	45.50
112	44.44	44.50	44.56	44.61	44.67	44.72	44.78	44.83	44.89	44.94
111	43.89	43.94	44.00	44.06	44.11	44.17	44.22	44.28	44.33	44.39
110	43.33	43.39	43.44	43.50	43.56	43.61	43.67	43.72	43.78	43.83
109	42.78	42.83	42.89	42.94	43.00	43.06	43.11	43.17	43.22	43.28
108	42.22	42.28	42.33	42.39	42.44	42.50	42.56	42.61	42.67	42.72
107	41.67	41.72	41.78	41.83	41.89	41.94	42.00	42.06	42.11	42.17
106	41.11	41.17	41.22	41.28	41.33	41.39	41.44	41.50	41.56	41.61
105	40.56	40.61	40.67	40.72	40.78	40.83	40.89	40.94	41.00	41.06
104	40.00	40.06	40.11	40.17	40.22	40.28	40.33	40.39	40.44	40.50
103	39.44	39.50	39.56	39.61	39.67	39.72	39.78	39.83	39.89	39.94
102	38.89	38.94	39.00	39.06	39.11	39.17	39.22	39.28	39.33	39.39
101	38.33	38.39	38.44	38.50	38.56	38.61	38.67	38.72	38.78	38.83
100	37.78	37.83	37.89	37.94	38.00	38.06	38.11	38.17	38.22	38.28
99	37.22	37.28	37.33	37.39	37.44	37.50	37.56	37.61	37.67	37.72
98	36.67	36.72	36.78	36.83	36.89	36.94	37.00	37.06	37.11	37.17
97	36.11	36.17	36.22	36.28	36.33	36.39	36.44	36.50	36.56	36.61
96	35.56	35.61	35.67	35.72	35.78	35.83	35.89	35.94	36.00	36.06
95	35.00	35.06	35.11	35.17	35.22	35.28	35.33	35.39	35.44	35.50
94	34.44	34.50	34.56	34.61	34.67	34.72	34.78	34.83	34.89	34.94
93	33.89	33.94	34.00	34.06	34.11	34.17	34.22	34.28	34.33	34.39
92	33.33	33.39	33.44	33.50	33.56	33.61	33.67	33.72	33.78	33.83
91	32.78	32.83	32.89	32.94	33.00	33.06	33.11	33.17	33.22	33.28
90	32.22	32.28	32.33	32.39	32.44	32.50	32.56	32.61	32.67	32.72
89	31.67	31.72	31.78	31.83	31.89	31.94	32.00	32.06	32.11	33.17
88	31.11	31.17	31.22	31.28	31.33	31.39	31.44	31.50	31.56	31.61
87	30.56	30.61	30.67	30.72	30.78	30.83	30.89	30.94	31.00	31.06
86	30.00	30.06	30.11	30.17	30.22	30.28	30.33	30.39	30.44	30.50
85	29.44	29.50	29.56	29.61	29.67	29.72	29.78	29.83	29.89	29.94
84	28.89	28.94	29.00	29.06	29.11	29.17	29.22	29.28	29.33	29.39
83	28.33	28.39	28.44	28.50	28.56	28.61	28.67	28.72	28.78	28.83
	0.	**1.**	**2.**	**3.**	**4.**	**5.**	**6.**	**7.**	**8.**	**9.**

Degrees of Fahren-heit.	Tenths of Degrees.									
	0.	**1.**	**2.**	**3.**	**4.**	**5.**	**6.**	**7.**	**8.**	**9.**
	Centig.	Centig.	Centig.	Centig.	Centig.	Centig.	Centig.	Centig.	Centig.	Centig.
+82	+27.78	+27.83	+27.89	+27.94	+28.00	+28.06	+28.11	+28.17	+28.22	+28.28
81	27.22	27.28	27.33	27.39	27.44	27.50	27.56	27.61	27.67	27.72
80	26.67	26.72	26.78	26.83	26.89	26.94	27.00	27.06	27.11	27.17
79	26.11	26.17	26.22	26.28	26.33	26.39	26.44	26.50	26.56	26.61
78	25.56	25.61	25.67	25.72	25.78	25.83	25.89	25.94	26.00	26.06
77	25.00	25.06	25.11	25.17	25.22	25.28	25.33	25.39	25.44	25.50
76	24.44	24.50	24.56	24.61	24.67	24.72	24.78	24.83	24.89	24.94
75	23.89	23.94	24.00	24.06	24.11	24.17	24.22	24.28	24.33	24.39
74	23.33	23.39	23.44	23.50	23.56	23.61	23.67	23.72	23.78	23.83
73	22.78	22.83	22.89	22.94	23.00	23.06	23.11	23.17	23.22	23.28
72	22.22	22.28	22.33	22.39	22.44	22.50	22.56	22.61	22.67	22.72
71	21.67	21.72	21.78	21.83	21.89	21.94	22.00	22.06	22.11	22.17
70	21.11	21.17	21.22	21.28	21.33	21.39	21.44	21.50	21.56	21.61
69	20.56	20.61	20.67	20.72	20.78	20.83	20.89	20.94	21.00	21.06
68	20.00	20.06	20.11	20.17	20.22	20.28	20.33	20.39	20.44	20.50
67	19.44	19.50	19.56	19.61	19.67	19.72	19.78	19.83	19.89	19.94
66	18.89	18.94	19.00	19.06	19.11	19.17	19.22	19.28	19.33	19.39
65	18.33	18.39	18.44	18.50	18.56	18.61	18.67	18.72	18.78	18.83
64	17.78	17.83	17.89	17.94	18.00	18.06	18.11	18.17	18.22	18.28
63	17.22	17.28	17.33	17.39	17.44	17.50	17.56	17.61	17.67	17.72
62	16.67	16.72	16.78	16.83	16.89	16.94	17.00	17.06	17.11	17.17
61	16.11	16.17	16.22	16.28	16.33	16.39	16.44	16.50	16.56	16.61
60	15.56	15.61	15.67	15.72	15.78	15.83	15.89	15.94	16.00	16.06
59	15.00	15.06	15.11	15.17	15.22	15.28	15.33	15.39	15.44	15.50
58	14.44	14.50	14.56	14.61	14.67	14.72	14.78	14.83	14.89	14.94
57	13.89	13.94	14.00	14.06	14.11	14.17	14.22	14.28	14.33	14.39
56	13.33	13.39	13.44	13.50	13.56	13.61	13.67	13.72	13.78	13.83
55	12.78	12.83	12.89	12.94	13.00	13.06	13.11	13.17	13.22	13.28
54	12.22	12.28	12.33	12.39	12.44	12.50	12.56	12.61	12.67	12.72
53	11.67	11.72	11.78	11.83	11.89	11.94	12.00	12.06	12.11	12.17
52	11.11	11.17	11.22	11.28	11.33	11.39	11.44	11.50	11.56	11.61
51	10.56	10.61	10.67	10.72	10.78	10.83	10.89	10.94	11.00	11.06
50	10.00	10.06	10.11	10.17	10.22	10.28	10.33	10.39	10.44	10.50
49	9.44	9.50	9.56	9.61	9.67	9.72	9.78	9.83	9.89	9.94
48	8.89	8.94	9.00	9.06	9.11	9.17	9.22	9.28	9.33	9.39
47	8.33	8.39	8.44	8.50	8.56	8.61	8.67	8.72	8.78	8.83
46	7.78	7.83	7.89	7.94	8.00	8.06	8.11	8.17	8.22	8.28
45	7.22	7.28	7.33	7.39	7.44	7.50	7.56	7.61	7.67	7.72
44	6.67	6.72	6.78	6.83	6.89	6.94	7.00	7.06	7.11	7.17
43	6.11	6.17	6.22	6.28	6.33	6.39	6.44	6.50	6.56	6.61
	0.	**1.**	**2.**	**3.**	**4.**	**5.**	**6.**	**7.**	**8.**	**9.**

Degrees of Fahrenheit.	Tenths of Degrees.									
	0.	1.	2.	3.	4.	5.	6.	7.	8.	9.
	Centig.	Centig.	Centig.	Centig.	Centig.	Centig.	Centig.	Centig.	Centig.	Centig.
+42	+5.56	+5.61	+5.67	+5.72	+5.78	+5.83	+5.89	+5.94	+6.00	+6.06
41	5.00	5.06	5.11	5.17	5.22	5.28	5.33	5.39	5.44	5.50
40	4.44	4.50	4.56	4.61	4.67	4.72	4.78	4.83	4.89	4.94
39	3.89	3.94	4.00	4.06	4.11	4.17	4.22	4.28	4.33	4.39
38	3.33	3.39	3.44	3.50	3.56	3.61	3.67	3.72	3.78	3.83
37	2.78	2.83	2.89	2.94	3.00	3.06	3.11	3.17	3.22	3.28
36	2.22	2.28	2.33	2.39	2.44	2.50	2.56	2.61	2.67	2.72
35	1.67	1.72	1.78	1.83	1.89	1.94	2.00	2.06	2.11	2.17
34	1.11	1.17	1.22	1.28	1.33	1.39	1.44	1.50	1.56	1.61
33	0.56	0.61	0.67	0.72	0.78	0.83	0.89	0.94	1.00	1.06
32	0.00	0.06	0.11	0.17	0.22	0.28	0.33	0.39	0.44	0.50
31	− 0.56	− 0.50	− 0.44	− 0.39	− 0.33	− 0.28	− 0.22	− 0.17	− 0.11	− 0.06
30	− 1.11	− 1.06	− 1.00	− 0.94	− 0.89	− 0.83	− 0.78	− 0.72	− 0.67	− 0.61
29	− 1.67	− 1.61	− 1.56	− 1.50	− 1.44	− 1.39	− 1.33	− 1.28	− 1.22	− 1.17
28	− 2.22	− 2.17	− 2.11	− 2.06	− 2.00	− 1.94	− 1.89	− 1.83	− 1.78	− 1.72
27	− 2.78	− 2.72	− 2.67	− 2.61	− 2.56	− 2.50	− 2.44	− 2.39	− 2.33	− 2.28
26	− 3.33	− 3.28	− 3.22	− 3.17	− 3.11	− 3.06	− 3.00	− 2.94	− 2.89	− 2.83
25	− 3.89	− 3.83	− 3.78	− 3.72	− 3.67	− 3.61	− 3.56	− 3.50	− 3.44	− 3.39
24	− 4.44	− 4.39	− 4.33	− 4.28	− 4.22	− 4.17	− 4.11	− 4.06	− 4.00	− 3.94
23	− 5.00	− 4.94	− 4.89	− 4.83	− 4.78	− 4.72	− 4.67	− 4.61	− 4.56	− 4.50
22	− 5.56	− 5.50	− 5.44	− 5.39	− 5.33	− 5.28	− 5.22	− 5.17	− 5.11	− 5.06
21	− 6.11	− 6.06	− 6.00	− 5.94	− 5.89	− 5.83	− 5.78	− 5.72	− 5.67	− 5.61
20	− 6.67	− 6.61	− 6.56	− 6.50	− 6.44	− 6.39	− 6.33	− 6.28	− 6.22	− 6.17
19	− 7.22	− 7.17	− 7.11	− 7.06	− 7.00	− 6.94	− 6.89	− 6.83	− 6.78	− 6.72
18	− 7.78	− 7.72	− 7.67	− 7.61	− 7.56	− 7.50	− 7.44	− 7.39	− 7.33	− 7.28
17	− 8.33	− 8.28	− 8.22	− 8.17	− 8.11	− 8.06	− 8.00	− 7.94	− 7.89	− 7.83
16	− 8.89	− 8.83	− 8.78	− 8.72	− 8.67	− 8.61	− 8.56	− 8.50	− 8.44	− 8.39
15	− 9.44	− 9.39	− 9.33	− 9.28	− 9.22	− 9.17	− 9.11	− 9.06	− 9.00	− 8.94
14	−10.00	− 9.94	− 9.89	− 9.83	− 9.78	− 9.72	− 9.67	− 9.61	− 9.56	− 9.50
13	−10.56	−11.50	−10.44	−10.39	−10.33	−10.28	−10.22	−10.17	−10.11	−10.06
12	−11.11	−11.06	−11.00	−10.94	−10.89	−10.83	−10.78	−10.72	−10.67	−10.61
11	−11.67	−11.61	−11.56	−11.50	−11.44	−11.39	−11.33	−11.28	−11.22	−11.17
10	−12.22	−12.17	−12.11	−12.06	−12.00	−11.94	−11.89	−11.83	−11.78	−11.72
9	−12.78	−12.72	−12.67	−12.61	−12.56	−12.50	−12.44	−12.39	−12.33	−12.28
8	−13.33	−13.28	−13.22	−13.17	−13.11	−13.06	−13.00	−12.94	−12.89	−12.83
7	−13.89	−13.83	−13.78	−13.72	−13.67	−13.61	−13.56	−13.50	−13.44	−13.39
6	−14.44	−14.39	−14.33	−14.28	−14.22	−14.17	−14.11	−14.06	−14.00	−13.94
5	−15.00	−14.94	−14.89	−14.83	−14.78	−14.72	−14.67	−14.61	−14.56	−14.50
4	−15.56	−15.50	−15.44	−15.39	−15.33	−15.28	−15.22	−15.17	−15.11	−15.06
3	−16.11	−16.06	−16.00	−15.94	−15.89	−15.83	−15.78	−15.72	−15.67	−15.61
	0.	1.	2.	3.	4.	5.	6.	7.	8.	9.

Degrees of Fahren-heit.	Tenths of Degrees.									
	0.	**1.**	**2.**	**3.**	**4.**	**5.**	**6.**	**7.**	**8.**	**9.**
	Centig.	Centig.	Centig.	Centig.	Centig.	Centig.	Centig.	Centig.	Centig.	Centig.
+ 2	−16.67	−16.61	−16.56	−16.50	−16.44	−16.39	−16.33	−16.28	−16.22	−16.17
1	−17.22	−17.17	−17.11	−17.06	−17.00	−16.94	−16.89	−16.83	−16.78	−16.72
0	−17.78	−17.72	−17.67	−17.61	−17.56	−17.50	−17.44	−17.39	−17.33	−17.28
− 0	−17.78	−17.83	−17.89	−17.94	−18.00	−18.06	−18.11	−18.17	−18.22	−18.28
− 1	−18.33	−18.39	−18.44	−18.50	−18.56	−18.61	−18.67	−18.72	−18.78	−18.83
− 2	−18.89	−18.94	−19.00	−19.06	−19.11	−19.17	−19.22	−19.28	−19.33	−19.39
− 3	−19.44	−19.50	−19.56	−19.61	−19.67	−19.72	−19.78	−19.83	−19.89	−19.94
− 4	−20.00	−20.06	−20.11	−20.17	−20.22	−20.28	−20.33	−20.39	−20.44	−20.50
− 5	−20.56	−20.61	−20.67	−20.72	−20.78	−20.83	−20.89	−20.94	−21.00	−21.06
− 6	−21.11	−21.17	−21.22	−21.28	−21.33	−21.39	−21.44	−21.50	−21.56	−21.61
− 7	−21.67	−21.72	−21.78	−21.83	−21.89	−21.94	−22.00	−22.06	−22.11	−22.17
− 8	−22.22	−22.28	−22.33	−22.39	−22.44	−22.50	−22.56	−22.61	−22.67	−22.72
− 9	−22.78	−22.83	−22.89	−22.94	−23.00	−23.06	−23.11	−23.17	−23.22	−23.28
−10	−23.33	−23.39	−23.44	−23.50	−23.56	−23.61	−23.67	−23.72	−23.78	−23.83
−11	−23.89	−23.94	−24.00	−24.06	−24.11	−24.17	−24.22	−24.28	−24.33	−24.39
−12	−24.44	−24.50	−24.56	−24.61	−24.67	−24.72	−24.78	−24.83	−24.89	−24.94
−13	−25.00	−25.06	−25.11	−25.17	−25.22	−25.28	−25.33	−25.39	−25.44	−25.50
−14	−25.56	−25.61	−25.67	−25.72	−25.78	−25.83	−25.89	−25.94	−26.00	−26.06
−15	−26.11	−26.17	−26.22	−26.28	−26.33	−26.39	−26.44	−26.50	−26.56	−26.61
−16	−26.67	−26.72	−26.78	−26.83	−26.89	−26.94	−27.00	−27.06	−27.11	−27.17
−17	−27.22	−27.28	−27.33	−27.39	−27.44	−27.50	−27.56	−27.61	−27.67	−27.72
−18	−27.78	−27.83	−27.89	−27.94	−28.00	−28.06	−28.11	−28.17	−28.22	−28.28
−19	−28.33	−28.39	−28.44	−28.50	−28.56	−28.61	−28.67	−28.72	−28.78	−28.83
−20	−28.89	−28.94	−29.00	−29.06	−29.11	−29.17	−29.22	−29.28	−29.33	−29.39
−21	−29.44	−29.50	−29.56	−29.61	−29.67	−29.72	−29.78	−29.83	−29.89	−29.94
−22	−30.00	−30.06	−30.11	−30.17	−30.22	−30.28	−30.33	−30.39	−30.44	−30.50
−23	−30.56	−30.61	−30.67	−30.72	−30.78	−30.83	−30.89	−30.94	−31.00	−31.06
−24	−31.11	−31.17	−31.22	−31.28	−31.33	−31.39	−31.44	−31.50	−31.56	−31.61
−25	−31.67	−31.72	−31.78	−31.83	−31.89	−31.94	−32.00	−32.06	−32.11	−32.17
−26	−32.22	−32.28	−32.33	−32.39	−32.44	−32.50	−32.56	−32.61	−32.67	−32.72
−27	−32.78	−32.83	−32.89	−32.94	−33.00	−33.06	−33.11	−33.17	−33.22	−33.28
−28	−33.33	−33.39	−33.44	−33.50	−33.56	−33.61	−33.67	−33.72	−33.78	−33.83
−29	−33.89	−33.94	−34.00	−34.06	−34.11	−34.17	−34.22	−34.28	−34.33	−34.39
−30	−34.44	−34.50	−34.56	−34.61	−34.67	−34.72	−34.78	−34.83	−34.89	−34.94
−31	−35.00	−35.06	−35.11	−35.17	−35.22	−35.28	−35.33	−35.39	−35.44	−35.50
−32	−35.56	−35.61	−35.67	−35.72	−35.78	−35.83	−35.89	−35.94	−36.00	−36.06
−33	−36.11	−36.17	−36.22	−36.28	−36.33	−36.39	−36.44	−36.50	−36.56	−36.61
−34	−36.67	−36.72	−36.78	−36.83	−36.89	−36.94	−37.00	−37.06	−37.11	−37.17
−35	−37.22	−37.28	−37.33	−37.39	−37.44	−37.50	−37.56	−37.61	−37.67	−37.72
−36	−37.78	−37.83	−37.89	−37.94	−38.00	−38.06	−38.11	−38.17	−38.22	−38.28
	0.	**1.**	**2.**	**3.**	**4.**	**5.**	**6.**	**7.**	**8.**	**9.**

Degrees of Fahrenheit.	Tenths of Degrees.									
	0.	**1.**	**2.**	**3.**	**4.**	**5.**	**6.**	**7.**	**8.**	**9.**
	Centig.	Centig.	Centig.	Centig.	Centig.	Centig.	Centig.	Centig.	Centig.	Centig.
−37	−38.33	−38.39	−38.44	−38.50	−38.56	−38.61	−38.67	−38.72	−38.78	−38.83
−38	−38.89	−38.94	−39.00	−39.06	−39.11	−39.17	−39.22	−39.28	−39.33	−39.39
−39	−39.44	−39.50	−39.56	−39.61	−39.67	−39.72	−39.78	−39.83	−39.89	−39.94
−40	−40.00	−40.06	−40.11	−40.17	−40.22	−40.28	−40.33	−40.39	−40.44	−40.50
−41	−40.56	−40.61	−40.67	−40.72	−40.78	−40.83	−40.89	−40.94	−41.00	−41.06
−42	−41.11	−41.17	−41.22	−41.28	−41.33	−41.39	−41.44	−41.50	−41.56	−41.61
−43	−41.67	−41.72	−41.78	−41.83	−41.89	−41.94	−42.00	−42.06	−42.11	−42.17
−44	−42.22	−42.28	−42.33	−42.39	−42.44	−42.50	−42.56	−42.61	−42.67	−42.72
−45	−42.78	−42.83	−42.89	−42.94	−43.00	−43.06	−43.11	−43.17	−43.22	−43.28
−46	−43.33	−43.39	−43.44	−43.50	−43.56	−43.61	−43.67	−43.72	−43.78	−43.83
−47	−43.89	−43.94	−44.00	−44.06	−44.11	−44.17	−44.22	−44.28	−44.33	−44.39
−48	−44.44	−44.50	−44.56	−44.61	−44.67	−44.72	−44.78	−44.83	−44.89	−44.94
−49	−45.00	−45.06	−45.11	−45.17	−45.22	−45.28	−45.33	−45.39	−45.44	−45.50
−50	−45.56	−45.61	−45.67	−45.72	−45.78	−45.83	−45.89	−45.94	−46.00	−46.06
−51	−46.11	−46.17	−46.22	−46.28	−46.33	−46.39	−46.44	−46.50	−46.56	−46.61
−52	−46.67	−46.72	−46.78	−46.83	−46.89	−46.94	−47.00	−47.06	−47.11	−47.17
−53	−47.22	−47.28	−47.33	−47.39	−47.44	−47.50	−47.56	−47.61	−47.67	−47.72
−54	−47.78	−47.83	−47.89	−47.94	−48.00	−48.06	−48.11	−48.17	−48.22	−48.28
−55	−48.33	−48.39	−48.44	−48.50	−48.56	−48.61	−48.67	−48.72	−48.78	−48.83
−56	−48.89	−48.94	−49.00	−49.06	−49.11	−49.17	−49.22	−49.28	−49.33	−49.39
−57	−49.44	−49.50	−49.56	−49.61	−49.67	−49.72	−49.78	−49.83	−49.89	−49.94
−58	−50.00	−50.06	−50.11	−50.17	−50.22	−50.28	−50.33	−50.39	−50.44	−50.50
−59	−50.56	−50.61	−50.67	−50.72	−50.78	−50.83	−50.89	−50.94	−51.00	−51.06
−60	−51.11	−51.17	−51.22	−51.28	−51.33	−51.39	−51.44	−51.50	−51.56	−51.61
−61	−51.67	−51.72	−51.78	−51.83	−51.89	−51.94	−52.00	−52.06	−52.11	−52.17
−62	−52.22	−52.28	−52.33	−52.39	−52.44	−52.50	−52.56	−52.61	−52.67	−52.72
−63	−52.78	−52.83	−52.89	−52.94	−53.00	−53.06	−53.11	−53.17	−53.22	−53.28
−64	−53.33	−53.39	−53.44	−53.50	−53.56	−53.61	−53.67	−53.72	−53.78	−53.83
−65	−53.89	−53.94	−54.00	−54.06	−54.11	−54.17	−54.22	−54.28	−54.33	−54.39
−66	−54.44	−54.50	−54.56	−54.61	−54.67	−54.72	−54.78	−54.83	−54.89	−54.94
−67	−55.00	−55.06	−55.11	−55.17	−55.22	−55.28	−55.33	−55.39	−55.44	−55.50
−68	−55.56	−55.61	−55.67	−55.72	−55.78	−55.83	−55.89	−55.94	−56.00	−56.06
−69	−56.11	−56.17	−56.22	−56.28	−56.33	−56.39	−56.44	−56.50	−56.56	−56.61
−70	−56.67	−56.72	−56.78	−56.83	−56.89	−56.94	−57.00	−57.06	−57.11	−57.17
−71	−57.22	−57.28	−57.33	−57.39	−57.44	−57.50	−57.56	−57.61	−57.67	−57.72
−72	−57.78	−57.83	−57.89	−57.94	−58.00	−58.06	−58.11	−58.16	−58.22	−58.28
−73	−58.33	−58.39	−58.44	−58.50	−58.56	−58.61	−58.67	−58.72	−58.78	−58.83
−74	−58.89	−58.94	−59.00	−59.06	−59.11	−59.17	−59.22	−59.28	−59.33	−59.39
−75	−59.44	−59.50	−59.56	−59.61	−59.67	−59.72	−59.78	−59.83	−59.89	−59.94
−76	−60.00	−60.06	−60.11	−60.17	−60.22	−60.28	−60.33	−60.39	−60.44	−60.50
	0.	**1.**	**2.**	**3.**	**4.**	**5.**	**6.**	**7.**	**8.**	**9.**

Degrees of Fahrenheit.	Tenths of a Degree.									
	0.	**1.**	**2.**	**3.**	**4.**	**5.**	**6.**	**7.**	**8.**	**9.**
	Reaumur.	Reaumur.	Reaumur.	Reaumur.	Reaumur.	Reaumur.	Reaumur.	Reaumur.	Reaumur.	Reaumur.
+122	+40.00	+40.04	+40.09	+40.13	+40.18	+40.22	+40.27	+40.31	+40.36	+40.40
121	39.56	39.60	39.64	39.69	39.73	39.78	39.82	39.87	39.91	39.96
120	39.11	39.16	39.20	39.24	39.29	39.33	39.38	39.42	39.47	39.51
119	38.67	38.71	38.76	38.80	38.84	38.89	38.93	38.98	39.02	39.07
118	38.22	38.27	38.31	38.36	38.40	38.44	38.49	38.53	38.58	38.62
117	37.78	37.82	37.87	37.91	37.96	38.00	38.04	38.09	38.13	38.18
116	37.33	37.38	37.42	37.47	37.51	37.56	37.60	37.64	37.69	37.73
115	36.89	36.93	36.98	37.02	37.07	37.11	37.16	37.20	37.24	37.29
114	36.44	36.49	36.53	36.58	36.62	36.67	36.71	36.76	36.80	36.84
113	36.00	36.04	36.09	36.13	36.18	36.22	36.27	36.31	36.36	36.40
112	35.56	35.60	35.64	35.69	35.73	35.78	35.82	35.87	35.91	35.96
111	35.11	35.16	35.20	35.24	35.29	35.33	35.38	35.42	35.47	35.51
110	34.67	34.71	34.76	34.80	34.84	34.89	34.93	34.98	35.02	35.07
109	34.22	34.27	34.31	34.36	34.40	34.44	34.49	34.53	34.58	34.62
108	33.78	33.82	33.87	33.91	33.96	34.00	34.04	34.09	34.13	34.18
107	33.33	33.38	33.42	33.47	33.51	33.56	33.60	33.64	33.69	33.73
106	32.89	32.93	32.98	33.02	33.07	33.11	33.16	33.20	33.24	33.29
105	32.44	32.49	32.53	32.58	32.62	32.67	32.71	32.76	32.80	32.84
104	32.00	32.04	32.09	32.13	32.18	32.22	32.27	32.31	32.36	32.40
103	31.56	31.60	31.64	31.69	31.73	31.78	31.82	31.87	31.91	31.96
102	31.11	31.16	31.20	31.24	31.29	31.33	31.38	31.42	31.47	31.51
101	30.67	30.71	30.76	30.80	30.84	30.89	30.93	30.98	31.02	31.07
100	30.22	30.27	30.31	30.36	30.40	30.44	30.49	30.53	30.58	30.62
99	29.78	29.82	29.87	29.91	29.96	30.00	30.04	30.09	30.13	30.18
98	29.33	29.38	29.42	29.47	29.51	29.56	29.60	29.64	29.69	29.73
97	28.89	28.93	28.98	29.02	29.07	29.11	29.16	29.20	29.24	29.29
96	28.44	28.49	28.53	28.58	28.62	28.67	28.71	28.76	28.80	28.84
95	28.00	28.04	28.09	28.13	28.18	28.22	28.27	28.31	28.36	28.40
94	27.56	27.60	27.64	27.69	27.73	27.78	27.82	27.87	27.91	27.96
93	27.11	27.16	27.20	27.24	27.29	27.33	27.38	27.42	27.47	27.51
92	26.67	26.71	26.76	26.80	26.84	26.89	26.93	26.98	27.02	27.07
91	26.22	26.27	26.31	26.36	26.40	26.44	26.49	26.53	26.58	26.62
90	25.78	25.82	25.87	25.91	25.96	26.00	26.04	26.09	26.13	26.18
89	25.33	25.38	25.42	25.47	25.51	25.56	25.60	25.64	25.69	25.73
88	24.89	24.93	24.98	25.02	25.07	25.11	25.16	25.20	25.24	25.29
87	24.44	24.49	24.53	24.58	24.62	24.67	24.71	24.76	24.80	24.84
86	24.00	24.04	24.09	24.13	24.18	24.22	24.27	24.31	24.36	24.40
85	23.56	23.60	23.64	23.69	23.73	23.78	23.82	23.87	23.91	23.96
84	23.11	23.16	23.20	23.24	23.29	23.33	23.38	23.42	23.47	23.51
83	22.67	22.71	22.76	22.80	22.84	22.89	22.93	22.98	23.02	23.07
82	22.22	22.27	22.31	22.36	22.40	22.44	22.49	22.53	22.58	22.62
	0.	**1.**	**2.**	**3.**	**4.**	**5.**	**6.**	**7.**	**8.**	**9.**

Degrees of Fahrenheit.	Tenths of a Degree.									
	0.	**1.**	**2.**	**3.**	**4.**	**5.**	**6.**	**7.**	**8.**	**9.**
	Reaumur.	Reaumur.	Reaumur.	Reaumur.	Reaumur.	Reaumur.	Reaumur.	Reaumur.	Reaumur.	Reaumur.
+81	+21.78	+21.82	+21.87	+21.91	+21.96	+22.00	+22.04	+22.09	+22.13	+22.18
80	21.33	21.38	21.42	21.47	21.51	21.56	21.60	21.64	21.69	21.73
79	20.89	20.93	20.98	21.02	21.07	21.11	21.16	21.20	21.24	21.29
78	20.44	20.49	20.53	20.58	20.62	20.67	20.71	20.76	20.80	20.84
77	20.00	20.04	20.09	20.13	20.18	20.22	20.27	20.31	20.36	20.40
76	19.56	19.60	19.64	19.69	19.73	19.78	19.82	19.87	19.91	19.96
75	19.11	19.16	19.20	19.24	19.29	19.33	19.38	19.42	19.47	19.51
74	18.67	18.71	18.76	18.80	18.84	18.89	18.93	18.98	19.02	19.07
73	18.22	18.27	18.31	18.36	18.40	18.44	18.49	18.53	18.58	18.62
72	17.78	17.82	17.87	17.91	17.96	18.00	18.04	18.09	18.13	18.18
71	17.33	17.38	17.42	17.47	17.51	17.56	17.60	17.64	17.69	17.73
70	16.89	16.93	16.98	17.02	17.07	17.11	17.16	17.20	17.24	17.29
69	16.44	16.49	16.53	16.58	16.62	16.67	16.71	16.76	16.80	16.84
68	16.00	16.04	16.09	16.13	16.18	16.22	16.27	16.31	16.36	16.40
67	15.56	15.60	15.64	15.69	15.73	15.78	15.82	15.87	15.91	15.96
66	15.11	15.16	15.20	15.24	15.29	15.33	15.38	15.42	15.47	15.51
65	14.67	14.71	14.76	14.80	14.84	14.89	14.93	14.98	15.02	15.07
64	14.22	14.27	14.31	14.36	14.40	14.44	14.49	14.53	14.58	14.62
63	13.78	13.82	13.87	13.91	13.96	14.00	14.04	14.09	14.13	14.18
62	13.33	13.38	13.42	13.47	13.51	13.56	13.60	13.64	13.69	13.73
61	12.89	12.93	12.98	13.02	13.07	13.11	13.16	13.20	13.24	13.29
60	12.44	12.49	12.53	12.58	12.62	12.67	12.71	12.76	12.80	12.84
59	12.00	12.04	12.09	12.13	12.18	12.22	12.27	12.31	12.36	12.40
58	11.56	11.60	11.64	11.69	11.73	11.78	11.82	11.87	11.91	11.96
57	11.11	11.16	11.20	11.24	11.29	11.33	11.38	11.42	11.47	11.51
56	10.67	10.71	10.76	10.80	10.84	10.89	10.93	10.98	11.02	11.07
55	10.22	10.27	10.31	10.36	10.40	10.44	10.49	10.53	10.58	10.62
54	9.78	9.82	9.87	9.91	9.96	10.00	10.04	10.09	10.13	10.18
53	9.33	9.38	9.42	9.47	9.51	9.56	9.60	9.64	9.69	9.73
52	8.89	8.93	8.98	9.02	9.07	9.11	9.16	9.20	9.24	9.29
51	8.44	8.49	8.53	8.58	8.62	8.67	8.71	8.76	8.80	8.84
50	8.00	8.04	8.09	8.13	8.18	8.22	8.27	8.31	8.36	8.40
49	7.56	7.60	7.64	7.69	7.73	7.78	7.82	7.87	7.91	7.96
48	7.11	7.16	7.20	7.24	7.29	7.33	7.38	7.42	7.47	7.51
47	6.67	6.71	6.76	6.80	6.84	6.89	6.93	6.98	7.02	7.07
46	6.22	6.27	6.31	6.36	6.40	6.44	6.49	6.53	6.58	6.62
45	5.78	5.82	5.87	5.91	5.96	6.00	6.04	6.09	6.13	6.18
44	5.33	5.38	5.42	5.47	5.51	5.56	5.60	5.64	5.69	5.73
43	4.89	4.93	4.98	5.02	5.07	5.11	5.16	5.20	5.24	5.29
42	4.44	4.49	4.53	4.58	4.62	4.67	4.71	4.76	4.80	4.84
41	4.00	4.04	4.09	4.13	4.18	4.22	4.27	4.31	4.36	4.40
	0.	**1.**	**2.**	**3.**	**4.**	**5.**	**6.**	**7.**	**8.**	**9.**

Degrees of Fahren-heit.	Tenths of a Degree.									
	0.	**1.**	**2.**	**3.**	**4.**	**5.**	**6.**	**7.**	**8.**	**9.**
	Reaumur.	Reaumur.	Reaumur.	Reaumur.	Reaumur.	Reaumur.	Reaumur.	Reaumur.	Reaumur.	Reaumur.
+40	+ 3.56	+ 3.60	+ 3.64	+ 3.69	+ 3.73	+ 3.78	+ 3.82	+ 3.87	+ 3.91	+ 3.96
39	3.11	3.16	3.20	3.24	3.29	3.33	3.38	3.42	3.47	3.51
38	2.67	2.71	2.76	2.80	2.84	2.89	2.93	2.98	3.02	3.07
37	2.22	2.27	2.31	2.36	2.40	2.44	2.49	2.53	2.58	2.62
36	1.78	1.82	1.87	1.91	1.96	2.00	2.04	2.09	2.13	2.18
35	1.33	1.38	1.42	1.47	1.51	1.56	1.60	1.64	1.69	1.73
34	0.89	0.93	0.98	1.02	1.07	1.11	1.16	1.20	1.24	1.29
33	0.44	0.49	0.53	0.58	0.62	0.67	0.71	0.76	0.80	0.84
32	0.00	0.04	0.09	0.13	0.18	0.22	0.27	0.31	0.36	0.40
31	– 0.44	– 0.40	– 0.36	– 0.31	– 0.27	– 0.22	– 0.18	– 0.13	– 0.09	– 0.04
30	– 0.89	– 0.84	– 0.80	– 0.76	– 0.71	– 0.67	– 0.62	– 0.58	– 0.53	– 0.49
29	– 1.33	– 1.29	– 1.24	– 1.20	– 1.16	– 1.11	– 1.07	– 1.02	– 0.98	– 0.93
28	– 1.78	– 1.73	– 1.69	– 1.64	– 1.60	– 1.56	– 1.51	– 1.47	– 1.42	– 1.38
27	– 2.22	– 2.18	– 2.13	– 2.09	– 2.04	– 2.00	– 1.96	– 1.91	– 1.87	– 1.82
26	– 2.67	– 2.62	– 2.58	– 2.53	– 2.49	– 2.44	– 2.40	– 2.36	– 2.31	– 2.27
25	– 3.11	– 3.07	– 3.02	– 2.98	– 2.93	– 2.89	– 2.84	– 2.80	– 2.76	– 2.71
24	– 3.56	– 3.51	– 3.47	– 3.42	– 3.38	– 3.33	– 3.29	– 3.24	– 3.20	– 3.16
23	– 4.00	– 3.96	– 3.91	– 3.87	– 3.82	– 3.78	– 3.73	– 3.69	– 3.64	– 3.60
22	– 4.44	– 4.40	– 4.36	– 4.31	– 4.27	– 4.22	– 4.18	– 4.13	– 4.09	– 4.04
21	– 4.89	– 4.84	– 4.80	– 4.76	– 4.71	– 4.67	– 4.62	– 4.58	– 4.53	– 4.49
20	– 5.33	– 5.29	– 5.24	– 5.20	– 5.16	– 5.11	– 5.07	– 5.02	– 4.98	– 4.93
19	– 5.78	– 5.73	– 5.69	– 5.64	– 5.60	– 5.56	– 5.51	– 5.47	– 5.42	– 5.38
18	– 6.22	– 6.18	– 6.13	– 6.09	– 6.04	– 6.00	– 5.96	– 5.91	– 5.87	– 5.82
17	– 6.67	– 6.62	– 6.58	– 6.53	– 6.49	– 6.44	– 6.40	– 6.36	– 6.31	– 6.27
16	– 7.11	– 7.07	– 7.02	– 6.98	– 6.93	– 6.89	– 6.84	– 6.80	– 6.76	– 6.71
15	– 7.56	– 7.51	– 7.47	– 7.42	– 7.38	– 7.33	– 7.29	– 7.24	– 7.20	– 7.16
14	– 8.00	– 7.96	– 7.91	– 7.87	– 7.82	– 7.78	– 7.73	– 7.69	– 7.64	– 7.60
13	– 8.44	– 8.40	– 8.36	– 8.31	– 8.27	– 8.22	– 8.18	– 8.13	– 8.09	– 8.04
12	– 8.89	– 8.84	– 8.80	– 8.76	– 8.71	– 8.67	– 8.62	– 8.58	– 8.53	– 8.49
11	– 9.33	– 9.29	– 9.24	– 9.20	– 9.16	– 9.11	– 9.07	– 9.02	– 8.98	– 8.93
10	– 9.78	– 9.73	– 9.69	– 9.64	– 9.60	– 9.56	– 9.51	– 9.47	– 9.42	– 9.38
9	–10.22	–10.18	–10.13	–10.09	–10.04	–10.00	– 9.96	– 9.91	– 9.87	– 9.82
8	–10.67	–10.62	–10.58	–10.53	–10.49	–10.44	–10.40	–10.36	–10.31	–10.27
7	–11.11	–11.07	–11.02	–10.98	–10.93	–10.89	–10.84	–10.80	–10.76	–10.71
6	–11.56	–11.51	–11.47	–11.42	–11.38	–11.33	–11.29	–11.24	–11.20	–11.16
5	–12.00	–11.96	–11.91	–11.87	–11.82	–11.78	–11.73	–11.69	–11.64	–11.60
4	–12.44	–12.40	–12.36	–12.31	–12.27	–12.22	–12.18	–12.13	–12.09	–12.04
3	–12.89	–12.84	–12.80	–12.76	–12.71	–12.67	–12.62	–12.58	–12.53	–12.49
2	–13.33	–13.29	–13.24	–13.20	–13.16	–13.11	–13.07	–12.02	–12.98	–12.93
1	–13.78	–13.73	–13.69	–13.64	–13.60	–13.56	–13.51	–13.47	–13.42	–13.38
+ 0	–14.22	–14.18	–14.13	–14.09	–14.04	–14.00	–13.96	–13.91	–13.87	–13.82
	0.	**1.**	**2.**	**3.**	**4.**	**5.**	**6.**	**7.**	**8.**	**9.**

Degrees of Fahrenheit.	Tenths of a Degree.									
	0.	**1.**	**2.**	**3.**	**4.**	**5.**	**6.**	**7.**	**8.**	**9.**
	Reaumur.	Reaumur.	Reaumur.	Reaumur.	Reaumur.	Reaumur.	Reaumur.	Reaumur.	Reaumur.	Reaumur.
- 0	-14.22	-14.27	-14.31	-14.36	-14.40	-14.44	-14.49	-14.53	-14.58	-14.62
- 1	-14.67	-14.71	-14.76	-14.80	-14.84	-14.89	-14.93	-14.98	-15.02	-15.07
- 2	-15.11	-15.16	-15.20	-15.24	-15.29	-15.33	-15.38	-15.42	-15.47	-15.51
- 3	-15.56	-15.60	-15.64	-15.69	-15.73	-15.78	-15.82	-15.87	-15.91	-15.96
- 4	-16.00	-16.04	-16.09	-16.13	-16.18	-16.22	-16.27	-16.31	-16.36	-16.40
- 5	-16.44	-16.49	-16.53	-16.58	-16.62	-16.67	-16.71	-16.76	-16.80	-16.84
- 6	-16.89	-17.93	-16.98	-17.02	-17.07	-17.11	-17.16	-17.20	-17.24	-17.29
- 7	-17.33	-17.38	-17.42	-17.47	-17.51	-17.56	-17.60	-17.64	-17.69	-17.73
- 8	-17.78	-18.82	-17.87	-17.91	-17.96	-18.00	-18.04	-18.09	-18.13	-18.18
- 9	-18.22	-18.27	-18.31	-18.36	-18.40	-18.44	-18.49	-18.53	-18.58	-18.62
-10	-18.67	-18.71	-18.76	-18.80	-18.84	-18.89	-18.93	-18.98	-19.02	-19.07
-11	-19.11	-19.16	-19.20	-19.24	-19.29	-19.33	-19.38	-19.42	-19.47	-19.51
-12	-19.56	-19.60	-19.64	-19.69	-19.73	-19.78	-19.82	-19.87	-19.91	-19.96
-13	-20.00	-20.04	-20.09	-20.13	-20.18	-20.22	-20.27	-20.31	-20.36	-20.40
-14	-20.44	-20.49	-20.53	-20.58	-20.62	-20.67	-20.71	-20.76	-20.80	-20.84
-15	-20.89	-20.93	-20.98	-21.02	-21.07	-21.11	-21.16	-21.20	-21.24	-21.29
-16	-21.33	-21.38	-21.42	-21.47	-21.51	-21.56	-21.60	-21.64	-21.69	-21.73
-17	-21.78	-21.82	-21.87	-21.91	-21.96	-22.00	-22.04	-22.09	-22.13	-22.18
-18	-22.22	-22.27	-22.31	-22.36	-22.40	-22.44	-22.49	-22.53	-22.58	-22.62
-19	-22.67	-22.71	-22.76	-22.80	-22.84	-22.89	-22.93	-22.98	-23.02	-23.07
-20	-23.11	-23.16	-23.20	-23.24	-23.29	-23.33	-23.38	-23.42	-23.47	-23.51
-21	-23.56	-23.60	-23.64	-23.69	-23.73	-23.78	-23.82	-23.87	-23.91	-23.96
-22	-24.00	-24.04	-24.09	-24.13	-24.18	-24.22	-24.27	-24.31	-24.36	-24.40
-23	-24.44	-24.49	-24.53	-24.58	-24.62	-24.67	-24.71	-24.76	-24.80	-24.84
-24	-24.89	-24.93	-24.98	-25.02	-25.07	-25.11	-25.16	-25.20	-25.24	-25.29
-25	-25.33	-25.38	-25.42	-25.47	-25.51	-25.56	-25.60	-25.64	-25.69	-25.73
-26	-25.78	-25.82	-25.87	-25.91	-25.96	-26.00	-26.04	-26.09	-26.13	-26.18
-27	-26.22	-26.27	-26.31	-26.36	-26.40	-26.44	-26.49	-26.53	-26.58	-26.62
-28	-26.67	-26.71	-26.76	-26.80	-26.84	-26.89	-26.93	-26.98	-27.02	-27.07
-29	-27.11	-27.16	-27.20	-27.24	-27.29	-27.33	-27.38	-27.42	-27.47	-27.51
-30	-27.56	-27.60	-27.64	-27.69	-27.73	-27.78	-27.82	-27.87	-27.91	-27.96
-31	-28.00	-28.04	-28.09	-28.13	-28.18	-28.22	-28.27	-28.31	-28.36	-28.40
-32	-28.44	-28.49	-28.53	-28.58	-28.62	-28.67	-28.71	-28.76	-28.80	-28.84
-33	-28.89	-28.93	-28.98	-29.02	-29.07	-29.11	-29.16	-29.20	-29.24	-29.29
-34	-29.33	-29.38	-29.42	-29.47	-29.51	-29.56	-29.60	-29.64	-29.69	-29.73
-35	-29.78	-29.82	-29.87	-29.91	-29.96	-30.00	-30.04	-30.09	-30.13	-30·18
-36	-30.22	-30.27	-30.31	-30.36	-30.40	-30.44	-30.49	-30.53	-30.58	-30.62
-37	-30.67	-30.71	-30.76	-30.80	-30.84	-30.89	-30.93	-30.98	-31.02	-31.07
-38	-31.11	--31.16	-31.20	-31.24	-31.29	-31.33	-31.38	-31.42	-31.47	-31.51
-39	-31.56	-31.60	-31.64	-31.69	-31.73	-31.78	-31.82	-31.87	-31.91	-31.96
-40	-32.00	-30.04	-30.09	-30.13	-30.18	-30.22	-30.27	-30.31	-30.36	-30.40
	0.	**1.**	**2.**	**3.**	**4.**	**5.**	**6.**	**7.**	**8.**	**9.**

VI.-VII.

COMPARISON

OF

THE CENTIGRADE THERMOMETER

WITH

THE THERMOMETERS OF FAHRENHEIT AND OF REAUMUR,

OR

TABLES

FOR CONVERTING CENTIGRADE DEGREES INTO DEGREES OF FAHRENHEIT AND OF REAUMUR;

GIVING THE CORRESPONDING VALUES FOR EACH TENTH OF A DEGREE, FROM +50° TO —54° CENTIGRADE.

Centigrade Degrees.	Tenths of Degrees.									
	0.	**1.**	**2.**	**3.**	**4.**	**5.**	**6.**	**7.**	**8.**	**9.**
	Fahren.	Fahren.	Fahren.	Fahren.	Fahren.	Fahren.	Fahren.	Fahren.	Fahren.	Fahren.
+50	+122.00	+122.18	+122.36	+122.54	+122.72	+122.90	+123.08	+123.26	+123.44	+123.62
49	120.20	120.38	120.56	120.74	120.92	121.10	121.28	121.46	121.64	121.82
48	118.40	118.58	118.76	118.94	119.12	119.30	119.48	119.66	119.84	120.02
47	116.60	116.78	116.96	117.14	117.32	117.50	117.68	117.86	118.04	118.22
46	114.80	114.98	115.16	115.34	115.52	115.70	115.88	116.06	116.24	116.42
45	113.00	113.18	113.36	113.54	113.72	113.90	114.08	114.26	114.44	114.62
44	111.20	111.38	111.56	111.74	111.92	112.10	112.28	112.46	112.64	112.82
43	109.40	109.58	109.76	109.94	110.12	110.30	110.48	110.66	110.84	111.02
42	107.60	107.78	107.96	108.14	108.32	108.50	108.68	108.86	109.04	109.22
41	105.80	105.98	106.16	106.34	106.52	106.70	106.88	107.06	107.24	107.42
40	104.00	104.18	104.36	104.54	104.72	104.90	105.08	105.26	105.44	105.62
39	102.20	102.38	102.56	102.74	102.92	103.10	103.28	103.46	103.64	103.82
38	100.40	100.58	100.76	100.94	101.12	101.30	101.48	101.66	101.84	102.02
37	98.60	98.78	98.96	99.14	99.32	99.50	99.68	99.86	100.04	100.22
36	96.80	96.98	97.16	97.34	97.52	97.70	97.88	98.06	98.24	98.42
35	95.00	95.18	95.36	95.54	95.72	95.90	96.08	96.26	96.44	96.62
34	93.20	93.38	93.56	93.74	93.92	94.10	94.28	94.46	94.64	94.82
33	91.40	91.58	91.76	91.94	92.12	92.30	92.48	92.66	92.84	93.02
32	89.60	89.78	89.96	90.14	90.32	90.50	90.68	90.86	91.04	91.22
31	87.80	87.98	88.16	88.34	88.52	88.70	88.88	89.06	89.24	89.42
30	86.00	86.18	86.36	86.54	86.72	86.90	87.08	87.26	87.44	87.62
29	84.20	84.38	84.56	84.74	84.92	85.10	85.28	85.46	85.64	85.82
28	82.40	82.58	82.76	82.94	83.12	83.30	83.48	83.66	83.84	84.02
27	80.60	80.78	80.96	81.14	81.32	81.50	81.68	81.86	82.04	82.22
26	78.80	78.98	79.16	79.34	79.52	79.70	79.88	80.06	80.24	80.42
25	77.00	77.18	77.36	77.54	77.72	77.90	78.08	78.26	78.44	78.62
24	75.20	75.38	75.56	75.74	75.92	76.10	76.28	76.46	76.64	76.82
23	73.40	73.58	73.76	73.94	74.12	74.30	74.48	74.66	74.84	75.02
22	71.60	71.78	71.96	72.14	72.32	72.50	72.68	72.86	73.04	73.22
21	69.80	69.98	70.16	70.34	70.52	70.70	70.88	71.06	71.24	71.42
20	68.00	68.18	68.36	68.54	68.72	68.90	69.08	69.26	69.44	69.62
19	66.20	66.38	66.56	66.74	66.92	67.10	67.28	67.46	67.64	67.82
18	64.40	64.58	64.76	64.94	65.12	65.30	65.48	65.66	65.84	66.02
17	62.60	62.78	62.96	63.14	63.32	63.50	63.68	63.86	64.04	64.22
16	60.80	60.98	61.16	61.34	61.52	61.70	61.88	62.06	62.24	62.42
15	59.00	59.18	59.36	59.54	59.72	59.90	60.08	60.26	60.44	60.62
14	57.20	57.38	57.56	57.74	57.92	58.10	58.28	58.46	58.64	58.82
13	55.40	55.58	55.76	55.94	56.12	56.30	56.48	56.66	56.84	57.02
12	53.60	53.78	53.96	54.14	54.32	54.50	54.68	54.86	55.04	55.22
11	51.80	51.98	52.16	52.34	52.52	52.70	52.88	53.06	53.24	53.42
	0.	**1.**	**2.**	**3.**	**4.**	**5.**	**6.**	**7.**	**8.**	**9.**

Centigrade Degrees.	Tenths of Degrees.									
	0.	**1.**	**2.**	**3.**	**4.**	**5.**	**6.**	**7.**	**8.**	**9.**
	Fahren.	Fahren.	Fahren.	Fahren.	Fahren.	Fahren.	Fahren.	Fahren.	Fahren.	Fahren.
+10	+50.00	+50.18	+50.36	+50.54	+50.72	+50.90	+51.08	+51.26	+51.44	+51.62
9	48.20	48.38	48.56	48.74	48.92	49.10	49.28	49.46	49.64	49.82
8	46.40	46.58	46.76	46.94	47.12	47.30	47.48	47.66	47.84	48.02
7	44.60	44.78	44.96	45.14	45.32	45.50	45.68	45.86	46.04	46.22
6	42.80	42.98	43.16	43.34	43.52	43.70	43.88	44.06	44.24	44.42
5	41.00	41.18	41.36	41.54	41.72	41.90	42.08	42.26	42.44	42.62
4	39.20	39.38	39.56	39.74	39.92	40.10	40.28	40.46	40.64	40.82
3	37.40	37.58	37.76	37.94	38.12	38.30	38.48	38.66	38.84	39.02
2	35.60	35.78	35.96	36.14	36.32	36.50	36.68	36.86	37.04	37.22
1	33.80	33.98	34.16	34.34	34.52	34.70	34.88	35.06	35.24	35.42
0	32.00	32.18	32.36	32.54	32.72	32.90	33.08	33.26	33.44	33.62
– 0	32.00	31.82	31.64	31.46	31.28	31.10	30.92	30.74	30.56	30.38
– 1	30.20	30.02	29.84	29.66	29.48	29.30	29.12	28.94	28.76	28.58
– 2	28.40	28.22	28.04	27.86	27.68	27.50	27.32	27.14	26.96	26.78
– 3	26.60	26.42	26.24	26.06	25.88	25.70	25.52	25.34	25.16	24.98
– 4	24.80	24.62	24.44	24.26	24.08	23.90	23.72	23.54	23.36	23.18
– 5	23.00	22.82	22.64	22.46	22.28	22.10	21.92	21.74	21.56	21.38
– 6	21.20	21.02	20.84	20.66	20.48	20.30	20.12	19.94	19.76	19.58
– 7	19.40	19.22	19.04	18.86	18.68	18.50	18.32	18.14	17.96	17.78
– 8	17.60	17.42	17.24	17.06	16.88	16.70	16.52	16.34	16.16	15.98
– 9	15.80	15.62	15.44	15.26	15.08	14.90	14.72	14.54	14.36	14.18
–10	14.00	13.82	13.64	13.46	13.28	13.10	12.92	12.74	12.56	12.38
–11	12.20	12.02	11.84	11.66	11.48	11.30	11.12	10.94	10.76	10.58
–12	10.40	10.22	10.04	9.86	9.68	9.50	9.32	9.14	8.96	8.78
–13	8.60	8.42	8.24	8.06	7.88	7.70	7.52	7.34	7.16	6.98
–14	6.80	6.62	6.44	6.26	6.08	5.90	5.72	5.54	5.36	5.18
–15	5.00	4.82	4.64	4.46	4.28	4.10	3.92	3.74	3.56	3.38
–16	3.20	3.02	2.84	2.66	2.48	2.30	2.12	1.94	1.76	1.58
–17	1.40	1.22	1.04	0.86	0.68	0.50	0.32	0.14	– 0.04	– 0.22
–18	– 0.40	– 0.58	– 0.76	– 0.94	– 1.12	– 1.30	– 1.48	– 1.66	– 1.84	– 2.02
–19	– 2.20	– 2.38	– 2.56	– 2.74	– 2.92	– 3.10	– 3.28	– 3.46	– 3.64	– 3,82
–20	– 4.00	– 4.18	– 4.36	– 4.54	– 4.72	– 4.90	– 5.08	– 5.26	– 5.44	– 5.62
–21	– 5.80	– 5.98	– 6.16	– 6.34	– 6.52	– 6.70	– 6.88	– 7.06	– 7.24	– 7.42
–22	– 7.60	– 7.78	– 7.96	– 8.14	– 8.32	– 8.50	– 8.68	– 8.86	– 9.04	– 9.22
–23	– 9.40	– 9.58	– 9.76	– 9.94	–10.12	–10.30	–10.48	–10.66	–10.84	–11.02
–24	–11.20	–11.38	–11.56	–11.74	–11.92	–12.10	–12.28	–12.46	–12.64	–12.82
–25	–13.00	–13.18	–13.36	–13.54	–13.72	–13.90	–14.08	–14.26	–14.44	–14.62
–26	–14.80	–14.98	–15.16	–15.34	–15.52	–15.70	–15.88	–16.06	–16.24	–16.42
–27	–16.60	–16.78	–16.96	–17.14	–17.32	–17.50	–17.68	–17.86	–18.04	–18.22
–28	–18.40	–18.58	–18.76	–18.94	–19.12	–19.30	–19.48	–19.66	–19.84	–20.02
–29	–20.20	–20.38	–20.56	–20.74	–20.92	–21.10	–21.28	–21.46	–21.64	–21.82
	0.	**1.**	**2.**	**3.**	**4.**	**5.**	**6.**	**7.**	**8.**	**9.**

Centigrade Degrees.	Tenths of Degrees.									
	0.	1.	2.	3.	4.	5.	6.	7.	8.	9.
	Fahren.	Fahren.	Fahren.	Fahren.	Fahren.	Fahren.	Fahren.	Fahren.	Fahren.	Fahren.
−30	−22.00	−22.18	−22.36	−22.54	−22.72	−22.90	−23.08	−23.26	−23.44	−23.62
−31	−23.80	−23.98	−24.16	−24.34	−24.52	−24.70	−24.88	−25.06	−25.24	−25.42
−32	−25.60	−25.78	−25.96	−26.14	−26.32	−26.50	−26.68	−26.86	−27.04	−27.22
−33	−27.40	−27.58	−27.76	−27.94	−28.12	−28.30	−28.48	−28.66	−28.84	−29.02
−34	−29.20	−29.38	−29.56	−29.74	−29.92	−30.10	−30.28	−30.46	−30.64	−30.82
−35	−31.00	−31.18	−31.36	−31.54	−31.72	−31.90	−32.08	−32.26	−32.44	−32.62
−36	−32.80	−32.98	−33.16	−33.34	−33.52	−33.70	−33.88	−34.06	−34.24	−34.42
−37	−34.60	−34.78	−34.96	−35.14	−35.32	−35.50	−35.68	−35.86	−36.04	−36.22
−38	−36.40	−36.58	−36.76	−36.94	−37.12	−37.30	−37.48	−37.66	−37.84	−38.02
−39	−38.20	−38.38	−38.56	−38.74	−38.92	−39.10	−39.28	−39.46	−39.64	−39.82
−40	−40.00	−40.18	−40.36	−40.54	−40.72	−40.90	−41.08	−41.26	−41.44	−41.62
−41	−41.80	−41.98	−42.16	−42.34	−42.52	−42.70	−42.88	−43.06	−43.24	−43.42
−42	−43.60	−43.78	−43.96	−44.14	−44.32	−44.50	−44.68	−44.86	−45.04	−45.22
−43	−45.40	−45.58	−45.76	−45.94	−46.12	−46.30	−46.48	−46.66	−46.84	−47.02
−44	−47.20	−47.38	−47.56	−47.74	−47.92	−48.10	−48.28	−48.46	−48.64	−48.82
−45	−49.00	−49.18	−49.36	−49.54	−49.72	−49.90	−50.08	−50.26	−50.44	−50.62
−46	−50.80	−50.98	−51.16	−51.34	−51.52	−51.70	−51.88	−52.06	−52.24	−52.42
−47	−52.60	−52.78	−52.96	−53.14	−53.32	−53.50	−53.68	−53.86	−54.04	−54.22
−48	−54.40	−54.58	−54.76	−54.94	−55.12	−55.30	−55.48	−55.66	−55.84	−56.02
−49	−56.20	−56.38	−56.56	−56.74	−56.92	−57.10	−57.28	−57.46	−57.64	−57.82
−50	−58.00	−58.18	−58.36	−58.54	−58.72	−58.90	−59.08	−59.26	−59.44	−59.62
−51	−59.80	−59.98	−60.16	−60.34	−60.52	−60.70	−60.88	−61.06	−61.24	−61.42
−52	−61.60	−61.78	−61.96	−62.14	−62.32	−62.50	−62.68	−62.86	−63.04	−63.22
−53	−63.40	−63.58	−63.76	−63.94	−64.12	−64.30	−64.48	−64.66	−64.84	−65.02
−54	−65.20	−65.38	−65.56	−65.74	−65.92	−66.10	−66.28	−66.46	−66.64	−66.82

TABLE FOR COMPARING THE CENTIGRADE AND FAHRENHEIT'S THERMOMETERS NEAR THE BOILING POINT.

Centigrade Degrees.	0.	1.	2.	3.	4.	5.	6.	7.	8.	9.
	Fahren.	Fahren.	Fahren.	Fahren.	Fahren.	Fahren.	Fahren.	Fahren.	Fahren.	Fahren.
100	212.00	212.18	212.36	212.54	212.72	212.90	213.08	213.26	213.44	213.62
99	210.20	210.38	210.56	210.74	210.92	211.10	211.28	211.46	211.64	211.82
98	208.40	208.58	208.76	208.94	209.12	209.30	209.48	209.66	209.84	210.02
97	206.60	206.78	206.96	207.14	207.32	207.50	207.68	207.86	208.04	208.22
96	204.80	204.98	205.16	205.34	205.52	205.70	205.88	206.06	206.24	206.42
95	203.00	203.18	203.36	203.54	203.72	203.90	204.08	204.26	204.44	204.62
94	201.20	201.38	201.56	201.74	201.92	202.10	202.28	202.46	202.64	202.82
93	199.40	199.58	199.76	199.94	200.12	200.30	200.48	200.66	200.84	201.02
92	197.60	197.78	197.96	198.14	198.32	198.50	198.68	198.86	199.04	199.22
91	195.80	195.98	196.16	196.34	196.52	196.70	196.88	197.06	197.24	197.42
90	194.00	194.18	194.36	194.54	194.72	194.90	195.08	195.26	195.44	195.62
89	192.20	192.38	192.56	192.74	192.92	193.10	193.28	193.46	193.64	193.82

Centigrade Degrees.	Tenths of Degrees.									
	0.	**1.**	**2.**	**3.**	**4.**	**5.**	**6.**	**7.**	**8.**	**9.**
	Reaum.	Reaum.	Reaum.	Reaum.	Reaum.	Reaum.	Reaum.	Reaum.	Reaum.	Reaum.
±40	±32.00	±32.08	±32.16	±32.24	±32.32	±32.40	±32.48	±32.56	±32.64	±32.72
39	31.20	31.28	31.36	31.44	31.52	31.60	31.68	31.76	31.84	31.92
38	30.40	30.48	30.56	30.64	30.72	30.80	30.88	30.96	31.04	31.12
37	29.60	29.68	29.76	29.84	29.92	30.00	30.08	30.16	30.24	30.32
36	28.80	28.88	28.96	29.04	29.12	29.20	29.28	29.36	29.44	29.52
35	28.00	28.08	28.16	28.24	28.32	28.40	28.48	28.56	28.64	28.72
34	27.20	27.28	27.36	27.44	27.52	27.60	27.68	27.76	27.84	27.92
33	26.40	26.48	26.56	26.64	26.72	26.80	26.88	26.96	27.04	27.12
32	25.60	25.68	25.76	25.84	25.92	26.00	26.08	26.16	26.24	26.32
31	24.80	24.88	24.96	25.04	25.12	25.20	25.28	25.36	25.44	25.52
30	24.00	24.08	24.16	24.24	25.32	24.40	24.48	24.56	24.64	24.72
29	23.20	23.28	23.36	23.44	23.52	23.60	23.68	23.76	23.84	23.92
28	22.40	22.48	22.56	22.64	22.72	22.80	22.88	22.96	23.04	23.12
27	21.60	21.68	21.76	21.84	21.92	22.00	22.08	22.16	22.24	22.32
26	20.80	20.88	20.96	21.04	21.12	21.20	21.28	21.36	21.44	21.52
25	20.00	20.08	20.16	20.24	20.32	20.40	20.48	20.56	20.64	20.72
24	19.20	19.28	19.36	19.44	19.52	19.60	19.68	19.76	19.84	19.92
23	18.40	18.48	18.56	18.64	18.72	18.80	18.88	18.96	19.04	19.12
22	17.60	17.68	17.76	17.84	17.92	18.00	18.08	18.16	18.24	18.32
21	16.80	16.88	16.96	17.04	17.12	17.20	17.28	17.36	17.44	17.52
20	16.00	16.08	16.16	16.24	16.32	16.40	16.48	16.56	16.64	16.72
19	15.20	15.28	15.36	15.44	15.52	15.60	15.68	15.76	15.84	15.92
18	14.40	14.48	14.56	14.64	14.72	14.80	14.88	14.96	15.04	15.12
17	13.60	13.68	13.76	13.84	13.92	14.00	14.08	14.16	14.24	14.32
16	12.80	12.88	12.96	13.04	13.12	13.20	13.28	13.36	13.44	13.52
15	12.00	12.08	12.16	12.24	12.32	12.40	12.48	12.56	12.64	12.72
14	11.20	11.28	11.36	11.44	11.52	11.60	11.68	11.76	11.84	11.92
13	10.40	10.48	10.56	10.64	10.72	10.80	10.88	10.96	11.04	11.12
12	9.60	9.68	9.76	9.84	9.92	10.00	10.08	10.16	10.24	10.32
11	8.80	8.88	8.96	9.04	9.12	9.20	9.28	9.36	9.44	9.52
10	8.00	8.08	8.16	8.24	8.32	8.40	8.48	8.56	8.64	8.72
9	7.20	7.28	7.36	7.44	7.52	7.60	7.68	7.76	7.84	7.92
8	6.40	6.48	6.56	6.64	6.72	6.80	6.88	6.96	7.04	7.12
7	5.60	5.68	5.76	5.84	5.92	6.00	6.08	6.16	6.24	6.32
6	4.80	4.88	4.96	5.04	5.12	5.20	5.28	5.36	5.44	5.52
5	4.00	4.08	4.16	4.24	4.32	4.40	4.48	4.56	4.64	4.72
4	3.20	3.28	3.36	3.44	3.52	3.60	3.68	3.76	3.84	3.92
3	2.40	2.48	2.56	2.64	2.72	2.80	2.88	2.96	3.04	3.12
2	1.60	1.68	1.76	1.84	1.92	2.00	2.08	2.16	2.24	2.32
1	0.80	0.88	0.96	1.04	1.12	1.20	1.28	1.36	1.44	1.52
°	°	°	°	°	°	°	°	°	°	°
0	0.00	0.08	0.16	0.24	0.32	0.40	0.48	0.56	0.64	0.72
	0.	**1.**	**2.**	**3.**	**4.**	**5.**	**6.**	**7.**	**8.**	**9.**

VIII.-IX.

COMPARISON

OF

REAUMUR'S THERMOMETER

WITH

THE THERMOMETER OF FAHRENHEIT AND THE CENTIGRADE THERMOMETER,

OR

TABLES

FOR CONVERTING DEGREES OF REAUMUR INTO DEGREES OF FAHRENHEIT AND INTO CENTIGRADE DEGREES;

GIVING THE CORRESPONDING VALUES FOR EACH TENTH OF A DEGREE, FROM +40° TO —40° REAUMUR.

Degrees of Reaumur.	Tenths of Degrees.									
	0.	**1.**	**2.**	**3.**	**4.**	**5.**	**6.**	**7.**	**8.**	**9.**
	Fahren.	Fahren.	Fahren.	Fahren.	Fahren.	Fahren.	Fahren.	Fahren.	Fahren.	Fahren.
+40	+122.00	+122.22	+122.45	+122.67	+122.90	+123.12	+123.35	+123.57	+123.80	+124.02
39	119.75	119.97	120.20	120.42	120.65	120.87	121.10	121.32	121.55	121.77
38	117.50	117.72	117.95	118.17	118.40	118.62	118.85	119.07	119.30	119.52
37	115.25	115.47	115.70	115.92	116.15	116.37	116.60	116.82	117.05	117.27
36	113.00	113.22	113.45	113.67	113.90	114.12	114.35	114.57	114.80	115.02
35	110.75	110.97	111.20	111.42	111.65	111.87	112.10	112.32	112.55	112.77
34	108.50	108.72	108.95	109.17	109.40	109.62	109.85	110.07	110.30	110.52
33	106.25	106.47	106.70	106.92	107.15	107.37	107.60	107.82	108.05	108.27
32	104.00	104.22	104.45	104.67	104.90	105.12	105.35	105.57	105.80	106.02
31	101.75	101.97	102.20	102.42	102.65	102.87	103.10	103.32	103.55	103.77
30	99.50	99.72	99.95	100.17	100.40	100.62	100.85	101.07	101.30	101.52
29	97.25	97.47	97.70	97.92	98.15	98.37	98.60	98.82	99.05	99.27
28	95.00	95.22	95.45	95.67	95.90	96.12	96.35	96.57	96.80	97.02
27	92.75	92.97	93.20	93.42	93.65	93.87	94.10	94.32	94.55	94.77
26	90.50	90.72	90.95	91.17	91.40	91.62	91.85	92.07	92.30	92.52
25	88.25	88.47	88.70	88.92	89.15	89.37	89.60	89.82	90.05	90.27
24	86.00	86.22	86.45	86.67	86.90	87.12	87.35	87.57	87.80	88.02
23	83.75	83.97	84.20	84.42	84.65	84.87	85.10	85.32	85.55	85.77
22	81.50	81.72	81.95	82.17	82.40	82.62	82.85	83.07	83.30	83.52
21	79.25	79.47	79.70	79.92	80.15	80.37	80.60	80.82	81.05	81.27
20	77.00	77.22	77.45	77.67	77.90	78.12	78.35	78.57	78.80	79.02
19	74.75	74.97	75.20	75.42	75.65	75.87	76.10	76.32	76.55	76.77
18	72.50	72.72	72.95	73.17	73.40	73.62	73.85	74.07	74.30	74.52
17	70.25	70.47	70.70	70.92	71.15	71.37	71.60	71.82	72.05	72.27
16	68.00	68.22	68.45	68.67	68.90	69.12	69.35	69.57	69.80	70.02
15	65.75	65.97	66.20	66.42	66.65	66.87	67.10	67.32	67.55	67.77
14	63.50	63.72	63.95	64.17	64.40	64.62	64.85	65.07	65.30	65.52
13	61.25	61.47	61.70	61.92	62.15	62.37	62.60	62.82	63.05	63.27
12	59.00	59.22	59.45	59.67	59.90	60.12	60.35	60.57	60.80	61.02
11	56.75	56.97	57.20	57.42	57.65	57.87	58.10	58.32	58.55	58.77
10	54.50	54.72	54.95	55.17	55.40	55.62	55.85	56.07	56.30	56.52
9	52.25	52.47	52.70	52.92	53.15	53.37	53.60	53.82	54.05	54.27
8	50.00	50.22	50.45	50.67	50.90	51.12	51.35	51.57	51.80	52.02
7	47.75	47.97	48.20	48.42	48.65	48.87	49.10	49.32	49.55	49.77
6	45.50	45.72	45.95	46.17	46.40	46.62	46.85	47.07	47.30	47.52
5	43.25	43.47	43.70	43.92	44.15	44.37	44.60	44.82	45.05	45.27
4	41.00	41.22	41.45	41.67	41.90	42.12	42.35	42.57	42.80	43.02
3	38.75	38.97	39.20	39.42	39.65	39.87	40.10	40.32	40.55	40.77
2	36.50	36.72	36.95	37.17	37.40	37.62	37.85	38.07	38.30	38.52
1	34.25	34.47	34.70	34.92	35.15	35.37	35.60	35.82	36.05	36.27
	0.	**1.**	**2.**	**3.**	**4.**	**5.**	**6.**	**7.**	**8.**	**9.**

Degrees of Reaumur.	Tenths of Degrees.									
	0.	**1.**	**2.**	**3.**	**4.**	**5.**	**6.**	**7.**	**8.**	**9.**
	Fahren.	Fahren.	Fahren.	Fahren.	Fahren.	Fahren.	Fahren.	Fahren.	Fahren.	Fahren.
+ 0	+32.00	+32.22	+32.45	+32.67	+32.90	+33.12	+33.35	+33.57	+33.80	+34.02
− 0	32.00	31.77	31.55	31.32	31.10	30.87	30.65	30.42	30.20	29.97
− 1	29.75	29.52	29.30	29.07	28.85	28.62	28.40	28.17	27.95	27.72
− 2	27.50	27.27	27.05	26.82	26.60	26.37	26.15	25.92	25.70	25.47
− 3	25.25	25.02	24.80	24.57	24.35	24.12	23.90	23.67	23.45	23.22
− 4	23.00	22.77	22.55	22.32	22.10	21.87	21.65	21.42	21.20	20.97
− 5	20.75	20.52	20.30	20.07	19.85	19.62	19.40	19.17	18.95	18.72
− 6	18.50	18.27	18.05	17.82	17.60	17.37	17.15	16.92	16.70	16.47
− 7	16.25	16.02	15.80	15.57	15.35	15.12	14.90	14.67	14.45	14.22
− 8	14.00	13.77	13.55	13.32	13.10	12.87	12.65	12.42	12.20	11.97
− 9	11.75	11.52	11.30	11.07	10.85	10.62	10.40	10.17	9.95	9.72
−10	9.50	9.27	9.05	8.82	8.60	8.37	8.15	7.92	7.70	7.47
−11	7.25	7.02	6.80	6.57	6.35	6.12	5.90	5.67	5.45	5.22
−12	5.00	4.77	4.55	4.32	4.10	3.87	3.65	3.42	3.20	2.97
−13	2.75	2.52	2.30	2.07	1.85	1.62	1.40	1.17	0.95	0.72
−14	0.50	0.27	0.05	− 0.17	− 0.40	− 0.62	− 0.85	− 1.07	− 1.30	− 1.52
−15	− 1.75	− 1.97	− 2.20	− 2.42	− 2.65	− 2.87	− 3.10	− 3.32	− 3.55	− 3.77
−16	− 4.00	− 4.22	− 4.45	− 4.67	− 4.90	− 5.12	− 5.35	− 5.57	− 5.80	− 6.02
−17	− 6.25	− 6.47	− 6.70	− 6.92	− 7.15	− 7.37	− 7.60	− 7.82	− 8.05	− 8.27
−18	− 8.50	− 8.72	− 8.95	− 9.17	− 9.40	− 9.62	− 9.85	−10.07	−10.30	−10.52
−19	−10.75	−10.97	−11.20	−11.42	−11.65	−11.87	−12.10	−12.32	−12.55	−12.77
−20	−13.00	−13.22	−13.45	−13.67	−13.90	−14.12	−14.35	−14.57	−14.80	−15.02
−21	−15.25	−15.47	−15.70	−15.92	−16.15	−16.37	−16.60	−16.82	−17.05	−17.27
−22	−17.50	−17.72	−17.95	−18.17	−18.40	−18.62	−18.85	−19.07	−19.30	−19.52
−23	−19.75	−19.97	−20.20	−20.42	−20.65	−20.87	−21.10	−21.32	−21.55	−21.77
−24	−22.00	−22.22	−22.45	−22.67	−22.90	−23.12	−23.35	−23.57	−23.80	−24.02
−25	−24.25	−24.47	−24.70	−24.92	−25.15	−25.37	−25.60	−25.82	−26.05	−26.27
−26	−26.50	−26.72	−26.95	−27.17	−27.40	−27.62	−27.85	−28.07	−28.30	−28.52
−27	−28.75	−28.97	−29.20	−29.42	−29.65	−29.87	−30.10	−30.32	−30.55	−30.77
−28	−31.00	−31.22	−31.45	−31.67	−31.90	−32.12	−32.35	−32.57	−32.80	−33.02
−29	−33.25	−33.47	−33.70	−33.92	−34.15	−34.37	−34.60	−34.82	−35.05	−35.27
−30	−35.50	−35.72	−35.95	−36.17	−36.40	−36.62	−36.85	−37.07	−37.30	−37.52
−31	−37.75	−37.97	−38.20	−38.42	−38.65	−38.87	−39.10	−39.32	−39.55	−39.77
−32	−40.00	−40.22	−40.45	−40.67	−40.90	−41.12	−41.35	−41.57	−41.80	−42.02
−33	−42.25	−42.47	−42.70	−42.92	−43.15	−43.37	−43.60	−43.82	−44.05	−44.27
−34	−44.50	−44.72	−44.95	−45.17	−45.40	−45.62	−45.85	−46.07	−46.30	−46.52
−35	−46.75	−46.97	−47.20	−47.42	−47.65	−47.87	−48.10	−48.32	−48.55	−48.77
−36	−49.00	−49.22	−49.45	−49.67	−49.90	−50.12	−50.35	−50.57	−50.80	−51.02
−37	−51.25	−51.47	−51.70	−51.92	−52.15	−52.37	−52.60	−52.82	−53.05	−53.27
−38	−53.50	−53.72	−53.95	−54.17	−54.40	−54.62	−54.85	−55.07	−55.30	−55.52
−39	−55.75	−55.97	−56.20	−56.42	−56.65	−56.87	−57.10	−57.32	−57.55	−57.77
	0.	**1.**	**2.**	**3.**	**4.**	**5.**	**6.**	**7.**	**8.**	**9.**

Degrees of Reaumur.	Tenths of Degrees.									
	0.	**1.**	**2.**	**3.**	**4.**	**5.**	**6.**	**7.**	**8.**	**9.**
	Centig.	Centig.	Centig.	Centig.	Centig.	Centig.	Centig.	Centig.	Centig.	Centig.
±40	±50.00	±50.13	±50.25	±50.38	±50.50	±50.63	±50.75	±50.88	±51.00	±51.13
39	48.75	48.88	49.00	49.13	49.25	49.38	49.50	49.63	49.75	49.88
38	47.50	47.63	47.75	47.88	48.00	48.13	48.25	48.38	48.50	48.63
37	46.25	46.38	46.50	46.63	46.75	46.88	47.00	47.13	47.25	47.38
36	45.00	45.13	45.25	45.38	45.50	45.63	45.75	45.88	46.00	46.13
35	43.75	43.88	44.00	44.13	44.25	44.38	44.50	44.63	44.75	44.88
34	42.50	42.63	42.75	42.88	43.00	43.13	43.25	43.38	43.50	43.63
33	41.25	41.38	41.50	41.63	41.75	41.88	42.00	42.13	42.25	42.38
32	40.00	40.13	40.25	40.38	40.50	40.63	40.75	40.88	41.00	41.13
31	38.75	38.88	39.00	39.13	39.25	39.38	39.50	39.63	39.75	39.88
30	37.50	37.63	37.75	37.88	38.00	38.13	38.25	38.38	38.50	38.63
29	36.25	36.38	36.50	36.63	36.75	36.88	37.00	37.13	37.25	37.38
28	35.00	35.13	35.25	35.38	35.50	35.63	35.75	35.88	36.00	36.13
27	33.75	33.88	34.00	34.13	34.25	34.38	34.50	34.63	34.75	34.88
26	32.50	32.63	32.75	32.88	33.00	33.13	33.25	33.38	33.50	33.63
25	31.25	31.38	31.50	31.63	31.75	31.88	32.00	32.13	32.25	32.38
24	30.00	30.13	30.25	30.38	30.50	30.63	30.75	30.88	31.00	31.13
23	28.75	28.88	29.00	29.13	29.25	29.38	29.50	29.63	29.75	29.88
22	27.50	27.63	27.75	27.88	28.00	28.13	28.25	28.38	28.50	28.63
21	26.25	26.38	26.50	26.63	26.75	26.88	27.00	27.13	27.25	27.38
20	25.00	25.13	25.25	25.38	25.50	25.63	25.75	25.88	26.00	26.13
19	23.75	23.88	24.00	24.13	24.25	24.38	24.50	24.63	24.75	24.88
18	22.50	22.63	22.75	22.88	23.00	23.13	23.25	23.38	23.50	23.63
17	21.25	21.38	21.50	21.63	21.75	21.88	22.00	22.13	22.25	22.38
16	20.00	20.13	20.25	20.38	20.50	20.63	20.75	20.88	21.00	21.13
15	18.75	18.88	19.00	19.13	19.25	19.38	19.50	19.63	19.75	19.88
14	17.50	17.63	17.75	17.88	18.00	18.13	18.25	18.38	18.50	18.63
13	16.25	16.38	16.50	16.63	16.75	16.88	17.00	17.13	17.25	17.38
12	15.00	15.13	15.25	15.38	15.50	15.63	15.75	15.88	16.00	16.13
11	13.75	13.88	14.00	14.13	14.25	14.38	14.50	14.63	14.75	14.88
10	12.50	12.63	12.75	12.88	13.00	13.13	13.25	13.38	13.50	13.63
9	11.25	11.38	11.50	11.63	11.75	11.88	12.00	12.13	12.25	12.38
8	10.00	10.13	10.25	10.38	10.50	10.63	10.75	10.88	11.00	11.13
7	8.75	8.88	9.00	9.13	9.25	9.38	9.50	9.63	9.75	9.88
6	7.50	7.63	7.75	7.88	8.00	8.13	8.25	8.38	8.50	8.63
5	6.25	6.38	6.50	6.63	6.75	6.88	7.00	7.13	7.25	7.38
4	5.00	5.13	5.25	5.38	5.50	5.63	5.75	5.88	6.00	6.13
3	3.75	3.88	4.00	4.13	4.25	4.38	4.50	4.63	4.75	4.88
2	2.50	2.63	2.75	2.88	3.00	3.13	3.25	3.38	3.50	3.63
1	1.25	1.38	1.50	1.63	1.75	1.88	2.00	2.13	2.25	2.38
0	0.00	0.13	0.25	0.38	0.50	0.63	0.75	0.88	1.00	1.13
	0.	**1.**	**2.**	**3.**	**4.**	**5.**	**6.**	**7.**	**8.**	**9.**

X.-XV.

TABLES

FOR

COMPARING THERMOMETRICAL DIFFERENCES

EXPRESSED IN DEGREES OF DIFFERENT SCALES,

IRRESPECTIVE OF THEIR ZERO POINT.

X. NUMBER OF DEGREES OF FAHRENHEIT = NUMBER OF CENTIGRADE DEGREES.

4° Reaumur = 5° Centigrade = 9° Fahrenheit.

Degrees of Fahrenheit.	Tenths of a Degree.									
	0.	**1.**	**2.**	**3.**	**4.**	**5.**	**6.**	**7.**	**8.**	**9.**
	Centig.	Centig.	Centig.	Centig.	Centig.	Centig.	Centig.	Centig.	Centig.	Centig.
0	0.00	0.06	0.11	0.17	0.22	0.28	0.33	0.39	0.44	0.50
1	0.56	0.61	0.67	0.72	0.78	0.83	0.89	0.94	1.00	1.06
2	1.11	1.17	1.22	1.28	1.33	1.39	1.44	1.50	1.56	1.61
3	1.67	1.72	1.78	1.83	1.89	1.94	2.00	2.06	2.11	2.22
4	2.22	2.28	2.33	2.39	2.44	2.50	2.56	2.61	2.67	2.72
5	2.78	2.83	2.89	2.94	3.00	3.06	3.11	3.17	3.22	3.28
6	3.33	3.39	3.44	3.50	3.56	3.61	3.67	4.72	3.78	3.83
7	3.89	3.94	4.00	4.06	4.11	4.17	4.22	4.28	4.33	4.39
8	4.44	4.50	4.56	4.61	4.67	4.72	4.78	4.83	4.89	4.94
9	5.00	5.06	5.11	5.17	5.22	5.28	5.32	5.39	5.44	5.50

XI. NUMBER OF DEGREES OF FAHRENHEIT = NUMBER OF DEGREES OF REAUMUR.

Degrees of Fahrenheit.	Tenths of a Degree.									
	0.	**1.**	**2.**	**3.**	**4.**	**5.**	**6.**	**7.**	**8.**	**9.**
	Reaumur.	Reaumur.	Reaumur.	Reaumur.	Reaumur.	Reaumur.	Reaumur.	Reaumur.	Reaumur.	Reaumur.
0	0.00	0.04	0.09	0.13	0.18	0.22	0.27	0.31	0.36	0.40
1	0.44	0.49	0.53	0.58	0.62	0.67	0.71	0.76	0.80	0.84
2	0.89	0.93	0.98	1.02	1.07	1.11	1.16	1.20	1.24	1.29
3	1.33	1.38	1.42	1.47	1.51	1.56	1.60	1.64	1.69	1.73
4	1.78	1.82	1.87	1.91	1.96	2.00	2.04	2.09	2.13	2.18
5	2.22	2.27	2.31	2.36	2.40	2.44	2.49	2.53	2.58	2.62
6	2.67	2.71	2.76	2.80	2.84	2.89	2.93	2.98	3.02	3.07
7	3.11	3.16	3.20	3.24	3.29	3.33	3.38	3.42	3.47	3.51
8	3.56	3.60	3.64	3.69	3.73	3.78	3.82	3.87	3.91	3.96
9	4.00	4.04	4.09	4.13	4.18	4.22	4.27	4.31	4.36	4.40

XII. NUMBER OF CENTIGRADE DEGREES = NUMBER OF DEGREES OF REAUMUR.

Centig. Degrees.	Tenths of a Degree.									
	0.	**1.**	**2.**	**3.**	**4.**	**5.**	**6.**	**7.**	**8.**	**9.**
	Reaumur.	Reaumur.	Reaumur.	Reaumur.	Reaumur.	Reaumur.	Reaumur.	Reaumur.	Reaumur.	Reaumur.
0	0.00	0.08	0.16	0.24	0.32	0.40	0.48	0.56	0.64	0.72
1	0.80	0.88	0.96	1.04	1.12	1.20	1.28	1.36	1.44	1.52
2	1.60	1.68	1.76	1.84	1.92	2.00	2.08	2.16	2.24	2.32
3	2.40	2.48	2.56	2.64	2.72	2.80	2.88	2.96	3.04	3.12
4	3.20	3.28	3.36	3.44	3.52	3.60	3.68	3.76	3.84	3.92
5	4.00	4.08	4.16	4.24	4.32	4.40	4.48	4.56	4.64	4.72
6	4.80	4.88	4.96	5.04	5.12	5.20	5.28	5.36	5.44	5.52
7	5.60	5.68	5.76	5.84	5.92	6.00	6.08	6.16	6.24	6.32
8	6.40	6.48	6.56	6.64	6.72	6.80	6.88	6.96	7.04	7.12
9	7.20	7.28	7.36	7.44	7.52	7.60	7.68	7.76	7.84	7.92

XIII. NUMBER OF CENTIGRADE DEGREES = NUMBER OF DEGREES OF FAHRENHEIT.

4° Reaumur = 5° Centigrade = 9° Fahrenheit.

Centig. Degrees.	Tenths of a Degree.									
	0.	**1.**	**2.**	**3.**	**4.**	**5.**	**6.**	**7.**	**8.**	**9.**
	Fahr.	Fahr.	Fahr.	Fahr.	Fahr.	Fahr.	Fahr.	Fahr.	Fahr.	Fahr.
0	0.00	0.18	0.36	0.54	0.72	0.90	1.08	1.26	1.44	1.62
1	1.80	1.98	2.16	2.34	2.52	2.70	2.88	3.06	3.24	3.42
2	3.60	3.78	3.96	4.14	4.32	4.50	4.68	4.86	5.04	5.22
3	5.40	5.58	5.76	5.94	6.12	6.30	6.48	6.66	6.84	7.02
4	7.20	7.38	7.56	7.74	7.92	8.10	8.28	8.46	8.64	8.82
5	9.00	9.18	9.36	9.54	9.72	9.90	10.08	10.26	10.44	10.62
6	10.80	10.98	11.16	11.34	11.52	11.70	11.88	12.06	12.24	12.42
7	12.60	12.78	12.96	13.14	13.32	13.50	13.68	13.86	14.04	14.22
8	14.40	14.58	14.76	14.94	15.12	15.30	15.48	15.66	15.84	16.02
9	16.20	16.38	16.56	16.74	16.92	17.10	17.28	17.46	17.64	17.82

XIV. NUMBER OF DEGREES OF REAUMUR = NUMBER OF CENTIGRADE DEGREES.

Degrees of Reaum.	Tenths of a Degree.									
	0.	**1.**	**2.**	**3.**	**4.**	**5.**	**6.**	**7.**	**8.**	**9.**
	Centig.	Centig.	Centig.	Centig.	Centig.	Centig.	Centig.	Centig.	Centig.	Centig.
0	0.00	0.12	0.25	0.37	0.50	0.62	0.75	0.87	1.00	1.12
1	1.25	1.37	1.50	1.62	1.75	1.87	2.00	2.12	2.25	2.37
2	2.50	2.62	2.75	2.87	3.00	3.12	3.25	3.37	3.50	3.62
3	3.75	3.87	4.00	4.12	4.25	4.37	4.50	4.62	4.75	4.87
4	5.00	5.12	5.25	5.37	5.50	5.62	5.75	5.87	6.00	6.12
5	6.25	6.37	6.50	6.62	6.75	6.87	7.00	7.12	7.25	7.37
6	7.50	7.62	7.75	7.87	8.00	8.12	8.25	8.37	8.50	8.62
7	8.75	8.87	9.00	9.12	9.25	9.37	9.50	9.62	9.75	9.87
8	10.00	10.12	10.25	10.37	10.50	10.62	10.75	10.87	11.00	11.12
9	11.25	11.37	11.50	11.62	11.75	11.87	12.00	12.12	12.25	12.37

XV. NUMBER OF DEGREES OF REAUMUR = NUMBER OF DEGREES OF FAHRENHEIT.

Degrees of Reaum.	Tenths of a Degree.									
	0.	**1.**	**2.**	**3.**	**4.**	**5.**	**6.**	**7.**	**8.**	**9.**
	Fahr.	Fahr.	Fahr.	Fahr.	Fahr.	Fahr.	Fahr.	Fahr.	Fahr.	Fahr.
0	0.00	0.22	0.45	0.67	0.90	1.12	1.35	1.57	1.80	2.02
1	2.25	2.47	2.70	2.92	3.15	3.37	3.60	3.82	4.05	4.27
2	4.50	4.72	4.95	5.17	5.40	5.62	5.85	6.07	6.30	6.52
3	6.75	6.97	7.20	7.42	7.65	7.87	8.10	8.32	8.55	8.77
4	9.00	9.22	9.45	9.67	9.90	10.12	10.35	10.57	10.80	11.02
5	11.25	11.47	11.70	11.92	12.15	12.37	12.60	12.82	13.05	13.27
6	13.50	13.72	13.95	14.17	14.40	14.62	14.85	15.07	15.30	15.52
7	15.75	15.97	16.20	16.42	16.65	16.87	17.10	17.32	17.55	17.77
8	18.00	18.22	18.45	18.67	18.90	19.12	19.35	19.57	19.80	20.02
9	20.25	20.47	20.70	20.92	21.15	21.37	21.60	21.82	22.05	22.27

METEOROLOGICAL TABLES.

II.

HYGROMETRICAL TABLES.

CONTENTS.

(The figures refer to the folio at the bottom of the page.)

PRACTICAL TABLES BASED ON REGNAULT'S HYGROMETRICAL CONSTANTS.

a. In French Measures.

b. In English Measures.

PRACTICAL TABLES BASED ON THE HYGROMETRICAL CONSTANTS ADOPTED IN THE GREENWICH OBSERVATIONS.

MISCELLANEOUS TABLES FOR COMPARISON.

APPENDIX.

FOR COMPARING QUANTITIES OF RAIN-WATER GIVEN IN DIFFERENT MEASURES.

HYGROMETRICAL TABLES.

HYGROMETERS, or instruments used for determining the amount of aqueous vapor present in the air, are of three classes. In the first, we find the hygrometers based on the absorption of moisture by hygroscopic substances, the best of which is Saussure's Hair-Hygrometer; in the second class, the Psychrometer, or wet-bulb thermometer, which gives the temperature of evaporation; in the third, the various instruments designed for ascertaining the temperature of the dew-point. From the data furnished by each of these instruments, and a table of the elastic forces of vapor at different temperatures, the humidity of the air can be deduced with more or less accuracy.

The use of the hygroscopic substances as hygrometers having been nearly given up on account of the inaccuracy of the results, the variability of the instruments, and the difficulty, if not impossibility, of making them comparable, the psychrometer and the dew-point instruments represent the two methods now usually employed in Meteorology. The following set, therefore, contains extensive tables, in French and English measures, for deducing the hygrometrical condition of the atmosphere from the indications of the Psychrometer and of the dew-point instruments, to which have been added tables of the weight of vapor, in a given space, at different temperatures, — an element often needed in Meteorology.

As, however, the results deduced from the same data furnished by the observations may considerably differ, according to the values of the elastic force of vapor, and the formulæ used in the computation, the tables have been arranged in two series.

The first series contains Regnault's table of the elastic forces of vapor, with tables of the three kinds above mentioned, together with a corresponding set in English measures. Tables V. to X. have been computed for this volume.

The second series gives the table of elastic forces of vapor deduced from Dalton's experiments, and adopted in the Greenwich Observations, together with the various tables based on it.

B

A third series of miscellaneous tables furnishes the means of comparing the different values of the elastic force and weight of vapor determined by various physicists, as well as the results of Saussure's Hair-Hygrometer, with those obtained by other methods.

An Appendix, containing tables for comparing the quantity of rain-water indicated in different measures, closes the set.

Though the first series of tables, based on Regnault's table of tensions, is recommended for ordinary use, as being derived from the determinations which seem to deserve the greatest degree of confidence, it was thought expedient to give also the Greenwich tables, which have been, and still are, so extensively used in England, in order to enable meteorologists to judge of the differences which exist between the results obtained by them and those deduced from the constants of Regnault and others.

PRACTICAL TABLES,

IN

FRENCH MEASURES,

BASED ON REGNAULT'S HYGROMETRICAL CONSTANTS.

TABLE

OF

THE ELASTIC FORCE OF AQUEOUS VAPOR,

EXPRESSED IN MILLIMETRES OF MERCURY FOR CENTIGRADE TEMPERATURES, BY REGNAULT.

This table contains the elastic forces of vapor corresponding to every tenth of a degree of temperature between —35° and +40° Centigrade, as determined by the experiments of V. Regnault, made by order of the French government, for the purpose of establishing the numerical value of the elements which enter into the computations concerning the steam-engine. These results are generally considered as the most accurate science possesses at present. They are published in the *Mémoires de l'Institut*, Tom. XXI.; and more correctly in Regnault's *Etudes sur l'Hygrométrie*, in the *Annales de Chimie et de Physique*. In Vol. XV. Regnault gives the table of elastic forces for every tenth of a degree from —10° to +35° Centigrade, which is reprinted in Table I. The numbers below —10° and above +35°, in the same table, have been taken from another table for every full degree, previously published in Vol. XI. p. 333 of the same periodical, and in the same volume of the *Mémoires de l'Institut*, extending from —32° to +230°.

It should be remarked, however, that the numbers below zero, in the two tables just mentioned, having been computed from different formulas of interpolation, slightly disagree. In order to establish a continuity, therefore, the numbers in Table I. corresponding to full degrees from —10° to —35° have been formed by starting from the value due to —10° in the larger table of Regnault, and subtracting from it the difference between —10° and —11° in the other table, in order to find the value of —11°, and so on, by subtracting successively the corresponding differences to —35°. For the fractions of degrees below —10°, the mean values have been adopted as sufficiently accurate for meteorological purposes.

I. ELASTIC FORCE OF AQUEOUS VAPOR,

EXPRESSED IN MILLIMETRES OF MERCURY FOR CENTIGRADE TEMPERATURES.

By REGNAULT.

Temperature Centigrade.	Tenths of Degrees.									
	0.	**1.**	**2.**	**3.**	**4.**	**5.**	**6.**	**7.**	**8.**	**9.**
°	Millim.	Millim.	Millim.	Millim.	Millim.	Millim.	Millim.	Millim.	Millim.	Millim.
−35	0.221	0.219	0.216	0.214	0.211	0.209	0.207	0.204	0.202	0.199
−34	0.247	0.244	0.242	0.249	0.237	0.234	0.231	0.229	0.226	0.224
−33	0.275	0.272	0.269	0.267	0.264	0.261	0.258	0.255	0.253	0.250
−32	0.305	0.302	0.299	0.296	0.293	0.290	0.287	0.284	0.281	0.278
−31	0.337	0.334	0.331	0.327	0.324	0.321	0.318	0.315	0.311	0.308
−30	0.371	0.368	0.364	0.361	0.357	0.354	0.351	0.347	0.344	0.340
−29	0.409	0.405	0.401	0.398	0.394	0.390	0.386	0.382	0.379	0.375
−28	0.449	0.445	0.441	0.437	0.433	0.429	0.425	0.421	0.417	0.413
−27	0.493	0.489	0.484	0.480	0.475	0.471	0.467	0.462	0.458	0.453
−26	0.540	0.535	0.531	0.526	0.521	0.516	0.512	0.507	0.502	0.498
−25	0.590	0.585	0.580	0.575	0.570	0.565	0.560	0.555	0.550	0.545
−24	0.645	0.639	0.634	0.628	0.623	0.617	0.612	0.606	0.601	0.595
−23	0.704	0.698	0.692	0.686	0.680	0.674	0.669	0.663	0.657	0.651
−22	0.768	0.762	0.755	0.749	0.742	0.736	0.730	0.723	0.717	0.710
−21	0.838	0.831	0.824	0.817	0.810	0.803	0.796	0.789	0.782	0.775
−20	0.912	0.905	0.897	0.890	0.882	0.875	0.868	0.860	0.853	0.845
−19	0.993	0.985	0.977	0.969	0.961	0.952	0.944	0.936	0.928	0.920
−18	1.080	1.071	1.063	1.054	1.045	1.036	1.028	1.019	1.010	1.002
−17	1.174	1.165	1.155	1.146	1.136	1.127	1.118	1.108	1.099	1.089
−16	1.275	1.265	1.255	1.245	1.235	1.224	1.214	1.204	1.194	1.184
−15	1.385	1.374	1.363	1.352	1.341	1.330	1.319	1.308	1.297	1.286
−14	1.503	1.491	1.479	1.468	1.456	1.444	1.432	1.420	1.409	1.397
−13	1.631	1.618	1.605	1.593	1.580	1.567	1.554	1.541	1.529	1.516
−12	1.768	1.754	1.741	1.727	1.713	1.699	1.686	1.672	1.658	1.645
−11	1.918	1.903	1.888	1.873	1.858	1.843	1.828	1.813	1.798	1.783
−10	2.078	2.062	2.046	2.030	2.014	1.998	1.982	1.966	1.950	1.934
− 9	2.261	2.242	2.223	2.204	2.186	2.168	2.150	2.132	2.114	2.096
− 8	2.456	2.436	2.416	2.396	2.376	2.356	2.337	2.318	2.299	2.280
− 7	2.666	2.645	2.624	2.603	2.582	2.561	2.540	2.519	2.498	2.477
− 6	2.890	2.867	2.844	2.821	2.798	2.776	2.754	2.732	2.710	2.688
− 5	3.131	3.106	3.082	3.058	3.034	3.010	2.986	2.962	2.938	2.914
− 4	3.387	3.361	3.335	3.309	3.283	3.257	3.231	3.206	3.181	3.156
− 3	3.662	3.634	3.606	3.578	3.550	3.522	3.495	3.468	3.441	3.414
− 2	3.955	3.925	3.895	3.865	3.836	3.807	3.778	3.749	3.720	3.691
− 1	4.267	4.235	4.203	4.171	4.140	4.109	4.078	4.047	4.016	3.985
− 0	4.600	4.565	4.531	4.497	4 463	4.430	4.397	4.364	4.331	4.299
	0.	**1.**	**2.**	**3.**	**4.**	**5.**	**6.**	**7.**	**8.**	**9.**

Centigrade Degrees.	Tenths of Degrees.									
	0.	**1.**	**2.**	**3.**	**4.**	**5.**	**6.**	**7.**	**8.**	**9.**
°	Millim.	Millim.	Millim.	Millim.	Millim.	Millim.	Millim.	Millim.	Millim.	Millim.
0	4.600	4.633	4.667	4.700	4.733	4.767	4.801	4.836	4.871	4.905
1	4.940	4.975	5.011	5.047	5.082	5.118	5.155	5.191	5.228	5.265
2	5.302	5.340	5.378	5.416	5.454	5.491	5.530	5.569	5.608	5.647
3	5.687	5.727	5.767	5.807	5.848	5.889	5.930	5.972	6.014	6.055
4	6.097	6.140	6.183	6.226	6.270	6.313	6.357	6.401	6.445	6.490
5	6.534	6.580	6.625	6.671	6.717	6.763	6.810	6.857	6.904	6.951
6	6.998	7.047	7.095	7.144	7.193	7.242	7.292	7.342	7.392	7.442
7	7.492	7.544	7.595	7.647	7.699	7.751	7.804	7.857	7.910	7.964
8	8.017	8.072	8.126	8.181	8.236	8.291	8.347	8.404	8.461	8.517
9	8.574	8.632	8.690	8.748	8.807	8.865	8.925	8.985	9.045	9.105
10	9.165	9.227	9.288	9.350	9.412	9.474	9.537	9.601	9.665	9.728
11	9.792	9.857	9.923	9.989	10.054	10.120	10.187	10.255	10.322	10.389
12	10.457	10.526	10.596	10.665	10.734	10.804	10.875	10.947	11.019	11.090
13	11.162	11.235	11.309	11.383	11.456	11.530	11.605	11.681	11.757	11.832
14	11.908	11.986	12.064	12.142	12.220	12.298	12.378	12.458	12.538	12.619
15	12.699	12.781	12.864	12.947	13.029	13.112	13.197	13.281	13.366	13.451
16	13.536	13.623	13.710	13.797	13.885	13.972	14.062	14.151	14.241	14.331
17	14.421	14.513	14.605	14.697	14.790	14.882	14.977	15.072	15.167	15.262
18	15.357	15.454	15.552	15.650	15.747	15.845	15.945	16.045	16.145	16.246
19	16.346	16.449	16.552	16.655	16.758	16.861	16.967	17.073	17.179	17.285
20	17.391	17.500	17.608	17.717	17.826	17.935	18.047	18.159	18.271	18.383
21	18.495	18.610	18.724	18.839	18.954	19.069	19.187	19.305	19.423	19.541
22	19.659	19.780	19.901	20.022	20.143	20.265	20.389	20.514	20.639	20.763
23	20.888	21.016	21.144	21.272	21.400	21.528	21.659	21.790	21.921	22.053
24	22.184	22.319	22.453	22.588	22.723	22.858	22.996	23.135	23.273	23.411
25	23.550	23.692	23.834	23.976	24.119	24.261	24.406	24.552	24.697	24.842
26	24.988	25.138	25.288	25.438	25.588	25.738	25.891	26.045	26.198	26.351
27	26.505	26.663	26.820	26.978	27.136	27.294	27.455	27.617	27.778	27.939
28	28.101	28.267	28.433	28.599	28.765	28.931	29.101	29.271	29.441	29.612
29	29.782	29.956	30.131	30.305	30.479	30.654	30.833	31.011	31.190	31.369
30	31.548	31.729	31.911	32.094	32.278	32.463	32.650	32.837	33.026	33.215
31	33.406	33.596	33.787	33.980	34.174	34.368	34.564	34.761	34.959	35.159
32	35.359	35.559	35.760	35.962	36.165	36.370	36.576	36.783	36.991	37.200
33	37.410	37.621	37.832	38.045	38.258	38.473	38.689	38.906	39.124	39.344
34	39.565	39.786	40.007	40.230	40.455	40.680	40.907	41.135	41.364	41.595
35	41.827	42.059	42.293	42.527	42.763	43.000	43.238	43.477	43.717	43.959
36	44.201	44.445	44.690	44.936	45.183	45.431	45.681	45.932	46.184	46.437
37	46.691	46.947	47.203	47.462	47.721	47.981	48.243	48.506	48.770	49.035
38	49.302	49.570	49.839	50.110	50.382	50.655	50.929	51.205	51.481	51.759
39	52.039	52.320	52.602	52.885	53.170	53.456	53.743	54.032	54.322	54.613
40	54.906	55.200	55.496	55.793	56.091	56.391	56.692	56.994	57.298	57.603
	0.	**1.**	**2.**	**3.**	**4.**	**5.**	**6.**	**7.**	**8.**	**9.**

II.

PSYCHROMETRICAL TABLES.

GIVING IMMEDIATELY THE FORCE OF AQUEOUS VAPOR AND THE RELATIVE HUMIDITY FROM THE INDICATIONS OF THE PSYCHROMETER.

CALCULATED BY M. T. HAEGHENS.

IN his *Etudes sur l'Hygrométrie,** M. V. Regnault discusses the theoretical bases of the formula of the Psychrometer, given by M. August, which was,

$$x = f' - \frac{0.568\,(t - t')}{640 - t'}\,h,$$

in which h represents the height of the barometer; t the temperature of the air given by the dry-bulb thermometer; t' the temperature of the wet-bulb thermometer; f' the force of aqueous vapor in the saturated air at a temperature equal to t'; x the elastic force of aqueous vapor which exists in the air at the time of the observation.

After having modified some of the numerical values, which form the coefficients, M. Regnault adopted this formula,

$$x = f' - \frac{0.429\,(t - t')}{610 - t'}\,h.$$

But comparative experiments, made by himself, showed that by substituting the coefficient 0.480 for that of 0.429, the calculated results, and those obtained by direct observation, agree perfectly in the fractions of saturation, which are greater than 0.40. This formula thus modified, or

$$x = f' - \frac{0.480\,(t - t')}{610 - t'}\,h,$$

has been used for calculating the following tables. In that part of the tables which supposes the wet-bulb to be covered with a film of ice, or below the freezing point, the value $610 - t'$, which represents the latent heat of aqueous vapor, has been changed into this: $610 + 79 - t' = 689 - t'$.

The only hypothesis made, is that of a mean barometric pressure h, equal to 755 millimetres. If we take into account the causes of errors inherent to the psychrometer, and to the tables of the force of vapor, by means of which the absolute force of vapor is calculated, as well as to the differences of these tensions, taken at temperatures differing only by *one* tenth of a degree, it will be obvious that the correction due to the variations of barometric pressure can almost always be neglected. Nevertheless, a separate table has been calculated, giving the *correction* to be applied to the numbers in the Psychrometrical Tables for the heights of the barometer between 650 and 800 millimetres. It will be found at the end of the tables.

The disposition of the tables is the following: —

The temperatures are noted in centigrade degrees; the elastic force of vapor in the air, or its pressure on the barometer, is expressed in millimetres of mercury; the rel-

---

* *Etudes sur l'Hygrométrie,* par M. V. Regnault. *Annales de Chimie et de Physique,* 3me Série, Tom XV., 1845.

ative humidity is indicated in per cent. of the full saturation of the air at the corresponding temperature of the dry-bulb thermometer t.

The first vertical column contains the indications of the wet-bulb thermometer t', beginning with the temperatures below the freezing point, when the bulb is covered with ice, from —35°, and continuing from the freezing point up to +35° centigrade, the bulb being simply wet.

The second column gives the differences of the force of vapor for each tenth (0°.1) of a degree, between each full degree of the first column. It enables the observer to find out the correction for any fraction of a degree of the wet-bulb thermometer.

The following double columns give immediately the force of vapor and the relative humidity, corresponding to each degree of the wet-bulb, placed in the first column, on the same horizontal line, and to differences of the two thermometers, or to $t — t'$, taken at every two tenths of a degree.

The horizontal column at the bottom indicates the mean difference, for each tenth of a degree, of the force of vapor contained in the same horizontal line. It gives the correction for the intermediate differences of the thermometers; 0.1, 0.3, 0.5, 0.7, 0.9, &c., &c.

To meet the wants arising from the extreme climate of North America, the tables of Mr. Haeghens have been extended from —15° to —35° centigrade, and from +30° to +35° of temperature of the wet-bulb, and to +40° of temperature of the dry-bulb thermometer. The forces of aqueous vapor of Regnault, as given in Table I., have been used for the calculations.

Use of the Tables.

Enter the tables with the difference of the two thermometers, or $t — t'$, and with the temperature of the wet-bulb thermometer t', taking the first three pages, when the temperature of the wet-bulb is below the freezing point; and the following ones when it is above the freezing point.

Seek first the column at the head of which you find the difference of the thermometers; go down as far as the horizontal line, at the beginning of which you see the temperature of the wet-bulb thermometer; there you find the force of vapor, and the relative humidity corresponding to your observation.

Two corrections for fractions may be required for a complete calculation of the force of vapor; one for the fractions of degrees of the wet-bulb thermometer; another for the intermediate differences of the two thermometers, viz. for 0.1, 0.3, 0.5, 0.7, &c.

The first correction for fractions of degrees of the wet-bulb thermometer is found by multiplying the decimal fraction by the number placed in the second vertical column next to the whole degree, which number is the value of a tenth of a degree. The product must be *added* to the value of the full degree given in the table, when the temperature of the wet-bulb is above the freezing point: it must be *subtracted* when the temperature is below the freezing point, and receives the sign —. This correction is too important to be neglected.

The second correction, less important, for the intermediate differences of the thermometers, which are greater by one tenth than those indicated in the tables, is given in the horizontal column at the bottom of the page. It is *constant* and always *subtractive.*

Examples of Calculation.

Difference of thermometers, or $t - t'$ $= 0°.8$.
Temperature of the wet-bulb thermometer, $t' = 11°.0$.
We find, page 18, for $t - t'$, fifth double column; and for t', first column,
The force of vapor in the air $= 9^{mm.}.31$.
Relative humidity, $= 90$.

Difference of thermometers, or $t - t'$, $= 7°.2$.
Wet-bulb thermometer, or t', $= 17°.9$.
We find, page 24, for $t - t'$, $= 7°.2$, and $t' = 17°.0$, force of vapor $10^{mm.}.02$.
Additive correction for fraction 0°.9, or $9 \times 0.09 =$ 0 .81.

Force of vapor in the air = 10 .83.
Relative humidity, 46

Difference of thermometers, $t - t' = 6°.5$.
Wet-bulb thermometer, $t' = 23°.6$.
We find, page 23, for $t' = 23°.0$, and $t - t'$, or difference, $= 6°.4$, force of vapor $16^{mm.}.94$; applying immediately the correction found at the bottom of the page for one tenth more difference, or $6°.4 + 0.1 = 6°.5$, we have,
Force of vapor $= 16^{mm.}.94 - 0.06$, or $16^{mm.}.88$.
Additive correction for fraction 0.6 of the wet-bulb, $6 \times 0.13 =$ 0 .78.

Force of vapor in the air = 17 .66.
Relative humidity, 56.

The wet-bulb thermometer covered with ice.

Difference of thermometers, $t - t' = 2°.8$.
Wet-bulb thermometer (ice), $t' = -8°.5$.
Page 17 gives for $t - t' = 2°.8$, and $t' = -8°.0$, force of vapor $= 1^{mm.}.0$.
Subtractive correction for fraction 0.5 of wet-bulb, $5 \times 0.019 =$ —0 .1.

Force of vapor in the air = 0 .9.
Relative humidity, 30.

Below the Freezing-Point; the Bulb covered with a Film of Ice.

Wet-Bulb Thermometer t' Centigrade Degrees.	Mean Vertical Difference for each 0°.1.	$t - t'$, Difference of Wet and Dry Bulb Thermometers.											
		0°.0		0°.2		0°.4		0°.6		0°.8		1°.0	
		Force of Vapor.	Relative Humidity.	Force of Vapor.	Relative Humidity.	Force of Vapor.	Relative Humidity.	Force of Vapor.	Relative Humidity.	Force of Vapor.	Relative Humidity.	Force of Vapor.	Relative Humidity.
°	Millim.	Millim.		Millim.		Millim.		Millim		Millim.		Millim.	
−35		0.22	100	0.12	53								
−34	0.003	0.25	100	0.15	58	0.05	18						
−33	0.003	0.27	100	0.17	62	0.07	26						
−32	0.003	0.30	100	0.20	66	0.10	33						
−31	0.003	0.34	100	0.24	69	0.14	39	0.03	10				
−30	0.004	0.37	100	0.27	71	0.17	44	0.07	17				
−29	0.004	0.41	100	0.31	74	0.21	46	0.11	25				
−28	0.004	0.45	100	0.35	76	0.25	53	0.15	31	0.04	9		
−27	0.004	0.49	100	0.39	78	0.29	57	0.19	36	0.09	17		
−26	0.005	0.54	100	0.44	80	0.34	60	0.24	41	0.13	23	0.03	6
−25	0.005	0.59	100	0.49	81	0.39	63	0.29	46	0.18	29	0.08	12
−24	0.005	0.64	100	0.54	82	0.44	66	0.34	50	0.24	34	0.14	19
−23	0.006	0.70	100	0.60	84	0.50	69	0.40	53	0.30	39	0.19	25
−22	0.006	0.77	100	0.67	85	0.56	71	0.46	57	0.36	44	0.26	31
−21	0.007	0.84	100	0.74	86	0.63	73	0.53	60	0.43	48	0.33	36
−20	0.008	0.91	100	0.81	87	0.71	75	0.61	63	0.50	51	0.40	40
−19	0.008	0.99	100	0.89	88	0.79	77	0.69	66	0.58	55	0.48	45
−18	0.008	1.08	100	0.98	89	0.87	78	0.77	68	0.67	58	0.57	48
−17	0.009	1.17	100	1.07	90	0.97	80	0.87	70	0.76	61	0.66	52
−16	0.010	1.27	100	1.17	90	1.07	81	1.97	72	0.86	63	0.76	55
−15	0.011	1.38	100	1.28	91	1.18	82	1.08	74	0.97	66	0.87	58
−14	0.012	1.50	100	1.40	92	1.30	83	1.19	76	1.09	68	0.99	61
−13	0.013	1.63	100	1.53	92	1.42	84	1.32	77	1.22	70	1.11	63
−12	0.014	1.77	100	1.66	93	1.56	85	1.46	78	1.35	71	1.25	65
−11	0.015	1.92	100	1.81	93	1.71	86	1.61	80	1.50	73	1.40	67
−10	0.016	2.08	100	1.97	94	1.87	87	1.77	81	1.66	75	1.56	69
− 9	0.019	2.26	100	2.16	94	2.05	88	1.95	82	1.85	76	1.74	71
− 8	0.021	2.46	100	2.35	94	2.25	89	2.14	83	2.04	78	1.94	73
− 7	0.023	2.67	100	2.56	94	2.46	89	2.35	84	2.25	79	2.15	74
− 6	0.024	2.89	100	2.79	95	2.68	90	2.58	85	2.47	80	2.37	76
− 5	0.025	3.13	100	3.03	95	2.92	90	2.82	86	2.71	81	2.61	77
− 4	0.028	3.39	100	3.28	95	3.18	91	3.07	87	2.97	82	2.86	78
− 3	0.029	3.66	100	3.56	96	3.45	92	3.35	87	3.24	83	3.14	79
− 2	0.031	3.96	100	3.85	96	3.75	92	3.64	88	3.54	84	3.43	80
− 1	0.033	4.27	100	4.16	96	4.06	92	3.95	89	3.85	85	3.74	81
− 0	0.034	4.60	100	4.50	96	4.40	93	4.29	89	4.19	86	4.08	82

Mean Horizontal Difference of Force of Vapor for each 0°.1 = 0.05 mm.

Below the Freezing-Point; the Bulb covered with a Film of Ice.

Wet-Bulb Thermometer, t′ Centigrade Degrees.	Mean Vertical Difference for each 0°.1.	t — t′, Difference of Wet and Dry Bulb Thermometers.											
		1°.2		1°.4		1°.6		1°.8		2°.0		2°.2	
		Force of Vapor.	Relative Humidity.	Force of Vapor.	Relative Humidity.	Force of Vapor.	Relative Humidity.	Force of Vapor.	Relative Humidity.	Force of Vapor.	Relative Humidity.	Force of Vapor.	Relative Humidity.
°	Millim.	Millim.		Millim.		Millim.		Millim.		Millim.		Millim.	
−35													
−34													
−33													
−32													
−31													
−30													
−29													
−28													
−27													
−26													
−25													
−24	0.006	0.04	5										
−23	0.006	0.09	12										
−22	0.007	0.16	18	0.05	6								
−21	0.007	0.23	24	0.12	13								
−20	0.008	0.30	30	0.20	18	0.09	9						
−19	0.008	0.38	34	0.28	25	0.17	15	0.07	6				
−18	0.009	0.46	39	0.36	30	0.26	21	0.16	13	0.05	4		
−17	0.010	0.56	43	0.46	35	0.35	26	0.25	18	0.15	11	0.04	3
−16	0.011	0.66	47	0.56	39	0.45	31	0.35	24	0.25	16	0.14	9
−15	0.013	0.77	50	0.66	43	0.56	36	0.46	29	0.36	22	0.25	15
−14	0.013	0.88	53	0.78	46	0.68	40	0.58	33	0.47	27	0.37	21
−13	0.015	1.01	56	0.91	50	0.80	43	0.70	37	0.60	31	0.50	25
−12	0.017	1.15	59	1.04	53	0.94	47	0.84	41	0.73	35	0.63	30
−11	0.018	1.30	61	1.19	55	1.09	50	0.99	44	0.88	39	0.78	34
−10	0.019	1.46	63	1.35	58	1.25	52	1.15	47	1.04	42	0.94	38
− 9	0.021	1.64	66	1.53	61	1.43	56	1.33	51	1.22	46	1.12	41
− 8	0.023	1.83	68	1.73	63	1.62	58	1.52	54	1.42	49	1.31	45
− 7	0.024	2.04	69	1.94	65	1.83	61	1.73	56	1.63	52	1.52	48
− 6	0.025	2.26	71	2.16	67	2.06	63	1.95	59	1.85	55	1.74	51
− 5	0.028	2.50	73	2.40	69	2.30	65	2.19	61	2.09	57	1.98	53
− 4	0.029	2.76	74	2.65	70	2.55	67	2.45	63	2.34	59	2.24	55
− 3	0.030	3.03	75	2.93	72	2.82	68	2.72	65	2.61	61	2.51	58
− 2	0.031	3.33	77	3.22	73	3.12	70	3.01	66	2.91	63	2.80	60
− 1		3.64	78	3.53	75	3.43	71	3.32	68	3.22	65	3.11	62

Mean Horizontal Difference of Force of Vapor for each 0°.1 = 0.05 mm.

Below the Freezing-Point; the Bulb covered with a Film of Ice.

Wet-Bulb Thermometer t′ Centigrade Degrees.	Mean Vertical Difference for each 0°.1.	t — t′, Difference of Wet and Dry Bulb Thermometers. 2°.4		2°.6		2°.8		3°.0		3°.2		3°.4	
		Force of Vapor.	Relative Humidity.	Force of Vapor.	Relative Humidity.	Force of Vapor.	Relative Humidity.	Force of Vapor.	Relative Humidity.	Force of Vapor.	Relative Humidity.	Force of Vapor.	Relative Humidity.
°	Millim.	Millim.		Millim.		Millim.		Millim.		Millim.		Millim.	
−15		0.15	9	0.05	3								
−14	0.011	0.27	15	0.16	9	0.06	4						
−13	0.013	0.39	20	0.29	14	0.19	9	0.08	4				
−12	0.013	0.53	25	0.42	19	0.32	14	0.22	10	0.11	5		
−11	0.015	0.68	29	0.57	24	0.47	19	0.36	15	0.26	10	0.16	6
	0.016												
−10		0.83	33	0.73	28	0.63	24	0.52	20	0.42	16	0.32	12
− 9	0.018	1.02	37	0.91	33	0.81	28	0.70	24	0.60	20	0.50	17
− 8	0.019	1.21	40	1.10	36	1.00	32	0.90	28	0.79	25	0.69	21
− 7	0.021	1.42	44	1.31	40	1.21	36	1.11	32	1.00	29	0.90	26
− 6	0.022	1.64	47	1.54	43	1.43	40	1.33	36	1.22	33	1.12	30
	0.024												
− 5		1.88	50	1.77	46	1.67	43	1.57	40	1.46	36	1.36	33
− 4	0.025	2.13	52	2.03	49	1.92	46	1.82	43	1.71	40	1.61	37
− 3	0.027	2.40	55	2.30	52	2.19	48	2.09	45	1.99	43	1.88	40
− 2	0.029	2.70	57	2.59	54	2.49	51	2.38	48	2.28	46	2.17	43
− 1	0.031	3.01	59	2.90	56	2.80	54	2.69	51	2.59	48	2.48	46

Wet-Bulb Thermometer t′	Mean Vertical Difference	3°.6		3°.8		4°.0		4°.2		4°.4		4°.6	
		Millim.		Millim.		Millim.		Millim.		Millim.		Millim.	
−15													
−14													
−13													
−12													
−11		0.05	2										
	0.016												
−10		0.21	8	0.11	4								
− 9	0.018	0.39	13	0.29	9	0.19	6	0.08	3				
− 8	0.019	0.58	18	0.48	14	0.38	11	0.27	8	0.17	5	0.06	2
− 7	0.021	0.79	22	0.69	19	0.59	16	0.48	13	0.38	10	0.27	7
− 6	0.022	1.01	26	0.91	23	0.81	20	0.70	17	0.60	15	0.49	12
	0.024												
− 5		1.25	30	1.15	27	1.04	24	0.94	22	0.83	19	0.73	16
− 4	0.025	1.50	34	1.40	31	1.30	28	1.19	26	1.09	23	0.98	20
− 3	0.027	1.78	37	1.67	34	1.57	32	1.46	29	1.36	27	1.25	24
− 2	0.029	2.07	40	1.96	37	1.86	35	1.75	33	1.65	30	1.54	28
− 1	0.031	2.38	43	2.27	40	2.17	38	2.06	36	1.96	34	1.85	31

Mean Horizontal Difference of Force of Vapor for each 0°.1 = 0.05 mm.

Wet-Bulb Thermometer. t′ Centigrade Degrees.	Mean Vertical Difference for each 0°.1.	t — t′, Difference of Wet and Dry-Bulb Thermometers.											
		0°.0		0°.2		0°.4		0°.6		0°.8		1°.0	
		Force of Vapor.	Relative Humidity.	Force of Vapor.	Relative Humidity.	Force of Vapor.	Relative Humidity.	Force of Vapor.	Relative Humidity.	Force of Vapor.	Relative Humidity.	Force of Vapor.	Relative Humidity.
°	Millim.	Millim.		Millim.		Millim.		Millim.		Millim.		Millim.	
0		4.60	100	4.48	96	4.36	92	4.24	88	4.12	85	4.01	81
	0.03												
1		4.94	100	4.82	96	4.70	93	4.58	89	4.46	85	4.35	82
	0.04												
2		5.30	100	5.18	96	5.06	93	4.94	89	4.83	86	4.71	83
	0.04												
3		5.69	100	5.57	97	5.45	93	5.33	90	5.21	87	5.09	83
	0.04												
4		6.10	100	5.98	97	5.86	93	5.74	90	5.62	87	5.50	84
	0.04												
5		6.53	100	6.41	97	6.29	94	6.17	91	6.05	88	5.94	85
	0.05												
6		7.00	100	6.88	97	6.76	94	6.64	91	6.52	88	6.40	85
	0.05												
7		7.49	100	7.37	97	7.25	94	7.13	91	7.01	89	6.89	86
	0.05												
8		8.02	100	7.90	97	7.78	94	7.66	92	7.54	89	7.42	86
	0.06												
9		8.57	100	8.45	97	8.33	95	8.21	92	8.09	89	7.97	86
	0.06												
10		9.17	100	9.04	97	8.92	95	8.80	93	8.68	90	8.56	87
	0.06												
11		9.79	100	9.67	97	9.55	95	9.43	93	9.31	90	9.19	88
	0.07												
12		10.46	100	10.34	98	10.21	95	10.09	93	9.97	90	9.85	88
	0.07												
13		11.16	100	11.04	98	10.92	95	10.80	93	10.68	91	10.56	89
	0.07												
14		11.91	100	11.79	98	11.66	95	11.54	93	11.42	91	11.30	89
	0.08												
15		12.70	100	12.58	98	12.46	96	12.33	93	12.21	91	12.09	89
	0.08												
16		13.54	100	13.41	98	13.29	96	13.17	94	13.05	92	12.93	90
	0.09												
17		14.42	100	14.30	98	14.18	96	14.05	94	13.93	92	13.81	90
	0.09												
18		15.36	100	15.23	98	15.11	96	14.99	94	14.87	92	14.75	90
	0.10												
19		16.35	100	16.22	98	16.10	96	15.98	94	15.86	92	15.73	91
	0.10												
20		17.39	100	17.27	98	17.15	96	17.02	94	16.90	92	16.78	91
	0.11												
21		18.50	100	18.37	98	18.25	96	18.13	94	18.00	92	17.88	91
	0.12												
22		19.66	100	19.54	98	19.41	96	19.29	95	19.17	93	19.04	91
	0.12												
23		20.89	100	20.76	98	20.64	96	20.52	95	20.39	93	20.27	91
	0.13												
24		22.18	100	22.06	98	21.94	97	21.81	95	21.69	93	21.57	92
	0.14												
25		23.55	100	23.43	98	23.30	97	23.18	95	23.05	93	22.93	92
	0.14												
26		24.99	100	24.86	98	24.74	97	24.62	95	24.49	93	24.37	92
	0.15												
27		26.51	100	26.38	98	26.26	97	26.13	95	26.01	93	25.88	92
	0.16												
28		28.10	100	27.98	98	27.85	97	27.73	95	27.60	93	27.48	92
	0.17												
29		29.78	100	29.66	98	29.53	97	29.41	95	29.28	94	29.16	92
	0.18												
30		31.55	100	31.42	98	31.30	97	31.17	95	30.05	94	30.92	93
	0.19												
31		33.40	100	33.28	98	33.15	97	33.03	96	32.90	94	32.78	93
	0.20												
32		35.36	100	35.23	99	35.11	97	34.98	96	34.86	94	34.73	93
	0.21												
33		37.41	100	37.28	99	37.16	98	37.03	96	36.91	94	36.78	93
	0.22												
34		39.56	100	39.43	99	39.31	98	39.18	96	39.06	94	38.93	93
	0.23												
35		41.83	100	41.70	99	41.58	98	41.45	96	41.33	95	41.20	93

Mean Horizontal Difference of Force of Vapor for each 0°.1 = 0.06 mm.

Wet-Bulb Thermometer. t′ Centigrade Degrees	Mean Vertical Difference for each 0°.1.	t — t, Difference of Wet and Dry-Bulb Thermometers.											
		1°.2		1°.4		1°.6		1°.8		2°.0		2°.2	
		Force of Vapor.	Relative Humidity.	Force of Vapor.	Relative Humidity.	Force of Vapor.	Relative Humidity.	Force of Vapor.	Relative Humidity.	Force of Vapor.	Relative Humidity.	Force of Vapor.	Relative Humidity.
°	Millim.	Millim.		Millim.		Millim.		Millim.		Millim.		Millim.	
0		3.89	78	3.77	74	3.65	71	3.53	67	3.41	64	3.29	61
1	0.03	4.23	79	4.11	75	3.99	72	3.87	69	3.75	66	3.63	63
2	0.04	4.59	80	4.47	76	4.35	73	4.23	70	4.11	67	3.99	65
3	0.04	4.97	80	4.85	77	4.73	74	4.61	71	4.49	69	4.37	66
4	0.04	5.38	81	5.26	78	5.14	75	5.02	73	4.90	70	4.78	67
5	0.04	5.82	82	5.70	79	5.58	77	5.46	74	5.34	71	5.22	69
6	0.05	6.28	83	6.16	80	6.04	77	5.92	75	5.80	72	5.68	70
7	0.05	6.77	83	6.65	81	6.53	78	6.41	76	6.29	73	6.17	71
8	0.05	7.29	84	7.17	81	7.05	79	6.93	76	6.81	74	6.69	72
9	0.06	7.85	84	7.73	82	7.61	80	7.49	77	7.37	75	7.25	73
10	0.06	8.44	85	8.32	83	8.20	80	8.08	78	7.96	76	7.84	74
11	0.06	9.07	86	8.95	83	8.82	81	8.70	79	8.58	77	8.46	75
12	0.07	9.73	86	9.61	84	9.49	82	9.37	80	9.25	78	9.12	76
13	0.07	10.43	86	10.31	84	10.19	82	10.07	80	9.95	78	9.83	76
14	0.08	11.18	87	11.06	85	10.94	83	10.81	81	10.69	79	10.57	77
15	0.08	11.97	87	11.85	85	11.73	83	11.60	81	11.48	80	11.36	78
16	0.08	12.80	88	12.68	86	12.56	84	12.44	82	12.32	80	12.19	78
17	0.09	13.69	88	13.57	86	13.44	84	13.32	83	13.20	81	13.08	79
18	0.09	14.62	88	14.50	87	14.38	85	14.26	83	14.13	81	14.01	80
19	0.10	15.61	89	15.49	87	15.37	85	15.24	83	15.12	82	15.00	80
20	0.11	16.65	89	16.53	87	16.41	86	16.29	84	16.16	82	16.04	81
21	0.11	17.76	89	17.63	88	17.51	86	17.39	84	17.27	83	17.14	81
22	0.12	18.92	90	18.80	88	18.67	86	18.55	85	18.43	83	18.30	82
23	0.12	20.15	90	20.02	88	19.90	87	19.78	85	19.65	83	19.53	82
24	0.13	21.44	90	21.32	88	21.20	87	21.07	85	20.95	84	20.82	82
25	0.14	22.81	90	22.68	89	22.56	87	22.44	86	22.31	84	22.19	83
26	0.14	24.24	90	24.12	89	23.99	87	23.87	86	23.75	85	23.62	83
27	0.15	25.76	91	25.63	89	25.51	88	25.39	86	25.26	85	25.14	83
28	0.16	27.35	91	27.23	89	27.10	88	26.98	87	26.86	85	26.73	84
29	0.17	29.03	91	28.91	90	28.78	88	28.66	87	28.53	85	28.41	84
30	0.18	30.80	91	30.67	90	30.55	89	30.42	87	30.30	86	30.17	84
31	0.19	32.65	91	32.53	90	32.40	89	32.28	87	32.15	86	32.03	85
32	0.20	34.61	91	34.48	90	34.36	89	34.23	88	34.11	86	33.98	85
33	0.21	36.66	92	36.53	90	36.41	89	36.28	88	36.16	86	36.03	85
34	0.22	38.81	92	38.68	90	38.56	89	38.43	88	38.31	87	38.18	85
35	0.23	41.07	92	40.94	91	40.82	89	40.69	88	40.57	87	40.44	86

Mean Horizontal Difference of Force of Vapor for each 0°.1 = 0.06 mm.

Wet-Bulb Thermometer. t′ Centigrade Degrees.	Mean Vertical Difference for each 0°.1.	t — t′, Difference of Wet and Dry-Bulb Thermometers.											
		2°.4		2°.6		2°.8		3°.0		3°.2		3°.4	
		Force of Vapor.	Relative Humidity.	Force of Vapor.	Relative Humidity.	Force of Vapor.	Relative Humidity.	Force of Vapor.	Relative Humidity.	Force of Vapor.	Relative Humidity.	Force of Vapor.	Relative Humidity.
°	Millim.	Millim.		Millim.		Millim.		Millim.		Millim.		Millim.	
0		3.17	58	3.06	55	2.94	52	2.82	50	2.70	47	2.58	44
1	0.03	3.51	60	3.39	57	3.27	54	3.16	52	3.04	49	2.92	47
2	0.04	3.87	62	3.75	59	3.63	56	3.51	54	3.39	51	3.28	49
3	0.04	4.25	63	4.13	61	4.02	58	3.90	56	3.78	53	3.66	51
4	0.04	4.66	65	4.54	62	4.42	60	4.30	57	4.18	55	4.06	53
5	0.04	5.10	66	4.98	64	4.86	61	4.74	59	4.62	57	4.50	55
6	0.05	5.56	67	5.44	65	5.32	63	5.20	61	5.08	58	4.96	56
7	0.05	6.05	69	5.93	66	5.81	64	5.69	62	5.57	60	5.45	58
8	0.05	6.57	70	6.45	68	6.33	65	6.21	63	6.09	61	5.97	59
9	0.06	7.13	71	7.01	69	6.89	67	6.77	65	6.64	63	6.52	61
10	0.06	7.72	72	7.59	70	7.47	68	7.35	66	7.23	64	7.11	62
11	0.06	8.34	73	8.22	71	8.10	69	7.98	67	7.86	65	7.74	63
12	0.07	9.00	74	8.88	72	8.76	70	8.64	68	8.52	66	8.40	64
13	0.07	9.71	75	9.58	73	9.46	71	9.34	69	9.22	67	9.10	66
14	0.07	10.45	75	10.33	73	10.21	72	10.08	70	9.96	68	9.84	67
15	0.08	11.24	76	11.12	74	10.99	72	10.87	71	10.75	69	10.63	67
16	0.08	12.07	77	11.95	75	11.83	73	11.71	72	11.58	70	11.46	68
17	0.09	12.95	77	12.83	76	12.71	74	12.59	72	12.47	71	12.34	69
18	0.09	13.89	78	13.77	76	13.64	75	13.52	73	13.40	72	13.28	70
19	0.10	14.87	78	14.75	77	14.63	75	14.51	74	14.38	72	14.26	71
20	0.10	15.92	79	15.79	77	15.67	76	15.55	74	15.43	73	15.30	72
21	0.11	17.02	80	16.90	78	16.77	77	16.65	75	16.53	74	16.40	72
22	0.12	18.18	80	18.06	79	17.93	77	17.81	76	17.69	74	17.56	73
23	0.12	19.41	80	19.28	79	19.16	78	19.04	76	18.91	75	18.79	73
24	0.13	20.70	81	20.58	79	20.45	78	20.33	77	20.21	75	20.08	74
25	0.14	22.06	81	21.94	80	21.82	79	21.69	77	21.57	76	21.45	75
26	0.14	23.50	82	23.37	80	23.25	79	23.13	78	23.00	77	22.88	75
27	0.15	25.01	82	24.89	81	24.76	79	24.64	78	24.51	77	24.39	76
28	0.16	26.61	83	26.48	81	26.36	80	26.23	79	26.11	77	25.98	76
29	0.17	28.28	83	28.16	81	28.03	80	27.91	79	27.69	77	27.76	76
30	0.18	30.05	83	29.92	82	29.80	81	29.67	79	29.55	78	29.42	77
31	0.19	31.90	83	31.78	82	31.65	81	31.53	80	31.40	78	31.28	77
32	0.20	33.86	84	33.73	82	33.61	81	33.48	80	33.36	79	33.23	78
33	0.21	35.90	84	35.77	83	35.65	81	35.52	80	35.40	79	35.27	78
34	0.22	38.06	84	37.93	83	37.81	82	37.68	81	37.56	80	37.43	78
35	0.23	40.31	84	40.18	83	40.06	82	39.93	81	39.81	80	39.68	79

Mean Horizontal Difference of Force of Vapor for each 0°.1 = 0.06 mm.

Wet-Bulb Thermometer. t′ Centigrade Degrees.	Mean Vertical Difference for each 0°.1.	t — t′, Difference of Wet and Dry-Bulb Thermometers. 3°.6 Force of Vapor.	3°.6 Relative Humidity.	3°.8 Force of Vapor.	3°.8 Relative Humidity.	4°.0 Force of Vapor.	4°.0 Relative Humidity.	4°.2 Force of Vapor.	4°.2 Relative Humidity.	4°.4 Force of Vapor.	4°.4 Relative Humidity.	4°.6 Force of Vapor.	4°.6 Relative Humidity.
°	Millim.	Millim.		Millim.		Millim.		Millim.		Millim.		Millim.	
0		2.46	41	2.34	39	2.22	36	2.11	34	1.99	32	1.87	29
1	0.03	2.80	44	2.68	42	2.56	39	2.44	37	2.32	35	2.20	32
2	0.04	3.16	46	3.04	44	2.92	42	2.80	39	2.68	37	2.56	35
3	0.04	3.54	49	3.42	46	3.30	44	3.18	42	3.06	40	2.94	38
4	0.04	3.94	51	3.82	48	3.71	46	3.59	44	3.47	42	3.35	40
5	0.04	4.38	52	4.26	50	4.14	48	4.02	46	3.90	44	3.78	42
6	0.05	4.84	54	4.72	52	4.60	50	4.48	48	4.36	46	4.24	44
7	0.05	5.33	56	5.21	54	5.09	52	4.97	50	4.85	48	4.73	46
8	0.05	5.85	57	5.73	56	5.61	54	5.49	52	5.37	50	5.25	48
9	0.06	6.40	59	6.28	57	6.16	55	6.04	53	5.92	52	5.80	50
10	0.06	6.99	60	6.87	58	6.75	57	6.63	55	6.51	53	6.39	52
11	0.06	7.61	61	7.49	60	7.37	58	7.25	56	7.13	55	7.01	53
12	0.07	8.28	62	8.15	61	8.03	59	7.91	58	7.79	56	7.67	55
13	0.07	8.98	64	8.85	63	8.73	61	8.61	59	8.49	57	8.37	56
14	0.07	9.72	65	9.60	63	9.48	62	9.35	60	9.23	59	9.11	57
15	0.08	10.51	66	10.38	64	10.26	63	10.14	61	10.02	60	9.90	58
16	0.08	11.34	67	11.22	65	11.10	64	10.97	62	10.85	61	10.73	59
17	0.09	12.22	68	12.10	67	11.98	65	11.85	63	11.73	62	11.61	61
18	0.09	13.15	69	13.03	67	12.91	66	12.79	64	12.66	63	12.54	62
19	0.10	14.14	69	14.02	68	13.89	66	13.77	65	13.65	64	13.53	62
20	0.11	15.18	70	15.06	69	14.94	67	14.81	66	14.69	65	14.57	63
21	0.11	16.28	71	16.16	69	16.04	68	15.91	67	15.79	65	15.67	64
22	0.12	17.44	71	17.32	70	17.20	69	17.07	67	16.95	66	16.83	65
23	0.12	18.67	72	18.54	71	18.42	69	18.30	68	18.17	67	18.05	66
24	0.13	19.96	73	19.84	71	19.71	70	19.59	69	19.46	68	19.34	66
25	0.14	21.32	73	21.20	72	21.07	71	20.95	70	20.83	68	20.70	67
26	0.14	22.75	74	22.63	73	22.50	71	22.38	70	22.26	69	22.13	68
27	0.15	24.27	74	24.14	73	24.02	72	23.89	71	23.77	70	23.64	68
28	0.16	25.86	75	25.73	74	25.61	72	25.48	71	25.36	70	25.24	69
29	0.17	27.44	75	27.31	74	27.29	73	27.16	72	27.04	71	26.91	70
30	0.18	29.30	76	29.17	75	29.05	73	28.92	72	28.80	71	28.67	70
31	0.19	31.15	76	31.03	75	30.90	74	30.78	73	30.65	72	30.53	71
32	0.20	33.10	77	32.97	76	32.85	75	32.72	73	32.60	72	32.47	71
33	0.21	35.15	77	35.02	76	34.90	75	34.77	74	34.65	73	34.52	72
34	0.22	37.30	77	37.17	76	37.05	75	36.92	74	36.80	73	36.67	72
35	0.23	39.56	78	39.43	77	39.31	76	39.18	74	39.06	73	38.93	72

Mean Horizontal Difference of Force of Vapor for each 0°.1 = 0.06 mm.

Wet-Bulb Thermometer. t' Centigrade Degrees.	Mean Vertical Difference for each 0°.1.	t — t', Difference of Wet and Dry-Bulb Thermometers.											
		4°.8		5°.0		5°.2		5°.4		5°.6		5°.8	
		Force of Vapor.	Relative Humidity.	Force of Vapor.	Relative Humidity.	Force of Vapor.	Relative Humidity.	Force of Vapor.	Relative Humidity.	Force of Vapor.	Relative Humidity.	Force of Vapor.	Relative Humidity.
°	Millim.	Millim.		Millim.		Millim.		Millim.		Millim.		Millim.	
0	0.03	1.75	27	1.63	25	1.51	23	1.39	21	1.27	19	1.15	17
1	0.04	2.08	30	1.97	28	1.85	26	1.73	24	1.61	22	1.49	20
2	0.04	2.44	33	2.32	31	2.20	29	2.08	27	1.96	25	1.85	23
3	0.04	2.82	36	2.70	34	2.58	32	2.46	30	2.34	28	2.22	26
4	0.04	3.23	38	3.11	36	2.99	34	2.87	33	2.75	31	2.63	29
5	0.05	3.66	40	3.54	39	3.42	37	3.30	35	3.18	33	3.06	32
6	0.05	4.12	43	4.00	41	3.88	39	3.76	37	3.64	36	3.52	34
7	0.05	4.61	45	4.49	43	4.37	41	4.25	40	4.13	38	4.01	36
8	0.06	5.13	47	5.01	45	4.89	43	4.77	42	4.65	40	4.53	39
9	0.06	5.68	48	5.56	47	5.44	45	5.32	44	5.20	42	5.08	41
10	0.06	6.27	50	6.15	48	6.02	47	5.90	45	5.78	44	5.66	42
11	0.07	6.89	52	6.77	50	6.65	49	6.53	47	6.40	46	6.28	44
12	0.07	7.55	53	7.43	52	7.31	50	7.18	49	7.06	47	6.94	46
13	0.07	8.25	55	8.13	53	8.01	52	7.88	50	7.76	49	7.64	47
14	0.08	8.99	56	8.87	54	8.75	53	8.62	51	8.50	50	8.38	49
15	0.08	9.78	57	9.65	55	9.53	54	9.41	53	9.29	51	9.17	50
16	0.09	10.61	58	10.49	57	10.36	55	10.24	54	10.12	53	10.00	51
17	0.09	11.49	59	11.37	58	11.24	56	11.12	55	11.00	54	10.88	53
18	0.10	12.42	60	12.30	59	12.17	58	12.05	56	11.93	55	11.81	54
19	0.11	13.40	61	13.28	60	13.16	59	13.04	57	12.91	56	12.79	55
20	0.11	14.44	62	14.32	61	14.20	60	14.08	58	13.95	57	13.83	56
21	0.12	15.54	63	15.42	62	15.30	60	15.17	59	15.05	58	14.93	57
22	0.12	16.70	64	16.58	63	16.46	61	16.33	60	16.21	59	16.09	58
23	0.13	17.93	65	17.80	63	17.68	62	17.56	61	17.43	60	17.31	59
24	0.14	19.22	65	19.09	64	18.97	63	18.85	62	18.72	61	18.60	60
25	0.14	20.58	66	20.46	65	20.33	64	20.21	63	20.08	62	19.96	60
26	0.15	22.01	67	21.88	65	21.76	64	21.63	63	21.51	62	21.39	61
27	0.16	23.52	67	23.40	66	23.27	65	23.15	64	23.02	63	22.90	62
28	0.17	25.11	68	24.99	67	24.86	66	24.74	65	24.61	64	24.49	63
29	0.18	26.79	68	26.66	67	26.54	66	26.41	65	26.29	64	26.16	63
30	0.19	28.55	69	28.42	68	28.30	67	28.17	66	28.05	65	27.92	64
31	0.20	30.40	70	30.28	69	30.15	68	30.03	67	29.90	66	29.78	65
32	0.21	32.35	70	32.22	69	32.10	68	31.97	67	31.85	66	31.72	65
33	0.22	34.40	71	34.27	70	34.15	69	34.02	68	33.90	67	33.77	66
34	0.23	36.55	71	36.42	70	36.30	69	36.17	68	36.05	67	35.92	66
35		38.80	71	38.68	70								

Mean Horizontal Difference of Force of Vapor for each 0°.1 = 0.06 mm.

Wet-Bulb Thermometer. t′ Centigrade Degrees.	Mean Vertical Difference for each 0°.1.	t − t′, Difference of Wet and Dry-Bulb Thermometers. 6°.0 Force of Vapor.	6°.0 Relative Humidity.	6°.2 Force of Vapor.	6°.2 Relative Humidity.	6°.4 Force of Vapor.	6°.4 Relative Humidity.	6°.6 Force of Vapor.	6°.6 Relative Humidity.	6°.8 Force of Vapor.	6°.8 Relative Humidity.	7°.0 Force of Vapor.	7°.0 Relative Humidity.
°	Millim.	Millim.		Millim.		Millim.		Millim.		Millim.		Millim.	
0		1.04	15	0.92	13	0.80	11	0.68	9	0.56	8	0.44	6
	0.03												
1		1.37	18	1.25	16	1.13	15	1.01	13	0.89	11	0.78	10
	0.04												
2		1.73	22	1.61	20	1.49	18	1.37	16	1.25	15	1.13	13
	0.04												
3		2.11	25	1.99	23	1.87	21	1.75	19	1.63	18	1.51	16
	0.04												
4		2.51	28	2.39	26	2.27	24	2.15	23	2.03	21	1.91	19
	0.04												
5		2.94	30	2.82	28	2.70	27	2.58	25	2.46	24	2.34	22
	0.05												
6		3.40	33	3.28	31	3.16	29	3.04	28	2.92	26	2.80	25
	0.05												
7		3.89	35	3.77	33	3.65	32	3.53	30	3.41	29	3.29	28
	0.05												
8		4.41	37	4.28	35	4.16	34	4.04	33	3.92	31	3.80	30
	0.06												
9		4.96	39	4.84	38	4.71	36	4.59	35	4.47	33	4.35	32
	0.06												
10		5.54	41	5.42	40	5.30	38	5.18	37	5.06	35	4.94	34
	0.06												
11		6.16	43	6.04	41	5.92	40	5.80	39	5.68	37	5.56	36
	0.07												
12		6.82	44	6.70	43	6.58	42	6.46	41	6.34	39	6.22	38
	0.07												
13		7.52	46	7.40	45	7.28	43	7.16	42	7.03	41	6.91	40
	0.07												
14		8.26	47	8.14	46	8.02	45	7.90	44	7.77	43	7.65	41
	0.08												
15		9.05	49	8.92	48	8.80	46	8.68	45	8.56	44	8.44	43
	0.08												
16		9.88	50	9.75	49	9.63	48	9.51	47	9.39	45	9.27	44
	0.09												
17		10.76	52	10.63	50	10.51	49	10.39	48	10.27	47	10.14	46
	0.09												
18		11.69	53	11.56	51	11.44	50	11.32	49	11.20	48	11.07	47
	0.10												
19		12.67	54	12.55	53	12.42	51	12.30	50	12.18	49	12.06	48
	0.11												
20		13.71	55	13.58	54	13.46	53	13.34	52	13.22	50	13.09	49
	0.11												
21		14.81	56	14.68	55	14.56	54	14.44	53	14.31	52	14.19	51
	0.12												
22		15.96	57	15.84	56	15.72	55	15.59	54	15.47	53	15.35	52
	0.12												
23		17.19	58	17.06	57	16.94	56	16.82	55	16.69	54	16.57	53
	0.13												
24		18.48	59	18.35	58	18.23	56	18.11	55	17.98	54	17.86	53
	0.14												
25		19.84	59	19.71	58	19.59	57	19.46	56	19.34	55	19.22	54
	0.14												
26		21.26	60	21.14	59	21.01	58	20.89	57	20.77	56	20.64	55
	0.15												
27		22.77	61	22.65	60	22.52	59	22.40	58	22.28	57	22.15	56
	0.16												
28		24.36	62	24.24	61	24.11	60	23.99	59	23.86	58	23.74	57
	0.17												
29		26.04	62	25.91	61	25.79	60	25.66	59	25.54	58	25.41	57
	0.18												
30		27.80	63	27.67	62	27.55	61	27.42	60	27.30	59	27.17	58
	0.19												
31		29.65	64	29.53	63	29.40	62	29.28	61	29.15	60	29.03	59
	0.20												
32		31.59	64	31.47	63	31.34	62	31.22	61	31.09	60	30.97	59
	0.21												
33		33.64	65	33.51	64	33.39	63	33.26	62	33.14	61	33.01	60
34													
35													

Mean Horizontal Difference of Force of Vapor for each 0°.1 = 0.06 mm.

Wet-Bulb Thermometer. t′ Centigrade Degrees.	Mean Vertical Difference for each 0°.1.	t — t′, Difference of Wet and Dry-Bulb Thermometers. 7°.2 Force of Vapor.	7°.2 Relative Humidity.	7°.4 Force of Vapor.	7°.4 Relative Humidity.	7°.6 Force of Vapor.	7°.6 Relative Humidity.	7°.8 Force of Vapor.	7°.8 Relative Humidity.	8°.0 Force of Vapor.	8°.0 Relative Humidity.	8°.2 Force of Vapor.	8°.2 Relative Humidity.
°	Millim.	Millim.		Millim.		Millim.		Millim.		Millim.		Millim.	
0		0.32	4	0.20	3	0.09	1						
1	0.03	0.66	8	0.54	7	0.42	5	0.30	4	0.18	2	0.06	1
2	0.04	1.01	12	0.89	10	0.77	9	0.65	7	0.53	6	0.41	4
3	0.04	1.39	15	1.27	13	1.15	12	1.03	11	0.91	9	0.79	8
4	0.04	1.79	18	1.67	16	1.55	15	1.43	14	1.31	13	1.19	11
5	0.04	2.22	21	2.10	19	1.98	18	1.86	17	1.74	16	1.62	14
	0.05												
6		2.78	24	2.66	23	2.44	21	2.32	20	2.20	18	2.08	17
7	0.05	3.16	26	3.04	25	2.92	24	2.80	22	2.68	21	2.56	20
8	0.05	3.68	29	3.56	27	3.44	26	3.32	25	3.20	24	3.08	22
9	0.06	4.23	31	4.11	30	3.99	28	3.87	27	3.75	26	3.63	25
10	0.06	4.82	33	4.70	32	4.57	30	4.45	29	4.33	28	4.21	27
	0.06												
11		5.44	35	5.32	34	5.19	32	5.07	31	4.95	30	4.83	29
12	0.07	6.09	37	5.97	36	5.85	34	5.73	33	5.61	32	5.49	31
13	0.07	6.79	39	6.67	37	6.55	36	6.43	35	6.31	34	6.18	33
14	0.07	7.53	40	7.41	39	7.29	38	7.17	37	7.04	36	6.92	35
15	0.08	8.31	42	8.19	41	8.07	40	7.95	39	7.83	37	7.71	36
	0.08												
16		9.14	43	9.02	42	8.90	41	8.78	40	8.66	39	8.53	38
17	0.09	10.02	45	9.90	44	9.78	43	9.66	42	9.53	40	9.41	39
18	0.09	10.95	46	10.83	45	10.71	44	10.58	43	10.46	42	10.34	41
19	0.10	11.93	47	11.81	46	11.69	45	11.56	44	11.44	43	11.32	42
20	0.10	12.97	48	12.85	47	12.72	46	12.60	45	12.48	44	12.36	43
	0.11												
21		14.07	50	13.94	49	13.82	48	13.70	47	13.58	46	13.45	45
22	0.12	15.22	51	15.10	50	14.98	49	14.85	48	14.73	47	14.61	46
23	0.12	16.45	52	16.32	51	16.20	50	16.08	49	15.95	48	15.83	47
24	0.13	17.73	52	17.61	52	17.49	51	17.36	50	17.24	49	17.12	48
25	0.14	19.09	53	18.97	52	18.85	52	18.72	51	18.60	50	18.47	49
	0.14												
26		20.52	54	20.39	53	20.27	52	20.14	51	20.02	51	19.90	50
27	0.15	22.03	55	21.90	54	21.78	53	21.65	52	21.53	51	21.41	51
28	0.16	23.61	55	23.49	54	23.36	53	23.24	53	23.11	52	22.99	51
29	0.17	25.29	56	25.16	55	25.04	54	24.91	54	24.79	53	24.66	52
30	0.18	27.05	57	26.92	56	26.80	55	26.67	55	26.55	54	26.42	53
	0.19												
31		28.90	58	28.78	57	28.65	56	28.53	55	28.40	55	28.27	54
32	0.20	30.85	59	30.72	58	30.60	57	30.47	56	30.35	56		
33													
34													
35													

Mean Horizontal Difference of Force of Vapor for each 0°.1 = 0.06 mm.

Wet-Bulb Thermometer. t′ Centigrade Degrees.	Mean Vertical Difference for each 0°.1.	t — t′, Difference of Wet and Dry-Bulb Thermometers.											
		8°.4		8°.6		8°.8		9°.0		9°.2		9°.4	
		Force of Vapor.	Relative Humidity.	Force of Vapor.	Relative Humidity.	Force of Vapor.	Relative Humidity.	Force of Vapor.	Relative Humidity.	Force of Vapor.	Relative Humidity.	Force of Vapor.	Relative Humidity.
°	Millim.	Millim.		Millim.		Millim.		Millim.		Millim.		Millim.	
0													
1													
2	0.04	0.30	3	0.18	2	0.06	1						
3	0.04	0.67	7	0.55	5	0.43	4	0.31	3	0.19	2	0.08	1
4	0.04	1.07	10	0.95	9	0.83	8	0.72	6	0.60	5	0.48	4
5	0.05	1.50	13	1.38	12	1.26	11	1.14	10	1.02	8	0.90	7
6	0.05	1.96	16	1.84	15	1.72	14	1.60	13	1.48	12	1.36	10
7	0.05	2.44	19	2.32	17	2.20	16	2.08	15	1.96	14	1.84	13
8	0.06	2.96	21	2.84	20	2.72	19	2.60	18	2.48	17	2.36	16
9	0.06	3.51	24	3.39	23	3.27	21	3.15	20	3.03	19	2.91	18
10	0.06	4.09	26	3.97	25	3.85	24	3.73	23	3.61	22	3.49	21
11	0.07	4.71	28	4.59	27	4.47	26	4.35	25	4.23	24	4.11	23
12	0.07	5.87	30	5.25	29	5.12	28	5.00	27	4.88	26	4.76	25
13	0.07	6.06	32	5.94	31	5.82	30	5.70	29	5.58	28	5.46	27
14	0.08	6.80	34	6.68	33	6.56	32	6.44	31	6.31	30	6.19	29
15	0.08	7.58	35	7.46	34	7.34	33	7.22	33	7.10	32	6.97	31
16	0.09	8.41	37	8.29	36	8.17	35	8.05	34	7.92	33	7.80	32
17	0.09	9.29	39	9.17	38	9.04	37	8.92	36	8.80	35	8.68	34
18	0.10	10.22	40	10.09	39	9.97	38	9.85	37	9.73	36	9.60	35
19	0.11	11.20	41	11.07	40	10.95	39	10.83	39	10.71	38	10.58	37
20	0.11	12.23	43	12.11	42	11.99	41	11.87	40	11.74	39	11.62	38
21	0.12	13.33	44	13.21	43	13.08	42	12.96	41	12.84	40	12.71	40
22	0.12	14.48	45	14.36	44	14.24	43	14.12	42	13.99	41	13.87	41
23	0.13	15.71	46	15.58	45	15.46	44	15.34	43	15.21	42	15.09	42
24	0.14	16.99	47	16.87	46	16.75	45	16.62	44	16.50	44	16.37	43
25	0.14	18.35	48	18.22	47	18.10	46	17.98	45	17.86	45	17.73	44
26	0.15	19.77	49	19.65	48	19.52	47	19.40	46	19.27	46	19.15	45
27	0.16	21.28	50	21.16	49	21.03	48	20.91	47	20.78	47	20.66	46
28	0.17	22.86	51	22.74	50	22.61	49	22.49	48	22.36	47	22.24	47
29	0.18	24.54	51	24.41	51	24.29	50	24.16	49	24.04	48	23.91	47
30	0.19	26.30	52	26.17	51	26.05	51	25.92	50	25.80	49	25.67	48
31		28.16	53	28.03	52	27.91	51	27.78	51				
32													
33													
34													
35													

Mean Horizontal Difference of Force of Vapor for each 0°.1 = 0.06 mm.

Wet-Bulb Thermometer. t′ Centigrade Degrees.	Mean Vertical Difference for each 0°.1.	t — t′, Difference of Wet and Dry-Bulb Thermometers.											
		9°.6		9°.8		10°.0		10°.2		10°.4		10°.6	
		Force of Vapor.	Relative Humidity.	Force of Vapor.	Relative Humidity.	Force of Vapor.	Relative Humidity.	Force of Vapor.	Relative Humidity.	Force of Vapor.	Relative Humidity.	Force of Vapor.	Relative Humidity.
°	Millim.	Millim.		Millim.		Millim.		Millim.		Millim.		Millim.	
0													
1													
2													
3													
4		0.36	3	0.24	2	0.12	1						
	0.04												
5		0.78	6	0.66	5	0.54	4	0.42	3	0.30	2	0.18	1
	0.05												
6		1.24	9	1.12	8	1.00	7	0.88	6	0.76	5	0.64	5
	0.05												
7		1.72	12	1.60	11	1.48	10	1.36	9	1.24	8	1.12	7
	0.05												
8		2.24	15	2.12	14	2.00	13	1.88	12	1.76	11	1.64	10
	0.06												
9		2.79	17	2.66	16	2.54	16	2.42	15	2.30	14	2.18	13
	0.06												
10		3.37	20	3.25	19	3.13	18	3.00	17	2.88	16	2.76	15
	0.06												
11		3.98	22	3.86	21	3.74	20	3.62	19	3.50	18	3.38	18
	0.07												
12		4.64	24	4.52	23	4.40	22	4.28	22	4.15	21	4.03	20
	0.07												
13		5.33	26	5.21	25	5.09	25	4.97	24	4.85	23	4.73	22
	0.07												
14		6.07	28	5.95	27	5.83	26	5.71	25	5.58	25	5.46	24
	0.08												
15		6.85	30	6.73	29	6.61	28	6.49	27	6.37	26	6.24	26
	0.08												
16		7.68	31	7.56	31	7.44	30	7.31	29	7.19	28	7.07	27
	0.09												
17		8.56	33	8.43	32	8.31	31	8.19	31	8.07	30	7.94	29
	0.09												
18		9.48	35	9.36	34	9.24	33	9.11	32	8.99	31	8.87	30
	0.10												
19		10.46	36	10.34	35	10.22	34	10.09	33	9.97	33	9.85	32
	0.11												
20		11.50	37	11.37	36	11.25	36	11.13	35	11.01	34	10.88	33
	0.11												
21		12.59	39	12.47	38	12.35	37	12.22	36	12.10	35	11.98	35
	0.12												
22		13.75	40	13.62	39	13.50	38	13.38	37	13.25	37	13.13	36
	0.12												
23		14.96	41	14.84	40	14.72	39	14.59	39	14.47	38	14.35	37
	0.13												
24		16.25	42	16.13	41	16.00	40	15.88	40	15.76	39	15.63	38
	0.14												
25		17.61	43	17.48	42	17.36	42	17.24	41	17.12	40	16.99	39
	0.14												
26		19.02	44	18.90	43	18.78	42	18.65	42	18.53	41	18.40	40
	0.15												
27		20.54	45	20.41	44	20.29	43	20.16	43	20.04	42	19.91	41
	0.16												
28		22.12	46	22.00	45	21.87	44	21.75	44	21.62	43	21.50	42
	0.17												
29		23.79	47	23.66	46	23.54	45	23.41	45	23.29	44	23.16	43
	0.18												
30		25.55	48	25.42	47	25.30	46						
31													
32													
33													
34													
35													

Mean Horizontal Difference of Force of Vapor for each 0°.1 = 0.06 mm.

Wet-Bulb Thermometer. t′ Centigrade Degrees.	Mean Vertical Difference for each 0°.1.	t — t′, Difference of Wet and Dry-Bulb Thermometers.											
		10°.8		11°.0		11°.2		11°.4		11°.6		11°.8	
		Force of Vapor.	Relative Humidity.	Force of Vapor.	Relative Humidity.	Force of Vapor.	Relative Humidity.	Force of Vapor.	Relative Humidity.	Force of Vapor.	Relative Humidity.	Force of Vapor.	Relative Humidity.
°	Millim.	Millim.		Millim.		Millim.		Millim.		Millim.		Millim.	
0													
1													
2													
3													
4													
5													
6	0.05	0.52	4	0.40	3	0.28	2	0.16	1				
7	0.05	1.00	7	0.88	6	0.76	5	0.64	4	0.52	3	0.40	2
8	0.06	1.52	9	1.40	9	1.27	8	1.15	7	1.03	6	0.91	5
9	0.06	2.06	12	1.94	11	1.82	10	1.70	10	1.58	9	1.46	8
10	0.06	2.64	14	2.52	14	2.40	13	2.28	12	2.16	11	2.04	11
11	0.07	3.26	17	3.14	16	3.02	15	2.90	14	2.77	14	2.65	13
12	0.07	3.91	19	3.79	18	3.67	17	3.55	17	3.43	16	3.31	15
13	0.07	4.61	21	4.49	20	4.36	19	4.24	19	4.12	18	4.00	17
14	0.08	5.34	23	5.22	22	5.10	21	4.98	21	4.86	20	4.73	19
15	0.08	6.12	25	6.00	24	5.88	23	5.76	22	5.63	22	5.51	21
16	0.09	6.95	27	6.83	26	6.70	25	6.58	24	6.46	23	6.34	22
17	0.09	7.82	28	7.70	27	7.58	27	7.46	26	7.33	25	7.21	24
18	0.10	8.75	29	8.63	29	8.50	28	8.38	27	8.26	27	8.14	26
19	0.10	9.73	31	9.60	30	9.48	30	9.36	29	9.24	28	9.11	28
20	0.11	10.76	33	10.64	32	10.51	31	10.39	30	10.27	30	10.15	29
21	0.12	11.85	34	11.73	33	11.61	32	11.48	32	11.36	31	11.24	30
22	0.12	13.01	35	12.88	34	12.76	34	12.64	33	12.51	32	12.39	32
23	0.13	14.22	36	14.10	36	13.98	35	13.85	34	13.73	34	13.61	33
24	0.14	15.51	38	15.39	37	15.27	36	15.15	35	15.02	35	14.90	34
25	0.14	16.87	39	16.74	38	16.62	37	16.49	36	16.37	36	16.24	35
26	0.15	18.28	39	18.16	39	18.03	38	17.91	37	17.78	37	17.66	36
27	0.16	19.79	40	19.67	40	19.54	39	19.42	38	19.29	38	19.17	37
28	0.17	21.37	41	21.25	41	21.12	40	21.00	39	20.87	39	20.75	38
29		23.04	42	22.91	42								
30													
31													
32													
33													
34													
35													

Mean Horizontal Difference of Force of Vapor for each 0°.1 = 0.06 mm.

Wet-Bulb Thermometer. t′ Centigrade Degrees.	Mean Vertical Difference for each 0°.1.	t — t′, Difference of Wet and Dry-Bulb Thermometers.											
		12°.0		12°.2		12°.4		12°.6		12°.8		13°.0	
		Force of Vapor.	Relative Humidity.	Force of Vapor.	Relative Humidity.	Force of Vapor.	Relative Humidity.	Force of Vapor.	Relative Humidity.	Force of Vapor.	Relative Humidity.	Force of Vapor.	Relative Humidity.
°	Millim.	Millim.		Millim.		Millim.		Millim.		Millim.		Millim.	
12		3.19	14	3.06	14	2.94	13	2.82	12	2.70	12	2.58	11
	0.07												
13		3.88	16	3.76	16	3.64	15	3.51	14	3.39	14	3.27	13
	0.07												
14		4.61	18	4.49	18	4.37	17	4.25	16	4.13	16	4.00	15
	0.08												
15		5.39	20	5.27	20	5.15	19	5.03	18	4.90	18	4.78	17
	0.08												
16		6.22	22	6.09	21	5.97	21	5.85	20	5.73	19	5.61	19
	0.09												
17		7.09	24	6.97	23	6.84	22	6.72	22	6.60	21	6.48	21
	0.09												
18		8.01	25	7.89	25	7.77	24	7.65	23	7.52	23	7.40	22
	0.10												
19		8.99	27	8.87	26	8.74	26	8.62	25	8.50	25	8.38	24
	0.10												
20		10.02	28	10.90	28	9.78	27	9.65	26	9.53	26	9.41	25
	0.11												
21		11.12	30	10.99	29	10.87	28	10.75	28	10.62	27	10.50	27
	0.12												
22		12.27	31	12.14	30	12.02	30	11.90	29	11.77	28	11.65	28
	0.12												
23		13.48	32	13.36	31	13.23	31	13.11	30	12.99	29	12.86	29
	0.13												
24		14.78	33	14.65	33	14.53	32	14.40	31	14.28	31	14.16	30
	0.14												
25		16.11	35	15.99	34	15.87	33	15.74	33	15.62	32	15.50	31
	0.14												
26		17.54	36	17.42	35	17.29	34	17.17	34	17.04	33	16.92	33
	0.15												
27		19.04	37	18.92	36	18.80	35	18.67	35	18.55	34	18.42	34
	0.16												
28		20.63	38										

Wet-Bulb Thermometer.	Mean Vertical Difference.	13°.2		13°.4		13°.6		13°.8		14°.0			
		Millim.		Millim.		Millim.		Millim.		Millim.		Millim.	
12		2.46	10	2.34	10	2.22	9	2.09	8	1.97	8		
	0.07												
13		3.15	12	3.03	12	2.91	11	2.79	11	2.66	10		
	0.07												
14		3.88	14	3.76	14	3.64	13	3.52	13	3.40	12		
	0.08												
15		4.66	16	4.54	16	4.42	15	4.29	15	4.17	14		
	0.08												
16		5.48	18	5.36	18	5.24	17	5.12	16	5.00	16		
	0.09												
17		6.36	20	6.23	19	6.11	19	5.99	18	5.87	17		
	0.09												
18		7.28	22	7.16	21	7.03	20	6.91	20	6.79	19		
	0.10												
19		8.25	23	8.13	22	8.01	22	7.89	21	7.76	21		
	0.10												
20		9.29	25	9.16	24	9.04	23	8.92	23	8.80	22		
	0.11												
21		10.38	26	10.25	25	10.13	25	10.01	24	9.89	24		
	0.12												
22		11.53	27	11.40	27	11.28	26	11.16	26	11.03	25		
	0.12												
23		12.74	28	12.62	28	12.49	27	12.37	27	12.25	26		
	0.13												
24		14.02	30	13.90	29	13.77	29	13.65	28	13.53	27		
	0.14												
25		15.37	31	15.25	30	15.12	30	15.00	29	14.88	29		
	0.14												
26		16.80	32	16.67	31	16.55	31	16.42	30	16.30	30		

Mean Horizontal Difference of Force of Vapor for each 0°.1 = 0.06 mm.

Correction for the Barometrical Height.

For the Barometrical Height below. Add.	Subtr'ct.	1°	2°	3°	4°	5°	6°	7°	8°	9°	10°	11°	12°	13°	14°
		Difference of Thermometers $t - t'$. Wet-Bulb above the Freezing Point.													
Millim.	Millim.	Milli.	Milli.	Milli.	Milli.	Milli.	Milli.	Milli.	Milli.	Milli.	Milli.	Milli.	Milli.	Milli.	Milli.
755	755	0.00	0.00	0.00	0.00	0.00	0.00	0.00	0.00	0.00	0.00	0.00	0.00	0.00	0.00
750	760	0.00	0.01	0.01	0.02	0.02	0.02	0.03	0.03	0.04	0.04	0.04	0.05	0.05	0.06
745	765	0.01	0.02	0.02	0.03	0.04	0.05	0.06	0.06	0.07	0.08	0.09	0.10	0.10	0.11
740	770	0.01	0.02	0.04	0.05	0.06	0.07	0.08	0.10	0.11	0.12	0.13	0.14	0.16	0.17
735	775	0.02	0.03	0.05	0.06	0.08	0.10	0.11	0.13	0.14	0.16	0.18	0.19	0.21	0.22
730	780	0.02	0.04	0.06	0.08	0.10	0.12	0.14	0.16	0.18	0.20	0.22	0.24	0.26	0.28
725	785	0.02	0.05	0.07	0.10	0.12	0.14	0.17	0.19	0.22	0.24	0.26	0.29	0.31	0.34
720	790	0.03	0.06	0.08	0.11	0.14	0.17	0.20	0.22	0.25	0.28	0.31	0.34	0.36	0.39
715	795	0.03	0.06	0.10	0.13	0.16	0.19	0.22	0.26	0.29	0.32	0.35	0.38	0.42	0.45
710	800	0.04	0.07	0.11	0.14	0.18	0.22	0.25	0.29	0.32	0.36	0.40	0.43	0.47	0.50
700	"	0.04	0.09	0.13	0.18	0.22	0.26	0.31	0.35	0.40	0.44	0.48	0.53	0.57	0.62
690	"	0.05	0.10	0.16	0.21	0.26	0.31	0.36	0.42	0.47	0.52	0.57	0.62	0.68	0.73
680	"	0.06	0.12	0.18	0.24	0.30	0.36	0.42	0.48	0.54	0.60	0.66	0.72	0.78	0.84
670	"	0.07	0.14	0.20	0.27	0.34	0.41	0.48	0.54	0.61	0.68	0.75	0.82	0.88	0.95
660	"	0.08	0.15	0.23	0.30	0.38	0.46	0.53	0.61	0.68	0.76	0.84	0.91	0.99	1.06
650	"	0.08	0.17	0.25	0.34	0.42	0.50	0.59	0.67	0.76	0.84	0.92	1.01	1.09	1.18

Add.	Subtr'ct.	1°	2°	3°	4°	5°
		Wet-bulb below the Freezing Point.				
755	755	0.00	0.00	0.00	0.00	0.00
750	760	0.00	0.01	0.01	0.01	0.02
745	765	0.01	0.01	0.02	0.03	0.04
740	770	0.01	0.02	0.03	0.04	0.05
735	775	0.01	0.03	0.04	0.06	0.07
730	780	0.02	0.04	0.05	0.07	0.09
725	785	0.02	0.04	0.06	0.08	0.11
720	790	0.02	0.05	0.07	0.10	0.12
715	795	0.03	0.06	0.08	0.11	0.14
710	800	0.03	0.06	0.09	0.13	0.16
700	"	0.04	0.08	0.12	0.15	0.19
690	"	0.05	0.09	0.14	0.18	0.23
680	"	0.05	0.11	0.16	0.21	0.26
670	"	0.06	0.12	0.18	0.24	0.30
660	"	0.07	0.13	0.20	0.27	0.33
650	"	0.07	0.15	0.22	0.29	0.36

EXAMPLE OF CALCULATION.

Wet-bulb above the Freezing Point.

$t' = 17^{\circ}.0$. $t - t' = 8^{\circ}.2$. $h = 710^{mm.}$

The tables give for mean barometrical height $755^{mm.}$ Force of vapor . . = 9.41 mm.

Additive correction for $710^{mm.}$ and $8^{\circ}.2$ = 0.30

Force of vapor . . = 9.71

The mean barometrical pressure, at a given place, being known, it is easy to make the above Psychrometrical Tables fitted for that place, by determining, by means of this last table, a *constant correction*, to be applied to the numbers in the tables, giving the force of vapor. This correction will be found by taking for $t - t'$, or the difference of thermometers, a mean value, the deviations of which will have little influence upon the accuracy of the results.

III.

TABLE

GIVING AT SIGHT THE RELATIVE HUMIDITY DEDUCED FROM THE INDICATIONS OF THE DEW POINT INSTRUMENTS.

BY M. T. HAEGHENS.

THIS table, which has been published in the *Annuaire Météorologique de France* for 1850, page 86, and following, has been calculated by Mr. Haeghens, using Regnault's Tables of Elastic Forces of Vapor. It gives directly the *relative humidity*, when the hygrometrical observations have been made by means of dew point instruments like those of Daniell, Regnault, Bache, and others.

These hygrometers are destined to find out the temperature of *the dew point*, that is the temperature to which it would be necessary to lower the temperature of the air, in order that this air be completely saturated by the aqueous vapor which it contained at the time of the observation.

The force of vapor contained in the air, or its *absolute humidity*, is thus the maximum of force of vapor which corresponds to the temperature of the dew point; it is given directly in the Table I. of the Elastic Forces of Vapor, by Regnault.

The ratio of that maximum of force of vapor at the temperature of the dew point to the force of vapor which corresponds, in the same table, to the temperature of the surrounding air at the time of the observation, is the *relative humidity*. This ratio is given in hundredths in the following table, which relieves the observer of the trouble of calculating it.

Let t = temperature of the air surrounding the instrument.

t' = temperature of the dew point.

$t - t'$ = the difference between these two temperatures.

The first column, on the left, contains the temperature of the air t, in centigrade degrees. The following ones, headed with the differences, $t - t'$, between the temperatures of the air and of the dew point, give the *relative humidity* corresponding to the two elements.

	Temp. of the Air = t.	Dew point = t'.	Difference $t - t'$.	Relative Humidity.
Example:	10°.0	4°.4	5°.6	68

Should the temperature of the air t', or the difference $t - t'$, fall between the numbers found in the columns, it is obvious, by glancing at the table, that an interpolation at sight will always be easy.

B

Temperature of the air. t=	t — t′ = Difference of Temperatures of the Dew Point and of the Air.														
	0°.0	0°.2	0°.4	0°.6	0°.8	1°.0	1°.2	1°.4	1°.6	1°.8	2°.0	2°.2	2°.4	2°.6	2°.8
Centig. −8	100	98	97	95	94	92	90	89	88	86	85	83	82	80	79
−7	100	98	97	95	94	92	91	89	88	86	85	83	82	81	79
−6	100	98	97	95	94	92	91	89	88	87	85	84	82	81	80
−5	100	98	97	95	94	92	91	89	88	87	85	84	82	81	80
−4	100	98	97	95	94	92	91	89	88	87	85	84	83	81	80
−3	100	98	97	95	94	92	91	90	88	87	85	84	83	81	80
−2	100	98	97	95	94	93	91	90	88	87	86	84	83	82	80
−1	100	98	97	95	94	93	91	90	89	87	86	85	83	82	81
0	100	98	97	96	94	93	91	90	89	87	86	85	83	82	81
+1	100	99	97	96	95	93	92	90	89	88	86	85	84	83	81
2	100	99	97	96	95	93	92	91	89	88	87	85	84	83	82
3	100	99	97	96	95	93	92	91	89	88	87	86	84	83	82
4	100	99	97	96	95	93	92	91	89	88	87	86	85	83	82
5	100	99	97	96	95	93	92	91	90	88	87	86	85	83	82
6	100	99	97	96	95	93	92	91	90	88	87	86	85	84	82
7	100	99	97	96	95	93	92	91	90	89	87	86	85	84	83
8	100	99	97	96	95	93	92	91	90	89	87	86	85	84	83
9	100	99	97	96	95	94	92	91	90	89	87	86	85	84	83
10	100	99	97	96	95	94	92	91	90	89	87	86	85	84	83
11	100	99	97	96	95	94	92	91	90	89	87	86	85	84	83
12	100	99	97	96	95	94	92	91	90	89	88	87	85	84	83
13	100	99	97	96	95	94	92	91	90	89	88	87	85	84	83
14	100	99	98	96	95	94	93	91	90	89	88	87	86	84	83
15	100	99	98	96	95	94	93	91	90	89	88	87	86	84	83
16	100	99	98	96	95	94	93	91	90	89	88	87	86	85	84
17	100	99	98	96	95	94	93	91	90	89	88	87	86	85	84
18	100	99	98	96	95	94	93	92	90	89	88	87	86	85	84
19	100	99	98	96	95	94	93	92	91	89	88	87	86	85	84
20	100	99	98	96	95	94	93	92	91	89	88	87	86	85	84
21	100	99	98	96	95	94	93	92	91	90	88	87	86	85	84
22	100	99	98	96	95	94	93	92	91	90	89	87	86	85	84
23	100	99	98	96	95	94	93	92	91	90	89	88	86	85	84
24	100	99	98	97	95	94	93	92	91	90	89	88	87	85	84
25	100	99	98	97	95	94	93	92	91	90	89	88	87	86	85
26	100	99	98	97	95	94	93	92	91	90	89	88	87	86	85
27	100	99	98	97	95	94	93	92	91	90	89	88	87	86	85
28	100	99	98	97	95	94	93	92	91	90	89	88	87	86	85
29	100	99	98	97	96	94	93	92	91	90	89	88	87	86	85
30	100	99	98	97	96	94	93	92	91	90	89	88	87	86	85
31	100	99	98	97	96	94	93	92	91	90	89	88	87	86	85
32	100	99	98	97	96	94	93	92	91	90	89	88	87	86	85
33	100	99	98	97	96	94	93	92	91	90	89	88	87	86	85
34	100	99	98	97	96	95	93	92	91	90	89	88	87	86	85
35	100	99	98	97	96	95	93	92	91	90	89	88	87	86	85

Temperature of the air. t=	t — t′ = Difference of Temperatures of the Dew Point and of the Air.														
	3°.0	3°.2	3.°4	3°.6	3°.8	4°.0	4°.2	4°.4	4°.6	4°.8	5°.0	5°.2	5°.4	5°.6	5°.8
Centig.															
−8	78	77	75	74	73	72	71	69	68	67	66	65	64	63	62
−7	78	77	75	74	73	72	71	69	68	67	66	65	64	63	62
−6	78	77	76	74	73	72	71	69	68	67	66	65	64	63	62
−5	79	77	76	75	73	72	71	70	68	67	66	65	64	63	62
−4	79	77	76	75	74	73	71	70	69	68	67	66	64	63	62
−3	79	77	76	75	74	73	72	70	69	68	67	66	65	64	63
−2	79	78	77	76	74	73	72	71	70	69	68	66	65	64	63
−1	79	78	77	76	75	73	72	71	70	69	68	67	66	65	64
0	80	78	77	76	75	74	73	71	70	69	68	67	66	65	64
+1	80	79	78	77	75	74	73	72	71	70	69	68	66	65	64
2	81	79	78	77	76	75	74	72	71	70	69	68	67	66	65
3	81	80	78	77	76	75	74	73	72	71	70	69	68	66	65
4	81	80	79	78	77	75	74	73	72	71	70	69	68	67	66
5	81	80	79	78	77	76	75	73	72	71	70	69	68	67	66
6	81	80	79	78	77	76	75	74	73	72	71	70	69	68	67
7	81	80	79	78	77	76	75	74	73	72	71	70	69	68	67
8	81	80	79	78	77	76	75	74	73	72	71	70	69	68	67
8	82	80	79	78	77	76	75	74	73	72	71	70	69	68	67
10	82	81	80	78	77	76	75	74	73	72	71	70	69	68	67
11	82	81	80	79	78	76	75	74	73	72	71	70	70	69	68
12	82	81	80	79	78	77	76	75	74	73	72	71	70	69	68
13	82	81	80	79	78	77	76	75	74	73	72	71	70	69	68
14	82	81	80	79	78	77	76	75	74	73	72	71	70	69	68
15	82	81	80	79	78	77	76	75	74	73	72	71	70	69	68
16	82	81	80	79	78	77	76	75	74	73	72	71	71	70	69
17	83	81	80	79	78	77	76	75	74	73	73	72	71	70	69
18	83	82	81	80	79	78	77	76	75	74	73	72	71	70	69
19	83	82	81	80	79	78	77	76	75	74	73	72	71	70	69
20	83	82	81	80	79	78	77	76	75	74	73	72	71	70	69
21	83	82	81	80	79	78	77	76	75	74	73	72	71	70	70
22	83	82	81	80	79	78	77	76	75	74	73	73	72	71	70
23	83	82	81	80	79	78	77	76	75	74	74	73	72	71	70
24	83	82	81	80	79	78	77	77	76	75	74	73	72	71	70
25	84	83	82	81	80	79	78	77	76	75	74	73	72	71	70
26	84	83	82	81	80	79	78	77	76	75	74	73	72	71	70
27	84	83	82	81	80	79	78	77	76	75	74	73	72	71	70
28	84	83	82	81	80	79	78	77	76	75	74	73	72	71	70
29	84	83	82	81	80	79	78	77	76	75	75	74	73	72	71
30	84	83	82	81	80	79	78	77	76	76	75	74	73	72	71
31	84	83	82	81	80	79	78	77	77	76	75	74	73	72	71
32	84	83	82	81	80	79	79	78	77	76	75	74	73	72	72
33	84	83	82	81	80	80	79	78	77	76	75	74	73	72	72
34	85	84	83	82	81	80	79	78	77	76	75	74	74	73	72
35	85	84	83	82	81	80	79	78	77	76	75	75	74	73	72

Temperature of the air. t=	t — t' = Difference of Temperatures of the Dew Point and of the Air.														
	6°0	6°.2	6°.4	6°.6	6°.8	7°.0	7°.2	7°.4	7°.6	7°.8	8°.0	8°.2	8°.4	8°.6	8°.8
Centig.															
−8															
−7															
−6	61	60	59	58	57	56									
−5	61	60	59	58	58	57	56	55	54	53	52				
−4	62	61	60	59	58	57	56	55	54	53	52				
−3	62	61	60	59	58	57	56	55	54	53	53	52	51	50	49
−2	62	61	60	60	59	58	57	56	55	54	53	52	51	50	49
−1	63	62	61	60	59	58	57	56	55	54	53	52	51	50	49
−0	63	62	61	60	59	58	57	56	55	54	53	53	52	51	50
+1	63	62	61	61	60	58	58	57	56	55	54	53	52	51	51
2	64	63	62	61	60	59	58	57	56	55	55	54	53	52	51
3	64	63	62	62	60	60	59	58	57	56	55	54	53	53	52
4	65	64	63	62	61	60	59	58	57	56	56	55	54	53	52
5	65	64	63	62	62	61	60	59	58	57	56	55	54	54	53
6	66	65	64	63	62	61	60	59	58	57	57	56	55	54	53
7	66	65	64	63	62	61	60	60	59	58	57	56	55	55	54
8	66	65	64	63	62	62	61	60	59	58	57	56	56	55	54
9	66	65	64	64	63	62	61	60	59	58	58	57	56	55	54
10	67	66	65	64	63	62	61	60	59	59	58	57	56	55	55
11	67	66	65	64	63	62	61	61	60	59	58	57	56	56	55
12	67	66	65	64	63	62	62	61	60	59	58	57	57	56	55
13	67	66	65	64	64	63	62	61	60	59	59	58	57	56	55
14	67	66	66	65	64	63	62	61	60	60	59	58	57	56	56
15	67	67	66	65	64	63	62	61	61	60	59	58	57	57	56
16	68	67	66	65	64	63	63	62	61	60	59	58	58	57	56
17	68	67	66	65	64	64	63	62	61	60	59	59	58	57	56
18	68	67	66	65	65	64	63	62	61	60	60	59	58	57	57
19	68	67	67	66	65	64	63	62	62	61	60	59	58	58	57
20	68	68	67	66	65	64	63	63	62	61	60	59	59	58	57
21	69	68	67	66	65	64	64	63	62	61	60	60	59	58	57
22	69	68	67	66	65	65	64	63	62	61	61	60	59	58	58
23	69	68	67	67	66	65	64	63	62	62	61	60	59	59	58
24	69	68	68	67	66	65	64	63	63	62	61	60	60	59	58
25	69	69	68	67	66	65	64	64	63	62	61	61	60	59	58
26	70	69	68	67	66	65	65	64	63	62	61	61	60	59	58
27	70	69	68	67	66	66	65	64	63	62	62	61	60	59	59
28	70	69	68	67	67	66	65	64	63	63	62	61	60	60	59
29	70	69	69	68	67	66	65	64	64	63	62	61	61	60	59
30	70	69	69	68	67	66	65	65	64	63	62	62	61	60	59
31	70	70	69	68	67	66	66	65	64	63	62	62	61	60	60
32	71	70	69	68	67	67	66	65	64	64	63	62	61	61	60
33	71	70	69	68	68	67	66	65	64	64	63	62	61	61	60
34	71	70	69	69	68	67	66	66	65	64	63	62	62	61	60
35	71	70	70	69	68	67	66	66	65	64	63	63	62	61	60

Temperature of the air. t=	t — t′ = Difference of Temperatures of the Dew Point and of the Air.														
	9°.0	9°.2	9°.4	9°.6	9°.8	10°.0	10°.2	10°.4	10°.6	10°.8	11°.0	11°.2	11°.4	11°.6	11°.8
Centig. −8															
−7															
−6															
−5															
−4															
−3															
−2															
−1															
0															
+1	50														
2	50	49	49	48	47	46									
3	51	50	49	48	48	47	46	45	45	44	43				
4	51	51	50	49	48	47	47	46	45	44	44	43	42	42	41
5	52	51	50	49	49	48	47	46	46	45	44	43	43	42	41
6	52	52	51	50	49	48	48	47	46	45	45	44	43	43	42
7	53	52	51	51	50	49	48	47	47	46	45	45	44	43	42
8	53	52	52	51	50	49	49	48	47	46	46	45	44	44	43
9	54	53	52	51	50	50	49	48	48	47	46	45	45	44	43
10	54	53	52	51	51	50	49	49	48	47	47	46	45	44	44
11	54	53	53	52	51	50	50	49	48	48	47	46	46	45	44
12	54	54	53	52	51	51	50	49	49	48	47	47	46	45	45
13	55	54	53	52	52	51	50	50	49	48	47	47	46	46	45
14	55	54	53	53	52	51	50	50	49	48	48	47	46	46	45
15	55	54	54	53	52	51	51	50	49	49	48	47	47	46	45
16	55	55	54	53	52	52	51	50	50	49	48	48	47	46	46
17	56	55	54	53	53	52	51	51	50	49	49	48	47	47	46
18	56	55	54	54	53	52	51	51	50	49	49	48	47	47	46
19	56	55	55	54	53	52	52	51	50	50	49	48	48	47	47
20	56	56	55	54	53	53	52	51	51	50	49	49	48	47	47
21	57	56	55	54	54	53	52	52	51	50	50	49	48	48	47
22	57	56	55	55	54	53	53	52	51	50	50	49	49	48	47
23	57	56	56	55	54	53	53	52	51	51	50	49	49	48	48
24	57	57	56	55	54	54	53	52	52	51	50	50	49	48	48
25	58	57	56	55	55	54	53	53	52	51	51	50	49	49	48
26	58	57	56	56	55	54	53	53	52	51	51	50	50	49	48
27	58	57	56	56	55	54	54	53	52	52	51	50	50	49	48
28	58	57	57	56	55	55	54	53	53	52	51	51	50	49	49
29	58	58	57	56	56	55	54	53	53	52	52	51	50	50	49
30	59	58	57	57	56	55	54	54	53	52	52	51	51	50	49
31	59	58	57	57	56	55	55	54	53	53	52	51	51	50	49
32	59	58	58	57	56	56	55	54	54	53	52	52	51	50	50
33	59	59	58	57	56	56	55	54	54	53	52	52	51	51	50
34	60	59	58	57	57	56	55	55	54	53	53	52	52	51	50
35	60	59	58	58	57	56	56	55	54	54	53	52	52	51	50

Temperature of the air. t =	t — t′ = Difference of Temperatures of the Dew Point and of the Air.														
	12°.0	12°.2	12°.4	12°.6	12°.8	13°.0	13°.2	13°.4	13°.6	13°.8	14°.0	14°.2	14°.4	14°.6	14°.8
Centig. −8															
−7															
−6															
−5															
−4															
−3															
−2															
−1															
0															
+1															
2															
3															
4	40	40	39	38	38	37									
5	41	40	39	39	38	38	37	36	36	35	35				
6	41	41	40	39	39	38	37	37	36	36	35	35	34	33	33
7	42	41	40	40	39	39	38	37	37	36	36	35	34	34	33
8	42	42	41	40	40	39	38	38	37	37	36	35	35	34	34
9	43	42	41	41	40	40	39	38	38	37	37	36	35	35	34
10	43	43	42	41	41	40	39	39	38	38	37	36	36	35	35
11	44	43	42	42	41	40	40	39	39	38	37	37	36	36	35
12	44	43	43	42	41	41	40	40	39	38	38	37	37	36	36
13	44	44	43	42	42	41	41	40	39	39	38	38	37	37	36
14	45	44	43	43	42	42	41	40	40	39	39	38	37	37	36
15	45	44	44	43	42	42	41	41	40	39	39	38	38	37	37
16	45	44	44	43	43	42	41	41	40	40	39	39	38	38	37
17	45	45	44	43	43	42	42	41	41	40	39	39	38	38	37
18	46	45	44	44	43	43	42	41	41	40	40	39	39	38	38
19	46	45	45	44	43	43	42	42	41	41	40	39	39	38	38
20	46	45	45	44	44	43	42	42	41	41	40	40	39	39	38
21	46	46	45	45	44	43	43	42	42	41	41	40	39	39	38
22	47	46	45	45	44	44	43	43	42	41	41	40	40	39	39
23	47	46	46	45	45	44	43	43	42	42	41	41	40	39	39
24	47	47	46	45	45	44	44	43	42	42	41	41	40	40	39
25	47	47	46	46	45	44	44	43	43	42	42	41	41	40	39
26	48	47	46	46	45	45	44	44	43	42	42	41	41	40	40
27	48	47	47	46	45	45	44	44	43	43	42	42	41	40	40
28	48	48	47	46	46	45	45	44	44	43	42	42	41	41	40
29	48	48	47	47	46	45	45	44	44	43	43	42	42	41	41
30	49	48	47	47	46	46	45	45	44	43	43	42	42	41	41
31	49	48	48	47	46	46	45	45	44	44	43	43	42	42	41
32	49	49	48	47	47	46	46	45	45	44	43	43	42	42	41
33	49	49	48	48	47	46	46	45	45	44	44	43	43	42	42
34	50	49	49	48	47	47	46	46	45	44	44	43	43	42	42
35	50	49	49	48	48	47	46	46	45	44	44	44	43	43	42

TABLE IV.

FACTOR $\frac{100}{F}$, FOR COMPUTING THE RELATIVE HUMIDITY, OR THE DEGREE OF MOISTURE OF THE AIR FROM ITS ABSOLUTE HUMIDITY, GIVEN IN MILLIMETRES.

BY HAEGHENS.

THE Relative Humidity, or the degree of moisture of the air, is the ratio of the quantity of vapor contained in the air to the quantity it could contain at the temperature observed, if fully saturated.

If we call

The force of vapor contained in the air $= f$,

The maximum of the force of vapor at the temperature of the air $=$ F,

The point of saturation $=$ 100,

we have the proportion,

$$\text{Relative Humidity} : 100 :: f : F,$$

and

$$\frac{f \times 100}{F} = \text{Relative Humidity in Hundredths.}$$

But as $\frac{f \times 100}{F} = f \times \frac{100}{F}$, it is obvious that the operation indicated by the former expression, viz. $\frac{f \times 100}{F}$, would be reduced to a simple multiplication, if we had a table of the factors $\frac{100}{F}$. Such a table is obtained by dividing the constant number 100 by each number in the Table of Elastic Forces of Vapor, and substituting the quotients to the tensions.

The following Table, taken from the *Annuairè Météorologique de la France*, for 1850, p. 79, gives the factor $\frac{100}{F}$ for every tenth of a degree from -10 to $+35°$ Centigrade, corresponding to the Forces of Vapor in Table I.

USE OF THE TABLE.

The force of vapor contained in the air being given in millimetres, multiply the number expressing it by the factor in the table corresponding to the temperature of the air at the time of the observation; the result will be the ***Relative Humidity in Hundredths.***

Examples.

1. Suppose the temperature of the air to be $=$ 24° Centigrade.
   " " force of vapor in the air to be $=$ 10.76 millimetres.

Opposite 24° is found in the table the factor 4.51.

Then $10.76 \times 4.51 = 48.5$, Relative Humidity in Hundredths.

---

2. Suppose the temperature of the air to be $=$ 16.7.
   " " force of vapor in the air to be $=$ 12.07.

Table gives for 16.7 the factor 7.07.

Then $12.07 \times 7.07 = 85.3$, Relative Humidity.

FACTOR $\frac{100}{F}$, TO COMPUTE THE RELATIVE HUMIDITY.

t = Temp. of Air, Centig.	Tenths of Degrees.									
	0.	1.	2.	3.	4.	5.	6.	7.	8.	9.
° −10	48.1	48.5	48.9	49.3	49.7	50.1	50.5	50.9	51.4	51.8
9	44.2	44.6	45.0	45.4	45.7	46.1	46.5	46.9	47.3	47.7
8	40.7	41.1	41.4	41.7	42.1	42.4	42.8	43.1	43.5	43.9
7	37.5	37.8	38.1	38.4	38.7	39.0	39.4	39.7	40.0	40.4
6	34.6	34.9	35.2	35.4	35.7	36.0	36.3	36.6	36.9	37.2
5	31.9	32.2	32.4	32.7	33.0	33.2	33.5	33.8	34.0	34.3
4	29.5	29.8	30.0	30.2	30.5	30.7	31.0	31.2	31.4	31.7
3	27.3	27.5	27.7	27.9	28.2	28.4	28.6	28.8	29.1	29.3
2	25.3	25.5	25.7	25.9	26.1	26.3	26.5	26.7	26.9	27.1
1	23.4	23.6	23.8	24.2	24.0	24.3	24.5	24.7	24.9	25.1
−0	21.7	21.9	22.1	22.2	22.4	22.6	22.8	22.9	23.1	23.3
+0	21.7	21.6	21.4	21.3	21.1	21.0	20.8	20.7	20.5	20.4
1	20.2	20.1	20.0	19.8	19.7	19.5	19.4	19.3	19.1	19.0
2	18.9	18.7	18.6	18.5	18.3	18.2	18.1	18.0	17.8	17.7
3	17.6	17.5	17.3	17.2	17.1	17.0	16.9	16.7	16.6	16.5
4	16.4	16.3	16.2	16.1	15.9	15.8	15.7	15.6	15.5	15.4
5	15.3	15.2	15.1	15.0	14.9	14.8	14.7	14.6	14.5	14.4
6	14.3	14.2	14.1	14.0	13.9	13.8	13.7	13.6	13.5	13.4
7	13.4	13.3	13.2	13.1	13.0	12.9	12.8	12.7	12.6	12.6
8	12.5	12.4	12.3	12.2	12.1	12.1	12.0	11.9	11.8	11.7
9	11.7	11.6	11.5	11.4	11.4	11.3	11.2	11.1	11.1	11.0
10	10.9	10.8	10.8	10.7	10.6	10.6	10.5	10.4	10.3	10.3
11	10.2	10.1	10.1	10.0	9.95	9.88	9.82	9.75	9.69	9.63
12	9.56	9.50	9.44	9.38	9.32	9.26	9.20	9.13	9.08	9.02
13	8.96	8.90	8.84	8.79	8.73	8.67	8.62	8.56	8.51	8.45
14	8.40	8.34	8.29	8.24	8.18	8.15	8.08	8.03	7.98	7.92
15	7.87	7.82	7.77	7.72	7.68	7.63	7.58	7.53	7.48	7.43
16	7.39	7.34	7.29	7.25	7.20	7.16	7.11	7.07	7.02	6.98
17	6.93	6.89	6.85	6.80	6.76	6.72	6.68	6.63	6.59	6.55
18	6.51	6.47	6.43	6.39	6.35	6.31	6.27	6.23	6.19	6.16
19	6.12	6.08	6.04	6.00	5.97	5.93	5.89	5.86	5.82	5.79
20	5.75	5.71	5.68	5.64	5.61	5.58	5.54	5.51	5.47	5.44
21	5.41	5.37	5.34	5.31	5.27	5.24	5.21	5.18	5.15	5.12
22	5.09	5.06	5.02	4.99	4.96	4.93	4.90	4.87	4.85	4.82
23	4.79	4.76	4.73	4.70	4.67	4.65	4.62	4.59	4.56	4.53
24	4.51	4.48	4.45	4.43	4.40	4.37	4.35	4.32	4.30	4.27
25	4.25	4.22	4.20	4.17	4.15	4.12	4.10	4.07	4.05	4.03
26	4.00	3.98	3.95	3.93	3.91	3.89	3.86	3.84	3.82	3.79
27	3.77	3.75	3.73	3.71	3.69	3.66	3.64	3.62	3.60	3.58
28	3.56	3.54	3.52	3.50	3.48	3.46	3.44	3.42	3.40	3.38
29	3.36	3.34	3.32	3.30	3.28	3.26	3.24	3.22	3.21	3.19
30	3.17	3.15	3.13	3.12	3.10	3.08	3.06	3.05	3.03	3.01
31	2.99	2.98	2.96	2.94	2.93	2.91	2.89	2.88	2.86	2.84
32	2.83	2.81	2.80	2.78	2.77	2.75	2.73	2.72	2.70	2.69
33	2.67	2.66	2.64	2.63	2.61	2.60	2.58	2.57	2.56	2.54
34	2.53	2.51	2.50	2.49	2.47	2.46	2.44	2.43	2.42	2.40
35	2.39	2.38	2.36	2.35	2.34	2.33	2.31	2.30	2.29	2.28

TABLE V.

WEIGHT OF VAPOR, IN GRAMMES,

CONTAINED IN A CUBIC METRE OF SATURATED AIR UNDER A BAROMETRIC PRESSURE OF 760 MILLIMETRES, AND AT TEMPERATURES BETWEEN —20° AND +40° CENTIGRADE.

THE theoretic density of aqueous vapor is very nearly 0.622, or $\frac{5}{8}$, of the density of the air at the same temperature and pressure. Regnault's experiments gave similar results. From this ratio the weight of the vapor contained in a given volume of air, the temperature and humidity of which are known, can be computed.

If we call

$t =$ the temperature of the air ;

$f =$ the elastic force of the vapor contained in the air at the time of the observation ;

$F =$ the maximum elastic force of vapor due to the temperature t, as given in the table ;

$p =$ the weight of the vapor contained in a litre of air at the temperature t, and with a force of vapor f;

$P =$ the weight of vapor in a litre of air at the temperature t, and at full saturation, or F.

Then,
$$p = 0.622 \frac{1.293223^{\text{gr.}}}{1 + 0.00367\,t} \cdot \frac{f}{760^{\text{mm.}}} .$$

In which 1.293223 grammes is the weight of a litre of dry air, at the temperature of zero Centigrade, and under a barometric pressure of 760 millimetres, according to the determination of Regnault ; 0.00367, the coefficient of the expansion of the air as found by the same ; 760 millimetres, the assumed normal barometric pressure.

The weight of a litre of air given by Regnault in the *Mémoires de l'Institut*, Tom. XXI. p. 157, is 1.293187 grammes ; but by correcting a slight error of computation (see E. Ritter, *Mémoires de la Société Physique de Genève*, Tom. XIII. p. 361), it becomes, as given above, 1.293223 grammes.

In order to obtain the weight of vapor in a cubic metre, or 1000 litres, of saturated air, the formula becomes,

$$P = 0.622 \frac{1293.223^{\text{gr.}}}{1 + 0.00367\,t} \cdot \frac{F}{760^{\text{mm.}}} .$$

From this formula Table V. has been computed. The tensions due to the temperatures in the first column are placed opposite the weights of vapor ; they are taken from Table I. It will be seen that, throughout the table, the number of grammes of vapor nearly corresponds to the number of millimetres of pressure expressing the tension.

The table of the weights of vapor given in Pouillet's *Eléments des Physique*, Tom. II. p. 707, being based on older values, gives results somewhat different. In that published by Becquerel, *Eléments de Physique Terrestre*, p. 354, Regnault's tensions and coefficient of expansion of the air have been used, but the value of the weight of vapor in a litre of air formerly determined by Biot and Arago, viz. 1.29954 grammes, has been retained.

V. WEIGHT OF VAPOR, IN GRAMMES,

CONTAINED IN A CUBIC METRE OF SATURATED AIR,

At Temperatures between —20° and +40° Centigrade.

Temperature of Dew-Point.	Force of Vapor.	Weight of Vapor.	Difference.	Temperature of Dew-Point.	Force of Vapor.	Weight of Vapor.	Difference.
Centigrade.	Millimetres.	Grammes.	Grammes.	Centigrade.	Millimetres.	Grammes.	Grammes.
−20°	0.912	1.042		+10°	9.165	9.357	
−19	0.993	1.130	0.088	11	9.792	9.962	0.605
−18	1.080	1.224	0.094	12	10.457	10.601	0.639
−17	1.174	1.325	0.101	13	11.162	11.276	0.675
−16	1.275	1.434	0.109	14	11.908	11.988	0.712
−15	1.385	1.551	0.118	15	12.699	12.739	0.751
−14	1.503	1.678	0.127	16	13.536	13.532	0.793
−13	1.631	1.813	0.134	17	14.421	14.367	0.835
−12	1.768	1.957	0.145	18	15.357	15.247	0.880
−11	1.918	2.114	0.157	19	16.346	16.173	0.926
−10	2.078	2.283	0.169	20	17.391	17.148	0.975
− 9	2.261	2.475	0.192	21	18.495	18.174	1.026
− 8	2.456	2.678	0.203	22	19.659	19.253	1.078
− 7	2.666	2.896	0.218	23	20.888	20.387	1.134
− 6	2.890	3.128	0.232	24	22.184	21.579	1.192
− 5	3.131	3.376	0.248	25	23.550	22.831	1.252
− 4	3.387	3.638	0.262	26	24.988	24.144	1.313
− 3	3.662	3.919	0.281	27	26.505	25.524	1.380
− 2	3.955	4.217	0.298	28	28.101	26.971	1.447
− 1	4.267	4.534	0.317	29	29.782	28.489	1.519
0	4.600	4.869	0.334	30	31.548	30.079	1.589
+ 1	4.940	5.209	0.341	31	33.405	31.744	1.666
2	5.302	5.571	0.361	32	35.359	33.491	1.747
3	5.687	5.953	0.383	33	37.410	35.317	1.827
4	6.097	6.360	0.406	34	39.565	37.230	1.913
5	6.534	6.791	0.431	35	41.827	39.231	2.001
6	6.998	7.247	0.456	36	44.201	41.323	2.092
7	7.492	7.731	0.484	37	46.691	43.510	2.187
8	8.017	8.243	0.512	38	49.302	45.795	2.285
9	8.574	8.785	0.541	39	52.039	48.182	2.387
+10	9.165	9.357	0.572	+40	54.906	50.674	2.492

PRACTICAL TABLES,

IN

ENGLISH MEASURES,

BASED ON REGNAULT'S HYGROMETRICAL CONSTANTS.

VI.

TABLE OF THE ELASTIC FORCE OF AQUEOUS VAPOR,

EXPRESSED IN ENGLISH INCHES OF MERCURY FOR TEMPERATURES OF FAHRENHEIT, REDUCED FROM REGNAULT'S TABLE.

THE values of the elastic force of vapor furnished by V. Regnault, which are found in Table I. of this Hygrometrical set, are derived from a series of experiments conducted, during several years, with great care, consummate skill, and all the means of precision which are at the disposal of modern science. The methods of investigation, and all the steps in each experiment, were minutely described and submitted to the judgment of the scientific, successively in separate papers in several volumes of the *Annales de Chimie et de Physique*, and collectively in his final Report to the Minister of Public Works, (see above, p. 9,) which fills Volume XXI. of the *Mémoires de l'Institut de France.* The confidence which has been deservedly granted to these determinations by nearly all scientific men, is increased by the fact that one of the best physicists and experimenters in Germany, Professor Magnus, came, about the same time, to results so little different, that both tables, for most purposes, may be considered identical. (Compare below, Table XXII.) It seems, therefore, that these values ought to be used in our hygrometrical tables, as has been done in France, in preference to the older and less reliable determinations on which they are based.

Though Regnault's table of the elastic force of vapor is considered, even, it is believed, by a majority of scientific men in England, as the most reliable which science now possesses, the author is not aware that any extensive reduction of it to English measures, such as is wanted for meteorological purposes, has been as yet published; still less a series of tables based on these values. Such a set of hygrometrical tables in English measures, corresponding to the preceding one in French measures, is offered here, which, it is hoped, supplies a real want felt by a large number of meteorologists.

Table VI. is Regnault's Table of the Elastic Force of Vapor as given in Table I., reduced to English measures, in which the fourth decimal is given in order to secure the third, and otherwise to facilitate the computations. From these values Tables VII. to X. have been computed.

VI. ELASTIC FORCE OF AQUEOUS VAPOR,

Expressed in English Inches of Mercury for Temperatures of Fahrenheit.

Reduced from Regnault's Table.

Temperature Fahrenheit.	Force of Vapor. Tenths of Degrees. 0	0.5	Temperature Fahrenheit.	Force of Vapor. Tenths of Degrees. 0	0.5	Temperature Fahrenheit.	Force of Vapor. Tenths of Degrees. 0	0.5	Temperature Fahrenheit.	Force of Vapor. Tenths of Degrees. 0	0.5
	Eng. In.	Eng. In.		Eng. In.	Eng. In.		Eng. In.	Eng. In.		Eng. In.	Eng. In.
−31	0.0087	0.0085	−19	0.0171	0.0167	− 8	0.0297	0.0290	+ 2	0.0476	0.0485
−30	0.0092	0.0090	−18	0.0181	0.0176	− 7	0.0312	0.0304	3	0.0498	0.0510
−29	0.0098	0.0095	−17	0.0190	0.0185	− 6	0.0327	0.0319	4	0.0521	0.0533
−28	0.0104	0.0101	−16	0.0200	0.0195	− 5	0.0343	0.0335	5	0.0545	0.0558
−27	0.0110	0.0107	−15	0.0210	0.0205	− 4	0.0359	0.0351	6	0.0570	0.0584
−26	0.0117	0.0114	−14	0.0221	0.0216	− 3	0.0376	0.0368	7	0.0597	0.0611
−25	0.0124	0.0120	−13	0.0232	0.0227	− 2	0.0395	0.0386	8	0.0625	0.0639
−24	0.0131	0.0127	−12	0.0244	0.0238	− 1	0.0414	0.0404	9	0.0654	0.0669
−23	0.0138	0.0135	−11	0.0257	0.0250	− 0	0.0434	0.0424	10	0.0684	0.0700
−22	0.0146	0.0142	−10	0.0270	0.0263	+ 0	0.0434	0.0444	11	0.0716	0.0732
−21	0.0154	0.0150	− 9	0.0283	0.0276	+ 1	0.0454	0.0465	12	0.0749	0.0766
−20	0.0163	0.0158	− 8	0.0297	0.0290	+ 2	0.0476	0.0487	+13	0.0783	0.0800

Temperature Fahrenheit.	Tenths of Degrees. 0.	1.	2.	3.	4.	5.	6.	7.	8.	9.
°	Eng. In.	Eng. In.	Eng. In.	Eng. In.	Eng. In.	Eng. In.	Eng. In.	Eng. In.	Eng. In.	Eng. In.
14	0.0818	0.0822	0.0826	0.0830	0.0834	0.0837	0.0841	0.0845	0.0849	0.0853
15	0.0857	0.0861	0.0865	0.0869	0.0873	0.0877	0.0881	0.0885	0.0889	0.0893
16	0.0898	0.0902	0.0906	0.0910	0.0914	0.0918	0.0923	0.0927	0.0931	0.0936
17	0.0940	0.0944	0.0949	0.0953	0.0958	0.0962	0.0967	0.0971	0.0975	0.0980
18	0.0984	0.0989	0.0993	0.0998	0.1002	0.1007	0.1012	0.1016	0.1021	0.1025
19	0.1030	0.1035	0.1040	0.1044	0.1049	0.1054	0.1059	0.1064	0.1068	0.1073
20	0.1078	0.1083	0.1088	0.1093	0.1098	0.1103	0.1108	0.1113	0.1118	0.1123
21	0.1128	0.1133	0.1138	0.1143	0.1148	0.1153	0.1159	0.1164	0.1169	0.1174
22	0.1179	0.1185	0.1190	0.1195	0.1200	0.1206	0.1211	0.1217	0.1222	0.1227
23	0.1233	0.1238	0.1244	0.1249	0.1255	0.1260	0.1266	0.1272	0.1277	0.1283
24	0.1289	0.1295	0.1300	0.1306	0.1312	0.1318	0.1324	0.1329	0.1335	0.1341
25	0.1347	0.1353	0.1359	0.1365	0.1371	0.1377	0.1383	0.1389	0.1395	0.1401
26	0.1407	0.1413	0.1419	0.1426	0.1432	0.1438	0.1444	0.1450	0.1457	0.1463
27	0.1469	0.1476	0.1482	0.1488	0.1495	0.1501	0.1508	0.1514	0.1521	0.1527
28	0.1534	0.1540	0.1547	0.1553	0.1560	0.1567	0.1573	0.1580	0.1587	0.1593
29	0.1600	0.1607	0.1613	0.1620	0.1627	0.1634	0.1641	0.1647	0.1654	0.1661
30	0.1668	0.1675	0.1682	0.1689	0.1696	0.1703	0.1710	0.1717	0.1724	0.1732
31	0.1739	0.1746	0.1753	0.1760	0.1767	0.1775	0.1782	0.1789	0.1796	0.1804
	0.	1.	2.	3.	4.	5.	6.	7.	8.	9.

EXPRESSED IN ENGLISH INCHES OF MERCURY FOR TEMPERATURES OF FAHRENHEIT.

Temperature of Fahrenheit.	Tenths of Degrees.									
	0.	**1.**	**2.**	**3.**	**4.**	**5.**	**6.**	**7.**	**8.**	**9.**
°	Eng. In.	Eng. In.	Eng. In.	Eng. In.	Eng. In.	Eng. In.	Eng. In.	Eng. In.	Eng. In.	Eng. In.
32	0.1811	0.1818	0.1825	0.1833	0.1840	0.1847	0.1854	0.1861	0.1869	0.1876
33	0.1883	0.1891	0.1898	0.1906	0.1913	0.1921	0.1928	0.1936	0.1944	0.1951
34	0.1959	0.1967	0.1974	0.1982	0.1990	0.1998	0.2006	0.2013	0.2021	0.2029
35	0.2037	0.2045	0.2053	0.2061	0.2070	0.2077	0.2086	0.2094	0.2102	0.2111
36	0.2119	0.2127	0.2135	0.2144	0.2152	0.2161	0.2169	0.2178	0.2186	0.2195
37	0.2204	0.2212	0.2221	0.2230	0.2238	0.2247	0.2256	0.2265	0.2273	0.2282
38	0.2291	0.2300	0.2309	0.2318	0.2327	0.2336	0.2345	0.2354	0.2364	0.2373
39	0.2382	0.2391	0.2400	0.2410	0.2419	0.2428	0.2438	0.2447	0.2457	0.2466
40	0.2476	0.2485	0.2495	0.2504	0.2514	0.2524	0.2533	0.2543	0.2553	0.2563
41	0.2572	0.2582	0.2592	0.2602	0.2612	0.2622	0.2632	0.2642	0.2652	0.2662
42	0.2672	0.2682	0.2692	0.2702	0.2713	0.2723	0.2733	0.2744	0.2754	0.2764
43	0.2775	0.2785	0.2796	0.2807	0.2817	0.2828	0.2839	0.2850	0.2860	0.2871
44	0.2882	0.2893	0.2904	0.2915	0.2926	0.2937	0.2948	0.2960	0.2971	0.2982
45	0.2993	0.3005	0.3016	0.3028	0.3039	0.3050	0.3062	0.3074	0.3085	0.3097
46	0.3108	0.3120	0.3132	0.3144	0.3156	0.3168	0.3179	0.3191	0.3203	0.3215
47	0.3228	0.3240	0.3252	0.3264	0.3276	0.3289	0.3301	0.3313	0.3326	0.3338
48	0.3351	0.3363	0.3376	0.3388	0.3401	0.3414	0.3426	0.3439	0.3452	0.3465
49	0.3477	0.3490	0.3503	0.3516	0.3529	0.3542	0.3556	0.3569	0.3582	0.3595
50	0.3608	0.3622	0.3635	0.3648	0.3661	0.3675	0.3688	0.3702	0.3715	0.3729
51	0.3743	0.3756	0.3770	0.3784	0.3798	0.3812	0.3826	0.3840	0.3854	0.3868
52	0.3882	0.3896	0.3911	0.3925	0.3939	0.3954	0.3968	0.3983	0.3997	0.4012
53	0.4027	0.4041	0.4056	0.4071	0.4086	0.4101	0.4116	0.4131	0.4146	0.4161
54	0.4176	0.4191	0.4207	0.4222	0.4237	0.4253	0.4268	0.4284	0.4299	0.4315
55	0.4331	0.4346	0.4362	0.4378	0.4394	0.4410	0.4426	0.4442	0.4458	0.4474
56	0.4490	0.4507	0.4523	0.4539	0.4556	0.4572	0.4589	0.4605	0.4622	0.4638
57	0.4655	0.4672	0.4689	0.4705	0.4722	0.4739	0.4756	0.4773	0.4791	0.4808
58	0.4825	0.4842	0.4859	0.4876	0.4894	0.4912	0.4929	0.4947	0.4964	0.4982
59	0.5000	0.5017	0.5035	0.5053	0.5071	0.5089	0.5107	0.5125	0.5143	0.5161
60	0.5179	0.5198	0.5216	0.5234	0.5253	0.5271	0.5290	0.5301	0.5328	0.5346
61	0.5365	0.5384	0.5403	0.5422	0.5441	0.5461	0.5480	0.5499	0.5519	0.5538
62	0.5558	0.5577	0.5597	0.5617	0.5636	0.5656	0.5676	0.5696	0.5716	0.5736
63	0.5756	0.5777	0.5797	0.5817	0.5838	0.5858	0.5879	0.5899	0.5920	0.5941
64	0.5962	0.5983	0.6004	0.6025	0.6046	0.6067	0.6088	0.6109	0.6131	0.6152
65	0.6173	0.6195	0.6217	0.6238	0.6260	0.6282	0.6304	0.6325	0.6347	0.6369
66	0.6392	0.6414	0.6436	0.6458	0.6481	0.6503	0.6525	0.6548	0.6571	0.6593
67	0.6616	0.6639	0.6662	0.6685	0.6708	0.6731	0.6754	0.6777	0.6800	0.6824
	0.	**1.**	**2.**	**3.**	**4.**	**5.**	**6.**	**7.**	**8.**	**9.**

Expressed in English Inches of Mercury for Temperatures of Fahrenheit.

Temperature of Fahrenheit.	Tenths of Degrees.									
	0.	**1.**	**2.**	**3.**	**4.**	**5.**	**6.**	**7.**	**8.**	**9.**
°	Eng. In.	Eng. In.	Eng. In.	Eng. In.	Eng. In.	Eng. In.	Eng. In.	Eng. In.	Eng. In.	Eng. In.
68	0.6847	0.6870	0.6894	0.6917	0.6941	0.6965	0.6989	0.7012	0.7036	0.7060
69	0.7084	0.7108	0.7133	0.7157	0.7181	0.7206	0.7230	0.7255	0.7280	0.7305
70	0.7329	0.7354	0.7379	0.7405	0.7430	0.7455	0.7480	0.7506	0.7531	0.7557
71	0.7583	0.7609	0.7634	0.7660	0.7686	0.7712	0.7739	0.7765	0.7791	0.7818
72	0.7844	0.7871	0.7897	0.7924	0.7951	0.7978	0.8005	0.8032	0.8059	0.8086
73	0.8113	0.8141	0.8168	0.8196	0.8223	0.8251	0.8279	0.8307	0.8335	0.8363
74	0.8391	0.8419	0.8447	0.8476	0.8504	0.8533	0.8561	0.8590	0.8619	0.8648
75	0.8676	0.8705	0.8735	0.8764	0.8793	0.8822	0.8852	0.8881	0.8911	0.8940
76	0.8970	0.9000	0.9030	0.9060	0.9090	0.9120	0.9150	0.9180	0.9211	0.9241
77	0.9272	0.9302	0.9333	0.9364	0.9395	0.9426	0.9457	0.9488	0.9519	0.9550
78	0.9582	0.9613	0.9645	0.9677	0.9709	0.9740	0.9773	0.9805	0.9837	0.9869
79	0.9902	0.9934	0.9967	1.0000	1.0033	1.0065	1.0099	1.0132	1.0165	1.0198
80	1.0232	1.0265	1.0299	1.0332	1.0366	1.0400	1.0434	1.0468	1.0503	1.0537
81	1.0572	1.0606	1.0641	1.0675	1.0710	1.0745	1.0780	1.0815	1.0851	1.0886
82	1.0922	1.0957	1.0993	1.1028	1.1064	1.1100	1.1136	1.1172	1.1209	1.1245
83	1.1281	1.1318	1.1354	1.1391	1.1428	1.1465	1.1502	1.1539	1.1576	1.1614
84	1.1651	1.1689	1.1726	1.1764	1.1802	1.1840	1.1878	1.1916	1.1954	1.1993
85	1.2031	1.2070	1.2108	1.2147	1.2186	1.2225	1.2264	1.2303	1.2342	1.2381
86	1.2421	1.2460	1.2500	1.2540	1.2580	1.2620	1.2660	1.2700	1.2740	1.2781
87	1.2821	1.2862	1.2903	1.2944	1.2985	1.3026	1.3068	1.3109	1.3151	1.3192
88	1.3234	1.3276	1.3318	1.3361	1.3403	1.3445	1.3488	1.3531	1.3573	1.3616
89	1.3659	1.3703	1.3746	1.3789	1.3833	1.3877	1.3920	1.3964	1.4008	1.4053
90	1.4097	1.4141	1.4186	1.4230	1.4275	1.4320	1.4365	1.4410	1.4456	1.4501
91	1.4546	1.4592	1.4638	1.4684	1.4730	1.4776	1.4822	1.4869	1.4915	1.4962
92	1.5008	1.5055	1.5102	1.5149	1.5197	1.5244	1.5291	1.5339	1.5387	1.5435
93	1.5482	1.5531	1.5579	1.5627	1.5676	1.5724	1.5773	1.5822	1.5871	1.5920
94	1.5969	1.6018	1.6068	1.6117	1.6167	1.6217	1.6267	1.6317	1.6367	1.6417
95	1.6468	1.6518	1.6569	1.6620	1.6671	1.6722	1.6773	1.6825	1.6876	1.6928
96	1.6980	1.7032	1.7084	1.7137	1.7189	1.7242	1.7295	1.7348	1.7401	1.7454
97	1.7508	1.7561	1.7615	1.7669	1.7723	1.7777	1.7831	1.7886	1.7940	1.7995
98	1.8050	1.8105	1.8160	1.8215	1.8271	1.8327	1.8382	1.8438	1.8494	1.8551
99	1.8607	1.8664	1.8720	1.8777	1.8834	1.8891	1.8949	1.9006	1.9064	1.9121
100	1.9179	1.9237	1.9295	1.9354	1.9412	1.9471	1.9530	1.9589	1.9648	1.9707
101	1.9766	1.9826	1.9885	1.9945	2.0005	2.0065	2.0126	2.0186	2.0247	2.0307
102	2.0368	2.0429	2.0490	2.0552	2.0613	2.0675	2.0737	2.0798	2.0861	2.0923
103	2.0985	2.1048	2.1110	2.1173	2.1236	2.1299	2.1362	2.1426	2.1489	2.1553
104	2.1617	2.1681	2.1745	2.1810	2.1874	2.1939	2.2004	2.2069	2.2135	2.2200
	0.	**1.**	**2.**	**3.**	**4.**	**5.**	**6.**	**7.**	**8.**	**9.**

VII.

PSYCHROMETRICAL TABLES,

GIVING, IN ENGLISH INCHES OF MERCURY, THE ELASTIC FORCE OF VAPOR CONTAINED IN THE AIR, AND ITS RELATIVE HUMIDITY IN HUNDREDTHS;

DERIVED FROM THE INDICATIONS OF THE WET AND DRY BULB THERMOMETERS, IN DEGREES OF FAHRENHEIT.

BY A. GUYOT.*

M. V. REGNAULT, in his *Etudes sur l'Hygrométrie Annales de Chimie et de Physique*, 3me série, Tom. XV. p. 129, after having discussed the theoretical bases of the psychrometric formula given by August, and modified the numerical values of some of its coefficients, adopts the formula

$$x = f - \frac{0.480\ (t - t')}{610 - t'}\ h$$

for temperatures above the freezing-point; and when the temperature of the wet thermometer is below the freezing-point, the bulb being covered with a film of ice,

$$x = f - \frac{0.480\ (t - t')}{689 - t'}\ h,$$

* While this table was going through the press, a similar one, prepared by Prof. T. H. Coffin for his private use, was published by the Smithsonian Institution, in order to meet an urgent demand from many quarters. Being based on the same formula, it gives the same results, except, perhaps, in degrees below 14° Fahrenheit, where the tables show slight discrepancies. These unimportant differences arise from the fact that Prof. Coffin's table was computed from Regnault's tensions, as given in the first edition of this collection, while the author's table is based on the table of tensions as given in this second edition, in which the values below 14° Fahrenheit have been somewhat modified, for reasons given above. The following table gives also the relative humidity with one more decimal, which makes the interpolations more easy; and a column of differences for finding the values for fractions of t'. A table for reducing the results to another barometric height is added at the end of the table.

in which

x represents the force of vapor in the air at the time of the observation;

t, the temperature of the air in Centigrade degrees, indicated by the dry thermometer;

t', the temperature of evaporation given by the wet thermometer;

f, the force of vapor in a saturated air at the temperature t';

h, the height of the barometer.

Substituting the Fahrenheit scale for the Centigrade, the formula, for temperatures above the freezing-point, reads

$$x = f - \frac{0.480 \times \frac{5}{9}(t - t')}{610 - \frac{5}{9}(t' - 32^\circ)} h = f - \frac{0.480\ (t - t')}{1130 - t'} h;$$

and below the freezing-point,

$$x = f - \frac{0.480 \times \frac{5}{9}(t - t')}{689 - \frac{5}{9}(t' - 32^\circ)} h = f - \frac{0.480\ (t - t')}{1240.2 - t'} h.$$

Making, further, $h = 29.7$ English inches, these formulæ become

$$x = f - \frac{0.480\ (t - t')}{1130 - t'} 29.7 = f - \frac{14.256\ (t - t')}{1130 - t'},$$

and

$$x = f - \frac{0.480\ (t - t')}{1240.2 - t'} 29.7 = f - \frac{14.256\ (t - t')}{1240.2 - t'}.$$

The mean barometric pressure for which the table has been computed, viz. 29.7 inches, is, within a small fraction, the same as that adopted in Haeghens's Tables, No. II., which is 755 millimetres = 29.725 Eng. inches. As that slight difference in the barometric pressure cannot cause, in the most extreme cases, a difference exceeding two thousandths of an inch in the elastic forces, the results in the two tables may be considered identical.

That barometric pressure, corresponding, in our latitudes, to a mean altitude of 250 to 300 feet above the sea, is likely to suit, without correction, the largest number of meteorological stations. Should the mean height of the barometer, in consequence of the elevation of the station, much differ from that adopted in the table, a constant correction can be determined, to be applied to the numbers in the table. At the end, page 72, will be found a table which furnishes that correction for barometric heights between 20 and 31 inches, and for values of $t - t'$ between 2° and 26° Fahrenheit.

The effect of the irregular variations of the barometer at the same station can, in most cases, be neglected; for the error due to that cause will scarcely ever exceed those which may arise from the uncertainty of the very elements on which the tables are based.

Arrangement of the Tables.

The same arrangement as is found in the Psychrometrical for the Centigrade scale has been adopted.

The first column at the left contains the indications of the wet-bulb thermometer, from —31° to 105° Fahrenheit.

The second column gives the differences of the force of vapor for each tenth of a degree, between each two consecutive full degrees in the first column. It enables the observer easily to find the values for the fractions of degrees of the wet thermometer.

The following double columns furnish the forces of vapor and the relative humidity corresponding to each full degree of the wet-bulb thermometer given in the first column in the same horizontal line, and to the difference of the two thermometers, or $t - t'$, found at the head of each column, for every half-degree from 0° to 26°.5. The relative humidity, or the fraction of saturation, is given in hundredths, which is near enough for meteorological purposes; but one decimal more has been added, though separated by a point, in order to facilitate the interpolations.

At the bottom of each page is found the mean difference, for each tenth of a degree, between the forces of vapor on the same line. It gives the means of finding the values for the intermediate differences of $t - t'$, not found in the tables.

Use of the Tables.

Enter the tables with the difference of the two thermometers, or $t - t'$, and the temperature of the wet-bulb thermometer, given by observation.

In the column headed by the observed difference of the thermometer, $t - t'$, and on the horizontal line headed by the observed temperature of the wet thermometer, t', are found the force of vapor, and the relative humidity corresponding to these temperatures.

For the fractions of degrees of the wet thermometer, multiply the decimal fraction by the number placed in the second column between the full degree and the next, and *add* the product if the temperature is above, and *subtract* it if it is below zero Fahrenheit.

The intermediate values of $t - t'$ not given in the table are found by *subtracting* the number in the line at the bottom of the page, multiplied by the number of additional tenths, from the value given in the table. This correction, being always very small, can usually be neglected.

For the relative humidity, interpolations at sight will generally suffice.

Examples.

1. Dry thermometer, $t = 50°$ F.
Wet thermometer, $t' = 43°$ F.
Difference, or $t - t' = 7°$ F.

Page 58, we find for $t - t' = 7°$ in the third double column, and for $t' = 43°$ in the first column

Force of vapor in the air $= 0.186$ inch.
Relative humidity in hundredths $= 51$

2.

Dry thermometer, t	$= 88^\circ.5$ F.		
Wet thermometer, t'	$= 76^\circ.3$ F.		
Difference, $t - t'$	$= 12^\circ.2$ F.		
Page **63**, Table gives for $t - t' = 12$ and $t' = 76^\circ$		=	0.735 inch.
Add for fraction of $t' = 0.3$,	0.003×3	=	0.009
Subtract for fraction of $t - t' = 0^\circ.2$,	$.0013 \times 2$	=	—0.003
	Force of vapor in the air	=	0.741
	Relative humidity	=	55

3.

Dry thermometer, t	$= - 4^\circ.5$ F.		
Wet thermometer, t'	$= 6^\circ.0$ F.		
Difference, $t - t'$	$= 1^\circ.5$ F.		
Page **50**, Table gives for $t - t' = 1^\circ.5$ and $t' = - 6^\circ$		=	0.016 inch.
Subtract for fraction of $t' = 0.5$,	0.0002×5	=	— 0.001
	Force of vapor in the air	=	0.015
	Relative humidity	=	45

Temperature, Fahrenheit. — Force of Vapor in English Inches. — Relative Humidity in Hundredths.

Wet-Bulb Thermometer t' Fahrenheit.	Mean Vertical Difference of Force of Vapor for each 0°.1.	t — t', or Difference of Wet and Dry Bulb Thermometers. t', below the Freezing-Point; the Bulb covered with a Film of Ice.											
		0°.0		0°.5		1°.0		1°.5		2°.0		2°.5	
		Force of Vapor.	Relative Humidity.	Force of Vapor.	Relative Humidity.	Force of Vapor.	Relative Humidity.	Force of Vapor.	Relative Humidity.	Force of Vapor.	Relative Humidity.	Force of Vapor.	Relative Humidity.
°		Eng. In.		Eng. In.		Eng. In.		Eng. In.		Eng. In.		Eng. In.	
−31		0.009	100	0.003	36.0								
−30	.00005	0.009	100	0.004	39.6								
−29	.00006	0.010	100	0.004	42.9								
−28	.00006	0.010	100	0.005	46.1								
−27	.00006	0.011	100	0.006	49.0								
−26	.00006	0.012	100	0.006	51.8								
−25	.00007	0.012	100	0.007	54.4								
−24	.00007	0.013	100	0.008	56.8								
−23	.00008	0.014	100	0.008	59.0								
−22	.00008	0.015	100	0.009	61.0								
−21	.00008	0.015	100	0.010	62.6	0.004	26.9						
−20	.00008	0.016	100	0.011	64.2	0.005	30.3						
−19	.00008	0.017	100	0.012	65.9	0.006	33.5						
−18	.00009	0.018	100	0.012	67.5	0.007	36.6						
−17	.0001	0.019	100	0.013	69.0	0.008	39.5						
−16	.0001	0.020	100	0.014	70.4	0.009	42.3						
−15	.0001	0.021	100	0.015	71.8	0.010	44.9	0.004	19.4				
−14	.0001	0.022	100	0.017	73.0	0.011	47.4	0.005	23.0				
−13	.0001	0.023	100	0.018	74.3	0.012	49.8	0.007	26.4				
−12	.0001	0.024	100	0.019	75.4	0.013	51.9	0.008	29.5				
−11	.0001	0.026	100	0.020	76.5	0.014	53.9	0.009	32.5				
−10	.0001	0.027	100	0.021	77.5	0.016	55.7	0.010	35.3	0.005	15.6		
− 9	.0001	0.028	100	0.023	78.5	0.017	57.7	0.012	38.3	0.006	19.1		
− 8	.0001	0.030	100	0.024	79.4	0.018	59.4	0.013	40.6	0.007	22.5		
− 7	.0001	0.031	100	0.026	80.3	0.020	61.1	0.014	43.0	0.009	25.7		
− 6	.0001	0.033	100	0.027	81.1	0.021	62.7	0.016	45.4	0.010	28.4	0.005	12.9
− 5	.0002	0.034	100	0.029	81.8	0.023	64.5	0.017	47.6	0.012	31.7	0.006	16.4
− 4	.0002	0.036	100	0.030	82.5	0.025	65.8	0.019	49.8	0.014	34.5	0.008	19.8
− 3	.0002	0.038	100	0.032	83.2	0.026	67.1	0.021	51.7	0.015	36.9	0.010	22.8
− 2	.0002	0.039	100	0.034	83.9	0.028	68.3	0.023	53.5	0.017	39.3	0.011	25.8
− 1	.0002	0.041	100	0.036	84.5	0.030	69.5	0.024	55.3	0.019	41.6	0.013	28.6
− 0	.0002	0.043	100	0.038	85.0	0.032	71.0	0.026	57.0	0.021	43.8	0.015	31.3

Mean Horizontal Difference of Force of Vapor for each 0°.1 = 0.0012.

Temperature, Fahrenheit. — Force of Vapor in English Inches. — Relative Humidity in Hundredths.

Wet-Bulb Thermometer t′ Fahrenheit.	Mean Vertical Difference of Force of Vapor for each 0°.1.	t — t′, or Difference of Wet and Dry Bulb Thermometers. t′, below the Freezing-Point, the Bulb covered with a Film of Ice. 0°.0 Force of Vapor.	0°.0 Relative Humidity.	0°.5 Force of Vapor.	0°.5 Relative Humidity.	1°.0 Force of Vapor.	1°.0 Relative Humidity.	1°.5 Force of Vapor.	1°.5 Relative Humidity.	2°.0 Force of Vapor.	2°.0 Relative Humidity.	2°.5 Force of Vapor.	2°.5 Relative Humidity.
°		Eng. In.		Eng. In.		Eng. In.		Eng. In.		Eng. In.		Eng. In.	
0	0.0002	0.043	100	0.038	85.0	0.032	70.7	0.026	57.0	0.021	43.8	0.015	31.3
1	.0002	0.045	100	0.040	85.6	0.034	71.8	0.028	58.6	0.023	46.0	0.017	33.9
2	.0002	0.047	100	0.042	86.2	0.036	73.0	0.031	60.2	0.025	48.0	0.019	36.4
3	.0002	0.050	100	0.044	86.7	0.038	74.0	0.033	61.8	0.027	50.0	0.022	38.8
4	.0002	0.052	100	0.046	87.2	0.041	75.0	0.035	63.3	0.030	52.0	0.024	41.2
5	.0002	0.055	100	0.049	87.7	0.043	76.0	0.038	64.7	0.032	53.8	0.026	43.4
6	.0003	0.057	100	0.051	88.2	0.046	76.9	0.040	66.0	0.034	55.3	0.029	45.2
7	.0003	0.059	100	0.054	88.6	0.048	77.7	0.043	67.1	0.037	56.8	0.031	47.0
8	.0003	0.062	100	0.057	89.0	0.051	78.4	0.045	68.2	0.040	58.2	0.034	48.8
9	.0003	0.065	100	0.059	89.4	0.054	79.1	0.048	69.2	0.043	59.6	0.037	50.5
10	.0003	0.068	100	0.062	89.8	0.057	79.7	0.051	70.1	0.046	61.0	0.040	52.2
11	.0003	0.071	100	0.066	90.1	0.061	80.4	0.054	71.1	0.049	62.3	0.043	53.8
12	.0003	0.075	100	0.069	90.4	0.063	81.0	0.058	72.1	0.052	63.5	0.046	55.3
13	.0004	0.078	100	0.072	90.7	0.067	81.6	0.061	73.0	0.056	64.8	0.050	56.8
14	.0004	0.082	100	0.076	91.0	0.071	82.3	0.065	73.9	0.059	65.9	0.054	58.2
15	.0004	0.086	100	0.080	91.3	0.074	82.9	0.069	74.8	0.063	67.1	0.057	59.7
16	.0004	0.090	100	0.084	91.6	0.078	83.4	0.073	75.7	0.067	68.2	0.061	61.0
17	.0004	0.094	100	0.088	91.9	0.083	84.0	0.077	76.5	0.071	69.2	0.066	62.3
18	.0005	0.098	100	0.093	92.1	0.087	84.5	0.081	77.2	0.076	70.2	0.070	63.5
19	.0005	0.103	100	0.097	92.4	0.092	85.0	0.086	78.0	0.080	71.2	0.075	64.7
20	.0005	0.108	100	0.102	92.6	0.096	85.5	0.091	78.7	0.085	72.1	0.079	65.8
21	.0005	0.113	100	0.107	92.9	0.101	86.0	0.096	79.4	0.090	73.0	0.084	66.9
22	.0005	0.118	100	0.112	93.1	0.107	86.4	0.101	80.0	0.095	73.8	0.089	68.0
23	.0006	0.123	100	0.118	93.3	0.112	86.8	0.106	80.7	0.100	74.6	0.095	68.9
24	.0006	0.129	100	0.123	93.6	0.117	87.2	0.112	81.2	0.106	75.4	0.100	69.9
25	.0006	0.135	100	0.129	93.8	0.123	87.6	0.118	81.8	0.112	76.1	0.106	70.7
26	.0006	0.141	100	0.135	94.0	0.129	88.0	0.123	82.4	0.117	76.8	0.112	71.6
27	.0006	0.147	100	0.141	94.1	0.136	88.3	0.130	82.9	0.124	77.5	0.118	72.5
28	.0007	0.153	100	0.148	94.3	0.142	88.7	0.136	83.4	0.130	78.2	0.125	73.3
29	.0007	0.160	100	0.154	94.5	0.149	89.0	0.143	83.9	0.137	78.8	0.131	74.0
30	.0007	0.167	100	0.161	94.7	0.155	89.3	0.150	84.3	0.144	79.4	0.138	74.8
31		0.174	100	0.168	94.8	0.162	89.6	0.157	84.8	0.151	80.0	0.145	75.6

Mean Horizontal Difference of Force of Vapor for each 0°.1 = 0.0012.

Temperature, Fahrenheit. — Force of Vapor in English Inches. — Relative Humidity in Hundredths.

Wet-Bulb Thermometer t' Fahrenheit.	Mean Vertical Difference of Force of Vapor for each 0°.1.	$t - t'$, or Difference of Wet and Dry Bulb Thermometers. t', below the Freezing-Point; the Bulb covered with a Film of Ice. 3°.0		3°.5		4°.0		4°.5		5°.0		5°.5	
		Force of Vapor.	Relative Humidity.	Force of Vapor.	Relative Humidity.	Force of Vapor.	Relative Humidity.	Force of Vapor.	Relative Humidity.	Force of Vapor.	Relative Humidity.	Force of Vapor.	Relative Humidity.
°		Eng. In.		Eng. In.		Eng. In.		Eng. In.		Eng. In.		Eng. In.	
0	0.0002	0.010	19.3	0.004	7.9								
1	.0002	0.012	22.3	0.006	11.3								
2	.0002	0.014	25.3	0.008	14.7								
3	.0002	0.016	28.1	0.010	17.8								
4	.0002	0.018	30.8	0.013	20.9	0.007	11.4						
5	.0002	0.021	33.4	0.015	23.8	0.010	14.6						
6	.0002	0.023	35.6	0.018	26.3	0.012	17.5	0.006	9.0				
7	.0003	0.026	37.7	0.020	28.8	0.014	20.2	0.009	12.0				
8	.0003	0.028	39.8	0.023	31.2	0.017	22.9	0.011	15.0				
9	.0003	0.031	41.8	0.026	33.5	0.020	25.5	0.014	17.9	0.009	10.6		
10	.0003	0.034	43.8	0.029	35.7	0.023	28.0	0.017	20.6	0.012	13.6		
11	.0003	0.037	45.7	0.032	37.9	0.026	30.4	0.020	23.3	0.014	16.4	0.009	9.9
12	.0003	0.041	47.5	0.035	40.0	0.029	32.7	0.024	25.8	0.018	19.2	0.012	12.9
13	.0004	0.044	49.2	0.039	42.0	0.033	35.0	0.027	28.3	0.022	21.9	0.016	15.8
14	.0004	0.048	50.9	0.042	43.9	0.037	37.1	0.031	30.7	0.025	24.5	0.020	18.5
15	.0004	0.052	52.5	0.046	45.7	0.040	39.2	0.035	32.9	0.029	26.9	0.023	21.2
16	.0004	0.056	54.1	0.050	47.5	0.044	41.2	0.039	35.1	0.033	29.3	0.027	23.7
17	.0004	0.060	55.6	0.054	49.2	0.049	43.1	0.043	37.2	0.037	31.6	0.032	26.2
18	.0004	0.065	57.0	0.059	50.9	0.053	44.9	0.047	39.2	0.042	33.7	0.036	28.5
19	.0005	0.069	58.4	0.063	52.5	0.058	46.7	0.052	41.2	0.046	35.8	0.040	30.7
20	.0005	0.074	59.8	0.068	54.0	0.062	48.3	0.057	43.0	0.050	37.8	0.045	32.9
21	.0005	0.079	61.0	0.073	55.4	0.067	50.0	0.062	44.7	0.056	39.7	0.050	34.9
22	.0005	0.084	62.2	0.078	56.8	0.072	51.5	0.067	46.4	0.061	41.5	0.055	36.8
23	.0006	0.089	63.4	0.083	58.1	0.078	52.9	0.072	48.0	0.066	43.3	0.061	38.6
24	.0006	0.095	64.4	0.089	59.3	0.083	54.3	0.077	49.6	0.072	44.9	0.066	40.5
25	.0006	0.100	65.5	0.095	60.5	0.089	55.6	0.083	51.0	0.078	46.5	0.072	42.2
26	.0006	0.106	66.5	0.101	61.7	0.095	56.9	0.089	52.4	0.083	48.0	0.078	43.9
27	.0006	0.113	67.5	0.107	62.8	0.101	58.2	0.095	53.8	0.090	49.6	0.084	45.5
28	.0007	0.119	68.5	0.113	63.9	0.108	59.4	0.102	55.2	0.096	51.0	0.090	47.0
29	.0007	0.126	69.4	0.120	64.9	0.114	60.6	0.108	56.4	0.103	52.4	0.097	48.5
30	.0007	0.132	70.3	0.127	65.9	0.121	61.7	0.115	57.7	0.109	53.7	0.104	49.9
31		0.139	71.2	0.134	66.9	0.128	62.8	0.122	58.8	0.116	55.0	0.111	51.2

Mean Horizontal Difference of Force of Vapor for each 0°.1 = 0.0012.

Temperature, Fahrenheit. — Force of Vapor in English Inches. — Relative Humidity in Hundredths.

$t - t'$, or Difference of Wet and Dry Bulb Thermometers.

t', below the Freezing-Point; the Bulb covered with a Film of Ice.

Wet-Bulb Thermometer t' Fahrenheit.	Mean Vertical Difference of Force of Vapor for each 0°.1.	6°.0		6°.5		7°.0		7°.5		8°.0		8°.5	
		Force of Vapor.	Relative Humidity.	Force of Vapor.	Relative Humidity.	Force of Vapor.	Relative Humidity.	Force of Vapor.	Relative Humidity.	Force of Vapor.	Relative Humidity.	Force of Vapor.	Relative Humidity.
°		Eng. In.		Eng. In.		Eng. In.		Eng. In.		Eng. In.		Eng. In.	
12	0.0003	0.007	6.8										
13	.0004	0.010	9.9										
14	.0004	0.014	12.8	0.008	7.5								
15	.0004	0.018	15.7	0.012	10.4	0.006	5.4						
16	.0004	0.022	18.4	0.016	13.3	0.010	8.4						
17	.0004	0.026	21.0	0.020	16.0	0.015	11.3	0.009	6.7				
18	.0005	0.030	23.5	0.025	18.6	0.019	14.0	0.013	9.6	0.008	5.3		
19	.0005	0.035	25.8	0.029	21.2	0.023	16.6	0.018	12.3	0.012	8.2	0.006	4.2
20	.0005	0.040	28.1	0.034	23.5	0.028	19.0	0.022	15.0	0.017	11.0	0.011	7.1
21	.0005	0.044	30.3	0.039	25.8	0.033	21.5	0.027	17.5	0.022	13.5	0.016	9.8
22	.0005	0.050	32.3	0.044	28.0	0.038	23.8	0.032	19.8	0.027	16.0	0.021	12.3
23	.0005	0.055	34.2	0.049	30.1	0.043	26.0	0.038	22.1	0.032	18.4	0.026	14.8
24	.0006	0.060	36.1	0.055	32.1	0.049	28.1	0.043	24.4	0.038	20.7	0.032	17.2
25	.0006	0.066	38.0	0.060	34.0	0.055	30.2	0.049	26.5	0.043	23.0	0.038	19.5
26	.0006	0.072	39.8	0.066	35.9	0.061	32.2	0.055	28.6	0.049	25.1	0.043	21.8
27	.0006	0.078	41.5	0.073	37.8	0.067	34.0	0.061	30.6	0.055	27.2	0.050	23.9
28	.0007	0.085	43.2	0.079	39.5	0.073	35.9	0.067	32.5	0.062	29.1	0.056	25.9
29	.0007	0.091	44.8	0.085	41.1	0.080	37.6	0.074	34.2	0.068	31.0	0.063	27.9
30	.0007	0.098	46.2	0.092	42.7	0.086	39.2	0.081	35.9	0.075	32.8	0.069	29.7
31		0.105	47.6	0.099	44.2	0.093	40.8	0.088	37.5	0.082	34.4	0.076	31.4

Wet-Bulb Thermometer t' Fahrenheit.	Mean Vertical Difference of Force of Vapor for each 0°.1.	9°.0		9°.5		10°.0		10°.5		11°.0		11°.5	
		Eng. In.		Eng. In.		Eng. In.		Eng. In.		Eng. In.		Eng. In.	
20	0.0005	0.005	3.4										
21	.0005	0.010	6.1	0.005	2.7								
22	.0005	0.015	8.8	0.010	5.4	0.004	2.2						
23	.0005	0.021	11.4	0.015	8.0	0.009	4.9						
24	.0006	0.026	13.9	0.020	10.6	0.015	7.5	0.009	4.5				
25	.0006	0.032	16.2	0.026	13.1	0.020	10.0	0.015	7.1	0.009	4.2		
26	.0006	0.038	18.5	0.032	15.4	0.026	12.4	0.021	9.5	0.015	6.8	0.009	4.1
27	.0006	0.044	20.7	0.038	17.7	0.032	14.7	0.027	11.9	0.021	9.2	0.015	6.5
28	.0007	0.050	22.8	0.045	19.9	0.039	16.9	0.033	14.2	0.027	11.5	0.022	8.9
29	.0007	0.057	24.9	0.051	21.9	0.045	19.0	0.040	16.3	0.034	13.7	0.028	11.1
30	.0007	0.064	26.7	0.058	23.8	0.052	21.0	0.046	18.4	0.041	15.8	0.035	13.3
31		0.071	28.5	0.065	25.7	0.059	22.9	0.053	20.3	0.048	17.8	0.042	15.3

Mean Horizontal Difference of Force of Vapor for each 0°.1 = 0.0012.

Temperature, Fahrenheit. — Force of Vapor in English Inches. — Relative Humidity in Hundredths.

Wet-Bulb Thermometer t' Fahrenheit.	Mean Vertical Difference of Force of Vapor for each 0°.1.	$t - t'$, or Difference of Wet and Dry Bulb Thermometers.											
		0°.0		0°.5		1°.0		1°.5		2°.0		2°.5	
		Force of Vapor.	Relative Humidity.	Force of Vapor.	Relative Humidity.	Force of Vapor.	Relative Humidity.	Force of Vapor.	Relative Humidity.	Force of Vapor.	Relative Humidity.	Force of Vapor.	Relative Humidity.
°		Eng. In.		Eng. In.		Eng. In.		Eng. In.		Eng. In.		Eng. In.	
32	0.0007	0.181	100	0.175	94.5	0.168	89.3	0.162	84.1	0.155	79.2	0.149	74.4
33	.0008	0.188	100	0.182	94.7	0.175	89.5	0.169	84.5	0.162	79.7	0.156	75.0
34	.0008	0.196	100	0.189	94.8	0.183	89.8	0.176	84.9	0.170	80.2	0.163	75.6
35	.0008	0.204	100	0.197	94.9	0.191	90.0	0.184	85.3	0.178	80.7	0.171	76.2
36	.0009	0.212	100	0.205	95.0	0.199	90.3	0.192	85.6	0.186	81.1	0.179	76.8
37	.0009	0.220	100	0.214	95.2	0.207	90.5	0.201	86.0	0.194	81.6	0.188	77.3
38	.0009	0.229	100	0.223	95.3	0.216	90.7	0.210	86.3	0.203	82.0	0.196	77.9
39	.0009	0.238	100	0.232	95.4	0.225	91.0	0.219	86.6	0.212	82.4	0.206	78.4
40	.0010	0.248	100	0.241	95.5	0.235	91.2	0.228	86.9	0.221	82.9	0.215	78.9
41	.0010	0.257	100	0.251	95.6	0.244	91.4	0.238	87.3	0.231	83.3	0.224	79.4
42	.0010	0.267	100	0.260	95.7	0.254	91.6	0.247	87.5	0.241	83.6	0.234	79.8
43	.0011	0.278	100	0.271	95.8	0.264	91.8	0.258	87.8	0.251	84.0	0.245	80.3
44	.0011	0.288	100	0.282	95.9	0.275	92.0	0.268	88.1	0.262	84.3	0.255	80.7
45	.0011	0.299	100	0.293	96.0	0.286	92.1	0.280	88.3	0.273	84.7	0.266	81.1
46	.0012	0.311	100	0.304	96.1	0.297	92.3	0.291	88.6	0.284	85.0	0.278	81.5
47	.0012	0.323	100	0.316	96.2	0.310	92.5	0.303	88.8	0.297	85.3	0.290	81.9
48	.0013	0.335	100	0.329	96.2	0.322	92.6	0.315	89.0	0.309	85.6	0.302	82.2
49	.0013	0.348	100	0.341	96.3	0.335	92.7	0.328	89.3	0.321	85.9	0.315	82.6
50	.0013	0.361	100	0.354	96.4	0.348	92.9	0.341	89.5	0.334	86.1	0.328	82.9
51	.0014	0.374	100	0.368	96.5	0.361	93.0	0.354	89.7	0.348	86.4	0.341	83.2
52	.0014	0.388	100	0.382	96.5	0.375	93.2	0.368	89.9	0.362	86.7	0.355	83.6
53	.0015	0.403	100	0.396	96.6	0.389	93.3	0.383	90.1	0.376	86.9	0.370	83.9
54	.0015	0.418	100	0.411	96.7	0.404	93.4	0.398	90.2	0.391	87.2	0.385	84.2
55	.0016	0.433	100	0.426	96.7	0.420	93.5	0.413	90.4	0.407	87.4	0.400	84.4
56	.0016	0.449	100	0.442	96.8	0.436	93.6	0.429	90.6	0.422	87.6	0.416	84.7
57	.0017	0.466	100	0.459	96.8	0.452	93.7	0.446	90.7	0.439	87.8	0.432	85.0
58	.0017	0.482	100	0.476	96.9	0.469	93.9	0.463	90.9	0.456	88.0	0.449	85.2
59	.0018	0.500	100	0.493	96.9	0.487	94.0	0.480	91.0	0.473	88.2	0.467	85.5
60	.0019	0.518	100	0.511	97.0	0.505	94.1	0.498	91.2	0.491	88.4	0.485	85.7
61	.0019	0.537	100	0.530	97.0	0.523	94.2	0.517	91.3	0.510	88.6	0.503	85.9
62	.0020	0.556	100	0.549	97.1	0.542	94.2	0.536	91.5	0.529	88.8	0.522	86.2
63	.0020	0.576	100	0.569	97.1	0.562	94.3	0.556	91.6	0.549	89.0	0.542	86.4
64	.0021	0.596	100	0.589	97.2	0.583	94.4	0.576	91.7	0.569	89.1	0.563	86.6
65	.0022	0.617	100	0.611	97.2	0.604	94.5	0.597	91.9	0.591	89.3	0.584	86.8
66	.0023	0.639	100	0.633	97.3	0.626	94.6	0.619	92.0	0.612	89.5	0.606	87.0
67		0.662	100	0.655	97.3	0.648	94.7	0.642	92.1	0.635	89.6	0.628	87.2

Mean Horizontal Difference of Force of Vapor for each 0°.1 = 0.0013.

Temperature, Fahrenheit. — Force of Vapor in English Inches. — Relative Humidity in Hundredths.

Wet-Bulb Thermometer t' Fahrenheit.	Mean Vertical Difference of Force of Vapor for each 0°.1.	$t - t'$, or Difference of Wet and Dry Bulb Thermometers.											
		0°.0		0°.5		1°.0		1°.5		2°.0		2°.5	
		Force of Vapor.	Relative Humidity.	Force of Vapor.	Relative Humidity.	Force of Vapor.	Relative Humidity.	Force of Vapor.	Relative Humidity.	Force of Vapor.	Relative Humidity.	Force of Vapor.	Relative Humidity.
°		Eng. In.		Eng. In.		Eng. In.		Eng. In.		Eng. In.		Eng. In.	
68	0.0023	0.685	100	0.678	97.3	0.671	94.7	0.665	92.2	0.658	89.8	0.651	87.3
69	.0024	0.708	100	0.702	97.4	0.695	94.8	0.688	92.3	0.682	89.9	0.675	87.5
70	.0025	0.733	100	0.726	97.4	0.720	94.9	0.713	92.4	0.706	90.0	0.699	87.7
71	.0026	0.758	100	0.752	97.5	0.745	95.0	0.738	92.5	0.731	90.2	0.725	87.9
72	.0027	0.784	100	0.778	97.5	0.771	95.0	0.764	92.7	0.757	90.3	0.751	88.0
73	.0028	0.811	100	0.805	97.5	0.798	95.1	0.791	92.7	0.784	90.4	0.778	88.2
74	.0028	0.839	100	0.832	97.6	0.826	95.2	0.819	92.8	0.812	90.6	0.805	88.3
75	.0029	0.868	100	0.861	97.6	0.854	95.2	0.847	92.9	0.841	90.7	0.834	88.5
76	.0030	0.897	100	0.890	97.6	0.883	95.3	0.877	93.0	0.870	90.8	0.863	88.6
77	.0031	0.927	100	0.920	97.7	0.914	95.4	0.907	93.1	0.900	90.9	0.893	88.8
78	.0032	0.958	100	0.951	97.7	0.945	95.4	0.938	93.2	0.931	91.0	0.924	88.9
79	.0033	0.990	100	0.983	97.7	0.977	95.5	0.970	93.3	0.963	91.1	0.956	89.0
80	.0034	1.023	100	1.016	97.7	1.010	95.5	1.003	93.4	0.996	91.2	0.989	89.2
81	.0035	1.057	100	1.050	97.8	1.044	95.6	1.037	93.4	1.030	91.3	1.023	89.3
82	.0036	1.092	100	1.085	97.8	1.079	95.6	1.072	93.5	1.065	91.4	1.058	89.4
83	.0037	1.128	100	1.121	97.8	1.115	95.7	1.108	93.6	1.101	91.5	1.094	89.5
84	.0038	1.165	100	1.158	97.8	1.152	95.7	1.145	93.6	1.138	91.6	1.131	89.6
85	.0039	1.203	100	1.196	97.9	1.189	95.8	1.183	93.7	1.176	91.7	1.169	89.7
86	.0040	1.242	100	1.235	97.9	1.228	95.8	1.222	93.8	1.215	91.8	1.208	89.8
87	.0041	1.282	100	1.275	97.9	1.268	95.9	1.263	93.8	1.256	91.9	1.249	90.0
88	.0042	1.323	100	1.317	97.9	1.310	95.9	1.303	93.9	1.296	92.0	1.289	90.1
89	.0044	1.366	100	1.359	97.9	1.352	95.9	1.345	94.0	1.339	92.0	1.332	90.2
90	.0045	1.410	100	1.403	98.0	1.396	96.0	1.389	94.0	1.382	92.1	1.375	90.3
91	.0046	1.455	100	1.448	98.0	1.441	96.0	1.434	94.1	1.427	92.2	1.420	90.3
92	.0048	1.501	100	1.494	98.0	1.487	96.1	1.480	94.1	1.473	92.3	1.466	90.4
93	.0049	1.548	100	1.541	98.0	1.535	96.1	1.528	94.2	1.521	92.4	1.514	90.5
94	.0050	1.597	100	1.590	98.1	1.583	96.1	1.576	94.3	1.569	92.4	1.562	90.6
95	.0051	1.647	100	1.640	98.1	1.633	96.2	1.626	94.3	1.619	92.5	1.612	90.7
96	.0053	1.698	100	1.691	98.1	1.684	96.2	1.677	94.4	1.670	92.6	1.664	90.8
97	.0054	1.751	100	1.744	98.1	1.739	96.2	1.730	94.4	1.723	92.6	1.716	90.9
98	.0056	1.805	100	1.798	98.1	1.791	96.3	1.784	94.5	1.777	92.7	1.770	90.9
99	.0057	1.861	100	1.854	98.1	1.847	96.3	1.840	94.5	1.833	92.8	1.826	91.0
100	.0059	1.918	100	1.911	98.2	1.904	96.3	1.897	94.6	1.890	92.8	1.883	91.1
101	.0060	1.977	100	1.970	98.2	1.963	96.4	1.956	94.6	1.949	92.9	1.942	91.2
102	.0062	2.037	100	2.030	98.2	2.023	96.4	2.016	94.7	2.009	92.9	2.002	91.2
103	.0063	2.098	100	2.092	98.2	2.085	96.4	2.078	94.7	2.071	93.0	2.064	91.3
104		2.162	100	2.155	98.2	2.148	96.5	2.141	94.7	2.134	93.1	2.127	91.4

Mean Horizontal Difference of Force of Vapor for each 0°.1 = 0.0013.

Temperature, Fahrenheit. — Force of Vapor in English Inches. — Relative Humidity in Hundredths.

Wet-Bulb Thermometer t′ Fahrenheit.	Mean Vertical Difference of Force of Vapor for each 0°.1.	t — t′, or Difference of Wet and Dry Bulb Thermometers.											
		3°.0		3°.5		4°.0		4°.5		5°.0		5°.5	
		Force of Vapor.	Relative Humidity.	Force of Vapor.	Relative Humidity.	Force of Vapor.	Relative Humidity.	Force of Vapor.	Relative Humidity.	Force of Vapor.	Relative Humidity.	Force of Vapor.	Relative Humidity.
°		Eng. In.		Eng. In.		Eng. In.		Eng. In.		Eng. In.		Eng. In.	
32	0.0007	0.142	69.8	0.136	65.3	0.129	61.0	0.123	56.8	0.116	52.7	0.110	48.8
33	.0007	0.149	70.5	0.143	66.1	0.136	61.9	0.130	57.7	0.123	53.7	0.117	50.0
34	.0008	0.157	71.2	0.150	66.9	0.144	62.8	0.137	58.6	0.131	54.7	0.124	51.2
35	.0008	0.165	71.9	0.158	67.7	0.152	63.6	0.145	59.5	0.139	55.7	0.132	52.3
36	.0008	0.173	72.6	0.166	68.5	0.160	64.5	0.153	60.5	0.147	56.7	0.140	53.4
37	.0009	0.181	73.2	0.175	69.2	0.168	65.3	0.162	61.4	0.155	57.7	0.149	54.5
38	.0009	0.190	73.8	0.183	69.9	0.177	66.1	0.170	62.3	0.164	58.7	0.157	55.5
39	.0010	0.199	74.4	0.192	70.6	0.186	66.9	0.179	63.2	0.173	59.7	0.166	56.5
40	.0010	0.208	75.0	0.202	71.3	0.195	67.7	0.189	64.1	0.182	60.7	0.176	57.5
41	.0010	0.218	75.6	0.211	72.0	0.205	68.4	0.198	65.0	0.192	61.7	0.185	58.5
42	.0010	0.228	76.2	0.221	72.6	0.215	69.1	0.208	65.7	0.202	62.4	0.195	59.4
43	.0011	0.238	76.7	0.232	73.2	0.225	69.8	0.219	66.3	0.212	63.1	0.205	60.2
44	.0011	0.249	77.2	0.242	73.7	0.236	70.4	0.229	67.0	0.223	63.8	0.216	61.1
45	.0011	0.260	77.7	0.253	74.3	0.247	71.0	0.240	67.6	0.234	64.6	0.227	61.8
46	.0012	0.271	78.1	0.265	74.8	0.258	71.6	0.252	68.3	0.245	65.3	0.238	62.6
47	.0012	0.283	78.6	0.277	75.3	0.270	72.2	0.264	68.9	0.257	66.0	0.250	63.3
48	.0013	0.296	79.0	0.289	75.8	0.282	72.7	0.276	69.6	0.269	66.7	0.263	64.0
49	.0013	0.308	79.4	0.302	76.3	0.295	73.3	0.288	70.2	0.282	67.4	0.275	64.7
50	.0013	0.321	79.8	0.315	76.7	0.308	73.8	0.301	70.9	0.295	68.1	0.288	65.4
51	.0014	0.335	80.2	0.328	77.2	0.321	74.3	0.315	71.4	0.308	68.7	0.302	66.0
52	.0014	0.349	80.5	0.342	77.6	0.335	74.7	0.329	71.9	0.322	69.2	0.315	66.6
53	.0015	0.363	80.9	0.356	78.0	0.350	75.2	0.343	72.5	0.336	69.8	0.330	67.2
54	.0015	0.378	81.2	0.371	78.4	0.365	75.6	0.358	72.9	0.351	70.3	0.345	67.8
55	.0016	0.393	81.6	0.387	78.8	0.380	76.1	0.373	73.4	0.367	70.8	0.360	68.3
56	.0016	0.409	81.9	0.403	79.1	0.396	76.5	0.389	73.9	0.383	71.3	0.376	68.9
57	.0017	0.426	82.2	0.419	79.5	0.412	76.9	0.406	74.3	0.399	71.8	0.392	69.4
58	.0017	0.443	82.5	0.436	79.8	0.429	77.2	0.423	74.8	0.416	72.3	0.409	69.9
59	.0018	0.460	82.8	0.453	80.2	0.447	77.6	0.440	75.1	0.433	72.7	0.427	70.3
60	.0019	0.478	83.1	0.471	80.5	0.465	78.0	0.458	75.5	0.451	73.1	0.445	70.8
61	.0019	0.497	83.3	0.490	80.8	0.483	78.3	0.477	75.9	0.470	73.5	0.463	71.3
62	.0020	0.516	83.6	0.509	81.1	0.502	78.6	0.496	76.3	0.489	74.0	0.482	71.7
63	.0020	0.536	83.8	0.529	81.4	0.522	79.0	0.516	76.6	0.509	74.3	0.502	72.1
64	.0021	0.556	84.1	0.549	81.7	0.543	79.3	0.536	77.0	0.529	74.7	0.523	72.5
65	.0022	0.577	84.3	0.570	81.9	0.564	79.6	0.557	77.3	0.550	75.1	0.544	72.9
66	.0023	0.599	84.6	0.592	82.2	0.586	79.9	0.579	77.6	0.572	75.4	0.566	73.3
67		0.622	84.8	0.615	82.4	0.608	80.2	0.601	78.0	0.595	75.8	0.588	73.7

Mean Horizontal Difference of Force of Vapor for each 0°.1 = 0.0013.

Temperature, Fahrenheit. — Force of Vapor in English Inches. — Relative Humidity in Hundredths.

Wet-Bulb Thermometer t' Fahrenheit.	Mean Vertical Difference of Force of Vapor for each 0°.1.	t — t', or Difference of Wet and Dry Bulb Thermometers.											
		3°.0		3°.5		4°.0		4°.5		5°.0		5°.5	
		Force of Vapor.	Relative Humidity.	Force of Vapor.	Relative Humidity.	Force of Vapor.	Relative Humidity.	Force of Vapor.	Relative Humidity.	Force of Vapor.	Relative Humidity.	Force of Vapor.	Relative Humidity.
°		Eng. In.		Eng. In.		Eng. In.		Eng. In.		Eng. In.		Eng. In.	
68	0.0024	0.644	85.0	0.638	82.7	0.631	80.4	0.624	78.3	0.618	76.1	0.611	74.0
69	.0024	0.668	85.2	0.661	82.9	0.655	80.7	0.648	78.6	0.641	76.4	0.635	74.4
70	.0025	0.693	85.4	0.686	83.2	0.679	81.0	0.672	78.8	0.666	76.8	0.659	74.7
71	.0026	0.718	85.6	0.711	83.4	0.704	81.2	0.698	79.1	0.691	77.1	0.684	75.1
72	.0027	0.744	85.8	0.737	83.6	0.731	81.5	0.724	79.4	0.718	77.4	0.710	75.4
73	.0028	0.771	86.0	0.764	83.8	0.757	81.7	0.751	79.7	0.744	77.6	0.737	75.7
74	.0028	0.799	86.2	0.792	84.0	0.785	81.9	0.778	79.9	0.772	77.9	0.765	76.0
75	.0029	0.827	86.3	0.820	84.2	0.814	82.2	0.807	80.2	0.800	78.2	0.793	76.3
76	.0030	0.856	86.5	0.850	84.4	0.843	82.4	0.836	80.4	0.829	78.4	0.823	76.6
77	.0031	0.887	86.7	0.880	84.6	0.873	82.6	0.866	80.6	0.860	78.7	0.853	76.8
78	.0032	0.918	86.8	0.911	84.8	0.904	82.8	0.897	80.8	0.890	78.9	0.884	77.1
79	.0033	0.949	87.0	0.943	85.0	0.936	83.0	0.929	81.1	0.922	79.2	0.916	77.4
80	.0034	0.982	87.1	0.976	85.1	0.969	83.2	0.962	81.3	0.955	79.4	0.949	77.6
81	.0035	1.016	87.3	1.010	85.3	1.003	83.4	0.996	81.5	0.989	79.7	0.982	77.9
82	.0036	1.051	87.4	1.045	85.5	1.038	83.6	1.031	81.7	1.024	79.9	1.017	78.1
83	.0037	1.087	87.5	1.080	85.6	1.074	83.7	1.067	81.9	1.060	80.1	1.053	78.3
84	.0038	1.124	87.7	1.117	85.8	1.111	83.9	1.104	82.1	1.096	80.3	1.090	78.5
85	.0039	1.162	87.8	1.155	85.9	1.148	84.1	1.142	82.3	1.135	80.5	1.128	78.8
86	.0040	1.201	87.9	1.194	86.1	1.187	84.2	1.181	82.4	1.174	80.7	1.167	79.0
87	.0041	1.242	88.1	1.235	86.2	1.228	84.4	1.222	82.6	1.215	80.9	1.208	79.2
88	.0042	1.282	88.2	1.276	86.3	1.269	84.6	1.262	82.8	1.255	81.1	1.248	79.4
89	.0044	1.325	88.3	1.318	86.5	1.311	84.7	1.304	83.0	1.297	81.3	1.291	79.6
90	.0045	1.369	88.4	1.362	86.6	1.355	84.9	1.348	83.1	1.341	81.4	1.334	79.8
91	.0046	1.413	88.5	1.407	86.7	1.400	85.0	1.393	83.3	1.386	81.6	1.379	80.0
92	.0047	1.460	88.6	1.453	86.9	1.446	85.1	1.439	83.4	1.432	81.8	1.425	80.2
93	.0049	1.507	88.7	1.500	87.0	1.493	85.3	1.486	83.6	1.480	82.0	1.473	80.3
94	.0050	1.556	88.8	1.549	87.1	1.542	85.4	1.535	83.8	1.528	82.1	1.521	80.5
95	.0051	1.606	88.9	1.599	87.2	1.592	85.5	1.585	83.9	1.578	82.3	1.571	80.7
96	.0052	1.657	89.0	1.650	87.3	1.643	85.7	1.636	84.0	1.629	82.4	1.622	80.9
97	.0054	1.709	89.1	1.702	87.5	1.696	85.8	1.688	84.2	1.682	82.6	1.675	81.0
98	.0055	1.764	89.2	1.757	87.6	1.750	85.9	1.743	84.3	1.736	82.7	1.729	81.2
99	.0057	1.819	89.3	1.812	87.7	1.805	86.0	1.798	84.4	1.792	82.9	1.785	81.3
100	.0058	1.876	89.4	1.869	87.8	1.863	86.2	1.856	84.6	1.849	83.0	1.842	81.5
101	.0060	1.935	89.5	1.928	87.9	1.921	86.3	1.914	84.7	1.907	83.2	1.900	81.6
102	.0062	1.995	89.6	1.988	88.0	1.981	86.4	1.974	84.8	1.967	83.3	1.961	81.8
103	.0063	2.057	89.7	2.050	88.1	2.043	86.5	2.036	84.9	2.029	83.4	2.022	81.9
104		2.120	89.8	2.113	88.2	2.106	86.6	2.099	85.1	2.092	83.5	2.085	82.1

Mean Horizontal Difference of Force of Vapor for each 0°.1 = 0.0013.

Temperature, Fahrenheit. — Force of Vapor in English Inches. — Relative Humidity in Hundredths.

Wet-Bulb Thermometer t′ Fahrenheit.	Mean Vertical Difference of Force of Vapor for each 0°.1.	t — t′, or Difference of Wet and Dry Bulb Thermometers.											
		6°.0		6°.5		7°.0		7°.5		8°.0		8°.5	
		Force of Vapor.	Relative Humidity.	Force of Vapor.	Relative Humidity.	Force of Vapor.	Relative Humidity.	Force of Vapor.	Relative Humidity.	Force of Vapor.	Relative Humidity.	Force of Vapor.	Relative Humidity.
°		Eng. In.		Eng. In.		Eng. In.		Eng. In.		Eng. In.		Eng. In.	
32	0.0007	0.103	45.0	0.097	41.4	0.090	37.9	0.084	34.5	0.077	31.2	0.071	28.0
33	.0007	0.110	46.3	0.104	42.7	0.097	39.3	0.091	36.0	0.084	32.8	0.078	29.6
34	.0008	0.118	47.6	0.111	44.1	0.105	40.7	0.098	37.4	0.092	34.3	0.085	31.2
35	.0008	0.126	48.8	0.119	45.3	0.113	42.0	0.106	38.8	0.100	35.7	0.093	32.8
36	.0009	0.134	50.0	0.127	46.6	0.121	43.3	0.114	40.2	0.108	37.2	0.101	34.3
37	.0009	0.142	51.1	0.136	47.8	0.129	44.6	0.123	41.6	0.116	38.6	0.109	35.7
38	.0009	0.151	52.2	0.144	49.0	0.138	45.9	0.131	42.9	0.125	40.0	0.118	37.2
39	.0009	0.160	53.3	0.153	50.1	0.147	47.1	0.140	44.1	0.134	41.3	0.127	38.5
40	.0010	0.169	54.3	0.163	51.3	0.156	48.3	0.149	45.4	0.143	42.6	0.136	39.9
41	.0010	0.179	55.4	0.172	52.3	0.166	49.4	0.159	46.6	0.153	43.9	0.146	41.2
42	.0010	0.189	56.3	0.182	53.4	0.175	50.5	0.169	47.7	0.162	45.0	0.156	42.4
43	.0011	0.199	57.2	0.192	54.3	0.186	51.5	0.179	48.8	0.173	46.1	0.166	43.6
44	.0011	0.209	58.1	0.203	55.3	0.196	52.5	0.190	49.8	0.183	47.2	0.177	44.7
45	.0011	0.220	59.0	0.214	56.2	0.207	53.5	0.201	50.8	0.194	48.3	0.188	45.8
46	.0012	0.232	59.8	0.225	57.0	0.219	54.4	0.212	51.8	0.206	49.3	0.198	46.9
47	.0012	0.244	60.6	0.237	57.9	0.231	55.2	0.224	52.7	0.217	50.2	0.211	47.9
48	.0013	0.256	61.3	0.249	58.7	0.243	56.1	0.236	53.6	0.230	51.2	0.223	48.8
49	.0013	0.269	62.0	0.262	59.4	0.255	56.9	0.249	54.5	0.242	52.1	0.236	49.7
50	.0013	0.282	62.7	0.275	60.2	0.268	57.7	0.262	55.3	0.255	52.9	0.249	50.6
51	.0014	0.295	63.4	0.288	60.9	0.282	58.4	0.275	56.1	0.269	53.7	0.262	51.5
52	.0014	0.309	64.1	0.302	61.6	0.296	59.2	0.289	56.8	0.282	54.6	0.276	52.3
53	.0015	0.323	64.7	0.317	62.3	0.310	59.9	0.303	57.6	0.297	55.3	0.290	53.2
54	.0015	0.338	65.3	0.332	62.9	0.325	60.6	0.318	58.3	0.312	56.1	0.305	53.9
55	.0016	0.354	65.9	0.347	63.5	0.340	61.2	0.334	59.0	0.327	56.8	0.320	54.9
56	.0017	0.369	66.5	0.363	64.1	0.356	61.9	0.349	59.7	0.343	57.5	0.336	55.4
57	.0017	0.386	67.0	0.379	64.7	0.373	62.5	0.366	60.3	0.359	58.2	0.353	56.1
58	.0017	0.403	67.5	0.396	65.3	0.389	63.1	0.383	60.9	0.376	58.8	0.369	56.8
59	.0018	0.420	68.0	0.413	65.8	0.407	63.6	0.400	61.5	0.393	59.5	0.387	57.5
60	.0018	0.438	68.5	0.431	66.3	0.425	64.2	0.418	62.1	0.411	60.1	0.405	58.1
61	.0019	0.457	69.0	0.450	66.9	0.443	64.7	0.436	62.7	0.430	60.7	0.423	58.7
62	.0020	0.476	69.5	0.469	67.4	0.462	65.3	0.456	63.2	0.449	61.3	0.442	59.3
63	.0021	0.495	70.0	0.489	67.8	0.482	65.8	0.475	63.8	0.469	61.8	0.462	59.9
64	.0021	0.516	70.4	0.509	68.3	0.503	66.3	0.496	64.3	0.489	62.4	0.483	60.5
65	.0022	0.537	70.8	0.530	68.8	0.524	66.8	0.517	64.8	0.510	62.9	0.504	61.0
66	.0023	0.559	71.2	0.552	69.2	0.545	67.2	0.539	65.3	0.532	63.4	0.525	61.6
67		0.581	71.6	0.575	69.6	0.568	67.7	0.561	65.7	0.554	63.9	0.549	62.1

Mean Horizontal Difference of Force of Vapor for each 0°.1 = 0.0013.

Temperature, Fahrenheit. — Force of Vapor in English Inches. — Relative Humidity in Hundredths.

Wet-Bulb Thermometer t′ Fahrenheit.	Mean Vertical Difference of Force of Vapor for each 0°.1.	t — t′, or Difference of Wet and Dry Bulb Thermometers.											
		6°.0		6°.5		7°.0		7°.5		8°.0		8°.5	
		Force of Vapor.	Relative Humidity.	Force of Vapor.	Relative Humidity.	Force of Vapor.	Relative Humidity.	Force of Vapor.	Relative Humidity.	Force of Vapor.	Relative Humidity.	Force of Vapor.	Relative Humidity.
°		Eng. In.		Eng. In.		Eng. In.		Eng. In.		Eng. In.		Eng. In.	
68	0.0024	0.604	72.0	0.597	70.0	0.591	68.1	0.584	66.2	0.577	64.4	0.571	62.6
69	.0024	0.628	72.4	0.621	70.4	0.614	68.5	0.608	66.6	0.601	64.8	0.594	63.0
70	.0025	0.652	72.7	0.646	70.8	0.639	68.9	0.632	67.1	0.625	65.3	0.619	63.5
71	.0026	0.678	73.1	0.671	71.2	0.664	69.3	0.657	67.5	0.651	65.7	0.644	64.0
72	.0027	0.704	73.4	0.697	71.5	0.690	69.7	0.683	67.9	0.677	66.1	0.670	64.4
73	.0028	0.730	73.8	0.724	71.9	0.717	70.1	0.710	68.3	0.703	66.5	0.697	64.8
74	.0029	0.758	74.1	0.751	72.2	0.745	70.4	0.738	68.7	0.731	66.9	0.724	65.3
75	.0030	0.787	74.4	0.780	72.6	0.773	70.8	0.766	69.0	0.760	67.3	0.753	65.7
76	.0030	0.816	74.7	0.809	72.9	0.802	71.1	0.796	69.4	0.789	67.7	0.782	66.1
77	.0031	0.846	75.0	0.839	73.2	0.832	71.4	0.826	69.7	0.819	68.1	0.812	66.4
78	.0032	0.877	75.3	0.870	73.5	0.863	71.8	0.857	70.1	0.850	68.4	0.843	66.8
79	.0033	0.909	75.6	0.902	73.8	0.895	72.1	0.888	70.4	0.882	68.8	0.875	67.2
80	.0034	0.942	75.8	0.935	74.1	0.928	72.4	0.921	70.7	0.915	69.1	0.908	67.5
81	.0035	0.976	76.1	0.969	74.4	0.962	72.7	0.955	71.0	0.948	69.4	0.942	67.9
82	.0036	1.011	76.4	1.004	74.6	0.997	73.0	0.990	71.3	0.983	69.8	0.977	68.2
83	.0037	1.046	76.6	1.040	74.9	1.033	73.3	1.026	71.6	1.019	70.1	1.012	68.5
84	.0038	1.083	76.8	1.077	75.2	1.070	73.5	1.063	71.9	1.056	70.4	1.049	68.8
85	.0038	1.121	77.1	1.114	75.4	1.108	73.8	1.101	72.2	1.094	70.7	1.087	69.1
86	.0039	1.160	77.3	1.153	75.7	1.147	74.1	1.140	72.5	1.133	70.9	1.126	69.4
87	.0040	1.201	77.5	1.194	75.9	1.187	74.3	1.181	72.7	1.174	71.2	1.167	69.7
88	.0042	1.241	77.7	1.235	76.1	1.228	74.6	1.221	73.0	1.214	71.5	1.207	70.0
89	.0044	1.284	78.0	1.277	76.4	1.270	74.8	1.263	73.3	1.256	71.8	1.250	70.3
90	.0045	1.327	78.2	1.321	76.6	1.314	75.0	1.307	73.5	1.300	72.0	1.293	70.6
91	.0046	1.372	78.4	1.365	76.8	1.359	75.3	1.352	73.7	1.345	72.3	1.338	70.8
92	.0047	1.418	78.6	1.412	77.0	1.405	75.5	1.398	74.0	1.391	72.5	1.384	71.1
93	.0049	1.466	78.8	1.459	77.2	1.452	75.7	1.445	74.2	1.438	72.8	1.431	71.3
94	.0050	1.514	79.0	1.507	77.4	1.501	75.9	1.494	74.4	1.487	73.0	1.480	71.6
95	.0051	1.564	79.1	1.557	77.6	1.550	76.1	1.544	74.7	1.537	73.2	1.530	71.8
96	.0052	1.615	79.3	1.608	77.8	1.602	76.3	1.595	74.9	1.588	73.4	1.581	72.1
97	.0054	1.668	79.5	1.661	78.0	1.654	76.5	1.647	75.1	1.640	73.7	1.633	72.3
98	.0056	1.722	79.7	1.715	78.2	1.708	76.7	1.701	75.3	1.694	73.9	1.688	72.5
99	.0057	1.778	79.8	1.771	78.4	1.764	76.9	1.757	75.5	1.750	74.1	1.743	72.7
100	.0059	1.835	80.0	1.828	78.5	1.821	77.1	1.814	75.7	1.807	74.3	1.800	72.9
101	.0060	1.893	80.2	1.887	78.7	1.880	77.3	1.873	75.9	1.866	74.5	1.859	73.2
102	.0061	1.954	80.3	1.947	78.9	1.940	77.4	1.933	76.1	1.926	74.7	1.919	73.4
103	.0063	2.015	80.5	2.008	79.0	2.001	77.6	1.994	76.2	1.987	74.9	1.980	73.6
104		2.078	80.6	2.071	79.2	2.064	77.8	2.057	76.4	2.051	75.1	2.044	73.8

Mean Horizontal Difference of Force of Vapor for each 0°.1 = 0.0013.

Temperature, Fahrenheit. — Force of Vapor in English Inches. — Relative Humidity in Hundredths.

Wet-Bulb Thermometer t′ Fahrenheit.	Mean Vertical Difference of Force of Vapor for each 0°.1.	t — t′, or Difference of Wet and Dry Bulb Thermometers.											
		9°.0		9°.5		10°.0		10°.5		11°.0		11°.5	
		Force of Vapor.	Relative Humidity.	Force of Vapor.	Relative Humidity.	Force of Vapor.	Relative Humidity.	Force of Vapor.	Relative Humidity.	Force of Vapor.	Relative Humidity.	Force of Vapor.	Relative Humidity.
°		Eng. In.		Eng. In.		Eng. In.		Eng. In.		Eng. In.		Eng. In	
32	0.0007	0.064	25.0	0.058	22.0	0.051	19.2	0.045	16.4	0.038	13.8	0.032	11.2
33	.0007	0.071	26.7	0.065	23.8	0.058	21.0	0.052	18.3	0.045	15.7	0.039	13.2
34	.0008	0.079	28.3	0.072	25.5	0.066	22.7	0.059	20.1	0.053	17.5	0.046	15.1
35	.0008	0.087	29.9	0.080	27.1	0.074	24.4	0.067	21.8	0.061	19.3	0.054	16.9
36	.0008	0.095	31.4	0.088	28.7	0.082	26.0	0.075	23.5	0.069	21.1	0.062	18.7
37	.0009	0.103	33.0	0.096	30.3	0.090	27.6	0.083	25.2	0.077	22.8	0.070	20.4
38	.0009	0.112	34.4	0.105	31.8	0.099	29.2	0.092	26.8	0.086	24.4	0.079	22.1
39	.0009	0.121	35.9	0.114	33.3	0.108	30.7	0.101	28.4	0.094	26.1	0.088	23.8
40	.0010	0.130	37.3	0.123	34.8	0.117	32.2	0.110	29.9	0.104	27.6	0.097	25.4
41	.0010	0.139	38.6	0.133	36.2	0.126	33.7	0.120	31.4	0.113	29.2	0.107	27.0
42	.0010	0.149	39.9	0.143	37.5	0.136	35.0	0.130	32.8	0.123	30.6	0.116	28.4
43	.0010	0.160	41.1	0.153	38.7	0.146	36.3	0.140	34.1	0.133	32.0	0.127	29.8
44	.0011	0.170	42.3	0.163	39.9	0.157	37.6	0.150	35.4	0.144	33.3	0.137	31.2
45	.0011	0.181	43.4	0.175	41.1	0.168	38.8	0.161	36.7	0.155	34.6	0.148	32.5
46	.0012	0.192	44.5	0.186	42.2	0.179	39.9	0.173	37.9	0.166	35.8	0.160	33.8
47	.0012	0.204	45.5	0.198	43.3	0.191	41.1	0.185	39.0	0.178	37.0	0.171	35.0
48	.0012	0.217	46.5	0.210	44.3	0.203	42.1	0.197	40.1	0.190	38.1	0.184	36.1
49	.0013	0.229	47.5	0.222	45.3	0.216	43.2	0.209	41.2	0.203	39.2	0.196	37.2
50	.0013	0.242	48.4	0.235	46.3	0.229	44.2	0.222	42.2	0.216	40.2	0.209	38.3
51	.0014	0.255	49.3	0.249	47.2	0.242	45.2	0.236	43.2	0.229	41.2	0.222	39.3
52	.0015	0.269	50.2	0.263	48.1	0.256	46.1	0.249	44.1	0.243	42.2	0.236	40.3
53	.0015	0.284	51.1	0.277	49.0	0.270	47.0	0.264	45.1	0.257	43.2	0.250	41.3
54	.0015	0.298	51.9	0.292	49.8	0.285	47.9	0.279	46.0	0.272	44.1	0.265	42.3
55	.0016	0.314	52.7	0.307	50.7	0.300	48.7	0.294	46.8	0.287	45.0	0.281	43.2
56	.0016	0.330	53.5	0.323	51.4	0.316	49.5	0.310	47.7	0.303	45.9	0.296	44.1
57	.0017	0.346	54.3	0.339	52.2	0.333	50.3	0.326	48.5	0.319	46.7	0.313	44.9
58	.0017	0.363	55.0	0.356	52.9	0.350	51.1	0.343	49.2	0.336	47.5	0.330	45.7
59	.0018	0.380	55.7	0.373	53.6	0.367	51.8	0.360	50.0	0.354	48.2	0.347	46.5
60	.0018	0.398	56.4	0.391	54.3	0.385	52.5	0.378	50.7	0.371	49.0	0.365	47.3
61	.0019	0.416	57.0	0.410	55.0	0.403	53.2	0.396	51.4	0.390	49.7	0.383	48.1
62	.0020	0.436	57.6	0.429	55.6	0.422	53.9	0.416	52.1	0.409	50.4	0.402	48.8
63	.0021	0.455	58.2	0.449	56.3	0.442	54.5	0.435	52.8	0.429	51.1	0.422	49.5
64	.0021	0.476	58.8	0.469	56.9	0.462	55.1	0.456	53.4	0.449	51.8	0.442	50.2
65	.0022	0.497	59.3	0.490	57.5	0.483	55.8	0.477	54.1	0.470	52.4	0.463	50.8
66	.0023	0.519	59.9	0.512	58.0	0.505	56.3	0.498	54.7	0.492	53.1	0.485	51.5
67		0.542	60.3	0.534	58.6	0.527	56.9	0.521	55.3	0.514	53.7	0.507	52.1

Mean Horizontal Difference of Force of Vapor for each 0°.1 = 0.0013.

Temperature, Fahrenheit. — Force of Vapor in English Inches. — Relative Humidity in Hundredths.

Wet-Bulb Thermometer t′ Fahrenheit.	Mean Vertical Difference of Force of Vapor for each 0°.1.	t — t′, or Difference of Wet and Dry Bulb Thermometers. 9°.0 Force of Vapor.	9°.0 Relative Humidity.	9°.5 Force of Vapor.	9°.5 Relative Humidity.	10°.0 Force of Vapor.	10°.0 Relative Humidity.	10°.5 Force of Vapor.	10°.5 Relative Humidity.	11°.0 Force of Vapor.	11°.0 Relative Humidity.	11°.5 Force of Vapor.	11°.5 Relative Humidity.
°		Eng. In.		Eng. In.		Eng. In.		Eng. In.		Eng. In.		Eng. In.	
68		0.564	60.8	0.557	59.1	0.550	57.4	0.544	55.8	0.537	54.2	0.530	52.7
69	0.0024	0.588	61.3	0.581	59.6	0.574	58.0	0.567	56.4	0.561	54.8	0.554	53.3
70	.0025	0.612	61.8	0.605	60.1	0.598	58.5	0.592	56.9	0.585	55.4	0.578	53.8
71	.0025	0.637	62.3	0.630	60.6	0.624	59.0	0.617	57.4	0.610	55.9	0.603	54.4
72	.0026	0.663	62.7	0.656	61.1	0.650	59.5	0.643	58.0	0.636	56.4	0.629	54.9
73	.0027	0.390	63.2	0.683	61.6	0.677	60.0	0.670	58.4	0.663	56.9	0.656	55.5
74	.0027	0.718	63.6	0.711	62.0	0.704	60.5	0.697	58.9	0.691	57.4	0.684	56.0
75	.0028	0.746	64.0	0.739	62.5	0.733	60.9	0.726	59.4	0.719	57.9	0.712	56.5
76	.0029	0.775	64.4	0.769	62.9	0.762	61.3	0.755	59.8	0.748	58.4	0.741	56.9
77	.0030	0.805	64.8	0.799	63.3	0.792	61.8	0.785	60.3	0.778	58.8	0.772	57.4
78	.0031	0.836	65.2	0.829	63.7	0.823	62.2	0.816	60.7	0.809	59.2	0.802	57.8
79	.0032	0.868	65.6	0.861	64.1	0.855	62.6	0.848	61.1	0.841	59.7	0.834	58.3
80	.0033	0.901	66.0	0.894	64.5	0.897	63.0	0.881	61.5	0.874	60.1	0.867	58.7
81	.0034	0.935	66.3	0.928	64.8	0.921	63.4	0.914	61.9	0.908	60.5	0.901	59.1
82	.0035	0.970	66.7	0.963	65.2	0.956	63.7	0.949	62.3	0.943	60.9	0.936	59.5
83	.0036	1.006	67.0	0.999	65.5	0.992	64.1	0.985	62.7	0.978	61.3	0.972	59.9
84	.0037	1.042	67.3	1.036	65.9	1.029	64.4	1.022	63.0	1.015	61.7	1.008	60.3
85	.0038	1.080	67.7	1.073	66.2	1.067	64.8	1.060	63.4	1.053	62.0	1.046	60.7
86	.0039	1.119	68.0	1.112	66.5	1.106	65.1	1.099	63.7	1.092	62.4	1.085	61.0
87	.0040	1.160	68.3	1.153	66.8	1.146	65.4	1.140	64.1	1.133	62.7	1.126	61.4
88	.0041	1.200	68.6	1.194	67.1	1.187	65.8	1.180	64.4	1.173	63.1	1.166	61.7
89	.0042	1.243	68.9	1.236	67.4	1.229	66.1	1.222	64.7	1.215	63.4	1.208	62.1
90	.0044	1.286	69.1	1.279	67.7	1.273	66.4	1.266	65.0	1.259	63.7	1.252	62.4
91	.0045	1.331	69.4	1.324	68.0	1.317	66.7	1.311	65.3	1.304	64.0	1.297	62.7
92	.0046	1.377	69.7	1.370	68.3	1.363	67.0	1.357	65.6	1.350	64.3	1.343	63.1
93	.0047	1.425	69.9	1.418	68.6	1.411	67.2	1.404	65.9	1.397	64.6	1.390	63.4
94	.0048	1.473	70.2	1.466	68.8	1.459	67.5	1.452	66.2	1.446	64.9	1.439	63.7
95	.0050	1.523	70.4	1.516	69.1	1.509	67.8	1.502	66.5	1.495	65.2	1.488	64.0
96	.0051	1.574	70.7	1.567	69.4	1.560	68.0	1.553	66.7	1.546	65.5	1.539	64.2
97	.0053	1.627	70.9	1.620	69.6	1.613	68.3	1.606	67.0	1.599	65.8	1.592	64.5
98	.0054	1.681	71.2	1.674	69.8	1.667	68.5	1.660	67.3	1.653	66.0	1.646	64.8
99	.0056	1.736	71.4	1.729	70.1	1.722	68.8	1.716	67.5	1.709	66.3	1.702	65.1
100	.0057	1.793	71.6	1.786	70.3	1.780	69.0	1.773	67.8	1.766	66.5	1.759	65.3
101	.0058	1.852	71.8	1.845	70.5	1.838	69.3	1.831	68.0	1.824	66.8	1.817	65.6
102	.0060	1.912	72.0	1.905	70.8	1.898	69.5	1.891	68.2	1.884	67.0	1.877	65.8
103	.0062	1.974	72.3	1.967	71.0	1.960	69.7	1.953	68.5	1.946	67.3	1.939	66.1
104	.0063	2.037	72.5	2.030	71.2	2.023	69.9	2.016	68.7	2.009	67.5	2.002	66.3

Mean Horizontal Difference of Force of Vapor for each 0°.1 = 0.0013.

Temperature, Fahrenheit. — Force of Vapor in English Inches. — Relative Humidity in Hundredths.

Wet-Bulb Thermometer t′ Fahrenheit.	Mean Vertical Difference of Force of Vapor for each 0°.1.	t — t′, or Difference of Wet and Dry Bulb Thermometers.											
		12°.0		12°.5		13°.0		13°.5		14°.0		14°.5	
		Force of Vapor.	Relative Humidity.	Force of Vapor.	Relative Humidity.	Force of Vapor.	Relative Humidity.	Force of Vapor.	Relative Humidity.	Force of Vapor.	Relative Humidity.	Force of Vapor.	Relative Humidity.
°		Eng. In.		Eng. In.		Eng. In.		Eng. In.		Eng. In.		Eng. In.	
32	0.0007	0.025	8.8	0.019	6.4	0.012	4.1						
33	.0007	0.032	10.8	0.026	8.4	0.019	6.2	0.013	4.0				
34	.0007	0.040	12.7	0.033	10.4	0.027	8.2	0.020	6.0	0.014	4.1		
35	.0008	0.048	14.6	0.041	12.3	0.034	10.1	0.028	8.0	0.021	6.1	0.015	4.2
36	.0008	0.056	16.4	0.049	14.2	0.042	12.0	0.036	10.0	0.029	8.1	0.023	6.2
37	.0009	0.064	18.2	0.057	16.0	0.051	13.9	0.044	11.9	0.038	10.0	0.031	8.2
38	.0009	0.072	19.9	0.066	17.8	0.059	15.7	0.053	13.7	0.046	11.9	0.040	10.1
39	.0009	0.081	21.6	0.075	19.5	0.068	17.5	0.062	15.5	0.055	13.7	0.049	11.9
40	.0010	0.091	23.3	0.084	21.2	0.078	19.2	0.071	17.2	0.064	15.4	0.058	13.6
41	.0010	0.100	24.9	0.094	22.8	0.087	20.8	0.081	18.9	0.074	17.1	0.067	15.3
42	.0010	0.110	26.4	0.103	24.3	0.097	22.4	0.090	20.5	0.084	18.6	0.077	16.8
43	.0011	0.120	27.8	0.114	25.8	0.107	23.9	0.100	22.0	0.095	20.1	0.087	18.3
44	.0011	0.131	29.2	0.124	27.2	0.118	25.3	0.111	23.5	0.104	21.5	0.098	19.8
45	.0011	0.142	30.5	0.135	28.6	0.129	26.7	0.122	24.9	0.115	22.9	0.109	21.2
46	.0012	0.153	31.8	0.146	30.0	0.140	28.1	0.133	26.3	0.127	24.3	0.119	22.7
47	.0012	0.165	33.0	0.158	31.2	0.152	29.3	0.145	27.6	0.138	25 7	0.132	24.0
48	.0013	0.177	34.2	0.170	32.4	0.164	30.6	0.157	28.8	0.151	27.0	0.144	25.4
49	.0013	0.190	35.3	0.183	33.5	0.176	31.7	0.170	30.0	0.163	28.3	0.157	26.7
50	.0014	0.202	36.4	0.196	34.6	0.189	32.9	0.183	31.2	0.176	29.5	0.169	27.9
51	.0014	0.216	37.5	0.209	35.7	0.202	34.0	0.196	32.3	0.189	30.7	0.183	29.1
52	.0014	0.229	38.5	0.223	36.8	0.216	35.1	0.210	33.4	0.203	31.8	0.196	30.2
53	.0015	0.244	39.5	0.237	37.8	0.231	36.1	0.224	34.5	0.217	32.9	0.211	31.4
54	.0015	0.259	40.5	0.252	38.8	0.245	37.1	0.239	35.5	0.232	34.0	0.226	32.4
55	.0016	0.274	41.5	0.267	39.8	0.261	38.1	0.254	36.5	0.247	35.0	0.241	33.5
56	.0016	0.290	42.4	0.283	40.7	0.276	39.1	0.270	37.5	0.263	35.9	0.257	34.4
57	.0017	0.306	43.2	0.299	41.6	0.293	40.0	0.286	38.4	0.280	36.9	0.273	35.4
58	.0017	0.323	44.1	0.316	42.4	0.310	40.8	0.303	39.3	0.296	37.8	0.290	36.3
59	.0018	0.340	44.9	0.334	43.3	0.327	41.7	0.320	40.1	0.314	38.7	0.307	37.2
60	.0018	0.358	45.7	0.351	44.1	0.345	42.5	0.338	41.0	0.331	39.5	0.325	38.1
61	.0019	0.376	46.4	0.370	44.9	0.363	43.3	0.356	41.8	0.350	40.3	0.343	38.9
62	.0020	0.396	47.2	0.389	45.6	0.382	44.1	0.376	42.6	0.369	41.2	0.362	39.8
63	.0021	0.415	47.9	0.409	46.4	0.402	44.8	0.395	43.4	0.389	41.9	0.382	40.6
64	.0021	0.436	48.6	0.429	47.1	0.422	45.6	0.416	44.1	0.409	42.7	0.402	41.3
65	.0022	0.457	49.3	0.450	47.8	0.443	46.3	0.437	44.8	0.431	43.4	0.423	42.1
66	.0023	0.478	49.9	0.472	48.4	0.465	47.0	0.458	45.5	0.452	44.1	0.445	42.8
67		0.501	50.6	0.494	49.1	0.487	47.6	0.481	46.2	0.474	44.8	0.467	43.5

Mean Horizontal Difference of Force of Vapor for each 0°.1 = 0.0013.

Temperature, Fahrenheit. — Force of Vapor in English Inches. — Relative Humidity in Hundredths.

Wet-Bulb Thermometer t′ Fahrenheit.	Mean Vertical Difference of Force of Vapor for each 0°.1.	t — t′, or Difference of Wet and Dry Bulb Thermometers.											
		12°.0		**12°.5**		**13°.0**		**13°.5**		**14°.0**		**14°.5**	
		Force of Vapor.	Relative Humidity.	Force of Vapor.	Relative Humidity.	Force of Vapor.	Relative Humidity.	Force of Vapor.	Relative Humidity.	Force of Vapor.	Relative Humidity.	Force of Vapor.	Relative Humidity.
°		Eng. In.		Eng. In.		Eng. In.		Eng. In.		Eng. In.		Eng. In.	
68	0.0024	0.524	51.2	0.517	49.7	0.510	48.3	0.503	46.9	0.497	45.5	0.490	44.1
69	.0024	0.547	51.8	0.541	50.3	0.534	48.9	0.527	47.5	0.520	46.1	0.514	44.8
70	.0025	0.572	52.4	0.565	50.9	0.558	49.5	0.551	48.1	0.545	46.8	0.538	45.5
71	.0026	0.597	52.9	0.590	51.5	0.583	50.1	0.577	48.7	0.570	47.4	0.563	46.1
72	.0026	0.623	53.5	0.616	52.1	0.609	50.7	0.603	49.3	0.596	48.0	0.589	46.7
73	.0027	0.650	54.0	0.643	52.6	0.636	51.3	0.629	49.9	0.623	48.6	0.616	47.3
74	.0028	0.677	54.5	0.670	53.2	0.664	51.8	0.657	50.5	0.650	49.2	0.643	47.9
75	.0029	0.705	55.0	0.699	53.7	0.692	52.3	0.685	51.0	0.678	49.7	0.672	48.4
76	.0030	0.735	55.5	0.728	54.2	0.721	52.9	0.714	51.5	0.708	50.3	0.701	48.9
77	.0031	0.765	56.0	0.759	54.7	0.752	53.4	0.745	52.1	0.739	50.8	0.731	49.5
78	.0032	0.796	56.5	0.782	55.2	0.782	53.8	0.775	52.5	0.768	51.3	0.762	50.0
79	.0033	0.827	56.9	0.821	55.6	0.814	54.3	0.807	53.0	0.800	51.8	0.794	50.5
80	.0034	0.860	57.3	0.853	56.1	0.847	54.8	0.840	53.5	0.833	52.2	0.826	51.0
81	.0035	0.894	57.8	0.887	56.5	0.880	55.2	0.874	53.9	0.867	52.7	0.860	51.4
82	.0036	0.929	58.2	0.922	56.9	0.915	55.6	0.909	54.4	0.902	53.2	0.895	51.9
83	.0037	0.965	58.6	0.958	57.3	0.951	56.1	0.944	54.8	0.937	53.6	0.931	52.4
84	.0038	1.002	59.0	0.995	57.7	0.988	56.5	0.981	55.2	0.974	54.0	0.968	52.8
85	.0039	1.039	59.4	1.033	58.1	1.026	56.8	1.019	55.6	1.012	54.4	1.005	53.2
86	.0040	1.078	59.7	1.071	58.5	1.065	57.2	1.058	56.0	1.051	54.8	1.044	53.6
87	.0041	1.119	60.1	1.112	58.8	1.105	57.6	1.099	56.4	1.092	55.2	1.085	54.0
88	.0042	1.159	60.5	1.152	59.2	1.146	58.0	1.139	56.8	1.132	55.6	1.125	54.4
89	.0044	1.202	60.9	1.195	59.6	1.188	58.3	1.181	57.1	1.174	56.0	1.167	54.8
90	.0045	1.245	61.3	1.238	59.9	1.231	58.7	1.225	57.5	1.218	56.3	1.211	55.2
91	.0046	1.290	61.6	1.283	60.2	1.276	59.0	1.269	57.9	1.263	66.7	1.256	55.6
92	.0047	1.336	61.9	1.329	60.6	1.322	59.4	1.315	58.2	1.309	57.0	1.302	55.9
93	.0049	1.383	62.2	1.376	60.9	1.370	59.7	1.363	58.5	1.356	57.4	1.349	56.3
94	.0050	1.432	62.5	1.425	61.2	1.418	60.0	1.411	58.9	1.404	57.7	1.397	56.6
95	.0051	1.482	62.7	1.475	61.5	1.468	60.4	1.461	59.2	1.454	58.1	1.447	57.0
96	.0052	1.533	63.0	1.526	61.8	1.519	60.7	1.512	59.5	1.505	58.4	1.498	57.3
97	.0054	1.585	63.3	1.578	62.1	1.571	61.0	1.564	59.8	1.558	58.7	1.551	57.6
98	.0056	1.639	63.6	1.632	62.4	1.625	61.3	1.618	60.1	1.612	59.0	1.605	57.9
99	.0057	1.695	63.9	1.688	62.7	1.681	61.6	1.674	60.4	1.667	59.3	1.660	58.2
100	.0059	1.752	64.2	1.745	63.0	1.738	62.0	1.731	60.7	1.724	59.6	1.717	58.5
101	.0060	1.810	64.4	1.803	63.2	1.797	62.3	1.790	61.0	1.783	59.9	1.776	58.8
102	.0062	1.870	64.7	1.863	63.5	1.857	62.6	1.850	61.3	1.843	60.2	1.836	59.1
103	.0063	1.932	64.9	1.925	63.8	1.918	62.9	1.911	61.5	1.904	60.4	1.897	59.4
104		1.995	65.2	1.988	64.0	1.981	63.2	1.974	61.8	1.967	60.7	1.960	59.6

Mean Horizontal Difference of Force of Vapor for each 0°.1 = 0.0013.

Temperature, Fahrenheit. — Force of Vapor in English Inches. — Relative Humidity in Hundredths.

Wet-Bulb Thermometer t′ Fahrenheit.	Mean Vertical Difference of Force of Vapor for each 0°.1.	t — t′, or Difference of Wet and Dry Bulb Thermometers.											
		15°.0		15°.5		16°.0		16°.5		17°.0		17°.5	
		Force of Vapor.	Relative Humidity.	Force of Vapor.	Relative Humidity.	Force of Vapor.	Relative Humidity.	Force of Vapor.	Relative Humidity.	Force of Vapor.	Relative Humidity.	Force of Vapor.	Relative Humidity.
°		Eng. In.		Eng. In.		Eng. In.		Eng. In.		Eng. In.		Eng. In.	
32													
33													
34													
35													
36		0.016	4.4										
37	0.0009	0.025	6.4	0.018	4.6								
38	.0009	0.033	8.3	0.027	6.5	0.020	4.8	0.014	3.2				
39	.0009	0.042	10.1	0.036	8.4	0.029	6.7	0.023	5.1	0.016	3.6	0.010	2.1
40	.0009	0.051	11.9	0.045	10.1	0.038	8.5	0.032	6.9	0.025	5.4	0.019	3.9
41	.0010	0.061	13.6	0.054	11.8	0.048	10.3	0.041	8.7	0.035	7.2	0.028	5.7
42	.0010	0.071	15.1	0.064	13.4	0.058	11.9	0.051	10.3	0.044	8.8	0.038	7.4
43	.0010	0.081	16.6	0.074	15.0	0.068	13.4	0.061	11.9	0.055	10.4	0.048	9.0
44	.0011	0.091	18.1	0.085	16.5	0.078	15.0	0.072	13.5	0.065	12.0	0.058	10.6
45	.0011	0.102	19.6	0.096	18.0	0.089	16.5	0.083	15.0	0.076	13.5	0.069	12.1
46	.0011	0.114	21.0	0.107	19.4	0.100	17.9	0.094	16.4	0.087	15.0	0.081	13.6
47	.0012	0.125	22.4	0.119	20.8	0.112	19.3	0.106	17.9	0.099	16.5	0.092	15.1
48	.0012	0.137	23.8	0.131	22.2	0.124	20.7	0.118	19.3	0.111	17.9	0.104	16.5
49	.0013	0.150	25.1	0.143	23.6	0.137	22.1	0.130	20.7	0.124	19.3	0.117	17.9
50	.0013	0.163	26.4	0.156	24.9	0.150	23.4	0.143	22.0	0.136	20.6	0.130	19.3
51	.0013	0.176	27.6	0.169	26.1	0.163	24.6	0.156	23.2	0.150	21.9	0.143	20.6
52	.0014	0.190	28.7	0.183	27.3	0.177	25.8	0.170	24.4	0.163	23.1	0.157	21.8
53	.0014	0.204	29.9	0.197	28.4	0.191	27.0	0.184	25.6	0.178	24.3	0.171	23.0
54	.0015	0.219	30.9	0.212	29.5	0.206	28.1	0.199	26.7	0.192	25.4	0.186	24.1
55	.0015	0.234	32.0	0.228	30.6	0.221	29.2	0.214	27.8	0.208	26.5	0.201	25.2
56	.0016	0.250	33.0	0.243	31.6	0.237	30.2	0.230	28.9	0.223	27.6	0.217	26.3
57	.0016	0.266	34.0	0.260	32.6	0.253	31.2	0.246	29.9	0.240	28.6	0.233	27.3
58	.0017	0.283	34.9	0.276	33.5	0.270	32.2	0.268	30.8	0.256	29.6	0.249	28.3
59	.0017	0.300	35.8	0.294	34.4	0.287	33.1	0.280	31.8	0.274	30.5	0.267	29.3
60	.0018	0.318	36.7	0.311	35.3	0.305	34.0	0.298	32.7	0.291	31.4	0.285	30.2
61	.0019	0.336	37.5	0.330	36.2	0.323	34.9	0.316	33.6	0.310	32.4	0.303	31.2
62	.0019	0.356	38.4	0.349	37.0	0.342	35.7	0.336	34.5	0.329	33.2	0.322	32.0
63	.0020	0.375	39.2	0.369	37.9	0.362	36.6	0.355	35.3	0.349	34.1	0.342	32.9
64	.0020	0.396	40.0	0.389	38.7	0.382	37.4	0.376	36.1	0.369	34.9	0.362	33.7
65	.0021	0.417	40.7	0.410	39.4	0.403	38.2	0.396	36.9	0.390	35.7	0.383	34.5
66	.0022	0.438	41.5	0.431	40.2	0.425	38.9	0.418	37.7	0.411	36.5	0.405	35.3
67	.0023	0.460	42.2	0.454	40.9	0.447	39.6	0.440	38.4	0.434	37.2	0.427	36.1

Mean Horizontal Difference of Force of Vapor for each 0°.1 = 0.0013.

Temperature, Fahrenheit. — Force of Vapor in English Inches. — Relative Humidity in Hundredths.

Wet-Bulb Thermometer t′ Fahrenheit.	Mean Vertical Difference of Force of Vapor for each 0°.1.	t — t′, or Difference of Wet and Dry Bulb Thermometers.											
		15°.0		15°.5		16°.0		16°.5		17°.0		17°.5	
		Force of Vapor.	Relative Humidity.	Force of Vapor.	Relative Humidity.	Force of Vapor.	Relative Humidity.	Force of Vapor.	Relative Humidity.	Force of Vapor.	Relative Humidity.	Force of Vapor.	Relative Humidity.
°		Eng. In.		Eng. In.		Eng. In.		Eng. In.		Eng. In.		Eng. In.	
68	0.0024	0.483	42.8	0.477	41.6	0.470	40.3	0.463	39.1	0.456	37.9	0.450	36.8
69	.0024	0.507	43.5	0.500	42.3	0.594	41.0	0.487	39.8	0.480	38.7	0.473	37.5
70	.0025	0.531	44.2	0.524	42.9	0.518	41.7	0.511	40.5	0.504	39.3	0.498	38.2
71	.0026	0.556	44.8	0.550	43.6	0.543	42.4	0.536	41.2	0.529	40.0	0.523	38.9
72	.0027	0.582	45.4	0.576	44.2	0.569	43.0	0.562	41.8	0.555	40.7	0.549	39.5
73	.0028	0.609	46.0	0.602	44.8	0.596	43.6	0.589	42.4	0.582	41.3	0.575	40.2
74	.0028	0.637	46.6	0.630	45.4	0.623	44.2	0.616	43.0	0.610	41.9	0.603	40.8
75	.0029	0.665	47.2	0.658	46.0	0.651	44.8	0.645	43.6	0.638	42.5	0.631	41.4
76	.0030	0.694	47.7	0.687	46.5	0.681	45.4	0.674	44.2	0.667	43.1	0.660	42.0
77	.0031	0.724	48.2	0.717	47.1	0.711	45.9	0.704	44.8	0.697	43.6	0.690	42.6
78	.0032	0.755	48.8	0.748	47.6	0.741	46.4	0.735	45.3	0.728	44.2	0.721	43.1
79	.0033	0.787	49.3	0.780	48.1	0.773	47.0	0.766	45.8	0.760	44.7	0.753	43.7
80	.0034	0.820	49.8	0.813	48.6	0.806	47.5	0.799	46.4	0.792	45.3	0.786	44.2
81	.0035	0.853	50.3	0.847	49.1	0.840	48.0	0.833	46.9	0.826	45.8	0.819	44.6
82	.0036	0.888	50.7	0.881	49.6	0.875	48.5	0.868	47.4	0.861	46.3	0.854	45.1
83	.0037	0.924	51.2	0.917	50.0	0.910	48.9	0.903	47.8	0.897	46.8	0.890	45.6
84	.0038	0.961	51.6	0.954	50.5	0.947	49.4	0.940	48.3	0.933	47.2	0.927	46.2
85	.0039	0.998	52.1	0.992	50.9	0.985	49.8	0.978	48.7	0.971	47.7	0.964	46.6
86	.0040	1.037	52.5	1.030	51.3	1.024	50.3	1.017	49.2	1.010	48.1	1.003	47.1
87	.0041	1.078	52.9	1.071	51.8	1.064	50.7	1.058	49.6	1.051	48.6	1.044	47.5
88	.0042	1.118	53.3	1.111	52.2	1.105	51.1	1.098	50.0	1.091	49.0	1.084	48.0
89	.0044	1.161	53.7	1.154	52.6	1.147	51.5	1.140	50.4	1.133	49.4	1.126	48.4
90	.0045	1.204	54.1	1.197	53.0	1.190	51.9	1.183	50.9	1.177	49.8	1.170	48.8
91	.0046	1.249	54.5	1.242	53.4	1.235	52.3	1.228	51.2	1.221	50.2	1.215	49.2
92	.0048	1.295	54.8	1.288	53.7	1.281	52.7	1.274	51.6	1.267	50.6	1.260	49.6
93	.0049	1.342	55.2	1.335	54.1	1.328	53.0	1.321	52.0	1.315	51.0	1.308	50.0
94	.0050	1.390	55.5	1.384	54.4	1.377	53.4	1.370	52.4	1.363	51.4	1.356	50.4
95	.0051	1.440	55.9	1.433	54.8	1.426	53.7	1.420	52.7	1.413	51.7	1.406	50.7
96	.0053	1.491	56.2	1.484	55.1	1.477	54.1	1.471	53.1	1.464	52.1	1.457	51.1
97	.0054	1.544	56.5	1.537	55.5	1.530	54.4	1.523	53.4	1.516	52.4	1.509	51.5
98	.0056	1.598	56.8	1.591	55.8	1.584	54.8	1.577	53.8	1.570	52.8	1.563	51.8
99	.0057	1.653	57.2	1.646	56.1	1.639	55.1	1.633	54.1	1.626	53.1	1.619	52.1
100	.0059	1.710	57.5	1.703	56.4	1.696	55.4	1.690	54.4	1.683	53.4	1.676	52.5
101	.0060	1.769	57.8	1.762	56.7	1.755	55.7	1.748	54.7	1.741	53.7	1.734	52.8
102	.0062	1.829	58.0	1.822	57.0	1.815	56.0	1.809	55.0	1.802	54.0	1.794	53.1
103	.0063	1.890	58.3	1.883	57.3	1.876	56.3	1.869	55.3	1.863	54.3	1.856	53.4
104		1.953	58.6	1.946	57.6	1.939	56.6	1.932	55.6	1.925	54.6	1.919	53.7

Mean Horizontal Difference of Force of Vapor for each 0°.1 = 0.0013.

Temperature, Fahrenheit. — Force of Vapor in English Inches. — Relative Humidity in Hundredths.

Wet-Bulb Thermometer t′ Fahrenheit.	Mean Vertical Difference of Force of Vapor for each 0°.1.	t — t′, or Difference of Wet and Dry Bulb Thermometers.											
		18°.0		18°.5		19°.0		19°.5		20°.0		20°.5	
		Force of Vapor.	Relative Humidity.	Force of Vapor.	Relative Humidity.	Force of Vapor.	Relative Humidity.	Force of Vapor.	Relative Humidity.	Force of Vapor.	Relative Humidity.	Force of Vapor.	Relative Humidity.
°		Eng. In.		Eng. In.		Eng. In.		Eng. In.		Eng. In.		Eng. In.	
32													
33													
34													
35													
36													
37													
38													
39													
40		0.012	2.5										
41	0.0010	0.022	4.3	0.015	3.0	0.009	1.6						
42	.0010	0.031	6.0	0.025	4.6	0.018	3.3	0.012	2.1				
43	.0010	0.041	7.6	0.035	6.3	0.028	5.0	0.022	3.7	0.015	2.6		
44	.0011	0.052	9.2	0.045	7.9	0.039	6.6	0.032	5.4	0.026	4.3	0.019	3.2
45	.0011	0.063	10.8	0.056	9.5	0.050	8.2	0.043	7.0	0.037	5.9	0.030	4.8
46	.0011	0.074	12.3	0.068	11.0	0.061	9.7	0.054	8.5	0.048	7.5	0.041	6.3
47	.0012	0.086	13.8	0.079	12.5	0.073	11.2	0.066	10.0	0.059	9.0	0.053	7.9
48	.0012	0.098	15.2	0.091	13.9	0.085	12.7	0.078	11.5	0.072	10.4	0.065	9.3
49	.0013	0.110	16.6	0.104	15.4	0.097	14.1	0.091	12.9	0.084	11.9	0.077	10.7
50	.0013	0.123	18.0	0.117	16.7	0.110	15.5	0.103	14.4	0.097	13.2	0.090	12.1
51	.0013	0.136	19.3	0.130	18.0	0.123	16.8	0.117	15.7	0.110	14.5	0.103	13.4
52	.0014	0.150	20.5	0.144	19.3	0.137	18.1	0.130	16.9	0.124	15.7	0.117	14.6
53	.0014	0.164	21.7	0.158	20.5	0.151	19.3	0.145	18.2	0.138	16.9	0.131	15.8
54	.0015	0.179	22.9	0.173	21.7	0.166	20.5	0.159	19.3	0.152	18.1	0.146	17.0
55	.0015	0.194	24.0	0.188	22.8	0.181	21.6	0.174	20.5	0.168	19.2	0.161	18.2
56	.0016	0.210	25.1	0.203	23.9	0.197	22.7	0.190	21.6	0.184	20.4	0.177	19.3
57	.0016	0.226	26.1	0.220	24.9	0.213	23.8	0.206	22.7	0.200	21.5	0.193	20.4
58	.0017	0.243	27.1	0.236	25.9	0.230	24.8	0.223	23.7	0.217	22.6	0.210	21.5
59	.0017	0.260	28.1	0.254	26.9	0.247	25.8	0.240	24.7	0.234	23.6	0.227	22.6
60	.0018	0.278	29.0	0.271	27.9	0.265	26.8	0.258	25.7	0.251	24.6	0.245	23.6
61	.0019	0.296	30.0	0.290	28.8	0.283	27.7	0.276	26.6	0.270	25.5	0.263	24.5
62	.0019	0.316	30.9	0.309	29.7	0.302	28.6	0.295	27.5	0.289	26.5	0.282	25.4
63	.0020	0.335	31.7	0.328	30.6	0.322	29.5	0.315	28.4	0.308	27.4	0.302	26.4
64	.0020	0.355	32.6	0.349	31.5	0.342	30.4	0.335	29.3	0.329	28.2	0.322	27.2
65	.0021	0.376	33.4	0.370	32.3	0.363	31.2	0.356	30.1	0.350	29.1	0.343	28.1
66	.0022	0.398	34.2	0.391	33.1	0.385	32.0	0.378	30.9	0.371	29.9	0.364	28.9
67	.0023	0.420	34.9	0.414	33.8	0.407	32.8	0.400	31.7	0.393	30.7	0.387	29.7

Mean Horizontal Difference of Force of Vapor for each 0°.1 = 0.0013.

Temperature, Fahrenheit. — Force of Vapor in English Inches. — Relative Humidity in Hundredths.

Wet-Bulb Thermometer t' Fahrenheit.	Mean Vertical Difference of Force of Vapor for each 0°.1.	t — t', or Difference of Wet and Dry Bulb Thermometers.											
		18°.0		18°.5		19°.0		19°.5		20°.0		20°.5	
		Force of Vapor.	Relative Humidity.	Force of Vapor.	Relative Humidity.	Force of Vapor.	Relative Humidity.	Force of Vapor.	Relative Humidity.	Force of Vapor.	Relative Humidity.	Force of Vapor.	Relative Humidity.
°		Eng. In.		Eng. In.		Eng. In.		Eng. In.		Eng. In.		Eng. In.	
68	0.0024	0.443	35.7	0.436	34.6	0.430	33.5	0.423	32.5	0.416	31.4	0.409	30.4
69	.0025	0.467	36.4	0.460	35.3	0.453	34.2	0.446	33.2	0.440	32.2	0.433	31.2
70	.0025	0.491	37.1	0.484	36.0	0.477	35.0	0.471	33.9	0.464	32.9	0.457	31.9
71	.0026	0.516	37.8	0.509	36.7	0.502	35.7	0.496	34.6	0.489	33.6	0.482	32.7
72	.0026	0.542	38.5	0.535	37.4	0.528	36.3	0.522	35.3	0.515	34.3	0.508	33.4
73	.0027	0.569	39.1	0.562	38.0	0.555	37.0	0.548	36.0	0.542	35.0	0.535	34.0
74	.0028	0.596	39.7	0.589	38.7	0.583	37.7	0.576	36.6	0.569	35.7	0.562	34.7
75	.0029	0.624	40.3	0.618	39.3	0.611	38.3	0.604	37.3	0.597	36.3	0.591	35.3
76	.0030	0.654	40.9	0.647	39.9	0.640	38.9	0.633	37.9	0.627	36.9	0.620	35.9
77	.0031	0.683	41.5	0.677	40.5	0.670	39.5	0.663	38.5	0.656	37.5	0.650	36.5
78	.0032	0.714	42.1	0.707	41.0	0.701	40.0	0.694	39.0	0.687	38.1	0.680	37.1
79	.0033	0.746	42.6	0.739	41.6	0.732	40.6	0.726	39.6	0.719	38.6	0.712	37.7
80	.0034	0.779	43.2	0.772	42.1	0.765	41.1	0.758	40.2	0.752	39.2	0.745	38.3
81	.0035	0.813	43.7	0.806	42.7	0.799	41.7	0.792	40.7	0.785	39.7	0.779	38.8
82	.0036	0.847	44.2	0.840	43.2	0.834	42.2	0.827	41.2	0.820	40.2	0.813	39.4
83	.0036	0.883	44.7	0.876	43.7	0.869	42.7	0.863	41.7	0.856	40.7	0.849	39.9
84	.0038	0.920	45.2	0.913	44.2	0.906	43.2	0.899	42.2	0.893	41.3	0.886	40.4
85	.0039	0.958	45.6	0.951	44.6	0.944	43.7	0.937	42.7	0.930	41.8	0.923	40.9
86	.0040	0.996	46.1	0.989	45.1	0.983	44.1	0.976	43.2	0.969	42.3	0.962	41.3
87	.0041	1.037	46.5	1.030	45.6	1.023	44.6	1.017	43.6	1.010	42.7	1.003	41.8
88	.0042	1.077	47.0	1.070	46.0	1.064	45.0	1.057	44.1	1.050	43.2	1.043	42.3
89	.0043	1.119	47.4	1.113	46.4	1.106	45.5	1.099	44.5	1.092	43.6	1.085	42.7
90	.0045	1.163	47.8	1.156	46.9	1.149	45.9	1.142	45.0	1.136	44.1	1.129	43.2
91	.0047	1.208	48.2	1.201	47.3	1.194	46.3	1.187	45.4	1.180	44.5	1.173	43.6
92	.0048	1.254	48.6	1.247	47.7	1.240	46.7	1.233	45.8	1.226	44.9	1.219	44.0
93	.0049	1.301	49.0	1.294	48.1	1.287	47.1	1.280	46.2	1.273	45.3	1.266	44.4
94	.0050	1.349	49.4	1.342	48.4	1.335	47.5	1.329	46.6	1.322	45.7	1.315	44.8
95	.0051	1.399	49.8	1.392	48.8	1.385	47.9	1.378	47.0	1.371	46.1	1.364	45.2
96	.0053	1.450	50.1	1.443	49.2	1.436	48.3	1.429	47.3	1.422	46.5	1.415	45.6
97	.0054	1.502	50.5	1.495	49.5	1.489	48.6	1.482	47.7	1.475	46.8	1.468	46.0
98	.0055	1.556	50.8	1.549	49.9	1.543	49.0	1.536	48.1	1.529	47.2	1.522	46.3
99	.0057	1.612	51.2	1.605	50.2	1.598	49.3	1.591	48.4	1.584	47.5	1.577	46.7
100	.0058	1.669	51.5	1.662	50.6	1.655	49.7	1.648	48.8	1.641	47.9	1.634	47.0
101	.0060	1.727	51.8	1.720	50.9	1.713	50.0	1.706	49.1	1.700	48.2	1.693	47.4
102	.0062	1.787	52.2	1.780	51.2	1.773	50.3	1.766	49.4	1.759	48.6	1.753	47.7
103	.0063	1.849	52.5	1.842	51.5	1.835	50.7	1.828	49.8	1.821	48.9	1.814	48.0
104		1.912	52.8	1.905	51.9	1.898	51.0	1.891	50.1	1.884	49.2	1.877	48.4

Mean Horizontal Difference of Force of Vapor for each 0°.1 = 0.0013.

Temperature, Fahrenheit. — Force of Vapor in English Inches. — Relative Humidity in Hundredths.

Wet-Bulb Thermometer t' Fahrenheit.	Mean Vertical Difference of Force of Vapor for each 0°.1.	$t - t'$, or Difference of Wet and Dry Bulb Thermometers.											
		21°.0		21°.5		22°.0		22°.5		23°.0		23°.5	
		Force of Vapor.	Relative Humidity.	Force of Vapor.	Relative Humidity.	Force of Vapor.	Relative Humidity.	Force of Vapor.	Relative Humidity.	Force of Vapor.	Relative Humidity.	Force of Vapor.	Relative Humidity.
°		Eng. In.		Eng. In.		Eng. In.		Eng. In.		Eng. In		Eng. In.	
32													
33													
34													
35													
36													
37													
38													
39													
40													
41													
42													
43													
44		0.013	2.0										
45	0.0011	0.023	3.7	0.017	2.6	0.010	1.6						
46	.0011	0.035	5.2	0.028	4.2	0.022	3.1	0.015	2.1				
47	.0012	0.046	6.8	0.040	5.7	0.033	4.7	0.027	3.7	0.020	2.7	0.013	1.8
48	.0012	0.058	8.2	0.052	7.2	0.045	6.2	0.039	5.2	0.032	4.2	0.025	3.3
49	.0013	0.071	9.7	0.064	8.6	0.058	7.6	0.051	6.6	0.044	5.7	0.038	4.7
50	.0013	0.084	11.0	0.077	10.0	0.070	9.0	0.064	8.0	0.057	7.1	0.051	6.1
51	.0013	0.097	12.3	0.090	11.3	0.084	10.3	0.077	9.3	0.070	8.3	0.064	7.4
52	.0014	0.110	13.5	0.104	12.5	0.097	11.5	0.091	10.6	0.084	9.6	0.077	8.7
53	.0014	0.125	14.8	0.118	13.7	0.111	12.8	0.105	11.8	0.098	10.9	0.092	9.9
54	.0015	0.139	16.0	0.133	14.9	0.126	14.0	0.120	13.0	0.113	12.1	0.106	11.2
55	.0015	0.155	17.1	0.148	16.1	0.141	15.1	0.135	14.2	0.128	13.3	0.121	12.4
56	.0016	0.170	18.2	0.164	17.2	0.157	16.3	0.150	15.3	0.144	14.4	0.137	13.5
57	.0016	0.186	19.4	0.180	18.4	0.173	17.4	0.167	16.5	0.160	15.6	0.153	14.7
58	.0017	0.203	20.5	0.197	19.5	0.190	18.5	0.183	17.6	0.177	16.7	0.170	15.8
59	.0017	0.220	21.5	0.214	20.6	0.207	19.6	0.200	18.7	0.194	17.7	0.187	16.9
60	.0018	0.238	22.5	0.231	21.6	0.225	20.6	0.218	19.6	0.211	18.7	0.205	17.8
61	.0019	0.256	23.4	0.250	22.5	0.243	21.5	0.236	20.6	0.230	19.7	0.223	18.8
62	.0019	0.275	24.4	0.269	23.5	0.262	22.4	0.255	21.5	0.249	20.6	0.242	19.7
63	.0020	0.295	25.3	0.288	24.4	0.282	23.3	0.275	22.4	0.268	21.5	0.262	20.7
64	.0020	0.315	26.1	0.309	25.3	0.302	24.2	0.295	23.3	0.289	22.4	0.282	21.6
65	.0021	0.336	27.0	0.330	26.1	0.323	25.1	0.316	24.2	0.309	23.3	0.303	22.4
66	.0022	0.358	27.9	0.351	27.0	0.344	26.0	0.338	25.1	0.331	24.2	0.324	23.3
67	.0023	0.380	28.7	0.373	27.8	0.367	26.8	0.360	25.9	0.353	25.0	0.346	24.2

Mean Horizontal Difference of Force of Vapor for each 0°.1 = 0.0013.

Temperature, Fahrenheit. — Force of Vapor in English Inches. — Relative Humidity in Hundredths.

Wet-Bulb Thermometer t' Fahrenheit.	Mean Vertical Difference of Force of Vapor for each 0°.1.	t — t', or Difference of Wet and Dry Bulb Thermometers. 21°.0 Force of Vapor.	21°.0 Relative Humidity.	21°.5 Force of Vapor.	21°.5 Relative Humidity.	22°.0 Force of Vapor.	22°.0 Relative Humidity.	22°.5 Force of Vapor.	22°.5 Relative Humidity.	23°.0 Force of Vapor.	23°.0 Relative Humidity.	23°.5 Force of Vapor.	23°.5 Relative Humidity.
°		Eng. In.		Eng. In.		Eng. In.		Eng. In.		Eng. In.		Eng. In.	
68		0.403	29.5	0.396	28.5	0.389	27.6	0.383	26.7	0.376	25.8	0.369	25.0
69	0.0024	0.426	30.2	0.420	29.3	0.413	28.4	0.406	27.5	0.399	26.6	0.393	25.8
70	.0024	0.451	31.0	0.444	30.1	0.437	29.1	0.430	28.2	0.424	27.4	0.417	26.5
71	.0025	0.476	31.7	0.469	30.8	0.462	29.9	0.455	29.0	0.449	28.1	0.442	27.3
72	.0026	0.501	32.4	0.495	31.5	0.488	30.6	0.481	29.7	0.475	28.8	0.468	28.0
73	.0027	0.528	33.1	0.521	32.2	0.515	31.3	0.508	30.4	0.501	29.5	0.494	28.7
74	.0028	0.556	33.8	0.549	32.8	0.542	31.9	0.535	31.1	0.529	30.2	0.522	29.4
75	.0028	0.584	34.4	0.577	33.5	0.570	32.6	0.564	31.7	0.557	30.9	0.550	30.0
76	.0029	0.613	35.0	0.606	34.1	0.599	33.2	0.593	32.3	0.586	31.5	0.579	30.7
77	.0030	0.643	35.6	0.636	34.7	0.629	33.8	0.623	33.0	0.616	32.1	0.609	31.3
78	.0031	0.674	36.2	0.667	35.3	0.660	34.4	0.653	33.6	0.647	32.7	0.640	31.9
79	.0032	0.705	36.8	0.699	35.9	0.692	35.0	0.685	34.2	0.678	33.3	0.671	32.5
80	.0033	0.738	37.4	0.731	36.5	0.724	35.6	0.718	34.7	0.711	33.9	0.704	33.1
81	.0034	0.772	37.9	0.765	37.0	0.758	36.1	0.751	35.3	0.745	34.5	0.738	33.5
82	.0035	0.806	38.4	0.800	37.6	0.793	36.7	0.786	35.8	0.779	35.0	0.772	34.2
83	.0036	0.842	39.0	0.835	38.1	0.829	37.2	0.822	36.4	0.815	35.5	0.808	34.7
84	.0037	0.879	39.5	0.872	38.6	0.865	37.7	0.858	36.9	0.852	36.1	0.845	35.2
85	.0038	0.917	40.0	0.910	39.1	0.903	38.2	0.896	37.4	0.889	36.6	0.882	35.8
86	.0039	0.955	40.4	0.948	39.6	0.942	38.7	0.935	37.9	0.928	37.1	0.921	36.3
87	.0040	0.995	40.9	0.988	40.1	0.981	39.2	0.975	38.4	0.968	37.5	0.961	36.7
88	.0041	1.036	41.4	1.029	40.5	1.022	39.7	1.016	38.8	1.009	38.0	1.002	37.2
89	.0042	1.078	41.8	1.071	41.0	1.065	40.1	1.058	39.3	1.051	38.5	1.044	37.7
90	.0044	1.122	42.3	1.115	41.4	1.108	40.6	1.101	39.7	1.094	38.9	1.088	38.1
91	.0045	1.166	42.7	1.160	41.9	1.153	41.0	1.146	40.2	1.139	39.4	1.132	38.6
92	.0046	1.212	43.1	1.206	42.3	1.199	41.4	1.192	40.6	1.185	39.8	1.178	39.0
93	.0048	1.260	43.5	1.253	42.7	1.246	41.9	1.239	41.0	1.232	40.2	1.225	39.4
94	.0049	1.308	43.9	1.301	43.1	1.294	42.3	1.287	41.4	1.280	40.6	1.274	39.9
95	.0050	1.358	44.3	1.351	43.5	1.344	42.7	1.337	41.8	1.330	41.0	1.323	40.3
96	.0051	1.408	44.7	1.402	43.9	1.395	43.0	1.388	42.2	1.381	41.4	1.374	40.7
97	.0053	1.461	45.1	1.454	44.3	1.447	43.4	1.440	42.6	1.433	41.8	1.426	41.1
98	.0054	1.515	45.5	1.508	44.6	1.501	43.8	1.494	43.0	1.487	42.2	1.480	41.4
99	.0056	1.570	45.8	1.563	45.0	1.556	44.2	1.550	43.4	1.543	42.6	1.536	41.8
100	.0057	1.627	46.2	1.620	45.4	1.613	44.5	1.607	43.7	1.600	43.0	1.593	42.2
101	.0059	1.686	46.5	1.679	45.7	1.672	44.9	1.665	44.1	1.658	43.3	1.651	42.5
102	.0060	1.746	46.8	1.739	46.0	1.732	45.2	1.725	44.4	1.718	43.7	1.711	42.9
103	.0062	1.807	47.2	1.800	46.4	1.793	45.6	1.786	44.8	1.779	44.0	1.772	43.2
104	.0063	1.870	47.5	1.863	46.7	1.856	45.9	1.849	45.1	1.842	44.3	1.835	43.6

Mean Horizontal Difference of Force of Vapor for each 0°.1 = 0.0013.

Temperature, Fahrenheit. — Force of Vapor in English Inches. — Relative Humidity in Hundredths.

Wet-Bulb Thermometer t' Fahrenheit.	Mean Vertical Difference of Force of Vapor for each 0°.1.	$t - t'$, or Difference of Wet and Dry Bulb Thermometers.											
		24°.0		**24°.5**		**25°.0**		**25°.5**		**26°.0**		**26°.5**	
		Force of Vapor.	Relative Humidity.	Force of Vapor.	Relative Humidity.	Force of Vapor.	Relative Humidity.	Force of Vapor.	Relative Humidity.	Force of Vapor.	Relative Humidity.	Force of Vapor.	Relative Humidity.
°		Eng. In.		Eng. In.		Eng. In.		Eng. In.		Eng. In.		Eng. In.	
32													
33													
34													
35													
36													
37													
38													
39													
40													
41													
42													
43													
44													
45													
46													
47													
48	0.0013	0.019	2.4	0.012	1.5								
49	.0013	0.031	3.9	0.025	3.0	0.018	2.2	0.011	1.3				
50	.0013	0.044	5.2	0.037	4.4	0.031	3.6	0.024	2.7	0.018	2.0	0.011	1.2
51	.0014	0.057	6.5	0.051	5.7	0.044	4.9	0.037	4.1	0.031	3.3	0.024	2.5
52	.0014	0.071	7.8	0.064	7.0	0.058	6.1	0.051	5.3	0.044	4.6	0.038	3.8
53	.0015	0.085	9.1	0.078	8.2	0.072	7.4	0.065	6.6	0.058	5.8	0.052	5.1
54	.0015	0.100	10.3	0.093	9.4	0.086	8.6	0.080	7.8	0.073	7.0	0.067	6.3
55	.0016	0.115	11.5	0.108	10.6	0.102	9.8	0.095	9.0	0.088	8.2	0.082	7.5
56	.0016	0.130	12.7	0.124	11.8	0.117	11.0	0.111	10.2	0.104	9.4	0.097	8.7
57	.0017	0.147	13.8	0.140	13.0	0.133	12.1	0.127	11.3	0.120	10.6	0.113	9.8
58	.0017	0.163	14.9	0.157	14.1	0.150	13.2	0.143	12.5	0.137	11.7	0.130	10.9
59	.0018	0.180	16.0	0.174	15.2	0.167	14.3	0.161	13.6	0.154	12.8	0.147	12.0
60	.0019	0.198	17.0	0.191	16.1	0.185	15.3	0.178	14.6	0.172	13.8	0.165	13.0
61	.0019	0.216	17.9	0.210	17.1	0.203	16.3	0.196	15.5	0.190	14.7	0.183	14.0
62	.0020	0.235	18.9	0.229	18.1	0.222	17.2	0.215	16.5	0.209	15.7	0.202	15.0
63	.0020	0.255	19.8	0.248	19.0	0.242	18.2	0.235	17.4	0.228	16.6	0.222	15.9
64	.0021	0.275	20.7	0.269	19.9	0.262	19.1	0.255	18.3	0.248	17.5	0.242	16.8
65	.0022	0.296	21.6	0.289	20.8	0.283	20.0	0.276	19.2	0.269	18.4	0.263	17.7
66	.0023	0.318	22.5	0.311	21.7	0.304	20.9	0.297	20.1	0.291	19.3	0.284	18.6
67		0.340	23.3	0.333	22.5	0.326	21.7	0.320	20.9	0.313	20.2	0.306	19.4

Mean Horizontal Difference of Force of Vapor for each 0°.1 = 0.0013.

Temperature, Fahrenheit. — Force of Vapor in English Inches. — Relative Humidity in Hundredths.

Wet-Bulb Thermometer t' Fahrenheit.	Mean Vertical Difference of Force of Vapor for each 0°.1.	$t - t'$, or Difference of Wet and Dry Bulb Thermometers. 24°.0		24°.5		25°.0		25°.5		26°.0		26°.5	
		Force of Vapor.	Relative Humidity.	Force of Vapor.	Relative Humidity.	Force of Vapor.	Relative Humidity.	Force of Vapor.	Relative Humidity.	Force of Vapor.	Relative Humidity.	Force of Vapor.	Relative Humidity.
°		Eng. In.		Eng. In.		Eng. In.		Eng. In.		Eng. In.		Eng In.	
68	0.0024	0.363	24.2	0.356	23.3	0.349	22.5	0.342	21.8	0.336	21.8	0.329	20.3
69	.0024	0.386	24.9	0.379	24.1	0.373	23.3	0.366	22.6	0.359	21.8	0.352	21.1
70	.0025	0.410	25.7	0.403	24.9	0.397	24.1	0.390	23.3	0.383	22.6	0.377	21.9
71	.0026	0.435	26.4	0.428	25.6	0.422	24.9	0.415	24.1	0.408	23.3	0.402	22.6
72	.0027	0.461	27.2	0.454	26.4	0.448	25.6	0.441	24.8	0.434	24.1	0.427	23.3
73	.0028	0.488	27.9	0.481	27.1	0.474	26.3	0.467	25.5	0.461	24.8	0.454	24.0
74	.0028	0.515	28.5	0.508	27.7	0.502	27.0	0.495	26.2	0.488	25.5	0.481	24.7
75	.0029	0.543	29.2	0.537	28.4	0.530	27.6	0.523	26.8	0.516	26.1	0.510	25.4
76	.0030	0.572	29.8	0.566	29.1	0.559	28.3	0.552	27.4	0.545	26.8	0.539	26.1
77	.0031	0.602	30.5	0.595	29.7	0.589	28.9	0.582	28.0	0.575	27.4	0.568	26.7
78	.0032	0.633	31.1	0.626	30.3	0.619	29.5	0.613	28.7	0.606	28.0	0.599	27.3
79	.0033	0.665	31.7	0.658	30.9	0.651	30.1	0.644	29.3	0.638	28.6	0.631	27.9
80	.0034	0.697	32.3	0.691	31.5	0.684	30.7	0.677	29.9	0.670	29.2	0.663	28.5
81	.0035	0.731	32.8	0.724	32.1	0.717	31.3	0.711	30.5	0.704	29.8	0.697	29.1
82	.0036	0.766	33.4	0.759	32.6	0.752	31.8	0.745	31.0	0.738	30.4	0.732	29.7
83	.0037	0.801	33.9	0.795	33.2	0.788	32.4	0.781	31.6	0.774	30.9	0.767	30.2
84	.0038	0.838	34.5	0.831	33.7	0.824	32.9	0.818	32.1	0.811	31.5	0.804	30.7
85	.0039	0.876	35.0	0.869	34.2	0.862	33.4	0.855	32.7	0.848	32.0	0.842	31.3
86	.0040	0.914	35.5	0.908	34.7	0.901	33.9	0.894	33.2	0.887	32.5	0.880	31.8
87	.0041	0.954	36.0	0.947	35.2	0.940	34.4	0.934	33.7	0.927	33.0	0.920	32.3
88	.0042	0.995	36.4	0.988	35.7	0.981	34.9	0.975	34.2	0.968	33.5	0.961	32.8
89	.0044	1.037	36.9	1.030	36.1	1.024	35.4	1.017	34.7	1.010	33.9	1.003	33.2
90	.0045	1.081	37.4	1.074	36.6	1.067	35.8	1.060	35.1	1.053	34.4	1.046	33.7
91	.0046	1.125	37.8	1.118	37.1	1.112	36.3	1.105	35.6	1.098	34.9	1.091	34.2
92	.0048	1.171	38.2	1.164	37.5	1.157	36.7	1.151	36.0	1.144	35.3	1.137	34.6
93	.0049	1.218	38.7	1.211	37.9	1.205	37.1	1.198	36.5	1.191	35.7	1.184	35.0
94	.0050	1.267	39.1	1.260	38.3	1.253	37.5	1.246	36.9	1.239	36.2	1.232	35.5
95	.0051	1.316	39.5	1.309	38.7	1.302	37.9	1.296	37.3	1.289	36.6	1.282	35.9
96	.0053	1.367	39.9	1.360	39.1	1.353	38.3	1.346	37.7	1.340	37.0	1.333	36.3
97	.0054	1.420	40.3	1.413	39.5	1.406	38.7	1.399	38.1	1.392	37.4	1.385	36.7
98	.0056	1.473	40.7	1.467	39.9	1.460	39.1	1.453	38.5	1.446	37.8	1.439	37.1
99	.0057	1.529	41.1	1.522	40.3	1.515	39.5	1.508	38.9	1.501	38.2	1.494	37.5
100	.0059	1.586	41.4	1.579	40.7	1.572	39.9	1.565	39.2	1.558	38.5	1.551	37.9
101	.0060	1.644	41.8	1.637	41.0	1.630	40.3	1.623	39.6	1.616	38.9	1.609	38.2
102	.0062	1.704	42.2	1.697	41.4	1.690	40.7	1.683	40.0	1.676	39.3	1.669	38.6
103	.0063	1.765	42.5	1.758	41.8	1.751	41.0	1.745	40.3	1.738	39.6	1.731	38.9
104		1.828	42.8	1.821	42.1	1.814	41.4	1.807	40.7	1.800	40.0	1.793	39.3

Mean Horizontal Difference of Force of Vapor for each 0°.1 = 0.0013.

Correction for Barometrical Height above or below the Normal Height of 29.7 inches.

For Baromet-rical Height.	Difference of Thermometers, or $t - t'$ Fahrenheit.												
	2°	**4°**	**6°**	**8°**	**10°**	**12°**	**14°**	**16°**	**18°**	**20°**	**22°**	**24°**	**26°**
	Wet Bulb above the Freezing-Point.												
Eng. In.	Inch.	Inch.	Inch.	Inch.	Inch.	Inch.	Inch.	Inch.	Inch.	Inch.	Inch.	Inch.	Inch.
31.0	–.001	–.002	–.003	–.005	–.006	–.007	–.008	–.009	–.010	–.012	–.013	–.014	–.015
30.5	.001	.001	.002	.003	.004	.004	.005	.006	.006	.007	.008	.009	.009
30.0	–.000	–.000	–.001	–.001	–.001	–.002	–.002	–.002	–.002	–.003	–.003	–.003	–.004
29.5	+.000	+.000	+.001	+.001	+.001	+.001	+.001	+.001	+.002	+.002	+.002	+.002	+.002
29.0	.001	.001	.002	.003	.003	.004	.004	.005	.006	.006	.007	.008	.008
28.5	.001	.002	.003	.004	.005	.006	.007	.009	.010	.011	.012	.013	.014
28.0	.001	.003	.005	.006	.008	.009	.011	.012	.014	.015	.017	.018	.020
27.5	.002	.004	.006	.007	.010	.012	.014	.016	.018	.020	.022	.024	.026
27.0	.002	.005	.007	.009	012	.014	.017	.019	.022	.024	.027	.029	.031
26.5	.003	.006	.008	.011	.014	.017	.020	.023	.026	.029	.031	.034	.037
26.0	.003	.006	.010	.013	.016	.020	.023	.026	.030	.033	.036	.040	.043
25.5	.004	.007	.011	.014	.019	.022	.025	.030	.034	.037	.041	.045	.049
25.0	.004	.008	.012	.016	.021	.025	.028	.033	.038	.042	.046	.050	.055
24.0	.005	.010	.015	.020	.025	.030	.034	.040	.046	.051	.056	.061	.066
23.0	.006	.012	.018	.023	.030	.035	.041	.047	.054	.060	.066	.072	.078
22.0	..007	.013	.020	.027	.034	.041	.047	.054	.062	.069	.076	.083	.090
21.0	.008	.015	.023	.030	.038	.046	.053	.062	.070	.077	.085	.093	.101
20.0	+.008	+.017	+.026	+.034	+.043	+.051	+.059	+.069	+.078	+.086	+.095	+.104	+.113

	Wet Bulb below the Freezing-Point.				
31.0	–.001	–.002	–.003	–.004	–.006
30.5	.001	.001	.002	.003	.003
30.0	–.000	–.000	–.001	–.001	–.001
29.5	+.000	+.000	+.000	+.001	+.001
29.0	.001	.001	.002	.002	.003
28.5	.001	.002	.003	.004	.005
28.0	.001	.003	.004	.005	.007
27.5	.002	.003	.005	.007	.009
27.0	.002	.004	.006	.008	.011
26.5	.002	.005	.007	.010	.013
26.0	.003	.006	.009	.012	.014
25.5	.003	.007	.010	.013	.016
25.0	.003	.007	.011	.015	.018
24.0	.004	.009	.013	.018	.022
23.0	.005	.010	.016	.021	.026
22.0	.006	.012	.018	.024	.030
21.0	.006	.014	.020	.027	.034
20.0	+.007	+.015	+.023	+.030	+.038

EXAMPLE OF CALCULATION.

Wet Bulb above the Freezing-Point.

$t' = 62°$ F. $t - t' = 10°$. Barom. = 26.5 in.

The large tables give for a mean barometrical height of 27.9 inches. Force of Vapor . . . . Inch. = 0.403

Additive correction, in this table, for B = 26.5 inches, and 10° . . = 0.014

Corrected Force of Vapor . = 0.417

The mean barometrical pressure, at a given place of observation, being known, the above Psychrometrical Tables may be fitted for that place, by determining, by means of this table, a *constant correction*, to be applied to the numbers in the tables, expressing the force of vapor. This correction will be found by taking for $t - t'$, or the difference of thermometers, a *mean value*, representing the mean moisture of the air. The errors arising from the deviations from that mean will little impair the accuracy of the results.

TABLE VIII.

FOR DEDUCING THE RELATIVE HUMIDITY OF THE AIR FROM THE INDICATIONS, IN ENGLISH MEASURES, OF THE DEW-POINT INSTRUMENTS.

THE object of every Dew-Point instrument is to ascertain, by causing a part of the apparatus to cool, the temperature at which the vapor contained in the air begins to condense, in the shape of light dew, on the cooled portion of the instrument. It is obvious that this is the temperature at which the atmosphere itself, if cooled likewise, would be fully saturated by the amount of vapor present in the air at the time of the observation.

The temperature of the dew-point being known, all the hygrometrical conditions of the air can be easily deduced from it.

The *Absolute Humidity*, or the total amount of vapor in the atmosphere, is expressed by the number, in the Tables of Elastic Forces of Vapor, due to that temperature.

The *Relative Humidity*, or the degree of moisture, being the ratio of the quantity of vapor actually contained in the air to the quantity it could contain if fully saturated, is expressed by the proportion

Relative Humidity : 1 : : Force of Vapor at Dew-Point : Maximum Force of Vapor.

Calling the

Force of Vapor at the Temperature of the Dew-Point, f;
Force of Vapor at the Temperature of the Air, F ;

then

$$\text{Relative Humidity} = \frac{f}{F}.$$

It is thus found by dividing the force of vapor due, in the Table of Elastic Forces, to the temperature of the dew-point, by the maximum of the force of vapor due, in the same table, to the temperature of the air at the time of the observation. F being always greater than f, when the air is not saturated, the Relative Humidity is expressed by a fraction, which is termed the *fraction of saturation*. Making the point of saturation = 100, in order to obtain this fraction in hundredths, we have

$$\text{Relative Humidity} = \frac{f \times 100}{F}.$$

Example.

Suppose the

Temperature of the Air, or t, to be	= 43° F.
Temperature of the Dew-Point, or t′, to be	= 35° F.
Difference between the two, or t — t′, to be	= 8° F.

Taking in Table VI. the Elastic Forces due to t and t′, we have

$$\frac{\text{Force of Vapor at } t'}{\text{Force of Vapor at } t} = \frac{.2037 \times 100}{.2775} = 73.4, \text{ Relative Humidity in Hundredths.}$$

The following Table VIII. gives, in hundredths, the fraction of saturation, or Relative Humidity, corresponding to each degree of t′, or of the temperature of the air, from 0° to 104° ; and for every half degree of t — t′, or of the difference between the temperature of the air and of the dew-point, from 0.°5 to 24.°5. Regnault's Table of Elastic Forces of Vapor, reduced to English measures, has been used in the computation.

Though the fraction of saturation expressed in hundredths indicates the Relative Humidity with sufficient accuracy, the thousandths have been added to facilitate, as remarked above in the preface to the Psychrometrical Tables, the interpolations for any number falling between those given in the table.

Use of the Table.

Example.

Temperature of Air, or t, being	= 62° F.
Temperature of the Dew-Point, or t′,	= 53° F.
Difference, or t — t′,	= 9° F.

Find out the Relative Humidity.

In the column of temperatures, the first on the left, find 62° ; on the same horizontal line, in the column headed 9°, is found 72.4, which is the Relative Humidity required.

Should it seem desirable to compute the Relative Humidity for values of t — t′ not contained in the table, the factors given below in Table IX. may be used. It may be seen, however, that an interpolation at sight will always suffice for meteorological purposes.

VIII.

FOR DEDUCING THE RELATIVE HUMIDITY OF THE AIR,

FROM THE INDICATIONS OF DEW-POINT INSTRUMENTS.

Relative Humidity expressed in Hundredths, full Saturation being = 100.

Temperature of Air, Fahrenheit.	$t - t'$ = Difference of Temperatures of the Air and of the Dew-Point. — Fahrenheit.									
	0.0	**0.5**	**1.0**	**1.5**	**2.0**	**2.5**	**3.0**	**3.5**	**4.0**	**4.5**
0°	100.	97.7	95.4	93.2	91.0	88.9	86.8	84.8	82.8	80.9
1	100.	97.7	95.5	93.3	91.1	89.0	86.9	84.9	82.9	81.0
2	100.	97.7	95.5	93.3	91.2	89.1	87.0	85.0	83.0	81.1
3	100.	97.8	95.5	93.4	91.2	89.2	87.1	85.1	83.1	81.2
4	100.	97.8	95.6	93.4	91.3	89.2	87.2	85.2	83.2	81.3
5	100.	97.8	95.6	93.5	91.4	89.3	87.3	85.3	83.3	81.4
6	100.	97.8	95.6	93.5	91.4	89.3	87.3	85.3	83.3	81.5
7	100.	97.8	95.6	93.5	91.4	89.3	87.3	85.3	83.4	81.5
8	100.	97.8	95.6	93.5	91.3	89.3	87.3	85.3	83.4	81.5
9	100.	97.8	95.6	93.5	91.3	89.3	87.3	85.3	83.4	81.5
10	100.	97.8	95.6	93.4	91.3	89.3	87.3	85.3	83.4	81.5
11	100.	97.8	95.6	93.4	91.3	89.3	87.3	85.3	83.4	81.6
12	100.	97.8	95.5	93.4	91.3	89.3	87.3	85.4	83.4	81.6
13	100.	97.8	95.5	93.4	91.3	89.3	87.3	85.4	83.5	81.6
14	100.	97.7	95.5	93.4	91.3	89.3	87.3	85.4	83.5	81.7
15	100.	97.7	95.5	93.4	91.3	89.4	87.4	85.5	83.5	81.7
16	100.	97.7	95.5	93.4	91.3	89.3	87.3	85.4	83.5	81.6
17	100.	97.7	95.5	93.4	91.3	89.3	87.3	85.3	83.4	81.6
18	100.	97.7	95.5	93.4	91.3	89.3	87.3	85.3	83.4	81.5
19	100.	97.8	95.5	93.4	91.3	89.3	87.2	85.2	83.3	81.4
	0.0	**0.5**	**1.0**	**1.5**	**2.0**	**2.5**	**3.0**	**3.5**	**4.0**	**4.5**

Temperature of Air, Fahrenheit.	t — t′ = Difference of Temperatures of the Air and of the Dew-Point.—Fahrenheit.									
	5.0	**5.5**	**6.0**	**6.5**	**7.0**	**7.5**	**8.0**	**8.5**	**9.0**	**9.5**
0°	79.0	77.2	75.4	73.6	71.9	70.1	68.5	66.9	65.3	63.7
1	79.1	77.3	75.5	73.7	72.0	70.2	68.6	67.0	65.4	63.8
2	79.2	77.4	75.6	73.8	72.1	70.3	68.7	67.1	65.5	64.0
3	79.3	77.5	75.7	73.9	72.2	70.5	68.8	67.2	65.6	64.1
4	79.4	77.6	75.8	74.0	72.3	70.6	68.9	67.3	65.7	64.2
5	79.5	77.7	75.9	74.1	72.4	70.7	69.1	67.4	65.8	64.4
6	79.6	77.8	76.0	74.2	72.5	70.8	69.2	67.6	66.0	64.5
7	79.6	77.8	76.0	74.3	72.6	70.9	69.3	67.7	66.1	64.6
8	79.6	77.9	76.1	74.4	72.7	71.0	69.4	67.8	66.2	64.7
9	79.7	77.9	76.1	74.4	72.7	71.1	69.5	67.9	66.3	64.8
10	79.7	77.9	76.2	74.5	72.8	71.2	69.6	68.0	66.4	64.9
11	79.7	78.0	76.2	74.5	72.8	71.2	69.6	68.0	66.5	64.9
12	79.8	78.0	76.2	74.5	72.9	71.2	69.6	68.0	66.5	65.0
13	79.8	78.0	76.3	74.6	72.9	71.3	69.6	68.1	66.5	65.0
14	79.8	78.1	76.3	74.6	72.9	71.3	69.6	68.1	66.5	65.1
15	79.8	78.1	76.3	74.6	72.9	71.3	69.7	68.1	66.6	65.1
16	79.8	78.0	76.2	74.5	72.9	71.2	69.6	68.1	66.5	65.1
17	79.7	77.9	76.1	74.5	72.8	71.2	69.6	68.0	66.5	65.0
18	79.6	77.8	76.1	74.4	72.7	71.1	69.5	68.0	66.5	65.0
19	79.6	77.8	76.0	74.3	72.7	71.1	69.5	68.0	66.4	65.0
	10.0	**10.5**	**11.0**	**11.5**	**12.0**	**12.5**	**13.0**	**13.5**	**14.0**	**14.5**
0°	62.1	60.7	59.2	57.7	56.3	54.9	53.6	52.3	51.0	49.8
1	62.3	60.8	59.3	57.9	56.5	55.1	53.7	52.5	51.2	50.0
2	62.4	61.0	59.5	58.1	56.6	55.3	53.9	52.7	51.4	50.1
3	62.6	61.1	59.6	58.2	56.8	55.5	54.1	52.8	51.5	50.3
4	62.7	61.3	59.8	58.4	57.0	55.7	54.3	53.0	51.7	50.5
5	62.9	61.4	60.0	58.6	57.2	55.8	54.5	53.2	51.9	50.7
6	63.0	61.5	60.1	58.7	57.3	55.9	54.6	53.3	52.0	50.8
7	63.1	61.7	60.2	58.8	57.4	56.0	54.7	53.4	52.1	50.9
8	63.2	61.8	60.3	58.9	57.5	56.2	54.8	53.5	52.3	51.0
9	63.3	61.9	60.4	59.0	57.6	56.3	54.9	53.6	52.4	51.2
10	63.4	62.1	60.5	59.1	57.7	56.4	55.0	53.8	52.5	51.3
11	63.5	62.1	60.6	59.2	57.8	56.5	55.1	53.9	52.6	51.4
12	63.5	62.1	60.6	59.3	57.9	56.6	55.2	54.0	52.7	51.5
13	63.5	62.2	60.7	59.3	58.0	56.6	55.3	54.1	52.8	51.6
14	63.6	62.3	60.8	59.4	58.1	56.7	55.4	54.2	52.9	51.7
15	63.6	62.3	60.8	59.5	58.1	56.8	55.5	54.3	53.0	51.8
16	63.6	62.3	60.8	59.5	58.1	56.8	55.5	54.3	53.0	51.8
17	63.6	62.2	60.8	59.4	58.1	56.7	55.5	54.2	53.0	51.8
18	63.5	62.2	60.7	59.4	58.0	56.7	55.4	54.2	53.0	51.8
19	63.5	62.1	60.7	59.3	58.0	56.6	55.4	54.2	52.9	51.8

Temperature of Air, Fahrenheit.	t — t' = Difference of Temperatures of the Air and of the Dew-Point.—Fahrenheit.									
	15.0	**15.5**	**16.0**	**16.5**	**17.0**	**17.5**	**18.0**	**18.5**	**19.0**	**19.5**
0°	48.5	47.3	46.1	45.0	43.9	42.8	41.6	40.6	39.5	38.5
1	48.7	47.5	46.3	45.2	44.0	42.9	41.8	40.8	39.7	38.7
2	48.9	47.7	46.5	45.4	44.2	43.1	42.0	41.0	39.9	38.9
3	49.1	47.9	46.7	45.5	44.4	43.3	42.2	41.2	40.2	39.2
4	49.3	48.1	46.9	45.7	44.6	43.5	42.4	41.4	40.4	39.4
5	49.4	48.2	47.1	45.9	44.8	43.7	42.6	41.6	40.6	39.6
6	49.6	48.4	47.2	46.1	44.9	43.9	42.8	41.8	40.7	39.8
7	49.7	48.5	47.3	46.2	45.1	44.0	42.9	41.9	40.9	39.9
8	49.8	48.7	47.5	46.4	45.3	44.2	43.1	42.1	41.1	40.1
9	50.0	48.8	47.6	46.5	45.4	44.3	43.3	42.2	41.2	40.2
10	50.1	48.9	47.8	46.7	45.6	44.5	43.4	42.4	41.4	40.4
11	50.2	49.0	47.9	46.8	45.7	44.6	43.5	42.5	41.5	40.5
12	50.3	49.1	48.0	46.9	45.8	44.7	43.6	42.6	41.6	40.6
13	50.4	49.2	48.1	47.0	45.9	44.8	43.7	42.7	41.7	40.7
14	50.5	49.3	48.2	47.1	46.0	44.9	43.8	42.8	41.8	40.8
15	50.6	49.4	48.3	47.2	46.1	45.0	43.9	42.9	41.9	40.9
16	50.6	49.5	48.3	47.2	46.1	45.0	44.0	43.0	41.9	41.0
17	50.6	49.5	48.3	47.2	46.1	45.0	44.0	43.0	42.0	41.0
18	50.6	49.5	48.3	47.2	46.2	45.0	44.1	43.1	42.0	41.1
19	50.6	49.5	48.3	47.3	46.2	45.1	44.1	43.1	42.1	41.1
	20.0	**20.5**	**21.0**	**21.5**	**22.0**	**22.5**	**23.0**	**23.5**	**24.0**	**24.5**
0°	37.5	36.5	35.5	34.6	33.7	32.8	31.9	31.0	30.2	29.3
1	37.7	36.8	35.8	34.8	33.9	33.0	32.1	31.3	30.4	29.6
2	37.9	37.0	36.0	35.1	34.2	33.3	32.4	31.5	30.7	29.9
3	38.2	37.2	36.2	35.3	34.4	33.5	32.6	31.8	30.9	30.1
4	38.4	37.4	36.5	35.6	34.6	33.8	32.9	32.0	31.2	30.4
5	38.6	37.7	36.7	35.8	34.9	34.0	33.1	32.3	31.4	30.6
6	38.8	37.8	36.9	36.0	35.0	34.2	33.3	32.5	31.6	30.8
7	38.9	38.0	37.0	36.1	35.2	34.3	33.5	32.6	31.8	31.0
8	39.1	38.1	37.2	36.3	35.4	34.5	33.6	32.8	32.1	31.2
9	39.2	38.3	37.3	36.4	35.5	34.7	33.8	33.0	32.3	31.4
10	39.4	38.4	37.5	36.6	35.7	34.8	34.0	33.1	32.5	31.6
11	39.5	38.6	37.6	36.7	35.8	35.0	34.1	33.3	32.6	31.7
12	39.6	38.7	37.8	36.9	36.0	35.1	34.2	33.4	32.7	31.8
13	39.8	38.8	37.9	37.0	36.1	35.2	34.4	33.6	32.8	32.0
14	39.9	39.0	38.0	37.1	36.2	35.4	34.5	33.7	32.9	32.1
15	40.0	39.1	38.2	37.3	36.4	35.5	34.7	33.9	33.0	32.2
16	40.0	39.1	38.2	37.3	36.4	35.6	34.7	33.9	33.1	32.3
17	40.1	39.2	38.2	37.4	36.5	35.6	34.8	34.0	33.1	32.4
18	40.1	39.2	38.3	37.4	36.5	35.7	34.8	34.0	33.2	32.4
19	40.2	39.3	38.3	37.5	36.6	35.7	34.9	34.1	33.2	32.5

Temperature of Air, Fahrenheit.	t — t' = Difference of Temperatures of the Air and of the Dew-Point. — Fahrenheit.									
	0.0	**0.5**	**1.0**	**1.5**	**2.0**	**2.5**	**3.0**	**3.5**	**4.0**	**4.5**
20°	100.	97.8	95.6	93.4	91.3	89.2	87.2	85.2	83.2	81.3
21	100.	97.8	95.6	93.4	91.3	89.3	87.3	85.3	83.3	81.5
22	100.	97.8	95.6	93.5	91.4	89.3	87.3	85.4	83.4	81.6
23	100.	97.8	95.6	93.5	91.4	89.4	87.4	85.5	83.5	81.7
24	100.	97.8	95.7	93.5	91.5	89.5	87.5	85.5	83.6	81.8
25	100.	97.8	95.7	93.6	91.5	89.5	87.6	85.6	83.7	81.9
26	100.	97.8	95.7	93.6	91.6	89.6	87.7	85.7	83.8	82.0
27	100.	97.9	95.8	93.7	91.7	89.7	87.8	85.9	84.0	82.1
28	100.	97.9	95.8	93.8	91.8	89.8	87.9	86.0	84.1	82.3
29	100.	97.9	95.9	93.8	91.8	89.9	88.0	86,1	84.2	82.4
30	100.	97.9	95.9	93.9	91.9	90.0	88.1	86.2	84.3	82.5
31	100.	98.0	96.0	94.0	92.0	90.1	88.2	86.4	84.5	82.7
32	100.	98.0	96.0	94.0	92.1	90.2	88.4	86.6	84.7	83.0
33	100.	98.0	96.1	94.1	92.2	90.4	88.6	86.7	84.9	83.2
34	100.	98.0	96.1	94.2	92.3	90.5	88.7	86.9	85.1	83.4
35	100.	98.0	96.1	94.3	92.4	90.6	88.9	87.1	85.3	83.6
36	100.	98.1	96.2	94.3	92.5	90.7	88.9	87.1	85.4	83.7
37	100.	98.1	96.2	94.3	92.5	90.7	88.9	87.2	85.4	83.7
38	100.	98.1	96.2	94.3	92.5	90.7	89.0	87.2	85.5	83.8
39	100.	98.1	96.2	94.3	92.5	90.7	89.0	87.2	85.5	83.9
40	100.	98.1	96.2	94.4	92.5	90.8	89.0	87.3	85.6	83.9
41	100.	98.1	96.2	94.4	92.6	90.8	89.1	87.3	85.7	84.0
42	100.	98.1	96.2	94.4	92.6	90.8	89.1	87.4	85.7	84.1
43	100.	98.1	96.3	94.4	92.6	90.9	89.2	87.5	85.8	84.2
44	100.	98.1	96.3	94.5	92.7	90.9	89.2	87.5	85.9	84.2
45	100.	98.1	96.3	94.5	92.7	91.0	89.3	87.6	85.9	84.3
46	100.	98.1	96.3	94.5	92.7	91.0	89.3	87.6	86.0	84.4
47	100.	98.1	96.3	94.5	92.8	91.0	89.3	87.7	86.0	84.4
48	100.	98.2	96.3	94.6	92.8	91.1	89.4	87.7	86.1	84.4
49	100.	98.2	96.4	94.6	92.8	91.1	89.4	87.7	86.1	84.5
50	100.	98.2	96.4	94.6	92.9	91.1	89.4	87.8	86.2	84.5
51	100.	98.2	96.4	94.6	92.9	91.2	89.5	87.8	86.2	84.6
52	100.	98.2	96.4	94.6	92.9	91.2	89.5	87.9	86.3	84.7
53	100.	98.2	96.4	94.7	92.9	91.2	89.6	87.9	86.3	84.7
54	100.	98.2	96.4	94.7	93.0	91.3	89.6	88.0	86.4	84.8
55	100.	98.2	96.5	94.7	93.0	91.3	89.7	88.0	86.4	84.8
56	100.	98.2	96.5	94.7	93.0	91.4	89.7	88.1	86.5	84.9
57	100.	98.2	96.5	94.8	93.1	91.4	89.7	88.1	86.5	85.0
58	100.	98.2	96.5	94.8	93.1	91.4	89.8	88.2	86.6	85.0
59	100.	98.2	96.5	94.8	93.1	91.5	89.8	88.2	86.6	85.1
60	100.	98.2	96.5	94.8	93.2	91.5	89.9	88.3	86.7	85.1
61	100.	98.3	96.5	94.9	93.2	91.5	89.9	88.3	86.7	85.2
62	100.	98.3	96.6	94.9	93.2	91.6	90.0	88.4	86.8	85.3
	0.0	**0.5**	**1.0**	**1.5**	**2.0**	**2.5**	**3.0**	**3.5**	**4.0**	**4.5**

Temperature of Air, Fahrenheit.	t — t′ = Difference of Temperatures of the Air and of the Dew-Point. — Fahrenheit.									
	0.0	**0.5**	**1.0**	**1.5**	**2.0**	**2.5**	**3.0**	**3.5**	**4.0**	**4.5**
62°	100.	98.3	96.6	94.9	93.2	91.6	90.0	88.4	86.8	85.3
63	100.	98.3	96.6	94.9	93.2	91.6	90.0	88.4	86.8	85.3
64	100.	98.3	96.6	94.9	93.3	91.6	90.0	88.5	86.9	85.3
65	100.	98.3	96.6	94.9	93.3	91.7	90.1	88.5	86.9	85.4
66	100.	98.3	96.6	94.9	93.3	91.7	90.1	88.5	87.0	85.4
67	100.	98.3	96.6	95.0	93.3	91.7	90.1	88.6	87.0	85.5
68	100.	98.3	96.6	95.0	93.4	91.8	90.2	88.6	87.1	85.5
69	100.	98.3	96.6	95.0	93.4	91.8	90.2	88.7	87.2	85.6
70	100.	98.3	96.7	95.0	93.4	91.8	90.3	88.7	87.2	85.7
71	100.	98.3	96.7	95.0	93.4	91.9	90.3	88.8	87.2	85.8
72	100.	98.3	96.7	95.1	93.5	91.9	90.3	88.8	87.3	85.8
73	100.	98.3	96.7	95.1	93.5	91.9	90.4	88.8	87.3	85.9
74	100.	98.3	96.7	95.1	93.5	91.9	90.4	88.9	87.4	85.9
75	100.	98.3	96.7	95.1	93.5	92.0	90.4	88.9	87.4	86.0
76	100.	98.3	96.7	95.1	93.6	92.0	90.5	89.0	87.5	86.0
77	100.	98.4	96.7	95.2	93.6	92.0	90.5	89.0	87.5	86.1
78	100.	98.4	96.7	95.2	93.6	92.1	90.5	89.1	87.6	86.1
79	100.	98.4	96.8	95.2	93.6	92.1	90.6	89.1	87.6	86.2
80	100.	98.4	96.8	95.2	93.6	92.1	90.6	89.1	87.7	86.2
81	100.	98.4	96.8	95.2	93.7	92.1	90.6	89.2	87.7	86.3
82	100.	98.4	96.8	95.2	93.7	92.2	90.7	89.2	87.8	86.3
83	100.	98.4	96.8	95.3	93.7	92.2	90.7	89.3	87.8	86.4
84	100.	98.4	96.8	95.3	93.7	92.2	90.8	89.3	87.8	86.4
85	100.	98.4	96.8	95.3	93.8	92.3	90.8	89.3	87.9	86.5
86	100.	98.4	96.8	95.3	93.8	92.3	90.8	89.4	87.9	86.5
87	100.	98.4	96.9	95.3	93.8	92.3	90.9	89.4	88.0	86.6
88	100.	98.4	96.9	95.3	93.8	92.3	90.9	89.4	88.0	86.6
89	100.	98.4	96.9	95.4	93.9	92.4	90.9	89.5	88.1	86.7
90	100.	98.4	96.9	95.4	93.9	92.4	91.0	89.5	88.1	86.7
91	100.	98.4	96.9	95.4	93.9	92.4	91.0	89.6	88.2	86.8
92	100.	98.5	96.9	95.4	93.9	92.5	91.0	89.6	88.2	86.8
93	100.	98.5	96.9	95.4	93.9	92.5	91.1	89.6	88.2	86.9
94	100.	98.5	96.9	95.4	94.0	92.5	91.1	89.7	88.3	86.9
95	100.	98.5	97.0	95.5	94.0	92.5	91.1	89.7	88.3	87.0
96	100.	98.5	97.0	95.5	94.0	92.6	91.2	89.7	88.4	87.0
97	100.	98.5	97.0	95.5	94.0	92.6	91.2	89.8	88.4	87.0
98	100.	98.5	97.0	95.5	94.1	92.6	91.2	89.8	88.4	87.1
99	100.	98.5	97.0	95.5	94.1	92.7	91.3	89.9	88.5	87.1
100	100.	98.5	97.0	95.6	94.1	92.7	91.3	89.9	88.5	87.2
101	100.	98.5	97.0	95.6	94.1	92.7	91.3	89.9	88.6	87.2
102	100.	98.5	97.0	95.6	94.2	92.7	91.4	90.0	88.6	87.3
103	100.	98.5	97.0	95.6	94.2	92.8	91.4	90.0	88.7	87.3
104	100.	98.5	97.0	95.6	94.2	92.8	91.4	90.0	88.7	87.4
	0.0	**0.5**	**1.0**	**1.5**	**2.0**	**2.5**	**3.0**	**3.5**	**4.0**	**4.5**

Temperature of Air, Fahrenheit.	t — t′ = Difference of Temperatures of the Air and of the Dew-Point. — Fahrenheit.									
	5.0	**5.5**	**6.0**	**6.5**	**7.0**	**7.5**	**8.0**	**8.5**	**9.0**	**9.5**
20°	79.5	77.7	75.9	74.2	72.6	71.0	69.4	67.9	66.4	64.9
21	79.6	77.8	76.0	74.3	72.7	71.1	69.5	68.0	66.4	65.0
22	79.7	77.9	76.1	74.4	72.8	71.2	69.6	68.0	66.5	65.0
23	79.8	78.0	76.2	74.6	72.9	71.3	69.6	68.1	66.5	65.0
24	79.9	78.1	76.4	74.7	73.0	71.4	69.7	68.1	66.6	65.1
25	80.0	78.2	76.5	74.8	73.1	71.5	69.8	68.2	66.6	65.1
26	80.2	78.4	76.6	74.9	73.2	71.7	70.0	68.4	66.8	65.3
27	80.3	78.5	76.8	75.1	73.4	71.8	70.1	68.6	67.0	65.5
28	80.5	78.7	76.9	75.2	73.6	72.0	70.3	68.8	67.2	65.7
29	80.6	78.8	77.1	75.4	73.7	72.1	70.5	68.9	67.4	65.9
30	80.7	78.9	77.2	75.6	73.9	72.3	70.7	69.1	67.6	66.1
31	81.0	79.2	77.5	75.8	74.2	72.6	71.0	69.4	67.9	66.4
32	81.2	79.4	77.7	76.1	74.4	72.8	71.3	69.7	68.2	66.7
33	81.4	79.7	78.0	76.4	74.7	73.1	71.5	70.0	68.5	67.0
34	81.7	79.9	78.3	76.6	75.0	73.4	71.8	70.3	68.8	67.3
35	81.9	80.2	78.5	76.9	75.3	73.7	72.1	70.6	69.1	67.6
36	82.0	80.3	78.6	77.0	75.4	73.9	72.3	70.8	69.3	67.8
37	82.0	80.4	78.8	77.2	75.6	74.0	72.5	71.0	69.5	68.1
38	82.1	80.5	78.9	77.3	75.8	74.2	72.7	71.2	69.8	68.3
39	82.2	80.6	79.0	77.4	75.9	74.4	72.9	71.5	70.0	68.6
40	82.3	80.7	79.1	77.6	76.1	74.6	73.2	71.7	70.2	68.8
41	82.4	80.8	79.2	77.7	76.2	74.7	73.2	71.8	70.3	68.9
42	82.5	80.9	79.3	77.8	76.3	74.8	73.3	71.9	70.5	69.0
43	82.5	80.9	79.4	77.9	76.4	74.9	73.4	72.0	70.6	69.2
44	82.6	81.0	79.5	78.0	76.5	75.0	73.5	72.1	70.7	69.3
45	82.7	81.1	79.6	78.0	76.5	75.1	73.6	72.2	70.8	69.4
46	82.8	81.2	79.6	78.1	76.6	75.1	73.7	72.3	70.9	69.5
47	82.8	81.2	79.7	78.2	76.7	75.2	73.8	72.4	71.0	69.6
48	82.9	81.3	79.8	78.2	76.8	75.3	73.9	72.5	71.1	69.7
49	82.9	81.3	79.8	78.3	76.8	75.4	74.0	72.6	71.2	69.8
50	83.0	81.4	79.9	78.4	76.9	75.5	74.0	72.7	71.3	69.9
51	83.0	81.5	80.0	78.5	77.0	75.5	74.1	72.8	71.4	70.0
52	83.1	81.5	80.0	78.5	77.1	75.6	74.2	72.8	71.5	70.1
53	83.2	81.6	80.1	78.6	77.2	75.7	74.3	72.9	71.6	70.2
54	83.2	81.7	80.2	78.7	77.2	75.8	74.4	73.0	71.7	70.3
55	83.3	81.8	80.3	78.8	77.3	75.9	74.5	73.1	71.8	70.4
56	83.4	81.8	80.3	78.9	77.4	76.0	74.6	73.2	71.9	70.5
57	83.4	81.9	80.4	78.9	77.5	76.1	74.7	73.3	72.0	70.6
58	83.5	82.0	80.5	79.0	77.6	76.2	74.8	73.4	72.1	70.7
59	83.6	82.0	80.6	79.1	77.7	76.2	74.9	73.5	72.2	70.9
60	83.6	82.1	80.6	79.2	77.7	76.3	75.0	73.6	72.3	71.0
61	83.7	82.2	80.7	79.2	77.8	76.4	75.0	73.7	72.4	71.0
62	83.7	82.2	80.8	79.3	77.9	76.5	75.1	73.8	72.4	71.1
	5.0	**5.5**	**6.0**	**6.5**	**7.0**	**7.5**	**8.0**	**8.5**	**9.0**	**9.5**

Temperature of Air, Fahrenheit.	$t - t'$ = Difference of Temperatures of the Air and of the Dew-Point. — Fahrenheit.									
	5.0	**5.5**	**6.0**	**6.5**	**7.0**	**7.5**	**8.0**	**8.5**	**9.0**	**9.5**
62°	83.7	82.2	80.8	79.3	77.9	76.5	75.1	73.8	72.4	71.1
63	83.8	82.3	80.8	79.4	78.0	76.6	75.2	73.9	72.5	71.2
64	83.9	82.4	80.9	79.5	78.1	76.7	75.3	74.0	72.6	71.3
65	83.9	82.4	81.0	79.6	78.1	76.8	75.4	74.0	72.7	71.4
66	84.0	82.5	81.1	79.6	78.2	76.8	75.5	74.1	72.8	71.5
67	84.0	82.6	81.1	79.7	78.3	76.9	75.6	74.2	72.9	71.6
68	84.1	82.6	81.2	79.8	78.4	77.0	75.7	74.3	73.0	71.7
69	84.2	82.7	81.3	79.9	78.5	77.1	75.7	74.4	73.1	71.8
70	84.2	82.8	81.3	79.9	78.5	77.2	75.8	74.5	73.2	71.9
71	84.3	82.8	81.4	80.0	78.6	77.3	75.9	74.6	73.3	72.0
72	84.3	82.9	81.5	80.1	78.7	77.3	76.0	74.7	73.4	72.1
73	84.4	83.0	81.5	80.1	78.7	77.4	76.1	74.8	73.5	72.2
74	84.5	83.0	81.6	80.2	78.8	77.5	76.2	74.9	73.6	72.3
75	84.5	83.1	81.7	80.3	78.9	77.6	76.2	74.9	73.7	72.4
76	84.6	83.1	81.7	80.4	78.9	77.7	76.3	75.0	73.7	72.5
77	84.6	83.2	81.8	80.4	79.0	77.7	76.4	75.1	73.8	72.6
78	84.7	83.3	81.9	80.5	79.1	77.8	76.5	75.2	73.9	72.7
79	84.7	83.3	81.9	80.6	79.1	77.9	76.6	75.3	74.0	72.8
80	84.8	83.4	82.0	80.6	79.2	78.0	76.7	75.4	74.1	72.9
81	84.9	83.5	82.1	80.7	79.3	78.0	76.7	75.5	74.2	73.0
82	84.9	83.5	82.1	80.8	79.4	78.1	76.8	75.5	74.3	73.0
83	85.0	83.6	82.2	80.8	79.4	78.2	76.9	75.6	74.4	73.1
84	85.0	83.6	82.3	80.9	79.5	78.3	77.0	75.7	74.5	73.2
85	85.1	83.7	82.3	81.0	79.6	78.4	77.1	75.8	74.6	73.3
86	85.1	83.7	82.4	81.1	79.7	78.4	77.1	75.9	74.6	73.4
87	85.2	83.8	82.5	81.1	79.8	78.5	77.2	76.0	74.7	73.5
88	85.2	83.9	82.5	81.2	79.9	78.6	77.3	76.1	74.8	73.6
89	85.3	83.9	82.6	81.3	79.9	78.7	77.4	76.1	74.9	73.7
90	85.3	84.0	82.6	81.3	80.0	78.7	77.5	76.2	75.0	73.8
91	85.4	84.0	82.7	81.4	80.1	78.8	77.5	76.3	75.1	73.9
92	85.4	84.1	82.8	81.5	80.2	78.9	77.6	76.4	75.2	74.0
93	85.5	84.2	82.8	81.5	80.2	79.0	77.7	76.5	75.2	74.0
94	85.6	84.2	82.9	81.6	80.3	79.0	77.8	76.6	75.3	74.1
95	85.6	84.3	83.0	81.7	80.4	79.1	77.9	76.6	75.4	74.2
96	85.7	84.3	83.0	81.7	80.4	79.2	77.9	76.7	75.5	74.3
97	85.7	84.4	83.1	81.8	80.5	79.3	78.0	76.8	75.6	74.4
98	85.8	84.4	83.1	81.9	80.6	79.3	78.1	76.9	75.7	74.5
99	85.8	84.5	83.2	81.9	80.7	79.4	78.2	77.0	75.8	74.6
100	85.9	84.6	83.3	82.0	80.7	79.5	78.3	77.0	75.8	74.7
101	85.9	84.6	83.3	82.0	80.8	79.6	78.3	77.1	75.9	74.8
102	86.0	84.7	83.4	82.1	80.9	79.6	78.4	77.2	76.0	74.9
103	86.0	84.7	83.4	82.2	80.9	79.7	78.5	77.3	76.1	74.9
104	86.1	84.8	83.5	82.2	81.0	79.8	78.6	77.4	76.2	75.0
	5.0	**5.5**	**6.0**	**6.5**	**7.0**	**7.5**	**8.0**	**8.5**	**9.0**	**9.5**

Temperature of Air, Fahrenheit.	t — t′ = Difference of Temperatures of the Air and of the Dew-Point.—Fahrenheit.									
	10.0	**10.5**	**11.0**	**11.5**	**12.0**	**12.5**	**13.0**	**13.5**	**14.0**	**14.5**
20°	63.5	62.1	60.6	59.3	58.0	56.6	55.4	54.1	52.9	51.7
21	63.5	62.1	60.7	59.3	58.0	56.6	55.4	54.2	53.0	51.8
22	63.5	62.1	60.7	59.4	58.0	56.7	55.5	54.2	53.0	51.8
23	63.6	62.1	60.7	59.4	58.0	56.7	55.5	54.3	53.0	51.9
24	63.6	62.1	60.7	59.4	58.1	56.8	55.5	54.3	53.1	51.9
25	63.6	62.1	60.7	59.4	58.1	56.8	55.6	54.4	53.1	52.0
26	63.8	62.3	60.9	59.6	58.3	57.0	55.7	54.5	53.3	52.1
27	64.0	62.5	61.1	59.8	58.5	57.2	55.9	54.6	53.4	52.2
28	64.2	62.7	61.3	60.0	58.6	57.3	56.0	54.8	53.5	52.3
29	64.4	63.0	61.5	60.2	58.8	57.5	56.2	54.9	53.7	52.4
30	64.6	63.2	61.8	60.4	59.0	57.7	56.3	55.1	53.8	52.6
31	64.9	63.5	62.1	60.7	59.3	58.0	56.6	55.4	54.1	52.9
32	65.2	63.8	62.4	61.0	59.6	58.3	57.0	55.7	54.4	53.2
33	65.5	64.1	62.7	61.3	59.9	58.6	57.3	56.0	54.7	53.5
34	65.8	64.4	63.0	61.6	60.2	58.9	57.6	56.3	55.0	53.8
35	66.1	64.7	63.3	61.9	60.5	59.2	57.9	56.6	55.4	54.1
36	66.4	64.9	63.5	62.1	60.8	59.5	58.2	56.9	55.6	54.4
37	66.6	65.2	63.8	62.4	61.1	59.8	58.5	57.2	55.9	54.7
38	66.9	65.5	64.1	62.7	61.4	60.1	58.8	57.5	56.2	55.0
39	67.1	65.7	64.4	63.0	61.7	60.3	59.1	57.8	56.5	55.3
40	67.4	66.0	64.6	63.3	62.0	60.6	59.4	58.1	56.8	55.6
41	67.5	66.1	64.8	63.5	62.1	60.9	59.6	58.3	57.1	55.9
42	67.7	66.3	65.0	63.6	62.3	61.1	59.8	58.6	57.3	56.1
43	67.8	66.4	65.1	63.8	62.5	61.3	60.0	58.8	57.6	56.4
44	67.9	66.6	65.3	64.0	62.7	61.5	60.3	59.0	57.8	56.6
45	68.1	66.7	65.4	64.2	62.9	61.7	60.5	59.3	58.1	56.9
46	68.2	66.9	65.6	64.3	63.0	61.8	60.6	59.4	58.2	57.0
47	68.3	67.0	65.7	64.4	63.2	61.9	60.7	59.5	58.3	57.2
48	68.4	67.1	65.8	64.5	63.3	62.0	60.8	59.6	58.5	57.3
49	68.5	67.2	65.9	64.6	63.4	62.1	61.0	59.8	58.6	57.4
50	68.6	67.3	66.0	64.7	63.5	62.2	61.1	59.9	58.7	57.6
51	68.7	67.4	66.1	64.9	63.6	62.4	61.2	60.0	58.9	57.7
52	68.8	67.5	66.2	65.0	63.7	62.5	61.3	60.1	59.0	57.8
53	68.9	67.6	66.4	65.1	63.9	62.6	61.4	60.3	59.1	58.0
54	69.0	67.7	66.5	65.2	64.0	62.7	61.6	60.4	59.2	58.1
55	69.1	67.8	66.6	65.3	64.1	62.9	61.7	60.5	59.4	58.2
56	69.2	67.9	66.7	65.4	64.2	63.0	61.8	60.6	59.5	58.4
57	69.3	68.1	66.8	65.6	64.3	63.1	61.9	60.8	59.6	58.5
58	69.5	68.2	66.9	65.7	64.4	63.2	62.1	60.9	59.8	58.6
59	69.6	68.3	67.0	65.8	64.6	63.4	62.2	61.0	59.9	58.8
60	69.7	68.4	67.1	65.9	64.7	63.5	62.3	61.2	60.0	58.9
61	69.8	68.5	67.2	66.0	64.8	63.6	62.4	61.3	60.1	59.0
62	69.9	68.6	67.4	66.1	64.9	63.7	62.6	61.4	60.3	59.1
	10.0	**10.5**	**11.0**	**11.5**	**12.0**	**12.5**	**13.0**	**13.5**	**14.0**	**14.5**

Temperature of Air, Fahrenheit.	t — t′ = Difference of Temperatures of the Air and of the Dew-Point. — Fahrenheit.									
	10.0	**10.5**	**11.0**	**11.5**	**12.0**	**12.5**	**13.0**	**13.5**	**14.0**	**14.5**
62°	69.9	68.6	67.4	66.1	64.9	63.7	62.6	61.4	60.3	59.1
63	70.0	68.7	67.5	66.2	65.0	63.8	62.7	61.5	60.4	59.3
64	70.1	68.8	67.6	66.3	65.1	64.0	62.8	61.6	60.5	59.4
65	70.2	68.9	67.7	66.5	65.3	64.1	62.9	61.8	60.6	59.5
66	70.3	69.0	67.8	66.6	65.4	64.2	63.0	61.9	60.8	59.7
67	70.4	69.1	67.9	66.7	65.5	64.3	63.2	62.0	60.9	59.8
68	70.5	69.2	68.0	66.8	65.6	64.4	63.3	62.1	61.0	59.9
69	70.6	69.3	68.1	66.9	65.7	64.5	63.4	62.3	61.1	60.0
70	70.7	69.4	68.2	67.0	65.8	64.7	63.5	62.4	61.3	60.2
71	70.8	69.5	68.3	67.1	65.9	64.8	63.6	62.5	61.4	60.3
72	70.9	69.6	68.4	67.2	66.0	64.9	63.7	62.6	61.5	60.4
73	71.0	69.7	68.5	67.3	66.2	65.0	63.9	62.7	61.6	60.5
74	71.1	69.8	68.6	67.4	66.3	65.1	64.0	62.8	61.7	60.7
75	71.1	69.9	68.7	67.5	66.4	65.2	64.1	63.0	61.9	60.8
76	71.2	70.0	68.8	67.6	66.5	65.3	64.2	63.1	62.0	60.9
77	71.3	70.1	68.9	67.8	66.6	65.5	64.3	63.2	62.1	61.0
78	71.4	70.2	69.0	67.9	66.7	65.6	64.4	63.3	62.2	61.1
79	71.5	70.3	69.1	68.0	66.8	65.7	64.5	63.4	62.3	61.3
80	71.6	70.4	69.2	68.1	66.9	65.8	64.7	63.6	62.5	61.4
81	71.7	70.5	69.3	68.2	67.0	65.9	64.8	63.7	62.6	61.5
82	71.8	70.6	69.4	68.3	67.1	66.0	64.9	63.8	62.7	61.6
83	71.9	70.7	69.5	68.4	67.2	66.1	65.0	63.9	62.8	61.8
84	72.0	70.8	69.6	68.5	67.3	66.2	65.1	64.0	62.9	61.9
85	72.1	70.9	69.7	68.6	67.4	66.3	65.2	64.1	63.0	62.0
86	72.2	71.0	69.8	68.7	67.5	66.4	65.3	64.2	63.2	62.1
87	72.3	71.1	69.9	68.8	67.7	66.5	65.4	64.4	63.3	62.2
88	72.4	71.2	70.0	68.9	67.8	66.6	65.5	64.5	63.4	62.3
89	72.5	71.3	70.1	69.0	67.9	66.8	65.7	64.6	63.5	62.5
90	72.6	71.4	70.2	69.1	68.0	66.9	65.8	64.7	63.6	62.6
91	72.7	71.4	70.3	69.2	68.1	67.0	65.9	64.8	63.7	62.7
92	72.8	71.5	70.4	69.3	68.2	67.1	66.0	64.9	63.9	62.8
93	72.9	71.6	70.5	69.4	68.3	67.2	66.1	65.0	64.0	62.9
94	72.9	71.7	70.6	69.5	68.4	67.3	66.2	65.1	64.1	63.0
95	73.0	71.8	70.7	69.6	68.5	67.4	66.3	65.2	64.2	63.2
96	73.1	71.9	70.8	69.7	68.6	67.5	66.4	65.4	64.3	63.3
97	73.2	72.0	70.9	69.8	68.7	67.6	66.5	65.5	64.4	63.4
98	73.3	72.1	71.0	69.9	68.8	67.7	66.6	65.6	64.5	63.5
99	73.4	72.3	71.1	70.0	68.9	67.8	66.7	65.7	64.6	63.6
100	73.5	72.4	71.2	70.1	69.0	67.9	66.8	65.8	64.8	63.7
101	73.6	72.5	71.3	70.2	69.1	68.0	67.0	65.9	64.9	63.9
102	73.7	72.6	71.4	70.3	69.2	68.1	67.1	66.0	65.0	64.0
103	73.8	72.7	71.5	70.4	69.3	68.2	67.2	66.1	65.1	64.1
104	73.9	72.8	71.6	70.5	69.4	68.3	67.3	66.2	65.2	64.2
	10.0	**10.5**	**11.0**	**11.5**	**12.0**	**12.5**	**13.0**	**13.5**	**14.0**	**14.5**

Temperature of Air, Fahrenheit.	t — t′ = Difference of Temperatures of the Air and of the Dew-Point.—Fahrenheit.									
	15.0	**15.5**	**16.0**	**16.5**	**17.0**	**17.5**	**18.0**	**18.5**	**19.0**	**19.5**
20°	50.6	49.5	48.4	47.3	46.2	45.1	44.1	43.1	42.1	41.2
21	50.6	49.5	48.4	47.3	46.2	45.1	44.2	43.2	42.2	41.2
22	50.7	49.5	48.4	47.4	46.3	45.2	44.2	43.2	42.2	41.3
23	50.7	49.6	48.5	47.4	46.3	45.2	44.2	43.3	42.3	41.3
24	50.7	49.6	48.5	47.4	46.4	45.3	44.3	43.3	42.3	41.4
25	50.8	49.7	48.5	47.5	46.4	45.4	44.3	43.3	42.4	41.4
26	50.9	49.8	48.6	47.6	46.5	45.4	44.4	43.4	42.4	41.5
27	51.0	49.9	48.7	47.7	46.6	45.5	44.5	43.5	42.5	41.6
28	51.1	50.0	48.8	47.7	46.7	45.6	44.6	43.6	42.6	41.6
29	51.2	50.1	48.9	47.8	46.8	45.7	44.7	43.7	42.7	41.7
30	51.4	50.2	49.0	47.9	46.8	45.8	44.7	43.7	42.7	41.8
31	51.7	50.5	49.4	48.2	47.1	46.1	45.0	44.0	43.0	42.0
32	52.0	50.8	49.7	48.5	47.4	46.4	45.3	44.3	43.3	42.3
33	52.3	51.1	50.0	48.8	47.7	46.6	45.6	44.5	43.5	42.5
34	52.6	51.4	50.3	49.1	48.0	46.9	45.9	44.8	43.8	42.8
35	52.9	51.7	50.6	49.4	48.3	47.2	46.1	45.1	44.1	43.0
36	53.2	52.0	50.9	49.7	48.6	47.5	46.4	45.4	44.4	43.3
37	53.5	52.3	51.2	50.0	48.9	47.8	46.7	45.7	44.7	43.6
38	53.8	52.6	51.5	50.3	49.2	48.1	47.0	46.0	45.0	43.9
39	54.1	52.9	51.8	50.6	49.5	48.4	47.3	46.3	45.3	44.2
40	54.4	53.2	52.1	50.9	49.8	48.7	47.6	46.6	45.6	44.5
41	54.7	53.5	52.3	51.2	50.1	49.0	47.9	46.9	45.8	44.8
42	54.9	53.8	52.6	51.5	50.4	49.3	48.2	47.2	46.1	45.1
43	55.2	54.0	52.9	51.8	50.7	49.6	48.5	47.5	46.4	45.4
44	55.5	54.3	53.2	52.1	50.9	49.9	48.8	47.7	46.7	45.7
45	55.7	54.6	53.4	52.3	51.2	50.2	49.1	48.0	47.0	46.0
46	55.9	54.7	53.6	52.5	51.4	50.4	49.3	48.3	47.2	46.2
47	56.0	54.9	53.8	52.7	51.6	50.6	49.5	48.5	47.5	46.5
48	56.2	55.0	54.0	52.9	51.8	50.8	49.8	48.7	47.7	46.7
49	56.3	55.2	54.1	53.1	52.0	51.0	50.0	49.0	47.9	47.0
50	56.5	55.4	54.3	53.2	52.2	51.2	50.2	49.2	48.2	47.2
51	56.6	55.5	54.4	53.4	52.3	51.3	50.3	49.3	48.3	47.4
52	56.7	55.6	54.6	53.5	52.5	51.5	50.5	49.5	48.5	47.5
53	56.9	55.8	54.7	53.6	52.6	51.6	50.6	49.6	48.6	47.7
54	57.0	55.9	54.8	53.8	52.7	51.7	50.7	49.8	48.8	47.8
55	57.1	56.0	55.0	53.9	52.9	51.9	50.9	49.9	48.9	48.0
56	57.3	56.2	55.1	54.1	53.0	52.0	51.0	50.0	49.1	48.1
57	57.4	56.3	55.2	54.2	53.2	52.2	51.2	50.2	49.2	48.3
58	57.5	56.4	55.4	54.3	53.3	52.3	51.3	50.3	49.4	48.4
59	57.7	56.6	55.5	54.5	53.4	52.4	51.4	50.5	49.5	48.6
60	57.8	56.7	55.6	54.6	53.6	52.6	51.6	50.6	49.7	48.7
61	57.9	56.8	55.8	54.7	53.7	52.7	51.7	50.8	49.8	48.9
62	58.0	57.0	55.9	54.9	53.8	52.8	51.9	50.9	49.9	49.0
	15.0	**15.5**	**16.0**	**16.5**	**17.0**	**17.5**	**18.0**	**18.5**	**19.0**	**19.5**

Temperature of Air, Fahrenheit.	t — t′ = Difference of Temperatures of the Air and of the Dew-Point. — Fahrenheit.									
	15.0	**15.5**	**16.0**	**16.5**	**17.0**	**17.5**	**18.0**	**18.5**	**19.0**	**19.5**
62°	58.0	57.0	55.9	54.9	53.8	52.8	51.9	50.9	49.9	49.0
63	58.2	57.1	56.0	55.0	54.0	53.0	52.0	51.0	50.1	49.1
64	58.3	57.2	56.2	55.1	54.1	53.1	52.1	51.2	50.2	49.3
65	58.4	57.4	56.3	55.3	54.3	53.3	52.3	51.3	50.4	49.4
66	58.6	57.5	56.4	55.4	54.4	53.4	52.4	51.5	50.5	49.6
67	58.7	57.6	56.6	55.5	54.5	53.5	52.6	51.6	50.6	49.7
68	58.8	57.8	56.7	55.7	54.7	53.7	52.7	51.7	50.8	49.9
69	59.0	57.9	56.8	55.8	54.8	53.8	52.8	51.9	50.9	50.0
70	59.1	58.0	57.0	55.9	54.9	53.9	53.0	52.0	51.1	50.1
71	59.2	58.2	57.1	56.1	55.1	54.1	53.1	52.1	51.2	50.3
72	59.3	58.3	57.2	56.2	55.2	54.2	53.2	52.3	51.3	50.4
73	59.5	58.4	57.4	56.3	55.3	54.3	53.4	52.4	51.5	50.6
74	59.6	58.5	57.5	56.5	55.5	54.5	53.5	52.6	51.6	50.7
75	59.7	58.7	57.6	56.6	55.6	54.6	53.6	52.7	51.7	50.8
76	59.8	58.8	57.8	56.7	55.7	54.7	53.8	52.8	51.9	51.0
77	60.0	58.9	57.9	56.9	55.9	54.9	53.9	53.0	52.0	51.1
78	60.1	59.1	58.0	57.0	56.0	55.0	54.0	53.1	52.2	51.2
79	60.2	59.2	58.1	57.1	56.1	55.1	54.2	53.2	52.3	51.4
80	60.3	59.3	58.3	57.3	56.3	55.3	54.3	53.4	52.4	51.5
81	60.5	59.4	58.4	57.4	56.4	55.4	54.5	53.5	52.6	51.7
82	60.6	59.6	58.5	57.5	56.5	55.5	54.6	53.6	52.7	51.8
83	60.7	59.7	58.6	57.6	56.6	55.7	54.7	53.8	52.8	51.9
84	60.8	59.8	58.8	57.8	56.8	55.8	54.8	53.9	53.0	52.1
85	60.9	59.9	58.9	57.9	56.9	55.9	55.0	54.0	53.1	52.2
86	61.1	60.0	59.0	58.0	57.0	56.1	55.1	54.2	53.2	52.3
87	61.2	60.2	59.1	58.1	57.2	56.2	55.2	54.3	53.4	52.5
88	61.3	60.3	59.3	58.3	57.3	56.3	55.4	54.4	53.5	52.6
89	61.4	60.4	59.4	58.4	57.4	56.5	55.5	54.6	53.7	52.7
90	61.6	60.5	59.5	58.5	57.6	56.6	55.6	54.7	53.8	52.9
91	61.7	60.7	59.6	58.7	57.7	56.7	55.8	54.8	53.9	53.0
92	61.8	60.8	59.8	58.8	57.8	56.9	55.9	55.0	54.1	53.2
93	61.9	60.9	59.9	58.9	57.9	57.0	56.0	55.1	54.2	53.3
94	62.0	61.0	60.0	59.0	58.1	57.1	56.2	55.2	54.3	53.4
95	62.1	61.1	60.1	59.2	58.2	57.2	56.3	55.4	54.5	53.6
96	62.3	61.3	60.3	59.3	58.3	57.4	56.4	55.5	54.6	53.7
97	62.4	61.4	60.4	59.4	58.4	57.5	56.5	55.6	54.7	53.8
98	62.5	61.5	60.5	59.5	58.6	57.6	56.7	55.8	54.9	54.0
99	62.6	61.6	60.6	59.6	58.7	57.7	56.8	55.9	55.0	54.1
100	62.7	61.7	60.7	59.8	58.8	57.9	56.9	56.0	55.1	54.2
101	62.8	61.9	60.9	59.9	58.9	58.0	57.1	56.2	55.3	54.4
102	63.0	62.0	61.0	60.0	59.1	58.1	57.2	56.3	55.4	54.5
103	63.1	62.1	61.1	60.1	59.2	58.3	57.3	56.4	55.5	54.6
104	63.2	62.2	61.2	60.3	59.3	58.4	57.5	56.6	55.7	54.8
	15.0	**15.5**	**16.0**	**16.5**	**17.0**	**17.5**	**18.0**	**18.5**	**19.0**	**19.5**

Temperature of Air, Fahrenheit.	t — t′ = Difference of Temperatures of the Air and of the Dew-Point.—Fahrenheit.									
	20.0	**20.5**	**21.0**	**21.5**	**22.0**	**22.5**	**23.0**	**23.5**	**24.0**	**24.5**
20°	40.2	39.3	38.4	37.5	36.6	35.8	34.9	34.1	33.3	32.5
21	40.3	39.4	38.4	37.6	36.7	35.8	35.0	34.2	33.4	32.6
22	40.3	39.4	38.5	37.6	36.8	35.9	35.1	34.3	33.5	32.7
23	40.4	39.5	38.6	37.7	36.8	36.0	35.2	34.4	33.6	32.8
24	40.4	39.6	38.6	37.8	36.9	36.1	35.2	34.4	33.6	32.9
25	40.5	39.6	38.7	37.8	37.0	36.2	35.3	34.5	33.7	33.0
26	40.5	39.7	38.8	37.9	37.0	36.2	35.4	34.6	33.8	33.1
27	40.6	39.7	38.8	38.0	37.1	36.3	35.5	34.7	33.9	33.1
28	40.7	39.8	38.9	38.0	37.2	36.3	35.5	34.7	34.0	33.2
29	40.8	39.9	38.9	38.1	37.2	36.4	35.6	34.8	34.0	33.3
30	40.8	39.9	39.0	38.1	37.3	36.5	35.7	34.9	34.1	33.4
31	41.1	40.2	39.2	38.4	37.5	36.7	35.9	35.1	34.3	33.6
32	41.3	40.4	39.5	38.6	37.7	37.0	36.1	35.3	34.5	33.8
33	41.6	40.6	39.7	38.8	38.0	37.2	36.3	35.5	34.7	34.0
34	41.8	40.9	39.9	39.1	38.2	37.4	36.5	35.7	34.9	34.2
35	42.1	41.1	40.2	39.3	38.4	37.7	36.7	35.9	35.1	34.4
36	42.3	41.4	40.4	39.6	38.7	37.9	37.0	36.2	35.4	34.6
37	42.6	41.7	40.7	39.8	38.9	38.2	37.2	36.4	35.6	34.8
38	42.8	42.0	41.0	40.1	39.2	38.4	37.5	36.6	35.8	35.0
39	43.1	42.3	41.3	40.4	39.5	38.6	37.7	36.9	36.0	35.2
40	43.3	42.6	41.6	40.7	39.8	38.9	38.0	37.1	36.3	35.4
41	43.7	42.9	41.9	41.0	40.0	39.1	38.3	37.4	36.5	35.7
42	44.0	43.2	42.2	41.2	40.3	39.4	38.5	37.7	36.8	36.0
43	44.3	43.4	42.5	41.5	40.6	39.7	38.8	38.0	37.1	36.3
44	44.7	43.7	42.8	41.8	40.9	40.0	39.1	38.2	37.4	36.6
45	45.0	44.0	43.1	42.1	41.2	40.3	39.4	38.5	37.7	36.8
46	45.2	44.3	43.3	42.4	41.4	40.5	39.7	38.8	37.9	37.1
47	45.5	44.5	43.6	42.6	41.7	40.8	39.9	39.1	38.2	37.4
48	45.7	44.8	43.8	42.9	42.0	41.1	40.2	39.3	38.5	37.6
49	46.0	45.0	44.1	43.2	42.2	41.3	40.5	39.6	38.7	37.9
50	46.2	45.3	44.3	43.4	42.5	41.6	40.7	39.9	39.0	37.2
51	46.4	45.4	44.5	43.6	42.7	41.8	40.9	40.1	39.2	38.4
52	46.6	45.5	44.7	43.8	42.9	42.0	41.2	40.3	39.5	38.6
53	46.7	45.8	44.9	44.0	43.1	42.2	41.4	40.5	39.7	38.9
54	46.9	46.0	45.1	44.2	43.3	42.4	41.6	40.8	39.9	39.1
55	47.0	46.1	45.2	44.4	43.5	42.6	41.8	41.0	40.1	39.3
56	47.2	46.3	45.4	44.5	43.6	42.8	42.0	41.1	40.3	39.5
57	47.3	46.4	45.5	44.7	43.8	42.9	42.1	41.3	40.5	39.6
58	47.5	46.6	45.7	44.8	43.9	43.1	42.3	41.4	40.6	39.8
59	47.6	46.7	45.8	45.0	44.1	43.2	42.4	41.6	40.8	40.0
60	47.8	46.9	46.0	45.1	44.2	43.4	42.5	41.7	40.9	40.1
61	47.9	47.0	46.1	45.3	44.4	43.5	42.7	41.9	41.1	40.3
62	48.1	47.2	46.3	45.4	44.5	43.7	42.8	42.0	41.2	40.4
	20.0	**20.5**	**21.0**	**21.5**	**22.0**	**22.5**	**23.0**	**23.5**	**24.0**	**24.5**

Temperature of Air, Fahrenheit.	t — t′ = Difference of Temperatures of the Air and of the Dew-Point. — Fahrenheit.									
	20.0	**20.5**	**21.0**	**21.5**	**22.0**	**22.5**	**23.0**	**23.5**	**24.0**	**24.5**
62°	48.1	47.2	46.3	45.4	44.5	43.7	42.8	42.0	41.2	40.4
63	48.2	47.3	46.4	45.5	44.7	43.8	43.0	42.2	41.4	40.6
64	48.4	47.5	46.6	45.7	44.8	44.0	43.1	42.3	41.5	40.7
65	48.6	47.6	46.7	45.8	45.0	44.1	43.3	42.5	41.7	40.9
66	48.7	47.8	46.9	46.0	45.1	44.3	43.4	42.6	41.8	41.0
67	48.8	47.9	47.0	46.1	45.3	44.4	43.6	42.8	42.0	41.2
68	48.9	48.0	47.2	46.3	45.4	44.6	43.7	42.9	42.1	41.3
69	49.1	48.2	47.3	46.4	45.6	44.7	43.9	43.1	42.3	41.5
70	49.2	48.3	47.4	46.6	45.7	44.9	44.0	43.2	42.4	41.6
71	49.4	48.5	47.6	46.7	45.9	45.0	44.2	43.4	42.6	41.8
72	49.5	48.6	47.7	46.9	46.0	45.2	44.3	43.5	42.7	41.9
73	49.6	48.8	47.9	47.0	46.1	45.3	44.5	43.7	42.9	42.1
74	49.8	48.9	48.0	47.1	46.3	45.4	44.6	43.8	43.0	42.2
75	49.9	49.0	48.2	47.3	46.4	45.6	44.8	44.0	43.1	42.4
76	50.1	49.2	48.3	47.4	46.6	45.7	44.9	44.1	43.3	42.5
77	50.2	49.3	48.5	47.6	46.7	45.9	45.1	44.2	43.4	42.6
78	50.3	49.5	48.6	47.7	46.9	46.0	45.2	44.4	43.6	42.8
79	50.5	49.6	48.7	47.8	47.0	46.2	45.3	44.5	43.7	43.0
80	50.6	49.7	48.9	48.0	47.2	46.3	45.5	44.7	43.9	43.1
81	50.8	49.9	49.0	48.1	47.3	46.5	45.6	44.8	44.0	43.2
82	50.9	50.0	49.2	48.3	47.4	46.6	45.8	45.0	44.2	43.4
83	51.0	50.1	49.3	48.4	47.6	46.8	45.9	45.1	44.3	43.5
84	51.2	50.3	49.4	48.6	47.7	46.9	46.1	45.3	44.5	43.7
85	51.3	50.4	49.6	48.7	47.9	47.0	46.2	45.4	44.6	43.8
86	51.4	50.6	49.7	48.8	48.0	47.2	46.4	45.6	44.8	44.0
87	51.6	50.7	49.8	49.0	48.1	47.3	46.5	45.7	44.9	44.1
88	51.7	50.8	50.0	49.1	48.3	47.5	46.6	45.8	45.0	44.3
89	51.9	51.0	50.1	49.3	48.4	47.6	46.8	46.0	45.2	44.4
90	52.0	51.1	50.3	49.4	48.6	47.7	46.9	46.1	45.3	44.6
91	52.1	51.3	50.4	49.5	48.7	47.9	47.1	46.3	45.5	44.7
92	52.3	51.4	50.5	49.7	48.8	48.0	47.2	46.4	45.6	44.8
93	52.4	51.5	50.7	49.8	49.0	48.2	47.4	46.6	45.8	45.0
94	52.5	51.7	50.8	50.0	49.1	48.3	47.5	46.7	45.9	45.1
95	52.7	51.8	50.9	50.1	49.3	48.4	47.6	46.8	46.1	45.3
96	52.8	51.9	51.1	50.2	49.4	48.6	47.8	47.0	46.2	45.4
97	52.9	52.1	51.2	50.4	49.5	48.7	47.9	47.1	46.3	45.6
98	53.1	52.2	51.4	50.5	49.7	48.9	48.1	47.3	46.5	45.7
99	53.2	52.3	51.5	50.6	49.8	49.0	48.2	47.4	46.6	45.9
100	53.4	52.5	51.6	50.8	50.0	49.1	48.3	47.5	46.8	46.0
101	53.5	52.6	51.8	50.9	50.1	49.3	48.5	47.7	46.9	46.2
102	53.6	52.8	51.9	51.1	50.2	49.4	48.6	47.8	47.1	46.3
103	53.8	52.9	52.0	51.2	50.4	49.6	48.8	48.0	47.2	46.4
104	53.9	53.0	52.2	51.3	50.5	49.7	48.9	48.1	47.3	46.6
	20.0	**20.5**	**21.0**	**21.5**	**22.0**	**22.5**	**23.0**	**23.5**	**24.0**	**24.5**

TABLE IX.

FACTOR $\frac{100}{F}$, FOR COMPUTING THE RELATIVE HUMIDITY, OR THE DEGREE OF MOISTURE OF THE AIR, EXPRESSED IN HUNDREDTHS, FROM ITS ABSOLUTE HUMIDITY GIVEN IN ENGLISH MEASURES.

THE Relative Humidity, or the degree of moisture of the air, is, as explained above, the ratio of the quantity of vapor contained in the air to the quantity it could contain at the temperature observed, if fully saturated.

If we call

The force of vapor contained in the air $= f$,

The maximum of the force of vapor at the temperature of the air $=$ F,

The point of saturation $=$ 100,

we have the proportion,

$$\text{Relative Humidity} : 100 :: f : F,$$

and

$$\frac{f \times 100}{F} = \text{Relative Humidity in Hundredths.}$$

But as $\frac{f \times 100}{F} = f \times \frac{100}{F}$, it is obvious that the operation indicated by the former expression, viz. $\frac{f \times 100}{F}$, would be reduced to a simple multiplication, if we had a table of the factors $\frac{100}{F}$. Such a table is obtained by dividing the constant number 100 by each number in the Table of Elastic Forces of Vapor, and substituting the quotients for the tensions, or forces of vapor.

The following Table gives the factor $\frac{100}{F}$ for every tenth of a degree from 0° to 104° Fahrenheit, corresponding to the Forces of Vapor in Table VI., or Regnault's table reduced to English measures.

USE OF THE TABLE.

The force of vapor contained in the air, or its absolute humidity, being given in English measures, multiply the number expressing it by the factor in the table corresponding to the temperature of the air at the time of the observation; the result will be the *Relative Humidity in Hundredths.*

Examples.

1. Suppose the temperature of the air to be = 60° Fahrenheit.
" " force of vapor in the air to be = .388 English inch.

Opposite 60° is found in the table the factor 193.1.

Then 0.388 × 193.1 = 74.9, Relative Humidity in Hundredths.

---

2. Suppose the temperature of the air to be = 74°.5 Fahrenheit.
" " force of vapor in the air to be = .650 English inch.

Table gives for 74°.5 the factor 117.2.

Then 0.650 × 117.2 = 76.2, Relative Humidity required.

IX. FACTOR $\frac{100}{F}$, FOR COMPUTING THE RELATIVE HUMIDITY, OR THE DEGREE OF MOISTURE OF THE AIR,

EXPRESSED IN HUNDREDTHS, FROM ITS ABSOLUTE HUMIDITY GIVEN IN ENGLISH INCHES.

Temperature of Air, Fahrenheit.	Tenths of Degrees.									
	0.	**1.**	**2.**	**3.**	**4.**	**5.**	**6.**	**7.**	**8.**	**9.**
0°	2306	2295	2285	2275	2264	2254	2243	2233	2222	2211
1	2201	2191	2181	2171	2162	2152	2142	2132	2122	2111
2	2101	2092	2083	2074	2064	2055	2045	2036	2026	2017
3	2007	1998	1990	1981	1972	1963	1954	1945	1936	1927
4	1918	1910	1901	1893	1885	1876	1868	1859	1851	1842
5	1834	1826	1818	1810	1802	1794	1786	1777	1769	1761
6	1753	1745	1738	1730	1722	1714	1707	1699	1691	1683
7	1675	1668	1660	1653	1646	1638	1631	1623	1616	1608
8	1600	1594	1587	1580	1572	1565	1558	1551	1544	1537
9	1529	1523	1516	1509	1503	1496	1489	1482	1475	1469
10	1462	1455	1449	1443	1436	1430	1423	1417	1410	1404
11	1397	1391	1385	1379	1373	1367	1361	1355	1348	1342
12	1336	1330	1324	1319	1313	1307	1301	1295	1289	1284
13	1278	1272	1267	1261	1255	1250	1244	1239	1233	1228
14	1222	1217	1211	1206	1200	1195	1189	1184	1178	1173
15	1167	1162	1157	1151	1146	1141	1136	1130	1125	1120
16	1114	1109	1104	1099	1094	1089	1084	1079	1074	1069
17	1064	1059	1055	1050	1045	1040	1035	1031	1026	1021
18	1016	1012	1007	1003	998.2	993.6	989.1	984.5	979.9	975.3
19	970.6	966.4	962.2	957.9	953.7	949.4	945.0	940.7	936.3	931.9
20	927.5	923.5	919.5	915.5	911.4	907.4	903.3	899.1	895.0	890.8
21	886.7	882.9	879.1	875.3	871.4	867.6	863.7	859.8	855.8	851.9
22	847.9	844.3	840.7	837.1	833.4	829.8	826.1	822.4	818.7	815.0
23	811.2	807.8	804.3	800.8	797.3	793.8	790.2	786.7	783.1	779.5
24	775.9	772.6	769.3	766.0	762.7	759.3	756.0	752.6	749.2	745.8
25	742.4	739.3	736.2	733.0	729.9	726.7	723.5	720.3	717.1	713.9
26	710.6	707.7	704.7	701.8	698.8	695.8	692.8	689.7	686.7	683.6
27	680.5	677.8	675.0	672.1	669.3	666.5	663.6	660.7	657.8	654.9
28	652.0	649.4	646.7	644.1	641.4	638.7	636.0	633.3	630.5	627.8
29	625.0	622.5	620.0	617.5	614.9	612.4	609.8	607.2	604.6	602.0
30	599.4	597.1	594.7	592.3	589.9	587.4	585.0	582.6	580.1	577.6
31	575.1	572.9	570.7	568.4	566.2	563.9	561.6	559.2	556.9	554.5
	0.	**1.**	**2.**	**3.**	**4.**	**5.**	**6.**	**7.**	**8.**	**9.**

Temperature of Air, Fahrenheit.	Tenths of Degrees.									
	0.	1.	2.	3.	4.	5.	6.	7.	8.	9.
32°	552.2	550.0	547.8	545.7	543.6	541.4	539.3	537.2	535.1	533.0
33	530.9	528.8	526.8	524.7	522.7	520.6	518.6	516.5	514.5	512.5
34	510.5	508.5	506.5	504.5	502.5	500.5	498.6	496.6	494.7	492.7
35	490.8	488.9	487.0	485.1	483.2	481.3	479.4	477.5	475.6	473.8
36	471.9	470.1	468.2	466.4	464.6	462.8	461.0	459.2	457.4	455.6
37	453.8	452.0	450.3	448.5	446.8	445.0	443.3	441.6	439.9	438.1
38	436.4	434.7	433.1	431.4	429.7	428.0	426.4	424.7	423.1	421.4
39	419.8	418.2	416.6	415.0	413.4	411.8	410.2	408.6	407.0	405.5
40	403.9	402.4	400.8	399.3	397.8	396.2	394.7	393.2	391.7	290.2
41	388.7	387.2	385.8	384.3	382.9	381.4	380.0	378.5	377.1	375.7
42	374.3	372.9	371.5	370.0	368.6	367.3	365.9	364.5	363.1	361.7
43	360.4	359.0	357.6	356.3	354.9	353.6	352.3	350.9	349.6	348.3
44	347.0	345.6	344.3	343.0	341.7	340.4	339.2	337.9	336.6	335.3
45	334.1	332.8	331.6	330.3	328.1	327.8	326.6	325.4	324.1	322.9
46	321.7	320.5	319.3	318.1	316.9	315.7	314.5	313.3	312.2	311.0
47	309.8	308.7	307.5	306.4	305.2	304.1	302.9	301.8	300.7	299.6
48	298.5	297.3	296.2	295.1	294.0	292.9	291.9	290.8	289.7	288.6
49	287.6	286.5	285.4	284.4	283.3	282.3	281.3	280.2	279.2	278.2
50	277.1	276.1	275.1	274.1	273.1	272.1	271.1	270.1	269.1	268.2
51	267.2	266.2	265.2	264.3	263.3	262.3	261.4	260.4	259.5	258.5
52	257.6	256.6	255.7	254.8	253.8	252.9	252.0	251.1	250.2	249.3
53	248.3	247.4	246.5	245.6	244.7	243.9	243.0	242.1	241.2	240.3
54	239.5	238.6	237.7	236.9	236.0	235.1	234.3	233.4	232.6	231.7
55	230.9	230.1	229.2	228.4	227.6	226.8	225.9	225.1	224.3	223.5
56	222.7	221.9	221.1	220.3	219.5	218.7	217.9	217.1	216.4	215.6
57	214.8	214.0	213.3	212.5	211.8	211.0	210.2	209.5	208.7	208.0
58	207.3	206.5	205.8	205.0	204.3	203.6	202.9	202.2	201.4	200.7
59	200.0	199.3	198.6	197.9	197.2	196.5	195.8	195.1	194.4	193.8
60	193.1	192.4	191.7	191.0	190.4	189.7	189.0	188.4	187.7	187.0
61	186.4	185.7	185.1	184.4	183.8	183.1	182.5	181.8	181.2	180.6
62	179.9	179.3	178.7	178.0	177.4	176.8	176.2	175.6	174.9	174.3
63	173.7	173.1	172.5	171.9	171.3	170.7	170.1	169.5	168.9	168.3
64	167.7	167.1	166.6	166.0	165.4	164.8	164.3	163.7	163.1	162.5
65	162.0	161.4	160.9	160.3	159.7	159.2	158.6	158.1	157.5	157.0
66	156.5	155.9	155.4	154.8	154.3	153.8	153.2	152.7	152.2	151.7
67	151.1	150.6	150.1	149.6	149.1	148.6	148.1	147.6	147.1	146.6
	0.	1.	2.	3.	4.	5.	6.	7.	8.	9.

Temperature of Air, Fahrenheit.	Tenths of Degrees.									
	0.	**1.**	**2.**	**3.**	**4.**	**5.**	**6.**	**7.**	**8.**	**9.**
68°	146.0	145.6	145.1	144.6	144.1	143.6	143.1	142.6	142.1	141.6
69	141.2	140.7	140.2	139.7	139.2	138.8	138.3	137.8	137.4	136.9
70	136.4	136.0	135.5	135.1	134.6	134.1	133.7	133.2	132.8	132.3
71	131.9	131.4	131.0	130.5	130.1	129.7	129.2	128.8	128.3	127.9
72	127.5	127.1	126.6	126.2	125.8	125.3	124.9	124.5	124.1	123.7
73	123.3	122.8	122.4	122.0	121.6	121.2	120.8	120.4	120.0	119.6
74	119.2	118.8	118.4	118.0	117.6	117.2	116.8	116.4	116.0	115.6
75	115.3	114.9	114.5	114.1	113.7	113.3	113.0	112.6	112.2	111.9
76	111.5	111.1	110.7	110.4	110.0	109.6	109.3	108.9	108.6	108.2
77	107.9	107.5	107.1	106.8	106.4	106.1	105.7	105.4	105.1	104.7
78	104.4	104.0	103.7	103.3	103.0	102.7	102.3	102.0	101.7	101.3
79	101.0	100.7	100.3	100.0	99.68	99.35	99.02	98.70	98.38	98.06
80	97.73	97.42	97.10	96.78	96.47	96.15	95.84	95.52	95.21	94.90
81	94.59	94.29	93.98	93.67	93.37	93.06	92.76	92.46	92.16	91.86
82	91.56	91.26	90.97	90.67	90.38	90.09	89.80	89.51	89.22	88.93
83	88.64	88.36	88.07	87.79	87.50	87.22	86.94	86.66	86.38	86.10
84	85.83	85.55	85.27	85.00	84.73	84.46	84.19	83.92	83.65	83.38
85	83.12	82.85	82.59	82.32	82.06	81.80	81.54	81.28	81.02	80.77
86	80.51	80.25	80.00	79.74	79.49	79.24	78.99	78.74	78.49	78.24
87	77.99	77.75	77.50	77.26	77.01	76.77	76.52	76.28	76.04	75.80
88	75.56	75.32	75.08	74.85	74.61	74.37	74.14	73.91	73.67	73.44
89	73.21	72.98	72.75	72.52	72.29	72.06	71.84	71.61	71.39	71.16
90	70.94	70.72	70.49	70.27	70.05	69.83	69.61	69.39	69.18	68.96
91	68.74	68.53	68.32	68.10	67.89	67.68	67.47	67.26	67.05	66.84
92	66.63	66.42	66.22	66.01	65.81	65.60	65.40	65.19	64.99	64.79
93	64.59	64.39	64.19	63.99	63.79	63.59	63.40	63.20	63.01	62.81
94	62.62	62.43	62.24	62.04	61.85	61.66	61.47	61.29	61.10	60.91
95	60.72	60.54	60.35	60.17	59.98	59.80	59.62	59.43	59.25	59.07
96	58.89	58.71	58.53	58.35	58.17	58.00	57.82	57.64	57.47	57.29
97	57.12	56.94	56.77	56.60	56.42	56.25	56.08	55.91	55.74	55.57
98	55.40	55.23	55.06	54.90	54.73	54.56	54.40	54.23	54.07	53.91
99	53.74	53.58	53.42	53.26	53.09	52.93	52.77	52.61	52.45	52.30
100	52.14	51.98	51.82	51.67	51.51	51.36	51.20	51.05	50.90	50.74
101	50.59	50.44	50.29	50.14	49.99	49.84	49.69	49.54	49.39	49.24
102	49.10	48.95	48.80	48.66	48.51	48.37	48.22	48.08	47.94	47.79
103	47.65	47.51	47.37	47.23	47.09	46.95	46.81	46.67	46.53	46.40
104	46.26	46.12	45.99	45.85	45.72	45.58	45.45	45.31	45.18	45.04
	0.	**1.**	**2.**	**3.**	**4.**	**5.**	**6.**	**7.**	**8.**	**9.**

TABLE X.

WEIGHT OF VAPOR, IN GRAINS TROY,

CONTAINED IN A CUBIC FOOT OF SATURATED AIR, UNDER A BAROMETRIC PRESSURE OF 30 ENGLISH INCHES, AT TEMPERATURES BETWEEN 0° AND 105° FAHRENHEIT.

THE weight of a litre of dry air at the temperature of zero Centigrade, or 32° Fahrenheit, and under a barometric pressure of 760 millimetres, as determined by the experiments of Regnault (*Mémoires de l'Institut,* Tom. XXI. p. 157), and corrected for a slight error of computation (see above, p. 38), is 1.293223 grammes. The co-efficient of expansion of the air, according to the same physicist, is 0.00367 for 1° Centigrade ; and the theoretic density of vapor is nearly 0.622, or $\frac{5}{8}$, of that of the air at the same temperature and pressure. From these elements the weight of the vapor contained in a determined volume of air, the temperature and humidity of which are known, can be deduced.

Reducing these values to English measures, 1 litre being = 61.02705 cubic inches, and 1 gramme = 15.43208 grains Troy, we have

$$1.293223 \text{ grammes} = 19.9571208 \text{ grains},$$

and

$$61.027051 \text{ cubic inches} : 19.9571208 \text{ grains} :: 1 \text{ cubic inch} : 0.32702 \text{ grain}.$$

Therefore, the weight of a cubic foot of dry air, at 32° Fahrenheit, under a pressure of 760 millimetres, or 29.918 English inches, is = 0.32702 grain × 1728 = 565.0923 grains Troy. Under a barometric pressure of 30 inches, it becomes

$$\frac{30}{29.918} \times 565.0923 = 566.5691 \text{ grains}.$$

The coefficient for the expansion of the air becomes 0.0020361 of its bulk for 1° Fahrenheit.

Now, if we call

t = the temperature of the air ;

W = the weight of vapor in a saturated air at the temperature t ;

F = the maximum of the force of vapor due to the temperature t, as given in the tables ;

then the weight of the vapor contained in a cubic foot of saturated air is given by the formula

$$W = 0.622 \frac{566.5691 \text{ grains}}{1 + 0.002036 \times (t - 32^\circ)} \cdot \frac{F}{30};$$

from which the values in Table X. have been computed. The forces of vapor due to the temperatures in the first column are those of Regnault, as given in Table VI.

It is evident, that, in order to find the weight of the vapor contained in the air at any state of humidity and pressure, it suffices to substitute for the normal values of $\frac{F}{30}$ the force of vapor and the barometric pressure given by the observation.

B

X. WEIGHT OF VAPOR, IN GRAINS TROY,

CONTAINED IN A CUBIC FOOT OF SATURATED AIR, AT TEMPERATURES BETWEEN 0° AND 105° FAHRENHEIT.

Temperature of Air, Fahren.	Force of Vapor in Eng. Inches.	Weight of Vapor in Grains.	Difference.	Temperature of Air, Fahren.	Force of Vapor in Eng. Inches.	Weight of Vapor in Grains.	Difference.	Temperature of Air, Fahren	Force of Vapor in Eng. Inches.	Weight of Vapor in Grains.	Difference.
0°	0.043	0.545	0.024	35°	0.204	2.379	0.090	70°	0.733	7.992	0.261
1	0.045	0.569	0.025	36	0.212	2.469	0.093	71	0.758	8.252	0.268
2	0.048	0.595	0.027	37	0.220	2.563	0.097	72	0.784	8.521	0.276
3	0.050	0.621	0.028	38	0.229	2.659	0.100	73	0.811	8.797	0.284
4	0.052	0.649	0.029	39	0.238	2.759	0.103	74	0.839	9.081	0.291
5	0.055	0.678	0.030	40	0.248	2.862	0.106	75	0.868	9.372	0.298
6	0.057	0.708	0.031	41	0.257	2.967	0.109	76	0.897	9.670	0.307
7	0.060	0.739	0.033	42	0.267	3.076	0.113	77	0.927	9.977	0.315
8	0.062	0.772	0.034	43	0.277	3.189	0.116	78	0.958	10.292	0.324
9	0.065	0.806	0.035	44	0.288	3.306	0.120	79	0.990	10.616	0.332
10	0.068	0.841	0.037	45	0.299	3.426	0.124	80	1.023	10.949	0.342
11	0.072	0.878	0.038	46	0.311	3.550	0.129	81	1.057	11.291	0.352
12	0.075	0.916	0.040	47	0.323	3.679	0.133	82	1.092	11.643	0.361
13	0.078	0.957	0.042	48	0.335	3.811	0.137	83	1.128	12.005	0.371
14	0.082	0.999	0.044	49	0.348	3.948	0.141	84	1.165	12.376	0.380
15	0.086	1.043	0.046	50	0.361	4.089	0.145	85	1.203	12.756	0.390
16	0.090	1.090	0.049	51	0.374	4.234	0.149	86	1.242	13.146	0.400
17	0.094	1.138	0.051	52	0.388	4.383	0.154	87	1.282	13.546	0.411
18	0.098	1.190	0.053	53	0.403	4.537	0.159	88	1.323	13.957	0.421
19	0.103	1.243	0.055	54	0.418	4.696	0.163	89	1.366	14.378	0.432
20	0.108	1.298	0.057	55	0.433	4.860	0.168	90	1.410	14.810	0.443
21	0.113	1.355	0.059	56	0.449	5.028	0.174	91	1.455	15.254	0.455
22	0.118	1.415	0.062	57	0.466	5.202	0.179	92	1.501	15.709	0.467
23	0.123	1.476	0.064	58	0.482	5.381	0.185	93	1.548	16.176	0.479
24	0.129	1.540	0.066	59	0.500	5.566	0.190	94	1.597	16.654	0.491
25	0.135	1.606	0.068	60	0.518	5.756	0.196	95	1.647	17.145	0.503
26	0.141	1.674	0.070	61	0.537	5.952	0.202	96	1.698	17.648	0.516
27	0.147	1.745	0.073	62	0.556	6.154	0.208	97	1.751	18.164	0.529
28	0.153	1.817	0.075	63	0.576	6.361	0.214	98	1.805	18.693	0.542
29	0.160	1.892	0.077	64	0.596	6.575	0.220	99	1.861	19.235	0.555
30	0.167	1.969	0.077	65	0.617	6.795	0.226	100	1.918	19.790	0.567
31	0.174	2.046	0.080	66	0.639	7.021	0.232	101	1.977	20.357	0.582
32	0.181	2.126	0.082	67	0.662	7.253	0.239	102	2.037	20.938	0.596
33	0.188	2.208	0.084	68	0.685	7.493	0.246	103	2.099	21.535	0.611
34	0.196	2.292	0.087	69	0.708	7.739	0.253	104	2.162	22.145	0.625
35	0.204	2.379		70	0.733	7.992		105	2.227	22.771	

PRACTICAL TABLES,

IN

ENGLISH MEASURES,

BASED ON THE HYGROMETRICAL CONSTANTS ADOPTED IN THE GREENWICH OBSERVATIONS.

TABLE

OF

THE ELASTIC FORCES OF AQUEOUS VAPOR,

UNDER A PRESSURE OF 30 INCHES, EXPRESSED IN ENGLISH INCHES OF MERCURY FOR TEMPERATURES OF FAHRENHEIT, ADOPTED IN THE GREENWICH OBSERVATIONS.

This table contains the values of the elastic force of vapor for temperatures from 0° to 90° Fahrenheit, derived from Dalton's experiments by Biot's formula, by Anderson, and published in *Edinburgh Encyclopædia*, Art. *Hygrometry*. It is republished, without the last decimal, in the volumes of the *Greenwich Magnetic and Meteorological Observations*, and on it are based the various hygrometrical tables published by Mr. Glaisher, either in the Greenwich volumes, or separately, most of which will be found below, Tables XII. to XVII.

Since Dalton published his experiments, numerous attempts have been made by various skilful physicists to determine with greater accuracy the elastic force of vapor. Dr. Ure in England, Regnault in France, and Magnus in Germany, deserve in this respect a special notice.

The last two experimenters having arrived simultaneously at results nearly identical, and their experiments having been conducted with all the care that modern science requires, and the means that it can secure, their determinations seem to command an especial confidence, and to deserve the preference over all others. It is, therefore, much to be regretted that the usefulness of the following otherwise so valuable tables, the formation of which involved so much labor, is in a measure impaired by the fact that they were computed from elements which cannot be regarded as the most reliable we now possess.

XI.

TABLE

OF THE

ELASTIC FORCE OF AQUEOUS VAPOR,

UNDER A BAROMETRIC PRESSURE OF 30 INCHES, EXPRESSED IN ENGLISH INCHES OF MERCURY FOR TEMPERATURES OF FAHRENHEIT.

FROM THE GREENWICH OBSERVATIONS.

Temperature Fahrenheit.	Tenths of Degrees.									
	0.	**1.**	**2.**	**3.**	**4.**	**5.**	**6.**	**7.**	**8.**	**9.**
°	Eng. In.	Eng. In.	Eng. In.	Eng. In.	Eng. In.	Eng. In.	Eng. In.	Eng. In.	Eng. In.	Eng. In.
0	0.061	0.061	0.062	0.062	0.062	0.062	0.063	0.063	0.063	0.063
1	0.064	0.064	0.064	0.064	0.065	0.065	0.065	0.065	0.066	0.066
2	0.066	0.066	0.067	0.067	0.067	0.067	0.068	0.068	0.068	0.068
3	0.069	0.069	0.069	0.069	0.070	0.070	0.070	0.071	0.071	0.071
4	0.071	0.072	0.072	0.072	0.072	0.073	0.073	0.073	0.073	0.074
5	0.074	0.074	0.075	0.075	0.075	0.075	0.076	0.076	0.076	0.077
6	0.077	0.077	0.077	0.078	0.078	0.078	0.079	0.079	0.079	0.080
7	0.080	0.080	0.080	0.081	0.081	0.081	0.082	0.082	0.082	0.083
8	0.083	0.083	0.083	0.084	0.084	0.084	0.085	0.085	0.085	0.086
9	0.086	0.086	0.087	0.087	0.087	0.088	0.088	0.088	0.089	0.089
10	0.089	0.090	0.090	0.090	0.091	0.091	0.091	0.092	0.092	0.092
11	0.093	0.093	0.093	0.094	0.094	0.094	0.095	0.095	0.096	0.096
12	0.096	0.097	0.097	0.097	0.098	0.098	0.098	0.099	0.099	0.099
13	0.100	0.100	0.101	0.101	0.101	0.102	0.102	0.102	0.103	0.103
14	0.104	0.104	0.104	0.105	0.105	0.106	0.106	0.106	0.107	0.107
15	0.108	0.108	0.108	0.109	0.109	0.110	0.110	0.110	0.111	0.111
16	0.112	0.112	0.112	0.113	0.113	0.114	0.114	0.115	0.115	0.115
17	0.116	0.116	0.117	0.117	0.118	0.118	0.118	0.119	0.119	0.120
18	0.120	0.121	0.121	0.121	0.122	0.122	0.123	0.123	0.124	0.124
19	0.125	0.125	0.126	0.126	0.126	0.127	0.127	0.128	0.128	0.129
20	0.129	0.130	0.130	0.131	0.131	0.132	0.132	0.133	0.133	0.134
	0.	**1.**	**2.**	**3.**	**4.**	**5.**	**6.**	**7.**	**8.**	**9.**

From the Greenwich Observations.

Temperature Fahrenheit.	Tenths of Degrees.									
	0.	1.	2.	3.	4.	5.	6.	7.	8.	9.
°	Eng. In.	Eng. In.	Eng. In.	Eng. In.	Eng. In.	Eng. In.	Eng. In.	Eng. In.	Eng. In.	Eng. In.
21	0.134	0.135	0.135	0.136	0.136	0.137	0.137	0.138	0.138	0.139
22	0.139	0.140	0.140	0.141	0.141	0.142	0.142	0.143	0.143	0.144
23	0.144	0.145	0.145	0.146	0.146	0.147	0.147	0.148	0.148	0.149
24	0.150	0.150	0.151	0.152	0.152	0.152	0.153	0.153	0.154	0.155
25	0.155	0.156	0.156	0.157	0.157	0.158	0.158	0.159	0.160	0.160
26	0.161	0.161	0.162	0.163	0.163	0.164	0.164	0.165	0.165	0.166
27	0.167	0.167	0.168	0.168	0.169	0.170	0.170	0.171	0.172	0.172
28	0.173	0.173	0.174	0.175	0.175	0.176	0.177	0.177	0.178	0.178
29	0.179	0.180	0.180	0.181	0.182	0.182	0.183	0.184	0.184	0.185
30	0.186	0.186	0.187	0.188	0.188	0.189	0.190	0.190	0.191	0.192
31	0.192	0.193	0.194	0.194	0.195	0.196	0.197	0.197	0.198	0.198
32	0.199	0.200	0.201	0.201	0.202	0.203	0.204	0.204	0.205	0.206
33	0.207	0.207	0.208	0.209	0.210	0.210	0.211	0.212	0.213	0.213
34	0.214	0.215	0.216	0.216	0.217	0.218	0.219	0.219	0.220	0.221
35	0.222	0.223	0.223	0.224	0.225	0.226	0.227	0.227	0.228	0.229
36	0.230	0.231	0.231	0.232	0.233	0.234	0.235	0.235	0.236	0.237
37	0.238	0.239	0.240	0.240	0.241	0.242	0.243	0.244	0.245	0.246
38	0.246	0.247	0.248	0.249	0.250	0.251	0.252	0.253	0.253	0.254
39	0.255	0.256	0.257	0.258	0.259	0.260	0.261	0.262	0.263	0.263
40	0.264	0.265	0.266	0.267	0.268	0.269	0.270	0.271	0.272	0.273
41	0.274	0.275	0.276	0.277	0.278	0.279	0.280	0.281	0.282	0.282
42	0.283	0.284	0.285	0.286	0.287	0.288	0.289	0.290	0.291	0.292
43	0.293	0.295	0.296	0.297	0.298	0.299	0.300	0.301	0.302	0.303
44	0.304	0.305	0.306	0.307	0.308	0.309	0.310	0.311	0.312	0.313
45	0.315	0.316	0.317	0.318	0.319	0.320	0.321	0.322	0.323	0.324
46	0.326	0.327	0.328	0.329	0.330	0.331	0.332	0.333	0.335	0.336
47	0.337	0.338	0.339	0.340	0.342	0.343	0.344	0.345	0.346	0.348
48	0.349	0.350	0.351	0.352	0.354	0.355	0.356	0.357	0.358	0.360
49	0.361	0.362	0.363	0.365	0.366	0.367	0.368	0.370	0.371	0.372
50	0.373	0.375	0.376	0.377	0.379	0.380	0.381	0.382	0.383	0.385
51	0.386	0.388	0.389	0.390	0.392	0.393	0.394	0.396	0.397	0.398
52	0.400	0.401	0.402	0.404	0.405	0.407	0.408	0.409	0.411	0.412
53	0.414	0.415	0.416	0.418	0.419	0.421	0.422	0.423	0.425	0.426
54	0.428	0.429	0.431	0.432	0.434	0.435	0.437	0.438	0.440	0.441
55	0.442	0.444	0.445	0.447	0.449	0.450	0.452	0.453	0.455	0.456
	0.	1.	2.	3.	4.	5.	6.	7.	8.	9.

From the Greenwich Observations.

Temperature Fahrenheit.	Tenths of Degrees.									
	0.	**1.**	**2.**	**3.**	**4.**	**5.**	**6.**	**7.**	**8.**	**9.**
°	Eng. In.	Eng. In.	Eng. In.	Eng. In.	Eng. In.	Eng. In.	Eng. In.	Eng. In.	Eng. In.	Eng. In.
56	0.458	0.459	0.461	0.462	0.464	0.465	0.467	0.469	0.470	0.472
57	0.473	0.475	0.476	0.478	0.480	0.481	0.483	0.485	0.486	0.488
58	0.489	0.491	0.493	0.494	0.496	0.498	0.499	0.501	0.503	0.504
59	0.506	0.508	0.509	0.511	0.513	0.515	0.516	0.518	0.520	0.521
60	0.523	0.525	0.527	0.528	0.530	0.532	0.534	0.536	0.537	0.539
61	0.541	0.543	0.544	0.546	0.548	0.550	0.552	0.554	0.555	0.557
62	0.559	0.561	0.563	0.565	0.567	0.568	0.570	0.572	0.574	0.576
63	0.578	0.580	0.582	0.584	0.586	0.588	0.590	0.591	0.593	0.595
64	0.597	0.599	0.601	0.603	0.605	0.607	0.609	0.611	0.613	0.615
65	0.617	0.619	0.621	0.623	0.626	0.628	0.630	0.632	0.634	0.636
66	0.638	0.640	0.642	0.644	0.646	0.648	0.651	0.653	0.655	0.657
67	0.659	0.661	0.664	0.666	0.668	0.670	0.672	0.674	0.677	0.679
68	0.681	0.684	0.686	0.688	0.690	0.692	0.695	0.697	0.699	0.701
69	0.704	0.706	0.708	0.711	0.713	0.715	0.717	0.720	0.722	0.725
70	0.727	0.729	0.732	0.734	0.736	0.739	0.741	0.744	0.746	0.748
71	0.751	0.753	0.756	0.758	0.761	0.763	0.766	0.768	0.771	0.773
72	0.776	0.778	0.781	0.783	0.785	0.787	0.790	0.792	0.795	0.797
73	0.801	0.803	0.806	0.809	0.811	0.814	0.817	0.819	0.822	0.824
74	0.827	0.830	0.832	0.835	0.838	0.840	0.843	0.846	0.849	0.851
75	0.854	0.857	0.860	0.862	0.865	0.868	0.871	0.873	0.876	0.879
76	0.882	0.885	0.887	0.890	0.893	0.896	0.899	0.902	0.905	0.908
77	0.910	0.913	0.916	0.919	0.922	0.925	0.928	0.931	0.934	0.937
78	0.940	0.943	0.946	0.949	0.952	0.955	0.958	0.961	0.964	0.967
79	0.970	0.973	0.976	0.979	0.983	0.986	0.989	0.992	0.995	0.998
80	1.001	1.005	1.008	1.011	1.014	1.017	1.021	1.024	1.027	1.030
81	1.034	1.037	1.040	1.043	1.047	1.050	1.053	1.057	1.060	1.063
82	1.067	1.069	1.073	1.077	1.080	1.083	1.087	1.090	1.094	1.097
83	1.101	1.104	1.108	1.111	1.114	1.118	1.121	1.125	1.129	1.132
84	1.136	1.139	1.143	1.146	1.150	1.153	1.157	1.160	1.164	1.167
85	1.171	1.175	1.178	1.182	1.186	1.190	1.193	1.197	1.201	1.205
86	1.209	1.212	1.216	1.220	1.224	1.228	1.232	1.235	1.239	1.243
87	1.247	1.251	1.255	1.258	1.262	1.266	1.270	1.274	1.278	1.282
88	1.286	1.290	1.294	1.298	1.302	1.306	1.310	1.314	1.318	1.322
89	1.326	1.330	1.335	1.339	1.343	1.347	1.351	1.355	1.359	1.364
90	1.368	1.372	1.376	1.381	1.385	1.389	1.393	1.397	1.402	1.406
	0.	**1.**	**2.**	**3.**	**4.**	**5.**	**6.**	**7.**	**8.**	**9.**

PSYCHROMETRICAL TABLE,

GIVING THE TEMPERATURE OF THE DEW-POINT, THE FORCE AND THE WEIGHT OF VAPOR IN THE ATMOSPHERE, AND ITS RELATIVE HUMIDITY, DEDUCED FROM THE INDICATIONS OF THE PSYCHROMETER, OR DRY AND WET BULB THERMOMETERS.

BY JAMES GLAISHER.

THIS elaborate table, first published in London, in 1847, in pamphlet form, by J. Glaisher, of the Royal Observatory at Greenwich, is based on the tables of elastic forces of vapor deduced from Dalton's experiments, and given above, Table XI.

The weight of a cubic foot of dry air at 32° Fahrenheit, and under the barometric pressure of 30 inches, which has been adopted by Glaisher, and from which the weight of vapor in a cubic foot of air is derived, is the mean of the determinations obtained by Shuckburgh and by Biot and Arago, which is 563.2154 grains Troy; 563 being the number actually used in the calculations. See Preface to the Table, p. 13, and also the *Greenwich Meteorological Observations* for 1842, p. xlvi.

The coefficient of the expansion of air which has been employed is that determined by the experiments of Gay-Lussac, according to which the air expands 0.00375 of its bulk for 1° Centigrade, or $\frac{1}{480}$ for 1° Fahrenheit.

All these values, as may be seen by comparing Tables VI. and XI. of the elastic forces, and also page 92, materially differ from those more recently determined with great care by Regnault, and on which are based the Psychrometrical Tables given above, page 50 *et seq*. This will account for the no inconsiderable differences often found between the results in the two tables derived from the same data. A few examples, taken from various parts of the tables, may be given here, in order to enable the meteorologist to judge of the amount of the discrepancies which may occur in the results when computed from different hygrometrical constants.

1. Suppose the temperature of the air indicated by the dry thermometer to be = 10° F.

The temperature of evaporation indicated by the wet thermometer = 9° F.

Difference 1° F.

Then, Glaisher's table gives,

The Force of Vapor = 0.065 inch.
The Relative Humidity = 0.730

Guyot's table gives,

The Force of Vapor = 0.054 inch.
The Relative Humidity = 0.791

2. By observation we have,

Dry Thermometer	=	50° F.
Wet Thermometer	=	40° F.
Difference	=	10° F.

Then, by Glaisher's table, we find,

Force of Vapor	=	0.186 inch.
Relative Humidity	=	0.495

And by Guyot's table, we find,

Force of Vapor	=	0.117 inch.
Relative Humidity	=	0.322

3. The reading of the

Dry Thermometer is	=	90° F.
Wet Thermometer is	=	70° F.
Difference	=	20° F.

By Glaisher's table we have,

Force of Vapor	=	0.523 inch.
Relative Humidity	=	0.381

And by Guyot's table,

Force of Vapor	=	0.464 inch.
Relative Humidity	=	0.329

The temperatures of the Dew-Point, given in Glaisher's tables, have been computed by means of the empirical factors given below, page 140, and in the manner there described. See Preface to the Table, page 11.

Arrangement of the Table.

In the first two columns, at the left, are found the indications, in degrees of Fahrenheit, of the dry and wet bulb thermometers. In the following columns, in their order, and opposite to each of the temperatures of the wet thermometer, are given the temperature of the dew-point; the force of vapor, in English inches; the weight of vapor, in grains, contained in a cubic foot of air; the amount of the same required for saturation; and the relative humidity in thousandths, corresponding to the difference of temperature between the two thermometers. The second half of the page, at the right, furnishes, in seven columns, the weight, in grains, of a cubic foot of air, under various barometric pressures from 28 to 31 inches, and in the different hygrometric conditions indicated by the differences of the two thermometers. These numbers have been computed in the manner described below, page 142.

The range of the table extends from 10° to 90° of the dry thermometer, or of the temperature of the air. From 10° to 34° Fahrenheit the results are calculated for every second, third, and fifth of a degree of the wet thermometer, and for extreme differences of the temperature of evaporation ranging from 2° to 5° below the temperature of the air. From 34° to 90° the results are given only for every full degree of the wet thermometer, and for extreme differences gradually increasing from 5° to 27°. This range falls short of the wants of the extreme climate of North America, where temperatures above 90° and far below 10° are of usual occurrence over a great portion of the continent. The same may be said of the range of the differences between the two thermometers in the first part of the table. The double interpolation for the fractions of degrees of both thermometers being rather too large to be neglected, its application becomes inconvenient.

Use of the Table.

Enter the table with the observed temperatures of the dry and wet bulb thermometers. On the same line as the last, and in their appropriate columns, the results deduced from these data will be found.

Example.

The observation has given,

Temperature of the air by the dry thermometer = 62° F.
Temperature of evaporation by the wet-bulb thermometer = 53° F.

Page 129, find in the first column, headed Reading of the Dry Thermometer, the temperature of 62°, and in the second, that of the wet, 53°. On the line beginning with 53° are found, in their respective columns, the results deduced from these data, viz.: —

The temperature of the Dew-point = 46°.7 F.
The force of vapor in the air = 0.333 inch.
The weight of vapor in a cubic foot of air = 3.72 grains.
The amount of vapor required for saturation = 2.53 grains.
The relative humidity in thousandths = 0.595

Reading of Thermometer, Fahr. Dry.	Reading of Thermometer, Fahr. Wet.	Temp. of Dew-Point, Fahr.	Force of Vapor in English Inches.	Weight of Vapor: In a Cubic Foot of Air.	Weight of Vapor: Reqd. for Sat'n. of a Cubic Ft. of Air.	Humidity, Saturation = 1.000.	Weight in Grains of a Cubic Foot of Air. Height of the Barometer in English Inches. in. **28.0**	in. **28.5**	in. **29.0**	in. **29.5**	in. **30.0**	in. **30.5**	in. **31.0**
°	°	°	in.	gr.	gr.		gr.	gr.	gr.	gr.	gr.	gr.	gr.
10	10.0	10.0	0.089	1.11	0.00	1.000	550.1	560.0	569.8	579.6	589.4	599.2	609.0
	9.8	8.3	0.084	1.05	0.06	0.946	550.2	560.1	569.9	579.7	589.5	599.3	609.1
	9.6	6.6	0.079	0.98	0.13	0.883	550.2	560.1	569.9	579.7	589.5	599.3	609.1
	9.4	4.9	0.074	0.92	0.19	0.829	550.2	560.1	569.9	579.7	589.5	599.3	609.1
	9.2	3.2	0.069	0.86	0.25	0.775	550.3	560.2	570.0	579.8	589.6	599.4	609.2
	9.0	1.5	0.065	0.81	0.30	0.730	550.3	560.3	570.0	579.8	589.6	599.4	609.3
11	11.0	11.0	0.093	1.15	0.00	1.000	548.9	558.7	568.5	578.3	588.1	597.9	607.7
	10.8	9.3	0.087	1.08	0.07	0.939	548.9	558.7	568.5	578.3	588.1	597.9	607.7
	10.6	7.6	0.082	1.02	0.13	0.887	549.0	558.8	568.6	578.4	588.2	598.0	607.8
	10.4	5.9	0.077	0.96	0.19	0.835	549.0	558.8	568.6	578.4	588.2	598.0	607.8
	10.2	4.2	0.072	0.90	0.25	0.783	549.0	558.8	568.6	578.4	588.2	598.0	607.8
	10.0	2.5	0.067	0.84	0.31	0.731	549.1	558.9	568.7	578.6	588.3	598.1	607.9
	9.8	0.8	0.063	0.78	0.37	0.679	549.1	558.9	568.7	578.6	588.3	598.1	607.9
12	12.0	12.0	0.096	1.19	0.00	1.000	547.7	557.5	567.2	577.0	586.8	596.6	606.4
	11.8	10.3	0.090	1.12	0.07	0.942	547.7	557.5	567.2	577.0	586.8	596.6	606.4
	11.6	8.6	0.085	1.05	0.14	0.883	547.8	557.6	567.3	577.1	586.9	596.7	606.5
	11.4	6.9	0.080	0.99	0.20	0.832	547.8	557.6	567.3	577.1	586.9	596.7	606.5
	11.2	5.2	0.075	0.93	0.26	0.782	547.8	557.6	567.3	577.1	586.9	596.7	606.5
	11.0	3.5	0.070	0.87	0.32	0.731	547.9	557.7	567.4	577.2	587.0	596.8	606.6
	10.8	1.8	0.066	0.81	0.38	0.681	547.9	557.7	567.4	577.2	587.0	596.8	606.6
	10.6	0.1	0.061	0.76	0.43	0.639	547.9	557.7	567.4	577.2	587.0	596.8	606.6
13	13.0	13.0	0.100	1.24	0.00	1.000	546.5	556.3	566.0	575.8	585.5	595.3	605.0
	12.8	11.3	0.094	1.16	0.08	0.936	546.5	556.3	566.0	575.8	585.5	595.3	605.0
	12.6	9.6	0.088	1.08	0.16	0.871	546.6	556.4	566.1	575.9	585.6	595.4	605.1
	12.4	7.9	0.083	1.02	0.22	0.823	546.7	556.5	566.2	576.0	585.7	595.5	605.2
	12.2	6.2	0.077	0.97	0.27	0.783	546.7	556.5	566.2	576.0	585.7	595.5	605.2
	12.0	4.5	0.073	0.91	0.33	0.734	546.7	556.5	566.2	576.0	585.7	595.5	605.2
	11.8	2.8	0.068	0.84	0.40	0.678	546.8	556.6	566.3	576.1	585.8	595.6	605.3
	11.6	1.1	0.064	0.79	0.45	0.637	546.8	556.6	566.3	576.1	585.8	595.6	605.3
14	14.0	14.0	0.104	1.28	0.00	1.000	545.3	555.0	564.7	574.4	584.2	594.0	603.7
	13.8	12.3	0.097	1.20	0.08	0.938	545.3	555.0	564.7	574.4	584.2	594.0	603.7
	13.6	10.6	0.091	1.12	0.16	0.875	545.4	555.1	564.8	574.5	584.3	594.1	603.8
	13.4	8.9	0.086	1.06	0.22	0.828	545.4	555.1	564.8	574.5	584.3	594.1	603.8
	13.2	7.2	0.080	1.00	0.28	0.782	545.4	555.1	564.8	574.5	584.3	594.1	603.8
	13.0	5.5	0.075	0.93	0.35	0.727	545.5	555.2	564.9	574.6	584.4	594.2	603.9
	12.8	3.8	0.071	0.87	0.41	0.680	545.5	555.2	564.9	574.6	584.4	594.2	603.9
	12.6	2.1	0.066	0.82	0.46	0.641	545.6	555.3	565.0	574.7	584.5	594.2	603.9
15	15.0	15.0	0.108	1.32	0.00	1.000	544.0	553.8	563.5	573.2	582.9	592.6	602.3
	14.8	13.3	0.101	1.24	0.08	0.940	544.0	553.8	563.5	573.2	582.9	592.6	602.3
	14.6	11.6	0.095	1.16	0.16	0.879	544.1	553.9	563.6	573.3	583.0	592.7	602.4
	14.4	9.9	0.089	1.10	0.22	0.833	544.1	553.9	563.6	573.3	583.0	592.7	602.4
	14.2	8.2	0.083	1.04	0.28	0.788	544.2	554.0	563.7	573.4	583.1	592.8	602.5
	14.0	6.5	0.078	0.97	0.35	0.735	544.2	554.0	563.7	573.4	583.1	592.8	602.5
	13.8	4.8	0.073	0.90	0.42	0.682	544.2	554.0	563.7	573.4	583.1	592.8	602.5
	13.6	3.1	0.069	0.85	0.47	0.644	544.3	554.1	563.8	573.5	583.2	592.9	602.6

Reading of Thermometer, Fahr. Dry.	Reading of Thermometer, Fahr. Wet.	Temp. of Dew-Point, Fahr.	Force of Vapor in English Inches.	Weight of Vapor In a Cubic Foot of Air.	Weight of Vapor Reqd. for Sat'n. of a Cubic Ft. of Air.	Humidity, Saturation = 1.000.	Weight in Grains of a Cubic Foot of Air. Height of the Barometer in English Inches. in. **28.0**	in. **28.5**	in. **29.0**	in. **29.5**	in. **30.0**	in. **30.5**	in. **31.0**
°	°	°	in.	gr.	gr.		gr.	gr.	gr.	gr.	gr.	gr.	gr.
16	16.0	16.0	0.112	1.37	0.00	1.000	542.8	552.5	562.2	571.9	581.6	591.3	601.0
	15.8	14.3	0.105	1.29	0.08	0.942	542.9	552.6	562.3	572.0	581.7	591.4	601.1
	15.6	12.6	0.098	1.21	0.16	0.883	542.9	552.6	562.3	572.0	581.7	591.4	601.1
	15.4	10.9	0.092	1.14	0.23	0.832	543.0	552.7	562.4	572.1	581.8	591.5	601.2
	15.2	9.2	0.087	1.07	0.30	0.781	543.0	552.7	562.4	572.1	581.8	591.5	601.2
	15.0	7.5	0.081	1.00	0.37	0.730	543.0	552.7	562.4	572.1	581.8	591.5	601.2
	14.8	5.8	0.076	0.94	0.43	0.686	543.1	552.8	562.5	572.1	581.9	591.6	601.3
	14.6	4.1	0.072	0.88	0.49	0.643	543.1	552.8	562.5	572.1	581.9	591.6	601.3
17	17.0	17.0	0.116	1.41	0.00	1.000	541.3	551.0	560.8	570.5	580.1	589.8	599.4
	16.8	15.3	0.109	1.33	0.08	0.943	541.3	551.0	560.8	570.5	580.1	589.8	599.4
	16.6	13.6	0.102	1.25	0.16	0.887	541.4	551.1	560.9	570.6	580.2	589.9	599.5
	16.4	11.9	0.096	1.17	0.24	0.830	541.4	551.1	560.9	570.6	580.2	589.9	599.5
	16.2	10.2	0.090	1.10	0.31	0.780	541.5	551.2	561.0	570.7	580.3	590.0	599.6
	16.0	8.5	0.084	1.03	0.38	0.730	541.5	551.2	561.0	570.7	580.3	590.0	599.6
	15.8	6.8	0.079	0.97	0.44	0.688	541.5	551.2	561.0	570.7	580.3	590.0	599.6
	15.6	5.1	0.074	0.91	0.50	0.646	541.6	551.3	561.1	570.8	580.4	590.1	599.7
18	18.0	18.0	0.120	1.47	0.00	1.000	540.5	550.2	559.8	569.5	579.1	588.8	598.4
	17.8	16.3	0.113	1.38	0.09	0.939	540.5	550.2	559.8	569.5	579.1	588.8	598.4
	17.6	14.6	0.106	1.29	0.18	0.878	540.6	550.3	559.9	569.6	579.2	588.9	598.5
	17.4	12.9	0.099	1.21	0.26	0.824	540.6	550.3	559.9	569.6	579.2	588.9	598.5
	17.2	11.2	0.093	1.14	0.33	0.776	540.7	550.4	560.0	569.7	579.3	589.0	598.6
	17.0	9.5	0.088	1.07	0.40	0.728	540.7	550.4	560.0	569.7	579.3	589.0	598.6
	16.8	7.8	0.082	1.01	0.46	0.688	540.7	550.5	560.1	569.8	579.3	589.0	598.6
	16.6	6.1	0.077	0.95	0.52	0.647	540.8	550.6	560.2	569.9	579.4	589.1	598.7
19	19.0	19.0	0.125	1.52	0.00	1.000	539.3	548.9	558.5	568.2	577.8	587.5	597.1
	18.8	17.3	0.117	1.43	0.09	0.941	539.3	548.9	558.5	568.2	577.8	587.5	597.1
	18.6	15.6	0.110	1.34	0.18	0.882	539.4	549.0	558.6	568.3	577.9	587.6	597.2
	18.4	13.9	0.103	1.26	0.26	0.829	539.4	549.0	558.6	568.3	577.9	587.6	597.2
	18.2	12.2	0.097	1.18	0.34	0.776	539.5	549.1	558.7	568.4	578.0	587.7	597.3
	18.0	10.5	0.091	1.11	0.41	0.730	539.5	549.1	558.7	568.4	578.0	587.7	597.3
	17.8	8.8	0.085	1.04	0.48	0.684	539.6	549.2	558.8	568.5	578.1	587.8	597.4
	17.6	7.1	0.080	0.98	0.54	0.645	539.6	549.2	558.8	568.5	578.1	587.8	597.4
20	20.0	20.0	0.129	1.58	0.00	1.000	538.1	547.7	557.3	566.9	576.5	586.1	595.7
	19.8	18.3	0.121	1.48	0.10	0.937	538.2	547.8	557.4	567.0	576.6	586.2	595.8
	19.6	16.6	0.114	1.38	0.20	0.874	538.3	547.9	557.5	567.1	576.7	586.3	595.9
	19.4	14.9	0.107	1.30	0.28	0.823	538.3	547.9	557.5	567.1	576.7	586.3	595.9
	19.2	13.2	0.101	1.23	0.35	0.779	538.3	547.9	557.5	567.1	576.7	586.3	595.9
	19.0	11.5	0.094	1.15	0.43	0.728	538.4	548.0	557.6	567.2	576.8	586.4	596.0
	18.8	9.8	0.089	1.08	0.50	0.684	538.4	548.0	557.6	567.2	576.8	586.4	596.0
	18.6	8.1	0.083	1.01	0.57	0.639	538.5	548.1	557.7	567.3	576.9	586.5	596.1
	18.4	6.4	0.078	0.95	0.63	0.601	538.5	548.1	557.7	567.3	576.9	586.5	596.1

Reading of Thermometer, Fahr.		Temp. of Dew-Point, Fahr.	Force of Vapor in English Inches.	Weight of Vapor		Humidity, Saturation = 1.000.	Weight in Grains of a Cubic Foot of Air. Height of the Barometer in English Inches.						
Dry.	Wet.			In a Cubic Foot of Air.	Reqd. for Sat'n. of a Cubic Ft. of Air.		in. **28.0**	in. **28.5**	in. **29.0**	in. **29.5**	in. **30.0**	in. **30.5**	in. **31.0**
°	°	°	in.	gr.	gr.		gr.	gr.	gr.	gr.	gr.	gr.	gr.
21	21.0	21.0	0.134	1.63	0.00	1.000	537.0	546.6	556.1	565.7	575.3	584.9	594.5
	20.8	19.3	0.126	1.53	0.10	0.939	537.0	546.6	556.1	565.7	575.3	584.9	594.5
	20.6	17.6	0.118	1.44	0.19	0.884	537.1	546.7	556.2	565.8	575.4	585.0	594.6
	20.4	15.9	0.111	1.36	0.27	0.835	537.1	546.7	556.2	565.8	575.4	585.0	594.6
	20.2	14.2	0.104	1.28	0.35	0.785	537.2	546.8	556.3	565.9	575.5	585.1	594.7
	20.0	12.5	0.098	1.20	0.43	0.736	537.2	546.8	556.3	565.9	575.5	585.1	594.7
	19.8	10.8	0.092	1.12	0.51	0.687	537.3	546.9	556.4	566.0	575.6	585.2	594.8
	19.6	9.1	0.086	1.05	0.58	0.644	537.3	546.9	556.4	566.0	575.6	585.2	594.8
	19.4	7.4	0.081	0.99	0.64	0.607	537.3	546.9	556.4	566.0	575.6	585.2	594.8
22	22.0	22.0	0.139	1.69	0.00	1.000	535.7	545.3	554.9	564.5	574.0	583.6	593.1
	21.8	20.3	0.131	1.59	0.10	0.941	535.8	545.4	555.0	564.6	574.1	583.7	593.2
	21.6	18.6	0.123	1.49	0.20	0.882	535.8	545.4	555.0	564.6	574.1	583.7	593.2
	21.4	16.9	0.115	1.40	0.29	0.828	535.9	545.5	555.1	564.7	574.2	583.8	593.3
	21.2	15.2	0.108	1.31	0.38	0.775	535.9	545.5	555.1	564.7	574.2	583.8	593.3
	21.0	13.5	0.102	1.23	0.46	0.728	536.0	545.6	555.2	564.8	574.3	583.9	593.4
	20.8	11.8	0.096	1.16	0.53	0.686	536.0	545.6	555.2	564.8	574.3	583.9	593.4
	20.6	10.1	0.090	1.09	0.60	0.645	536.1	545.7	555.3	564.9	574.4	584.0	593.5
	20.4	8.4	0.084	1.02	0.67	0.604	536.1	545.7	555.3	564.9	574.4	584.0	593.5
	20.2	6.7	0.079	0.96	0.73	0.568	536.1	545.7	555.3	564.9	574.4	584.0	593.5
23	23.0	23.0	0.144	1.75	0.00	1.000	534.6	544.2	553.7	563.3	572.8	582.4	591.9
	22.8	21.3	0.136	1.65	0.10	0.943	534.6	544.2	553.7	563.3	572.8	582.4	591.9
	22.6	19.6	0.127	1.55	0.20	0.886	534.7	544.3	553.8	563.4	572.9	582.5	592.0
	22.4	17.9	0.120	1.45	0.30	0.829	534.7	544.3	553.8	563.4	572.9	582.5	592.0
	22.2	16.2	0.112	1.36	0.39	0.777	534.8	544.4	553.9	563.5	573.0	582.6	592.1
	22.0	14.5	0.106	1.28	0.47	0.731	534.8	544.4	553.9	563.5	573.0	582.6	592.1
	21.8	12.8	0.099	1.21	0.54	0.691	534.9	544.5	554.0	563.6	573.1	582.7	592.2
	21.6	11.1	0.093	1.13	0.62	0.646	534.9	544.5	554.0	563.6	573.1	582.7	592.2
	21.4	9.4	0.087	1.06	0.69	0.606	535.0	544.6	554.1	563.7	573.2	582.8	592.3
	21.2	7.7	0.082	1.00	0.75	0.571	535.0	544.6	554.1	563.7	573.2	582.8	592.3
24	24.0	24.0	0.150	1.81	0.00	1.000	533.4	542.9	552.4	562.0	571.5	581.1	590.6
	23.8	22.5	0.142	1.72	0.09	0.951	533.5	543.0	552.5	562.1	571.6	581.2	590.7
	23.6	21.1	0.135	1.63	0.18	0.901	533.5	543.1	552.5	562.1	571.6	581.2	590.7
	23.4	19.6	0.127	1.55	0.26	0.856	533.6	543.2	552.6	562.2	571.7	581.3	590.8
	23.2	18.2	0.121	1.46	0.35	0.807	533.6	543.2	552.6	562.2	571.7	581.3	590.8
	23.0	16.7	0.115	1.38	0.43	0.762	533.7	543.3	552.7	562.3	571.8	581.4	590.9
	22.8	15.2	0.108	1.31	0.50	0.724	533.7	543.3	552.7	562.3	571.8	581.4	590.9
	22.6	13.8	0.103	1.24	0.57	0.685	533.7	543.3	552.7	562.3	571.8	581.4	590.9
	22.4	12.3	0.097	1.18	0.63	0.652	533.8	543.4	552.8	562.4	571.9	581.5	591.0
	22.2	10.8	0.091	1.12	0.69	0.634	533.8	543.4	552.8	562.4	571.9	581.5	591.0

Reading of Thermometer, Fahr. Dry.	Reading of Thermometer, Fahr. Wet.	Temp. of Dew-Point, Fahr.	Force of Vapor in English Inches.	Weight of Vapor: In a Cubic Foot of Air.	Weight of Vapor: Reqd. for Sat'n. of a Cubic Ft. of Air.	Humidity, Saturation = 1.000.	Weight in Grains of a Cubic Foot of Air. Height of the Barometer in English Inches. in. 28.0	in. 28.5	in. 29.0	in. 29.5	in. 30.0	in. 30.5	in. 31.0
°	°	°	in.	gr.	gr.		gr.	gr.	gr.	gr.	gr.	gr.	gr.
25	25.0	25.0	0.155	1.87	0.00	1.000	532.3	541.8	551.3	560.8	570.3	579.8	589.3
	24.8	23.7	0.148	1.78	0.09	0.952	532.3	541.8	551.3	560.8	570.3	579.8	589.3
	24.6	22.4	0.141	1.70	0.17	0.909	532.4	541.9	551.4	560.9	570.4	579.9	589.4
	24.4	21.2	0.135	1.62	0.25	0.867	532.4	541.9	551.4	560.9	570.4	579.9	589.4
	24.2	19.9	0.129	1.55	0.32	0.829	532.4	541.9	551.4	560.9	570.4	579.9	589.4
	24.0	18.6	0.123	1.48	0.49	0.791	532.5	542.0	551.5	561.0	570.5	580.0	589.5
	23.8	17.3	0.117	1.41	0.46	0.754	532.5	542.0	551.5	561.0	570.5	580.0	589.5
	23.6	16.0	0.112	1.34	0.53	0.717	532.6	542.1	551.6	561.1	570.6	580.1	589.6
	23.4	14.8	0.107	1.28	0.59	0.685	532.6	542.1	551.6	561.1	570.6	580.1	589.6
	23.2	13.5	0.102	1.22	0.65	0.653	532.6	542.1	551.6	561.1	570.6	580.1	589.6
26	26.0	26.0	0.161	1.93	0.00	1.000	531.1	540.6	550.0	559.5	569.0	578.5	588.0
	25.8	24.8	0.154	1.85	0.08	0.959	531.2	540.7	550.1	559.6	569.1	578.6	588.1
	25.6	23.6	0.147	1.78	0.15	0.923	531.2	540.7	550.1	559.6	569.1	578.6	588.1
	25.4	22.3	0.141	1.70	0.23	0.881	531.2	540.7	550.1	559.6	569.1	578.6	588.1
	25.2	21.2	0.135	1.62	0.31	0.839	531.3	540.8	550.2	559.7	569.2	578.7	588.2
	25.0	19.9	0.129	1.55	0.38	0.804	531.3	540.8	550.2	559.7	569.2	578.7	588.2
	24.8	18.7	0.123	1.48	0.45	0.767	531.4	540.9	550.3	559.8	569.3	578.8	588.3
	24.6	17.5	0.118	1.41	0.52	0.731	531.4	540.9	550.3	559.8	569.3	578.8	588.3
	24.4	16.2	0.112	1.35	0.58	0.700	531.4	540.9	550.3	559.8	569.3	578.8	588.3
	24.2	15.0	0.108	1.29	0.64	0.668	531.5	541.0	550.4	559.9	569.4	578.9	588.4
27	27.0	27.0	0.167	2.00	0.00	1.000	529.9	539.4	548.9	558.4	567.8	577.3	586.7
	26.7	25.2	0.156	1.88	0.12	0.940	529.9	539.4	548.9	558.4	567.8	577.4	586.8
	26.4	23.3	0.146	1.76	0.24	0.880	530.0	539.5	549.0	558.5	567.9	577.5	586.9
	26.1	21.5	0.137	1.64	0.36	0.820	530.1	539.6	549.1	558.6	568.0	577.6	587.0
	25.8	19.7	0.128	1.53	0.47	0.765	530.1	539.6	549.1	558.6	568.0	577.6	587.0
	25.5	17.8	0.119	1.43	0.57	0.715	530.2	539.7	549.2	558.7	568.1	577.7	587.1
	25.2	16.0	0.112	1.34	0.66	0.670	530.3	539.8	549.3	558.8	568.2	577.8	587.2
	24.9	14.2	0.104	1.26	0.74	0.630	530.3	539.8	549.3	558.8	568.2	577.8	587.2
	24.6	12.4	0.098	1.17	0.83	0.585	530.4	539.9	549.4	558.9	568.3	577.9	587.3
	24.3	10.5	0.091	1.09	0.91	0.545	530.5	540.0	549.5	559.0	568.3	577.9	587.3
28	28.0	28.0	0.173	2.07	0.00	1.000	528.7	538.1	547.6	557.0	566.5	575.9	585.4
	27.7	26.3	0.163	1.95	0.12	0.942	528.8	538.2	547.7	557.1	566.6	576.0	585.5
	27.4	24.6	0.153	1.84	0.23	0.889	528.9	538.3	547.8	557.2	566.7	576.1	585.6
	27.1	22.9	0.144	1.73	0.34	0.836	528.9	538.3	547.8	557.2	566.7	576.1	585.6
	26.8	21.2	0.135	1.62	0.45	0.783	529.0	538.4	547.9	557.3	566.8	576.2	585.7
	26.5	19.4	0.126	1.52	0.55	0.734	529.1	538.5	548.0	557.4	566.9	576.3	585.8
	26.2	17.7	0.119	1.42	0.65	0.686	529.1	538.5	548.0	557.4	566.9	576.3	585.8
	25.9	16.0	0.112	1.34	0.73	0.648	529.2	538.6	548.1	557.5	567.0	576.4	585.9
	25.6	14.3	0.105	1.26	0.82	0.604	529.2	538.6	548.1	557.5	567.0	576.4	585.9
	25.3	12.6	0.098	1.18	0.89	0.571	529.2	538.6	548.1	557.5	567.0	576.4	585.9

Reading of Thermometer, Fahr. Dry.	Reading of Thermometer, Fahr. Wet.	Temp. of Dew-Point, Fahr.	Force of Vapor in English Inches.	Weight of Vapor: In a Cubic Foot of Air.	Weight of Vapor: Reqd. for Sat'n. of a Cubic Ft. of Air.	Humidity, Saturation = 1.000.	Weight in Grains of a Cubic Foot of Air. Height of the Barometer in English Inches. in. **28.0**	in. **28.5**	in. **29.0**	in. **29.5**	in. **30.0**	in. **30.5**	in. **31.0**
°	°	°	in.	gr.	gr.		gr.	gr.	gr.	gr.	gr.	gr.	gr.
29	29.0	29.0	0.179	2.14	0.00	1.000	527.6	537.0	546.5	555.9	565.3	574.7	584.1
	28.7	27.5	0.170	2.03	0.11	0.949	527.7	537.1	546.6	556.0	565.4	574.8	584.2
	28.4	26.0	0.161	1.92	0.22	0.898	527.7	537.1	546.6	556.0	565.4	574.8	584.2
	28.1	24.5	0.152	1.82	0.32	0.851	527.8	537.2	546.7	556.1	565.5	574.9	584.3
	27.8	23.0	0.144	1.73	0.41	0.809	527.8	537.2	546.7	556.1	565.5	574.9	584.3
	27.5	21.5	0.137	1.64	0.50	0.766	527.9	537.3	546.7	556.2	565.6	575.0	584.5
	27.2	20.0	0.129	1.55	0.59	0.725	528.0	537.4	546.8	556.2	565.7	575.1	584.6
	26.9	18.5	0.122	1.47	0.67	0.687	528.0	537.4	546.8	556.3	565.7	575.2	584.6
	26.6	17.0	0.116	1.38	0.76	0.645	528.1	537.5	546.9	556.4	565.8	575.3	584.7
	26.3	15.5	0.110	1.30	0.84	0.617	528.1	537.5	546.9	556.4	565.8	575.3	584.7
30	30.0	30.0	0.186	2.21	0.00	1.000	526.5	535.9	545.3	554.7	564.1	573.5	582.9
	29.7	28.6	0.177	2.10	0.11	0.951	526.5	535.9	545.3	554.7	564.1	573.5	582.9
	29.4	27.2	0.168	2.00	0.21	0.905	526.6	536.0	545.4	554.8	564.2	573.6	583.0
	29.1	25.9	0.160	1.91	0.30	0.864	526.7	536.1	545.5	554.9	564.3	573.7	583.1
	28.8	24.5	0.152	1.82	0.39	0.824	526.7	536.1	545.5	554.9	564.3	573.7	583.1
	28.5	23.1	0.145	1.73	0.48	0.783	526.8	536.2	545.6	555.0	564.4	573.8	583.2
	28.2	21.7	0.138	1.64	0.57	0.742	526.8	536.2	545.6	555.0	564.4	573.8	583.2
	27.9	20.3	0.131	1.56	0.65	0.706	526.9	536.3	545.7	555.1	564.5	573.9	583.3
	27.6	19.0	0.125	1.49	0.72	0.674	526.9	536.3	545.7	555.1	564.5	573.9	583.3
	27.3	17.6	0.118	1.42	0.79	0.643	527.0	536.4	545.8	555.2	564.6	574.0	583.4
31	31.0	31.0	0.192	2.29	0.00	1.000	525.4	534.7	544.1	553.5	562.9	572.3	581.7
	30.7	29.9	0.185	2.20	0.09	0.961	525.4	534.7	544.1	553.5	562.9	572.3	581.7
	30.4	28.8	0.178	2.12	0.17	0.926	525.5	534.8	544.2	553.6	563.0	572.4	581.8
	30.1	27.7	0.171	2.04	0.25	0.891	525.5	534.8	544.2	553.6	563.0	572.4	581.8
	29.8	26.6	0.164	1.95	0.34	0.852	525.6	534.9	544.3	553.7	563.1	572.5	581.9
	29.5	25.5	0.158	1.87	0.42	0.817	525.6	534.9	544.3	553.7	563.1	572.5	581.9
	29.2	24.4	0.152	1.80	0.49	0.786	525.6	534.9	544.3	553.7	563.1	572.5	581.9
	28.9	23.4	0.146	1.73	0.56	0.756	525.7	535.0	544.4	553.8	563.2	572.6	582.0
	28.6	22.3	0.141	1.67	0.62	0.729	525.7	535.0	544.4	553.8	563.2	572.6	582.0
	28.3	21.2	0.135	1.60	0.69	0.699	525.7	535.0	544.4	553.8	563.2	572.6	582.0
32	32.0	32.0	0.199	2.37	0.00	1.000	524.2	533.5	542.9	552.3	561.6	570.9	580.3
	31.6	30.8	0.191	2.27	0.10	0.958	524.3	533.6	543.0	552.4	561.7	571.0	580.4
	31.2	29.5	0.182	2.17	0.20	0.916	524.4	533.7	543.1	552.5	561.8	571.1	580.5
	30.8	28.3	0.175	2.07	0.30	0.874	524.4	533.7	543.1	552.5	561.8	571.1	580.6
	30.4	27.0	0.167	1.98	0.39	0.836	524.5	533.8	543.2	552.6	561.9	571.2	580.6
	30.0	25.8	0.160	1.90	0.47	0.802	524.5	533.8	543.2	552.6	561.9	571.2	580.6
	29.6	24.6	0.153	1.82	0.55	0.768	524.6	533.9	543.3	552.7	562.0	571.3	580.7
	29.2	23.3	0.146	1.74	0.63	0.735	524.6	533.9	543.3	552.7	562.0	571.3	580.7
	28.8	22.1	0.140	1.67	0.70	0.705	524.6	533.9	543.3	552.7	562.0	571.3	580.7
	28.4	20.8	0.133	1.60	0.77	0.675	524.7	534.0	543.4	552.8	562.1	571.4	580.8

Reading of Thermometer, Fahr.		Temp. of Dew-Point, Fahr.	Force of Vapor in English Inches.	Weight of Vapor		Humidity, Saturation = 1.000.	Weight in Grains of a Cubic Foot of Air.						
							Height of the Barometer in English Inches.						
Dry.	Wet.			In a Cubic Foot of Air.	Reqd. for Sat'n. of a Cubic Ft. of Air.		in. **28.0**	in. **28.5**	in. **29.0**	in. **29.5**	in. **30.0**	in. **30.5**	in. **31.0**
°	°	°	in.	gr.	gr.		gr.	gr.	gr.	gr.	gr.	gr.	gr.
33	33.0	33.0	0.207	2.45	0.00	1.000	523.0	532.3	541.7	551.1	560.4	569.7	579.1
	32.5	31.6	0.197	2.33	0.12	0.951	523.1	532.5	541.8	551.2	560.5	569.8	579.2
	32.0	30.2	0.187	2.22	0.23	0.906	523.2	532.6	541.9	551.3	560.6	569.9	579.3
	31.5	28.8	0.178	2.11	0.34	0.862	523.3	532.7	542.0	551.4	560.7	570.0	579.4
	31.0	27.4	0.169	2.01	0.44	0.821	523.3	532.7	542.0	551.4	560.7	570.0	579.4
	30.5	26.0	0.161	1.91	0.54	0.780	523.4	532.8	542.1	551.5	560.8	570.1	579.5
	30.0	24.6	0.153	1.82	0.63	0.743	523.4	532.8	542.1	551.5	560.8	570.1	579.5
	29.5	23.2	0.145	1.74	0.71	0.711	523.5	532.9	542.2	551.6	560.9	570.2	579.6
	29.0	21.8	0.138	1.65	0.80	0.674	523.5	532.9	542.2	551.6	560.9	570.2	579.6
	28.5	20.4	0.131	1.57	0.88	0.641	523.6	533.0	542.3	551.7	561.0	570.3	579.7
34	34.0	34.0	0.214	2.53	0.00	1.000	521.9	531.2	540.6	549.9	559.2	568.5	577.8
	33.5	32.7	0.204	2.42	0.11	0.957	522.0	531.4	540.7	550.0	559.3	568.6	577.9
	33.0	31.4	0.195	2.31	0.22	0.913	522.0	531.4	540.7	550.0	559.3	568.6	577.9
	32.5	30.1	0.186	2.21	0.32	0.874	522.1	531.5	540.8	550.1	559.4	568.7	578.0
	32.0	28.8	0.178	2.11	0.42	0.834	522.1	531.5	540.8	550.1	559.4	568.7	578.0
	31.5	27.5	0.170	2.01	0.52	0.795	522.2	531.6	540.9	550.2	559.5	568.8	578.1
	31.0	26.2	0.162	1.91	0.62	0.755	522.3	531.7	541.0	550.3	559.6	568.9	578.2
	30.5	24.9	0.155	1.83	0.70	0.724	522.3	531.7	541.0	550.3	559.6	568.9	578.2
	30.0	23.6	0.147	1.75	0.78	0.692	522.4	531.8	541.1	550.4	559.7	569.0	578.3
	29.5	22.3	0.141	1.67	0.86	0.660	522.4	531.8	541.1	550.4	559.7	569.0	578.3
	29.0	21.0	0.134	1.59	0.94	0.629	522.5	531.9	541.2	550.5	559.8	569.1	578.4
35	35	35.0	0.222	2.62	0.00	1.000	520.8	530.1	539.4	548.7	558.0	567.3	576.6
	34	32.5	0.203	2.40	0.22	0.916	520.9	530.2	539.5	548.8	558.1	567.4	576.7
	33	30.0	0.186	2.19	0.43	0.836	521.0	530.3	539.6	548.9	558.3	567.5	576.8
	32	27.5	0.170	2.00	0.62	0.764	521.1	530.4	539.7	549.0	558.4	567.6	576.9
	31	25.0	0.155	1.83	0.79	0.698	521.2	530.5	539.8	549.1	558.5	567.7	577.0
	30	22.5	0.142	1.68	0.94	0.641	521.3	530.6	539.9	549.2	558.6	567.8	577.1
	29	20.0	0.129	1.53	1.09	0.584	521.3	530.7	540.0	549.3	558.6	567.9	577.2
	28	17.5	0.117	1.39	1.23	0.531	521.4	530.8	540.1	549.4	558.7	568.0	577.3
	27	15.0	0.108	1.27	1.35	0.485	521.5	530.9	540.2	549.5	558.7	568.1	577.4
36	36	36.0	0.230	2.71	0.00	1.000	519.7	529.0	538.3	547.5	556.8	566.1	575.4
	35	33.5	0.210	2.48	0.23	0.915	519.8	529.1	538.4	547.6	556.9	566.2	575.5
	34	31.0	0.192	2.27	0.44	0.838	519.9	529.2	538.5	547.7	557.0	566.3	575.6
	33	28.5	0.176	2.07	0.64	0.764	520.0	529.3	538.6	547.8	557.1	566.4	575.7
	32	26.0	0.161	1.89	0.82	0.698	520.1	529.4	538.7	547.9	557.2	566.5	575.8
	31	23.5	0.147	1.74	0.97	0.642	520.2	529.5	538.8	548.0	557.3	566.6	575.9
	30	21.0	0.134	1.58	1.13	0.583	520.3	529.6	538.9	548.1	557.4	566.7	576.0
	29	18.5	0.122	1.45	1.26	0.535	520.4	529.7	539.0	548.2	557.5	566.8	576.1
	28	16.0	0.112	1.32	1.39	0.487	520.5	529.8	539.1	548.3	557.6	566.9	576.2

Reading of Thermometer, Fahr.		Temp. of Dew-Point, Fahr.	Force of Vapor in English Inches.	Weight of Vapor		Humidity, Saturation = 1.000.	Weight in Grains of a Cubic Foot of Air. Height of the Barometer in English Inches.						
Dry.	Wet.			In a Cubic Foot of Air.	Reqd. for Sat'n. of a Cubic Ft. of Air.		in. **28.0**	in. **28.5**	in. **29.0**	in. **29.5**	in. **30.0**	in. **30.5**	in. **31.0**
°	°	°	in.	gr.	gr.		gr.	gr.	gr.	gr.	gr.	gr.	gr.
37	37	37.0	0.238	2.80	0.00	1.000	518.6	527.8	537.1	546.3	555.6	564.8	574.1
	36	34.5	0.218	2.56	0.24	0.914	518.7	527.9	537.2	546.4	555.7	564.9	574.2
	35	32.0	0.199	2.35	0.45	0.839	518.8	528.0	537.3	546.5	555.8	565.0	574.3
	34	29.5	0.182	2.14	0.66	0.764	518.9	528.1	537.4	546.6	555.9	565.1	574.4
	33	27.0	0.167	1.96	0.84	0.700	519.0	528.2	537.5	546.7	556.0	565.2	574.5
	32	24.5	0.152	1.79	1.01	0.640	519.1	528.3	537.6	546.8	556.1	565.3	574.6
	31	22.0	0.139	1.64	1.16	0.586	519.2	528.4	537.7	546.9	556.2	565.4	574.7
	30	19.5	0.127	1.50	1.30	0.536	519.3	528.5	537.8	547.1	556.3	565.5	574.8
	29	17.0	0.116	1.37	1.43	0.489	519.4	528.6	537.9	547.2	556.4	565.6	574.9
38	38	38.0	0.246	2.89	0.00	1.000	517.4	526.6	535.9	545.1	554.4	563.6	572.9
	37	35.5	0.226	2.65	0.24	0.917	517.5	526.7	536.0	545.2	554.5	563.7	573.0
	36	33.0	0.207	2.43	0.46	0.841	517.6	526.8	536.1	545.3	554.6	563.8	573.1
	35	30.5	0.189	2.22	0.67	0.768	517.7	526.9	536.2	545.4	554.7	563.9	573.2
	34	28.0	0.173	2.03	0.86	0.703	517.8	527.0	536.3	545.5	554.8	564.0	573.3
	33	25.5	0.158	1.85	1.04	0.640	517.9	527.1	536.4	545.6	554.9	564.1	573.4
	32	23.0	0.144	1.70	1.19	0.588	518.0	527.2	536.5	545.7	555.0	564.2	573.5
	31	20.5	0.132	1.54	1.35	0.533	518.1	527.3	536.6	545.8	555.1	564.3	573.6
	30	18.0	0.120	1.39	1.50	0.481	518.2	527.4	536.7	545.9	555.2	564.4	573.7
39	39	39.0	0.255	2.99	0.00	1.000	516.3	525.5	534.7	543.9	553.2	562.4	571.6
	38	36.5	0.234	2.74	0.25	0.917	516.4	525.6	534.8	544.0	553.3	562.5	571.7
	37	34.0	0.214	2.51	0.48	0.840	516.5	525.7	534.9	544.1	553.4	562.6	571.8
	36	31.5	0.196	2.30	0.69	0.769	516.6	525.8	535.0	544.2	553.5	562.7	571.9
	35	29.0	0.179	2.10	0.89	0.703	516.7	525.9	535.1	544.3	553.6	562.8	572.1
	34	26.5	0.164	1.91	1.08	0.639	516.8	526.0	535.2	544.4	553.7	562.9	572.2
	33	24.0	0.150	1.76	1.23	0.589	516.9	526.1	535.3	544.5	553.8	563.0	572.3
	32	21.5	0.137	1.60	1.39	0.535	517.0	526.2	535.4	544.6	553.9	563.1	572.4
	31	19.0	0.125	1.46	1.53	0.488	517.1	526.3	535.6	544.8	554.1	563.3	572.6
	30	16.5	0.114	1.32	1.67	0.442	517.2	526.4	535.7	544.9	554.2	563.4	572.7
40	40	40.0	0.264	3.09	0.00	1.000	515.2	524.4	533.6	542.8	552.0	561.2	570.4
	39	37.8	0.245	2.86	0.23	0.926	515.3	524.5	533.7	542.9	552.1	561.3	570.5
	38	35.6	0.227	2.65	0.44	0.858	515.4	524.6	533.8	543.0	552.2	561.4	570.6
	37	33.4	0.210	2.45	0.64	0.793	515.5	524.7	533.9	543.1	552.3	561.5	570.7
	36	31.2	0.194	2.27	0.82	0.734	515.6	524.8	534.0	543.2	552.4	561.6	570.8
	35	29.0	0.179	2.09	1.00	0.676	515.7	524.9	534.1	543.3	552.5	561.7	570.9
	34	26.8	0.165	1.94	1.15	0.628	515.8	525.0	534.2	543.4	552.6	561.8	571.0
	33	24.6	0.153	1.79	1.30	0.579	515.9	525.1	534.3	543.5	552.7	561.9	571.1
	32	22.4	0.141	1.65	1.44	0.534	516.0	525.2	534.4	543.6	552.8	562.0	571.2
	31	20.2	0.130	1.53	1.56	0.495	516.1	525.3	534.5	543.7	552.9	562.1	571.3
	30	18.0	0.120	1.42	1.67	0.459	516.1	525.3	534.5	543.8	553.0	562.2	571.4

Reading of Thermometer, Fahr. Dry.	Wet.	Temp. of Dew-Point, Fahr.	Force of Vapor in English Inches.	Weight of Vapor: In a Cubic Foot of Air.	Weight of Vapor: Reqd. for Sat'n. of a Cubic Ft. of Air.	Humidity, Saturation = 1.000.	Weight in Grains of a Cubic Foot of Air. Height of the Barometer in English Inches. in. **28.0**	in. **28.5**	in. **29.0**	in. **29.5**	in. **30.0**	in. **30.5**	in. **31.0**
°	°	°	in.	gr.	gr.		gr.	gr.	gr.	gr.	gr.	gr.	gr.
41	41	41.0	0.274	3.19	0.00	1.000	514.1	523.3	532.5	541.6	550.8	560.0	569.2
	40	38.8	0.253	2.96	0.23	0.928	514.2	523.4	532.6	541.7	550.9	560.1	569.3
	39	36.6	0.235	2.74	0.45	0.859	514.3	523.5	532.7	541.8	551.0	560.2	569.4
	38	34.4	0.217	2.54	0.65	0.796	514.4	523.6	532.8	541.9	551.1	560.3	569.5
	37	32.2	0.201	2.35	0.84	0.737	514.5	523.7	532.9	542.0	551.2	560.4	569.6
	36	30.0	0.186	2.16	1.03	0.677	514.6	523.8	533.0	542.1	551.3	560.5	569.7
	35	27.8	0.172	2.01	1.18	0.630	514.7	523.9	533.1	542.2	551.4	560.6	569.8
	34	25.6	0.158	1.85	1.34	0.580	514.8	524.0	533.2	542.3	551.5	560.7	569.9
	33	23.4	0.146	1.71	1.48	0.536	514.9	524.1	533.3	542.4	551.6	560.8	570.0
	32	21.2	0.135	1.58	1.61	0.495	514.9	524.1	533.3	542.5	551.7	560.9	570.1
	31	19.0	0.125	1.46	1.73	0.458	515.0	524.2	533.4	542.6	551.8	561.0	570.2
42	42	42.0	0.283	3.30	0.00	1.000	513.0	522.2	531.3	540.5	549.6	558.8	567 9
	41	39.8	0.263	3.06	0.24	0.927	513.1	522.3	531.4	540.6	549.7	558.9	568.0
	40	37.6	0.243	2.83	0.47	0.858	513.2	522.4	531.5	540.7	549.9	559.0	568.1
	39	35.4	0.225	2.63	0.67	0.797	513.3	522.5	531.6	540.8	550.0	559.1	568.2
	38	33.2	0.208	2.43	0.87	0.736	513.4	522.6	531.7	540.9	550.1	559.2	568.3
	37	31.0	0.192	2.24	1.06	0.679	513.5	522.7	531.8	541.0	550.2	559.3	568.4
	36	28.8	0.178	2.08	1.22	0.631	513.6	522.8	531.9	541.1	550.3	559.4	568.5
	35	26.6	0.164	1.91	1.39	0.579	513.7	522.9	532.0	541.2	550.4	559.5	568.6
	34	24.4	0.152	1.77	1.53	0.536	513.8	523.0	532.1	541.3	550.5	559.6	568.7
	33	22.2	0.140	1.63	1.67	0.494	513.9	523.1	532.2	541.4	550.6	559.7	568.8
	32	20.0	0.129	1.51	1.79	0.458	513.9	523.1	532.3	541.5	550.6	559.8	569.0
43	43	43.0	0.293	3.41	0.00	1.000	511.8	520.9	530.1	539.3	548.4	557.5	566.7
	42	40.8	0.272	3.16	0.25	0.927	511.9	521.0	530.2	539.4	548.6	557.7	566.9
	41	38.6	0.252	2.93	0.48	0.859	512.0	521.1	530.3	539.5	548.7	557.8	567.0
	40	36.4	0.233	2.71	0.70	0.795	512.1	521.2	530.4	539.6	548.8	557.9	567.1
	39	34.2	0.216	2.51	0.90	0.736	512.2	521.3	530.5	539.7	548.9	558.0	567.2
	38	32.0	0.199	2.32	1.09	0.680	512.3	521.4	530.7	539.8	549.0	558.1	567.3
	37	29.8	0.184	2.15	1.26	0.630	512.4	521.5	530.8	539.9	549.1	558.2	567.4
	36	27.6	0.170	1.98	1.43	0.581	512.5	521.6	530.9	540.0	549.2	558.3	567.5
	35	25.4	0.157	1.82	1.59	0.534	512.6	521.7	531.0	540.1	549.3	558.4	567.6
	34	23.2	0.145	1.69	1.72	0.495	512.7	521.8	531.1	540.2	549.4	558.5	567.7
	33	21.0	0.134	1.56	1.85	0.458	512.9	522.0	531.2	540.3	549.5	558.6	567.8
44	44	44.0	0.304	3.52	0.00	1.000	510.8	519.9	529.0	538.1	547.3	556.4	565.5
	43	41.8	0.282	3.27	0.25	0.929	510.9	520.0	529.1	538.2	547.5	556.5	565.7
	42	39.6	0.261	3.02	0.50	0.858	511.0	520.1	529.2	538.3	547.6	556.6	565.8
	41	37.4	0.241	2.80	0.72	0.796	511.1	520.2	529.3	538.4	547.7	556.7	565.9
	40	35.2	0.223	2.60	0.92	0.739	511.2	520.3	529.4	538.5	547.8	556.8	566.0
	39	33.0	0.207	2.40	1.12	0.682	511.3	520.4	529.5	538.6	547.9	556.9	566.1
	38	30.8	0.191	2.22	1.30	0.631	511.4	520.5	529.6	538.7	548.0	557.0	566.2
	37	28.6	0.177	2.05	1.47	0.582	511.5	520.6	529.7	538.8	548.1	557.1	566.3
	36	26.4	0.163	1.89	1.63	0.537	511.6	520.7	529.8	538.9	548.2	557.2	566.4
	35	24.2	0.151	1.75	1.77	0.497	511.7	520.8	529.9	539.0	548.3	557.3	566.5
	34	22.0	0.139	1.62	1.90	0.460	511.7	520.8	530.0	539.1	548.3	557.4	566.6

Reading of Thermometer, Fahr.		Temp. of Dew-Point, Fahr.	Force of Vapor in English Inches.	Weight of Vapor		Humidity, Saturation = 1.000.	Weight in Grains of a Cubic Foot of Air.						
							Height of the Barometer in English Inches.						
Dry.	Wet.			In a Cubic Foot of Air.	Reqd. for Sat'n. of a Cubic Ft. of Air.		in. **28.0**	in. **28.5**	in. **29.0**	in. **29.5**	in. **30.0**	in. **30.5**	in. **31.0**
°	°	°	in.	gr.	gr.		gr.	gr.	gr.	gr.	gr.	gr.	gr.
45	45	45.0	0.315	3.64	0.00	1.000	509.7	518.8	527.9	537.0	546.1	555.2	564.3
	44	42.9	0.292	3.39	0.25	0.931	509.8	518.9	528.0	537.1	546.3	555.3	564.5
	43	40.8	0.272	3.14	0.50	0.863	509.9	519.0	528.1	537.2	546.4	555.4	564.6
	42	38.7	0.253	2.92	0.72	0.802	510.0	519.1	528.2	537.3	546.5	555.5	564.7
	41	36.6	0.235	2.70	0.94	0.742	510.1	519.2	528.3	537.4	546.6	555.6	564.8
	40	34.5	0.218	2.52	1.12	0.692	510.2	519.3	528.4	537.5	546.7	555.7	564.9
	39	32.4	0.202	2.34	1.30	0.643	510.3	519.4	528.5	537.6	546.8	555.8	565.0
	38	30.3	0.188	2.16	1.48	0.593	510.4	519.5	528.6	537.7	546.9	555.9	565.1
	37	28.2	0.174	2.01	1.63	0.552	510.5	519.6	528.7	537.8	547.0	556.0	565.2
	36	26.1	0.161	1.87	1.77	0.514	510.6	519.7	528.8	537.9	547.1	556.1	565.3
	35	24.0	0.150	1.73	1.91	0.475	510.7	519.8	528.9	538.0	547.2	556.3	565.4
46	46	46.0	0.326	3.76	0.00	1.000	508.6	517.7	526.7	535.8	544.9	554.0	563.1
	45	43.9	0.303	3.50	0.26	0.931	508.7	517.8	526.8	535.9	545.0	554.1	563.2
	44	41.8	0.282	3.25	0.51	0.864	508.8	517.9	526.9	536.0	545.1	554.2	563.3
	43	39.7	0.262	3.02	0.74	0.803	508.9	518.0	527.0	536.1	545.2	554.3	563.4
	42	37.6	0.243	2.80	0.96	0.745	509.0	518.1	527.2	536.3	545.4	554.5	563.6
	41	35.5	0.226	2.61	1.15	0.694	509.1	518.2	527.3	536.4	545.5	554.6	563.7
	40	33.4	0.210	2.42	1.34	0.643	509.2	518.3	527.4	536.5	545.6	554.7	563.8
	39	31.3	0.194	2.24	1.52	0.596	509.3	518.4	527.5	536.6	545.7	554.8	563.9
	38	29.2	0.180	2.08	1.68	0.553	509.4	518.5	527.6	536.7	545.8	554.9	564.0
	37	27.1	0.167	1.93	1.83	0.514	509.5	518.6	527.7	536.8	545.9	555.0	564.1
	36	25.0	0.155	1.79	1.97	0.476	509.5	518.6	527.7	536.8	545.9	555.0	564.1
47	47	47.0	0.337	3.88	0.00	1.000	507.5	516.5	525.6	534.7	543.8	552.8	561.9
	46	44.9	0.313	3.62	0.26	0.933	507.6	516.6	525.7	534.8	543.9	552.9	562.0
	45	42.8	0.291	3.36	0.52	0.866	507.8	516.7	525.9	535.0	544.1	553.1	562.2
	44	40.7	0.271	3.12	0.76	0.804	507.9	516.8	526.0	535.1	544.2	553.2	562.3
	43	38.6	0.252	2.90	0.98	0.747	508.0	516.9	526.1	535.2	544.3	553.3	562.4
	42	36.5	0.234	2.70	1.18	0.696	508.1	517.0	526.2	535.3	544.4	553.4	562.5
	41	34.4	0.217	2.51	1.37	0.647	508.2	517.1	526.3	535.4	544.5	553.5	562.6
	40	32.3	0.201	2.32	1.56	0.598	508.3	517.2	526.4	535.5	544.6	553.6	562.7
	39	30.2	0.187	2.16	1.72	0.557	508.4	517.3	526.5	535.6	544.7	553.7	562.8
	38	28.1	0.173	2.00	1.88	0.515	508.5	517.4	526.6	535.7	544.8	553.8	562.9
	37	26.0	0.161	1.85	2.03	0.477	508.5	517.6	526.7	535.8	544.9	554.0	563.1
48	48	48.0	0.349	4.01	0.00	1.000	506.4	515.4	524.5	533.5	542.6	551.6	560.7
	47	45.9	0.324	3.73	0.28	0.930	506.5	515.5	524.6	533.7	542.8	551.8	560.9
	46	43.8	0.302	3.47	0.54	0.865	506.6	515.6	524.7	533.8	542.9	551.9	561.0
	45	41.7	0.281	3.23	0.78	0.805	506.7	515.7	524.8	533.9	543.0	552.0	561.1
	44	39.6	0.261	3.00	1.01	0.748	506.8	515.8	524.9	534.0	543.1	552.1	561.2
	43	37.5	0.242	2.79	1.22	0.696	506.9	515.9	525.0	534.1	543.2	552.2	561.3
	42	35.4	0.225	2.60	1.41	0.648	507.0	516.0	525.1	534.2	543.3	552.3	561.4
	41	33.3	0.209	2.40	1.61	0.598	507.1	516.1	525.2	534.4	543.5	552.5	561.5
	40	31.2	0.194	2.24	1.77	0.558	507.2	516.2	525.3	534.5	543.5	552.5	561.6
	39	29.1	0.180	2.07	1.94	0.516	507.3	516.3	525.4	534.6	543.6	552.6	561.6
	38	27.0	0.167	1.92	2.09	0.479	507.4	516.4	525.5	534.7	543.6	552.7	561.7
	37	24.9	0.155	1.77	2.24	0.441	507.4	516.4	525.6	534.7	543.7	552.8	561.8

Reading of Thermometer, Fahr.		Temp. of Dew-Point, Fahr.	Force of Vapor in English Inches.	Weight of Vapor		Humidity, Saturation = 1.000.	Weight in Grains of a Cubic Foot of Air. Height of the Barometer in English Inches.						
Dry.	Wet.			In a Cubic Foot of Air.	Reqd. for Sat'n. of a Cubic Ft. of Air.		in. **28.0**	in. **28.5**	in. **29.0**	in. **29.5**	in. **30.0**	in. **30.5**	in. **31.0**
°	°	°	in.	gr.	gr.		gr.	gr.	gr.	gr.	gr.	gr.	gr.
49	49	49.0	0.361	4.14	0.00	1.000	505.3	514.3	523.3	532.3	541.4	550.4	559.4
	48	46.9	0.336	3.85	0.29	0.930	505.4	514.4	523.4	532.4	541.5	550.5	559.5
	47	44.8	0.312	3.59	0.55	0.867	505.6	514.6	523.6	532.6	541.7	550.7	559.7
	46	42.7	0.290	3.34	0.80	0.807	505.7	514.7	523.7	532.7	541.8	550.8	559.8
	45	40.6	0.270	3.10	1.04	0.749	505.9	514.9	523.8	532.9	542.0	551.0	560.0
	44	38.5	0.251	2.88	1.26	0.696	506.0	515.0	523.9	533.0	542.1	551.1	560.1
	43	36.4	0.233	2.68	1.46	0.647	506.1	515.1	524.0	533.1	542.2	551.2	560.2
	42	34.3	0.216	2.49	1.65	0.601	506.2	515.2	524.1	533.2	542.3	551.3	560.3
	41	32.2	0.201	2.32	1.82	0.560	506.3	515.3	524.2	533.3	542.4	551.4	560.4
	40	30.1	0.186	2.14	2.00	0.517	506.3	515.3	524.3	533.4	542.5	551.5	560.5
	39	28.0	0.173	1.99	2.15	0.481	506.4	515.4	524.4	533.5	542.6	551.6	560.6
	38	25.9	0.160	1.84	2.30	0.444	506.4	515.4	524.4	533.5	542.6	551.6	560.6
50	50	50.0	0.373	4.28	0.00	1.000	504.1	513.1	522.1	531.1	540.2	549.2	558.2
	49	48.0	0.349	3.99	0.29	0.932	504.2	513.2	522.2	531.2	540.3	549.3	558.3
	48	46.0	0.326	3.73	0.55	0.871	504.4	513.4	522.4	531.4	540.5	549.5	558.5
	47	44.0	0.304	3.48	0.80	0.813	504.5	513.5	522.5	531.5	540.6	549.6	558.6
	46	42.0	0.283	3.25	1.03	0.759	504.6	513.6	522.6	531.6	540.7	549.7	558.7
	45	40.0	0.264	3.03	1.25	0.708	504.8	513.8	522.8	531.8	540.9	549.9	558.9
	44	38.0	0.246	2.82	1.46	0.659	504.9	513.9	522.9	532.0	541.0	550.0	559.0
	43	36.0	0.230	2.63	1.65	0.614	505.1	514.1	523.1	532.1	541.2	550.2	559.2
	42	34.0	0.214	2.45	1.83	0.572	505.2	514.2	523.2	532.2	541.3	550.3	559.3
	41	32.0	0.199	2.28	2.00	0.533	505.3	514.3	523.3	532.3	541.4	550.4	559.4
	40	30.0	0.186	2.12	2.16	0.495	505.4	514.4	523.4	532.4	541.5	550.5	559.5
	39	28.0	0.173	1.97	2.31	0.460	505.5	514.5	523.5	532.5	541.6	550.6	559.6
51	51	51.0	0.386	4.42	0.00	1.000	503.1	512.1	521.1	530.0	539.0	548.0	557.0
	50	49.0	0.361	4.12	0.30	0.932	503.2	512.2	521.2	530.1	539.1	548.1	557.1
	49	47.0	0.337	3.85	0.57	0.871	503.3	512.3	521.3	530.3	539.3	548.3	557.3
	48	45.0	0.315	3.60	0.82	0.814	503.4	512.4	521.4	530.4	539.4	548.4	557.4
	47	43.0	0.293	3.36	1.06	0.760	503.5	512.5	521.5	530.5	539.5	548.5	557.5
	46	41.0	0.274	3.13	1.29	0.708	503.7	512.7	521.7	530.7	539.7	548.7	557.7
	45	39.0	0.255	2.92	1.50	0.661	503.8	512.8	521.8	530.8	539.8	548.8	557.8
	44	37.0	0.238	2.72	1.70	0.615	503.9	512.9	521.9	530.9	539.9	548.9	557.9
	43	35.0	0.222	2.54	1.88	0.575	504.0	513.0	522.0	531.0	540.0	549.0	558.0
	42	33.0	0.207	2.36	2.06	0.534	504.1	513.1	522.1	531.1	540.1	549.1	558.1
	41	31.0	0.192	2.20	2.22	0.498	504.2	513.2	522.2	531.2	540.3	549.3	558.3
	40	29.0	0.179	2.05	2.37	0.464	504.3	513.3	522.3	531.3	540.4	549.4	558.4
52	52	52.0	0.400	4.56	0.00	1.000	502.1	511.0	520.0	528.9	537.9	546.8	555.8
	51	50.0	0.373	4.26	0.30	0.934	502.2	511.1	520.1	529.0	538.0	546.9	555.9
	50	48.0	0.349	3.98	0.58	0.873	502.4	511.3	520.3	529.2	538.2	547.1	556.1
	49	46.0	0.326	3.72	0.84	0.816	502.5	511.4	520.4	529.3	538.3	547.2	556.2
	48	44.0	0.304	3.47	1.09	0.761	502.6	511.5	520.5	529.4	538.4	547.3	556.3
	47	42.0	0.283	3.23	1.33	0.709	502.8	511.7	520.7	529.6	538.6	547.5	556.5
	46	40.0	0.264	3.02	1.54	0.662	502.9	511.8	520.8	529.7	538.7	547.6	556.6
	45	38.0	0.246	2.81	1.75	0.616	502.9	511.9	520.9	529.8	538.8	547.8	556.8
	44	36.0	0.230	2.63	1.93	0.577	503.1	512.0	521.0	529.9	539.0	548.0	557.0
	43	34.0	0.214	2.44	2.12	0.535	503.2	512.1	521.1	530.0	539.1	548.1	557.1
	42	32.0	0.199	2.28	2.28	0.500	503.3	512.3	521.3	530.2	539.2	548.2	557.2
	41	30.0	0.186	2.13	2.43	0.467	503.4	512.4	521.4	530.3	539.3	548.3	557.3

Reading of Thermometer, Fahr.		Temp. of Dew-Point, Fahr.	Force of Vapor in English Inches.	Weight of Vapor		Humidity, Saturation = 1.000.	Weight in Grains of a Cubic Foot of Air.						
				In a Cubic Foot of Air.	Reqd. for Sat'n. of a Cubic Ft. of Air.		Height of the Barometer in English Inches.						
Dry.	Wet.						in. **28.0**	in. **28.5**	in. **29.0**	in. **29.5**	in. **30.0**	in. **30.5**	in. **31.0**
°	°	°	in.	gr.	gr.		gr.	gr.	gr.	gr.	gr.	gr.	gr.
53	53	53.0	0.414	4.71	0.00	1.000	500.9	509.8	518.8	527.7	536.7	545.6	554.6
	52	51.0	0.386	4.40	0.31	0.934	501.1	510.0	519.0	527.9	536.9	545.8	554.8
	51	49.0	0.361	4.11	0.60	0.873	501.2	510.1	519.1	528.0	537.0	545.9	554.9
	50	47.0	0.337	3.84	0.87	0.815	501.4	510.3	519.3	528.2	537.2	546.1	555.1
	49	45.0	0.315	3.58	1.13	0.760	501.5	510.4	519.4	528.3	537.3	546.2	555.2
	48	43.0	0.293	3.34	1.37	0.709	501.6	510.5	519.5	528.4	537.4	546.3	555.3
	47	41.0	0.274	3.12	1.59	0.662	501.7	510.6	519.6	528.5	537.5	546.4	555.4
	46	39.0	0.255	2.91	1.80	0.618	501.8	510.7	519.7	528.6	537.6	546.5	555.5
	45	37.0	0.238	2.71	2.00	0.575	502.0	510.9	519.9	528.8	537.8	546.7	555.7
	44	35.0	0.222	2.53	2.18	0.537	502.1	511.0	520.0	528.9	537.9	546.8	555.8
	43	33.0	0.207	2.35	2.36	0.499	502.1	511.0	520.0	528.9	538.0	546.9	555.9
	42	31.0	0.192	2.18	2.53	0.463	502.2	511.1	520.1	529.0	538.1	547.0	556.0
54	54	54.0	0.428	4.86	0.00	1.000	499.9	508.8	517.8	526.7	535.6	544.5	553.5
	53	52.0	0.400	4.54	0.32	0.934	500.0	508.9	517.9	526.8	535.7	544.6	553.6
	52	50.0	0.373	4.25	0.61	0.875	500.2	509.1	518.1	527.0	535.9	544.8	553.8
	51	48.0	0.349	3.96	0.90	0.815	500.3	509.2	518.2	527.1	536.0	544.9	553.9
	50	46.0	0.326	3.70	1.16	0.761	500.4	509.3	518.3	527.2	536.1	545.0	554.0
	49	44.0	0.304	3.45	1.41	0.709	500.6	509.5	518.5	527.4	536.3	545.2	554.2
	48	42.0	0.283	3.23	1.63	0.665	500.7	509.6	518.6	527.5	536.4	545.3	554.3
	47	40.0	0.264	3.01	1.85	0.619	500.8	509.7	518.7	527.6	536.5	545.4	554.4
	46	38.0	0.246	2.80	2.06	0.576	500.9	509.8	518.8	527.7	536.7	545.6	554.6
	45	36.0	0.230	2.61	2.25	0.537	501.0	509.9	518.9	527.8	536.8	545.7	554.7
	44	34.0	0.214	2.43	2.43	0.500	501.1	510.0	519.0	527.9	536.9	545.8	554.8
	43	32.0	0.199	2.27	2.59	0.467	501.2	510.1	519.1	528.0	537.0	545.9	554.9
	42	30.0	0.186	2.10	2.76	0.432	501.3	510.2	519.2	528.1	537.1	546.0	555.0
	41	28.0	0.173	1.96	2.90	0.403	501.4	510.3	519.3	528.2	537.2	546.1	555.1
	40	26.0	0.161	1.82	3.04	0.375	501.5	510.4	519.4	528.3	537.3	546.2	555.2
55	55	55.0	0.442	5.02	0.00	1.000	498.8	507.7	516.6	525.5	534.4	543.3	552.2
	54	53.3	0.418	4.74	0.28	0.944	499.0	507.9	516.8	525.7	534.6	543.5	552.4
	53	51.6	0.394	4.46	0.56	0.888	499.1	508.0	516.9	525.8	534.7	543.6	552.5
	52	49.9	0.372	4.23	0.79	0.843	499.3	508.2	517.1	526.0	534.9	543.8	552.7
	51	48.2	0.351	3.98	1.04	0.793	499.4	508.3	517.2	526.1	535.0	543.9	552.8
	50	46.5	0.331	3.76	1.26	0.749	499.5	508.4	517.3	526.2	535.1	544.0	552.9
	49	44.8	0.312	3.55	1.47	0.707	499.7	508.6	517.5	526.3	535.3	544.2	553.1
	48	43.1	0.295	3.34	1.68	0.665	499.8	508.7	517.6	526.5	535.4	544.3	553.3
	47	41.4	0.278	3.14	1.88	0.626	499.8	508.7	517.6	526.6	535.5	544.4	553.4
	46	39.7	0.262	2.97	2.05	0.591	499.9	508.8	517.7	526.7	535.6	544.5	553.5
	45	38.0	0.246	2.79	2.23	0.556	500.0	508.9	517.9	526.8	535.7	544.6	553.6
	44	36.3	0.232	2.64	2.38	0.526	500.1	509.0	518.0	526.9	535.8	544.7	553.7
	43	34.6	0.219	2.47	2.55	0.492	500.2	509.1	518.1	527.0	535.9	544.8	553.8
	42	32.9	0.206	2.32	2.70	0.462	500.3	509.2	518.2	527.1	536.0	544.9	553.9
	41	31.2	0.194	2.20	2.82	0.438	500.4	509.3	518.3	527.1	536.0	544.9	554.0
	40	29.5	0.182	2.07	2.95	0.412	500.5	509.3	518.4	527.2	536.1	545.0	554.1
	39	27.8	0.172	1.95	3.07	0.388	500.6	509.4	518.5	527.3	536.2	545.1	554.2
	38	26.1	0.161	1.83	3.19	0.365	500.7	509.5	518.6	527.4	536.2	545.1	554.2

Reading of Thermometer, Fahr.		Temp. of Dew-Point, Fahr.	Force of Vapor in English Inches.	Weight of Vapor		Humidity, Saturation = 1.000.	Weight in Grains of a Cubic Foot of Air. Height of the Barometer in English Inches.						
Dry.	Wet.			In a Cubic Foot of Air.	Reqd. for Sat'n. of a Cubic Ft. of Air.		in. **28.0**	in. **28.5**	in. **29.0**	in. **29.5**	in. **30.0**	in. **30.5**	in. **31.0**
°	°	°	in.	gr.	gr.		gr.	gr.	gr.	gr.	gr.	gr.	gr.
56	56	56.0	0.458	5.18	0.00	1.000	497.7	506.6	515.5	524.4	533.2	542.1	551.0
	55	54.3	0.432	4.89	0.29	0.944	497.9	506.8	515.7	524.6	533.4	542.3	551.2
	54	52.6	0.408	4.61	0.57	0.890	498.0	506.9	515.8	524.7	533.5	542.4	551.3
	53	50.9	0.385	4.37	0.81	0.844	498.2	507.1	516.0	524.9	533.7	542.6	551.5
	52	49.2	0.363	4.11	1.07	0.793	498.3	507.2	516.1	525.0	533.8	542.7	551.6
	51	47.5	0.343	3.87	1.31	0.747	498.4	507.3	516.2	525.1	533.9	542.8	551.7
	50	45.8	0.323	3.66	1.52	0.706	498.6	507.5	516.4	525.3	534.1	543.0	551.9
	49	44.1	0.305	3.45	1.73	0.666	498.6	507.5	516.4	525.3	534.2	543.1	552.0
	48	42.4	0.287	3.25	1.93	0.627	498.7	507.6	516.5	525.4	534.3	543.2	552.1
	47	40.7	0.271	3.07	2.11	0.593	498.8	507.7	516.6	525.5	534.4	543.3	552.2
	46	39.0	0.255	2.89	2.29	0.558	498.9	507.8	516.7	525.6	534.5	543.4	552.3
	45	37.3	0.240	2.73	2.45	0.527	499.0	507.9	516.8	525.7	534.6	543.5	552.4
	44	35.6	0.227	2.56	2.62	0.494	499.1	508.0	516.9	525.8	534.7	543.6	552.5
	43	33.9	0.213	2.41	2.77	0.465	499.2	508.1	517.0	525.9	534.8	543.7	552.6
	42	32.2	0.201	2.27	2.91	0.438	499.3	508.2	517.1	526.0	534.9	543.8	552.7
	41	30.5	0.189	2.14	3.04	0.413	499.4	508.3	517.2	526.1	535.0	543.9	552.8
	40	28.8	0.178	2.01	3.17	0.388	499.5	508.4	517.3	526.2	535.1	544.1	552.9
	39	27.1	0.167	1.89	3.29	0.365	499.5	508.4	517.3	526.2	535.1	544.1	552.9
57	57	57.0	0.473	5.34	0.00	1.000	496.6	505.5	514.4	523.2	532.1	540.9	549.8
	56	55.3	0.447	5.05	0.29	0.946	496.8	505.7	514.6	523.4	532.3	541.1	550.0
	55	53.6	0.422	4.76	0.58	0.891	496.9	505.8	514.7	523.5	532.4	541.2	550.1
	54	51.9	0.398	4.50	0.84	0.843	497.1	506.0	514.9	523.7	532.6	541.4	550.3
	53	50.2	0.376	4.25	1.09	0.796	497.2	506.1	515.0	523.8	532.7	541.5	550.4
	52	48.5	0.355	4.00	1.34	0.749	497.3	506.2	515.1	523.9	532.8	541.6	550.5
	51	46.8	0.335	3.78	1.56	0.709	497.5	506.4	515.3	524.1	533.0	541.8	550.7
	50	45.1	0.316	3.56	1.78	0.667	497.6	506.5	515.4	524.2	533.1	541.9	550.8
	49	43.4	0.298	3.36	1.98	0.629	497.7	506.6	515.5	524.3	533.2	542.0	550.9
	48	41.7	0.281	3.17	2.17	0.594	497.8	506.7	515.6	524.4	533.3	542.1	551.0
	47	40.0	0.264	2.99	2.35	0.560	497.9	506.8	515.7	524.5	533.4	542.2	551.2
	46	38.3	0.249	2.81	2.53	0.526	498.0	506.9	515.8	524.6	533.5	542.3	551.3
	45	36.6	0.235	2.65	2.69	0.496	498.1	507.0	515.9	524.7	533.6	542.4	551.4
	44	34.9	0.221	2.50	2.84	0.468	498.2	507.1	516.0	524.8	533.7	542.5	551.5
	43	33.2	0.208	2.35	2.99	0.440	498.3	507.2	516.1	524.9	533.8	542.6	551.6
	42	31.5	0.196	2.21	3.13	0.414	498.3	507.2	516.1	524.9	533.8	542.6	551.6
	41	29.8	0.184	2.08	3.26	0.390	498.4	507.3	516.2	525.1	533.9	542.7	551.7
	40	28.1	0.173	1.96	3.38	0.367	498.5	507.4	516.3	525.2	534.0	542.8	551.8

Reading of Thermometer, Fahr.		Temp. of Dew-Point, Fahr.	Force of Vapor in English Inches.	Weight of Vapor		Humidity, Saturation = 1.000.	Weight in Grains of a Cubic Foot of Air. Height of the Barometer in English Inches.						
Dry.	Wet.			In a Cubic Foot of Air.	Reqd. for Sat'n. of a Cubic Ft. of Air.		in. **28.0**	in. **28.5**	in. **29.0**	in. **29.5**	in. **30.0**	in. **30.5**	in. **31.0**
°	°	°	in.	gr.	gr.		gr.	gr.	gr.	gr.	gr.	gr.	gr.
58	58	58.0	0.489	5.51	0.00	1.000	495.5	504.3	513.2	522.0	530.9	539.7	548.6
	57	56.3	0.462	5.21	0.30	0.946	495.7	504.5	513.4	522.2	531.1	539.9	548.8
	56	54.6	0.437	4.92	0.59	0.893	495.8	504.6	513.5	522.3	531.2	540.0	548.9
	55	52.9	0.412	4.64	0.87	0.842	496.0	504.8	513.7	522.5	531.4	540.2	549.1
	54	51.2	0.389	4.39	1.12	0.797	496.1	504.9	513.8	522.7	531.6	540.4	549.3
	53	49.5	0.367	4.14	1.37	0.751	496.2	505.0	513.9	522.8	531.7	540.5	549.4
	52	47.8	0.346	3.90	1.61	0.708	496.4	505.2	514.1	523.0	531.9	540.7	549.6
	51	46.1	0.327	3.68	1.83	0.668	496.5	505.3	514.2	523.1	532.0	540.8	549.7
	50	44.4	0.308	3.48	2.03	0.632	496.6	505.4	514.3	523.2	532.1	540.9	549.8
	49	42.7	0.290	3.28	2.23	0.595	496.7	505.5	514.4	523.3	532.2	541.0	549.9
	48	41.0	0.274	3.08	2.43	0.559	496.8	505.6	514.5	523.4	532.3	541.1	550.0
	47	39.3	0.258	2.91	2.60	0.528	496.9	505.7	514.6	523.5	532.4	541.2	550.1
	46	37.6	0.243	2.74	2.77	0.497	497.0	505.8	514.7	523.6	532.5	541.3	550.2
	45	35.9	0.229	2.58	2.93	0.469	497.1	505.9	514.8	523.7	532.6	541.4	550.3
	44	34.2	0.216	2.43	3.08	0.441	497.2	506.0	514.9	523.8	532.7	541.5	550.4
	43	32.5	0.203	2.29	3.22	0.416	497.3	506.1	515.1	523.9	532.8	541.6	550.5
	42	30.8	0.191	2.15	3.36	0.390	497.4	506.2	515.2	524.1	532.9	541.7	550.6
	41	29.1	0.180	2.03	3.48	0.368	497.5	506.3	515.3	524.2	533.0	541.8	550.7
	40	27.4	0.169	1.91	3.60	0.347	497.5	506.3	515.3	524.2	533.0	541.8	550.7
59	59	59.0	0.506	5.69	0.00	1.000	494.5	503.3	512.2	521.0	529.8	538.6	547.5
	58	57.3	0.478	5.37	0.32	0.944	494.6	503.4	512.3	521.1	529.9	538.7	547.6
	57	55.6	0.452	5.08	0.61	0.893	494.7	503.5	512.4	521.2	530.0	538.8	547.7
	56	53.9	0.426	4.79	0.90	0.842	494.8	503.6	512.5	521.3	530.1	538.9	547.8
	55	52.2	0.402	4.53	1.16	0.796	494.9	503.7	512.6	521.4	530.3	539.1	548.0
	54	50.5	0.380	4.28	1.41	0.752	495.1	503.9	512.8	521.6	530.5	539.3	548.2
	53	48.8	0.358	4.03	1.66	0.708	495.3	504.1	513.0	521.8	530.7	539.5	548.4
	52	47.1	0.338	3.80	1.89	0.668	495.4	504.2	513.1	521.9	530.8	539.6	548.5
	51	45.4	0.319	3.60	2.09	0.633	495.5	504.3	513.2	522.0	530.9	539.7	548.6
	50	43.7	0.301	3.39	2.30	0.596	495.7	504.5	513.4	522.2	531.1	539.9	548.8
	49	42.0	0.283	3.19	2.50	0.561	495.8	504.6	513.4	522.3	531.2	540.0	548.9
	48	40.3	0.267	3.01	2.68	0.529	495.9	504.7	513.5	522.4	531.3	540.1	549.0
	47	38.6	0.252	2.84	2.85	0.499	496.0	504.8	513.6	522.5	531.4	540.2	549.1
	46	36.9	0.237	2.67	3.02	0.469	496.1	504.9	513.7	522.6	531.5	540.3	549.2
	45	35.2	0.223	2.51	3.18	0.441	496.2	505.0	513.8	522.7	531.6	540.4	549.3
	44	33.5	0.210	2.37	3.32	0.417	496.3	505.1	513.9	522.8	531.7	540.5	549.4
	43	31.8	0.198	2.23	3.46	0.392	496.4	505.2	514.1	522.9	531.8	540.6	549.5
	42	30.1	0.186	2.09	3.60	0.367	496.5	505.3	514.2	523.0	531.9	540.7	549.6
	41	28.4	0.175	1.97	3.72	0.346	496.6	505.4	514.3	523.1	532.0	540.8	549.7
	40	26.7	0.165	1.85	3.84	0.325	496.6	505.4	514.3	523.1	532.0	540.8	549.7

Reading of Thermometer, Fahr.		Temp. of Dew-Point, Fahr.	Force of Vapor in English Inches.	Weight of Vapor		Humidity, Saturation = 1.000.	Weight in Grains of a Cubic Foot of Air. Height of the Barometer in English Inches.						
Dry.	Wet.			In a Cubic Foot of Air.	Reqd. for Sat'n. of a Cubic Ft. of Air.		in. **28.0**	in. **28.5**	in. **29.0**	in. **29.5**	in. **30.0**	in. **30.5**	in. **31.0**
°	°	°	in.	gr.	gr.		gr.	gr.	gr.	gr.	gr.	gr.	gr.
60	60	60.0	0.523	5.87	0.00	1.000	493.4	502.2	511.0	519.8	528.6	537.4	546.2
	59	58.3	0.494	5.54	0.33	0.944	493.6	502.4	511.2	520.0	528.8	537.6	546.4
	58	56.6	0.467	5.24	0.63	0.893	493.7	502.5	511.3	520.1	528.9	537.7	546.5
	57	54.9	0.441	4.95	0.92	0.843	493.8	502.6	511.4	520.2	529.0	537.8	546.6
	56	53.2	0.416	4.68	1.19	0.797	494.0	502.8	511.6	520.4	529.2	538.0	546.8
	55	51.5	0.393	4.41	1.46	0.751	494.2	503.0	511.8	520.6	529.4	538.2	547.0
	54	49.8	0.371	4.17	1.70	0.710	494.4	503.2	512.0	520.8	529.6	538.4	547.2
	53	48.1	0.350	3.92	1.95	0.668	494.5	503.3	512.1	520.9	529.7	538.5	547.4
	52	46.4	0.330	3.70	2.17	0.630	494.7	503.4	512.3	521.1	529.9	538.7	547.6
	51	44.7	0.311	3.49	2.38	0.595	494.8	503.5	512.4	521.2	530.0	538.8	547.7
	50	43.0	0.293	3.29	2.58	0.561	494.8	503.6	512.5	521.3	530.1	538.9	547.8
	49	41.3	0.277	3.10	2.77	0.528	494.9	503.7	512.6	521.4	530.2	539.0	547.9
	48	39.6	0.261	2.93	2.94	0.499	495.0	503.8	512.7	521.5	530.3	539.1	548.0
	47	37.9	0.246	2.75	3.12	0.468	495.1	503.9	512.8	521.6	530.4	539.2	548.1
	46	36.2	0.231	2.60	3.27	0.443	495.2	504.0	512.9	521.7	530.5	539.3	548.2
	45	34.5	0.218	2.45	3.42	0.417	495.3	504.1	513.0	521.8	530.6	539.4	548.3
	44	32.8	0.205	2.31	3.56	0.394	495.4	504.2	513.1	521.9	530.7	539.5	548.4
	43	31.1	0.193	2.17	3.70	0.370	495.5	504.3	513.2	522.0	530.8	539.6	548.5
	42	29.4	0.182	2.04	3.83	0.348	495.6	504.4	513.3	522.1	530.9	539.7	548.6
	41	27.7	0.171	1.92	3.95	0.327	495.6	504.4	513.3	522.1	530.9	539.7	548.7
61	61	61.0	0.541	6.06	0.00	1.000	492.3	501.1	509.9	518.7	527.5	536.3	545.1
	60	59.3	0.511	5.72	0.34	0.944	492.5	501.3	510.1	518.9	527.7	536.5	545.3
	59	57.6	0.483	5.40	0.66	0.891	492.6	501.4	510.2	519.0	527.8	536.6	545.4
	58	55.9	0.456	5.11	0.95	0.843	492.8	501.6	510.4	519.2	528.0	536.8	545.6
	57	54.2	0.431	4.83	1.23	0.797	493.0	501.8	510.6	519.4	528.2	537.0	545.8
	56	52.5	0.407	4.55	1.51	0.751	493.1	501.9	510.7	519.5	528.3	537.1	545.9
	55	50.8	0.383	4.30	1.76	0.710	493.3	502.1	510.9	519.7	528.5	537.3	546.1
	54	49.1	0.362	4.05	2.01	0.668	493.4	502.2	511.0	519.8	528.6	537.4	546.2
	53	47.4	0.342	3.83	2.23	0.632	493.5	502.3	511.1	519.9	528.7	537.5	546.3
	52	45.7	0.322	3.61	2.45	0.596	493.6	502.4	511.2	520.0	528.8	537.6	546.4
	51	44.0	0.304	3.40	2.66	0.561	493.8	502.6	511.4	520.2	529.0	537.8	546.6
	50	42.3	0.286	3.21	2.85	0.530	493.9	502.7	511.5	520.3	529.1	537.9	546.7
	49	40.6	0.270	3.02	3.04	0.498	494.0	502.8	511.6	520.4	529.2	538.0	546.8
	48	38.9	0.254	2.85	3.21	0.470	494.1	502.9	511.7	520.5	529.3	538.1	546.9
	47	37.2	0.240	2.69	3.37	0.444	494.2	503.0	511.8	520.6	529.4	538.2	547.0
	46	35.5	0.226	2.53	3.53	0.417	494.3	503.1	511.9	520.7	529.5	538.3	547.1
	45	33.8	0.213	2.38	3.68	0.393	494.4	503.2	512.0	520.8	529.6	538.4	547.2
	44	32.1	0.200	2.24	3.82	0.370	494.5	503.3	512.1	520.9	529.7	538.5	547.3
	43	30.4	0.188	2.11	3.95	0.348	494.6	503.4	512.2	521.0	529.8	538.6	547.4
	42	28.7	0.177	1.99	4.07	0.328	494.7	503.5	512.3	521.1	529.9	538.7	547.5
	41	27.0	0.167	1.87	4.19	0.309	494.7	503.5	512.3	521.1	529.9	538.7	547.5

Reading of Thermometer, Fahr.		Temp. of Dew-Point, Fahr.	Force of Vapor in English Inches.	Weight of Vapor.		Humidity, Saturation = 1.000.	Weight in Grains of a Cubic Foot of Air.						
							Height of the Barometer in English Inches.						
Dry.	Wet.			In a Cubic Foot of Air.	Reqd. for Sat'n. of a Cubic Ft. of Air.		in. **28.0**	in. **28.5**	in. **29.0**	in. **29.5**	in. **30.0**	in. **30.5**	in. **31.0**
°	°	°	in.	gr.	gr.		gr.	gr.	gr.	gr.	gr.	gr.	gr.
62	62	62.0	0.559	6.25	0.00	1.000	491.2	499.9	508.7	517.5	526.3	535.1	543.9
	61	60.3	0.528	5.91	0.34	0.946	491.4	500.1	508.9	517.7	526.5	535.3	544.1
	60	58.6	0.499	5.58	0.67	0.893	491.5	500.2	509.0	517.8	526.6	535.4	544.2
	59	56.9	0.472	5.27	0.98	0.843	491.7	500.4	509.2	518.0	526.8	535.6	544.4
	58	55.2	0.445	4.99	1.26	0.798	491.9	500.6	509.4	518.2	527.0	535.8	544.6
	57	53.5	0.421	4.70	1.55	0.752	492.0	500.7	509.5	518.3	527.1	535.9	544.7
	56	51.8	0.397	4.44	1.81	0.710	492.1	500.7	509.5	518.4	527.3	536.1	544.9
	55	50.1	0.375	4.19	2.06	0.670	492.2	500.9	509.7	518.5	527.4	536.2	545.0
	54	48.4	0.354	3.95	2.30	0.632	492.4	501.1	509.9	518.7	527.6	536.4	545.2
	53	46.7	0.333	3.72	2.53	0.595	492.5	501.3	510.1	518.9	527.7	536.5	545.3
	52	45.0	0.315	3.52	2.73	0.563	492.7	501.5	510.3	519.1	527.9	536.7	545.5
	51	43.3	0.297	3.31	2.94	0.530	492.8	501.6	510.4	519.2	528.0	536.8	545.6
	50	41.6	0.280	3.13	3.12	0.501	492.9	501.7	510.5	519.3	528.1	536.9	545.7
	49	39.9	0.263	2.95	3.30	0.472	493.0	501.8	510.6	519.4	528.2	537.0	545.8
	48	38.2	0.248	2.77	3.48	0.443	493.1	501.9	510.7	519.5	528.3	537.1	545.9
	47	36.5	0.234	2.61	3.64	0.418	493.2	502.0	510.8	519.6	528.4	537.2	546.0
	46	34.8	0.220	2.47	3.78	0.395	493.3	502.1	510.9	519.7	528.5	537.3	546.1
	45	33.1	0.207	2.32	3.93	0.371	493.3	502.1	511.0	519.7	528.6	537.3	546.1
	44	31.4	0.195	2.18	4.07	0.349	493.4	502.2	511.0	519.8	528.6	537.4	546.2
	43	29.7	0.184	2.06	4.19	0.330	493.4	502.2	511.1	519.8	528.6	537.4	546.2
	42	28.0	0.173	1.94	4.31	0.311	493.5	502.3	511.2	519.9	528.7	537.5	546.3
	41	26.3	0.163	1.83	4.42	0.293	493.6	502.4	511.3	520.0	528.8	537.6	546.4
63	63	63.0	0.578	6.45	0.00	1.000	490.2	498.9	507.7	516.4	525.2	533.9	542.7
	62	61.3	0.546	6.10	0.35	0.946	490.4	499.1	507.9	516.6	525.4	534.1	542.9
	61	59.6	0.516	5.76	0.69	0.893	490.5	499.2	508.0	516.7	525.5	534.2	543.0
	60	57.9	0.488	5.44	1.01	0.843	490.7	499.4	508.2	516.9	525.7	534.4	543.2
	59	56.2	0.461	5.15	1.30	0.798	490.9	499.6	508.4	517.1	525.9	534.6	543.4
	58	54.5	0.435	4.86	1.59	0.753	491.0	499.7	508.5	517.2	526.0	534.7	543.5
	57	52.8	0.411	4.59	1.86	0.712	491.1	499.8	508.6	517.3	526.2	534.9	543.7
	56	51.1	0.388	4.33	2.12	0.671	491.2	499.9	508.7	517.4	526.3	535.0	543.8
	55	49.4	0.366	4.09	2.36	0.634	491.3	500.0	508.8	517.5	526.4	535.1	543.9
	54	47.7	0.345	3.85	2.60	0.597	491.5	500.2	509.0	517.7	526.6	535.3	544.1
	53	46.0	0.326	3.63	2.82	0.563	491.7	500.4	509.2	518.0	526.8	535.5	544.3
	52	44.3	0.307	3.43	3.02	0.532	491.8	500.5	509.3	518.1	526.9	535.6	544.4
	51	42.6	0.289	3.24	3.21	0.502	491.9	500.6	509.4	518.2	527.0	535.7	544.5
	50	40.9	0.273	3.05	3.40	0.473	492.0	500.7	509.5	518.3	527.1	535.8	544.6
	49	39.2	0.257	2.07	3.58	0.445	492.1	500.8	509.6	518.4	527.2	535.9	544.7
	48	37.5	0.242	2.71	3.74	0.420	492.2	500.9	509.7	518.5	527.3	536.0	544.8
	47	35.8	0.228	2.56	3.89	0.397	492.3	501.0	509.8	518.6	527.4	536.1	544.9
	46	34.1	0.215	2.41	4.04	0.374	492.4	501.1	509.9	518.7	527.5	536.2	545.0
	45	32.4	0.202	2.26	4.19	0.351	492.5	501.2	510.0	518.8	527.6	536.3	545.1
	44	30.7	0.190	2.13	4.32	0.330	492.5	501.2	510.0	518.8	527.6	536.3	545.1
	43	29.0	0.179	2.00	4.45	0.310	492.6	501.3	510.1	518.9	527.7	536.4	545.2
	42	27.3	0.168	1.87	4.58	0.290	492.7	501.4	510.2	519.0	527.8	536.5	545.3

Reading of Thermometer, Fahr.		Temp. of Dew-Point, Fahr.	Force of Vapor in English Inches.	Weight of Vapor.		Humidity, Saturation = 1.000.	Weight in Grains of a Cubic Foot of Air.						
							Height of the Barometer in English Inches.						
Dry.	Wet.			In a Cubic Foot of Air.	Reqd. for Sat'n. of a Cubic Ft. of Air.		in. **28.0**	in. **28.5**	in. **29.0**	in. **29.5**	in. **30.0**	in. **30.5**	in. **31.0**
°	°	°	in.	gr.	gr.		gr.	gr.	gr.	gr.	gr.	gr.	gr.
64	64	64.0	0.597	6.65	0.00	1.000	489.1	497.8	506.6	515.3	524.0	532.7	541.5
	63	62.3	0.565	6.29	0.36	0.946	489.3	498.0	506.8	515.5	524.2	532.9	541.7
	62	60.6	0.534	5.94	0.71	0.893	489.5	498.2	507.0	515.7	524.4	533.1	541.9
	61	58.9	0.504	5.61	1.04	0.843	489.7	498.4	507.2	515.9	524.6	533.3	542.1
	60	57.2	0.476	5.31	1.34	0.798	489.9	498.6	507.4	516.1	524.8	533.5	542.3
	59	55.5	0.450	5.01	1.64	0.753	490.0	498.7	507.5	516.2	524.9	533.6	542.4
	58	53.8	0.425	4.73	1.92	0.711	490.1	498.8	507.6	516.3	525.1	533.8	542.6
	57	52.1	0.401	4.47	2.18	0.672	490.2	498.9	507.7	516.4	525.2	533.9	542.7
	56	50.4	0.379	4.23	2.42	0.636	490.4	499.1	507.9	516.6	525.4	534.1	542.9
	55	48.7	0.357	3.98	2.67	0.598	490.5	499.2	508.0	516.7	525.5	534.2	543.0
	54	47.0	0.337	3.75	2.90	0.564	490.7	499.4	508.2	516.9	525.7	534.4	543.2
	53	45.3	0.318	3.55	3.10	0.534	490.8	499.5	508.3	517.0	525.8	534.5	543.3
	52	43.6	0.300	3.34	3.31	0.502	490.9	499.6	508.4	517.1	525.9	534.6	543.4
	51	41.9	0.282	3.15	3.50	0.473	491.0	499.7	508.5	517.2	526.0	534.7	543.5
	50	40.2	0.266	2.96	3.69	0.445	491.2	499.9	508.7	517.4	526.1	534.9	543.7
	49	38.5	0.251	2.79	3.86	0.419	491.3	500.0	508.8	517.5	526.2	535.0	543.8
	48	36.8	0.236	2.63	4.02	0.396	491.4	500.1	508.9	517.6	526.3	535.1	543.9
	47	35.1	0.223	2.47	4.18	0.372	491.5	500.2	509.0	517.7	526.4	535.2	544.0
	46	33.4	0.210	2.33	4.32	0.351	491.6	500.3	509.1	517.8	526.5	535.3	544.1
	45	31.7	0.197	2.19	4.46	0.330	491.7	500.4	509.2	517.9	526.6	535.4	544.2
	44	30.0	0.186	2.06	4.59	0.310	491.7	500.4	509.2	517.9	526.6	535.4	544.2
	43	28.3	0.175	1.94	4.71	0.292	491.8	500.5	509.3	518.0	526.7	535.5	544.3
	42	26.6	0.164	1.83	4.82	0.275	491.9	500.6	509.4	518.1	526.8	535.6	544.4
65	65	65.0	0.617	6.87	0.00	1.000	488.1	496.8	505.5	514.2	522.9	531.6	540.3
	64	63.4	0.586	6.51	0.36	0.947	488.3	497.0	505.7	514.4	523.1	531.8	540.5
	63	61.8	0.555	6.17	0.70	0.898	488.5	497.2	505.9	514.6	523.3	532.0	540.7
	62	60.2	0.527	5.85	1.02	0.851	488.7	497.4	506.1	514.8	523.5	532.2	540.9
	61	58.6	0.499	5.55	1.32	0.808	488.9	497.6	506.3	515.0	523.7	532.4	541.1
	60	57.0	0.473	5.25	1.62	0.765	489.0	497.7	506.5	515.2	523.9	532.6	541.3
	59	55.4	0.449	4.98	1.89	0.725	489.1	497.8	506.6	515.3	524.0	532.7	541.5
	58	53.8	0.425	4.72	2.15	0.687	489.3	498.0	506.8	515.5	524.2	532.9	541.7
	57	52.2	0.402	4.47	2.40	0.651	489.4	498.1	506.9	515.6	524.3	533.0	541.8
	56	50.6	0.381	4.23	2.64	0.616	489.6	498.3	507.1	515.8	524.5	533.2	542.0
	55	49.0	0.361	4.01	2.86	0.584	489.7	498.4	507.2	515.9	524.6	533.3	542.1
	54	47.4	0.342	3.79	3.08	0.552	489.8	498.5	507.3	516.0	524.7	533.4	542.2
	53	45.8	0.323	3.60	3.27	0.524	489.9	498.6	507.4	516.1	524.8	533.5	542.3
	52	44.2	0.306	3.39	3.48	0.493	490.0	498.7	507.5	516.2	524.9	533.6	542.4
	51	42.6	0.289	3.22	3.65	0.469	490.1	498.8	507.6	516.3	525.0	533.7	542.5
	50	41.0	0.274	3.04	3.83	0.442	490.2	498.9	507.7	516.4	525.1	533.8	542.6
	49	39.4	0.259	2.87	4.00	0.418	490.3	499.0	507.8	516.5	525.2	533.9	542.7
	48	37.8	0.245	2.72	4.15	0.396	490.3	499.0	507.8	516.5	525.2	533.9	542.7
	47	36.2	0.231	2.57	4.30	0.374	490.4	499.1	507.9	516.6	525.3	534.0	542.8
	46	34.6	0.219	2.43	4.44	0.354	490.5	499.2	508.0	516.7	525.4	534.1	542.9
	45	33.0	0.207	2.31	4.56	0.336	490.6	499.3	508.1	516.8	525.5	534.2	543.0
	44	31.4	0.195	2.17	4.70	0.316	490.7	499.4	508.2	516.9	525.6	534.3	543.1
	43	29.8	0.184	2.05	4.82	0.299	490.7	499.4	508.2	516.9	525.6	534.3	543.1
	42	28.2	0.174	1.94	4.93	0.283	490.8	499.5	508.3	517.0	525.7	534.4	543.2

Reading of Thermometer, Fahr.		Temp. of Dew-Point, Fahr.	Force of Vapor in English Inches.	Weight of Vapor		Humidity, Saturation = 1.000.	Weight in Grains of a Cubic Foot of Air. Height of the Barometer in English Inches.						
Dry.	Wet.			In a Cubic Foot of Air.	Reqd. for Sat'n. of a Cubic Ft. of Air.		in. **28.0**	in. **28.5**	in. **29.0**	in. **29.5**	in. **30.0**	in. **30.5**	in. **31.0**
°	°	°	in.	gr.	gr.		gr.	gr.	gr.	gr.	gr.	gr.	gr.
66	66	66.0	0.638	7.08	0.00	1.000	487.0	495.7	504.4	513.1	521.8	530.5	539.2
	65	64.4	0.605	6.72	0.36	0.949	487.2	495.9	504.6	513.3	522.0	530.7	539.4
	64	62.8	0.574	6.35	0.73	0.897	487.3	496.0	504.7	513.4	522.1	530.8	539.5
	63	61.2	0.544	6.04	1.04	0.853	487.5	496.2	504.9	513.6	522.3	531.0	539.7
	62	59.6	0.516	5.72	1.36	0.808	487.7	496.4	505.1	513.8	522.5	531.2	539.9
	61	58.0	0.489	5.42	1.66	0.766	487.9	496.6	505.3	514.0	522.7	531.4	540.1
	60	56.4	0.464	5.14	1.94	0.726	488.0	496.7	505.4	514.1	522.8	531.5	540.2
	59	54.8	0.440	4.88	2.20	0.689	488.1	496.8	505.5	514.2	523.0	531.7	540.4
	58	53.2	0.416	4.62	2.46	0.652	488.2	496.9	505.6	514.3	523.1	531.8	540.5
	57	51.6	0.394	4.37	2.71	0.619	488.4	497.1	505.8	514.5	523.3	532.0	540.7
	56	50.0	0.373	4.15	2.93	0.586	488.5	497.2	505.9	514.6	523.4	532.1	540.8
	55	48.4	0.354	3.92	3.16	0.553	488.6	497.3	506.1	514.8	523.5	532.2	541.0
	54	46.8	0.335	3.72	3.36	0.525	488.8	497.5	506.3	515.0	523.7	532.4	541.2
	53	45.2	0.317	3.51	3.57	0.496	488.9	497.6	506.4	515.1	523.8	532.5	541.3
	52	43.6	0.300	3.33	3.75	0.470	489.0	497.7	506.5	515.2	523.9	532.6	541.4
	51	42.0	0.283	3.14	3.94	0.443	489.1	497.8	506.6	515.3	524.0	532.7	541.5
	50	40.4	0.268	2.97	4.11	0.419	489.2	497.9	506.7	515.4	524.1	532.8	541.6
	49	38.8	0.253	2.81	4.27	0.397	489.3	498.0	506.8	515.5	524.2	532.9	541.7
	48	37.2	0.240	2.66	4.42	0.376	489.4	498.1	506.9	515.6	524.3	533.0	541.8
	47	35.6	0.227	2.51	4.57	0.355	489.4	498.1	506.9	515.6	524.3	533.0	541.8
	46	34.0	0.214	2.37	4.71	0.335	489.5	498.2	507.0	515.7	524.4	533.1	541.9
	45	32.4	0.202	2.24	4.84	0.316	489.6	498.3	507.1	515.8	524.5	533.2	542.0
	44	30.8	0.191	2.12	4.96	0.299	489.7	498.4	507.2	515.9	524.6	533.3	542.1
	43	29.2	0.180	2.00	5.08	0.283	489.7	498.4	507.2	515.9	524.6	533.3	542.1
67	67	67.0	0.659	7.30	0.00	1.000	485.9	494.6	503.3	512.0	520.6	529.3	538.0
	66	65.4	0.626	6.93	0.37	0.949	486.1	494.8	503.5	512.2	520.8	529.5	538.2
	65	63.8	0.593	6.55	0.75	0.897	486.3	495.0	503.7	512.4	521.0	529.7	538.4
	64	62.2	0.563	6.23	1.07	0.853	486.5	495.2	503.9	512.6	521.2	529.9	538.6
	63	60.6	0.534	5.91	1.39	0.810	486.7	495.4	504.1	512.8	521.4	530.1	538.8
	62	59.0	0.506	5.60	1.70	0.767	486.8	495.5	504.2	512.9	521.6	530.3	539.0
	61	57.4	0.480	5.31	1.99	0.728	486.9	495.6	504.3	513.0	521.7	530.4	539.1
	60	55.8	0.455	5.04	2.26	0.691	487.1	495.8	504.5	513.2	521.9	530.6	539.3
	59	54.2	0.431	4.77	2.53	0.653	487.2	495.9	504.6	513.3	522.0	530.7	539.4
	58	52.6	0.408	4.52	2.78	0.619	487.3	496.0	504.7	513.4	522.1	530.8	539.5
	57	51.0	0.386	4.28	3.02	0.586	487.5	496.2	504.9	513.6	522.3	531.0	539.7
	56	49.4	0.366	4.05	3.25	0.555	487.6	496.3	505.0	513.7	522.4	531.1	539.8
	55	47.8	0.346	3.83	3.47	0.524	487.8	496.5	505.1	513.8	522.6	531.2	549.9
	54	46.2	0.328	3.62	3.68	0.496	487.9	496.6	505.2	513.9	522.7	531.3	540.0
	53	44.6	0.310	3.43	3.87	0.470	488.0	496.7	505.3	514.0	522.8	531.4	540.1
	52	43.0	0.293	3.25	4.05	0.445	488.1	496.8	504.4	514.1	522.9	531.5	540.2
	51	41.4	0.278	3.08	4.22	0.422	488.2	496.9	505.5	514.2	523.0	531.6	540.3
	50	39.8	0.263	2.91	4.39	0.399	488.4	497.1	505.7	514.4	523.1	531.8	540.5
	49	38.2	0.248	2.75	4.55	0.377	488.5	497.2	505.8	514.5	523.2	531.9	540.6

Reading of Thermometer, Fahr. Dry.	Reading of Thermometer, Fahr. Wet.	Temp. of Dew-Point, Fahr.	Force of Vapor in English Inches.	Weight of Vapor In a Cubic Foot of Air.	Weight of Vapor Reqd. for Sat'n. of a Cubic Ft. of Air.	Humidity, Saturation = 1.000.	Weight in Grains of a Cubic Foot of Air. Height of the Barometer in English Inches. in. **28.0**	in. **28.5**	in. **29.0**	in. **29.5**	in. **30.0**	in. **30.5**	in. **31.0**
°	°	°	in.	gr.	gr.		gr.	gr.	gr.	gr.	gr.	gr.	gr.
67	49	38.2	0.248	2.75	4.55	0.377	488.5	497.2	505.8	514.5	523.2	531.9	540.6
	48	36.6	0.235	2.60	4.70	0.356	488.6	497.3	505.9	514.6	523.3	532.0	540.7
	47	35.0	0.222	2.46	4.84	0.337	488.7	497.4	505.9	514.7	523.4	532.1	540.8
	46	33.4	0.210	2.32	4.98	0.318	488.7	497.4	506.0	514.7	523.4	532.1	540.8
	45	31.8	0.198	2.19	5.11	0.301	488.8	497.5	506.1	514.8	523.5	532.2	540.9
	44	30.2	0.187	2.07	5.23	0.284	488.9	497.6	506.2	514.9	523.6	532.3	541.0
68	68	68.0	0.681	7.53	0.00	1.000	484.9	493.5	502.2	510.8	519.5	528.1	536.8
	67	66.4	0.646	7.15	0.38	0.949	485.1	493.8	502.5	511.1	519.7	528.4	537.1
	66	64.8	0.613	6.77	0.76	0.899	485.3	494.0	502.6	511.2	519.9	528.6	537.3
	65	63.2	0.582	6.43	1.10	0.854	485.5	494.2	502.8	511.4	520.1	528.8	537.5
	64	61.6	0.552	6.10	1.43	0.810	485.7	494.4	503.0	511.6	520.3	529.0	537.7
	63	60.0	0.523	5.78	1.75	0.768	485.8	494.5	503.1	511.8	520.5	529.2	537.9
	62	58.4	0.496	5.47	2.06	0.726	485.9	494.6	503.3	512.0	520.7	529.4	538.1
	61	56.8	0.470	5.20	2.33	0.691	486.0	494.7	503.4	512.1	520.8	529.5	538.3
	60	55.2	0.445	4.93	2.60	0.655	486.2	494.9	503.6	512.3	521.0	529.7	538.5
	59	53.6	0.422	4.67	2.86	0.620	486.3	495.0	503.7	512.4	521.1	529.8	538.6
	58	52.0	0.400	4.42	3.11	0.587	486.4	495.1	503.8	512.5	521.2	529.9	538.6
	57	50.4	0.379	4.19	3.34	0.556	486.6	495.3	504.0	512.7	521.4	530.1	538.8
	56	48.8	0.358	3.96	3.57	0.526	486.7	495.4	504.1	512.8	521.5	530.2	538.9
	55	47.2	0.339	3.75	3.78	0.498	486.8	495.5	504.2	512.9	521.6	530.3	539.0
	54	45.6	0.321	3.54	3.99	0.470	486.9	495.6	504.3	513.0	521.7	530.4	539.1
	53	44.0	0.304	3.35	4.18	0.445	487.0	495.7	504.4	513.1	521.8	530.5	539.2
	52	42.4	0.287	3.17	4.36	0.421	487.1	495.8	504.5	513.2	521.9	530.6	539.3
	51	40.8	0.272	3.00	4.53	0.399	487.2	495.9	504.6	513.3	522.0	530.7	539.4
	50	39.2	0.257	2.84	4.69	0.377	487.3	496.0	504.7	513.4	522.1	530.8	539.5
	49	37.6	0.243	2.68	4.85	0.356	487.4	496.1	504.8	513.5	522.2	530.9	539.6
	48	36.0	0.230	2.54	4.99	0.337	487.5	496.2	504.9	513.6	522.3	531.0	539.7
	47	34.4	0.217	2.40	5.13	0.319	487.6	496.3	505.0	513.7	522.4	531.1	539.8
	46	32.8	0.205	2.27	5.26	0.302	487.6	496.3	505.0	513.7	522.4	531.1	539.8
	45	31.2	0.194	2.15	5.38	0.286	487.7	496.4	505.1	513.8	522.5	531.2	539.9
	44	29.6	0.183	2.04	5.49	0.271	487.8	496.5	505.2	513.9	522.6	531.3	540.0
69	69	69.0	0.704	7.76	0.00	1.000	483.8	492.4	501.1	509.7	518.3	527.0	535.6
	68	67.4	0.668	7.37	0.39	0.950	484.0	492.6	501.3	509.9	518.5	527.2	535.8
	67	65.8	0.634	7.00	0.76	0.902	484.2	492.8	501.5	510.1	518.7	527.4	536.0
	66	64.2	0.601	6.63	1.13	0.854	484.4	493.0	501.7	510.3	518.9	527.6	536.2
	65	62.6	0.570	6.29	1.47	0.810	484.6	493.2	501.9	510.5	519.1	527.8	536.4
	64	61.0	0.541	5.97	1.79	0.769	484.8	493.4	502.1	510.7	519.3	528.0	536.6
	63	59.4	0.513	5.65	2.11	0.728	485.0	493.6	502.3	510.9	519.5	528.2	536.8
	62	57.8	0.486	5.37	2.39	0.693	485.1	493.7	502.4	511.0	519.6	528.3	536.9
	61	56.2	0.461	5.09	2.67	0.657	485.1	493.7	502.6	511.2	519.8	528.5	537.1
	60	54.6	0.437	4.82	2.94	0.621	485.2	493.9	502.7	511.3	519.9	528.6	537.3
	59	53.0	0.414	4.57	3.19	0.589	485.4	494.1	502.8	511.5	520.1	528.8	537.5
	58	51.4	0.392	4.33	3.43	0.558	485.5	494.2	502.9	511.6	520.2	528.9	537.6

Reading of Thermometer, Fahr.		Temp. of Dew-Point, Fahr.	Force of Vapor in English Inches.	Weight of Vapor		Humidity, Saturation = 1.000.	Weight in Grains of a Cubic Foot of Air.						
				In a Cubic Foot of Air.	Reqd. for Sat'n. of a Cubic Ft. of Air.		Height of the Barometer in English Inches.						
Dry.	Wet.						in. **28.0**	in. **28.5**	in. **29.0**	in. **29.5**	in. **30.0**	in. **30.5**	in. **31.0**
°	°	°	in.	gr.	gr.		gr.	gr.	gr.	gr.	gr.	gr.	gr.
69	58	51.4	0.392	4.33	3.43	0.558	485.5	494.2	502.9	511.6	520.2	528.9	537.6
	57	49.8	0.371	4.09	3.67	0.527	485.7	494.4	503.1	511.8	520.4	529.1	537.8
	56	48.2	0.351	3.87	3.89	0.499	485.8	494.5	503.2	511.9	520.5	529.2	537.9
	55	46.6	0.332	3.66	4.10	0.472	485.9	494.6	503.3	512.0	520.6	529.3	538.0
	54	45.0	0.315	3.47	4.29	0.447	486.0	494.7	503.4	512.1	520.7	529.4	538.1
	53	43.4	0.298	3.29	4.47	0.424	486.1	494.8	503.5	512.2	520.8	529.5	538.2
	52	41.8	0.282	3.11	4.65	0.401	486.2	494.9	503.6	512.3	520.9	529.6	538.3
	51	40.2	0.266	2.94	4.82	0.379	486.3	495.0	503.7	512.4	521.0	529.7	538.4
	50	38.6	0.252	2.78	4.98	0.358	486.4	495.1	503.8	512.5	521.1	529.8	538.5
	49	37.0	0.238	2.63	5.13	0.339	486.5	495.2	503.9	512.6	521.2	529.9	538.6
	48	35.4	0.225	2.49	5.27	0.321	486.6	495.3	504.0	512.7	521.3	530.0	538.7
	47	33.8	0.213	2.34	5.42	0.302	486.7	495.4	504.1	512.8	521.4	530.1	538.8
	46	32.2	0.201	2.20	5.56	0.284	486.8	495.5	504.2	512.9	521.5	530.2	538.9
	45	30.6	0.190	2.06	5.70	0.266	486.8	495.5	504.2	512.9	521.5	530.2	538.9
70	70	70.0	0.727	8.00	0.00	1.000	482.8	491.4	500.0	508.6	517.2	525.8	534.4
	69	68.5	0.692	7.62	0.38	0.953	483.0	491.6	500.2	508.8	517.4	526.0	534.6
	68	67.0	0.659	7.26	0.74	0.907	483.2	491.8	500.4	509.0	517.6	526.2	534.8
	67	65.5	0.628	6.91	1.09	0.865	483.3	491.9	500.5	509.1	517.7	526.3	534.9
	66	64.0	0.597	6.57	1.43	0.822	483.5	492.1	500.7	509.3	517.9	526.5	535.1
	65	62.5	0.568	6.25	1.75	0.781	483.7	492.3	500.9	509.5	518.1	526.7	535.3
	64	61.0	0.541	5.95	2.05	0.744	483.8	492.4	501.0	509.6	518.3	526.9	535.5
	63	59.5	0.515	5.66	2.34	0.708	484.0	492.6	501.2	509.8	518.5	527.1	535.7
	62	58.0	0.489	5.38	2.62	0.672	484.2	492.8	501.4	510.0	518.7	527.3	535.9
	61	56.5	0.465	5.12	2.88	0.640	484.3	492.9	501.5	510.1	518.8	527.4	536.0
	60	55.0	0.442	4.87	3.13	0.609	484.4	493.0	501.6	510.2	518.9	527.5	536.1
	59	53.5	0.421	4.62	3.38	0.578	484.6	493.2	501.8	510.4	519.1	527.7	536.3
	58	52.0	0.400	4.40	3.60	0.550	484.7	493.3	501.9	510.5	519.2	527.8	536.4
	57	50.5	0.380	4.18	3.82	0.522	484.8	493.4	502.0	510.6	519.3	527.9	536.5
	56	49.0	0.361	3.96	4.04	0.495	484.9	493.5	502.1	510.7	519.4	528.0	536.6
	55	47.5	0.343	3.76	4.24	0.470	485.1	493.7	502.3	510.9	519.6	528.2	536.8
	54	46.0	0.326	3.57	4.43	0.446	485.2	493.8	502.4	511.0	519.7	528.3	536.9
	53	44.5	0.309	3.40	4.60	0.425	485.3	493.9	502.5	511.1	519.8	528.4	537.0
	52	43.0	0.292	3.23	4.77	0.404	485.4	494.0	502.6	511.2	519.9	528.5	537.1
	51	41.5	0.279	3.07	4.93	0.384	485.5	494.1	502.7	511.3	520.0	528.6	537.2
	50	40.0	0.264	2.81	5.19	0.351	485.5	494.1	502.7	511.3	520.0	528.6	537.2
	49	38.5	0.251	2.76	5.24	0.345	485.6	494.2	502.8	511.4	520.1	528.7	537.3
	48	37.0	0.238	2.63	5.37	0.329	485.7	494.3	502.9	511.5	520.2	528.8	537.4
	47	35.5	0.226	2.50	5.50	0.313	485.8	494.4	503.0	511.6	520.3	528.9	537.5
	46	34.0	0.214	2.37	5.63	0.296	485.8	494.4	503.0	511.6	520.3	528.9	537.5
	45	32.5	0.203	2.24	5.76	0.280	485.9	494.5	503.1	511.7	520.4	529.0	537.6
	44	31.0	0.192	2.12	5.88	0.265	486.0	494.6	503.2	511.8	520.5	529.1	537.7
	43	29.5	0.182	2.01	5.99	0.251	486.1	494.7	503.3	511.9	520.6	529.2	537.8

Reading of Thermometer, Fahr.		Temp. of Dew-Point, Fahr.	Force of Vapor in English Inches.	Weight of Vapor		Humidity, Saturation = 1.000.	Weight in Grains of a Cubic Foot of Air. Height of the Barometer in English Inches.						
Dry.	Wet.			In a Cubic Foot of Air.	Reqd. for Sat'n. of a Cubic Ft. of Air.		in. **28.0**	in. **28.5**	in. **29.0**	in. **29.5**	in. **30.0**	in. **30.5**	in. **31.0**
°	°	°	in.	gr.	gr.		gr.	gr.	gr.	gr.	gr.	gr	gr.
71	71	71.0	0.751	8.25	0.00	1.000	481.6	490.2	498.8	507.4	516.0	524.6	533.2
	70	69.5	0.715	7.86	0.39	0.953	481.8	490.4	499.0	507.6	516.2	524.8	533.4
	69	68.0	0.681	7.48	0.77	0.907	482.0	490.6	499.2	507.8	516.4	525.0	533.6
	68	66.5	0.648	7.13	1.12	0.865	482.2	490.8	499.4	508.0	516.6	525.2	533.8
	67	65.0	0.617	6.79	1.46	0.823	482.4	491.0	499.6	508.2	516.8	525.4	534.0
	66	63.5	0.588	6.45	1.80	0.782	482.6	491.2	499.8	508.4	517.0	525.6	534.2
	65	62.0	0.559	6.14	2.11	0.744	482.8	491.4	500.0	508.6	517.2	525.8	534.4
	64	60.5	0.532	5.85	2.40	0.709	483.0	491.6	500.2	508.8	517.4	526.0	534.6
	63	59.0	0.506	5.56	2.69	0.674	483.1	491.7	500.3	508.9	517.5	526.1	534.7
	62	57.5	0.481	5.28	2.97	0.640	483.2	491.8	500.4	509.0	517.7	526.3	534.9
	61	56.0	0.458	5.03	3.22	0.609	483.3	491.9	500.5	509.1	517.8	526.4	535.0
	60	54.5	0.435	4.78	3.47	0.579	483.5	492.1	500.7	509.3	518.0	526.6	535.1
	59	53.0	0.414	4.54	3.71	0.550	483.6	492.2	500.8	509.4	518.1	526.7	535.2
	58	51.5	0.393	4.31	3.94	0.522	483.8	492.4	501.0	509.6	518.3	526.9	535.4
	57	50.0	0.373	4.10	4.15	0.497	483.9	492.5	501.1	509.7	518.4	527.0	535.5
	56	48.5	0.355	3.89	4.36	0.471	484.0	492.6	501.2	509.9	518.5	527.1	535.6
	55	47.0	0.337	3.69	4.56	0.447	484.1	492.7	501.3	510.0	518.6	527.2	535.7
	54	45.5	0.320	3.51	4.74	0.425	484.2	492.8	501.4	510.1	518.7	527.3	535.8
	53	44.0	0.304	3.33	4.92	0.404	484.3	492.9	501.5	510.2	518.8	527.4	535.9
	52	42.5	0.288	3.16	5.09	0.383	484.4	493.0	501.6	510.3	518.9	527.5	536.0
	51	41.0	0.274	3.00	5.25	0.364	484.5	493.1	501.7	510.4	519.0	527.6	536.1
	50	39.5	0.260	2.85	5.40	0.345	484.6	493.2	501.8	510.5	519.1	527.7	536.2
	49	38.0	0.246	2.70	5.55	0.327	484.7	493.3	501.9	510.6	519.2	527.8	536.3
	48	36.5	0.234	2.57	5.68	0.312	484.7	493.3	501.9	510.6	519.2	527.8	536.3
	47	35.0	0.222	2.44	5.81	0.296	484.8	493.4	502.0	510.7	519.3	527.9	536.4
	46	33.5	0.210	2.31	5.94	0.280	484.9	493.5	502.1	510.8	519.4	528.0	536.5
	45	32.0	0.199	2.19	6.06	0.265	485.0	493.6	502.2	510.9	519.5	528.1	536.6
	44	30.5	0.189	2.08	6.17	0.252	485.0	493.6	502.2	510.9	519.5	528.1	536.6
72	72	72.0	0.776	8.50	0.00	1.000	480.6	489.2	497.8	506.4	514.9	523.5	532.1
	71	70.5	0.739	8.10	0.40	0.953	480.8	489.4	498.0	506.5	515.1	523.7	532.3
	70	69.0	0.704	7.71	0.79	0.907	481.0	489.6	498.2	506.7	515.3	523.9	532.5
	69	67.5	0.670	7.35	1.15	0.865	481.2	489.8	498.4	506.9	515.5	524.1	532.7
	68	66.0	0.638	7.00	1.50	0.824	481.4	490.0	498.5	507.1	515.7	524.3	532.9
	67	64.5	0.607	6.66	1.84	0.784	481.6	490.2	498.7	507.3	515.9	524.5	533.1
	66	63.0	0.578	6.33	2.17	0.745	481.7	490.3	498.8	507.4	516.1	524.7	533.3
	65	61.5	0.550	6.03	2.47	0.710	481.8	490.4	499.0	507.6	516.2	524.8	533.4
	64	60.0	0.523	5.73	2.77	0.674	482.0	490.6	499.2	507.8	516.4	525.0	533.6
	63	58.5	0.498	5.45	3.05	0.641	482.1	490.7	499.3	507.9	516.5	525.1	533.7
	62	57.0	0.473	5.18	3.32	0.610	482.3	490.9	499.5	508.1	516.7	525.3	533.9
	61	55.5	0.450	4.93	3.57	0.580	482.5	491.1	499.7	508.3	516.9	525.5	534.1
	60	54.0	0.428	4.68	3.82	0.551	482.6	491.2	499.8	508.4	517.0	525.6	534.2
	59	52.5	0.407	4.45	4.05	0.523	482.8	491.4	500.0	508.6	517.2	525.8	534.4

Reading of Thermometer, Fahr. Dry.	Wet.	Temp. of Dew-Point, Fahr.	Force of Vapor in English Inches.	Weight of Vapor: In a Cubic Foot of Air.	Weight of Vapor: Reqd. for Sat'n. of a Cubic Ft. of Air.	Humidity, Saturation = 1.000.	Weight in Grains of a Cubic Foot of Air. Height of the Barometer in English Inches. in. 28.0	in. 28.5	in. 29.0	in. 29.5	in. 30.0	in. 30.5	in. 31.0
°	°	°	in.	gr.	gr.		gr.	gr.	gr.	gr.	gr.	gr.	gr.
72	59	52.5	0.407	4.45	4.05	0.523	482.8	491.4	500.0	508.6	517.2	525.8	534.4
	58	51.0	0.386	4.23	4.27	0.498	482.9	491.5	500.1	508.7	517.3	525.9	534.5
	57	49.5	0.367	4.02	4.48	0.473	483.0	491.6	500.2	508.8	517.4	526.0	534.6
	56	48.0	0.349	3.82	4.68	0.449	483.1	491.7	500.3	508.9	517.5	526.1	534.7
	55	46.5	0.331	3.63	4.87	0.427	483.2	491.8	500.4	509.0	517.6	526.2	534.8
	54	45.0	0.315	3.45	5.05	0.406	483.3	491.9	500.5	509.1	517.7	526.2	534.9
	53	43.5	0.299	3.28	5.22	0.386	483.3	492.0	500.6	509.2	517.8	526.3	535.0
	52	42.0	0.283	3.11	5.39	0.366	483.5	492.1	500.7	509.3	517.9	526.4	535.1
	51	40.5	0.269	2.95	5.55	0.347	483.6	492.2	500.8	509.4	518.0	526.5	535.2
	50	39.0	0.255	2.80	5.70	0.329	483.7	492.3	500.9	509.5	518.1	526.6	535.3
	49	37.5	0.242	2.66	5.84	0.313	483.8	492.4	501.0	509.6	518.2	526.7	535.4
	48	36.0	0.230	2.52	5.98	0.296	483.8	492.4	501.0	509.6	518.2	526.7	535.4
	47	34.5	0.218	2.39	6.11	0.281	483.9	492.5	501.2	509.7	518.3	526.8	535.5
	46	33.0	0.207	2.27	6.23	0.267	484.0	492.6	501.3	509.8	518.4	526.9	535.6
	45	31.5	0.196	2.16	6.34	0.254	484.1	492.7	501.3	509.9	518.5	527.1	535.7
73	73	73.0	0.801	8.76	0.00	1.000	479.6	488.1	496.7	505.2	513.8	522.3	530.9
	72	71.5	0.736	8.35	0.41	0.953	479.8	488.3	496.9	505.4	514.0	522.5	531.1
	71	70.0	0.727	7.95	0.81	0.908	480.0	488.5	497.1	505.6	514.2	522.7	531.3
	70	68.5	0.692	7.57	1.19	0.864	480.2	488.7	497.3	505.8	514.4	522.9	531.5
	69	67.0	0.659	7.21	1.55	0.823	480.4	488.9	497.5	506.0	514.6	523.1	531.7
	68	65.5	0.628	6.87	1.89	0.784	480.5	489.0	497.6	506.1	514.8	523.3	531.9
	67	64.0	0.597	6.53	2.23	0.745	480.7	489.2	497.8	506.3	515.0	523.5	532.1
	66	62.5	0.568	6.22	2.54	0.710	480.8	489.3	497.9	506.4	515.1	523.6	532.2
	65	61.0	0.541	5.92	2.84	0.676	481.0	489.5	498.1	506.6	515.3	523.8	532.4
	64	59.5	0.515	5.63	3.13	0.643	481.1	489.6	498.2	506.8	515.4	524.0	532.6
	63	58.0	0.489	5.34	3.42	0.610	481.2	489.8	498.4	507.0	515.6	524.2	532.8
	62	56.5	0.465	5.09	3.67	0.581	481.4	490.0	498.6	507.2	515.8	524.4	533.0
	61	55.0	0.442	4.84	3.92	0.553	481.6	490.2	498.8	507.4	516.0	524.6	533.2
	60	53.5	0.421	4.59	4.17	0.524	481.7	490.3	498.9	507.5	516.1	524.7	533.3
	59	52.0	0.400	4.37	4.39	0.499	481.8	490.4	499.0	507.6	516.2	524.8	533.4
	58	50.5	0.380	4.16	4.60	0.475	482.0	490.6	499.2	507.8	516.4	525.0	533.6
	57	49.0	0.361	3.94	4.82	0.450	482.1	490.7	499.3	507.9	516.5	525.1	533.7
	56	47.5	0.343	3.74	5.02	0.427	482.2	490.8	499.4	508.0	516.6	525.2	533.8
	55	46.0	0.326	3.56	5.20	0.406	482.3	490.9	499.5	508.1	516.7	525.3	533.9
	54	44.5	0.309	3.38	5.38	0.386	482.4	491.0	499.6	508.2	516.8	525.4	534.0
	53	43.0	0.293	3.21	5.55	0.366	482.5	491.1	499.7	508.3	516.9	525.5	534.1
	52	41.5	0.279	3.05	5.71	0.348	482.6	491.2	499.8	508.4	517.0	525.6	534.2
	51	40.0	0.264	2.89	5.87	0.330	482.7	491.3	499.9	508.5	517.1	525.7	534.3
	50	38.5	0.251	2.74	6.02	0.313	482.8	491.4	500.0	508.6	517.2	525.8	534.4
	49	37.0	0.238	2.60	6.16	0.297	482.9	491.5	500.0	508.6	517.2	525.8	534.4
	48	35.5	0.226	2.47	6.29	0.282	483.0	491.6	500.1	508.7	517.3	525.9	534.5
	47	34.0	0.214	2.34	6.42	0.267	483.1	491.7	500.2	508.8	517.4	526.0	534.6
	46	32.5	0.203	2.22	6.54	0.253	433.3	491.9	500.4	509.1	517.6	526.2	534.8

Reading of Thermometer, Fahr.		Temp. of Dew-Point, Fahr.	Force of Vapor in English Inches.	Weight of Vapor		Humidity, Saturation = 1.000.	Weight in Grains of a Cubic Foot of Air.						
							Height of the Barometer in English Inches.						
Dry.	Wet.			In a Cubic Foot of Air.	Reqd. for Sat'n. of a Cubic Ft. of Air.		in. **28.0**	in. **28.5**	in. **29.0**	in. **29.5**	in. **30.0**	in. **30.5**	in. **31.0**
°	°	°	in.	gr.	gr.		gr.	gr.	gr.	gr.	gr.	gr.	gr.
74	74	74.0	0.827	9.04	0.00	1.000	478.4	486.9	495.5	504.0	512.6	521.1	529.7
	73	72.5	0.787	8.60	0.44	0.951	478.6	487.1	495.7	504.2	512.8	521.3	529.9
	72	71.0	0.751	8.20	0.84	0.907	478.8	487.3	495.9	504.4	513.0	521.5	530.1
	71	69.5	0.715	7.81	1.23	0.864	479.0	487.5	496.1	504.6	513.2	521.7	530.3
	70	68.0	0.681	7.44	1.60	0.823	479.2	487.7	496.3	504.8	513.4	521.9	530.5
	69	66.5	0.648	7.08	1.96	0.783	479.4	487.9	496.5	505.0	513.6	522.1	530.7
	68	65.0	0.617	6.75	2.29	0.747	479.6	488.1	496.7	505.2	513.8	522.3	530.9
	67	63.5	0.588	6.41	2.63	0.709	479.8	488.3	496.9	505.4	514.0	522.5	531.1
	66	62.0	0.559	6.10	2.94	0.675	480.0	488.5	497.1	505.6	514.2	522.7	531.3
	65	60.5	0.532	5.81	3.23	0.643	480.1	488.7	497.3	505.9	514.4	522.9	531.5
	64	59.0	0.506	5.52	3.52	0.611	480.3	488.9	497.5	506.1	514.6	523.2	531.8
	63	57.5	0.481	5.24	3.80	0.580	480.5	489.1	497.7	506.3	514.8	523.4	532.0
	62	56.0	0.458	4.99	4.05	0.552	480.6	489.2	497.8	506.4	514.9	523.5	532.1
	61	54.5	0.435	4.75	4.29	0.525	480.7	489.3	497.9	506.5	515.0	523.6	532.2
	60	53.0	0.414	4.52	4.52	0.500	480.9	489.5	498.1	506.7	515.2	523.8	532.4
	59	51.5	0.393	4.29	4.75	0.475	481.0	489.6	498.2	506.8	515.3	523.9	532.5
	58	50.0	0.373	4.08	4.96	0.451	481.1	489.7	498.3	506.9	515.4	524.0	532.6
	57	48.5	0.355	3.86	5.18	0.427	481.2	489.8	498.4	507.0	515.5	524.1	532.7
	56	47.0	0.337	3.66	5.38	0.405	481.3	489.9	498.5	507.1	515.6	524.2	532.8
	55	45.5	0.320	3.48	5.56	0.385	481.4	490.0	498.6	507.2	515.7	524.3	532.9
	54	44.0	0.304	3.32	5.72	0.367	481.5	490.1	498.7	507.3	515.8	524.4	533.0
	53	42.5	0.288	3.15	5.89	0.348	481.6	490.2	498.8	507.4	515.9	524.5	533.1
	52	41.0	0.274	2.99	6.05	0.331	481.7	490.3	498.9	507.5	516.0	524.6	533.2
	51	39.5	0.260	2.83	6.21	0.313	481.8	490.4	499.0	507.6	516.1	524.7	533.3
	50	38.0	0.246	2.69	6.35	0.298	481.9	490.5	499.1	507.7	516.2	524.8	533.4
	49	36.5	0.234	2.55	6.49	0.282	481.9	490.5	499.1	507.7	516.2	524.8	533.4
	48	35.0	0.222	2.42	6.62	0.268	482.0	490.6	499.2	507.8	516.3	524.9	533.5
	47	33.5	0.210	2.30	6.74	0.254	482.1	490.7	499.2	507.9	516.4	525.0	533.6
75	75	75.0	0.854	9.31	0.00	1.000	477.4	485.9	494.4	502.9	511.5	520.0	528.5
	74	73.5	0.814	8.87	0.44	0.953	477.6	486.1	494.6	503.1	511.7	520.2	528.7
	73	72.0	0.776	8.45	0.86	0.908	477.8	486.3	494.8	503.3	511.9	520.4	528.9
	72	70.5	0.739	8.05	1.26	0.865	478.0	486.5	495.0	503.5	512.1	520.6	529.1
	71	69.0	0.704	7.67	1.64	0.824	478.2	486.7	495.2	503.7	512.3	520.8	529.3
	70	67.5	0.670	7.30	2.01	0.784	478.3	486.8	495.3	503.8	512.5	521.0	529.5
	69	66.0	0.638	6.95	2.36	0.746	478.5	487.0	495.5	504.0	512.7	521.2	529.7
	68	64.5	0.607	6.62	2.69	0.711	478.7	487.2	495.7	504.2	512.9	521.4	529.9
	67	63.0	0.578	6.30	3.01	0.677	478.9	487.4	495.9	504.4	513.1	521.6	530.1
	66	61.5	0.550	5.99	3.32	0.643	479.1	487.6	496.1	504.6	513.3	521.8	530.3
	65	60.0	0.523	5.69	3.62	0.611	479.3	487.8	496.4	504.9	513.5	522.0	530.6
	64	58.5	0.498	5.42	3.89	0.582	479.5	488.0	496.6	505.1	513.7	522.2	530.8
	63	57.0	0.473	5.15	4.16	0.553	479.6	488.1	496.7	505.2	513.8	522.3	530.9
	62	55.5	0.450	4.90	4.41	0.526	479.7	484.2	496.8	505.3	513.9	522.4	531.0

Reading of Thermometer, Fahr.		Temp. of Dew-Point, Fahr.	Force of Vapor in English Inches.	Weight of Vapor		Humidity, Saturation = 1.000.	Weight in Grains of a Cubic Foot of Air. Height of the Barometer in English Inches.						
Dry.	Wet.			In a Cubic Foot of Air.	Reqd. for Sat'n. of a Cubic Ft. of Air.		in. **28.0**	in. **28.5**	in. **29.0**	in. **29.5**	in. **30.0**	in. **30.5**	in. **31.0**
°	°	°	in.	gr.	gr.		gr.	gr.	gr.	gr.	gr.	gr.	gr.
75	62	55.5	0.450	4.90	4.41	0.526	479.7	488.2	496.8	505.3	513.9	522.4	531.0
	61	54.0	0.428	4.66	4.65	0.501	479.9	488.4	497.0	505.5	514.1	522.6	531.2
	60	52.5	0.407	4.43	4.88	0.476	480.0	488.5	497.1	505.6	514.2	522.7	531.3
	59	51.0	0.386	4.21	5.10	0.452	480.1	488.6	497.2	505.7	514.3	522.8	531.4
	58	49.5	0.367	4.00	5.31	0.429	480.3	488.8	497.4	505.9	514.5	523.0	531.6
	57	48.0	0.349	3.79	5.52	0.407	480.4	488.9	497.5	506.0	514.6	523.1	531.7
	56	46.5	0.331	3.60	5.71	0.387	480.5	489.0	497.6	506.1	514.7	523.2	531.8
	55	45.0	0.315	3.42	5.89	0.367	480.6	489.1	497.7	506.2	514.8	523.3	531.9
	54	43.5	0.299	3.25	6.06	0.349	480.7	489.2	497.8	506.3	514.9	523.4	532.0
	53	42.0	0.283	3.09	6.22	0.332	480.8	489.3	497.9	506.4	515.0	523.5	532.1
	52	40.5	0.269	2.93	6.38	0.315	480.8	489.3	497.9	506.4	515.0	523.5	532.1
	51	39.0	0.255	2.78	6.53	0.299	480.9	489.4	498.0	506.5	515.1	523.6	532.2
	50	37.5	0.242	2.64	6.67	0.284	481.0	489.5	498.1	506.6	515.2	523.7	532.3
	49	36.0	0.230	2.51	6.80	0.270	481.1	489.6	498.2	506.7	515.3	523.8	532.4
	48	34.5	0.218	2.39	6.92	0.257	481.2	489.7	498.3	506.8	515.4	523.9	532.5
76	76	76.0	0.882	9.60	0.00	1.000	476.3	484.8	493.3	501.8	510.3	518.8	527.3
	75	74.5	0.840	9.14	0.46	0.952	476.6	485.1	493.6	502.1	510.6	519.1	527.6
	74	73.0	0.801	8.71	0.89	0.907	476.8	485.3	493.8	502.3	510.8	519.3	527.8
	73	71.5	0.763	8.30	1.30	0.865	477.0	485.5	494.0	502.6	511.1	519.6	528.1
	72	70.0	0.727	7.90	1.70	0.823	477.2	485.7	494.3	502.8	511.3	519.8	528.3
	71	68.5	0.692	7.53	2.07	0.784	477.4	485.9	494.5	503.0	511.5	520.0	528.5
	70	67.0	0.659	7.17	2.43	0.747	477.6	486.1	494.7	503.2	511.7	520.2	528.7
	69	65.5	0.628	6.83	2.77	0.711	477.8	486.3	494.9	503.4	511.9	520.4	528.9
	68	64.0	0.597	6.49	3.11	0.676	477.9	486.4	495.0	503.6	512.1	520.6	529.2
	67	62.5	0.568	6.16	3.44	0.642	478.1	486.6	465.2	503.8	512.3	520.8	529.4
	66	61.0	0.541	5.88	3.72	0.613	478.2	486.7	495.3	503.9	512.4	520.9	529.5
	65	59.5	0.515	5.59	4.01	0.582	478.3	486.8	495.4	504.0	512.5	521.0	529.6
	64	58.0	0.489	5.31	4.29	0.553	478.5	487.0	495.6	504.2	512.7	521.2	529.8
	63	56.5	0.465	5.06	4.54	0.527	478.6	487.1	495.7	504.3	512.8	521.3	529.9
	62	55.0	0.442	4.81	4.79	0.501	478.8	487.3	495.9	504.5	513.0	521.5	530.1
	61	53.5	0.421	4.57	5.03	0.476	479.0	487.5	496.1	504.7	513.2	521.7	530.3
	60	52.0	0.400	4.34	5.26	0.452	479.1	487.6	496.2	504.8	513.3	521.8	530.4
	59	50.5	0.380	4.13	5.47	0.430	499.2	487.7	496.3	504.9	513.4	521.9	530.5
	58	49.0	0.361	3.92	5.68	0.408	499.3	487.8	496.4	505.0	513.5	522.0	530.6
	57	47.5	0.343	3.73	5.87	0.389	499.4	487.9	496.5	505.1	513.6	522.1	530.7
	56	46.0	0.326	3.54	6.06	0.369	499.5	488.0	496.6	505.2	513.7	522.2	530.8
	55	44.5	0.309	3.36	6.24	0.351	499.6	488.1	496.7	505.3	513.8	522.3	530.9
	54	43.0	0.293	3.19	6.41	0.332	499.7	488.2	496.8	505.4	513.9	522.4	531.0
	53	41.5	0.279	3.03	6.57	0.316	499.8	488.3	496.9	505.5	514.0	522.5	531.1
	52	40.0	0.264	2.88	6.72	0.301	499.9	488.4	497.0	505.6	514.1	522.6	531.2
	51	38.5	0.251	2.73	6.87	0.284	500.0	488.5	497.1	505.7	514.2	522.7	531.3
	50	37.0	0.238	2.59	7.01	0.269	500.1	488.6	497.2	505.8	514.3	522.8	531.4
	49	35.5	0.226	2.46	7.14	0.256	500.2	488.7	497.3	505.9	514.4	522.9	531.5

Reading of Thermometer, Fahr.		Temp. of Dew-Point, Fahr.	Force of Vapor in English Inches.	Weight of Vapor		Humidity, Saturation = 1.000.	Weight in Grains of a Cubic Foot of Air.						
							Height of the Barometer in English Inches.						
Dry.	Wet.			In a Cubic Foot of Air.	Reqd. for Sat'n. of a Cubic Ft. of Air.		in. **28.0**	in. **28.5**	in. **29.0**	in. **29.5**	in. **30.0**	in. **30.5**	in. **31.0**
°	°	°	in.	gr.	gr.		gr.	gr.	gr.	gr.	gr.	gr.	gr.
77	77	77.0	0.910	9.89	0.00	1.000	475.3	483.8	492.3	500.8	509.2	517.7	526.2
	76	75.5	0.868	9.42	0.47	0.953	475.5	484.0	492.5	501.0	509.4	517.9	526.4
	75	74.0	0.827	8.99	0.90	0.909	475.7	484.2	492.7	501.2	509.6	518.1	526.6
	74	72.5	0.787	8.57	1.32	0.867	475.9	484.4	492.9	501.4	509.9	518.4	526.9
	73	71.0	0.751	8.15	1.74	0.824	476.1	484.6	493.1	501.6	510.1	518.6	527.1
	72	69.5	0.715	7.77	2.12	0.786	476.3	484.8	493.3	501.8	510.3	518.8	527.3
	71	68.0	0.681	7.40	2.49	0.748	476.5	485.0	493.5	502.0	510.5	519.0	527.5
	70	66.5	0.648	7.04	2.85	0.712	476.7	485.2	493.7	502.2	510.7	519.2	527.7
	69	65.0	0.617	6.71	3.18	0.678	476.9	485.4	493.9	502.4	510.9	519.4	527.9
	68	63.5	0.588	6.37	3.52	0.644	477.0	485.6	494.1	502.6	511.1	519.6	528.1
	67	62.0	0.559	6.06	3.83	0.613	477.2	485.8	494.3	502.8	511.3	519.8	528.3
	66	60.5	0.532	5.77	4.12	0.583	477.4	486.0	494.5	503.0	511.5	520.0	528.5
	65	59.0	0.506	5.49	4.40	0.556	477.5	486.1	494.6	503.1	511.6	520.1	528.6
	64	57.5	0.481	5.21	4.68	0.527	477.7	486.3	494.8	503.3	511.8	520.3	528.8
	63	56.0	0.458	4.96	4.93	0.501	477.9	486.5	495.0	503.5	512.0	520.5	529.0
	62	54.5	0.435	4.70	5.19	0.476	478.0	486.6	495.1	503.7	512.1	520.6	529.1
	61	53.0	0.414	4.49	5.40	0.454	478.0	486.6	495.1	503.7	512.2	520.7	529.3
	60	51.5	0.393	4.26	5.63	0.431	478.1	486.7	495.2	503.8	512.3	520.8	529.4
	59	50.0	0.373	4.05	5.84	0.410	478.2	486.8	495.3	503.9	512.4	520.9	529.5
	58	48.5	0.355	3.85	6.04	0.389	478.3	486.9	495.4	504.0	512.5	521.0	529.6
	57	47.0	0.337	3.65	6.24	0.369	478.5	487.1	495.6	504.1	512.7	521.2	529.8
	56	45.5	0.320	3.47	6.42	0.351	478.6	487.2	495.7	504.2	512.8	521.3	529.9
	55	44.0	0.304	3.29	6.60	0.333	478.7	487.3	495.8	504.3	512.9	521.4	530.0
	54	42.5	0.288	3.13	6.76	0.317	478.8	487.4	495.9	504.4	513.0	521.5	530.1
	53	41.0	0.274	2.97	6.92	0.301	478.9	487.5	496.0	504.5	513.1	521.6	530.2
	52	39.5	0.260	2.82	7.07	0.285	479.0	487.6	496.1	504.6	513.2	521.7	530.3
	51	38.0	0.246	2.67	7.22	0.270	479.1	487.7	496.2	504.7	513.3	521.8	530.4
	50	36.5	0.234	2.53	7.36	0.256	479.1	487.7	496.2	504.7	513.3	521.8	530.4
78	78	78.0	0.940	10.19	0.00	1.000	474.1	482.5	491.0	499.4	508.0	516.4	524.9
	77	76.5	0.896	9.72	0.47	0.954	474.4	482.9	491.4	499.9	508.3	516.7	525.2
	76	75.0	0.854	9.25	0.94	0.908	474.7	483.2	491.6	500.1	508.6	517.1	525.6
	75	73.5	0.814	8.82	1.37	.0.865	474.9	483.4	491.8	500.3	508.8	517.3	525.8
	74	72.0	0.776	8.40	1.79	0.824	475.2	483.7	492.1	500.6	509.1	517.6	526.1
	73	70.5	0.739	8.00	2.19	0.785	475.4	483.9	492.3	500.8	500.3	517.8	526.3
	72	69.0	0.704	7.62	2.57	0.748	475.6	484.1	492.5	501.0	509.5	518.0	526.5
	71	67.5	0.670	7.25	2.94	0.711	475.8	484.3	492.7	501.2	509.7	518.2	526.7
	70	66.0	0.638	6.91	3.28	0.678	475.9	484.4	492.9	501.4	509.9	518.4	526.9
	69	64.5	0.607	6.58	3.61	0.646	476.1	484.6	493.1	501.6	510.1	518.6	527.1
	68	63.0	0.578	6.26	3.93	0.614	476.3	484.8	493.3	501.8	510.3	518.8	527.3
	67	61.5	0.550	5.96	4.23	0.585	476.4	484.9	493.4	501.9	510.4	518.9	527.4
	66	60.0	0.523	5.66	4.53	0.555	476.6	485.1	493.6	502.1	510.6	519.1	527.6
	65	58.5	0.498	5.38	4.81	0.528	476.8	485.3	493.8	502.3	510.8	519.3	527.8

Reading of Thermometer, Fahr.		Temp. of Dew-Point, Fahr.	Force of Vapor in English Inches.	Weight of Vapor		Humidity, Saturation = 1.000.	Weight in Grains of a Cubic Foot of Air. Height of the Barometer in English Inches.						
Dry.	Wet.			In a Cubic Foot of Air.	Reqd. for Sat'n. of a Cubic Ft. of Air.		in. 28.0	in. 28.5	in. 29.0	in. 29.5	in. 30.0	in. 30.5	in. 31.0
°	°	°	in.	gr.	gr.		gr.	gr.	gr.	gr.	gr.	gr.	gr.
78	65	58.5	0.498	5.38	4.81	0.528	476.8	485.3	493.8	502.3	510.8	519.3	527.8
	64	57.0	0.473	5.12	5.07	0.502	476.8	485.3	493.9	502.4	510.9	519.4	527.9
	63	55.5	0.450	4.88	5.31	0.479	476.9	485.4	494.0	502.5	511.0	519.5	528.0
	62	54.0	0.428	4.63	5.56	0.454	477.1	485.6	494.2	502.7	511.2	519.7	528.2
	61	52.5	0.407	4.40	5.79	0.432	477.2	485.7	494.3	502.8	511.3	519.8	528.3
	60	51.0	0.386	4.18	6.01	0.409	477.3	485.8	494.4	502.9	511.4	519.9	528.4
	59	49.5	0.367	3.98	6.21	0.391	477.4	485.9	494.5	503.0	511.5	520.0	528.5
	58	48.0	0.349	3.78	6.41	0.371	477.5	486.0	494.6	503.1	511.6	520.1	528.6
	57	46.5	0.331	3.59	6.60	0.352	477.6	486.1	494.7	503.2	511.7	520.2	528.7
	56	45.0	0.315	3.41	6.78	0.335	477.8	486.3	494.8	503.3	511.9	520.4	528.9
	55	43.5	0.299	3.24	6.95	0.318	477.9	486.4	494.9	503.4	512.0	520.5	529.0
	54	42.0	0.283	3.07	7.12	0.301	478.0	486.5	495.0	503.5	512.1	520.6	529.1
	53	40.5	0.269	2.92	7.27	0.287	478.1	486.5	495.0	503.5	512.1	520.6	529.1
	52	39.0	0.255	2.77	7.42	0.272	478.2	486.6	495.1	503.6	512.2	520.7	529.2
	51	37.5	0.242	2.63	7.56	0.258	478.3	486.7	495.2	503.7	512.3	520.8	529.3
79	79	79.0	0.970	10.50	0.00	1.000	473.1	481.5	490.0	498.4	506.9	515.3	523.8
	78	77.5	0.925	10.01	0.49	0.953	473.4	481.8	490.3	498.7	507.2	515.6	524.1
	77	76.0	0.882	9.54	0.96	0.909	473.7	482.1	490.6	499.0	507.5	515.9	524.4
	76	74.5	0.840	9.10	1.40	0.867	473.8	482.2	490.7	499.2	507.7	516.2	524.7
	75	73.0	0.801	8.66	1.84	0.825	474.0	482.4	490.9	499.4	507.9	516.4	524.9
	74	71.5	0.763	8.25	2.25	0.786	474.3	482.7	491.2	499.7	508.2	516.7	525.2
	73	70.0	0.727	7.86	2.64	0.749	474.5	482.9	491.4	499.9	508.4	516.9	525.4
	72	68.5	0.692	7.48	3.02	0.712	474.7	483.1	491.6	500.1	508.6	517.1	525.6
	71	67.0	0.659	7.12	3.38	0.678	474.9	483.4	491.9	500.4	508.8	517.3	525.8
	70	65.5	0.628	6.79	3.71	0.647	475.1	483.6	462.1	500.6	509.0	517.5	526.0
	69	64.0	0.597	6.45	4.05	0.614	475.3	483.8	492.3	500.8	509.2	517.7	526.2
	68	62.5	0.568	6.14	4.36	0.585	475.4	483.9	492.4	500.9	509.3	517.8	526.3
	67	61.0	0.541	5.84	4.66	0.556	475.6	484.1	492.6	501.1	509.5	518.0	526.5
	66	59.5	0.515	5.55	4.95	0.529	475.7	484.2	492.7	501.2	509.6	518.1	526.6
	65	58.0	0.489	5.28	5.22	0.503	475.8	484.3	492.8	501.3	509.8	518.3	526.8
	64	56.5	0.465	5.02	5.48	0.478	476.0	484.5	493.0	501.5	510.0	518.5	527.0
	63	55.0	0.442	4.78	5.72	0.455	476.1	484.6	493.1	501.6	510.1	518.6	527.1
	62	53.5	0.421	4.54	5.96	0.432	476.3	484.8	493.3	501.8	510.3	518.8	527.3
	61	52.0	0.400	4.31	6.19	0.410	476.4	484.9	493.4	501.9	510.4	518.9	527.4
	60	50.5	0.380	4.10	6.40	0.390	476.5	485.0	493.5	502.0	510.5	519.0	527.5
	59	49.0	0.361	3.90	6.60	0.371	476.6	485.1	493.6	502.1	510.6	519.1	527.6
	58	47.5	0.343	3.71	6.79	0.353	476.7	485.2	493.7	502.2	510.7	519.2	527.7
	57	46.0	0.326	3.52	6.98	0.335	476.8	485.3	493.8	502.3	510.8	519.3	527.8
	56	44.5	0.309	3.34	7.16	0.318	476.9	485.4	493.9	502.4	510.9	519.4	527.9
	55	43.0	0.293	3.17	7.33	0.301	477.0	485.5	494.0	502.5	511.0	519.5	528.0
	54	41.5	0.279	3.01	7.49	0.287	477.1	485.6	494.1	502.6	511.1	519.6	528.1
	53	40.0	0.264	2.86	7.64	0.272	477.2	485.7	494.2	502.7	511.2	519.7	528.2
	52	38.5	0.251	2.72	7.78	0.260	477.3	485.8	494.3	502.8	511.3	519.8	528.3

Reading of Thermometer, Fahr.		Temp. of Dew-Point, Fahr.	Force of Vapor in English Inches.	Weight of Vapor		Humidity, Saturation = 1.000.	Weight in Grains of a Cubic Foot of Air. Height of the Barometer in English Inches.						
Dry.	Wet.			In a Cubic Foot of Air.	Reqd. for Sat'n. of a Cubic Ft. of Air.		in. 28.0	in. 28.5	in. 29.0	in. 29.5	in. 30.0	in. 30.5	in. 31.0
°	°	°	in.	gr.	gr.		gr.	gr.	gr.	gr.	gr.	gr.	gr
80	80	80.0	1.001	10.81	0.00	1.000	472.0	480.4	488.9	497.3	505.7	514.1	522.6
	79	78.5	0.955	10.31	0.50	0.954	472.3	480.7	489.1	497.5	506.0	514.4	522.9
	78	77.0	0.910	9.83	0.98	0.909	472.5	480.9	489.4	497.9	506.3	514.7	523.2
	77	75.5	0.868	9.37	1.44	0.867	472.7	481.1	489.6	498.1	506.5	514.9	523.4
	76	74.0	0.827	8.93	1.88	0.826	473.0	481.4	489.9	498.4	506.8	515.2	523.7
	75	72.5	0.787	8.50	2.31	0.786	473.2	481.6	490.1	498.6	507.0	515.4	523.9
	74	71.0	0.751	8.11	2.70	0.750	473.4	481.8	490.3	498.8	507.2	515.6	524.1
	73	69.5	0.715	7.71	3.10	0.713	473.6	482.1	490.6	499.1	507.5	515.9	524.4
	72	68.0	0.681	7.35	3.46	0.680	473.8	482.3	490.8	499.3	507.7	516.1	524.6
	71	66.5	0.648	6.99	3.82	0.647	474.0	482.5	491.0	499.5	507.9	516.3	524.8
	70	65.0	0.617	6.66	4.15	0.616	474.2	482.7	491.2	499.7	508.1	516.5	525.0
	69	63.5	0.588	6.33	4.48	0.586	474.4	482.9	491.4	499.9	508.3	516.7	525.2
	68	62.0	0.559	6.03	4.78	0.558	474.5	483.0	491.5	500.0	508.4	516.8	525.3
	67	60.5	0.532	5.74	5.07	0.531	474.7	483.2	491.7	500.2	508.6	517.0	525.5
	66	59.0	0.506	5.45	5.36	0.504	474.9	483.4	491.9	500.4	508.8	517.2	525.7
	65	57.5	0.481	5.18	5.63	0.479	475.0	483.5	492.0	500.5	508.9	517.3	525.8
	64	56.0	0.458	4.93	5.96	0.456	475.2	483.7	492.2	500.7	509.1	517.5	526.0
	63	54.5	0.435	4.69	6.12	0.434	475.3	483.8	492.3	500.8	509.2	517.6	526.1
	62	53.0	0.414	4.46	6.35	0.413	475.4	483.9	492.4	500.9	509.3	517.7	526.2
	61	51.5	0.393	4.23	6.58	0.391	475.5	484.0	492.5	501.0	509.4	517.8	526.3
	60	50.0	0.373	4.02	6.79	0.372	475.6	484.1	492.6	501.1	509.5	517.9	526.4
	59	48.5	0.355	3.82	6.99	0.353	475.7	484.2	492.7	501.2	509.6	518.0	526.5
	58	47.0	0.337	3.63	7.18	0.336	475.9	484.4	492.9	501.4	509.8	518.2	526.7
	57	45.5	0.320	3.45	7.36	0.319	476.0	484.5	493.1	501.5	509.9	518.3	526.8
	56	44.0	0.304	3.27	7.54	0.302	476.1	484.6	493.2	501.6	510.0	518.4	526.9
	55	42.5	0.288	3.11	7.70	0.288	476.2	484.7	493.3	501.7	510.1	518.5	527.0
	54	41.0	0.274	2.96	7.85	0.274	476.3	484.8	493.4	501.8	510.2	518.6	527.1
	53	39.5	0.260	2.82	7.99	0.261	476.3	484.8	493.4	501.8	510.2	518.6	527.1
81	81	81.0	1.034	11.14	0.00	1.000	471.0	479.4	487.8	496.2	504.6	513.0	521.4
	80	79.5	0.986	10.62	0.52	0.953	471.3	479.7	488.1	496.5	504.9	513.3	521.7
	79	78.0	0.940	10.13	1.01	0.910	471.5	479.9	488.4	496.8	505.2	513.6	522.1
	78	76.5	0.896	9.65	1.49	0.866	471.7	480.1	488.6	497.0	505.4	513.8	522.3
	77	75.0	0.854	9.20	1.94	0.826	472.0	480.4	488.9	497.3	505.7	514.1	522.6
	76	73.5	0.814	8.77	2.37	0.787	472.2	480.6	489.1	497.5	505.9	514.3	522.8
	75	72.0	0.776	8.35	2.79	0.750	472.5	480.9	489.4	497.8	506.2	514.6	523.1
	74	70.5	0.739	7.95	3.19	0.713	472.6	481.0	489.5	497.9	506.4	514.8	523.3
	73	69.0	0.704	7.57	3.57	0.680	472.8	481.2	489.7	498.1	506.6	515.0	523.5
	72	67.5	0.670	7.21	3.93	0.647	473.0	481.4	489.9	498.3	506.8	515.2	523.7
	71	66.0	0.638	6.87	4.27	0.617	473.2	481.6	490.1	498.5	507.0	515.4	523.9
	70	64.5	0.607	6.54	4.60	0.587	473.4	481.8	490.3	498.7	507.2	515.6	524.1
	69	63.0	0.578	6.22	4.92	0.558	473.6	482.0	490.5	498.9	507.4	515.8	524.3
	68	61.5	0.550	5.92	5.22	0.531	473.7	482.2	490.7	499.1	507.6	516.0	524.5

Reading of Thermometer, Fahr.		Temp. of Dew-Point, Fahr.	Force of Vapor in English Inches.	Weight of Vapor		Humidity, Saturation = 1.000.	Weight in Grains of a Cubic Foot of Air.						
							Height of the Barometer in English Inches.						
Dry.	Wet.			In a Cubic Foot of Air.	Reqd. for Sat'n. of a Cubic Ft. of Air.		in. 28.0	in. 28.5	in. 29.0	in. 29.5	in. 30.0	in. 30.5	in. 31.0
°	°	°	in.	gr.	gr.		gr.	gr.	gr.	gr.	gr.	gr.	gr.
81	68	61.5	0.550	5.92	5.22	0.531	473.7	482.2	490.7	499.1	507.6	516.0	524.5
	67	60.0	0.523	5.62	5.52	0.505	473.8	482.3	490.8	499.2	507.7	516.1	524.6
	66	58.5	0.498	5.31	5.83	0.477	474.0	482.5	491.0	499.4	507.9	516.3	524.8
	65	57.0	0.473	5.08	6.06	0.456	474.1	482.6	491.1	499.5	508.0	516.4	524.9
	64	55.5	0.450	4.84	6.30	0.434	474.3	482.8	491.3	499.7	508.2	516.6	525.1
	63	54.0	0.428	4.60	6.54	0.413	474.4	482.9	491.4	499.8	508.3	516.7	525.2
	62	52.5	0.407	4.37	6.77	0.392	474.5	483.0	491.5	499.9	508.4	516.8	525.3
	61	51.0	0.386	4.15	6.99	0.373	474.6	483.1	491.6	500.0	508.5	516.9	525.4
	60	49.5	0.367	3.95	7.19	0.355	474.7	483.2	491.7	500.1	508.6	517.0	525.5
	59	48.0	0.349	3.75	7.39	0.337	474.9	483.4	491.9	500.3	508.8	517.2	525.7
	58	46.5	0.331	3.56	7.58	0.320	475.0	483.5	492.0	500.4	508.9	517.3	525.8
	57	45.0	0.315	3.38	7.76	0.303	475.1	483.6	492.1	500.5	509.0	517.4	525.9
	56	43.5	0.299	3.21	7.93	0.289	475.2	483.7	492.2	500.6	509.1	517.5	526.0
	55	42.0	0.283	3.05	8.09	0.274	475.3	483.8	492.3	500.7	509.2	517.6	526.1
	54	40.5	0.269	2.90	8.24	0.260	475.3	483.8	492.3	500.7	509.2	517.6	526.1
82	82	82.0	1.067	11.47	0.00	1.000	470.0	478.4	486.8	495.2	503.5	511.9	520.3
	81	80.5	1.017	10.94	0.53	0.954	470.3	478.7	487.0	495.4	503.8	512.2	520.6
	80	79.0	0.970	10.44	1.03	0.910	470.6	479.0	487.3	495.7	504.1	512.5	520.9
	79	77.5	0.925	9.95	1.52	0.868	470.7	479.1	487.5	495.9	504.3	512.7	521.1
	78	76.0	0.882	9.49	1.98	0.827	471.0	479.4	487.8	496.2	504.6	513.0	521.4
	77	74.5	0.840	9.03	2.44	0.787	471.2	479.6	488.0	496.4	504.8	513.2	521.6
	76	73.0	0.801	8.60	2.87	0.750	471.5	479.9	488.3	496.7	505.1	513.5	521.9
	75	71.5	0.763	8.19	3.28	0.714	471.6	480.0	488.5	496.9	505.3	513.7	522.1
	74	70.0	0.727	7.81	3.66	0.681	471.8	480.2	488.6	497.1	505.5	513.9	522.4
	73	68.5	0.692	7.43	4.04	0.648	472.0	480.4	488.8	497.3	505.7	514.1	522.6
	72	67.0	0.659	7.08	4.39	0.618	472.2	480.6	489.0	497.5	505.9	514.3	522.8
	71	65.5	0.628	6.75	4.72	0.588	472.4	480.8	489.2	497.7	506.1	514.5	523.0
	70	64.0	0.597	6.41	5.06	0.559	472.5	480.8	489.4	497.9	506.3	514.7	523.2
	69	62.5	0.568	6.10	5.37	0.532	472.6	481.0	489.5	498.0	506.4	514.8	523.3
	68	61.0	0.541	5.81	5.66	0.507	472.8	481.2	489.7	498.2	506.6	515.0	523.5
	67	59.5	0.515	5.52	5.95	0.481	473.0	481.4	489.9	498.4	506.8	515.2	523.7
	66	58.0	0.489	5.25	6.22	0.458	473.1	481.5	490.0	498.5	506.9	515.3	523.8
	65	56.5	0.465	4.99	6.48	0.435	473.2	481.6	490.1	498.6	507.0	515.4	523.9
	64	55.0	0.442	4.75	6.72	0.414	473.4	481.8	490.3	498.8	507.2	515.6	524.1
	63	53.5	0.421	4.51	6.96	0.393	473.5	482.0	490.5	499.0	507.4	515.8	524.3
	62	52.0	0.400	4.29	7.18	0.374	473.6	482.1	490.6	499.1	507.5	515.9	524.4
	61	50.5	0.380	4.08	7.39	0.356	473.7	482.2	490.7	499.2	507.6	516.0	524.4
	60	49.0	0.361	3.87	7.60	0.337	473.8	482.3	490.8	499.3	507.7	516.1	524.5
	59	47.5	0.343	3.68	7.79	0.320	473.9	482.4	490.9	499.4	507.8	516.2	524.6
	58	46.0	0.326	3.50	7.97	0.305	474.0	482.5	491.0	499.5	507.9	516.3	524.7
	57	44.5	0.309	3.32	8.15	0.289	474.1	482.6	491.1	499.6	508.0	516.4	524.8
	56	43.0	0.293	3.15	8.32	0.274	474.2	482.7	491.2	499.7	508.1	516.5	524.9
	55	41.5	0.279	2.99	8.48	0.260	474.3	482.8	491.3	499.8	508.2	516.6	525.1

Reading of Thermometer, Fahr.		Temp. of Dew-Point, Fahr.	Force of Vapor in English Inches.	Weight of Vapor		Humidity, Saturation = 1.000.	Weight in Grains of a Cubic Foot of Air. Height of the Barometer in English Inches.						
Dry.	Wet.			In a Cubic Foot of Air.	Reqd. for Sat'n. of a Cubic Ft. of Air.		in. **28.0**	in. **28.5**	in. **29.0**	in. **29.5**	in. **30.0**	in. **30.5**	in. **31.0**
°	°	°	in.	gr.	gr.		gr.	gr.	gr.	gr.	gr.	gr.	gr.
83	83	83.0	1.101	11.82	0.00	1.000	468.8	477.2	485.5	493.9	502.3	510.6	519.0
	82	81.5	1.050	11.27	0.55	0.953	469.1	477.5	485.8	494.2	502.6	511.0	519.4
	81	80.0	1.001	10.75	1.07	0.909	469.4	477.8	486.1	494.5	502.9	511.3	519.7
	80	78.5	0.955	10.25	1.57	0.868	469.7	478.1	486.4	494.8	503.2	511.6	520.0
	79	77.0	0.910	9.78	2.04	0.828	470.0	478.4	486.7	495.1	503.5	511.9	520.3
	78	75.5	0.868	9.31	2.51	0.786	470.3	478.7	487.0	495.4	503.8	512.2	520.6
	77	74.0	0.827	8.88	2.94	0.751	470.5	478.9	487.2	495.6	504.0	512.4	520.8
	76	72.5	0.787	8.45	3.37	0.715	470.6	479.0	487.4	495.8	504.2	512.6	521.0
	75	71.0	0.751	8.05	3.77	0.681	470.8	479.2	487.6	496.0	504.4	512.8	521.2
	74	69.5	0.715	7.66	4.16	0.647	471.0	479.4	487.8	496.2	504.6	513.0	521.4
	73	68.0	0.681	7.30	4.52	0.618	471.2	479.6	488.0	496.4	504.8	513.2	521.6
	72	66.5	0.648	6.95	4.87	0.588	471.4	479.8	488.2	496.6	505.0	513.4	521.8
	71	65.0	0.617	6.62	5.20	0.560	471.6	480.0	488.4	496.8	505.2	513.6	522.0
	70	63.5	0.588	6.29	5.53	0.533	471.7	480.1	488.5	497.0	505.4	513.8	522.3
	69	62.0	0.559	5.99	5.83	0.507	471.9	480.3	488.7	497.2	505.6	514.0	522.5
	68	60.5	0.532	5.70	6.12	0.482	472.0	480.4	488.8	497.3	505.7	514.1	522.6
	67	59.0	0.506	5.42	6.40	0.459	472.2	480.6	489.0	497.5	505.9	514.3	522.8
	66	57.5	0.481	5.15	6.67	0.435	472.4	480.8	489.2	497.7	506.1	514.5	523.0
	65	56.0	0.458	4.90	6.92	0.414	472.4	480.8	489.3	497.8	506.2	514.6	523.1
	64	54.5	0.435	4.66	7.18	0.394	472.5	480.9	489.4	497.9	506.3	514.7	523.2
	63	53.0	0.414	4.43	7.39	0.375	472.7	481.1	489.6	498.1	506.5	514.9	523.4
	62	51.5	0.393	4.21	7.61	0.356	472.8	481.2	489.7	498.2	506.6	515.0	523.5
	61	50.0	0.373	4.00	7.82	0.339	472.9	481.3	489.8	498.3	506.7	515.1	523.6
	60	48.5	0.355	3.80	8.02	0.322	473.1	481.4	489.9	498.4	506.8	515.2	523.7
	59	47.0	0.337	3.60	8.22	0.305	473.2	481.5	490.0	498.5	506.9	515.3	523.8
	58	45.5	0.320	3.42	8.40	0.289	473.3	481.6	490.1	498.6	507.0	515.4	523.9
	57	44.0	0.304	3.25	8.57	0.276	473.4	481.7	490.2	498.7	507.1	515.5	524.0
	56	42.5	0.288	3.09	8.73	0.261	473.5	481.8	490.3	498.8	507.2	515.6	524.1
84	84	84.0	1.136	12.17	0.00	1.000	467.8	476.2	484.5	492.7	501.2	509.6	517.9
	83	82.5	1.083	11.61	0.56	0.954	468.1	476.4	484.8	493.2	501.5	509.8	518.2
	82	81.0	1.034	11.07	1.10	0.910	468.4	476.7	485.1	493.5	501.8	510.1	518.5
	81	79.5	0.986	10.55	1.62	0.867	468.6	476.9	485.4	493.7	502.1	510.5	518.8
	80	78.0	0.940	10.07	2.10	0.827	468.9	477.3	485.7	494.0	502.4	510.8	519.1
	79	76.5	0.896	9.59	2.58	0.788	469.1	477.5	485.9	494.2	502.6	511.0	519.3
	78	75.0	0.854	9.14	3.03	0.751	469.4	477.8	486.1	494.5	502.9	511.3	519.7
	77	73.5	0.814	8.71	3.46	0.716	469.6	478.0	486.3	494.7	503.1	511.5	519.9
	76	72.0	0.776	8.30	3.87	0.682	469.8	478.2	486.5	494.9	503.3	511.7	520.1
	75	70.5	0.739	7.90	4.27	0.649	470.1	478.5	486.8	495.2	503.6	512.0	520.4
	74	69.0	0.704	7.53	4.64	0.619	470.3	478.7	487.0	495.4	503.8	512.2	520.6
	73	67.5	0.670	7.17	5.00	0.589	470.5	478.9	487.2	495.6	504.0	512.4	520.8
	72	66.0	0.638	6.83	5.34	0.561	470.6	479.0	487.4	495.8	504.2	512.6	521.0
	71	64.5	0.607	6.50	5.67	0.534	470.7	479.1	487.5	495.9	504.3	512.7	521.1

Reading of Thermometer, Fahr.		Temp. of Dew-Point, Fahr.	Force of Vapor in English Inches.	Weight of Vapor		Humidity, Saturation = 1.000.	Weight in Grains of a Cubic Foot of Air.						
							Height of the Barometer in English Inches.						
Dry.	Wet.			In a Cubic Foot of Air.	Reqd. for Sat'n. of a Cubic Ft. of Air.		in. **28.0**	in. **28.5**	in. **29.0**	in. **29.5**	in. **30.0**	in. **30.5**	in. **31.0**
°	°	°	in.	gr.	gr.		gr.	gr.	gr.	gr.	gr.	gr.	gr.
84	71	64.5	0.607	6.50	5.67	0.534	470.7	479.1	487.5	495.9	504.3	512.7	521.1
	70	63.0	0.578	6.18	5.99	0.508	470.9	479.3	487.7	496.1	504.5	512.9	521.3
	69	61.5	0.550	5.87	6.30	0.482	471.1	479.5	487.9	496.3	504.7	513.1	521.5
	68	60.0	0.523	5.59	6.58	0.459	471.2	479.6	488.0	496.4	504.8	513.2	521.6
	67	58.5	0.498	5.31	6.86	0.436	471.4	479.8	488.2	496.6	505.0	513.4	521.8
	66	57.0	0.473	5.05	7.12	0.415	471.6	480.0	488.3	496.7	505.2	513.6	522.1
	65	55.5	0.450	4.81	7.36	0.395	471.6	480.0	488.4	496.8	505.3	513.7	522.2
	64	54.0	0.428	4.57	7.60	0.375	471.7	480.1	488.5	496.9	505.4	513.8	522.3
	63	52.5	0.407	4.35	7.82	0.357	471.8	480.2	488.6	497.0	505.5	513.9	522.4
	62	51.0	0.386	4.13	8.04	0.339	471.9	480.4	488.8	497.2	505.7	514.0	522.5
	61	49.5	0.367	3.93	8.24	0.323	472.1	480.5	488.9	497.3	505.8	514.1	522.6
	60	48.0	0.349	3.73	8.44	0.306	472.2	480.6	489.0	497.4	505.9	514.2	522.7
	59	46.5	0.331	3.55	8.62	0.292	472.3	480.7	489.1	497.5	506.0	514.3	522.8
	58	45.0	0.315	3.37	8.80	0.277	472.4	480.8	489.2	497.6	506.1	514.4	522.9
	57	43.5	0.299	3.20	8.97	0.263	472.5	480.9	489.3	497.7	506.2	514.5	523.0
85	85	85.0	1.171	12.53	0.00	1.000	466.8	475.2	483.5	491.8	500.1	508.5	516.8
	84	83.5	1.118	11.95	0.58	0.954	467.1	475.4	483.7	492.1	500.4	508.7	517.1
	83	82.0	1.067	11.40	1.13	0.910	467.3	475.6	484.0	492.4	500.7	509.0	517.4
	82	80.5	1.017	10.87	1.66	0.868	467.6	475.9	484.3	492.7	501.0	509.3	517.7
	81	79.0	0.970	10.38	2.15	0.829	467.8	476.1	484.5	492.9	501.2	509.5	517.9
	80	77.5	0.925	9.89	2.64	0.789	468.1	476.4	484.8	493.2	501.5	509.8	518.2
	79	76.0	0.882	9.43	3.10	0.753	468.4	476.7	485.1	493.5	501.8	510.1	518.5
	78	74.5	0.840	8.98	3.55	0.717	468.6	476.9	485.3	493.7	502.0	510.3	518.7
	77	73.0	0.801	8.55	3.98	0.682	468.7	477.1	485.5	493.9	502.2	510.5	518.9
	76	71.5	0.763	8.15	4.38	0.650	469.0	477.4	485.8	494.2	502.5	510.8	519.2
	75	70.0	0.727	7.76	4.77	0.619	469.2	477.6	486.0	494.4	502.7	511.0	519.4
	74	68.5	0.692	7.39	5.14	0.589	469.4	477.8	486.2	494.6	502.9	511.2	519.6
	73	67.0	0.659	7.04	5.49	0.562	469.7	478.1	486.5	494.9	503.2	511.5	519.9
	72	65.5	0.628	6.71	5.82	0.536	469.9	478.3	486.7	495.1	503.4	511.7	520.1
	71	64.0	0.597	6.37	6.16	0.508	470.1	478.5	486.9	495.3	503.6	511.9	520.3
	70	62.5	0.568	6.07	6.46	0.484	470.3	478.7	487.1	495.5	503.8	512.1	520.5
	69	61.0	0.541	5.77	6.76	0.460	470.5	478.9	487.2	495.6	504.0	512.4	520.8
	68	59.5	0.515	5.48	7.05	0.437	470.6	479.0	487.3	495.7	504.1	513.5	520.9
	67	58.0	0.489	5.21	7.32	0.415	470.6	479.0	487.4	495.8	504.2	512.6	521.0
	66	56.5	0.465	4.96	7.57	0.396	470.7	479.1	487.5	495.9	504.3	512.7	521.1
	65	55.0	0.442	4.72	7.81	0.377	470.8	479.2	487.6	496.0	504.4	512.8	521.2
	64	53.5	0.421	4.49	8.04	0.359	470.9	479.3	487.7	496.1	504.5	512.9	521.3
	63	52.0	0.400	4.26	8.27	0.340	471.1	479.5	487.9	496.3	504.7	513.1	521.5
	62	50.5	0.380	4.05	8.48	0.323	471.2	479.6	488.1	496.4	504.8	513.2	521.6
	61	49.0	0.361	3.85	8.68	0.307	471.3	479.7	488.2	496.5	504.9	513.3	521.7
	60	47.5	0.343	3.66	8.87	0.292	471.4	479.8	488.3	496.6	505.0	513.4	521.8
	59	46.0	0.326	3.48	9.05	0.278	471.5	479.9	488.4	496.7	505.1	513.5	521.9
	58	44.5	0.309	3.31	9.22	0.264	471.6	480.1	488.5	496.8	505.2	513.6	522.1

Reading of Thermometer, Fahr. Dry.	Wet.	Temp. of Dew-Point, Fahr.	Force of Vapor in English Inches.	Weight of Vapor: In a Cubic Foot of Air.	Weight of Vapor: Reqd. for Sat'n. of a Cubic Ft. of Air.	Humidity, Saturation = 1.000.	Weight in Grains of a Cubic Foot of Air. Height of the Barometer in English Inches. in. **28.0**	in. **28.5**	in. **29.0**	in. **29.5**	in. **30.0**	in. **30.5**	in. **31.0**
°	°	°	in.	gr.	gr.		gr.	gr.	gr.	gr.	gr.	gr.	gr.
86	86	86.0	1.209	12.91	0.00	1.000	465.7	474.0	482.3	490.6	498.9	507.2	515.5
	85	84.5	1.153	12.31	0.60	0.954	466.0	474.3	482.6	490.9	499.2	507.5	515.8
	84	83.0	1.101	11.75	1.16	0.910	466.3	474.6	482.9	491.2	499.5	507.8	516.1
	83	81.5	1.050	11.20	1.71	0.868	466.5	474.8	483.2	491.5	499.8	508.1	516.5
	82	80.0	1.001	10.69	2.22	0.828	466.8	475.1	483.5	491.8	500.1	508.4	516.8
	81	78.5	0.955	10.19	2.72	0.789	467.1	475.4	483.8	492.1	500.4	508.7	517.1
	80	77.0	0.910	9.71	3.20	0.752	467.3	475.6	484.0	492.3	500.7	509.0	517.4
	79	75.5	0.868	9.25	3.66	0.717	467.5	475.8	484.2	492.5	500.9	509.2	517.6
	78	74.0	0.827	8.82	4.09	0.683	467.8	476.1	484.5	492.8	501.2	509.5	517.9
	77	72.5	0.787	8.40	4.51	0.651	468.0	476.3	484.7	493.0	501.4	509.7	518.1
	76	71.0	0.751	8.00	4.91	0.619	468.2	476.5	484.9	493.2	501.6	509.9	518.3
	75	69.5	0.715	7.62	5.29	0.590	468.3	476.6	485.0	493.4	501.8	510.2	518.6
	74	68.0	0.681	7.26	5.65	0.562	468.5	476.8	485.2	493.6	502.0	510.4	518.8
	73	66.5	0.648	6.91	6.00	0.535	468.8	477.1	485.5	493.9	502.2	510.6	519.0
	72	65.0	0.617	6.58	6.33	0.509	468.9	477.2	485.6	494.0	502.4	510.8	519.2
	71	63.5	0.588	6.26	6.65	0.485	469.1	477.4	485.8	494.2	502.6	511.0	519.4
	70	62.0	0.559	5.95	6.96	0.461	469.2	477.5	485.9	494.3	502.7	511.1	519.5
	69	60.5	0.532	5.66	7.25	0.438	469.4	477.7	486.1	494.5	502.9	511.3	519.7
	68	59.0	0.506	5.38	7.53	0.417	469.6	477.9	486.3	494.7	503.1	511.5	519.9
	67	57.5	0.481	5.11	7.80	0.396	469.8	478.1	486.5	494.9	503.3	511.7	520.1
	66	56.0	0.458	4.87	8.04	0.377	469.9	478.2	486.6	495.0	503.4	511.8	520.2
	65	54.5	0.435	4.63	8.28	0.359	470.0	478.3	486.7	495.1	503.5	511.9	520.3
	64	53.0	0.414	4.40	8.51	0.341	470.1	478.4	486.8	495.1	503.6	512.0	520.4
	63	51.5	0.393	4.19	8.72	0.325	470.2	478.5	486.9	495.2	503.7	512.1	520.5
	62	50.0	0.373	3.98	8.93	0.308	470.4	478.7	487.1	495.4	503.9	512.2	520.7
	61	48.5	0.355	3.78	9.13	0.293	470.5	478.8	487.2	495.5	504.0	512.3	520.8
	60	47.0	0.337	3.59	9.32	0.278	470.6	478.9	487.3	495.6	504.1	512.4	520.9
	59	45.5	0.320	3.40	9.51	0.263	470.7	479.0	487.4	495.7	504.2	512.5	521.0
87	87	87.0	1.247	13.29	0.00	1.000	464.5	472.8	481.1	489.4	497.7	506.0	514.3
	86	85.5	1.190	12.68	0.61	0.954	464.8	473.1	481.4	489.7	498.0	506.3	514.6
	85	84.0	1.136	12.10	1.19	0.910	465.1	473.4	481.7	490.0	498.3	506.6	514.9
	84	82.5	1.083	11.54	1.75	0.868	465.4	473.7	482.0	490.3	498.6	506.9	515.2
	83	81.0	1.034	11.01	2.28	0.828	465.7	474.0	482.3	490.6	498.9	507.2	515.5
	82	79.5	0.986	10.49	2.80	0.789	466.0	474.3	482.6	490.9	499.2	507.5	515.8
	81	78.0	0.940	10.01	3.28	0.753	466.3	474.6	482.9	491.2	499.5	507.8	516.1
	80	76.5	0.896	9.54	3.75	0.718	466.5	474.8	483.1	491.4	499.8	508.1	516.5
	79	75.0	0.854	9.09	4.20	0.684	466.8	475.1	483.5	491.8	500.1	508.4	516.8
	78	73.5	0.814	8.66	4.63	0.652	467.0	475.3	483.7	492.0	500.3	508.6	517.0
	77	72.0	0.776	8.24	5.05	0.620	467.2	475.5	483.9	492.2	500.5	508.8	517.2
	76	70.5	0.739	7.85	5.44	0.591	467.3	475.6	484.0	492.3	500.7	509.0	517.4
	75	69.0	0.704	7.48	5.81	0.563	467.5	475.8	484.2	492.5	500.9	509.2	517.6
	74	67.5	0.670	7.12	6.17	0.536	467.7	476.0	484.4	492.7	501.1	509.4	517.8

Reading of Thermometer, Fahr.		Temp. of Dew-Point, Fahr.	Force of Vapor in English Inches.	Weight of Vapor		Humidity, Saturation = 1.000.	Weight in Grains of a Cubic Foot of Air. Height of the Barometer in English Inches.						
Dry.	Wet.			In a Cubic Foot of Air.	Reqd. for Sat'n. of a Cubic Ft. of Air.		in. **28.0**	in. **28.5**	in. **29.0**	in. **29.5**	in. **30.0**	in. **30.5**	in. **31.0**
°	°	°	in.	gr.	gr.		gr.	gr.	gr.	gr.	gr.	gr.	gr.
87	74	67.5	0.670	7.12	6.17	0.536	467.7	476.0	484.4	492.7	501.1	509.4	517.8
	73	66.0	0.638	6.78	6.51	0.510	467.9	476.2	484.6	492.9	501.3	509.6	518.0
	72	64.5	0.607	6.46	6.83	0.486	468.1	476.4	484.8	493.1	501.5	509.8	518.2
	71	63.0	0.578	6.14	7.15	0.462	468.3	476.6	485.0	493.3	501.7	510.1	518.5
	70	61.5	0.550	5.85	7.44	0.440	468.4	476.7	485.1	493.5	501.9	510.3	518.7
	69	60.0	0.523	5.56	7.73	0.418	468.5	476.9	485.3	493.7	502.0	510.4	518.8
	68	58.5	0.498	5.28	8.01	0.397	468.7	477.1	485.5	493.9	502.2	510.6	519.0
	67	57.0	0.473	5.02	8.27	0.378	468.8	477.2	485.6	494.0	502.3	510.7	519.1
	66	55.5	0.450	4.77	8.52	0.359	468.9	477.3	485.7	494.1	502.4	510.7	519.2
	65	54.0	0.428	4.54	8.75	0.342	469.1	477.5	485.9	494.3	502.6	510.9	519.4
	64	52.5	0.407	4.33	8.96	0.326	469.2	477.6	486.1	494.4	502.7	511.0	519.5
	63	51.0	0.386	4.12	9.17	0.310	469.3	477.7	486.2	494.5	502.8	511.1	519.6
	62	49.5	0.367	3.91	9.38	0.294	469.4	477.8	486.3	494.6	502.9	511.2	519.7
	61	48.0	0.349	3.71	9.58	0.279	469.6	477.9	486.5	494.8	503.1	511.4	519.9
	60	46.5	0.331	3.51	9.78	0.264	469.7	478.1	486.6	494.9	503.2	511.5	520.0
88	88	88.0	1.286	13.68	0.00	1.000	463.5	471.7	480.0	488.3	496.6	504.8	513.1
	87	86.5	1.228	13.06	0.62	0.954	463.8	472.0	480.3	488.6	496.9	505.1	513.4
	86	85.0	1.171	12.46	1.22	0.911	464.2	472.4	480.7	489.0	497.3	505.6	513.9
	85	83.5	1.118	11.88	1.80	0.868	464.4	472.7	481.0	489.3	497.6	505.9	514.2
	84	82.0	1.067	11.34	2.34	0.829	464.7	473.0	481.3	489.6	497.9	506.2	514.5
	83	80.5	1.017	10.81	2.87	0.790	465.0	473.3	481.6	489.9	498.2	506.5	514.8
	82	79.0	0.970	10.31	3.37	0.754	465.2	473.5	481.8	490.1	498.4	506.7	515.0
	81	77.5	0.925	9.83	3.85	0.718	465.5	473.8	482.1	490.4	498.7	507.0	515.3
	80	76.0	0.882	9.37	4.31	0.685	465.8	474.1	482.4	490.7	499.0	507.3	515.6
	79	74.5	0.840	8.93	4.75	0.653	466.1	474.4	482.7	491.0	499.3	507.6	515.9
	78	73.0	0.801	8.50	5.18	0.621	466.3	474.6	482.9	491.2	499.5	507.8	516.2
	77	71.5	0.763	8.09	5.59	0.591	466.4	474.7	483.0	491.3	499.7	508.0	516.4
	76	70.0	0.727	7.71	5.97	0.563	466.6	474.9	483.2	491.5	499.9	508.2	516.6
	75	68.5	0.692	7.34	6.34	0.537	466.8	475.1	483.4	491.7	500.1	508.4	516.8
	74	67.0	0.659	6.99	6.69	0.511	467.0	475.3	483.6	491.9	500.3	508.6	517.0
	73	65.5	0.628	6.66	7.02	0.487	467.2	475.5	483.8	492.1	500.5	508.8	517.2
	72	64.0	0.597	6.33	7.35	0.463	467.4	475.7	484.0	492.3	500.7	509.0	517.4
	71	62.5	0.568	6.03	7.65	0.441	467.4	475.7	484.0	492.4	500.8	509.1	517.5
	70	61.0	0.541	5.74	7.94	0.420	467.6	475.9	484.2	492.6	501.0	509.3	517.7
	69	59.5	0.515	5.45	8.23	0.398	467.7	476.0	484.3	492.7	501.2	509.4	517.8
	68	58.0	0.489	5.18	8.50	0.378	467.9	476.2	484.5	492.9	501.3	509.6	518.0
	67	56.5	0.465	4.93	8.75	0.359	468.1	476.4	484.7	493.1	501.5	509.8	518.2
	66	55.0	0.442	4.69	8.99	0.342	468.2	476.5	484.8	493.2	501.6	509.9	518.3
	65	53.5	0.421	4.47	9.21	0.326	468.3	476.6	484.9	493.3	501.7	510.0	518.4
	64	52.0	0.400	4.25	9.43	0.310	468.4	476.7	485.1	493.4	501.8	510.1	518.5
	63	50.5	0.380	4.04	9.64	0.295	468.6	476.9	485.3	493.6	502.0	510.3	518.7
	62	49.0	0.361	3.83	9.85	0.280	468.7	477.1	485.4	493.7	502.1	510.4	518.8
	61	47.5	0.343	3.62	10.06	0.265	468.8	477.2	485.5	493.8	502.2	510.5	518.9

Reading of Thermometer, Fahr. Dry.	Reading of Thermometer, Fahr. Wet.	Temp. of Dew-Point, Fahr.	Force of Vapor in English Inches.	Weight of Vapor: In a Cubic Foot of Air.	Weight of Vapor: Reqd. for Sat'n. of a Cubic Ft. of Air.	Humidity, Saturation = 1.000.	Weight in Grains of a Cubic Foot of Air. Height of the Barometer in English Inches. in. 28.0	in. 28.5	in. 29.0	in. 29.5	in. 30.0	in. 30.5	in. 31.0
°	°	°	in.	gr.	gr		gr.	gr.	gr.	gr.	gr.	gr.	gr.
89	89	89.0	1.326	14.08	0.00	1.000	462.4	470.6	478.9	487.1	495.4	503.6	511.9
	88	87.5	1.266	13.44	0.64	0.954	462.7	470.9	479.2	487.4	495.7	503.9	512.2
	87	86.0	1.209	12.84	1.24	0.912	463.0	471.2	479.5	487.8	496.1	504.4	512.7
	86	84.5	1.153	12.24	1.84	0.869	463.3	471.5	479.8	488.1	496.4	504.7	513.0
	85	83.0	1.101	11.68	2.40	0.830	463.6	471.8	480.1	488.4	496.7	505.0	513.3
	84	81.5	1.050	11.13	2.95	0.791	464.0	472.2	480.5	488.8	497.1	505.4	513.7
	83	80.0	1.001	10.62	3.46	0.754	464.2	472.5	480.8	489.1	497.4	505.7	514.0
	82	78.5	0.955	10.13	3.95	0.719	464.4	472.7	481.0	489.3	497.6	505.9	514.2
	81	77.0	0.910	9.66	4.42	0.686	464.7	473.0	481.3	489.6	497.9	506.2	514.5
	80	75.5	0.868	9.20	4.88	0.653	464.9	473.2	481.5	489.8	498.1	506.4	514.7
	79	74.0	0.827	8.77	5.31	0.623	465.2	473.5	481.8	490.1	498.4	506.7	515.0
	78	72.5	0.787	8.35	5.73	0.593	465.4	473.7	482.0	490.3	498.6	506.9	515.2
	77	71.0	0.751	7.96	6.12	0.565	465.6	473.9	482.2	490.5	498.8	507.1	515.4
	76	69.5	0.715	7.57	6.51	0.537	465.8	474.1	482.4	490.7	499.0	507.3	515.7
	75	68.0	0.681	7.21	6.87	0.512	466.0	474.3	482.6	490.9	499.2	507.5	515.8
	74	66.5	0.648	6.87	7.21	0.488	466.2	474.5	482.8	491.1	499.4	507.7	516.0
	73	65.0	0.617	6.54	7.54	0.465	466.3	474.6	482.9	491.2	499.6	507.9	516.3
	72	63.5	0.588	6.22	7.86	0.442	466.5	474.8	483.1	491.4	499.8	508.1	516.5
	71	62.0	0.559	5.91	8.17	0.420	466.7	475.0	483.3	491.7	500.0	508.3	516.7
	70	60.5	0.532	5.62	8.46	0.399	466.8	475.1	483.4	491.8	500.1	508.4	516.8
	69	59.0	0.506	5.35	8.73	0.380	467.0	475.3	483.6	492.0	500.3	508.6	517.0
	68	57.5	0.481	5.08	9.00	0.361	467.1	475.4	483.7	492.1	500.4	508.7	517.1
	67	56.0	0.458	4.84	9.24	0.343	467.2	475.5	483.8	492.2	500.5	508.8	517.2
	66	54.5	0.435	4.61	9.47	0.327	467.4	475.7	483.9	492.4	500.7	509.1	517.4
	65	53.0	0.414	4.39	9.69	0.312	467.5	475.8	484.1	492.5	500.8	509.2	517.5
	64	51.5	0.393	4.17	9.91	0.296	467.6	475.9	484.2	492.6	500.9	509.3	517.6
	63	50.0	0.373	3.96	10.12	0.281	467.7	476.1	484.3	492.7	501.0	509.4	517.7
	62	48.5	0.355	3.76	10.32	0.267	467.8	476.2	484.4	492.8	501.1	509.5	517.8
90	90	90.0	1.368	14.50	0.00	1.000	461.3	469.5	477.8	486.0	494.3	502.5	510.8
	89	88.5	1.306	13.84	0.66	0.954	461.6	469.8	478.1	486.3	494.6	502.8	511.1
	88	87.0	1.247	13.22	1.28	0.910	462.0	470.2	478.5	486.7	495.0	503.2	511.5
	87	85.5	1.190	12.61	1.89	0.870	462.3	470.5	478.8	487.0	495.3	503.5	511.8
	86	84.0	1.136	12.03	2.47	0.830	462.7	470.9	479.2	487.4	495.7	503.9	512.1
	85	82.5	1.083	11.47	3.03	0.791	463.0	471.2	479.5	487.7	496.0	504.2	512.5
	84	81.0	1.034	10.94	3.56	0.755	463.2	471.5	479.8	488.0	496.3	504.5	512.8
	83	79.5	0.986	10.43	4.07	0.719	463.4	471.7	480.0	488.2	496.5	504.7	513.0
	82	78.0	0.940	9.95	4.55	0.686	463.7	472.0	480.3	488.5	496.8	505.0	513.3
	81	76.5	0.896	9.48	5.02	0.653	464.0	472.3	480.6	488.8	497.1	505.3	513.6
	80	75.0	0.854	9.03	5.47	0.622	464.2	472.5	480.7	488.9	497.3	505.5	513.9
	79	73.5	0.814	8.61	5.89	0.594	464.3	472.6	480.9	489.1	497.5	505.7	514.1
	78	72.0	0.776	8.20	6.30	0.565	464.5	472.8	481.1	489.3	497.7	505.9	514.3
	77	70.5	0.739	7.80	6.70	0.538	464.7	473.0	481.3	489.5	497.9	506.1	514.5

Reading of Thermometer, Fahr.		Temp. of Dew-Point, Fahr.	Force of Vapor in English Inches.	Weight of Vapor		Humidity, Saturation = 1.000.	Weight in Grains of a Cubic Foot of Air.						
				In a Cubic Foot of Air.	Reqd. for Sat'n. of a Cubic Ft. of Air.		Height of the Barometer in English Inches.						
Dry.	Wet.						in. **28.0**	in. **28.5**	in. **29.0**	in. **29.5**	in. **30.0**	in. **30.5**	in. **31.0**
°	°	°	in.	gr.	gr.		gr.	gr.	gr.	gr.	gr.	gr.	gr.
90	77	70.5	0.739	7.80	6.70	0.538	464.7	473.0	481.3	489.5	497.9	506.1	514.5
	76	69.0	0.704	7.43	7.07	0.512	465.0	473.3	481.6	489.8	498.2	506.4	514.8
	75	67.5	0.670	7.08	7.42	0.488	465.2	473.5	481.8	490.0	498.4	506.6	515.0
	74	66.0	0.638	6.74	7.76	0.465	465.4	473.7	482.0	490.2	498.6	506.8	515.2
	73	64.5	0.607	6.42	8.08	0.443	465.6	473.9	482.2	490.4	498.8	507.0	515.4
	72	63.0	0.578	6.10	8.40	0.421	465.7	474.0	482.3	490.5	498.9	507.1	515.5
	71	61.5	0.550	5.81	8.69	0.400	465.9	474.2	482.5	490.7	499.1	507.3	515.7
	70	60.0	0.523	5.52	8.98	0.381	466.1	474.4	482.8	491.0	499.3	507.5	515.9
	69	58.5	0.498	5.25	9.25	0.362	466.2	474.5	482.9	491.1	499.4	507.6	516.0
	68	57.0	0.473	4.99	9.51	0.344	466.4	474.7	483.1	491.3	499.6	507.8	516.2
	67	55.5	0.450	4.74	9.76	0.327	466.5	474.8	483.2	491.4	499.7	507.9	516.3
	66	54.0	0.428	4.52	9.98	0.312	466.6	474.9	483.3	491.5	499.8	508.0	516.4
	65	52.5	0.407	4.30	10.20	0.297	466.7	475.0	483.4	491.6	499.9	508.1	516.5
	64	51.0	0.386	4.09	10.41	0.282	466.9	475.2	483.6	491.8	500.1	508.3	516.6
	63	49.5	0.367	3.90	10.60	0.269	467.0	475.3	483.7	491.9	500.2	508.4	516.7

TABLE XIII.

FACTORS FOR COMPUTING THE FORCE OF VAPOR, FROM THE READINGS OF THE PSYCHROMETER, BY APJOHN'S FORMULA.

DR. APJOHN'S formula for deducing the force of vapor, and the temperature of the dew-point, from the readings of the Psychrometer, as given in the Proceedings of the Royal Irish Academy for 1840, is

$$f'' = f' - \frac{d}{88} \times \frac{h}{30},$$

when the readings of the wet-bulb thermometer are *above* 32° Fahr., in which formula

f'' = the force of vapor at the temperature of the dew-point in degrees of Fahr.,
f' = the force of vapor at the temperature of evaporation given by the wet-bulb thermometer,
d = the difference between the readings of the dry and wet thermometers,
h = the height of the barometer in English inches at the time of the observation.

When the readings of the wet-bulb thermometer are *below* 32° Fahr., and the bulb is covered with ice, the formula becomes

$$f'' = f' - \frac{d}{96} \times \frac{h}{30}.$$

The factors in the following table, which is taken from the Greenwich Observations for 1843, represent $\frac{d}{88} \times \frac{1}{30}$ and $\frac{d}{96} \times \frac{1}{30}$, computed for all differences between the wet and dry bulb thermometers, or values of d, from 0° to 21°.

USE OF THE TABLE.

To find out the force of vapor in the air, and the temperature of the dew-point, by means of these factors, let the factor corresponding to d, or the difference between the wet and dry thermometer in the first column, be multiplied into the observed height of the barometer, and subtract the result from the force of vapor, in Table XI., due to the temperature of evaporation, indicated by the wet-bulb thermometer; the rest is the force of vapor in the air at the time of the observation; and the temperature of the dew-point is that which is due to it in Table XI.

EXAMPLE.

The observation gives,

Dry-bulb thermometer	= 79°	Fahr., or the temperature of the air.
Wet-bulb "	= 69°	" or temperature of evaporation.
Difference	10°	
Height of barometer	29.7	English inches.

In the Table, 2d part, is found, — factor for a difference of 10° = 0.00379 × 29.7, or height of barometer = 0.113, which, subtracted from the force of vapor due to 69°, in Table XI., = 0.704 — 0.113, gives force of vapor in the air = 0.591 inches, and temperature of the dew-point 62°.5.

When the temperature of the wet bulb is *below* 32° Fahrenheit, the factors in the first part of the Table must be used.

XIII. FACTOR $\frac{a}{96} \times \frac{1}{30}$, FOR COMPUTING THE FORCE OF VAPOR BY APJOHN'S FORMULA.

Below 32° Fahrenheit; the Wet Bulb covered with a Film of Ice.

d, or Difference of Wet and Dry Bulb Therm.	Tenths of Degrees.									
	0.	1.	2.	3.	4.	5.	6.	7.	8.	9.
° 0	0.00000	0.00003	0.00007	0.00010	0.00014	0.00017	0.00020	0.00024	0.00027	0.00030
1	.00034	.00037	.00041	.00044	.00047	.00051	.00054	.00058	.00061	.00064
2	.00068	.00071	.00075	.00078	.00081	.00085	.00088	.00092	.00095	.00099
3	.00102	.00105	.00109	.00112	.00116	.00119	.00122	.00126	.00129	.00133
4	.00136	.00139	.00143	.00146	.00150	.00153	.00156	.00160	.00163	.00167
5	.00170	.00173	.00177	.00180	.00184	.00187	.00190	.00194	.00198	.00201
6	.00204	.00207	.00211	.00214	.00218	.00221	.00224	.00228	.00231	.00235
7	.00238	.00241	.00245	.00248	.00252	.00255	.00258	.00262	.00265	.00269
8	.00272	.00275	.00279	.00282	.00285	.00289	.00292	.00296	.00299	.00302
9	.00306	.00309	.00313	.00316	.00319	.00323	.00326	.00330	.00333	.00337
10	.00340	.00343	.00347	.00350	.00354	.00357	.00360	.00364	.00367	.00370

FACTOR $\frac{d}{88} \times \frac{1}{30}$.

Reading of Wet-Bulb Thermometer above 32° Fahrenheit.

d, or Difference of Wet and Dry Bulb Therm.	Tenths of Degrees.									
	0.	1.	2.	3.	4.	5.	6.	7.	8.	9.
° 0	0.00000	0.00004	0.00008	0.00011	0.00015	0.00019	0.00023	0.00027	0.00030	0.00034
1	.00038	.00042	.00046	.00049	.00053	.00057	.00061	.00064	.00068	.00072
2	.00076	.00080	.00083	.00087	.00091	.00095	.00098	.00102	.00106	.00110
3	.00114	.00118	.00121	.00125	.00129	.00132	.00137	.00140	.00144	.00148
4	.00151	.00135	.00159	.00163	.00167	.00171	.00174	.00178	.00182	.00186
5	.00189	.00193	.00197	.00201	.00205	.00209	.00212	.00216	.00220	.00224
6	.00228	.00231	.00235	.00239	.00242	.00246	.00250	.00254	.00258	.00261
7	.00265	.00269	.00273	.00277	.00280	.00284	.00288	.00292	.00295	.00299
8	.00303	.00307	.00311	.00315	.00318	.00322	.00326	.00330	.00333	.00337
9	.00341	.00345	.00349	.00352	.00356	.00360	.00364	.00368	.00371	.00375
10	.00379	.00383	.00386	.00390	.00394	.00398	.00401	.00405	.00409	.00412
11	.00416	.00420	.00424	.00428	.00432	.00436	.00439	.00443	.00447	.00451
12	.00454	.00458	.00462	.00466	.00470	.00474	.00477	.00481	.00485	.00489
13	.00493	.00496	.00500	.00504	.00508	.00511	.00515	.00519	.00522	.00526
14	.00530	.00534	.00538	.00541	.00545	.00549	.00553	.00556	.00560	.00564
15	.00568	.00572	.00576	.00580	.00584	.00587	.00591	.00595	.00598	.00602
16	.00606	.00610	.00614	.00618	.00622	.00625	.00629	.00633	.00636	.00640
17	.00644	.00648	.00652	.00655	.00659	.00663	.00666	.00670	.00674	.00678
18	.00682	.00686	.00690	.00693	.00697	.00701	.00704	.00708	.00712	.00716
19	.00720	.00724	.00728	.00731	.00735	.00739	.00742	.00746	.00750	.00754
20	.00758	.00761	.00765	.00769	.00773	.00777	.00780	.00784	.00788	.00792

TABLE XIV.

In the *Greenwich Magnetic and Meteorological Observations* for 1842 and 1843, Mr. Glaisher discussed the relation between the temperature of evaporation given by the Wet-bulb Thermometer and the temperature of the Dew-Point as given by Daniell's Hygrometer. Comparing the observations taken simultaneously every six hours with the Psychrometer, and with Daniell's Dew-Point Hygrometer, and dividing the average difference between the temperatures of the Wet and Dry bulb by the average difference of the temperature of the Dew-Point and of the Air, he obtained the empirical factors given in the following Table.

The observations from which they are deduced are those taken at the Observatory in the years 1841 to 1845, for the temperatures below 35° F., and in the years 1841 to 1843, for the temperatures above 35° F.

The observations made at Toronto Observatory, Canada West, in similar circumstances, in the years 1840 to 1842, were also compared in the same manner, and the factors derived from them showed a very close accordance for temperatures above 30° F., but were found smaller at temperatures below 30° F.

The errors in the temperature of the Dew-Point, which may result by using the Greenwich factors, though frequently within half a degree, often amount, however, to ± 2 or 3 degrees, and, in extreme cases, to ± 4 or 5 degrees, as shown in the volume of the *Greenwich Observations* for 1842, p. 60 of the *Abstracts*.

Use of the Table.

Multiply the difference between the Wet-bulb and Dry-bulb Thermometers by the factor standing in the Table opposite the reading of the Dry-bulb, and subtract the product from the reading of the Dry-bulb; the remainder will be the temperature of the Dew-Point.

Example. — Dry-bulb = 62° F.; Wet-bulb = 55°; Difference = 7°.

Opposite 62°, in the first column, stands the factor 1.7, which multiplied by 7°, the difference, gives 11°.9, to be subtracted from the Dry-bulb; or 62° — 11°.9 = 50°.1, temperature of the Dew-Point.

XIV. FACTORS TO FIND OUT THE TEMPERATURE OF THE DEW-POINT FROM THE READINGS OF THE PSYCHROMETER. — GLAISHER.

Dry-Bulb Therm. Fahren.	Factors.	Dry-Bulb Therm. Fahren.	Factors.	Dry-Bulb Therm. Fahren.	Factors.	Dry-Bulb Therm. Fahren.	Factors.	Dry-Bulb Therm. Fahren.	Factors.
21°	8.5	35°	2.6	49°	2.2	63°	1.7	77°	1.5
22	8.5	36	2.6	50	2.1	64	1.7	78	1.5
23	8.5	37	2.5	51	2.1	65	1.7	79	1.5
24	7.3	38	2.5	52	2.0	66	1.6	80	1.5
25	6.4	39	2.5	53	2.0	67	1.6	81	1.5
26	6.1	40	2.4	54	2.0	68	1.6	82	1.5
27	5.9	41	2.4	55	2.0	69	1.5	83	1.5
28	5.7	42	2.4	56	1.9	70	1.5	84	1.5
29	5.0	43	2.4	57	1.9	71	1.5	85	1.5
30	4.6	44	2.3	58	1.9	72	1.5	86	1.5
31	3.6	45	2.3	59	1.8	73	1.5	87	1.5
32	3.1	46	2.3	60	1.8	74	1.5	88	1.5
33	2.8	47	2.2	61	1.8	75	1.5	89	1.5
34	2.6	48	2.2	62	1.7	76	1.5	90	1.5

XV. WEIGHT OF VAPOR, IN GRAINS TROY, CONTAINED IN A CUBIC FOOT OF SATURATED AIR, AT TEMPERATURES BETWEEN 0° AND 94° FAHRENHEIT.

From the Greenwich Observations.

Temperature of Air, Fahren.	Weight of Vapor, in Grains.	Temperature of Air, Fahren.	Weight of Vapor, in Grains.	Temperature of Air, Fahren.	Weight of Vapor, in Grains.	Temperature of Air, Fahren.	Weight of Vapor, in Grains.	Temperature of Air, Fahren.	Weight of Vapor, in Grains.
0°	0.78	19°	1.52	38°	2.89	57°	5.34	76°	9.60
1	0.81	20	1.58	39	2.99	58	5.51	77	9.89
2	0.84	21	1.63	40	3.09	59	5.69	78	10.19
3	0.87	22	1.69	41	3.19	60	5.87	79	10.50
4	0.90	23	1.75	42	3.30	61	6.06	80	10.81
5	0.93	24	1.81	43	3.41	62	6.25	81	11.14
6	0.97	25	1.87	44	3.52	63	6.45	82	11.47
7	1.00	26	1.93	45	3.64	64	6.65	83	11.82
8	1.04	27	2.00	46	3.76	65	6.87	84	12.17
9	1.07	28	2.07	47	3.88	66	7.08	85	12.53
10	1.11	29	2.14	48	4.01	67	7.30	86	12.91
11	1.15	30	2.21	49	4.14	68	7.53	87	13.29
12	1.19	31	2.29	50	4.28	69	7.76	88	13.68
13	1.24	32	2.37	51	4.42	70	8.00	89	14.08
14	1.28	33	2.45	52	4.56	71	8.25	90	14.50
15	1.32	34	2.53	53	4.71	72	8.50	91	14.91
16	1.37	35	2.62	54	4.86	73	8.76	92	15.33
17	1.41	36	2.71	55	5.02	74	9.04	93	15.76
18	1.47	37	2.80	56	5.18	75	9.31	94	16.22

XVI. FACTORS TO DEDUCE THE WEIGHT OF VAPOR CONTAINED IN A CUBIC FOOT OF AIR, AT THE TIME OF A GIVEN OBSERVATION, FROM THE INDICATIONS OF DEW-POINT INSTRUMENTS. — GREENW. OBS.

t = Temperature of Air; **t″** = Temperature of Dew-Point.

Difference or t — t″.	Factors.	Difference or t — t″.	Factors.	Difference or t — t″.	Factors.	Difference or t — t″.	Factors.	Difference or t — t″.	Factors.
1	0.999	9	0.982	17	0.966	25	0.951	33	0.935
2	0.996	10	0.980	18	0.964	26	0.949	34	0.934
3	0.994	11	0.978	19	0.962	27	0.947	35	0.932
4	0.992	12	0.976	20	0.960	28	0.945	36	0.930
5	0.990	13	0.974	21	0.958	29	0.943	37	0.929
6	0.988	14	0.972	22	0.956	30	0.942	38	0.927
7	0.986	15	0.970	23	0.954	31	0.939	39	0.925
8	0.984	16	0.968	24	0.952	32	0.937	40	0.923

USE OF TABLE XVI. — The difference between the temperatures of the air and of the Dew-Point being known, multiply the factor in the Table corresponding to that difference into the weight of a cubic foot of vapor at the temperature of the Dew-Point, as given in Table XV., and the product will be the weight of vapor in a cubic foot of air at the time of the observation.

Example. — Temperature of air = 60° F.; Dew-Point = 52°; Diff. = 8°.

Table gives for a difference of 8°, factor 0.984; Table XV. gives weight of a cubic foot of vapor at temperature 52° = 4.$^{gr.}$56.

Hence, 0.984 × 4.56 = 4$^{gr.}$.49, the weight of vapor required.

TABLE XVII.

FOR COMPARING THE WEIGHT OF A CUBIC FOOT OF DRY AND OF SATURATED AIR.

THIS table is composed of two tables found in the *Greenwich Meteorological Observations* for 1842, pages xlvi. and li.; the first containing the weight of a cubic foot of dry air, under a barometric pressure of 30 inches, at temperatures between 0° and 90° F.; the other giving the weight of a cubic foot of saturated air under the same barometric pressure and temperature, together with the excess of the first above the last.

The weight of a cubic foot of dry air, on which the tables are based, is assumed to be 563 grains Troy, being a mean value, in round numbers, between the determinations of Shuckburgh, which is 557.7295 grains, and that of Biot and Arago, 568.7013. The true mean is 563.2154, but 563 is the number used in the calculations.

The coefficient of the expansion of the air is that of Gay-Lussac, viz. 0.00375 for 1° Centigrade, or 0.002083 of its bulk for 1° Fahrenheit.

Use of the Table.

This table shows the amount of buoyancy imparted to the air by the addition of moisture; and from it, the temperature and the relative humidity of the air being known, the weight of a cubic foot of air, in the actual condition of the atmosphere at the time of an observation, can be deduced.

It suffices to take in the fourth column, headed "Excess," the quantity corresponding to the temperature of the air in the first, multiply it into the given Relative Humidity, and subtract the product from the number in the second column. The result will be the weight of a cubic foot of air at the existing temperature and moisture, under a barometric pressure of 30 inches.

This result will be reduced to its true value, under the barometric pressure given by the observation, by multiplying it by $\frac{\text{Height of Barometer}}{30}$.

Example.

The temperature of the air is 60° F.; the relative humidity, 0.852; the barometer reads 29 inches.

The table gives, for temperature of air, 60°; excess, $3.35 \times 0.852 = 2.85$, which, subtracted from 531.91 in the second column, = 529.12, weight of a cubic foot of air under 30 inches of pressure; and $529.12 \times \frac{29 \text{ inch}}{30} = 511.48$, the weight of a cubic foot of air in the given conditions of temperature, moisture, and barometric pressure.

XVII. FOR COMPARING THE WEIGHT OF A CUBIC FOOT OF DRY AND OF SATURATED AIR,

AT TEMPERATURES BETWEEN 0° AND 90° FAHRENHEIT.

From the Greenwich Observations.

Temperature Fahren.	Weight of a cubic foot of Dry Air.	Weight of a cubic foot of Saturated Air.	Excess of Dry Air.	Temperature Fahren.	Weight of a cubic foot of Dry Air.	Weight of a cubic foot of Saturated Air.	Excess of Dry Air.	Temperature Fahren.	Weight of a cubic foot of Dry Air	Weight of a cubic foot of Saturated Air.	Excess of Dry Air.
°	Grains.	Grains.	Grains	°	Grains.	Grains.	Grains.	°	Grains.	Grains.	Grains.
0	603.21	602.77	0.44	30	565.35	564.08	1.27	60	531.97	528.62	3.35
1	601.87	601.40	0.47	31	564.17	562.86	1.31	61	530.93	527.48	3.45
2	600.52	600.03	0.49	32	563.00	561.64	1.36	62	529.88	526.32	3.56
3	599.20	598.69	0.51	33	561.84	560.42	1.42	63	528.84	525.17	3.67
4	597.87	597.34	0.53	34	560.67	559.20	1.47	64	527.81	524.03	3.78
5	596.55	596.01	0.54	35	559.51	558.01	1.50	65	526.88	522.90	3.88
6	595.24	594.69	0.55	36	558.35	556.79	1.56	66	525.76	521.75	4.01
7	593.94	593.36	0.58	37	557.21	555.61	1.60	67	524.75	520.61	4.14
8	592.63	592.04	0.59	38	556.05	554.40	1.65	68	523.72	519.46	4.26
9	591.33	590.72	0.61	39	554.91	553.20	1.71	69	522.70	518.29	4.41
10	590.04	589.40	0.64	40	553.77	552.00	1.77	70	521.70	517.17	4.53
11	588.75	588.07	0.68	41	552.65	550.80	1.84	71	520.70	516.02	4.68
12	587.48	586.78	0.70	42	551.52	549.63	1.89	72	519.69	514.87	4.82
13	586.21	585.49	0.72	43	550.39	548.44	1.95	73	518.70	513.75	4.95
14	584.94	584.18	0.75	44	549.27	547.26	2.01	74	517.70	512.61	5.09
15	583.67	582.89	0.78	45	548.16	546.06	2.10	75	516.71	511.46	5.25
16	582.41	581.61	0.80	46	547.05	544.88	2.17	76	515.73	510.32	5.41
17	581.15	580.33	0.82	47	545.97	543.75	2.22	77	514.74	509.18	5.56
18	579.91	579.06	0.85	48	544.85	542.55	2.30	78	513.77	508.04	5.73
19	578.67	577.79	0.88	49	543.75	541.36	2.39	79	512.80	506.91	5.89
20	577.44	576.54	0.90	50	542.65	540.21	2.44	80	511.82	505.74	6.08
21	576.21	575.27	0.94	51	541.55	539.04	2.51	81	510.87	504.61	6.26
22	574.98	574.01	0.97	52	540.48	537.87	2.61	82	509.89	503.45	6.44
23	573.76	572.76	1.00	53	539.41	536.71	2.70	83	508.93	502.32	6.61
24	572.55	571.50	1.05	54	538.33	535.55	2.78	84	507.97	501.16	6.81
25	571.33	570.26	1.07	55	537.27	534.39	2.88	85	507.03	500.05	6.98
26	570.13	569.01	1.12	56	536.19	533.22	2.97	86	506.07	498.87	7.20
27	568.92	567.77	1.15	57	535.12	532.06	3.06	87	505.11	497.71	7.40
28	567.73	566.53	1.20	58	534.07	530.92	3.15	88	504.19	496.58	7.61
29	566.54	565.31	1.23	59	533.03	529.77	3.26	89	503.25	495.44	7.81
30	565.35	564.08	1.27	60	531.97	528.62	3.35	90	502.32	494.28	8.04

MISCELLANEOUS TABLES,

FOR

COMPARING THE HYGROMETRICAL RESULTS OBTAINED BY DIFFERENT AUTHORITIES.

MISCELLANEOUS TABLES.

The object of these Tables is to afford the means of comparing the different determinations of the hygrometrical elements which have been obtained, or adopted, by various physicists, especially the values of the elastic forces of vapor given in other tables than those contained in the preceding pages.

Table XVIII., giving the elastic forces of vapor, expressed in millimetres of mercury, for Centigrade temperatures, was calculated by August from Dalton's experiments, and reduced to French measures in the translation of Kaemtz's *Meteorology*, by Chas. Martins, page 70, from which it has been taken. On these values are based the first psychrometrical tables published by August, in Berlin, 1825.

Table XIX. is the table computed by Kaemtz from his own experiments. It is found, reduced to French measures, in the same volume, page 68.

Table XX. furnishes the results of the experiments made by Professor Magnus, in Berlin, and published in Poggendorf's *Annalen*, Tom. LXI. p. 226, and also in the *Annales de Chimie et de Physique*, 3me série, Tom. XII. p. 88, from which this table was copied.

Table XXI. has been published by the Committee of Physics and Meteorology of the Royal Society, in their *Report on the Objects of Scientific Inquiry in these Sciences*, London, 1840, p. 89. The values which it contains are not derived from new experiments, but are probably computed from those existing at that time.

Table XXII. furnishes a synoptic view of the differences in the values of the force of vapor adopted by various authorities, prepared with the view of facilitating their comparison. A reference to their respective origin will be found below, page 152.

Table XXIII., showing the weight, in grammes, of the vapor contained in a cubic metre of saturated air, at different temperatures, is taken from Pouillet's *Eléments de Physique*, Tom. II. p. 707.

Table XXIV. gives the weights as derived from August's experiments, in Kaemtz's *Vorlesungen über Meteorologie*. The table is copied from the French translation, by Martins, page 73. The tensions have been added, opposite the weights, and are extracted from August's table.

Table XXV. is found in Biot's *Traité de Physique*, Tom. I. p. 533.

XVIII. ELASTIC FORCE OF AQUEOUS VAPOR,

EXPRESSED IN MILLIMETRES OF MERCURY FOR EVERY TENTH OF A CENTIGRADE DEGREE.

Calculated by AUGUST.

Centigrade Degrees.	Tenths of Degrees.									
	0.	**1.**	**2.**	**3.**	**4.**	**5.**	**6.**	**7.**	**8.**	**9.**
°	Millim.	Millim.	Millim.	Millim.	Millim.	Millim.	Millim.	Millim.	Millim.	Millim.
−31	0.45	0.45	0.45	0.44	0.44	0.43	0.43	0.42	0.42	0.41
−30	0.50	0.49	0.49	0.48	0.48	0.47	0.47	0.46	0.46	0.45
−29	0.54	0.54	0.54	0.53	0.53	0.52	0.52	0.51	0.51	0.50
−28	0.59	0.58	0.58	0.57	0.57	0.56	0.56	0.55	0.55	0.54
−27	0.63	0.63	0.63	0.62	0.62	0.61	0.61	0.60	0.60	0.59
−26	0.70	0.69	0.68	0.68	0.67	0.66	0.66	0.65	0.64	0.64
−25	0.77	0.76	0.75	0.75	0.74	0.73	0.73	0.72	0.71	0.71
−24	0.83	0.83	0.82	0.82	0.81	0.80	0.80	0.79	0.78	0.78
−23	0.90	0.89	0.88	0.88	0.87	0.86	0.86	0.85	0.84	0.84
−22	0.99	0.98	0.97	0.96	0.95	0.95	0.94	0.93	0.92	0.91
−21	1.06	1.05	1.04	1.04	1.03	1.02	1.02	1.01	1.00	1.00
−20	1.15	1.14	1.13	1.12	1.11	1.11	1.10	1.09	1.08	1.07
−19	1.26	1.25	1.24	1.23	1.22	1.21	1.20	1.18	1.17	1.16
−18	1.33	1.32	1.31	1.31	1.30	1.29	1.29	1.28	1.27	1.27
−17	1.44	1.43	1.42	1.41	1.40	1.39	1.38	1.36	1.35	1.34
−16	1.56	1.54	1.53	1.52	1.51	1.50	1.49	1.47	1.46	1.45
−15	1.69	1.68	1.67	1.65	1.64	1.63	1.61	1.60	1.59	1.57
−14	1.80	1.79	1.78	1.77	1.76	1.75	1.74	1.72	1.71	1.70
−13	1.96	1.94	1.93	1.91	1.89	1.88	1.86	1.85	1.83	1.82
−12	2.12	2.10	2.09	2.07	2.05	2.04	2.02	2.01	1.99	1.98
−11	2.30	2.28	2.26	2.25	2.23	2.21	2.19	2.17	2.16	2.14
−10	2.48	2.46	2.44	2.43	2.41	2.39	2.37	2.35	2.34	2.32
− 9	2.66	2.64	2.62	2.61	2.59	2.57	2.55	2.53	2.52	2.50
− 8	2.86	2.84	2.82	2.80	2.78	2.76	2.74	2.72	2.70	2.68
− 7	3.09	3.06	3.04	3.02	3.00	2.97	2.95	2.93	2.91	2.88
− 6	3.32	3.29	3.27	3.25	3.23	3.20	3.18	3.16	3.14	3.11
− 5	3.56	3.56	3.54	3.51	3.48	3.46	3.43	3.40	3.37	3.35
− 4	3.83	3.80	3.78	3.75	3.72	3.70	3.67	3.64	3.61	3.59
− 3	4.11	4.07	4.05	4.02	3.99	3.97	3.94	3.91	3.88	3.86
− 2	4.40	4.37	4.34	4.32	4.29	4.26	4.23	4.20	4.17	4.14
− 1	4.71	4.68	4.65	4.62	4.59	4.56	4.53	4.49	4.46	4.43
− 0	5.05	5.01	4.98	4.95	4.91	4.88	4.85	4.81	4.78	4.74
+ 0	5.05	5.09	5.12	5.16	5.19	5.23	5.27	5.30	5.34	5.37
	0.	**1.**	**2.**	**3.**	**4.**	**5.**	**6.**	**7.**	**8.**	**9.**

Centigrade Degrees.	Tenths of Degrees.									
	0.	**1.**	**2.**	**3.**	**4.**	**5.**	**6.**	**7.**	**8.**	**9.**
°	Millim.	Millim.	Millim.	Millim.	Millim.	Millim.	Millim.	Millim.	Millim.	Millim.
1	5.41	5.45	5.49	5.52	5.56	5.60	5.64	5.68	5.72	5.75
2	5.80	5.84	5.88	5.92	5.96	6.00	6.04	6.08	6.13	6.17
3	6.20	6.24	6.29	6.33	6.37	6.41	6.46	6.50	6.54	6.59
4	6.63	6.68	6.72	6.77	6.81	6.86	6.90	6.95	6.99	7.04
5	7.08	7.13	7.18	7.23	7.28	7.33	7.38	7.43	7.48	7.53
6	7.58	7.63	7.68	7.74	7.79	7.84	7.89	7.94	7.99	8.05
7	8.10	8.15	8.21	8.26	8.32	8.37	8.43	8.48	8.53	8.59
8	8.64	8.70	8.76	8.82	8.87	8.93	8.99	9.05	9.11	9.17
9	9.23	9.30	9.36	9.43	9.50	9.57	9.63	9.70	9.77	9.84
10	9.90	9.96	10.02	10.08	10.14	10.20	10.25	10.31	10.37	10.43
11	10.49	10.56	10.63	10.69	10.76	10.83	10.90	10.96	11.03	11.10
12	11.17	11.24	11.31	11.38	11.45	11.52	11.59	11.66	11.73	11.80
13	11.86	11.94	12.02	12.10	12.18	12.26	12.34	12.42	12.50	12.58
14	12.66	12.74	12.82	12.90	12.98	13.05	13.13	13.21	13.29	13.37
15	13.44	13.52	13.61	13.69	13.77	13.86	13.94	14.02	14.11	14.19
16	14.28	14.37	14.47	14.56	14.65	14.74	14.84	14.93	15.02	15.11
17	15.20	15.29	15.38	15.46	15.55	15.64	15.73	15.82	15.90	15.99
18	16.08	16.17	16.27	16.36	16.45	16.54	16.64	16.73	16.82	16.91
19	17.01	17.13	17.25	17.37	17.49	17.61	17.73	17.85	17.97	18.09
20	18.20	18.31	18.43	18.54	18.65	18.76	18.88	18.99	19.10	19.21
21	19.33	19.45	19.56	19.68	19.80	19.92	20.03	20.15	20.27	20.39
22	20.51	20.63	20.76	20.88	21.01	21.13	21.25	21.38	21.50	21.63
23	21.75	21.88	22.00	22.13	22.26	22.38	22.51	22.63	22.76	22.89
24	23.01	23.13	23.24	23.36	23.48	23.60	23.71	23.83	23.95	24.07
25	24.18	24.34	24.50	24.67	24.83	24.99	25.15	25.32	25.48	25.64
26	25.81	25.97	26.13	26.28	26.44	26.60	26.76	26.92	27.07	27.23
27	27.39	27.55	27.71	27.86	28.02	28.18	28.34	28.50	28.65	28.81
28	28.96	29.13	29.29	29.46	29.63	29.79	29.96	30.13	30.30	30.46
29	30.63	30.81	30.98	31.16	31.33	31.51	31.69	31.86	32.04	32.21
30	32.39	32.57	32.76	32.94	33.13	33.31	33.50	33.68	33.87	34.05
31	34.24	34.43	34.63	34.82	35.02	35.21	35.40	35.60	35.79	35.99
32	36.18	36.38	36.59	36.79	36.99	37.20	37.40	37.60	37.80	38.01
33	38.21	38.43	38.64	38.86	39.08	39.29	39.51	39.73	39.94	40.16
34	40.38	40.60	40.82	41.04	41.26	41.49	41.71	41.93	42.15	42.37
35	42.59	42.82	43.05	43.28	43.51	43.74	43.97	44.20	44.43	44.66
	0.	**1.**	**2.**	**3.**	**4.**	**5.**	**6.**	**7.**	**8.**	**9.**

XIX. ELASTIC FORCE OF AQUEOUS VAPOR,

EXPRESSED IN MILLIMETRES OF MERCURY, FOR CENTIGRADE TEMPERATURES.

By KAEMTZ.

Temperature Centi-Grade.	Force of Vapor.	Temperature Centi-grade.	Force of Vapor.	Temperature Centi-grade.	Force of Vapor.	Temperature Centi-grade.	Force of Vapor.	Temperature Centi-grade.	Force of Vapor.
°	Millim.	°	Millim.	°	Millim.	°	Millim.	°	Millim.
–25	0.68	–12	1.92	0	4.58	12	10.24	24	21.43
–24	0.72	–11	2.05	1	4.92	13	10.91	25	22.74
–23	0.79	–10	2.21	2	5.26	14	11.62	26	24.16
–22	0.86	– 9	2.39	3	5.64	15	12.38	27	25.56
–21	0.92	– 8	2.57	4	6.01	16	13.17	28	27.07
–20	1.01	– 7	2.78	5	6.45	17	14.03	29	28.67
–19	1.10	– 6	2.98	6	6.90	18	14.93	30	30.36
–18	1.20	– 5	3.20	7	7.38	19	15.86	31	32.17
–17	1.29	– 4	3.45	8	7.89	20	16.87	32	33.95
–16	1.40	– 3	3.70	9	8.41	21	17.91	33	35.95
–15	1.51	– 2	3.97	10	9.00	22	19.04	34	37.99
–14	1.62	– 1	4.26	11	9.58	23	20.21	35	40.15
–13	1.76	0	4.58	12	10.24	24	21.43	36	42.40

XX. ELASTIC FORCE OF AQUEOUS VAPOR,

EXPRESSED IN MILLIMETRES OF MERCURY, FOR CENTIGRADE TEMPERATURES.

By MAGNUS.

Temperature Centi-grade.	Force of Vapor.	Temperature Centi-grade.	Force of Vapor.	Temperature Centi-grade.	Force of Vapor.	Temperature Centi-grade.	Force of Vapor.	Temperature Centi-grade.	Force of Vapor.
°	Millim.	°	Millim.	°	Millim.	°	Millim.	°	Millim.
–20	0.916	–7	2.671	6	6.939	19	16.345	32	35.419
–19	0.999	–6	2.886	7	7.436	20	17.396	33	37.473
–18	1.089	–5	3.115	8	7.964	21	18.505	34	39.630
–17	1.186	–4	3.361	9	8.525	22	19.675	35	41.893
–16	1.290	–3	3.624	10	9.126	23	20.909	36	44.268
–15	1.403	–2	3.905	11	9.751	24	22.211	37	46.758
–14	1.525	–1	4.205	12	10.421	25	23.582	38	49.368
–13	1.655	0	4.525	13	11.130	26	25.026	39	52.103
–12	1.796	+1	4.867	14	11.882	27	26.547	40	54.964
–11	1.947	2	5.231	15	12.677	28	28.148	41	57.969
–10	2.109	3	5.619	16	13.519	29	29.832	42	61.109
– 9	2.284	4	6.032	17	14.409	30	31.602	43	64.396
– 8	2.471	5	6.471	18	15.351	31	33.464	44	67.833

XXI. ELASTIC FORCE OF AQUEOUS VAPOR,

EXPRESSED IN ENGLISH INCHES OF MERCURY, FOR TEMPERATURES OF FAHRENHEIT.

From the Royal Society's Report.

Temperature of Air.	Force of Vapor.	Temperature of Air.	Force of Vapor.	Temperature of Air.	Force of Vapor.	Temperature of Air.	Force of Vapor.
Fahrenheit.	Eng. Inches.	Fahrenheit.	Eng. Inches.	Fahrenheit.	Eng. Inches.	Fahrenheit.	Eng. Inches.
0°	0.051	31°	0.179	62°	0.551	93°	1.514
1	0.053	32	0.186	63	0.570	94	1.562
2	0.056	33	0.193	64	0.590	95	1.610
3	0.058	34	0.200	65	0.611	96	1.660
4	0.060	35	0.208	66	0.632	97	1.712
5	0.063	36	0.216	37	0.654	98	1.764
6	0.066	37	0.224	68	0.676	99	1.819
7	0.069	38	0.233	69	0.699	100	1.874
8	0.071	39	0.242	70	0.723	101	1.931
9	0.074	40	0.251	71	0.748	102	1.990
10	0.078	41	0.260	72	0.773	103	2.050
11	0.081	42	0.270	73	0.799	104	2.112
12	0.084	43	0.280	74	0.826	105	2.176
13	0.088	44	0.291	75	0.854	106	2.241
14	0.092	45	0.302	76	0.882	107	2.307
15	0.095	46	0.313	77	0.911	108	2.376
16	0.099	47	0.324	78	0.942	109	2.447
17	0.103	48	0.336	79	0.973	110	2.519
18	0.107	49	0.349	80	1.005	111	2.593
19	0.112	50	0.361	81	1.036	112	2.669
20	0.116	51	0.375	82	1.072	113	2.747
21	0.121	52	0.389	83	1.106	114	2.826
22	0.126	53	0.402	84	1.142	115	2.908
23	0.131	54	0.417	85	1.179	116	2.992
24	0.136	55	0.432	86	1.217	117	3.078
25	0.142	56	0.447	87	1.256	118	3.166
26	0.147	57	0.463	88	1.296	119	3.257
27	0.153	58	0.480	89	1.337	120	3.349
28	0.159	59	0.497	90	1.380	121	3.444
29	0.165	60	0.514	91	1.423	122	3.542
30	0.172	61	0.532	92	1.468	123	3.641
31	0.179	62	0.551	93	1.514	124	3.743

TABLE XXII.

FOR SHOWING THE DIFFERENCES IN THE VALUES OF THE ELASTIC FORCE OF AQUEOUS VAPOR ADOPTED BY DIFFERENT AUTHORITIES.

THE following synoptic view of the values of the elastic force of vapor adopted by various authorities, furnishes the means of readily comparing them, and of appreciating the amount of the differences which they exhibit. The values are given both in English and in French measures.

Dalton's values are copied from the *Edinburgh Encyclopædia*, Art. *Hygrometry.* Those adopted in the *Greenwich Observations* are found in the same article, and also in the volumes published annually by that Observatory. Biot's table of tensions is, in fact, the same, computed by Pouillet from Dalton's results, by Biot's formula, and published in Biot's *Traité de Physique*, Tom. I. p. 531. Dr. Ure's results are taken from his Memoir in the *Philosophical Transactions* for 1818, p. 347. In the column headed "Daniell" are given the forces of vapor as found in the table published in his *Meteorological Essays*, 2d edition, p. 596, a table computed by Galbraith, from Dr. Ure's experiments, by the formula of Ivory.

For the columns headed Royal Society, August, Kaemtz, Magnus, and Regnault, see above, p. 147.

XXII. FOR SHOWING THE DIFFERENCES IN THE VALUES OF THE ELASTIC FORCE OF AQUEOUS VAPOR, ADOPTED BY DIFFERENT AUTHORITIES.

FORCE OF VAPOR EXPRESSED IN ENGLISH INCHES FOR TEMPERATURES OF FAHRENHEIT.

Temperature of Air, Fahrenheit.	Force of Vapor according to									Temperature of Air, Fahrenheit.
	Dalton.	Greenwich Observations.	Ure.	Daniell.	Royal Society.	August.	Kaemtz.	Magnus.	Regnault.	
°	Eng. In.	Eng. In.	Eng. In.	Eng. In.	Eng. In.	Eng. In.	Eng. In.	Eng. In.	Eng. In.	°
0	0.064	0.061	....	0.068	0.051	0.053	0.048	0.044	0.043	0
10	0.090	0.089	....	0.098	0.078	0.082	0.074	0.070	0.068	10
20	0.129	0.129	....	0.140	0.116	0.124	0.112	0.108	0.108	20
30	0.186	0.186	....	0.200	0.172	0.184	0.166	0.164	0.167	30
32	0.200	0.199	0.200	0.216	0.186	0.199	0.180	0.178	0.181	32
40	0.263	0.264	0.250	0.280	0.251	0.269	0.244	0.245	0.248	40
50	0.375	0.373	0.360	0.400	0.361	0.390	0.354	0.359	0.361	50
60	0.524	0.523	0.516	0.560	0.516	0.547	0.505	0.517	0.518	60
70	0.721	0.727	0.726	0.770	0.723	0.766	0.710	0.733	0.733	70
80	1.000	1.001	1.010	1.060	1.005	1.058	0.988	1.025	1.023	80
90	1.360	1.368	1.360	1.430	1.380	1.442	1.354	1.412	1.410	90
95	1.580	1.594	1.640	1.636	1.562	1.677	1.581	1.649	1.647	95
100	1.860	1.852	1.860	....	1.874	....	....	1.921	1.918	100

FORCE OF VAPOR EXPRESSED IN MILLIMETRES FOR CENTIGRADE TEMPERATURES.

Temperature of Air, Centigrade.	Force of Vapor according to									Temperature of Air, Centigrade.
	Dalton.	Greenwich Observations.	Biot.	Daniell.	Royal Society.	August.	Kaemtz.	Magnus.	Regnault.	
°	Millim.	Millim.	Millim.	Millim.	Millim.	Millim.	Millim.	Millim.	Millim.	°
−20	....	....	1.33	....	....	1.15	1.01	0.91	0.91	−20
−15	1.93	1.88	1.88	2.11	1.60	1.69	1.51	1.40	1.38	−15
−10	2.64	2.62	2.63	2.92	2.34	2.48	2.21	2.11	2.08	−10
− 5	3.66	3.66	3.66	4.01	3.33	3.56	3.20	3.11	3.13	− 5
0	5.08	5.06	5.06	5.49	4.72	5.05	4.58	4.52	4.60	0
+ 5	6.93	6.95	6.95	7.42	6.60	7.08	6.45	6.47	6.53	+ 5
10	9.52	9.48	9.47	10.16	9.17	9.90	9.00	9.13	9.16	10
15	12.88	12.85	12.84	13.79	12.62	13.44	12.38	12.68	12.70	15
20	17.17	17.30	17.31	18.34	17.17	18.20	16.87	17.40	17.39	20
25	23.11	23.12	23.09	24.54	23.14	24.18	22.74	23.58	23.55	25
30	30.73	30.70	30.64	32.33	30.91	32.39	30.36	31.60	31.55	30
35	40.13	40.47	40.40	41.55	40.89	42.59	40.15	41.89	41.83	35
40	....	....	53.00	....	53.64	....	....	54.96	54.91	40

XXIII. WEIGHT OF VAPOR, IN GRAMMES, CONTAINED IN A CUBIC METRE OF SATURATED AIR, AT TEMPERATURES BETWEEN —20° AND +40° CENTIGRADE.—POUILLET.

Temperature of Dew-Point.	Force of Vapor.	Weight of Vapor.	Temperature of Dew-Point.	Force of Vapor.	Weight of Vapor.	Temperature of Dew-Point.	Force of Vapor.	Weight of Vapor.
Centigrade.	Millim.	Grammes.	Centigrade.	Millim.	Grammes.	Centigrade.	Millim.	Grammes.
−20°	1.3	1.5	11°	10.1	10.3	26°	24.4	23.8
−15	1.9	2.1	12	10.7	10.9	27	25.9	25.1
−10	2.6	2.9	13	11.4	11.6	28	27.4	26.4
− 5	3.7	4.0	14	12.1	12.2	29	29.0	27.9
0	5.0	5.4	15	12.8	13.0	30	30.6	29.4
+ 1	5.4	5.7	16	13.6	13.7	31	32.4	31.0
2	5.7	6.1	17	14.5	14.5	32	34.3	32.6
3	6.1	6.5	18	15.4	15.3	33	36.2	34.3
4	6.5	6.9	19	16.3	16.2	34	38.3	36.2
5	6.9	7.3	20	17.3	17.1	35	40.4	38.1
6	7.4	7.7	21	18.3	18.1	36	42.7	40.2
7	7.9	8.2	22	19.4	19.1	37	45.0	42.2
8	8.4	8.7	23	20.6	20.2	38	47.6	44.4
9	8.9	9.2	24	21.8	21.3	39	50.1	46.7
10	9.5	9.7	25	23.1	22.5	40	53.0	49.2

XXIV. WEIGHT OF VAPOR, IN GRAMMES, CONTAINED IN A CUBIC METRE OF SATURATED AIR, AT TEMPERATURES BETWEEN —25° AND +36° CENTIGR.—KAEMTZ.

Temperature of Dew-Point.	Force of Vapor.	Weight of Vapor.	Temperature of Dew-Point.	Force of Vapor.	Weight of Vapor.	Temperature of Dew-Point.	Force of Vapor.	Weight of Vapor.
Centigrade.	Millim.	Grammes.	Centigrade.	Millim.	Grammes.	Centigrade.	Millim.	Grammes.
−25°	0.77	0.93	−4°	3.83	4.37	16°	14.28	14.97
−24	0.83	1.01	−3	4.11	4.70	17	15.20	15.84
−23	0.90	1.10	−2	4.40	5.01	18	16.08	16.76
−22	0.99	1.19	−1	4.71	5.32	19	17.01	17.75
−21	1.06	1.26	0	5.05	5.66	20	18.20	18.77
−20	1.15	1.38	+1	5.41	6.00	21	19.33	19.82
−19	1.26	1.47	2	5.80	6.42	22	20.51	20.91
−18	1.33	1.60	3	6.20	6.84	23	21.75	22.09
−17	1.44	1.74	4	6.63	7.32	24	23.01	23.36
−16	1.56	1.84	5	7.08	7.77	25	24.18	24.61
−15	1.69	2.00	6	7.58	8.25	26	25.81	25.96
−14	1.80	2.14	7	8.10	8.79	26	27.39	27.34
−13	1.96	2.33	8	8.64	9.30	28	28.96	28.81
−12	2.12	2.48	9	9.23	9.86	29	30.63	30.35
−11	2.30	2.63	10	9.90	10.57	30	32.39	31.93
−10	2.48	2.87	11	10.49	11.18	31	34.24	33.65
− 9	2.66	3.08	12	11.17	11.83	32	36.18	35.45
− 8	2.86	3.30	13	11.86	12.57	33	38.21	37.20
− 7	3.09	3.53	14	12.66	13.33	34	40.38	39.12
− 6	3.32	3.80	15	13.44	14.17	35	42.59	41.13
− 5	3.56	4.08	16	14.28	14.97	36	44.96	43.17

XXV. FORCES OF VAPOR AND RELATIVE HUMIDITY,

CORRESPONDING TO THE DEGREES OF SAUSSURE'S HAIR-HYGROMETER, AT THE TEMPERATURE OF 10° CENTIGRADE.

From the Experiments of Gay-Lussac.

The force of vapor is expressed in hundredths, the tension at full saturation being represented by 100.

Degrees of Hair-Hygrometer.	Force of Vapor.	Relative Humidity in Thousandths.	Degrees of Hair-Hygrometer.	Force of Vapor.	Relative Humidity in Thousandths.	Degrees of Hair-Hygrometer.	Force of Vapor.	Relative Humidity in Thousandths.
0°	0.00	0.000	34°	17.10		67°	43.73	
1	0.45		35	17.68	0.177	68	44.89	
2	0.90		36	18.30		69	46.04	
3	1.35		37	18.92		70	47.19	0.472
4	1.80		38	19.54		71	48.51	
5	2.25	0.022	39	20.16		72	49.82	0.500
6	2.71		40	20.78	0.208	73	51.14	
7	3.18		41	21.45		74	52.45	
8	3.64		42	22.12		75	53.76	0.538
9	4.10		43	22.79		76	55.25	
10	4.57	0.046	44	23.46		77	56.74	
11	5.05		45	24.13	0.241	78	58.24	
12	5.52		46	24.86		79	59.73	
13	6.00		47	25.59		80	61.22	0.612
14	6.48		48	26.32		81	62.89	
15	6.96	0.070	49	27.06		82	64.57	
16	7.46		50	27.79	0.278	83	66.24	
17	7.95		51	28.58		84	67.92	
18	8.45		52	29.38		85	69.59	0.696
19	8.95		53	30.17		86	71.49	
20	9.45	0.094	54	30.97		87	73.39	
21	9.97		55	31.76	0.318	88	75.29	
22	10.49		56	32.66		89	77.19	
23	11.01		57	33.57		90	79.09	0.791
24	11.53		58	34.47		91	81.09	
25	12.05	0.120	59	35.37		92	83.08	
26	12.59		60	36.28	0.363	93	85.08	
27	13.14		61	37.31		94	87.07	
28	13.69		62	38.34		95	89.06	0.891
29	14.23		63	39.36		96	91.25	
30	14.78	0.148	64	40.39		97	93.44	
31	15.36		65	41.42	0.414	98	95.63	
32	15.94		66	42.58		99	97.81	
33	16.52		67	43.73		100	100.00	1.000

XXVI.

TABLE

FOR

DEDUCING THE RELATIVE HUMIDITY IN HUNDREDTHS, FROM THE INDICATIONS OF SAUSSURE'S HÁIR-HYGROMETER;

Calculated from the Experiments of Melloni.

BY M. T. HAEGHENS.

THE Hair-Hygrometer of Saussure having been formerly used for long series of observations, and being still employed by some meteorologists, notwithstanding the imperfection of this instrument, on account of its giving directly the relative humidity without calculation, it was desirable to ascertain the correspondence of the degrees of that hygrometer with the relative humidity expressed in hundredths, as in the preceding table. Though these instruments compared with each other, show very often great discrepancies in their indications, yet a large number of them agree sufficiently well with the experiments of Melloni, August, and others, to allow the following table of comparison to be constructed, which table may be considered as giving good approximations. For the calculation of it, Mr. Haeghens used the results of Melloni, which agree also satisfactorily with a series of observations very carefully made by M. Delcros. See *Annuaire Météorologique de la France, pour* 1850.

RELATIVE HUMIDITY IN HUNDREDTHS.

Degrees of Saussure's Hygrometer. Tens.	Degrees of Saussure's Hygrometer. Units.									
	0.	**1.**	**2.**	**3.**	**4.**	**5.**	**6.**	**7.**	**8.**	**9.**
	Humidity	Humidity	Humidity	Humidity	Humidity	Humidity	Humidity	Humidity	Humidity	Humidity
0	0	0	1	1	2	3	3	4	4	5
1	5	6	6	7	8	8	9	10	11	11
2	12	12	13	14	15	16	17	18	18	19
3	19	20	21	22	23	24	24	25	26	26
4	27	27	28	28	29	30	31	32	33	34
5	35	36	37	37	38	39	40	41	42	43
6	44	45	46	47	49	50	51	52	53	55
7	56	57	58	59	61	62	63	65	66	68
8	69	70	72	73	75	77	78	79	81	82
9	83	85	87	88	90	91	93	95	97	98
10	100	.	.	.	.	.	.	.	.	.

TABLE XXVII.

THE following Table shows the Relative Humidity, in hundredths, corresponding to the degrees of Saussure's Hair-Hygrometer, as determined by various physicists. It is found in Kaemtz, *Vorlesungen über Meteorologie*, page 100; also in the French translation by Martins, *Cours de Météorologie*, page 80.

XXVI. RELATIVE HUMIDITY, CORRESPONDING TO THE DEGREES OF SAUSSURE'S HAIR-HYGROMETER.

Saturation = 100.

Degrees of Hair-Hygrometer.	Relative Humidity according to				Degrees of Hair-Hygrometer.
	Gay-Lussac.	Prinsep.	August.	Melloni.	
100°	100.0	100.0	100.0	100.0	100°
95	89.1	88.7	94.0	90.8	95
90	79.1	78.2	86.0	83.1	90
85	69.6	68.3	79.0	76.5	85
80	61.2	59.2	71.0	68.9	80
75	53.8	50.6	64.0	62.0	75
70	47.2	43.6	56.0	55.6	70
65	41.4	37.2	48.0	49.6	65
60	36.3	31.5	41.0	44.0	60
55	31.8	26.3	36.0	39.1	55
50	27.8	21.8	31.0	34.6	50
45	24.1	17.7	27.0	29.8	45
40	20.8	14.3	23.0	27.0	40
35	17.7	11.4	19.0	23.8	35
30	14.8	9.1	16.0	19.0	30
25	12.0	7.1	13.0	16.4	25
20	9.4	4.9	10.0	11.7	20
15	7.0	3.0	7.0	8.3	15
10	4.6	1.6	4.0	5.0	10
5	2.2	0.6	2.0	2.6	5
0	0.0	0.0	0.0	0.0	0

APPENDIX

TO

THE HYGROMETRICAL TABLES.

TABLES

FOR

COMPARING THE QUANTITIES OF RAIN-WATER.

The three kinds of measures which are most in use for noting the quantities of rain and melted snow, are the Centimetres and Millimetres in France, the Paris or French inches and lines in Germany, and the English inches and decimals in England, America, and also in Russia, the Russian foot being the same as the English foot. The following tables will facilitate the comparison of these various measures with each other.

A glance at the tables will show that the first column on the left contains the numbers to be converted, and the heads of the following columns the fractions of these numbers, or units, each of which is one tenth of those in the first column. Shorter tables, at the bottom, give, when necessary, the value of proportional parts still smaller than those found in the larger tables.

Example.

Let 13 Centimetres be converted into French inches and lines.

Take, in Table II., the line beginning with 10 Centimetres in the first column, follow that line as far as the column headed 3 Centimetres, and there will be found the number of 4 inches 9.63 lines, which is the corresponding value in French inches of 10 + 3, or 13 Centimetres.

If the number is followed by a fraction, as for instance, 13.5 Centimetres, or 135 Millimetres, we find, —

			French Inches. Lines.
In the larger table	13	Centimetres =	4 .9,63
In the smaller table at the bottom	5	Millimetres =	.2,216
	Or 13.5	Centimetres =	4.11,846

When the measures which are to be compared are both subdivided into decimal parts, the equivalents of the numbers greater than 9.9 may be found by moving the decimal point.

Example.

Let 346.7 Centimetres be converted into English inches.

In Table I., in the column headed 4, on the fourth line,

we find	3.4 Centimetres =	1.3386 English inches.
Moving the decimal point by two places we have	340 Centimetres =	133.86 English inches.
Then, in the column headed 7, on the line beginning with 6, we find	6.7 Centimetres =	2.64
Making together	346.7 Centimetres =	136.50 English inches.

I. CONVERSION OF CENTIMETRES INTO ENGLISH INCHES AND DECIMALS.

1 Centimetre = 0.3937079 English Inch.

Centi-metres.	Millimetres.									
	0.	**1.**	**2.**	**3.**	**4.**	**5.**	**6.**	**7.**	**8.**	**9.**
	Eng.Inch.	Eng.Inch.	Eng.Inch.	Eng.Inch.	Eng.Inch.	Eng.Inch.	Eng.Inch.	Eng.Inch.	Eng.Inch	Eng.Inch.
0	0.0000	0.0394	0.0787	0.1181	0.1575	0.1969	0.2362	0.2756	0.3150	0.3543
1	0.3937	0.4331	0.4724	0.5118	0.5512	0.5906	0.6299	0.6693	0.7087	0.7480
2	0.7874	0.8268	0.8662	0.9055	0.9449	0.9843	1.0236	1.0630	1.1024	1.1418
3	1.1811	1.2205	1.2599	1.2992	1.3386	1.3780	1.4173	1.4567	1.4961	1.5355
4	1.5748	1.6142	1.6536	1.6929	1.7323	1.7717	1.8111	1.8504	1.8898	1.9292
5	1.9685	2.0079	2.0473	2.0867	2.1260	2.1654	2.2048	2.2441	2.2835	2.3229
6	2.3622	2.4016	2.4410	2.4804	2.5197	2.5591	2.5985	2.6378	2.6772	2.7166
7	2.7560	2.7953	2.8347	2.8741	2.9134	2.9528	2.9922	3.0316	3.0709	3.1103
8	3.1497	3.1890	3.2284	3.2678	3.3071	3.3465	3.3859	3.4253	3.4646	3.5040
9	3.5434	3.5827	3.6221	3.6615	3.7009	3.7402	3.7796	3.8190	3.8583	3.8977

II. CONVERSION OF CENTIMETRES INTO FRENCH INCHES, LINES, AND DECIMALS.

1 Centimetre = 0. inches 4.43296 Paris lines.

Centi-metres.	Units.									
	0.	**1.**	**2.**	**3.**	**4.**	**5.**	**6.**	**7.**	**8.**	**9.**
	Fr.In. Lin.	Fr.In. Lin.	Fr.In. Lin.	Fr.In. Lin.	Fr.In. Lin.	Fr.In. Lin.	Fr.In. Lin.	Fr.In. Lin.	Fr.In. Lin.	Fr.In. Lin.
0	0. 0,00	0. 4,43	0. 8,87	1. 1,30	1. 5,73	1.10,16	2. 2,60	2. 7,03	2.11,46	3. 3,90
10	3. 8,33	4. 0,76	4. 5,20	4. 9,63	5. 2,06	5. 6,49	5.10,93	6. 3,36	6. 7,79	7. 0,23
20	7. 4,66	7. 9,09	8. 1,53	8. 5,96	8.10,39	9. 2,82	9. 7,26	9.11,69	10. 4,12	10. 8,56
30	11, 0,99	11. 5,42	11. 9,85	12. 2,29	12. 6,72	12.11,15	13. 3,59	13. 8,02	14. 0,45	14. 4,89
40	14. 9,32	15. 1,75	15. 6,18	15.10,62	16. 3,05	16. 7,48	16.11,92	17. 4,35	17. 8,78	18. 1,22
50	18. 5,65	18.10,08	19. 2,51	19. 6,95	19.11,38	20. 3,81	20. 8,25	21. 0,68	21. 5,11	21. 9,54
60	22. 1,98	22. 6,41	22.10,84	23. 3,28	23. 7,71	24. 0,14	24. 4,58	24. 9,01	25. 1,44	25. 5,87
70	25.10,31	26. 2,74	26. 7,17	26.11,61	27. 4,04	27. 8,47	28. 0,90	28. 5,34	28. 9,77	29. 2,20
80	29. 6,64	29.11,07	30. 3,50	30. 7,93	31. 0,37	31. 4,80	31. 9,23	32. 1,67	32. 6,10	32.10,53
90	33. 2,97	33. 7,40	33.11,83	34. 4,26	34. 8,70	35. 1,13	35. 5,56	35.10,00	36. 2,43	36 6,86
	Centim.	Fr.In. Lin.	Centim.	Fr.In. Lin.	Centim.	Fr.In. Lin.	Centim.	Fr.In. Lin.	Centim.	Fr.In. Lin.
	100	36.11,30	200	73.10.59	300	110.9,89	400	147.9,18	500	184.8,48

CONVERSION OF CENTIMETRES INTO FRENCH LINES AND DECIMALS.

Centi-metres.	Units.									
	0.	**1.**	**2.**	**3.**	**4.**	**5.**	**6.**	**7.**	**8.**	**9.**
	Fr. Lines.	Fr. Lines.	Fr. Lines.	Fr. Lines.	Fr. Lines.	Fr. Lines.	Fr. Lines.	Fr. Lines.	Fr. Lines.	Fr. Lines.
0	0.00	4.43	8.87	13.30	17.73	22.16	26.60	31.03	35.46	39.90
10	44.33	48.76	53.20	57.63	62.06	66.49	70.93	75.36	79.79	84.23
20	88.66	93.09	97.53	101.96	106.39	110.82	115.26	119.69	124.12	128.56
30	132.99	137.42	141.85	146.29	150.72	155.15	159.59	164.02	168.45	172.89
40	177.32	181.75	186.18	190.62	195.05	199.48	203.92	208.35	212.78	217.22
50	221.65	226.08	230.51	234.95	239.38	243.81	248.25	252.68	257.11	261.54
60	265.98	270.41	274.84	279.28	283.71	288.14	292.58	297.01	301.44	305.87
70	310.31	314.74	319.17	323.61	328.04	332.47	336.90	341.34	345.77	350.20
80	354.64	359.07	363.50	367.93	372.37	376.80	381.23	385.67	390.10	394.53
90	398.97	403.40	407.83	412.26	416.70	421.13	425.56	430.00	434.43	438.86

CONVERSION OF MILLIMETRES INTO FRENCH LINES AND DECIMALS.

0.	**1.**	**2.**	**3.**	**4.**	**5.**	**6.**	**7.**	**8.**	**9.**
Fr. Lines.	Fr. Lines.	Fr. Lines.	Fr. Lines.	Fr. Lines	Fr. Lines.	Fr. Lines.	Fr. Lines.	Fr. Lines.	Fr. Lines.
0.0	0.443	0.887	1.330	1.773	2.216	2.660	3.103	3.546	3.990

III. CONVERSION OF ENGLISH INCHES INTO CENTIMETRES.

1 English Inch = 2.53995 Centimetres.

English Inches.	Units.									
	0.	**1.**	**2.**	**3.**	**4.**	**5.**	**6.**	**7.**	**8.**	**9.**
	Centim.	Centim.	Centim.	Centim.	Centim.	Centim.	Centim.	Centim.	Centim.	Centim.
0	0.00	2.54	5.08	7.62	10.16	12.70	15.24	17.78	20.32	22.86
10	25.40	27.94	30.48	33.02	35.56	38.10	40.64	43.18	45.72	48.26
20	50.80	53.34	55.88	58.42	60.96	63.50	66.04	68.58	71.12	73.66
30	76.20	78.74	81.28	83.82	86.36	88.90	91.44	93.98	96.52	99.06
40	101.60	104.14	106.68	109.22	111.76	114.30	116.84	119.38	121.92	124.46
50	127.00	129.54	132.08	134.62	137.16	139.70	142.24	144.78	147.32	149.86
60	152.40	154.94	157.48	160.02	162.56	165.10	167.64	170.18	172.72	175.26
70	177.80	180.34	182.88	185.42	187.96	190.50	193.04	195.58	198.12	200.66
80	203.20	205.74	208.28	210.82	213.36	215.90	218.44	220.98	223.52	226.06
90	228.60	231.14	233.68	236.22	238.76	241.30	243.84	246.38	248.92	251.46
100	254.00	256.54	259.08	261.62	264.16	266.70	269.24	271.78	274.32	276.85
110	279.39	281.93	284.47	287.01	289.55	292.09	294.63	297.17	299.71	302.25
120	304.79	307.33	309.87	312.41	314.95	317.49	320.03	322.57	325.11	327.65
130	330.19	332.73	335.27	337.81	340.35	342.89	345.43	347.97	350.51	353.05
140	355.59	358.13	360.67	363.21	365.75	368.29	370.83	373.37	375.91	378.45
150	380.99	383.53	386.07	388.61	391.15	393.69	396.23	398.77	401.31	403.85
160	406.39	408.93	411.47	414.01	416.55	419.09	421.63	424.17	426.71	429.25
170	431.79	434.33	436.87	439.41	441.95	444.49	447.03	449.57	452.11	454.65
180	457.19	459.73	462.27	464.81	467.35	469.89	472.43	474.97	477.51	480.05
190	482.59	485.13	487.67	490.21	492.75	495.29	497.83	500.37	502.91	505.45
200	507.99	510.53	513.07	515.61	518.15	520.69	523.23	525.77	528.31	530.85
	Tenths of an Inch.									
	0.	**1.**	**2.**	**3.**	**4.**	**5.**	**6.**	**7.**	**8.**	**9.**
	Centim.	Centim.	Centim.	Centim.	Centim.	Centim.	Centim.	Centim.	Centim.	Centim.
	0.000	0.254	0.508	0.762	1.016	1.270	1.524	1.778	2.032	2.286

IV. CONVERSION OF ENGLISH INCHES INTO FRENCH INCHES AND LINES.

1 English Inch = 0. inches 11.2595 Paris lines.

Eng. Inches.	Units.									
	0.	**1.**	**2.**	**3.**	**4.**	**5.**	**6.**	**7.**	**8.**	**9.**
	Fr.In. Lin.	Fr.In. Lin.	Fr.In. Lin.	Fr.In. Lin.	Fr.In. Lin.	Fr.In. Lin.	Fr.In. Lin.	Fr.In. Lin.	Fr.In. Lin.	Fr.In. Lin.
0	0. 0,00	0.11,26	1.10,52	2. 9,78	3. 9,04	4. 8,30	5. 7,56	6. 6,82	7. 6,08	8. 5,34
10	9. 4,59	10. 3,85	11. 3,11	12. 2,37	13. 1,63	14. 0,89	15. 0,15	15.11,41	16.10,67	17. 9,93
20	18. 9,19	19. 8,45	20. 7,71	21. 6,97	22. 6,23	23. 5,49	24. 4,75	25. 4,01	26. 3,27	27 2,53
30	28. 1,78	29. 1,04	30. 0,30	30.11,56	31.10,82	32.10,03	33. 9,34	34. 8,60	35. 7,86	36. 7,12
40	37. 6,38	38. 5,64	39. 4,90	40. 4,16	41. 3,42	42. 2,68	43. 1,94	44. 1,20	45. 0,46	45.11,72
50	46.10,97	47.10,23	48. 9,49	49. 8,75	50. 8,01	51. 7,27	52. 6,53	53. 5,79	54. 5,05	55. 4,31
60	56. 3,57	57. 2,83	58. 2,09	59. 1,35	60. 0,61	60.11,87	61.11,13	62.10,39	63. 9,65	64. 8,91
70	65. 8,16	66. 7,42	67. 6,68	68. 5,94	69. 5,20	70. 4,46	71. 3,72	72. 2,98	73. 2,24	74. 1,50
80	75. 0,76	76. 0,02	76.11,28	77.10,54	78. 9,80	79. 9,06	80. 8,32	81. 7,58	82. 6,84	83. 6,10
90	84. 5,35	85. 4,61	86. 3,87	87. 3,13	88. 2,39	89. 1,65	90. 0,91	91. 0,17	91.11,43	92.10,69
	Eng. Inch.	Fr.In. Lin.	Eng.Inch.	Fr.In. Lin.	Eng.Inch.	Fr.In. Lin.	Eng.Inch.	Fr.In. Lin.	Eng.Inch.	Fr.In. Lin.
	100	93.9,95	200	187.7,90	300	281.5,85	400	375.3,80	500	469.1,75
	Tenths of an Inch.									
	0.	**1.**	**2.**	**3.**	**4.**	**5.**	**6.**	**7.**	**8.**	**9.**
	Fr.In. Lin.	Fr.In. Lin.	Fr.In. Lin.	Fr.In. Lin.	Fr.In. Lin.	Fr.In. Lin.	Fr.In. Lin.	Fr.In. Lin.	Fr.In. Lin.	Fr.In. Lin.
	0.0,00	0.1,13	0.2,25	0.3,38	0.4,50	0.5,63	0.6,76	0.7,88	0.9,01	0.10,13

V. CONVERSION OF FRENCH INCHES INTO CENTIMETRES.

1 French Inch = 2.7070 Centimetres.

French Inches.	Units.									
	0.	**1.**	**2.**	**3.**	**4.**	**5.**	**6.**	**7.**	**8.**	**9.**
	Centim.	Centim.	Centim.	Centim.	Centim.	Centim.	Centim.	Centim.	Centim.	Centim.
0	0.00	2.71	5.41	8.12	10.83	13.53	16.24	18.95	21.66	24.36
10	27.07	29.78	32.48	35.19	37.90	40.60	43.31	46.02	48.73	51.43
20	54.14	56.85	59.55	62.26	64.97	67.67	70.38	73.09	75.80	78.50
30	81.21	83.92	86.62	89.33	92.04	94.74	97.45	100.16	102.87	105.57
40	108.28	110.99	113.69	116.40	119.11	121.81	124.52	127.23	129.94	132.64
50	135.35	138.06	140.76	143.47	146.18	148.88	151.59	154.30	157.01	159.71
60	162.42	165.13	167.83	170.54	172.25	175.95	178.66	181.37	184.08	186.78
70	189.49	192.20	194.90	197.61	200.32	203.02	205.73	208.44	211.15	213.85
80	216.56	219.27	221.97	224.68	227.39	230.09	232.80	235.51	238.22	240.92
90	243.63	246.34	249.04	251.75	254.46	257.16	259.87	262.58	265.29	267.99
100	270.70	273.41	276.11	278.82	281.53	284.23	286.94	289.65	292.36	295.06
110	297.77	300.48	303.18	305.89	308.60	311.30	314.01	316.72	319.42	322.13
120	324.84	327.55	330.25	332.96	335.67	338.37	341.08	343.79	346.49	349.20
130	351.91	354.62	357.32	360.03	362.74	365.44	368.15	370.86	373.56	376.27
140	378.98	381.69	384.39	387.10	389.81	392.51	395.22	397.93	400.63	403.34
150	406.05	408.76	411.46	414.17	416.88	419.58	422.29	425.00	427.70	430.41
160	433.12	435.83	438.53	441.24	443.95	446.65	449.36	452.07	454.77	457.48
170	460.19	462.90	465.60	468.31	471.02	473.72	476.43	479.14	481.84	484.55
180	487.26	489.97	492.67	495.38	498.09	500.79	503.50	506.21	508.91	511.62
190	514.33	517.04	519.74	522.45	525.16	527.86	530.57	533.28	535.98	538.69
200	541.40	544.11	546.81	549.52	552.23	554.93	557.64	560.35	563.05	565.76

CONVERSION OF FRENCH LINES INTO CENTIMETRES.

1 French Line = 0.22558 Centimetre.

French Lines.	Tenths of a Line.									
	0.	**1.**	**2.**	**3.**	**4.**	**5.**	**6.**	**7.**	**8.**	**9.**
	Centim.	Centim.	Centim.	Centim.	Centim.	Centim.	Centim.	Centim.	Centim.	Centim.
0	0.000	0.023	0.045	0.068	0.090	0.113	0.135	0.158	0.180	0.203
1	0.226	0.248	0.271	0.293	0.316	0.338	0.361	0.383	0.406	0.429
2	0.451	0.474	0.496	0.519	0.541	0.564	0.587	0.609	0.632	0.654
3	0.677	0.699	0.722	0.744	0.767	0.790	0.812	0.835	0.857	0.880
4	0.902	0.925	0.947	0.970	0.993	1.015	1.038	1.060	1.083	1.105
5	1.128	1.150	1.173	1.196	1.218	1.241	1.263	1.286	1.308	1.331
6	1.353	1.376	1.399	1.421	1.444	1.466	1.489	1.511	1.534	1.557
7	1.579	1.602	1.624	1.647	1.669	1.692	1.714	1.737	1.760	1.782
8	1.805	1.827	1.850	1.872	1.895	1.917	1.940	1.963	1.985	2.008
9	2.030	2.053	2.075	2.098	2.120	2.143	2.166	2.188	2.211	2.233
10	2.256	2.278	2.301	2.324	2.346	2.369	2.391	2.414	2.436	2.459
11	2.481	2.504	2.527	2.549	2.572	2.594	2.617	2.639	2.662	2.684
12	2.707	2.730	2.752	2.775	2.797	2.820	2.842	2.865	2.887	2.910

VI. CONVERSION OF FRENCH INCHES INTO ENGLISH INCHES AND DECIMALS.

1 French Inch = 1.065765 English Inch.

French Inches.	Units.									
	0.	**1.**	**2.**	**3.**	**4.**	**5.**	**6.**	**7.**	**8.**	**9.**
	Eng Inch.	Eng.Inch.	Eng.Inch.	Eng.Inch.	Eng.Inch.	Eng.Inch.	Eng.Inch.	Eng.Inch.	Eng.Inch.	Eng.Inch.
0	0.000	1.066	2.132	3.197	4.263	5.329	6.395	7.460	8.526	9.592
10	10.658	11.723	12.789	13.855	14.921	15.986	17.052	18.118	19.184	20.250
20	21.315	22.381	23.447	24.513	25.578	26.644	27.710	28.776	29.841	30.907
30	31.973	33.039	34.104	35.170	36.236	37.302	38.368	39.433	40.499	41.565
40	42.631	43.696	44.762	45.828	46.894	47.959	49.025	50.091	51.157	52.222
50	53.288	54.354	55.420	56.486	57.551	58.617	59.683	60.749	61.814	62.880
60	63.946	65.012	66.077	67.143	68.209	69.275	70.340	71.407	72.472	73.538
70	74.604	75.669	76.735	77.801	78.867	79.932	80.998	82.064	83.130	84.195
80	85.261	86.327	87.393	88.458	89.524	90.590	91.656	92.722	93.787	94.853
90	95.919	96.985	98.050	99.116	100.182	101.248	102.314	103.379	104.445	105.511
100	106.576	107.642	108.708	109.774	110.840	111.905	112.971	114.037	115.103	116.168
110	117.234	118.300	119.366	120.431	121.497	122.563	123.629	124.695	125.760	126.826
120	127.892	128.958	130.023	131.089	132.155	133.221	134.286	135.352	136.418	137.484
130	138.549	139.615	140.681	141.747	142.813	143.878	144.944	146.010	147.076	148.141
140	149.207	150.273	151.339	152.404	153.470	154.536	155.602	156.667	157.733	158.799
150	159.865	160.931	161.996	163.062	164.128	165.194	166.259	167.325	168.391	169.457
160	170.522	171.588	172.654	173.720	174.785	175.851	176.917	177.983	179.049	180.114
170	181.180	182.246	183.312	184.377	185.443	186.509	187.575	188.640	189.706	190.772
180	191.838	192.903	193.969	195.035	196.101	197.167	198.232	199.298	200.364	201.430
190	202.495	203.561	204.627	205.693	206.758	207.824	208.890	209.956	211.021	212.087
200	213.153	214.219	215.285	216.350	217.416	218.482	219.548	220.613	221.679	222.745

CONVERSION OF FRENCH LINES INTO ENGLISH INCHES.

1 French Line = 0.088814 English Inch.

French Lines.	Tenths of a Line.									
	0.	**1.**	**2.**	**3.**	**4.**	**5.**	**6.**	**7.**	**8.**	**9.**
	Eng.Inch.	Eng.Inch.	Eng.Inch.	Eng.Inch.	Eng Inch.	Eng.Inch.	Eng Inch.	Eng.Inch	Eng.Inch.	Eng Inch.
0	0.0000	0.0089	0.0178	0.0266	0.0355	0.0444	0.0533	0.0622	0.0711	0.0799
1	0.0888	0.0977	0.1066	0.1155	0.1243	0.1332	0.1421	0.1510	0.1599	0.1687
2	0.1776	0.1865	0.1954	0.2043	0.2132	0.2220	0.2309	0.2398	0.2487	0.2576
3	0.2664	0.2753	0.2842	0.2931	0.3020	0.3108	0.3197	0.3286	0.3375	0.3464
4	0.3553	0.3641	0.3730	0.3819	0.3908	0.3997	0.4085	0.4174	0.4263	0.4352
5	0.4441	0.4530	0.4618	0.4707	0.4796	0.4885	0.4974	0.5062	0.5151	0.5240
6	0.5329	0.5418	0.5506	0.5595	0.5684	0.5773	0.5862	0.5951	0.6039	0.6128
7	0.6217	0.6306	0.6395	0.6483	0.6572	0.6661	0.6750	0.6839	0.6927	0.7016
8	0.7105	0.7194	0.7283	0.7372	0.7460	0.7549	0.7638	0.7727	0.7816	0.7904
9	0.7993	0.8082	0.8171	0.8260	0.8349	0.8437	0.8526	0.8615	0.8704	0.8793
10	0.8881	0.8970	0.9059	0.9148	0.9237	0.9325	0.9414	0.9503	0.9592	0.9681
11	0.9770	0.9858	0.9947	1.0036	1.0125	1.0214	1.0302	1.0391	1.0480	1.0569
12	1.0658	1.0746	1.0835	1.0924	1.1013	1.1102	1.1191	1.1279	1.1368	1.1457

METEOROLOGICAL TABLES.

III.

BAROMETRICAL TABLES.

CONTENTS.

(The figures refer to the folio at the bottom of the page.)

For Comparing the different Barometrical Scales.

For Comparing Barometrical Differences.

For Reducing Barometrical Observations to the Freezing Point.

For Correcting Barometrical Observations for Capillary Action.

COMPARISON

OF

THE BAROMETRICAL SCALES,

OR

TABLES

FOR CONVERTING THE INDICATIONS OF THE ENGLISH, METRICAL, OLD FRENCH, AND RUSSIAN BAROMETERS INTO EACH OTHER.

COMPARISON

OF

THE BAROMETRICAL SCALES.

THE following tables are intended for converting into each other the four most important Barometrical Scales. They are sufficiently detailed to save the labor of any calculation or even of interpolation for the ordinary wants of Meteorology. But before making use of them, for comparing the observations taken with barometers of different scales, it is necessary *to reduce the observed heights to the temperature of the freezing point*, or to any other temperature, provided it be the same for all, by means of the tables calculated for this purpose, and which will be found below. The reason of it may be readily understood.

The length of the bars of metal, or of other substances, which represent the standard measures of length which obtain among different nations, varying with the temperature, it was necessary to determine a fixed point of temperature at which they really ought to have the length adopted as the standard unit of measure. This temperature is the *normal* temperature of the standard, and the length of the standard-bar, at this temperature, is the *true* length of it.

If the normal temperature of the various standards used for dividing Barometrical Scales were the same, the heights of the barometrical column, taken with these scales, could be compared directly, provided the scales be made of the same substance, brass, for instance, because their variations above or below this normal temperature would remain parallel with each other. But unfortunately it is not so. The English Yard is a standard at the temperature of 62° Fahrenheit; the Old French Toise, at 13° Reaumur; the Metre, at the freezing point, or zero Centigrade. Thus metallic rods intended to represent these various units of measure give the true or standard length only when at these respective temperatures; at any other temperature they are longer or shorter than the standard, and their subdivisions, inches, lines, or millimetres, partake of the error.

It is obvious, therefore, that the barometrical heights, taken with different scales, cannot be compared *directly* by means of the following tables, which give the relation between these scales at their respective *normal* temperatures. For suppose the temperature of the three barometers to be the freezing point, or 32° Fahrenheit, the scale of the Metrical Barometer alone will actually represent the standard length, and the millimeters will have the true length; while the inches and lines of the Old French and of the English Barometers will be too short, causing thus the barometrical column to appear too high. If the temperature of the instruments be 62° Fahrenheit, the divisions of the English Barometer will have the true standard length, and those of the Old French Barometer nearly so; but the millimeters of the Metrical Barometer will be too long, causing the barometrical column to appear too low. It is to neutralize the effect of those inequalities arising from the expansion of the scale, that it is necessary, *before* comparing the observations taken with the three barometers, to reduce them to the same temperature. This is done by means of the tables above mentioned, for reducing the barometer to the freezing point, which suppose the scales to be of brass from top to bottom, and which take into account the expansion or contraction they undergo by the variations of temperature.

But in doing so, we must be aware that the accuracy of the comparison depends in part upon the correctness of the indications of the attached thermometers, which determine the amount of the correction to be applied for reducing the barometers to the freezing point. If the thermometers do not agree, an error is introduced which will affect the height of the reduced columns, and the final comparison. Therefore the correction of the attached thermometers ought to be ascertained and applied to them *before* the reduction is made; or if this correction is unknown, it will be well to place the instruments to be compared in the most favorable conditions for taking the same temperature, and then to take the temperature given by *one* of the thermometers to reduce both barometers. If the correction of the attached thermometer has not been applied before the reduction, it will be contained, *after* the reduction, in the total correction of the instrument. If it be so, this circumstance must be indicated.

In computing the following tables, the value of the Metre, as determined by Capt. Kater, (Philosoph. Transact. for 1818, p. 109, and Baily's Astronomical Tables, p. 192,) has been adopted, viz. 1 Metre, at 0° Centigrade = 39.37079 English inches, at 62° Fahrenheit. The relation of the Metre (legal) to the Old French system of measures is known to be 1 Metre = 443.296 French or Paris lines. From these equations are derived the elements used in the computations, which are found at the head of each table.

Besides the larger Tables I. – VIII., a set of smaller ones, Tables IX. – XVI., has been added, which will be found useful for comparing Barometrical differences, such as ranges, amount of variation in a given time, &c., expressed in measures of different scales, in which only small quantities occur that are not found in the large tables.

I.–II.

COMPARISON

OF

THE ENGLISH BAROMETER

WITH

THE METRICAL AND THE OLD FRENCH BAROMETERS,

OR

TABLES

FOR CONVERTING ENGLISH INCHES INTO MILLIMETRES, AND INTO FRENCH OR PARIS LINES AND DECIMALS;

GIVING THE VALUES CORRESPONDING TO EVERY TENTH OF AN INCH, FROM 9 TO 19 INCHES; AND TO EVERY HUNDREDTH, FROM 19 TO 31.5 ENGLISH INCHES.

USE OF TABLE I.

Example.

THE English Barometer reads 20.657 inches. What would be the corresponding height in the Metrical Barometer?

In Table I., first column on the left, look out the line of 20 inches 6 tenths; on that line, in the sixth column, headed 5 hundredths, is found the value in millimetres for

	20.65	inches	=	524.50	millimetres.
At the bottom of the page, for	0.007	"	=	0.18	"
Or for	20.657	"	=	524.68	"

which would be the reading of the Metrical Barometer.

This example may serve for all tables, throughout the volume, which are constructed on the same plan.

1 English Inch = 25.39954 Millimetres.

English Inches.	Tenths of an Inch. 0.	1.	2.	3.	4.	5.	6.	7.	8.	9.
	Millim.	Millim.	Millim.	Millim.	Millim.	Millim.	Millim.	Millim.	Millim.	Millim.
9	228.60	231.14	233.68	236.22	238.76	241.30	243.84	246.38	248.92	251.46
10	254.00	256.54	259.08	261.62	264.16	266.70	269.24	271.78	274.32	276.85
11	279.39	281.93	284.47	287.01	289.55	292.09	294.63	297.17	299.71	302.25
12	304.79	307.33	309.87	312.41	314.95	317.49	320.03	322.57	325.11	327.65
13	330.19	332.73	335.27	337.81	340.35	342.89	345.43	347.97	350.51	353.05
14	355.59	358.13	360.67	363.21	365.75	368.29	370.83	373.37	375.91	378.45
15	380.99	383.53	386.07	388.61	391.15	393.69	396.23	398.77	401.31	403.85
16	406.39	408.93	411.47	414.01	416.55	419.09	421.63	424.17	426.71	429.25
17	431.79	434.33	436.87	439.41	441.95	444.49	447.03	449.57	452.11	454.65
18	457.19	459.73	462.27	464.81	467.35	469.89	472.43	474.97	477.51	480.05

English Inches and tenths.	Hundredths of an Inch. 0.	1.	2.	3.	4.	5.	6.	7.	8.	9.
	Millim.	Millim.	Millim.	Millim.	Millim.	Millim.	Millim.	Millim.	Millim.	Millim.
19.0	482.59	482.85	483.10	483.35	483.61	483.86	484.12	484.37	484.62	484.88
1	485.13	485.39	485.64	485.89	486.15	486.40	486.66	486.91	487.16	487.42
2	487.67	487.93	488.18	488.43	488.69	488.94	489.20	489.45	489.70	489.96
3	490.21	490.47	490.72	490.97	491.23	491.48	491.74	491.99	492.24	492.50
4	492.75	493.01	493.26	493.51	493.77	494.02	494.28	494.53	494.78	495.04
5	495.29	495.55	495.80	496.05	496.31	496.56	496.81	497.07	497.32	497.58
6	497.83	498.08	498.34	498.59	498.85	499.10	499.35	499.61	499.86	500.12
7	500.37	500.62	500.88	501.13	501.39	501.64	501.89	502.15	502.40	502.66
8	502.91	503.16	503.42	503.67	503.93	504.18	504.43	504.69	504.94	505.20
9	505.45	505.70	505.96	506.21	506.47	506.72	506.97	507.23	507.48	507.74
20.0	507.99	508.24	508.50	508.75	509.01	509.26	509.51	509.77	510.02	510.28
1	510.53	510.78	511.04	511.29	511.55	511.80	512.05	512.31	512.56	512.82
2	513.07	513.32	513.58	513.83	514.09	514.34	514.59	514.85	515.10	515.36
3	515.61	515.86	516.12	516.37	516.63	516.88	517.13	517.39	517.64	517.90
4	518.15	518.40	518.66	518.91	519.17	519.42	519.67	519.93	520.18	520.44
5	520.69	520.94	521.20	521.45	521.71	521.96	522.21	522.47	522.72	522.98
6	523.23	523.48	523.74	523.99	524.25	524.50	524.75	525.01	525.26	525.52
7	525.77	526.02	526.28	526.53	526.79	527.04	527.29	527.55	527.80	528.06
8	528.31	528.56	528.82	529.07	529.33	529.58	529.83	530.09	530.34	530.60
9	530.85	531.10	531.36	531.61	531.87	532.12	532.37	532.63	532.88	533.14

Thousandths of an Inch.

0.	1.	2.	3.	4.	5.	6.	7.	8.	9.
0.0	0.03	0.05	0.08	0.10	0.13	0.15	0.18	0.20	0.23

English Inches and tenths.	Hundredths of an Inch.									
	0.	**1.**	**2.**	**3.**	**4.**	**5.**	**6.**	**7.**	**8.**	**9.**
	Millim.	Millim.	Millim.	Millim.	Millim.	Millim.	Millim.	Millim.	Millim.	Millim.
21.0	533.39	533.64	533.90	534.15	534.41	534.66	534.91	535.17	535.42	535.68
1	535.93	536.18	536.44	536.69	536.95	537.20	537.45	537.71	537.96	538.22
2	538.47	538.72	538.98	539.23	539.49	539.74	539.99	540.25	540.50	540.76
3	541.01	541.26	541.52	541.77	542.03	542.28	542.53	542.79	543.04	543.30
4	543.55	543.80	544.06	544.31	544.57	544.82	545.07	545.33	545.58	545.84
5	546.09	546.34	546.60	546.85	547.11	547.36	547.61	547.87	548.12	548.38
6	548.63	548.88	549.14	549.39	549.65	549.90	550.15	550.41	550.66	550.92
7	551.17	551.42	551.68	551.93	552.19	552.44	552.69	552.95	553.20	553.46
8	553.71	553.96	554.22	554.47	554.73	554.98	555.23	555.49	555.74	556.00
9	556.25	556.50	556.76	557.01	557.27	557.52	557.77	558.03	558.28	558.54
22.0	558.79	559.04	559.30	559.55	559.81	560.06	560.31	560.57	560.82	561.08
1	561.33	561.58	561.84	562.09	562.35	562.60	562.85	563.11	563.36	563.62
2	563.87	564.12	564.38	564.63	564.89	565.14	565.39	565.65	565.90	566.16
3	566.41	566.66	566.92	567.17	567.43	567.68	567.93	568.19	568.44	568.70
4	568.95	569.20	569.46	569.71	569.97	570.22	570.47	570.73	570.98	571.24
5	571.49	571.74	572.00	572.25	572.51	572.76	573.01	573.27	573.52	573.78
6	574.03	574.28	574.54	574.79	575.05	575.30	575.55	575.81	576.06	576.32
7	576.57	576.82	577.08	577.33	577.59	577.84	578.09	578.35	578.60	578.86
8	579.11	579.36	579.62	579.87	580.13	580.38	580.63	580.89	581.14	581.40
9	581.65	581.90	582.16	582.41	582.67	582.92	583.17	583.43	583.68	583.94
23.0	584.19	584.44	584.70	584.95	585.21	585.46	585.71	585.97	586.22	586.48
1	586.73	586.98	587.24	587.49	587.75	588.00	588.25	588.51	588.76	589.02
2	589.27	589.52	589.78	590.03	590.29	590.54	590.79	591.05	591.30	591.56
3	591.81	592.06	592.32	592.57	592.83	593.08	593.33	593.59	593.84	594.10
4	594.35	594.60	594.86	595.11	595.37	595.62	595.87	596.13	596.38	596.64
5	596.89	597.14	597.40	597.65	597.91	598.16	598.41	598.67	598.92	599.18
6	599.43	599.68	599.94	600.19	600.45	600.70	600.95	601.21	601.46	601.72
7	601.97	602.22	602.48	602.73	602.99	603.24	603.49	603.75	604.00	604.26
8	604.51	604.76	605.02	605.27	605.53	605.78	606.03	606.29	606.54	606.79
9	607.05	607.30	607.56	607.81	608.06	608.32	608.57	608.83	609.08	609.33
24.0	609.59	609.84	610.10	610.35	610.60	610.86	611.11	611.37	611.62	611.87
1	612.13	612.38	612.64	612.89	613.14	613.40	613.65	613.91	614.16	614.41
2	614.67	614.92	615.18	615.43	615.68	615.94	616.19	616.45	616.70	616.95
3	617.21	617.46	617.72	617.97	618.22	618.48	618.73	618.99	619.24	619.49
4	619.75	620.00	620.26	620.51	620.76	621.02	621.27	621.53	621.78	622.03

Thousandths of an Inch.									
0.	**1.**	**2.**	**3.**	**4.**	**5.**	**6.**	**7.**	**8.**	**9.**
0.0	0.03	0.05	0.08	0.10	0.13	0.15	0.18	0.20	0.23

English Inches and tenths.	Hundredths of an Inch.									
	0.	**1.**	**2.**	**3.**	**4.**	**5.**	**6.**	**7.**	**8.**	**9.**
	Millim.	Millim.	Millim.	Millim.	Millim.	Millim.	Millim.	Millim.	Millim.	Millim.
24.5	622.29	622.54	622.80	623.05	623.30	623.56	623.81	624.07	624.32	624.57
6	624.83	625.08	625.34	625.59	625.84	626.10	626.35	626.61	626.86	627.11
7	627.37	627.62	627.88	628.13	628.38	628.64	628.89	629.15	629.40	629.65
8	629.91	630.16	630.42	630.67	630.92	631.18	631.43	631.69	631.94	632.19
9	632.45	632.70	632.96	633.21	633.46	633.72	633.97	634.23	634.48	634.73
25.0	634.99	635.24	635.50	637.75	636.00	636.26	636.51	636.77	637.02	637.27
1	637.53	637.78	638.04	638.29	638.54	638.80	639.05	639.31	639.56	639.81
2	640.07	640.32	640.58	640.83	641.08	641.34	641.59	641.85	642.10	642.35
3	642.61	642.86	643.12	643.37	643.62	643.88	644.13	644.39	644.64	644.89
4	645.15	645.40	645.66	645.91	646.16	646.42	646.67	646.93	647.18	647.43
5	647.69	647.94	648.20	648.45	648.70	648.96	649.21	649.47	649.72	649.97
6	650.23	650.48	650.74	650.99	651.24	651.50	651.75	652.01	652.26	652.51
7	652.77	653.02	653.28	653.53	653.78	654.04	654.29	654.55	654.80	655.05
8	655.31	655.56	655.82	656.07	656.32	656.58	656.83	657.09	657.34	657.59
9	657.85	658.10	658.36	658.61	658.86	659.12	659.37	659.63	659.88	660.13
26.0	660.39	660.64	660.90	661.15	661.40	661.66	661.91	662.17	662.42	662.67
1	662.93	663.18	663.44	663.69	663.94	664.20	664.45	664.71	664.96	665.21
2	665.47	665.72	665.98	666.23	666.48	666.74	666.99	667.25	667.50	667.75
3	668.01	668.26	668.52	668.77	669.02	669.28	669.53	669.79	670.04	670.29
4	670.55	670.80	671.06	671.31	671.56	671.82	672.07	672.33	672.58	672.83
5	673.09	673.34	673.60	673.85	674.10	674.36	674.61	674.87	675.12	675.37
6	675.63	675.88	676.14	676.39	676.64	676.90	677.15	677.41	677.66	677.91
7	678.17	678.42	678.68	678.93	679.18	679.44	679.69	679.95	680.20	680.45
8	680.71	680.96	681.22	681.47	681.72	681.98	682.23	682.49	682.74	682.99
9	683.25	683.50	683.76	684.01	684.26	684.52	684.77	685.03	685.28	685.53
27.0	685.79	686.04	686.30	686.55	686.80	687.06	687.31	687.57	687.82	688.07
1	688.33	688.58	688.84	689.09	689.34	689.60	689.85	690.11	690.36	690.61
2	690.87	691.12	691.38	691.63	691.88	692.14	692.39	692.65	692.90	693.15
3	693.41	693.66	693.92	694.17	694.42	694.68	694.93	695.19	695.44	695.69
4	695.95	696.20	696.46	696.71	696.96	697.22	697.47	697.73	697.98	698.23
5	698.49	698.74	699.00	699.25	699.50	699.76	700.01	700.27	700.52	700.77
6	701.03	701.28	701.54	701.79	702.04	702.30	702.55	702.81	703.06	703.31
7	703.57	703.82	704.08	704.33	704.58	704.84	705.09	705.35	705.60	705.85
8	706.11	706.36	706.62	706.87	707.12	707.38	707.63	707.89	708.14	708.39
9	708.65	708.90	709.16	709.41	709.66	709.92	710.17	710.43	710.68	710.93

Thousandths of an Inch.									
0.	**1.**	**2.**	**3.**	**4.**	**5.**	**6.**	**7.**	**8.**	**9.**
0.0	0.03	0.05	0.08	0.10	0.13	0.15	0.18	0.20	0.23

English Inches and tenths.	Hundredths of an Inch.									
	0.	**1.**	**2.**	**3.**	**4.**	**5.**	**6.**	**7.**	**8.**	**9.**
	Millim.	Millim.	Millim.	Millim.	Millim.	Millim.	Millim.	Millim.	Millim.	Millim.
28.0	711.19	711.44	711.70	711.95	.712.20	712.46	712.71	712.97	713.22	713.47
1	713.73	713.98	714.24	714.49	714.74	715.00	715.25	715 51	715.76	716.01
2	716.27	716.52	716.78	717.03	717.28	717.54	717.79	718.04	718.30	718.55
3	718.81	719.06	719.31	719.57	719.82	720.08	720.33	720.58	720.84	721.09
4	721.35	721.60	721.85	722.11	722.36	722.62	722.87	723.12	723.38	723.63
5	723.89	724.14	724.39	724.65	724.90	725.16	725.41	725.66	725.92	726.17
6	726.43	726.68	726.93	727.19	727.44	727.70	727.95	728.20	728.46	728.71
7	728.97	729.22	729.47	729.73	729.98	730.24	730.49	730.74	731.00	731.25
8	731.51	731.76	732.01	732.27	732.52	732.78	733.03	733.28	733.54	732.79
9	734.05	734.30	734.55	734.81	735.06	735.32	735.57	735.82	736.08	736.33
29.0	736.59	736.84	737.09	737.35	737.60	737.86	738.11	738.36	738.62	738.87
1	739.13	739.38	739.63	739.89	740.14	740.40	740.65	740.90	741.16	741.41
2	741.67	741.92	742.17	742.43	742.68	742.94	743.19	743.44	743.70	743.95
3	744.21	744.46	744.71	744.97	745.22	745.48	745.73	745.98	746.24	746.49
4	746.75	747.00	747.25	747.51	747.76	748.02	748.27	748.52	748.78	749.03
5	749.29	749.54	749 79	750.05	750.30	750.56	750.81	751.06	751.32	751.57
6	751.83	752.08	752.33	752.59	752.84	753.10	753.35	753.60	753.86	754.11
7	754.37	754.62	754.87	755.13	755.38	755.64	755.89	756.14	756.40	756.65
8	756.91	757.16	757.41	757.67	757.92	758.18	758.43	758.68	758.94	759.19
9	759.45	759.70	759.95	760.21	760.46	760.72	760.97	761.22	761.48	761.73
30.0	761.99	762.24	762.49	762.75	763.00	763.26	763.51	763.76	764.02	764.27
1	764.53	764.78	765.03	765.29	765.54	765.80	766.05	766.30	766.56	766.81
2	767.07	767.32	767.57	767.83	768.08	768.34	768.59	768.84	769.10	769.35
3	769.61	769.86	770.11	770.37	770.62	770.88	771.13	771.38	771.64	771.89
4	772.15	772.40	772.65	772.91	773.16	773.42	773.67	773.92	774.18	774.43
5	774.69	774.94	775.19	775.45	775.70	775.96	776.21	776.46	776.72	776.97
6	777.23	777.48	777.73	777.99	778.24	778.50	778.75	779.00	779.26	779.51
7	779.77	780.02	780.27	780.53	780.78	781.04	781.29	781.54	781.80	782.05
8	782.31	782.56	782.81	783.07	783.32	783.58	783.83	784.08	784.34	784.59
9	784.85	785.10	785.35	785.61	785.86	786.12	786.37	786.62	786.88	787.13
31.0	787.39	787.64	787.89	788.15	788.40	788.66	788.91	789.16	789.42	789.67
1	789.93	790.18	790.43	790.69	790.94	791.20	791.45	791.70	791.96	792.21
2	792.47	792.72	792.97	793.23	793.48	793.74	793.99	794.24	794.50	794.75
3	795.01	795.26	795.51	795.77	796.02	796.28	796.53	796.78	797.04	797.29
4	797.55	797.80	798.05	798.31	798.56	798.82	799.07	799.32	799.58	799.83

Thousandths of an Inch.									
0.	**1.**	**2.**	**3.**	**4.**	**5.**	**6.**	**7.**	**8.**	**9.**
0.0	0.03	0.05	0.08	0.10	0.13	0.15	0.18	0.20	0.23

1 English Inch = 11.2595 French or Paris Lines.

English Inches.	Tenths of an Inch.									
	0.	1.	2.	3.	4.	5.	6.	7.	8.	9.
	Par.lines.	Par.lines.	Par.lines.	Par.lines.	Par.lines.	Par.lines.	Par.lines.	Par.lines.	Par.lines.	Par.lines.
11	123.85	124.98	126.11	127.23	128.36	129.48	130.61	131.74	132.86	133.99
12	135.11	136.24	137.37	138.49	139.62	140.74	141.87	143.00	144.12	145.25
13	146.37	147.50	148.63	149.75	150.88	152.00	153.13	154.26	155.38	156.51
14	157.63	158.76	159.88	161.01	162.14	163.26	164.39	165.51	166.64	167.77
15	168.89	170.02	171.14	172.27	173.40	174.52	175.65	176.77	177.90	179.03
16	180.15	181.28	182.40	183.53	184.66	185.78	186.91	188.03	189.16	190.29

Hundredths of an Inch.									
0.	1.	2.	3.	4.	5.	6.	7.	8.	9.
0.000	0.113	0.225	0.338	0.450	0.563	0.676	0.788	0.901	1.013

English Inches and Tenths.	Hundredths of an Inch.									
	0.	1.	2.	3.	4.	5.	6.	7.	8.	9.
	Par.lines.	Par.lines.	Par.lines.	Par.lines.	Par.lines.	Par.lines.	Par.lines.	Par.lines.	Par.lines.	Par.lines.
17.0	191.41	191.52	191.64	191.75	191.86	191.97	192.09	192.20	192.31	192.42
1	192.54	192.65	192.76	192.88	192.99	193.10	193.21	193.33	193.44	193.55
2	193.66	193.78	193.89	194.00	194.11	194.23	194.34	194.45	194.56	194.68
3	194.79	194.90	195.01	195.13	195.24	195.35	195.46	195.58	195.69	195.80
4	195.92	196.03	196.14	196.25	196.37	196.48	196.59	196.70	196.82	196.93
5	197.04	197.15	197.27	197.38	197.49	197.60	197.72	197.83	197.94	198.05
6	198.17	198.28	198.39	198.50	198.62	198.73	198.84	198.96	199.07	199.18
7	199.29	199.41	199.52	199.63	199.74	199.86	199.97	200.08	200.19	200.31
8	200.42	200.53	200.64	200.76	200.87	200.98	201.09	201.21	201.32	201.43
9	201.55	201.66	201.77	201.88	202.00	202.11	202.22	202.33	202.45	202.56
18.0	202.67	202.78	202.90	203.01	203.12	203.23	203.35	203.46	203.57	203.68
1	203.80	203.91	204.02	204.13	204.25	204.36	204.47	204.59	204.70	204.81
2	204.92	205.04	205.15	205.26	205.37	205.49	205.60	205.71	205.82	205.94
3	206.05	206.16	206.27	206.39	206.50	206.61	206.72	206.84	206.95	207.06
4	207.17	207.29	207.40	207.51	207.63	207.74	207.85	207.96	208.08	208.19
5	208.30	208.41	208.53	208.64	208.75	208.86	208.98	209.09	209.20	209.31
6	209.43	209.54	209.65	209.76	209.88	209.99	210.10	210.21	210.33	210.44
7	210.55	210.67	210.78	210.89	211.00	211.12	211.23	211.34	211.45	211.57
8	211.68	211.79	211.90	212.02	212.13	212.24	212.35	212.47	212.58	212.69
9	212.80	212.92	213.03	213.14	213.25	213.37	213.48	213.59	213.71	213.82
19.0	213.93	214.04	214.16	214.27	214.38	214.49	214.61	214.72	214.83	214.94
1	215.06	215.17	215.28	215.39	215.51	215.62	215.73	215.84	215.96	216.07
2	216.18	216.29	216.41	216.52	216.63	216.75	216.86	216.97	217.08	217.20
3	217.31	217.42	217.53	217.65	217.76	217.87	217.98	218.10	218.21	218.32
4	218.43	218.55	218.66	218.77	218.88	219.00	219.11	219.22	219.34	219.45
5	219.56	219.67	219.79	219.90	220.01	220.12	220.24	220.35	220.46	220.57
6	220.69	220.80	220.91	221.02	221.14	221.25	221.36	221.47	221.59	221.70
7	221.81	221.92	222.04	222.15	222.26	222.38	222.49	222.60	222.71	222.83
8	222.94	223.05	223.16	223.28	223.39	223.50	223.61	223.73	223.84	223.95
9	224.06	224.18	224.29	224.40	224.51	224.63	224.74	224.85	224.96	225.08

1 English Inch = 11.2595 French or Paris Lines.

English Inches and Tenths.	Hundredths of an Inch.									
	0.	**1.**	**2.**	**3.**	**4.**	**5.**	**6.**	**7.**	**8.**	**9.**
	Par.lines.	Par.lines.	Par.lines.	Par.lines.	Par.lines.	Par.lines.	Par.lines.	Par.lines.	Par.lines.	Par.lines.
20.0	225.19	225.30	225.42	225.53	225.64	225.75	225.87	225.98	226.09	226.20
1	226.32	226.43	226.54	226.65	226.77	226.88	226.99	227.10	227.22	227.33
2	227.44	227.55	227.67	227.78	227.89	228.00	228.12	228.23	228.34	228.46
3	228.57	228.68	228.79	228.91	229.02	229.13	229.24	229.36	229.47	229.58
4	229.69	229.81	229.92	230.03	230.14	230.26	230.37	230.48	230.59	230.71
5	230.82	230.93	231.04	231.16	231.27	231.38	231.50	231.61	231.72	231.83
6	231.95	232.06	232.17	232.28	232.40	232.51	232.62	232.73	232.85	232.96
7	233.07	233.18	233.30	233.41	233.52	233.63	233.75	233.86	233.97	234.09
8	234.20	234.31	234.42	234.54	234.65	234.76	234.87	234.99	235.10	235.21
9	235.32	235.44	235.55	235.66	235.77	235.89	236.00	236.11	236.22	236.34
21.0	236.45	236.56	236.67	236.79	236.90	237.01	237.13	237.24	237.35	237.46
1	237.58	237.69	237.80	237.91	238.03	238.14	238.25	238.36	238.48	238.59
2	238.70	238.81	238.93	239.04	239.15	239.26	239.38	239.49	239.60	239.71
3	239.83	239.94	240.05	240.17	240.28	240.39	240.50	240.62	240.73	240.84
4	240.95	241.07	241.18	241.29	241.40	241.52	241.63	241.74	241.85	241.97
5	242.08	242.19	242.30	242.42	242.53	242.64	242.75	242.87	242.98	243.09
6	243.21	243.32	243.43	243.54	243.66	243.77	243.88	243.99	244.11	244.22
7	244.33	244.44	244.56	244.67	244.78	244.89	245.01	245.12	245.23	245.34
8	245.46	245.57	245.68	245.79	245.91	246.02	246.13	246.25	246.36	246.47
9	246.58	246.70	246.81	246.92	247.03	247.15	247.26	247.37	247.48	247.60
22.0	247.71	247.82	247.93	248.05	248.16	248.27	248.38	248.50	248.61	248.72
1	248.83	248.95	249.06	249.17	249.29	249.40	249.51	249.62	249.74	249.85
2	249.96	250.07	250.19	250.30	250.41	250.52	250.64	250.75	250.86	250.97
3	251.09	251.20	251.31	251.42	251.54	251.65	251.76	251.88	251.99	252.10
4	252.21	252.33	252.44	252.55	252.66	252.78	252.89	253.00	253.11	253.23
5	253.34	253.45	253.56	253.68	253.79	253.90	254.01	254.13	254.24	254.35
6	254.46	254.58	254.69	254.80	254.92	255.03	255.14	255.25	255.37	255.48
7	255.59	255.70	255.82	255.93	256.04	256.15	256.27	256.38	256.49	256.60
8	256.72	256.83	256.94	257.05	257.17	257.28	257.39	257.50	257.62	257.73
9	257.84	257.96	258.07	258.18	258.29	258.41	258.52	258.63	258.74	258.86
23.0	258.97	259.08	259.19	259.31	259.42	259.53	259.64	259.76	259.87	259.98
1	260.09	260.21	260.32	260.43	260.54	260.66	260.77	260.88	261.00	261.11
2	261.22	261.33	261.45	261.56	261.67	261.78	261.90	262.01	262.12	262.23
3	262.35	262.46	262.57	262.68	262.80	262.91	263.02	263.13	263.25	263.36
4	263.47	263.58	263.70	263.81	263.92	264.04	264.15	264.26	264.37	264.49
5	264.60	264.71	264.82	264.94	265.05	265.16	265.27	265.39	265.50	265.61
6	265.72	265.84	265.95	266.06	266.17	266.29	266.40	266.51	266.62	266.74
7	266.85	266.96	267.08	267.19	267.30	267.41	267.53	267.64	267.75	267.86
8	267.98	268.09	268.20	268.31	268.43	268.54	268.65	268.76	268.88	268.99
9	269.10	269.21	269.33	269.44	269.55	269.67	269.78	269.89	270.00	270.12
	0.	**1.**	**2.**	**3.**	**4.**	**5.**	**6.**	**7.**	**8.**	**9.**

1 English Inch = 11.2595 French or Paris Lines.

English Inches and Tenths.	Hundredths of an Inch.									
	0.	**1.**	**2.**	**3.**	**4.**	**5.**	**6.**	**7.**	**8.**	**9.**
	Par.lines.	Par.lines.	Par.lines.	Par.lines.	Par.lines.	Par.lines.	Par.lines.	Par.lines.	Par.lines.	Par.lines.
24.0	270.23	270.34	270.45	270.57	270.68	270.79	270.90	271.02	271.13	271.24
1	271.35	271.47	271.58	271.69	271.80	271.92	272.03	272.14	272.25	272.37
2	272.48	272.59	272.71	272.82	272.93	273.04	273.16	273.27	273.38	273.49
3	273.61	273.72	273.83	273.94	274.06	274.17	274.28	274.39	274.51	274.62
4	274.73	274.84	274.96	275.07	275.18	275.29	275.41	275.52	275.63	275.75
5	275.86	275.97	276.08	276.20	276.31	276.42	276.53	276.65	276.76	276.87
6	276.98	277.10	277.21	277.32	277.43	277.55	277.66	277.77	277.88	278.00
7	278.11	278.22	278.33	278.45	278.56	278.67	278.79	278.90	279.01	279.12
8	279.24	279.35	279.46	279.57	279.69	279.80	279.91	280.02	280.14	280.25
9	280.36	280.47	280.59	280.70	280.81	280.92	281.04	281.15	281.26	281.38
25.0	281.49	281.60	281.71	281.83	281.94	282.05	282.16	282.28	282.39	282.50
1	282.61	282.73	282.84	282.95	283.06	283.18	283.29	283.40	283.51	283.63
2	283.74	283.85	283.96	284.08	284.19	284.30	284.41	284.53	284.64	284.75
3	284.87	284.98	285.09	285.20	285.32	285.43	285.54	285.65	285.77	285.88
4	285.99	286.10	286.22	286.33	286.44	286.55	286.67	286.78	286.89	287.00
5	287.12	287.23	287.34	287.46	287.57	287.68	287.79	287.91	288.02	288.13
6	288.24	288.36	288.47	288.58	288.69	288.81	288.92	289.03	289.14	289.26
7	289.37	289.48	289.59	289.71	289.82	289.93	290.04	290.16	290.27	290.38
8	290.50	290.61	290.72	290.83	290.95	291.06	291.17	291.28	291.40	291.51
9	291.62	291.73	291.85	291.96	292.07	292.18	292.30	292.41	292.52	292.63
26.0	292.75	292.86	292.97	293.08	293.20	293.31	293.42	293.54	293.65	293.76
1	293.87	293.99	294.10	294.21	294.32	294.44	294.55	294.66	294.77	294.89
2	295.00	295.11	295.22	295.34	295.45	295.56	295.67	295.79	295.90	296.01
3	296.12	296.24	296.35	296.46	296.58	296.69	296.80	296.91	297.03	297.14
4	297.25	297.36	297.48	297.59	297.70	297.81	297.93	298.04	298.15	298.26
5	298.38	298.49	298.60	298.71	298.83	298.94	299.05	299.17	299.28	299.39
6	299.50	299.62	299.73	299.84	299.95	300.07	300.18	300.29	300.40	300.52
7	300.63	300.74	300.85	300.97	301.08	301.19	301.30	301.42	301.53	301.64
8	301.75	301.87	301.98	302.09	302.20	302.32	302.43	302.54	302.66	302.77
9	302.88	302.99	303.11	303.22	303.33	303.44	303.56	303.67	303.78	303.89
27.0	304.01	304.12	304.23	304.34	304.46	304.57	304.68	304.79	304.91	305.02
1	305.13	305.25	305.36	305.47	305.58	305.70	305.81	305.92	306.03	306.15
2	306.26	306.37	306.48	306.60	306.71	306.82	306.93	307.05	307.16	307.27
3	307.38	307.50	307.61	307.72	307.83	307.95	308.06	308.17	308.29	308.40
4	308.51	308.62	308.74	308.85	308.96	309.07	309.19	309.30	309.41	309.52
5	309.64	309.75	309.86	309.97	310.09	310.20	310.31	310.42	310.54	310.65
6	310.76	310.87	310.99	311.10	311.21	311.33	311.44	311.55	311.66	311.78
7	311.89	312.00	312.11	312.23	312.34	312.45	312.56	312.68	312.79	312.90
8	313.01	313.13	313.24	313.35	313.46	313.58	313.69	313.80	313.91	314.03
9	314.14	314.25	314.37	314.48	314.59	314.70	314.82	314.93	315.04	315.15
	0.	**1.**	**2.**	**3.**	**4.**	**5.**	**6.**	**7.**	**8.**	**9.**

1 English Inch = 11.2595 French or Paris Lines.

English Inches and Tenths.	Hundredths of an Inch.									
	0.	**1.**	**2.**	**3.**	**4.**	**5.**	**6.**	**7.**	**8.**	**9.**
	Par.lines.	Par.lines.	Par.lines.	Par.lines.	Par.lines.	Par.lines.	Par.lines.	Par.lines.	Par.lines.	Par.lines.
28.0	315.27	315.38	315.49	315.60	315.72	315.83	315.94	316.05	316.17	316.28
1	316.39	316.50	316.62	316.73	316.84	316.95	317.07	317.18	317.29	317.41
2	317.52	317.63	317.74	317.86	317.97	318.08	318.19	318.31	318.42	318.53
3	318.64	318.76	318.87	318.98	319.09	319.21	319.32	319.43	319.54	319.66
4	319.77	319.88	319.99	320.11	320.22	320.33	320.45	320.56	320.67	320.78
5	320.90	321.01	321.12	321.23	321.35	321.46	321.57	321.68	321.80	321.91
6	322.02	322.13	322.25	322.36	322.47	322.58	322.70	322.81	322.92	323.04
7	323.15	323.26	323.37	323.49	323.60	323.71	323.82	323.94	324.05	324.16
8	324.27	324.39	324.50	324.61	324.72	324.84	324.95	325.06	325.17	325.29
9	325.40	325.51	325.62	325.74	325.85	325.96	326.08	326.19	326.30	326.41
29.0	326.53	326.64	326.75	326.86	326.98	327.09	327.20	327.31	327.43	327.54
1	327.65	327.76	327.88	327.99	328.10	328.21	328.33	328.44	328.55	328.66
2	328.78	328.89	329.00	329.12	329.23	329.34	329.45	329.57	329.68	329.79
3	329.90	330.02	330.13	330.24	330.35	330.47	330.58	330.69	330.80	330.92
4	331.03	331.14	331.25	331.37	331.48	331.59	331.70	331.82	331.93	332.04
5	332.16	332.27	332.38	332.49	332.61	332.72	332.83	332.94	333.06	333.17
6	333.28	333.39	333.51	333.62	333.73	333.84	333.96	334.07	334.18	334.29
7	334.41	334.52	334.63	334.74	334.86	334.97	335.08	335.20	335.31	335.42
8	335.53	335.65	335.76	335.87	335.98	336.10	336.21	336.32	336.43	336.55
9	336.66	336.77	336.88	337.00	337.11	337.22	337.33	337.45	337.56	337.67
30.0	337.78	337.90	338.01	338.12	338.24	338.35	338.46	338.57	338.69	338.80
1	338.91	339.02	339.14	339.25	339.36	339.47	339.59	339.70	339.81	339.92
2	340.04	340.15	340.26	340.37	340.49	340.60	340.71	340.83	340.94	341.05
3	341.16	341.28	341.39	341.50	341.61	341.73	341.84	341.95	342.06	342.18
4	342.29	342.40	342.51	342.63	342.74	342.85	342.96	343.08	343.19	343.30
5	343.41	343.53	343.64	343.75	343.87	343.98	344.09	344.20	344.32	344.43
6	344.54	344.65	344.77	344.88	344.99	345.10	345.22	345.33	345.44	345.55
7	345.67	345.78	345.89	346.00	346.12	346.23	346.34	346.45	346.57	346.68
8	346.79	346.91	347.02	347.13	347.24	347.36	347.47	347.58	347.69	347.81
9	347.92	348.03	348.14	348.26	348.37	348.48	348.59	348.71	348.82	348.93
31.0	349.04	349.16	349.27	349.38	349.49	349.61	349.72	349.83	349.95	350.06
1	350.17	350.28	350.40	350.51	350.62	350.73	350.85	350.96	351.07	351.18
2	351.30	351.41	351.52	351.63	351.75	351.86	351.97	352.08	352.20	352.31
3	352.42	352.53	352.65	352.76	352.87	352.99	353.10	353.21	353.32	353.44
4	353.55	353.66	353.77	353.89	354.00	354.11	354.22	354.34	354.45	354.56
5	354.67	354.79	354.90	355.01	355.12	355.24	355.35	355.46	355.57	355.69
6	355.80	355.91	356.03	356.14	356.25	356.36	356.48	356.59	356.70	356.81

Thousandths of an Inch.									
0.	**1.**	**2.**	**3.**	**4.**	**5.**	**6.**	**7.**	**8.**	**9.**
0.000	0.011	0.023	0.034	0.045	0.056	0.068	0.079	0.090	0.101

III.–IV.

COMPARISON

OF

THE METRICAL BAROMETER

WITH

THE ENGLISH AND THE OLD FRENCH BAROMETERS,

OR

TABLES

FOR CONVERTING MILLIMETRES INTO ENGLISH INCHES AND DECIMALS, AND INTO FRENCH OR PARIS LINES;

GIVING THE VALUES CORRESPONDING TO EVERY MILLIMETRE FROM 250 TO 600; AND TO EVERY TENTH OF A MILLIMETRE FROM 600 TO 800 MILLIMETRES.

1 Metre = 39.37079 English Inches.

Millimetres. Tens.	Millimetres. Units.									
	0.	**1.**	**2.**	**3.**	**4.**	**5.**	**6.**	**7.**	**8.**	**9.**
	Eng. In.	Eng. In.	Eng. In.	Eng. In.	Eng. In.	Eng. In.	Eng. In.	Eng. In	Eng. In.	Eng. In.
250	9.843	9.882	9.921	9.961	10.000	10.040	10.079	10.118	10.158	10.197
260	10.236	10.276	10.315	10.355	10.394	10.433	10.473	10.512	10.551	10.591
270	10.630	10.669	10.709	10.748	10.788	10.827	10.866	10.906	10.945	10.984
280	11.024	11.063	11.103	11.142	11.181	11.221	11.260	11.299	11.339	11.378
290	11.418	11.457	11.496	11.536	11.575	11.614	11.654	11.693	11.732	11.772
300	11.811	11.851	11.890	11.929	11.969	12.008	12.047	12.087	12.126	12.166
310	12.205	12.244	12.284	12.323	12.362	12.402	12.441	12.481	12.520	12.559
320	12.599	12.638	12.677	12.717	12.756	12.795	12.835	12.874	12.914	12.953
330	12.992	13.032	13.071	13.110	13.150	13.189	13.229	13.268	13.307	13.347
340	13.386	13.425	13.465	13.504	13.544	13.583	13.622	13.662	13.701	13.740
350	13.780	13.819	13.859	13.898	13.937	13.977	14.016	14.055	14.095	14.134
360	14.173	14.213	14.252	14.292	14.331	14.370	14.410	14.449	14.488	14.528
370	14.567	14.607	14.646	14.685	14.725	14.764	14.803	14.843	14.882	14.922
380	14.961	15.000	15.040	15.079	15.118	15.158	15.197	15.236	15.276	15.315
390	15.355	15.494	15.433	15.473	15.512	15.551	15.591	15.630	15.670	15.709
400	15.748	15.788	15.827	15.866	15.906	15.945	15.985	16.024	16.063	16.103
410	16.142	16.181	16.221	16.260	16.300	16.339	16.378	16.418	16.458	16.496
420	16.536	16.575	16.614	16.654	16.693	16.733	16.772	16.811	16.851	16.890
430	16.929	16.969	17.008	17.048	17.087	17.126	17.166	17.205	17.244	17.284
440	17.323	17.362	17.402	17.441	17.481	17.520	17.559	17.599	17.638	17.677
450	17.717	17.756	17.796	17.835	17.874	17.914	17.953	17.992	18.032	18.071
460	18.111	18.150	18.189	18.229	18.268	18.307	18.347	18.386	18.426	18.465
470	18.504	18.544	18.583	18.622	18.662	18.701	18.740	18.780	18.819	18.859
480	18.898	18.937	18.977	19.016	19.055	19.095	19.134	19.174	19.213	19.252
490	19.292	19.331	19.370	19.410	19.449	19.489	19.528	19.567	19.607	19.646
500	19.685	19.725	19.764	19.804	19.843	19.882	19.922	19.961	20.000	20.040
510	20.079	20.118	20.158	20.197	20.237	20.276	20.315	20.355	20.394	20.433
520	20.473	20.512	20.552	20.591	20.630	20.670	20.709	20.748	20.788	20.827
530	20.867	20.906	20.945	20.985	21.024	21.063	21.103	21.142	21.181	21.221
540	21.260	21.300	21.339	21.378	21.418	21.457	21.496	21.536	21.575	21.615
550	21.654	21.693	21.733	21.772	21.811	21.851	21.890	21.930	21.969	22.008
560	22.048	22.087	22.126	22.166	22.205	22.244	22.284	22.323	22.363	22.402
570	22.441	22.481	22.520	22.559	22.599	22.638	22.678	22.717	22.756	22.796
580	22.835	22.874	22.914	22.953	22.993	23.032	23.071	23.111	23.150	23.189
590	23.229	23.268	23.308	23.347	23.386	23.426	23.465	23.504	23.544	23.583

Tenths of Millimetres.

0.	**1.**	**2.**	**3.**	**4.**	**5.**	**6.**	**7.**	**8.**	**9.**
0.000	0.004	0.008	0.012	0.016	0.020	0.024	0.028	0.031	0.035

1 Metre = 39.37079 English Inches.

Millimetres.	Tenths of Millimetres.									
	0.	**1.**	**2.**	**3.**	**4.**	**5.**	**6.**	**7.**	**8.**	**9.**
	Eng. In.	Eng. In.	Eng. In.	Eng. In.	Eng. In.	Eng. In.	Eng. In.	Eng. In.	Eng. In.	Eng. In.
600	23.622	23.626	23.630	23.634	23.638	23.642	23.646	23.650	23.654	23.658
601	23.662	23.666	23.670	23.674	23.678	23.682	23.685	23.689	23.693	23.697
602	23.701	23.705	23.709	23.713	23.717	23.721	23.725	23.729	23.733	23.737
603	23.741	23.745	23.748	23.752	23.756	23.760	23.764	23.768	23.772	23.776
604	23.780	23.784	23.788	23.792	23.796	23.800	23.804	23.808	23.811	23.815
605	23.819	23.823	23.827	23.831	23.835	23.839	23.843	23.847	23.851	23.855
606	23.859	23.863	23.867	23.871	23.874	23.878	23.882	23.886	23.890	23.894
607	23.898	23.902	23.906	23.910	23.914	23.918	23.922	23.926	23.930	23.934
608	23.937	23.941	23.945	23.949	23.953	23.957	23.961	23.965	23.969	23.973
609	23.977	23.981	23.985	23.989	23.993	23.996	24.000	24.004	24.008	24.012
610	24.016	24.020	24.024	24.028	24.032	24.036	24.040	24.044	24.048	24.052
611	24.056	24.059	24.063	24.067	24.071	24.075	24.079	24.083	24.087	24.091
612	24.095	24.099	24.103	24.107	24.111	24.115	24.119	24.122	24.126	24.130
613	24.134	24.138	24.142	24.146	24.150	24.154	24.158	24.162	24.166	24.170
614	24.174	24.178	24.182	24.185	24.189	24.193	24.197	24.201	24.205	24.209
615	24.213	24.217	24.221	24.225	24.229	24.233	24.237	24.241	24.245	24.248
616	24.252	24.256	24.260	24.264	24.268	24.272	24.276	24.280	24.284	24.288
617	24.292	24.296	24.300	24.304	24.308	24.311	24.315	24.319	24.323	24.327
618	24.331	24.335	24.339	24.343	24.347	24.351	24.355	24.359	24.363	24.367
619	24.371	24.374	24.378	24.382	24.386	24.390	24.394	24.398	24.402	24.406
620	24.410	24.414	24.418	24.422	24.426	24.430	24.434	24.437	24.441	24.445
621	24.449	24.453	24.457	24.461	24.465	24.469	24.473	24.477	24.481	24.485
622	24.489	24.493	24.497	24.500	24.504	24.508	24.512	24.516	24.520	24.524
623	24.528	24.532	24.536	24.540	24.544	24.548	24.552	24.556	24.559	24.563
624	24.567	24.571	24.575	24.579	24.583	24.587	24.591	24.595	24.599	24.603
625	24.607	24.611	24.615	24.619	24.622	24.626	24.630	24.634	24.638	24.642
626	24.646	24.650	24.654	24.658	24.662	24.666	24.670	24.674	24.678	24.682
627	24.685	24.689	24.693	24.697	24.701	24.705	24.709	24.713	24.717	24.721
628	24.725	24.729	24.733	24.737	24.741	24.745	24.748	24.752	24.756	24.760
629	24.764	24.768	24.772	24.776	24.780	24.784	24.788	24.792	24.796	24.800
630	24.804	24.808	24.811	24.815	24.819	24.823	24.827	24.831	24.835	24.839
631	24.843	24.847	24.851	24.855	24.859	24.863	24.867	24.871	24.874	24.878
632	24.882	24.886	24.890	24.894	24.898	24.902	24.906	24.910	24.914	24.918
633	24.922	24.926	24.930	24.934	24.937	24.941	24.945	24.949	24.953	24.957
634	24.961	24.965	24.969	24.973	24.977	24.981	24.985	24.989	24.993	24.997
635	25.000	25.004	25.008	25.012	25.016	25.020	25.024	25.028	25.032	25.036
636	25.040	25.044	25.048	25.052	25.056	25.060	25.063	25.067	25.071	25.075
637	25.079	25.083	25.087	25.091	25.095	25.099	25.103	25.107	25.111	25.115
638	25.119	25.123	25.126	25.130	25.134	25.138	25.142	25.146	25.150	25.154
639	25.158	25.162	25.166	25.170	25.174	25.178	25.182	25.185	25.189	25.193
	0.	**1.**	**2.**	**3.**	**4.**	**5.**	**6.**	**7.**	**8.**	**9.**

1 Metre = 39.37079 English Inches.

Millimetres.	Tenths of Millimetres.									
	0.	1.	2.	3.	4.	5.	6.	7.	8.	9.
	Eng. In.	Eng. In.	Eng. In.	Eng. In.	Eng. In.	Eng. In.	Eng. In.	Eng. In.	Eng. In.	Eng. In.
640	25.197	25.201	25.205	25.209	25.213	25.217	25.221	25.225	25.229	25.233
641	25.237	25.241	25.245	25.248	25.252	25.256	25.260	25.264	25.268	25.272
642	25.276	25.280	25.284	25.288	25.292	25.296	25.300	25.304	25.308	25.311
643	25.315	25.319	25.323	25.327	25.331	25.335	25.339	25.343	25.347	25.351
644	25.355	25.359	25.363	25.367	25.371	25.374	25.378	25.382	25.386	25.390
645	25.394	25.398	25.402	25.406	25.410	25.414	25.418	25.422	25.426	25.430
646	25.434	25.437	25.441	25.445	25.449	25.453	25.457	25.461	25.465	25.469
647	25.473	25.477	25.481	25.485	25.489	25.493	25.497	25.500	25.504	25.508
648	25.512	25.516	25.520	25.524	25.528	25.532	25.536	25.540	25.544	25.548
649	25.552	25.556	25.560	25.563	25.567	25.571	25.575	25.579	25.583	25.587
650	25.591	25.595	25.599	25.603	25.607	25.611	25.615	25.619	25.623	25.626
651	25.630	25.634	25.638	25.642	25.646	25.650	25.654	25.658	25.662	25.666
652	25.670	25.674	25.678	25.682	25.686	25.689	25.693	25.697	25.701	25.705
653	25.709	25.713	25.717	25.721	25.725	25.729	25.733	25.737	25.741	25.745
654	25.748	25.752	25.756	25.760	25.764	25.768	25.772	25.776	25.780	25.784
655	25.788	25.792	25.796	25.800	25.804	25.808	25.811	25.815	25.819	25.823
656	25.827	25.831	25.835	25.839	25.843	25.847	25.851	25.855	25.859	25.863
657	25.867	25.871	25.874	25.878	25.882	25.886	25.890	25.894	25.898	25.902
658	25.906	25.910	25.914	25.918	25.922	25.926	25.930	25.934	25.937	25.941
659	25.945	25.949	25.953	25.957	25.961	25.965	25.969	25.973	25.977	25.981
660	25.985	25.989	25.993	25.997	26.000	26.004	26.008	26.012	26.016	26.020
661	26.024	26.028	26.032	26.036	26.040	26.044	26.048	26.052	26.056	26.060
662	26.063	26.067	26.071	26.075	26.079	26.083	26.087	26.091	26.095	26.099
663	26.103	26.107	26.111	26.115	26.119	26.123	26.126	26.130	26.134	26.138
664	26.142	26.146	26.150	26.154	26.158	26.162	26.166	26.170	26.174	26.178
665	26.182	26.186	26.189	26.193	26.197	26.201	26.205	26.209	26.213	26.217
666	26 221	26.225	26.229	26.233	26.237	26.241	26.245	26.249	26.252	26.256
667	26.260	26.264	26.268	26.272	26.276	26.280	26.284	26.288	26.292	26.296
668	26.300	26.304	26.308	26.311	26.315	26.319	26.323	26.327	26.331	26.335
669	26.339	26.343	26.347	26.351	26.355	26.359	26.363	26.367	26.371	26.374
670	26.378	26.382	26.386	26.390	26.394	26.398	26.402	26.406	26.410	26.414
671	26.418	26.422	26.426	26.430	26.434	26.437	26.441	26.445	26.449	26.453
672	26.457	26.461	26.465	26.469	26.473	26.477	26.481	26.485	26.489	26.493
673	26.497	26.500	26.504	26.508	26.512	26.516	26.520	26.524	26.528	26.532
674	26.536	26.540	26.544	26.548	26.552	26.556	26.560	26.563	26.567	26.571
675	26.575	26.579	26.583	26.587	26.591	26.595	26.599	26.603	26.607	26.611
676	26.615	26.619	26.623	26.626	26.630	26.634	26.638	26.642	26.646	26.650
677	26.654	26.658	26.662	26.666	26.670	26.674	26.678	26.682	26.686	26.689
678	26.693	26.697	26.701	26.705	26.709	26.713	26.717	26.721	26.725	26.729
679	26.733	26.737	26.741	26.745	26.749	26.752	26.756	26.760	26.764	26.768
	0.	1.	2.	3.	4.	5.	6.	7.	8.	9.

1 Metre = 39.37079 English Inches.

Millimetres.	Tenths of Millimetres.									
	0.	**1.**	**2.**	**3.**	**4.**	**5.**	**6.**	**7.**	**8.**	**9.**
	Eng. In.	Eng. In.	Eng. In.	Eng. In.	Eng. In.	Eng. In.	Eng. In.	Eng. In.	Eng. In.	Eng. In.
680	26.772	26.776	26.780	26.784	26.788	26.792	26.796	26.800	26.804	26.808
681	26.812	26.815	26.819	26.823	26.827	26.831	26.835	26.839	26.843	26.847
682	26.851	26.855	26.859	26.863	26.867	26.871	26.875	26.878	26.882	26.886
683	26.890	26.894	26.898	26.902	26.906	26.910	26.914	26.918	26.922	26.926
684	26.930	26.934	26.937	26.941	26.945	26.949	26.953	26.957	26.961	26.965
685	26.969	26.973	26.977	26.981	26.985	26.989	26.993	26.997	27.000	27.004
686	27.008	27.012	27.016	27.020	27.024	27.028	27.032	27.036	27.040	27.044
687	27.048	27.052	27.056	27.060	27.063	27.067	27.071	27.075	27.079	27.083
688	27.087	27.091	27.095	27.099	27.103	27.107	27.111	27.115	27.119	27.123
689	27.126	27.130	27.134	27.138	27.142	27.146	27.150	27.154	27.158	27.162
690	27.166	27.170	27.174	27.178	27.182	27.186	27.189	27.193	27.197	27.201
691	27.205	27.209	27.213	27.217	27.221	27.225	27.229	27.233	27.237	27.241
692	27.245	27.249	27.252	27.256	27.260	27.264	27.268	27.272	27.276	27.280
693	27.284	27.288	27.292	27.296	27.300	27.304	27.308	27.312	27.315	27.319
694	27.323	27.327	27.331	27.335	27.339	27.343	27.347	27.351	27.355	27.359
695	27.363	27.367	27.371	27.375	27.378	27.382	27.386	27.390	27.394	27.398
696	27.402	27.406	27.410	27.414	27.418	27.422	27.426	27.430	27.434	27.438
697	27.441	27.445	27.449	27.453	27.457	27.461	27.465	27.469	27.473	27.477
698	27.481	27.485	27.489	27.493	27.497	27.500	27.504	27.508	27.512	27.516
699	27.520	27.524	27.528	27.532	27.536	27.540	27.544	27.548	27.552	27.556
700	27.560	27.563	27.567	27.571	27.575	27.579	27.583	27.587	27.591	27.595
701	27.599	27.603	27.607	27.611	27.615	27.619	27.623	27.626	27.630	27.634
702	27.638	27.642	27.646	27.650	27.654	27.658	27.662	27.666	27.670	27.674
703	27.678	27.682	27.686	27.689	27.693	27.697	27.701	27.705	27.709	27.713
704	27.717	27.721	27.725	27.729	27.733	27.737	27.741	27.745	27.749	27.752
705	27.756	27.760	27.764	27.768	27.772	27.776	27.780	27.784	27.788	27.792
706	27.796	27.800	27.804	27.808	27.812	27.815	27.819	27.823	27.827	27.831
707	27.835	27.839	27.843	27.847	27.851	27.855	27.859	27.863	27.867	27.871
708	27.875	27.878	27.882	27.886	27.890	27.894	27.898	27.902	27.906	27.910
709	27.914	27.918	27.922	27.926	27.930	27.934	27.938	27.941	27.945	27.949
710	27.953	27.957	27.961	27.965	27.969	27.973	27.977	27.981	27.985	27.989
711	27.993	27.997	28.001	28.004	28.008	28.012	28.016	28.020	28.024	28.028
712	28.032	28.036	28.040	28.044	28.048	28.052	28.056	28.060	28.063	28.067
713	28.071	28.075	28.079	28.083	28.087	28.091	28.095	28.099	28.103	28.107
714	28.111	28.115	28.119	28.123	28.126	28.130	28.134	28.138	28.142	28.146
715	28.150	28.154	28.158	28.162	28.166	28.170	28.174	28.178	28.182	28.186
716	28.189	28.193	28.197	28.201	28.205	28.209	28.213	28.217	28.221	28.225
717	28.229	28.233	28.237	28.241	28.245	28.249	28.252	28.256	28.260	28.264
718	28.268	28.272	28.276	28.280	28.284	28.288	28.292	28.296	28.300	28.304
719	28.308	28.312	28.315	28.319	28.323	28.327	28.331	28.335	28.339	28.343
	0.	**1.**	**2.**	**3.**	**4.**	**5.**	**6.**	**7.**	**8.**	**9.**

1 Metre = 39.37079 English Inches.

Millimetres.	Tenths of Millimetres.									
	0.	1.	2.	3.	4.	5.	6.	7.	8.	9.
	Eng. In.	Eng. In.	Eng. In.	Eng. In.	Eng. In.	Eng. In.	Eng. In.	Eng. In.	Eng. In.	Eng. In.
720	28.347	28.351	28.355	28.359	28.363	28.367	28.371	28.375	28.378	28.382
721	28.386	28.390	28.394	28.398	28.402	28.406	28.410	28.414	28.418	28.422
722	28.426	28.430	28.434	28.438	28.441	28.445	28.449	28.453	28.457	28.461
723	28.465	28.469	28.473	28.477	28.481	28.485	28.489	28.493	28.497	28.501
724	28.504	28.508	28.512	28.516	28.520	28.524	28.528	28.532	28.536	28.540
725	28.544	28.548	28.552	28.556	28.560	28.564	28.567	28.571	28.575	28.579
726	28.583	28.587	28.591	28.595	28.599	28.603	28.607	28.611	28.615	28.619
727	28.623	28.627	28.630	28.634	28.638	28.642	28.646	28.650	28.654	28.658
728	28.662	28.666	28.670	28.674	28.678	28.682	28.686	28.689	28.693	28.697
729	28.701	28.705	28.709	28.713	28.717	28.721	28.725	28.729	28.733	28.737
730	28.741	28.745	28.749	28.752	28.756	28.760	28.764	28.768	28.772	28.776
731	28.780	28.784	28.788	28.792	28.796	28.800	28.804	28.808	28.812	28.815
732	28.819	28.823	28.827	28.831	28.835	28.839	28.843	28.847	28.851	28.855
733	28.859	28.863	28.867	28.871	28.875	28.878	28.882	28.886	28.890	28.894
734	28.898	28.902	28.906	28.910	28.914	28.918	28.922	28.926	28.930	28.934
735	28.938	28.941	28.945	28.949	28.953	28.957	28.961	28.965	28.969	28.973
736	28.977	28.981	28.985	28.989	28.993	28.997	29.001	29.004	29.008	29.012
737	29.016	29.020	29.024	29.028	29.032	29.036	29.040	29.044	29.048	29.052
738	29.056	29.060	29.064	29.067	29.071	29.075	29.079	29.083	29.087	29.091
739	29.095	29.099	29.103	29.107	29.111	29.115	29.119	29.123	29.127	29.130
740	29.134	29.138	29.142	29.146	29.150	29.154	29.158	29.162	29.166	29.170
741	29.174	29.178	29.182	29.186	29.190	29.193	29.197	29.201	29.205	29.209
742	29.213	29.217	29.221	29.225	29.229	29.233	29.237	29.241	29.245	29.249
743	29.252	29.256	29.260	29.264	29.268	29.272	29.276	29.280	29.284	29.288
744	29.292	29.296	29.300	29.304	29.308	29.312	29.315	29.319	29.323	29.327
745	29.331	29.335	29.339	29.343	29.347	29.351	29.355	29.359	29.363	29.367
746	29.371	29.375	29.378	29.382	29.386	29.390	29.394	29.398	29.402	29.406
747	29.410	29.414	29.418	29.422	29.426	29.430	29.434	29.438	29.441	29.445
748	29.449	29.453	29.457	29.461	29.465	29.469	29.473	29.477	29.481	29.485
749	29.489	29.493	29.497	29.501	29.504	29.508	29.512	29.516	29.520	29.524
750	29.528	29.532	29.536	29.540	29.544	29.548	29.552	29.556	29.560	29.564
751	29.567	29.571	29.575	29.579	29.583	29.587	29.591	29.595	29.599	29.603
752	29.607	29.611	29.615	29.619	29.623	29.627	29.630	29.634	29.638	29.642
753	29.646	29.650	29.654	29.658	29.662	29.666	29.670	29.674	29.678	29.682
754	29.686	29.690	29.693	29.697	29.701	29.705	29.709	29.713	29.717	29.721
755	29.725	29.729	29.733	29.737	29.741	29.745	29.749	29.753	29.756	29.760
756	29.764	29.768	29.772	29.776	29.780	29.784	29.788	29.792	29.796	29.800
757	29.804	29.808	29.812	29.815	29.819	29.823	29.827	29.831	29.835	29.839
758	29.843	29.847	29.851	29.855	29.859	29.863	29.867	29.871	29.875	29.878
759	29.882	29.886	29.890	29.894	29.898	29.902	29.906	29.910	29.914	29.918
	0.	1.	2.	3.	4.	5.	6.	7.	8.	9.

1 Metre = 39.37079 English Inches.

Millimetres.	Tenths of Millimetres.									
	0.	**1.**	**2.**	**3.**	**4.**	**5.**	**6.**	**7.**	**8.**	**9.**
	Eng. In.	Eng. In.	Eng. In.	Eng. In.	Eng. In.	Eng. In.	Eng. In.	Eng. In	Eng. In.	Eng. In.
760	29.922	29.926	29.930	29.934	29.938	29.941	29.945	29.949	29.953	29.957
761	29.961	29.965	29.969	29.973	29.977	29.981	29.985	29.989	29.993	29.997
762	30.001	30.004	30.008	30.012	30.016	30.020	30.024	30.028	30.032	30.036
763	30.040	30.044	30.048	30.052	30.056	30.060	30.064	30.067	30.071	30.075
764	30.079	30.083	30.087	30.091	30.095	30.099	30.103	30.107	30.111	30.115
765	30.119	30.123	30.127	30.130	30.134	30.138	30.142	30.146	30.150	30.154
766	30.158	30.162	30.166	30.170	30.174	30.178	30.182	30.186	30.190	30.193
767	30.197	30.201	30.205	30.209	30.213	30.217	30.221	30.225	30.229	30.233
768	30.237	30.241	30.245	30.249	30.253	30.256	30.260	30.264	30.268	30.272
769	30.276	30.280	30.284	30.288	30.292	30.296	30.300	30.304	30.308	30.312
770	30.316	30.319	30.323	30.327	30.331	30.335	30.339	30.343	30.347	30.351
771	30.355	30.359	30.363	30.367	30.371	30.375	30.379	30.382	30.386	30.390
772	30.394	30.398	30.402	30.406	30.410	30.414	30.418	30.422	30.426	30.430
773	30.434	30.438	30.441	30.445	30.449	30.453	30.457	30.461	30.465	30.469
774	30.473	30.477	30.481	30.485	30.489	30.493	30.497	30.501	30.504	30.508
775	30.512	30.516	30.520	30.524	30.528	30.532	30.536	30.540	30.544	30.548
776	30.552	30.556	30.560	30.564	30.567	30.571	30.575	30.579	30.583	30.587
777	30.591	30.595	30.599	30.603	30.607	30.611	30.615	30.619	30.623	30.627
778	30.630	30.634	30.638	30.642	30.646	30.650	30.654	30.658	30.662	30.666
779	30.670	30.674	30.678	30.682	30.686	30.690	30.693	30.697	30.701	30.705
780	30.709	30.713	30.717	30.721	30.725	30.729	30.733	30.737	30.741	30.745
781	30.749	30.753	30.756	30.760	30.764	30.768	30.772	30.776	30.780	30.784
782	30.788	30.792	30.796	30.800	30.804	30.808	30.812	30.816	30.819	30.823
783	30.827	30.831	30.835	30.839	30.843	30.847	30.851	30.855	30.859	30.863
784	30.867	30.871	30.875	30.879	30.882	30.886	30.890	30.894	30.898	30.902
785	30.906	30.910	30.914	30.918	30.922	30.926	30.930	30.934	30.938	30.942
786	30.945	30.949	30.953	30.957	30.961	30.965	30.969	30.973	30.977	30.981
787	30.985	30.989	30.993	30.997	31.001	31.004	31.008	31.012	31.016	31.020
788	31.024	31.028	31.032	31.036	31.040	31.044	31.048	31.052	31.056	31.060
789	31.064	31.067	31.071	31.075	31.079	31.083	31.087	31.091	31.095	31.099
790	31.103	31.107	31.111	31.115	31.119	31.123	31.127	31.130	31.134	31.138
791	31.142	31.146	31.150	31.154	31.158	31.162	31.166	31.170	31.174	31.178
792	31.182	31.186	31.190	31.193	31.197	31.201	31.205	31.209	31.213	31.217
793	31.221	31.225	31.229	31.233	31.237	31.241	31.245	31.249	31.253	31.256
994	31.260	31.264	31.268	31.272	31.276	31.280	31.284	31.288	31.292	31.296
795	31.300	31.304	31.308	31.312	31.316	31.319	31.323	31.327	31.331	31.335
796	31.339	31.343	31.347	31.351	31.355	31.359	31.363	31.367	31.371	31.375
797	31.379	31.382	31.386	31.390	31.394	31.398	31.402	31.406	31.410	31.414
798	31.418	31.422	31.426	31.430	31.434	31.438	31.442	31.445	31.449	31.453
799	31.457	31.461	31.465	31.469	31.473	31.477	31.481	31.485	31.489	31.493
800	31.497	31.501	31.505	31.508	31.512	31.516	31.520	31.524	31.528	31.532

Hundredths of Millimetres.									
0.	**1.**	**2.**	**3.**	**4.**	**5.**	**6.**	**7.**	**8.**	**9.**
.0000	.0004	.0008	.0012	.0016	.0020	.0024	.0028	.0031	.0035

1 Millimetre = 0.443296 French or Paris Line.

Millimetres. Tens.	Millimetres. Units. 0.	1.	2.	3.	4.	5.	6.	7.	8.	9.
	Par.lines.	Par.lines.	Par.lines.	Par.lines.	Par.lines.	Par.lines.	Par.lines.	Par.lines.	Par.lines.	Par.lines.
300	132.99	133.43	133.88	134.32	134.76	135.21	135.65	136.09	136.54	136.98
310	137.42	137.87	138.31	138.75	139.19	139.64	140.08	140.52	140.97	141.41
320	141.85	142.30	142.74	143.18	143.63	144.07	144.51	144.96	145.40	145.84
330	146.29	146.73	147.17	147.62	148.06	148.50	148.95	149.39	149.83	150.28
340	150.72	151.16	151.61	152.05	152.49	152.94	153.38	153.82	154.27	154.71
350	155.15	155.60	156.04	156.48	156.93	157.37	157.81	158.26	158.70	159.14
360	159.59	160.03	160.47	160.92	161.36	161.80	162.25	162.69	163.13	163.58
370	164.02	164.46	164.91	165.35	165.79	166.24	166.68	167.12	167.57	168.01
380	168.45	168.90	169.34	169.78	170.23	170.67	171.11	171.56	172.00	172.44
390	172.89	173.33	173.77	174.22	174.66	175.10	175.55	175.99	176.43	176.88
400	177.32	177.76	178.20	178.65	179.09	179.53	179.98	180.42	180.86	181.31
410	181.75	182.19	182.64	183.08	183.52	183.97	184.41	184.85	185.30	185.74
420	186.18	186.63	187.07	187.51	187.96	188.40	188.84	189.29	189.73	190.17
430	190.62	191.06	191.50	191.95	192.39	192.83	193.28	193.72	194.16	194.61
440	195.05	195.49	195.94	196.38	196.82	197.27	197.71	198.15	198.60	199.04
450	199.48	199.93	200.37	200.81	201.26	201.70	202.14	202.59	203.03	203.47
460	203.92	204.36	204.80	205.25	205.69	206.13	206.58	207.02	207.46	207.91
470	208.35	208.79	209.24	209.68	210.12	210.57	211.01	211.45	211.90	212.34
480	212.78	213.23	213.67	214.11	214.56	215.00	215.44	215.88	216.33	216.77
490	217.22	217.66	218.10	218.54	218.99	219.43	219.87	220.32	220.76	221.20
500	221.65	222.09	222.53	222.98	223.42	223.86	224.31	224.75	225.19	225.64
510	226.08	226.52	226.97	227.41	227.85	228.30	228.74	229.18	229.63	230.07
520	230.51	230.96	231.40	231.84	232.29	232.73	233.17	233.62	234.06	234.50
530	234.95	235.39	235.83	236.28	236.72	237.16	237.61	238.05	238.49	238.94
540	239.38	239.82	240.27	240.71	241.15	241.60	242.04	242.48	242.93	243.37
550	243.81	244.26	244.70	245.14	245.59	246.03	246.47	246.92	247.36	247.80
560	248.25	248.69	249.13	249.57	250.01	250.46	250.91	251.35	251.79	252.24
570	252.68	253.12	253.57	254.01	254.45	254.90	255.34	255.78	256.23	256.67
580	257.11	257.55	258.00	258.44	258.88	259.32	259.77	260.21	260.66	261.10
590	261.54	261.99	262.43	262.87	263.32	263.76	264.20	264.65	265.09	265.53

Tenths of Millimetres.

0.	1.	2.	3.	4.	5.	6.	7.	8.	9.
0.000	0.044	0.089	0.133	0.177	0.222	0.266	0.310	0.355	0.399

Hundredths of Millimetres.

0.	1.	2.	3.	4.	5.	6.	7.	8.	9.
0.000	0.004	0.009	0.013	0.018	0.022	0.027	0.031	0.035	0.040

1 Millimetre = 0.443296 French Line.

Millimetres.	Tenths of Millimetres.									
	0.	**1.**	**2.**	**3.**	**4.**	**5.**	**6.**	**7.**	**8.**	**9.**
	Par.lines.	Par.lines.	Par.lines.	Par.lines.	Par.lines.	Par.lines.	Par.lines.	Par.lines.	Par.lines.	Par.lines.
600	265.98	266.02	266.07	266.11	266.15	266.20	266.24	266.29	266.33	266.38
601	266.42	266.47	266.51	266.55	266.60	266.64	266.69	266.73	266.78	266.82
602	266.86	266.91	266.95	267.00	267.04	267.09	267.13	267.17	267.22	267.26
603	267.31	267.35	267.40	267.44	267.48	267.53	267.57	267.62	267.66	267.71
604	267.75	267.80	267.84	267.88	267.93	267.97	268.02	268.06	268.11	268.15
605	268.19	268.24	268.28	268.33	268.37	268.42	268.46	268.50	268.55	268.59
606	268.64	268.68	268.73	268.77	268.81	268.86	268.90	268.95	268.99	269.04
607	269.08	269.13	269.17	269.21	269.26	269.30	269.35	269.39	269.44	269.48
608	269.52	269.57	269.61	269.66	269.70	269.75	269.79	269.83	269.88	269.92
609	269.97	270.01	270.06	270.10	270.14	270.19	270.23	270.28	270.32	270.37
610	270.41	270.45	270.50	270.54	270.59	270.63	270.68	270.72	270.77	270.81
611	270.85	270.90	270.94	270.99	271.03	271.08	271.12	271.16	271.21	271.25
612	271.30	271.34	271.39	271.43	271.47	271.52	271.56	271.61	271.65	271.70
613	271.74	271.78	271.83	271.87	271.92	271.96	272.01	272.05	272.10	272.14
614	272.18	272.23	272.27	272.32	272.36	272.41	272.45	272.49	272.54	272.58
615	272.63	272.67	272.72	272.76	272.80	272.85	272.89	272.94	272.98	273.03
616	273.07	273.11	273.16	273.20	273.25	273.29	273.34	273.38	273.42	273.47
617	273.51	273.56	273.60	273.65	273.69	273.74	273.78	273.82	273.87	273.91
618	273.96	274.00	274.05	274.09	274.13	274.18	274.22	274.27	274.31	274.36
619	274.40	274.44	274.49	274.53	274.58	274.62	274.67	274.71	274.75	274.80
620	274.84	274.89	274.93	274.98	275.02	275.07	275.11	275.15	275.20	275.24
621	275.29	275.33	275.38	275.42	275.46	275.51	275.55	275.60	275.64	275.69
622	275.73	275.77	275.82	275.86	275.91	275.95	276.00	276.04	276.08	276.13
623	276.17	276.22	276.26	276.31	276.35	276.38	276.44	276.48	276.53	276.57
624	276.62	276.66	276.71	276.75	276.79	276.84	276.88	276.93	276.97	277.02
625	277.06	277.10	277.15	277.19	277.24	277.28	277.33	277.37	277.41	277.46
626	277.50	277.55	277.59	277.64	277.58	277.72	277.77	277.81	277.86	277.90
627	277.95	277.99	278.04	278.08	278.12	278.17	278.21	278.26	278.30	278.35
628	278.39	278.43	278.48	278.52	278.57	278.61	278.66	278.70	278.74	278.79
629	278.83	278.88	278.92	278.97	279.01	279.05	279.10	279.14	279.19	279.23
630	279.28	279.32	279.37	279.41	279.45	279.50	279.54	279.59	279.63	279.68
631	279.72	279.76	279.81	279.85	279.90	279.94	279.99	280.03	280.07	280.12
632	280.16	280.21	280.25	280.30	280.34	280.38	280.43	280.47	280.52	280.56
633	280.61	280.65	280.70	280.74	280.78	280.83	280.87	280.92	280.96	281.01
634	281.05	281.09	281.14	281.18	281.23	281.27	281.32	281.36	281.40	281.45
635	281.49	281.54	281.58	281.63	281.67	281.71	281.76	281.80	281.85	281.89
636	281.94	281.98	282.02	282.07	282.11	282.16	282.20	282.25	282.29	282.34
637	282.38	282.42	282.47	282.51	282.56	282.60	282.65	282.69	282.73	282.78
638	282.82	282.87	282.91	282.96	283.00	283.04	283.09	283.13	283.18	283.22
639	283.27	283.31	283.35	283.40	283.44	283.49	283.53	283.58	283.62	283.67
	0.	**1.**	**2.**	**3.**	**4.**	**5.**	**6.**	**7.**	**8.**	**9.**

1 Millimetre = 0.443296 French Line.

Millimetres.	Tenths of Millimetres.									
	0.	1.	2.	3.	4.	5.	6.	7.	8.	9.
	Par.lines.	Par.lines.	Par.lines.	Par.lines.	Par.lines.	Par.lines.	Par.lines.	Par.lines.	Par.lines.	Par.lines.
640	283.71	283.75	283.80	283.84	283.89	283.93	283.98	284.02	284.06	284.11
641	284.15	284.20	284.24	284.29	284.33	284.37	284.42	284.46	284.51	284.55
642	284.60	284.64	284.68	284.73	284.77	284.82	284.86	284.91	284.95	284.99
643	285.04	285.08	285.13	285.17	285.22	285.26	285.31	285.35	285.39	285.44
644	285.48	285.53	285.57	285.62	285.66	285.70	285.75	285.79	285.84	285.88
645	285.93	285.97	286.01	286.06	286.10	286.15	286.19	286.24	286.28	286.32
646	286.37	286.41	286.46	286.50	286.55	286.59	286.64	286.68	286.72	286.77
647	286.81	286.86	286.90	286.95	286.99	287.03	287.08	287.12	287.17	287.21
648	287.26	287.30	287.34	287.39	287.43	287.48	287.52	287.57	287.61	287.65
649	287.70	287.74	287.79	287.83	287.88	287.92	287.96	288.01	288.05	288.10
650	288.14	288.19	288.23	288.28	288.32	288.36	288.41	288.45	288.50	288.54
651	288.59	288.63	288.67	288.72	288.76	288.81	288.85	288.90	288.94	288.98
652	289.03	289.07	289.12	289.16	289.21	289.25	289.29	289.34	289.38	289.43
653	289.47	289.52	289.56	289.61	289.65	289.69	289.74	289.78	289.83	289.87
654	289.92	289.96	290.00	290.05	290.09	290.14	290.18	290.23	290.27	290.31
655	290.36	290.40	290.45	290.49	290.54	290.58	290.62	290.67	290.71	290.76
656	290.80	290.85	290.89	290.94	290.98	291.02	291.07	291.11	291.16	291.20
657	291.25	291.29	291.33	291.38	291.42	291.47	291.51	291.56	291.60	291.64
658	291.69	291.73	291.78	291.82	291.87	291.91	291.95	292.00	292.04	292.09
659	292.13	292.18	292.22	292.26	292.31	292.35	292.40	292.44	292.49	292.53
660	292.58	292.62	292.66	292.71	292.75	292.80	292.84	292.89	292.93	292.97
661	293.02	293.06	293.11	293.15	293.20	293.24	293.28	293.33	293.37	293.42
662	293.46	293.51	293.55	293.59	293.64	293.68	293.73	293.77	293.82	293.86
663	293.91	293.95	293.99	294.04	294.08	294.13	294.17	294.22	294.26	294.30
664	294.35	294.39	294.44	294.48	294.53	294.57	294.61	294.66	294.70	294.75
665	294.79	294.84	294.88	294.92	294.97	295.01	295.06	295.10	295.15	295.19
666	295.24	295.28	295.32	295.37	295.41	295.46	295.50	295.55	295.59	295.63
667	295.68	295.72	295.77	295.81	295.86	295.90	295.94	295.99	296.03	296.08
668	296.12	296.17	296.21	296.25	296.30	296.34	296.39	296.43	296.48	296.52
669	296.56	296.61	296.65	296.70	296.74	296.79	296.83	296.88	296.92	296.96
670	297.01	297.05	297.10	297.14	297.19	297.23	297.27	297.32	297.36	297.41
671	297.45	297.50	297.54	297.58	297.63	297.67	297.72	297.76	297.81	297.85
672	297.89	297.94	297.98	298.03	298.07	298.12	298.16	298.21	298.25	298.29
673	298.34	298.38	298.43	298.47	298.52	298.56	298.60	298.65	298.69	298.74
674	298.78	298.83	298.87	298.91	298.96	299.00	299.05	299.09	299.14	299.18
675	299.22	299.27	299.31	299.36	299.40	299.45	299.49	299.54	299.58	299.62
676	299.67	299.71	299.76	299.80	299.85	299.89	299.93	299.98	300.02	300.07
677	300.11	300.16	300.20	300.24	300.29	300.33	300.38	300.42	300.47	300.51
678	300.55	300.60	300.64	300.69	300.73	300.78	300.82	300.86	300.91	300.95
679	301.00	301.04	301.09	301.13	301.18	301.22	301.26	301.31	301.35	301.40
	0.	1.	2.	3.	4.	5.	6.	7.	8.	9.

1 Millimetre = 0.443296 French Line.

Millimetres.	Tenths of Millimetres.									
	0.	**1.**	**2.**	**3.**	**4.**	**5.**	**6.**	**7.**	**8.**	**9.**
	Par.lines.	Par.lines.	Par.lines.	Par.lines.	Par.lines.	Par.lines.	Par.lines.	Par.lines.	Par.lines.	Par.lines.
680	301.44	301.49	301.53	301.57	301.62	301.66	301.71	301.75	301.80	301.84
681	301.88	301.93	301.97	302.02	302.06	302.11	302.15	302.19	302.24	302.28
682	302.33	302.37	302.42	302.46	302.51	302.55	302.59	302.64	302.68	302.73
683	302.77	302.82	302.86	302.90	302.95	302.99	303.04	303.08	303.13	303.17
684	303.21	303.26	303.30	303.35	303.39	303.44	303.48	303.52	303.57	303.61
685	303.66	303.70	303.75	303.79	303.83	303.88	303.92	303.97	304.01	304.06
686	304.10	304.15	304.19	304.23	304.28	304.32	304.37	304.41	304.46	304.50
687	304.54	304.59	304.63	304.68	304.72	304.77	304.81	304.85	304.90	304.94
688	304.99	305.03	305.08	305.12	305.16	305.21	305.25	305.30	305.34	305.39
689	305.43	305.48	305.52	305.56	305.61	305.65	305.70	305.74	305.79	305.83
690	305.87	305.92	305.96	306.01	306.05	306.10	306.14	306.18	306.23	306.27
691	306.32	306.36	306.41	306.45	306.49	306.54	306.58	306.63	306.67	306.72
692	306.76	306.81	306.85	306.89	306.94	306.98	307.03	307.07	307.12	307.16
693	307.20	307.25	307.29	307.34	307.38	307.43	307.47	307.51	307.56	307.60
694	307.65	307.69	307.74	307.78	307.82	307.87	307.91	307.96	308.00	308.05
695	308.09	308.13	308.18	308.22	308.27	308.31	308.36	308.40	308.45	308.49
696	308.53	308.58	308.62	308.67	308.71	308.76	308.80	308.84	308.89	308.93
697	308.98	309.02	309.07	309.11	309.15	309.20	309.24	309.29	309.33	309.38
698	309.42	309.46	309.51	309.55	309.60	309.64	309.69	309.73	309.78	309.82
699	309.86	309.91	309.95	310.00	310.04	310.09	310.13	310.17	310.22	310.26
700	310.31	310.35	310.40	310.44	310.48	310.53	310.57	310.62	310.66	310.71
701	310.75	310.79	310.84	310.88	310.93	310.97	311.02	311.06	311.11	311.15
702	311.19	311.24	311.28	311.33	311.37	311.42	311.46	311.50	311.55	311.59
703	311.64	311.68	311.73	311.77	311.81	311.86	311.90	311.95	311.99	312.04
704	312.08	312.12	312.17	312.21	312.26	312.30	312.35	312.39	312.43	312.48
705	312.52	312.57	312.61	312.66	312.70	312.75	312.79	312.83	312.88	312.92
706	312.97	313.01	313.06	313.10	313.14	313.19	313.23	313.28	313.32	313.37
707	313.41	313.45	313.50	313.54	313.59	313.63	313.68	313.72	313.76	313.81
708	313.85	313.90	313.94	313.99	314.03	314.08	314.12	314.16	314.21	314.25
709	314.30	314.34	314.39	314.43	314.47	314.52	314.56	314.61	314.65	314.70
710	314.74	314.78	314.83	314.87	314.92	314.96	315.01	315.05	315.09	315.14
711	315.18	315.23	315.27	315.32	315.36	315.41	315.45	315.49	315.54	315.58
712	315.63	315.67	315.72	315.76	315.80	315.85	315.89	315.94	315.98	316.03
713	316.07	316.11	316.16	316.20	316.25	316.29	316.34	316.38	316.42	316.47
714	316.51	316.56	316.60	316.65	316.69	316.73	316.78	316.82	316.87	316.91
715	316.96	317.00	317.05	317.09	317.13	317.18	317.22	317.27	317.31	317.36
716	317.40	317.44	317.49	317.53	317.58	317.62	317.67	317.71	317.75	317.80
717	317.84	317.89	317.93	317.98	318.02	318.06	318.11	318.15	318.20	318.24
718	318.29	318.33	318.38	318.42	318.46	318.51	318.55	318.60	318.64	318.69
719	318.73	318.77	318.82	318.86	318.91	318.95	319.00	319.04	319.08	319.13
	0.	**1.**	**2.**	**3.**	**4.**	**5.**	**6.**	**7.**	**8.**	**9.**

1 Millimetre = 0.443296 French Line.

Millimetres.	Tenths of Millimetres.									
	0.	**1.**	**2.**	**3.**	**4.**	**5.**	**6.**	**7.**	**8.**	**9.**
	Par.lines.	Par.lines.	Par.lines.	Par.lines.	Par.lines.	Par.lines.	Par.lines.	Par.lines.	Par.lines.	Par.lines.
720	319.17	319.22	319.26	319.31	319.35	319.39	319.44	319.48	319.53	319.57
721	319.62	319.66	319.70	319.75	319.79	319.84	319.88	319.93	319.97	320.02
722	320.06	320.10	320.15	320.19	320.24	320.28	320.33	320.37	320.41	320.46
723	320.50	320.55	320.59	320.64	320.68	320.72	320.77	320.81	320.86	320.90
724	320.95	320.99	321.03	321.08	321.12	321.17	321.21	321.26	321.30	321.35
725	321.39	321.43	321.48	321.52	321.57	321.61	321.66	321.70	321.74	321.79
726	321.83	321.88	321.92	321.97	322.01	322.05	322.10	322.14	322.19	322.23
727	322.28	322.32	322.36	322.41	322.45	322.50	322.54	322.59	322.63	322.68
728	322.72	322.76	322.81	322.85	322.90	322.94	322.99	323.03	323.07	323.12
729	323.16	323.21	323.25	323.30	323.34	323.38	323.43	323.47	323.52	323.56
730	323.61	323.65	323.69	323.74	323.78	323.83	323.87	323.92	323.96	324.00
731	324.05	324.09	324.14	324.18	324.23	324.27	324.32	324.36	324.40	324.45
732	324.49	324.54	324.58	324.63	324.67	324.71	324.76	324.80	324.85	324.89
733	324.94	324.98	325.02	325.07	325.11	325.16	325.20	325.25	325.29	325.33
734	325.38	325.42	325.47	325.51	325.56	325.60	325.65	325.69	325.73	325.78
735	325.82	325.87	325.91	325.96	326.00	326.04	326.09	326.13	326.18	326.22
736	326.27	326.31	326.35	326.40	326.44	326.49	326.53	326.58	326.62	326.66
737	326.71	326.75	326.80	326.84	326.89	326.93	326.98	327.02	327.06	327.11
738	327.15	327.20	327.24	327.29	327.33	327.37	327.42	327.46	327.51	327.55
739	327.60	327.64	327.68	327.73	327.77	327.82	327.86	327.91	327.95	327.99
740	328.04	328.08	328.13	328.17	328.22	328.26	328.30	328.35	328.39	328.44
741	328.48	328.53	328.57	328.62	328.66	328.70	328.75	328.79	328.84	328.88
742	328.93	328.97	329.01	329.06	329.10	329.15	329.19	329.24	329.28	329.32
743	329.37	329.41	329.46	329.50	329.55	329.59	329.63	329.68	329.72	329.77
744	329.81	329.86	329.90	329.95	329.99	330.03	330.08	330.12	330.17	330.21
745	330.26	330.30	330.34	330.39	330.43	330.48	330.52	330.57	330.61	330.65
746	330.70	330.74	330.79	330.83	330.88	330.92	330.96	331.01	331.05	331.10
747	331.14	331.19	331.23	331.28	331.32	331.36	331.41	331.45	331.50	331.54
748	331.59	331.63	331.67	331.72	331.76	331.81	331.85	331.90	331.94	331.98
749	332.03	332.07	332.12	332.16	332.21	332.25	332.29	332.34	332.38	332.43
750	332.47	332.52	332.56	332.60	332.65	332.69	332.74	332.78	332.83	332.87
751	332.92	332.96	333.00	333.05	333.09	333.14	333.18	333.23	333.27	333.31
752	333.36	333.40	333.45	333.49	333.54	333.58	333.62	333.67	333.71	333.76
753	333.80	333.85	333.89	333.93	333.98	334.02	334.07	334.11	334.16	334.20
754	334.25	334.29	334.33	334.38	334.42	334.47	334.51	334.56	334.60	334.64
755	334.69	334.73	334.78	334.82	334.87	334.91	334.95	335.00	335.04	335.09
756	335.13	335.18	335.22	335.26	335.31	335.35	335.40	335.44	335.49	335.53
757	335.58	335.62	335.66	335.71	335.75	335.80	335.84	335.89	335.93	335.97
758	336.02	336.06	336.11	336.15	336.20	336.24	336.28	336.33	336.37	336.42
759	336.46	336.51	336.55	336.59	336.64	336.68	336.73	336.77	336.82	336.86
	0.	**1.**	**2.**	**3.**	**4.**	**5.**	**6.**	**7.**	**8.**	**9.**

1 Millimetre = 0.443296 French Line.

Millimetres.	Tenths of Millimetres.									
	0.	1.	2.	3.	4.	5.	6.	7.	8.	9.
	Par.lines.	Par.lines.	Par.lines.	Par.lines.	Par.lines.	Par.lines.	Par.lines.	Par.lines.	Par.lines.	Par.lines.
760	336.90	336.95	336.99	337.04	337.08	337.13	337.17	337.22	337.26	337.30
761	337.35	337.39	337.44	337.48	337.53	337.57	337.61	337.66	337.70	337.75
762	337.79	337.84	337.88	337.92	337.97	338.01	338.06	338.10	338.15	338.19
763	338.23	338.28	338.32	338.37	338.41	338.46	338.50	338.55	338.59	338.63
764	338.68	338.72	338.77	338.81	338.86	338.90	338.94	338.99	339.03	339.08
765	339.12	339.17	339.21	339.25	339.30	339.34	339.39	339.43	339.48	339.52
766	339.56	339.61	339.65	339.70	339.74	339.79	339.83	339.87	339.92	339.96
767	340.01	340.05	340.10	340.14	340.19	340.23	340.27	340.32	340.36	340.41
768	340.45	340.50	340.54	340.58	340.63	340.67	340.72	340.76	340.81	340.85
769	340.89	340.94	340.98	341.03	341.07	341.12	341.16	341.20	341.25	341.29
770	341.34	341.38	341.43	341.47	341.52	341.56	341.60	341.65	341.69	341.74
771	341.78	341.83	341.87	341.91	341.96	342.00	342.05	342.09	342.14	342.18
772	342.22	342.27	342.31	342.36	342.40	342.45	342.49	342.53	342.58	342.62
773	342.67	342.71	342.76	342.80	342.85	342.89	342.93	342.98	343.02	343.07
774	343.11	343.16	343.20	343.24	343.29	343.33	343.38	343.42	343.47	343.51
775	343.55	343.60	343.64	343.69	343.73	343.78	343.82	343.86	343.91	343.95
776	344.00	344.04	344.09	344.13	344.17	344.22	344.26	344.31	344.35	344.40
777	344.44	344.49	344.53	344.57	344.62	344.66	344.71	344.75	344.80	344.84
778	344.88	344.93	344.97	345.02	345.06	345.11	345.15	345.19	345.24	345.28
779	345.33	345.37	345.42	345.46	345.50	345.55	345.59	345.64	345.68	345.73
780	345.77	345.82	345.86	345.90	345.95	345.99	346.04	346.08	346.13	346.17
781	346.21	346.26	346.30	346.35	346.39	346.44	346.48	346.52	346.57	346.61
782	346.66	346.70	346.75	346.79	346.83	346.88	346.92	346.97	347.01	347.06
783	347.10	347.15	347.19	347.23	347.28	347.32	347.37	347.41	347.46	347.50
784	347.54	347.59	347.63	347.68	347.72	347.77	347.81	347.85	347.90	347.94
785	347.99	348.03	348.08	348.12	348.16	348.21	348.25	348.30	348.34	348.39
786	348.43	348.47	348.52	348.56	348.61	348.65	348.70	348.74	348.79	348.83
787	348.87	348.92	348.96	349.01	349.05	349.10	349.14	349.18	349.23	349.27
788	349.32	349.36	349.41	349.45	349.49	349.54	349.58	349.63	349.67	349.72
789	349.76	349.80	349.85	349.89	349.94	349.98	350.03	350.07	350.12	350.16
790	350.20	350.25	350.29	350.34	350.38	350.43	350.47	350.51	350.56	350.60
791	350.65	350.69	350.74	350.78	350.82	350.87	350.91	350.96	351.00	351.05
792	351.09	351.13	351.18	351.22	351.27	351.31	351.36	351.40	351.44	351.49
793	351.53	351.58	351.62	351.67	351.71	351.76	351.80	351.84	351.89	351.93
794	351.98	352.02	352.07	352.11	352.15	352.20	352.24	352.29	352.33	352.38
795	352.42	352.46	352.51	352.55	352.60	352.64	352.69	352.73	352.77	352.82
796	352.86	352.91	352.95	353.00	353.04	353.09	353.13	353.17	353.22	353.26
797	353.31	353.35	353.40	353.44	353.48	353.53	353.57	353.62	353.66	353.71
798	353.75	353.79	353.84	353.88	353.93	353.97	354.02	354.06	354.10	354.15
799	354.19	354.24	354.28	354.33	354.37	354.42	354.46	354.50	354.55	354.59
800	354.64	354.68	354.73	354.77	354.81	354.86	354.90	354.95	354.99	355.04
	0.	1.	2.	3.	4.	5.	6.	7.	8.	9.

V.-VI.

COMPARISON

OF

THE OLD FRENCH BAROMETER

WITH

THE ENGLISH AND THE METRICAL BAROMETERS,

OR

TABLES

FOR CONVERTING FRENCH OR PARIS LINES INTO ENGLISH INCHES AND DECIMALS, AND INTO MILLIMETRES;

GIVING THE VALUES CORRESPONDING TO EVERY PARIS LINE FROM 120 TO 216 LINES, OR FROM 10 TO 18 INCHES; AND TO EVERY TENTH OF A LINE FROM 216 TO 348 LINES, OR FROM 18 TO 29 FRENCH INCHES.

TABLE V.

MM. J. J. Pohl and J. Schabus have published, in the number for March, 1852, of the *Proceedings of the Imperial Academy of Vienna, Class of Mathematics and Natural Philosophy*, a set of short Thermometrical and Barometrical Reduction Tables, among which is found a table for the reduction of the Old French Barometrical Scale into the English. As this table shows slight discrepancies from the one given in the following pages, it may not be out of place to state that they arise from an accidental error in the equation used by MM. Pohl and Schabus in computing their table. Adopting, as they do, Bird's value of the metre, viz.

1 metre = 39.37062 English inches,

the value of the Paris line is

1 Paris line = 0.088813 English inches.

But the table seems to have been computed by using the equation

1 Paris line = 0.088823 English inches,

which gives, at the end of the table,

348 lines × .088823 = 30.9104 English inches,

instead of

348 " × .088813 = 30.9069 " "

thus causing an error = 0.0035 " "

which, of course, gradually diminishes in lower numbers.

1 Paris Line = 0.088814 English Inch.

French or Paris Lines. Tens.	Units.									
	0.	**1.**	**2.**	**3.**	**4.**	**5.**	**6.**	**7.**	**8.**	**9.**
10 Inch.	Eng. In.	Eng. In.	Eng. In.	Eng. In.	Eng. In.	Eng. In.	Eng. In.	Eng. In.	Eng. In.	Eng. In.
120	10.658	10.746	10.835	10.924	11.013	11.102	11.191	11.279	11.368	11.457
130	11.546	11.635	11.723	11.812	11.901	11.990	12.079	12.168	12.256	12.345
140	12.434	12.523	12.612	12.700	12.789	12.878	12.967	13.056	13.144	13.233
150	13.322	13.411	13.500	13.589	13.677	13.766	13.855	13.944	14.033	14.121
160	14.210	14.299	14.388	14.477	14.565	14.654	14.743	14.832	14.921	15.010
170	15.098	15.187	15.276	15.365	15.454	15.542	15.631	15.720	15.809	15.898
180	15.987	16.075	16.164	16.253	16.342	16.431	16.519	16.608	16.697	16.786
190	16.875	16.963	17.052	17.141	17.230	17.319	17.408	17.496	17.585	17.674
200	17.763	17.852	17.940	18.029	18.118	18.207	18.296	18.384	18.473	18.562
210	18.651	18.740	18.829	18.917	19.006	19.095	19.184	19.273	19.361	19.450

Paris Lines.	Tenths.									
	0.	**1.**	**2.**	**3.**	**4.**	**5.**	**6.**	**7.**	**8.**	**9.**
18 Inch.	Eng. In.	Eng. In.	Eng. In.	Eng. In.	Eng. In.	Eng. In.	Eng. In.	Eng. In.	Eng. In.	Eng. In.
216	19.184	19.193	19.202	19.210	19.219	19.228	19.237	19.246	19.255	19.264
217	19.273	19.282	19.290	19.299	19.308	19.317	19.326	19.335	19.344	19.353
218	19.361	19.370	19.379	19.388	19.397	19.406	19.415	19.424	19.433	19.441
219	19.450	19.459	19.468	19.477	19.486	19.495	19.504	19.512	19.521	19.530
220	19.539	19.548	19.557	19.566	19.575	19.583	19.592	19.601	19.610	19.619
221	19.628	19.637	19.646	19.655	19.663	19.672	19.681	19.690	19.699	19.708
222	19.717	19.726	19.734	19.743	19.752	19.761	19.770	19.779	19.788	19.797
223	19.806	19.814	19.823	19.832	19.840	19.850	19.859	19.868	19.877	19.885
224	19.894	19.903	19.912	19.921	19.930	19.939	19.948	19.957	19.965	19.974
225	19.983	19.992	20.001	20.010	20.019	20.028	20.036	20.045	20.054	20.063
226	20.072	20.081	20.090	20.099	20.107	20.116	20.125	20.134	20.143	20.152
227	20.161	20.170	20.179	20.187	20.196	20.205	20.214	20.223	20.232	20.241
19 Inch.										
228	20.250	20.258	20.267	20.276	20.285	20.294	20.303	20.312	20.321	20.330
229	20.338	20.347	20.356	20.365	20.374	20.383	20.392	20.401	20.409	20.418
230	20.427	20.436	20.445	20.454	20.463	20.472	20.481	20.489	20.498	20.507
231	20.516	20.525	20.534	20.543	20.552	20.560	20.569	20.578	20.587	20.596
232	20.605	20.614	20.623	20.631	20.640	20.649	20.658	20.667	20.676	20.685
233	20.694	20.703	20.711	20.720	20.729	20.738	20.747	20.756	20.765	20.774
234	20.782	20.791	20.800	20.809	20.818	20.827	20.836	20.845	20.854	20.862
235	20.871	20.880	20.889	20.898	20.907	20.916	20.925	20.933	20.942	20.951
236	20.960	20.969	20.978	20.987	20.996	21.005	21.013	21.022	21.031	21.040
237	21.049	21.058	21.067	21.076	21.084	21.093	21.102	21.111	21.120	21.129
238	21.138	21.147	21.155	21.164	21.173	21.182	21.191	21.200	21.209	21.218
239	21.227	21.235	21.244	21.253	21.262	21.271	21.280	21.289	21.298	21.306

Hundredths of a Line.									
0.	**1.**	**2.**	**3.**	**4.**	**5.**	**6.**	**7.**	**8.**	**9.**
.000	.001	.002	.003	.004	.004	.005	.006	.007	.008

1 Paris Line = 0.088814 English Inch.

French or Paris Lines.	Tenths of a Line.									
	0.	**1.**	**2.**	**3.**	**4.**	**5.**	**6.**	**7.**	**8.**	**9.**
20 Inches.	Eng. In.	Eng. In.	Eng. In.	Eng. In.	Eng. In.	Eng. In.	Eng. In.	Eng. In.	Eng. In.	Eng. In.
240	21.315	21.324	21.333	21.342	21.351	21.360	21.369	21.378	21.386	21.395
241	21.404	21.413	21.422	21.431	21.440	21.449	21.457	21.466	21.475	21.484
242	21.493	21.502	21.511	21.520	21.529	21.537	21.546	21.555	21.564	21.573
243	21.582	21.591	21.600	21.608	21.617	21.626	21.635	21.644	21.653	21.662
244	21.671	21.679	21.688	21.697	21.706	21.715	21.724	21.733	21.742	21.751
245	21.759	21.768	21.777	21.786	21.795	21.804	21.813	21.822	21.830	21.839
246	21.848	21.857	21.866	21.875	21.884	21.893	21.902	21.910	21.919	21.928
247	21.937	21.946	21.955	21.964	21.973	21.981	21.990	21.999	22.008	22.017
248	22.026	22.035	22.044	22.053	22.061	22.070	22.079	22.088	22.097	22.106
249	22.115	22.124	22.132	22.141	22.150	22.159	22.168	22.177	22.186	22.195
250	22.203	22.212	22.221	22.230	22.239	22.248	22.257	22.266	22.275	22.283
251	22.292	22.301	22.310	22.319	22.328	22.337	22.346	22.354	22.363	22.372
21 In. =										
252	22.381	22.390	22.399	22.408	22.417	22.426	22.434	22.443	22.452	22.461
253	22.470	22.479	22.488	22.497	22.505	22.514	22.523	22.532	22.541	22.550
254	22.559	22.568	22.577	22.585	22.594	22.603	22.612	22.621	22.630	22.639
255	22.648	22.656	22.665	22.674	22.683	22.692	22.701	22.710	22.719	22.728
256	22.736	22.745	22.754	22.763	22.772	22.781	22.790	22.799	22.807	22.816
257	22.825	22.834	22.843	22.852	22.861	22.870	22.878	22.887	22.896	22.905
258	22.914	22.923	22.932	22.941	22.950	22.958	22.967	22.976	22.985	22.994
259	23.003	23.012	23.021	23.029	23.038	23.047	23.056	23.065	23.074	23.083
260	23.092	23.101	23.109	23.118	23.127	23.136	23.145	23.154	23.163	23.172
261	23.180	23.189	23.198	23.207	23.216	23.225	23.234	23.243	23.252	23.260
262	23.269	23.278	23.287	23.296	23.305	23.314	23.323	23.331	23.340	23.349
263	23.358	23.367	23.376	23.385	23.394	23.402	23.411	23.420	23.429	23.438
22 In. =										
264	23.447	23.456	23.465	23.474	23.482	23.491	23.500	23.509	23.518	23.527
265	23.536	23.545	23.553	23.562	23.571	23.580	23.589	23.598	23.607	23.616
266	23.625	23.633	23.642	23.651	23.660	23.669	23.678	23.687	23.696	23.704
267	23.713	23.722	23.731	23.740	23.749	23.758	23.767	23.776	23.784	23.793
268	23.802	23.811	23.820	23.829	23.838	23.847	23.855	23.864	23.873	23.882
269	23.891	23.900	23.909	23.918	23.926	23.935	23.944	23.953	23.962	23.971
270	23.980	23.989	23.998	24.006	24.015	24.024	24.033	24.042	24.051	24.060
271	24.069	24.077	24.086	24.095	24.104	24.113	24.122	24.131	24.140	24.149
272	24.157	24.166	24.175	24.184	24.193	24.202	24.211	24.220	24.228	24.237
273	24.246	24.255	24.264	24.273	24.282	24.291	24.300	24.308	24.317	24.326
274	24.335	24.344	24.353	24.362	24.371	24.379	24.388	24.397	24.406	24.415
275	24.424	24.433	24.442	24.450	24.459	24.468	24.477	24.486	24.495	24.504

Hundredths of a Line.									
0.	**1.**	**2.**	**3.**	**4.**	**5.**	**6.**	**7.**	**8.**	**9.**
.0000	.0009	.0018	.0027	.0036	.0044	.0053	.0062	.0071	.0080

1 Paris Line = 0.088814 English Inch.

French or Paris Lines.	Tenths of a Line.									
	0.	**1.**	**2.**	**3.**	**4.**	**5.**	**6.**	**7.**	**8.**	**9.**
23 Inches.	Eng. In.	Eng. In.	Eng. In.	Eng. In.	Eng. In.	Eng. In.	Eng. In.	Eng. In.	Eng. In.	Eng. In.
276	24.513	24.522	24.530	24.539	24.548	24.557	24.566	24.575	24.584	24.593
277	24.601	24.610	24.619	24.628	24.637	24.646	24.655	24.664	24.673	24.681
278	24.690	24.699	24.708	24.717	24.726	24.735	24.744	24.752	24.761	24.770
279	24.779	24.788	24.797	24.806	24.815	24.824	24.832	24.841	24.850	24.859
280	24.868	24.877	24.886	24.895	24.903	24.912	24.921	24.930	24.939	24.948
281	24.957	24.966	24.974	24.983	24.992	25.001	25.010	25.019	25.028	25.037
282	25.046	25.054	25.063	25.072	25.081	25.090	25.099	25.108	25.117	25.125
283	25.134	25.143	25.152	25.161	25.170	25.179	25.188	25.197	25.205	25.214
284	25.223	25.232	25.241	25.250	25.259	25.268	25.276	25.285	25.294	25.303
285	25.312	25.321	25.330	25.339	25.348	25.356	25.365	25.374	25.383	25.392
286	25.401	25.410	25.419	25.427	25.436	25.445	25.454	25.463	25.472	25.481
287	25.490	25.498	25.507	25.516	25.525	25.534	25.543	25.552	25.561	25.570
24 In. =										
288	25.578	25.587	25.596	25.605	25.614	25.623	25.632	25.641	25.649	25.658
289	25.667	25.676	25.685	25.694	25.703	25.712	25.721	25.729	25.738	25.747
290	25.756	25.765	25.774	25.783	25.792	25.800	25.809	25.818	25.827	25.836
291	25.845	25.854	25.863	25.872	25.880	25.889	25.898	25.907	25.916	25.925
292	25.934	25.943	25.951	25.960	25.969	25.978	25.987	25.996	26.005	26.014
293	26.023	26.031	26.040	26.049	26.058	26.067	26.076	26.085	26.094	26.102
294	26.111	26.120	26.129	26.138	26.147	26.156	26.165	26.173	26.182	26.191
295	26.200	26.209	26.218	26.227	26.236	26.245	26.253	26.262	26.271	26.280
296	26.289	26.298	26.307	26.316	26.324	26.333	26.342	26.351	26.360	26.369
297	26.378	26.387	26.396	26.404	26.413	26.422	26.431	26.440	26.449	26.458
298	26.467	26.475	26.484	26.493	26.502	26.511	26.520	26.529	26.538	26.547
299	26.555	26.564	26.573	26.582	26.591	26.600	26.609	26.618	26.626	26.635
25 In. =										
300	26.644	26.653	26.662	26.671	26.680	26.689	26.697	26.706	26.715	26.724
301	26.733	26.742	26.751	26.760	26.769	26.777	26.786	26.795	26 804	26.813
302	26.822	26.831	26.840	26.848	26.857	26.866	26.875	26.884	26.893	26.902
303	26.911	26.920	26.928	26.937	26.946	26.955	26.964	26.973	26.982	26.991
304	26.999	27.008	27.017	27.026	27.035	27.044	27.053	27.062	27.071	27.079
305	27.088	27.097	27.106	27.115	27.124	27.133	27.142	27.150	27.159	27.168
306	27.177	27.186	27.195	27.204	27.213	27.221	27.230	27.239	27.248	27.257
307	27.266	27.275	27.284	27.293	27.301	27.310	27.319	27.328	27.337	27.346
308	27.355	27.364	27.372	27.381	27.390	27.399	27.408	27.417	27.426	27.435
309	27.444	27.452	27.461	27.470	27.479	27.488	27.497	27.506	27.515	27.523
310	27.532	27.541	27.550	27.559	27.568	27.577	27.586	27.595	27.603	27.612
311	27.621	27.630	27.639	27.648	27.657	27.666	27.674	27.683	27.692	27.701

Hundredths of a Line.

0.	**1.**	**2.**	**3.**	**4.**	**5.**	**6.**	**7.**	**8.**	**9.**
.0000	.0009	.0018	.0027	.0036	.0044	.0053	.0062	.0071	.0080

1 Paris Line = 0.088814 English Inch.

French or Paris Lines.	Tenths of a Line.									
	0.	**1.**	**2.**	**3.**	**4.**	**5.**	**6.**	**7.**	**8.**	**9.**
26 Inches.	Eng. In.	Eng. In.	Eng. In.	Eng. In.	Eng. In.	Eng. In.	Eng. In.	Eng. In.	Eng. In.	Eng. In.
312	27.710	27.719	27.728	27.737	27.745	27.754	27.763	27.772	27.781	27.790
313	27.799	27.808	27.817	27.825	27.834	27.843	27.852	27.861	27.870	27.879
314	27.888	27.896	27.905	27.914	27.923	27.932	27.941	27.950	27.959	27.968
315	27.976	27.985	27.994	28.003	28.012	28.021	28.030	28.039	28.047	28.056
316	28.065	28.074	28.083	28.092	28.101	28.110	28.119	28.127	28.136	28.145
317	28.154	28.163	28.172	28.181	28.190	28.198	28.207	28.216	28.225	28.234
318	28.243	28.252	28.261	28.269	28.278	28.287	28.296	28.305	28.314	28.323
319	28.332	28.341	28.349	28.358	28.367	28.376	28.385	28.394	28.403	28.412
320	28.420	28.429	28.438	28.447	28.456	28.465	28.474	28.483	28.492	28.500
321	28.509	28.518	28.527	28.536	28.545	28.554	28.563	28.571	28.580	28.589
322	28.598	28.607	28.616	28.625	28.634	28.643	28.651	28.660	28.669	28.678
323	28.687	28.696	28.705	28.714	28.722	28.731	28.740	28.749	28.758	28.767
27 In. =										
324	28.776	28.785	28.793	28.802	28.811	28.820	28.829	28.838	28.847	28.856
325	28.865	28.873	28.882	28.891	28.900	28.909	28.918	28.927	28.936	28.944
326	28.953	28.962	28.971	28.980	28.989	28.998	29.007	29.016	29.024	29.033
327	29.042	29.051	29.060	29.069	29.078	29.087	29.095	29.104	29.113	29.122
328	29.131	29.140	29.149	29.158	29.167	29.175	29.184	29.193	29.202	29.211
329	29.220	29.229	29.238	29.246	29.255	29.264	29.273	29.282	29.291	29.300
330	29.309	29.318	29.326	29.335	29.344	29.353	29.362	29.371	29.380	29.389
331	29.397	29.406	29.415	29.424	29.433	29.442	29.451	29.460	29.468	29.477
332	29.486	29.495	29.504	29.513	29.522	29.531	29.540	29.548	29.557	29.566
333	29.575	29.584	29.593	29.602	29.611	29.619	29.628	29.637	29.646	29.655
334	29.664	29.673	29.682	29.691	29.699	29.708	29.717	29.726	29.735	29.744
335	29.753	29.762	29.770	29.779	29.788	29.797	29.806	29.815	29.824	29.833
28 In. =										
336	29.842	29.850	29.859	29.868	29.877	29.886	29.895	29.904	29.913	29.921
337	29.930	29.939	29.948	29.957	29.966	29.975	29.984	29.992	30.001	30.010
338	30.019	30.028	30.037	30.046	30.055	30.064	30.072	30.081	30.090	30.099
339	30.108	30.117	30.126	30.135	30.143	30.152	30.161	30.170	30.179	30.188
340	30.197	30.206	30.215	30.223	30.232	30.241	30.250	30.259	30.268	30.277
341	30.286	30.294	30.303	30.312	30.321	30.330	30.339	30.348	30.357	30.366
342	30.374	30.383	30.392	30.401	30.410	30.419	30.428	30.437	30.445	30.454
343	30.463	30.472	30.481	30.490	30.499	30.508	30.516	30.525	30.534	30.543
344	30.552	30.561	30.570	30.579	30.588	30.596	30.605	30.614	30.623	30.632
345	30.641	30.650	30.659	30.667	30.676	30.685	30.694	30.703	30.712	30.721
346	30.730	30.739	30.747	30.756	30.765	30.774	30.783	30.792	30.801	39.810
347	30.818	30.827	30.836	30.845	30.854	30.863	30.872	30.881	30.890	30.898
29 In. =										
348	30.907	30.916	30.925	30.934	30.943	30.952	30.961	30.969	30.978	30.987

Hundredths of a Line.									
0.	**1.**	**2.**	**3.**	**4.**	**5.**	**6.**	**7.**	**8.**	**9.**
.0000	.0009	.0018	.0027	.0036	.0044	.0053	.0062	.0071	.0080

1 Paris Line = 2.255829 Millimetres.

French or Paris Lines.	Units.									
Tens.	**0.**	**1.**	**2.**	**3.**	**4.**	**5.**	**6.**	**7.**	**8.**	**9.**
10 Inch.	Millim.	Millim.	Millim.	Millim.	Millim.	Millim.	Millim.	Millim.	Millim.	Millim.
120	270.70	272.96	275.21	277.47	279.72	281.98	284.23	286.49	288.75	291.00
130	293.26	295.51	297.77	300.03	302.28	304.54	306.79	309.05	311.30	313.56
140	315.82	318.07	320.33	322.58	324.84	327.10	329.35	331.61	333.86	336.12
150	338.37	340.63	342.89	345.14	347.40	349.65	351.91	354.17	356.42	358.68
160	360.93	363.19	365.44	367.70	369.96	372.21	374.47	376.72	378.98	381.24
170	383.49	385.75	388.00	390.26	392.51	394.77	397.03	399.28	401.54	403.79
180	406.05	408.30	410.56	412.82	415.07	417.33	419.58	421.84	424.10	426.35
190	428.61	430.86	433.12	435.37	437.63	439.89	442.14	444.40	446.65	448.91
200	451.17	453.42	455.68	457.93	460.19	462.44	464.70	466.96	469.21	471.47
210	473.72	475.98	478.24	480.49	482.75	485.00	487.26	489.51	491.77	494.03

Paris Lines.	Tenths of a Line.									
	0.	**1.**	**2.**	**3.**	**4.**	**5.**	**6.**	**7.**	**8.**	**9.**
18 Inch.	Millim.	Millim.	Millim.	Millim.	Millim.	Millim.	Millim.	Millim.	Millim.	Millim.
216	487.26	487.48	487.71	487.94	488.16	488.39	488.61	488.84	489.06	489.29
217	489.51	489.74	489.97	490.19	490.42	490.64	490.87	491.09	491.32	491.55
218	491.77	492.00	492.22	492.45	492.67	492.90	493.12	493.35	493.58	493.80
219	494.03	494.25	494.48	494.70	494.93	495.15	495.38	495.61	495.83	496.06
220	496.28	496.51	496.73	496.96	497.18	497.41	497.64	497.86	498.09	498.31
221	498.54	498.76	498.99	499.21	499.44	499.67	499.89	500.12	500.34	500.57
222	500.79	501.02	501.25	501.47	501.70	501.92	502.15	502.37	502.60	502.82
223	503.05	503.28	503.50	503.73	503.95	504.18	504.40	504.63	504.85	505.08
224	505.31	505.53	505.76	505.98	506.21	506.43	506.66	506.88	507.11	507.34
225	507.56	507.79	508.01	508.24	508.46	508.69	508.91	509.14	509.37	509.59
226	509.82	510.04	510.27	510.49	510.72	510.95	511.17	511.40	511.62	511.85
227	512.07	512.30	512.52	512.75	512.98	513.20	513.43	513.65	513.88	514.10
19 Inch.										
228	514.33	514.55	514.78	515.01	515.23	515.46	515.68	515.91	516.13	516.36
229	516.58	516.81	517.04	517.26	517.49	517.71	517.94	518.16	518.39	518.61
230	518.84	519.07	519.29	519.52	519.74	519.97	520.19	520.42	520.65	520.87
231	521.10	521.32	521.55	521.77	522.00	522.22	522.45	522.68	522.90	523.13
232	523.35	523.58	523.80	524.03	524.25	524.48	524.71	524.93	525.16	525.38
233	525.61	525.83	526.06	526.28	526.51	526.74	526.96	527.19	527.41	527.64
234	527.86	528.09	528.32	528.54	528.77	528.99	529.22	529.44	529.67	529.89
235	530.12	530.35	530.57	530.80	531.02	531.25	531.47	531.70	531.92	532.15
236	532.38	532.60	532.83	533.05	533.28	533.50	533.73	533.95	534.18	534.41
237	534.63	534.86	535.08	535.31	535.53	535.76	535.98	536.21	536.44	536.66
238	536.89	537.11	537.34	537.56	537.79	538.02	538.24	538.47	538.69	538.92
239	539.14	539.37	539.59	539.82	540.05	540.27	540.50	540.72	540.95	541.17

Tenths of a Line.

0.	**1.**	**2.**	**3.**	**4.**	**5.**	**6.**	**7.**	**8.**	**9.**
0.00	0.23	0.45	0.68	0.90	1.13	1.35	1.58	1.80	2.03

1 Paris Line = 2.255829 Millimetres.

Paris or French Lines.	Tenths of a Line. 0.	1.	2.	3.	4.	5.	6.	7.	8.	9.
20 Inches.	Millim.	Millim.	Millim.	Millim.	Millim.	Millim.	Millim.	Millim.	Millim.	Millim.
240	541.40	541.62	541.85	542.08	542.30	542.53	542.75	542.98	543.20	543.43
241	543.65	543.88	544.11	544.33	544.56	544.78	545.01	545.23	545.46	545.69
242	545.91	546.14	546.36	546.59	546.81	547.04	547.26	547.49	547.72	547.94
243	548.17	548.39	548.62	548.84	549.07	549.29	549.52	549.75	549.97	550.20
244	550.42	550.65	550.87	551.10	551.32	551.55	551.78	552.00	552.23	552.45
245	552.68	552.90	553.13	553.35	553.58	553.81	554.03	554.26	554.48	554.71
246	554.93	555.16	555.39	555.61	555.84	556.06	556.29	556.51	556.74	556.96
247	557.19	557.42	557.64	557.87	558.09	558.32	558.54	558.77	558.99	559.22
248	559.45	559.67	559.90	560.12	560.35	560.57	560.80	561.02	561.25	561.48
249	561.70	561.93	562.15	562.38	562.60	562.83	563.05	563.28	563.51	563.73
250	563.96	564.18	564.41	564.63	564.86	565.09	565.31	565.54	565.76	565.99
251	566.21	566.44	566.66	566.89	567.12	567.34	567.57	567.79	568.02	568.24
21 Inches.										
252	568.47	568.69	568.92	569.15	569.37	569.60	569.82	570.05	570.27	570.50
253	570.72	570.95	571.18	571.40	571.63	571.85	572.08	572.30	572.53	572.75
254	572.98	573.21	573.43	573.66	573.88	574.11	574.33	574.56	574.79	575.01
255	575.24	575.46	575.69	575.91	576.14	576.36	576.59	576.82	577.04	577.27
256	577.49	577.72	577.94	578.17	578.39	578.62	578.85	579.07	579.30	579.52
257	579.75	579.97	580.20	580.42	580.65	580.88	581.10	581.33	581.55	581.78
258	582.00	582.23	582.46	582.68	582.91	583.13	583.36	583.58	583.81	584.03
259	584.26	584.49	584.71	584.94	585.16	585.39	585.61	585.84	586.06	586.29
260	586.52	586.74	586.97	587.19	587.42	587.64	587.87	588.09	588.32	588.55
261	588.77	589.00	589.22	589.45	589.67	589.90	590.12	590.35	590.58	590.80
262	591.03	591.25	591.48	591.70	591.93	592.16	592.38	592.61	592.83	593.06
263	593.28	593.51	593.73	593.96	594.19	594.41	594.64	594.86	595.09	595.31
22 Inches.										
264	595.54	595.76	595.99	596.22	596.44	596.67	596.89	597.12	597.34	597.57
265	597.79	598.02	598.25	598.47	598.70	598.92	599.15	599.37	599.60	599.82
266	600.05	600.28	600.50	600.73	600.95	601.18	601.40	601.63	601.86	602.08
267	602.31	602.53	602.76	602.98	603.21	603.43	603.66	603.89	604.11	604.34
268	604.56	604.79	605.01	605.24	605.46	605.69	605.92	606.14	606.37	606.59
269	606.82	607.04	607.27	607.49	607.72	607.95	608.17	608.40	608.62	608.85
270	609.07	609.30	609.52	609.75	609.98	610.20	610.43	610.65	610.88	611.10
271	611.33	611.56	611.78	612.01	612.23	612.46	612.68	612.91	613.13	613.36
272	613.59	613.81	614.04	614.26	614.49	614.71	614.94	615.16	615.39	615.62
273	615.84	616.07	616.29	616.52	616.74	616.97	617.19	617.42	617.65	617.87
274	618.10	618.32	618.55	618.77	619.00	619.23	619.45	619.68	619.90	620.13
275	620.35	620.58	620.80	621.03	621.26	621.48	621.71	621.93	622.16	622.38

Hundredths of a Line.

0.	1.	2.	3.	4.	5.	6.	7.	8.	9.
0.000	0.023	0.045	0.068	0.090	0.113	0.135	0.158	0.180	0.203

1 Paris Line = 2.255829 Millimetres.

Paris or French Lines.	Tenths of a Line.									
	0.	**1.**	**2.**	**3.**	**4.**	**5.**	**6.**	**7.**	**8.**	**9.**
23 Inches.	Millim.	Millim.	Millim.	Millim.	Millim.	Millim.	Millim.	Millim.	Millim.	Millim.
276	622.61	622.83	623.06	623.29	623.51	623.74	623.96	624.19	624.41	624.64
277	624.86	625.09	625.32	625.54	625.77	625.99	626.22	626.44	626.67	626.89
278	627.12	627.35	627.57	627.80	628.02	628.25	628.47	628.70	628.93	629.15
279	629.38	629.60	629.83	630.05	630.28	630.50	630.73	630.96	631.18	631.41
280	631.63	631.86	632.08	632.31	632.53	632.76	632.99	633.21	633.44	633.66
281	633.89	634.11	634.34	634.56	634.79	635.02	635.24	635.47	635.69	635.92
282	636.14	636.37	636.59	636.82	637.05	637.27	637.50	637.72	637.95	638.17
283	638.40	638.63	638.85	639.08	639.30	639.53	639.75	639.98	640.20	640.43
284	640.66	640.88	641.11	641.33	641.56	641.78	642.01	642.23	643.46	642.69
285	642.91	643.14	643.36	643.59	643.81	644.04	644.26	644.49	644.72	644.94
286	645.17	645.39	645.62	645.84	646.07	646.30	646.52	646.75	646.97	647.20
287	647.42	647.65	647.87	648.10	648.33	648.55	648.78	649.00	649.23	649.45
24 Inches.										
288	649.68	649.90	650.13	650.36	650.58	650.81	651.03	651.26	651.48	651.71
289	651.93	652.16	652.39	652.61	652.84	653.06	653.29	653.51	653.74	653.96
290	654.19	654.42	654.64	654.87	655.09	655.32	655.54	655.77	656.00	656.22
291	656.45	656.67	656.90	657.12	657.35	657.57	657.80	658.03	658.25	658.48
292	658.70	658.93	659.15	659.38	659.60	659.83	660.06	660.28	660.51	660.73
293	660.96	661.18	661.41	661.63	661.86	662.09	662.31	662.54	662.76	662.99
294	663.21	663.44	663.66	663.89	664.12	664.34	664.57	664.79	665.02	665.24
295	665.47	665.70	665.92	666.15	666.37	666.60	666.82	667.05	667.27	667.50
296	667.73	667.95	668.18	668.40	668.63	668.85	669.08	669.30	669.53	669.76
297	669.98	670.21	670.43	670.66	670.88	671.11	671.33	671.56	671.79	672.01
298	672.24	672.46	672.69	672.91	673.14	673.36	673.59	673.82	674.04	674.27
299	674.49	674.72	674.94	675.17	675.40	675.62	675.85	676.07	676.30	676.52
25 Inches.										
300	676.75	676.97	677.20	677.43	677.65	677.88	678.10	678.33	678.55	678.78
301	679.00	679.23	679.46	679.68	679.91	680.13	680.36	680.58	680.81	681.03
302	681.26	681.49	681.71	681.94	682.16	682.39	682.61	682.84	683.07	683.29
303	683.52	683.74	683.97	684.19	684.42	684.64	684.87	685.10	685.32	685.55
304	685.77	686.00	686.22	686.45	686.67	686.90	687.13	687.35	687.58	687.80
305	688.03	688.25	688.48	688.70	688.93	689.16	689.38	689.61	689.83	690.06
306	690.28	690.51	690.73	690.96	691.19	691.41	691.64	691.86	692.09	692.31
307	692.54	692.77	692.99	693.22	693.44	693.67	693.89	694.12	694.34	694.57
308	694.80	695.02	695.25	695.47	695.70	695.92	696.15	696.37	696.60	696.83
309	697.05	697.28	697.50	697.73	697.95	698.18	698.40	698.63	698.86	699.08
310	699.31	699.53	699.76	699.98	700.21	700.43	700.66	700.89	701.11	701.34
311	701.56	701.79	702.01	702.24	702.47	702.69	702.92	703.14	703.37	703.59

Hundredths of a Line.									
0.	**1.**	**2.**	**3.**	**4.**	**5.**	**6.**	**7.**	**8.**	**9.**
0.000	0.023	0.045	0.068	0.090	0.113	0.135	0.158	0.180	0.203

1 Paris Line = 2.255829 Millimetres.

Paris or French Lines.	Tenths of a Line.									
	0.	**1.**	**2.**	**3.**	**4.**	**5.**	**6.**	**7.**	**8.**	**9.**
26 Inches.	Millim.	Millim.	Millim.	Millim.	Millim.	Millim.	Millim.	Millim.	Millim.	Millim.
312	703.82	704.04	704.27	704.50	704.72	704.95	705.17	705.40	705.62	705.85
313	706.07	706.30	706.53	706.75	706.98	707.20	707.43	707.65	707.88	708.10
314	708.33	708.56	708.78	709.01	709.23	709.46	709.68	709.91	710.13	710.36
315	710.59	710.81	711.04	711.26	711.49	711.71	711.94	712.17	712.39	712.62
316	712.84	713.07	713.29	713.52	713.74	713.97	714.20	714.42	714.65	714.87
317	715.10	715.32	715.55	715.77	716.00	716.23	716.45	716.68	716.90	717.13
318	717.35	717.58	717.80	718.03	718.26	718.48	718.71	718.93	719.16	719.38
319	719.61	719.84	720.06	720.29	720.51	720.74	720.96	721.19	721.41	721.64
320	721.87	722.09	722.32	722.54	722.77	722.99	723.22	723.44	723.67	723.90
321	724.12	724.35	724.57	724.80	725.02	725.25	725.47	725.70	725.93	726.15
322	726.38	726.60	726.83	727.05	727.28	727.50	727.73	727.96	728.18	728.41
323	728.63	728.86	729.08	729.31	729.54	729.76	729.99	730.21	730.44	730.66
27 Inches.										
324	730.89	731.11	731.34	731.57	731.79	732.02	732.24	732.47	732.69	732.92
325	733.14	733.37	733.60	733.82	734.05	734.27	734.50	734.72	734.95	735.17
326	735.40	735.63	735.85	736.08	736.30	736.53	736.75	736.98	737.20	737.43
327	737.66	737.88	738.11	738.33	738.56	738.78	739.01	739.24	739.46	739.69
328	739.91	740.14	740.36	740.59	740.81	741.04	741.27	741.49	741.72	741.94
329	742.17	742.39	742.62	742.84	743.07	743.30	743.52	743.75	743.97	744.20
330	744.42	744.65	744.87	745.10	745.33	745.55	745.78	746.00	746.23	746.45
331	746.68	746.90	747.13	747.36	747.58	747.81	748.03	748.26	748.48	748.71
332	748.94	749.16	749.39	749.61	749.84	750.06	750.29	750.51	750.74	750.97
333	751.19	751.42	751.64	751.87	752.09	752.32	752.54	752.77	753.00	753.22
334	753.45	753.67	753.90	754.12	754.35	754.57	754.80	755.03	755.25	755.48
335	755.70	755.93	756.15	756.38	756.61	756.83	757.06	757.28	757.51	757.73
28 Inches.										
336	757.96	758.18	758.41	758.64	758.86	759.09	759.31	759.54	759.76	759.99
337	760.21	760.44	760.67	760.89	761.12	761.34	761.57	761.79	762.02	762.24
338	762.47	762.70	762.92	763.15	763.37	763.60	763.82	764.05	764.27	764.50
339	764.73	764.95	765.18	765.40	765.63	765.85	766.08	766.31	766.53	766.76
340	766.98	767.21	767.43	767.66	767.88	768.11	768.34	768.56	768.79	769.01
341	769.24	769.46	769.69	769.91	770.14	770.37	770.59	770.82	771.04	771.27
342	771.49	771.72	771.94	772.17	772.40	772.62	772.85	773.07	773.30	773.52
343	773.75	773.97	774.20	774.43	774.65	774.88	775.10	775.33	775.55	775.78
344	776.01	776.23	776.46	776.68	776.91	777.13	777.36	777.58	777.81	778.04
345	778.26	778.49	778.71	778.94	779.16	779.39	779.61	779.84	780.07	780.29
346	780.52	780.74	780.97	781.19	781.42	781.64	781.87	782.10	782.32	782.55
347	782.77	783.00	783.22	783.45	783.67	783.90	784.13	784.35	784.58	784.80
29 Inches.										
348	785.03	785.25	785.48	785.71	785.93	786.16	786.38	786.61	786.83	787.06

Hundredths of a Line.

0.	**1.**	**2.**	**3.**	**4.**	**5.**	**6.**	**7.**	**8.**	**9.**
0.000	0.023	0.045	0.068	0.090	0.113	0.135	0.158	0.180	0.203

VII. - VIII.

COMPARISON

OF

THE RUSSIAN BAROMETER

WITH

THE METRICAL AND THE OLD FRENCH BAROMETERS,

OR

TABLES

FOR CONVERTING RUSSIAN HALF-LINES INTO MILLIMETRES,
AND INTO FRENCH OR PARIS LINES;

GIVING THE VALUES CORRESPONDING TO EVERY HALF-LINE FROM 440 TO 540,
OR FROM 22 TO 27 INCHES; AND TO EVERY TENTH, FROM 540 TO 610
HALF-LINES, OR FROM 27 TO 30.5 ENGLISH INCHES.

RUSSIAN BAROMETER.

A REGULAR system of Meteorological Observations has been established by order of the Russian government throughout the extensive regions placed under its sway, and a vast amount of observations made in Europe, in Asia, and in North America have already been published. The scale of the barometer employed in this system is divided in units, each of which is equal to one half of a Russian, or English decimal line, that is, 1 = 0.05 of an inch, 600 half-lines of the Russian Barometer being = 30 inches of the English Barometer.

The conversion of this scale, which is the English scale, slightly modified in its form, is easy. It suffices to divide the Russian heights by two, and to put back, by one figure, the decimal point, in order to have them converted into English inches and decimals. This transformation is so easy to effect, that a peculiar table for it would seem superfluous.

The normal temperature of the standard being the same as that of the English, that is, 13°$\frac{1}{3}$ Reaumur, or 62° Fahrenheit, the reduction of the Russian Barometer to the freezing point can be made by means of the table for reducing the English Barometers. But the attached thermometer being that of Reaumur, its indications must be first converted into degrees of Fahrenheit.

Tables VII. and VIII., which follow, have been computed in order to render more easy the comparison and the use of the Barometrical Observations recorded in the large collection, published annually by order of the Emperor of Russia, under the name of *Annuaire Météorologique et Magnétique du Corps des Ingénieurs des Mines*.

1 Russian Half-Line = 1.269977 Millimetres.

Russian Half-Lines.	Units or Russian Half-Lines.									
	0.	**1.**	**2.**	**3.**	**4.**	**5.**	**6.**	**7.**	**8.**	**9.**
22 Inch.	Millim.	Millim.	Millim.	Millim.	Millim.	Millim.	Millim.	Millim.	Millim.	Millim.
440	558.79	560.06	561.33	562.60	563.87	565.14	566.41	567.68	568.95	570.22
450	571.49	572.76	574.03	575.30	576.57	577.84	579.11	580.38	581.65	582.92
460	584.19	585.46	586.73	588.00	589.27	590.54	591.81	593.08	594.35	595.62
470	596.89	598.16	599.43	600.70	601.97	603.24	604.51	605.78	607.05	608.32
480	609.59	610.86	612.13	613.40	614.67	615.94	617.21	618.48	619.75	621.02
24.5 In.										
490	622.29	623.56	624.83	626.10	627.37	628.64	629.91	631.18	632.45	633.72
500	634.99	636.26	637.53	638.80	640.07	641.34	642.61	643.88	645.15	646.42
510	647.69	648.96	650.23	651.50	652.77	654.04	655.31	656.58	657.85	659.12
520	660.39	661.66	662.93	664.20	665.47	666.74	668.01	669.28	670.55	671.82
530	673.09	674.36	675.63	676.90	678.17	679.44	680.71	681.98	683.25	684.52

Russian Half-Lines.	Tenths.									
	0.	**1.**	**2.**	**3.**	**4.**	**5.**	**6.**	**7.**	**8.**	**9.**
27 Inch.	Millim.	Millim.	Millim.	Millim.	Millim.	Millim.	Millim.	Millim.	Millim.	Millim.
540	685.79	685.91	686.04	686.17	686.30	686.42	686.55	686.68	686.80	686.93
541	687.06	687.18	687.31	687.44	687.57	687.69	687.82	687.95	688.07	688.20
542	688.33	688.45	688.58	688.71	688.84	688.96	689.09	689.22	689.34	689.47
543	689.60	689.72	689.85	689.98	690.11	690.23	690.36	690.49	690.61	690.74
544	690.87	690.99	691.12	691.25	691.38	691.50	691.63	691.76	691.88	692.01
545	692.14	692.26	692.39	692.52	692.65	692.77	692.90	693.03	693.15	693.28
546	693.41	693.53	693.66	693.79	693.91	694.04	694.17	694.30	694.42	694.55
547	694.68	694.80	694.93	695.06	695.19	695.31	695.44	695.57	695.69	695.82
548	695.95	696.07	696.20	696.33	696.46	696.58	696.71	696.84	696.96	697.09
549	697.22	697.34	697.47	697.60	697.73	697.85	697.98	698.11	698.23	698.36
27.5 In.										
550	698.49	698.61	698.74	698.87	699.00	699.12	699.25	699.38	699.50	699.63
551	699.76	699.88	700.01	700.14	700.27	700.39	700.52	700.65	700.77	700.90
552	701.03	701.15	701.28	701.41	701.54	701.66	701.79	701.92	702.04	702.17
553	702.30	702.42	702.55	702.68	702.81	702.93	703.06	703.19	703.31	703.44
554	703.57	703.69	703.82	703.95	704.08	704.20	704.33	704.46	704.58	704.71
555	704.84	704.96	705.09	705.22	705.35	705.47	705.60	705.73	705.85	705.98
556	706.11	706.23	706.36	706.49	706.62	706.74	706.87	707.00	707.12	707.25
557	707.38	707.50	707.63	707.76	707.89	708.01	708.14	708.27	708.39	708.52
558	708.65	708.77	708.90	709.03	709.16	709.28	709.41	709.54	709.66	709.79
559	709.92	710.14	710.27	710.40	710.53	710.65	710.78	710.81	710.93	711.06
28 Inch.										
560	711.19	711.31	711.44	711.57	711.70	711.82	711.95	712.08	712.20	712.33
561	712.46	712.58	712.71	712.84	712.97	713.09	713.22	713.35	713.47	713.60
562	713.73	713.85	713.98	714.11	714.24	714.36	714.49	714.62	714.74	714.87
563	715.00	715.12	715.25	715.38	715.51	715.63	715.76	715.89	716.01	716.14
564	716.27	716.39	716.52	716.65	716.78	716.90	717.03	717.16	717.28	717.41
565	717.54	717.66	717.79	717.92	718.04	718.17	718.30	718.43	718.55	718.68
566	718.81	718.93	719.06	719.19	719.31	719.44	719.57	719.70	719.82	719.95
567	720.08	720.20	720.33	720.46	720.58	720.71	720.84	720.97	721.09	721.22
568	721.35	721.47	721.60	721.73	721.85	721.98	722.11	722.24	722.36	722.49
569	722.62	722.74	722.87	723.00	723.12	723.25	723.38	723.51	723.63	723.76

1 Russian Half-Line = 1.269977 Millimetre.

Russian Half-Lines.	Tenths.									
	0.	**1.**	**2.**	**3.**	**4.**	**5.**	**6.**	**7.**	**8.**	**9.**
28.5 Inch.	Millim.	Millim.	Millim.	Millim.	Millim.	Millim.	Millim.	Millim.	Millim.	Millim.
570	723.89	724.01	724.14	724.27	724.39	724.52	724.65	724.78	724.90	725.03
571	725.16	725.28	725.41	725.54	725.66	725.79	725.92	726.05	726.17	726.30
572	726.43	726.55	726.68	726.81	726.93	727.06	727.19	727.32	727.44	727.57
573	727.70	727.82	727.95	728.08	728.20	728.33	728.46	728.59	728.71	728.84
574	728.97	729.08	729.21	729.34	729.46	729.59	729.73	729.85	729.97	730.11
575	730.24	730.36	730.49	730.62	730.74	730.87	731.00	731.13	731.25	731.38
576	731.51	731.63	731.76	731.89	732.01	732.14	732.27	732.40	732.52	732.65
577	732.78	732.90	733.03	733.16	733.28	733.41	733.54	733.67	733.79	733.92
578	734.05	734.17	734.30	734.43	734.55	734.68	734.81	734.94	735.06	735.19
579	735.32	735.44	735.57	735.70	735.82	735.95	736.08	736.21	736.33	736.46
29 Inch.										
580	736.59	736.71	736.84	736.97	737.09	737.22	737.35	737.48	737.60	737.73
581	737.86	737.98	738.11	738.24	738.36	738.49	738.62	738.75	738.87	739.00
582	739.13	739.25	739.38	739.51	739.63	739.76	739.89	740.02	740.14	740.27
583	740.40	740.52	740.65	740.78	740.90	741.03	741.16	741.29	741.41	741.54
584	741.67	741.79	741.92	742.05	742.17	742.30	742.43	742.56	742.68	742.81
585	742.94	743.06	743.19	743.32	743.44	743.57	743.70	743.83	743.95	744.08
586	744.21	744.33	744.46	744.59	744.71	744.84	744.97	745.10	745.22	745.35
587	745.48	745.60	745.73	745.86	745.98	746.11	746.24	746.37	746.49	746.62
588	746.75	746.87	747.00	747.13	747.25	747.38	747.51	747.64	747.76	747.89
589	748.02	748.14	748.27	748.40	748.52	748.65	748.78	748.91	749.03	749.16
29.5 In.										
590	749.29	749.41	749.54	749.67	749.79	749.92	750.05	750.18	750.30	750.43
591	750.56	750.68	750.81	750.94	751.06	751.19	751.32	751.45	751.57	751.70
592	751.83	751.95	752.08	752.21	752.33	752.46	752.59	752.72	752.84	752.97
593	753.10	753.22	753.35	753.48	753.60	753.73	753.86	753.99	754.11	754.24
594	754.37	754.49	754.62	754.75	754.87	755.00	755.13	755.26	755.38	755.51
595	755.64	755.76	755.89	756.02	756.14	756.27	756.40	756.53	756.65	756.78
596	756.91	757.03	757.16	757.29	757.41	757.54	757.67	757.80	757.92	758.05
597	758.18	758.30	758.43	758.56	758.68	758.81	758.94	759.07	759.19	759.32
598	759.45	759.57	759.70	759.84	759.96	760.09	760.21	760.34	760.46	760.59
599	760.72	760.84	760.97	761.10	761.22	761.35	761.48	761.61	761.73	761.86
30 Inch.										
600	761.99	762.11	762.24	762.37	762.49	762.62	762.75	762.88	763.00	763.13
601	763.26	763.38	763.51	763.64	763.76	763.89	764.02	764.15	764.27	764.40
602	764.53	764.65	764.78	764.91	765.03	765.16	765.29	765.42	765.54	765.67
603	765.80	765.92	766.05	766.18	766.30	766.43	766.56	766.69	766.81	766.95
604	767.07	767.19	767.32	767.45	767.57	767.70	767.83	767.96	768.08	768.21
605	768.34	768.46	768.59	768.72	768.84	768.97	769.10	769.23	769.35	769.48
606	769.61	769.73	769.85	769.99	770.11	770.24	770.37	770.50	770.62	770.75
607	770.88	771.00	771.13	771.26	771.38	771.51	771.64	771.77	771.89	772.02
608	772.15	772.27	772.40	772.53	772.65	772.78	772.91	773.03	773.16	773.29
609	773.42	773.54	773.67	773.80	773.92	774.05	774.18	774.30	774.43	774.56

Hundredths.									
0.000	0.013	0.025	0.038	0.051	0.063	0.076	0.089	0.102	0.114

1 Russian Half-Line = 0.562976 Paris Line.

Russian Half-Lines.	Units or Russian Half-Lines.									
	0.	**1.**	**2.**	**3.**	**4.**	**5.**	**6.**	**7.**	**8.**	**9.**
22 Inch.	Par. line.	Par. line.	Par. line.	Par. line.	Par line.	Par line.	Par. line.	Par. line	Par line.	Par. line.
440	247.71	248.27	248.84	249.40	249.96	250.52	251.09	251.65	252.21	252.78
450	253.34	253.90	254.47	255.03	255.59	256.15	256.72	257.28	257.84	258.41
460	258.97	259.53	260.09	260.66	261.22	261.78	262.35	262.91	263.47	264.04
470	264.60	265.16	265.72	266.29	266.85	267.41	267.98	268.54	269.10	269.67
480	270.23	270.79	271.35	271.92	272.48	273.04	273.61	274.17	274.73	275.30
24.5 In.										
490	275.86	276.42	276.98	277.55	278.11	278.67	279.24	279.80	280.36	280.93
500	281.49	282.05	282.61	283.18	283.74	284.30	284.87	285.43	285.99	286.55
510	287.12	287.68	288.24	288.81	289.37	289.93	290.50	291.06	291.62	292.18
520	292.75	293.31	293.87	294.44	295.00	295.56	296.13	296.69	297.25	297.81
530	298.38	298.94	299.50	300.07	300.63	301.19	301.76	302.32	302.88	303.44

Russian Half-Lines.	Tenths.									
	0.	**1.**	**2.**	**3.**	**4.**	**5.**	**6.**	**7.**	**8.**	**9.**
27 Inch.	Par. line.	Par. line.	Par. line.	Par. line.	Par. line.	Par. line.	Par. line.	Par. line.	Par. line.	Par. line.
540	304.01	304.06	304.12	304.18	304.23	304.29	304.34	304.40	304.46	304.51
541	304.57	304.63	304.68	304.74	304.80	304.85	304.91	304.96	305.02	305.08
542	305.13	305.19	305.25	305.30	305.36	305.41	305.47	305.53	305.58	305.64
543	305.70	305.75	305.81	305.86	305.92	305.98	306.03	306.09	306.15	306.20
544	306.26	306.32	306.37	306.43	306.48	306.54	306.60	306.65	306.71	306.77
545	306.82	306.88	306.93	306.99	307.05	307.10	307.16	307.22	307.27	307.33
546	307.38	307.44	307.50	307.55	307.61	307.67	307.72	307.78	307.84	307.89
547	307.95	308.00	308.06	308.12	308.17	308.23	308.29	308.34	308.40	308.45
548	308.51	308.57	308.62	308.68	308.74	308.79	308.85	308.90	308.96	309.02
549	309.07	309.13	309.19	309.24	309.30	309.36	309.41	309.47	309.52	309.58
27.5 In.										
550	309.64	309.69	309.75	309.81	309.86	309.92	309.97	310.03	310.09	310.14
551	310.20	310.26	310.31	310.37	310.42	310.48	310.54	310.59	310.65	310.71
552	310.76	310.82	310.88	310.93	310.99	311.04	311.10	311.16	311.21	311.27
553	311.33	311.38	311.44	311.49	311.55	311.61	311.66	311.72	311.78	311.83
554	311.89	311.95	312.00	312.06	312.11	312.17	312.23	312.28	312.34	312.40
555	312.45	312.51	312.56	312.62	312.68	312.73	312.79	312.85	312.90	312.96
556	313.01	313.07	313.13	313.18	313.24	313.30	313.35	313.41	313.47	313.52
557	313.56	313.63	313.69	313.75	313.80	313.86	313.92	313.97	314.03	314.08
558	314.14	314.20	314.25	314.31	314.37	314.42	314.48	314.53	314.59	314.65
559	314.70	314.76	314.82	314.87	314.93	314.99	315.04	315.10	315.15	315.21
28 Inch.										
560	315.27	315.32	315.38	315.44	315.49	315.55	315.60	315.66	315.72	315.77
561	315.83	315.89	315.94	316.00	316.05	316.11	316.17	316.22	316.28	316.34
562	316.39	316.45	316.51	316.56	316.62	316.67	316.73	316.79	316.84	316.90
563	316.96	317.01	317.07	317.12	317.18	317.24	317.29	317.35	317.41	317.46
564	317.52	317.57	317.63	317.69	317.74	317.80	317.86	317.91	317.97	318.03
565	318.08	318.14	318.19	318.25	318.31	318.36	318.42	318.48	318.53	318.59
566	318.64	318.70	318.76	318.81	318.87	318.93	318.98	319.04	319.09	319.15
567	319.21	319.26	319.32	319.38	319.43	319.49	319.55	319.60	319.66	319.71
568	319.77	319.83	319.88	319.94	320.00	320.05	320.11	320.16	320.22	320.28
569	320.33	320.39	320.45	320.50	320.56	320.61	320.67	320.73	320.78	320.84

1 Russian Half-Line = 0.562976 Paris Line.

Russian Half-Lines.	Tenths.									
	0.	**1.**	**2.**	**3.**	**4.**	**5.**	**6.**	**7.**	**8.**	**9.**
28.5 Inch.	Par. line.	Par. line.	Par. line.	Par. line.	Par. line.	Par. line.	Par. line.	Par. line.	Par. line.	Par. line.
570	320.90	320.95	321.01	321.07	321.12	321.18	321.23	321.29	321.35	321.40
571	321.46	321.52	321.57	321.63	321.68	321.74	321.80	321.85	321.91	321.97
572	322.02	322.08	322.13	322.19	322.25	322.30	322.36	322.42	322.47	322.53
573	322.59	322.64	322.70	322.75	322.81	322.87	322.92	322.98	323.04	323.09
574	323.15	323.20	323.26	323.32	323.37	323.43	323.49	323.54	323.60	323.65
575	323.71	323.77	323.82	323.88	323.94	323.99	324.05	324.11	324.16	324.22
576	324.27	324.33	324.39	324.44	324.50	324.56	324.61	324.67	324.72	324.78
577	324.84	324.89	324.95	325.01	325.06	325.12	325.17	325.23	325.29	325.34
578	325.40	325.46	325.51	325.57	325.63	325.68	325.74	325.79	325.85	325.91
579	325.96	326.02	326.08	326.13	326.19	326.24	326.30	326.36	326.41	326.47
29 Inch.										
580	326.53	326.58	326.64	326.69	326.75	326.81	326.86	326.92	326.98	327.03
581	327.09	327.15	327.20	327.26	327.31	327.37	327.43	327.48	327.54	327.60
582	327.65	327.71	327.76	327.82	327.88	327.93	327.99	328.05	328.10	328.16
583	328.22	328.27	328.33	328.38	328.44	328.50	328.55	328.61	328.67	328.72
584	328.78	328.83	328.89	328.95	329.00	329.06	329.12	329.17	329.23	329.28
585	329.34	329.40	329.45	329.51	329.57	329.62	329.68	329.74	329.79	329.85
586	329.90	329.96	330.02	330.07	330.13	330.19	330.24	330.30	330.35	330.41
587	330.47	330.52	330.58	330.64	330.69	330.75	330.80	330.86	330.92	330.97
588	331.03	331.09	331.14	331.20	331.26	331.31	331.37	331.42	331.48	331.54
589	331.59	331.65	331.71	331.76	331.82	331.87	331.93	331.99	332.04	332.10
29.5 In.										
590	332.16	332.21	332.27	332.32	332.38	332.44	332.49	332.55	332.61	332.66
591	332.72	332.78	332.83	332.89	332.94	333.00	333.06	333.11	333.17	333.23
592	333.28	333.34	333.39	333.45	333.51	333.56	333.62	333.68	333.73	333.79
593	333.84	333.90	333.96	334.01	334.07	334.13	334.18	384.24	334.30	334.35
594	334.41	334.46	334.52	334.58	334.63	334.69	334.75	334.80	334.86	334.91
595	334.97	335.03	335.08	335.14	335.20	335.25	335.31	335.36	335.42	335.48
596	335.53	335.59	335.65	335.70	335.76	335.82	335.87	335.93	335.98	336.04
597	336.10	336.15	336.21	336.27	336.32	336.38	336.43	336.49	336.55	336.60
598	336.66	336.72	336.77	336.83	336.88	336.94	337.00	337.05	337.11	337.17
599	337.22	337.28	337.34	337.39	337.45	337.50	337.56	337.62	337.67	337.73
30 Inch.										
600	337.79	337.84	337.90	337.95	338.01	338.07	338.12	338.18	338.24	338.29
601	338.35	338.40	338.46	338.52	338.57	338.63	338.69	338.74	338.80	338.86
602	338.91	338.97	339.02	339.08	339.14	339.19	339.25	339.31	339.36	339.42
603	339.47	339.53	339.59	339.64	339.70	339.76	339.81	339.87	339.92	339.98
604	340.04	340.09	340.15	340.21	340.26	340.32	340.38	340.43	340.49	340.54
605	340.60	340.66	340.71	340.77	340.83	340.88	340.94	340.99	341.05	341.11
606	341.16	341.22	341.28	341.33	341.39	341.44	341.50	341.56	341.61	341.67
607	341.73	341.78	341.84	341.90	341.95	342.01	342.06	342.12	342.18	342.23
608	342.29	342.35	342.40	342.46	342.51	342.57	342.63	342.68	342.74	342.80
609	342.85	342.91	342.96	343.02	343.08	343.13	343.19	343.25	343.30	343.36

Hundredths.

0.000	0.006	0.011	0.017	0.022	0.028	0.034	0.039	0.045	0.051

IX.-XVI.

COMPARISON

OF

BAROMETRICAL DIFFERENCES

EXPRESSED IN MEASURES OF DIFFERENT SCALES,

OR

TABLES

FOR CONVERTING ENGLISH INCHES, MILLIMETRES, PARIS LINES, AND RUSSIAN HALF-LINES INTO EACH OTHER.

IX. CONVERSION OF ENGLISH INCHES INTO MILLIMETRES.

1 English Inch = 25.39954 Millimetres.

English Inches and Tenths.	Hundredths of an Inch.									
	0.	**1.**	**2.**	**3.**	**4.**	**5.**	**6.**	**7.**	**8.**	**9.**
	Millim.	Millim.	Millim.	Millim.	Millim.	Millim.	Millim.	Millim.	Millim.	Millim.
0.0	0.000	0.254	0.508	0.762	1.016	1.270	1.524	1.778	2.032	2.286
0.1	2.540	2.794	3.048	3.302	3.556	3.810	4.064	4.318	4.572	4.826
0.2	5.080	5.334	5.588	5.842	6.096	6.350	6.604	6.858	7.112	7.366
0.3	7.620	7.874	8.128	8.382	8.636	8.890	9.144	9.398	9.652	9.906
0.4	10.160	10.414	10.668	10.922	11.176	11.430	11.684	11.938	12.192	12.446
0.5	12.700	12.954	13.208	13.462	13.716	13.970	14.224	14.478	14.732	14.986
0.6	15.240	15.494	15.748	16.002	16.256	16.510	16.764	17.018	17.272	17.526
0.7	17.780	18.034	18.288	18.542	18.796	19.050	19.304	19.558	19.812	20.066
0.8	20.320	20.574	20.828	21.082	21.336	21.590	21.844	22.098	22.352	22.606
0.9	22.860	23.114	23.368	23.622	23.876	24.130	24.384	24.638	24.892	25.146
1.0	25.400	25.654	25.908	26.162	26.416	26.670	26.924	27.178	27.432	27.685
1.1	27.939	28.193	28.447	28.701	28.955	29.209	29.463	29.717	29.971	30.225
1.2	30.479	30.733	30.987	31.241	31.495	31.749	32.003	32.257	32.511	32.765
1.3	33.019	33.273	33.527	33.781	34.035	34.289	34.543	34.797	35.051	35.305
1.4	35.559	35.813	36.067	36.321	36.575	36.829	37.083	37.337	37.591	37.845
1.5	38.099	38.353	38.607	38.861	39.115	39.369	39.623	39.877	40.131	40.385
1.6	40.639	40.893	41.147	41.401	41.655	41.909	42.163	42.417	42.671	42.925
1.7	43.179	43.433	43.687	43.941	44.195	44.449	44.703	44.957	45.211	45.465
1.8	45.719	45.973	46.227	46.481	46.735	46.989	47.243	47.497	47.751	48.005

X. CONVERSION OF ENGLISH INCHES INTO FRENCH OR PARIS LINES.

1 English Inch = 11.259515 Paris Lines.

English Inches and Tenths.	Hundredths of an Inch.									
	0.	**1.**	**2.**	**3.**	**4.**	**5.**	**6.**	**7.**	**8.**	**9.**
	Par. line.	Par. line.	Par. line.	Par. line.	Par. line.	Par. line.	Par. line.	Par. line.	Par. line.	Par. line.
0.0	0.000	0.113	0.225	0.338	0.450	0.563	0.676	0.788	0.901	1.013
0.1	1.126	1.239	1.351	1.464	1.576	1.689	1.802	1.914	2.027	2.139
0.2	2.252	2.364	2.477	2.590	2.702	2.815	2.927	3.040	3.153	3.265
0.3	3.378	3.490	3.603	3.716	3.828	3.941	4.053	4.166	4.279	4.391
0.4	4.504	4.616	4.729	4.842	4.954	5.067	5.179	5.292	5.405	5.517
0.5	5.630	5.742	5.855	5.968	6.080	6.193	6.305	6.418	6.531	6.643
0.6	6.756	6.868	6.981	7.093	7.206	7.319	7.431	7.544	7.656	7.769
0.7	7.882	7.994	8.107	8.219	8.332	8.445	8.557	8.670	8.782	8.895
0.8	9.008	9.120	9.233	9.345	9.458	9.571	9.683	9.796	9.908	10.021
0.9	10.134	10.246	10.359	10.471	10.584	10.697	10.809	10.922	11.034	11.147
1.0	11.260	11.372	11.485	11.597	11.710	11.822	11.935	12.048	12.160	12.273
1.1	12.385	12.498	12.611	12.723	12.836	12.948	13.061	13.174	13.286	13.399
1.2	13.511	13.624	13.737	13.849	13.962	14.074	14.187	14.300	14.412	14.525
1.3	14.637	14.750	14.863	14.975	15.088	15.200	15.313	15.426	15.538	15.651
1.4	15.763	15.876	15.988	16.101	16.214	16.326	16.439	16.551	16.664	16.777
1.5	16.889	17.002	17.114	17.227	17.340	17.452	17.565	17.677	17.790	17.903
1.6	18.015	18.128	18.240	18.353	18.466	18.578	18.691	18.803	18.916	19.029
1.7	19.141	19.254	19.366	19.479	19.592	19.704	19.817	19.929	20.042	20.155
1.8	20.267	20.380	20.492	20.605	20.717	20.830	20.943	21.055	21.168	21.280

1 Metre = 39.37079 English Inches.

Millimetres.	Tenths of a Millimetre.									
	0.	**1.**	**2.**	**3.**	**4.**	**5.**	**6.**	**7.**	**8.**	**9.**
	Eng. In.	Eng. In.	Eng. In.	Eng. In.	Eng. In.	Eng. In.	Eng. In.	Eng. In.	Eng. In.	Eng. In.
0	0.0000	0.0039	0.0079	0.0118	0.0157	0.0197	0.0236	0.0276	0.0315	0.0354
1	0.0394	0.0433	0.0472	0.0512	0.0551	0.0591	0.0630	0.0669	0.0709	0.0748
2	0.0787	0.0827	0.0866	0.0906	0.0945	0.0984	0.1024	0.1063	0.1102	0.1142
3	0.1181	0.1220	0.1260	0.1299	0.1339	0.1378	0.1417	0.1457	0.1496	0.1535
4	0.1575	0.1614	0.1654	0.1693	0.1732	0.1772	0.1811	0.1850	0.1890	0.1929
5	0.1969	0.2008	0.2047	0.2087	0.2126	0.2165	0.2205	0.2244	0.2283	0.2323
6	0.2362	0.2402	0.2441	0.2480	0.2520	0.2559	0.2598	0.2638	0.2677	0.2717
7	0.2756	0.2795	0.2835	0.2874	0.2913	0.2953	0.2992	0.3032	0.3071	0.3110
8	0.3150	0.3189	0.3228	0.3268	0.3307	0.3347	0.3386	0.3425	0.3465	0.3504
9	0.3543	0.3583	0.3622	0.3661	0.3701	0.3740	0.3780	0.3819	0.3858	0.3898
10	0.3937	0.3976	0.4016	0.4055	0.4095	0.4134	0.4173	0.4213	0.4252	0.4291
11	0.4331	0.4370	0.4410	0.4449	0.4488	0.4528	0.4567	0.4606	0.4646	0.4685
12	0.4724	0.4764	0.4803	0.4843	0.4882	0.4921	0.4961	0.5000	0.5039	0.5079
13	0.5118	0.5158	0.5197	0.5236	0.5276	0.5315	0.5354	0.5394	0.5433	0.5473
14	0.5512	0.5551	0.5591	0.5630	0.5669	0.5709	0.5748	0.5788	0.5827	0.5866
15	0.5906	0.5945	0.5984	0.6024	0.6063	0.6102	0.6142	0.6181	0.6221	0.6260
16	0.6299	0.6339	0.6378	0.6417	0.6457	0.6496	0.6536	0.6575	0.6614	0.6654
17	0.6693	0.6732	0.6772	0.6811	0.6851	0.6890	0.6929	0.6969	0.7008	0.7047
18	0.7087	0.7126	0.7165	0.7205	0.7244	0.7284	0.7323	0.7362	0.7402	0.7441
19	0.7480	0.7520	0.7559	0.7599	0.7638	0.7677	0.7717	0.7756	0.7795	0.7835
20	0.7874	0.7914	0.7953	0.7992	0.8032	0.8071	0.8110	0.8150	0.8189	0.8228
21	0.8268	0.8307	0.8347	0.8386	0.8425	0.8465	0.8504	0.8543	0.8583	0.8622
22	0.8662	0.8701	0.8740	0.8780	0.8819	0.8858	0.8898	0.8937	0.8977	0.9016
23	0.9055	0.9095	0.9134	0.9173	0.9213	0.9252	0.9292	0.9331	0.9370	0.9410
24	0.9449	0.9488	0.9528	0.9567	0.9606	0.9646	0.9685	0.9725	0.9764	0.9803
25	0.9843	0.9882	0.9921	0.9961	1.0000	1.0040	1.0079	1.0118	1.0158	1.0197
26	1.0236	1.0276	1.0315	1.0355	1.0394	1.0433	1.0473	1.0512	1.0551	1.0591
27	1.0630	1.0669	1.0709	1.0748	1.0788	1.0827	1.0866	1.0906	1.0945	1.0984
28	1.1024	1.1063	1.1103	1.1142	1.1181	1.1221	1.1260	1.1299	1.1339	1.1378
29	1.1418	1.1457	1.1496	1.1536	1.1575	1.1614	1.1654	1.1693	1.1732	1.1772
30	1.1811	1.1851	1.1890	1.1929	1.1969	1.2008	1.2047	1.2087	1.2126	1.2166
31	1.2205	1.2244	1.2284	1.2323	1.2362	1.2402	1.2441	1.2481	1.2520	1.2559
32	1.2599	1.2638	1.2677	1.2717	1.2756	1.2796	1.2835	1.2874	1.2914	1.2953
33	1.2992	1.3032	1.3071	1.3110	1.3150	1.3189	1.3229	1.3268	1.3307	1.3347
34	1.3386	1.3425	1.3465	1.3504	1.3544	1.3583	1.3622	1.3662	1.3701	1.3740
35	1.3780	1.3819	1.3859	1.3898	1.3937	1.3977	1.4016	1.4055	1.4095	1.4134
36	1.4173	1.4213	1.4252	1.4292	1.4331	1.4370	1.4410	1.4449	1.4488	1.4528
37	1.4567	1.4607	1.4646	1.4685	1.4725	1.4764	1.4803	1.4843	1.4882	1.4922
38	1.4961	1.5000	1.5040	1.5079	1.5118	1.5158	1.5197	1.5236	1.5276	1.5315
39	1.5355	1.5394	1.5433	1.5473	1.5512	1.5551	1.5591	1.5630	1.5670	1.5709
40	1.5748	1.5788	1.5827	1.5866	1.5906	1.5945	1.5985	1.6024	1.6063	1.6103
	0.	**1.**	**2.**	**3.**	**4.**	**5.**	**6.**	**7.**	**8.**	**9.**

XII. CONVERSION OF MILLIMETRES INTO FRENCH OR PARIS LINES.

1 Millimetre = 0.443296 Paris Line.

Millimetres.	Tenths of a Millimetre.									
	0.	**1.**	**2.**	**3.**	**4.**	**5.**	**6.**	**7.**	**8.**	**9.**
	Par. line.	Par. line.	Par. line.	Par. line.	Par. line.	Par. line.	Par. line.	Par. line.	Par. line.	Par. line.
0	0.000	0.044	0.089	0.133	0.177	0.222	0.266	0.310	0.355	0.399
1	0.443	0.488	0.532	0.576	0.621	0.665	0.709	0.754	0.798	0.842
2	0.887	0.931	0.975	1.020	1.064	1.108	1.153	1.197	1.241	1.286
3	1.330	1.374	1.419	1.463	1.507	1.552	1.596	1.640	1.685	1.729
4	1.773	1.818	1.862	1.906	1.950	1.995	2.039	2.083	2.128	2.172
5	2.216	2.261	2.305	2.349	2.394	2.438	2.482	2.527	2.571	2.615
6	2.660	2.704	2.748	2.793	2.837	2.881	2.926	2.970	3.014	3.059
7	3.103	3.147	3.192	3.236	3.280	3.325	3.369	3.413	3.458	3.502
8	3.546	3.591	3.635	3.679	3.724	3.768	3.812	3.857	3.901	3.945
9	3.990	4.034	4.078	4.123	4.167	4.211	4.256	4.300	4.344	4.389
10	4.433	4.477	4.522	4.566	4.610	4.655	4.699	4.743	4.788	4.832
11	4.876	4.921	4.965	5.009	5.054	5.098	5.142	5.187	5.231	5.275
12	5.320	5.364	5.408	5.453	5.497	5.541	5.586	5.630	5.674	5.719
13	5.763	5.807	5.851	5.896	5.940	5.984	6.029	6.073	6.117	6.162
14	6.206	6.250	6.295	6.339	6.383	6.428	6.472	6.516	6.561	6.605
15	6.649	6.694	6.738	6.782	6.827	6.871	6.915	6.960	7.004	7.048
16	7.093	7.137	7.181	7.226	7.270	7.314	7.359	7.403	7.447	7.492
17	7.536	7.580	7.625	7.669	7.713	7.758	7.802	7.846	7.891	7.935
18	7.979	8.024	8.068	8.112	8.157	8.201	8.245	8.290	8.334	8.378
19	8.423	8.467	8.511	8.556	8.600	8.644	8.689	8.733	8.777	8.822
20	8.866	8.910	8.955	8.999	9.043	9.088	9.132	9.176	9.221	9.265
21	9.309	9.354	9.398	9.442	9.487	9.531	9.575	9.620	9.664	9.708
22	9.753	9.797	9.841	9.886	9.930	9.974	10.018	10.063	10.107	10.151
23	10.196	10.240	10.284	10.329	10.373	10.417	10.462	10.506	10.550	10.595
24	10.639	10.683	10.728	10.772	10.816	10.861	10.905	10.949	10.994	11.038
25	11.082	11.127	11.171	11.215	11.260	11.304	11.348	11.393	11.437	11.481
26	11.526	11.570	11.614	11.659	11.703	11.747	11.792	11.836	11.880	11.925
27	11.969	12.013	12.058	12.102	12.146	12.191	12.235	12.279	12.324	12.368
28	12.412	12.457	12.501	12.545	12.590	12.634	12.678	12.723	12.767	12.811
29	12.856	12.900	12.944	12.989	13.033	13.077	13.122	13.166	13.210	13.255
30	13.299	13.343	13.388	13.432	13.476	13.521	13.565	13.609	13.654	13.698
31	13.742	13.786	13.831	13.875	13.919	13.964	14.008	14.052	14.097	14.141
32	14.185	14.230	14.274	14.318	14.363	14.407	14.451	14.496	14.540	14.584
33	14.629	14.673	14.717	14.762	14.806	14.850	14.895	14.939	14.983	15.028
34	15.072	15.116	15.161	15.205	15.249	15.294	15.338	15.382	15.427	15.471
35	15.515	15.560	15.604	15.648	15.693	15.737	15.781	15.826	15.870	15.914
36	15.959	16.003	16.047	16.092	16.136	16.180	16.225	16.269	16.313	16.358
37	16.402	16.446	16.491	16.535	16.579	16.624	16.668	16.712	16.757	16.801
38	16.845	16.890	16.934	16.978	17.023	17.067	17.111	17.156	17.200	17.244
39	17.289	17.333	17.377	17.422	17.466	17.510	17.555	17.599	17.643	17.688
40	17.732	17.776	17.820	17.865	17.909	17.953	17.998	18.042	18.086	18.131
	0.	**1.**	**2.**	**3.**	**4.**	**5.**	**6.**	**7.**	**8.**	**9.**

1 Paris Line = 2.255829 Millimetres.

Paris Lines.	Tenths of a Line.									
	0.	1.	2.	3.	4.	5.	6.	7.	8.	9.
	Millim.	Millim.	Millim.	Millim.	Millim.	Millim.	Millim.	Millim.	Millim.	Millim.
0	0.000	0.226	0.451	0.677	0.902	1.128	1.353	1.579	1.805	2.030
1	2.256	2.481	2.707	2.933	3.158	3.384	3.609	3.835	4.060	4.286
2	4.512	4.737	4.963	5.188	5.414	5.640	5.865	6.091	6.316	6.542
3	6.767	6.993	7.219	7.444	7.670	7.895	8.121	8.347	8.572	8.798
4	9.023	9.249	9.474	9.700	9.926	10.151	10.377	10.602	10.828	11.054
5	11.279	11.505	11.730	11.956	12.181	12.407	12.633	12.858	13.084	13.309
6	13.535	13.761	13.986	14.212	14.437	14.663	14.888	15.114	15.340	15.565
7	15.791	16.016	16.242	16.468	16.693	16.919	17.144	17.370	17.595	17.821
8	18.047	18.272	18.498	18.723	18.949	19.175	19.400	19.626	19.851	20.077
9	20.302	20.528	20.754	20.979	21.205	21.430	21.656	21.882	22.107	22.333
10	22.558	22.784	23.009	23.235	23.461	23.686	23.912	24.137	24.363	24.589
11	24.814	25.040	25.265	25.491	25.716	25.942	26.168	26.393	26.619	26.844
12	27.070	27.296	27.521	27.747	27.972	28.198	28.423	28.649	28.875	29.100
13	29.326	29.551	29.777	30.003	30.228	30.454	30.679	30.905	31.130	31.356
14	31.582	31.807	32.033	32.258	32.485	32.711	32.936	33.162	33.387	33.613
15	33.837	34.063	34.289	34.514	34.740	34.965	35.191	35.417	35.642	35.868
16	36.093	36.319	36.544	36.770	36.996	37.221	37.447	37.672	37.898	38.124
17	38.349	38.575	38.800	39.026	39.251	39.477	39.703	39.928	40.154	40.379
18	40.605	40.831	41.056	41.282	41.507	41.733	41.958	42.184	42.410	42.635

XIV. CONVERSION OF FRENCH OR PARIS LINES INTO ENGLISH INCHES.

1 Paris Line = 0.088814 English Inch.

Paris Lines.	Tenths of a Line.									
	0.	1.	2.	3.	4.	5.	6.	7.	8.	9.
	Eng. In.	Eng. In.	Eng. In.	Eng. In.	Eng. In.	Eng. In.	Eng. In.	Eng. In.	Eng. In.	Eng. In.
0	0.0000	0.0089	0.0178	0.0266	0.0355	0.0444	0.0533	0.0622	0.0711	0.0799
1	0.0888	0.0977	0.1066	0.1155	0.1243	0.1332	0.1421	0.1510	0.1599	0.1687
2	0.1776	0.1865	0.1954	0.2043	0.2132	0.2220	0.2309	0.2398	0.2487	0.2576
3	0.2664	0.2753	0.2842	0.2931	0.3020	0.3108	0.3197	0.3286	0.3375	0.3464
4	0.3553	0.3641	0.3730	0.3819	0.3908	0.3997	0.4085	0.4174	0.4263	0.4352
5	0.4441	0.4530	0.4618	0.4707	0.4796	0.4885	0.4974	0.5062	0.5151	0.5240
6	0.5329	0.5418	0.5506	0.5595	0.5684	0.5773	0.5862	0.5951	0.6039	0.6128
7	0.6217	0.6306	0.6395	0.6483	0.6572	0.6661	0.6750	0.6839	0.6927	0.7016
8	0.7105	0.7194	0.7283	0.7372	0.7460	0.7549	0.7638	0.7727	0.7816	0.7904
9	0.7993	0.8082	0.8171	0.8260	0.8349	0.8437	0.8526	0.8615	0.8704	0.8793
10	0.8881	0.8970	0.9059	0.9148	0.9237	0.9325	0.9414	0.9503	0.9592	0.9681
11	0.9770	0.9858	0.9947	1.0036	1.0125	1.0214	1.0302	1.0391	1.0480	1.0569
12	1.0658	1.0746	1.0835	1.0924	1.1013	1.1102	1.1191	1.1279	1.1368	1.1457
13	1.1546	1.1635	1.1723	1.1812	1.1901	1.1990	1.2079	1.2168	1.2256	1.2345
14	1.2434	1.2523	1.2612	1.2700	1.2789	1.2878	1.2967	1.3056	1.3144	1.3233
15	1.3322	1.3411	1.3500	1.3589	1.3677	1.3766	1.3855	1.3944	1.4033	1.4121
16	1.4210	1.4299	1.4388	1.4477	1.4565	1.4654	1.4743	1.4832	1.4921	1.5010
17	1.5098	1.5187	1.5276	1.5365	1.5454	1.5542	1.5631	1.5720	1.5809	1.5898
18	1.5987	1.6075	1.6164	1.6253	1.6342	1.6431	1.6519	1.6608	1.6697	1.6786

XV. CONVERSION OF RUSSIAN HALF-LINES INTO MILLIMETRES.

1 Russian Half-Line = 1.269977 Millimetres.

Russian Half-Lines.	Tenths.									
	0.	**1.**	**2.**	**3.**	**4.**	**5.**	**6.**	**7.**	**8.**	**9.**
	Millim.	Millim.	Millim.	Millim.	Millim.	Millim.	Millim.	Millim.	Millim.	Millim.
0	0.000	0.127	0.254	0.381	0.508	0.635	0.762	0.889	1.016	1.143
1	1.270	1.397	1.524	1.651	1.778	1.905	2.032	2.159	2.286	2.413
2	2.540	2.667	2.794	2.921	3.048	3.175	3.302	3.429	3.556	3.683
3	3.810	3.937	4.064	4.191	4.318	4.445	4.572	4.699	4.826	4.953
4	5.080	5.207	5.334	5.461	5.588	5.715	5.842	5.969	6.096	6.223
5	6.350	6.477	6.604	6.731	6.858	6.985	7.112	7.239	7.366	7.493
6	7.620	7.747	7.874	8.001	8.128	8.255	8.382	8.509	8.636	8.763
7	8.890	9.017	9.144	9.271	9.398	9.525	9.652	9.779	9.906	10.033
8	10.160	10.287	10.414	10.541	10.668	10.795	10.922	11.049	11.176	11.303
9	11.430	11.557	11.684	11.811	11.938	12.065	12.192	12.319	12.446	12.573
10	12.700	12.827	12.954	13.081	13.208	13.335	13.462	13.589	13.716	13.843
11	13.970	14.097	14.224	14.351	14.478	14.605	14.732	14.859	14.986	15.113
12	15.240	15.367	15.494	15.621	15.748	15.875	16.002	16.129	16.256	16.383
13	16.510	16.637	16.764	16.891	17.018	17.145	17.272	17.399	17.526	17.653
14	17.780	17.907	18.034	18.161	18.288	18.415	18.542	18.669	18.796	18.923
15	19.050	19.177	19.304	19.431	19.558	19.685	19.812	19.939	20.066	20.193
16	20.320	20.447	20.574	20.701	20.828	20.955	21.082	21.209	21.336	21.463
17	21.590	21.717	21.844	21.971	22.098	22.225	22.352	22.479	22.606	22.733
18	22.860	22.987	23.114	23.241	23.368	23.495	23.622	23.749	23.876	24.003

XVI. CONVERSION OF RUSSIAN HALF-LINES INTO PARIS LINES.

1 Russian Half-Line = 0.562976 Paris Line.

Russian Half-Lines.	Tenths.									
	0.	**1.**	**2.**	**3.**	**4.**	**5.**	**6.**	**7.**	**8.**	**9.**
	Par. line.	Par. line.	Par. line.	Par. line.	Par. line.	Par. line.	Par. line.	Par. line.	Par. line.	Par. line.
0	0.000	0.056	0.113	0.169	0.225	0.281	0.338	0.394	0.450	0.507
1	0.563	0.619	0.676	0.732	0.788	0.844	0.901	0.957	1.013	1.070
2	1.126	1.182	1.239	1.295	1.351	1.407	1.464	1.520	1.576	1.633
3	1.689	1.745	1.802	1.858	1.914	1.970	2.027	2.083	2.139	2.196
4	2.252	2.308	2.364	2.421	2.477	2.533	2.590	2.646	2.702	2.759
5	2.815	2.871	2.927	2.984	3.040	3.096	3.153	3.209	3.265	3.322
6	3.378	3.434	3.490	3.547	3.603	3.659	3.716	3.772	3.828	3.885
7	3.941	3.997	4.053	4.110	4.166	4.222	4.279	4.335	4.391	4.448
8	4.504	4.560	4.616	4.673	4.729	4.785	4.842	4.898	4.954	5.010
9	5.067	5.123	5.179	5.236	5.292	5.348	5.405	5.461	5.517	5.573
10	5.630	5.686	5.742	5.799	5.855	5.911	5.968	6.024	6.080	6.136
11	6.193	6.249	6.305	6.362	6.418	6.474	6.531	6.587	6.643	6.699
12	6.756	6.812	6.868	6.925	6.981	7.037	7.093	7.150	7.206	7.262
13	7.319	7.375	7.431	7.488	7.544	7.600	7.656	7.713	7.769	7.825
14	7.882	7.938	7.994	8.051	8.107	8.163	8.219	8.276	8.332	8.388
15	8.445	8.501	8.557	8.614	8.670	8.726	8.782	8.839	8.895	8.951
16	9.008	9.064	9.120	9.177	9.233	9.289	9.345	9.402	9.458	9.514
17	9.571	9.627	9.683	9.739	9.796	9.852	9.908	9.965	10.021	10.077
18	10.134	10.190	10.246	10.302	10.359	10.415	10.471	10.528	10.584	10.640

TABLES

FOR

REDUCING BAROMETRICAL OBSERVATIONS,

TAKEN AT ANY TEMPERATURE,

TO THE TEMPERATURE OF THE FREEZING POINT.

TABLES

FOR

REDUCING THE BAROMETRICAL OBSERVATIONS TAKEN AT ANY TEMPERATURE TO THE TEMPERATURE OF THE FREEZING POINT.

THE variations of the mercurial column in a stationary barometer are due to two causes, the changes of atmospheric pressure and the variations of temperature of the mercury, which affect the length of the column by changing its density. The variations of atmospheric pressure, which alone the barometer is destined to ascertain, are therefore hidden, and their observation falsified by the expansion or contraction of the mercury due to changes of temperature. For, supposing that, while the atmospheric pressure remains the same, the temperature of the instrument becomes lower, the mercurial column will become shorter, and the barometer will appear to fall; if the pressure becomes less, but the temperature increases, the expansion of the mercury will tend to compensate the diminution of pressure, and the barometer may remain stationary, or even may rise, while it ought to be falling; in other cases the action of temperature will tend to increase the amount of the changes of the barometrical height. It is therefore evident that successive observations, with the same barometer, do not give *directly* the actual changes of atmospheric pressure, unless they have been taken exactly at the same temperature, a case which, in practice, seldom occurs. Likewise simultaneous observations, taken with various barometers, do not give *directly* the actual differences of the absolute pressure of the atmosphere above the instruments. To obtain the true barometrical heights, that is, the action of the atmospheric pressure alone, the influence of the temperature must first be eliminated from the observed heights. This is done by reducing, by means of the following Tables, the various barometrical columns to the length they would have at a given temperature, which is the same for all. For the sake of convenient comparison, the freezing point has been almost universally adopted as the standard temperature to which all observations are to be reduced.

CONSTRUCTION OF THE TABLES.

In all the following Tables the barometers are supposed to be furnished with brass scales, extending from the surface of the mercury in the cistern to the top of the mercurial column. The correction to be applied is therefore composed of two elements: the correction for the expansion of the mercury, and that for the expansion of the scale; both of which ought to be, and have been, taken into account.

Indeed, the correction for the expansion of mercury is not sufficient to reduce the readings to the height which the barometer would indicate, under the same pressure, at the temperature of the freezing point. For when the temperature rises the mercurial column expands; but then the scale also grows longer, and this will tend to lower the reading of the height. The correction for the expansion of the mercury must thus be diminished by the amount of that of the scale, that is, by nearly $\frac{1}{10}$, this being the proportion between the expansion of brass and that of mercury.

It is also the expansion of the scale which causes an apparent anomaly in the Tables for the Reduction of the English and Old French Barometers. It can be seen, that, though the observations are to be reduced to the freezing point, or to 32° Fahrenheit and zero Reaumur, the Tables give still a correction for observations taken at that temperature. The reason of it is, that the normal length of the English and Old French standards has not been determined at the temperature of the freezing point, as is the case with the metre, but respectively at the temperatures of 62° Fahrenheit and 13° Reaumur. It is thus *only at these temperatures* that the scales graduated with these standards have their true length. Above and below, the inches of the scales are longer or shorter than the inches of the standards. At the freezing point, therefore, the correction for the expansion of the mercury is null, but that for the expansion of the scale is not. The scale being too short, the reading will be too high, and a *subtractive* correction must still be applied, which will be gradually compensated at lower temperatures by the now *additive* correction of the mercurial column. Thus the point of no correction will occur at 28°.5 Fahrenheit, instead of 32°, in the English Barometer, and at —1°.5 Reaumur, instead of zero, in the Old French.

Schumacher has calculated and published in his *Collection of Tables*, &c., and in his *Jahrbuch* for 1836, 1837, and 1838, extensive tables for the reduction of the English, Old French, and Metrical Barometers, using the following general formula :—

Let h = observed height.
" t = temperature of the attached thermometer.
" T = temperature to which the observed height is to be reduced.
" m = expansion, in volume, of mercury.
" l = linear expansion of brass.
" ϑ = normal temperature of the standard scale.

The reduction to the freezing point will be given by the formula,—

$$h \,.\, \frac{m\,(t-T)-l\,(t-\vartheta)}{1+m\,(t-T)}$$

The following tables, which may be found more convenient for ordinary use, have been calculated after the same formula. Table XVII., published in the Instructions of the Royal Society of London, is mostly abstracted from the table of Schumacher. It gives the reduction of the English Barometer, adopting the following values :—

Let h = observed height in English inches.
" t = temperature of attached thermometer in degrees of Fahrenheit.
" m = expansion, in volume, of mercury for one degree Fahrenheit = 0.0001001.
" l = linear expansion of brass for one degree Fahrenheit = 0.0000104344.

The normal temperature of standard being = 62°.

The reduction to 32° Fahrenheit will be given then by the formula,

$$-\,h \,.\, \frac{m\,(t-32)-l\,(t-62)}{1+m\,(t-32)}\,.$$

The elements for the other tables are found at the head of each.

XVII.

ENGLISH BAROMETER.

TABLE

GIVING THE CORRECTION TO BE APPLIED TO ENGLISH BAROMETERS,

WITH BRASS SCALES EXTENDING FROM THE CISTERN TO THE TOP OF THE MERCURIAL COLUMN, FOR REDUCING THE OBSERVATIONS TO THIRTY-TWO DEGREES FAHRENHEIT.

Table XVII.

The following Table, calculated after that of Schumacher, has been adopted by the Committee of Physics and Meteorology of the Royal Society of London. It gives immediately the correction for every degree of Fahrenheit, and for every half-inch from 20 up to 31 inches. The scale of the barometer is supposed to be of brass, extending from the cistern to the top of the mercurial column. The difference of expansion of brass and mercury is taken into account. The standard temperature of the yard being 62° Fahr., and not 32° Fahr., the difference of expansion of the scale and of the mercurial column carries the point of no correction down to 29° Fahr. Therefore, from 29° up the correction must be *subtracted* from, from 29° down it must be *added* to, the observed height.

Examples of Calculation.

Barometer, observed height, ... 30.231
Attached thermometer 82° Fahr.

See in the last page the column of 30 inches; go down as far as the horizontal line corresponding with 82° in the first vertical column, which contains the temperatures; you will find there the correction —.143. We have thus: —

Barometer, observed height, ...	30.231
Subtractive correction for 82° Fahr., . . .	—0.143
Barometer at 32° Fahr., . . .	30.088

Barometer, observed height, ... 29.743
Attached thermometer 25° Fahr.

The column of 29.5 inches opposite to 25° Fahr. gives an *additive* correction of, ... +0.009

Barometer at 32° Fahr., . . . 29.752

It will be easy to apply also the correction for fractions of a degree Fahrenheit; for example: —

Barometer, observed height, ... 28.358
Attached thermometer 71.3

In the column of 28.5 inches, we find that the difference between the correction for 71° and that for 72° is .003; dividing this difference proportionally to the fraction, we have for three tenths of a degree a correction of —.001, which added to —.108, the correction for 71°, makes a total correction of, ... —.109

And barometer at 32° Fahr., ... 28.249

Degrees of Fahrenheit.	English Inches.								Degrees of Fahrenheit.
	20	20.5	21	21.5	22	22.5	23	23.5	
0°	+.051	+.053	+.054	+.055	+.056	+.058	+.059	+.060	0°
1	.049	.051	.052	.053	.054	.056	.057	.058	1
2	.048	.049	.050	.051	.052	.054	.055	.056	2
3	.046	.047	.048	.049	.050	.052	.053	.054	3
4	.044	.045	.046	.047	.048	.050	.051	.052	4
5	.042	.043	.044	.045	.046	.048	.049	.050	5
6	+.040	+.042	+.042	+.044	+.044	+.046	+.047	+.048	6
7	.039	.040	.041	.042	.042	.044	.044	.046	7
8	.037	.038	.039	.040	.041	.041	.042	.043	8
9	.035	.036	.037	.038	.039	.039	.040	.041	9
10	.033	.034	.035	.036	.037	.037	.038	.039	10
11	+.031	+.032	+.033	+.034	+.035	+.035	+.036	+.037	11
12	.030	.030	.031	.032	.033	.033	.034	.035	12
13	.028	.029	.029	.030	.031	.031	.032	.033	13
14	.026	.027	.027	.028	.029	.029	.030	.031	14
15	.024	.025	.026	.026	.027	.027	.028	.029	15
16	+.022	+.023	+.024	+.024	+.025	+.025	+.026	+.026	16
17	.021	.021	.022	.022	.023	.023	.024	.024	17
18	.019	.019	.020	.020	.021	.021	.022	.022	18
19	.017	.018	.018	.018	.019	.019	.020	.020	19
20	.015	.016	.016	.016	.017	.017	.018	.018	20
21	+.014	+.014	+.014	+.015	+.015	+.015	+.015	+.016	21
22	.012	.012	.012	.013	.013	.013	.013	.014	22
23	.010	.010	.010	.011	.011	.011	.011	.012	23
24	.008	.008	.009	.009	.009	.009	.009	.010	24
25	.006	.007	.007	.007	.007	.007	.007	.007	25
26	+.005	+.005	+.005	+.005	+.005	+.005	+.005	+.005	26
27	.003	.003	.003	.003	.003	.003	.003	.003	27
28	.001	.001	.001	.001	.001	.001	.001	.001	28
29	−.001	−.001	−.001	−.001	−.001	−.001	−.001	−.001	29
30	.003	.003	.003	.003	.003	.003	.003	.003	30
31	−.005	−.005	−.005	−.005	−.005	−.005	−.005	−.005	31
32	.006	.006	.007	.007	.007	.007	.007	.007	32
33	.008	.008	.008	.009	.009	.009	.009	.010	33
34	.010	.010	.010	.011	.011	.011	.011	.012	34
35	.012	.012	.012	.013	.013	.013	.013	.014	35
36	−.013	−.014	−.014	−.014	−.015	−.015	−.016	−.016	36
37	.015	.016	.016	.016	.017	.017	.018	.018	37
38	.017	.017	.018	.018	.019	.019	.020	.020	38
39	.019	.019	.020	.020	.021	.021	.022	.022	39
40	.021	.021	.022	.022	.023	.023	.024	.024	40
41	−.022	−.023	−.024	−.024	−.025	−.025	−.026	−.026	41
42	.024	.025	.025	.026	.027	.027	.028	.028	42
43	.026	.027	.027	.028	.029	.029	.030	.031	43
44	.028	.029	.029	.030	.031	.031	.032	.033	44
45	.030	.030	.031	.032	.033	.033	.034	.035	45
46	−.031	−.032	−.033	−.034	−.035	−.035	−.036	−.037	46
47	.033	.034	.035	.036	.036	.037	.038	.039	47
48	.035	.036	.037	.038	.038	.039	.040	.041	48
49	.037	.038	.039	.040	.040	.041	.042	.043	49
50	.038	.039	.040	.041	.042	.043	.044	.045	50

Degrees of Fahrenheit.	English Inches.								Degrees of Fahrenheit.
	20	20.5	21	21.5	22	22.5	23	23.5	
° 51	−.040	−.041	−.042	−.043	−.044	−.045	−.046	−.047	° 51
52	.042	.043	.044	.045	.046	.047	.048	.049	52
53	.044	.045	.046	.047	.048	.049	.050	.052	53
54	.046	.047	.048	.049	.050	.051	.052	.054	54
55	.047	.049	.050	.051	.052	.053	.055	.056	55
56	−.049	−.050	−.052	−.053	−.054	−.055	−.057	−.058	56
57	.051	.052	.054	.055	.056	.057	.059	.060	57
58	.053	.054	.055	.057	.058	.059	.061	.062	58
59	.055	.056	.057	.059	.060	.061	.063	.064	59
60	.056	.058	.059	.061	.062	.063	.065	.066	60
61	−.058	−.060	−.061	−.062	−.064	−.065	−.067	−.068	61
62	.060	.061	.063	.064	.066	.067	.069	.070	62
63	.062	.063	.065	.066	.068	.069	.071	.072	63
64	.063	.065	.067	.068	.070	.071	.073	.075	64
65	.065	.067	.068	.070	.072	.073	.075	.077	65
66	−.067	−.069	−.070	−.072	−.074	−.075	−.077	−.079	66
67	.069	.071	.072	.074	.076	.077	.079	.081	67
68	.071	.072	.074	.076	.078	.079	.081	.083	68
69	.072	.074	.076	.078	.080	.081	.083	.085	69
70	.074	.076	.078	.080	.082	.083	.085	.087	70
71	−.076	−.078	−.080	−.082	−.083	−.085	−.087	−.089	71
72	.078	.080	.082	.084	.085	.087	.089	.091	72
73	.079	.081	.083	.085	.087	.089	.091	.093	73
74	.081	.083	.085	.087	.089	.091	.093	.095	74
75	.083	.085	.087	.089	.091	.093	.095	.098	75
76	−.085	−.087	−.089	−.091	−.093	−.095	−.097	−.100	76
77	.087	.089	.091	.093	.095	.097	.100	.102	77
78	.088	.091	.093	.095	.097	.099	.102	.104	78
79	.090	.092	.095	.097	.099	.101	.104	.106	79
80	.092	.094	.096	.099	.101	.103	.106	.108	80
81	−.094	−.096	−.098	−.101	−.103	−.105	−.108	−.110	81
82	.095	.098	.100	.103	.105	.107	.110	.112	82
83	.097	.100	.102	.104	.107	.109	.112	.114	83
84	.099	.101	.104	.106	.109	.111	.114	.116	84
85	.101	.103	.106	.108	.111	.113	116	.118	85
86	−.103	−.105	−.108	−.110	−.113	−.115	−.118	−.120	86
87	.104	.107	.109	.112	.115	.117	.120	.123	87
88	.106	.109	.111	.114	.117	.119	.122	.125	88
89	.108	.111	.113	.116	.119	.121	.124	.127	89
90	.110	.112	.115	.118	.121	.123	.126	.129	90
91	−.111	−.114	−.117	−.120	−.122	−.125	−.128	−.131	91
92	.113	.116	.119	.122	.124	.127	.130	.133	92
93	.115	.118	.121	.124	.126	.129	.132	.135	93
94	.117	.120	.122	.125	.128	.131	.134	.137	94
95	.118	.121	.124	.127	.130	.133	.136	.139	95
96	−.120	−.123	−.126	−.129	−.132	−.135	−.138	−.141	96
97	.122	.125	.128	.131	.134	.137	.140	.143	97
98	.124	.127	.130	.133	.136	.139	.142	.145	98
99	.125	.129	.132	.135	.138	.141	.144	.147	99
100	.127	.130	.134	.137	140	.143	.146	.150	100

Degrees of Fahrenheit.	English Inches.								Degrees of Fahrenheit.
	24	24.5	25	25.5	26	26.5	27	27.5	
° 0	+.061	+.063	+.064	+.065	+.067	+.068	+.069	+.071	° 0
1	.059	.061	.062	.063	.064	.065	.067	.068	1
2	.057	.058	.060	.061	.062	.063	.064	.066	2
3	.055	.056	.057	.059	.060	.061	.062	.063	3
4	.053	.054	.055	.056	.057	.058	.059	.061	4
5	.051	.052	.053	.054	.055	.056	.057	.058	5
6	+.049	+.050	+.051	+.052	+.053	+.054	+.055	+.056	6
7	.046	.047	.048	.049	.050	.051	.052	.053	7
8	.044	.045	.046	.047	.048	.049	.050	.051	8
9	.042	.043	.044	.045	.046	.046	.047	.048	9
10	.040	.041	.042	.042	.043	.044	.045	.046	10
11	+.038	+.039	+.039	+.040	+.041	+.042	+.042	+.043	11
12	.036	.036	.037	.038	.039	.039	.040	.041	12
13	.033	.034	.035	.036	.036	.037	.038	.038	13
14	.031	.032	.033	.033	.034	.035	.035	.036	14
15	.029	.030	.030	.031	.032	.032	.033	.033	15
16	+.027	+.028	+.028	+.029	+.029	+.030	+.030	+.031	16
17	.025	.025	.026	.026	.027	.027	.028	.028	17
18	.023	.023	.024	.024	.025	.025	.025	.026	18
19	.021	.021	.021	.022	.022	.023	.023	.024	19
20	.018	.019	.019	.020	.020	.020	.021	.021	20
21	+.016	+.017	+.017	+.017	+.018	+.018	+.018	+.019	21
22	.014	.014	.015	.015	.015	.016	.016	.016	22
23	.012	.012	.012	.013	.013	.013	.013	.014	23
24	.010	.010	.010	.010	.011	.011	.011	.011	24
25	.008	.008	.008	.008	.008	.008	.009	.009	25
26	+.005	+.006	+.006	+.006	+.006	+.006	+.006	+.006	26
27	.003	.003	.003	.003	.004	.004	.004	.004	27
28	.001	.001	.001	.001	.001	.001	.001	.001	28
29	−.001	−.001	−.001	−.001	−.001	−.001	−.001	−.001	29
30	.003	.003	.003	.004	.004	.004	.004	.004	30
31	−.005	−.006	−.006	−.006	−.006	−.006	−.006	−.006	31
32	.008	.008	.008	.008	.008	.008	.008	.009	32
33	.010	.010	.010	.010	.011	.011	.011	.011	33
34	.012	.012	.012	.013	.013	.013	.013	.014	34
35	.014	.014	.015	.015	.015	.015	.016	.016	35
36	−.016	−.017	−.017	−.017	−.017	−.018	−.018	−.019	36
37	.018	.019	.019	.019	.020	.020	.021	.021	37
38	.020	.021	.021	.022	.022	.023	.023	.023	38
39	.023	.023	.024	.024	.024	.025	.025	.026	39
40	.025	.025	.026	.026	.027	.027	.028	.028	40
41	−.027	−.027	−.028	−.029	−.029	−.030	−.030	−.031	41
42	.029	.030	.030	.031	.031	.032	.033	.033	42
43	.031	.032	.032	.033	.034	.034	.035	.036	43
44	.033	.034	.035	.035	.036	.037	.037	.038	44
45	.035	.036	.037	.038	.038	.039	.040	.041	45
46	−.038	−.038	−.039	−.040	−.041	−.042	−.042	−.043	46
47	.040	.041	.041	.042	.043	.044	.045	.046	47
48	.042	.043	.044	.045	.045	.046	.047	.048	48
49	.044	.045	.046	.047	.048	.049	.050	.050	49
50	.046	.047	.048	.049	.050	.051	.052	.053	50

Degrees of Fahrenheit.	24	24.5	25	25.5	26	26.5	27	27.5	Degrees of Fahrenheit.
	English Inches.								
° 51	−.048	−.049	−.050	−.051	−.052	−.053	−.054	−.055	° 51
52	.050	.052	.053	.054	.055	.056	.057	.058	52
53	.053	.054	.055	.056	.057	.058	.059	.060	53
54	.055	.056	.057	.058	.059	.060	.062	.063	54
55	.057	.058	.059	.060	.062	.063	.064	.065	55
56	−.059	−.060	−.061	−.063	−.064	−.065	−.066	−.068	56
57	.061	.062	.064	.065	.066	.068	.069	.070	57
58	.063	.065	.066	.067	.069	.070	.071	.073	58
59	.065	.067	.068	.070	.071	.072	.074	.075	59
60	.068	.069	.070	.072	.073	.075	.076	.077	60
61	−.070	−.071	−.073	−.074	−.075	−.077	−.078	−.080	61
62	.072	.073	.075	.076	.078	.079	.081	.082	62
63	.074	.076	.077	.079	.080	.082	.083	.085	63
64	.076	.078	.079	.081	.082	.084	.086	.087	64
65	.078	.080	.082	.083	.085	.086	.088	.090	65
66	−.080	−.082	−.084	−.085	−.087	−.089	−.090	−.092	66
67	.083	.084	.086	.088	.089	.091	.093	.095	67
68	.085	.086	.088	.090	.092	.094	.095	.097	68
69	.087	.089	.090	.092	.094	.096	.098	.100	69
70	.089	.091	.093	.095	.096	.098	.100	.102	70
71	−.091	−.093	−.095	−.097	−.099	−.101	−.102	−.104	71
72	.093	.095	.097	.099	.101	.103	.105	.107	72
73	.095	.097	.099	.101	.103	.105	.107	.109	73
74	.097	.099	.102	.104	.106	.108	.110	.112	74
75	.100	.102	.104	.106	.108	.110	.112	.114	75
76	−.102	−.104	−.106	−.108	−.110	−.112	−.114	−.117	76
77	.104	.106	.108	.110	.112	.115	117	.119	77
78	.106	.108	.110	.113	.115	.117	.119	.122	78
79	.108	.110	.113	.115	.117	.119	.122	.124	79
80	.110	.113	.115	.117	.119	.122	.124	.126	80
81	−.112	−.115	−117	−.119	−.122	−.124	−.126	−.129	81
82	.114	.117	.119	.122	.124	.126	.129	.131	82
83	.117	.119	.121	.124	.126	.129	.131	.134	83
84	.119	.121	.124	.126	.129	.131	.134	.136	84
85	.121	.123	.126	.128	.131	.133	.136	.139	85
86	−.123	−.126	−.128	−.131	−.133	−.136	−.138	−.141	86
87	.125	.128	.130	.133	.136	.138	.141	.143	87
88	.127	.130	.133	.135	.138	.141	.143	.146	88
89	.129	.132	.135	.137	.140	.143	.146	.148	89
90	.131	.134	.137	.140	.142	.145	.148	.151	90
91	−.134	−.136	−.139	−.142	−.145	−.148	−.150	−.153	91
92	.136	.139	.141	.144	.147	.150	.153	.156	92
93	.138	.141	.144	.147	.149	.152	.155	.158	93
94	.140	.143	.146	.149	.152	.155	.157	.161	94
95	.142	.145	.148	.151	.154	.157	.160	.163	95
96	−.144	−.147	−.150	−.153	−.156	−.159	−.162	−.165	96
97	.146	.149	.152	.156	.159	.162	.165	.168	97
98	.148	.152	.155	.158	.161	.164	.167	.170	98
99	.151	.154	.157	.160	.163	.166	.169	.173	99
100	.153	.156	.159	.162	.165	.169	.172	.175	100

Degrees of Fahrenheit.	English Inches.							Degrees of Fahrenheit.
	28	**28.5**	**29**	**29.5**	**30**	**30.5**	**31**	
0°	+.072	+.073	+.074	+.076	+.077	+.078	+.080	0°
1	.069	.071	.072	.073	.074	.076	.077	1
2	.067	.068	.069	.070	.072	.073	.074	2
3	.064	.065	.067	.068	.069	.070	.071	3
4	.062	.063	.064	.065	.066	.067	.068	4
5	.059	.060	.061	.062	.063	.065	.066	5
6	+.057	+.058	+.059	+.060	+.061	+.062	+.063	6
7	.054	.055	.056	.057	.058	.059	.060	7
8	.052	.053	.054	.054	.055	.056	.057	8
9	.049	.050	.051	.052	.053	.054	.054	9
10	.047	.047	.048	.049	.050	.051	.052	10
11	+.044	+.045	+.046	+.046	+.047	+.048	+.049	11
12	.042	.042	.043	.044	.045	.045	.046	12
13	.039	.040	.040	.041	.042	.043	.043	13
14	.037	.037	.038	.038	.039	.040	.040	14
15	.034	.035	.035	.036	.036	.037	.038	15
16	+.032	+.032	+.033	+.033	+.034	+.034	+.035	16
17	.029	.030	.030	.031	.031	.032	.032	17
18	.026	.027	.027	.028	.028	.029	.029	18
19	.024	.024	.025	.025	.026	.026	.027	19
20	.021	.022	.022	.023	.023	.023	.024	20
21	+.019	+.019	+.020	+.020	+.020	+.021	+.021	21
22	.016	.017	.017	.017	.018	.018	.018	22
23	.014	.014	.014	.015	.015	.015	.015	23
24	.011	.012	.012	.012	.012	.012	.013	24
25	.009	.009	.009	.009	.009	.010	.010	25
26	+.006	+.006	+.007	+.007	+.007	+.007	+.007	26
27	.004	.004	.004	.004	.004	.004	.004	27
28	.001	.001	.001	.001	.001	.001	.001	28
29	−.001	−.001	−.001	−.001	−.001	−.001	−.001	29
30	.004	.004	.004	.004	.004	.004	.004	30
31	−.006	−.006	−.007	−.007	−.007	−.007	−.007	31
32	.009	.009	.009	.009	.009	.010	.010	32
33	.011	.012	.012	.012	.012	.012	.012	33
34	.014	.014	.014	.015	.015	.015	.015	34
35	.016	.017	.017	.017	.018	.018	.018	35
36	−.019	−.019	−.020	−.020	−.020	−.021	−.021	36
37	.021	.022	.022	.022	.023	.023	.024	37
48	.024	.024	.025	.025	.026	.026	.026	38
39	.026	.027	.027	.028	.028	.029	.029	39
40	.029	.029	.030	.030	.031	.031	.032	40
41	−.031	−.032	−.033	−.033	−.034	−.034	−.035	41
42	.034	.034	.035	.036	.036	.037	.037	42
43	.036	.037	.038	.038	.039	.040	.040	43
44	.039	.040	.040	.041	.042	.042	.043	44
45	.041	.042	.043	.044	.044	.045	.046	45
46	−.044	−.045	−.045	−.046	−.047	−.048	−.049	46
47	.046	.047	.048	.049	.050	.051	.051	47
48	.049	.050	.051	.052	.052	.053	.054	48
49	.051	.052	.053	.054	.055	.056	.057	49
50	.054	.055	.056	.057	.058	.059	.060	50

Degrees of Fahrenheit.	English Inches.							Degrees of Fahrenheit.
	28	**28.5**	**29**	**29.5**	**30**	**30.5**	**31**	
° 51	−.056	−.057	−.058	−.059	−.060	−.061	−.062	° 51
52	.059	.060	.061	.062	.063	.064	.065	52
53	.061	.063	.064	.065	.066	.067	.068	53
54	.064	.065	.066	.067	.068	.070	.071	54
55	.066	.068	.069	.070	.071	.072	.073	55
56	−.069	−.070	−.071	−.073	−.074	−.075	−.076	56
57	.071	.073	.074	.075	.076	.078	.079	57
58	.074	.075	.077	.078	.079	.081	.082	58
59	.076	.078	.079	.080	.082	.083	.085	59
60	.079	.080	.082	.083	.085	.086	.087	60
61	−.081	−.083	−.084	−.086	−.087	−.089	−.090	61
62	.084	.085	.087	.088	.090	.091	.093	62
63	.086	.088	.089	.091	.093	.094	.096	63
64	.089	.090	.092	.094	.095	.097	.098	64
65	.091	.093	.095	.096	.098	.100	.101	65
66	−.094	−.096	−.097	−.099	−.101	−.102	−.104	66
67	.096	.098	.100	.102	.103	.105	.107	67
68	.099	.101	.102	.104	.106	.108	.109	68
69	.101	.103	.105	.107	.109	.110	.112	69
70	.104	.106	.108	.109	.111	.113	.115	70
71	−.106	−.108	−.110	−.112	−.114	−.116	−.118	71
72	.109	.111	.113	.115	.117	.119	.120	72
73	.111	.113	.115	.117	.119	.121	.123	73
74	.114	.116	.118	.120	.122	.124	.126	74
75	.116	.118	.120	.122	.125	.127	.129	75
76	−.119	−.121	−.123	−.125	−.127	−.129	−.131	76
77	.121	.123	.126	.128	.130	.132	.134	77
78	.124	.126	.128	.130	.133	.135	.137	78
79	.126	.128	.131	.133	.135	.137	.140	79
80	.129	.131	.133	.136	.138	.140	.143	80
81	−.131	−.134	−.136	−.138	−.141	−.143	−.145	81
82	.134	.136	.138	.141	.143	.146	.148	82
83	.136	.139	.141	.143	.146	.148	.151	83
84	.139	.141	.144	.146	.149	.151	.154	84
85	.141	.144	.146	.149	.151	.154	.156	85
86	−.144	−.146	−.149	−.151	−.154	−.156	−.159	86
87	.146	.149	.151	.154	.157	.159	.162	87
88	.149	.151	.154	.157	.159	.162	.165	88
89	.151	.154	.156	.159	.162	.165	.167	89
90	.153	.156	.159	.162	.164	.167	.170	90
91	−.156	−.159	−.162	−.165	−.167	−.170	−.173	91
92	.158	.161	.164	.167	.170	.172	.175	92
93	.161	.164	.167	.170	.172	.175	.178	93
94	.163	.166	.169	.172	.175	.177	.180	94
95	.166	.169	.172	.175	.178	.180	.183	95
96	−.168	−.171	−.174	−.178	−.181	−.183	−.186	96
97	.171	.174	.177	.180	.183	.186	.189	97
98	.173	.176	.179	.183	.186	.188	.191	98
99	.176	.179	.182	.185	.188	.191	.194	99
100	.178	.181	.184	.188	.191	.194	.197	100

TABLE XVIII.

In most of the common barometers the scale is engraved upon a short plate of brass, fixed upon the wooden frame of the instrument. In such a case, the compound expansion of the two substances can only be guessed at, and the correction to be applied to the observations for reducing them to the freezing point cannot be determined with precision. As a near approximation for such imperfect instruments, the following table, taken from the Instructions of the Royal Society of London, may be used. In this table, the expansion of glass, which is less than that of brass, and greater than that of wood, has been substituted to that of brass, as an approximate value for a scale composed of these last two substances.

CORRECTION TO BE APPLIED TO ENGLISH BAROMETERS, THE SCALES OF WHICH ARE ENGRAVED ON *GLASS*, TO REDUCE THE OBSERVATIONS TO THE FREEZING POINT.

Degrees of Fahrenheit.	Inches.							Degrees of Fahrenheit.
	28.0	**28.5**	**29.0**	**29.5**	**30.0**	**30.5**	**31.0**	
25	+.017	+.017	+.017	+.018	+.018	+.018	+.019	25
30	+.005	+.005	+.005	+.005	+.005	+.005	+.005	30
35	−.007	−.007	−.007	−.008	−.008	−.008	−.008	35
40	−.019	−.020	−.020	−.020	−.021	−.021	−.021	40
45	−.031	−.032	−.032	−.033	−.033	−.034	−.035	45
50	−.043	−.044	−.045	−.046	−.046	−.047	−.048	50
55	−.055	−.056	−.057	−.058	−.059	−.060	−.061	55
60	−.067	−.068	−.069	−.071	−.072	−.074	−.075	60
65	−.079	−.081	−.082	−.083	−.085	−.086	−.088	65
70	−.091	−.093	−.094	−.096	−.098	−.100	−.101	70
75	−.103	−.105	−.106	−.109	−.111	−.114	−.116	75
80	−.115	−.117	−.119	−.121	−.124	−.127	−.130	80
85	−.127	−.129	−.131	−.134	−.137	−.140	−.144	85

XIX.

METRICAL BAROMETER.

TABLE

FOR

REDUCING TO THE FREEZING POINT THE BAROMETRICAL COLUMN,

MEASURED BY BRASS SCALES, EXTENDING FROM THE CISTERN TO THE TOP; CALCULATED FROM 260 TO 865 MILLIMETRES, AND FOR EACH DEGREE CENTIGRADE.

BY M. T. DELCROS.

TABLE XIX.

THIS table has been calculated by using the following coefficients of dilatation: —

Brass, linear dilatation, from Laplace and Lavoisier for 100° C. = 0.0018782.
Mercury, dilatation in volume, from Dulong and Petit for 100° C. = 0.0180180.
Dilatation of the mercurial column for 100° C. . . . = 0.0161398.
Dilatation of the mercurial column for 1° C. . . . = 0.0001614.
Observed height reduced to freezing point,

$$H = h - h\ (0.0001614). \quad T = h - h\left(\frac{T}{6196}\right).$$

The second term of this last formula is given by the table, when the temperature T and the height h of the barometer are known; this correction must be *subtracted* from the observed height h, when the temperature is above freezing point; it is to be *added* when the temperature is below zero, or freezing point.

This table allows the barometrical heights taken at the highest summits, and in the deepest mines, to be corrected.

Examples of Calculation.

		mm.
Barometer, observed height, ...		567.49
Temperature of the barometer, +12°.7.		
Second page,	for 10°.0 = 0.912 mm.	
	for 2.0 = 0.182	
	for 0.7 = 0.064	
	Total, = 1.158	
Subtractive correction, ...		— 1.16
	Barometer at zero,	566.33

		mm.
Barometer, observed height, ...		454.17
Temperature of the barometer, —7°.8.		
First page,	for 7°.0 = 0.514 mm.	
	for 0.8 = 0.059	
	Total, = 0.573	
Additive correction, ...		+0.57
	Barometer at zero,	454.74

Height of the Barometer.	TEMPERATURE CENTIGRADE.								
	1°	2°	3°	4°	5°	6°	7°	8°	9°
Millim.	Millim.	Millim.	Millim.	Millim.	Millim.	Millim.	Millim.	Millim.	Millim.
260	0.042	0.084	0.126	0.168	0.210	0.252	0.294	0.336	0.378
265	0.043	0.086	0.128	0.171	0.214	0.257	0.299	0.342	0.385
270	0.044	0.087	0.131	0.174	0.218	0.261	0.305	0.349	0.392
275	0.044	0.089	0.133	0.178	0.222	0.266	0.311	0.355	0.399
280	0.045	0.090	0.136	0.181	0.226	0.271	0.316	0.362	0.407
285	0.046	0.092	0.138	0.184	0.230	0.276	0.322	0.368	0.414
290	0.047	0.094	0.140	0.187	0.234	0.281	0.328	0.374	0.421
295	0.048	0.095	0.143	0.190	0.238	0.286	0.333	0.381	0.428
300	0.048	0.097	0.145	0.194	0.242	0.291	0.339	0.387	0.436
305	0.049	0.098	0.148	0.197	0.246	0.295	0.345	0.394	0.443
310	0.050	0.100	0.150	0.200	0.250	0.300	0.350	0.400	0.450
315	0.051	0.102	0.152	0.203	0.254	0.305	0.356	0.407	0.458
320	0.052	0.103	0.155	0.207	0.258	0.310	0.361	0.413	0.465
325	0.052	0.105	0.157	0.210	0.262	0.315	0.367	0.420	0.472
330	0.053	0.106	0.160	0.213	0.266	0.320	0.374	0.426	0.479
335	0.054	0.108	0.162	0.216	0.270	0.324	0.379	0.432	0.487
340	0.055	0.110	0.165	0.219	0.274	0.329	0.384	0.439	0.494
345	0.056	0.111	0.167	0.223	0.278	0.334	0.390	0.445	0.501
350	0.056	0.113	0.169	0.226	0.282	0.339	0.395	0.452	0.508
355	0.057	0.115	0.172	0.229	0.286	0.344	0.401	0.458	0.516
360	0.058	0.116	0.174	0.232	0.290	0.349	0.407	0.465	0.523
365	0.059	0.118	0.177	0.236	0.294	0.353	0.412	0.471	0.530
370	0.060	0.119	0.179	0.239	0.299	0.358	0.418	0.478	0.537
375	0.060	0.121	0.182	0.242	0.303	0.363	0.424	0.484	0.545
380	0.061	0.123	0.184	0.245	0.307	0.368	0.429	0.491	0.552
385	0.062	0.124	0.186	0.249	0.311	0.373	0.435	0.497	0.559
390	0.063	0.126	0.189	0.252	0.315	0.378	0.441	0.504	0.566
395	0.064	0.127	0.191	0.255	0.319	0.382	0.446	0.510	0.574
400	0.065	0.129	0.194	0.258	0.323	0.387	0.452	0.516	0.581
405	0.065	0.131	0.196	0.261	0.327	0.392	0.457	0.523	0.588
410	0.066	0.132	0.198	0.265	0.331	0.397	0.463	0.529	0.596
415	0.067	0.134	0.201	0.268	0.335	0.402	0.469	0.536	0.603
420	0.068	0.136	0.203	0.271	0.339	0.407	0.474	0.542	0.610
425	0.068	0.137	0.206	0.274	0.343	0.411	0.480	0.549	0.617
430	0.069	0.139	0.208	0.278	0.347	0.416	0.486	0.555	0.625
435	0.070	0.140	0.211	0.281	0.351	0.421	0.491	0.562	0.632
440	0.071	0.142	0.213	0.284	0.355	0.426	0.497	0.568	0.639
445	0.072	0.144	0.215	0.287	0.359	0.431	0.503	0.574	0.646
450	0.073	0.145	0.218	0.290	0.363	0.436	0.508	0.581	0.654
455	0.073	0.147	0.220	0.294	0.367	0.441	0.514	0.587	0.661
	1°	2°	3°	4°	5°	6°	7°	8°	9°

Height of the Barometer.	TEMPERATURE CENTIGRADE.								
	1°	2°	3°	4°	5°	6°	7°	8°	9°
Millim.	Millim.	Millim.	Millim.	Millim.	Millim.	Millim.	Millim.	Millim.	Millim.
460	0.0742	0.1485	0.2227	0.2970	0.371	0.445	0.520	0.594	0.668
465	0.0750	0.1501	0.2251	0.3002	0.375	0.450	0.525	0.600	0.675
470	0.0759	0.1517	0.2276	0.3034	0.379	0.455	0.531	0.607	0.683
475	0.0767	0.1533	0.2300	0.3066	0.383	0.460	0.537	0.613	0.690
480	0.0775	0.1549	0.2324	0.3099	0.387	0.465	0.542	0.620	0.697
485	0.0783	0.1565	0.2348	0.3131	0.391	0.470	0.548	0.626	0.704
490	0.0791	0.1582	0.2373	0.3163	0.395	0.474	0.554	0.633	0.712
495	0.0800	0.1598	0.2397	0.3195	0.399	0.479	0.559	0.639	0.719
500	0.0807	0.1614	0.2421	0.3228	0.403	0.484	0.565	0.646	0.726
505	0.0815	0.1630	0.2445	0.3260	0.407	0.489	0.570	0.652	0.734
510	0.0823	0.1646	0.2469	0.3293	0.412	0.494	0.576	0.658	0.741
515	0.0831	0.1662	0.2493	0.3325	0.416	0.499	0.582	0.665	0.748
520	0.0839	0.1679	0.2518	0.3357	0.420	0.504	0.587	0.671	0.755
525	0.0847	0.1695	0.2542	0.3389	0.424	0.508	0.593	0.678	0.763
530	0.0855	0.1711	0.2566	0.3422	0.428	0.513	0.599	0.684	0.770
535	0.0863	0.1727	0.2590	0.3454	0.432	0.518	0.604	0.691	0.777
540	0.0872	0.1743	0.2615	0.3486	0.436	0.523	0.610	0.697	0.784
545	0.0879	0.1759	0.2639	0.3518	0.440	0.528	0.616	0.704	0.792
550	0.0888	0.1775	0.2663	0.3551	0.444	0.533	0.621	0.710	0.799
555	0.0896	0.1791	0.2687	0.3583	0.448	0.537	0.627	0.717	0.806
560	0.0904	0.1808	0.2712	0.3615	0.452	0.542	0.633	0.723	0.813
565	0.0912	0.1824	0.2736	0.3647	0.456	0.547	0.638	0.730	0.821
570	0.0920	0.1840	0.2760	0.3680	0.460	0.552	0.644	0.736	0.828
575	0.0928	0.1856	0.2784	0.3712	0.464	0.557	0.650	0.742	0.835
580	0.0936	0.1872	0.2808	0.3744	0.468	0.562	0.655	0.749	0.842
585	0.0944	0.1888	0.2833	0.3777	0.472	0.566	0.661	0.755	0.850
590	0.0952	0.1904	0.2857	0.3809	0.476	0.571	0.667	0.762	0.857
595	0.0960	0.1921	0.2881	0.3841	0.480	0.576	0.672	0.768	0.864
600	0.0968	0.1937	0.2905	0.3874	0.484	0.581	0.678	0.775	0.872
605	0.0976	0.1953	0.2929	0.3906	0.488	0.586	0.683	0.781	0.879
610	0.0985	0.1969	0.2954	0.3938	0.492	0.591	0.689	0.788	0.886
615	0.0993	0.1985	0.2978	0.3970	0.496	0.595	0.695	0.794	0.893
620	0.1001	0.2001	0.3002	0.4003	0.500	0.600	0.700	0.800	0.901
625	0.1009	0.2017	0.3026	0.4035	0.504	0.605	0.706	0.807	0.908
630	0.1017	0.2034	0.3050	0.4067	0.508	0.610	0.712	0.813	0.915
635	0.1025	0.2050	0.3074	0.4099	0.512	0.615	0.717	0.820	0.922
640	0.1033	0.2066	0.3099	0.4132	0.516	0.620	0.723	0.826	0.930
645	0.1041	0.2082	0.3123	0.4164	0.520	0.625	0.729	0.833	0.937
650	0.1049	0.2098	0.3147	0.4196	0.524	0.629	0.734	0.839	0.944
655	0.1057	0.2114	0.3172	0.4229	0.529	0.634	0.740	0.846	0.951
660	0.1065	0.2130	0.3196	0.4261	0.533	0.639	0.746	0.852	0.959
	1°	2°	3°	4°	5°	6°	7°	8°	9°

Height of the Barometer.	TEMPERATURE CENTIGRADE.								
	1°	2°	3°	4°	5°	6°	7°	8°	9°
Millim.	Millim.	Millim.	Millim.	Millim.	Millim.	Millim.	Millim.	Millim.	Millim.
665	0.1073	0.2146	0.3220	0.4293	0.537	0.644	0.751	0.859	0.966
670	0.1081	0.2163	0.3244	0.4326	0.541	0.649	0.757	0.865	0.973
675	0.1089	0.2179	0.3268	0.4358	0.545	0.654	0.763	0.871	0.980
680	0.1097	0.2195	0.3292	0.4390	0.549	0.658	0.768	0.878	0.988
685	0.1106	0.2211	0.3317	0.4423	0.553	0.663	0.774	0.884	0.995
690	0.1114	0.2227	0.3341	0.4455	0.557	0.668	0.780	0.891	1.002
695	0.1122	0.2233	0.3365	0.4487	0.561	0.673	0.785	0.897	1.010
700	0.1130	0.2260	0.3389	0.4520	0.565	0.678	0.791	0.904	1.017
705	0.1138	0.2276	0.3414	0.4552	0.569	0.683	0.797	0.910	1.024
710	0.1146	0.2292	0.3438	0.4584	0.573	0.688	0.802	0.917	1.031
715	0.1154	0.2308	0.3462	0.4616	0.577	0.691	0.808	0.923	1.039
720	0.1162	0.2324	0.3486	0.4648	0.581	0.697	0.813	0.930	1.046
725	0.1170	0.2340	0.3510	0.4680	0.585	0.702	0.819	0.936	1.053
730	0.1178	0.2356	0.3535	0.4713	0.589	0.707	0.825	0.943	1.060
735	0.1186	0.2372	0.3559	0.4745	0.593	0.712	0.830	0.949	1.068
740	0.1104	0.2389	0.3583	0.4777	0.597	0.717	0.836	0.955	1.075
745	0.1202	0.2405	0.3607	0.4809	0.601	0.721	0.842	0.962	1.082
750	0.1210	0.2421	0.3631	0.4842	0.605	0.726	0.847	0.968	1.089
755	0.1218	0.2437	0.3655	0.4874	0.609	0.731	0.853	0.975	1.097
760	0.1227	0.2453	0.3680	0.4906	0.613	0.736	0.859	0.981	1.104
765	0.1235	0.2469	0.3704	0.4939	0.617	0.741	0.864	0.988	1.111
770	0.1243	0.2486	0.3728	0.4971	0.621	0.746	0.870	0.994	1.118
775	0.1251	0.2502	0.3752	0.5003	0.625	0.750	0.876	1.001	1.126
780	0.1259	0.2518	0.3777	0.5036	0.629	0.755	0.881	1.007	1.133
785	0.1267	0.2534	0.3801	0.5068	0.633	0.760	0.888	1.014	1.140
790	0.1275	0.2550	0.3825	0.5100	0.637	0.765	0.893	1.020	1.148
795	0.1283	0.2566	0.3849	0.5132	0.641	0.770	0.898	1.026	1.155
800	0.1291	0.2582	0.3874	0.5165	0.646	0.775	0.904	1.033	1.162
805	0.1299	0.2598	0.3898	0.5197	0.650	0.780	0.909	1.039	1.169
810	0.1307	0.2615	0.3922	0.5230	0.654	0.784	0.915	1.046	1.177
815	0.1315	0.2621	0.3946	0.5262	0.658	0.789	0.921	1.052	1.184
820	0.1323	0.2647	0.3970	0.5294	0.662	0.794	0.926	1.059	1.191
825	0.1331	0.2653	0.3994	0.5326	0.666	0.799	0.932	1.065	1.198
830	0.1340	0.2679	0.4019	0.5358	0.670	0.804	0.938	1.072	1.206
835	0.1348	0.2695	0.4043	0.5391	0.674	0.809	0.943	1.078	1.213
840	0.1356	0.2712	0.4067	0.5423	0.678	0.813	0.949	1.085	1.220
845	0.1364	0.2728	0.4091	0.5455	0.682	0.818	0.955	1.091	1.227
850	0.1372	0.2744	0.4116	0.5488	0.686	0.823	0.960	1.097	1.235
855	0.1380	0.2760	0.4140	0.5520	0.690	0.828	0.966	1.104	1.242
860	0.1388	0.2776	0.4164	0.5552	0.694	0.833	0.972	1.110	1.249
865	0.1396	0.2792	0.4188	0.5584	0.698	0.838	0.977	1.117	1.256
	1°	2°	3°	4°	5°	6°	7°	8°	9°

XX.

METRICAL BAROMETER.

TABLE

FOR

REDUCING TO THE FREEZING POINT THE BAROMETRICAL COLUMN,

MEASURED BY BRASS SCALES, EXTENDING FROM THE CISTERN TO THE TOP; CALCULATED FOR THE HEIGHTS BETWEEN 605 AND 800 MILLIMETRES, AND FOR EVERY TENTH OF A DEGREE, FROM 0° TO + AND — 35° CENTIGRADE.

BY M. T. HAEGHENS.

TABLE XX.

This table has been calculated by using the same coefficients of dilatation as in the preceding table, viz.: —

Brass, linear dilatation, from Laplace and Lavoisier for 100° C. = 0.0018782.

Mercury, dilatation in volume, from Dulong and Petit for 100° C. = 0.0180180.

Dilatation of the mercurial column for 100° C. . . . = 0.0161398.

Dilatation of the mercurial column for 1° C. . . . = 0.0001614.

This table, calculated for the reduction of long series of meteorological observations, gives immediately the value of the correction for each tenth of a degree up to 35° C. above, and down to 35° C. below, the freezing point, and for mercurial columns extending from 605 to 800 millimetres.

Examples of Calculation.

	mm.
Barometer, observed height, ...	754.17
Temperature of the attached thermometer, +17°.8.	

For finding the correction, seek in the horizontal column, headed *barometer*, at the head of the pages, the corresponding height of the barometer; it will be found, p. 31, barometer 755$^{mm.}$ (from 752.50 to 757.50); next seek in the first vertical column, containing the temperatures, 17°, follow then horizontally this line as far as the column of 8 tenths, and you find there 2.17 millimetres, which is the correction, or the quantity to be subtracted for reducing the observed height to zero. We have thus: —

	mm.
Observed height, ...	754.17
Subtractive correction for +17°.8 = ...	— 2.17
Barometer at zero,	752.00

If the temperature is below zero, the correction will be additive.

	mm.
Observed height, ...	729.72
Temperature of the attached thermometer, —8°.4.	
Additive correction, ...	+0.99
Barometer at zero,	730.71

Centigrade Degrees.	BAROMETER: 605mm. (from 602.51 to 607.50).									
	Tenths of Degrees.									
	0.	**1.**	**2.**	**3.**	**4.**	**5.**	**6.**	**7.**	**8.**	**9.**
°	Millim.	Millim.	Millim.	Millim.	Millim.	Millim.	Millim.	Millim.	Millim.	Millim.
0	0.00	0.01	0.02	0.03	0.04	0.05	0.06	0.07	0.08	0.09
1	0.10	0.11	0.12	0.13	0.14	0.15	0.16	0.17	0.18	0.19
2	0.20	0.21	0.21	0.22	0.23	0.24	0.25	0.26	0.27	0.28
3	0.29	0.30	0.31	0.32	0.33	0.34	0.35	0.36	0.37	0.38
4	0.39	0.40	0.41	0.42	0.43	0.44	0.45	0.46	0.47	0.48
5	0.49	0.50	0.51	0.52	0.53	0.54	0.55	0.56	0.57	0.58
6	0.59	0.60	0.61	0.62	0.63	0.63	0.64	0.65	0.66	0.67
7	0.68	0.69	0.70	0.71	0.72	0.73	0.74	0.75	0.76	0.77
8	0.78	0.79	0.80	0.81	0.82	0.83	0.84	0.85	0.86	0.87
9	0.88	0.89	0.90	0.91	0.92	0.93	0.94	0.95	0.96	0.97
10	0.98	0.99	1.00	1.01	1.02	1.03	1.04	1.05	1.05	1.06
11	1.07	1.08	1.09	1.10	1.11	1.12	1.13	1.14	1.15	1.16
12	1.17	1.18	1.19	1.20	1.21	1.22	1.23	1.24	1.25	1.26
13	1.27	1.28	1.29	1.30	1.31	1.32	1.33	1.34	1.35	1.36
14	1.37	1.38	1.39	1.40	1.41	1.42	1.43	1.44	1.45	1.46
15	1.46	1.47	1.48	1.49	1.50	1.51	1.52	1.53	1.54	1.55
16	1.56	1.57	1.58	1.59	1.60	1.61	1.62	1.63	1.64	1.65
17	1.66	1.67	1.68	1.69	1.70	1.71	1.72	1.73	1.74	1.75
18	1.76	1.77	1.78	1.79	1.80	1.81	1.82	1.83	1.84	1.85
19	1.86	1.87	1.87	1.88	1.89	1.90	1.91	1.92	1.93	1.94
20	1.95	1.96	1.97	1.98	1.99	2.00	2.01	2.02	2.03	2.04
21	2.05	2.06	2.07	2.08	2.09	2.10	2.11	2.12	2.13	2.14
22	2.15	2.16	2.17	2.18	2.19	2.20	2.21	2.22	2.23	2.24
23	2.25	2.26	2.27	2.28	2.29	2.29	2.30	2.31	2.32	2.33
24	2.34	2.35	2.36	2.37	2.38	2.39	2.40	2.41	2.42	2.43
25	2.44	2.45	2.46	2.47	2.48	2.49	2.50	2.51	2.52	2.53
26	2.54	2.55	2.56	2.57	2.58	2.59	2.60	2.61	2.62	2.63
27	2.64	2.65	2.66	2.67	2.68	2.69	2.70	2.71	2.71	2.72
28	2.73	2.74	2.75	2.76	2.77	2.78	2.79	2.80	2.81	2.82
29	2.83	2.84	2.85	2.86	2.87	2.88	2.89	2.90	2.91	2.92
30	2.93	2.94	2.95	2.96	2.97	2.98	2.99	3.00	3.01	3.02
31	3.03	3.04	3.05	3.06	3.07	3.08	3.09	3.10	3.11	3.12
32	3.12	3.13	3.14	3.15	3.16	3.17	3.18	3.19	3.20	3.21
33	3.22	3.23	3.24	3.25	3.26	3.27	3.28	3.29	3.30	3.31
34	3.32	3.33	3.34	3.35	3.36	3.37	3.38	3.39	3.40	3.41
35	3.42	3.43	3.44	3.45	3.46	3.47	3.48	3.49	3.50	3.51
	0.	**1.**	**2.**	**3.**	**4.**	**5.**	**6.**	**7.**	**8.**	**9.**

Centi-grade Degrees.	BAROMETER: 610mm. (from 607.51 to 612.50).									
	Tenths of Degrees.									
	0.	**1.**	**2.**	**3.**	**4.**	**5.**	**6.**	**7.**	**8.**	**9.**
°	Millim.	Millim.	Millim.	Millim.	Millim.	Millim.	Millim.	Millim.	Millim.	Millim.
0	0.00	0.01	0.02	0.03	0.04	0.05	0.06	0.07	0.08	0.09
1	0.10	0.11	0.12	0.13	0.14	0.15	0.16	0.17	0.18	0.19
2	0.20	0.21	0.22	0.23	0.24	0.25	0.26	0.27	0.28	0.29
3	0.30	0.31	0.32	0.32	0.33	0.34	0.35	0.36	0.37	0.38
4	0.39	0.40	0.41	0.42	0.43	0.44	0.45	0.46	0.47	0.48
5	0.49	0.50	0.51	0.52	0.53	0.54	0.55	0.56	0.57	0.58
6	0.59	0.60	0.61	0.62	0.63	0.64	0.65	0.66	0.67	0.68
7	0.69	0.70	0.71	0.72	0.73	0.74	0.75	0.76	0.77	0.78
8	0.79	0.80	0.81	0.82	0.83	0.84	0.85	0.86	0.87	0.88
9	0.89	0.90	0.91	0.92	0.93	0.94	0.95	0.96	0.96	0.97
10	0.98	0.99	1.00	1.01	1.02	1.03	1.04	1.05	1.06	1.07
11	1.08	1.09	1.10	1.11	1.12	1.13	1.14	1.15	1.16	1.17
12	1.18	1.19	1.20	1.21	1.22	1.23	1.24	1.25	1.26	1.27
13	1.28	1.29	1.30	1.31	1.32	1.33	1.34	1.35	1.36	1.37
14	1.38	1.39	1.40	1.41	1.42	1.43	1.44	1.45	1.46	1.47
15	1.48	1.49	1.50	1.51	1.52	1.53	1.54	1.55	1.56	1.57
16	1.58	1.59	1.59	1.60	1.61	1.62	1.63	1.64	1.65	1.66
17	1.67	1.68	1.69	1.70	1.71	1.72	1.73	1.74	1.75	1.76
18	1.77	1.78	1.79	1.80	1.81	1.82	1.83	1.84	1.85	1.86
19	1.87	1.88	1.89	1.90	1.91	1.92	1.93	1.94	1.95	1.96
20	1.97	1.98	1.99	2.00	2.01	2.02	2.03	2.04	2.05	2.06
21	2.07	2.08	2.09	2.10	2.11	2.12	2.13	2.14	2.15	2.16
22	2.17	2.18	2.19	2.20	2.21	2.22	2.23	2.23	2.24	2.25
23	2.26	2.27	2.28	2.29	2.30	2.31	2.32	2.33	2.34	2.35
24	2.36	2.37	2.38	2.39	2.40	2.41	2.42	2.43	2.44	2.45
25	2.46	2.47	2.48	2.49	2.50	2.51	2.52	2.53	2.54	2.55
26	2.56	2.57	2.58	2.59	2.60	2.61	2.62	2.63	2.64	2.65
27	2.66	2.67	2.68	2.69	2.70	2.71	2.72	2.73	2.74	2.75
28	2.76	2.77	2.78	2.79	2.80	2.81	2.82	2.83	2.84	2.85
29	2.86	2.86	2.87	2.88	2.89	2.90	2.91	2.92	2.93	2.94
30	2.95	2.96	2.97	2.98	2.99	3.00	3.01	3.02	3.03	3.04
31	3.05	3.06	3.07	3.08	3.09	3.10	3.11	3.12	3.13	3.14
32	3.15	3.16	3.17	3.18	3.19	3.20	3.21	3.22	3.23	3.24
33	3.25	3.26	3.27	3.28	3.29	3.30	3.31	3.32	3.33	3.34
34	3.35	3.36	3.37	3.38	3.39	3.40	3.41	3.42	3.43	3.44
35	3.45	3.46	3.47	3.48	3.49	3.50	3.51	3.52	3.53	3.54
	0.	**1.**	**2.**	**3.**	**4.**	**5.**	**6.**	**7.**	**8.**	**9.**

BAROMETER: 615$^{mm.}$ (from 612.51 to 617.50).

Centigrade Degrees.	Tenths of Degrees.									
	0.	**1.**	**2.**	**3.**	**4.**	**5.**	**6.**	**7.**	**8.**	**9.**
°	Millim.	Millim.	Millim.	Millim.	Millim.	Millim.	Millim.	Millim.	Millim.	Millim.
0	0.00	0.01	0.02	0.03	0.04	0.05	0.06	0.07	0.08	0.09
1	0.10	0.11	0.12	0.13	0.14	0.15	0.16	0.17	0.18	0.19
2	0.20	0.21	0.22	0.23	0.24	0.25	0.26	0.27	0.28	0.29
3	0.30	0.31	0.32	0.33	0.34	0.35	0.36	0.37	0.38	0.39
4	0.40	0.41	0.42	0.43	0.44	0.45	0.46	0.47	0.48	0.49
5	0.50	0.51	0.52	0.53	0.54	0.55	0.56	0.57	0.58	0.59
6	0.60	0.61	0.62	0.63	0.64	0.65	0.66	0.67	0.68	0.68
7	0.69	0.70	0.71	0.72	0.73	0.74	0.75	0.76	0.77	0.78
8	0.79	0.80	0.81	0.82	0.83	0.84	0.85	0.86	0.87	0.88
9	0.89	0.90	0.91	0.92	0.93	0.94	0.95	0.96	0.97	0.98
10	0.99	1.00	1.01	1.02	1.03	1.04	1.05	1.06	1.07	1.08
11	1.09	1.10	1.11	1.12	1.13	1.14	1.15	1.16	1.17	1.18
12	1.19	1.20	1.21	1.22	1.23	1.24	1.25	1.26	1.27	1.28
13	1.29	1.30	1.31	1.32	1.33	1.34	1.35	1.36	1.37	1.38
14	1.39	1.40	1.41	1.42	1.43	1.44	1.45	1.46	1.47	1.48
15	1.49	1.50	1.51	1.52	1.53	1.54	1.55	1.56	1.57	1.58
16	1.59	1.60	1.61	1.62	1.63	1.64	1.65	1.66	1.67	1.68
17	1.69	1.70	1.71	1.72	1.73	1.74	1.75	1.76	1.77	1.78
18	1.79	1.80	1.81	1.82	1.83	1.84	1.85	1.86	1.87	1.88
19	1.89	1.90	1.91	1.92	1.93	1.94	1.95	1.96	1.97	1.98
20	1.99	2.00	2.01	2.01	2.02	2.03	2.04	2.05	2.06	2.07
21	2.08	2.09	2.10	2.11	2.12	2.13	2.14	2.15	2.16	2.17
22	2.18	2.19	2.20	2.21	2.22	2.23	2.24	2.25	2.26	2.27
23	2.28	2.29	2.30	2.31	2.32	2.33	2.34	2.35	2.36	2.37
24	2.38	2.39	2.40	2.41	2.42	2.43	2.44	2.45	2.46	2.47
25	2.48	2.49	2.50	2.51	2.52	2.53	2.54	2.55	2.56	2.57
26	2.58	2.59	2.60	2.61	2.62	2.63	2.64	2.65	2.66	2.67
27	2.68	2.69	2.70	2.71	2.72	2.73	2.74	2.75	2.76	2.77
28	2.78	2.79	2.80	2.81	2.82	2.83	2.84	2.85	2.86	2.87
29	2.88	2.89	2.90	2.91	2.92	2.93	2.94	2.95	2.96	2.97
30	2.98	2.99	3.00	3.01	3.02	3.03	3.04	3.05	3.06	3.07
31	3.08	3.09	3.10	3.11	3.12	3.13	3.14	3.15	3.16	3.17
32	3.18	3.19	3.20	3.21	3.22	3.23	3.24	3.25	3.26	3.27
33	3.28	3.29	3.30	3.31	3.32	3.33	3.34	3.35	3.36	3.36
34	3.37	3.38	3.39	3.40	3.41	3.42	3.43	3.44	3.45	3.46
35	3.47	3.48	3.49	3.50	3.51	3.52	3.53	3.54	3.55	3.56
	0.	**1.**	**2.**	**3.**	**4.**	**5.**	**6.**	**7.**	**8.**	**9.**

BAROMETER: 620mm. (from 617.51 to 622.50).

Tenths of Degrees.

Centigrade Degrees.	0.	1.	2.	3.	4.	5.	6.	7.	8.	9.
°	Millim.	Millim.	Millim.	Millim.	Millim.	Millim.	Millim.	Millim.	Millim	Millim.
0	0.00	0.01	0.02	0.03	0.04	0.05	0.06	0.07	0.08	0.09
1	0.10	0.11	0.12	0.13	0.14	0.15	0.16	0.17	0.18	0.19
2	0.20	0.21	0.22	0.23	0.24	0.25	0.26	0.27	0.28	0.29
3	0.30	0.31	0.32	0.33	0.34	0.35	0.36	0.37	0.38	0.39
4	0.40	0.41	0.42	0.43	0.44	0.45	0.46	0.47	0.48	0.49
5	0.50	0.51	0.52	0.53	0.54	0.55	0.56	0.57	0.58	0.59
6	0.60	0.61	0.62	0.63	0.64	0.65	0.66	0.67	0.68	0.69
7	0.70	0.71	0.72	0.73	0.74	0.75	0.76	0.77	0.78	0.79
8	0.80	0.81	0.82	0.83	0.84	0.85	0.86	0.87	0.88	0.89
9	0.90	0.91	0.92	0.93	0.94	0.95	0.96	0.97	0.98	0.99
10	1.00	1.01	1.02	1.03	1.04	1.05	1.06	1.07	1.08	1.09
11	1.10	1.11	1.12	1.13	1.14	1.15	1.16	1.17	1.18	1.19
12	1.20	1.21	1.22	1.23	1.24	1.25	1.26	1.27	1.28	1.29
13	1.30	1.31	1.32	1.33	1.34	1.35	1.36	1.37	1.38	1.39
14	1.40	1.41	1.42	1.43	1.44	1.45	1.46	1.47	1.48	1.49
15	1.50	1.51	1.52	1.53	1.54	1.55	1.56	1.57	1.58	1.59
16	1.60	1.61	1.62	1.63	1.64	1.65	1.66	1.67	1.68	1.69
17	1.70	1.71	1.72	1.73	1.74	1.75	1.76	1.77	1.78	1.79
18	1.80	1.81	1.82	1.83	1.84	1.85	1.86	1.87	1.88	1.89
19	1.90	1.91	1.92	1.93	1.94	1.95	1.96	1.97	1.98	1.99
20	2.00	2.01	2.02	2.03	2.04	2.05	2.06	2.07	2.08	2.09
21	2.10	2.11	2.12	2.13	2.14	2.15	2.16	2.17	2.18	2.19
22	2.20	2.21	2.22	2.23	2.24	2.25	2.26	2.27	2.28	2.29
23	2.30	2.31	2.32	2.33	2.34	2.35	2.36	2.37	2.38	2.39
24	2.40	2.41	2.42	2.43	2.44	2.45	2.46	2.47	2.48	2.49
25	2.50	2.51	2.52	2.53	2.54	2.55	2.56	2.57	2.58	2.59
26	2.60	2.61	2.62	2.63	2.64	2.65	2.66	2.67	2.68	2.69
27	2.70	2.71	2.72	2.73	2.74	2.75	2.76	2.77	2.78	2.79
28	2.80	2.81	2.82	2.83	2.84	2.85	2.86	2.87	2.88	2.89
29	2.90	2.91	2.92	2.93	2.94	2.95	2.96	2.97	2.98	2.99
30	3.00	3.01	3.02	3.03	3.04	3.05	3.06	3.07	3.08	3.09
31	3.10	3.11	3.12	3.13	3.14	3.15	3.16	3.17	3.18	3.19
32	3.20	3.21	3.22	3.23	3.24	3.25	3.26	3.27	3.28	3.29
33	3.30	3.31	3.32	3.33	3.34	3.35	3.36	3.37	3.38	3.39
34	3.40	3.41	3.42	3.43	3.44	3.45	3.46	3.47	3.48	3.49
35	3.50	3.51	3.52	3.53	3.54	3.55	3.56	3.57	3.58	3.59
	0.	1.	2.	3.	4.	5.	6.	7.	8.	9.

Centigrade Degrees.	BAROMETER: 625mm. (from 622.51 to 627.50). Tenths of Degrees.									
	0.	1.	2.	3.	4.	5.	6.	7.	8.	9.
°	Millim.	Millim.	Millim.	Millim.	Millim.	Millim.	Millim.	Millim.	Millim	Millim.
0	0.00	0.01	0.02	0.03	0.04	0.05	0.06	0.07	0.08	0.09
1	0.10	0.11	0.12	0.13	0.14	0.15	0.16	0.17	0.18	0.19
2	0.20	0.21	0.22	0.23	0.24	0.25	0.26	0.27	0.28	0.29
3	0.30	0.31	0.32	0.33	0.34	0.35	0.36	0.37	0.38	0.39
4	0.40	0.41	0.42	0.43	0.44	0.45	0.46	0.47	0.48	0.49
5	0.50	0.51	0.52	0.53	0.54	0.55	0.56	0.58	0.59	0.60
6	0.61	0.62	0.63	0.64	0.65	0.66	0.67	0.68	0.69	0.70
7	0.71	0.72	0.73	0.74	0.75	0.76	0.77	0.78	0.79	0.80
8	0.81	0.82	0.83	0.84	0.85	0.86	0.87	0.88	0.89	0.90
9	0.91	0.92	0.93	0.94	0.95	0.96	0.97	0.98	0.99	1.00
10	1.01	1.02	1.03	1.04	1.05	1.06	1.07	1.08	1.09	1.10
11	1.11	1.12	1.13	1.14	1.15	1.16	1.17	1.18	1.19	1.20
12	1.21	1.22	1.23	1.24	1.25	1.26	1.27	1.28	1.29	1.30
13	1.31	1.32	1.33	1.34	1.35	1.36	1.37	1.38	1.39	1.40
14	1.41	1.42	1.43	1.44	1.45	1.46	1.47	1.48	1.49	1.50
15	1.51	1.52	1.53	1.54	1.55	1.56	1.57	1.58	1.59	1.60
16	1.61	1.62	1.63	1.64	1.65	1.66	1.67	1.68	1.69	1.70
17	1.71	1.73	1.74	1.75	1.76	1.77	1.78	1.79	1.80	1.81
18	1.82	1.83	1.84	1.85	1.86	1.87	1.88	1.89	1.90	1.91
19	1.92	1.93	1.94	1.95	1.96	1.97	1.98	1.99	2.00	2.01
20	2.02	2.03	2.04	2.05	2.06	2.07	2.08	2.09	2.10	2.11
21	2.12	2.13	2.14	2.15	2.16	2.17	2.18	2.19	2.20	2.21
22	2.22	2.23	2.24	2.25	2.26	2.27	2.28	2.29	2.30	2.31
23	2.32	2.33	2.34	2.35	2.36	2.37	2.38	2.39	2.40	2.41
24	2.42	2.43	2.44	2.45	2.46	2.47	2.48	2.49	2.50	2.51
25	2.52	2.53	2.54	2.55	2.56	2.57	2.58	2.59	2.60	2.61
26	2.62	2.63	2.64	2.65	2.66	2.67	2.68	2.69	2.70	2.71
27	2.72	2.73	2.74	2.75	2.76	2.77	2.78	2.79	2.80	2.81
28	2.82	2.83	2.84	2.85	2.87	2.88	2.89	2.90	2.91	2.92
29	2.93	2.94	2.95	2.96	2.97	2.98	2.99	3.00	3.01	3.02
30	3.03	3.04	3.05	3.06	3.07	3.08	3.09	3.10	3.11	3.12
31	3.13	3.14	3.15	3.16	3.17	3.18	3.19	3.20	3.21	3.22
32	3.23	3.24	3.25	3.26	3.27	3.28	3.29	3.30	3.31	3.32
33	3.33	3.34	3.35	3.36	3.37	3.38	3.39	3.40	3.41	3.42
34	3.43	3.44	3.45	3.46	3.47	3.48	3.49	3.50	3.51	3.52
35	3.53	3.54	3.55	3.56	3.57	3.58	3.59	3.60	3.61	3.62
	0.	1.	2.	3.	4.	5.	6.	7.	8.	9.

BAROMETER: 630$^{mm.}$ (from 627.51 to 632.50).

Centigrade Degrees.	Tenths of Degrees.									
	0.	**1.**	**2.**	**3.**	**4.**	**5.**	**6.**	**7.**	**8.**	**9.**
°	Millim.	Millim.	Millim.	Millim.	Millim.	Millim.	Millim.	Millim.	Millim.	Millim.
0	0.00	0.01	0.02	0.03	0.04	0.05	0.06	0.07	0.08	0.09
1	0.10	0.11	0.12	0.13	0.14	0.15	0.16	0.17	0.18	0.19
2	0.20	0.21	0.22	0.23	0.24	0.25	0.26	0.27	0.28	0.29
3	0.31	0.32	0.33	0.34	0.35	0.36	0.37	0.38	0.39	0.40
4	0.41	0.42	0.43	0.44	0.45	0.46	0.47	0.48	0.49	0.50
5	0.51	0.52	0.53	0.54	0.55	0.56	0.57	0.58	0.59	0.60
6	0.61	0.62	0.63	0.64	0.65	0.66	0.67	0.68	0.69	0.70
7	0.71	0.72	0.73	0.74	0.75	0.76	0.77	0.78	0.79	0.80
8	0.81	0.82	0.83	0.84	0.85	0.86	0.87	0.88	0.89	0.90
9	0.92	0.93	0.94	0.95	0.96	0.97	0.98	0.99	1.00	1.01
10	1.02	1.03	1.04	1.05	1.06	1.07	1.08	1.09	1.10	1.11
11	1.12	1.13	1.14	1.15	1.16	1.17	1.18	1.19	1.20	1.21
12	1.22	1.23	1.24	1.25	1.26	1.27	1.28	1.29	1.30	1.31
13	1.32	1.33	1.34	1.35	1.36	1.37	1.38	1.39	1.40	1.41
14	1.42	1.43	1.44	1.45	1.46	1.47	1.48	1.49	1.50	1.52
15	1.53	1.54	1.55	1.56	1.57	1.58	1.59	1.60	1.61	1.62
16	1.63	1.64	1.65	1.66	1.67	1.68	1.69	1.70	1.71	1.72
17	1.73	1.74	1.75	1.76	1.77	1.78	1.79	1.80	1.81	1.82
18	1.83	1.84	1.85	1.86	1.87	1.88	1.89	1.90	1.91	1.92
19	1.93	1.94	1.95	1.96	1.97	1.98	1.99	2.00	2.01	2.02
20	2.03	2.04	2.05	2.06	2.07	2.08	2.09	2.10	2.11	2.13
21	2.14	2.15	2.16	2.17	2.18	2.19	2.20	2.21	2.22	2.23
22	2.24	2.25	2.26	2.27	2.28	2.29	2.30	2.31	2.32	2.33
23	2.34	2.35	2.36	2.37	2.38	2.39	2.40	2.41	2.42	2.43
24	2.44	2.45	2.46	2.47	2.48	2.49	2.50	2.51	2.52	2.53
25	2.54	2.55	2.56	2.57	2.58	2.59	2.60	2.61	2.62	2.63
26	2.64	2.65	2.66	2.67	2.68	2.69	2.70	2.71	2.73	2.74
27	2.75	2.76	2.77	2.78	2.79	2.80	2.81	2.82	2.83	2.84
28	2.85	2.86	2.87	2.88	2.89	2.90	2.91	2.92	2.93	2.94
29	2.95	2.96	2.97	2.98	2.99	3.00	3.01	3.02	3.03	3.04
30	3.05	3.06	3.07	3.08	3.09	3.10	3.11	3.12	3.13	3.14
31	3.15	3.16	3.17	3.18	3.19	3.20	3.21	3.22	3.23	3.24
32	3.25	3.26	3.27	3.28	3.29	3.30	3.31	3.32	3.34	3.35
33	3.36	3.37	3.38	3.39	3.40	3.41	3.42	3.43	3.44	3.45
34	3.46	3.47	3.48	3.49	3.50	3.51	3.52	3.53	3.54	3.55
35	3.56	3.57	3.58	3.59	3.60	3.61	3.62	3.63	3.64	3.65
	0.	**1.**	**2.**	**3.**	**4.**	**5.**	**6.**	**7.**	**8.**	**9.**

Centigrade Degrees.	BAROMETER: 635mm. (from 632.51 to 637.50). Tenths of Degrees.									
	0.	1.	2.	3.	4.	5.	6.	7.	8.	9.
°	Millim.	Millim.	Millim.	Millim.	Millim.	Millim.	Millim.	Millim.	Millim.	Millim.
0	0.00	0.01	0.02	0.03	0.04	0.05	0.06	0.07	0.08	0.09
1	0.10	0.11	0.12	0.13	0.14	0.15	0.16	0.17	0.18	0.19
2	0.20	0.22	0.23	0.24	0.25	0.26	0.27	0.28	0.29	0.30
3	0.31	0.32	0.33	0.34	0.35	0.36	0.37	0.38	0.39	0.40
4	0.41	0.42	0.43	0.44	0.45	0.46	0.47	0.48	0.49	0.50
5	0.51	0.52	0.53	0.54	0.55	0.56	0.57	0.58	0.59	0.60
6	0.61	0.63	0.64	0.65	0.66	0.67	0.68	0.69	0.70	0.71
7	0.72	0.73	0.74	0.75	0.76	0.77	0.78	0.79	0.80	0.81
8	0.82	0.83	0.84	0.85	0.86	0.87	0.88	0.89	0.90	0.91
9	0.92	0.93	0.94	0.95	0.96	0.97	0.98	0.99	1.00	1.01
10	1.02	1.04	1.05	1.06	1.07	1.08	1.09	1.10	1.11	1.12
11	1.13	1.14	1.15	1.16	1.17	1.18	1.19	1.20	1.21	1.22
12	1.23	1.24	1.25	1.26	1.27	1.28	1.29	1.30	1.31	1.32
13	1.33	1.34	1.35	1.36	1.37	1.38	1.39	1.40	1.41	1.42
14	1.43	1.45	1.46	1.47	1.48	1.49	1.50	1.51	1.52	1.53
15	1.54	1.55	1.56	1.57	1.58	1.59	1.60	1.61	1.62	1.63
16	1.64	1.65	1.66	1.67	1.68	1.69	1.70	1.71	1.72	1.73
17	1.74	1.75	1.76	1.77	1.78	1.79	1.80	1.81	1.82	1.83
18	1.84	1.86	1.87	1.88	1.89	1.90	1.91	1.92	1.93	1.94
19	1.95	1.96	1.97	1.98	1.99	2.00	2.01	2.02	2.03	2.04
20	2.05	2.06	2.07	2.08	2.09	2.10	2.11	2.12	2.13	2.14
21	2.15	2.16	2.17	2.18	2.19	2.20	2.21	2.22	2.23	2.24
22	2.25	2.27	2.28	2.29	2.30	2.31	2.32	2.33	2.34	2.35
23	2.36	2.37	2.38	2.39	2.40	2.41	2.42	2.43	2.44	2.45
24	2.46	2.47	2.48	2.49	2.50	2.51	2.52	2.53	2.54	2.55
25	2.56	2.57	2.58	2.59	2.60	2.61	2.62	2.63	2.64	2.65
26	2.66	2.67	2.69	2.70	2.71	2.72	2.73	2.74	2.75	2.76
27	2.77	2.78	2.79	2.80	2.81	2.82	2.83	2.84	2.85	2.86
28	2.87	2.88	2.89	2.90	2.91	2.92	2.93	2.94	2.95	2.96
29	2.97	2.98	2.99	3.00	3.01	3.02	3.03	3.04	3.05	3.06
30	3.07	3.08	3.10	3.11	3.12	3.13	3.14	3.15	3.16	3.17
31	3.18	3.19	3.20	3.21	3.22	3.23	3.24	3.25	3.26	3.27
32	3.28	3.29	3.30	3.31	3.32	3.33	3.34	3.35	3.36	3.37
33	3.38	3.39	3.40	3.41	3.42	3.43	3.44	3.45	3.46	3.47
34	3.48	3.49	3.51	3.52	3.53	3.54	3.55	3.56	3.57	3.58
35	3.59	3.60	3.61	3.62	3.63	3.64	3.65	3.66	3.67	3.68
	0.	1.	2.	3.	4.	5.	6.	7.	8.	9.

BAROMETER: 640mm. (from 637.51 to 642.50).

Centigrade Degrees.	Tenths of Degrees. 0.	1.	2.	3.	4.	5.	6.	7.	8.	9.
°	Millim.	Millim.	Millim.	Millim.	Millim.	Millim.	Millim.	Millim.	Millim.	Millim.
0	0.00	0.01	0.02	0.03	0.04	0.05	0.06	0.07	0.08	0.09
1	0.10	0.11	0.12	0.13	0.14	0.15	0.17	0.18	0.19	0.20
2	0.21	0.22	0.23	0.24	0.25	0.26	0.27	0.28	0.29	0.30
3	0.31	0.32	0.33	0.34	0.35	0.36	0.37	0.38	0.39	0.40
4	0.41	0.42	0.43	0.44	0.45	0.46	0.48	0.49	0.50	0.51
5	0.52	0.53	0.54	0.55	0.56	0.57	0.58	0.59	0.60	0.61
6	0.62	0.63	0.64	0.65	0.66	0.67	0.68	0.69	0.70	0.71
7	0.72	0.73	0.74	0.75	0.76	0.77	0.78	0.80	0.81	0.82
8	0.83	0.84	0.85	0.86	0.87	0.88	0.89	0.90	0.91	0.92
9	0.93	0.94	0.95	0.96	0.97	0.98	0.99	1.00	1.01	1.02
10	1.03	1.04	1.05	1.06	1.07	1.08	1.09	1.11	1.12	1.13
11	1.14	1.15	1.16	1.17	1.18	1.19	1.20	1.21	1.22	1.23
12	1.24	1.25	1.26	1.27	1.28	1.29	1.30	1.31	1.32	1.33
13	1.34	1.35	1.36	1.37	1.38	1.39	1.40	1.42	1.43	1.44
14	1.45	1.46	1.47	1.48	1.49	1.50	1.51	1.52	1.53	1.54
15	1.55	1.56	1.57	1.58	1.59	1.60	1.61	1.62	1.63	1.64
16	1.65	1.66	1.67	1.68	1.69	1.70	1.71	1.72	1.74	1.75
17	1.76	1.77	1.78	1.79	1.80	1.81	1.82	1.83	1.84	1.85
18	1.86	1.87	1.88	1.89	1.90	1.91	1.92	1.93	1.94	1.95
19	1.96	1.97	1.98	1.99	2.00	2.01	2.02	2.03	2.05	2.06
20	2.07	2.08	2.09	2.10	2.11	2.12	2.13	2.14	2.15	2.16
21	2.17	2.18	2.19	2.20	2.21	2.22	2.23	2.24	2.25	2.26
22	2.27	2.28	2.29	2.30	2.31	2.32	2.33	2.34	2.36	2.37
23	2.38	2.39	2.40	2.41	2.42	2.43	2.44	2.45	2.46	2.47
24	2.48	2.49	2.50	2.51	2.52	2.53	2.54	2.55	2.56	2.57
25	2.58	2.59	2.60	2.61	2.62	2.63	2.64	2.65	2.66	2.68
26	2.69	2.70	2.71	2.72	2.73	2.74	2.75	2.76	2.77	2.78
27	2.79	2.80	2.81	2.82	2.83	2.84	2.85	2.86	2.87	2.88
28	2.89	2.90	2.91	2.92	2.93	2.94	2.95	2.96	2.97	2.99
29	3.00	3.01	3.02	3.03	3.04	3.05	3.06	3.07	3.08	3.09
30	3.10	3.11	3.12	3.13	3.14	3.15	3.16	3.17	3.18	3.19
31	3.20	3.21	3.22	3.23	3.24	3.25	3.26	3.27	3.28	3.30
32	3.31	3.32	3.33	3.34	3.35	3.36	3.37	3.38	3.39	3.40
33	3.41	3.42	3.43	3.44	3.45	3.46	3.47	3.48	3.49	3.50
34	3.51	3.52	3.53	3.54	3.55	3.56	3.57	3.58	3.59	3.60
35	3.62	3.63	3.64	3.65	3.66	3.67	3.68	3.69	3.70	3.71
	0.	1.	2.	3.	4.	5.	6.	7.	8.	9.

Centi-grade Degrees.	BAROMETER: 645mm. (from 642.51 to 647.50). Tenths of Degrees.									
	0.	**1.**	**2.**	**3.**	**4.**	**5.**	**6.**	**7.**	**8.**	**9.**
°	Millim.	Millim.	Millim.	Millim.	Millim.	Millim.	Millim.	Millim.	Millim.	Millim.
0	0.00	0.01	0.02	0.03	0.04	0.05	0.06	0.07	0.08	0.09
1	0.10	0.11	0.12	0.14	0.15	0.16	0.17	0.18	0.19	0.20
2	0.21	0.22	0.23	0.24	0.25	0.26	0.27	0.28	0.29	0.30
3	0.31	0.32	0.33	0.34	0.35	0.36	0.37	0.39	0.40	0.41
4	0.42	0.43	0.44	0.45	0.46	0.47	0.48	0.49	0.50	0.51
5	0.52	0.53	0.54	0.55	0.56	0.57	0.58	0.59	0.60	0.61
6	0.62	0.64	0.65	0.66	0.67	0.68	0.69	0.70	0.71	0.72
7	0.73	0.74	0.75	0.76	0.77	0.78	0.79	0.80	0.81	0.82
8	0.83	0.84	0.85	0.86	0.87	0.88	0.90	0.91	0.92	0.93
9	0.94	0.95	0.96	0.97	0.98	0.99	1.00	1.01	1.02	1.03
10	1.04	1.05	1.06	1.07	1.08	1.09	1.10	1.11	1.12	1.13
11	1.15	1.16	1.17	1.18	1.19	1.20	1.21	1.22	1.23	1.24
12	1.25	1.26	1.27	1.28	1.29	1.30	1.31	1.32	1.33	1.34
13	1.35	1.36	1.37	1.38	1.39	1.41	1.42	1.43	1.44	1.45
14	1.46	1.47	1.48	1.49	1.50	1.51	1.52	1.53	1.54	1.55
15	1.56	1.57	1.58	1.59	1.60	1.61	1.62	1.63	1.64	1.66
16	1.67	1.68	1.69	1.70	1.71	1.72	1.73	1.74	1.75	1.76
17	1.77	1.78	1.79	1.80	1.81	1.82	1.83	1.84	1.85	1.86
18	1.87	1.88	1.89	1.91	1.92	1.93	1.94	1.95	1.96	1.97
19	1.98	1.99	2.00	2.01	2.02	2.03	2.04	2.05	2.06	2.07
20	2.08	2.09	2.10	2.11	2.12	2.13	2.14	2.15	2.17	2.18
21	2.19	2.20	2.21	2.22	2.23	2.24	2.25	2.26	2.27	2.28
22	2.29	2.30	2.31	2.32	2.33	2.34	2.35	2.36	2.37	2.38
23	2.39	2.40	2.42	2.43	2.44	2.45	2.46	2.47	2.48	2.49
24	2.50	2.51	2.52	2.53	2.54	2.55	2.56	2.57	2.58	2.59
25	2.60	2.61	2.62	2.63	2.64	2.65	2.66	2.67	2.69	2.70
26	2.71	2.72	2.73	2.74	2.75	2.76	2.77	2.78	2.79	2.80
27	2.81	2.82	2.83	2.84	2.85	2.86	2.87	2.88	2.89	2.90
28	2.92	2.93	2.94	2.95	2.96	2.97	2.98	2.99	3.00	3.01
29	3.02	3.03	3.04	3.05	3.06	3.07	3.08	3.09	3.10	3.11
30	3.12	3.13	3.14	3.15	3.16	3.18	3.19	3.20	3.21	3.22
31	3.23	3.24	3.25	3.26	3.27	3.28	3.29	3.30	3.31	3.32
32	3.33	3.34	3.35	3.36	3.37	3.38	3.39	3.40	3.41	3.42
33	3.44	3.45	3.46	3.47	3.48	3.49	3.50	3.51	3.52	3.53
34	3.54	3.55	3.56	3.57	3.58	3.59	3.60	3.61	3.62	3.63
35	3.64	3.65	3.66	3.67	3.68	3.69	3.70	3.71	3.72	3.73
	0.	**1.**	**2.**	**3.**	**4.**	**5.**	**6.**	**7.**	**8.**	**9.**

BAROMETER : 650mm. (from 647.51 to 652.50).

Centigrade Degrees.	Tenths of Degrees.									
	0.	**1.**	**2.**	**3.**	**4.**	**5.**	**6.**	**7.**	**8.**	**9.**
°	Millim.	Millim.	Millim.	Millim.	Millim.	Millim.	Millim.	Millim.	Millim.	Millim.
0	0.00	0.01	0.02	0.03	0.04	0.05	0.06	0.07	0.08	0.09
1	0.11	0.12	0.13	0.14	0.15	0.16	0.17	0.18	0.19	0.20
2	0.21	0.22	0.23	0.24	0.25	0.26	0.27	0.28	0.29	0.30
3	0.32	0.33	0.34	0.35	0.36	0.37	0.38	0.39	0.40	0.41
4	0.42	0.43	0.44	0.45	0.46	0.47	0.48	0.49	0.50	0.51
5	0.53	0.54	0.55	0.56	0.57	0.58	0.59	0.60	0.61	0.62
6	0.63	0.64	0.65	0.66	0.67	0.68	0.69	0.70	0.71	0.72
7	0.73	0.75	0.76	0.77	0.78	0.79	0.80	0.81	0.82	0.83
8	0.84	0.85	0.86	0.87	0.88	0.89	0.90	0.91	0.92	0.93
9	0.94	0.96	0.97	0.98	0.99	1.00	1.01	1.02	1.03	1.04
10	1.05	1.06	1.07	1.08	1.09	1.10	1.11	1.12	1.13	1.14
11	1.15	1.17	1.18	1.19	1.20	1.21	1.22	1.23	1.24	1.25
12	1.26	1.27	1.28	1.29	1.30	1.31	1.32	1.33	1.34	1.35
13	1.36	1.37	1.39	1.40	1.41	1.42	1.43	1.44	1.45	1.46
14	1.47	1.48	1.49	1.50	1.51	1.52	1.53	1.54	1.55	1.56
15	1.57	1.58	1.60	1.61	1.62	1.63	1.64	1.65	1.66	1.67
16	1.68	1.69	1.70	1.71	1.72	1.73	1.74	1.75	1.76	1.77
17	1.78	1.79	1.81	1.82	1.83	1.84	1.85	1.86	1.87	1.88
18	1.89	1.90	1.91	1.92	1.93	1.94	1.95	1.96	1.97	1.98
19	1.99	2.00	2.01	2.03	2.04	2.05	2.06	2.07	2.08	2.09
20	2.10	2.11	2.12	2.13	2.14	2.15	2.16	2.17	2.18	2.19
21	2.20	2.21	2.22	2.24	2.25	2.26	2.27	2.28	2.29	2.30
22	2.31	2.32	2.33	2.34	2.35	2.36	2.37	2.38	2.39	2.40
23	2.41	2.42	2.43	2.44	2.46	2.47	2.48	2.49	2.50	2.51
24	2.52	2.53	2.54	2.55	2.56	2.57	2.58	2.59	2.60	2.61
25	2.62	2.63	2.64	2.65	2.67	2.68	2.69	2.70	2.71	2.72
26	2.73	2.84	2.75	2.76	2.77	2.78	2.79	2.80	2.81	2.82
27	2.83	2.84	2.85	2.86	2.88	2.89	2.90	2.91	2.92	2.93
28	2.94	2.95	2.96	2.97	2.98	2.99	3.00	3.01	3.02	3.03
29	3.04	3.05	3.06	3.07	3.08	3.10	3.11	3.12	3.13	3.14
30	3.15	3.16	3.17	3.18	3.19	3.20	3.21	3.22	3.23	3.24
31	3.25	3.26	3.27	3.28	3.29	3.31	3.32	3.33	3.34	3.35
32	3.36	3.37	3.38	3.39	3.40	3.41	3.42	3.43	3.44	3.45
33	3.46	3.47	3.48	3.49	3.50	3.52	3.53	3.54	3.55	3.56
34	3.57	3.58	3.59	3.60	3.61	3.62	3.63	3.64	3.65	3.66
35	3.67	3.68	3.69	3.70	3.71	3.72	3.74	3.75	3.76	3.77
	0.	**1.**	**2.**	**3.**	**4.**	**5.**	**6.**	**7.**	**8.**	**9.**

Centigrade Degrees.	BAROMETER: 655$^{mm.}$ (from 652.51 to 657.50).									
	Tenths of Degrees.									
	0.	**1.**	**2.**	**3.**	**4.**	**5.**	**6.**	**7.**	**8.**	**9.**
°	Millim.	Millim.	Millim.	Millim.	Millim.	Millim.	Millim.	Millim.	Millim.	Millim.
0	0.00	0.01	0.02	0.03	0.04	0.05	0.06	0.07	0.09	0.10
1	0.11	0.12	0.13	0.14	0.15	0.16	0.17	0.18	0.19	0.20
2	0.21	0.22	0.23	0.24	0.25	0.26	0.28	0.29	0.30	0.31
3	0.32	0.33	0.34	0.35	0.36	0.37	0.38	0.39	0.40	0.41
4	0.42	0.43	0.44	0.46	0.47	0.48	0.49	0.50	0.51	0.52
5	0.53	0.54	0.55	0.56	0.57	0.58	0.59	0.60	0.61	0.62
6	0.63	0.65	0.66	0.67	0.68	0.69	0.70	0.71	0.72	0.73
7	0.74	0.75	0.76	0.77	0.78	0.79	0.80	0.81	0.83	0.84
8	0.85	0.86	0.87	0.88	0.89	0.90	0.91	0.92	0.93	0.94
9	0.95	0.96	0.97	0.98	0.99	1.00	1.02	1.03	1.04	1.05
10	1.06	1.07	1.08	1.09	1.10	1.11	1.12	1.13	1.14	1.15
11	1.16	1.17	1.18	1.20	1.21	1.22	1.23	1.24	1.25	1.26
12	1.27	1.28	1.29	1.30	1.31	1.32	1.33	1.34	1.35	1.36
13	1.37	1.39	1.40	1.41	1.42	1.43	1.44	1.45	1.46	1.47
14	1.48	1.49	1.50	1.51	1.52	1.53	1.54	1.55	1.57	1.58
15	1.59	1.60	1.61	1.62	1.63	1.64	1.65	1.66	1.67	1.68
16	1.69	1.70	1.71	1.72	1.73	1.74	1.76	1.77	1.78	1.79
17	1.80	1.81	1.82	1.83	1.84	1.85	1.86	1.87	1.88	1.89
18	1.90	1.91	1.92	1.94	1.95	1.96	1.97	1.98	1.99	2.00
19	2.01	2.02	2.03	2.04	2.05	2.06	2.07	2.08	2.09	2.10
20	2.11	2.13	2.14	2.15	2.16	2.17	2.18	2.19	2.20	2.21
21	2.22	2.23	2.24	2.25	2.26	2.27	2.28	2.29	2.31	2.32
22	2.33	2.34	2.35	2.36	2.37	2.38	2.39	2.40	2.41	2.42
23	2.43	2.44	2.45	2.46	2.47	2.48	2.50	2.51	2.52	2.53
24	2.54	2.55	2.56	2.57	2.58	2.59	2.60	2.61	2.62	2.63
25	2.64	2.65	2.66	2.68	2.69	2.70	2.71	2.72	2.73	2.74
26	2.75	2.76	2.77	2.78	2.79	2.80	2.81	2.82	2.83	2.84
27	2.85	2.87	2.88	2.89	2.90	2.91	2.92	2.93	2.94	2.95
28	2.96	2.97	2.98	2.99	3.00	3.01	3.02	3.03	3.05	3.06
29	3.07	3.08	3.09	3.10	3.11	3.12	3.13	3.14	3.15	3.16
30	3.17	3.18	3.19	3.20	3.21	3.22	3.24	3.25	3.26	3.27
31	3.28	3.29	3.30	3.31	3.32	3.33	3.34	3.35	3.36	3.37
32	3.38	3.39	3.40	3.42	3.43	3.44	3.45	3.46	3.47	3.48
33	3.49	3.50	3.51	3.52	3.53	3.54	3.55	3.56	3.57	3.58
34	3.59	3.61	3.62	3.63	3.64	3.65	3.66	3.67	3.68	3.69
35	3.70	3.71	3.72	3.73	3.74	3.75	3.76	3.77	3.79	3.80
	0.	**1.**	**2.**	**3.**	**4.**	**5.**	**6.**	**7.**	**8.**	**9.**

Centigrade Degrees.	BAROMETER: 660mm. (from 657.51 to 662.50).									
	Tenths of Degrees.									
	0.	**1.**	**2.**	**3.**	**4.**	**5.**	**6.**	**7.**	**8.**	**9.**
°	Millim.	Millim.	Millim.	Millim.	Millim.	Millim.	Millim.	Millim.	Millim.	Millim.
0	0.00	0.01	0.02	0.03	0.04	0.05	0.06	0.08	0.09	0.10
1	0.11	0.12	0.13	0.14	0.15	0.16	0.17	0.18	0.19	0.20
2	0.21	0.22	0.23	0.25	0.26	0.27	0.28	0.29	0.30	0.31
3	0.32	0.33	0.34	0.35	0.36	0.37	0.38	0.39	0.41	0.42
4	0.43	0.44	0.45	0.46	0.47	0.48	0.49	0.50	0.51	0.52
5	0.53	0.54	0.55	0.57	0.58	0.59	0.60	0.61	0.62	0.63
6	0.64	0.65	0.66	0.67	0.68	0.69	0.70	0.71	0.72	0.74
7	0.75	0.76	0.77	0.78	0.79	0.80	0.81	0.82	0.83	0.84
8	0.85	0.86	0.87	0.88	0.90	0.91	0.92	0.93	0.94	0.95
9	0.96	0.97	0.98	0.99	1.00	1.01	1.02	1.03	1.04	1.06
10	1.07	1.08	1.09	1.10	1.11	1.12	1.13	1.14	1.15	1.16
11	1.17	1.18	1.19	1.20	1.21	1.23	1.24	1.25	1.26	1.27
12	1.28	1.29	1.30	1.31	1.32	1.33	1.34	1.35	1.36	1.37
13	1.39	1.40	1.41	1.42	1.43	1.44	1.45	1.46	1.47	1.48
14	1.49	1.50	1.51	1.52	1.53	1.55	1.56	1.57	1.58	1.59
15	1.60	1.61	1.62	1.63	1.64	1.65	1.66	1.67	1.68	1.69
16	1.70	1.72	1.73	1.74	1.75	1.76	1.77	1.78	1.79	1.80
17	1.81	1.82	1.83	1.84	1.85	1.86	1.88	1.89	1.90	1.91
18	1.92	1.93	1.94	1.95	1.96	1.97	1.98	1.99	2.00	2.01
19	2.02	2.04	2.05	2.06	2.07	2.08	2.09	2.10	2.11	2.12
20	2.13	2.14	2.15	2.16	2.17	2.18	2.19	2.21	2.22	2.23
21	2.24	2.25	2.26	2.27	2.28	2.29	2.30	2.31	2.32	2.33
22	2.34	2.35	2.37	2.38	2.39	2.40	2.41	2.42	2.43	2.44
23	2.45	2.46	2.47	2.48	2.49	2.50	2.51	2.53	2.54	2.55
24	2.56	2.57	2.58	2.59	2.60	2.61	2.62	2.63	2.64	2.65
25	2.66	2.67	2.68	2.70	2.71	2.72	2.73	2.74	2.75	2.76
26	2.77	2.78	2.79	2.80	2.81	2.82	2.83	2.84	2.86	2.87
27	2.88	2.89	2.90	2.91	2.92	2.93	2.94	2.95	2.96	2.97
28	2.98	2.99	3.00	3.02	3.03	3.04	3.05	3.06	3.07	3.08
29	3.09	3.10	3.11	3.12	3.13	3.14	3.15	3.16	3.17	3.19
30	3.20	3.21	3.22	3.23	3.24	3.25	3.26	3.27	3.28	3.29
31	3.30	3.31	3.32	3.33	3.35	3.36	3.37	3.38	3.39	3.40
32	3.41	3.42	3.43	3.44	3.45	3.46	3.47	3.48	3.49	3.51
33	3.52	3.53	3.54	3.55	3.56	3.57	3.58	3.59	3.60	3.61
34	3.62	3.63	3.64	3.65	3.66	3.68	3.69	3.70	3.71	3.72
35	3.73	3.74	3.75	3.76	3.77	3.78	3.79	3.80	3.81	3.82
	0.	**1.**	**2.**	**3.**	**4.**	**5.**	**6.**	**7.**	**8.**	**9.**

Centigrade Degrees.	BAROMETER: 665mm. (from 662.51 to 667.50).									
	Tenths of Degrees.									
	0.	1.	2.	3.	4.	5.	6.	7.	8.	9.
°	Millim.	Millim.	Millim.	Millim.	Millim.	Millim.	Millim.	Millim.	Millim.	Millim.
0	0.00	0.01	0.02	0.03	0.04	0.05	0.06	0.08	0.09	0.10
1	0.11	0.12	0.13	0.14	0.15	0.16	0.17	0.18	0.19	0.20
2	0.22	0.23	0.24	0.25	0.26	0.27	0.28	0.29	0.30	0.31
3	0.32	0.33	0.34	0.35	0.37	0.38	0.39	0.40	0.41	0.42
4	0.43	0.44	0.45	0.46	0.47	0.48	0.49	0.51	0.52	0.53
5	0.54	0.55	0.56	0.57	0.58	0.59	0.60	0.61	0.62	0.63
6	0.64	0.66	0.67	0.68	0.69	0.70	0.71	0.72	0.73	0.74
7	0.75	0.76	0.77	0.78	0.79	0.81	0.82	0.83	0.84	0.85
8	0.86	0.87	0.88	0.89	0.90	0.91	0.92	0.93	0.95	0.96
9	0.97	0.98	0.99	1.00	1.01	1.02	1.03	1.04	1.05	1.06
10	1.07	1.08	1.10	1.11	1.12	1.13	1.14	1.15	1.16	1.17
11	1.18	1.19	1.20	1.21	1.22	1.23	1.25	1.26	1.27	1.28
12	1.29	1.30	1.31	1.32	1.33	1.34	1.35	1.36	1.37	1.39
13	1.40	1.41	1.42	1.43	1.44	1.45	1.46	1.47	1.48	1.49
14	1.50	1.51	1.52	1.54	1.55	1.56	1.57	1.58	1.59	1.60
15	1.61	1.62	1.63	1.64	1.65	1.66	1.67	1.69	1.70	1.71
16	1.72	1.73	1.74	1.75	1.76	1.77	1.78	1.79	1.80	1.81
17	1.83	1.84	1.85	1.86	1.87	1.88	1.89	1.90	1.91	1.92
18	1.93	1.94	1.95	1.96	1.98	1.99	2.00	2.01	2.02	2.03
19	2.04	2.05	2.06	2.07	2.08	2.09	2.10	2.11	2.13	2.14
20	2.15	2.16	2.17	2.18	2.19	2.20	2.21	2.22	2.23	2.24
21	2.25	2.27	2.28	2.29	2.30	2.31	2.32	2.33	2.34	2.35
22	2.36	2.37	2.38	2.39	2.40	2.42	2.43	2.44	2.45	2.46
23	2.47	2.48	2.49	2.50	2.51	2.52	2.53	2.54	2.56	2.57
24	2.58	2.59	2.60	2.61	2.62	2.63	2.64	2.65	2.66	2.67
25	2.68	2.69	2.71	2.72	2.73	2.74	2.75	2.76	2.77	2.78
26	2.79	2.80	2.81	2.82	2.83	2.84	2.86	2.87	2.88	2.89
27	2.90	2.91	2.92	2.93	2.94	2.95	2.96	2.97	2.98	3.00
28	3.01	3.02	3.03	3.04	3.05	3.06	3.07	3.08	3.09	3.10
29	3.11	3.12	3.13	3.15	3.16	3.17	3.18	3.19	3.20	3.21
30	3.22	3.23	3.24	3.25	3.26	3.27	3.28	3.30	3.31	3.32
31	3.33	3.34	3.35	3.36	3.37	3.38	3.39	3.40	3.41	3.42
32	3.44	3.45	3.46	3.47	3.48	3.49	3.50	3.51	3.52	3.53
33	3.54	3.55	3.56	3.57	3.59	3.60	3.61	3.62	3.63	3.64
34	3.65	3.66	3.67	3.68	3.69	3.70	3.71	3.72	3.74	3.75
35	3.76	3.77	3.78	3.79	3.80	3.81	3.82	3.83	3.84	3.85
	0.	1.	2.	3.	4.	5.	6.	7.	8.	9.

Centi-grade Degrees.	BAROMETER: 670mm. (from 667.51 to 672.50.)									
	Tenths of Degrees.									
	0.	**1.**	**2.**	**3.**	**4.**	**5.**	**6.**	**7.**	**8.**	**9.**
°	Millim.	Millim.	Millim.	Millim.	Millim.	Millim.	Millim.	Millim.	Millim.	Millim.
0	0.00	0.01	0.02	0.03	0.04	0.05	0.07	0.08	0.09	0.10
1	0.11	0.12	0.13	0.14	0.15	0.16	0.17	0.18	0.20	0.21
2	0.22	0.23	0.24	0.25	0.26	0.27	0.28	0.29	0.30	0.31
3	0.32	0.34	0.35	0.36	0.37	0.38	0.39	0.40	0.41	0.42
4	0.43	0.44	0.45	0.47	0.48	0.49	0.50	0.51	0.52	0.53
5	0.54	0.55	0.56	0.57	0.58	0.60	0.61	0.62	0.63	0.64
6	0.65	0.66	0.67	0.68	0.69	0.70	0.71	0.73	0.74	0.75
7	0.76	0.77	0.78	0.79	0.80	0.81	0.82	0.83	0.84	0.85
8	0.87	0.88	0.89	0.90	0.91	0.92	0.93	0.94	0.95	0.96
9	0.97	0.98	1.00	1.01	1.02	1.03	1.04	1.05	1.06	1.07
10	1.08	1.09	1.10	1.11	1.13	1.14	1.15	1.16	1.17	1.18
11	1.19	1.20	1.21	1.22	1.23	1.24	1.25	1.27	1.28	1.29
12	1.30	1.31	1.32	1.33	1.34	1.35	1.36	1.37	1.38	1.40
13	1.41	1.42	1.43	1.44	1.45	1.46	1.47	1.48	1.49	1.50
14	1.51	1.53	1.54	1.55	1.56	1.57	1.58	1.59	1.60	1.61
15	1.62	1.63	1.64	1.66	1.67	1.68	1.69	1.70	1.71	1.72
16	1.73	1.74	1.75	1.76	1.77	1.78	1.80	1.81	1.82	1.83
17	1.84	1.85	1.86	1.87	1.88	1.89	1.90	1.91	1.92	1.94
18	1.95	1.96	1.97	1.98	1.99	2.00	2.01	2.02	2.03	2.04
19	2.06	2.07	2.08	2.09	2.10	2.11	2.12	2.13	2.14	2.15
20	2.16	2.17	2.18	2.20	2.21	2.22	2.23	2.24	2.25	2.26
21	2.27	2.28	2.29	2.30	2.31	2.33	2.34	2.35	2.36	2.37
22	2.38	2.39	2.40	2.41	2.42	2.43	2.44	2.46	2.47	2.48
23	2.49	2.50	2.51	2.52	2.53	2.54	2.55	2.56	2.57	2.59
24	2.60	2.61	2.62	2.63	2.64	2.65	2.66	2.67	2.68	2.69
25	2.70	2.71	2.73	2.74	2.75	2.76	2.77	2.78	2.79	2.80
26	2.81	2.82	2.83	2.84	2.86	2.87	2.88	2.89	2.90	2.91
27	2.92	2.93	2.94	2.95	2.96	2.97	2.99	3.00	3.01	3.02
28	3.03	3.04	3.05	3.06	3.07	3.08	3.09	3.10	3.11	3.13
29	3.14	3.15	3.16	3.17	3.18	3.19	3.20	3.21	3.22	3.23
30	3.24	3.26	3.27	3.28	3.29	3.30	3.31	3.32	3.33	3.34
31	3.35	3.36	3.37	3.39	3.40	3.41	3.42	3.43	3.44	3.45
32	3.46	3.47	3.48	3.49	3.50	3.52	3.53	3.54	3.55	3.56
33	3.57	3.58	3.59	3.60	3.61	3.62	3.63	3.64	3.66	3.67
34	3.68	3.69	3.70	3.71	3.72	3.73	3.74	3.75	3.76	3.77
35	3.79	3.80	3.81	3.82	3.83	3.84	3.85	3.86	3.87	3.88
	0.	**1.**	**2.**	**3.**	**4.**	**5.**	**6.**	**7.**	**8.**	**9.**

Centigrade Degrees.	BAROMETER: 675mm. (from 672.51 to 677.50).									
	Tenths of Degrees.									
	0.	**1.**	**2.**	**3.**	**4.**	**5.**	**6.**	**7.**	**8.**	**9.**
°	Millim.	Millim.	Millim.	Millim.	Millim.	Millim.	Millim.	Millim.	Millim.	Millim.
0	0.00	0.01	0.02	0.03	0.04	0.05	0.07	0.08	0.09	0.10
1	0.11	0.12	0.13	0.14	0.15	0.16	0.17	0.19	0.20	0.21
2	0.22	0.23	0.24	0.25	0.26	0.27	0.28	0.29	0.31	0.32
3	0.33	0.34	0.35	0.36	0.37	0.38	0.39	0.40	0.41	0.42
4	0.44	0.45	0.46	0.47	0.48	0.49	0.50	0.51	0.52	0.53
5	0.54	0.56	0.57	0.58	0.59	0.60	0.61	0.62	0.63	0.64
6	0.65	0.66	0.68	0.69	0.70	0.71	0.72	0.73	0.74	0.75
7	0.76	0.77	0.78	0.80	0.81	0.82	0.83	0.84	0.85	0.86
8	0.87	0.88	0.89	0.90	0.92	0.93	0.94	0.95	0.96	0.97
9	0.98	0.99	1.00	1.01	1.02	1.03	1.05	1.06	1.07	1.08
10	1.09	1.10	1.11	1.12	1.13	1.14	1.15	1.17	1.18	1.19
11	1.20	1.21	1.22	1.23	1.24	1.25	1.26	1.27	1.29	1.30
12	1.31	1.32	1.33	1.34	1.35	1.36	1.37	1.38	1.39	1.41
13	1.42	1.43	1.44	1.45	1.46	1.47	1.48	1.49	1.50	1.51
14	1.53	1.54	1.55	1.56	1.57	1.58	1.59	1.60	1.61	1.62
15	1.63	1.65	1.66	1.67	1.68	1.69	1.70	1.71	1.72	1.73
16	1.74	1.75	1.76	1.78	1.79	1.80	1.81	1.82	1.83	1.84
17	1.85	1.86	1.87	1.88	1.90	1.91	1.92	1.93	1.94	1.95
18	1.96	1.97	1.98	1.99	2.00	2.02	2.03	2.04	2.05	2.06
19	2.07	2.08	2.09	2.10	2.11	2.12	2.14	2.15	2.16	2.17
20	2.18	2.19	2.20	2.21	2.22	2.23	2.24	2.26	2.27	2.28
21	2.29	2.30	2.31	2.32	2.33	2.34	2.35	2.36	2.38	2.39
22	2.40	2.41	2.42	2.43	2.44	2.45	2.46	2.47	2.48	2.49
23	2.51	2.52	2.53	2.54	2.55	2.56	2.57	2.58	2.59	2.60
24	2.61	2.63	2.64	2.65	2.66	2.67	2.68	2.69	2.70	2.71
25	2.72	2.73	2.75	2.76	2.77	2.78	2.79	2.80	2.81	2.82
26	2.83	2.84	2.85	2.87	2.88	2.89	2.90	2.91	2.92	2.93
27	2.94	2.95	2.96	2.97	2.99	3.00	3.01	3.02	3.03	3.04
28	3.05	3.06	3.07	3.08	3.09	3.10	3.12	3.13	3.14	3.15
29	3.16	3.17	3.18	3.19	3.20	3.21	3.22	3.24	3.25	3.26
30	3.27	3.28	3.29	3.30	3.31	3.32	3.33	3.34	3.36	3.37
31	3.38	3.39	3.40	3.41	3.42	3.43	3.44	3.45	3.46	3.48
32	3.49	3.50	3.51	3.52	3.53	3.54	3.55	3.56	3.57	3.58
33	3.60	3.61	3.62	3.63	3.64	3.65	3.66	3.67	3.68	3.69
34	3.70	3.72	3.73	3.74	3.75	3.76	3.77	3.78	3.79	3.80
35	3.81	3.82	3.83	3.85	3.86	3.87	3.88	3.89	3.90	3.91
	0.	**1.**	**2.**	**3.**	**4.**	**5.**	**6.**	**7.**	**8.**	**9.**

Centigrade Degrees.	BAROMETER : 680mm. (from 677.51 to 682.50).									
	Tenths of Degrees.									
	0.	1.	2.	3.	4.	5.	6.	7.	8.	9.
°	Millim.	Millim.	Millim.	Millim.	Millim.	Millim.	Millim.	Millim.	Millim.	Millim.
0	0.00	0.01	0.02	0.03	0.04	0.05	0.07	0.08	0.09	0.10
1	0.11	0.12	0.13	0.14	0.15	0.16	0.18	0.19	0.20	0.21
2	0.22	0.23	0.24	0.25	0.26	0.27	0.29	0.30	0.31	0.32
3	0.33	0.34	0.35	0.36	0.37	0.38	0.40	0.41	0.42	0.43
4	0.44	0.45	0.46	0.47	0.48	0.49	0.50	0.52	0.53	0.54
5	0.55	0.56	0.57	0.58	0.59	0.60	0.61	0.63	0.64	0.65
6	0.66	0.67	0.68	0.69	0.70	0.71	0.72	0.74	0.75	0.76
7	0.77	0.78	0.79	0.80	0.81	0.82	0.83	0.85	0.86	0.87
8	0.88	0.89	0.90	0.91	0.92	0.93	0.94	0.95	0.97	0.98
9	0.99	1.00	1.01	1.02	1.03	1.04	1.05	1.06	1.08	1.09
10	1.10	1.11	1.12	1.13	1.14	1.15	1.16	1.17	1.19	1.20
11	1.21	1.22	1.23	1.24	1.25	1.26	1.27	1.28	1.30	1.31
12	1.32	1.33	1.34	1.35	1.36	1.37	1.38	1.39	1.40	1.42
13	1.43	1.44	1.45	1.46	1.47	1.48	1.49	1.50	1.51	1.53
14	1.54	1.55	1.56	1.57	1.58	1.59	1.60	1.61	1.62	1.64
15	1.65	1.66	1.67	1.68	1.69	1.70	1.71	1.72	1.73	1.75
16	1.76	1.77	1.78	1.79	1.80	1.81	1.82	1.83	1.84	1.85
17	1.87	1.88	1.89	1.90	1.91	1.92	1.93	1.94	1.95	1.96
18	1.98	1.99	2.00	2.01	2.02	2.03	2.04	2.05	2.06	2.07
19	2.09	2.10	2.11	2.12	2.13	2.14	2.15	2.16	2.17	2.18
20	2.20	2.21	2.22	2.23	2.24	2.25	2.26	2.27	2.28	2.29
21	2.30	2.32	2.33	2.34	2.35	2.36	2.37	2.38	2.39	2.40
22	2.41	2.43	2.44	2.45	2.46	2.47	2.48	2.49	2.50	2.51
23	2.52	2.54	2.55	2.56	2.57	2.58	2.59	2.60	2.61	2.62
24	2.63	2.65	2.66	2.67	2.68	2.69	2.70	2.71	2.72	2.73
25	2.74	2.75	2.77	2.78	2.79	2.80	2.81	2.82	2.83	2.84
26	2.85	2.86	2.88	2.89	2.90	2.91	2.92	2.93	2.94	2.95
27	2.96	2.97	2.99	3.00	3.01	3.02	3.03	3.04	3.05	3.06
28	3.07	3.08	3.10	3.11	3.12	3.13	3.14	3.15	3.16	3.17
29	3.18	3.19	3.20	3.22	3.23	3.24	3.25	3.26	3.27	3.28
30	3.29	3.30	3.31	3.33	3.34	3.35	3.36	3.37	3.38	3.39
31	3.40	3.41	3.42	3.44	3.45	3.46	3.47	3.48	3.49	3.50
32	3.51	3.52	3.53	3.54	3.56	3.57	3.58	3.59	3.60	3.61
33	3.62	3.63	3.64	3.65	3.67	3.68	3.69	3.70	3.71	3.72
34	3.73	3.74	3.75	3.76	3.78	3.79	3.80	3.81	3.82	3.83
35	3.84	3.85	3.86	3.87	3.89	3.90	3.91	3.92	3.93	3.94
	0.	1.	2.	3.	4.	5.	6.	7.	8.	9.

BAROMETER: 685$^{mm.}$ (from 682.51 to 687.50).

Centigrade Degrees.	Tenths of Degrees.									
	0.	**1.**	**2.**	**3.**	**4.**	**5.**	**6.**	**7.**	**8.**	**9.**
°	Millim.	Millim.	Millim.	Millim.	Millim.	Millim.	Millim.	Millim.	Millim.	Millim.
0	0.00	0.01	0.02	0.03	0.04	0.06	0.07	0.08	0.09	0.10
1	0.11	0.12	0.13	0.14	0.15	0.17	0.18	0.19	0.20	0.21
2	0.22	0.23	0.24	0.25	0.27	0.28	0.29	0.30	0.31	0.32
3	0.33	0.34	0.35	0.36	0.38	0.39	0.40	0.41	0.42	0.43
4	0.44	0.45	0.46	0.48	0.49	0.50	0.51	0.52	0.53	0.54
5	0.55	0.56	0.57	0.59	0.60	0.61	0.62	0.63	0.64	0.65
6	0.66	0.67	0.69	0.70	0.71	0.72	0.73	0.74	0.75	0.76
7	0.77	0.78	0.80	0.81	0.82	0.83	0.84	0.85	0.86	0.87
8	0.88	0.90	0.91	0.92	0.93	0.94	0.95	0.96	0.97	0.98
9	1.00	1.01	1.02	1.03	1.04	1.05	1.06	1.07	1.08	1.09
10	1.11	1.12	1.13	1.14	1.15	1.16	1.17	1.18	1.19	1.21
11	1.22	1.23	1.24	1.25	1.26	1.27	1.28	1.29	1.30	1.32
12	1.33	1.34	1.35	1.36	1.37	1.38	1.39	1.40	1.42	1.43
13	1.44	1.45	1.46	1.47	1.48	1.49	1.50	1.51	1.53	1.54
14	1.55	1.56	1.57	1.58	1.59	1.60	1.61	1.63	1.64	1.65
15	1.66	1.67	1.68	1.69	1.70	1.71	1.72	1.74	1.75	1.76
16	1.77	1.78	1.79	1.80	1.81	1.82	1.84	1.85	1.86	1.87
17	1.88	1.89	1.90	1.91	1.92	1.93	1.95	1.96	1.97	1.98
18	1.99	2.00	2.01	2.02	2.03	2.05	2.06	2.07	2.08	2.09
19	2.10	2.11	2.12	2.13	2.14	2.16	2.17	2.18	2.19	2.20
20	2.21	2.22	2.23	2.24	2.26	2.27	2.28	2.29	2.30	2.31
21	2.32	2.33	2.34	2.35	2.37	2.38	2.39	2.40	2.41	2.42
22	2.43	2.44	2.45	2.47	2.48	2.49	2.50	2.51	2.52	2.53
23	2.54	2.55	2.56	2.58	2.59	2.60	2.61	2.62	2.63	2.64
24	2.65	2.66	2.68	2.69	2.70	2.71	2.72	2.73	2.74	2.75
25	2.76	2.78	2.79	2.80	2.81	2.82	2.83	2.84	2.85	2.86
26	2.87	2.89	2.90	2.91	2.92	2.93	2.94	2.95	2.96	2.97
27	2.99	3.00	3.01	3.02	3.03	3.04	3.05	3.06	3.07	3.08
28	3.10	3.11	3.12	3.13	3.14	3.15	3.16	3.17	3.18	3.20
29	3.21	3.22	3.23	3.24	3.25	3.26	3.27	3.28	3.29	3.31
30	3.32	3.33	3.34	3.35	3.36	3.37	3.38	3.39	3.41	3.42
31	3.43	3.44	3.45	3.46	3.47	3.48	3.49	3.50	3.52	3.53
32	3.54	3.55	3.56	3.57	3.58	3.59	3.60	3.62	3.63	3.64
33	3.65	3.66	3.67	3.68	3.69	3.70	3.71	3.73	3.74	3.75
34	3.76	3.77	3.78	3.79	3.80	3.81	3.83	3.84	3.85	3.86
35	3.87	3.88	3.89	3.90	3.91	3.92	3.94	3.95	3.96	3.97
	0.	**1.**	**2.**	**3.**	**4.**	**5.**	**6.**	**7.**	**8.**	**9.**

BAROMETER : 690mm. (from 687.51 to 692.50).

Centigrade Degrees.	Tenths of Degrees.									
	0.	**1.**	**2.**	**3.**	**4.**	**5.**	**6.**	**7.**	**8.**	**9.**
°	Millim.	Millim.	Millim.	Millim.	Millim.	Millim.	Millim.	Millim.	Millim.	Millim.
0	0.00	0.01	0.02	0.03	0.04	0.06	0.07	0.08	0.09	0.10
1	0.11	0.12	0.13	0.14	0.16	0.17	0.18	0.19	0.20	0.21
2	0.22	0.23	0.25	0.26	0.27	0.28	0.29	0.30	0.31	0.32
3	0.33	0.35	0.36	0.37	0.38	0.39	0.40	0.41	0.42	0.43
4	0.45	0.46	0.47	0.48	0.49	0.50	0.51	0.52	0.53	0.55
5	0.56	0.57	0.58	0.59	0.60	0.61	0.62	0.63	0.65	0.66
6	0.67	0.68	0.69	0.70	0.71	0.72	0.74	0.75	0.76	0.77
7	0.78	0.79	0.80	0.81	0.82	0.84	0.85	0.86	0.87	0.88
8	0.89	0.90	0.91	0.92	0.94	0.95	0.96	0.97	0.98	0.99
9	1.00	1.01	1.02	1.04	1.05	1.06	1.07	1.08	1.09	1.10
10	1.11	1.12	1.14	1.15	1.16	1.17	1.18	1.19	1.20	1.21
11	1.23	1.24	1.25	1.26	1.27	1.28	1.29	1.30	1.31	1.33
12	1.34	1.35	1.36	1.37	1.38	1.39	1.40	1.41	1.43	1.44
13	1.45	1.46	1.47	1.48	1.49	1.50	1.51	1.53	1.54	1.55
14	1.56	1.57	1.58	1.59	1.60	1.61	1.63	1.64	1.65	1.66
15	1.67	1.68	1.69	1.70	1.72	1.73	1.74	1.75	1.76	1.77
16	1.78	1.79	1.80	1.82	1.83	1.84	1.85	1.86	1.87	1.88
17	1.89	1.90	1.92	1.93	1.94	1.95	1.96	1.97	1.98	1.99
18	2.00	2.02	2.03	2.04	2.05	2.06	2.07	2.08	2.09	2.10
19	2.12	2.13	2.14	2.15	2.16	2.17	2.18	2.19	2.21	2.22
20	2.23	2.24	2.25	2.26	2.27	2.28	2.29	2.31	2.32	2.33
21	2.34	2.35	2.36	2.37	2.38	2.39	2.41	2.42	2.43	2.44
22	2.45	2.46	2.47	2.48	2.49	2.51	2.52	2.53	2.54	2.55
23	2.56	2.57	2.58	2.59	2.61	2.62	2.63	2.64	2.65	2.66
24	2.67	2.68	2.70	2.71	2.72	2.73	2.74	2.75	2.76	2.77
25	2.78	2.80	2.81	2.82	2.83	2.84	2.85	2.86	2.87	2.88
26	2.90	2.91	2.92	2.93	2.94	2.95	2.96	2.97	2.98	3.00
27	3.01	3.02	3.03	3.04	3.05	3.06	3.07	3.08	3.10	3.11
28	3.12	3.13	3.14	3.15	3.16	3.17	3.19	3.20	3.21	3.22
29	3.23	3.24	3.25	3.26	3.27	3.29	3.30	3.31	3.32	3.33
30	3.34	3.35	3.36	3.37	3.39	3.40	3.41	3.42	3.43	3.44
31	3.45	3.46	3.47	3.49	3.50	3.51	3.52	3.53	3.54	3.55
32	3.56	3.57	3.59	3.60	3.61	3.62	3.63	3.64	3.65	3.66
33	3.68	3.69	3.70	3.71	3.72	3.73	3.74	3.75	3.76	3.78
34	3.79	3.80	3.81	3.82	3.83	3.84	3.85	3.86	3.88	3.89
35	3.90	3.91	3.92	3.93	3.94	3.95	3.96	3.98	3.99	4.00
	0.	**1.**	**2.**	**3.**	**4.**	**5.**	**6.**	**7.**	**8.**	**9.**

BAROMETER : 695$^{mm.}$ (from 692.51 to 697.50).

Centigrade Degrees.	Tenths of Degrees.									
	0.	**1.**	**2.**	**3.**	**4.**	**5.**	**6.**	**7.**	**8.**	**9.**
°	Millim.	Millim.	Millim.	Millim.	Millim.	Millim.	Millim.	Millim.	Millim.	Millim.
0	0.00	0.01	0.02	0.03	0.04	0.06	0.07	0.08	0.09	0.10
1	0.11	0.12	0.13	0.15	0.16	0.17	0.18	0.19	0.20	0.21
2	0.22	0.24	0.25	0.26	0.27	0.28	0.29	0.30	0.31	0.33
3	0.34	0.35	0.36	0.37	0.38	0.39	0.40	0.42	0.43	0.44
4	0.45	0.46	0.47	0.48	0.49	0.50	0.52	0.53	0.54	0.55
5	0.56	0.57	0.58	0.59	0.61	0.62	0.63	0.64	0.65	0.66
6	0.67	0.68	0.70	0.71	0.72	0.73	0.74	0.75	0.76	0.77
7	0.79	0.80	0.81	0.82	0.83	0.84	0.85	0.86	0.87	0.89
8	0.90	0.91	0.92	0.93	0.94	0.95	0.96	0.98	0.99	1.00
9	1.01	1.02	1.03	1.04	1.05	1.07	1.08	1.09	1.10	1.11
10	1.12	1.13	1.14	1.16	1.17	1.18	1.19	1.20	1.21	1.22
11	1.23	1.25	1.26	1.27	1.28	1.29	1.30	1.31	1.32	1.33
12	1.35	1.36	1.37	1.38	1.39	1.40	1.41	1.42	1.44	1.45
13	1.46	1.47	1.48	1.49	1.50	1.51	1.52	1.54	1.55	1.56
14	1.57	1.58	1.59	1.60	1.61	1.63	1.64	1.65	1.66	1.67
15	1.68	1.69	1.71	1.72	1.73	1.74	1.75	1.76	1.77	1.78
16	1.79	1.81	1.82	1.83	1.84	1.85	1.86	1.87	1.88	1.90
17	1.91	1.92	1.93	1.94	1.95	1.96	1.97	1.99	2.00	2.01
18	2.02	2.03	2.04	2.05	2.06	2.08	2.09	2.10	2.11	2.12
19	2.13	2.14	2.15	2.16	2.18	2.19	2.20	2.21	2.22	2.23
20	2.24	2.25	2.27	2.28	2.29	2.30	2.31	2.32	2.33	2.34
21	2.36	2.37	2.38	2.39	2.40	2.41	2.42	2.43	2.45	2.46
22	2.47	2.48	2.49	2.50	2.51	2.52	2.53	2.55	2.56	2.57
23	2.58	2.59	2.60	2.61	2.62	2.64	2.65	2.66	2.67	2.68
24	2.69	2.70	2.71	2.73	2.74	2.75	2.76	2.77	2.78	2.79
25	2.80	2.82	2.83	2.84	2.85	2.86	2.87	2.88	2.89	2.91
26	2.92	2.93	2.94	2.95	2.96	2.97	2.98	3.00	3.01	3.02
27	3.03	3.04	3.05	3.06	3.07	3.08	3.10	3.11	3.12	3.13
28	3.14	3.15	3.16	3.17	3.19	3.20	3.21	3.22	3.23	3.24
29	3.25	3.26	3.28	3.29	3.30	3.31	3.32	3.33	3.34	3.35
30	3.37	3.38	3.39	3.40	3.41	3.42	3.43	3.44	3.45	3.47
31	3.48	3.49	3.50	3.51	3.52	3.53	3.54	3.56	3.57	3.58
32	3.59	3.60	3.61	3.62	3.63	3.65	3.66	3.67	3.68	3.69
33	3.70	3.71	3.72	3.74	3.75	3.76	3.77	3.78	3.79	3.80
34	3.81	3.83	3.84	3.85	3.86	3.87	3.88	3.89	3.90	3.91
35	3.93	3.94	3.95	3.96	3.97	3.98	3.99	4.00	4.02	4.03
	0.	**1.**	**2.**	**3.**	**4.**	**5.**	**6.**	**7.**	**8.**	**9.**

Centigrade Degrees.	BAROMETER: 700$^{mm.}$ (from 697.51 to 702.50).									
	Tenths of Degrees.									
	0.	**1.**	**2.**	**3.**	**4.**	**5.**	**6.**	**7.**	**8.**	**9.**
°	Millim.	Millim.	Millim.	Millim.	Millim.	Millim.	Millim.	Millim.	Millim.	Millim.
0	0.00	0.01	0.02	0.03	0.05	0.06	0.07	0.08	0.09	0.10
1	0.11	0.12	0.14	0.15	0.16	0.17	0.18	0.19	0.20	0.21
2	0.23	0.24	0.25	0.26	0.27	0.28	0.29	0.31	0.32	0.33
3	0.34	0.35	0.36	0.37	0.38	0.40	0.41	0.42	0.43	0.44
4	0.45	0.46	0.47	0.49	0.50	0.51	0.52	0.53	0.54	0.55
5	0.56	0.58	0.59	0.60	0.61	0.62	0.63	0.64	0.66	0.67
6	0.68	0.69	0.70	0.71	0.72	0.73	0.75	0.76	0.77	0.78
7	0.79	0.80	0.81	0.82	0.84	0.85	0.86	0.87	0.88	0.89
8	0.90	0.92	0.93	0.94	0.95	0.96	0.97	0.98	0.99	1.01
9	1.02	1.03	1.04	1.05	1.06	1.07	1.08	1.10	1.11	1.12
10	1.13	1.14	1.15	1.16	1.17	1.19	1.20	1.21	1.22	1.23
11	1.24	1.25	1.27	1.28	1.29	1.30	1.31	1.32	1.33	1.34
12	1.36	1.37	1.38	1.39	1.40	1.41	1.42	1.43	1.45	1.46
13	1.47	1.48	1.49	1.50	1.51	1.53	1.54	1.55	1.56	1.57
14	1.58	1.59	1.60	1.62	1.63	1.64	1.65	1.66	1.67	1.68
15	1.69	1.71	1.72	1.73	1.74	1.75	1.76	1.77	1.79	1.80
16	1.81	1.82	1.83	1.84	1.85	1.86	1.88	1.89	1.90	1.91
17	1.92	1.93	1.94	1.95	1.97	1.98	1.99	2.00	2.01	2.02
18	2.03	2.04	2.06	2.07	2.08	2.09	2.10	2.11	2.12	2.14
19	2.15	2.16	2.17	2.18	2.19	2.20	2.21	2.23	2.24	2.25
20	2.26	2.27	2.28	2.29	2.30	2.32	2.33	2.34	2.35	2.36
21	2.37	2.38	2.40	2.41	2.42	2.43	2.44	2.45	2.46	2.47
22	2.49	2.50	2.51	2.52	2.53	2.54	2.55	2.56	2.58	2.59
23	2.60	2.61	2.62	2.63	2.64	2.66	2.67	2.68	2.69	2.70
24	2.71	2.72	2.73	2.75	2.76	2.77	2.78	2.79	2.80	2.81
25	2.82	2.84	2.85	2.86	2.87	2.88	2.89	2.90	2.91	2.93
26	2.94	2.95	2.96	2.97	2.98	2.99	3.01	3.02	3.03	3.04
27	3.05	3.06	3.07	3.08	3.10	3.11	3.12	3.13	3.14	3.15
28	3.16	3.17	3.19	3.20	3.21	3.22	3.23	3.24	3.25	3.27
29	3.28	3.29	3.30	3.31	3.32	3.33	3.34	3.36	3.37	3.38
30	3.39	3.40	3.41	3.42	3.43	3.45	3.46	3.47	3.48	3.49
31	3.50	3.51	3.52	3.54	3.55	3.56	3.57	3.58	3.59	3.60
32	3.62	3.63	3.64	3.65	3.66	3.67	3.68	3.69	3.71	3.72
33	3.73	3.74	3.75	3.76	3.77	3.78	3.80	3.81	3.82	3.83
34	3.84	3.85	3.86	3.88	3.89	3.90	3.91	3.92	3.93	3.94
35	3.95	3.97	3.98	3.99	4.00	4.01	4.02	4.03	4.04	4.06
	0.	**1.**	**2.**	**3.**	**4.**	**5.**	**6.**	**7.**	**8.**	**9.**

Centigrade Degrees.	BAROMETER : 705mm. (from 702.51 to 707.50).									
	Tenths of Degrees.									
	0.	**1.**	**2.**	**3.**	**4.**	**5.**	**6.**	**7.**	**8.**	**9.**
°	Millim.	Millim.	Millim.	Millim.	Millim.	Millim.	Millim.	Millim.	Millim.	Millim.
0	0.00	0.01	0.02	0.03	0.05	0.06	0.07	0.08	0.09	0.10
1	0.11	0.13	0.14	0.15	0.16	0.17	0.18	0.19	0.20	0.22
2	0.23	0.24	0.25	0.26	0.27	0.28	0.30	0.31	0.32	0 33
3	0.34	0.35	0.36	0.38	0.39	0.40	0.41	0.42	0.43	0.44
4	0.46	0.47	0.48	0.49	0.50	0.51	0.52	0.53	0.55	0.56
5	0.57	0.58	0.59	0.60	0.61	0.63	0.64	0.65	0.66	0.67
6	0.68	0.69	0.71	0.72	0.73	0.74	0.75	0.76	0.77	0.79
7	0.80	0.81	0.82	0.83	0.84	0.85	0.86	0.88	0.89	0.90
8	0.91	0.92	0.93	0.94	0.96	0.97	0.98	0.99	1.00	1.01
9	1.02	1.04	1.05	1.06	1.07	1.08	1.09	1.10	1.12	1.13
10	1.14	1.15	1.16	1.17	1.18	1.19	1.21	1.22	1.23	1.24
11	1.25	1.26	1.27	1.29	1.30	1.31	1.32	1.33	1.34	1.35
12	1.37	1.38	1.39	1.40	1.41	1.42	1.43	1.45	1.46	1.47
13	1.48	1.49	1.50	1.51	1.52	1.54	1.55	1.56	1.57	1.58
14	1.59	1.60	1.62	1.63	1.64	1.65	1.66	1.67	1.68	1.70
15	1.71	1.72	1.73	1.74	1.75	1.76	1.78	1.79	1.80	1.81
16	1.82	1.83	1.84	1.85	1.87	1.88	1.89	1.90	1.91	1.92
17	1.93	1.95	1.96	1.97	1.98	1.99	2.00	2.01	2.03	2.04
18	2.05	2.06	2.07	2.08	2.09	2.11	2.12	2.13	2.14	2.15
19	2.16	2.17	2.18	2.20	2.21	2.22	2.23	2.24	2.25	2.26
20	2.28	2.29	2.30	2.31	2.32	2.33	2.34	2.36	2.37	2.38
21	2.39	2.40	2.41	2.42	2.44	2.45	2.46	2.47	2.48	2.49
22	2.50	2.51	2.53	2.54	2.55	2.56	2.57	2.58	2.59	2.61
23	2.62	2.63	2.64	2.65	2.66	2.67	2.69	2.70	2.71	2.72
24	2.73	2.74	2.75	2.77	2.78	2.79	2.80	2.81	2.82	2.83
25	2.84	2.86	2.87	2.88	2.89	2.90	2.91	2.92	2.94	2.95
26	2.96	2.97	2.98	2.99	3.00	3.02	3.03	3.04	3.05	3.06
27	3.07	3.08	3.10	3.11	3.12	3.13	3.14	3.15	3.16	3.17
28	3.19	3.20	3.21	3.22	3.23	3.24	3.25	3.27	3.28	3.29
29	3.30	3.31	3.32	3.33	3.35	3.36	3.37	3.38	3.39	3.40
30	3.41	3.42	3.44	3.45	3.46	3.47	3.48	3.49	3.50	3.52
31	3.53	3.54	3.55	3.56	3.57	3.58	3.60	3.61	3.62	3.63
32	3.64	3.65	3.66	3.68	3.69	3.70	3.71	3.72	3.73	3.74
33	3.75	3.77	3.78	3.79	3.80	3.81	3.82	3.83	3.85	3.86
34	3.87	3.88	3.89	3.90	3.91	3.93	3.94	3.95	3.96	3.97
35	3.98	3.99	4.01	4.02	4.03	4.04	4.05	4.06	4.07	4.08
	0.	**1.**	**2.**	**3.**	**4.**	**5.**	**6.**	**7.**	**8.**	**9.**

BAROMETER: 710mm. (from 707.51 to 712.50).

Centigrade Degrees.	Tenths of Degrees.									
	0.	**1.**	**2.**	**3.**	**4.**	**5.**	**6.**	**7.**	**8.**	**9.**
°	Millim.	Millim.	Millim.	Millim.	Millim.	Millim.	Millim.	Millim.	Millim.	Millim.
0	0.00	0.01	0.02	0.03	0.05	0.06	0.07	0.08	0.09	0.10
1	0.11	0.13	0.14	0.15	0.16	0.17	0.18	0.19	0.21	0.22
2	0.23	0.24	0.25	0.26	0.28	0.29	0.30	0.31	0.32	0.33
3	0.34	0.36	0.37	0.38	0.39	0.40	0.41	0.42	0.44	0.45
4	0.46	0.47	0.48	0.49	0.50	0.52	0.53	0.54	0.55	0.56
5	0.57	0.58	0.60	0.61	0.62	0.63	0.64	0.65	0.66	0.68
6	0.69	0.70	0.71	0.72	0.73	0.74	0.76	0.77	0.78	0.79
7	0.80	0.81	0.83	0.84	0.85	0.86	0.87	0.88	0.89	0.91
8	0.92	0.93	0.94	0.95	0.96	0.97	0.99	1.00	1.01	1.02
9	1.03	1.04	1.05	1.07	1.08	1.09	1.10	1.11	1.12	1.13
10	1.15	1.16	1.17	1.18	1.19	1.20	1.21	1.23	1.24	1.25
11	1.26	1.27	1.28	1.29	1.31	1.32	1.33	1.34	1.35	1.36
12	1.38	1.39	1.40	1.41	1.42	1.43	1.44	1.46	1.47	1.48
13	1.49	1.50	1.51	1.52	1.54	1.55	1.56	1.57	1.58	1.59
14	1.60	1.62	1.63	1.64	1.65	1.66	1.67	1.68	1.70	1.71
15	1.72	1.73	1.74	1.75	1.76	1.78	1.79	1.80	1.81	1.82
16	1.83	1.84	1.86	1.87	1.88	1.89	1.90	1.91	1.93	1.94
17	1.95	1.96	1.97	1.98	1.99	2.01	2.02	2.03	2.04	2.05
18	2.06	2.07	2.09	2.10	2.11	2.12	2.13	2.14	2.15	2.17
19	2.18	2.19	2.20	2.21	2.22	2.23	2.25	2.26	2.27	2.28
20	2.29	2.30	2.31	2.33	2.34	2.35	2.36	2.37	2.38	2.40
21	2.41	2.42	2.43	2.44	2.45	2.46	2.48	2.49	2.50	2.51
22	2.52	2.53	2.54	2.56	2.57	2.58	2.59	2.60	2.61	2.62
23	2.64	2.65	2.66	2.67	2.68	2.69	2.70	2.72	2.73	2.74
24	2.75	2.76	2.77	2.78	2.80	2.81	2.82	2.83	2.84	2.85
25	2.86	2.88	2.89	2.90	2.91	2.92	2.93	2.95	2.96	2.97
26	2.98	2.99	3.00	3.01	3.03	3.04	3.05	3.06	3.07	3.08
27	3.09	3.11	3.12	3.13	3.14	3.15	3.16	3.17	3.19	3.20
28	3.21	3.22	3.23	3.24	3.25	3.27	3.28	3.29	3.30	3.31
29	3.32	3.33	3.35	3.36	3.37	3.38	3.39	3.40	3.41	3.43
30	3.44	3.45	3.46	3.47	3.48	3.50	3.51	3.52	3.53	3.54
31	3.55	3.56	3.58	3.59	3.60	3.61	3.62	3.63	3.64	3.66
32	3.67	3.68	3.69	3.70	3.71	3.72	3.74	3.75	3.76	3.77
33	3.78	3.79	3.80	3.82	3.83	3.84	3.85	3.86	3.87	3.88
34	3.90	3.91	3.92	3.93	3.94	3.95	3.96	3.98	3.99	4.00
35	4.01	4.02	4.03	4.05	4.06	4.07	4.08	4.09	4.10	4.11
	0.	**1.**	**2.**	**3.**	**4.**	**5.**	**6.**	**7.**	**8.**	**9.**

Centigrade Degrees.	BAROMETER : 715mm. (from 712.51 to 717.50).									
	Tenths of Degrees.									
	0.	1.	2.	3.	4.	5.	6.	7.	8.	9.
°	Millim.	Millim.	Millim.	Millim.	Millim.	Millim.	Millim.	Millim.	Millim.	Millim.
0	0.00	0.01	0.02	0.04	0.05	0.06	0.07	0.08	0.09	0.10
1	0.12	0.13	0.14	0.15	0.16	0.17	0.18	0.20	0.21	0.22
2	0.23	0.24	0.25	0.27	0.28	0.29	0.30	0.31	0.32	0.33
3	0.35	0.36	0.37	0.38	0.39	0.40	0.42	0.43	0.44	0.45
4	0.46	0.47	0.48	0.50	0.51	0.52	0.53	0.54	0.55	0.57
5	0.58	0.59	0.60	0.61	0.62	0.63	0.65	0.66	0.67	0.68
6	0.69	0.70	0.72	0.73	0.74	0.75	0.76	0.77	0.78	0.80
7	0.81	0.82	0.83	0.84	0.85	0.87	0.88	0.89	0.90	0.91
8	0.92	0.93	0.95	0.96	0.97	0.98	0.99	1.00	1.02	1.03
9	1.04	1.05	1.06	1.07	1.08	1.10	1.11	1.12	1.13	1.14
10	1.15	1.17	1.18	1.19	1.20	1.21	1.22	1.23	1.25	1.26
11	1.27	1.28	1.29	1.30	1.32	1.33	1.34	1.35	1.36	1.37
12	1.38	1.40	1.41	1.42	1.43	1.44	1.45	1.47	1.48	1.49
13	1.50	1.51	1.52	1.53	1.55	1.56	1.57	1.58	1.59	1.60
14	1.62	1.63	1.64	1.65	1.66	1.67	1.68	1.70	1.71	1.72
15	1.73	1.74	1.75	1.77	1.78	1.79	1.80	1.81	1.82	1.83
16	1.85	1.86	1.87	1.88	1.89	1.90	1.92	1.93	1.94	1.95
17	1.96	1.97	1.98	2.00	2.01	2.02	2.03	2.04	2.05	2.07
18	2.08	2.09	2.10	2.11	2.12	2.13	2.15	2.16	2.17	2.18
19	2.19	2.20	2.22	2.23	2.24	2.25	2.26	2.27	2.28	2.30
20	2.31	2.32	2.33	2.34	2.35	2.37	2.38	2.39	2.40	2.41
21	2.42	2.43	2.45	2.46	2.47	2.48	2.49	2.50	2.52	2.53
22	2.54	2.55	2.56	2.57	2.58	2.60	2.61	2.62	2.63	2.64
23	2.65	2.67	2.68	2.69	2.70	2.71	2.72	2.74	2.75	2.76
24	2.77	2.78	2.79	2.80	2.82	2.83	2.84	2.85	2.86	2.87
25	2.89	2.90	2.91	2.92	2.93	2.94	2.95	2.97	2.98	2.99
26	3.00	3.01	3.02	3.04	3.05	3.06	3.07	3.08	3.09	3.10
27	3.12	3.13	3.14	3.15	3.16	3.17	3.19	3.20	3.21	3.22
28	3.23	3.24	3.26	3.27	3.28	3.29	3.30	3.31	3.32	3.34
29	3.35	3.36	3.37	3.38	3.39	3.40	3.42	3.43	3.44	3.45
30	3.46	3.47	3.49	3.50	3.51	3.52	3.53	3.54	3.55	3.57
31	3.58	3.59	3.60	3.61	3.62	3.64	3.65	3.66	3.67	3.68
32	3.69	3.70	3.72	3.73	3.74	3.75	3.76	3.77	3.79	3.80
33	3.81	3.82	3.83	3.84	3.85	3.87	3.88	3.89	3.90	3.91
34	3.92	3.94	3.95	3.96	3.97	3.98	3.99	4.00	4.02	4.03
35	4.04	4.05	4.06	4.07	4.09	4.10	4.11	4.12	4.13	4.14
	0.	1.	2.	3.	4.	5.	6.	7.	8.	9.

BAROMETER: 720mm. (from 717.51 to 722.50).

Centigrade Degrees.	Tenths of Degrees. 0.	1.	2.	3.	4.	5.	6.	7.	8.	9.
°	Millim.	Millim.	Millim.	Millim.	Millim.	Millim.	Millim.	Millim.	Millim.	Millim.
0	0.00	0.01	0.02	0.03	0.05	0.06	0.07	0.08	0.09	0.10
1	0.12	0.13	0.14	0.15	0.16	0.17	0.19	0.20	0.21	0.22
2	0.23	0.24	0.26	0.27	0.28	0.29	0.30	0.31	0.33	0.34
3	0.35	0.36	0.37	0.38	0.40	0.41	0.42	0.43	0.44	0.45
4	0.46	0.48	0.49	0.50	0.51	0.52	0.53	0.55	0.56	0.57
5	0.58	0.59	0.60	0.62	0.63	0.64	0.65	0.66	0.67	0.69
6	0.70	0.71	0.72	0.73	0.74	0.76	0.77	0.78	0.79	0.80
7	0.81	0.83	0.84	0.85	0.86	0.87	0.88	0.89	0.91	0.92
8	0.93	0.94	0.95	0.96	0.98	0.99	1.00	1.01	1.02	1.03
9	1.05	1.06	1.07	1.08	1.09	1.10	1.12	1.13	1.14	1.15
10	1.16	1.17	1.19	1.20	1.21	1.22	1.23	1.24	1.26	1.27
11	1.28	1.29	1.30	1.31	1.32	1.34	1.35	1.36	1.37	1.38
12	1.39	1.41	1.42	1.43	1.44	1.45	1.46	1.48	1.49	1.50
13	1.51	1.52	1.53	1.55	1.56	1.57	1.58	1.59	1.60	1.62
14	1.63	1.64	1.65	1.66	1.67	1.69	1.70	1.71	1.72	1.73
15	1.74	1.75	1.77	1.78	1.79	1.80	1.81	1.82	1.84	1.85
16	1.86	1.87	1.88	1.89	1.91	1.92	1.93	1.94	1.95	1.96
17	1.98	1.99	2.00	2.01	2.02	2.03	2.05	2.06	2.07	2.08
18	2.09	2.10	2.11	2.13	2.14	2.15	2.16	2.17	2.18	2.20
19	2.21	2.22	2.23	2.24	2.25	2.27	2.28	2.29	2.30	2.31
20	2.32	2.34	2.35	2.36	2.37	2.38	2.39	2.41	2.42	2.43
21	2.44	2.45	2.46	2.48	2.49	2.50	2.51	2.52	2.53	2.54
22	2.56	2.57	2.58	2.59	2.60	2.61	2.63	2.64	2.65	2.66
23	2.67	2.68	2.70	2.71	2.72	2.73	2.74	2.75	2.77	2.78
24	2.79	2.80	2.81	2.82	2.84	2.85	2.86	2.87	2.88	2.89
25	2.91	2.92	2.93	2.94	2.95	2.96	2.97	2.99	3.00	3.01
26	3.02	3.03	3.04	3.06	3.07	3.08	3.09	3.10	3.11	3.13
27	3.14	3.15	3.16	3.17	3.18	3.20	3.21	3.22	3.23	3.24
28	3.25	3.27	3.28	3.29	3.30	3.31	3.32	3.34	3.35	3.36
29	3.37	3.38	3.39	3.40	3.42	3.43	3.44	3.45	3.46	3.47
30	3.49	3.50	3.51	3.52	3.53	3.54	3.56	3.57	3.58	3.59
31	3.60.	3.61	3.63	3.64	3.65	3.66	3.67	3.68	3.70	3.71
32	3.72	3.73	3.74	3.75	3.77	3.78	3.79	3.80	3.81	3.82
33	3.83	3.85	3.86	3.87	3.88	3.89	3.90	3.92	3.93	3.94
34	3.95	3.96	3.97	3.99	4.00	4.01	4.02	4.03	4.04	4.06
35	4.07	4.08	4.09	4.10	4.11	4.13	4.14	4.15	4.16	4.17
	0.	1.	2.	3.	4.	5.	6.	7.	8.	9.

Centigrade Degrees.	BAROMETER : 725mm. (from 722.51 to 727.50).									
	Tenths of Degrees.									
	0.	**1.**	**2.**	**3.**	**4.**	**5.**	**6.**	**7.**	**8.**	**9.**
°	Millim.	Millim.	Millim.	Millim.	Millim.	Millim.	Millim.	Millim.	Millim.	Millim.
0	0.00	0.01	0.02	0.04	0.05	0.06	0.07	0.08	0.09	0.11
1	0.12	0.13	0.14	0.15	0.16	0.18	0.19	0.20	0.21	0.22
2	0.23	0.25	0.26	0.27	0.28	0.29	0.30	0.32	0.33	0.34
3	0.35	0.36	0.37	0.39	0.40	0.41	0.42	0.43	0.44	0.46
4	0.47	0.48	0.49	0.50	0.51	0.53	0.54	0.55	0.56	0.57
5	0.59	0.60	0.61	0.62	0.63	0.64	0.66	0.67	0.68	0.69
6	0.70	0.71	0.73	0.74	0.75	0.76	0.77	0.78	0.80	0.81
7	0.82	0.83	0.84	0.85	0.87	0.88	0.89	0.90	0.91	0.92
8	0.94	0.95	0.96	0.97	0.98	0.99	1.01	1.02	1.03	1.04
9	1.05	1.06	1.08	1.09	1.10	1.11	1.12	1.14	1.15	1.16
10	1.17	1.18	1.19	1.21	1.22	1.23	1.24	1.25	1.26	1.28
11	1.29	1.30	1.31	1.32	1.33	1.35	1.36	1.37	1.38	1.39
12	1.40	1.42	1.43	1.44	1.45	1.46	1.47	1.49	1.50	1.51
13	1.52	1.53	1.54	1.56	1.57	1.58	1.59	1.60	1.61	1.63
14	1.64	1.65	1.66	1.67	1.69	1.70	1.71	1.72	1.73	1.74
15	1.76	1.77	1.78	1.79	1.80	1.81	1.83	1.84	1.85	1.86
16	1.87	1.88	1.90	1.91	1.92	1.93	1.94	1.95	1.97	1.98
17	1.99	2.00	2.01	2.02	2.04	2.05	2.06	2.07	2.08	2.09
18	2.11	2.12	2.13	2.14	2.15	2.16	2.18	2.19	2.20	2.21
19	2.22	2.23	2.25	2.26	2.27	2.28	2.29	2.31	2.32	2.33
20	2.34	2.35	2.36	2.38	2.39	2.40	2.41	2.42	2.43	2.45
21	2.46	2.47	2.48	2.49	2.50	2.52	2.53	2.54	2.55	2.56
22	2.57	2.59	2.60	2.61	2.62	2.63	2.64	2.66	2.67	2.68
23	2.69	2.70	2.71	2.73	2.74	2.75	2.76	2.77	2.78	2.80
24	2.81	2.82	2.83	2.84	2.86	2.87	2.88	2.89	2.90	2.91
25	2.93	2.94	2.95	2.96	2.97	2.98	3.00	3.01	3.02	3.03
26	3.04	3.05	3.07	3.08	3.09	3.10	3.11	3.12	3.14	3.15
27	3.16	3.17	3.18	3.19	3.21	3.22	3.23	3.24	3.25	3.26
28	3.28	3.29	3.30	3.31	3.32	3.33	3.35	3.36	3.37	3.38
29	3.39	3.41	3.42	3.43	3.44	3.45	3.46	3.48	3.49	3.50
30	3.51	3.52	3.53	3.55	3.56	3.57	3.58	3.59	3.60	3.62
31	3.63	3.64	3.65	3.66	3.67	3.69	3.70	3.71	3.72	3.73
32	3.74	3.76	3.77	3.78	3.79	3.80	3.81	3.83	3.84	3.85
33	3.86	3.87	3.88	3.90	3.91	3.92	3.93	3.94	3.96	3.97
34	3.98	3.99	4.00	4.01	4.03	4.04	4.05	4.06	4.07	4.08
35	4.10	4.11	4.12	4.13	4.14	4.15	4.17	4.18	4.19	4.20
	0.	**1.**	**2.**	**3.**	**4.**	**5.**	**6.**	**7.**	**8.**	**9.**

BAROMETER : $730^{mm.}$ (from 727.51 to 732.50).

Centigrade Degrees.	Tenths of Degrees.									
	0.	**1.**	**2.**	**3.**	**4.**	**5.**	**6.**	**7.**	**8.**	**9.**
°	Millim.	Millim.	Millim.	Millim.	Millim.	Millim.	Millim.	Millim.	Millim.	Millim.
0	0.00	0.01	0.02	0.04	0.05	0.06	0.07	0.08	0.09	0.11
1	0.12	0.13	0.14	0.15	0.16	0.18	0.19	0.20	0.21	0.22
2	0.24	0.25	0.26	0.27	0.28	0.29	0.31	0.32	0.33	0.34
3	0.35	0.37	0.38	0.39	0.40	0.41	0.42	0.44	0.45	0.46
4	0.47	0.48	0.49	0.51	0.52	0.53	0.54	0.55	0.57	0.58
5	0.59	0.60	0.61	0.62	0.64	0.65	0.66	0.67	0.68	0.70
6	0.71	0.72	0.73	0.74	0.75	0.77	0.78	0.79	0.80	0.81
7	0.82	0.84	0.85	0.86	0.87	0.88	0.90	0.91	0.92	0.93
8	0.94	0.95	0.97	0.98	0.99	1.00	1.01	1.03	1.04	1.05
9	1.06	1.07	1.08	1.10	1.11	1.12	1.13	1.14	1.15	1.17
10	1.18	1.19	1.20	1.21	1.23	1.24	1.25	1.26	1.27	1.28
11	1.30	1.31	1.32	1.33	1.34	1.35	1.37	1.38	1.39	1.40
12	1.41	1.43	1.44	1.45	1.46	1.47	1.48	1.50	1.51	1.52
13	1.53	1.54	1.56	1.57	1.58	1.59	1.60	1.61	1.63	1.64
14	1.65	1.66	1.67	1.68	1.70	1.71	1.72	1.73	1.74	1.76
15	1.77	1.78	1.79	1.80	1.81	1.83	1.84	1.85	1.86	1.87
16	1.89	1.90	1.91	1.92	1.93	1.94	1.96	1.97	1.98	1.99
17	2.00	2.01	2.03	2.04	2.05	2.06	2.07	2.09	2.10	2.11
18	2.12	2.13	2.14	2.16	2.17	2.18	2.19	2.20	2.22	2.23
19	2.24	2.25	2.26	2.27	2.29	2.30	2.31	2.32	2.33	2.34
20	2.36	2.37	2.38	2.39	2.40	2.42	2.43	2.44	2.45	2.46
21	2.47	2.49	2.50	2.51	2.52	2.53	2.54	2.56	2.57	2.58
22	2.59	2.60	2.62	2.63	2.64	2.65	2.66	2.67	2.69	2.70
23	2.71	2.72	2.73	2.75	2.76	2.77	2.78	2.79	2.80	2.82
24	2.83	2.84	2.85	2.86	2.87	2.89	2.90	2.91	2.92	2.93
25	2.95	2.96	2.97	2.98	2.99	3.01	3.02	3.03	3.04	3.05
26	3.06	3.08	3.09	3.10	3.11	3.12	3.13	3.15	3.16	3.17
27	3.18	3.19	3.20	3.22	3.23	3.24	3.25	3.26	3.28	3.29
28	3.30	3.31	3.32	3.33	3.35	3.36	3.37	3.38	3.39	3.41
29	3.42	3.43	3.44	3.45	3.46	3.48	3.49	3.50	3.51	3.52
30	3.53	3.55	3.56	3.57	3.58	3.59	3.61	3.62	3.63	3.64
31	3.65	3.66	3.68	3.69	3.70	3.71	3.72	3.73	3.75	3.76
32	3.77	3.78	3.79	3.81	3.82	3.83	3.84	3.85	3.86	3.88
33	3.89	3.90	3.91	3.92	3.94	3.95	3.96	3.97	3.98	3.99
34	4.01	4.02	4.03	4.04	4.05	4.06	4.07	4.09	4.10	4.11
35	4.12	4.14	4.15	4.16	4.17	4.18	4.19	4.21	4.22	4.23
	0.	**1.**	**2.**	**3.**	**4.**	**5.**	**6.**	**7.**	**8.**	**9.**

BAROMETER: 735mm. (from 732.51 to 737.50).

Centigrade Degrees.	Tenths of Degrees.									
	0.	**1.**	**2.**	**3.**	**4.**	**5.**	**6.**	**7.**	**8.**	**9.**
°	Millim.	Millim.	Millim.	Millim.	Millim.	Millim.	Millim.	Millim.	Millim.	Millim.
0	0.00	0.01	0.02	0.04	0.05	0.06	0.07	0.08	0.09	0.11
1	0.12	0.13	0.14	0.15	0.17	0.18	0.19	0.20	0.21	0.23
2	0.24	0.25	0.26	0.27	0.28	0.30	0.31	0.32	0.33	0.34
3	0.36	0.37	0.38	0.39	0.40	0.42	0.43	0.44	0.45	0.46
4	0.47	0.49	0.50	0.51	0.52	0.53	0.55	0.56	0.57	0.58
5	0.59	0.61	0.62	0.63	0.64	0.65	0.66	0.68	0.69	0.70
6	0.71	0.72	0.74	0.75	0.76	0.77	0.78	0.79	0.81	0.82
7	0.83	0.84	0.85	0.87	0.88	0.89	0.90	0.91	0.93	0.94
8	0.95	0.96	0.97	0.98	1.00	1.01	1.02	1.03	1.04	1.06
9	1.07	1.08	1.09	1.10	1.12	1.13	1.14	1.15	1.16	1.17
10	1.19	1.20	1.21	1.22	1.23	1.25	1.26	1.27	1.28	1.29
11	1.30	1.32	1.33	1.34	1.35	1.36	1.37	1.39	1.40	1.41
12	1.42	1.44	1.45	1.46	1.47	1.48	1.49	1.51	1.52	1.53
13	1.54	1.55	1.57	1.58	1.59	1.60	1.61	1.63	1.64	1.65
14	1.66	1.67	1.69	1.70	1.71	1.72	1.73	1.74	1.76	1.77
15	1.78	1.79	1.80	1.82	1.83	1.84	1.85	1.86	1.87	1.89
16	1.90	1.91	1.92	1.93	1.95	1.96	1.97	1.98	1.99	2.00
17	2.02	2.03	2.04	2.05	2.06	2.08	2.09	2.10	2.11	2.12
18	2.14	2.15	2.16	2.17	2.18	2.19	2.21	2.22	2.23	2.24
19	2.25	2.27	2.28	2.29	2.30	2.31	2.33	2.34	2.35	2.36
20	2.37	2.38	2.40	2.41	2.42	2.43	2.44	2.46	2.47	2.48
21	2.49	2.50	2.51	2.53	2.54	2.55	2.56	2.57	2.59	2.60
22	2.61	2.62	2.63	2.65	2.66	2.67	2.68	2.69	2.70	2.72
23	2.73	2.74	2.75	2.76	2.78	2.79	2.80	2.81	2.82	2.84
24	2.85	2.86	2.87	2.88	2.89	2.91	2.92	2.93	2.94	2.95
25	2.97	2.98	2.99	3.00	3.01	3.03	3.04	3.05	3.06	3.07
26	3.08	3.10	3.11	3.12	3.13	3.14	3.16	3.17	3.18	3.19
27	3.20	3.21	3.23	3.24	3.25	3.26	3.27	3.29	3.30	3.31
28	3.32	3.33	3.35	3.36	3.37	3.38	3.39	3.40	3.42	3.43
29	3.44	3.45	3.46	3.48	3.49	3.50	3.51	3.52	3.54	3.55
30	3.56	3.57	3.58	3.59	3.61	3.62	3.63	3.64	3.65	3.67
31	3.68	3.69	3.70	3.71	3.72	3.74	3.75	3.76	3.77	3.78
32	3.80	3.81	3.82	3.83	3.84	3.86	3.87	3.88	3.89	3.90
33	3.91	3.93	3.94	3.95	3.96	3.97	3.99	4.00	4.01	4.02
34	4.03	3.05	4.06	4.07	4.08	4.09	4.10	4.12	4.13	4.14
35	4.15	4.16	4.18	4.19	4.20	4.21	4.22	4.24	4.25	4.26
	0.	**1.**	**2.**	**3.**	**4.**	**5.**	**6.**	**7.**	**8.**	**9.**

BAROMETER: 740mm. (from 737.51 to 742.50).

Centigrade Degrees.	Tenths of Degrees. 0.	1.	2.	3.	4.	5.	6.	7.	8.	9.
°	Millim.	Millim.	Millim.	Millim.	Millim.	Millim.	Millim.	Millim.	Millim.	Millim.
0	0.00	0.01	0.02	0.04	0.05	0.06	0.07	0.08	0.09	0.11
1	0.12	0.13	0.14	0.16	0.17	0.18	0.19	0.20	0.21	0.23
2	0.24	0.25	0.26	0.27	0.29	0.30	0.31	0.32	0.33	0.35
3	0.36	0.37	0.38	0.39	0.41	0.42	0.43	0.44	0.45	0.47
4	0.48	0.49	0.50	0.51	0.53	0.54	0.55	0.56	0.57	0.59
5	0.60	0.61	0.62	0.63	0.64	0.66	0.67	0.68	0.69	0.70
6	0.72	0.73	0.74	0.75	0.76	0.78	0.79	0.80	0.81	0.82
7	0.84	0.85	0.86	0.87	0.88	0.90	0.91	0.92	0.93	0.94
8	0.96	0.97	0.98	0.99	1.00	1.02	1.03	1.04	1.05	1.06
9	1.07	1.09	1.10	1.11	1.12	1.13	1.15	1.16	1.17	1.18
10	1.19	1.21	1.22	1.23	1.24	1.25	1.27	1.28	1.29	1.30
11	1.31	1.33	1.34	1.35	1.36	1.37	1.39	1.40	1.41	1.42
12	1.43	1.45	1.46	1.47	1.48	1.49	1.50	1.52	1.53	1.54
13	1.55	1.56	1.58	1.59	1.60	1.61	1.62	1.64	1.65	1.66
14	1.67	1.68	1.70	1.71	1.72	1.73	1.74	1.76	1.77	1.78
15	1.79	1.80	1.82	1.83	1.84	1.85	1.86	1.88	1.89	1.90
16	1.91	1.92	1.93	1.95	1.96	1.97	1.98	1.99	2.01	2.02
17	2.03	2.04	2.05	2.07	2.08	2.09	2.10	2.11	2.13	2.14
18	2.15	2.16	2.17	2.19	2.20	2.21	2.22	2.23	2.25	2.26
19	2.27	2.28	2.29	2.31	2.32	2.33	2.34	2.35	2.36	2.38
20	2.39	2.40	2.41	2.42	2.44	2.45	2.46	2.47	2.48	2.50
21	2.51	2.52	2.53	2.54	2.56	2.57	2.58	2.59	2.60	2.62
22	2.63	2.64	2.65	2.66	2.68	2.69	2.70	2.71	2.72	2.74
23	2.75	2.76	2.77	2.78	2.79	2.81	2.82	2.83	2.84	2.85
24	2.87	2.88	2.89	2.90	2.91	2.93	2.94	2.95	2.96	2.97
25	2.99	3.00	3.01	3.02	3.03	3.05	3.06	3.07	3.08	3.09
26	3.11	3.12	3.13	3.14	3.15	3.17	3.18	3.19	3.20	3.21
27	3.22	3.24	3.25	3.26	3.27	3.28	3.30	3.31	3.32	3.33
28	3.34	3.36	3.37	3.38	3.39	3.40	3.42	3.43	3.44	3.45
29	3.46	3.48	3.49	3.50	3.51	3.52	3.54	3.55	3.56	3.57
30	3.58	3.60	3.61	3.62	3.63	3.64	3.65	3.67	3.68	3.69
31	3.70	3.71	3.73	3.74	3.75	3.76	3.77	3.79	3.80	3.81
32	3.82	3.83	3.85	3.86	3.87	3.88	3.89	3.91	3.92	3.93
33	3.94	3.95	3.97	3.98	3.99	4.00	4.01	4.02	4.04	4.05
34	4.06	4.07	4.08	4.10	4.11	4.12	4.13	4.14	4.16	4.17
35	4.18	4.19	4.20	4.22	4.23	4.24	4.25	4.26	4.28	4.29
	0.	1.	2.	3.	4.	5.	6.	7.	8.	9.

BAROMETER : 745mm. (from 742.51 to 747.50).

Centigrade Degrees.	Tenths of Degrees.									
	0.	1.	2.	3.	4.	5.	6.	7.	8.	9.
°	Millim.	Millim.	Millim.	Millim.	Millim.	Millim.	Millim.	Millim.	Millim.	Millim.
0	0.00	0.01	0.02	0.04	0.05	0.06	0.07	0.08	0.10	0.11
1	0.12	0.13	0.14	0.16	0.17	0.18	0.19	0.20	0.22	0.23
2	0.24	0.25	0.26	0.28	0.29	0.30	0.31	0.32	0.34	0.35
3	0.36	0.37	0.38	0.40	0.41	0.42	0.43	0.44	0.46	0.47
4	0.48	0.49	0.51	0.52	0.53	0.54	0.55	0.57	0.58	0.59
5	0.60	0.61	0.63	0.64	0.65	0.66	0.67	0.69	0.70	0.71
6	0.72	0.73	0.75	0.76	0.77	0.78	0.79	0.81	0.82	0.83
7	0.84	0.85	0.87	0.88	0.89	0.90	0.91	0.93	0.94	0.95
8	0.96	0.97	0.99	1.00	1.01	1.02	1.03	1.05	1.06	1.07
9	1.08	1.09	1.11	1.12	1.13	1.14	1.15	1.17	1.18	1.19
10	1.20	1.21	1.23	1.24	1.25	1.26	1.27	1.29	1.30	1.31
11	1.32	1.33	1.35	1.36	1.37	1.38	1.39	1.41	1.42	1.43
12	1.44	1.45	1.47	1.48	1.49	1.50	1.52	1.53	1.54	1.55
13	1.56	1.58	1.59	1.60	1.61	1.62	1.64	1.65	1.66	1.67
14	1.68	1.70	1.71	1.72	1.73	1.74	1.76	1.77	1.78	1.79
15	1.80	1.82	1.83	1.84	1.85	1.86	1.88	1.89	1.90	1.91
16	1.92	1.94	1.95	1.96	1.97	1.98	2.00	2.01	2.02	2.03
17	2.04	2.06	2.07	2.08	2.09	2.10	2.12	2.13	2.14	2.15
18	2.16	2.18	2.19	2.20	2.21	2.22	2.24	2.25	2.26	2.27
19	2.28	2.30	2.31	2.32	2.33	2.34	2.36	2.37	2.38	2.39
20	2.40	2.42	2.43	2.44	2.45	2.46	2.48	2.49	2.50	2.51
21	2.53	2.54	2.55	2.56	2.57	2.59	2.60	2.61	2.62	2.63
22	2.65	2.66	2.67	2.68	2.69	2.71	2.72	2.73	2.74	2.75
23	2.77	2.78	2.79	2.80	2.81	2.83	2.84	2.85	2.86	2.87
24	2.89	2.90	2.91	2.92	2.93	2.95	2.96	2.97	2.98	2.99
25	3.01	3.02	3.03	3.04	3.05	3.07	3.08	3.09	3.10	3.11
26	3.13	3.14	3.15	3.16	3.17	3.19	3.20	3.21	3.22	3.23
27	3.25	3.26	3.27	3.28	3.29	3.31	3.32	3.33	3.34	3.35
28	3.37	3.38	3.39	3.40	3.41	3.43	3.44	3.45	3.46	3.48
29	3.49	3.50	3.51	3.52	3.54	3.55	3.56	3.57	3.58	3.60
30	3.61	3.62	3.63	3.64	3.66	3.67	3.68	3.69	3.70	3.72
31	3.73	3.74	3.75	3.76	3.78	3.79	3.80	3.81	3.82	3.84
32	3.85	3.86	3.87	3.88	3.90	3.91	3.92	3.93	3.94	3.96
33	3.97	3.98	3.99	4.00	4.02	4.03	4.04	4.05	4.06	4.08
34	4.09	4.10	4.11	4.12	4.14	4.15	4.16	4.17	4.18	4.20
35	4.21	4.22	4.23	4.24	4.26	4.27	4.28	4.29	4.30	4.32
	0.	1.	2.	3.	4.	5.	6.	7.	8.	9.

BAROMETER: 750$^{mm.}$ (from 747.51 to 752.50).

Centigrade Degrees.	Tenths of Degrees.									
	0.	**1.**	**2.**	**3.**	**4.**	**5.**	**6.**	**7.**	**8.**	**9.**
°	Millim.	Millim.	Millim.	Millim.	Millim.	Millim.	Millim.	Millim.	Millim.	Millim.
0	0.00	0.01	0.02	0.04	0.05	0.06	0.07	0.08	0.10	0.11
1	0.12	0.13	0.15	0.16	0.17	0.18	0.19	0.21	0.22	0.23
2	0.24	0.25	0.27	0.28	0.29	0.30	0.31	0.33	0.34	0.35
3	0.36	0.38	0.39	0.40	0.41	0.42	0.44	0.45	0.46	0.47
4	0.48	0.50	0.51	0.52	0.53	0.55	0.56	0.57	0.58	0.59
5	0.61	0.62	0.63	0.64	0.65	0.67	0.68	0.69	0.70	0.71
6	0.73	0.74	0.75	0.76	0.77	0.79	0.80	0.81	0.82	0.84
7	0.85	0.86	0.87	0.88	0.90	0.91	0.92	0.93	0.94	0.96
8	0.97	0.98	0.99	1.00	1.02	1.03	1.04	1.05	1.07	1.08
9	1.09	1.10	1.11	1.13	1.14	1.15	1.16	1.17	1.19	1.20
10	1.21	1.22	1.23	1.25	1.26	1.27	1.28	1.30	1.31	1.32
11	1.33	1.34	1.36	1.37	1.38	1.39	1.40	1.42	1.43	1.44
12	1.45	1.46	1.48	1.49	1.50	1.51	1.53	1.54	1.55	1.56
13	1.57	1.59	1.60	1.61	1.62	1.63	1.65	1.66	1.67	1.68
14	1.69	1.71	1.72	1.73	1.74	1.76	1.77	1.78	1.79	1.80
15	1.82	1.83	1.84	1.85	1.86	1.88	1.89	1.90	1.91	1.92
16	1.94	1.95	1.96	1.97	1.99	2.00	2.01	2.02	2.03	2.05
17	2.06	2.07	2.08	2.09	2.11	2.12	2.13	2.14	2.15	2.17
18	2.18	2.19	2.20	2.21	2.23	2.24	2.25	2.26	2.28	2.29
19	2.30	2.31	2.32	2.34	2.35	2.36	2.37	2.38	2.40	2.41
20	2.42	2.43	2.45	2.46	2.47	2.48	2.49	2.51	2.52	2.53
21	2.54	2.55	2.57	2.58	2.59	2.60	2.61	2.63	2.64	2.65
22	2.66	2.68	2.69	2.70	2.71	2.72	2.73	2.75	2.76	2.77
23	2.78	2.80	2.81	2.82	2.83	2.84	2.86	2.87	2.88	2.89
24	2.91	2.92	2.93	2.94	2.95	2.97	2.98	2.99	3.00	3.01
25	3.03	3.04	3.05	3.06	3.07	3.09	3.10	3.11	3.12	3.14
26	3.15	3.16	3.17	3.18	3.20	3.21	3.22	3.23	3.24	3.26
27	3.27	3.28	3.29	3.30	3.32	3.33	3.34	3.35	3.37	3.38
28	3.39	3.40	3.41	3.43	3.44	3.45	3.46	3.47	3.49	3.50
29	3.51	3.52	3.54	3.55	3.56	3.57	3.58	3.60	3.61	3.62
30	3.63	3.64	3.66	3.67	3.68	3.69	3.70	3.72	3.73	3.74
31	3.75	3.76	3.78	3.79	3.80	3.81	3.83	3.84	3.85	3.86
32	3.87	3.89	3.90	3.91	3.92	3.93	3.95	3.96	3.97	3.98
33	3.99	4.01	4.02	4.03	4.04	4.06	4.07	4.08	4.09	4.10
34	4.12	4.13	4.14	4.15	4.16	4.18	4.19	4.20	4.21	4.22
35	4.24	4.25	4.26	4.27	4.29	4.30	4.31	4.32	4.33	4.35
	0.	**1.**	**2.**	**3.**	**4.**	**5.**	**6.**	**7.**	**8.**	**9.**

Centigrade Degrees.	BAROMETER : 755$^{mm.}$ (from 752.51 to 757.50).									
	Tenths of Degrees.									
	0.	**1.**	**2.**	**3.**	**4.**	**5.**	**6.**	**7.**	**8.**	**9.**
°	Millim.	Millim.	Millim.	Millim.	Millim.	Millim.	Millim.	Millim.	Millim.	Millim.
0	0.00	0.01	0.02	0.04	0.05	0.06	0.07	0.09	0.10	0.11
1	0.12	0.13	0.15	0.16	0.17	0.18	0.19	0.21	0.22	0.23
2	0.24	0.26	0.27	0.28	0.29	0.30	0.32	0.33	0.34	0.35
3	0.37	0.38	0.39	0.40	0.41	0.43	0.44	0.45	0.46	0.48
4	0.49	0.50	0.51	0.52	0.54	0.55	0.56	0.57	0.58	0.60
5	0.61	0.62	0.63	0.65	0.66	0.67	0.68	0.69	0.71	0.72
6	0.73	0.74	0.76	0.77	0.78	0.79	0.80	0.82	0.83	0.84
7	0.85	0.87	0.88	0.89	0.90	0.91	0.93	0.94	0.95	0.96
8	0.97	0.99	1.00	1.01	1.02	1.04	1.05	1.06	1.07	1.08
9	1.10	1.11	1.12	1.13	1.15	1.16	1.17	1.18	1.19	1.21
10	1.22	1.23	1.24	1.26	1.27	1.28	1.29	1.30	1.32	1.33
11	1.34	1.35	1.36	1.38	1.39	1.40	1.41	1.43	1.44	1.45
12	1.46	1.47	1.49	1.50	1.51	1.52	1.54	1.55	1.56	1.57
13	1.58	1.60	1.61	1.62	1.63	1.65	1.66	1.67	1.68	1.69
14	1.71	1.72	1.73	1.74	1.75	1.77	1.78	1.79	1.80	1.82
15	1.83	1.84	1.85	1.86	1.88	1.89	1.90	1.91	1.93	1.94
16	1.95	1.96	1.97	1.99	2.00	2.01	2.02	2.04	2.05	2.06
17	2.07	2.08	2.10	2.11	2.12	2.13	2.14	2.16	2.17	2.18
18	2.19	2.21	2.22	2.23	2.24	2.25	2.27	2.28	2.29	2.30
19	2.32	2.33	2.34	2.35	2.36	2.38	2.39	2.40	2.41	2.42
20	2.44	2.45	2.46	2.47	2.49	2.50	2.51	2.52	2.53	2.55
21	2.56	2.57	2.58	2.60	2.61	2.62	2.63	2.64	2.66	2.67
22	2.68	2.69	2.71	2.72	2.73	2.74	2.75	2.77	2.78	2.79
23	2.80	2.81	2.83	2.84	2.85	2.86	2.88	2.89	2.90	2.91
24	2.92	2.94	2.95	2.96	2.97	2.99	3.00	3.01	3.02	3.03
25	3.05	3.06	3.07	3.08	3.10	3.11	3.12	3.13	3.14	3.16
26	3.17	3.18	3.19	3.20	3.22	3.23	3.24	3.25	3.27	3.28
27	3.29	3.30	3.31	3.33	3.34	3.35	3.36	3.38	3.39	3.40
28	3.41	3.42	3.44	3.45	3.46	3.47	3.49	3.50	3.51	3.52
29	3.53	3.55	3.56	3.57	3.58	3.59	3.61	3.62	3.63	3.64
30	3.66	3.67	3.68	3.69	3.70	3.72	3.73	3.74	3.75	3.77
31	3.78	3.79	3.80	3.81	3.83	3.84	3.85	3.86	3.88	3.89
32	3.90	3.91	3.92	3.94	3.95	3.96	3.97	3.98	4.00	4.01
33	4.02	4.03	4.05	4.06	4.07	4.08	4.09	4.11	4.12	4.13
34	4.14	4.16	4.17	4.18	4.19	4.20	4.22	4.23	4.24	4.25
35	4.26	4.28	4.29	4.30	4.31	4.33	4.34	4.35	4.36	4.37
	0.	**1.**	**2.**	**3.**	**4.**	**5.**	**6.**	**7.**	**8.**	**9.**

BAROMETER: 760$^{mm.}$ (from 757.51 to 762.50).

Centigrade Degrees.	Tenths of Degrees.									
	0.	**1.**	**2.**	**3.**	**4.**	**5.**	**6.**	**7.**	**8.**	**9.**
°	Millim.	Millim.	Millim.	Millim.	Millim.	Millim.	Millim.	Millim.	Millim.	Millim.
0	0.00	0.01	0.02	0.04	0.05	0.06	0.07	0.09	0.10	0.11
1	0.12	0.13	0.15	0.16	0.17	0.18	0.20	0.21	0.22	0.23
2	0.25	0.26	0.27	0.28	0.29	0.31	0.32	0.33	0.34	0.36
3	0.37	0.38	0.39	0.40	0.42	0.43	0.44	0.45	0.47	0.48
4	0.49	0.50	0.52	0.53	0.54	0.55	0.56	0.58	0.59	0.60
5	0.61	0.63	0.64	0.65	0.66	0.67	0.69	0.70	0.71	0.72
6	0.74	0.75	0.76	0.77	0.79	0.80	0.81	0.82	0.83	0.85
7	0.86	0.87	0.88	0.90	0.91	0.92	0.93	0.94	0.96	0.97
8	0.98	0.99	1.01	1.02	1.03	1.04	1.05	1.07	1.08	1.09
9	1.10	1.12	1.13	1.14	1.15	1.17	1.18	1.19	1.20	1.21
10	1.23	1.24	1.25	1.26	1.28	1.29	1.30	1.31	1.32	1.34
11	1.35	1.36	1.37	1.39	1.40	1.41	1.42	1.44	1.45	1.46
12	1.47	1.48	1.50	1.51	1.52	1.53	1.55	1.56	1.57	1.58
13	1.59	1.61	1.62	1.63	1.64	1.66	1.67	1.68	1.69	1.71
14	1.72	1.73	1.74	1.75	1.77	1.78	1.79	1.80	1.82	1.83
15	1.84	1.85	1.86	1.88	1.89	1.90	1.91	1.93	1.94	1.95
16	1.96	1.97	1.99	2.00	2.01	2.02	2.04	2.05	2.06	2.07
17	2.09	2.10	2.11	2.12	2.13	2.15	2.16	2.17	2.18	2.20
18	2.21	2.22	2.23	2.24	2.26	2.27	2.28	2.29	2.31	2.32
19	2.33	2.34	2.36	2.37	2.38	2.39	2.40	2.42	2.43	2.44
20	2.45	2.47	2.48	2.49	2.50	2.51	2.53	2.54	2.55	2.56
21	2.58	2.59	2.60	2.61	2.63	2.64	2.65	2.66	2.67	2.69
22	2.70	2.71	2.72	2.74	2.75	2.76	2.77	2.78	2.80	2.81
23	2.82	2.83	2.85	2.86	2.87	2.88	2.89	2.91	2.92	2.93
24	2.94	2.96	2.97	2.98	2.99	3.01	3.02	3.03	3.04	3.05
25	3.07	3.08	3.09	3.10	3.12	3.13	3.14	3.15	3.16	3.18
26	3.19	3.20	3.21	3.23	3.24	3.25	3.26	3.28	3.29	3.30
27	3.31	3.32	3.34	3.35	3.36	3.37	3.39	3.40	3.41	3.42
28	3.43	3.45	3.46	3.47	3.48	3.50	3.51	3.52	3.53	3.54
29	3.56	3.57	3.58	3.59	3.61	3.62	3.63	3.64	3.66	3.67
30	3.68	3.69	3.70	3.72	3.73	3.74	3.75	3.77	3.78	3.79
31	3.80	3.81	3.83	3.84	3.85	3.86	3.88	3.89	3.90	3.91
32	3.93	3.94	3.95	3.96	3.97	3.99	4.00	4.01	4.02	4.04
33	4.05	4.06	4.07	4.08	4.10	4.11	4.12	4.13	4.15	4.16
34	4.17	4.18	4.20	4.21	4.22	4.23	4.24	4.26	4.27	4.28
35	4.29	4.31	4.32	4.33	4.34	4.35	4.37	4.38	4.39	4.40
	0.	**1.**	**2.**	**3.**	**4.**	**5.**	**6.**	**7.**	**8.**	**9.**

Centigrade Degrees.	BAROMETER: 765mm. (from 762.51 to 767.50).									
	Tenths of Degrees.									
	0.	**1.**	**2.**	**3.**	**4.**	**5.**	**6.**	**7.**	**8.**	**9.**
°	Millim.	Millim.	Millim.	Millim.	Millim.	Millim.	Millim.	Millim.	Millim.	Millim.
0	0.00	0.01	0.02	0.04	0.05	0.06	0.07	0.09	0.10	0.11
1	0.12	0.14	0.15	0.16	0.17	0.19	0.20	0.21	0.22	0.23
2	0.25	0.26	0.27	0.28	0.30	0.31	0.32	0.33	0.35	0.36
3	0.37	0.38	0.40	0.41	0.42	0.43	0.44	0.46	0.47	0.48
4	0.49	0.51	0.52	0.53	0.54	0.56	0.57	0.58	0.59	0.61
5	0.62	0.63	0.64	0.65	0.67	0.68	0.69	0.70	0.72	0.73
6	0.74	0.75	0.77	0.78	0.79	0.80	0.82	0.83	0.84	0.85
7	0.86	0.88	0.89	0.90	0.91	0.93	0.94	0.95	0.96	0.98
8	0.99	1.00	1.01	1.02	1.04	1.05	1.06	1.07	1.09	1.10
9	1.11	1.12	1.14	1.15	1.16	1.17	1.19	1.20	1.21	1.22
10	1.23	1.25	1.26	1.27	1.28	1.30	1.31	1.32	1.33	1.35
11	1.36	1.37	1.38	1.40	1.41	1.42	1.43	1.44	1.46	1.47
12	1.48	1.49	1.51	1.52	1.53	1.54	1.56	1.57	1.58	1.59
13	1.61	1.62	1.63	1.64	1.65	1.67	1.68	1.69	1.70	1.72
14	1.73	1.74	1.75	1.77	1.78	1.79	1.80	1.82	1.83	1.84
15	1.85	1.86	1.88	1.89	1.90	1.91	1.93	1.94	1.95	1.96
16	1.98	1.99	2.00	2.01	2.02	2.04	2.05	2.06	2.07	2.09
17	2.10	2.11	2.12	2.14	2.15	2.16	2.17	2.19	2.20	2.21
18	2.22	2.23	2.25	2.26	2.27	2.28	2.30	2.31	2.32	2.33
19	2.35	2.36	2.37	2.38	2.40	2.41	2.42	2.43	2.44	2.46
20	2.47	2.48	2.49	2.51	2.52	2.53	2.54	2.56	2.57	2.58
21	2.59	2.61	2.62	2.63	2.64	2.65	2.67	2.68	2.69	2.70
22	2.72	2.73	2.74	2.75	2.77	2.78	2.79	2.80	2.82	2.83
23	2.84	2.85	2.86	2.88	2.89	2.90	2.91	2.93	2.94	2.95
24	2.96	2.98	2.99	3.00	3.01	3.03	3.04	3.05	3.06	3.07
25	3.09	3.10	3.11	3.12	3.14	3.15	3.16	3.17	3.19	3.20
26	3.21	3.22	3.23	3.25	3.26	3.27	3.28	3.30	3.31	3.32
27	3.33	3.35	3.36	3.37	3.38	3.40	3.41	3.42	3.43	3.44
28	3.46	3.47	3.48	3.49	3.51	3.52	3.53	3.54	3.56	3.57
29	3.58	3.59	3.61	3.62	3.63	3.64	3.65	3.67	3.68	3.69
30	3.70	3.72	3.73	3.74	3.75	3.77	3.78	3.79	3.80	3.82
31	3.83	3.84	3.85	3.86	3.88	3.89	3.90	3.91	3.93	3.94
32	3.95	3.96	3.98	3.99	4.00	4.01	4.03	4.04	4.05	4.06
33	4.07	4.09	4.10	4.11	4.12	4.14	4.15	4.16	4.17	4.19
34	4.20	4.21	4.22	4.24	4.25	4.26	4.27	4.28	4.30	4.31
35	4.32	4.33	4.35	4.36	4.37	4.38	4.40	4.41	4.42	4.43
	0.	**1.**	**2.**	**3.**	**4.**	**5.**	**6.**	**7.**	**8.**	**9.**

Centigrade Degrees.	BAROMETER : 770mm. (from 767.51 to 772.50).									
	Tenths of Degrees.									
	0.	**1.**	**2.**	**3.**	**4.**	**5.**	**6.**	**7.**	**8.**	**9.**
°	Millim.	Millim.	Millim.	Millim.	Millim.	Millim.	Millim.	Millim.	Millim.	Millim.
0	0.00	0.01	0.02	0.04	0.05	0.06	0.07	0.09	0.10	0.11
1	0.12	0.14	0.15	0.16	0.17	0.19	0.20	0.21	0.22	0.24
2	0.25	0.26	0.27	0.29	0.30	0.31	0.32	0.34	0.35	0.36
3	0.37	0.39	0.40	0.41	0.42	0.43	0.45	0.46	0.47	0.48
4	0.50	0.51	0.52	0.53	0.55	0.56	0.57	0.58	0.60	0.61
5	0.62	0.63	0.65	0.66	0.67	0.68	0.70	0.71	0.72	0.73
6	0.75	0.76	0.77	0.78	0.80	0.81	0.82	0.83	0.85	0.86
7	0.87	0.88	0.89	0.91	0.92	0.93	0.94	0.96	0.97	0.98
8	0.99	1.01	1.02	1.03	1.04	1.06	1.07	1.08	1.09	1.11
9	1.12	1.13	1.14	1.16	1.17	1.18	1.19	1.21	1.22	1.23
10	1.24	1.26	1.27	1.28	1.29	1.30	1.32	1.33	1.34	1.35
11	1.37	1.38	1.39	1.40	1.42	1.43	1.44	1.45	1.47	1.48
12	1.49	1.50	1.52	1.53	1.54	1.55	1.57	1.58	1.59	1.60
13	1.62	1.63	1.64	1.65	1.67	1.68	1.69	1.70	1.72	1.73
14	1.74	1.75	1.76	1.78	1.79	1.80	1.81	1.83	1.84	1.85
15	1.86	1.88	1.89	1.90	1.91	1.93	1.94	1.95	1.96	1.98
16	1.99	2.00	2.01	2.03	2.04	2.05	2.06	2.08	2.09	2.10
17	2.11	2.13	2.14	2.15	2.16	2.17	2.19	2.20	2.21	2.22
18	2.24	2.25	2.26	2.27	2.29	2.30	2.31	2.32	2.34	2.35
19	2.36	2.37	2.39	2.40	2.41	2.42	2.44	2.45	2.46	2.47
20	2.49	2.50	2.51	2.52	2.54	2.55	2.56	2.57	2.58	2.60
21	2.61	2.62	2.63	2.65	2.66	2.67	2.68	2.70	2.71	2.72
22	2.73	2.75	2.76	2.77	2.78	2.80	2.81	2.82	2.83	2.85
23	2.86	2.87	2.88	2.90	2.91	2.92	2.93	2.95	2.96	2.97
24	2.98	3.00	3.01	3.02	3.03	3.04	3.06	3.07	3.08	3.09
25	3.11	3.12	3.13	3.14	3.16	3.17	3.18	3.19	3.21	3.22
26	3.23	3.24	3.26	3.27	3.28	3.29	3.31	3.32	3.33	3.34
27	3.36	3.37	3.38	3.39	3.41	3.42	3.43	3.44	3.45	3.47
28	3.48	3.49	3.50	3.52	3.53	3.54	3.55	3.57	3.58	3.59
29	3.60	3.62	3.63	3.64	3.65	3.67	3.68	3.69	3.70	3.72
30	3.73	3.74	3.75	3.77	3.78	3.79	3.80	3.82	3.83	3.84
31	3.85	3.87	3.88	3.89	3.90	3.91	3.93	3.94	3.95	3.96
32	3.98	3.99	4.00	4.01	4.03	4.04	4.05	4.06	4.08	4.09
33	4.10	4.11	4.13	4.14	4.15	4.16	4.18	4.19	4.20	4.21
34	4.23	4.24	4.25	4.26	4.28	4.29	4.30	4.31	4.32	4.34
35	4.35	4.36	4.37	4.39	4.40	4.41	4.42	4.44	4.45	4.46
	0.	**1.**	**2.**	**3.**	**4.**	**5.**	**6.**	**7.**	**8.**	**9.**

BAROMETER : 775mm. (from 772.51 to 777.50).

Tenths of Degrees.

Centigrade Degrees.	0.	1.	2.	3.	4.	5.	6.	7.	8.	9.
°	Millim.	Millim.	Millim.	Millim.	Millim.	Millim.	Millim.	Millim.	Millim.	Millim.
0	0.00	0.01	0.03	0.04	0.05	0.06	0.08	0.09	0.10	0.11
1	0.13	0.14	0.15	0.16	0.18	0.19	0.20	0.21	0.23	0.24
2	0.25	0.26	0.28	0.29	0.30	0.31	0.33	0.34	0.35	0.36
3	0.38	0.39	0.40	0.41	0.43	0.44	0.45	0.46	0.48	0.49
4	0.50	0.51	0.53	0.54	0.55	0.56	0.58	0.59	0.60	0.61
5	0.63	0.64	0.65	0.66	0.68	0.69	0.70	0.71	0.73	0.74
6	0.75	0.76	0.78	0.79	0.80	0.81	0.83	0.84	0.85	0.86
7	0.88	0.89	0.90	0.91	0.93	0.94	0.95	0.96	0.98	0.99
8	1.00	1.01	1.03	1.04	1.05	1.06	1.08	1.09	1.10	1.11
9	1.13	1.14	1.15	1.16	1.18	1.19	1.20	1.21	1.23	1.24
10	1.25	1.26	1.28	1.29	1.30	1.31	1.33	1.34	1.35	1.36
11	1.38	1.39	1.40	1.41	1.43	1.44	1.45	1.46	1.48	1.49
12	1.50	1.51	1.53	1.54	1.55	1.56	1.58	1.59	1.60	1.61
13	1.63	1.64	1.65	1.66	1.68	1.69	1.70	1.71	1.73	1.74
14	1.75	1.76	1.78	1.79	1.80	1.81	1.83	1.84	1.85	1.86
15	1.88	1.89	1.90	1.91	1.93	1.94	1.95	1.96	1.98	1.99
16	2.00	2.01	2.03	2.04	2.05	2.06	2.08	2.09	2.10	2.11
17	2.13	2.14	2.15	2.16	2.18	2.19	2.20	2.21	2.23	2.24
18	2.25	2.26	2.28	2.29	2.30	2.31	2.33	2.34	2.35	2.36
19	2.38	2.39	2.40	2.41	2.43	2.44	2.45	2.46	2.48	2.49
20	2.50	2.51	2.53	2.54	2.55	2.56	2.58	2.59	2.60	2.61
21	2.63	2.64	2.65	2.66	2.68	2.69	2.70	2.71	2.73	2.74
22	2.75	2.76	2.78	2.79	2.80	2.81	2.83	2.84	2.85	2.86
23	2.88	2.89	2.90	2.91	2.93	2.94	2.95	2.96	2.98	2.99
24	3.00	3.01	3.03	3.04	3.05	3.06	3.08	3.09	3.10	3.11
25	3.13	3.14	3.15	3.16	3.18	3.19	3.20	3.21	3.23	3.24
26	3.25	3.26	3.28	3.29	3.30	3.31	3.33	3.34	3.35	3.36
27	3.38	3.39	3.40	3.41	3.43	3.44	3.45	3.46	3.48	3.49
28	3.50	3.51	3.53	3.54	3.55	3.56	3.58	3.59	3.60	3.61
29	3.63	3.64	3.65	3.66	3.68	3.69	3.70	3.72	3.73	3.74
30	3.75	3.77	3.78	3.79	3.80	3.82	3.83	3.84	3.85	3.87
31	3.88	3.89	3.90	3.92	3.93	3.94	3.95	3.97	3.98	3.99
32	4.00	4.02	4.03	4.04	4.05	4.07	4.08	4.09	4.10	4.12
33	4.13	4.14	4.15	4.17	4.18	4.19	4.20	4.22	4.23	4.24
34	4.25	4.27	4.28	4.29	4.30	4.32	4.33	4.34	4.35	4.37
35	4.38	4.39	4.40	4.42	4.43	4.44	4.45	4.47	4.48	4.49
	0.	1.	2.	3.	4.	5.	6.	7.	8.	9.

BAROMETER: 780mm. (from 777.51 to 782.50).

Centigrade Degrees.	Tenths of Degrees.									
	0.	**1.**	**2.**	**3.**	**4.**	**5.**	**6.**	**7.**	**8.**	**9.**
°	Millim.	Millim.	Millim.	Millim.	Millim.	Millim.	Millim.	Millim.	Millim.	Millim.
0	0.00	0.01	0.03	0.04	0.05	0.06	0.08	0.09	0.10	0.11
1	0.13	0.14	0.15	0.16	0.18	0.19	0.20	0.21	0.23	0.24
2	0.25	0.26	0.28	0.29	0.30	0.31	0.33	0.34	0.35	0.37
3	0.38	0.39	0.40	0.42	0.43	0.44	0.45	0.47	0.48	0.49
4	0.50	0.52	0.53	0.54	0.55	0.57	0.58	0.59	0.60	0.62
5	0.63	0.64	0.65	0.67	0.68	0.69	0.70	0.72	0.73	0.74
6	0.76	0.77	0.78	0.79	0.81	0.82	0.83	0.84	0.86	0.87
7	0.88	0.89	0.91	0.92	0.93	0.94	0.96	0.97	0.98	0.99
8	1.01	1.02	1.03	1.04	1.06	1.07	1.08	1.10	1.11	1.12
9	1.13	1.15	1.16	1.17	1.18	1.20	1.21	1.22	1.23	1.25
10	1.26	1.27	1.28	1.30	1.31	1.32	1.33	1.35	1.36	1.37
11	1.38	1.40	1.41	1.42	1.44	1.45	1.46	1.47	1.49	1.50
12	1.51	1.52	1.54	1.55	1.56	1.57	1.59	1.60	1.61	1.62
13	1.64	1.65	1.66	1.67	1.69	1.70	1.71	1.72	1.74	1.75
14	1.76	1.78	1.79	1.80	1.81	1.83	1.84	1.85	1.86	1.88
15	1.89	1.90	1.91	1.93	1.94	1.95	1.96	1.98	1.99	2.00
16	2.01	2.03	2.04	2.05	2.06	2.08	2.09	2.10	2.11	2.13
17	2.14	2.15	2.17	2.18	2.19	2.20	2.22	2.23	2.24	2.25
18	2.27	2.28	2.29	2.30	2.32	2.33	2.34	2.35	2.37	2.38
19	2.39	2.40	2.42	2.43	2.44	2.45	2.47	2.48	2.49	2.51
20	2.52	2.53	2.54	2.56	2.57	2.58	2.59	2.61	2.62	2.63
21	2.64	2.66	2.67	2.68	2.69	2.71	2.72	2.73	2.74	2.76
22	2.77	2.78	2.79	2.81	2.82	2.83	2.85	2.86	2.87	2.88
23	2.90	2.91	2.92	2.93	2.95	2.96	2.97	2.98	3.00	3.01
24	3.02	3.03	3.05	3.06	3.07	3.08	3.10	3.11	3.12	3.14
25	3.15	3.16	3.17	3.19	3.20	3.21	3.22	3.24	3.25	3.26
26	3.27	3.29	3.30	3.31	3.32	3.34	3.35	3.36	3.37	3.39
27	3.40	3.41	3.42	3.44	3.45	3.46	3.47	3.49	3.50	3.51
28	3.52	3.54	3.55	3.56	3.58	3.59	3.60	3.61	3.63	3.64
29	3.65	3.66	3.68	3.69	3.70	3.71	3.73	3.74	3.75	3.76
30	3.78	3.79	3.80	3.81	3.83	3.84	3.85	3.86	3.88	3.89
31	3.90	3.92	3.93	3.94	3.95	3.97	3.98	3.99	4.00	4.02
32	4.03	4.04	4.05	4.07	4.08	4.09	4.10	4.12	4.13	4.14
33	4.15	4.17	4.18	4.19	4.20	4.22	4.23	4.24	4.26	4.27
34	4.28	4.29	4.31	4.32	4.33	4.34	4.36	4.37	4.38	4.39
35	4.41	4.42	4.43	4.44	4.46	4.47	4.48	4.49	4.51	4.52
	0.	**1.**	**2.**	**3.**	**4.**	**5.**	**6.**	**7.**	**8.**	**9.**

BAROMETER: 785mm. (from 782.51 to 787.50).

Centigrade Degrees.	Tenths of Degrees.									
	0.	1.	2.	3.	4.	5.	6.	7.	8.	9.
°	Millim.	Millim.	Millim.	Millim.	Millim.	Millim.	Millim.	Millim.	Millim.	Millim.
0	0.00	0.01	0.03	0.04	0.05	0.06	0.08	0.09	0.10	0.11
1	0.13	0.14	0.15	0.16	0.18	0.19	0.20	0.22	0.23	0.24
2	0.25	0.27	0.28	0.29	0.30	0.32	0.33	0.34	0.35	0.37
3	0.38	0.39	0.41	0.42	0.43	0.44	0.46	0.47	0.48	0.49
4	0.51	0.52	0.53	0.54	0.56	0.57	0.58	0.60	0.61	0.62
5	0.63	0.65	0.66	0.67	0.68	0.70	0.71	0.72	0.73	0.75
6	0.76	0.77	0.79	0.80	0.81	0.82	0.84	0.85	0.86	0.87
7	0.89	0.90	0.91	0.92	0.94	0.95	0.96	0.98	0.99	1.00
8	1.01	1.03	1.04	1.05	1.06	1.08	1.09	1.10	1.11	1.13
9	1.14	1.15	1.17	1.18	1.19	1.20	1.22	1.23	1.24	1.25
10	1.27	1.28	1.29	1.30	1.32	1.33	1.34	1.36	1.37	1.38
11	1.39	1.41	1.42	1.43	1.44	1.46	1.47	1.48	1.50	1.51
12	1.52	1.53	1.55	1.56	1.57	1.58	1.60	1.61	1.62	1.63
13	1.65	1.66	1.67	1.69	1.70	1.71	1.72	1.74	1.75	1.76
14	1.77	1.79	1.80	1.81	1.82	1.84	1.85	1.86	1.88	1.89
15	1.90	1.91	1.93	1.94	1.95	1.96	1.98	1.99	2.00	2.01
16	2.03	2.04	2.05	2.07	2.08	2.09	2.10	2.12	2.13	2.14
17	2.15	2.17	2.18	2.19	2.20	2.22	2.23	2.24	2.26	2.27
18	2.28	2.29	2.31	2.32	2.33	2.34	2.36	2.37	2.38	2.39
19	2.41	2.42	2.43	2.45	2.46	2.47	2.48	2.50	2.51	2.52
20	2.53	2.55	2.56	2.57	2.58	2.60	2.61	2.62	2.64	2.65
21	2.66	2.67	2.69	2.70	2.71	2.72	2.74	2.75	2.76	2.77
22	2.79	2.80	2.81	2.83	2.84	2.85	2.86	2.88	2.89	2.90
23	2.91	2.93	2.94	2.95	2.96	2.98	2.99	3.00	3.02	3.03
24	3.04	3.05	3.07	3.08	3.09	3.10	3.12	3.13	3.14	3.15
25	3.17	3.18	3.19	3.21	3.22	3.23	3.24	3.26	3.27	3.28
26	3.29	3.31	3.32	3.33	3.34	3.36	3.37	3.38	3.40	3.41
27	3.42	3.43	3.45	3.46	3.47	3.48	3.50	3.51	3.52	3.53
28	3.55	3.56	3.57	3.59	3.60	3.61	3.62	3.64	3.65	3.66
29	3.67	3.69	3.70	3.71	3.72	3.74	3.75	3.76	3.78	3.79
30	3.80	3.81	3.83	3.84	3.85	3.86	3.88	3.89	3.90	3.91
31	3.93	3.94	3.95	3.97	3.98	3.99	4.00	4.02	4.03	4.04
32	4.05	4.07	4.08	4.09	4.11	4.12	4.13	4.14	4.16	4.17
33	4.18	4.19	4.21	4.22	4.23	4.24	4.26	4.27	4.28	4.30
34	4.31	4.32	4.33	4.35	4.36	4.37	4.38	4.40	4.41	4.42
35	4.43	4.45	4.46	4.47	4.49	4.50	4.51	4.52	4.54	4.55
	0.	1.	2.	3.	4.	5.	6.	7.	8.	9.

Centigrade Degrees.	BAROMETER : 790mm. (from 787.51 to 792.50). Tenths of Degrees.									
	0.	**1.**	**2.**	**3.**	**4.**	**5.**	**6.**	**7.**	**8.**	**9.**
°	Millim.	Millim.	Millim.	Millim.	Millim.	Millim.	Millim.	Millim.	Millim.	Millim.
0	0.00	0.01	0.03	0.04	0.05	0.06	0.08	0.09	0.10	0.11
1	0.13	0.14	0.15	0.17	0.18	0.19	0.20	0.22	0.23	0.24
2	0.26	0.27	0.28	0.29	0.31	0.32	0.33	0.34	0.36	0.37
3	0.38	0.40	0.41	0.42	0.43	0.45	0.46	0.47	0.48	0.50
4	0.51	0.52	0.54	0.55	0.56	0.57	0.59	0.60	0.61	0.62
5	0.64	0.65	0.66	0.68	0.69	0.70	0.71	0.73	0.74	0.75
6	0.77	0.78	0.79	0.80	0.82	0.83	0.84	0.85	0.87	0.88
7	0.89	0.91	0.92	0.93	0.94	0.96	0.97	0.98	0.99	1.01
8	1.02	1.03	1.05	1.06	1.07	1.08	1.10	1.11	1.12	1.13
9	1.15	1.16	1.17	1.19	1.20	1.21	1.22	1.24	1.25	1.26
10	1.28	1.29	1.30	1.31	1.33	1.34	1.35	1.36	1.38	1.39
11	1.40	1.42	1.43	1.44	1.45	1.47	1.48	1.49	1.50	1.52
12	1.53	1.54	1.56	1.57	1.58	1.59	1.61	1.62	1.63	1.64
13	1.66	1.67	1.68	1.70	1.71	1.72	1.73	1.75	1.76	1.77
14	1.79	1.80	1.81	1.82	1.84	1.85	1.86	1.87	1.89	1.90
15	1.91	1.93	1.94	1.95	1.96	1.98	1.99	2.00	2.01	2.03
16	2.04	2.05	2.07	2.08	2.09	2.10	2.12	2.13	2.14	2.15
17	2.17	2.18	2.19	2.21	2.22	2.23	2.24	2.26	2.27	2.28
18	2.30	2.31	2.32	2.33	2.35	2.36	2.37	2.38	2.40	2.41
19	2.42	2.44	2.45	2.46	2.47	2.49	2.50	2.51	2.52	2.54
20	2.55	2.56	2.58	2.59	2.60	2.61	2.63	2.64	2.65	2.66
21	2.68	2.69	2.70	2.72	2.73	2.74	2.75	2.77	2.78	2.79
22	2.81	2.82	2.83	2.84	2.86	2.87	2.88	2.89	2.91	2.92
23	2.93	2.95	2.96	2.97	2.98	3.00	3.01	3.02	3.03	3.05
24	3.06	3.07	3.09	3.10	3.11	3.12	3.14	3.15	3.16	3.17
25	3.19	3.20	3.21	3.23	3.24	3.25	3.26	3.28	3.29	3.30
26	3.32	3.33	3.34	3.35	3.37	3.38	3.39	3.40	3.42	3.43
27	3.44	3.46	3.47	3.48	3.49	3.51	3.52	3.53	3.54	3.56
28	3.57	3.58	3.60	3.61	3.62	3.63	3.65	3.66	3.67	3.68
29	3.70	3.71	3.72	3.74	3.75	3.76	3.77	3.79	3.80	3.81
30	3.83	3.84	3.85	3.86	3.88	3.89	3.90	3.91	3.93	3.94
31	3.95	3.97	3.98	3.99	4.00	4.02	4.03	4.04	4.05	4.07
32	4.08	4.09	4.11	4.12	4.13	4.14	4.16	4.17	4.18	4.19
33	4.21	4.22	4.23	4.25	4.26	4.27	4.28	4.30	4.31	4.32
34	4.34	4.35	4.36	4.37	4.39	4.40	4.41	4.42	4.44	4.45
35	4.46	4.48	4.49	4.50	4.51	4.53	4.54	4.55	4.56	4.58
	0.	**1.**	**2.**	**3.**	**4.**	**5.**	**6.**	**7.**	**8.**	**9.**

BAROMETER : 795$^{mm.}$ (from 792.51 to 797.50).

Centi-grade Degrees.	Tenths of Degrees.									
	0.	1.	2.	3.	4.	5.	6.	7.	8.	9.
°	Millim.	Millim.	Millim.	Millim.	Millim.	Millim.	Millim.	Millim.	Millim.	Millim.
0	0.00	0.01	0.03	0.04	0.05	0.06	0.08	0.09	0.10	0.12
1	0.13	0.14	0.15	0.17	0.18	0.19	0.21	0.22	0.23	0.24
2	0.26	0.27	0.28	0.30	0.31	0.32	0.33	0.35	0.36	0.37
3	0.38	0.40	0.41	0.42	0.44	0.45	0.46	0.47	0.49	0.50
4	0.51	0.53	0.54	0.55	0.56	0.58	0.59	0.60	0.62	0.63
5	0.64	0.65	0.67	0.68	0.69	0.71	0.72	0.73	0.74	0.76
6	0.77	0.78	0.80	0.81	0.82	0.83	0.85	0.86	0.87	0.89
7	0.90	0.91	0.92	0.94	0.95	0.96	0.98	0.99	1.00	1.01
8	1.03	1.04	1.05	1.06	1.08	1.09	1.10	1.12	1.13	1.14
9	1.15	1.17	1.18	1.19	1.21	1.22	1.23	1.24	1.26	1.27
10	1.28	1.30	1.31	1.32	1.33	1.35	1.36	1.37	1.39	1.40
11	1.41	1.42	1.44	1.45	1.46	1.48	1.49	1.50	1.51	1.53
12	1.54	1.55	1.57	1.58	1.59	1.60	1.62	1.63	1.64	1.66
13	1.67	1.68	1.69	1.71	1.72	1.73	1.75	1.76	1.77	1.78
14	1.80	1.81	1.82	1.83	1.85	1.86	1.87	1.89	1.90	1.91
15	1.92	1.94	1.95	1.96	1.98	1.99	2.00	2.01	2.03	2.04
16	2.05	2.07	2.08	2.09	2.10	2.12	2.13	2.14	2.16	2.17
17	2.18	2.19	2.21	2.22	2.23	2.25	2.26	2.27	2.28	2.30
18	2.31	2.32	2.34	2.35	2.36	2.37	2.39	2.40	2.41	2.43
19	2.44	2.45	2.46	2.48	2.49	2.50	2.51	2.53	2.54	2.55
20	2.57	2.58	2.59	2.60	2.62	2.63	2.64	2.66	2.67	2.68
21	2.69	2.71	2.72	2.73	2.75	2.76	2.77	2.78	2.80	2.81
22	2.82	2.84	2.85	2.86	2.87	2.89	2.90	2.91	2.93	2.94
23	2.95	2.96	2.98	2.99	3.00	3.02	3.03	3.04	3.05	3.07
24	3.08	3.09	3.11	3.12	3.13	3.14	3.16	3.17	3.18	3.19
25	3.21	3.22	3.23	3.25	3.26	3.27	3.28	3.30	3.31	3.32
26	3.34	3.35	3.36	3.37	3.39	3.40	3.41	3.43	3.44	3.45
27	3.46	3.48	3.49	3.50	3.52	3.53	3.54	3.55	3.57	3.58
28	3.59	3.61	3.62	3.63	3.64	3.66	3.67	3.68	3.70	3.71
29	3.72	3.73	3.75	3.76	3.77	3.79	3.80	3.81	3.82	3.84
30	3.85	3.86	3.88	3.89	3.90	3.91	3.93	3.94	3.95	3.96
31	3.98	3.99	4.00	4.02	4.03	4.04	4.05	4.07	4.08	4.09
32	4.11	4.12	4.13	4.14	4.16	4.17	4.18	4.20	4.21	4.22
33	4.23	4.25	4.26	4.27	4.29	4.30	4.31	4.32	4.34	4.35
34	4.36	4.38	4.39	4.40	4.41	4.43	4.44	4.45	4.47	4.48
35	4.49	4.50	4.52	4.53	4.54	4.56	4.57	4.58	4.59	4.61
	0.	1.	2.	3.	4.	5.	6.	7.	8.	9.

BAROMETER : 800mm. (from 797.51 to 802.50).

Centigrade Degrees.	Tenths of Degrees.									
	0.	**1.**	**2.**	**3.**	**4.**	**5.**	**6.**	**7.**	**8.**	**9.**
°	Millim.	Millim.	Millim.	Millim.	Millim.	Millim.	Millim.	Millim.	Millim.	Millim.
0	0.00	0.01	0.03	0.04	0.05	0.06	0.08	0.09	0.10	0.12
1	0.13	0.14	0.15	0.17	0.18	0.19	0.21	0.22	0.23	0.25
2	0.26	0.27	0.28	0.30	0.31	0.32	0.34	0.35	0.36	0.37
3	0.39	0.40	0.41	0.43	0.44	0.45	0.46	0.48	0.49	0.50
4	0.52	0.53	0.54	0.56	0.57	0.58	0.59	0.61	0.62	0.63
5	0.65	0.66	0.67	0.68	0.70	0.71	0.72	0.74	0.75	0.76
6	0.77	0.79	0.80	0.81	0.83	0.84	0.85	0.87	0.88	0.89
7	0.90	0.92	0.93	0.94	0.96	0.97	0.98	0.99	1.01	1.02
8	1.03	1.05	1.06	1.07	1.08	1.10	1.11	1.12	1.14	1.15
9	1.16	1.17	1.19	1.20	1.21	1.23	1.24	1.25	1.27	1.28
10	1.29	1.30	1.32	1.33	1.34	1.36	1.37	1.38	1.39	1.41
11	1.42	1.43	1.45	1.46	1.47	1.48	1.50	1.51	1.52	1.54
12	1.55	1.56	1.58	1.59	1.60	1.61	1.63	1.64	1.65	1.67
13	1.68	1.69	1.70	1.72	1.73	1.74	1.76	1.77	1.78	1.79
14	1.81	1.82	1.83	1.85	1.86	1.87	1.89	1.90	1.91	1.92
15	1.94	1.95	1.96	1.98	1.99	2.00	2.01	2.03	2.04	2.05
16	2.07	2.08	2.09	2.10	2.12	2.13	2.14	2.16	2.17	2.18
17	2.20	2.21	2.22	2.23	2.25	2.26	2.27	2.29	2.30	2.31
18	2.32	2.34	2.35	2.36	2.38	2.39	2.40	2.41	2.43	2.44
19	2.45	2.47	2.48	2.49	2.50	2.52	2.53	2.54	2.56	2.57
20	2.58	2.60	2.61	2.62	2.63	2.65	2.66	2.67	2.69	2.70
21	2.71	2.72	2.74	2.75	2.76	2.78	2.79	2.80	2.81	2.83
22	2.84	2.85	2.87	2.88	2.89	2.91	2.92	2.93	2.94	2.96
23	2.97	2.98	3.00	3.01	3.02	3.03	3.05	3.06	3.07	3.09
24	3.10	3.11	3.12	3.14	3.15	3.16	3.18	3.19	3.20	3.22
25	3.23	3.24	3.25	3.27	3.28	3.29	3.31	3.32	3.33	3.34
26	3.36	3.37	3.38	3.40	3.41	3.42	3.43	3.45	3.46	3.47
27	3.49	3.50	3.51	3.52	3.54	3.55	3.56	3.58	3.59	3.60
28	3.62	3.63	3.64	3.65	3.67	3.68	3.69	3.71	3.72	3.73
29	3.74	3.76	3.77	3.78	3.80	3.81	3.82	3.83	3.85	3.86
30	3.87	3.89	3.90	3.91	3.93	3.94	3.95	3.96	3.98	3.99
31	4.00	4.02	4.03	4.04	4.05	4.07	4.08	4.09	4.11	4.12
32	4.13	4.14	4.16	4.17	4.18	4.20	4.21	4.22	4.24	4.25
33	4.26	4.27	4.29	4.30	4.31	4.33	4.34	4.35	4.36	4.38
34	4.39	4.40	4.42	4.43	4.44	4.45	4.47	4.48	4.49	4.51
35	4.52	4.53	4.55	4.56	4.57	4.58	4.60	4.61	4.62	4.64
	0.	**1.**	**2.**	**3.**	**4.**	**5.**	**6.**	**7.**	**8.**	**9.**

XXI.

OLD FRENCH BAROMETER.

TABLE

FOR

REDUCING TO THE FREEZING POINT THE OBSERVATIONS TAKEN WITH OLD FRENCH BAROMETERS,

PROVIDED WITH BRASS SCALES, EXTENDING FROM THE CISTERN TO THE TOP OF THE MERCURIAL COLUMN; CALCULATED FROM 240 TO 345 LINES, OR FROM 23 INCHES 4 LINES TO 28 INCHES 9 LINES.

BY KAEMTZ.

TABLE XXI.

This table is taken from KAEMTZ's *Lehrbuch der Météorologie*, Vol. II. p. 236. To render it more useful, the first page, giving the corrections for Barometrical Heights between 240 and 280 Paris lines, has been added.

The values adopted by Kaemtz for reducing the Old French Barometer are the following: —

Let h = observed height in French lines.
" t = temperature of attached thermometer in degrees of Reaumur.
" m = expansion of mercury between 0 and 80° Reaumur = 0.018018.
" l = linear expansion of brass between 0 and 80° Reaumur = 0.0018782.
The normal temperature of standard being = 13° Reaumur.
And the formula becomes, —

$$-h \cdot \frac{m \times t - l\,(t - 13)}{1 + m \times t}$$

The Table gives the corrections only for full degrees and for every fifth line; but the intermediate values can easily be found by an interpolation at sight.

Example of Reduction.

Observed height . . . . = 325.32 lines.
Attached thermometer . . . . . . = 12.5 Reaumur.

In the line beginning with 12°, and in the vertical column headed 325 lines, we find,

Correction for	12°	= —0.89	lines.
Interpolation for	0°.5	= —0.03	"
Correction for	12°.5	= —0.92	"

And we have,

Observed height,	325.32	"
Correction for 12°.5,	—0.92	"
Height at the freezing point =	324.40	lines.

Normal Temperature of the Scale = 13° Reaumur.

Attached Thermometer. Degrees of Reaumur.	Barometer in Paris Lines. 240	245	250	255	260	265	270	275	Attached Thermometer. Degrees of Reaumur.
°	Par. Lines.	Par. Lines.	Par. Lines.	Par. Lines.	Par. Lines.	Par. Lines.	Par. Lines.	Par. Lines.	°
−15	+0.65	+0.66	+0.68	+0.69	+0.70	+0.72	+0.73	+0.75	−15
−14	0.60	0.61	0.63	0.64	0.65	0.67	0.68	0.69	−14
−13	0.55	0.57	0.58	0.59	0.60	0.61	0.62	0.64	−13
−12	0.51	0.52	0.53	0.54	0.55	0.56	0.57	0.58	−12
−11	0.46	0.47	0.48	0.49	0.50	0.51	0.52	0.52	−11
−10	0.41	0.42	0.43	0.44	0.44	0.45	0.46	0.47	−10
− 9	+0.36	+0.37	+0.38	+0.38	+0.39	+0.40	+0.41	+0.41	− 9
− 8	0.31	0.32	0.33	0.33	0.34	0.35	0.35	0.36	− 8
− 7	0.27	0.27	0.28	0.28	0.29	0.29	0.30	0.30	− 7
− 6	0.22	0.22	0.23	0.23	0.24	0.24	0.24	0.25	− 6
− 5	0.17	0.17	0.18	0.18	0.18	0.19	0.19	0.19	− 5
− 4	+0.12	+0.12	+0.13	+0.13	+0.13	+0.13	+0.14	+0.14	− 4
− 3	0.07	0.07	0.08	0.08	0.08	0.08	0.08	0.08	− 3
− 2	+0.02	+0.03	+0.03	+0.03	+0.03	+0.03	+0.03	+0.03	− 2
− 1	−0.02	−0.03	−0.03	−0.03	−0.03	−0.03	−0.03	−0.03	− 1
0	−0.07	−0.07	−0.08	−0.08	−0.08	−0.08	−0.08	−0.08	0
+ 1	−0.12	−0.12	−0.13	−0.13	−0.13	−0.13	−0.14	−0.14	+ 1
2	0.17	0.17	0.18	0.18	0.18	0.19	0.19	0.19	2
3	0.22	0.22	0.23	0.23	0.24	0.24	0.24	0.25	3
4	0.27	0.27	0.28	0.28	0.29	0.29	0.30	0.30	4
5	0.31	0.32	0.33	0.33	0.34	0.35	0.35	0.36	5
+ 6	−0.36	−0.37	−0.38	−0.38	−0.39	−0.40	−0.41	−0.41	+ 6
7	0.41	0.42	0.43	0.44	0.44	0.45	0.46	0.47	7
8	0.46	0.47	0.48	0.49	0.50	0.51	0.52	0.52	8
9	0.51	0.52	0.53	0.54	0.55	0.56	0.57	0.58	9
10	0.55	0.57	0.58	0.59	0.60	0.61	0.62	0.64	10
+11	−0.60	−0.61	−0.63	−0.64	−0.65	−0.67	−0.68	−0.69	+11
12	0.65	0.66	0.68	0.69	0.70	0.72	0.73	0.75	12
13	0.70	0.71	0.73	0.74	0.76	0.77	0.79	0.80	13
14	0.75	0.76	0.78	0.79	0.81	0.82	0.84	0.86	14
15	0.80	0.81	0.83	0.84	0.86	0.88	0.89	0.91	15
+16	−0.84	−0.86	−0.88	−0.90	−0.91	−0.93	−0.95	−0.97	+16
17	0.89	0.91	0.93	0.95	0.97	0.98	1.00	1.02	17
18	0.94	0.96	0.98	1.00	1.02	1.04	1.06	1.08	18
19	0.99	1.01	1.03	1.05	1.07	1.09	1.11	1.13	19
20	1.04	1.06	1.08	1.10	1.12	1.14	1.17	1.19	20
+21	−1.08	−1.11	−1.13	−1.15	−1.17	−1.20	−1.22	−1.24	+21
22	1.13	1.16	1.18	1.20	1.23	1.25	1.27	1.30	22
23	1.18	1.20	1.23	1.25	1.28	1.30	1.33	1.35	23
24	1.23	1.25	1.28	1.31	1.33	1.36	1.38	1.41	24
25	1.28	1.30	1.33	1.36	1.38	1.41	1.44	1.46	25

Normal Temperature of the Scale = 13° Reaumur.

Attached Thermometer. Degrees of Reaumur.	Barometer in Paris Lines. 280	285	290	295	300	305	310	Attached Thermometer. Degrees of Reaumur.
°	Par. Lines.	Par. Lines.	Par. Lines.	Par. Lines.	Par. Lines.	Par. Lines.	Par. Lines.	°
−15	+0.77	+0.78	+0.79	+0.81	+0.82	+0.84	+0.85	−15
−14	0.71	0.73	0.74	0.75	0.76	0.77	0.79	−14
−13	0.65	0.67	0.68	0.69	0.70	0.71	0.72	−13
−12	0.60	0.61	0.62	0.63	0.64	0.65	0.66	−12
−11	0.54	0.55	0.56	0.57	0.58	0.59	0.60	−11
−10	0.48	0.49	0.50	0.51	0.52	0.53	0.54	−10
− 9	+0.43	+0.44	+0.44	+0.45	+0.46	+0.46	+0.47	− 9
− 8	0.37	0.38	0.38	0.39	0.40	0.40	0.41	− 8
− 7	0.31	0.32	0.32	0.33	0.34	0.34	0.35	− 7
− 6	0.26	0.26	0.26	0.27	0.27	0.28	0.28	− 6
− 5	0.20	0.20	0.21	0.21	0.21	0.22	0.22	− 5
− 4	+0.14	+0.15	+0.15	+0.15	+0.15	+0.16	+0.16	− 4
− 3	0.09	0.09	0.09	0.09	0.09	0.09	0.09	− 3
− 2	+0.03	+0.03	+0.03	+0.03	+0.03	+0.03	+0.03	− 2
− 1	−0.03	−0.03	−0.03	−0.03	−0.03	−0.03	−0.03	− 1
0	−0.08	−0.09	−0.09	−0.09	−0.09	−0.09	−0.09	0
+ 1	−0.14	−0.14	−0.15	−0.15	−0.15	−0.15	−0.16	+ 1
2	0.20	0.20	0.21	0.21	0.21	0.22	0.22	2
3	0.26	0.26	0.27	0.27	0.27	0.28	0.28	3
4	0.31	0.32	0.32	0.33	0.33	0.34	0.35	4
5	0.37	0.37	0.38	0.39	0.40	0.40	0.41	5
+ 6	−0.43	−0.43	−0.44	−0.45	−0.46	−0.46	−0.47	+ 6
7	0.48	0.49	0.50	0.51	0.52	0.53	0.53	7
8	0.54	0.55	0.56	0.57	0.58	0.59	0.60	8
9	0.60	0.61	0.62	0.63	0.64	0.65	0.66	9
10	0.65	0.66	0.68	0.69	0.70	0.71	0.72	10
+11	−0.71	−0.72	−0.74	−0.75	−0.76	−0.77	−0.79	+11
12	0.77	0.78	0.80	0.81	0.82	0.84	0.85	12
13	0.82	0.84	0.85	0.87	0.88	0.90	0.91	13
14	0.88	0.90	0.91	0.93	0.94	0.96	0.98	14
15	0.94	0.95	0.97	0.99	1.00	1.02	1.04	15
+16	−0.99	−1.01	−1.03	−1.05	−1.07	−1.08	−1.10	+16
17	1.05	1.07	1.09	1.11	1.13	1.15	1.16	17
18	1.11	1.13	1.15	1.17	1.19	1.21	1.23	18
19	1.16	1.18	1.21	1.23	1.25	1.27	1.29	19
20	1.22	1.24	1.27	1.29	1.31	1.33	1.35	20
+21	−1.28	−1.30	−1.33	−1.35	−1.37	−1.39	−1.42	+21
22	1.34	1.36	1.38	1.41	1.43	1.45	1.48	22
23	1.39	1.41	1.44	1.47	1.49	1.52	1.54	23
24	1.45	1.47	1.50	1.53	1.55	1.58	1.60	24
25	1.50	1.53	1.56	1.59	1.61	1.64	1.67	25

Normal Temperature of the Scale = 13° Reaumur.

Attached Thermometer. Degrees of Reaumur.	Barometer in Paris Lines. 315	320	325	330	335	340	345	Attached Thermometer. Degrees of Reaumur.
°	Par. Lines.	Par. Lines.	Par. Lines.	Par. Lines.	Par. Lines.	Par. Lines.	Par. Lines.	°
−15	+0.86	+0.88	+0.89	+0.90	+0.92	+0.93	+0.95	−15
−14	0.80	0.81	0.83	0.84	0.85	0.86	0.88	−14
−13	0.74	0.75	0.76	0.78	0.78	0.79	0.81	−13
−12	0.67	0.68	0.69	0.70	0.71	0.73	0.74	−12
−11	0.61	0.62	0.63	0.64	0.65	0.66	0.67	−11
−10	0.54	0.55	0.56	0.57	0.58	0.59	0.60	−10
− 9	+0.48	+0.49	+0.50	+0.50	+0.51	+0.52	+0.53	− 9
− 8	0.42	0.42	0.43	0.44	0.44	0.45	0.46	− 8
− 7	0.35	0.36	0.36	0.37	0.37	0.38	0.39	− 7
− 6	0.29	0.29	0.30	0.30	0.31	0.31	0.32	− 6
− 5	0.22	0.23	0.23	0.24	0.24	0.24	0.25	− 5
− 4	+0.16	+0.16	+0.17	+0.17	+0.17	+0.17	+0.18	− 4
− 3	0.10	0.10	0.10	0.10	0.10	0.10	0.11	− 3
− 2	+0.03	+0.03	+0.03	+0.03	+0.03	+0.03	+0.04	− 2
− 1	−0.03	−0.03	−0.03	−0.03	−0.03	−0.03	−0.03	− 1
0	−0.10	−0.10	−0.10	−0.10	−0.10	−0.10	−0.10	0
+ 1	−0.16	−0.16	−0.16	−0.17	−0.17	−0.17	−0.17	+ 1
2	0.22	0.23	0.23	0.23	0.24	0.24	0.24	2
3	0.29	0.29	0.30	0.30	0.31	0.31	0.31	3
4	0.35	0.36	0.36	0.37	0.37	0.38	0.38	4
5	0.42	0.42	0.43	0.44	0.44	0.45	0.45	5
+ 6	−0.48	−0.49	−0.49	−0.50	−0.51	−0.52	−0.53	+ 6
7	0.54	0.55	0.56	0.57	0.58	0.59	0.60	7
8	0.61	0.62	0.63	0.64	0.65	0.66	0.67	8
9	0.67	0.68	0.69	0.70	0.71	0.72	0.74	9
10	0.74	0.75	0.76	0.77	0.78	0.79	0.81	10
+11	−0.80	−0.81	−0.82	−0.84	−0.85	−0.86	−0.88	+11
12	0.86	0.88	0.89	0.90	0.92	0.93	0.95	12
13	0.93	0.94	0.96	0.97	0.99	1.00	1.02	13
14	0.99	1.01	1.02	1.04	1.05	1.07	1.09	14
15	1.05	1.07	1.09	1.10	1.12	1.14	1.16	15
+16	−1.12	−1.14	−1.15	−1.17	−1.19	−1.21	−1.23	+16
17	1.18	1.20	1.22	1.24	1.26	1.28	1.30	17
18	1.25	1.27	1.29	1.31	1.33	1.35	1.37	18
19	1.31	1.33	1.35	1.37	1.39	1.41	1.44	19
20	1.37	1.40	1.42	1.44	1.46	1.48	1.51	20
+21	−1.44	−1.46	−1.48	−1.51	−1.53	−1.55	−1.58	+21
22	1.50	1.53	1.55	1.57	1.60	1.62	1.65	22
23	1.57	1.59	1.62	1.64	1.67	1.69	1.72	23
24	1.63	1.66	1.68	1.71	1.73	1.76	1.79	24
25	1.69	1.72	1.75	1.78	1.80	1.83	1.86	25

TABLES

FOR CORRECTING THE

DEPRESSION OF THE BAROMETRICAL COLUMN

DUE TO CAPILLARY ACTION.

CORRECTION FOR CAPILLARY ACTION.

It is known that the effects of capillary action are not the same in different liquids. In a tube plunged in water, the liquid in the tube rises *higher* than the level of the water in the vessel, and terminates by a concave surface, which is called a *concave meniscus.* In a tube plunged in mercury the liquid in the tube stands *lower* than the mercury in the vessel, and terminates by a convex surface, or a *convex meniscus.* It is thus evident that the mercurial column in the tube of a Barometer does not rise to its true height, and that it needs to be corrected for the depression due to capillarity, before it indicates the real pressure of the atmosphere.

La Place, in the *Mécanique Céleste*, Tom. IV., has shown that the value of that correction depends upon the form of the meniscus, and gave a formula to compute it. As this form varies in tubes of different bores, so does the depression, which diminishes as the diameter of the tube increases. The form of the meniscus, however, was supposed to be the same in tubes of the same diameter, and constant in the same tube; and on this supposition the tables generally used for correcting the capillary action have been computed. But more accurate observations have proved that, owing to various causes not yet all well understood, the form of the meniscus is often different in tubes of the same diameter, and that it is even variable in the tube of the same instrument.

It thus became necessary to construct new tables, taking into consideration, in a given case, both the diameter of the tube and the form of the meniscus. Such tables, with a double entry, have been given by Schleiermacher, in the *Bibliothèque Universelle de Genève*, Tom. VIII.; by Bravais, in the *Annales de Physique et de Chimie*, Tom. V. p. 508; and by Delcros. The numbers in these tables agree very closely; but as Delcros's table is more extended than that of Schleiermacher, and in a more convenient form than that of Bravais, it is given below, together with a reduction of it to English measures, for the ordinary use.

The other tables may serve for comparison.

Table XXII., from the *Report of the Committee of Physics and Meteorology* of the Royal Society of London, 1840, gives the correction to be applied to English barometers for capillary action in boiled and unboiled tubes. It takes into account the diameter of the tube, but not the variations of the height of the meniscus, or of the convexity which terminates the barometrical column. This last element is supposed to be in its *normal state*, and *constant.*

Tables XXIII. and XXIV., by Delcros, in the *Annuaire Météorologique de France*, for 1849, give the means of finding the true correction to be applied to metrical barometers for capillary action.

The first shows the normal height of the meniscus when in contact with the air (as is the case in the inferior branch of a siphon barometer), and in the barometric vacuum at the top of the column, in tubes of different bores. It enables the observer to judge better of its variations.

Table XXIV. has been calculated by Delcros after the formulas of Schleiermacher, making the constant x equal to $6^{mm.}.5278$, being the mean value between that of Gay-Lussac $= 6^{mm.}.5262$, and that of Schleiermacher $= 6^{mm.}.5295$. It gives the amount of the capillary action in millimetres of mercury, taking into account both the size of the bore, or the internal radius of the tube, which will be found in the vertical argument, and the height of the meniscus, given in the horizontal argument. The internal radius of the tube is supposed to be known; the height of the meniscus, or the vertical distance from the base, that is, from the sharp line where the mercury ceases to be in contact with the walls of the tube, to the very top of the convexity, can be ascertained by measuring it several times by means of the vernier.

Example: — Suppose the internal radius of the tube to be $3^{mm.}.2$, and the height of the meniscus to be $0^{mm.}.8$; seek in the first vertical column the number $3^{mm.}.2$; follow then the horizontal line as far as the vertical column headed $0^{mm.}.8$, you find there the number $0^{mm.}.776$, which is the amount of the depression due to capillary action, or the value of the correction to be *added* to the observation.

Table XXV. is taken from Pouillet's *Eléments de Physique*, Vol. II. p. 698 (1853).

Table XXVI. is found in Gehler's *Physicalische Wörterbuch*, and in Schubarth, *Physicalische Tabellen*, p. 21.

Table XXVII., which is Delcros's table reduced into English measures, gives the means of correcting with more accuracy the indications of the English barometers. For its use, see, above, the explanation to Table XXIV.

Table XXVIII. is from Baily's *Astronomical Tables*.

XXII. Table for the Correction to be added to English Barometers for Capillary Action.

Diameter of Tube.	Correction for Unboiled Tubes.	Correction for Boiled Tubes.
Inch.	Inch.	Inch.
0.60	0.004	0.002
0.50	0.007	0.003
0.45	0.010	0.005
0.40	0.014	0.007
0.35	0.020	0.010
0.30	0.028	0.014
0.25	0.040	0.020
0.20	0.060	0.029
0.15	0.088	0.044
0.10	0.142	0.070

XXIII. Table of the Height of the Meniscus of the Barometrical Column.

Internal Radius of the Tube in Millimetres.	Normal Height of the Meniscus in Millimetres. In the Air.	Normal Height of the Meniscus in Millimetres. In the Vacuum.
1	0.427	0.34
2	0.795	0.64
3	1.079	0.86
4	1.287	1.03
5	1.413	1.13
6	1.488	1.19
7	1.524	1.22

VERTICAL ARGUMENT = INTERNAL RADIUS OF TUBE. HORIZONTAL ARGUMENT = HEIGHT OF MENISCUS IN MILLIMETRES.

Radius of the Tube in Millimetres.	Height of the Meniscus in Millimetres.																		Radius of the Tube in Millimetres.
	0.1	**0.2**	**0.3**	**0.4**	**0.5**	**0.6**	**0.7**	**0.8**	**0.9**	**1.0**	**1.1**	**1.2**	**1.3**	**1.4**	**1.5**	**1.6**	**1.7**	**1.8**	
	Millim.	Millim.	Millim.	Millim.	Millim.	Millim.	Millim.	Millim.	Millim.	Millim.	Millim.	Millim.	Millim.	Millim.	Millim.	Millim.	Millim.	Millim.	
1.0	1.268	2.460	3.516	4.396	5.085	"	"	"	"	"	"	"	"	"	"	"	"	"	1.0
1.2	0.876	1.715	2.484	3.162	3.728	4.190	"	"	"	"	"	"	"	"	"	"	"	"	1.2
1.4	0.638	1.256	1.836	2.363	2.825	3.218	3.542	"	"	"	"	"	"	"	"	"	"	"	1.4
1.6	0.484	0.955	1.404	1.820	2.196	2.528	2.812	3.050	"	"	"	"	"	"	"	"	"	"	1.6
1.8	0.378	0.747	1.103	1.437	1.746	2.024	2.270	2.483	2.662	"	"	"	"	"	"	"	"	"	1.8
2.0	0.302	0.598	0.885	1.158	1.413	1.648	1.859	2.046	2.209	2.348	"	"	"	"	"	"	"	"	2.0
2.2	0.245	0.487	0.723	0.948	1.161	1.360	1.541	1.705	1.851	1.978	2.087	"	"	"	"	"	"	"	2.2
2.4	0.203	0.403	0.599	0.787	0.966	1.135	1.292	1.436	1.565	1.680	1.780	1.866	"	"	"	"	"	"	2.4
2.6	0.170	0.337	0.502	0.661	0.813	0.958	1.093	1.218	1.332	1.436	1.528	1.608	1.676	"	"	"	"	"	2.6
2.8	0.143	0.285	0.425	0.560	0.691	0.815	0.932	1.041	1.142	1.235	1.318	1.392	1.456	1.511	"	"	"	"	2.8
3.0	0.122	0.243	0.362	0.478	0.591	0.698	0.800	0.896	0.985	1.068	1.143	1.210	1.270	1.322	1.368	"	"	"	3.0
3.2	0.105	0.209	0.312	0.412	0.509	0.602	0.691	0.776	0.855	0.928	0.995	1.057	1.112	1.161	1.203	1.238	"	"	3.2
3.4	0.091	0.181	0.269	0.356	0.441	0.523	0.601	0.675	0.745	0.810	0.871	0.926	0.976	1.021	1.061	1.095	"	"	3.4
3.6	0.079	0.157	0.234	0.310	0.384	0.455	0.524	0.590	0.652	0.710	0.764	0.814	0.860	0.901	0.938	0.970	"	"	3.6
3.8	0.069	0.137	0.205	0.271	0.336	0.399	0.459	0.517	0.572	0.624	0.673	0.718	0.760	0.797	0.831	0.861	0.887	"	3.8
4.0	0.060	0.120	0.180	0.238	0.295	0.350	0.404	0.455	0.504	0.551	0.594	0.635	0.673	0.707	0.738	0.766	0.790	"	4.0
4.2	0.053	0.106	0.158	0.210	0.260	0.309	0.356	0.402	0.446	0.487	0.526	0.563	0.597	0.628	0.657	0.682	0.705	"	4.2
4.4	0.047	0.094	0.140	0.185	0.230	0.273	0.315	0.356	0.395	0.432	0.467	0.500	0.531	0.559	0.585	0.609	0.630	"	4.4
4.6	0.042	0.083	0.124	0.164	0.204	0.242	0.280	0.316	0.351	0.384	0.416	0.445	0.473	0.499	0.522	0.544	0.563	"	4.6
4.8	0.037	0.074	0.110	0.146	0.181	0.215	0.249	0.281	0.312	0.342	0.370	0.397	0.422	0.445	0.467	0.486	0.504	"	4.8
5.0	0.033	0.065	0.098	0.130	0.161	0.192	0.221	0.250	0.278	0.305	0.330	0.354	0.377	0.398	0.418	0.436	0.452	"	5.0
5.2	0.029	0.058	0.087	0.116	0.144	0.171	0.198	0.224	0.248	0.272	0.295	0.317	0.337	0.356	0.374	0.390	0.405	0.418	5.2
5.4	0.026	0.052	0.078	0.103	0.128	0.153	0.177	0.200	0.222	0.244	0.264	0.284	0.302	0.319	0.336	0.350	0.364	0.376	5.4
5.6	0.023	0.047	0.070	0.092	0.115	0.137	0.158	0.179	0.199	0.218	0.237	0.255	0.271	0.287	0.301	0.315	0.327	0.338	5.6
5.8	0.021	0.042	0.062	0.083	0.103	0.122	0.142	0.160	0.178	0.196	0.213	0.228	0.243	0.257	0.271	0.283	0.294	0.304	5.8
6.0	0.019	0.037	0.056	0.074	0.092	0.110	0.127	0.144	0.160	0.176	0.191	0.205	0.219	0.231	0.243	0.254	0.264	0.273	6.0
6.2	0.017	0.034	0.050	0.067	0.083	0.099	0.114	0.129	0.144	0.158	0.172	0.185	0.197	0.208	0.219	0.229	0.238	0.246	6.2
6.4	0.015	0.030	0.045	0.060	0.074	0.089	0.103	0.116	0.130	0.142	0.154	0.166	0.177	0.187	0.197	0.206	0.214	0.221	6.4
6.6	0.014	0.027	0.041	0.054	0.067	0.080	0.093	0.105	0.117	0.128	0.139	0.150	0.160	0.169	0.178	0.186	0.193	0.200	6.6
6.8	0.012	0.024	0.037	0.049	0.061	0.072	0.084	0.095	0.105	0.116	0.126	0.135	0.144	0.153	0.160	0.168	0.174	0.180	6.8
7.0	0.011	0.022	0.033	0.044	0.055	0.065	0.075	0.085	0.095	0.105	0.114	0.122	0.130	0.138	0.145	0.152	0.158	0.163	7.0

FROM POUILLET.

Internal Diameter of Tube.	Depression.	Differences.	Internal Diameter of Tube.	Depression.	Differences.	Internal Diameter of Tube.	Depression.	Differences.
Millimetres.	Millimetres.	Millimet.	Millimetres.	Millimetres.	Millimet.	Millimetres.	Millimetres.	Millimet.
2.00	4.579	0.985	8.50	0.604	0.070	15.00	0.127	0.015
2.50	3.595	0.692	9.00	0.534	0.061	15.50	0.112	0.013
3.00	2.902	0.487	9.50	0.473	0.054	16.00	0.099	0.012
3.50	2.415	0.362	10.00	0.419	0.047	16.50	0.087	0.010
4.00	2.053	0.301	10.50	0.372	0.042	17.00	0.077	0.009
4.50	1.752	0.245	11.00	0.330	0.037	17.50	0.068	0.008
5.00	1.507	0.201	11.50	0.293	0.033	18.00	0.060	0.007
5.50	1.306	0.170	12.00	0.260	0.030	18.50	0.053	0.006
6.00	1.136	0.141	12.50	0.230	0.026	19.00	0.047	0.006
6.50	0.995	0.118	13.00	0.204	0.023	19.50	0.041	0.005
7.00	0.877	0.102	13.50	0.181	0.020	20.00	0.036	0.004
7.50	0.775	0.091	14.00	0.161	0.018	20.50	0.032	0.004
8.00	0.684	0.080	14.50	0.143	0.016	21.00	0.028	

XXVI. DEPRESSION OF THE BAROMETRICAL COLUMN DUE TO CAPILLARY ACTION.

Internal Diameter of Tube.	Depression according to				Internal Diameter of Tube.	Depression according to			
	La Place.	Young.	Ivory.	Cavendish.		La Place.	Young.	Ivory.	Cavendish.
Millimetres.	Millim.	Millim.	Millim.	Millim.	Millimetres.	Millim.	Millim.	Millim.	Millim.
2.00	4.454	4.887	4.888	4.472	11.50	0.315			
2.50	3.568				12.00	0.281	0.242	0.253	0.200
3.00	2.918	2.986	2.988	3.054	12.50	0.250			
3.50	2.442				13.00	0.223	0.188	0.196	0.170
4.00	2.068	2.063	2.066	2.187	13.50	0.198			
4.50	1.774				14.00	0.176	0.144	0.152	0.150
5.00	1.534	1.510	1.513	1.735	14.50	0.156			
5.50	1.337				15.00	0.137	0.111	0.118	0.131
6.00	1.171	1.139	1.134	1.377	15.50	0.121			
6.50	1.030				16.00	0.107	0.088	0.087	
7.00	0.909	0.869	0.868	1.073	16.50	0.094			
7.50	0.803				17.00	0.083	0.068	0.071	
8.00	0.712	0.669	0.673	0.820	17.50	0.073			
8.50	0.632				18.00	0.064	0.053	0.054	
9.00	0.562	0.517	0.521	0.608	18.50	0.056			
9.50	0.500				19.00	0.049	0.041	0.042	
10.00	0.445	0.402	0.406	0.406	19.50	0.043			
10.50	0.397				20.00	0.038	0.031	0.031	
11.00	0.354	0.311	0.316	0.270	20.50	0.034			
11.50	0.315				21.00	0.030	0.024	0.024	

XXVII. DEPRESSION OF THE BAROMETRICAL COLUMN DUE TO CAPILLARY ACTION, REDUCED INTO ENGLISH INCHES FROM DELCROS'S TABLE.

Internal Diameter of Tube.	Height of Meniscus in Thousandths of an English Inch.													
	5	**10**	**15**	**20**	**25**	**30**	**35**	**40**	**45**	**50**	**55**	**60**	**65**	**70**
Eng. In.	Inch.	Inch.	Inch.	Inch.	Inch.	Inch.	Inch.	Inch.	Inch.	Inch.	Inch.	Inch.	Inch.	Inch.
0.10	0.040	0.076	0.109	0.136	0.155									
0.12	.027	.053	.076	.097	.114									
0.14	.019	.038	.056	.071	.085	0.097								
0.16	.015	.029	.042	.055	.066	.076	0.084							
0.18	.011	.022	.033	.043	.052	.060	.067	0.073						
0.20	.009	.018	.026	.034	.042	.049	.055	.060	0.064					
0.22	.007	.014	.021	.028	.034	.040	.045	.049	.053	0.057				
0.24	.006	.012	.017	.023	.028	.033	.037	.041	.045	.048	0.050			
0.26	.005	.010	.014	.019	.023	.027	.031	.035	.038	.040	.043	0.045		
0.28	.004	.008	.012	.016	.019	.023	.026	.029	.032	.034	.036	.038		
0.30	.003	.007	.010	.013	.016	.019	.022	.025	.027	.029	.031	.033	0.034	
0.32	.003	.006	.009	.011	.014	.016	.019	.021	.023	.025	.027	.028	.030	
0.34	.002	.005	.007	.010	.012	.014	.016	.018	.020	.022	.023	.024	.026	
0.36	.002	.004	.006	.008	.010	.012	.014	.016	.017	.019	.020	.021	.022	
0.38	.002	.004	.005	.007	.009	.010	.012	.013	.015	.016	.017	.018	.019	
0.40	.002	.003	.005	.006	.008	.009	.010	.012	.013	.014	.015	.016	.017	
0.42	.001	.003	.004	.005	.007	.008	.009	.010	.011	.012	.013	.014	.015	0.015
0.44	.001	.002	.004	.005	.006	.007	.008	.009	.010	.011	.011	.012	.013	.013
0.46	.001	.002	.003	.004	.005	.006	.007	.008	.008	.009	.010	.011	.011	.012
0.48	.001	.002	.003	.004	.004	.005	.006	.007	.007	.008	.009	.009	.010	.010
0.50	.001	.002	.002	.003	.004	.004	.005	.006	.006	.007	.008	.008	.008	.009
0.52	.001	.001	.002	.003	.003	.004	.005	.005	.006	.006	.007	.007	.007	.008
0.54	.001	.001	.002	.002	.003	.003	.004	.004	.005	.005	.006	.006	.006	.007
	5	**10**	**15**	**20**	**25**	**30**	**35**	**40**	**45**	**50**	**55**	**60**	**65**	**70**

XXVIII. DEPRESSION OF THE BAROMETRICAL COLUMN DUE TO CAPILLARY ACTION, EXPRESSED IN ENGLISH INCHES. — BAILY.

Diameter of Tube.	Depression according to			Diameter of Tube.	Depression according to		
	Ivory.	Young.	La Place.		Ivory.	Young.	La Place.
Eng. Inch.	Eng. Inch.	Eng. Inch.	Eng. Inch.	Eng. Inch.	Eng. Inch.	Eng. Inch.	Eng. Inch.
0.05	0.2949	0.2964	0. . . .	0.35	0.0212	0.0196	0.0216
0.10	.1404	.1424	.1394	0.40	.0154	.0139	.0159
0.15	.0865	.0880	.0854	0.45	.0112	.0100	.0117
0.20	.0583	.0589	.0580	0.50	.0082	.0074	.0087
0.25	.0409	.0404	.0412	0.60	.0043	.0045	.0046
0.30	.0293	.0280	.0296	0.70	.0023	. . . .	.0024
0.35	0.0212	0.0196	0.0216	0.80	0.0012	0. . . .	0.0013

METEOROLOGICAL TABLES.

IV.

HYPSOMETRICAL TABLES.

CONTENTS.

(The figures refer to the folio at the bottom of the page.)

1. Barometrical Measurement of Heights.

For converting English Yards and Feet into different Measures.

BAROMETRICAL

MEASUREMENT OF HEIGHTS,

OR

TABLES

FOR COMPUTING DIFFERENCES OF ELEVATIÓN FROM BAROMETRICAL OBSERVATIONS.

HYPSOMETRICAL TABLES

FOR

COMPUTING DIFFERENCES OF ELEVATION FROM BAROMETRICAL OBSERVATIONS.

Numerous determinations of altitude are one of the great desiderata of physical science, and no more ready means for obtaining them is at the disposal of the scientific man than the Barometer. A traveller, furnished with the improved and convenient instruments we can now command, and with some experience in using them, can take a large number of barometric observations for determining heights, at the cost of little trouble or time. It is, however, quite otherwise with the computations by which the results are obtained. The prospect of that tedious and time-robbing labor not only too often cools the zeal of the observer, but a vast amount of data actually collected remain of no avail from the want of having been computed.

The object of this much enlarged set of Hypsometrical Tables is to facilitate the task of the computer. It contains practical tables adapted to the three usual barometrical scales, and, among them, No. I., II., and V. are so disposed as to dispense with the use of logarithms, and to reduce the computation to the simplest arithmetical operations. The others suppose the use of logarithms, a method which may still be preferred by some observers.

As these various tables represent the development of the principal formulæ which have been proposed, the computer is enabled to compare the results obtained by each of them, and to select that which he most approves.

These formulæ may be referred to two classes, the respective types of which are Laplace's and Bessel's formulæ.

Laplace, in the *Mécanique Céleste*, Tom. IV. p. 292, gave a complete solution of the problem, and proposed a formula which soon superseded the older and less accurate formulæ of De Luc, Shuckburgh, and others. The coefficients which enter in it were derived from the best determinations of the needed physical constants which science could then furnish, the most important of which are the relative weight of the air and of the mercury, and the rate of expansion of air by heat. The first was assumed to be $\frac{1}{10467}$, according to the experiments of Biot and Arago; and the barometrical coefficiency deduced from it, 18317 metres. This coefficient was, however, empirically increased to 18336 metres, in order to adjust the results of the formula to those furnished by the careful trigonometrical measurements made by Ramond for the purpose of testing its correctness. It becomes 18393 metres when including the correction due to the effect of the decrease of gravity with the height on the density of the mercurial column and of the air. The coefficient expressing the expansion of the air by heat, as determined by Gay-Lussac, viz. 0.00375 of its bulk for one Centigrade degree, was adopted, but Laplace increased it to 0.004, in order to take into the account the effect of the greater expansive power of the vapors contained in the atmosphere.

These values have been retained in the different formulæ proposed later by Gauss, in Schumacher's *Jahrbuch* for **1840**, by Schmidt, *Mathem. und Physische Geographie*, II. p. **205**, and by Baily, *Astronomical Tables*, p. **183**, which, therefore, only change the form without changing the results. D'Aubuisson, in his formula and tables, *Traité de Géognosie*, p. **488**, only reduced the barometrical coefficient to its theoretical value, which he determined to be **18365** metres, leaving unchanged the other coefficients of Laplace's formula.

Bessel first introduced, in his formula, *Astronomische Nachrichten*, No. **356**, a separate correction for the effect of moisture. The correction for the temperature of the air is computed in his tables for two values of the coefficient, that of Gay-Lussac, **0.00375**, and that of Rudberg, **0.00365**. Laplace's barometrical coefficient is retained, but the correction for the decrease of gravity is considerably modified.

In Elie Ritter's formula, in the *Mémoires de la Societé de Physique de Genève*, Tom. XIII. p. **343**, the corrections for temperature and moisture are also separated; but other values of the barometrical and thermometrical coefficients, derived from Regnault's determinations, are used, and a new method is proposed for applying the correction due to the expansion of air, which is made proportional to the square of the difference between the observed temperatures at each station.

Baeyer's formula, recently published in Poggendorf's *Annalen der Physik und Chemie*, Tom. XCVIII. p. **371**, does not belong to either of the two classes just mentioned; for while it keeps Laplace's barometrical and thermometrical coefficients, it corrects the effect of temperature by a method analogous to that of Ritter, and it entirely neglects the effect of aqueous vapor.

In the following set the tables of Delcros, Guyot, and Loomis develop the formula of Laplace. The much larger tables of Delcros render unnecessary those of Oltmanns, which are yearly reprinted in the *Annuaire du Bureau des Longitudes.* Instead of Gauss's tables will be found the tables of Dippe, which are computed from the same formula, but are more extended. Baily's tables close the first series. The tables of Plantamour, computed from Bessel's formula, are given here in preference to Bessel's tables, because Plantamour substituted for Laplace's barometrical coefficient that derived from the probably more accurate determination of the relative weight of the air and mercury by Regnault, viz. **18404.8** metres. E. Ritter's tables, computed from his own formula, give perhaps, in extreme cases, better results; but as, in ordinary circumstances, the altitudes obtained do not much differ from those furnished by the less complicated tables of Plantamour, they were not reprinted here.

The miscellaneous tables which follow furnish useful materials for solving several questions connected with the barometrical measurements.

Regnault's table of Barometric Pressures corresponding to Temperatures of the Boiling Point of Water, revised by Moritz, and its reduction to English measures, will be found a valuable addition for thermometrical measurements of height.

The Appendix to the Hypsometrical Tables now offers, in a new form, a complete series of tables for the comparison of the different measures of length generally used for indicating altitudes, the convenience of which will be fully appreciated by those who have attempted to collect and to use the abundant contributions furnished by all civilized nations to that branch of geographical science.

I.

TABLES

FOR

DETERMINING DIFFERENCES OF LEVEL BY MEANS OF BAROMETRICAL OBSERVATIONS,

COMPUTED FROM THE COMPLETE FORMULA OF LAPLACE,

By M. T. Delcros.

Construction of the Tables.

If we take z = difference of level of the two barometers,
a = earth's mean radius = 6366200 metres,
L = mean latitude between the two stations,

and further:—

At Station.
- Lower.
  - h = observed height of the barometer,
  - T = temperature of the barometer,
  - t = temperature of the air,
- Upper.
  - h' = observed height of the barometer,
  - T′ = temperature of the barometer,
  - t' = temperature of the air,

and if we make finally $\mathrm{H} = h + h'.\ \left(\frac{\mathrm{T}-\mathrm{T}'}{6196}\right)$,

we shall have, according to Laplace, the following general and complete equation:—

$$z = 18336 \text{ metres} \times \left\{\begin{array}{l} \left(1 + \frac{2.\,(t+t')}{1000}\right) \\ (1 + 0.0028371 \cos. 2.\,\mathrm{L}) \\ \left((1 + \frac{z}{a}).\ \ \mathrm{Log.}\left(\frac{h}{\mathrm{H}}\right) + \frac{z}{a}\ 0.868589\right) \end{array}\right\},$$

after the proper transformations this equation becomes:—

$$z = \mathrm{Log.}\left(\frac{h}{\mathrm{H}}\right) 18336 \text{ metres} \times \left\{\begin{array}{l} \left(1 + \frac{2.\,(t+t')}{1000}\right) \\ (1 + 0.0028371 \cos. 2.\,\mathrm{L}) \\ \left(1 + \dfrac{\left(\log.\left(\frac{h}{\mathrm{H}}\right) + 0.868589\right).\ \frac{z}{a}}{\mathrm{Log.}\left(\frac{h}{\mathrm{H}}\right)}\right) \end{array}\right\}$$

introducing into this expression the value in metres of a, the earth's mean radius, making $z = \text{Log.} \left(\frac{h}{\text{H}}\right)$ 18336 and $\text{Log.} \left(\frac{h}{\text{H}}\right) = \left(\frac{z}{18336}\right)$, which can be done without sensible error, the above formula takes the following form, sufficiently accurate for practical purposes : —

$$z = \text{Log.} \left(\frac{h}{\text{H}}\right). \ 18336 \text{ metres} \times \left\{ \begin{array}{l} \left(1 + \frac{2. \ (t + t')}{1000}\right) \\ (1 + 0.0028371 \ \cos. \ 2. \ \text{L}) \\ \left(1 + \frac{z + 15926}{6366200}\right) \end{array} \right\}$$

the four factors of which can easily be developed in tables, as has been done by Mr. Oltmanns. But though this *savant* chose to develop also the second factor, I found better not to do so, partly because the calculation of it is very easy, and also on account of the great extent it would have been necessary to give to this table, in order to avoid troublesome interpolations.

In the calculation of $h'. \left(\frac{\text{T} - \text{T}'}{6196}\right)$, Mr. Oltmanns used the constant coefficient of the absolute expansion of the mercurial column ; I took that of the relative expansion of the mercury and of the brass scale. It is obvious, therefore, that if the scale of the barometer employed was of wood, glass, iron, or of another substance, it would be necessary to make use of as many different coefficients, and the Table II. could not be used. Moreover, Oltmanns combined the last two factors of the general formula in one single table with double entry. This table I have calculated, extending it sufficiently to avoid a double interpolation ; but as it seemed to me much too extensive, I substituted for it Tables III. and IV., which are more condensed, without rendering any troublesome interpolation necessary.

I carried the calculation of these tables beyond the limits at which Oltmanns chose to stop, in order that they may answer for the most extreme cases.

At the head of each table will be found the factor of which it is the development; this makes any other explanation superfluous.

All these tables give, at sight, the numbers wanted; only when very great precision is desired, a slight interpolation, at sight, and very easy to apply, may be required. My principal object was to relieve the computer of the troublesome and annoying labor of interpolations.

I added to these four tables the small Table V., taken from the *Annuaire du Bureau des Longitudes* of Paris. It will be seldom used.

When calculating differences of level, in the same order, with the tables, and by the complete formula of Laplace, the results thus obtained never differ by more than one decimetre in the most extreme cases. The following example will illustrate this statement. I take the observation made in a balloon, by Gay-Lussac, at Paris, as an extreme case, which is very well adapted to manifest the errors of the tables, if there were any, by comparing the results obtained by means of them with those of the direct calculation according to the complete formula of Laplace, from which they are derived.

Example of Calculation by the complete Formula of Laplace and by the Tables Height of the Balloon of Gay-Lussac.

The observation gave:—

Balloon	$h' = 328.80$ mm.	$T' = -\ 9^\circ.5$	$t' = -\ 9^\circ.5$
Paris	$h = 765.68$	$T = +\ 30.8$	$t = +\ 30.8$

$T - T' = +\ 40.3 \quad (t + t') = +\ 21.3$ et $2\ (t + t') = 42^\circ.6$

With these data the formula of Laplace gives the following calculation:—

Log. h'. = 328.80		= 2.5169318	
Log. $(T - T') = +\ 40.3$		= 1.6053050	
Log. dilat. coefficient = 0.0001614		= 6.2079035	
Corr. a = +	Milli. 2.14 log.	= 0.3301403	
h' =	328.80		
H =	330.94 log.		= 2.5197480
	log. h = 765.68		= 2.8840473
(Log. h — Log. H) = Difference of Log.			= 0.3642993
Log. of (Log. h — log. H) =			9.5614583
Log. general coefficient = 18336 =			4.2633046
Log. $\left(\left(\frac{h}{H}\right) 18336 \right) = (A + a) =$			3.8247629

Corresponding number = 6679.79 = $(A + a)$

Log. cos. 2 L = 97° 40′ = —	9.1251872
Log. constant = 0.0028371 = +	7.4528746
Log. $(A + a)$ = 6679.79. = +	3.8247629
Log. $\left((0.0028371.\ \text{Cos. } 2\ L) \times (A + a) \right) = -$	0.4028247

Corresponding number =	Milli. — 2.53		
	6679.79		
$(A + a + \beta) =$	6677.26		
Corr. temp. air $= v =$	284.45		= Metres. (6.677×42.6)
$(A + a + \beta + v) =$	6961.71		
Constant =. +	15926		
	22887.71	... Log. ...	= 4.3596022
Comp^t. log. a =	6366200	... Log. ...	= 3.1961197
$(A + a + \beta + v) =$	6961.71	... Log. ...	= 3.8427153
$\delta = +$	25.03	Log.	= + 1.3984372
$(A + a + \beta + v + \delta) =$	6986.74		
Altitude barom. Paris =	48.70		
Altitude of balloon =	7035.44	by the formula of Laplace.	

Now let us calculate by the tables, placing side by side the corresponding results given by the formula of Laplace.

	Millim.		
Balloon	$h' = 328.80$	$T' = -\ 9^{\circ}.5$	$t' = -\ 9^{\circ}.5$
Paris	$h = 765.68$	$T = +\ 30.8$	$t = +\ 30.8$

		Metres.	
with $\left\{ \begin{array}{l} h' = 328.80 \\ h = 765.68 \end{array} \right\}$	Table I. gives	$\left\{ \begin{array}{l} 1478.4 \\ 8209.8 \end{array} \right.$	By the formula of Laplace we found :
	$A =$	6731.4	
with $(T' - T) = -\ 40^{\circ}.3$, Table II. gives	$\alpha =$	-52.0	Metres.
	$(A + \alpha) =$	6679.4	6679.79
with $L = 48^{\circ}\ 50'$, Table III. gives α	$= -$	2.3	$-$ 2.53
	$(A + \alpha + \beta) =$	6677.1	6677.26
with $2\ (t + t')$ direct calculation gives v	$= +$	284.5	$+$ 284.45
	$(A + \alpha + \beta + v) =$	6961.6	6961.71
with 6960, Table IV. gives δ	$= +$	25.1	$+$ 25.03
	$(A + \alpha + \beta + v + \delta) =$	6986.7	6986.74
Altitude of barometer at Paris	$= +$	48.7	$+$ 48.70
Therefore altitude of balloon	$=$	7035.4	7035.44

Two results which are sensibly identical. This ought not to astonish us; the tables being the exact development of the formula, they ought to give the same results, provided in both cases nothing has been neglected, and the four factors have been calculated in the same relative order.

DELCROS.

Disposition and Use of the Tables.

The disposition of the tables is the following : —

In Table I., the first column on the left contains the height of the barometer in millimetres, corrected for the error of the instrument.

The second column headed N (number), gives in metres the first two figures of the number corresponding to each height of the barometer in the first column ; the third column, headed 0.0, gives the remaining figures for the full number of millimetres ; the following columns give the remaining figures for the same number of millimetres and each decimal fraction of a millimetre which may follow it. The value of the hundredths is to be found in the last column.

Example : — Height of Barometer = 761.00.

We look out in the first column for the number 761, and we find on the same line in the second column, 81 ; in the third column, headed 0.0, or full number, 61.1. The corresponding number is thus 8161.1 metres.

Height of barometer = 761.35.

The second column gives 81; the column headed 0.3 gives, on the same line, 64.2. The corresponding number is then 8164.2. Adding the value of five hundredths of millim., being $0^{m}.5$, as indicated in the last column, we have 8164.7 metres, corresponding to 761.35 millim.

The other four tables need no further explanation.

To calculate, by means of the tables, a difference of level from two barometrical observations, proceed in the following manner: —

1. Take the height of the barometer at the lower station, or h, and seek in Table I. the number corresponding to this height. Seek likewise the number corresponding to the height of the barometer at the upper station. Subtract the second from the first. The remainder is the approximate difference of level between the two stations. Then apply the following corrections.

2. Correction to be applied for the temperature of the barometers.

If T' be the temperature of the attached thermometer at the upper station, and T that of the attached thermometer at the lower station, take the difference, or $T' - T$, and seek in Table II. the number corresponding to this difference.

When T' is smaller than T, that is, when the temperature of the attached thermometer of the upper station is lower than that of the lower station, the correction is to be *subtracted* from the approximate height; when T' is greater than T, it is to be *added*.

3. Correction for the temperature of the air.

The first correction having been applied, multiply the number obtained, or N, by the double sum of the temperatures of the air at both stations, and divide the product by 1000; the number thus found, or the quantity expressed by $\frac{N}{1000} \cdot 2\,(t + t')$ is the correction in metres which is to be *added* to the preceding number N.

4. Tables III. and IV. give two corrections; the first due to the decrease of gravitation in latitude, which is to be *added* when the mean latitude of the places of observation is between the 45th parallel and the equator; and to be *subtracted* when it is between the same parallel and the poles, as indicated at the head of the columns. The second correction, due to the decrease of gravitation on the vertical line, is always *additive*.

5. Table V. gives another small correction to be added in the case of the lower station being very elevated above the level of the ocean.

Examples of Calculation.

Measurement of the Height of Guanaxuato. By M. de Humboldt.

Barometer at the upper station,	$h' = 600^{mm}.95$	$T' = 21^{\circ}.3$	$t' = 21^{\circ}.3$
Barometer at the level of the sea,	$h = 763.15$	$T = 25.3$	$t = 25.3$

Table I. gives the corresponding numbers,	$\begin{cases} h = 8183.5 \\ h' = 6280.8 \end{cases}$
Difference,	1902.7
Table II. gives for T′ — T,	— 5.2
Difference,	1897.5 = N
$\frac{N}{1000} \cdot 2\,(t + t') = 1.897 \times 93.2,$	+ 176.8
Sum,	2074.3
Table III. gives for mean latitude of 21°,	+ 4.3
Table IV. gives for decrease of gravitation in the vertical line,	+ 6.0
Hence altitude of Guanaxuato above the ocean,	2084.6

Measurement of the height of Mont Blanc, August 29, 1844. *By MM. Bravais and Martins.*

Barometer at one metre below the summit,	$h' = 424.05$ mm.	T′ = — 4.2°	t' = — 7.6°
Barometer of the Observatory of Geneva,	$h = 729.65$	T = 18.6	t = 19.3

Table I. gives for numbers corresponding to	$\begin{cases} h = 7826.0 \\ h' = 3504.4 \end{cases}$
Difference,	4321.6
Table II. gives for T′ — T,	— 29.3
Difference,	4292.3 = N
$\frac{N}{1000} \cdot 2\,(t + t') = 4292 \times 23.4 =$	+ 100.4
Sum,	4392.7
Table III. gives for the mean latitude of 46°,	— 0.4
Difference,	4392.3
Table IV. for decrease of gravitation in the vertical line	+ 13.7
Table V. for the elevation of the lower station,	+ 0.5
Sum,	4406.5
Elevation of the lower barometer above the ocean,	407.0
Hence elevation of upper barometer above the ocean,	4813.5
Finally, height of the summit of Mont Blanc above the ocean,	4814.5

TABLE I. — Giving A = 18336 × log. H or *h*...., argument H or *h* in Millimetres.

Barometer *H* or h.	N.	0.0	0.1	0.2	0.3	0.4	0.5	0.6	0.7	0.8	0.9	Parts for each 0.01mm.
Milli.	Metr.	Metres.	Metres.	Metres.	Metres.	Metres.	Metres.	Metres.	Metres.	Metres.	Metres.	Metr.
288	4	23.4	26.2	28.9	31.7	34.4	37.2	40.0	42.7	45.5	48.2	1 0.3
289	4	51.0	53.8	56.5	59.3	62.0	64.8	67.5	70.3	73.0	75.8	2 0.5
290	4	78.5	81.3	84.0	86.7	89.5	92.2	95.0	97.7			3 0.8
290	5									00.4	03.2	4 1.1
291	5	05.9	08.7	11.4	14.1	16.8	19.6	22.3	25.0	27.8	30.5	5 1.4
292	5	33.2	36.0	38.7	41.4	44.1	46.8	49.6	52.3	55.0	57.7	6 1.6
293	5	60.5	63.2	65.9	68.6	71.3	74.0	76.7	79.5	82.2	84.9	7 1.9
294	5	87.6	90.3	93.0	95.7	98.4						8 2.2
294	6						01.1	03.8	06.5	09.2	11.9	9 2.4
295	6	14.6	17.3	20.0	22.7	25.4	28.1	30.8	33.5	36.2	38.9	
296	6	41.6	44.3	47.0	49.6	52.3	55.0	57.7	60.4	63.1	65.8	
297	6	68.4	71.1	73.8	76.5	79.1	81.8	84.5	87.2	89.9	92.5	
298	6	95.2	97.9									
298	7			00.5	03.2	05.9	08.6	11.2	13.9	16.6	19.2	
299	7	21.9	24.5	27.2	29.9	32.5	35.2	37.8	40.5	43.2	45.8	
300	7	48.5	51.1	53.8	56.4	59.1	61.7	64.4	67.0	69.7	72.3	
301	7	75.0	77.6	80.3	82.9	85.5	88.2	90.8	93.5	96.1	98.7	
302	8	01.4	04.0	06.6	09.3	11.9	14.5	17.2	19.8	22.4	25.1	
303	8	27.7	30.3	33.0	35.6	38.2	40.8	43.5	46.1	48.6	51.3	
304	8	54.0	56.6	59.2	61.8	64.4	67.0	69.6	72.3	74.9	77.5	
305	8	80.1	82.7	85.3	87.9	90.5	93.1	95.7	98.3			
305	9									01.0	03.6	
306	9	06.2	08.8	11.4	14.0	16.6	19.2	21.8	24.4	27.0	29.6	1 0.3
307	9	32.1	34.7	37.3	39.9	42.5	45.1	47.7	50.3	52.9	55.5	2 0.5
308	9	58.0	60.6	63.2	65.8	68.4	70.9	73.5	76.1	78.7	81.3	3 0.8
309	9	83.9	86.4	89.0	91.6	94.1	96.7	99.3				4 1.0
309	10								01.9	04.4	07.0	5 1.3
310	10	09.6	12.1	14.7	17.3	19.8	22.4	25.0	27.5	30.1	32.7	6 1.5
311	10	35.2	37.8	40.3	42.9	45.5	48.0	50.6	53.1	55.7	58.2	7 1.8
312	10	60.8	63.3	65.9	68.4	71.0	73.5	76.1	78.6	81.2	83.7	8 2.1
313	10	86.3	88.8	91.4	93.9	96.4	99.0					9 2.3
313	11							01.5	04.1	06.6	09.1	
314	11	11.7	14.2	16.7	19.3	21.8	24.3	26.9	29.4	31.9	34.5	
315	11	37.0	39.5	42.0	44.6	47.1	49.6	52.1	54.7	57.2	59.7	
316	11	62.2	64.8	67.3	69.8	72.3	74.8	77.3	79.9	82.4	84.9	
317	11	87.4	89.9	92.4	94.9	97.4	99.9					
317	12							02.4	05.0	07.5	10.0	
318	12	12.5	15.0	17.5	20.0	22.5	25.0	27.5	30.0	32.5	35.0	
319	12	37.5	40.0	42.5	45.0	47.5	50.0	52.4	54.9	57.4	59.9	
320	12	62.4	64.9	67.4	69.9	72.3	74.8	77.3	79.8	82.3	84.8	
321	12	87.2	89.7	92.2	94.7	92.1	99.6					
321	13							02.1	04.6	07.1	09.5	
322	13	12.0	14.5	17.0	19.4	21.9	24.4	26.8	29.3	31.8	34.2	
323	13	36.7	39.2	41.6	44.1	46.6	49.0	51.5	53.9	56.4	58.9	
324	13	61.3	63.8	66.2	68.7	71.1	73.6	76.1	78.5	81.0	83.4	
325	13	85.9	88.3	90.8	93.2	95.7	98.1					
325	14							00.5	03.0	05.4	07.9	
Barometer *H* or h.	N.	0.0	0.1	0.2	0.3	0.4	0.5	0.6	0.7	0.8	0.9	Parts for each 0.01mm.

326 to 364mm.

Barometer *H* or h.	N.	0.0	0.1	0.2	0.3	0.4	0.5	0.6	0.7	0.8	0.9	Parts for each 0.01mm.
Milli.	Metr.	Metres.	Metres.	Metres.	Metres.	Metres.	Metres.	Metres.	Metres.	Metres.	Metres.	Metr.
326	14	10.3	12.8	15.2	17.6	20.1	22.5	25.0	27.4	29.8	32.3	1 0.2
327	14	34.7	37.2	39.6	42.0	44.5	46.9	49.3	51.7	54.2	56.6	2 0.5
328	14	59.0	61.5	63.9	66.3	68.7	71.2	73.6	76.0	78.4	80.9	3 0.7
329	14	83.3	85.7	88.1	90.5	92.9	95.4	97.8				4 1.0
329	15								00.2	02.6	05.0	5 1.2
330	15	07.4	09.9	12.3	14.7	17.1	19.5	21.9	24.3	26.7	29.1	6 1.5
331	15	31.5	33.9	36.3	38.7	41.2	43.6	46.0	48.4	50.8	53.2	7 1.7
332	15	55.6	58.0	60.4	62.8	65.1	67.5	69.9	72.3	74.7	77.1	8 2.0
333	15	79.5	81.9	84.3	86.7	89.1	91.4	93.8	96.2	98.6		9 2.2
333	16										01.0	
334	16	03.4	05.8	08.1	10.5	12.9	15.3	17.7	20.0	22.4	24.8	
335	16	27.2	29.6	31.9	34.3	36.7	39.1	41.4	43.8	46.2	48.8	
336	16	50.9	53.3	55.7	58.0	60.4	62.8	65.1	67.5	69.9	72.2	1 0.2
337	16	74.6	77.0	79.3	81.7	84.0	86.4	88.8	91.1	93.5	95.8	2 0.4
338	16	98.2										3 0.7
338	17		00.5	02.9	05.2	07.6	10.0	12.3	14.7	17.0	19.4	4 1.0
339	17	21.7	24.1	26.4	28.8	31.1	33.4	35.8	38.1	40.5	42.8	5 1.2
340	17	45.2	47.5	49.8	52.2	54.5	56.9	59.2	61.5	63.9	66.2	6 1.5
341	17	68.6	70.9	73.2	75.6	77.9	80.2	82.6	84.9	87.2	89.5	7 1.7
342	17	91.9	94.2	96.5	98.9							8 1.9
342	18					01.2	03.5	05.8	08.2	10.5	12.8	9 2.2
343	18	15.1	17.4	19.8	22.1	24.4	26.7	29.0	31.4	33.7	36.0	
344	18	38.3	40.6	42.9	45.2	47.6	49.9	52.2	54.5	56.8	59.1	
345	18	61.4	63.7	66.0	68.3	70.6	73.0	75.3	77.6	79.9	82.2	
346	18	84.5	86.8	89.1	91.4	93.7	96.0	98.3				
346	19								00.6	02.9	05.2	
347	19	07.5	09.6	12.0	14.3	16.6	18.9	21.2	23.5	25.8	28.1	
348	19	30.4	32.7	34.9	37.2	39.5	41.8	44.1	46.4	48.6	50.9	
349	19	53.2	55.5	57.8	60.1	62.3	64.6	66.9	69.2	71.5	73.7	
350	19	76.0	78.3	80.6	82.8	85.1	87.4	89.6	91.9	94.2	96.5	1 0.2
351	19	98.7										2 0.4
351	20		01.0	03.3	05.5	07.8	10.1	12.3	14.6	16.8	19.1	3 0.7
352	20	21.4	23.6	25.9	28.2	30.4	32.7	34.9	37.2	39.5	41.7	4 0.9
353	20	44.0	46.2	48.5	50.7	53.0	55.2	57.5	59.7	62.0	64.2	5 1.1
354	20	66.5	68.7	71.0	73.2	75.5	77.7	80.0	82.2	84.5	86.7	6 1.3
355	20	89.0	91.2	93.4	95.7	97.9						7 1.6
355	21						00.2	02.4	04.6	06.9	09.1	8 1.8
356	21	11.4	13.6	15.8	18.1	20.3	22.5	24.8	27.0	29.2	31.5	9 2.1
357	21	33.7	35.9	38.2	40.4	42.6	44.8	47.1	49.3	51.5	53.7	
358	21	56.0	58.2	60.4	62.6	64.9	67.1	69.3	71.5	73.7	76.0	
359	21	78.2	80.4	82.6	84.8	87.0	89.3	91.5	93.7	95.9	98.1	
360	22	00.3	02.5	04.8	07.0	09.2	11.4	13.6	15.8	18.0	20.2	
361	22	22.4	24.6	26.8	29.0	31.2	33.4	35.6	37.9	40.1	42.3	
362	22	44.5	46.7	48.9	51.0	53.2	55.4	57.6	59.8	62.0	64.2	
363	22	66.4	68.6	70.8	73.0	75.2	77.4	79.6	81.8	83.9	86.1	
364	22	88.3	90.5	92.7	94.9	97.1	99.3					
364	23							01.4	03.6	05.8	08.0	
Barometer *H* or h.	N.	0.0	0.1	0.2	0.3	0.4	0.5	0.6	0.7	0.8	0.9	Parts for each 0.01mm.

365 to 403mm.

Barometer H or h.	N.	0.0	0.1	0.2	0.3	0.4	0.5	0.6	0.7	0.8	0.9	Parts for each 0.01mm.
Milli.	Metr.	Metres.	Metres.	Metres.	Metres.	Metres.	Metres.	Metres.	Metres.	Metres.	Metres.	Metr.
365	23	10.2	12.4	14.5	16.7	18.9	21.1	23.2	25.4	27.6	29.8	1 0.2
366	23	32.0	34.1	36.3	38.5	40.7	42.8	45.0	47.2	49.3	51.5	2 0.4
367	23	53.7	55.9	58.0	60.2	62.4	64.5	66.7	68.9	71.0	73.2	3 0.6
368	23	75.4	77.5	79.7	81.8	84.0	86.2	88.3	90.5	92.6	94.8	4 0.9
369	23	97.0	99.1									5 1.1
369	24			01.3	03.4	05.6	07.7	09.9	12.1	14.2	16.4	6 1.3
370	24	18.5	20.6	22.8	24.9	27.1	29.2	31.4	33.5	35.7	37.8	7 1.5
371	24	40.0	42.1	44.3	46.4	48.6	50.7	52.9	55.0	57.2	59.3	8 1.7
372	24	61.5	63.6	65.8	67.9	70.1	72.2	74.3	76.5	78.6	80.8	9 1.9
373	24	82.9	85.0	87.2	89.3	91.4	93.6	95.7	97.8	99.9		
373	25										02.1	
374	25	04.2	06.3	08.4	10.6	12.7	14.8	16.9	19.0	21.2	23.3	
375	25	25.4	27.5	29.6	31.8	33.9	36.0	38.1	40.2	42.4	44.5	
376	25	46.6	48.7	50.8	53.0	55.1	57.2	59.3	61.4	63.6	65.7	
377	25	67.8	69.9	72.0	74.1	76.2	78.3	80.5	82.6	84.7	86.8	
378	25	88.9	91.0	93.1	95.2	97.3	99.4					
378	26							01.5	03.6	05.7	07.8	
379	26	09.9	12.0	14.1	16.2	18.3	20.4	22.5	24.6	26.7	28.8	
380	26	30.9	33.0	35.1	37.2	39.3	41.3	43.4	45.5	47.6	49.7	
381	26	51.8	53.9	56.0	58.1	60.2	62.2	64.3	66.4	68.5	70.6	
382	26	72.7	74.8	76.9	78.9	81.0	83.1	85.2	87.3	89.3	91.4	
383	26	93.5	95.6	97.7	99.7							
383	27					01.8	03.9	06.0	08.1	10.1	12.2	1 0.2
384	27	14.3	16.4	18.4	20.5	22.6	24.6	26.7	28.8	30.9	32.9	2 0.4
385	27	35.0	37.1	39.1	41.2	43.2	45.3	47.4	49.4	51.5	53.5	3 0.6
386	27	55.6	57.7	59.7	61.8	63.8	65.9	68.0	70.0	72.1	74.1	4 0.9
387	27	76.2	78.3	80.3	82.4	84.4	86.5	88.6	90.6	92.7	94.7	5 1.1
388	27	96.8	98.8									6 1.3
388	28			00.9	02.9	05.0	07.0	09.1	11.1	13.2	15.2	7 1.5
389	28	17.3	19.3	21.4	23.4	25.5	27.5	29.6	31.6	33.7	35.7	8 1.7
390	28	37.8	39.8	41.9	43.9	46.0	48.0	50.0	52.1	54.1	56.2	9 1.9
391	28	58.2	60.2	62.3	64.3	66.3	68.3	70.4	72.4	74.4	76.5	
392	28	78.5	80.5	82.6	84.6	86.6	88.6	90.7	92.7	94.7	96.8	
393	28	98.8										
393	29		00.8	02.8	04.9	06.9	08.9	10.9	12.9	15.0	17.0	
394	29	19.0	21.0	23.0	25.1	27.1	29.1	31.1	33.1	35.2	37.2	
395	29	39.2	41.2	43.2	45.2	47.2	49.2	51.3	53.3	55.3	57.3	
396	29	59.3	61.3	63.3	65.3	67.3	69.3	71.4	73.4	75.4	77.4	
397	29	79.4	81.4	83.4	85.4	87.4	89.4	91.5	93.5	95.5	97.5	
398	29	99.5										
398	30		01.5	03.5	05.5	07.5	09.5	11.5	13.5	15.5	17.5	
399	30	19.5	21.5	23.5	25.5	27.5	29.4	31.4	33.4	35.4	37.4	
400	30	39.4	41.4	43.4	45.4	47.4	49.4	51.3	53.3	55.3	57.3	
401	30	59.3	61.3	63.3	65.2	67.2	69.2	71.2	73.2	75.1	77.1	
402	30	79.1	81.1	83.1	85.0	87.0	89.0	91.0	93.0	94.9	96.9	
403	30	98.9										
Barometer H or h.	N.	0.0	0.1	0.2	0.3	0.4	0.5	0.6	0.7	0.8	0.9	Parts for each 0.01mm.

403 to 442mm.

Barometer H or h.	N.	0.0	0.1	0.2	0.3	0.4	0.5	0.6	0.7	0.8	0.9	Parts for each 0.01mm.	
Milli.	Metr.	Metres.	Metres.	Metres.	Metres.	Metres	Metres.	Metres.	Metres.	Metres.	Metres.		Metr.
403	31		00.9	02.8	04.8	06.8	08.7	10.7	12.7	14.7	16.6	1	0.2
404	31	18.6	20.6	22.5	24.5	26.5	28.4	30.4	32.4	34.4	36.3	2	0.4
405	31	38.3	40.3	42.2	44.2	46.1	48.1	50.1	52.0	54.0	55.9	3	0.6
406	31	57.9	59.9	61.8	63.8	65.7	67.7	69.7	71.6	73.6	75.5	4	0.8
407	31	77.5	79.5	81.4	83.4	85.3	87.3	89.3	91.2	93.2	95.1	5	1.0
408	31	97.1	99.0									6	1.2
408	32			01.0	02.9	04.9	06.8	08.8	10.7	12.7	14.6	7	1.4
409	32	16.6	18.5	20.5	22.4	24.4	26.3	28.2	30.2	32.1	34.1	8	1.6
410	32	36.0	37.9	39.9	41.8	43.8	45.7	47.6	49.6	51.5	53.5	9	1.8
411	32	55.4	57.3	59.3	61.2	63.2	65.1	67.0	69.0	70.9	72.9		
412	32	74.8	76.7	78.7	80.6	82.5	84.4	86.4	88.3	90.2	92.2		
413	32	94.1	96.0	97.9	99.9								
413	33					01.8	03.7	05.6	07.5	09.5	11.4		
414	33	13.3	15.2	17.1	19.1	21.0	22.9	24.8	26.7	28.7	30.6		
415	33	32.5	34.4	36.3	38.3	40.2	42.1	44.0	45.9	47.9	49.8		
416	33	51.7	53.6	55.5	57.4	59.3	61.2	63.2	65.1	67.0	68.9		
417	33	70.8	72.7	74.6	76.5	78.4	80.3	82.3	84.2	86.1	88.0		
418	33	89.9	91.8	93.7	95.6	97.5	99.4						
418	34							01.3	03.2	05.1	07.0		
419	34	08.9	10.8	12.7	14.6	16.5	18.4	20.3	22.2	24.1	26.0		
420	34	27.9	29.8	31.7	33.6	35.5	37.3	39.2	41.1	43.0	44.9		
421	34	46.8	48.7	50.6	52.5	54.4	56.2	58.1	60.0	61.9	63.8		
422	34	65.7	67.6	69.5	71.4	73.3	75.1	77.0	78.9	80.8	82.7	1	0.2
423	34	84.6	86.5	88.4	90.2	92.1	94.0	95.9	97.8	99.6		2	0.4
423	35										01.5	3	0.6
424	35	03.4	05.3	07.2	09.0	10.9	12.8	14.7	16.6	18.4	20.3	4	0.8
												5	1.0
425	35	22.2	24.1	25.9	27.8	29.6	31.5	33.4	35.2	37.1	38.9	6	1.2
426	35	40.8	42.7	44.5	46.4	48.3	50.1	52.0	53.9	55.8	57.6	7	1.4
427	35	59.5	61.4	63.2	65.1	67.0	68.8	70.7	72.6	74.5	76.3	8	1.6
428	35	78.2	80.1	81.9	83.8	85.6	87.5	89.4	91.2	93.1	94.9	9	1.8
429	35	96.8	98.6										
429	36			00.5	02.3	04.2	06.0	07.9	09.7	11.6	13.4		
430	36	15.3	17.1	19.0	20.8	22.7	24.6	26.4	28.2	30.1	31.9		
431	36	33.8	35.6	37.5	39.3	41.2	43.0	44.8	46.7	48.5	50.4		
432	36	52.2	54.0	55.9	57.7	59.6	61.4	63.2	65.1	66.9	68.8		
433	36	70.6	72.4	74.3	76.1	78.0	79.8	81.6	83.5	85.3	87.2		
434	36	89.0	90.8	92.7	94.5	96.3	98.1						
434	37							00.0	01.8	03.6	05.5		
435	37	07.3	09.1	11.0	12.8	14.6	16.4	18.3	20.1	21.9	23.8		
436	37	25.6	27.4	29.2	31.1	32.9	34.7	36.5	38.3	40.2	42.0		
437	37	43.8	45.6	47.5	49.3	51.1	52.9	54.8	56.6	58.4	60.3		
438	37	62.1	63.9	65.7	67.6	69.4	71.2	73.0	74.8	76.7	78.5		
439	37	80.3	82.1	83.9	85.7	87.5	89.3	91.2	93.0	94.8	96.6		
440	37	98.4											
440	38		00.2	02.0	03.8	05.6	07.5	09.3	11.1	12.9	14.7		
441	38	16.5	18.3	20.1	21.9	23.7	25.5	27.3	29.1	30.9	32.7		
442	38	34.5	36.3	38.1	39.9	41.7	43.5	45.3	47.1	48.9	50.7		
Barometer H or h.	N.	0.0	0.1	0.2	0.3	0.4	0.5	0.6	0.7	0.8	0.9	Parts for each 0.01mm.	

443 to 482mm.

Barometer H or h.	N.	0.0	0.1	0.2	0.3	0.4	0.5	0.6	0.7	0.8	0.9	Parts for each 0.01mm.
		Tenth of Millimetre.										
Milli.	Metr.	Metres.	Metres.	Metres.	Metres.	Metres.	Metres.	Metres.	Metres.	Metres.	Metres.	Metr.
443	38	52.5	54.3	56.1	57.9	59.7	61.4	63.2	65.0	66.8	68.6	
444	38	70.4	72.2	74.0	75.8	77.6	79.3	81.1	82.9	84.7	86.5	
445	38	88.3	90.1	91.9	93.7	95.5	97.2	99.0				
445	39								00.8	02.6	04.4	
446	39	06.2	08.0	09.8	11.5	13.3	15.1	16.9	18.7	20.4	22.2	
447	39	24.0	25.8	27.6	29.3	31.1	32.9	34.7	36.5	38.2	40.0	
448	39	41.8	43.6	45.4	47.1	48.9	50.7	52.5	54.3	56.0	57.8	
449	39	59.6	61.4	63.1	64.9	66.7	68.4	70.2	72.0	73.8	75.5	
450	39	77.3	79.1	80.8	82.6	84.3	86.1	87.9	89.6	91.4	93.1	
451	39	94.9	96.7	98.4								
451	40				00.2	02.0	03.7	05.5	07.3	09.1	10.8	
452	40	12.6	14.4	16.1	17.9	19.6	21.4	23.2	24.9	26.7	28.4	
453	40	30.2	32.0	33.7	35.5	37.2	39.0	40.8	42.5	44.3	46.0	
454	40	47.8	49.5	51.3	53.0	54.8	56.5	58.3	60.0	61.8	63.5	
455	40	65.3	67.0	68.8	70.5	72.3	74.0	75.8	77.5	79.3	81.0	1 0.2
456	40	82.8	84.5	86.3	88.0	89.8	91.5	93.2	95.0	96.7	98.5	2 0.3
457	41	00.2	01.9	03.7	05.4	07.2	08.9	10.6	12.4	14.1	15.9	3 0.5
458	41	17.6	19.3	21.1	22.8	24.6	26.3	28.0	29.8	31.5	33.3	4 0.7
459	41	35.0	36.7	38.5	40.2	41.9	43.6	45.4	47.1	48.8	50.6	5 0.9
460	41	52.3	54.0	55.8	57.5	59.2	60.9	62.7	64.4	66.1	67.9	6 1.0
461	41	69.6	71.3	73.1	74.8	76.5	78.2	80.0	81.7	83.4	85.2	7 1.2
462	41	86.9	88.6	90.3	92.1	93.8	95.5	97.2	98.9			8 1.4
462	42									00.7	02.3	9 1.6
463	42	04.1	05.8	07.5	09.3	11.0	12.7	14.4	16.1	17.9	19.6	
464	42	21.3	23.0	24.7	26.4	28.1	29.8	31.6	33.3	35.0	36.7	
465	42	38.4	40.1	41.8	43.5	45.2	46.9	48.7	50.4	52.1	53.8	
466	42	55.5	57.2	58.9	60.6	62.3	64.0	65.8	67.5	69.2	70.9	
467	42	72.6	74.3	76.0	77.7	79.4	81.1	82.8	84.5	86.2	87.9	
468	42	89.6	91.3	93.0	94.7	96.4	98.1	99.8				
468	43								01.5	03.2	04.9	
469	43	06.6	08.3	10.0	11.7	13.4	15.1	16.8	18.5	20.2	21.9	
470	43	23.6	25.3	27.0	28.7	30.4	32.0	33.7	35.4	37.1	38.8	
471	43	40.5	42.2	43.9	45.6	47.3	48.9	50.6	52.3	54.0	55.7	
472	43	57.4	59.1	60.8	62.5	64.2	65.8	67.5	69.2	70.9	72.6	
473	43	74.3	76.0	77.7	79.3	81.0	82.7	84.4	86.1	87.7	89.4	
474	43	91.1	92.8	94.5	96.1	97.8	99.5					
474	44							01.2	02.9	04.5	06.2	
475	44	07.9	09.6	11.2	12.9	14.6	16.2	17.9	19.6	21.3	22.9	
476	44	24.6	26.3	27.9	29.6	31.3	33.9	35.6	37.3	39.0	40.6	
477	44	41.3	43.0	44.6	46.3	48.0	49.6	51.3	53.0	54.7	56.3	
478	44	58.0	59.7	61.3	63.0	64.7	66.3	68.0	69.7	71.4	73.0	
479	44	74.7	76.4	78.0	79.7	81.3	83.0	84.7	86.3	88.0	89.6	
480	44	91.3	93.0	94.6	96.3	97.9	99.6					
480	45							01.3	02.9	04.6	06.2	
481	45	07.9	09.5	11.2	12.8	14.5	16.1	17.7	19.4	21.0	22.7	
482	45	24.3	25.9	27.6	29.2	30.9	32.5	34.2	35.8	37.5	39.1	
Barometer H or h.	N.	0.0	0.1	0.2	0.3	0.4	0.5	0.6	0.7	0.8	0.9	Parts for each 0.01mm.

483 to 524^{mm}.

Barometer H or h.	N.	0.0	0.1	0.2	0.3	0.4	0.5	0.6	0.7	0.8	0.9	Parts for each 0.01mm.	
Milli.	Metr.	Metres.	Metres.	Metres.	Metres.	Metres	Metres.	Metres.	Metres.	Metres.	Metres.		Metr.
483	45	40.8	42.4	44.1	45.7	47.4	49.0	50.7	52.3	54.0	55.6	1	0.2
484	45	57.3	58.9	60.6	62.2	63.9	65 5	67.1	68.8	70.4	72.1	2	0.3
485	45	73.7	75.3	77.0	78.6	80.3	81.9	83.6	85.2	86.9	88.5	3	0.5
486	45	90.2	91.8	93.5	95.1	96.8	98.4					4	0.6
486	46							00.0	01.7	03.3	05.0	5	0.8
487	46	06.6	08.2	09.9	11.5	13.1	14.7	16.4	18.0	19.6	21.3	6	1.0
488	46	22.9	24.5	26.2	27.8	29.4	31.0	32.7	34.3	35.9	37.6	7	1.1
489	46	39.2	40.8	42.4	44.1	45.7	47.3	48.9	50.5	52.2	53.8	8	1.3
490	46	55.4	57.0	58.6	60.3	61.9	63.5	65.1	66.7	68.4	70.0	9	1.4
491	46	71.6	73.2	74.9	76.5	78.1	79.7	81.4	83.0	84.6	86.3		
492	46	87.9	89.5	91.1	92.8	94.4	96.0	97.6	99.2				
492	47									00.9	02.5		
493	47	04.1	05.7	07.3	08.9	10.5	12.1	13.8	15.4	17.0	18.6		
494	47	20.2	21.8	23.4	25.0	26.6	28.2	29.9	31.5	33.1	34.7		
495	47	36.3	37.9	39.5	41.1	42.7	44.3	45.9	47.5	49.1	50.7		
496	47	52.3	53.9	55.5	57.1	58.7	60.3	61.9	63.5	65.1	66.7		
497	47	68.3	69.9	71.5	73.1	74.7	76.3	78.0	79.6	81.2	82.8		
498	47	84.4	86.0	87.6	89.2	90.8	92.4	94.0	95.6	97.2	98.8		
499	48	00.4	02.0	03.6	05.2	06.8	08.3	09.9	11.5	13.1	14.7		
500	48	16.3	17.9	19.5	21.1	22.7	24.2	25.8	27.4	89.0	30.6		
501	48	32.2	33.8	35.4	37.0	38.6	40.1	41.7	43.3	44.9	46.5		
502	48	48.1	49.7	51.3	52.9	54.5	56.0	57.6	59.2	60.8	62.4		
503	48	64.0	65.6	67.2	68.7	70.3	71.9	73.5	75.1	76.6	78.2		
504	48	79.8	81.4	83.0	84.5	86.1	87.7	89.3	90.9	92.4	94.0		
505	48	95.6	97.2	98.7									
505	49				00.3	01.9	03.4	05.0	06.6	08.2	09.7		
506	49	11.3	12.9	14.4	16.0	17.6	19.1	20.7	22.3	23.9	25.4		
507	49	27.0	28.6	30.1	31.7	33.3	34.8	36.4	38.0	39.6	41.1		
508	49	42.7	44.3	45.8	47.4	49.0	50.5	52.1	53.7	55.3	56.8		
509	49	58.4	60.0	61.5	63.1	64.6	66.2	67.8	69.3	70.9	72.4		
510	49	74.0	75.6	77.1	78.7	80.2	81.8	83.4	84.9	86.5	88.0		
511	49	89.6	91.2	92.7	94.3	95.8	97.4	99.0					
511	50								00.5	02.1	03.6		
512	50	05.2	06.7	08.3	09.8	11.4	12.9	14.5	16.0	17.6	19.1		
513	50	20.7	22.2	23.8	25.3	26.9	28.4	30.0	31.5	33.1	34.6		
514	50	36.2	37.7	39.3	40.8	42.4	43.9	45.5	46.0	48.6	50.1		
515	50	51.7	53.2	54.8	56.3	57.9	59.4	61.0	62.5	64.1	65.6		
516	50	67.2	68.7	70.3	71.8	73.4	74.9	76.4	78.0	79.5	81.1		
517	50	82.6	84.1	85.7	87.2	88.7	90.2	91.8	93.3	94.8	96.4		
518	50	97.9	99.4										
518	51			01.0	02.5	04.1	05.6	07.1	08.7	10.2	11.8		
519	51	13.3	14.8	16.4	17.9	19.4	20.9	22.5	24.0	25.5	27.1		
520	51	28.6	30.1	31.7	33.2	34.7	36.2	37.8	39.3	40.8	42.4		
521	51	43.9	45.4	47.0	48.5	50.0	51.5	53.1	54.6	56.1	57.7		
522	51	59.2	60.7	62.2	63.8	65.3	66.8	68.3	69.8	71.4	72.9		
523	51	74.4	75.9	77.5	79.0	80.5	82.0	83.6	85.1	86.6	88.2		
524	51	89.7	91.2	92.7	94.3	95.8	97.3	98.8					
Barometer H or h.	N.	0.0	0.1	0.2	0.3	0.4	0.5	0.6	0.7	0.8	0.9	Parts for each 0.01mm.	

524 to 565mm.

Barometer H or h.	N.	0.0	0.1	0.2	0.3	0.4	0.5	0.6	0.7	0.8	0.9	Parts for each 0.01mm.
Milli.	Metr.	Metres.	Metres.	Metres.	Metres.	Metres	Metres.	Metres.	Metres.	Metres.	Metres.	Metr.
524	52								00.3	01.9	03.4	
525	52	04.9	06.4	07.9	09.4	10.9	12.4	14.0	15.5	17.0	18.5	
526	52	20.0	21.5	23.0	24.5	26.0	27.5	29.1	30.6	32.1	33.6	
527	52	35.1	36.6	38.1	39.6	41.1	42.6	44.2	45.7	47.2	48.7	
528	52	50.2	51.7	53.2	54.7	56.2	57.7	59.3	60.8	62.3	63.8	
529	52	65.3	66.8	68.3	69.8	71.3	72.8	74.3	75.8	77.3	78.8	1 0.1
530	52	80.3	81.8	83.3	84.8	86.3	87.8	89.3	90.8	92.3	93.8	2 0.3
531	52	95.3	96.8	98.3	99.8							3 0.4
531	53					01.3	02.8	04.3	05.8	07.3	08.8	4 0.6
532	53	10.3	11.8	13.3	14.8	16.3	17.8	19.3	20.8	22.3	23.8	5 0.7
533	53	25.3	26.8	28.3	29.8	31.3	32.7	34.2	35.7	37.2	38.7	6 0.9
534	53	40.2	41.7	43.2	44.7	46.2	47.6	49.1	50.6	52.1	53.6	7 1.0
535	53	55.1	56.5	58.1	59.6	61.1	62.5	64.0	65.5	67.0	68.5	8 1.2
536	53	70.0	71.5	73.0	74.4	75.9	77.4	78.9	80.4	81.8	83.3	9 1.3
537	53	84.8	86.3	87.8	89.2	90.7	92.2	93.7	95.2	96.6	98.1	
538	53	99.6										
538	54		01.1	02.6	04.0	05.5	07.0	08.5	10.0	11.4	12.9	
539	54	14.4	15.9	17.4	18.8	20.3	21.8	23.3	24.8	26.2	27.7	
540	54	29.2	30.7	32.1	33.6	35.1	36.5	38.0	39.5	41.0	42.4	
541	54	43.9	45.4	46.8	48.3	49.8	51.2	52.7	54.2	55.7	57.1	
542	54	58.6	60.1	61.5	63.0	64.5	66.0	67.4	68.9	70.4	71.8	
543	54	73.3	74.8	76.2	77.7	79.1	80.6	82.1	83.5	85.0	86.4	
544	54	87.9	89.4	90.8	92.3	93.7	95.2	96.7	98.1	99.6		
544	55										01.0	
545	55	02.5	04.0	05.4	06.9	08.4	09.8	11.3	12.8	14.3	15.7	
546	55	17.2	18.7	20.1	21.6	23.0	24.5	26.0	27.4	28.9	30.3	
547	55	31.8	33.3	34.7	36.1	37.6	39.0	40.5	41.9	43.4	44.8	
548	55	46.3	47.7	49.2	50.6	52.1	53.5	55.0	56.4	57.9	59.3	
549	55	60.8	62.2	63.7	65.1	66.6	68.0	69.5	70.9	72.4	73.8	
550	55	75.3	76.7	78.2	79.6	81.1	82.5	84.0	85.4	86.9	88.3	
551	55	89.8	91.2	92.7	94.1	95.6	97.0	98.4	99.9			
551	56									01.3	02.8	1 0.1
552	56	04.2	05.6	07.1	08.5	10.0	11.4	12.8	14.3	15.7	17.2	2 0.3
553	56	18.6	20.0	21.5	22.9	24.4	25.8	27.2	28.7	30.1	31.6	3 0.4
554	56	33.0	34.4	35.9	37.3	38.8	40.2	41.6	43.1	44.5	46.0	4 0.6
555	56	47.4	48.8	50.3	51.7	53.1	54.5	56.0	57.4	58.8	60.3	5 0.7
556	56	61.7	63.1	64.6	66.0	67.4	68.8	70.3	71.7	73.1	74.6	6 0.9
557	56	76.0	77.4	78.9	80.3	81.7	83.1	84.6	86.0	87.4	88.9	7 1.0
558	57	90.3	91.7	93.2	94.6	96.0	97.4	98.9				8 1.2
558	57								00.3	01.7	03.2	9 1.3
559	57	04.6	06.0	07.4	08.9	10.3	11.7	13.1	14.5	16.0	17.4	
560	57	18.8	20.2	21.6	23.1	24.5	25.9	27.3	28.7	30.2	31.6	
561	57	33.0	34.4	35.8	37.3	38.7	40.1	41.5	42.9	44.4	45.8	
562	57	47.2	48.6	50.0	51.4	52.8	54.2	55.7	57.1	58.5	59.9	
563	57	61.3	62.7	64.1	65.5	66.9	68.3	69.8	71.2	72.6	74.0	
564	57	75.4	76.8	78.2	79.6	81.0	82.4	83.9	85.3	86.7	88.1	
565	57	89.5	90.9	92.4	93.8	95.2	96.6	98.0	99.4			
Barometer H or h	N.	0.0	0.1	0.2	0.3	0.4	0.5	0.6	0.7	0.8	0.9	Parts for each 0.01mm.

565 to 605mm.

Barometer H or h.	N.	0.0	0.1	0.2	0.3	0.4	0.5	0.6	0.7	0.8	0.9	Parts for each 0.01mm.
Milli.	Metr.	Metres.	Metres.	Metres.	Metres.	Metres.	Metres.	Metres.	Metres.	Metres.	Metres.	Metr.
565	58									00.8	02.2	
566	58	03.6	05.0	06.4	07.8	09.2	10.6	12.1	13.5	14.9	16.3	
567	58	17.7	19.1	20.5	21.9	23.3	24.7	26.1	27.5	28.9	30.3	
568	58	31.7	33.1	34.5	35.9	37.3	38.7	40.1	41.5	42.9	44.3	
569	58	45.7	47.1	48.5	49.9	51.3	52.7	54.1	55.5	56.9	58.3	
570	58	59.7	61.1	62.5	63.9	65.3	66.7	68.1	69.5	70.9	72.3	
571	58	73.7	75.1	76.5	77.9	79.3	80.6	82.0	83.4	84.8	86.2	
572	58	87.6	89.0	90.4	91.8	93.2	94.5	95.9	97.3	98.7		
572	59										00.1	
573	59	01.5	02.9	04.3	05.7	07.1	08.4	09.8	11.2	12.6	14.0	
574	59	15.4	16.8	18.2	19.6	21.0	22.3	23.7	25.1	26.5	27.9	
575	59	29.3	30.7	32.1	33.4	34.8	36.2	37.6	39.0	40.3	41.7	
576	59	43.1	44.5	45.9	47.2	48.6	50.0	51.4	52.8	54.1	55.5	1 0.1
577	59	56.9	58.3	59.7	61.0	62.4	63.8	65.2	66.6	67.9	69.3	2 0.3
578	59	70.7	72.1	73.5	74.8	76.2	77.6	79.0	80.4	81.7	83.1	3 0.4
579	59	84.5	85.9	87.2	88.6	90.0	91.3	92.7	94.1	95.5	96.8	4 0.5
580	59	98.2	99.6									5 0.7
580	60			00.9	02.3	03.7	05.0	06.4	07.8	09.2	10.5	6 0.8
581	60	11.9	13.3	14.6	16.0	17.4	18.7	20.1	21.5	22.9	24.2	7 1.0
582	60	25.6	27.0	28.3	29.7	31.1	32.4	33.8	35.2	36.6	37.9	8 1.1
583	60	39.3	40.7	42.0	43.4	44.7	46.1	47.5	48.8	50.2	51.5	9 1.2
584	60	52.9	54.3	55.6	57.0	58.4	59.7	61.1	62.5	63.9	65.2	
585	60	66.6	68.0	69.3	70.7	72.0	73.4	74.8	76.1	77.5	78.8	
586	60	80.2	81.6	82.9	84.3	85.6	87.0	88.4	89.7	91.1	92.4	
587	60	93.8	95.1	96.5	97.8	99.2						
587	61						00.5	01.9	03.2	04.6	05.9	
588	61	07.3	08.6	10.0	11.3	12.7	14.0	15.4	16.7	18.1	19.4	
589	61	20.8	22.1	23.5	24.8	26.2	27.5	28.9	30.2	31.6	32.9	
590	61	34.3	35.6	37.0	38.3	39.7	41.0	42.4	43.7	45.1	46.4	
591	61	47.8	49.1	50.5	51.8	53.2	54.5	55.9	57.2	58.6	59.9	
592	61	61.3	62.6	64.0	65.3	66.7	68.0	69.3	70.7	72.0	73.4	
593	61	74.7	76.0	77.4	78.7	80.1	81.4	82.7	84.1	85.4	86.8	
594	61	88.1	89.4	90.8	92.1	93.5	94.8	96.1	97.5	98.8		
594	62										00.2	
595	62	01.5	02.8	04.2	05.5	06.9	08.2	09.5	10.9	12.2	13.6	
596	62	14.9	16.2	17.6	18.9	20.2	21.5	22.9	24.2	25.5	26.9	
597	62	28.2	29.5	30.9	32.2	33.6	34.9	36.2	37.6	38.9	40.3	
598	62	41.6	42.9	44.3	45.6	46.9	48.2	49.6	50.9	52.2	53.6	
599	62	54.9	56.2	57.6	58.9	60.2	61.5	62.9	64.2	65.5	66.9	
600	62	68.2	69.5	70.8	72.2	73.5	74.8	76.1	77.4	78.8	80.1	
601	62	81.4	82.7	84.1	85.4	86.7	88.0	89.4	90.7	92.0	93.4	
602	62	94.7	96.0	97.3	98.7							
602	63					00.0	01.3	02.6	03.9	05.3	06.6	
603	63	07.9	09.2	10.5	11.9	13.2	14.5	15.8	17.1	18.5	19.8	
604	63	21.1	22.4	23.7	25.1	26.4	27.7	29.0	30.3	31.7	33.0	
605	63	34.3	35.6	36.9	38.2	39.5	40.8	42.2	43.5	44.8	46.1	
Barometer H or h.	N.	0.0	0.1	0.2	0.3	0.4	0.5	0.6	0.7	0.8	0.9	Parts for each 0.01mm.

606 to 647mm.

Barometer H or h.	N.	0.0	0.1	0.2	0.3	0.4	0.5	0.6	0.7	0.8	0.9	Parts for each 0.01mm.
Milli.	Metr.	Metres.	Metres.	Metres.	Metres.	Metres.	Metres.	Metres.	Metres.	Metres.	Metres.	Metr.
606	63	47.4	48.7	50.0	51.3	52.6	53.9	55.3	56.6	57.9	59.2	
607	63	60.5	61.8	63.1	64.5	65.8	67.1	68.4	69.7	71.1	72.4	
608	63	73.7	75.0	76.3	77.6	78.9	80.2	81.5	82.8	84.1	85.4	
609	63	86.7	88.0	89.3	90.6	91.9	93.2	94.6	95.9	97.2	98.5	
610	63	99.8										
610	64		01.1	02.4	03.7	05.0	06.3	07.6	08.9	10.2	11.5	
611	64	12.8	14.1	15.4	16.7	18.0	19.3	20.7	22.0	23.3	24.6	
612	64	25.9	27.2	28.5	29.8	31.1	32.4	33.7	35.0	36.3	37.6	
613	64	38.9	40.2	41.5	42.8	44.1	45.4	46.7	48.0	49.3	50.6	
614	64	51.9	53.2	54.5	55.8	57.1	58.3	59.6	60.9	62.2	63.5	
615	64	64.8	66.1	67.4	68.7	70.0	71.2	72.5	73.8	75.1	76.4	
616	64	77.7	79.0	80.3	81.6	82.9	84.2	85.5	86.8	88.1	89.4	
617	64	90.7	92.0	93.3	94.6	95.9	97.1	98.4	99.7			
617	65									01.0	02.3	
618	65	03.6	04.9	06.2	07.4	08.7	10.0	11.3	12.6	13.8	15.1	
619	65	16.4	17.7	19.0	20.3	21.6	22.8	24.1	25.4	26.7	28.0	
620	65	29.3	30.6	31.9	33.1	34.4	35.7	37.0	38.3	39.5	40.8	
621	65	42.1	43.4	44.7	45.9	47.2	48.5	49.8	51.1	52.3	53.6	1 0.1
622	65	54.9	56.2	57.5	58.7	60.0	61.3	62.6	63.9	65.1	66.4	2 0.2
623	65	67.7	69.0	70.3	71.5	72.8	74.1	75.4	76.7	77.9	79.2	3 0.4
624	65	80.5	81.8	83.0	84.3	85.6	86.8	88.1	89.4	90.7	91.9	4 0.5
												5 0.6
625	65	93.2	94.5	95.8	97.0	98.3	99.6					6 0.8
625	66							00.9	02.2	03.4	04.7	7 0.9
626	66	06.0	07.3	08.5	09.8	11.1	12.3	13.6	14.9	16.2	17.4	8 1.0
627	66	18.7	20.0	21.2	22.5	23.8	25.0	26.3	27.6	28.9	30.1	9 1.1
628	66	31.4	32.7	33.9	36.2	56.4	37.7	39.0	40.2	41.5	42.7	
629	66	44.0	45.3	46.5	47.8	49.1	50.3	51.6	52.9	54.2	55.4	
630	66	56.7	58.0	59.2	60.5	61.7	63.0	64.3	65.5	66.8	68.0	
631	66	69.3	70.6	71.8	73.1	74.4	75.6	76.9	78.2	79.5	80.7	
632	66	82.0	83.2	84.5	85.7	87.0	88.2	89.5	90.7	92.0	93.2	
633	66	94.5	95.8	97.0	98.3	99.5						
633	67						00.8	02.1	03.3	04.6	05.8	
634	67	07.1	08.4	09.6	10.9	12.1	13.4	14.7	15.9	17.2	18.4	
635	67	19.7	20.9	22.2	23.4	24.7	25.9	27.2	28.4	29.7	30.9	
636	67	32.2	33.4	34.7	35.9	37.2	38.4	39.7	40.9	42.2	43.4	
637	67	44.7	45.9	47.2	48.4	49.7	50.9	52.2	53.4	54.7	55.9	
638	67	57.2	58.4	59.7	60.9	62.2	63.4	64.7	65.9	67.2	68.4	
639	67	69.7	70.9	72.2	73.4	74.7	75.9	77.1	78.4	79.6	80.9	
640	67	82.1	83.3	84.6	85.8	87.1	88.3	89.6	90.8	92.1	93.3	
641	67	94.6	95.8	97.1	98.3	99.6						
641	68						00.8	02.0	03.3	04.5	05.8	
642	68	07.0	08.2	09.5	10.7	12.0	13.2	14.4	15.7	16.9	18.2	
643	68	19.4	20.6	21.9	23.1	24.3	25.5	26.8	28.0	29.2	30.5	
644	68	31.7	32.9	34.2	35.4	36.7	37.9	39.1	40.4	41.6	42.9	
645	68	44.1	45.3	46.6	47.8	49.0	50.2	51.5	52.7	53.9	55.2	
646	68	56.4	57.6	58.9	60.1	61.3	62.5	63.8	65.0	66.2	67.5	
647	68	68.7	69.9	71.2	72.4	73.6	74.8	76.1	77.3	78.5	79.8	
Barometer H or h.	N.	0.0	0.1	0.2	0.3	0.4	0.5	0.6	0.7	0.8	0.9	Parts for each 0.01mm.

648 to 689mm.

Barometer H or h.	N.	0.0	0.1	0.2	0.3	0.4	0.5	0.6	0.7	0.8	0.9	Parts for each 0.01mm.
Milli.	Metr.	Metres.	Metres.	Metres.	Metres.	Metres.	Metres.	Metres.	Metres.	Metres.	Metres.	Metr.
648	68	81.0	82.2	83.5	84.7	85.9	87.1	88.4	89.6	90.8	92.1	
649	68	93.3	94.5	95.8	97.0	98.2	99.4					
649	69							00.7	01.9	03.1	04.4	
650	69	05.6	06.8	08.0	09.3	10.5	11.7	12.9	14.1	15.4	16.6	
651	69	17.8	19.0	20.2	21.5	22.7	23.9	25.1	26.3	27.6	28.8	
652	69	30.0	31.2	32.4	33.7	34.9	36.1	37.3	38.5	39.8	41.0	
653	69	42.2	43.4	44.6	45.9	47.1	48.3	49.5	50.7	52.0	53.2	
654	69	54.4	55.6	56.8	58.1	59.3	60.5	61.7	62.9	64.2	65.4	
655	69	66.6	67.8	69.0	70.2	71.4	72.6	73.9	75.1	76.3	77.5	
656	69	78.7	79.9	81.1	82.4	83.6	84.8	86.0	87.2	88.5	89.7	
657	69	90.9	92.1	93.3	94.5	95.7	96.9	98.2	99.4			
657	70									00.6	01.8	
658	70	03.0	04.2	05.4	06.6	07.8	09.0	10.3	11.5	12.7	13.9	
659	70	15.1	16.3	17.5	18.7	19.9	21.1	22.4	23.6	24.8	26.0	
660	70	27.2	28.4	29.6	30.8	32.0	33.2	34.4	35.6	36.8	38.0	1 0.1
661	70	39.2	40.4	41.6	42.8	44.0	45.2	46.4	47.6	48.8	50.0	2 0.2
662	70	51.2	52.4	53.6	54.8	56.0	57.2	58.5	59.7	60.9	62.1	3 0.4
663	70	63.3	64.5	65.7	66.9	68.1	69.3	70.5	71.7	72.9	74.1	4 0.5
664	70	75.3	76.5	77.7	78.9	80.1	81.2	82.4	83.6	84.8	86.0	5 0.6
665	70	87.2	88.4	89.6	90.8	92.0	93.2	94.4	95.6	96.8	98.0	6 0.7
666	70	99.2										7 0.8
666	71		00.4	01.6	02.8	04.0	05.2	06.4	07.6	08.8	10.0	8 1.0
667	71	11.2	12.4	13.6	14.8	16.0	17.1	18.3	19.5	20.7	21.9	9 1.1
668	71	23.1	24.3	25.5	26.7	27.9	29.0	30.2	31.4	32.6	33.8	
669	71	35.0	36.2	37.4	38.6	39.8	40.9	42.1	43.3	44.5	45.7	
670	71	46.9	48.1	49.3	50.5	51.7	52.8	54.0	55.2	56.4	57.6	
671	71	58.8	60.0	61.2	62.3	63.5	64.7	65.9	67.1	68.2	69.4	
672	71	70.6	71.8	73.0	74.2	75.4	76.5	77.7	78.9	80.1	81.3	
673	71	82.5	83.7	84.9	86.0	87.2	88.4	89.6	90.8	91.9	93.1	
674	71	94.3	95.5	96.7	97.8	99.0						
674	72						00.2	01.4	02.6	03.7	04.9	
675	72	06.1	07.3	08.5	09.6	10.8	12.0	13.2	14.4	15.5	16.7	
676	72	17.9	19.1	20.3	21.4	22.6	23.8	25.0	26.2	27.3	28.5	
677	72	29.7	30.9	32.0	33.2	34.4	35.5	36.7	37.9	39.1	40.2	
678	72	41.4	42.6	43.8	44.9	46.1	47.3	48.5	49.7	50.8	52.0	
679	72	53.2	54.4	55.5	56.7	57.9	59.0	60.2	61.4	62.6	63.7	
680	72	64.9	66.1	67.2	68.4	69.6	70.7	71.9	73.1	74.3	75.4	
681	72	76.6	77.8	78.9	80.1	81.3	82.4	83.6	84.8	86.0	87.1	1 0.1
682	72	88.3	89.5	90.6	91.8	93.0	94.1	95.3	96.5	97.7	98.8	2 0.2
683	73	00.0	01.2	02.3	03.5	04.6	05.8	07.0	08.1	09.3	10.4	3 0.3
684	73	11.6	12.8	13.9	15.1	16.2	17.4	18.6	19.7	20.9	22.0	4 0.5
685	73	23.2	24.4	25.5	26.7	27.8	29.0	30.2	31.3	32.5	33.6	5 0.6
686	73	34.8	36.0	37.1	38.3	39.4	40.6	41.8	42.9	44.1	45.2	6 0.7
687	73	46.4	47.6	48.7	49.9	51.0	52.2	53.4	54.5	55.7	56.8	7 0.8
688	73	58.0	59.2	60.3	61.5	62.6	63.8	65.0	66.1	67.3	68.4	8 0.9
689	73	69.6	70.7	71.9	73.0	74.2	75.3	76.5	77.6	78.8	79.9	9 1.1
Barometer H or h.	N.	0.0	0.1	0.2	0.3	0.4	0.5	0.6	0.7	0.8	0.9	Parts for each 0.01mm.

690 to 730mm.

Barom-eter H or h.	N.	0.0	0.1	0.2	0.3	0.4	0.5	0.6	0.7	0.8	0.9	Parts for each 0.01mm.
Milli.	Metr.	Metres.	Metres.	Metres.	Metres.	Metres.	Metres.	Metres.	Metres.	Metres.	Metres.	Metr.
690	73	81.1	82.3	83.4	84.6	85.7	86.9	88.1	89.2	90.4	91.5	
691	73	92.7	93.8	95.0	96.1	97.3	98.4	99.6				
691	74								00.7	01.9	03.0	
692	74	04.2	05.3	06.5	07.6	08.8	09.9	11.1	12.2	13.4	14.5	
693	74	15.7	16.8	18.0	19.1	20.3	21.4	22.6	23.7	24.9	26.0	
694	74	27.2	28.3	29.5	30.6	31.8	32.9	34.1	35.2	36.4	37.5	
695	74	38.7	39.8	41.0	42.1	43.3	44.4	45.5	46.7	47.8	49.0	
696	74	50.1	51.2	52.4	53.5	54.7	55.8	56.9	58.1	59.2	60.4	
697	74	61.5	62.6	63.8	64.9	66.1	67.2	68.3	69.5	70.6	71.8	
698	74	72.9	74.0	75.2	76.3	77.5	78.6	79.7	80.9	82.0	83.2	
699	74	84.3	85.4	86.6	87.7	88.9	90.0	91.1	92.3	93.4	94.6	
700	74	95.7	96.8	98.0	99.1							
700	75					00.3	01.4	02.5	03.7	04.8	06.0	
701	75	07.1	08.2	09.4	10.5	11.6	12.7	13.9	15.0	16.1	17.3	
702	75	18.4	19.5	20.7	21.8	23.0	24.1	25.2	26.4	27.5	28.7	
703	75	29.8	30.9	32.1	33.2	34.3	35.4	36.6	37.7	38.8	40.0	
704	75	41.1	42.2	43.4	44.5	45.6	46.7	47.9	49.0	50.1	51.3	
705	75	52.4	53.5	54.7	55.8	56.9	58.0	59.2	60.3	61.4	62.6	
706	75	63.7	64.8	66.0	67.1	68.2	69.3	70.5	71.6	72.7	73.9	
707	75	75.0	76.1	77.2	78.4	79.5	80.6	81.7	82.8	84.0	85.1	
708	75	86.2	87.3	88.5	89.6	90.7	91.8	93.0	94.1	95.2	96.4	
709	75	97.5	98.6	99.7								
709	76				00.9	02.0	03.1	04.2	05.3	06.5	07.6	
710	76	08.7	09.8	10.9	12.1	13.2	14.3	15.4	16.5	17.7	18.8	
711	76	19.9	21.0	22.1	23.3	24.4	25.5	26.6	27.7	28.9	30.0	
712	76	31.1	32.2	33.3	34.4	35.5	36.6	37.8	38.9	40.0	41.1	1 | 0.1
713	76	42.2	43.3	44.4	45.6	46.7	47.8	48.9	50.0	51.2	52.3	2 | 0.2
714	76	53.4	54.5	55.6	56.8	57.9	59.0	60.1	61.2	62.4	63.5	3 | 0.3
715	76	64.6	65.7	66.8	67.9	69.0	70.1	71.3	72.4	73.5	74.6	4 | 0.4
716	76	75.7	76.8	77.9	79.0	80.1	81.2	82.4	83.5	84.6	85.7	5 | 0.5
717	76	86.8	87.9	89.0	90.1	91.2	92.3	93.5	94.6	95.7	96.8	6 | 0.7
718	76	97.9	99.0									7 | 0.8
718	77			00.1	01.2	02.3	03.4	04.6	05.7	06.8	07.9	8 | 0.9
719	77	09.0	10.1	11.2	12.3	13.4	14.5	15.7	16.8	17.9	19.0	9 | 1.0
720	77	20.1	21.2	22.3	23.4	24.5	25.6	26.7	27.8	28.9	30.0	
721	77	31.1	32.2	33.3	34.4	35.5	36.6	37.7	38.8	39.9	41.0	
722	77	42.1	43.2	44.3	45.4	46.5	47.6	48.7	49.8	50.9	52.0	
723	77	53.1	54.2	55.3	56.4	57.5	58.6	59.8	60.9	62.0	63.1	
724	77	64.2	65.3	66.4	67.5	68.6	69.6	70.7	71.8	72.9	74.0	
725	77	75.1	76.2	77.3	78.4	79.5	80.6	81.7	82.8	83.9	85.0	
726	77	86.1	87.2	88.3	89.4	90.5	91.6	92.7	93.8	94.9	96.0	
727	77	97.1	98.2	99.3								
727	78				00.4	01.5	02.5	03.6	04.7	05.8	06.9	
728	78	08.0	09.1	10.2	11.3	12.4	13.5	14.6	15.7	16.8	17.9	
729	78	19.0	20.1	21.2	22.3	23.4	24.4	25.5	26.6	27.7	28.8	
730	78	29.9	31.0	32.1	33.3	34.3	35.3	36.4	37.5	38.6	39.7	
Barom-eter H or h.	N.	0.0	0.1	0.2	0.3	0.4	0.5	0.6	0.7	0.8	0.9	Parts for each 0.01mm.

731 to 770mm.

Barometer H or h.	N.	0.0	0.1	0.2	0.3	0.4	0.5	0.6	0.7	0.8	0.9	Parts for each 0.01mm.
Milli.	Metr.	Metres.	Metres.	Metres.	Metres.	Metres.	Metres.	Metres.	Metres.	Metres.	Metres.	Metr.
731	78	40.8	41.9	43.0	44.1	45.2	46.2	47.3	48.4	49.5	50.6	
732	78	51.7	52.8	53.9	54.9	56.0	57.0	58.2	59.3	60.3	61.4	
733	78	62.5	63.6	64.7	65.8	66.9	67.9	69.0	70.1	71.2	72.3	
734	78	73.4	74.5	75.6	76.6	77.7	78.8	79.9	81.0	82.0	83.1	
735	78	84.2	85.3	86.4	87.5	88.6	89.6	90.7	91.8	92.9	94.0	
736	78	95.1	96.2	97.3	98.3	99.4						
736	79						00.5	01.6	02.7	03.7	04.8	
737	79	05.9	07.0	08.1	09.1	10.2	11.3	12.4	13.5	14.5	15.6	
738	79	16.7	17.8	18.9	19.9	21.0	22.1	23.2	24.3	25.3	26.4	
739	79	27.5	28.6	29.6	30.7	31.8	32.8	33.9	35.0	36.1	37.1	
740	79	38.2	39.3	40.4	41.4	42.5	43.6	44.7	45.8	46.8	47.9	
741	79	49.0	50.1	51.1	52.2	53.3	54.3	55.4	56.5	57.6	58.6	
742	79	59.7	60.8	61.8	62.9	64.0	65.0	66.1	67.2	68.3	69.3	
743	79	70.4	71.5	72.6	73.6	74.7	75.8	76.9	78.0	79.0	80.1	
744	79	81.2	82.3	83.3	84.4	85.5	86.5	87.6	88.7	89.8	90.8	
745	79	91.9	93.0	94.0	95.1	96.1	97.2	98.3	99.3			
745	80									00.4	01.4	
746	80	02.5	03.6	04.6	05.7	06.8	07.8	08.9	10.0	11.1	12.3	
747	80	13.2	14.3	15.3	16.4	17.4	18.5	19.6	20.6	21.7	22.7	
748	80	23.8	24.9	25.9	27.0	28.0	29.1	30.2	31.2	32.3	33.3	
749	80	34.4	35.5	36.5	37.6	38.7	39.7	40.8	41.9	43.0	44.0	
750	80	45.1	46.2	47.3	48.4	49.4	50.5	51.6	52.6	53.7	54.7	
751	80	55.7	56.8	57.8	58.9	59.9	61.0	62.1	63.1	64.2	65.2	
752	80	66.3	67.4	68.4	69.5	70.5	71.6	72.7	73.7	74.8	75.8	
753	80	76.9	78.0	79.0	80.1	81.1	82.2	83.3	84.3	85.4	86.4	
754	80	87.5	88.5	89.6	90.6	91.7	92.7	93.8	94.8	95.9	96.9	1 0.1
755	80	98.0	99.1									2 0.2
755	81			00.1	01.2	02.2	03.3	04.4	05.4	06.5	07.5	3 0.3
756	81	08.6	09.6	10.7	11.7	12.8	13.8	14.9	15.9	17.0	18.0	4 0.4
757	81	19.1	20.1	21.2	22.2	23.3	24.3	25.4	26.4	27.5	28.5	5 0.5
758	81	29.6	30.6	31.7	32.7	33.8	34.8	35.9	36.9	38.0	39.0	6 0.6
759	81	40.1	41.1	42.2	43.2	44.3	45.3	46.4	47.4	48.5	49.5	7 0.7
												8 0.8
760	81	50.6	51.6	52.7	53.7	54.8	55.8	56.9	57.9	59.0	60.0	9 0.9
761	81	61.1	62.1	63.2	64.2	65.3	66.3	67.3	68.4	69.4	70.5	
762	81	71.5	72.5	73.6	74.6	75.7	76.7	77.8	78.8	79.9	80.9	
763	81	82.0	83.0	84.1	85.1	86.2	87.2	88.2	89.3	90.3	91.4	
764	81	92.4	93.4	94.5	95.5	96.6	97.6	98.6	99.7			
764	82									00.7	01.8	
765	82	02.8	03.8	04.9	05.9	07.0	08.0	09.0	10.1	11.1	12.2	
766	82	13.2	14.2	15.3	16.3	17.4	18.4	19.4	20.5	21.5	22.6	
767	82	23.6	24.6	25.7	26.7	27.8	28.8	29.8	30.9	31.9	33.0	
768	82	34.0	35.0	36.1	37.1	38.2	39.2	40.2	41.3	42.3	43.4	
769	82	44.4	45.4	46.5	47.5	48.5	49.5	50.6	51.6	52.6	53.7	
770	82	54.7	55.7	56.8	57.8	58.8	59.8	60.9	61.9	62.9	64.0	
Barometer H or h.	N.	0.0	0.1	0.2	0.3	0.4	0.5	0.6	0.7	0.8	0.9	Parts for each 0.01mm.

771 to 810mm.

Barometer H or h.	N.	0.0	0.1	0.2	0.3	0.4	0.5	0.6	0.7	0.8	0.9	Parts for each 0.01mm.
Milli.	Metr.	Metres.	Metres.	Metres.	Metres.	Metres.	Metres.	Metres.	Metres.	Metres.	Metres.	Metr.
771	82	65.0	66.0	67.1	68.1	69.2	70.2	71.2	72.3	73.3	74.4	
772	82	75.4	76.4	77.5	78.5	79.5	80.5	81.6	82.6	83.6	84.7	
773	82	85.7	86.7	87.8	88.8	89.8	90.8	91.9	92.9	93.9	95.0	
774	82	96.0	97.0	98.0	99.1							
774	83					00.1	01.1	02.1	03.1	04.2	05.2	
775	83	06.2	07.2	08.3	09.3	10.3	11.3	12.4	13.4	14.4	15.5	
776	83	16.5	17.5	18.5	19.6	20.6	21.6	22.6	23.6	24.7	25.7	
777	83	26.7	27.7	28.8	29.8	30.8	31.8	32.9	33.9	34.9	36.0	
778	83	37.0	38.0	39.0	40.1	41.1	42.1	43.1	44.1	45.2	46.2	
779	83	47.2	48.2	49.2	50.3	51.3	52.3	53.3	54.3	55.4	56.4	
780	83	57.4	58.4	59.4	60.5	61.5	62.5	63.5	64.5	65.6	66.6	
781	83	67.6	68.6	69.6	70.7	71.7	72.7	73.7	74.7	75.8	76.8	
782	83	77.8	78.8	79.8	80.9	81.9	82.9	83.9	84.9	86.0	87.0	
783	83	88.0	89.0	90.0	91.1	92.1	93.1	94.1	95.1	96.2	97.2	
784	83	98.2	99.2									
784	84			00.2	01.2	02.2	03.2	04.3	05.3	06.3	07.3	
785	84	08.3	09.3	10.3	11.4	12.4	13.4	14.4	15.4	16.5	17.5	
786	84	18.5	19.5	20.5	21.5	22.5	23.5	24.6	25.6	26.6	27.6	
787	84	28.6	29.6	30.6	31.6	32.6	33.6	34.7	35.7	36.7	37.7	
788	84	38.7	39.7	40.7	41.7	42.7	43.7	44.8	45.8	46.8	47.8	
789	84	48.8	49.8	50.8	51.8	52.8	53.8	54.9	55.9	56.9	57.9	
790	84	58.9	59.9	60.9	61.9	62.9	63.9	65.0	66.0	67.0	68.0	
791	84	68.9	69.9	70.9	71.9	72.9	73.9	75.0	76.0	77.0	78.0	1 0.1
792	84	79.0	80.0	81.0	82.0	83.0	84.0	85.0	86.0	87.0	88.0	2 0.2
793	84	89.0	90.0	91.0	92.0	93.0	94.0	95.1	96.1	97.1	98.1	3 0.3
794	84	99.1										4 0.4
794	85		00.1	01.1	02.1	03.1	04.1	05.1	06.1	07.1	08.1	5 0.5
795	85	09.1	10.1	11.1	12.1	13.1	14.1	15.1	16.1	17.1	18.1	6 0.6
796	85	19.1	20.1	21.1	22.1	23.1	24.1	25.1	26.1	27.1	28.1	7 0.7
797	85	29.1	30.1	31.1	32.1	33.1	34.1	35.1	36.1	37.1	38.1	8 0.8
798	85	39.1	40.1	41.1	42.1	43.1	44.1	45.1	46.1	47.1	48.1	9 0.9
799	85	49.1	50.1	51.1	52.0	53.0	54.1	55.0	56.0	57.0	58.0	
800	85	59.0	60.0	61.0	62.0	63.0	64.0	65.0	66.0	67.0	68.0	
801	85	69.0	70.0	70.9	71.9	72.9	73.9	74.9	75.9	76.9	77.9	
802	85	78.9	79.9	80.9	81.9	82.9	83.9	84.9	85.8	86.8	87.8	
803	85	88.8	89.8	90.8	91.8	92.8	93.8	94.8	95.8	96.7	97.7	
804	85	98.7	99.7									
804	86			00.7	01.7	02.7	03.7	04.7	05.7	06.6	07.6	
805	86	08.6	09.6	10.6	11.6	12.6	13.6	14.6	15.5	16.5	17.5	
806	86	18.5	19.5	20.5	21.5	22.5	23.4	24.4	25.4	26.4	27.4	
807	86	28.4	29.4	30.4	31.3	32.3	33.3	34.3	35.3	36.3	37.3	
808	86	38.3	39.2	40.2	41.2	42.2	43.2	44.2	45.1	46.1	47.1	
809	86	48.1	49.1	50.1	51.1	52.0	53.0	54.0	55.0	56.0	57.0	
810	86	57.9	58.9	59.9	60.9	61.9	62.8	63.8	64.8	65.8	66.8	
Barometer H or h.	N.	0.0	0.1	0.2	0.3	0.4	0.5	0.6	0.7	0.8	0.9	Parts for each 0.01mm.

TABLE II. Correction for Difference of Temperature of Attached Thermometers.

Temperature of Barometers at Station { Upper = T′ / Lower = T.

T′ — T Centig.	Correct. Metres.	T′ — T Centigrade.	Correct. Metres.	T′ — T Centigrade.	Correct. Metres.	T′ — T Centigrade.	Correct. Metres.	T′ — T Centigrade.	Correct. Metres.
0.0	0.0	8.0	10.3	16.0	20.6	24.0	30.9	32.0	41.3
0.2	0.3	8.2	10.6	16.2	20.9	24.2	31.2	32.2	41.5
0.4	0.5	8.4	10.8	16.4	21.1	24.4	31.5	32.4	41.8
0.6	0.8	8.6	11.1	16.6	21.4	24.6	31.7	32.6	42.0
0.8	1.0	8.8	11.3	16.8	21.7	24.8	32.0	32.8	42.3
1.0	1.3	9.0	11.6	17.0	21.9	25.0	32.2	33.0	42.5
1.2	1.5	9.2	11.9	17.2	22.2	25.2	32.5	33.2	42.8
1.4	1.8	9.4	12.1	17.4	22.4	25.4	32.7	33.4	43.1
1.6	2.1	9.6	12.4	17.6	22.7	25.6	33.0	33.6	43.3
1.8	2.3	9.8	12.6	17.8	22.9	25.8	33.3	33.8	43.6
2.0	2.6	10.0	12.9	18.0	23.2	26.0	33.5	34.0	43.8
2.2	2.8	10.2	13.1	18.2	23.5	26.2	33.8	34.2	44.1
2.4	3.1	10.4	13.4	18.4	23.7	26.4	34.0	34.4	44.3
2.6	3.4	10.6	13.7	18.6	24.0	26.6	34.3	34.6	44.6
2.8	3.6	10.8	13.9	18.8	24.2	26.8	34.6	34.8	44.9
3.0	3.9	11.0	14.2	19.0	24.5	27.0	34.8	35.0	45.1
3.2	4.1	11.2	14.5	19.2	24.8	27.2	35.1	35.2	45.4
3.4	4.4	11.4	14.7	19.4	25.0	27.4	35.3	35.4	45.6
3.6	4.6	11.6	15.0	19.6	25.3	27.6	35.6	35.6	45.9
3.8	4.9	11.8	15.2	19.8	25.5	27.8	35.8	35.8	46.2
4.0	5.2	12.0	15.5	20.0	25.8	28.0	36.1	36.0	46.4
4.2	5.4	12.2	15.8	20.2	26.0	28.2	36.4	36.2	46.7
4.4	5.7	12.4	16.0	20.4	26.3	28.4	36.6	36.4	46.9
4.6	5.9	12.6	16.3	20.6	26.6	28.6	36.9	36.6	47.2
4.8	6.2	12.8	16.5	20.8	26.8	28.8	37.1	36.8	47.4
5.0	6.4	13.0	16.8	21.0	27.1	29.0	37.4	37.0	47.7
5.2	6.7	13.2	17.0	21.2	27.3	29.2	37.6	37.2	48.0
5.4	7.0	13.4	17.3	21.4	27.6	29.4	37.9	37.4	48.2
5.6	7.2	13.6	17.5	21.6	27.8	29.6	38.2	37.6	48.5
5.8	7.5	13.8	17.8	21.8	28.1	29.8	38.4	37.8	48.7
6.0	7.7	14.0	18.0	22.0	28.4	30.0	38.7	38.0	49.0
6.2	8.0	14.2	18.3	22.2	28.6	30.2	38.9	38.2	49.2
6.4	8.3	14.4	18.5	22.4	28.9	30.4	39.2	38.4	49.5
6.6	8.5	14.6	18.8	22.6	29.1	30.6	39.5	38.6	49.8
6.8	8.8	14.8	19.0	22.8	29.4	30.8	39.7	38.8	50.0
7.0	9.0	15.0	19.3	23.0	29.7	31.0	40.0	39.0	50.3
7.2	9.3	15.2	19.6	23.2	29.9	31.2	40.2	39.2	50.5
7.4	9.5	15.4	19.8	23.4	30.2	31.4	40.5	39.4	50.8
7.6	9.8	15.6	20.1	23.6	30.4	31.6	40.7	39.6	51.1
7.8	10.1	15.8	20.3	23.8	30.7	31.8	41.0	39.8	51.3
8.0	10.3	16.0	20.6	24.0	30.9	32.0	41.3	40.0	51.6

This Table supposes the scale to be of *brass* from the top to the cistern. If it were of glass or of wood, the argument T′ — T ought to be diminished at the ratio of 54 to 62.

In calculating the formula of Laplace, we begin by reducing the barometers to the same temperature by means of the following formula: $H = h' + h'\left(\frac{T' - T}{6196}\right)$. Table II. saves this trouble, and gives, in metres, the correction due to the difference of temperature of the barometers.

TABLE III. Correction for Decrease of Gravitation in Latitude.

$\beta = (0.0028371 \text{ cosin. } 2 \text{ L}). \quad (A + \alpha + \beta).$

The Argument is the Mean Latitude between the two Stations.

LATITUDE.		Correction, in metres, for								
Correction. Added.	Subtr'ct	1000	2000	3000	4000	5000	6000	7000	8000	9000
°	°									
0	90	2.8	5.7	8.5	11.3	14.2	17.0	19.9	22.7	25.7
1	89	2.8	5.7	8.5	11.3	14.2	17.0	19.8	22.7	25.6
2	88	2.8	5.7	8.5	11.3	14.1	17.0	19.8	22.6	25.5
3	87	2.8	5.6	8.5	11.3	14.1	16.9	19.7	22.6	25.4
4	86	2.8	5.6	8.4	11.2	14.0	16.9	19.7	22.5	25.3
5	85	2.8	5.6	8.4	11.2	14.0	16.8	19.6	22.3	25.1
6	84	2.8	5.5	8.3	11.1	13.9	16.6	19.4	22.2	25.0
7	83	2.7	5.5	8.2	11.0	13.8	16.5	19.3	22.0	24.8
8	82	2.7	5.4	8.2	10.9	13.6	16.4	19.1	21.8	24.5
9	81	2.7	5.4	8.1	10.8	13.5	16.2	18.9	21.6	24.3
10	80	2.7	5.3	8.0	10.7	13.3	16.0	18.7	21.3	24.0
11	79	2.6	5.2	7.9	10.5	13.1	15.8	18.4	21.0	23.7
12	78	2.6	5.2	7.8	10.4	13.0	15.5	18.1	20.7	23.3
13	77	2.5	5.1	7.6	10.2	12.7	15.3	17.8	20.4	22.9
14	76	2.5	5.0	7.5	10.0	12.5	15.0	17.5	20.0	22.5
15	75	2.5	4.9	7.4	9.8	12.3	14.7	17.2	19.7	22.1
16	74	2.4	4.8	7.2	9.6	12.0	14.4	16.8	19.2	21.6
17	73	2.4	4.7	7.0	9.4	11.8	14.1	16.5	18.8	21.2
18	72	2.3	4.6	6.9	9.2	11.5	13.8	16.1	18.4	20.7
19	71	2.2	4.5	6.7	8.9	11.2	13.4	15.6	17.9	20.1
20	70	2.2	4.3	6.5	8.7	10.9	13.0	15.2	17.4	19.6
21	69	2.1	4.2	6.3	8.4	10.5	12.6	14.7	16.9	19.0
22	68	2.0	4.1	6.1	8.2	10.2	12.2	14.3	16.3	18.4
23	67	2.0	3.9	5.9	7.9	9.8	11.8	13.8	15.8	17.7
24	66	1.9	3.8	5.7	7.6	9.5	11.4	13.3	15.2	17.1
25	65	1.8	3.6	5.5	7.3	9.1	10.9	12.8	14.6	16.4
26	64	1.7	3.5	5.2	7.0	8.7	10.5	12.2	14.0	15.7
27	63	1.7	3.3	5.0	6.7	8.3	10.0	11.7	13.3	15.0
28	62	1.6	3.2	4.8	6.3	7.9	9.5	11.1	12.7	14.3
29	61	1.5	3.0	4.5	6.0	7.5	9.0	10.5	12.0	13.5
30	60	1.4	2.8	4.3	5.7	7.1	8.5	9.9	11.3	12.8
31	59	1.3	2.7	4.0	5.3	6.6	8.0	9.3	10.6	12.0
32	58	1.2	2.5	3.7	5.0	6.2	7.5	8.7	9.9	11.2
33	57	1.1	2.3	3.5	4.6	5.8	6.9	8.1	9.2	10.4
34	56	1.1	2.1	3.2	4.2	5.3	6.4	7.4	8.5	9.6
35	55	1.0	1.9	2.9	3.9	4.8	5.8	6.8	7.8	8.7
36	54	0.9	1.7	2.6	3.5	4.4	5.3	6.1	7.0	7.9
37	53	0.8	1.6	2.3	3.1	3.9	4.7	5.5	6.2	7.0
38	52	0.7	1.4	2.1	2.7	3.4	4.1	4.8	5.5	6.2
39	51	0.6	1.2	1.8	2.4	2.9	3.5	4.1	4.7	5.3
40	50	0.5	1.0	1.5	2.0	2.5	3.0	3.4	3.9	4.4
41	49	0.4	0.8	1.2	1.6	2.0	2.4	2.8	3.2	3.5
42	48	0.3	0.6	0.9	1.2	1.5	1.8	2.1	2.4	2.7
43	47	0.2	0.4	0.6	0.8	1.0	1.2	1.4	1.6	1.8
44	46	0.1	0.2	0.3	0.4	0.5	0.6	0.7	0.8	0.9
45	45	0.0	0.0	0.0	0.0	0.0	0.0	0.0	0.0	0.0

TABLE IV. Correction for Decrease of Gravitation on a Vertical Line.

$$\delta = \left(\frac{A + \alpha + \beta + \upsilon + 15296}{6366200}\right) \times A\ (+ \alpha + \beta + \upsilon).$$

Argument = (A + α + β + υ).

Approximate Difference of Level.	Correspond. Correction Positive.	Approximate Difference of Level.	Correspond. Correction Positive.	Approximate Difference of Level.	Correspond. Correction Positive.	Approximate Difference of Level.	Correspond. Correction Positive.
Metres.	Metres.	Metres.	Metres.	Metres.	Metres.	Metres.	Metres.
100	0.2	2100	6.0	4100	12.9	6100	21.1
200	0.5	2200	6.3	4200	13.3	6200	21.6
300	0.8	2300	6.6	4300	13.7	6300	22.0
400	1.0	2400	6.9	4400	14.1	6400	22.5
500	1.3	2500	7.3	4500	14.5	6500	22.9
600	1.6	2600	7.6	4600	14.9	6600	23.4
700	1.8	2700	7.9	4700	15.3	6700	23.9
800	2.1	2800	8.3	4800	15.7	6800	24.3
900	2.4	2900	8.6	4900	16.1	6900	24.8
1000	2.7	3000	8.9	5000	16.5	7000	25.3
1100	2.9	3100	9.3	5100	16.9	7100	25.7
1200	3.2	3200	9.6	5200	17.3	7200	26.2
1300	3.5	3300	10.0	5300	17.7	7300	26.7
1400	3.8	3400	10.3	5400	18.1	7400	27.2
1500	4.1	3500	10.7	5500	18.5	7500	27.7
1600	4.4	3600	11.1	5600	19.0	7600	28.1
1700	4.7	3700	11.4	5700	19.4	7700	28.6
1800	5.0	3800	11.8	5800	19.8	7800	29.1
1900	5.3	3900	12.2	5900	20.3	7900	29.6
2000	5.6	4000	12.5	6000	20.7	8000	30.1

TABLE V. Correction for the Elevation of the Lower Station above Ocean.

Argument = Height of Barometer at Lower Station.

Approximate Difference of Level.	Height of Barometer at Lower Station in Millimetres.							
	400	450	500	550	600	650	700	750
Metres.	Metres.	Metres.	Metres.	Metres.	Metres.	Metres.	Metres.	Metres.
1000	1.7	1.4	1.1	0.9	0.6	0.4	0.2	0.0
2000	3.4	2.8	2.2	1.7	1.3	0.8	0.4	0.1
3000	5.1	4.2	3.3	2.6	1.9	1.3	0.7	0.1
4000	6.8	5.6	4.4	3.4	2.5	1.7	0.9	0.1
5000	8.5	6.9	5.5	4.3	3.1	2.1	1.1	0.1
6000	10.3	8.3	6.7	5.2	3.8	2.5	1.3	0.2
7000	12.0	9.7	7.8	6.0	4.4	2.9	1.5	0.2
8000	13.7	11.1	8.9	6.9	5.0	3.4	1.8	0.2
9000	15.4	12.5	10.0	7.7	5.7	3.8	2.0	0.3

II.

TABLES

FOR COMPUTING DIFFERENCES OF ELEVATION FROM BAROMETRICAL OBSERVATIONS.

BY A. GUYOT.

TABLES which, like the preceding ones by Delcros, in metrical measures, are sufficiently extensive to save the necessity of interpolations, relieve the computer of most of his trouble, and considerably reduce the chances of error in the computations. They thus render to science itself a real service, by inducing observers to determine a larger number of points, and to secure the accuracy of the results by repeating their observations at the same point in various atmospheric circumstances, both of which they can do without fear of being overwhelmed by the labor of the computation.

Similar tables are here offered to the observers who use instruments graduated to English measures. Like those of Delcros, the new tables are based on Laplace's formula, with a slight modification of only one constant. They dispense with the use of logarithms, and give the differences of level corresponding to every thousandth of an inch from 12 to 31 inches by means of the simplest arithmetical operations, so that the data being prepared and corrected, the computation of an elevation takes but a few minutes, and is done with scarcely any chance of error.

Laplace's formula and constants were adopted for the computation of the tables in preference to others found in the following sets for reasons which a few words will explain.

It has been remarked, page 9, that, in consequence of Laplace's constants having been retained in Gauss's, Schmidt's, and Baily's formulæ, they all give similar results ; but that Bessel's formula differs in separating the correction due to the moisture of the air from that due to its temperature, while in Laplace's, and in the formulæ just mentioned, both are united. To introduce a separate correction for the expansion of aqueous vapor is, in the writer's view, a doubtful improvement. The laws of the distribution and transmission of moisture through the atmosphere are too little known, and its amount, especially in mountain regions, is too variable, and depends too much upon local winds and local condensation, to allow a reasonable hope of obtaining the mean humidity of the layer of air between the two stations by means of hygrometrical observations taken at each of them. These doubts are confirmed by the experience of the author and of many other observers, which shows that, on an average, Laplace's method works not only as well as the other, but more uniformly well. At any rate, the gain, if there is any, is not clear enough to compensate for the undesirable complication of the formula.

Though the several co-efficients of Laplace's formula need perhaps to be modified according to more recent and probably more accurate determinations of the physical constants on which they depend, as has been proposed by Plantamour, E. Ritter, and lately by the writer himself in a paper read before the American Association for the Advancement of Science at their meeting in Montreal, they have been retained in preparing the following tables, partly because it was found that the errors due to the various co-efficients nearly compensate each other; partly on the ground that, until a severe test, by means of actual comparative measurements made for the purpose, has shown the expediency of these modifications, it seemed desirable to adhere to the old constants, and thus to preserve a uniformity in the results with the tables of Oltmans, Delcros, Gauss, Baily, and others, which have already been extensively used. The substitution of the co-efficient 0.00260, expressing, according to Schmidt's computation (*Mathem. und Physic. Geogr.*, II. p. 202), the variation of gravity in latitude, for the value 0.002837, does not sensibly alter the altitudes obtained.

The close agreement of the determinations furnished by Laplace's formula, in barometrical measurements carefully conducted, made in favorable circumstances, and during the warm season, with those obtained from repeated trigonometrical observations, or by the spirit-level, strongly testifies in favor of its general correctness. A few striking examples will suffice to show it.

The altitude of Mont Blanc, measured by the barometer, by MM. Bravais and Martins, on the 29th of August, 1844, and computed by Delcros, by means of nine corresponding stations situated on all sides of the mountain (see *Annuaire Météorologique de France*, for 1851, p. 274), was found to be 4810 metres. The altitude of the same point, being the mean of seven of the most elaborate and reliable geodetic measurements, which cost nearly twenty years of labor, is 4809.6 metres.

For smaller elevations the formula seems to answer equally well.

The barometrical measurement of Mount Washington, in New Hampshire, by the author, on the 8th and 9th of August, 1851, gave, by Delcros's Tables, for the mean of eight observations, taken at different hours of the day, 5466.7 English feet above Gorham, N. H., 6285.7 above high tide, and 6291.7 feet above the mean level of the ocean in Portland harbor. In August, 1852, W. A. Goodwin, Civil Engineer, starting from Gorham Railroad Station, found, by the spirit-level, Mount Washington to be 6285.5 feet above mean tide. In September, 1853, Captain T. J. Cram, of the Topographical Engineers, executed, in behalf of the Coast Survey, a careful measurement with the spirit-level, on the same line, for the purpose of testing the various methods of measuring altitudes, and found Mount Washington to be 6293 English feet above the mean level of the ocean.

In lower latitudes the formula showed equally good results. By a barometrical measurement in July, 1856, the altitude of the highest peak of the Black Mountain, North Carolina, about Lat. 36°, was found by the author to be 6701 English feet; and that of the highest Mountain House 5248 feet. In September, 1857, Major T. C. Turner, Chief Engineer of the Morganton Railroad, ran a line of levels from the same point which was used as the lower station for the barometrical measurement, to the top of the highest peak, and found its altitude to be 6711 English feet, and that of the Mountain House 5246 feet. Other points on the line agreed equally well.

Such an agreement, in so considerable elevations, is all that can be desired.

These figures show conclusively, that, when the errors which may arise from the great variability of the data furnished by the instruments have been removed by a repetition, in various states of the atmosphere, and by a proper combination of simultaneous observations at stations not too distant from each other, those which remain and may be attributed to the formula cannot be considerable. But, on the other hand, we have no right to expect such results from single observations, taken, perhaps, in unsettled weather, without paying any regard to the time of the day at which they were made, to the distance or the non-simultaneity of the corresponding observations, or to other unfavorable circumstances. It is too well known that in such cases large errors may and do actually occur; but for these the formula ought not to be held responsible.

Arrangement of the Tables.

If we call

h = the observed height of the barometer, τ = the temperature of the barometer, t = the temperature of the air — at the lower station;

h' = the observed height of the barometer, τ' = the temperature of the barometer, t' = the temperature of the air — at the upper station.

If we make, further,

Z = the difference of level between the two barometers;

L = the mean latitude between the two stations;

H = the height of the barometer at the upper station reduced to the temperature of the barometer at the lower station; or,

$H = h' \{1 + 0.00008967 (\tau - \tau')\}$;

The expansion of the mercurial column, measured by a brass scale, for 1° Fahrenheit = 0.00008967;

The increase of gravity from the equator to the poles = 0.00520048, or 0.00260 to the 45th degree of latitude;

The earth's mean radius = 20,886,860 English feet;

Then, Laplace's formula, reduced to English measures, reads as follows:

$$Z = \log \frac{h}{H} \times 60158.6 \text{ English feet} \left\{ \begin{array}{l} \left(1 + \frac{t + t' - 64}{900}\right). \\ (1 + 0.00260 \cos 2\,L). \\ \left(1 + \frac{z + 52252}{20886860} + \frac{h}{10443430}\right). \end{array} \right.$$

Table I. gives, in English feet, the value of log H or $h \times 60158.6$ for every hundredth of an inch, from 12 to 31 inches in the barometer, together with the value of the additional thousandths, in a separate column. These values have been diminished by a constant, which does not alter the difference required.

Table II. gives the correction 2.343 feet $\times (\tau - \tau')$ for the difference of the temperatures of the barometers at the two stations, or $\tau - \tau'$. As the temperature at the upper station is generally lower, $\tau - \tau'$ is usually positive, and the correction *negative*. It becomes *positive* when the temperature of the upper barometer is higher, and $\tau - \tau'$ negative. When the heights of the barometers have been reduced to the same temperature, or to the freezing point, this table will not be used.

Table IV. shows the correction $D' \frac{z + 52252}{20886860}$ to be applied to the approximate altitude for the decrease of gravity on a vertical acting on the density of the mercurial column. It is always *additive*.

Table V. furnishes the small correction $\frac{h}{10443430}$ for the decrease of gravity on a vertical acting on the density of the air; the height of the barometer h at the lower station representing its approximate altitude. Like the preceding correction, it is always *additive.*

Use of the Tables.

In Table I. find first the numbers corresponding to the observed heights of the barometer h and h'. Suppose, for instance, $h = 29.345$ in.; find in the first column on the left the number 29.3; on the same horizontal line, in the column headed .04, is given the number corresponding to 29.34 = 28121.7; in the last column but one on the right, we find for .005 = 4.5, or for 29.345 = 28126.2. Take likewise the value of h', and find the difference.

If the barometrical heights have not been previously reduced to the same temperature, or to the freezing point, apply to the difference the correction found in Table II. opposite the number representing $\tau - \tau'$; we thus obtain the approximate difference of level, D.

For computing the correction due to the expansion of the air according to its temperature, or $D \times \left(\frac{t + t' - 64}{900}\right)$, make the sum of the temperatures, subtract from that sum 64; multiply the rest into the approximate difference D, and divide the product by 900. This correction is of the same sign as $(t + t' - 64)$. By applying it, we obtain a second approximate difference of level, D'.

In Table III., with D' and the mean latitude of the stations, find the correction for variation of gravity in latitude, and add it to D', paying due attention to the sign.

In Table IV. with D', and in Table V. with D' and the height of the barometer at the lower station, take the corrections for the decrease of gravity on a vertical, and add them to the approximate difference of level.

The sum thus found is the true difference of level between the two stations, or Z; by adding the elevation of the lower station above the level of the sea, when known, we obtain the *altitude* of the upper station.

The use of the small table, VI., by means of which approximate differences of level can be obtained by a single multiplication, is explained below, page 90.

Example 1.

Measurement of Mount Washington, New Hampshire, by A. Guyot, August 8th, 1851, 4 P. M.; the barometer at the lower station being at 825 English feet above the mean level of the sea; at the upper station at one foot below the summit.

The observation gave,

	Barometer.	Attached Thermometer.	Temperature of Air.
Gorham,	$h = 29.272$ in.	$\tau = 70°.70$ F.	$t = 72°.05$ F.
Mount Washington,	$h' = 24.030$ "	$\tau' = 54°.52$ F.	$t' = 50°.54$ F.
		$\tau - \tau' = 16°.38$ F.	122°.59 F.
			— 64°
			$t + t' - 64 = 58°.59$ F.

Table I. gives for h = 29.272 inches,	. . .	28,061.00
" " for h' = 24.030 "	. . .	22,905.60
Difference,	. .	5,155.40
Table II. gives for $\tau - \tau'$ = 16°.38	. . .	— 37.64
Approximate difference of level, D =		5,117.76
$\frac{D \times (t + t' - 64)}{900} = \frac{5118 \times 58.6}{900} =$		333.19
Second approximate difference, D' =		5,450.95
Table III. gives for D' = 5450 and Lat. 44°	. .	0.50
Table IV. gives for D' = 5450	. . . .	14.94
Table V. gives for h = 29.27	. . . .	0.00
Barometer below summit,	. .	— 1.00
Mount Washington above Gorham, or	. . Z =	5,465.39
Barometer at Gorham above sea level	. .	825.00
Mount Washington above the sea, or altitude,	. .	6,290.39 Eng. ft.

Example 2.

Measurement of the highest peak of the Black Mountain, in North Carolina, July 11th, 1856, by A. Guyot.

By observation we have at,

	Barometer.	Attached Thermometer.	Temperature of Air.
Mountain House,	h = 24.934 in.	τ = 64°.58 F.	t = 61°.34 F.
Highest Peak,	h' = 23.662 "	τ' = 61°.88 F.	t' = 59°.36 F.
		$\tau - \tau'$ = 2°.70 F.	120°.70 F.
			— 64°
			$t + t' - 64$ = 56°.7 F.

Table I. gives for h = 24.934	. . . . . .	23,870.4
" " for h' = 23.662	. . . . .	22,502.4
Difference,	. . . . .	1,368.0
Table II. gives for $\tau - \tau'$ = 2.7	. . . . .	— 6.3
Approximate difference, D =		1,361.7
$\frac{D \times (t + t' - 64)}{900} = \frac{1362 \times 56.7}{900} =$		85.8
Second approximate difference, D' =		1,447.5
Table III. gives for D' = 1448 and Lat. 36°	. .	1.2
Table IV. gives for D' = 1448	. . . .	3.8
Table V. gives for D' = 1448 and h = 25	. .	0.7
Highest peak above Mountain House, or	. . Z =	1,453.2
Mountain House above the sea	. . . . .	5,248.4
Black Mountain, highest peak above the sea, or altitude,		6,701.6 Eng. ft.

II.

TABLES

FOR COMPUTING THE DIFFERENCE IN THE HEIGHT OF TWO PLACES FROM BAROMETRICAL OBSERVATIONS.

I. $D = 60158.58 \times \log H$ or h. Argument, the observed Height of the Barometer at either Station.

Barometer in Eng. Inch.	Hundredths of an Inch. .00	.01	.02	.03	.04	.05	.06	.07	.08	.09	Thousandths of an Inch.		Barometer in Eng. Inch.
	Eng. Feet.	Eng. Feet.	Eng. Feet.	Eng. Feet.	Eng. Feet.	Eng. Feet.	Eng. Feet.	Eng. Feet.	Eng. Feet.	Eng. Feet.			
12.0	4763.4	4785.2	4806.9	4828.7	4850.4	4872.1	4893.7	4915.4	4937.0	4958.6			12.0
12.1	4980.2	5001.8	5023.4	5044.9	5066.4	5087.9	5109.4	5130.9	5152.4	5173.8			12.1
12.2	5195.2	5216.6	5238.0	5259.4	5280.7	5302.1	5323.4	5344.7	5367.0	5387.2		Feet.	12.2
12.3	5408.5	5429.8	5452.0	5472.2	5493.4	5514.5	5535.7	5556.8	5578.9	5599.0	1	2.1	12.3
12.4	5620.1	5641.2	5662.2	5683.2	5704.3	5725.3	5746.2	5767.2	5788.1	5809.0	2	4.2	12.4
12.5	5829.9	5850.8	5871.7	5892.6	5913.4	5934.2	5955.0	5975.8	5996.6	6017.4	3	6.2	12.5
12.6	6038.1	6058.8	6079.6	6100.2	6120.9	6141.6	6162.2	6182.8	6203 5	6224.0	4	8.3	12.6
12.7	6244.6	6265.2	6285.8	6306.3	6326.8	6347.3	6367.8	6388.3	6408.8	6429.2	5	10.4	12.7
12.8	6449.6	6470.0	6490.4	6510.8	6531.1	6551.5	6571.8	6592.1	6612.4	6632.7	6	12.5	12.8
12.9	6652.9	6673.2	6693.4	6713.6	6733.8	6754.0	6774.1	6794.3	6814.4	6834.5	7	14.6	12.9
13.0	6854.7	6874.7	6894.8	6914.9	6934.9	6955.0	6975.0	6995.0	7014.9	7034.9	8	16.6	13.0
13.1	7054.9	7074.8	7094.7	7114.6	7134.5	7154.4	7174.3	7194.1	7213.9	7233.8	9	18.7	13.1
13.2	7253.6	7273.3	7293.1	7312.9	7332.6	7352.3	7372.1	7391.8	7411.4	7431.1			13.2
13.3	7450.8	7470.4	7490.0	7509.6	7529.2	7548.8	7568.4	7587.9	7607.4	7627.0			13.3
13.4	7646.5	7666.0	7685.4	7704.9	7724.4	7743.8	7763.2	7782.6	7802.0	7821.4			13.4

Barometer in Eng. Inch.	Hundredths of an Inch. .00	.01	.02	.03	.04	.05	.06	.07	.08	.09	Thousandths of an Inch.		Barometer in Eng. Inch.
	Eng. Feet.	Eng. Feet.	Eng. Feet.	Eng. Feet.	Eng. Feet.	Eng. Feet.	Eng. Feet.	Eng. Feet.	Eng. Feet.	Eng. Feet.		Feet.	
13.5	7840.8	7860.1	7879.4	7898.7	7918.0	7937.3	7956.6	7975.8	7995.1	8014.3			13.5
13.6	8033.6	8052.8	8071.9	8091.1	8110.3	8129.4	8148.6	8167.7	8186.8	8205.9			13.6
13.7	8225.0	8244.0	8263.1	8282.1	8301.1	8320.1	8339.1	8358.1	8377.1	8396.0	1	1.9	13.7
13.8	8415.0	8433.9	8452.8	8471.7	8490.6	8509.4	8528.3	8547.1	8565.9	8584.8	2	3.8	13.8
13.9	8603.6	8622.3	8641.1	8659.9	8678.6	8697.4	8716.1	8734.8	8753.5	8772.2	3	5.6	13.9
14.0	8790.8	8809.5	8828.2	8846.8	8865.4	8884.0	8902.6	8921.2	8939.7	8958.3	4	7.5	14.0
14.1	8976.8	8995.4	9013.9	9032.4	9050.8	9069.3	9087.8	9106.2	9124.6	9143.0	5	9.4	14.1
14.2	9161.4	9179.8	9198.2	9216.6	9234.9	9253.3	9271.6	9289.9	9308.2	9326.5	6	11.3	14.2
14.3	9344.7	9363.0	9381.3	9399.5	9417.7	9436.0	9454.2	9472.3	9490.5	9508.7	7	13.2	14.3
14.4	9526.8	9545.0	9563.1	9581.2	9599.3	9617.4	9635.5	9653.5	9671.6	9689.6	8	15.0	14.4
14.5	9707.6	9725.7	9743.7	9761.7	9779.6	9797.6	9815.6	9833.5	9851.4	9869.3	9	17.0	14.5
14.6	9887.2	9905.1	9923.0	9940.9	9958.7	9976.5	9994.4	10012.2	10030.0	10047.8			14.6
14.7	10065.5	10083.3	10101.1	10118.8	10136.6	10154.3	10172.0	10189.7	10207.4	10225.1			14.7
14.8	10242.7	10260.4	10278.0	10295.7	10313.3	10330.9	10348.5	10366.1	10383.6	10401.2	1	1.7	14.8
14.9	10418.7	10436.3	10453.8	10471.3	10488.8	10506.3	10523.7	10541.2	10558.6	10576.0	2	3.4	14.9
15.0	10593.4	10610.8	10628.2	10645.6	10662.9	10680.3	10697.6	10715.0	10732.3	10749.6	3	5.1	15.0
15.1	10766.9	10784.1	10801.5	10818.7	10836.0	10853.2	10870.5	10887.7	10904.9	10922.1	4	6.8	15.1
15.2	10939.3	10956.5	10973.6	10990.8	11008.0	11025.1	11042.2	11059.3	11076.4	11093.5	5	8.5	15.2
15.3	11110.6	11127.7	11144.7	11161.8	11178.8	11195.8	11212.8	11229.8	11246.8	11263.8	6	10.2	15.3
15.4	11280.8	11297.8	11314.7	11331.6	11348.6	11365.5	11382.4	11399.3	11416.2	11433.0	7	11.9	15.4
15.5	11449.9	11466.7	11483.6	11500.4	11517.2	11534.0	11550.8	11567.6	11584.4	11601.1	8	13.6	15.5
15.6	11617.9	11634.6	11651.4	11668.1	11684.8	11701.5	11718.2	11734.9	11751.6	11768.2	9	15.3	15.6
15.7	11784.9	11801.5	11818.2	11834.8	11851.4	11868.0	11884.6	11901.1	11917.7	11934.3			15.7
15.8	11950.8	11967.3	11983.8	12000.4	12016.9	12033.3	12049.8	12066.3	12082.7	12099.2			15.8
15.9	12115.6	12132.0	12148.4	12164.8	12181.2	12197.6	12214.0	12230.4	12246.7	12263.1			15.9

Barometer in Eng. Inch.	Hundredths of an Inch.										Thousandths of an Inch.		Barometer in Eng. Inch.
	.00	.01	.02	.03	.04	.05	.06	.07	.08	.09			
	Eng. Feet.	Eng. Feet.	Eng. Feet.	Eng. Feet.	Eng. Feet.	Eng. Feet.	Eng. Feet.	Eng Feet.	Eng. Feet.	Eng Feet.		Feet.	
16.0	12279.6	12295.9	12312.2	12328.5	12344.8	12361.1	12377.4	12393.6	12409.9	12426.1			16.0
16.1	12442.4	12458.6	12474.8	12491.0	12507.2	12523.4	12539.6	12555.7	12571.9	12588.0			16.1
16.2	12604.2	12620.3	12636.4	12652.5	12668.6	12684.7	12700.8	12716.8	12732.9	12748.9	1	1.6	16.2
16.3	12765.0	12781.0	12797.0	12813.0	12829.0	12845.0	12861.0	12876.9	12892.9	12908.8	2	3.1	16.3
16.4	12924.8	12940.7	12956.6	12972.5	12988.4	13004.3	13020.2	13036.0	13051.9	13067.7	3	4.7	16.4
16.5	13083.6	13099.4	13115.2	13131.0	13146.8	13162.6	13178.4	13194.2	13210.0	13225.7	4	6.3	16.5
16.6	13241.5	13257.2	13272.9	13288.6	13304.3	13320.0	13335.7	13351.4	13367.1	13382.7	5	7.8	16.6
16.7	13398.4	13414.0	13429.6	13445.2	13460.8	13476.4	13492.0	13507.6	13523.2	13538.7	6	9.4	16.7
16.8	13554.3	13569.8	13585.4	13600.9	13616.4	13631.9	13647.4	13662.9	13678.4	13693.9	7	11.0	16.8
16.9	13709.4	13724.8	13740.3	13755.7	13771.1	13786.5	13801.9	13817.3	13832.7	13848.1	8	12.5	16.9
17.0	13863.5	13878.8	13894.2	13909.6	13924.9	13940.2	13955.6	13970.9	13986.2	14001.5	9	14.1	17.0
17.1	14016.8	14032.0	14047.3	14062.6	14077.8	14093.0	14108.3	14123.5	14138.7	14153.9			17.1
17.2	14169.1	14184.3	14199.4	14214.6	14229.8	14244.9	14260.1	14275.2	14290.3	14305.5			17.2
17.3	14320.6	14335.7	14350.8	14365.8	14380.9	14396.0	14411.0	14426.1	14441.1	14456.2			17.3
17.4	14471.2	14486.2	14501.2	14516.2	14531.2	14546.1	14561.1	14576.1	14591.0	14605.9	1	1.5	17.4
17.5	14620.9	14635.8	14650.7	14665.6	14680.5	14695.4	14710.3	14725.2	14740.1	14754.9	2	2.9	17.5
17.6	14769.8	14784.6	14799.4	14814.3	14829.1	14843.9	14858.7	14873.5	14888.2	14903.0	3	4.4	17.6
17.7	14917.8	14932.5	14947.3	14962.0	14976.8	14991.5	15006.2	15020.9	15035.6	15050.3	4	5.8	17.7
17.8	15065.0	15079.6	15094.3	15109.0	15123.6	15138.2	15152.9	15167.5	15182.1	15196.7	5	7.3	17.8
17.9	15211.3	15225.9	15240.5	15255.0	15269.6	15284.2	15298.7	15313.3	15327.8	15342.4	6	8.8	17.9
18.0	15356.8	15371.3	15385.8	15400.3	15414.8	15429.3	15443.7	15458.2	15472.7	15487.1	7	10.2	18.0
18.1	15501.5	15516.0	15530.4	15544.8	15559.2	15573.6	15588.0	15602.4	15616.8	15631.2	8	11.7	18.1
18.2	15645.5	15659.9	15674.2	15688.5	15702.9	15717.2	15731.5	15745.8	15760.1	15774.4	9	13.1	18.2
18.3	15788.6	15802.9	15817.2	15831.4	15845.7	15859.9	15874.2	15888.4	15902.6	15916.8			18.3
18.4	15931.0	15945.2	15959.4	15973.6	15987.8	16001.9	16016.1	16030.2	16044.4	16058.5			18.4

Barometer in Eng. Inch.	Hundredths of an Inch.										Thousandths of an Inch.		Barometer in Eng. Inch.
	.00	.01	.02	.03	.04	.05	.06	.07	.08	.09			
	Eng. Feet.	Eng. Feet.	Eng. Feet.	Eng. Feet.	Eng. Feet.	Eng. Feet.	Eng. Feet.	Eng. Feet.	Eng. Feet.	Eng. Feet.		Feet.	
18.5	16072.6	16086.8	16100.9	16115.0	16129.1	16143.2	16157.3	16171.3	16185.4	16199.5			18.5
18.6	16213.5	16227.6	16211.6	16255.6	16269.7	16283.7	16297.7	16311.7	16325.7	16339.6			18.6
18.7	16353.5	16367.5	16381.5	16395.4	16409.4	16423.3	16437.2	16451.2	16465.1	16479.0	1	1.4	18.7
18.8	16492.9	16506.8	16520.7	16534.6	16548.5	16562.3	16576.2	16590.0	16603.9	16617.8	2	2.7	18.8
18.9	16631.5	16645.4	16659.2	16673.0	16686.8	16700.6	16714.4	16728.1	16741.9	16755.7	3	4.1	18.9
19.0	16769.4	16783.2	16796.9	16810.6	16824.3	16838.1	16851.8	16865.5	16879.2	16892.8	4	5.4	19.0
19.1	16906.5	16920.2	16933.9	16947.5	16961.2	16974.9	16988.5	17002.1	17015.8	17029.4	5	6.8	19.1
19.2	17043.0	17056.6	17070.2	17083.8	17097.4	17110.9	17124.5	17138.1	17151.6	17165.2	6	8.1	19.2
19.3	17178.7	17192.2	17205.8	17219.3	17232.8	17246.3	17259.8	17273.3	17286.8	17300.3	7	9.5	19.3
19.4	17313.7	17327.2	17340.6	17354.1	17367.5	17380.9	17394.4	17407.8	17421.2	17434.6	8	10.9	19.4
19.5	17448.0	17461.4	17474.8	17488.2	17501.6	17515.0	17528.3	17541.7	17555.0	17568.4	9	12.2	19.5
19.6	17581.7	17595.0	17608.3	17621.7	17635.0	17648.2	17661.5	17674.8	17688.1	17701.4			19.6
19.7	17714.6	17727.9	17741.1	17754.4	17767.6	17780.8	17794.1	17807.3	17820.5	17833.7			19.7
19.8	17846.9	17860.1	17873.3	17886.5	17899.6	17912.8	17926.0	17939.1	17952.2	17965.4			19.8
19.9	17978.5	17991.6	18004.8	18017.9	18031.0	18044.1	18057.2	18070.3	18083.4	18096.4	1	1.3	19.9
20.0	18109.5	18122.6	18135.6	18148.7	18161.7	18174.8	18187.8	18200.8	18213.8	18226.8	2	2.6	20.0
20.1	18239.8	18252.8	18265.8	18278.8	18291.8	18304.8	18317.7	18330.7	18343.6	18356.6	3	3.9	20.1
20.2	18369.5	18332.5	18395.4	18408.3	18421.2	18434.1	18447.0	18459.9	18472.8	18485.7	4	5.1	20.2
20.3	18498.5	18511.4	18524.3	18537.1	18550.0	18562.8	18575.7	18588.5	18601.3	18614.1	5	6.4	20.3
20.4	18626.9	18639.7	18652.5	18665.3	18678.1	18690.9	18703.6	18716.4	18729.1	18741.9	6	7.7	20.4
20.5	18754.6	18767.4	18780.1	18792.9	18805.6	18818.3	18831.0	18843.7	18856.4	18869.1	7	9.0	20.5
20.6	18881.8	18894.5	18907.2	18919.9	18932.5	18945.2	18957.8	18970.5	18983.1	18995.7	8	10.3	20.6
20.7	19008.3	19021.0	19033.6	19046.2	19053.8	19071.4	19083.9	19096.5	19109.1	19121.7	9	11.6	20.7
20.8	19134.2	19146.8	19159.3	19171.9	19184.4	19196.9	19209.5	19222.0	19234.5	19247.0			20.8
20.9	19259.5	19272.0	19284.5	19297.1	19309.5	19322.0	19334.4	19346.9	19359.3	19371.8			20.9

Barometer in Eng. Inch.	Hundredths of an Inch. .00	.01	.02	.03	.04	.05	.06	.07	.08	.09	Thousandths of an Inch.		Barometer in Eng. Inch.
	Eng. Feet.	Eng. Feet.	Eng. Feet.	Eng. Feet.	Eng. Feet.	Eng. Feet.	Eng. Feet.	Eng. Feet.	Eng. Feet.	Eng. Feet.		Feet.	
21.0	19384.3	19396.7	19409.1	19421.5	19434.0	19446.4	19458.8	19471.2	19483.6	19496.0			21.0
21.1	19508.4	19520.8	19533.1	19545.5	19557.9	19570.2	19582.6	19594.9	19607.3	19619.6	1	1.2	21.1
21.2	19632.0	19644.3	19656.6	19668.9	19681.2	19693.5	19705.8	19718.0	19730.3	19742.6	2	2.4	21.2
21.3	19754.9	19767.1	19779.4	19791.6	19803.9	19816.1	19828.4	19840.6	19852.8	19865.0	3	3.6	21.3
21.4	19877.3	19889.5	19901.7	19913.9	19926.0	19938.2	19950.4	19962.6	19974.7	19986.9	4	4.8	21.4
21.5	19999.1	20011.2	20023.3	20035.5	20047.6	20059.7	20071.8	20083.9	20096.1	20108.2	5	6.0	21.5
21.6	20120.3	20132.3	20144.4	20156.5	20168.6	20180.7	20192.7	20204.8	20216.9	20228.9	6	7.2	21.6
21.7	20241.0	20253.0	20265.0	20277.0	20289.1	20301.1	20313.1	20325.1	20337.1	20349.1	7	8.4	21.7
21.8	20361.1	20373.0	20385.0	20397.0	20409.0	20420.9	20432.9	20444.8	20456.8	20468.7	8	9.7	21.8
21.9	20480.7	20492.6	20504.5	20516.4	20528.3	20540.2	20552.1	20564.0	20575.9	20587.8	9	10.9	21.9
22.0	20599.7	20611.5	20623.4	20635.3	20647.1	20659.0	20670.8	20682.7	20694.5	20706.3			22.0
22.1	20718.2	20732.0	20741.8	20753.6	20765.4	20777.2	20789.0	20801.8	20812.6	20824.4			22.1
22.2	20836.2	20847.9	20859.7	20871.4	20883.2	20894.9	20906.7	20918.4	20930.1	20941.9			22.2
22.3	20953.6	20965.3	20977.0	20988.7	21000.4	21012.1	21023.8	21035.4	21047.1	21058.8	1	1.1	22.3
22.4	21070.5	21082.1	21093.8	21105.4	21117.1	21128.7	21140.4	21152.0	21163.6	21175.3	2	2.3	22.4
22.5	21186.9	21198.5	21210.1	21221.6	21233.2	21244.8	21256.4	21268.0	21279.5	21291.1	3	3.4	22.5
22.6	21302.6	21314.2	21325.8	21337.3	21348.9	21360.4	21371.9	21383.5	21395.0	21406.5	4	4.6	22.6
22.7	21418.1	21429.6	21441.1	21452.5	21464.0	21475.5	21487.0	21498.5	21509.9	21521.4	5	5.7	22.7
22.8	21532.9	21544.3	21555.8	21567.2	21578.7	21590.1	21601.6	21613.0	21624.4	21635.8	6	6.8	22.8
22.9	21647.3	21658.7	21670.1	21681.4	21692.8	21704.2	21715.6	21727.0	21738.3	21749.7	7	8.0	22.9
23.0	21761.0	21772.4	21783.7	21795.1	21806.4	21817.7	21829.1	21840.4	21851.7	21863.0	8	9.1	23.0
23.1	21874.3	21885.6	21897.0	21908.3	21919.6	21930.8	21942.1	21953.4	21964.7	21976.0	9	10.2	23.1
23.2	21987.2	21998.5	22009.8	22021.0	22032.3	22043.5	22054.7	22066.0	22077.2	22088.4			23.2
23.3	22099.6	22110.8	22122.1	22133.3	22144.5	22155.6	22166.8	22178.0	22189.2	22200.4			23.3
23.4	22211.5	22222.7	22233.9	22245.0	22256.2	22267.3	22278.4	22289.6	22300.7	22311.8			23.4

Barometer in Eng. Inch.	Hundredths of an Inch. .00	.01	.02	.03	.04	.05	.06	.07	.08	.09	Thousandths of an Inch.		Barometer in Eng. Inch.
	Eng. Feet.	Eng. Feet.	Eng. Feet.	Eng. Feet.	Eng. Feet.	Eng. Feet.	Eng Feet.	Eng Feet.	Eng Feet	Eng. Feet			
23.5	22322.9	22334.0	22345.2	22356.3	22367.4	22378.4	22389.5	22400.6	22411.7	22422.8			23.5
23.6	22433.8	22444.9	22456.0	22467.0	22478.1	22489.1	22500.2	22511.2	22522.3	22533.3		Feet	23.6
23.7	22544.3	22555.4	22566.4	22577.4	22588.4	22599.4	22610.4	22621.4	22632.4	22643.4			23.7
23.8	22654.3	22665.3	22676.3	22687.2	22698.2	22709.1	22720.1	22731.0	22742.0	22752.9	1	1.1	23.8
23.9	22763.8	22774.8	22785.7	22796.6	22807.5	22818.4	22829.4	22840.3	22851.2	22862.0	2	2.2	23.9
24.0	22873.0	22883.9	22894.7	22905.6	22916.5	22927.4	22938.2	22949.1	22960.0	22970.8	3	3.2	24.0
24.1	22981.7	22992.5	23003.3	23014.2	23025.0	23035.8	23046.6	23057.5	23068.3	23079.1	4	4.3	24.1
24.2	23089.9	23100.7	23111.4	23122.2	23133.0	23143.8	23154.5	23165.3	23176.1	23186.8	5	5.4	24.2
24.3	23197.6	23208.3	23219.1	23229.8	23240.5	23251.3	23262.0	23272.7	23283.4	23294.2	6	6.5	24.3
24.4	23304.9	23315.6	23326.3	23337.0	23347.6	23358.3	23369.0	23379.7	23390.3	23401.0	7	7.5	24.4
24.5	23411.7	23422.3	23433.0	23443.7	23454.3	23464.9	23475.6	23486.2	23496.8	23507.4	8	8.6	24.5
24.6	23518.1	23528.7	23539.3	23549.9	23560.5	23571.1	23581.7	23592.3	23602.9	23613.5	9	9.7	24.6
24.7	23624.1	23634.6	23645.2	23655.8	23666.3	23676.9	23687.5	23698.0	23708.6	23719.1			24.7
24.8	23729.7	23740.2	23750.7	23761.2	23771.7	23782.3	23792.8	23803.3	23813.8	23824.3			24.8
24.9	23834.8	23845.3	23855.7	23866.2	23876.7	23887.2	23897.7	23908.2	23918.6	23929.1	1	1.0	24.9
25.0	23939.5	23949.9	23960.4	23970.8	23981.3	23991.7	24002.1	24012.5	24023.0	24033.4	2	2.1	25.0
25.1	24043.8	24054.2	24064.6	24075.0	24085.4	24095.7	24106.1	24116.5	24126.9	24137.2	3	3.1	25.1
25.2	24147.6	24158.0	24168.3	24178.7	24189.0	24199.4	24209.7	24220.1	24230.4	24240.8	4	4.1	25.2
25.3	24251.1	24261.4	24271.8	24282.1	24292.4	24302.7	24313.0	24323.3	24333.6	24343.9	5	5.1	25.3
25.4	24354.2	24364.5	24374.7	24385.0	24395.3	24405.5	24415.8	24426.1	24436.3	24446.6	6	6.2	25.4
25.5	24456.8	24467.0	24477.3	24487.5	24497.8	24508.0	24518.2	24528.4	24538.7	24548.9	7	7.2	25.5
25.6	24559.1	24569.3	24579.5	24589.7	24599.9	24610.0	24620.2	24630.4	24640.6	24650.7	8	8.2	25.6
25.7	24660.9	24671.1	24681.2	24691.4	24701.5	24711.7	24721.8	24732.0	24742.1	24752.3	9	9.2	25.7
25.8	24762.4	24772.5	24782.6	24792.8	24802.9	24813.0	24823.1	24833.2	24843.3	24853.4			25.8
25.9	24863.5	24873.6	24883.7	24893.7	24903.8	24913.9	24924.0	24934.0	24944.1	24954.1			25.9

Barometer in Eng. Inch.	Hundredths of an Inch. .00	.01	.02	.03	.04	.05	.06	.07	.08	.09	Thousandths of an Inch.		Barometer in Eng. Inch.
	Eng. Feet.	Eng. Feet.	Eng. Feet.	Eng. Feet.	Eng. Feet.	Eng. Feet.	Eng. Feet.	Eng. Feet.	Eng. Feet.	Eng. Feet.		Feet.	
26.0	24964.2	24974.2	24984.3	24994.3	25004.4	25014.4	25024.4	25034.4	25044.5	25054.5			26.0
26.1	25064.5	25074.5	25084.5	25094.5	25104.5	25114.5	25124.5	25134.5	25144.4	25154.4			26.1
26.2	25164.4	25174.4	25184.3	25194.3	25204.2	25214.2	25224.1	25234.1	25244.0	25254.0	1	1.0	26.2
26.3	25263.9	25273.8	25283.8	25293.7	25303.6	25313.5	25323.4	25333.3	25343.2	25353.1	2	2.0	26.3
26.4	25363.0	25372.9	25382.8	25392.7	25402.6	25412.4	25422.3	25432.2	25442.1	25451.9	3	2.9	26.4
26.5	25461.8	25471.7	25481.5	25491.4	25501.2	25511.0	25520.9	25530.7	25540.5	25550.4	4	3.9	26.5
26.6	25560.2	25570.0	25579.8	25589.7	25599.5	25609.3	25619.1	25628.9	25638.7	25648.5	5	4.9	26.6
26.7	25658.3	25668.1	25677.8	25687.6	25697.4	25707.1	25716.9	25726.7	25736.4	25746.2	6	5.9	26.7
26.8	25755.9	25765.6	25775.4	25785.1	25794.8	25804.6	25814.3	25824.0	25833.8	25843.5	7	6.9	26.8
26.9	25853.2	25862.9	25872.6	25882.3	25892.0	25901.7	25911.4	25921.1	25930.8	25940.5	8	7.8	26.9
27.0	25950.2	25959.9	25969.6	25979.2	25988.9	25998.6	26008.2	26017.9	26027.5	26037.2	9	8.8	27.0
27.1	26046.8	26056.5	26066.1	26075.7	26085.3	26095.0	26104.6	26114.2	26123.8	26133.4			27.1
27.2	26143.0	26152.6	26162.2	26171.8	26181.4	26191.0	26200.6	26210.2	26219.8	26229.3			27.2
27.3	26238.9	26248.5	26258.0	26267.6	26277.2	26286.7	26296.3	26305.8	26315.3	26324.9			27.3
27.4	26334.4	26344.0	26353.5	26363.0	26372.5	26382.1	26391.6	26401.1	26410.6	26420.1	1	0.9	27.4
27.5	26429.6	26439.1	26448.6	26458.1	26467.6	26477.1	26486.5	26496.0	26505.5	26514.9	2	1.9	27.5
27.6	26524.4	26533.9	26543.3	26552.8	26562.3	26571.7	26581.2	26590.6	26600.0	26609.5	3	2.8	27.6
27.7	26618.9	26628.4	26637.8	26647.2	26656.7	26666.1	26675.5	26684.9	26694.3	26703.7	4	3.7	27.7
27.8	26713.1	26722.5	26731.9	26741.3	26750.7	26760.1	26769.5	26778.8	26788.2	26797.6	5	4.7	27.8
27.9	26806.9	26816.3	26825.6	26835.0	26844.3	26853.7	26863.0	26872.3	26881.7	26891.0	6	5.6	27.9
28.0	26900.4	26909.7	26919.0	26928.4	26937.7	26947.0	26956.3	26965.6	26975.0	26984.3	7	6.5	28.0
28.1	26993.6	27002.9	27012.2	27021.5	27030.7	27040.0	27049.3	27058.6	27067.8	27077.1	8	7.5	28.1
28.2	27086.4	27095.6	27104.9	27114.2	27123.4	27132.7	27141.9	27151.2	27160.4	27169.6	9	8.4	28.2
28.3	27178.9	27188.1	27197.3	27206.5	27215.7	27225.0	27234.2	27243.4	27252.6	27261.8			28.3
28.4	27271.0	27280.2	27289.4	27298.6	27307.8	27317.0	27326.2	27335.3	27344.5	27353.7			28.4

Barometer in Eng. Inch.	Hundredths of an Inch. .00	.01	.02	.03	.04	.05	.06	.07	.08	.09	Thousandths of an Inch.		Barometer in Eng. Inch.
	Eng. Feet.	Eng. Feet.	Eng. Feet.	Eng. Feet.	Eng. Feet.	Eng. Feet.	Eng. Feet.	Eng. Feet.	Eng. Feet.	Eng. Feet.		Feet.	
28.5	27362.9	27372.0	27381.2	27390.4	27399.5	27408.7	27417.8	27427.0	27436.1	27445.2			28.5
28.6	27454.4	27463.5	27472.6	27481.8	27490.9	27500.0	27509.1	27518.2	27527.4	27536.5			28.6
28.7	27545.6	27554.7	27563.8	27572.9	27582.0	27591.1	27600.2	27609.3	27618.3	27627.4	1	0.9	28.7
28.8	27636.5	27645.5	27654.6	27663.7	27672.7	27681.8	27690.8	27699.9	27708.9	27717.9	2	1.8	28.8
28.9	27727.0	27736.0	27745.1	27754.1	27763.1	27772.2	27781.2	27790.2	27799.2	27808.3	3	2.7	28.9
29.0	27817.2	27826.2	27835.2	27844.2	27853.2	27862.2	27871.2	27880.2	27889.1	27898.1	4	3.6	29.0
29.1	27907.1	27916.1	27925.0	27934.0	27943.0	27951.9	27960.9	27969.8	27978.8	27987.7	5	4.5	29.1
29.2	27996.7	28005.6	28014.6	28023.5	28032.4	28041.4	28050.3	28059.2	28068.2	28077.1	6	5.4	29.2
29.3	28086.0	28094.9	28103.8	28112.8	28121.7	28130.6	28139.5	28148.4	28157.3	28166.2	7	6.3	29.3
29.4	28175.1	28184.0	28192.9	28201.7	28210.6	28219.5	28228.4	28237.2	28246.1	28254.9	8	7.2	29.4
29.5	28263.8	28272.6	28281.5	28290.3	28299.2	28308.0	28316.9	28325.7	28334.5	28343.4	9	8.1	29.5
29.6	28352.2	28361.0	28369.8	28378.7	28387.5	28396.3	28405.1	28413.9	28422.7	28431.5			29.6
29.7	28440.3	28449.1	28457.9	28466.7	28475.4	28484.2	28493.0	28501.8	28510.6	28519.3			29.7
29.8	28528.1	28536.9	28545.6	28554.4	28563.2	28571.9	28580.7	28589.4	28598.2	28606.9			29.8
29.9	28615.7	28624.4	28633.2	28641.9	28650.6	28659.3	28668.1	28676.8	28685.5	28694.2	1	8.6	29.9
30.0	28702.9	28711.6	28720.3	28729.0	28737.7	28746.4	28755.1	28763.8	28772.5	28781.1	2	1.7	30.0
30.1	28789.8	28798.5	28807.2	28815.9	28824.5	28833.2	28841.9	28850.5	28859.2	28867.9	3	2.6	30.1
30.2	28876.5	28885.2	28893.8	28902.5	28911.1	28919.8	28928.4	28937.0	28945.7	28954.3	4	3.4	30.2
30.3	28962.9	28971.5	28980.1	28988.8	28997.4	29006.0	29014.6	29023.2	29031.7	29040.3	5	4.3	30.3
30.4	29048.9	29057.5	29066.1	29074.7	29083.3	29091.8	29100.4	29109.0	29117.6	29126.2	6	5.2	30.4
30.5	29134.7	29143.3	29151.9	29160.4	29169.0	29177.6	29186.1	29194.7	29203.2	29211.8	7	6.0	30.5
30.6	29220.3	29228.9	29237.4	29245.9	29254.4	29262.9	29271.5	29280.0	29288.5	29297.0	8	6.9	30.6
30.7	29305.5	29314.0	29322.5	29331.1	29339.6	29348.1	29356.6	29365.1	29373.5	29382.0	9	7.7	30.7
30.8	29390.5	29399.0	29407.5	29416.0	29424.4	29432.9	29441.4	29449.8	29458.3	29466.8			30.8
30.9	29475.2	29483.7	29492.1	29500.6	29509.0	29517.5	29525.9	29534.3	29542.8	29551.2			30.9

II. Correction for $\tau - \tau'$, or Difference of the Temperature of the Barometers at the Two Stations.

This Correction is *negative* when the attached Thermometer at the Upper Station is lowest; *positive*, when the attached Thermometer at the Upper Station is highest.

$\tau - \tau'$ Fahrenheit.	Correction in Eng. Feet.	$\tau - \tau'$ Fahrenheit.	Correction in Eng. Feet.	$\tau - \tau'$ Fahrenheit.	Correction in Eng. Feet.	$\tau - \tau'$ Fahrenheit.	Correction in Eng. Feet.	$\tau - \tau'$ Fahrenheit.	Correction in Eng. Feet.	$\tau - \tau'$ Fahrenheit.	Correction in Eng. Feet.	$\tau - \tau'$ Fahrenheit.	Correction in Eng. Feet.	$\tau - \tau'$ Fahrenheit.	Correction in Eng. Feet.	$\tau - \tau'$ Fahrenheit.	Correction in Eng. Feet.	$\tau - \tau'$ Fahrenheit.	Correction in Eng. Feet.
°		°		°		°		°		°		°		°		°		°	
1.0	2.3	11.0	25.8	21.0	49.2	31.0	72.6	41.0	96.0	51.0	119.5	61.0	142.9	71.0	166.3	81.0	189.7	91.0	213.2
1.5	3.5	11.5	26.9	21.5	50.4	31.5	73.8	41.5	97.2	51.5	120.6	61.5	144.1	71.5	167.5	81.5	190.9	91.5	214.3
2.0	4.7	12.0	28.1	22.0	51.5	32.0	75.0	42.0	98.4	52.0	121.8	62.0	145.2	72.0	168.7	82.0	192.1	92.0	215.5
2.5	5.9	12.5	29.3	22.5	52.7	32.5	76.1	42.5	99.6	52.5	123.0	62.5	146.4	72.5	169.8	82.5	193.3	92.5	216.7
3.0	7.0	13.0	30.5	23.0	53.9	33.0	77.3	43.0	100.7	53.0	124.2	63.0	147.6	73.0	171.0	83.0	194.4	93.0	217.9
3.5	8.2	13.5	31.6	23.5	55.1	33.5	78.5	43.5	101.9	53.5	125.3	63.5	148.8	73.5	172.2	83.5	195.6	93.5	219.0
4.0	9.4	14.0	32.8	24.0	56.2	34.0	79.6	44.0	103.1	54.0	126.5	64.0	149.9	74.0	173.4	84.0	196.8	94.0	220.2
4.5	10.5	14.5	34.0	24.5	57.4	34.5	80.8	44.5	104.2	54.5	127.7	64.5	151.1	74.5	174.5	84.5	197.9	94.5	221.4
5.0	11.7	15.0	35.1	25.0	58.6	35.0	82.0	45.0	105.4	55.0	128.8	65.0	152.3	75.0	175.7	85.0	199.1	95.0	222.5
5.5	12.9	15.5	36.3	25.5	59.7	35.5	83.2	45.5	106.6	55.5	130.0	65.5	153.4	75.5	176.9	85.5	200.3	95.5	223.7
6.0	14.1	16.0	37.5	26.0	60.9	36.0	84.3	46.0	107.8	56.0	131.2	66.0	154.6	76.0	178.0	86.0	201.5	96.0	224.9
6.5	15.2	16.5	38.7	26.5	62.1	36.5	85.5	46.5	108.9	56.5	132.4	66.5	155.8	76.5	179.2	86.5	202.6	96.5	226.1
7.0	16.4	17.0	39.8	27.0	63.2	37.0	86.7	47.0	110.1	57.0	133.5	67.0	157.0	77.0	180.4	87.0	203.8	97.0	227.2
7.5	17.6	17.5	41.0	27.5	64.4	37.5	87.8	47.5	111.3	57.5	134.7	67.5	158.1	77.5	181.6	87.5	205.0	97.5	228.4
8.0	18.7	18.0	42.2	28.0	65.6	38.0	89.0	48.0	112.4	58.0	135.9	68.0	159.3	78.0	182.7	88.0	206.1	98.0	229.6
8.5	19.9	18.5	43.3	28.5	66.8	38.5	90.2	48.5	113.6	58.5	137.0	68.5	160.5	78.5	183.9	88.5	207.3	98.5	230.7
9.0	21.1	19.0	44.5	29.0	67.9	39.0	91.4	49.0	114.8	59.0	138.2	69.0	161.6	79.0	185.1	89.0	208.5	99.0	231.9
9.5	22.3	19.5	45.7	29.5	69.1	39.5	92.5	49.5	116.0	59.5	139.4	69.5	162.8	79.5	186.2	89.5	209.7	99.5	233.1
10.0	23.4	20.0	46.9	30.0	70.3	40.0	93.7	50.0	117.1	60.0	140.6	70.0	164.0	80.0	187.4	90.0	210.8	100.0	234.3
10.5	24.6	20.5	48.0	30.5	71.4	40.5	94.9	50.5	118.3	60.5	141.7	70.5	165.2	80.5	188.6	90.5	212.0	100.5	235.4

… OF GRAVITY IN VARIOUS LATITUDES.

Correction *positive* from Latitude 0° to 45°. *Negative* from 45° to 90°.

Approximate Difference of Level.	Latitude.																							Approximate Difference of Level.
	0° 90°	2° 88°	4° 86°	6° 84°	8° 82°	10° 80°	12° 78°	14° 76°	16° 74°	18° 72°	20° 70°	22° 68°	24° 66°	26° 64°	28° 62°	30° 60°	32° 58°	34° 56°	36° 54°	38° 52°	40° 50°	42° 48°	44° 46°	45°	
Eng. Feet.	Feet.	Feet.	Feet.	Feet.	Feet.	Feet.	Feet.	Feet.	Feet.	Feet.	Feet.	Feet.	Feet.	Feet.	Feet.	Feet.	Feet.	Feet.	Feet.	Feet.	Feet.	Feet.	Feet.	Feet.	Eng. Feet.
1000	2.6	2.6	2.6	2.5	2.5	2.4	2.4	2.3	2.2	2.1	2.0	1.9	1.7	1.6	1.5	1.3	1.1	1.0	0.8	0.6	0.5	0.3	0.1	0	1000
2000	5.2	5.2	5.1	5.1	5.0	4.9	4.7	4.6	4.4	4.2	4.0	3.7	3.5	3.2	2.9	2.6	2.3	1.9	1.6	1.3	0.9	0.5	0.2	0	2000
3000	7.8	7.8	7.7	7.6	7.5	7.3	7.1	6.9	6.6	6.3	6.0	5.6	5.2	4.8	4.4	3.9	3.4	2.9	2.4	1.9	1.4	0.8	0.3	0	3000
4000	10.4	10.4	10.3	10.2	10.0	9.8	9.5	9.2	8.8	8.4	8.0	7.5	7.0	6.4	5.8	5.2	4.6	3.9	3.2	2.5	1.8	1.1	0.4	0	4000
5000	13.0	13.0	12.9	12.7	12.5	12.2	11.9	11.5	11.0	10.5	10.0	9.4	8.7	8.0	7.3	6.5	5.7	4.9	4.0	3.1	2.3	1.4	0.5	0	5000
6000	15.6	15.6	15.4	15.3	15.0	14.7	14.3	13.8	13.2	12.6	11.9	11.2	10.4	9.6	8.7	7.8	6.8	5.8	4.8	3.8	2.7	1.6	0.5	0	6000
7000	18.2	18.2	18.0	17.8	17.5	17.1	16.6	16.1	15.4	14.7	13.9	13.1	12.2	11.2	10.2	9.1	8.0	6.8	5.6	4.4	3.2	1.9	0.6	0	7000
8000	20.8	20.7	20.6	20.3	20.0	19.5	19.0	18.4	17.6	16.8	15.9	15.0	13.9	12.8	11.6	10.4	9.1	7.8	6.4	5.0	3.6	2.2	0.7	0	8000
9000	23.4	23.3	23.2	22.9	22.5	22.0	21.4	20.7	19.8	18.9	17.9	16.8	15.7	14.4	13.1	11.7	10.3	8.8	7.2	5.7	4.1	2.4	0.8	0	9000
10000	26.0	25.9	25.7	25.4	25.0	24.4	23.8	23.0	22.0	21.0	19.9	18.7	17.4	16.0	14.5	13.0	11.4	9.7	8.0	6.3	4.5	2.7	0.9	0	10000
11000	28.6	28.5	28.3	28.0	27.5	26.9	26.1	25.3	24.3	23.1	21.9	20.6	19.1	17.6	16.0	14.3	12.5	10.7	8.8	6.9	5.0	3.0	1.0	0	11000
12000	31.2	31.1	30.9	30.5	30.0	29.3	28.5	27.5	26.5	25.2	23.9	22.4	20.9	19.2	17.4	15.6	13.7	11.7	9.6	7.5	5.4	3.3	1.1	0	12000
13000	33.8	33.7	33.5	33.1	32.5	31.8	30.9	29.8	28.7	27.3	25.9	24.3	22.6	20.8	18.9	16.9	14.8	12.7	10.4	8.2	5.9	3.5	1.2	0	13000
14000	36.4	36.3	36.0	35.6	35.0	34.2	33.3	32.1	30.9	29.4	27.9	26.2	24.4	22.4	20.4	18.2	16.0	13.6	11.2	8.8	6.3	3.8	1.3	0	14000
15000	39.0	38.9	38.6	38.1	37.5	36.6	35.6	34.4	33.1	31.6	29.9	28.1	26.1	24.0	21.8	19.5	17.1	14.6	12.1	9.4	6.8	4.1	1.4	0	15000
16000	41.6	41.5	41.2	40.7	40.0	39.1	38.0	36.7	35.3	33.7	31.9	29.9	27.8	25.6	23.3	20.8	18.2	15.6	12.9	10.1	7.2	4.3	1.5	0	16000
17000	44.2	44.1	43.8	43.2	42.5	41.5	40.4	39.0	37.5	35.8	33.9	31.8	29.6	27.2	24.7	22.1	19.4	16.6	13.7	10.7	7.7	4.6	1.5	0	17000
18000	46.8	46.7	46.3	45.8	45.0	44.0	42.8	41.3	39.7	37.9	35.8	33.7	31.3	28.8	26.2	23.4	20.5	17.5	14.5	11.3	8.1	4.9	1.6	0	18000
19000	49.4	49.3	48.9	48.3	47.5	46.4	45.1	43.6	41.9	40.0	37.8	35.5	33.1	30.4	27.6	24.7	21.7	18.5	15.3	12.0	8.6	5.2	1.7	0	19000
20000	52.0	51.9	51.5	50.9	50.0	48.9	47.5	45.9	44.1	42.1	39.8	37.4	34.8	32.0	29.1	26.0	22.8	19.5	16.1	12.6	9.0	5.4	1.8	0	20000
21000	54.6	54.5	54.1	53.4	52.5	51.3	49.9	48.2	46.3	44.2	41.8	39.3	36.5	33.6	30.5	27.3	23 9	20.5	16.9	13.2	9.5	5.7	1.9	0	21000
22000	57.2	57.1	56.6	55.9	55.0	53.7	52.3	50.5	48.5	46.3	43.8	41.1	38.3	35.2	32.0	28.6	25.1	21.4	17.7	13.8	9.9	6.0	2.0	0	22000
23000	59.8	59.7	59.2	58.5	57.5	56.2	54.6	52.8	50.7	48.4	45.8	43.0	40.0	36.8	33.4	29.9	26.2	22.4	18.5	14.5	10.4	6.2	2.1	0	23000
24000	62.4	62.2	61.8	61.0	60.0	58.6	57.0	55.1	52.9	50.5	47.8	44.9	41.8	38.4	34.9	31.2	27.4	23.4	19.3	15.1	10.8	6.5	2.2	0	24000
25000	65.0	64.8	64.4	63.6	62.5	61.1	59.4	57.4	55.1	52.6	49.8	46.8	43.5	40.0	36.3	32.5	28.5	24.3	20.1	15.7	11.3	6.8	2.3	0	25000

IV. Correction for Decrease of Gravity on a Vertical. — V. Correction for the Height of the Lower Station. — Positive.

Approximate Difference of Level.	IV. Decrease of Gravity on a Vertical. *Positive.* 0	+500	V. Height of the Barometer, in English Inches, at Lower Station. 16	18	20	22	24	26	28
Eng Feet.	Feet.	Feet.	Feet.	Feet.	Feet.	Feet.	Feet.	Feet.	Feet.
1000	2.5	3.9	1.6	1.3	1.0	0.8	0.6	0.4	0.2
2000	5.2	6.6	3.1	2.5	2.0	1.5	1.1	0.7	0.3
3000	7.9	9.3	4.7	3.8	3.0	2.3	1.7	1.1	0.5
4000	10.8	12.2	6.3	5.1	4.0	3.1	2.2	1.4	0.7
5000	13.7	15.2	7.8	6.4	5.0	3.8	2.8	1.8	0.8
6000	16.7	18.3	9.4	7.6	6.0	4.6	3.3	2.1	1.0
7000	19.9	21.5	11.0	8.9	7.1	5.4	3.9	2.5	1.2
8000	23.1	24.7	12.5	10.2	8.1	6.2	4.4	2.8	1.3
9000	26.4	28.1	14.1	11.4	9.1	6.9	5.0	3.2	1.5
10000	29.8	31.5	15.7	12.7	10.1	7.7	5.5	3.5	1.7
11000	33.3	35.1	17.2	14.0	11.1	8.5	6.1	3.9	1.8
12000	36.9	38.7	18.8	15.3	12.1	9.2	6.6	4.2	2.0
13000	40.6	42.5	20.4	16.5	13.1	10.0	7.2	4.6	2.2
14000	44.4	46.3	21.9	17.8	14.1	10.8	7.7	4.9	2.3
15000	48.3	50.3	23.5	19.1	15.1	11.5	8.3	5.3	2.5
16000	52.3	54.3	25.1	20.3	16.1	12.3	8.8	5.6	2.7
17000	56.4	58.4	26.6	21.6	17.1	13.1	9.4	6.0	2.8
18000	60.5	62.6	28.2	22.9	18.1	13.8	9.9	6.3	3.0
19000	64.8	67.0	29.8	24.1	19.2	14.6	10.5	6.7	3.2
20000	69.2	71.4	31.3	25.4	20.2	15.4	11.0	7.0	3.3
21000	73.6	75.9	32.9	26.7	21.2	16.1	11.6	7.4	3.5
22000	78.2	80.5	34.5	28.0	22.2	16.9	12.1	7.7	3.7
23000	82.9	85.2	36.0	29.2	23.2	17.7	12.7	8.1	3.8
24000	87.6	90.0	37.6	30.5	24.2	18.5	13.2	8.4	4.0
25000	92.5	94.9	39.1	31.8	25.2	19.2	13.8	8.8	4.1

VI. Height of a Column of Air, corresponding to One Tenth of an Inch in the Barometer.

Barometer Reading in English Inches.	Temperature of the Air, Fahrenheit, being 40°	45°	50°	55°	60°	65°	70°	75°	80°	85°
	Feet.	Feet.	Feet.	Feet.	Feet.	Feet.	Feet.	Feet.	Feet.	Feet.
18.5	144.5	146.1	147.7	149.3	150.8	152.4	154.0	155.6	157.1	158.7
19.0	140.8	142.3	143.8	145.4	146.9	148.4	150.0	151.5	153.1	154.6
19.5	137.1	138.6	140.1	141.6	143.1	144.6	146.1	147.6	149.1	150.6
20.0	133.7	135.2	136.7	138.1	139.6	141.0	142.5	144.0	145.4	146.9
20.5	130.4	131.8	133.2	134.6	136.1	137.5	138.9	140.3	141.8	143.2
21.0	127.3	128.7	130.1	131.5	132.9	134.3	135.7	137.0	138.4	139.8
21.5	124.0	125.3	126.7	128.0	129.4	130.7	132.1	133.4	134.8	136.1
22.0	121.5	122.8	124.2	125.5	126.8	128.2	129.5	130.8	132.1	133.5
22.5	118.9	120.2	121.5	122.8	124.1	125.4	126.7	128.0	129.3	130.6
23.0	116.2	117.5	118.8	120.0	121.3	122.6	123.8	125.1	126.4	127.7
23.5	113.8	115.0	116.3	117.5	118.8	120.0	121.2	122.5	123.7	125.0
24.0	111.3	112.6	113.8	115.0	116.2	117.4	118.6	119.9	121.1	122.3
24.5	109.0	110.2	111.4	112.6	113.8	114.9	116.1	117.3	118.5	119.7
25.0	107.0	108.1	109.3	110.5	111.6	112.8	114.0	115.1	116.3	117.5
25.5	104.7	105.9	107.0	108.2	109.3	110.4	111.6	112.7	113.9	115.0
26.0	102.8	103.9	105.0	106.2	107.3	108.4	109.5	110.6	111.8	112.9
26.5	100.9	102.0	103.1	104.2	105.3	106.4	107.5	108.6	109.7	110.8
27.0	99.0	100.1	101.2	102.3	103.3	104.4	105.5	106.6	107.7	108.7
27.5	96.9	97.9	99.0	100.1	101.1	102.2	103.2	104.3	105.4	106.4
28.0	95.8	96.8	97.9	98.9	100.0	101.0	102.0	103.1	104.1	105.2
28.5	93.8	94.9	95.9	96.9	97.9	99.0	100.0	101.0	102.0	103.1
29.0	92.3	93.3	94.3	95.3	96.3	97.4	98.4	99.4	100.4	101.4
29.5	90.6	91.6	92.6	93.5	94.5	95.5	96.5	97.5	98.5	99.5
30.0	89.1	90.0	91.0	92.0	92.9	93.9	94.9	95.9	96.8	97.8
30.5	86.6	87.5	88.5	89.4	90.4	91.3	92.3	93.2	94.2	95.1

III.

TABLE

FOR

COMPUTING THE DIFFERENCE IN THE HEIGHTS OF TWO PLACES BY MEANS OF THE BAROMETER.

By Prof. Elias Loomis.

This table was computed from the formula of Laplace, modified in accordance with the results of more recent determinations.

Suppose that we have observed

$$\text{At the lower station.}\left\{\begin{array}{l}H\text{, the height of the barometer,}\\ T\text{, the temperature of the barometer,}\\ t\text{, the temperature of the air,}\end{array}\right.$$

$$\text{At the upper station.}\left\{\begin{array}{l}h'\text{, the height of the barometer,}\\ T'\text{, the temperature of the barometer,}\\ t'\text{, the temperature of the air.}\end{array}\right.$$

Represent by s the height of the lower station above the level of the sea, by L the latitude of the place, and by h the observed height h' reduced to the temperature T.

The difference of level x between the two stations is given by the formula,

$$x = 60158.\ 6\text{ ft.} \times \log. \tfrac{H}{h} \times \left\{\begin{array}{l}\left(1 + \frac{t + t' - 64}{900}\right)\\ (1 + 0.00265 \cos. 2\text{ L})\\ \left(1 + \frac{x + 52251}{20888629} + \frac{s}{10444315}\right)\end{array}\right\}$$

But h represents the height h' reduced from the temperature T′ to the temperature T. The expansion of mercury for 1° Fahr. is 0.0001000; that of the brass which forms the scale of the barometer is 0.0000104; the difference is 0.0000896. Hence we have $h = h' \{1 + 0.0000896\ (T - T')\}$.

Therefore,

$$60158.\ 6\text{ ft. log. } \tfrac{H}{h} = 60158.6\text{ ft. log. } \tfrac{H}{h'} - 2.3409\text{ ft. } (T - T').$$

Part I. of the accompanying Table furnishes in English feet the value of the expression 60158.6 log. H for heights of the barometer from 11 to 31 inches; only they have all been diminished by the constant 27541.5 feet which does not change the difference

$$60158.6\text{ log. H} - 60158.6\text{ log. } h.$$

Part II. furnishes the correction $-$ 2.3409 (T $-$ T′) depending upon the difference T $-$ T′ of the temperatures of the barometers at the two stations. This correction is generally negative. It would be positive if T — T′ were negative; that is, if the temperature T′ of the barometer at the upper station exceeded the temperature T at the lower station.

Part III. gives the correction $A \times 0.00265 \cos. 2L$, to be applied to the approximate altitude A, and which arises from the variation of gravity from the latitude of 45 degrees, to the latitude L of the place of observation. This correction has the same sign as cos. 2 L; that is, it is positive from the equator to 45 degrees, and negative from 45 degrees to the pole.

Part IV. gives the correction $A \times \frac{A + 52251}{20888629}$, which is always to be added to the approximate height A, and which is due to the diminution of gravity on the vertical.

Part V. furnishes for the approximate difference of level A the small correction $A \times \frac{s}{10444315}$ corresponding to several values of the height s of the lower station. But in place of s there has been substituted as the argument of the table, the height H of the barometer at this station.

Method of Computation.

Take from Part I. the two numbers corresponding to the observed barometric heights H and h'. From their difference subtract the correction 2.3409 (T — T′) found in Part II. with the difference T — T′ of the thermometers attached to the barometers. We thus obtain an approximate altitude a.

We then calculate the correction $a\,\frac{t + t' - 64}{900}$ for the temperature of the air, by multiplying the nine-hundredth part of a by the sum of the temperatures t and t' diminished by 64. This correction is of the same sign as $t + t' - 64$. We thus obtain a second approximate altitude A.

With A and the latitude of the place L, we seek in Part III. the correction $A \times 0.00265 \cos. 2L$ arising from the variation of gravity with the latitude.

For the approximate height A, Part IV. gives the correction $A \times \frac{A + 52251}{20888629}$ arising from the diminution of gravity on a vertical. This correction is always additive.

Finally, when the height s of the lower station is considerable, the small correction $A \times \frac{s}{10444315}$ may be found in Part V. This correction is always additive.

Example 1.

M. Humboldt made the following observations on the mountain of Guanaxuato, in Mexico, in Latitude 21°, viz.

	Upper station.	Lower station near the sea.
Thermometer in open air,	$t' = 70°.3$	$t = 77°.5$
Thermometer to barometer,	$T' = 70°.3$	$T = 77°.5$
Barometer,	$h' = 23.66$	$H = 30.046$

Required the difference in the height of the two stations.

Part I. gives { for H = 30.046 inches	27649.7
Part I. gives { for h = 23.66 inches	21406.9
Difference	6242.8
Part II. gives for T — T′ = 7°.2,	—16.9
Approximate altitude a,	6225.9
$\frac{a}{900}(t + t' - 64) = 6.918 \times 83.8$,	+579.7
Second approximate altitude A,	6805.6
Part III. gives for A = 6806, and L = 21°,	+13.3
Part IV. gives for 6806,	+19.3
Height above the sea,	6838.2 feet.

Example 2.

M. Gay Lussac in his celebrated balloon ascent in 1805, found his barometer to indicate 12.945 English inches, the temperature being 14°.9 Fahrenheit. The barometer at Paris at the same time indicated 30.145 English inches with a temperature of 87°.44 Fahrenheit. Required the elevation of the balloon above Paris.

Part I. gives { for H = 30.145 inches,	27735.6
Part I. gives { for h' = 12.945 inches,	5650.4
Difference,	22085.2
Part II. gives for T — T′ = 72°.54,	—169.9
Approximate altitude a,	21915.3
$\frac{a}{900}(t + t' - 64) = 24.35 \times 38.34$,	+933.6
Second approximate altitude A,	22848.9
Part III. gives for A = 22848, and L = 48° 50′	—8.2
Part IV. gives for 22848,	+82.1
Height of balloon above Paris,	22922.8 feet.

PART I.

Argument, the observed Height of the Barometer at either Station.

Inches.	Feet.	Diff.	Inches.	Feet.	Diff.	Inches.	Feet.	Diff.	Inches.	Feet.	Diff.
11.0	1396.9	236.4	16.0	11186.3	162.8	21.0	18291.0	124.1	26.0	23871.0	100.3
11.1	1633.3	234.3	16.1	11349.1	161.8	21.1	18415.1	123.6	26.1	23971.3	99.9
11.2	1867.6	232.3	16.2	11510.9	160.8	21.2	18538.7	122.9	26.2	24071.2	99.5
11.3	2099.9	230.2	16.3	11671.7	159.8	21.3	18661.6	122.4	26.3	24170.7	99.1
11.4	2330.1	228.2	16.4	11831.5	158.8	21.4	18784.0	121.8	26.4	24269.8	98.8
11.5	2558.3	226.2	16.5	11990.3	157.9	21.5	18905.8	121.2	26.5	24368.6	98.4
11.6	2784.5	224.2	16.6	12148.2	156.9	21.6	19027.0	120.7	26.6	24467.0	98.1
11.7	3008.7	222.4	16.7	12305.1	155.9	21.7	19147.7	120.1	26.7	24565.1	97.6
11.8	3231.1	220.5	16.8	12461.0	155.1	21.8	19267.8	119.6	26.8	24662.7	97.3
11.9	3451.6	218.6	16.9	12616.1	154.1	21.9	19387.4	119.0	26.9	24760.0	97.0
12.0	3670.2	216.8	17.0	12770.2	153.3	22.0	19506.4	118.5	27.0	24857.0	96.6
12.1	3887.0	215.0	17.1	12923.5	152.3	22.1	19624.9	118.0	27.1	24953.6	96.2
12.2	4102.0	213.3	17.2	13075.8	151.5	22.2	19742.9	117.4	27.2	25049.8	95.9
12.3	5315.3	211.6	17.3	13227.3	150.6	22.3	19860.3	116.9	27.3	25145.7	95.5
12.4	4526.9	209.8	17.4	13377.9	149.7	22.4	19977.2	116.4	27.4	25241.2	95.2
12.5	4736.7	208.2	17.5	13527.6	148.9	22.5	20093.6	115.8	27.5	25336.4	94.8
12.6	4944.9	206.5	17.6	13676.5	148.0	22.6	20209.4	115.4	27.6	25431.2	94.5
12.7	5151.4	205.0	17.7	13824.5	147.2	22.7	20324.8	114.8	27.7	25525.7	94.2
12.8	5356.4	203.3	17.8	13971.7	146.3	22.8	20439.6	114.4	27.8	25619.9	93.8
12.9	5559.7	201.7	17.9	14118.0	145.6	22.9	20554.0	113.8	27.9	25713.7	93.4
13.0	5761.4	200.2	18.0	14263.6	144.7	23.0	20667.8	113.3	28.0	25807.1	93.2
13.1	5961.6	198.7	18.1	14408.3	144.0	23.1	20781.1	112.9	28.1	25900.3	92.8
13.2	6160.3	197.2	18.2	14552.3	143.1	23.2	20894.0	112.4	28.2	25993.1	92.5
13.3	6357.5	195.7	18.3	14695.4	142.4	23.3	21006.4	111.9	28.3	26085.6	92.1
13.4	6553.2	194.3	18.4	14837.8	141.6	23.4	21118.3	111.4	28.4	26177.7	91.9
13.5	6747.5	192.8	18.5	14979.4	140.9	23.5	21229.7	110.9	28.5	26269.6	91.5
13.6	6940.3	191.4	18.6	15120.3	140.0	23.6	21340.6	110.5	28.6	26361.1	91.2
13.7	7131.7	190.0	18.7	15260.3	139.4	23.7	21451.1	110.0	28.7	26452.3	90.9
13.8	7321.7	188.6	18.8	15399.7	138.6	23.8	21561.1	109.5	28.8	26543.2	90.5
13.9	7510.3	187.3	18.9	15538.3	137.9	23.9	21670.6	109.1	28.9	26633.7	90.3
14.0	7697.6	186.0	19.0	15676.2	137.1	24.0	21779.7	108.7	29.0	26724.0	89.9
14.1	7883.6	184.6	19.1	15813.3	136.5	24.1	21888.4	108.2	29.1	26813.9	89.6
14.2	8068.2	183.3	19.2	15949.8	135.7	24.2	21996.6	107.7	29.2	26903.5	89.3
14.3	8251.5	182.1	19.3	16085.5	135.0	24.3	22104.3	107.3	29.3	26992.8	89.1
14.4	8433.6	180.8	19.4	16220.5	134.3	24.4	22211.6	106.8	29.4	27081.9	88.7
14.5	8614.4	179.6	19.5	16354.8	133.7	24.5	22318.4	106.4	29.5	27170.6	88.4
14.6	8794.0	178.3	19.6	16488.5	132.9	24.6	22424.8	106.0	29.6	27259.0	88.1
14.7	8972.3	177.2	19.7	16621.4	132.3	24.7	22530.8	105.6	29.7	27347.1	87.8
14.8	9149.5	176.0	19.8	16753.7	131.6	24.8	22636.4	105.1	29.8	27434.9	87.6
14.9	9325.5	174.8	19.9	16885.3	131.0	24.9	22741.5	104.8	29.9	27522.5	87.2
15.0	9500.3	173.5	20.0	17016.3	130.3	25.0	22846.3	104.3	30.0	27609.7	86.9
15.1	9673.8	172.4	20.1	17146.6	129.7	25.1	22950.6	103.8	30.1	27696.6	86.7
15.2	9846.2	171.3	20.2	17276.3	129.0	25.2	23054.4	103.5	30.2	27783.3	86.4
15.3	10017.5	170.2	20.3	17405.3	128.4	25.3	23157.9	103.1	30.3	27869.7	86.0
15.4	10187.7	169.1	20.4	17533.7	127.7	25.4	23261.0	102.6	30.4	27955.7	85.8
15.5	10356.8	168.0	20.5	17661.4	127.2	25.5	23363.6	102.3	30.5	28041.5	85.6
15.6	10524.8	167.0	20.6	17788.6	126.5	25.6	23465.9	101.8	30.6	28127.1	85.2
15.7	10691.8	165.9	20.7	17915.1	125.9	25.7	23567.7	101.5	30.7	28212.3	85.0
15.8	10857.7	164.8	20.8	18041.0	125.3	25.8	23669.2	101.1	30.8	28297.3	84.7
15.9	11022.5	163.8	20.9	18166.3	124.7	25.9	23770.3	100.7	30.9	28382.0	84.4
16.0	11186.3		21.0	18291.0		26.0	23871.0		31.0	28466.4	

PART II.

Correction due to T — T′, or the Difference of the Temperatures of the Barometers at the two Stations.

This Correction is Negative when the Temperature at the Upper Station is lowest, and vice versâ.

T — T′.	Correction.	T — T′.	Correction.	T — T′.	Correction.	T — T′.	Correction.	T — T′.	Correction.	T — T′.	Correction.
Fah't.	Feet.	Fah't.	Feet.	Fah't.	Feet.	Fah't.	Feet.	Fah't.	Feet.	Fah't.	Feet.
°		°		°		°		°		°	
1	2.3	14	32.8	27	63.2	40	93.6	53	124.1	66	154.5
2	4.7	15	35.1	28	65.5	41	96.0	54	126.4	67	156.8
3	7.0	16	37.5	29	67.9	42	98.3	55	128.7	68	159.2
4	9.4	17	39.8	30	70.2	43	100.7	56	131.1	69	161.5
5	11.7	18	42.1	31	72.6	44	103.0	57	133.4	70	163.9
6	14.0	19	44.5	32	74.9	45	105.3	58	135.8	71	166.2
7	16.4	20	46.8	33	77.3	46	107.7	59	138.1	72	168.6
8	18.7	21	49.2	34	79.6	47	110.0	60	140.4	73	170.9
9	21.1	22	51.5	35	81.9	48	112.4	61	142.8	74	173.3
10	23.4	23	53.8	36	84.3	49	114.7	62	145.1	75	175.6
11	25.8	24	56.2	37	86.6	50	117.0	63	147.5	76	177.9
12	28.1	25	58.5	38	89.0	51	119.4	64	149.8	77	180.3
13	30.4	26	60.9	39	91.3	52	121.7	65	152.2	78	182.6

PART III.

Correction due to the Change of Gravity from the Latitude of 45° to the Latitude of the Place of Observation.

Positive from Lat. 0° to 45°; Negative from Lat. 45° to 90°.

PART IV.

Correction for Decrease of Gravity on a Vertical.

Always Positive.

PART V.

Correction due to the Height of the Lower Station.

Always Positive.

App. Alt.	Part III. Latitude. 0° / 90°	10° / 80°	20° / 70°	30° / 60°	40° / 50°	45°	Part IV.	Part V. Height of Barometer at Lower Station. 16 in.	18 in.	20 in.	22 in.	24 in.	26 in.	28 in.	App. Alt.
Feet.	Feet.	Feet.	Feet.	Feet.	Feet.	Feet.	Feet.	Feet.	Feet.	Feet.	Feet.	Feet.	Feet.	Feet.	Feet.
1000	2.6	2.5	2.0	1.3	0.5	0	2.5	1.6	1.3	1.0	0.8	0.6	0.4	0.2	1000
2000	5.3	5.0	4.1	2.6	0.9	0	5.2	3.1	2.5	2.0	1.5	1.1	0.7	0.3	2000
3000	7.9	7.5	6.1	4.0	1.4	0	7.9	4.7	3.8	3.0	2.3	1.7	1.1	0.5	3000
4000	10.6	10.0	8.1	5.3	1.8	0	10.8	6.3	5.1	4.0	3.1	2.2	1.4	0.7	4000
5000	13.2	12.4	10.1	6.6	2.3	0	13.7	7.8	6.4	5.0	3.8	2.8	1.8	0.8	5000
6000	15.9	14.9	12.2	7.9	2.8	0	16.7	9.4	7.6	6.0	4.6	3.3	2.1	1.0	6000
7000	18.5	17.4	14.2	9.3	3.2	0	19.9	11.0	8.9	7.1	5.4	3.9	2.5	1.2	7000
8000	21.2	19.9	16.2	10.6	3.7	0	23.1	12.5	10.2	8.1	6.2	4.4	2.8	1.3	8000
9000	23.8	22.4	18.3	11.9	4.1	0	26.4	14.1	11.4	9.1	6.9	5.0	3.2	1.5	9000
10000	26.5	24.9	20.3	13.2	4.6	0	29.8	15.7	12.7	10.1	7.7	5.5	3.5	1.7	10000
11000	29.1	27.4	22.3	14.6	5.1	0	33.3	17.2	14.0	11.1	8.5	6.1	3.9	1.8	11000
12000	31.8	29.9	24.4	15.9	5.5	0	36.9	18.8	15.3	12.1	9.2	6.6	4.2	2.0	12000
13000	34.4	32.4	26.4	17.2	6.0	0	40.6	20.4	16.5	13.1	10.0	7.2	4.6	2.2	13000
14000	37.1	34.9	28.4	18.5	6.4	0	44.4	21.9	17.8	14.1	10.8	7.7	4.9	2.3	14000
15000	39.7	37.3	30.4	19.9	6.9	0	48.3	23.5	19.1	15.1	11.5	8.3	5.3	2.5	15000
16000	42.4	39.8	32.5	21.2	7.4	0	52.3	25.1	20.3	16.1	12.3	8.8	5.6	2.7	16000
17000	45.0	42.3	34.5	22.5	7.8	0	56.4	26.6	21.6	17.1	13.1	9.4	6.0	2.8	17000
18000	47.7	44.8	36.5	23.8	8.3	0	60.5	28.2	22.9	18.1	13.8	9.9	6.3	3.0	18000
19000	50.3	47.3	38.6	25.2	8.7	0	64.8	29.8	24.1	19.2	14.6	10.5	6.7	3.2	19000
20000	53.0	49.8	40.6	26.5	9.2	0	69.2	31.3	25.4	20.2	15.4	11.0	7.0	3.3	20000
21000	55.6	52.3	42.6	27.8	9.7	0	73.6	32.9	26.7	21.2	16.1	11.6	7.4	3.5	21000
22000	58.3	54.8	44.7	29.1	10.1	0	78.2	34.5	28.0	22.2	16.9	12.1	7.7	3.7	22000
23000	60.9	57.3	46.7	30.5	10.6	0	82.9	36.0	29.2	23.2	17.7	12.7	8.1	3.8	23000
24000	63.6	59.8	48.7	31.8	11.0	0	87.6	37.6	30.5	24.2	18.5	13.2	8.4	4.0	24000
25000	66.2	62.2	50.7	33.1	11.5	0	92.5	39.1	31.8	25.2	19.2	13.8	8.8	4.1	25000

IV.

TABLES

FOR REDUCING BAROMETRICAL OBSERVATIONS TO THE LEVEL OF THE SEA, OR TO ANY OTHER LEVEL, AND FOR COMPUTING DIFFERENCES OF ELEVATION MEASURED BY THE BAROMETER, BY M. C. DIPPE.

THE following tables, published by M. C. DIPPE, in the *Astronomische Nachrichten*, No. 1056, November, 1856, are a modification and extension of Gauss's tables, published in Schumacher's *Jahrbuch*, for 1836 and the following years, which are based on the formula of Laplace. In this new form they answer a double purpose. They give the means of solving a problem which often occurs in Meteorology, viz.: The difference of elevation between two stations, and the temperature of the air at both, being known, to reduce the height of the barometer at one of the stations to the height it would have at the other. They are likewise adapted to the computation of heights from barometrical observations.

The formula of Laplace, which has been used, the Metres being reduced to Toises, and the Centigrade degrees to degrees of Reaumur, reads as follows:

$$h = 9407\overset{\text{T.}}{.}73 \left(1 + \frac{t + t'}{400}\right) (1 + a \cos 2\phi) \left(1 + \frac{h}{r}\right) \left\{\log \frac{b}{b'} + 2 \log \left(1 + \frac{h}{r}\right)\right\}.$$

Where t and t' = the temperatures of the air, in degrees of Reaumur, at the lower and upper station,

b and b' = the height of the barometer, in any scale, reduced to the freezing point, at the lower and upper station,

h = the difference of level, in toises, between the two stations,

r = the distance, in toises, of the lower station to the centre of the Earth,

ϕ = the latitude of the place of observation,

a = the increase of gravity from the equator to the poles.

Making, besides, m = the modulus of the common logarithms, the formula becomes, with sufficient accuracy,

$$\log b - \log b' = h \left\{\frac{1}{9407.73} \cdot \frac{1}{1 + \frac{t + t'}{400}} - \frac{2\,m}{r}\right\} \cdot \frac{1}{1 + a \cos 2\phi} \cdot \frac{1}{1 + \frac{h}{r}}.$$

Assuming r, or the radius of the Earth, at 45° latitude = 3266631 toises, and a = 0.002595, instead of 0.002845 adopted in Gauss's tables, and making

$$u = \log b - \log b',$$

$$a = \log \left(\frac{1}{9407.73} \cdot \frac{1}{1 + \frac{t + t'}{400}} - \frac{2\,m}{r}\right),$$

$$c = -\, m\, a \cos 2\phi,$$

$$c' = -\frac{m\,h}{r},$$

then the reduction of *the height of the barometer to another level* is given by the formula,

1. $\log u = \log h + a + c + c'$;
2. $\log b = \log b' + u$.

Table I. contains the values of a for the argument $t + t'$; 10 units are to be subtracted from the characteristic.

Table II. gives the values of c for the argument ϕ, or the correction for the change of gravity in latitude, which is *negative* from 0° to 45°, *positive* from 45° to 90°.

Table III. furnishes the values of c' for the argument h in toises, or the correction for the decrease of gravity on the vertical. Both in Tables II. and III. the values of c and c' are given in units of the fifth decimal place.

The *difference of elevation of the two stations* is given by the formula,

1. $u = \log b - \log b'$,
2. $\log h = \log u + A + c + c'$,

in which A is the arithmetical complement of a, and the corrections c and c' receive *contrary signs*. For the sake of convenience, the values of A have been placed in Table I., and in Table III. the correction for A is found in another column, with the more convenient argument $v = \log u + A$.

If the heights of the barometers have not been reduced to the freezing point, then, B and B′ being the unreduced heights of the barometers, and T and T′ the temperature of the attached thermometer *in degrees of Reaumur*,

$$b : b' = \frac{B}{1 + \frac{T}{4440}} : \frac{B'}{1 + \frac{T'}{4440}},$$

and making $\frac{m}{4440} = \beta$,

$$u = \log b - \log b' = (\log B - \beta T) - (\log B' - \beta T').$$

Instead of $\beta = 0.000098$, we can write with sufficient accuracy 0.00010.

Use of the Tables.

These tables can be used in any latitude, and for any barometrical scale; but the indications of the barometers *must be reduced to the freezing point;* and the temperatures of the air *must be given in degrees of Reaumur.* The tables suppose the use of logarithms with 5 decimals, such as those of Lalande, and give the results in toises.

I. *For Reducing Barometrical Observations to another Level.*

Given h in toises, t, t', ϕ, and b or b'.
To find b or b'.

In Table I. with the argument $t + t'$, take a,
In Table II. with the argument ϕ, take c,
In Table III. with the argument h, take c',

the last two corrections being given in units of the fifth decimal, making

$$\log h + a + c + c' - 10 \text{ (whole units)} = \log u.$$

Then we have

for a level lower by h toises, $\log b = \log b' + u$;
for a level higher by h toises, $\log b' = \log b - u$.

If h, or the difference of elevation, is given in metres, take c', which is always negative, from Table III. (for A) with the argument $v = \log h + 9.71$, and write

$$\log u = 9.71018 + \log h + a + c + c' - 10 \text{ (whole units)}.$$

Then again is $\log b = \log b' + u$.

Example 1.

Suppose the height of the barometer, reduced to the freezing point, to be $b' =$ 295.39 Paris lines; the temperature of the air $t' = 11°.8$ Reaumur, and the latitude $\phi = 51° \ 48'$; the increase of heat downwards being 1° Reaumur for 100 toises. What is the height of the barometer, reduced to the freezing point, at a station lower by $h = 498.2$ toises?

In th' case $t = t' + 4°.98 = 16°.78$, and $t + t' = 28°.58$.

Then

	$\log h =$	2.69740
Table I. for 28°.58 gives	$a =$	5.99538
Table II. for 51° 48′ gives	$c =$	+ 0.00026
Table III. for 498 toises gives	$c' =$	— 0.00007
	$\log u =$	8.69297 — 10
	$u =$	0.04931
	$\log b' =$	2.47040
	$\log b =$	2.51971
Barometer at the lower station	$b =$	330.90 Paris lines.

Example 2.

Suppose the reduced barometer $b' = 598.6$ millimetres; the temperature of the air $t' = 18°.0$ Centigrade $= 14°.4$ Reaumur; the difference of elevation $h = 2217$ metres; $\phi = 3°$. The temperature of the air at the lower station $t = 27°.5$ Centigrade $= 22°.0$ Reaumur, and $t + t' = 36°.4$ Reaumur.

Then	$\log h = \{$	$\log 2217 =$	3.34577
		$+$	9.71018
			3.05595 $v = 3.06$
		$a =$	5.98750
		$c =$	— 0.00112
		$c' =$	— 0.00015
		$\log u =$	9.04218 — 10
		$u =$	0.11020
		$\log b' =$	9.77714
		$\log b =$	9.88734
Barometer at the lower station		$b =$	771.5 millimetres.

2. *For Computing Differences of Elevation from Barometrical Observations.*

Given the unreduced height of the barometer at the lower and upper station, B and B′; the temperatures of the attached thermometers, T and T′; the temperatures of the air, t and t'; and the latitude, ϕ.

To find h, or the difference of elevation between the two stations.

Subtract (log B′ — 10 T′) from (log B — 10 T), paying due attention to the nature of the signs of T and T′, and taking the numbers 10 T and 10 T′ as units of the fifth decimal. Calling then (log B — 10 T) — (log B′ — 10 T′) $= u$, or if the heights of the Barometers are reduced to the freezing point, $\log b - \log b' = u$, take,

In Table I., A with the argument $t + t'$, and make $v = \log u + A$.
In Table II., with the argument ϕ, take c reversing the sign.

In Table III., for A, with the argument v, take c', which, in this case, is always ***positive*** ; then, remembering that the values of c and c' are given in units of the fifth decimal, we have,

$$v + c + c' = \log h \text{ in toises,}$$
$$v + c + c' + 0.28982 = \log h \text{ in metres,}$$
$$v + c + c' + 0.80584 = \log h \text{ in English feet.}$$

Example 1.

L. station B $= 329.013$ Paris lines ; $T = +15\overset{\circ}{.}88$ R. ; $t = +15\overset{\circ}{.}96$ R. ; $\phi = 45^\circ\, 32'$.
U. station B′ $= 268.215$ Paris lines ; $T' = +\ 8.40$ R. ; $t = +\ 7.92$ R.

$$t + t' = 23.88 \text{ R.}$$

$$\log B = 2.51722 - 10 \times 15.88 = 2.51563$$
$$\log B' = 2.42848 - 10 \times 8.4 = 2.42764$$
$$u = 0.08799$$
$$\log u = 8.94443$$
$$A = 3.99982$$
$$v = 2.94425$$
$$c = -0.00002$$
$$c' = +0.00012$$
$$\log h = 2.94435$$
$$h = 879.74 \text{ toises.}$$

Example 2.

L. station B $= 763.15$ millimetres ; $T = t = 25\overset{\circ}{.}3$ Cent. $= 20\overset{\circ}{.}24$ R. ; $\phi = 21^\circ$.
U. station B′ $= 600.95$ millimetres ; $T' = t' = 21.3$ Cent. $= 17.04$ R.

$$t + t' = 37.28 \text{ R.}$$

$$\log B = 9.88261 - 10 \times 20.24 = 9.88059$$
$$\log B' = 9.77884 - 10 \times 17.04 = 9.77714$$
$$u = 0.10345$$
$$\log u = 9.01473$$
$$A = 4.01337$$
$$v = 3.02810$$
$$c = +0.00084$$
$$c' = +0.00014$$
$$\log h = 3.02908 \text{ for toises.}$$
$$0.28982$$
$$\log h = 3.31890 \text{ for metres.}$$
$$\log h = 3.02908 \text{ for toises.}$$
$$0.30584$$
$$\log h = 3.83492 \text{ for English feet.}$$

$$h = 1069.3 \text{ toises} = 2084.0 \text{ metres} = 6837.9 \text{ English feet.}$$

I. Argument: Sum of the Temperatures of the Air in Degrees of Reaumur.

$t+t'$	Correction for			$t+t'$	Correction for		
Reaumur.	a	Difference.	A	Reaumur.	a	Difference.	A
−60°	6.09617	128	3.90383	−20°	6.04776	115	3.95224
−59	6.09489	127	3.90511	−19	6.04661	114	3.95339
−58	6.09362	127	3.90638	−18	6.04547	113	3.95453
−57	6.09235	127	3.90765	−17	6.04434	114	3.95566
−56	6.09108	126	3.90892	−16	6.04320	113	3.95680
−55	6.08982	126	3.91018	−15	6.04207	113	3.95793
−54	6.08856	126	3.91144	−14	6.04094	113	3.95906
−53	6.08730	125	3.91270	−13	6.03981	112	3.96019
−52	6.08605	125	3.91395	−12	6.03869	112	3.96131
−51	6.08480	124	3.91520	−11	6.03757	112	3.96243
−50	6.08356	125	3.91644	−10	6.03645	112	3.96355
−49	6.08231	123	3.91769	− 9	6.03533	111	3.96467
−48	6.08108	124	3.91892	− 8	6.03422	111	3.96578
−47	6.07984	123	3.92016	− 7	6.03311	110	3.96689
−46	6.07861	123	3.92139	− 6	6.03201	111	3.96799
−45	6.07738	122	3.92262	− 5	6.03090	110	3.96910
−44	6.07616	122	3.92384	− 4	6.02980	109	3.97020
−43	6.07494	122	3.92506	− 3	6.02871	110	3.97129
−42	6.07372	122	3.92628	− 2	6.02761	109	3.97239
−41	6.07250	121	3.92750	− 1	6.02652	109	3.97348
−40	6.07129	120	3.92871	0	6.02543	109	3.97457
−39	6.07009	121	3.92991	+ 1	6.02434	108	3.97566
−38	6.06888	120	3.93112	2	6.02326	109	3.97674
−37	6.06768	120	3.93232	3	6.02217	108	3.97783
−36	6.06648	119	3.93352	4	6.02109	107	3.97891
−35	6.06529	119	3.93471	5	6.02002	107	3.97998
−34	6.06410	119	3.93590	6	6.01895	108	3.98105
−33	6.06291	118	3.93709	7	6.01787	107	3.98213
−32	6.06173	118	3.93827	8	6.01680	106	3.98320
−31	6.06055	118	3.93945	9	6.01574	106	3.98426
−30	6.05937	118	3.94063	10	6.01468	106	3.98532
−29	6.05819	117	3.94181	11	6.01362	106	3.98638
−28	6.05702	117	3.94298	12	6.01256	106	3.98744
−27	6.05585	116	3.94415	13	6.01150	105	3.98850
−26	6.05469	117	3.94531	14	6.01045	105	3.98955
−25	6.05352	116	3.94648	15	6.00940	105	3.99060
−24	6.05236	115	3.94764	16	6.00835	104	3.99165
−23	6.05121	116	3.94879	17	6.00731	105	3.99269
−22	6.05005	115	3.94995	18	6.00626	104	3.99374
−21	6.04890	114	3.95110	19	6.00522	104	3.99478
−20	6.04776		3.95224	+20	6.00418		3.99582

(*Continued.*)

$t+t'$ Reaumur.	Correction for a	Difference.	Correction for A	$t+t'$ Reaumur.	Correction for a	Difference.	Correction for A
+20°	6.00418	103	3.99582	+40°	5.98393	99	4.01607
21	6.00315	103	3.99685	41	5.98294	99	4.01706
22	6.00212	104	3.99788	42	5.98195	98	4.01805
23	6.00108	102	3.99892	43	5.98097	99	4.01903
24	6.00006	103	3.99994	44	5.97998	98	4.02002
25	5.99903	102	4.00097	45	5.97900	97	4.02100
26	5.99801	102	4.00199	46	5.97803	98	4.02197
27	5.99699	102	4.00301	47	5.97705	97	4.02295
28	5.99597	102	4.00403	48	5.97608	97	4.02392
29	5.99495	101	4.00505	49	5.97511	97	4.02489
30	5.99394	101	4.00606	50	5.97414	97	4.02586
31	5.99293	101	4.00707	51	5.97317	96	4.02683
32	5.99192	101	4.00808	52	5.97221	97	4.02779
33	5.99091	100	4.00909	53	5.97124	96	4.02876
34	5.98991	101	4.01009	54	5.97028	95	4.02972
35	5.98890	100	4.01110	55	5.96933	96	4.03067
36	5.98790	99	4.01210	56	5.96837	95	4.03163
37	5.98691	100	4.01309	57	5.96742	96	4.03258
38	5.98591	99	4.01409	58	5.96646	95	4.03354
39	5.98492		4.01508	59	5.96551		4.03449

II. Latitude. — Correction for a.

For A reverse the Signs of c.

ϕ	c	ϕ	ϕ	c	ϕ	ϕ	c	ϕ
0°	−113+	90°	15°	−98+	75°	30°	−56+	60°
1	113	89	16	96	74	31	53	59
2	112	88	17	93	73	32	49	58
3	112	87	18	91	72	33	46	57
4	112	86	19	89	71	34	42	56
5	111	85	20	86	70	35	39	55
6	110	84	21	84	69	36	35	54
7	109	83	22	81	68	37	31	53
8	108	82	23	78	67	38	27	52
9	107	81	24	75	66	39	23	51
10	106	80	25	72	65	40	20	50
11	104	79	26	69	64	41	16	49
12	103	78	27	66	63	42	12	48
13	101	77	28	63	62	43	8	47
14	100	76	29	60	61	44	4	46
15	−98+	75	30	−56+	60	45	−0+	45

III. Decrease of Gravity on the Vertical. — Correction

For a, argument h, in Toises, c' always Negative.

h	c'	h	c'
100	1	1600	21
200	3	1700	23
300	4	1800	24
400	5	1900	25
500	7	2000	27
600	8	2100	28
700	9	2200	29
800	11	2300	31
900	12	2400	32
1000	13	2500	33
1100	15	2600	35
1200	16	2700	36
1300	17	2800	37
1400	19	2900	39
1500	20	3000	40
1600	21	3500	47

For A, arg. v, c' always Positive.

v	c'
1.8	1
1.9	1
2.0	1
2.1	2
2.2	2
2.3	3
2.4	3
2.5	4
2.6	5
2.7	7
2.8	8
2.9	11
3.0	13
3.1	17
3.2	21
3.3	27
3.4	33
3.5	42
3.6	53

V.

TABLES

FOR REDUCING BAROMETRICAL OBSERVATIONS TO ANOTHER LEVEL, AND FOR COMPUTING DIFFERENCES OF ELEVATION MEASURED BY THE BAROMETER, BY M. C. DIPPE.

In No. 1088 of the *Astronomische Nachrichten*, published in June, 1857, Dr. Dippe gives the following set of Tables for reducing barometrical observations to another level, and for computing heights. These tables, being based, as the preceding ones (IV.), on the formula of Laplace, and computed with the same constants, give results nearly identical, but dispense with the use of logarithms.

Use of the Tables.

The tables suppose the height of the barometer to be expressed in French inches or Paris lines, and the temperature in degrees of Reaumur; they give the differences of level in French toises.

The signs used have the following signification: —

At Lower Station.
- B = Observed Height of Barometer in Paris lines.
- T = Attached Thermometer in degrees of Reaumur.
- b = Barometer reduced to the freezing point.
- t = Temperature of the air, detached Thermometer.

At Upper Station.
- B′ = Observed Height of Barometer.
- T′ = Attached Thermometer.
- b' = Barometer at the freezing point.
- t' = Temperature of the air.

ϕ = Latitude of the place.
h = Difference of elevation between the two stations.

I. *For Reducing Barometrical Observations to another Level.*

Given, h in toises, t, t', ϕ, and b or b'.
To find b or b'.

Make first $2\tau = \frac{t+t}{2}$ and τ, and

In Table I., with the argument 2τ, take τ';
In Table III., with the arguments h and τ, take C;
In Table IV., with the arguments h and ϕ, take C';

Make, further,

$$u = h + C + C' \text{ and } \frac{u}{100}\tau';$$

And if b' be given, and b required,

In Table II., with the argument b, take H;

then is
$$H = H' + (u - \frac{u}{100}\tau'),$$

and the height of the barometer, in Table II., due to H, is b required.

If b be given, and b' required for a level higher by h toises, then,

In Table II., with the argument b, take H'.

Make, further,

$$H' = H - (u - \frac{u}{100}\tau'),$$

and b' is the height of the barometer in Table II., corresponding to H'.

Example 1.

Suppose the height of the barometer reduced to the freezing point to be $b' =$ 295.39 Paris lines; the temperature of the air $t' = 11°.8$ Reaumur; and the latitude $\phi = 51°.48$; the increase of heat downwards being 1° Reaumur for 100 toises. What is the height of the barometer reduced to the freezing point, at a station lower by $h = 498.2$ toises?

In this case, $t' = 11°.8$; $t = 11°.8 + 4°.98$; $t + t' = 28°.58$;

$$2\tau = \frac{t+t}{2} = 14°.29; \; \tau = 7°.15;$$

and according to Table I. $\tau' = +6.67$.

With h and τ, in Table III., we find $C = -\;1.4$
With h and ϕ, in Table IV., we find $C' = +\;0.3$
We add $h = 498.2$
and we have $u = 497.1$;
$-\frac{u}{100}\tau' = -33.15$
463.95
With b', in Table II., we find $H' = 367.86$
$H = 831.81$

$\frac{u}{100} = 4.971$
$\tau' = +6.67$
29.83
2.98
.34
$\frac{u}{100}\tau' = +33.15$

Finally, with H, in Table II., we find $b = 330.91$ Paris lines, which is the required height of the barometer at the lower station. Gauss's tables (IV.) would give $b =$ 330.90 lines.

Example 2.

Suppose $b' = 330.46$ Paris lines; $t' = -12°.3$ Reaumur; $h' = 92.7$ toises; $\phi = 62°$.

In this case, assuming $t = t'$,

$$2\tau = \frac{t + t'}{2} = -12°.3; \quad \tau = -6.15;$$

and according to Table I. $\tau' = -6.55$.

With h and τ, in Table III., take $C = - \ 0.2$
With h and ϕ, in Table IV., take $C' = + \ 0.1$
Add $h = \ 92.7$
We have $u = \ 92.6$
$-\frac{u}{100}\tau' = + \ 6.07$
98.67
With b', in Table II., take $H' = 826.22$
$H = 924.89$

$\frac{u}{100} = 0.926$
$\tau' = -6.55$
5.56
$.46$
$.05$
$\frac{u}{100}\tau' = -6.07$

With H, in Table II., we find $b = 338.53$ Paris lines. Gauss's tables (IV.) would give $b = 338.54$ lines.

II. *For Computing Differences of Elevation from Barometrical Observations.*

Suppose to be given B, B′, T, T′, t, t', ϕ; required h.

Make first $\tau = \frac{t + t'}{4}$ and $T - T'$.

Then in Table II., with the argument $\begin{cases} \text{B take H,} \\ \text{B}' \text{ take H}', \end{cases}$

and make

$$u = (H - H') + \frac{H - H'}{100}\tau - (T - T'),$$

in which each full degree of $T - T'$ corresponds to a toise.

Further, in Table III., with u and τ, take C reversing the sign;
in Table IV., with u and ϕ, take C′ reversing the sign;
in Table V., with $T - T'$ and τ, take C″ with the signs of $T - T'$.

Then the difference of elevation required is

$$h = u + C + C' + C''.$$

If the heights of the barometer, reduced to the freezing point, or b and b', are given,

then in Table II., with the argument, $\begin{cases} b \text{ take H} \\ b' \text{ take H}', \end{cases}$

and make

$$u = H - H' + \frac{H - H'}{100}\tau.$$

Further, in Table III., take C reversing the sign;
in Table IV., take C′ reversing the sign;

and

$$h = u + C + C'.$$

D

Example 1.

Suppose to be given,

$B = 333.6$ Paris lines; $T = +17°.0$ Reaumur; $t = +19°.0$ R.; $\phi = 48°$.
$B' = 289.9$ Paris lines; $T' = +16°.3$ Reaumur; $t' = +15°.2$ R.

$$T - T' = 0°.7 \qquad t + t' = +34°.2$$
$$\tau = +\ 8.55$$

In Table II. with B take $H = 864.9$
" with B′ take $H' = 291.2$

$$H - H' = 573.7$$
$$\frac{H - H'}{100}\tau = 49.06$$
$$-(T - T') = -0.7$$
$$u = 622.06$$

$$\frac{H - H'}{100} = 5.73$$
$$\tau = +8.55$$
$$45\ 90$$
$$2.87$$
$$.29$$
$$\frac{H - H'}{100}\tau = 49.06$$

In Table III., with u and τ, take $C = +1.8$
In Table IV., with u and ϕ, take $C' = -0.2$
In Table V., with $T - T'$ and τ take $C'' = 0.0$

Difference of elevation, or $h = 623.66$ toises.

Gauss's Tables give 623.64 toises.

Example 2.

Suppose to be given,

$b = 342.68$ Paris lines; $t = -10°.38$ Reaumur; $\phi = 65°$.
$b' = 285.47$ Paris lines; $t' = -14°.94$ Reaumur; $T - T' = 0°$. R.

$$t + t' = -25°.32$$
$$\tau = -\ 6.33$$

In Table II. with b take $H = 974.58$
" with b' take $H' = 228.28$

$$H - H' = 746.30$$
$$\frac{H - H'}{100}\tau = -47.24$$
$$u = 699.06$$

$$\frac{H - H'}{100} = 7\ 463$$
$$\tau = -6.33$$
$$44.78$$
$$2.24$$
$$.22$$
$$\frac{H - H'}{100}\tau - 47.24$$

In Table III., with u and τ, take $C = +1.8$
In Table IV., with u and ϕ, take $C' = -1.2$

$$h = 699.66$$

Gauss's Tables give $h = 699.72$ toises.

V.

TABLES

FOR REDUCING BAROMETRICAL OBSERVATIONS TO ANOTHER LEVEL, AND FOR COMPUTING DIFFERENCES OF ELEVATION, BY M. C. DIPPE.

TABLE I. — *Argument,* the observed Height of the Barometer at either Station.

Barometer in Paris Lines.	Tenths of a Line.									
	0	**1**	**2**	**3**	**4**	**5**	**6**	**7**	**8**	**9**
B or B′	H or H′ in Toises =									
270	0.7	2.2	3.7	5.2	6.7	8.2	9.7	11.2	12.8	14.3
271	15.8	17.3	18.8	20.3	21.8	23.3	24.8	26.3	27.8	29.3
272	30.8	32.3	33.8	35.3	36.8	38.3	39.8	41.3	42.8	44.3
273	45.8	47.3	48.8	50.3	51.8	53.3	54.8	56.3	57.8	59.3
274	60.8	62.2	63.7	65.2	66.7	68.2	69.7	71.2	72.7	74.1
275	75.6	77.1	78.6	80.1	81.6	83.1	84.5	86.0	87.5	89.0
23 Inch.										
276	90.5	91.9	93.4	94.9	96.4	97.9	99.3	100.8	102.3	103.8
277	105.2	106.7	108.2	109.7	111.1	112.6	114.1	115.6	117.0	118.5
278	120.0	121.4	122.9	124.4	125.8	127.3	128.8	130.2	131.7	133.2
279	134.6	136.1	137.6	139.0	140.5	142.0	143.4	144.9	146.3	147.8
280	149.3	150.7	152.2	153.6	155.1	156.5	158.0	159.5	160.9	162.4
281	163.8	165.3	166.7	168.2	169.6	171.1	172.5	174.0	175.4	176.9
282	178.3	179.8	181.2	182.7	184.1	185.6	187.0	188.5	189.9	191.4
283	192.8	194.2	195.7	197.1	198.6	200.0	201.4	202.9	204.3	205.8
284	207.2	208.6	210.1	211.5	213.0	214.4	215.8	217.3	218.7	220.1
285	221.6	223.0	224.4	225.9	227.3	228.7	230.2	231.6	233.0	234.5
286	235.9	237.3	238.7	240.2	241.6	243.0	244.4	245.9	247.3	248.7
287	250.1	251.6	253.0	254.4	255.8	257.3	258.7	260.1	261.5	262.9
24 Inch.										
288	264.4	265.8	267.2	268.6	270.0	271.4	272.9	274.3	275.7	277.1
289	278.5	279.9	281.3	282.8	284.2	285.6	287.0	288.4	289.8	291.2
290	292.6	294.0	295.4	296.8	298.3	299.7	301.1	302.5	303.9	305.3
291	306.7	308.1	309.5	310.9	312.3	313.7	315.1	316.5	317.9	319.3
292	320.7	322.1	323.5	324.9	326.3	327.7	329.1	330.5	331.9	333.3
293	334.7	336.1	337.5	338.9	340.2	341.6	343.0	344.4	345.8	347.2
294	348.6	350.0	351.4	352.8	354.2	355.5	356.9	358.3	359.7	361.1
295	362.5	363.9	365.2	366.6	368.0	369.4	370.8	372.2	373.5	374.9
296	376.3	377.7	379.1	380.4	381.8	383.2	384.6	385.9	387.3	388.7
297	390.1	391.5	392.8	394.2	395.6	397.0	398.3	399.7	401.1	402.4
298	403.8	405.2	406.5	407.9	409.3	410.7	412.0	413.4	414.8	416.1
299	417.5	418.9	420.2	421.6	423.0	424.3	425.7	427.1	428.4	429.8
25 Inch.										
300	431.1	432.5	433.9	435.2	436.6	437.9	439.3	440.7	442.0	443.4
301	444.7	446.1	447.5	448.8	450.2	451.5	452.9	454.2	455.6	456.9
302	458.3	459.6	461.0	462.3	463.7	465.0	466.4	467.8	469.1	470.5
303	471.8	473.1	474.5	475.8	477.2	478.5	479.9	481.2	482.6	483.9
304	485.3	486.6	487.9	489.3	490.6	492.0	493.3	494.7	496.0	497.3
305	498.7	500.0	501.4	502.7	504.0	505.4	506.7	508.0	509.4	510.7
306	512.0	513.4	514.7	516.0	517.4	518.7	520.1	521.4	522.7	524.0

TABLE I. *Continued.*

Barometer in Paris Lines.	Tenths of a Line.									
	0	**1**	**2**	**3**	**4**	**5**	**6**	**7**	**8**	**9**
306	512.0	513.4	514.7	516.0	517.4	518.7	520.1	521.4	522.7	524.0
307	525.4	526.7	528.0	529.4	530.7	532.0	533.4	534.7	536.0	537.4
308	538.7	540.0	541.3	542.6	544.0	545.3	546.6	547.9	549.3	550.6
309	551.9	553.2	554.6	555.9	557.2	558.5	559.8	561.2	562.5	563.8
310	565.1	566.4	567.8	569.1	570.4	571.7	573.0	574.3	575.6	576.9
311	578.3	579.6	580.9	582.2	583.5	584.8	586.1	587.5	588.8	590.1
26 Inch.										
312	591.4	592.7	594.0	595.3	596.6	597.9	599.2	600.6	601.9	603.2
313	604.5	605.8	607.1	608.4	609.7	611.0	612.3	613.6	614.9	616.2
314	617.5	618.8	620.1	621.4	622.7	624.0	625.3	626.6	627.9	629.2
315	630.5	631.8	633.1	634.4	635.7	637.0	638.3	639.5	640.8	642.1
316	643.4	644.7	646.0	647.3	648.6	649.9	651.2	652.5	653.8	655.1
317	656.3	657.6	658.9	660.2	661.5	662.8	664.1	665.4	666.6	667.9
318	669.2	670.5	671.8	673.1	674.3	675.6	676.9	678.2	679.5	680.8
319	682.0	683.3	684.6	685.9	687.2	688.4	689.7	691.0	692.3	693.6
320	694.8	696.1	697.4	698.7	699.9	701.2	702.5	703.8	705.0	706.3
321	707.6	708.9	710.1	711.4	712.7	713.9	715.2	716.5	717.7	719.0
322	720.3	721.6	722.8	724.1	725.4	726.6	727.9	729.2	730.4	731.7
323	733.0	734.2	735.5	736.7	738.0	739.3	740.5	741.8	743.1	744.3
27 Inch.										
324	745.6	746.8	748.1	749.4	750.6	751.9	753.2	754.4	755.7	756.9
325	758.2	759.4	760.7	761.9	763.2	764.5	765.7	767.0	768.2	769.5
326	770.7	772.0	773.2	774.5	775.7	777.0	778.2	779.5	780.7	782.0
327	783.2	784.5	785.7	787.0	788.2	789.5	790.7	792.0	793.2	794.5
328	795.7	797.0	798.2	799.4	800.7	801.9	803.2	804.4	805.7	806.9
329	808.2	809.4	810.6	811.9	813.1	814.4	815.6	816.8	818.1	819.3
330	820.6	821.8	823.0	824.3	825.5	826.7	828.0	829.2	830.4	831.7
331	832.9	834.2	835.4	836.6	837.9	839.1	840.3	841.6	842.8	844.0
332	845.2	846.5	847.7	848.9	850.2	851.4	852.6	853.9	855.1	856.3
333	857.5	858.8	860.0	861.2	862.4	863.7	864.9	866.1	867.3	868.6
334	869.8	871.0	872.2	873.4	874.7	875.9	877.1	878.3	879.6	880.8
335	882.0	883.2	884.4	885.7	886.9	888.1	889.3	890.5	891.7	893.0
28 Inch.										
336	894.2	895.4	896.6	897.8	899.0	900.3	901.5	902.7	903.9	905.1
337	906.3	907.5	908.7	909.9	911.2	912.4	913.6	914.8	916.0	917.2
338	918.4	919.6	920.8	922.0	923.3	924.5	925.7	926.9	928.1	929.3
339	930.5	931.7	932.9	934.1	935.3	936.5	937.7	938.9	940.1	941.3
340	942.5	943.7	944.9	946.1	947.3	948.5	949.7	950.9	952.1	953.3
341	954.5	955.7	956.9	958.1	959.3	960.5	961.7	962.9	964.1	965.3
342	966.5	967.7	968.9	970.1	971.3	972.5	973.7	974.8	976.0	977.2
343	978.4	979.6	980.8	982.0	983.2	984.4	985.6	986.8	987.9	989.1
344	990.3	991.5	992.7	993.9	995.1	996.2	997.4	998.6	999.8	1001.0
345	1002.2	1003.4	1004.5	1005.7	1006.9	1008.1	1009.3	1010.5	1011.6	1012.8
346	1014.0	1015.2	1016.4	1017.5	1018.7	1019.9	1021.1	1022.3	1023.4	1024.6
347	1025.8	1027.0	1028.1	1029.3	1030.5	1031.7	1032.8	1034.0	1035.2	1036.4
29 Inch.										
348	1037.5	1038.7	1039.9	1041.1	1042.2	1043.4	1044.6	1045.8	1046.9	1048.1

Table II.

CORRECTION FOR THE TEMPERATURE OF THE AIR.

Argument, $2\tau = \frac{t + t'}{2}$.

2τ	τ'	Diff.	2τ	τ'	Diff.	2τ	τ'	Diff.	2τ	τ'	Diff.
−25	−14.29	0.65	−12	−6.38	0.56	+ 1	+0.50	0.49	+14	+ 6.54	0.44
−24	−13.64	0.64	−11	−5.82	0.56	2	0.99	0.49	15	6.98	0.43
−23	−13.00	0.64	−10	−5.26	0.55	3	1.48	0.48	16	7.41	0.42
−22	−12.36	0.63	− 9	−4.71	0.54	4	1.96	0.48	17	7.83	0.43
−21	−11.73	0.62	− 8	−4.17	0.54	5	2.44	0.47	18	8.26	0.42
−20	−11.11	0.61	− 7	−3.63	0.54	6	2.91	0.47	19	8.68	0.41
−19	−10.50	0.61	− 6	−3.09	0.53	7	3.38	0.47	20	9.09	0.41
−18	− 9.89	0.60	− 5	−2.56	0.52	8	3.85	0.46	21	9.50	0.41
−17	− 9.29	0.59	− 4	−2.04	0.52	9	4.31	0.45	22	9.91	0.40
−16	− 8.70	0.59	− 3	−1.52	0.51	10	4.76	0.45	23	10.31	0.40
−15	− 8.11	0.58	− 2	−1.01	0.51	11	5.21	0.45	24	10.71	0.40
−14	− 7.53	0.58	− 1	−0.50	0.50	12	5.66	0.44	25	11.11	0.39
−13	− 6.95	0.57	0	0.00	0.50	13	6.10	0.44	26	11.50	0.39
−12	− 6.38		+ 1	+0.50		+14	+6.54		+27	+11.89	

Table III. for *C*.

Arguments, *h* and τ.

In computing Heights *reverse* the signs of *C*. — Arguments, τ and *u*.

h, (*u*) Toises.	τ, in Degrees of Reaumur =								
	−16°	**−12°**	**−8°**	**−4°**	**0°**	**+4°**	**+8°**	**+12°**	**+16°**
50	0.1	0.1	0.1	0.1	0.1	0.1	0.1	0.1	0.1
100	0.2	0.2	0.2	0.2	0.3	0.3	0.3	0.3	0.3
150	0.3	0.3	0.4	0.4	0.4	0.4	0.4	0.4	0.4
200	0.4	0.5	0.5	0.5	0.5	0.5	0.6	0.6	0.6
250	0.5	0.6	0.6	0.6	0.6	0.7	0.7	0.7	0.7
300	0.7	0.7	0.7	0.7	0.8	0.8	0.8	0.9	0.9
350	0.8	0.8	0.8	0.9	0.9	0.9	1.0	1.0	1.1
400	0.9	0.9	1.0	1.0	1.0	1.1	1.1	1.2	1.2
450	1.0	1.1	1.1	1.1	1.2	1.2	1.3	1.3	1.4
500	1.1	1.2	1.2	1.3	1.3	1.4	1.4	1.5	1.5
550	1.2	1.3	1.4	1.4	1.5	1.5	1.6	1.6	1.7
600	1.4	1.4	1.5	1.6	1.6	1.7	1.7	1.8	1.9
650	1.5	1.6	1.6	1.7	1.8	1.8	1.9	1.9	2.0
700	1.6	1.7	1.8	1.8	1.9	2.0	2.0	2.1	2.2
750	1.7	1.8	1.9	2.0	2.0	2.1	2.2	2.3	2.3
800	1.9	2.0	2.0	2.1	2.2	2.3	2.4	2.4	2.5
850	2.0	2.1	2.2	2.3	2.3	2.4	2.5	2.6	2.7
900	2.1	2.2	2.3	2.4	2.5	2.6	2.7	2.8	2.9
950	2.3	2.4	2.5	2.6	2.7	2.7	2.9	3.0	3.1
1000	2.4	2.5	2.6	2.7	2.8	2.9	3.1	3.2	3.3

Table IV. for C′.

CORRECTION IN TOISES FOR THE CHANGE OF GRAVITY IN LATITUDE.

In computing Heights, reverse the signs of C′. Arguments φ and u.

Latitude. —	Latitude. +	Approximate Difference of Level, in Toises. 100	200	300	400	500	600	700	800	900	1000
0	90	0.3	0.5	0.8	1.0	1.3	1.6	1.8	2.1	2.3	2.6
5	85	0.3	0.5	0.8	1.0	1.3	1.5	1.8	2.0	2.3	2.6
10	80	0.2	0.5	0.7	1.0	1.2	1.5	1.7	2.0	2.2	2.4
15	75	0.2	0.4	0.7	0.9	1.1	1.3	1.6	1.8	2.0	2.3
20	70	0.2	0.4	0.6	0.8	1.0	1.2	1.4	1.6	1.8	2.0
25	65	0.2	0.3	0.5	0.7	0.8	1.0	1.2	1.3	1.5	1.7
30	60	0.1	0.3	0.4	0.5	0.6	0.8	0.9	1.0	1.2	1.3
35	55	0.1	0.2	0.3	0.4	0.4	0.5	0.6	0.7	0.8	0.9
36	54	0.1	0.2	0.2	0.3	0.4	0.5	0.6	0.6	0.7	0.8
37	53	0.1	0.1	0.2	0.3	0.4	0.5	0.5	0.6	0.6	0.7
38	52	0.1	0.1	0.2	0.3	0.3	0.4	0.4	0.5	0.6	0.6
39	51	0.1	0.1	0.2	0.2	0.3	0.3	0.4	0.4	0.5	0.5
40	50	0.1	0.1	0.1	0.2	0.2	0.3	0.3	0.4	0.4	0.5
41	49	0.0	0.1	0.1	0.1	0.2	0.2	0.3	0.3	0.3	0.4
42	48	0.0	0.1	0.1	0.1	0.1	0.2	0.2	0.2	0.2	0.3
43	47	0.0	0.0	0.1	0.1	0.1	0.1	0.1	0.1	0.2	0.2
44	46	0.0	0.0	0.0	0.0	0.0	0.1	0.1	0.1	0.1	0.1
45	45	0.0	0.0	0.0	0.0	0.0	0.0	0.0	0.0	0.0	0.0

Table V. for C″.

Arguments τ and T — T′. To be used only in computing Heights.

T — T′ Reaumur.	Correction for T — T′, in Toises, with the *same* sign; τ = −12°	−10°	−8°	−6°	−4°	−2°	0°	+2°	+4°	+6°
0	0.0	0.0	0.0	0.0	0.0	0.0	0.0	0.0	0.0	0.0
1	0.2	0.2	0.2	0.1	0.1	0.1	0.1	0.1	0.0	0.0
2	0.4	0.3	0.3	0.3	0.2	0.2	0.2	0.1	0.1	0.0
3	0.6	0.5	0.5	0.4	0.4	0.3	0.2	0.2	0.1	0.1
4	0.8	0.7	0.6	0.5	0.5	0.4	0.3	0.2	0.2	0.1
5	1.0	0.9	0.8	0.7	0.6	0.5	0.4	0.3	0.2	0.1
6	1.1	1.0	0.9	0.8	0.7	0.6	0.5	0.4	0.3	0.1
7	1.3	1.2	1.1	0.9	0.8	0.7	0.6	0.4	0.3	0.2
8	1.5	1.4	1.2	1.1	0.9	0.8	0.6	0.5	0.3	0.2
9	1.7	1.6	1.4	1.2	1.1	0.9	0.7	0.6	0.4	0.2
10	1.9	1.7	1.5	1.4	1.2	1.0	0.8	0.6	0.4	0.2

Correction for T — T′ with *contrary* sign; τ =

T — T′	+8°	+10°	+12°	+14°	T — T′	+8°	+10°	+12°	+14°
1	0.0	0.0	0.0	0.0	6	0.0	0.1	0.2	0.3
2	0.0	0.0	0.1	0.1	7	0.0	0.1	0.2	0.3
3	0.0	0.0	0.1	0.1	8	0.0	0.1	0.2	0.4
4	0.0	0.0	0.1	0.2	9	0.0	0.1	0.2	0.4
5	0.0	0.1	0.2	0.2	10	0.0	0.1	0.3	0.4

VI.

LAPLACE'S FORMULA FOR COMPUTING DIFFERENCES OF ELEVATION FROM BAROMETRICAL OBSERVATIONS, MODIFIED BY BABINET.

In the *Comptes Rendus de l'Académie des Sciences* for March, 1851, M. Babine proposes the following modification of Laplace's formula, the object of which is to dispense both with the use of logarithms and with tables of any kind.

Laplace's formula is,

$$z = 18393 \text{ metres } (\log H - \log h) \left[1 + \frac{2\,(T+t)}{1000}\right],$$

z being the difference of level between the two stations,
H, the height of barometer at the lower station,
h, the height of barometer at the upper station,
T, temperature of air at the lower station,
t, temperature of air at the upper station.

The two barometers are supposed to be reduced to the same temperature. The small correction for the latitude is omitted.

For elevations less than 1000 metres, and even for much greater elevations, if approximate results only are needed, the formula may be transformed into the following:

$$z = 16000 \text{ metres } \frac{H-h}{H+h} \left[1 + \frac{2\,(T+t)}{1000}\right].$$

Example 1.

Suppose,
at lower station, barometer at zero Cent. = $755^{mm.}$; temperature of air 15° Cent.
at upper station, barometer at zero Cent. = $745^{mm.}$; temperature of air 10° Cent.

$H - h = 10^{mm.}$ $\qquad T + t = 25°$ Cent.
$H + h = 1500^{mm.}$ $\qquad 2\,(T + t) = \frac{50}{1000} = .05.$

Then $\qquad z = 16000\frac{10}{1500} \times (1.05) = 112$ metres.

Laplace's formula, by Delcros's tables, would give 111.6 metres.

Example 2.

Suppose,
at lower station, barometer at zero Cent. = $730^{mm.}$; temperature of air 20° Cent.
at upper station, barometer at zero Cent. = $635^{mm.}$; temperature of air 15° Cent.

$H - h = 95^{mm.}$ $\qquad T + t = 35°$ Cent.
$H + h = 1365^{mm.}$ $\qquad 2\,(T + t) = \frac{70}{1000} = .07.$

Then $\qquad z = 16000\frac{95}{1365} \times (1.07) = 1191.5$ metres.

Laplace's formula, by Delcros's tables, would give 1191.1 metres.

For greater elevations an intermediate station may be supposed.

Babinet's formula reduced to English measures becomes,

$$z = 52494 \text{ English feet } \frac{H-h}{H+h} \left[1 + \frac{(T+t-64)}{900}\right];$$

but as, in this form, it loses the simplicity of its coefficient, it will be found, on trial, that its use requires rather more computing than the author's tables (II.), p. 38, which give more accurate results.

VII.

TABLES

FOR COMPUTING THE DIFFERENCE IN THE HEIGHTS OF TWO PLACES BY MEANS OF THE BAROMETER. — BAILY.

BAILY, in his *Astronomical Tables and Formulæ*, page **111**, gives the following final formula:

$$x = 60345.51 \{1 + .0011111 (t + t' - 64^\circ)\}$$
$$\times \log \text{ of } \left\{ \frac{\beta}{\beta'} \times \frac{1}{1 + .0001 (\tau - \tau')} \right\} \times \{1 + .002695 \cos 2 \phi\}.$$

Where ϕ = the latitude of the place,

β = the height of the barometer, τ = the temperature, Fahrenheit, of the mercury, t = the temperature, Fahrenheit, of the air,	at the lower station.
β' = the height of the barometer, τ' = the temperature, Fahrenheit, of the mercury, t' = the temperature, Fahrenheit, of the air.	at the upper station.

The numerical values assumed are as follows: —

The constant barometrical coefficient	= 60158.53 English feet.
The expansion of *moist* air for 1° Fahrenheit	= .0022222.
The expansion of mercury for 1° Fahrenheit	= .0001001.
The increase of gravitation from Equator to Poles	= .00539.
The radius of the Earth at ϕ	= 20898240 English feet.
The height of lower station assumed	= 4000 English feet.

Make A = the log of the first term, in English feet.
B = the log of $1 + .0001 (\tau - \tau')$.
C = the log of the last term.
D = $\log \beta - (\log \beta' + B)$.

Then, by the tables which follow, the logarithm of the difference of altitude in English feet

$$= A + C + \log D.$$

These tables are given as printed in Lee's *Collection of Tables and Formulæ, for the Use of the Topographical Engineers, U. S. Army*, 2d edition, pp. 84, 85.

I. Thermometers in the Open Air.

$t+t'$	A	$t+t'$	A	$t+t'$	A	$t+t'$	A	$t+t'$	A
°		°		°		°		°	
1	4.74914	37	4.76742	73	4.78497	109	4.80183	145	4.81807
2	4.74966	38	4.76792	74	4.78544	110	4.80229	146	4.81851
3	4.75017	39	4.76842	75	4.78592	111	4.80275	147	4.81895
4	4.75069	40	4.76891	76	4.78640	112	4.80321	148	4.81939
5	4.75120	41	4.76941	77	4.78688	113	4.80367	149	4.81983
6	4.75172	42	4.76990	78	4.78735	114	4.80412	150	4.82027
7	4.75223	43	4.77039	79	4.78783	115	4.80458	151	4.82071
8	4.75274	44	4.77089	80	4.78830	116	4.80504	152	4.82115
9	4.75326	45	4.77138	81	4.78878	117	4.80550	153	4.82159
10	4.75377	46	4.77187	82	4.78925	118	4.80595	154	4.82203
11	4.75428	47	4.77236	83	4.78972	119	4.80641	155	4.82247
12	4.75479	48	4.77285	84	4.79019	120	4.80687	156	4.82291
13	4.75531	49	4.77334	85	4.79066	121	4.80732	157	4.82335
14	4.75582	50	4.77383	86	4.79113	122	4.80777	158	4.82379
15	4.75633	51	4.77432	87	4.79160	123	4.80822	159	4.82423
16	4.75684	52	4.77481	88	4.79207	124	4.80867	160	4.82466
17	4.75735	53	4.77530	89	4.79254	125	4.80912	161	4.82510
18	4.75786	54	4.77579	90	4.79301	126	4.80957	162	4.82553
19	4.75837	55	4.77628	91	4.79348	127	4.81002	163	4.82596
20	4.75888	56	4.77677	92	4.79395	128	4.81047	164	4.82640
							4.81092	165	4.82683
21	4.75938	57	4.77726	93	4.79442	129			
22	4.75989	58	4.77774	94	4.79488	130	4.81137	166	4.82727
23	4.76039	59	4.77823	95	4.79535	131	4.81182	167	4.82770
24	4.76090	60	4.77871	96	4.79582	132	4.81227	168	4.82813
25	4.76140	61	4.77919	97	4.79629	133	4.81272	169	4.82857
26	4.76190	62	4.77968	98	4.79675	134	4.81317	170	4.82900
27	4.76241	63	4.78016	99	4.79722	135	4.81362	171	4.82943
28	4.76291	64	4.78065	100	4.79768	136	4.81407	172	4.82986
29	4.76342	65	4.78113	101	4.79814	137	4.81452	173	4.83030
30	4.76392	66	4.78161	102	4.79860	138	4.81496	174	4.83073
31	4.76442	67	4.78209	103	4.79907	139	4.81541	175	4.83116
32	4.76492	68	4.78257	104	4.79953	140	4.81585	176	4.83159
33	4.76542	69	4.78305	105	4.79999	141	4.81630	177	4.83201
34	4.76592	70	4.78352	106	4.80045	142	4.81675	178	4.83244
35	4.76642	71	4.78400	107	4.80091	143	4.81719	179	4.83287
36	4.76692	72	4.78449	108	4.80137	144	4.81763	180	4.83329

II. Attached Thermometer.						III. Latitude of the Place.	
$\tau - \tau'$	B	$\tau - \tau'$	B	$\tau - \tau'$	B	ϕ	C
° 0	0.00000	° 20	0.00087	° 40	0.00174	° 0	0.00117
1	0.00004	21	0.00091	41	0.00178	5	0.00115
2	0.00009	22	0.00096	42	0.00182	10	0.00110
3	0.00013	23	0.00100	43	0.00187	15	0.00100
4	0.00017	24	0.00104	44	0.00191	20	0.00090
5	0.00022	25	0.00109	45	0.00195	25	0.00075
6	0.00026	26	0.00113	46	0.00200	30	0.00058
7	0.00030	27	0.00117	47	0.00204	35	0.00040
8	0.00035	28	0.00122	48	0.00208	40	0.00020
9	0.00039	29	0.00126	49	0.00213	45	0.00000
10	0.00043	30	0.00130	50	0.00217	50	9.99980
11	0.00048	31	0.00135	51	0.00221	55	9.99960
12	0.00052	32	0.00139	52	0.00226	60	9.99942
13	0.00056	33	0.00143	53	0.00230	65	9.99925
14	0.00061	34	0.00148	54	0.00234	70	9.99910
15	0.00065	35	0.00152	55	0.00239	75	9.99900
16	0.00069	36	0.00156	56	0.00243	80	9.99890
17	0.00074	37	0.00161	57	0.00247	85	9.99885
18	0.00078	38	0.00165	58	0.00252	90	9.99883
19	0.00083	39	0.00169	59	0.00256		

Example.

	Upper Station.	Lower Station.
Thermometer in open air,	$t' = 70.4$,	$t = 77.6$.
Attached Thermometer,	$\tau' = 70.4$,	$\tau = 77.6$.
Barometer,	$\beta' = 23.66$ inches,	$\beta = 30.05$ inches.
Latitude of the place	$\phi = 21°$.	

$B = 0.00031$	$\log D = 9.01502$
$\log \beta' = 1.37401$	$C = 0.00087$
1.37432	$A = 4.81939$
$\log \beta = 1.47784$	3.83528
$D = 0.10352$	$= 6843.7$ English feet.

VIII.

TABLES

FOR COMPUTING DIFFERENCES OF ELEVATION FROM BAROMETRICAL OBSERVATIONS, BASED ON BESSEL'S FORMULA.

By E. PLANTAMOUR.

[These Tables, computed by Professor E. PLANTAMOUR, Director of the Observatory at Geneva, Switzerland, are found in Vol. XIII. Part 1, of the *Mémoires de la Société de Physique, &c. de Genève*, p. 63, together with the following explanations.]

IN No. 356 of the *Astronomische Nachrichten*, Bessel published a paper on the measurement of heights by means of the barometer, in which he deduces a formula which contains a factor depending on the humidity of the air. This formula is:

$$\log \frac{P}{P'} = \frac{(g) \cdot H' - H}{l(1 + KT)} \left[1 - a \frac{0.002561}{\sqrt{PP'}} \cdot 10^{\,0.0279712\,T - 0.0000625826\,T^2}\right],$$

where the various quantities have the following signification: —

h being the elevation of the lower station, and
h' the elevation of the upper station above the level of the sea,
a $=$ the radius of the Earth,

$H = \frac{a\,h}{a + h}$,

$H' = \frac{a\,h'}{a + h'}$;

P $=$ the weight of the atmosphere at the lower station,
P′ $=$ the weight of the atmosphere at the upper station,
the unit of weight assumed being the pressure of a column of mercury

of 336.905 Paris lines, at the temperature of the freezing point, or zero Reaumur, and under the 45th degree of latitude.

(g) = the gravity, at the level of the sea, in the mean latitude between the two places of observation.

Therefore, calling ϕ the latitude,

$(g) = 1 - 0.0026257 \cos \phi$,

l = the constant barometrical coefficient depending on the relative density of the mercury and of the air,

k = the coefficient of the dilatation of the air,

T = the mean temperature of the layer of air between the lower and upper station,

a = the fraction of saturation of the same layer.

The second term in the parenthesis, destined to take into account the aqueous vapor in the air, was obtained by assuming that the elastic force of vapor for a temperature T is represented, in unit of weight, by the expression,

$$p = 0.0067407 \times 10^{\,0.0279712\,T - 0.0000625826\,T^2}.$$

Multiplying the second member by 336.905 we find the expression of the elastic force of vapor that Laplace deduced from Dalton's experiments. Substituting, in the computation, Regnault's results, the numerical value of these coefficients is somewhat changed, and we find then

$$p = 0.0060527 \times 10^{\,0.0301975\,T - 0.000080170\,T}.$$

Bessel's tables give the difference of elevation in toises. The logarithm of the difference is obtained by the sum of four logarithms. The same form is preserved in the following tables; but the differences of elevation are given in metres.

The term due to the expansion of the air is computed in Bessel's tables for two values of the coefficient, viz. that of Gay-Lussac, 0.00375, and that of Rudberg, 0.003648; in the new tables it is only computed for that of Regnault, 0.003665.

The relative density of dry air at the freezing point, under a barometrical pressure of $0^{m}.76$, and at the 45th degree of latitude, and of mercury in the same circumstances, adopted by Bessel, is that determined by the experiments of Biot and Arago, viz. $\frac{1}{10466.8}$. The value of that constant derived from Regnault's experiments has been substituted. Regnault found the weight of a litre of dry air, at zero Centigrade, under a pressure of $0^{m}.76$, and at the latitude of Paris, to be 1.293187 grammes, which, reduced to the gravity of the 45th degree of latitude, becomes 1.292732 grammes. The weight of a litre of mercury, at zero Centigrade, he found to be 13596 grammes; the ratio is thus:

$$D = \frac{1}{10517.3},$$

or about $\frac{1}{205}$ smaller than the value adopted by Bessel. If the constant coefficient L is expressed by $L = \frac{0^{m.}.76}{D \cdot \mu}$, μ being the modulus of the common logarithms, its numerical value becomes

$$L = 18404^{m.}.8.$$

In order to reduce the formula into tables, Bessel caused it to undergo several modifications, which we have followed, introducing the values of the constants above mentioned.

Let b and b' be the heights of the barometer, expressed in the metrical scale, at the two stations; t and t', the temperatures of the mercury measured with a brass scale; we have,

$$P = \frac{b}{0^{m.}.76} \cdot (g) \cdot \left(\frac{a}{a+h}\right)^2 \frac{(1 + 0.00001879\, t)}{(1 + 0.00018018\, t)},$$

and

$$P' = \frac{b'}{0^{m.}.76} \cdot (g) \cdot \left(\frac{a}{a+h'}\right)^2 \frac{(1 + 0.00001879\, t')}{(1 + 0.00018018\, t')}.$$

Therefore,

$$\log P = \log b + \log (g) - \log 0^{m.}.76 - \frac{2\,H\,\mu}{a} - \mu\, t\, [0.00018018 - 0.00001879],$$

$$\log P' = \log b' + \log (g) - \log 0^{m.}.76 - \frac{2\,H'\,\mu}{a} - \mu\, t'\, [0.00018018 - 0.00001879].$$

If we call B, B′ the heights of the barometer reduced to the freezing point, which we obtain by making

$$\log B = \log b - t \,.\, 0.000070095\,; \qquad \log B' = \log b' - t' \,.\, 0.000070095,$$

$$\log \frac{P}{P'} = \log B - \log B' + \frac{H' - H}{7329755},$$

and with sufficient accuracy,

$$\sqrt{P\,P'} = \frac{\sqrt{B\,B'}}{0^{m.}.76}.$$

Substituting these expressions in the formula, it becomes,

$$\log B - \log B' =$$

$$\frac{(g) \,.\, H' - H}{L\,(1 + K\,T)} \left[1 - \frac{L\,(1 + K\,T)}{(g) \,.\, 7329755} - \frac{a \,.\, 0.001748}{\sqrt{B\,B'}} \,.\, 10^{\,0.0301975\,T - 0.000080170\,T^2}\right].$$

If we set instead of a the half sum $\frac{a + a}{2}$ of the fraction of saturation observed at both stations, we find, after some transformations,

$$\log B - \log B' = \frac{(g)\,(H' - H)\,(397.25 - KT)}{398.25\,.\,L\,(1 + KT)} \times$$

$$\left[1 - \frac{(a + a')\,.\,0.34807}{(397.25 - KT)\,\sqrt{BB'}}\,.\,10^{\,0.0301975\,T - 0.000080170\,T^2}\right].$$

Making further,

$$V = \frac{398.25}{397.25 - KT}\,L\,(1 + KT),$$

$$W = \frac{0.34807}{397.25 - KT}\,.\,10^{\,0.0301975\,T - 0.000080170\,T^2},$$

we shall have for the logarithm of the approximate difference of level between the two stations $H' - H$,

$$\log (H' - H) = \log [\log B - \log B']$$

$$+ \log V + \log \frac{1}{1 - W\frac{a + a'}{\sqrt{BB'}}} + \log \frac{(g)}{1}.$$

Table I. gives the values of log V and log W, both of which only depend on the temperature; the argument is the sum of the temperature of the air, τ and τ', observed at both stations, supposing $\tau + \tau' = 2\,T$.

Table II. gives the factor depending on the humidity of the air; with the argument

$$W\,.\,\log \frac{(a + a')}{\sqrt{BB'}},$$

we obtain

$$\log \frac{1}{1 - W\frac{(a + a')}{\sqrt{BB'}}} = \log V'.$$

Table III. gives the factor depending on the latitude for every degree, viz.

$$\log G' = \log \frac{1}{(g)}.$$

The logarithm of the approximate difference is thus given by the sum of four logarithms. To obtain the exact elevation, the small correction found in Table IV. must be added to the number corresponding to that logarithm. For we have, with the necessary accuracy,

$$h' - h = H' - H + \frac{H'^2}{a} - \frac{H^2}{a}.$$

Table IV. gives, for every 200 metres, the quantity $\frac{H^2}{a}$; the number in the table corresponding to $\frac{H'^2}{a}$ must be *added* to the approximate elevation; and the number corresponding to $\frac{H^2}{a}$ must be subtracted from the same.

Use of the Tables.

Reduce first the observed height of the barometer at both stations to the freezing point by means of the usual tables, or by the logarithmic formula,

$$\log B = \log b - t \, . \, 0.00007, \qquad \log B' = \log b' - t' \, 0.00007 \, ;$$

b and b' being, in fractions of metre, the observed heights at the temperatures t and t' marked by the attached thermometers ; and B and B′ the reduced height at the lower and upper station.

Take the difference of log B and log B′, and find, in the tables of the common logarithms, the logarithm of that difference, viz. log (log B — log B′) ; find also the logarithm of the product $\sqrt{B\,B'}$, or

$$\log \sqrt{B\,B'} = \frac{\log B + \log B'}{2}.$$

Make further the sum $\tau + \tau'$ of the temperature of the air at both stations, and likewise the sum of $a + a'$ of the fraction of saturation.

Then, in Table I., with argument $\tau + \tau'$, take log V and log W ; further, to log W add log $(a + a')$, and subtract log $\sqrt{B\,B'}$; and with the logarithm thus obtained as argument, take in Table II. log V′.

Table III. with the mean latitude of the stations gives log G′.

H′ — H being the approximate difference of level between the two stations, we have

$$\log (H' - H) = \log (\log B - \log B') + \log V + \log V' + \log G'.$$

The altitude of the lower station being known, we deduce from H′ — H the approximate altitude, H′, of the upper station ; h', the exact altitude, or $h' - h$, the difference of elevation, is given by the formula,

$$h' - h = H' - H + \frac{H'^2}{a} - \frac{H^2}{a}.$$

Table IV. gives the values of $\frac{H'^2}{a}$ and $\frac{H^2}{a}$ for the values of H′ or H for every 200 metres.

Example 1.

Computing the height of St. Bernard, taking Geneva, 407 metres above the level of the sea, as the lower station. The observation gives,

$B = 726.43$ millimetres	$B' = 563.64$ millimetres	
$\tau = + 8°.97$ Centigrade	$\tau' = - 1°.89$ Centig.	$\tau + \tau' = + 7°.08$
$a = 0.77$	$a' = 0.80$	$a + a' = 1.57$

$\log B = 9.86119$	$\log \sqrt{(B\,B')} = 9.8061$
$\log B' = 9.75100$	Table I. $\log W = 7.0511$
$\log B - \log B' = 0.11019$	$\log (a + a') = 0.1959$
	$\log \frac{(a + a')}{\sqrt{B\,B'}} \, . \, W = 7.4409$

D

$\log [\log B - \log B'] = \quad 9.04215$

In Table I. argt. $\tau + \tau' = +7.08$, $\log V = \quad 4.27164$

In Table II. argt. 7.4409, $\log V' = \quad 0.00120$

In Table III. argt. 46°, $\log G' = -0.00004$

$\log (H' - H) = \quad 3.31495$

$H' - H = \quad 2065.1$ metres.

In Table IV. $\frac{H'^2}{a} - \frac{H^2}{a} = + \quad 0.9$

$h' - h = \quad 2066.0$

Geneva altitude $h = \quad 407.0$

St. Bernard above the level of the sea $h' = \quad 2473.0$ metres.

Example 2.

Computing the height of Mont Blanc from the observations of Bravais and Martins, on the 29th of August, 1844, taking St. Bernard (2473.0 metres) as the lower station. The observation gives,

$B = 568.03$ millimetres	$B' = 424.29$ millimetres
$\tau = +7°.6$ Centigrade	$\tau' = -9°.1$ Centig. $\tau + \tau' = -1°.5$
$a = \quad 0.59$	$a' = \quad 0.57 \qquad a + a' = \quad 1.16$

$\log B = 9.75437$	$\log \sqrt{B B'} = -9.6910$
$\log B' = 9.62766$	Table I. $\log W = \quad 6.9183$
$\log B - \log B' = 0.12671$	$\log (a + a') = \quad 0.0648$
	$\log \frac{(a + a')}{\sqrt{B B'}} \cdot W = \quad 7.2921$

$\log [\log B - \log B'] = \quad 9.10281$

In Table I. argt. — 1°.5, $\log V = \quad 4.26483$

In Table II. argt. 7.2921, $\log V' = \quad 0.00087$

In Table III. argt. 46°, $\log G' = -0.00004$

$\log (H' - H) = \quad 3.36847$

$H' - H = \quad 2336.0$ metres.

In Table IV. { with argument 4800 $+ \frac{H'^2}{a} = + \quad 3.6$
with argument 2473 $- \frac{H^2}{a} = - \quad 0.9$ }

$h' - h = \quad 2338.7$

St. Bernard altitude, $h = \quad 2473.0$

Mont Blanc above the sea, $h' = \quad 4811.7$ metres.

TABLE I.

Argument = $\tau + \tau'$. Centigrade Degrees.

$\tau+\tau'$.	log. V.	log. W.	$\tau+\tau'$.	log. V.	log. W.	$\tau+\tau'$.	log. V.	log. W.
°			°			°		
−24	4.24644	6.5362	+15	4.27783	7.1692	+54	4.30711	7.7033
−23	4.24728	6.5441	+16	4.27861	7.1839	+55	4.30784	7.7160
−22	4.24811	6.5620	+17	4.27938	7.1985	+56	4.30856	7.7287
−21	4.24894	6.5797	+18	4.28016	7.2131	+57	4.30929	7.7413
−20	4.24977	6.5974	+19	4.28093	7.2275	+58	4.31001	7.7539
−19	4.25059	6.6157	+20	4.28170	7.2420	+59	4.31073	7.7664
−18	4.25142	6.6341	+21	4.28247	7.2564	+60	4.31145	7.7789
−17	4.25225	6.6521	+22	4.28323	7.2708	+61	4.31217	7.7914
−16	4.25307	6.6700	+23	4.28400	7.2850	+62	4.31288	7.8038
−15	4.25389	6.6879	+24	4.28477	7.2993	+63	4.31360	7.8161
−14	4.25471	6.7057	+25	4.28553	7.3135	+64	4.31432	7.8285
−13	4.25553	6.7232	+26	4.28629	7.3276	+65	4.31503	7.8407
−12	4.25634	6.7407	+27	4.28705	7.3417	+66	4.31574	7.8530
−11	4.25716	6.7581	+28	4.28781	7.3557			
−10	4.25797	6.7755	+29	4.28857	7.3697			
− 9	4.25878	6.7926	+30	4.28933	7.3837			
− 8	4.25959	6.8096	+31	4.29008	7.3975			
− 7	4.26040	6.8266	+32	4.29084	7.4114			
− 6	4.26121	6.8436	+33	4.29159	7.4252			
− 5	4.26202	6.8603	+34	4.29234	7.4389			
− 4	4.26282	6.8770	+35	4.29319	7.4526			
− 3	4.26362	6.8935	+36	4.29384	7.4662			
− 2	4.26443	6.9100	+37	4.29459	7.4798			
− 1	4.26523	6.9263	+38	4.29534	7.4933			
0	4.26603	6.9426	+39	4.29608	7.5068			
+ 1	4.26682	6.9581	+40	4.29683	7.5202			
+ 2	4.26762	6.9736	+41	4.29757	7.5336			
+ 3	4.26841	6.9889	+42	4.29831	7.5470			
+ 4	4.26921	7.0043	+43	4.29905	7.5602			
+ 5	4.27000	7.0195	+44	4.29979	7.5735			
+ 6	4.27079	7.0347	+45	4.30053	7.5867			
+ 7	4.27157	7.0499	+46	4.30127	7.5999			
+ 8	4.27236	7.0650	+47	4.30200	7.6130			
+ 9	4.27315	7.0800	+48	4.30273	7.6260			
+10	4.27393	7.0950	+49	4.30347	7.6390			
+11	4.27471	7.1099	+50	4.30420	7.6519			
+12	4.27550	7.1248	+51	4.30493	7.6648			
+13	4.27628	7.1397	+52	4.30566	7.6777			
+14	4.27705	7.1545	+53	4.30639	7.6905			
+15	4.27783	7.1692	+54	4.30711	7.7033			

TABLE IV.

Arg't. = Height.

H'. H.	$\pm$
Metres.	Metres.
200	0.01
400	0.03
600	0.06
800	0.10
1000	0.16
1200	0.23
1400	0.31
1600	0.40
1800	0.51
2000	0.63
2200	0.76
2400	0.90
2600	1.06
2800	1.23
3000	1.41
3200	1.61
3400	1.82
3600	2.04
3800	2.27
4000	2.51
4200	2.77
4400	3.04
4600	3.32
4800	3.62
5000	3.93
5200	4.25
5400	4.58
5600	4.93
5800	5.28
6000	5.65
6200	6.04
6400	6.43
6600	6.84
6800	7.26
7000	7.70
7200	8.14
7400	8.60

TABLE II.

Argument = log. W. $\frac{(\varkappa + \varkappa')}{\sqrt{B B'}}$.

Argum't.	log. V'.	Argum't.	log. V'.	Argum't.	log. V'.
6.5	0.00014	7.70	0.00218	8.09	0.00538
6.6	0.00017	7.71	0.00223	8.10	0.00550
6.7	0.00022	7.72	0.00229	8.11	0.00563
6.8	0.00027	7.73	0.00234	8.12	0.00576
6.9	0.00034	7.74	0.00239	8.13	0.00590
7.0	0.00043	7.75	0.00245	8.14	0.00604
7.1	0.00055	7.76	0.00251	8.15	0.00618
7.2	0.00069	7.77	0.00256	8.16	0.00632
7.3	0.00087	7.78	0.00262	8.17	0.00647
7.4	0.00109	7.79	0.00269	8.18	0.00662
7.41	0.00112	7.80	0.00275	8.19	0.00678
7.42	0.00114	7.81	0.00281	8.20	0.00694
7.43	0.00117	7.82	0.00288	8.21	0.00710
7.44	0.00120	7.83	0.00295	8.22	0.00727
7.45	0.00123	7.84	0.00302	8.23	0.00744
7.46	0.00125	7.85	0.00309	8.24	0.00761
7.47	0.00128	7.86	0.00316	8.25	0.00779
7.48	0.00131	7.87	0.00323	8.26	0.00798
7.49	0.00134	7.88	0.00331	8.27	0.00816
7.50	0.00138	7.89	0.00338	8.28	0.00835
7.51	0.00141	7.90	0.00346	8.29	0.00855
7.52	0.00144	7.91	0.00354	8.30	0.00875
7.53	0.00147	7.92	0.00363	8.31	0.00896
7.54	0.00151	7.93	0.00371	8.32	0.00917
7.55	0.00154	7.94	0.00380	8.33	0.00939
7.56	0.00158	7.95	0.00389	8.34	0.00961
7.57	0.00162	7.96	0.00398	8.35	0.00983
7.58	0.00165	7.97	0.00407		
7.59	0.00169	7.98	0.00417		
7.60	0.00173	7.99	0.00427		
7.61	0.00177	8.00	0.00437		
7.62	0.00181	8.01	0.00447		
7.63	0.00186	8.02	0.00457		
7.64	0.00190	8.03	0.00468		
7.65	0.00194	8.04	0.00479		
7.66	0.00199	8.05	0.00490		
7.67	0.00204	8.06	0.00502		
7.68	0.00208	8.07	0.00513		
7.69	0.00213	8.08	0.00525		
7.70	0.00218	8.09	0.00538		

TABLE III.

Argument = Latitude.

φ.	log. G'.	φ.	log. G'.
0°	+0.00114	40°	+0.00020
1	+0.00114	41	+0.00016
2	+0.00114	42	+0.00012
3	+0.00114	43	+0.00008
4	+0.00113	44	+0.00004
5	+0.00112	45	0.00000
6	+0.00112	46	−0.00004
7	+0.00111	47	−0.00008
8	+0.00110	48	−0.00012
9	+0.00109	49	−0.00016
10	+0.00107	50	−0.00020
11	+0.00106	51	−0.00024
12	+0.00104	52	−0.00028
13	+0.00103	53	−0.00031
14	+0.00101	54	−0.00035
15	+0.00099	55	−0.00039
16	+0.00097	56	−0.00043
17	+0.00095	57	−0.00046
18	+0.00092	58	−0.00050
19	+0.00090	59	−0.00054
20	+0.00087	60	−0.00057
21	+0.00085	61	−0.00060
22	+0.00082	62	−0.00064
23	+0.00079	63	−0.00067
24	+0.00076	64	−0.00070
25	+0.00073	65	−0.00073
26	+0.00070	66	−0.00076
27	+0.00067	67	−0.00079
28	+0.00064	68	−0.00082
29	+0.00060	69	−0.00085
30	+0.00057	70	−0.00087
31	+0.00054	71	−0.00090
32	+0.00050	72	−0.00092
33	+0.00046	73	−0.00094
34	+0.00043	74	−0.00097
35	+0.00039	75	−0.00099
36	+0.00035	76	−0.00101
37	+0.00031	77	−0.00102
38	+0.00028	78	−0.00104
39	+0.00024	79	−0.00106
40	+0.00020	80	−0.00107

CORRECTION

FOR THE HOUR OF THE DAY AND THE SEASON OF THE YEAR AT WHICH THE OBSERVATIONS HAVE BEEN TAKEN.

In all the preceding tables, the mean temperature of the layer of air between the two stations is assumed to be given by the half-sum of the temperatures observed at each station, or by $\frac{t+t'}{2}$. Experience, however, has proved that this assumption is not true under all meteorological circumstances, and that, not to speak of more irregular influences, the temperature expressed by $\frac{t+t'}{2}$ differs in + or — from the true mean temperature by a quantity which considerably varies with the hour of the day, the season of the year, and the elevation at which the observations are taken. The amount of the correction for the temperature of the air, as given by the various formulas, thus needs to be modified accordingly. In the absence of the data necessary for establishing the law of the decrease of heat on the vertical in the various layers of the atmosphere, at the different periods of the day and of the year, and at different latitudes, which alone would furnish the means of determining the true value of this correction in these various circumstances, the following empirical tables enable us to form a judgment of the importance of that correction.

Tables IX. and X. are taken from Berghaus, *Grundriss der Geographie*, p. 91, and in the Tables accompanying the same work, p. 71. The correction to be applied for the hour of the day at which the observations have been taken, is found by multiplying the approximate height obtained by the factors in Table IX., giving to the correction the sign of the factor. This table and the following are calculated to be used in the climate of Germany, and for elevations not much exceeding 5,000 feet. The influence of the seasons on the correction is not taken into the account; judging from Table IX., the correction may be, perhaps, too small for the summer months, and may better answer for the autumn. Using these factors, we obtain for the differences of level, in toises, placed at the head of each column, in Table X., the correction corresponding to each hour, from 6 A. M. to 10 P. M.

TABLE IX.

CORRECTION FOR THE HOUR OF THE DAY.

Hour.	Factor.	Hour.	Factor.	Hour.	Factor.
A. M. 6	+0.0075	Noon.	−0.0054	P. M. 5	−0.0011
7	+0.0050	P. M. 1	−0.0057	6	+0.0013
8	+0.0025	2	−0.0059	7	+0.0022
9	−0.0005	3	−0.0045	8	+0.0032
10	−0.0035	4	−0.0031	9	+0.0043
11	−0.0044	5	−0.0011	10	+0.0054

TABLE IX.

CORRECTION FOR THE HOUR OF THE DAY.

ARGUMENT, THE HOUR, AND THE APPROXIMATE HEIGHT IN TOISES.

Hour.	Correction, in Toises, for									Hour.
	100	**200**	**300**	**400**	**500**	**600**	**700**	**800**	**900**	
A. M. 6	+0.7	+1.5	+2.2	+3.0	+3.7	+4.5	+5.2	+6.0	+6.7	6 A. M.
7	+0.5	+1.0	+1.5	+2.0	+2.5	+3.0	+3.5	+4.0	+4.5	7
8	+0.2	+0.5	+0.7	+1.0	+1.2	+1.5	+1.8	+2.0	+2.3	8
9	−0.0	−0.1	−0.1	−0.2	−0.2	−0.3	−0.3	−0.4	−0.4	9
10	−0.3	−0.7	−1.0	−1.4	−2.1	−2.4	−2.8	−3.1	−3.5	10
11	−0.4	−0.9	−1.3	−1.8	−2.2	−2.7	−3.1	−3.6	−4.0	11
Noon.	−0.5	−1.1	−1.6	−2.2	−2.7	−3.3	−3.8	−4.4	−4.9	Noon.
P. M. 1	−0.6	−1.1	−1.7	−2.3	−2.8	−3.4	−4.0	−4.5	−5.1	1 P. M.
2	−0.6	−1.2	−1.8	−2.4	−3.0	−3.5	−4.1	−4.7	−5.3	2
3	−0.4	−0.9	−1.3	−1.8	−2.2	−2.7	−3.1	−3.6	−4.0	3
4	−0.3	−0.6	−0.9	−1.2	−1.5	−1.8	−2.1	−2.4	−2.7	4
5	−0.1	−0.2	−0.3	−0.4	−0.5	−0.6	−0.7	−0.8	−0.9	5
6	+0.1	+0.2	+0.4	+0.5	+0.5	+0.8	+0.9	+1.0	+1.1	6
7	+0.2	+0.4	+0.7	+0.9	+1.1	+1.3	+1.6	+1.8	+2.0	7
8	+0.3	+0.6	+0.9	+1.3	+1.6	+1.9	+2.2	+2.5	+2.9	8
9	+0.4	+0.8	+1.3	+1.7	+2.1	+2.6	+3.0	+3.4	+3.8	9
10	+0.5	+1.1	+1.6	+2.1	+2.7	+3.2	+2.8	+4.3	+4.8	10

Table XI. is found in the *Résumé des Observations Thermométrique et Barométriques faites à Genève et au Grand St. Bernard pendant les dix années* 1841 *à* 1850, a very elaborate paper by Professor E. Plantamour, Director of the Observatory at Geneva, published in Vol. XIII. of the *Mémoires de la Société de Physique de Genève.* The author, after having determined the difference of elevation between Geneva (407.0 metres above the level of the sea) and the Great St. Bernard, by means of the corresponding observations, made during these 10 years, and using his own tables given above, reversed the problem. Assuming the difference of level thus found, viz. 2066 metres, to be the true height of the layer of air between the two stations, and its weight being given by the barometrical observations, he deduced from these data its mean density, and from the density its mean temperature at every even hour in every month of the year. Comparing these mean temperatures with those given at the same hours by the half-sum of the temperatures taken at the upper and the lower station, he found the differences contained in Table XI., which are the corrections to be applied to the half-sums of the temperatures to obtain, in this particular case, the true mean temperatures. The second part of the table has been computed by multiplying each temperature in the first by 7.5 metres, in order to show the value of that correction in barometrical measurements.

TABLE XI.

CORRECTION TO BE APPLIED TO THE HALF-SUMS OF THE TEMPERATURES OF THE AIR, OBSERVED AT GENEVA AND AT THE GREAT ST. BERNARD, TO OBTAIN THE TRUE MEAN TEMPERATURE OF THE AIR BETWEEN THE TWO STATIONS.

	Correction, in Centigrade Degrees, for												
Hour.	Jan.	Feb.	March.	April.	May.	June.	July.	Aug.	Sept.	Oct.	Nov.	Dec.	Year.
	°	°	°	°	°	°	°	°	°	°	°	°	°
Noon.	−0.5	−1.7	−3.0	−3.9	−4.1	−4.4	−4.4	−3.8	−2.7	−1.6	−0.4	+0.7	−2.5
2	−0.2	−1.5	−2.8	−3.7	−4.0	−4.4	−4.4	−3.8	−2.6	−1.5	−0.2	+0.7	−2.3
4	+0.4	−0.6	−1.6	−2.5	−2.7	−3.4	−3.6	−2.9	−1.7	−0.7	+0.4	+1.3	−1.5
6	+1.2	+0.7	−0.2	−0.9	−1.3	−2.1	−2.2	−1.6	−0.5	+0.4	+1.3	+2.1	−0.3
8	+1.5	+1.4	+0.6	0.0	0.0	−0.6	−0.7	−0.5	+0.3	+1.3	+1.7	+2.6	+0.6
10	+1.7	+1.5	+1.2	+0.6	+0.7	+0.5	−0.1	+0.1	+0.8	+1.7	+1.8	+2.6	+1.1
Midnight.	+1.9	+1.8	+1.9	+1.3	+1.8	+1.6	+0.9	+1.2	+1.3	+2.3	+2.1	+2.5	+1.7
2	+2.0	+2.2	+2.5	+1.9	+2.2	+2.0	+1.5	+2.0	+1.9	+2.5	+2.4	+2.6	+2.2
4	+2.3	+2.5	+2.6	+1.8	+1.7	+1.4	+1.1	+1.8	+2.1	+2.5	+2.7	+2.9	+2.1
6	+2.0	+2.0	+1.7	+0.7	+0.4	+0.1	0.0	+0.7	+1.5	+1.7	+2.3	+2.9	+1.3
8	+1.5	+1.1	0.0	−1.3	−2.0	−2.2	−2.4	−1.7	−0.4	+0.6	+1.7	+2.5	−0.3
10	+0.4	−0.4	−2.0	−3.1	−3.5	−3.8	−3.7	−3.1	−2.0	−1.0	+0.3	+1.3	−1.7
Mean,	+1.2	+0.8	+0.1	−0.8	−0.9	−1.2	−1.5	−0.9	−0.2	+0.7	+1.3	+2.1	0.0

	Correction, in Metres, for												
Hour.	Jan.	Feb.	March.	April.	May.	June.	July.	Aug.	Sept.	Oct.	Nov.	Dec.	Year.
Noon.	− 3.7	−12.7	−22.5	−29.2	−30.7	−33.0	−33.0	−28.5	−20.2	−12.0	− 3.0	+ 5.2	−18.7
2	− 1.5	−11.2	−21.0	−27.7	−30.0	−33.0	−33.0	−28.5	−19.5	−11.2	− 1.5	+ 5.2	−17.2
4	+ 3.0	− 4.5	−12.0	−18.7	−20.2	−25.5	−27.0	−21.7	−12.7	− 5.2	+ 3.0	+ 9.7	−11.2
6	+ 9.0	+ 5.2	− 1.5	− 6.7	− 9.7	−15.7	−16.5	−12.0	− 3.7	+ 3.0	+ 9.7	+15.7	− 2.2
8	+11.2	+10.5	+ 4.5	0.0	0.0	− 4.5	− 5.2	− 3.7	+ 2.2	+ 9.7	+12.7	+19.5	+ 4.5
10	+12.7	+11.2	+ 9.0	+ 4.5	+ 5.2	+ 3.7	− 0.7	+ 0.7	+ 6.0	+12.7	+13.5	+19.5	+ 8.5
Midnight.	+14.5	+13.5	+14.5	+ 9.7	+13.5	+12.0	+ 6.7	+ 9.0	+ 9.7	+17.2	+15.7	+18.7	+12.7
2	+15.0	+16.5	+18.7	+14.2	+16.5	+15.0	+11.2	+15.0	+14.2	+18.7	+18.0	+19.5	+16.5
4	+17.2	+18.7	+19.5	+13.5	+12.7	+10.5	+ 8.2	+13.5	+15.7	+18.7	+20.2	+21.7	+15.7
6	+15.0	+15.0	+12.7	+ 5.2	+ 3.0	+ 0.7	0.0	+ 5.2	+11.2	+12.7	+17.2	+21.7	+ 9.7
8	+11.2	+ 8.2	0.0	− 9.7	−15.0	−16.5	−18.0	−12.7	− 3.0	+ 4.5	+12.7	+18.7	− 2.2
10	+ 3.0	− 3.0	−15.0	−23.2	−26.2	−28.5	−27.7	−23.2	−15.0	− 7.5	+ 2.2	+ 9.7	−12.7
Mean,	+9.0	+6.0	+0.7	−6.0	−6.7	−9.0	−11.2	−6.7	−1.5	+5.2	+9.7	+15.7	0.0

THE elevation of a place in the interior of a continent where regular meteorological observations are made, may be ascertained by taking the yearly means of the barometer reduced to the freezing point, and of the temperature of the air, as data for the upper station, and the yearly means of the reduced barometer and of the free thermometer at the level of the sea, as the data for the lower station. The Hypsometric Tables then will give the difference of level. As observation, however, has shown that the mean height of the barometer at the level of the sea is not the same in all latitudes, it is necessary to take for such a comparison the mean height of the barometer which belongs to the latitude of the station the elevation of which is to be computed, or that which is nearest to it.

Table XII., published by Schouw, in Poggendorf's *Annalen*, and in the *Comptes Rendus de l'Académie des Sciences*, Tom. III. p. 573, gives in Paris lines the mean height of the barometer in various latitudes. The reduction into millimetres is from Martins's French translation of Kaemtz's *Meteorology*, p. 278 ; the corresponding values in English inches, and the new stations, Savannah, Ga., Philadelphia, Pa., and Cambridge, Mass., have been added. The mean heights last mentioned have been derived from three years of observations at Savannah, by Dr. John F. Posey, from June, 1853, to June, 1856, published in the American Almanac; from four years of hourly observations at Girard College, Philadelphia, by Prof. A. D. Bache ; and from ten years of observations at Cambridge Observatory. They have been reduced to a common absolute standard and to mean tide-water at the respective places.

These mean barometric heights, corrected for the variation of gravity in latitude, according to the proposition of Poggendorf, by the formula $b = b\,45\ (1 - 0.0025935 \cos 2\,\phi)$, where b is the height of the barometer in latitude ϕ, and $b\,45$ the corresponding height at the forty-fifth degree of latitude, are found in another column. For computing the elevations, the uncorrected heights are to be used.

The mean barometric pressure, as shown by Table XIII. from Kaemtz's *Précis de Météorologie*, French translation, p. 281, is not the same in all seasons, and the monthly means differ by a quantity which also varies with the latitude. If, therefore, the height of an inland station is to be ascertained from the barometrical means of one or more months only, the computation must be made with the mean pressure in the corresponding months at the level of the sea ; or if this is not known, the yearly means taken from Table XII. must be corrected for the difference between the monthly means of the given month, or months, and the annual mean in the same latitude, as derived from the comparison of the numbers in Table XIII.

Example.

Suppose an inland station, in latitude 40° N. ; the mean barometric pressure for July is 26.30 inches, and its elevation is to be computed from it.

Table XII. gives for latitude 40°, at Philadelphia, reduced to the level of the sea, 30.053 inches. Table XIII. gives for the mean for July, at the same place, 759.80 millimetres, and for the year, 760.25 millimetres (both not reduced to the level of the sea), difference — 0.45 millimetres = — 0.017 English inches, which is to be subtracted from the annual mean, 30.053, to reduce it to the mean of July ; or

30.053 — 0.017 = 30.036. This last number is to be used in the computation, with the mean temperature of July at both stations.

Towards the tropical regions, the irregular or non-periodic variations of the barometer, which in high and middle latitudes are so considerable as to render simultaneous observations indispensable for the measurement of heights, gradually decrease and nearly cease to exist, while the monthly and daily periodic variations, which are small in high latitudes, considerably increase. Within the tropics, therefore, the oscillations of the barometer being far more uniform, observations made during a short period of time, or even single observations, may be used for computing heights, without corresponding observations, by referring them to the mean pressure at the level of the sea as to a constant, provided this last has been corrected for the monthly and daily periodic variation at the place.

Table XIII. furnishes the means of applying the correction for the monthly variation, as described above. Table XIV., which gives the mean height of the barometer at all hours of the day in various latitudes, enables the observer to correct the data according to the hour at which the observations have been taken. This table is from Kaemtz's *Vorlesungen über Meteorologie*, French translation, p. 249. The column Bossekop is from the observations of the French Scientific Expedition in the North; the column Philadelphia, from the observations at Girard College, has been added.

The correction for the hourly variation is found by taking the difference between the mean of the hour of observation and the daily mean, and correcting accordingly, with due regard to the signs, either the yearly mean at the sea level, or the observation at the upper station.

Example.

The barometer at Caracas, latitude 10° 30′ N., on the 20th of August, at 4 o'clock P. M., reads 680.57 millimetres.

In Table XII. the mean height of the barometer at La Guayra, lat. 10° N.	=	760.17 millimetres,
By Table XIII. we find for August a correction . .	=	— 2.95
Mean barometer in August	=	757.22
In Table XIV. daily mean — mean at 4 P. M. gives for 4 P. M. a correction	=	— 1.17
Mean barometer at La Guayra in August, at 4 P. M.	=	756.05 millimetres,

which is the number to be used for the computation of the height of Caracas. In this case, however, the monthly correction, being derived from a higher latitude, may be too small. Both corrections can of course be applied, with contrary signs, to the observation at Caracas, leaving then the mean height at the level of the sea as a constant.

TABLE XII.

MEAN HEIGHT OF THE BAROMETER,

IN VARIOUS LATITUDES, REDUCED TO THE LEVEL OF THE SEA, AND TO THE FREEZING POINT.

Places.	Latitude.	In Millimetres.		In English Inches.		In Paris Lines.	
		Observed.	Corrected for Gravity.	Observed.	Corrected for Gravity.	Observed.	Corrected for Gravity.
	° ′						
Cape of Good Hope,	33 S.	763.01	762.20	30.041	30.008	338.24	337.88
Rio Janeiro, Brazil,	23 S.	764.03	762.65	30.080	30.026	338.69	338.08
Christiansborg, Guinea,	5 30N.	760.10	758.16	29.925	29.850	336.95	336.09
La Guayra, Venezuela,	10	760.17	758.32	29.928	29.855	336.98	336.16
St. Thomas, W. Indies,	19	760.51	758.95	29.942	29.881	337.13	336.44
Macao, China,	23	762.99	761.61	30.040	29.986	338.23	337.62
Teneriffe, Canary Isles,	28	764.21	763.10	30.087	30.044	338.77	338.28
Savannah, Georgia,	32	764.59	763.74	30.102	30.070	338.93	338.57
Funchal, Madeira,	32 30	765.18	764.34	30.126	30.093	339.20	338.83
Tripoli, Northern Africa,	33	767.41	766.60	30.214	30.182	340.19	339.83
Palermo, Sicily,	38	762.95	762.47	30.038	30.019	338.21	338.00
Philadelphia, Penn.	40	763.35	763.00	30.053	30.040	338.38	338.23
Naples, Italy,	41	762.34	762.06	30.014	30.003	337.94	337.82
Cambridge, Mass.	42	762.44	762.24	30.018	30.010	337.99	337.90
Florence, Italy,	43 30	761.93	761.81	29.997	29.993	337.76	337.71
Avignon, France,	44	762.02	761.95	30.001	29.998	337.80	337.77
Bologna, Italy,	44 30	762.18	762.13	30.007	30.005	337.87	337.85
Padua, Italy,	45	762.18	762.18	30.007	30.007	337.87	337.87
Paris, France,	49	761.41	761.68	29.978	29.988	337.53	337.65
London, England,	51 30	760.96	761.41	29.960	29.978	337.33	337.53
Altona, Denmark,	53 30	760.42	761.01	29.938	29.961	337.09	337.35
Dantzig, Prussia,	54 30	760.10	760.76	29.925	29.952	336.95	337.24
Königsberg, Prussia,	54 30	760.49	761.14	29.941	29.967	337.12	337.41
Apenrade, Denmark,	55	759.58	760.71	29.906	29.950	336.72	337.22
Edinburgh, Scotland,	56	758.25	759.00	29.853	29.882	336.13	336.46
Christiania, Norway,	60	758.64	759.63	29.868	29.908	336.30	336.74
Hardanger, Norway,	60	756.94	757.04	29.801	29.841	335.55	335.99
Bergen, Norway,	60	757.01	758.00	29.804	29.844	335.58	336.02
Reikiavig, Iceland,	64	752.00	753.20	29.607	29.654	333.36	333.89
Godthaab, S. Greenland,	64	751.94	753.13	29.605	29.651	333.33	333.86
Eyafiord, Iceland,	66	753.58	754.89	29.669	29.721	334.06	334.64
Godhavn, Disco, Greenl.	68	753.76	755.16	29.677	29.731	334.14	334.76
Upernavik, N. Greenl.	73	755.18	756.80	29.732	29.796	334.77	335.49
Melville Isl., Arct. Amer.	74 30	757.08	758.75	29.807	29.872	335.61	336.35
Spitzbergen,	75 30	756.76	758.48	29.794	29.862	335.47	336.23

XIII. MEAN HEIGHT OF THE BAROMETER, IN ALL MONTHS OF THE YEAR, IN VARIOUS LATITUDES.

Not reduced to the Level of the Sea.

Places,	HAVANA.	CALCUTTA.	MACAO.	CAIRO.	SAVANNAH.	PHILADELPHIA.	CAMBRIDGE.	PARIS.	ST. PETERSBURG.
Latitude,	23° 9′	22° 33′	22° 11′	30° 2′	32° 5′	39° 58′	42° 23′	48° 50′	59° 56′
Jan.	765.24	764.57	767.93	762.40	762.80	760.97	761.37	758.86	762.54
Feb.	760.15	758.86	767.01	"	763.76	759.63	760.90	759.09	763.10
March,	760.98	756.24	766.08	759.43	763.05	760.51	759.09	756.33	760.76
April,	759.58	753.83	761.93	760.10	763.10	760.05	759.37	755.18	761.19
May,	758.19	750.81	761.64	758.23	763.39	759.09	759.63	755.61	760.94
June,	760.67	748.10	757.31	754.42	764.37	759.22	758.91	757.28	759.83
July,	760.67	747.54	757.91	753.90	764.02	759.80	760.34	756.52	758.25
Aug.	757.33	748.53	757.91	754.06	765.54	760.54	761.11	756.74	759.94
Sept.	757.46	751.83	762.22	756.70	763.36	761.25	761.83	756.61	761.19
Oct.	758.19	755.25	763.37	759.70	763.13	760.68	761.07	754.42	760.82
Nov.	761.25	758.37	766.17	760.76	763.41	760.49	760.85	755.75	758.05
Dec.	763.62	760.59	768.65	761.82	761.12	760.82	760.80	755.09	760.22
Year,	760.28	754.54	763.18	758.32	763.41	760.25	760.44	756.46	760.57

XIV. MEAN HEIGHT OF THE BAROMETER, AT ALL HOURS OF THE DAY, IN VARIOUS LATITUDES.

Not reduced to the Level of the Sea.

Places,	PACIFIC OCEAN.	CUMANA.	LA GUAYRA.	CALCUTTA.	PHILADELPHIA.	PADUA.	HALLE	ST. PETERSBURG	BOSSEKOP.
Latitude,	0° 0′	10° 28′N.	10° 36′N.	22° 35′N.	39° 58′N.	45° 24′N.	51° 29′N.	59° 56′N.	69° 58′N.
Observers,	Horner.	Humboldt.	Boussingault.	Balfour.	Bache.	Ciminello.	Kaemtz.	Kupffer.	Bravais.
	Millim.	Millim.	Millim.	Millim.	Millim.	Millim.	Millim.	Millim.	Millim.
Midnight,	752.47	756.86	759.64	758.80	760.49	757.01	753.23	759.35	754.90
1	752.20	756.53	759.34	758.62	760.46	756.90	753.14	"	"
2	751.77	756.21	759.05	758.57	760.41	756.84	753.05	759.32	754.79
3	751.63	755.89	758.81	758.49	760.34	756.78	752.99	"	"
4	751.32	755.66	758.68	758.47	760.39	756.74	752.99	759.32	754.70
5	751.65	755.79	758.85	758.44	760.49	756.75	753.34	"	"
6	751.95	756.18	759.32	758.68	760.75	756.79	753.12	759.39	754.68
7	752.48	756.58	759.94	759.16	761.00	756.89	753.24	"	"
8	752.95	756.98	760.50	759.88	761.15	757.01	753.37	759.49	754.75
9	753.16	757.31	759.63	760.11	761.22	757.08	753.44	"	"
10	753.15	757.32	760.50	760.19	761.17	757.14	753.46	759.51	754.96
11	752.80	757.01	759.99	760.09	760.97	757.07	753.40	"	"
Noon,	752.35	756.57	759.41	759.61	760.56	757.02	753.29	759.47	755.01
1	751.87	755.99	758.91	759.22	760.13	756.85	753.11	"	"
2	751.55	755.47	758.41	758.39	759.83	756.67	752.99	759.38	754.96
3	751.15	755.14	758.12	758.12	759.65	756.54	752.89	"	"
4	751.02	754.96	758.05	757.91	759.65	756.47	753.84	759.32	754.82
5	751.31	755.14	758.10	757.93	759.70	756.46	752.86	"	"
6	751.71	755.41	758.40	758.01	759.85	756.50	752.91	759.31	754.87
7	751.93	755.81	758.90	758.02	760.08	756.63	753.02	"	"
8	752.35	756.21	759.19	758.54	760.31	756.79	753.14	759.32	754.89
9	752.74	756.59	759.69	759.24	760.49	756.92	753.24	"	"
10	752.85	756.87	759.93	759.33	760.59	757.02	753.31	759.36	754.92
11	752.86	757.15	759.98	759.09	760.72	757.02	753.29	"	"
Mean,	752.13	756.33	759.22	758.87	760.43	756.83	753.19	759.38	754.85

TABLE XIV. shows that, after all irregular variations of the barometer have been eliminated, there remains a double period of rise and fall within the twenty-four hours, and that the amplitude of these daily oscillations is greatest within the tropics, and goes on diminishing towards the polar regions.

According to Kaemtz, the mean time of the daily maxima and minima, or the mean tropic hours for the northern hemisphere, are as follows:—

	h.
The minimum of the afternoon is reached at	4.05 P. M.
The maximum of the evening is reached at	10.11 P. M.
The minimum of the night is reached at	3.45 A. M.
The maximum of the morning is reached at	9.37 A. M.

Even in temperate and high latitudes these diurnal variations, though small, must be taken into account, if great accuracy is required, in reducing corresponding observations made at a somewhat different hour to the time of the observation at the station the height of which is to be determined. But in so doing, it must be remembered that the times of the minima and maxima change with the seasons, as is shown by Table XV. from Kaemtz, p. 251 of the French translation.

XV. TROPIC HOURS OF THE DAILY VARIATION OF THE BAROMETER AT HALLE. LAT. 51° 30′ N.

Month.	Minimum, P. M.	Maximum, P. M.	Minimum, A. M.	Maximum, A. M.	Month.	Minimum, P. M.	Maximum, P. M.	Minimum, A. M.	Maximum, A. M.
	h.	h.	h.	h.		h.	h.	h.	h.
Jan.	2.81	9.17	4.91	9.91	July,	5.21	11.04	3.04	8.73
Feb.	3.43	9.46	3.86	9.66	Aug.	4.86	10.66	3.06	8.96
March,	3.82	9.80	3.87	10.10	Sept.	4.55	10.45	3.45	9.71
April,	4.46	10.27	3.53	9.53	Oct.	4.17	10.24	3.97	10.07
May,	5.43	10.93	3.03	9.13	Nov.	3.52	9.85	4.68	10.08
June,	5.20	10.93	2.83	8.73	Dec.	3.15	9.11	4.91	10.18

This shifting of the times of maxima and minima with the seasons diminishes with the latitude, and tends to disappear towards the equator, with the inequality of the days and nights. The elevation above the level of the sea also causes a change in the tropic hours of the daily variation which is not yet sufficiently studied.

Table XIV. gives evidence that the amplitude of the hourly oscillation is greatest under the equator, and gradually decreases towards the pole. Kaemtz computes its mean value in various latitudes and at the level of the sea, as follows:—

XV′. AMPLITUDE OF DAILY VARIATIONS IN VARIOUS LATITUDES.

Latitude.	Variation.	Latitude.	Variation.	Latitude.	Variation.	Latitude	Variation.
° ′	Millim.	° ′	Millim.	° ′	Millim.	° ′	Millim.
0 0	2.28	23 55	1.80	39 4	1.13	52 33	0.45
5 26 N.	2.26	29 28	1.58	43 34	0.90	57 17	0.23
17 52	2.03	34 26	1.35	48 1	0.67	62 25	0.00

The amplitude also decreases with the elevation, at least in our latitudes; it was found to be on the Faulhorn, in Switzerland, 9000 feet above the sea level, 0.27 millimetres, while it was 0.90 millimetres at Geneva.

TABLES

FOR REDUCING BAROMETRICAL OBSERVATIONS TO THE LEVEL OF THE SEA, OR TO ANOTHER LEVEL.

To reduce barometric means taken at a given elevation to the height they would have if taken at the level of the sea, or barometric observations made at different elevations to a common level, in order to eliminate the influence of altitude in the comparison of barometric pressures, is a problem, the solution of which is often needed in meteorology.

For a complete and accurate reduction, embracing all cases, Tables IV. and V., by Dippe, given above, pages 54 *et seq.*, may be used. But when the difference of height between the two stations, or above the sea-level, does not exceed a few hundred feet, the small tables XVI. to XIX., in three different scales, will be found more convenient.

Tables XVI. and XVII. have been computed from the constants of Laplace's formula, the barometric coefficient, including the correction for the decrease of gravity on a vertical, being respectively 60,345.51 English feet and 56,621.83 Paris feet; and the coefficient for expansion of moist air 0.004.

In Table XVIII. the coefficient 18,420 metres, deduced from Regnault's experiments (see *Proceedings of the Amer. Assoc. for Adv. of Science*, 1857), and his coefficient for expansion of dry air, 0.003665, increased to 0.0039, in order to include the action of moisture, have been used.

Use of the Tables.

The correction for reducing the barometer to the level of the sea is found by the formula

$$C = \frac{f}{N} \times \frac{h'}{h},$$

where C is the correction required; f, the elevation of the station; N, the number in the tables; h', the reading of the barometer; h, the normal height of barometer at the sea-level.

Example.

At Cambridge Observatory, Massachusetts, at 71.34 English feet above mean tide, the mean barometer is = 29.939 inches; the mean temperature 47°.3 Fahrenheit; what would be the height at the level of the sea?

In Table XVI. we take for 47°.3 = 90.49, or, in order to get the correction in a fraction of an inch, 904.9.

Then

$$C = \frac{71.34}{904.9} \times \frac{29.939}{30} = 0.079, \text{ correction required};$$

and

$$29.939 + 0.079 = 30.018 \text{ inches, height of the barometer at the level of the sea.}$$

It will be seen that the quantity represented by the second member can be neglected without causing a sensible error in the correction. In this case the error does not amount to .001; it scarcely would reach .002 for 250 feet of elevation; so that the reduction can be made in most cases by a simple division; viz. $\frac{f}{N}$.

XVI. HEIGHT, IN ENGLISH FEET, OF A COLUMN OF AIR CORRESPONDING TO A TENTH OF AN ENGLISH INCH IN THE BAROMETER, AT TEMPERATURES BETWEEN 32° AND 100° FAHRENHEIT,

The Barometric Pressure at the Lower Station being = 30 English Inches.

Temperature of Air, Fahren.	Height in English Feet.	Temperature of Air, Fahren.	Height in English Feet.	Temperature of Air, Fahren.	Height in English Feet.	Temperature of Air, Fahren.	Height in English Feet.	Temperature of Air, Fahren.	Height in English Feet.
32°	87.51	46°	90.23	60°	92.95	74°	95.67	87°	98.20
33	87.70	47	90.42	61	93.15	75	95.87	88	98.40
34	87.90	48	90.62	62	93.34	76	96.06	89	98.59
35	88.09	49	90.81	63	93.53	77	96.26	90	98.79
36	88.28	50	91.01	64	93.73	78	96.45	91	98.98
37	88.48	51	91.20	65	93.92	79	96.65	92	99.17
38	88.67	52	91.40	66	94.12	80	96.84	93	99.37
39	88.87	53	91.59	67	94.31	81	97.04	94	99.56
40	89.06	54	91.78	68	94.51	82	97.23	95	99.76
41	89.26	55	91.98	69	94.70	83	97.42	96	99.95
42	89.45	56	92.17	70	94.90	84	97.62	97	100.15
43	89.65	57	92.37	71	95.09	85	97.81	98	100.34
44	89.84	58	92.56	72	95.29	86	98.01	99	100.54
45	90.03	59	92.76	73	95.48	87	98.20	100	100.73

XVII. HEIGHT, IN FRENCH FEET, OF A COLUMN OF AIR CORRESPONDING TO A PARIS LINE IN THE BAROMETER, AT TEMPERATURES OF THE AIR BETWEEN 0° AND 34° REAUMUR,

The Barometric Pressure at the Lower Station being = 337 Paris Lines.

Temperature of Air, Reaumur.	Height in French Feet.	Temperature of Air, Reaumur.	Height in French Feet.	Temperature of Air, Reaumur.	Height in French Feet.	Temperature of Air, Reaumur.	Height in French Feet.	Temperature of Air, Reaumur.	Height in French Feet.
0°	73.08	7°	75.63	14°	78.19	21°	80.75	28°	83.31
1	73.44	8	76.00	15	78.56	22	81.11	29	83.67
2	73.81	9	76.36	16	78.92	23	81.48	30	84.04
3	74.17	10	76.73	17	79.29	24	81.85	31	84.40
4	74.54	11	77.10	18	79.65	25	82.21	32	84.77
5	74.90	12	77.46	19	80.02	26	82.58	33	85.13
6	75.27	13	77.83	20	80.38	27	82.94	34	85.50

XVIII. HEIGHT, IN METRES, OF A COLUMN OF AIR CORRESPONDING TO A MILLIMETRE IN THE BAROMETER, AT TEMPERATURES BETWEEN 0° AND 39° CENTIGRADE,

The Barometric Pressure at the Lower Station being = 760 Millimetres.

Temperature of Air, Centigr	Height in Metres.	Temperature of Air, Centigr.	Height in Metres.	Temperature of Air, Centigr.	Height in Metres.	Temperature of Air, Centigr.	Height in Metres.	Temperature of Air, Centigr.	Height in Metres.
0°	10.54	8°	10.86	16°	11.19	24°	11.52	32°	11.85
1	10.58	9	10.91	17	11.23	25	11.56	33	11.89
2	10.62	10	10.95	18	11.28	26	11.60	34	11.93
3	10.66	11	10.99	19	11.32	27	11.64	35	11.97
4	10.70	12	11.03	20	11.36	28	11.69	36	12.01
5	10.74	13	11.07	21	11.40	29	11.73	37	12.06
6	10.78	14	11.11	22	11.44	30	11.77	38	12.10
7	10.82	15	11.15	23	11.48	31	11.81	39	12.14

Table XIX. gives, in metrical measure, the values of a millimetre in the barometer at different elevations and Centigrade temperatures. The values are derived from Laplace's constants, as in Tables XVI. and XVII.

This table may be used, as the preceding ones, for reducing barometrical observations to the level of the sea, and also to any other level by a similar process.

Example.

Suppose the barometer to read 700 millimetres at the altitude of 750 metres, the temperature of air being = 16° Centigrade ; what would be the reading at a station lower by 350 metres, assuming the temperature of the air downwards to increase at the rate of 1° Centigrade for 185 metres ?

The temperature of air at lower station will be 16° + 1°.9 = 17°.9

The approximate height of barometer about 73 centimetres.

Then, in Table XIX. we find for 16° and 70 centimetres,	12.15
" " " for 17°.9 and 73 centimetres,	11.73
Mean	11.94

And

$$\frac{350}{11.94} = 29.31, \text{ or barometer at lower station } 700 + 29.31 = 729.31 \text{ millimetres.}$$

Delcros's tables, with these data, would give for the difference of level 349.76, instead of 350 metres ; the corresponding error in the height of the barometrical column does not exceed 0.08 millimetre, and thus remains within the limits of error which may be expected in an ordinary observation.

The principal object of this table, however, is to furnish the scientific traveller with the means of readily computing on the spot approximate differences of level, by simply multiplying the difference between the readings of the barometer at each station by the half sum of the numbers in the table corresponding to the data given by the observations.

Example.

Suppose the barometer at the lower station to read 732.5, and at the upper station 703.2 millimetres ; the temperature of the air being respectively 18° and 16° Centigrade.

The difference of the barometers, supposed to be reduced to the same temperature, is 29.3 millimetres.

Then, Table XIX. gives for 18° Centigrade and 73 centimetres,	11.73
" " for 16° Centigrade and 70 centimetres,	12.15
Half sum, or mean,	11.94

And, 29.3 × 11.94 = 349.8 metres = difference of level required.

By the large tables of Delcros, we find for the same data 350.1 metres.

This table can be considered as a complement to Delcros's tables, and may save the traveller the trouble of carrying the larger tables.

A similar table in English measures is found above, at the end of the author's larger tables (II. Table VI.), page 48 of this series, the use of which is explained by the examples just given.

XIX. HEIGHT, IN METRES, OF A COLUMN OF AIR, CORRESPONDING TO A MILLIMETRE IN THE BAROMETER, AT DIFFERENT TEMPERATURES AND ELEVATIONS.

Temperature of Air, Centig.	Barometer at the Lower Station, Reading in Centimetres.									
	76	**75**	**74**	**73**	**72**	**71**	**70**	**69**	**68**	**67**
°	Metres.	Metres.	Metres.	Metres.	Metres.	Metres.	Metres.	Metres.	Metres.	Metres.
0	10.52	10.66	10.80	10.94	11.10	11.26	11.42	11.59	11.75	11.93
2	10.60	10.74	10.89	11.03	11.19	11.35	11.51	11.68	11.85	12.03
4	10.69	10.83	10.97	11.12	11.28	11.44	11.60	11.77	11.94	12.13
6	10.77	10.91	11.06	11.20	11.37	11.53	11.69	11.86	12.04	12.22
8	10.85	11.00	11.15	11.29	11.46	11.62	11.78	11.96	12.13	12.32
10	10.94	11.08	11.23	11.38	11.55	11.71	11.87	12.05	12.22	12.41
12	11.02	11.17	11.32	11.47	11.63	11.80	11.97	12.14	12.32	12.51
14	11.11	11.25	11.41	11.55	11.72	11.89	12.06	12.23	12.41	12.60
16	11.19	11.34	11.49	11.64	11.81	11.98	12.15	12.33	12.51	12.70
18	11.27	11.43	11.58	11.73	11.90	12.07	12.24	12.42	12.60	12.79
20	11.36	11.51	11.67	11.82	11.99	12.16	12.33	12.51	12.69	12.89
22	11.44	11.60	11.75	11.90	12.08	12.25	12.42	12.61	12.79	12.99
24	11.53	11.68	11.84	11.99	12.17	12.34	12.51	12.70	12.88	13.08
26	11.61	11.77	11.93	12.08	12.26	12.43	12.61	12.79	12.98	13.18
28	11.70	11.85	12.01	12.17	12.35	12.52	12.70	12.88	13.07	13.27
30	11.78	11.94	12.10	12.25	12.43	12.61	12.79	12.98	13.16	13.37
32	11.86	12.02	12.18	12.34	12.52	12.70	12.88	13.07	13.26	13.46
34	11.95	12.11	12.27	12.43	12.61	12.79	12.97	13.16	13.35	13.56
36	12.03	12.19	12.36	12.52	12.70	12.88	13.06	13.25	13.45	13.65
38	12.12	12.28	12.44	12.60	12.79	12.97	13.15	13.35	13.54	13.75

Temperature of Air, Centig.	Barometer in Centimetres.									
	66	**65**	**64**	**63**	**62**	**61**	**60**	**59**	**58**	**57**
°	Metres.	Metres.	Metres.	Metres.	Metres.	Metres.	Metres.	Metres.	Metres.	Metres.
0	12.11	12.30	12.49	12.69	12.89	13.10	13.32	13.55	13.78	14.03
2	12.21	12.40	12.59	12.79	13.00	13.21	13.43	13.66	13.89	14.14
4	12.31	12.50	12.69	12.89	13.10	13.31	13.54	13.77	14.00	14.25
6	12.40	12.60	12.79	13.00	13.20	13.42	13.64	13.88	14.11	14.36
8	12.50	12.69	12.89	13.10	13.31	13.52	13.75	13.98	14.22	14.47
10	12.60	12.79	12.99	13.20	13.41	13.63	13.86	14.09	14.34	14.59
12	12.69	12.89	13.09	13.30	13.51	13.73	13.96	14.20	14.45	14.70
14	12.79	12.99	13.19	13.40	13.62	13.84	14.07	14.31	14.56	14.81
16	12.89	13.09	13.29	13.50	13.72	13.94	14.18	14.42	14.67	14.92
18	12.98	13.19	13.39	13.61	13.82	14.05	14.28	14.53	14.78	15.04
20	13.08	13.28	13.49	13.71	13.93	14.15	14.39	14.63	14.89	15.15
22	13.18	13.38	13.59	13.81	14.03	14.26	14.50	14.74	15.00	15.26
24	13.27	13.48	13.69	13.91	14.13	14.36	14.60	14.85	15.11	15.37
26	13.37	13.58	13.79	14.01	14.24	14.47	14.71	14.96	15.22	15.48
28	13.47	13.68	13.89	14.11	14.34	14.57	14.82	15.07	15.33	15.60
30	13.57	13.78	13.99	14.22	14.44	14.68	14.92	15.18	15.44	15.71
32	13.66	13.87	14.09	14.32	14.55	14.78	15.03	15.28	15.55	15.82
34	13.76	13.97	14.19	14.44	14.65	14.89	15.14	15.39	15.66	15.93
36	13.86	14.07	14.29	14.52	14.75	14.99	15.24	15.50	15.77	16.05

When the Barometrical means to be used have been derived from observations taken at such hours of the day as, if combined, do not give the true mean pressure, they must be reduced to the true means by using the Tables XX. and XXI. These tables give the corrections to be applied to the hourly means, in each month, for reducing them to the means which would have been given by observations made at each of the twenty-four hours. The correction for any given set of hours is found by taking the mean of the corrections due to each of the combined hours, paying due attention to the signs. Table XX. has been computed from the hourly observations made under the superintendence of Professor A. D. Bache, at Girard College, Philadelphia. Table XXI. is from the Greenwich Observations, by Glaisher.

XX.

North America. — Philadelphia. *Lat.* 39° 58′ N. *Long.* 75° 11′ W. *Greenw.*

Corrections to be applied to the Means of the Hours of Observation to obtain the true Mean Barometric Pressure of the respective Days, Months, and of the Year.

Barometer in English Inches.

Hour.	Jan.	Feb.	March.	April.	May.	June.	July.	Aug.	Sept.	Oct.	Nov.	Dec.	Year.
	Inch.	Inch.	Inch.	Inch.	Inch	Inch.	Inch.	Inch.	Inch.	Inch.	Inch.	Inch.	Inch.
Midnight.	+.002	−.009	−.007	−.004	−.002	+.003	−.007	−.003	−.002	+.007	+.003	−.010	−.0024
1	+.001	−.007	−.002	−.001	+.003	+.007	+.001	−.001	+.005	+.007	+.007	−.011	+.0007
2	−.007	−.003	−.001	+.006	+.007	+.010	+.004	+.004	+.010	+.011	+.011	−.016	+.0030
3	−.008	+.002	+.009	+.005	+.007	+.007	+.003	+.005	+.009	+.011	+.007	−.014	+.0036
4	−.003	+.003	+.009	+.002	+.003	+.002	.000	+.004	+.005	+.007	+.003	−.010	+.0038
5	−.003	.000	+.002	−.007	−.006	−.007	−.010	−.005	−.006	−.003	−.006	−.008	−.0050
6	−.009	−.004	−.011	−.020	−.019	−.022	−.019	−.017	−.016	−.012	−.012	−.015	−.0147
7	−.021	−.013	−.020	−.029	−.026	−.024	−.025	−.023	−.023	−.021	−.019	−.023	−.0222
8	−.032	−.023	−.028	−.034	−.031	−.029	−.028	−.026	−.029	−.030	−.028	−.029	−.0290
9	−.040	−.026	−.028	−.035	−.028	−.027	−.027	−.033	−.031	−.029	−.034	−.030	−.0307
10	−.041	−.026	−.025	−.033	−.024	−.025	−.026	−.030	−.029	−.026	−.038	−.032	−.0296
11	−.023	−.019	−.016	−.023	−.018	−.019	−.019	−.022	−.021	−.014	−.017	−.011	−.0185
Noon.	+.006	−.004	−.002	−.008	−.006	−.010	−.012	−.012	−.009	+.001	+.006	+.005	−.0037
1	+.028	+.017	+.014	+.006	+.005	.000	.000	.000	+.005	+.006	+.023	+.024	+.0107
2	+.037	+.032	+.031	+.021	+.017	+.011	+.011	+.012	+.020	+.028	+.033	+.034	+.0240
3	+.034	+.034	+.034	+.034	+.028	+.019	+.020	+.022	+.024	+.028	+.033	+.034	+.0287
4	+.031	+.032	+.034	+.042	+.032	+.027	+.027	+.027	+.030	+.028	+.027	+.030	+.0306
5	+.024	+.024	+.025	+.036	+.034	+.030	+.028	+.029	+.027	+.021	+.018	+.026	+.0267
6	+.015	+.014	+.016	+.031	+.027	+.023	+.028	+.028	+.023	+.012	+.005	+.021	+.0202
7	+.003	+.006	+.007	+.022	+.016	+.018	+.021	+.018	+.016	+.001	−.002	+.018	+.0123
8	+.003	.000	−.003	+.009	+.002	+.010	+.014	+.008	+.007	−.009	−.006	+.013	+.0040
9	−.002	−.008	−.010	+.001	−.010	.000	+.003	+.003	−.001	−.013	−.007	+.012	−.0027
10	−.003	−.012	−.011	−.003	−.018	−.003	−.004	−.001	−.005	−.016	−.010	+.008	−.0065
11	+.002	−.011	−.017	−.010	−.019	−.005	−.002	−.002	−.004	−.009	−.003	+.005	−.0064
6, 2, 10	+.008	+.005	+.003	−.001	−.007	−.005	−.004	−.002	.000	.000	+.004	+.009	+ .001
7, 2, 9	+.008	+.004	.000	−.002	−.006	−.004	−.004	−.003	−.001	−.002	+.002	+.008	.000
9, 12, 3, 9	.000	−.001	−.001	−.002	−.004	−.004	−.004	−.004	−.004	−.003	−.001	+.005	− .002

XXI. England. — Greenwich. *Lat.* 51° 29′ N.; *Long.* 0° 0′.

Corrections to be applied to the Means of the Hours of Observation, or Sets of Hours, to obtain the true Mean *Barometric Pressure* for the respective Months. — Glaisher.

English Inches.

Hours.	Jan.	Feb.	March	April.	May.	June.	July.	Aug.	Sept.	Oct.	Nov.	Dec.	Mean.
	Inch.	Inch.	Inch.	Inch.	Inch.	Inch.	Inch	Inch.	Inch.	Inch.	Inch	Inch.	Inch.
Midn. . .	.000	–.001	–.002	–.008	–.005	.000	–.006	–.010	–.005	–.005	–.011	–.004	–.005
1	.001	.004	.013	.000	.002	.004	.000	.000	.000	.004	–.005	.001	.002
2	.002	.008	.020	.007	.004	.005	.003	.007	.005	.010	.003	.006	.007
3	.005	.012	.023	.010	.005	.004	.005	.011	.010	.015	.008	.010	.009
4	.011	.014	.022	.011	.005	.001	.005	.014	.012	.020	.013	.012	.012
5	.015	.015	.019	.011	.006	–.002	.006	.011	.014	.022	.016	.014	.012
6	.015	.012	.012	.006	.006	–.006	.002	.005	.010	.018	.015	.011	.009
7	.010	.007	.005	–.003	.006	–.010	–.004	.000	.001	.008	.010	.006	.003
8	.003	.000	–.004	–.008	.003	–.012	–.008	–.007	–.006	–.003	.003	.004	–.003
9	–.008	–.008	–.010	–.011	–.007	–.012	–.010	–.008	–.011	–.009	–.005	–.010	–.009
10	–.010	–.015	–.015	–.014	–.009	–.011	–.010	–.009	–.013	–.014	–.007	–.015	–.012
11	–.014	–.016	–.015	–.011	–.006	–.009	–.009	–.008	–.010	–.014	–.005	–.015	–.011
Noon. . .	–.005	–.012	–.010	–.008	–.002	–.006	–.006	–.005	–.005	–.010	.002	–.009	–.006
1	.002	–.006	–.005	–.004	.000	–.003	–.003	.000	.000	–.003	.007	.003	–.001
2	.005	.003	.000	.003	.003	.003	.001	.003	.004	.004	.011	.008	.004
3	.004	.006	003	.009	.006	.007	.005	.005	.008	.005	.010	.010	.006
4	.002	.008	.005	.004	.010	.013	.009	.009	.010	.003	.008	.009	.007
5	.000	.006	.004	.014	.014	.017	.013	.011	.011	.000	.004	.006	.008
6	–.003	.002	.000	.011	.015	.017	.013	.011	.006	–.005	.000	.002	.006
7	–.005	–.004	–.006	–.007	.010	.014	.010	.005	.000	–.008	–.006	–.003	.000
8	–.006	–.006	–.012	–.005	.000	.008	.004	–.005	–.005	–.011	–.012	–.006	–.005
9	–.007	–.008	–.015	–.009	–.006	.003	–.001	–.010	–.009	–.014	–.017	–.019	–.008
10	–.005	–.007	–.012	–.012	–.008	–.002	–.005	–.015	–.011	–.012	–.019	–.010	–.010
11	–.004	–.005	–.010	–.012	–.008	–.002	–.012	–.015	–.011	–.009	–.017	–.009	–.009
6. 6	.006	.007	.006	.008	.011	.005	.008	.008	.008	.006	.007	.006	.008
7. 7	.002	.002	.000	–.005	.008	.002	.003	.002	.000	.000	.002	.002	.001
8. 8	–.002	–.003	–.008	–.006	.002	–.002	–.002	–.006	–.006	–.007	–.004	–.001	–.004
9. 9	–.007	–.008	–.013	–.010	–.006	–.004	–.005	–.009	–.010	–.012	–.011	–.009	–.009
10.10	–.007	–.011	–.014	–.013	–.009	–.006	–.007	–.012	–.012	–.013	–.013	–.012	–.011
7. 2. 9	.003	.001	–.003	–.003	.001	–.001	–.001	–.002	–.001	–.001	.001	.002	.001
6. 2. 8	.005	.003	.000	.001	.003	.002	.002	.001	.003	.004	.005	.004	.003
6. 2.10	.005	.003	.000	–.001	.000	–.002	–.001	–.002	.001	.003	.002	.003	.001
6. 2. 6	.006	.006	.004	.007	.008	–.005	.005	.006	.007	.006	.009	.007	.006
7. 2	.007	.005	.003	.000	.004	–.004	–.001	.002	.002	.006	.010	.007	.003
8. 2	.004	.002	–.002	–.002	.003	–.004	–.003	–.002	–.001	.000	.007	.006	.001
8. 1	.002	–.003	–.004	–.006	.001	–.007	–.006	–.003	–.003	–.003	.005	.003	–.002
7. 1	.006	.001	.000	–.003	.003	–.006	–.003	.000	.000	.002	.008	.004	.001
9.12.3.9	–.004	–.005	–.008	–.005	–.002	–.002	–.003	–.004	–.004	–.007	–.002	–.004	–.004

The numbers without sign must be added; those with the sign — must be subtracted.

D

XXII. TABLE TO REDUCE, BY INTERPOLATION,

THE OBSERVATIONS TO THE SAME ABSOLUTE TIME.

DECIMALS OF AN HOUR.

Min.	Decimal.	Min.	Decimal.	Min.	Decimal.	Min.	Decimal.	Min.	Decimal.	Min.	Decimal.
1	.017	11	.183	21	.350	31	.517	41	.683	51	.850
2	.033	12	.200	22	.367	32	.533	42	.700	52	.867
3	.050	13	.217	23	.383	33	.550	43	.717	53	.883
4	.067	14	.233	24	.400	34	.567	44	.733	54	.900
5	.083	15	.250	25	.417	35	.583	45	.750	55	.917
6	.100	16	.267	26	.433	36	.600	46	.767	56	.933
7	.117	17	.283	27	.450	37	.617	47	.783	57	.950
8	.133	18	.300	28	.467	38	.633	48	.800	58	.967
9	.150	19	.317	29	.483	39	.650	49	.817	59	.983
10	.167	20	.333	30	.500	40	.667	50	.833	60	1.000

Table for Correction of Curvature and Refraction.

From a mountain, when furnished with a barometer, or with an apparatus for determining the temperature of boiling water, and a pocket level, an observer can find the elevations of distant points, which are in sight, but lower than the mountain itself on which he stands. He has only to seek, with the level, the point on the slope of the mountain which corresponds to the point at a distance that he wishes to determine, and to take there a barometrical, or a boiling point observation. This observation is to be calculated in the usual way, but the result must be corrected for the curvature of the surface of the globe, and for the atmospheric refraction, by means of the following Table.

This method, which furnishes the means of multiplying, without much trouble, the measurements of heights, gives approximations which are sufficient for most of the purposes of Physical Geography. It may even seem preferable to direct measurements for determining the mean elevation of certain physical lines, which are best estimated when seen from a distance; such as the upper limit of the growth of trees, the limits of different kinds of vegetation, that of permanent snow, that of the mean elevation of the crest of a mountain range, &c.

Table XXIII. is taken from Captain Lee's *Collection of Tables and Formulæ*, 2d edit., page 81.

Showing the Difference of the Apparent and True Level, in feet and decimals, for Distances in feet and miles.

Distances in Feet.	Correction in Feet.			Distances in Miles.	Correction in Feet.		
	For Curvature.	For Refraction.	For Curvature and Refraction.		For Curvature.	For Refraction.	For Curvature and Refraction.
100	.00024	.00004	.00020	¼	.0417	.0060	.0357
150	.00054	.00008	.00046	½	.1668	.0238	.1430
200	.00094	.00013	.00083	¾	.3752	.0536	.3216
250	.00149	.00021	.00128	1	.6670	.0953	.5717
300	.00215	.00031	.00184	1½	1.5008	.2144	1.2864
350	.00293	.00042	.00251	2	2.6680	.3811	2.2869
400	.00383	.00055	.00328	2½	4.1688	.5955	3.5733
450	.00484	.00069	.00415	3	6.0030	.8561	5.1469
500	.00598	.00085	.00513	3½	8.1708	1.1673	7.0035
550	.00724	.00103	.00621	4	10.6720	1.5246	9.1474
600	.00861	.00123	.00738	4½	13.5468	1.9295	11.5773
650	.01010	.00144	.00866	5	16.6750	2.3821	14.2929
700	.01172	.00167	.01005	5½	20.1769	2.8824	17.2945
750	.01345	.00192	.01153	6	24.0120	3.4303	20.5817
800	.01531	.00219	.01312	6½	28.1809	4.0258	24.1551
850	.01728	.00247	.01481	7	32.6830	4.6690	28.0143
900	.01938	.00277	.01661	7½	37.5190	5.3599	32.1591
950	.02159	.00308	.01851	8	42.6880	6.0997	36.5883
1000	.02392	.00333	.02059	8½	48.1910	6.8844	41.3066
1050	.02638	.00377	.02261	9	54.0270	7.7181	46.3089
1100	.02895	.00414	.02481	9½	60.1971	8.5996	51.5975
1150	.03164	.00452	.02712	10	66.7000	9.5286	57.1714
1200	.03445	.00492	.02953	11	80.7070	11.5296	69.1774
1250	.03738	.00534	.03204	12	96.0480	13.7211	82.3269
1300	.04043	.00578	.03465	13	112.7230	16.1033	96.6197
1350	.04361	.00623	.03738	14	130.7320	18.6760	112.0560
1400	.04689	.00670	.04019	15	150.0750	21.4393	128.6357
1450	.05030	.00719	.04311	16	170.7520	24.3931	146.3589
1500	.05383	.00769	.04614	17	192.7630	27.5376	165.2254
1550	.05748	.00821	.04927	18	216.1086	30.8727	185.2359
1600	.06125	.00875	.05250	19	240.7870	34.3981	206.3889
1650	.06514	.00931	.05583	20	266.8000	38.1143	228.6857
1700	.06914	.00988	.05926				
1750	.07327	.01047	.06280				
1800	.07752	.01107	.06645				
1850	.08188	.01170	.07018				
1900	.08637	.01234	.07403				
1950	.09098	.01300	.07798				
2000	.09570	.01367	.08203				

THERMOMETRICAL

MEASUREMENT OF HEIGHTS,

OR

TABLES

FOR DEDUCING DIFFERENCES OF LEVEL FROM OBSERVATIONS OF THE TEMPERATURE OF BOILING WATER.

THERMOMETRICAL MEASUREMENT OF HEIGHTS.

TABLES

FOR DEDUCING DIFFERENCES OF LEVEL FROM THE TEMPERATURE OF THE BOILING POINT OF WATER.

WHEN water is heated in the open air, the elastic force of the vapors produced from it gradually increases, until it becomes equal to the incumbent weight of the atmosphere. Then, the pressure of the atmosphere being overcome, the steam escapes rapidly in large bubbles, and the water boils. The temperature at which, in the open air, water boils, thus depends upon the weight of the atmospheric column above it, and under a less barometric pressure the water will boil at a lower temperature than under a greater pressure. Now, as the weight of the atmosphere decreases with the elevation, it is obvious that, in ascending a mountain, the *higher* the station where an observation is taken, the *lower* the temperature at which water boils at that station will be.

The difference of elevation between two places, therefore, can be deduced from the temperature of boiling water observed at each station. It is only necessary to find the barometric pressures which correspond to those temperatures, and, the atmospheric pressures at both places being known, to compute the difference of level by a formula, or by the tables given above for computing heights from barometrical observations.

From the above, it may be seen that the heights determined by means of the temperature of boiling water are less reliable than those deduced from barometrical observations. Both derive the difference of altitude from the difference of atmospheric pressure. But the temperature of boiling water gives only *indirectly* the atmospheric pressure, which is given *directly* by the barometer. This method is thus liable to all the chances of error which may affect the measurements by means of the barometer, besides adding to them new ones peculiar to itself, the principal of which, not to speak of the differences exhibited in the various tables of the force of vapor, is the difficulty of ascertaining with the necessary accuracy the true temperature of boiling water. In the present state of thermometry it would hardly be safe, indeed, to answer, in the most favorable circumstances, for quantities so small as hundredths of degrees, even when the thermometer has been constructed with the utmost care ; moreover, the quality of the glass of the instrument, the form and the substance of the vessel containing the water, the nature of the water itself, the place at which the bulb of the thermometer is placed, whether in the current of steam or in the water, — all these circumstances cause no inconsiderable variations to take place in the indications of thermometers observed under the same atmospheric pressure. Owing to these various causes, an observation of the boiling point, differing by one tenth of a degree from the true temperature, ought to be still admitted as a good one. Now, as the tables show, an error of one tenth of a degree Centigrade in the temperature of boiling water would cause an error of 2 millimetres in the barometric pressure, or of from 70 to 80 feet in the final result, while with a good barometer the error of pressure will hardly ever exceed one tenth of a millimetre, making a difference of 3 feet in altitude.

Notwithstanding these imperfections, the hypsometric thermometer, or thermobarometer, is of the greatest utility to travellers traversing distant or rough countries, on account of its being more conveniently transported, and much less liable to accidents than the mercurial barometer. The best form for it is that contrived and described by Regnault in the *Annales de Chimie et de Physique*, Tom. XIV. p. 202. It consists of an accurate thermometer with long degrees, subdivided into tenths, whose bulb is placed, about 2 or 3 centimetres above the surface of the water, in the steam arising from distilled water in a cylindrical vessel, the water being made to boil by a spirit-lamp. The whole instrument when closed is about 6 inches long; when drawn out for observation, about 14 inches.

Table XXIV. of barometric pressures corresponding to temperatures of boiling water, has been calculated by Regnault from his Tables of Forces of Vapor, and published in the *Annales de Chimie et de Physique*, Tom. XIV. p. 206. It gives, in millimetres of mercury, the barometric pressures corresponding to every tenth of a Centigrade degree; for greater convenience, the values for every hundredth have been added.

The accuracy of this table has been tested by direct observation by Mr. Wisse, a traveller competent in such matters, who noted down simultaneously the temperatures of the boiling point of water and the height of the barometer, in various parts of the Andes, up to the summit of the volcano of Pichincha, including in his observations barometrical pressures ranging from 752 to 430 millimetres of mercury. The agreement between the barometric pressures given here by Regnault and those found by Wisse are very satisfactory, the differences never exceeding a few tenths of a millimetre. See *Annales de Chimie et de Physique*, Tom. XXVIII. p. 123.

Table XXV. is the same table, revised by A. Moritz, who, in a communication to the Académie des Sciences, in October, 1856, called the attention to some slight errors of computation in Regnault's table, and gave the corrected numbers for every whole degree from 40° to 102° Centigrade. Those numbers are given here from 80° upwards, as published in the *Journal de l'Institut;* the values for every tenth of a degree, and their differences, have been computed to fit the table for practical use. The comparison of the two tables will show that the corrections mostly amount to a few hundredths, and never exceed one tenth of a millimetre.

Table XXVI. is table XXV. reduced to English measures.

Centig. Degrees.	Hundredths of a Degree.									
	0.	**1.**	**2.**	**3.**	**4.**	**5.**	**6.**	**7.**	**8.**	**9.**
°	Millim.	Millim.	Millim.	Millim.	Millim.	Millim.	Millim.	Millim.	Millim.	Millim.
85.0	433.04	433.21	433.38	433.55	433.72	433.89	434.07	434.24	434.41	434.58
85.1	434.75	434.92	435.09	435.26	435.43	435.60	435.78	435.95	436.12	436.29
85.2	436.46	436.63	436.80	436.97	437.14	437.31	437.49	437.66	437.83	438.00
85.3	438.17	438.34	438.51	438.69	438.86	439.03	439.20	439.37	439.55	439.72
85.4	439.89	440.06	440.23	440.41	440.58	440.75	440.93	441.10	441.27	441.45
85.5	441.62	441.79	441.97	442.14	442.31	442.48	442.66	442.83	443.00	443.18
85.6	443.35	443.52	443.70	443.87	444.05	444.22	444.39	444.57	444.74	444.92
85.7	445.09	445.26	445.44	445.61	445.79	445.96	446.14	446.31	446.49	446.67
85.8	446.84	447.01	447.19	447.36	447.54	447.71	447.89	448.06	448.24	448.41
85.9	448.59	448.76	448.94	449.11	449.29	449.46	449.64	449.81	449.99	450.16
86.0	450.34	450.52	450.69	450.87	451.04	451.22	451.40	451.57	451.75	451.92
86.1	452.10	452.28	452.45	452.63	452.81	452.98	453.16	453.34	453.52	453.69
86.2	453.87	454.05	454.22	454.40	454.58	454.75	454.93	455.11	455.29	455.46
86.3	455.64	455.82	456.00	456.17	456.35	456.53	456.71	456.89	457.06	457.24
86.4	457.42	457.60	457.78	457.96	458.14	458.31	458.49	458.67	458.85	459.03
86.5	459.21	459.39	459.57	459.75	459.93	460.10	460.28	460.46	460.64	460.82
86.6	461.00	461.18	461.36	461.54	461.72	461.90	462.08	462.26	462.44	462.62
86.7	462.80	462.98	463.16	463.34	463.52	463.70	463.88	464.06	464.24	464.42
86.8	464.60	464.78	464.96	465.14	465.32	465.50	465.69	465.87	466.05	466.23
86.9	466.41	466.59	466.77	466.95	467.13	467.31	467.50	467.68	467.86	468.04
87.0	468.22	468.40	468.58	468.77	468.95	469.13	469.31	469.49	469.68	469.86
87.1	470.04	470.22	470.41	470.59	470.77	470.95	471.14	471.32	471.50	471.69
87.2	471.87	472.05	472.24	472.42	472.60	472.78	472.97	473.15	473.33	473.52
87.3	473.70	473.88	474.07	474.25	474.44	474.62	474.80	474.99	475.17	475.36
87.4	475.54	475.72	475.91	476.09	476.28	476.46	476.64	476.83	477.01	477.20
87.5	477.38	477.56	477.75	477.93	478.12	478.30	478.49	478.67	478.86	479.04
87.6	479.23	479.41	479.60	479.78	479.97	480.15	480.34	480.52	480.71	480.89
87.7	481.08	481.27	481.45	481.64	481.82	482.01	482.20	482.38	482.57	482.75
87.8	482.94	483.13	483.31	483.50	483.69	483.87	484.06	484.25	484.44	484.62
87.9	484.81	485.00	485.19	485.37	485.56	485.75	485.94	486.13	486.31	486.50
88.0	486.69	486.88	487.07	487.25	487.44	487.63	487.82	488.01	488.19	488.38
88.1	488.57	488.76	488.95	489.13	489.32	489.51	489.70	489.89	490.07	490.26
88.2	490.45	490.64	490.83	491.02	491.21	491.39	491.58	491.77	491.96	492.15
88.3	492.34	492.53	492.72	492.91	493.10	493.29	493.48	493.67	493.86	494.05
88.4	494.24	494.43	494.62	494.81	495.00	495.19	495.39	495.58	495.77	495.96
88.5	496.15	496.34	496.53	496.72	496.91	497.10	497.30	497.49	497.68	497.87
88.6	498.06	498.25	498.44	498.64	498.83	499.02	499.21	499.40	499.60	499.79
88.7	499.98	500.17	500.36	500.56	500.75	500.94	501.13	501.32	501.52	501.71
88.8	501.90	502.09	502.28	502.48	502.67	502.86	503.05	503.24	503.44	503.63
88.9	503.82	504.01	504.21	504.40	504.60	504.79	504.98	505.18	505.37	505.57
	0.	**1.**	**2.**	**3.**	**4.**	**5.**	**6.**	**7.**	**8.**	**9.**

Centig. Degrees.	Hundredths of a Degree.									
	0.	1.	2.	3.	4.	5.	6.	7.	8.	9.
°	Millim.	Millim.	Millim.	Millim.	Millim.	Millim.	Millim.	Millim.	Millim.	Millim.
89.0	505.76	505.95	506.15	506.34	506.54	506.73	506.92	507.12	507.31	507.51
89.1	507.70	507.89	508.09	508.28	508.48	508.67	508.87	509.06	509.26	509.45
89.2	509.65	509.84	510.04	510.23	510.43	510.62	510.82	511.01	511.21	511.40
89.3	511.60	511.80	511.99	512.19	512.38	512.58	512.78	512.97	513.17	513.36
89.4	513.56	513.76	513.95	514.15	514.35	514.54	514.74	514.94	515.14	515.33
89.5	515.53	515.73	515.92	516.12	516.32	516.51	516.71	516.91	517.11	517.30
89.6	517.50	517.70	517.90	518.09	518.29	518.49	518.69	518.89	519.08	519.28
89.7	519.48	519.68	519.88	520.07	520.27	520.47	520.67	520.87	521.06	521.26
89.8	521.46	521.66	521.86	522.06	522.26	522.46	522.66	522.86	523.05	523.25
89.9	523.45	523.65	523.85	524.05	524.25	524.45	524.65	524.85	525.05	525.25
90.0	525.45	525.65	525.85	526.05	526.25	526.45	526.65	526.85	527.05	527.25
90.1	527.45	527.65	527.85	528.05	528.25	528.45	528.66	528.86	529.06	529.26
90.2	529.46	529.66	529.86	530.07	530.27	530.47	530.67	530.87	531.08	531.28
90.3	531.48	531.68	531.88	532.09	532.29	532.49	532.69	532.89	533.10	533.30
90.4	533.50	533.70	533.91	534.11	534.31	534.51	534.72	534.92	535.12	535.33
90.5	535.53	535.73	535.94	536.14	536.35	536.55	536.75	536.96	537.16	537.37
90.6	537.57	537.77	537.98	538.18	538.39	538.59	538.79	539.00	539.20	539.41
90.7	539.61	539.81	540.02	540.22	540.43	540.63	540.84	541.04	541.25	541.45
90.8	541.66	541.87	542.07	542.28	542.48	542.69	542.90	543.10	543.31	543.51
90.9	543.72	543.93	544.13	544.34	544.54	544.75	544.96	545.16	545.37	545.57
91.0	545.78	545.99	546.19	546.40	546.61	546.81	547.03	547.23	547.44	547.64
91.1	547.85	548.06	548.26	548.47	548.68	548.88	549.09	549.30	549.51	549.71
91.2	549.92	550.13	550.34	550.54	550.75	550.96	551.17	551.38	551.58	551.79
91.3	552.00	552.21	552.42	552.63	552.84	553.04	553.25	553.46	553.67	553.88
91.4	554.09	554.30	554.51	554.72	554.93	555.14	555.35	555.56	555.77	555.98
91.5	556.19	556.40	556.61	556.82	557.03	557.24	557.45	557.66	557.87	558.08
91.6	558.29	558.50	558.71	558.92	559.13	559.34	559.55	559.76	559.97	560.18
91.7	560.39	560.60	560.81	561.03	561.24	561.45	561.66	561.87	562.09	562.30
91.8	562.51	562.72	562.93	563.15	563.36	563.57	563.78	563.99	564.21	564.42
91.9	564.63	564.86	565.06	565.27	565.48	565.69	565.91	566.12	566.33	566.55
92.0	566.76	566.97	567.19	567.40	567.61	567.85	568.04	568.25	568.46	568.68
92.1	568.89	569.10	569.32	569.53	569.75	569.96	570.17	570.39	570.60	570.82
92.2	571.03	571.24	571.46	571.67	571.89	572.10	572.32	572.53	572.75	572.96
92.3	573.18	573.40	573.61	573.83	574.04	574.26	574.48	574.69	574.91	575.12
92.4	575.34	575.56	575.77	575.99	576.20	576.42	576.64	576.85	577.07	577.28
92.5	577.50	577.72	577.93	578.15	578.37	578.58	578.80	579.02	579.24	579.45
92.6	579.67	579.89	580.10	580.32	580.54	580.75	580.97	581.19	581.41	581.62
92.7	581.84	582.06	582.28	582.49	582.71	582.93	583.15	583.37	583.58	583.80
92.8	584.02	584.24	584.46	584.68	584.90	585.11	585.33	585.55	585.77	585.99
92.9	586.21	586.43	586.65	586.87	587.09	587.31	587.53	587.75	587.97	588.19
	0.	1.	2.	3.	4.	5.	6.	7.	8.	9.

Centig. Degrees.	Hundredths of a Degree.									
	0.	**1.**	**2.**	**3.**	**4.**	**5.**	**6.**	**7.**	**8.**	**9.**
°	Millim.	Millim.	Millim.	Millim.	Millim.	Millim.	Millim.	Millim.	Millim.	Millim.
93.0	588.41	588.63	588.85	589.07	589.29	589.51	589.73	589.95	590.17	590.39
93.1	590.61	590.83	591.05	591.27	591.49	591.71	591.94	592.16	592.38	592.60
93.2	592.82	593.04	593.26	593.49	593.71	593.93	594.15	594.37	594.60	594.82
93.3	595.04	595.26	595.48	595.71	595.93	596.15	596.37	596.59	596.82	597.04
93.4	597.26	597.48	597.71	597.93	598.15	598.37	598.60	598.82	599.04	599.27
93.5	599.49	599.71	599.94	600.16	600.38	600.60	600.83	601.05	601.27	601.50
93.6	601.72	601.94	602.17	602.39	602.62	602.84	603.07	603.29	603.52	603.74
93.7	603.97	604.19	604.42	604.64	604.87	605.09	605.32	605.54	605.77	605.99
93.8	606.22	606.45	606.67	606.90	607.12	607.35	607.58	607.80	608.03	608.25
93.9	608.48	608.71	608.93	609.16	609.38	609.61	609.84	610.06	610.29	610.51
94.0	610.74	610.97	611.19	611.42	611.65	611.87	612.10	612.33	612.56	612.78
94.1	613.01	613.24	613.47	613.69	613.92	614.15	614.38	614.61	614.83	615.06
94.2	615.29	615.52	615.75	615.97	616.21	616.43	616.66	616.89	617.12	617.35
94.3	617.58	617.81	618.04	618.27	618.50	618.72	618.95	619.18	619.41	619.64
94.4	619.87	620.10	620.33	620.56	620.79	621.02	621.25	621.48	621.71	621.94
94.5	622.17	622.40	622.63	622.86	623.09	623.32	623.56	623.79	624.02	624.25
94.6	624.48	624.71	624.94	625.17	625.40	625.63	625.87	626.10	626.33	626.56
94.7	626.79	627.02	627.25	627.49	627.72	627.95	628.18	628.41	628.65	628.88
94.8	629.11	629.34	629.58	629.81	630.04	630.27	630.51	630.74	630.97	631.21
94.9	631.44	631.67	631.91	632.14	632.38	632.61	632.84	633.08	633.31	633.55
95.0	633.78	634.01	634.25	634.48	634.72	634.95	635.18	635.42	635.65	635.89
95.1	636.12	636.35	636.59	636.82	637.06	637.29	637.53	637.76	638.00	638.23
95.2	638.47	638.71	638.94	639.18	639.41	639.65	639.89	640.12	640.36	640.59
95.3	640.83	641.07	641.30	641.54	641.77	642.01	642.25	642.48	642.72	642.95
95.4	643.19	643.43	643.67	643.90	644.14	644.38	644.62	644.86	645.09	645.33
95.5	645.57	645.81	646.05	646.28	646.52	646.76	647.00	647.24	647.47	647.71
95.6	647.95	648.19	648.43	648.67	648.91	649.14	649.38	649.62	649.86	650.10
95.7	650.34	650.58	650.82	651.06	651.30	651.53	651.77	652.01	652.25	652.49
95.8	652.73	652.97	653.21	653.45	653.69	653.93	654.17	654.41	654.65	654.89
95.9	655.13	655.37	655.61	655.85	656.09	656.33	656.58	656.82	657.06	657.30
96.0	657.54	657.78	658.02	658.26	658.50	658.74	658.99	659.23	659.47	659.71
96.1	659.95	660.19	660.43	660.68	660.92	661.16	661.40	661.64	661.89	662.13
96.2	662.37	662.61	662.86	663.10	663.34	663.58	663.83	664.07	664.31	664.56
96.3	664.80	665.04	665.29	665.53	665.78	666.02	666.26	666.51	666.75	667.00
96.4	667.24	667.48	667.73	667.97	668.22	668.46	668.71	668.95	669.20	669.44
96.5	669.69	669.93	670.18	670.42	670.67	670.91	671.16	671.40	671.65	671.99
96.6	672.14	672.39	672.63	672.88	673.12	673.37	673.62	673.86	674.11	674.35
96.7	674.60	674.85	675.09	675.34	675.59	675.83	676.08	676.33	676.58	676.82
96.8	677.07	677.32	677.57	677.81	678.06	678.31	678.56	678.81	679.05	679.30
96.9	679.55	679.80	680.05	680.29	680.54	680.79	681.04	681.29	681.53	681.78
	0.	**1.**	**2.**	**3.**	**4.**	**5.**	**6.**	**7.**	**8.**	**9.**

Centig. Degrees.	Hundredths of a Degree.									
	0.	**1.**	**2.**	**3.**	**4.**	**5.**	**6.**	**7.**	**8.**	**9.**
°	Millim.	Millim.	Millim.	Millim.	Millim.	Millim.	Millim.	Millim.	Millim.	Millim.
97.0	682.03	682.28	682.53	682.78	683.03	683.27	683.52	683.77	684.02	684.27
97.1	684.52	684.77	685.02	685.27	685.52	685.77	686.02	686.27	686.52	686.77
97.2	687.02	687.27	687.52	687.77	688.02	688.27	688.53	688.78	689.03	689.28
97.3	689.53	689.78	690.03	690.28	690.53	690.78	691.04	691.29	691.54	691.79
97.4	692.04	692.29	692.54	692.80	693.05	693.30	693.55	693.80	694.06	694.31
97.5	694.56	694.81	695.06	695.32	695.57	695.82	696.07	696.32	696.58	696.83
97.6	697.08	697.33	697.59	697.84	698.09	698.34	698.60	698.85	699.10	699.36
97.7	699.61	699.86	700.12	700.37	700.63	700.88	701.13	701.39	701.64	701.90
97.8	702.15	702.40	702.66	702.91	703.17	703.42	703.68	703.93	704.19	704.44
97.9	704.70	704.96	705.21	705.47	705.72	705.98	706.24	706.49	706.75	707.00
98.0	707.26	707.52	707.77	708.03	708.28	708.54	708.80	709.05	709.31	709.56
98.1	709.82	710.08	710.33	710.59	710.85	711.10	711.36	711.62	711.88	712.13
98.2	712.39	712.65	712.91	713.16	713.42	713.68	713.94	714.20	714.45	714.71
98.3	714.97	715.22	715.49	715.75	716.01	716.26	716.52	716.78	717.04	717.30
98.4	717.56	717.82	718.08	718.34	718.60	718.85	719.11	719.37	719.63	719.89
98.5	720.15	720.41	720.67	720.93	721.19	721.45	721.71	721.97	722.23	722.49
98.6	722.75	723.01	723.27	723.53	723.79	724.05	724.31	724.57	724.83	725.09
98.7	725.35	725.61	725.87	726.13	726.39	726.65	726.92	727.18	727.44	727.70
98.8	727.96	728.22	728.48	728.75	729.01	729.27	729.53	729.79	730.06	730.32
98.9	730.58	730.84	731.11	731.37	731.63	731.89	732.16	732.42	732.68	732.95
99.0	733.21	733.47	733.74	734.00	734.27	734.53	734.79	735.06	735.32	735.59
99.1	735.85	736.11	736.38	736.64	736.91	737.17	737.44	737.70	737.97	738.23
99.2	738.50	738.77	739.03	739.30	739.56	739.83	740.10	740.36	740.63	740.89
99.3	741.16	741.43	741.69	741.96	742.23	742.49	742.76	743.03	743.30	743.56
99.4	743.83	744.10	744.36	744.63	744.90	745.16	745.43	745.70	745.97	746.23
99.5	746.50	746.77	747.04	747.30	747.57	747.84	748.11	748.38	748.64	748.91
99.6	749.18	749.45	749.72	749.99	750.26	750.52	750.79	751.06	751.33	751.60
99.7	751.87	752.14	752.41	752.68	752.95	753.22	753.49	753.76	754.03	754.30
99.8	754.57	754.84	755.11	755.38	755.65	755.92	756.20	756.47	756.74	757.01
99.9	757.28	757.55	757.82	758.10	758.37	758.64	758.91	759.18	759.46	759.73
100.0	760.00	760.27	760.55	760.82	761.09	761.36	761.64	761.91	762.18	762.46
100.1	762.73	763.00	763.28	763.55	763.82	764.09	764.37	764.64	764.91	765.19
100.2	765.46	765.73	766.01	766.28	766.56	766.83	767.10	767.38	767.65	767.93
100.3	768.20	768.47	768.75	769.02	769.30	769.57	769.85	770.12	770.40	770.67
100.4	770.95	771.23	771.50	771.78	772.05	772.33	772.61	772.88	773.16	773.43
100.5	773.71	773.99	774.26	774.54	774.82	775.09	775.37	775.65	775.93	776.20
100.6	776.48	776.76	777.04	777.31	777.59	777.87	778.15	778.43	778.70	778.98
100.7	779.26	779.54	779.82	780.09	780.37	780.65	780.93	781.21	781.48	781.76
100.8	782.04	782.32	782.60	782.88	783.16	783.43	783.71	783.99	784.27	784.55
100.9	784.83	785.11	785.39	785.67	785.95	786.23	786.51	786.79	787.07	787.35
101.0	787.63	787.91	788.19	788.47	788.75	789.03	789.31	789.59	789.87	790.15
	0.	**1.**	**2.**	**3.**	**4.**	**5.**	**6.**	**7.**	**8.**	**9.**

TABLE XXV.

BAROMETRIC PRESSURES CORRESPONDING TO TEMPERATURES OF THE BOILING POINT OF WATER,

EXPRESSED IN MILLIMETRES OF MERCURY FOR CENTIGRADE TEMPERATURES.

By Regnault, revised by Moritz.

Boiling Point, Centigrade.	Barometer in Millimetres.	Differ-ence.	Boiling Point, Centigrade.	Barometer in Millimetres.	Differ-ence.	Boiling Point, Centigrade.	Barometer in Millimetres.	Differ-ence.
80.0	354.62	1.44	83.0	400.07	1.60	86.0	450.30	1.76
80.1	356.06	1.45	83.1	401.66	1.60	86.1	452.06	1.77
80.2	357.50	1.45	83.2	403.26	1.61	86.2	453.83	1.77
80.3	358.96	1.46	83.3	404.87	1.61	86.3	455.60	1.78
80.4	360.41	1.46	83.4	406.48	1.62	86.4	457.38	1.78
80.5	361.87	1.47	83.5	408.10	1.62	86.5	459.17	1.79
80.6	363.34	1.47	83.6	409.72	1.63	86.6	460.96	1.80
80.7	364.81	1.48	83.7	411.35	1.63	86.7	462.75	1.80
80.8	366.29	1.48	83.8	412.98	1.64	86.8	464.55	1.81
80.9	367.77	1.49	83.9	414.62	1.64	86.9	466.36	1.81
81.0	369.26	1.49	84.0	416.26	1.65	87.0	468.17	1.82
81.1	370.75	1.50	84.1	417.91	1.66	87.1	469.99	1.83
81.2	372.25	1.50	84.2	419.57	1.66	87.2	471.82	1.83
81.3	373.75	1.51	84.3	421.23	1.67	87.3	473.65	1.84
81.4	375.25	1.51	84.4	422.89	1.67	87.4	475.49	1.84
81.5	376.77	1.52	84.5	424.56	1.68	87.5	477.33	1.85
81.6	378.28	1.52	84.6	426.24	1.68	87.6	479.18	1.86
81.7	379.81	1.53	84.7	427.92	1.69	87.7	481.04	1.86
81.8	381.33	1.53	84.8	429.61	1.69	87.8	482.90	1.87
81.9	382.87	1.54	84.9	431.30	1.70	87.9	484.76	1.87
82.0	384.40	1.54	85.0	433.00	1.70	88.0	486.64	1.88
82.1	385.95	1.55	85.1	434.71	1.71	88.1	488.52	1.89
82.2	387.49	1.55	85.2	436.42	1.72	88.2	490.40	1.89
82.3	389.05	1.56	85.3	438.13	1.72	88.3	492.29	1.90
82.4	390.61	1.56	85.4	439.85	1.73	88.4	494.19	1.90
82.5	392.17	1.57	85.5	441.58	1.73	88.5	496.09	1.91
82.6	393.74	1.57	85.6	443.31	1.74	88.6	498.00	1.92
82.7	395.31	1.58	85.7	445.05	1.74	88.7	500.92	1.92
82.8	396.89	1.58	85.8	446.80	1.75	88.8	501.84	1.93
82.9	398.48	1.59	85.9	448.55	1.76	88.9	503.77	1.93
83.0	400.07		86.0	450.30		89.0	505.70	

Boiling Point, Centigrade.	Barometer in Millimetres.	Difference.	Boiling Point, Centigrade.	Barometer in Millimetres.	Difference.	Boiling Point, Centigrade.	Barometer in Millimetres.	Difference.
° 89.0	505.70	1.94	° 93.0	588.33	2.20	° 97.0	681.93	2.49
89.1	507.65	1.95	93.1	590.53	2.21	97.1	684.42	2.50
89.2	509.59	1.95	93.2	592.74	2.22	97.2	686.92	2.51
89.3	511.54	1.96	93.3	594.96	2.22	97.3	689.42	2.51
89.4	513.50	1.97	93.4	597.18	2.23	97.4	691.94	2.52
89.5	515.47	1.97	93.5	599.41	2.24	97.5	694.46	2.53
89.6	517.44	1.98	93.6	601.65	2.24	97.6	696.98	2.54
89.7	519.42	1.98	93.7	603.89	2.25	97.7	699.52	2.54
89.8	521.40	1.99	93.8	606.14	2.26	97.8	702.06	2.55
89.9	523.39	2.00	93.9	608.40	2.26	97.9	704.62	2.56
90.0	525.39	2.00	94.0	610.66	2.27	98.0	707.17	2.57
90.1	527.40	2.01	94.1	612.93	2.28	98.1	709.74	2.57
90.2	529.41	2.02	94.2	615.21	2.29	98.2	712.31	2.58
90.3	531.42	2.02	94.3	617.50	2.29	98.3	714.90	2.59
90.4	533.44	2.03	94.4	619.79	2.30	98.4	717.49	2.60
90.5	535.47	2.04	94.5	622.09	2.31	98.5	720.08	2.61
90.6	537.51	2.04	94.6	624.39	2.31	98.6	722.69	2.61
90.7	539.55	2.05	94.7	626.71	2.32	98.7	725.30	2.62
90.8	541.60	2.05	94.8	629.93	2.33	98.8	727.93	2.63
90.9	543.65	2.06	94.9	631.36	2.33	98.9	730.55	2.64
91.0	545.71	2.07	95.0	633.69	2.34	99.0	733.19	2.64
91.1	547.78	2.07	95.1	636.03	2.35	99.1	735.84	2.65
91.2	549.86	2.08	95.2	638.38	2.36	99.2	738.49	2.66
91.3	551.94	2.09	95.3	640.74	2.36	99.3	741.15	2.67
91.4	554.03	2.09	95.4	643.10	2.37	99.4	743.82	2.68
91.5	556.12	2.10	95.5	645.48	2.38	99.5	746.50	2.68
91.6	558.22	2.11	95.6	647.86	2.39	99.6	749.18	2.69
91.7	560.33	2.11	95.7	650.24	2.39	99.7	751.87	2.70
91.8	562.44	2.12	95.8	652.63	2.40	99.8	754.57	2.71
91.9	564.56	2.13	95.9	655.04	2.41	99.9	757.28	2.72
92.0	566.69	2.13	96.0	657.44	2.42	100.0	760.00	2.73
92.1	568.82	2.14	96.1	659.86	2.42	100.1	762.73	2.73
92.2	570.96	2.15	96.2	662.28	2.43	100.2	765.46	2.74
92.3	573.11	2.15	96.3	664.71	2.44	100.3	768.20	2.75
92.4	575.27	2.16	96.4	667.15	2.44	100.4	770.95	2.76
92.5	577.43	2.17	96.5	669.59	2.45	100.5	773.71	2.77
92.6	579.59	2.17	96.6	672.05	2.46	100.6	776.47	2.77
92.7	581.77	2.18	96.7	674.51	2.47	100.7	779.25	2.78
92.8	583.95	2.19	96.8	676.97	2.47	100.8	782.03	2.79
92.9	586.14	2.19	96.9	679.45	2.48	100.9	784.82	2.80
93.0	588.33		97.0	681.93		101.0	787.62	

TABLE XXVI.

BAROMETRIC PRESSURES CORRESPONDING TO TEMPERATURES OF THE BOILING POINT OF WATER,

EXPRESSED IN ENGLISH INCHES FOR TEMPERATURES OF FAHRENHEIT.

Reduced from Regnault's Table, revised by Moritz.

Boiling Point, Fahren.	Barometer in English Inches.	Difference.	Boiling Point, Fahren.	Barometer in English Inches.	Difference.	Boiling Point, Fahren.	Barometer in English Inches.	Difference.	Boiling Point, Fahren.	Barometer in English Inches.	Difference.
°			°			°			°		
185.0	17.048	0.037	188.0	18.195	0.039	191.0	19.407	0.042	194.0	20.685	0.044
185.1	17.085	.037	188.1	18.235	.039	191.1	19.448	.042	194.1	20.729	.044
185.2	17.122	.037	188.2	18.274	.039	191.2	19.490	.042	194.2	20.773	.044
185.3	17.160	.037	188.3	18.314	.040	191.3	19.532	.042	194.3	20.817	.044
185.4	17.197	.038	188.4	18.353	.040	191.4	19.573	.042	194.4	20.861	.044
185.5	17.235	.038	188.5	18.393	.040	191.5	19.615	.042	194.5	20.905	.044
185.6	17.272	.038	188.6	18.432	.040	191.6	19.657	.042	194.6	20.949	.044
185.7	17.310	.038	188.7	18.472	.040	191.7	19.699	.042	194.7	20.993	.044
185.8	17.348	.038	188.8	18.512	.040	191.8	19.741	.042	194.8	21.038	.044
185.9	17.385	.038	188.9	18.552	.040	191.9	19.783	.042	194.9	21.082	.044
186.0	17.423	.038	189.0	18.592	.040	192.0	19.825	.042	195.0	21.126	.045
186.1	17.461	.038	189.1	18.632	.040	192.1	19.868	.042	195.1	21.171	.045
186.2	17.499	.038	189.2	18.672	.040	192.2	19.910	.042	195.2	21.216	.045
186.3	17.537	.038	189.3	18.712	.040	192.3	19.952	.042	195.3	21.260	.045
186.4	17.575	.038	189.4	18.753	.040	192.4	19.995	.043	195.4	21.305	.045
186.5	17.614	.038	189.5	18.793	.040	192.5	20.037	.043	195.5	21.350	.045
186.6	17.652	.038	189.6	18.833	.040	192.6	20.080	.043	195.6	21.395	.045
186.7	17.690	.038	189.7	18.874	.041	192.7	20.123	.043	195.7	21.440	.045
186.8	17.729	.038	189.8	18.914	.041	192.8	20.166	.043	195.8	21.485	.045
186.9	17.767	.039	189.9	18.955	.041	192.9	20.208	.043	195.9	21.530	.045
187.0	17.806	.039	190.0	18.996	.041	193.0	20.251	.043	196.0	21.576	.045
187.1	17.844	.039	190.1	19.036	.041	193.1	20.294	.043	196.1	21.621	.045
187.2	17.883	.039	190.2	19.077	.041	193.2	20.338	.043	196.2	21.666	.046
187.3	17.922	.039	190.3	19.118	.041	193.3	20.381	.043	196.3	21.712	.046
187.4	17.961	.039	190.4	19.159	.041	193.4	20.424	.043	196.4	21.758	.046
187.5	18.000	.039	190.5	19.200	.041	193.5	20.467	.043	196.5	21.803	.046
187.6	18.039	.039	190.6	19.241	.041	193.6	20.511	.043	196.6	21.849	.046
187.7	18.078	.039	190.7	19.283	.041	193.7	20.554	.044	196.7	21.895	.046
187.8	18.117	.039	190.8	19.324	.041	193.8	20.598	.044	196.8	21.941	.046
187.9	18.156	0.039	190.9	19.365	0.041	193.9	20.641	0.044	196.9	21.987	0.046
188.0	18.195		191.0	19.407		194.0	20.685		197.0	22.033	

Boiling Point, Fahren.	Barometer in English Inches.	Difference.	Boiling Point, Fahren.	Barometer in English Inches.	Difference.	Boiling Point, Fahren.	Barometer in English Inches.	Difference.	Boiling Point, Fahren.	Barometer in English Inches.	Difference.
°			°			°			°		
197.0	22.033	0.046	201.0	23.943	0.049	205.0	25.990	0.053	209.0	28.180	0.057
197.1	22.079	.046	201.1	23.993	.050	205.1	26.043	.053	209.1	28.237	.057
197.2	22.125	.046	201.2	24.042	.050	205.2	26.096	.053	209.2	28.293	.057
197.3	22.172	.046	201.3	24.092	.050	205.3	26.149	.053	209.3	28.350	.057
197.4	22.218	.046	201.4	24.142	.050	205.4	26.202	.053	209.4	28.407	.057
197.5	22.264	.047	201.5	24.191	.050	205.5	26.255	.053	209.5	28.464	.057
197.6	22.311	.047	201.6	24.241	.050	205.6	26.309	.054	209.6	28.521	.057
197.7	22.358	.047	201.7	24.291	.050	205.7	26.362	.054	209.7	28.579	.057
197.8	22.404	.047	201.8	24.341	.050	205.8	26.416	.054	209.8	28.636	.057
197.9	22.451	.047	201.9	24.391	.050	205.9	26.470	.054	209.9	28.693	.058
198.0	22.498	.047	202.0	24.442	.050	206.0	26.523	.054	210.0	28.751	.058
198.1	22.545	.047	202.1	24.492	.050	206.1	26.577	.054	210.1	28.809	.058
198.2	22.592	.047	202.2	24.542	.050	206.2	26.631	.054	210.2	28.866	.058
198.3	22.639	.047	202.3	24.593	.051	206.3	26.685	.054	210.3	28.924	.058
198.4	22.686	.047	202.4	24.644	.051	206.4	26.740	.054	210.4	28.982	.058
198.5	22.734	.047	202.5	24.694	.051	206.5	26.794	.054	210.5	29.040	.058
198.6	22.781	.047	202.6	24.745	.051	206.6	26.848	.054	210.6	29.098	.058
198.7	22.829	.048	202.7	24.796	.051	206.7	26.903	.055	210.7	29.156	.058
198.8	22.876	.048	202.8	24.847	.051	206.8	26.957	.055	210.8	29.215	.058
198.9	22.924	.048	202.9	24.898	.051	206.9	27.012	.055	210.9	29.273	.059
199.0	22.971	.048	203.0	24.949	.051	207.0	27.066	.055	211.0	29.331	.059
199.1	23.019	.048	203.1	25.000	.051	207.1	27.121	.055	211.1	29.390	.059
199.2	23.067	.048	203.2	25.051	.051	207.2	27.176	.055	211.2	29.449	.059
199.3	23.115	.048	203.3	25.103	.051	207.3	27.231	.055	211.3	29.508	.059
199.4	23.163	.048	203.4	25.154	.052	207.4	27.286	.055	211.4	29.566	.059
199.5	23.211	.048	203.5	25.206	.052	207.5	27.341	.055	211.5	29.625	.059
199.6	23.259	.048	203.6	25.257	.052	207.6	27.397	.055	211.6	29.684	.059
199.7	23.308	.048	203.7	25.309	.052	207.7	27.452	.055	211.7	29.744	.059
199.8	23.356	.048	203.8	25.361	.052	207.8	27.507	.056	211.8	29.803	.059
199.9	23.405	.049	203.9	25.413	.052	207.9	27.563	.056	211.9	29.862	.059
200.0	23.453	.049	204.0	25.465	.052	208.0	27.618	.056	212.0	29.922	.060
200.1	23.502	.049	204.1	25.517	.052	208.1	27.674	.056	212.1	29.981	.060
200.2	23.550	.049	204.2	25.569	.052	208.2	27.730	.056	212.2	30.041	.060
200.3	23.599	.049	204.3	25.621	.052	208.3	27.786	.056	212.3	30.101	.060
200.4	23.648	.049	204.4	25.674	.052	208.4	27.842	.056	212.4	30.161	.060
200.5	23.697	.049	204.5	25.726	.053	208.5	27.898	.056	212.5	30.221	.060
200.6	23.746	.049	204.6	25.779	.053	208.6	27.954	.056	212.6	30.281	.060
200.7	23.795	.049	204.7	25.831	.053	208.7	28.011	.056	212.7	30.341	.060
200.8	23.845	.049	204.8	25.884	.053	208.8	28.067	.056	212.8	30.401	.060
200.9	23.894	0.049	204.9	25.937	0.053	208.9	28.123	0.057	212.9	30.461	0.060
201.0	23.943		205.0	25.990		209.0	28.180		213.0	30.522	

APPENDIX

TO

THE HYPSOMETRICAL TABLES.

COMPARISON OF THE DIFFERENT MEASURES OF LENGTH MOST GENERALLY USED FOR INDICATING ALTITUDES.

COMPARISON

OF THE MEASURES OF LENGTH MOST GENERALLY USED FOR INDICATING ALTITUDES.

It is too well known that the measures used in scientific researches among civilized nations are not uniform, as the convenience of all would require. In France the metre is employed; in England and North America, the yard and its third part, the English foot; in Germany, most commonly, the Old French or Paris foot, the sixth part of the French toise called the *Toise du Perou;* at the same time, however, though not so extensively, the Rhine foot, in Denmark and Holland, and especially in Prussia, where it has been declared, under the name of Prussian foot, the legal measure in that kingdom; in Austria, the klafter of Vienna and its sixth part, the foot of Vienna; in Switzerland, the Swiss or federal foot, which has been adjusted to the metrical system, and is three tenths of a metre; and so on.

The numerous altitudes ascertained, either by private efforts, or in connection with the public works, and quite especially with the extensive geodetic operations carried on by the governments of these various countries for the survey of a regular map, are expressed in the measures respectively adopted by each of them. These heights, however, before they can be compared, require to be uniformly reduced to any one of these measures. Their relation to each other, therefore, is given here, together with numerous reduction tables, designed to save both the useless expenditure of time and the almost unavoidable errors arising from so numerous reductions.

The exact relation of the standard measures above mentioned is not easily ascertained, and the numbers given by the best authorities by no means always agree; for the manufacture of exact copies of a standard scale, and the accurate comparison of it, require considerable skill, and belong to the most delicate operations of physics. The numbers used for computing the following tables have been adopted, after a careful review of the authorities, as the most reliable. A few words on the most important original legal standards of measures may not be unwelcome. For further details on the subject the reader is referred principally to Dove's work, *Maas und Messen*, 2d edition, Berlin, 1835.

The principal original, legal standards are the following: —

1. *The Toise du Perou*, the old French standard, made in 1735, in Paris, by Langlois, under the direction of Godin, is a bar of iron which has its standard length at the temperature of 13° Reaumur. It is known as the Toise du Perou, because it was used by the French Academicians Bouguer and La Condamine in their measurement

of an arc of the meridian in Peru. What follows will show that it may almost be called the only common standard, to which all the others are referred for comparison.

2. *The Metre* is a standard bar of platina, made by Lenoir in Paris, which has its normal length at the temperature of zero Centigrade, or the freezing point. Its length is intended to make it a natural standard, and to represent the ten-millionth part of the terrestrial arc comprised between the equator and the pole, or of a quarter of the meridian. The length of this arc given by the measurement ordered for the purpose by the Assemblée Nationale, of the arc of the meridian between Barcelona, through France, to Dunkirk, combined with the measurements previously made in Peru and in Lapland, gave for the distance of the equator from the pole 5,130,740 toises, with an ellipticity of $\frac{1}{334}$, and for the length of the metre 443.29596 lines of the toise du Perou, assumed to be 443.296 lines, or 3 feet 11.296 lines. This last quantity was declared in 1799 to be the length of the *legal* metre, and *vrai et définitif*, and is the length of Lenoir's platina standard. Later and more extensive measurements in various parts of the globe, however, seem to indicate that this quantity is somewhat too small. The latest and most exact results we now possess, combined and computed by Bessel, would make the quarter of the meridian 10,000,856 metres, and the metre = 443.29979 Paris lines; Schmidt's computation would make it 443.29977 lines, and both numbers are confirmed by Airy's results. The legal metre is thus, in fact, as Dove remarks, a legalized part of the toise du Perou, and this last remains the primitive standard. But it must be added that a natural standard, in the absolute sense of the word, is a Utopian one, which ever-changing Nature never will give us. The metre is, for all practical purposes, what it was intended to be, a natural standard; though it must be confessed that, in practice, the question is not whether and how far a standard is a natural or a conventional one, but how readily and accurately it can be obtained, or recovered when lost.

3. *The English Standard Yard* is a brass bar, made by Bird in 1760, which was declared, by act of Parliament, 1st May, 1825, the legal measure of length when at the temperature of 62° Fahrenheit, under the name of *Imperial Standard.* Another standard, sometimes also called Parliamentary Standard, was made by Bird in 1758. Sir George Schuckburgh found both to be nearly identical, at least within 0.0002 of an inch. (*Philos. Trans. for* 1798, p. 170.)

Another scale of brass, however, made by Troughton for Sir George Schuckburgh, described in the *Philosophical Transactions for* 1798, and known as Schuckburgh's scale, obtained among scientific men, perhaps, a higher degree of authority, on account of the great accuracy of its division, and of its apparatus, devised by Troughton, for delicate comparisons. That scale was used by Captain Kater, in 1818, in his researches for determining the length of the pendulum beating a second at London, and also the length of the metre, expressed in English inches of the imperial standard. (*Phil. Trans. for* 1818.)

Numerous attempts to determine the relation between the English and the French measures show no inconsiderable discrepancies in their results. Omitting the older comparisons with the toise, we give here the value of the metre in English imperial inches, as resulting from the most reliable comparisons.

A standard scale made and divided by Troughton, and in all particulars identical with Shuckburgh's scale, was brought to France in 1801 by Pictet. The comparison of it with the standard metre, made by Prony, Legendre, and Méchain, gave, after due reduction of the two standards to their respective normal temperatures,

1 metre at 32° Fahr. = 39.371 English imperial inches at 62° Fahr.

This determination was adopted for all reductions in Kelly's *Universal Cambist*, and in the French translation of the work, published in Paris in 1823.

A new comparison was made with great care by Captain H. Kater, in 1818. (See *Philos. Trans. for* 1818, p. 103.) The standards used were a brass scale metre, by Fortin, terminated with parallel planes (*mètre à bouts*), and a bar of platina on which the length of the metre was marked by two very fine lines (*mètre à traits*). Both were compared with Shuckburgh's scale, and a double series of experiments gave as the mean result:

Brass metre at 32° Fahr. =	39.37076	inches of	Shuckburgh's	scale	at 62°	Fahr.
Platina metre at 32° Fahr. =	39.37081	"	"	"	"	"
Mean	39.37079	"	"	"	"	"

On this value of the metre are based the reduction tables by Matthieu, published yearly in the *Annuaire du Bureau des Longitudes;* and it has come into general use, both in Europe and in this country.

Captain Kater gives besides, in the same paper, p. 109, note, the value of the metre compared with Bird's Parliamentary standard as being

1 metre at 32° F. = 39.37062 imp. inches of Bird's Parliamentary standard at 62° F.

This value has been adopted by Dove, as being the legal one, in his reduction tables in his work, *Maas und Messen*, p. 175, &c., and by many German authorities.

According to Baily's experiments, made in 1835, when engaged in constructing a new standard for the Royal Astronomical Society (*Memoirs R. Ast. Soc.*, Vol. IX.), the value of the metre is (Lee, *Collection of Tables and Formulæ*, p. 62)

1 metre at 32° F. = 39.370092 imperial standard inches at 62° F.

The original legal standards having been lost in the fire which destroyed, several years ago, the Parliament Houses, an act of Parliament provided for the construction of new ones; but as the report of the committee having charge of the construction of the new British standard has not yet been published, the discussion of the subject must be postponed.

The value adopted in the following tables, is that determined by Captain Kater, viz. 1 metre = 39.37079 English inches.

It may not be out of place to remark that Schumacher, in the first edition of his *Sammlung von Hülfstafeln*, used the value 1 metre = 39.3827 English inches, as given in the *Base du Système Métrique;* but this number, which expresses the relation of both standards when at the freezing point, becomes 39.37079 when they are respectively reduced to their normal temperatures. Schumacher's tables, therefore, must be corrected accordingly.

4. The *actual standard of length of the United States* is a brass scale of eighty-two inches in length, prepared for the Coast Survey of the United States, by Troughton of London, meant to be identical with the English Imperial Standard, and deposited in the office of weights and measures. The temperature at which it is a standard is 62° Fahrenheit, and the yard measure is between the 27th and 63d inches of the

scale. (See *Report on the Construction and Distribution of Weights and Measures*, by Prof. A. D. Bache, 1857.)

Hassler, first Superintendent of the United States Coast Survey, made an elaborate comparison of eleven different standard metres with the brass scale of eighty-two inches, by Troughton. Three of the standard metres, certified to be correct by high authorities, seem to deserve especial confidence : — 1. An iron metre, presented to Mr. Hassler by Tralles, which was one of the three that Tralles had made by Lenoir at the same time with those distributed to the committee on the weights and measures. 2. Another metre of iron, also by Lenoir, verified by Bouvard and Arago, and declared by them to be identical with the original. 3. A platina standard by Fortin, verified by Arago, and found to be $\frac{11}{1000}$ of a millimetre too long, for which error allowance was made. Their comparison with the Troughton scale at the temperature of the freezing point gave :

1. Iron metre of Tralles = 39.3809171 inches of the Troughton scale.
2. Iron metre of Lenoir = 39.3799487 " " "
3. Platina metre of Fortin = 39.3804194 " " "

Or, correcting for expansion, and reducing them to their respective standard temperatures :

1. Iron metre of Tralles at 32° F.	= 39.36850	English inches of the Troughton scale of 82 inches at 62° F.
2. Iron metre of Lenoir at 32° F.	= 39.36754	
3. Platina metre of Fortin at 32° F.	= 39.36789	

Hassler, in his Report to Congress on Weights and Measures, in 1832, adopts the first value, viz. :

1 metre at 32° F. = 39.3809171 inches of the Troughton scale at 32° F ;

and the Troughton scale was declared the United States standard, from which copies were to be made.

This value materially differs from those given by other careful comparisons, while, on the other hand, the close accordance of the numbers corresponding to the various standard metres proves the accuracy of Hassler's method and comparison. It is, therefore, difficult not to ascribe, with Baily, this discrepancy to some inaccuracy in the length of the Troughton scale of 82 inches. But as that scale has been declared the standard of length of the United States, it seems better to call it, as is done in the Coast Survey Reports, the *American yard*, and its subdivisions the *American foot* and *inch*, and to consider it as a new standard, similar to, but not identical with, the English imperial standard. The value of the metre expressed in American standard inches is given in the Coast Survey Report for 1853, as

1 metre at 32° F. = 39.36850535 United States standard inches at 62° F.

We learn from the *Report on Weights and Measures*, by Prof. A. D. Bache, 1857, p. 18, that two copies of the new British standards, now in progress of construction, viz. a bronze standard, No. 11, and a malleable iron standard, No. 57, have been presented by the British government to the United States. A series of careful comparisons, made in 1856, by Mr. Saxton, under the direction of Prof. A. D. Bache, of the British bronze standard, No. 11, with the Troughton scale of eighty-two inches, showed that the British bronze standard yard is shorter than the American yard by 0.00087 inch.

Comparisons of the American standards with new French standards, recently presented to the United States by the French government, are still in progress.

For the present, however, it seems best to adhere to the value of the metre, expressed in American standard inches, adopted by the Coast Survey as given above. From this value the separate tables, which will be found below, for the reduction of the American yard and foot, were computed.

5. The *Klafter of Vienna* is a silver line let into a prismatic bar of iron, on which the length of the klafter was engraved by Voigtländer. It has its normal length at 13° Reaumur, and was declared by law, in 1816, the standard klafter of Vienna. On the same silver line the French toise is marked, from the standard toise sent, in 1760, by La Caille and La Condamine to the Observatory of Vienna. According to Berghaus's Tables, in his *Grundriss der Geographie*, the value of the klafter of Vienna is 840.76134 French lines of the toise du Pérou.

6. The *Prussian Foot* is marked on a standard iron bar, 3 feet long, made by Pistor in Berlin; it is a standard at the temperature of 13° Reaumur. The length of the Prussian foot was declared by law to be = 139.13 lines of the toise du Pérou.

7. A *Mexican Vara*, the standard length, brought from Mexico at the close of the war, by Major Turnbull of the Topographical Engineers, was presented to the Office of Weights and Measures. This standard was made by soldering sheet-brass upon the tinned surface of an iron bar. A careful comparison of its length with the American standard was made under the direction of Prof. Bache, which gave its length to be = 32.9682 inches at 58°.7 Fahrenheit, or 32.9680 when reduced to 62° Fahrenheit.

The relation of that particular Mexican standard to the Spanish standard not being known, it was thought better to adopt, for the present, the value of the Spanish Vara, and of its third part, the Castilian foot, found in Thionville, *Traité des Poids et Mesures*, &c., in Balbi's *Abrégé de Géographie*, viz. 1 vara = 0.847965 metre.

From the fundamental equations indicated above have been derived all those which have been used for computing the reduction tables given in the Appendix. At the head of each table will be found the value from which it was computed.

The tables are so arranged as to give *directly* the reduction of any whole number not exceeding three or four figures, and larger numbers within the limits needed for altitudes, by means of a *single* addition.

Example.

Reduce 25,351 English feet into metres.

In Table XVI., on the line beginning with 25,000 and in the column headed 300, take for 25,300 = 7711.30 metres.

In the second part of the table, on the line beginning with 50, and in column headed 1, take for 51 = 15.54 "

English feet 25,351 = 7726.84 "

The fractions, which seldom occur, are treated as whole numbers, taking care only properly to move the decimal point.

Tables XL. to XLIV. will be found convenient for converting fractional parts of a toise or of a foot into each other.

TO CONVERT

FRENCH TOISES

INTO DIFFERENT MEASURES OF LENGTH.

I. CONVERSION OF FRENCH TOISES INTO METRES.

1 Toise = 1.94903659 Metre.

Toises. Tens.	Units. 0.	1.	2.	3.	4.	5.	6.	7.	8.	9.
	Metres.	Metres.	Metres.	Metres.	Metres.	Metres.	Metres.	Metres.	Metres.	Metres.
0	0.000	1.949	3.898	5.847	7.796	9.745	11.694	13.643	15.592	17.541
10	19.490	21.439	23.388	25.337	27.287	29.236	31.185	33.124	35.073	37.022
20	38.981	40.930	42.879	44.828	46.777	48.726	50.675	52.624	54.573	56.522
30	58.471	60.420	62.369	64.318	66.267	68.216	70.165	72.114	74.063	76.012
40	77.961	79.911	81.860	83.809	85.758	87.707	89.656	91.605	93.554	95.503
50	97.452	99.401	101.350	103.299	105.248	107.197	109.146	111.095	113.044	114.993
60	116.942	118.891	120.840	122.789	124.738	126.687	128.636	130.585	132.534	134.484
70	136.433	138.382	140.331	142.280	144.229	146.178	148.127	150.076	152.025	153.974
80	155.923	157.872	159.821	161.770	163.719	165.668	167.617	169.566	171.515	173.464
90	175.413	177.362	179.311	181.260	183.209	185.158	187.108	189.057	191.006	192.955

Thousands.	Hundreds. 0.	100.	200.	300.	400.	500.	600.	700.	800.	900.
	Metres.	Metres.	Metres.	Metres.	Metres.	Metres.	Metres.	Metres.	Metres.	Metres.
0	0.00	194.90	389.81	584.71	779.61	974.52	1169.42	1364.33	1559.23	1754.13
1000	1949.04	2143.94	2338.84	2533.75	2728.65	2923.55	3118.46	3312.36	3507.27	3702.17
2000	3898.07	4092.98	4287.88	4482.78	4677.69	4872.59	5067.50	5262.40	5457.30	5652.21
3000	5847.11	6042.01	6236.92	6431.82	6626.72	6821.63	7016.53	7211.44	7406.34	7601.24
4000	7796.15	7991.05	8185.95	8380.86	8575.76	8770.66	8965.57	9160.47	9355.38	9550.28
5000	9745.18	9940.09	10135.0	10329.9	10524.8	10719.7	10914.6	11109.5	11304.4	11499.3

II. CONVERSION OF TOISES INTO FRENCH OR PARIS FEET.

1 Toise = 6 French Feet.

Toises. Tens.	Units. 0.	1.	2.	3.	4.	5.	6.	7.	8.	9.
	Par Feet	Par.Feet.	Par.Feet.	Par.Feet.	Par.Feet.	Par Feet.	Par.Feet.	Par.Feet	Par Feet	Par Feet.
0	0.00	6	12	18	24	30	36	42	48	54
10	60	66	72	78	84	90	96	102	108	114
20	120	126	132	138	144	150	156	162	168	174
30	180	186	192	198	204	210	216	222	228	234
40	240	246	252	258	264	270	276	282	288	294
50	300	306	312	318	324	330	336	342	348	354
60	360	366	372	378	384	390	396	402	408	414
70	420	426	432	438	444	450	456	462	468	474
80	480	486	492	498	504	510	516	522	528	534
90	540	546	552	558	564	570	576	582	588	594

III. CONVERSION OF FRENCH TOISES INTO ENGLISH FEET AND DECIMALS.

1 Toise = 6.394590 English Feet.

Toises. Tens.	Units. 0.	1.	2.	3.	4.	5.	6.	7.	8.	9.
	Eng. feet	Eng. feet.	Eng feet.	Eng feet.	Eng. feet.	Eng feet.	Eng. feet.	Eng feet.	Eng. feet.	Eng. feet.
0	0.000	6.395	12.789	19.184	25.578	31.973	38.368	44.762	51.157	57.551
10	63.946	70.340	76.735	83.130	89.524	95.919	102.313	108.708	115.103	121.497
20	127.892	134.286	140.681	147.076	153.470	159.865	166.259	172.654	179.049	185.443
30	191.838	198.232	204.627	211.021	217.416	223.811	230.205	236.600	242.994	249.389
40	255.784	262.178	268.573	274.967	281.362	287.757	294.151	300.546	306.940	313.335
50	319.729	326.124	332.519	338.913	345.308	351.702	358.097	364.492	370.886	377.281
60	383.675	390.070	396.465	402.859	409.254	415.648	422.043	428.438	434.832	441.227
70	447.621	454.016	460.410	466.805	473.200	479.594	485.989	492.383	498.778	505.173
80	511.567	517.962	524.356	530.751	537.146	543.540	549.935	556.329	562.724	569.119
90	575.513	581.908	588.302	594.697	601.091	607.486	613.881	620.275	626.670	633.064

Thousands.	Hundreds. 0.	100.	200.	300.	400.	500.	600.	700.	800.	900.
	Eng. feet.	Eng. feet.	Eng. feet.	Eng. feet.	Eng. feet.	Eng. feet.	Eng. feet.	Eng. feet	Eng feet.	Eng. feet.
0	0.0	639.5	1278.9	1918.4	2557.8	3197.3	3836.8	4476.2	5115.7	5755.1
1000	6394.6	7034.0	7673.5	8313.0	8952.4	9591.9	10231.3	10870.8	11510.3	12149.7
2000	12789.2	13428.6	14068.1	14707.6	15347.0	15986.5	16625.9	17265.4	17904.9	18544.3
3000	19183.8	19823.2	20462.7	21102.1	21741.6	22381.1	23020.5	23660.0	24299.4	24938.9
4000	25578.4	26217.8	26857.3	27496.7	28136.2	28775.7	29415.1	30054.6	30694.0	31333.5
5000	31972.9	32612.4	33251.9	33891.3	34530.8	35170.2	35809.7	36449.2	37088.6	37728.1

IV. CONVERSION OF FRENCH TOISES INTO AMERICAN FEET.

1 Toise = 6.394219 American Feet.

Toises. Tens.	Units. 0.	1.	2.	3.	4.	5.	6.	7.	8.	9.
	Am. Feet.	Am. Feet	Am. Feet.	Am. Feet.	Am. Feet.	Am. Feet	Am. Feet	Am Feet.	Am. Feet.	Am. Feet.
0	0.000	6.394	12.788	19.183	25.577	31.971	38.365	44.760	51.154	57.548
10	63.942	70.336	76.731	83.125	89.519	95.913	102.308	108.702	115.096	121.490
20	127.884	134.279	140.673	147.067	153.461	159.855	166.250	172.644	179.038	185.432
30	191.827	198.221	204.615	211.009	217.403	223.798	230.192	236.586	242.980	249.375
40	255.769	262.163	268.557	274.951	281.346	287.740	294.134	300.528	306.923	313.317
50	319.711	326.105	332.499	338.894	345.288	351.682	358.076	364.470	370.865	377.259
60	383.653	390.047	396.442	402.836	409.230	415.624	422.018	428.413	434.807	441.201
70	447.595	453.990	460.384	466.778	473.172	479.566	485.961	492.355	498.749	505.143
80	511.538	517.932	524.326	530.720	537.114	543.509	549.903	556.297	562.691	569.085
90	575.480	581.874	588.268	594.662	601.057	607.451	613.845	620.239	626.633	633.028

Thousands.	Hundreds. 0.	100.	200.	300.	400.	500.	600.	700.	800.	900.
	Am. Feet	Am Feet	Am Feet	Am Feet	Am. Feet.	Am Feet	Am Feet.	Am Feet	Am Feet	Am. Feet.
0	0.0	639.4	1278.8	1918.3	2557.7	3197.1	3836.5	4476.0	5115.4	5754.8
1000	6394.2	7033.6	7673.1	8312.5	8951.9	9591.3	10230.8	10870.2	11509.6	12149.0
2000	12788.4	13427.9	14067.3	14706.7	15346.1	15985.5	16625.0	17264.4	17903.8	18543.2
3000	19182.7	19822.1	20461.5	21100.9	21740.3	22379.8	23019.2	23658.6	24298.0	24937.5
4000	25576.9	26216.3	26855.7	27495.1	28134.6	28774.0	29413.4	30052.8	30692.3	31331.7
5000	31971.1	32610.5	33249.9	33889.4	34528.8	35168.2	35807.6	36447.0	37086.5	37725.9

TO CONVERT

METRES

INTO DIFFERENT MEASURES OF LENGTH.

1 LEGAL METRE = 443.296 FRENCH OR PARIS LINES.

V. CONVERSION OF METRES INTO TOISES AND DECIMALS.

1 Metre = 0.513074074 Toise.

Metres. Thousands.	Hundreds.									
	0.	**100.**	**200.**	**300.**	**400.**	**500.**	**600.**	**700.**	**800.**	**900.**
	Toises.	Toises.	Toises.	Toises.	Toises.	Toises.	Toises.	Toises.	Toises.	Toises.
0	0.00	51.31	102.61	153.92	205.23	256.54	307.84	359.15	410.46	461.77
1000	513.07	564.38	615.69	667.00	718.30	769.61	820.92	872.23	923.53	974.84
2000	1026.15	1077.46	1128.76	1180.07	1231.38	1282.69	1333.99	1385.30	1436.61	1487.91
3000	1539.22	1590.53	1641.84	1693.14	1744.45	1795.76	1847.07	1898 37	1949.68	2000.99
4000	2052.30	2103.60	2154.91	2206.22	2257.53	2308.83	2360.14	2411.45	2462.76	2514.06
5000	2565.37	2616.68	2667.98	2719.29	2770.60	2821.91	2873.21	2924.52	2975.83	3027.14
6000	3078.44	3129.75	3181.06	3232.37	3283.67	3334.98	3386.29	3437.60	3488.90	3540.21
7000	3591.52	3642.83	3694.13	3745.44	3796.75	3848.06	3899.36	3950.67	4001.98	4053.28
8000	4104.59	4155.90	4207.21	4258.51	4309.82	4361.13	4412.44	4463.74	4515.05	4566.36
9000	4617.67	4668.97	4720.28	4771.59	4822.90	4874.20	4925.51	4976.82	5028.13	5079.43

Metres. Tens.	Units.									
	0.	**1.**	**2.**	**3.**	**4.**	**5.**	**6.**	**7.**	**8.**	**9.**
	Toises.	Toises.	Toises.	Toises.	Toises.	Toises.	Toises.	Toises.	Toises.	Toises.
0	0.000	0.513	1.026	1.539	2.052	2.565	3.078	3.592	4.105	4.618
10	5.131	5.644	6.157	6.670	7.183	7.696	8.209	8.722	9.235	9.748
20	10.261	10.775	11.288	11.801	12.314	12.827	13.340	13.853	14.366	14.879
30	15.392	15.905	16.418	16.931	17.445	17.958	18.471	18.984	19.497	20.010
40	20.523	21.036	21.549	22.062	22.575	23.088	23.601	24.114	24.628	25.141
50	25.654	26.167	26.680	27.193	27.706	28.219	28.732	29.245	29.758	30.271
60	30.784	31.298	31.811	32.324	32.837	33.350	33.863	34.376	34.889	35.402
70	35.915	36.428	36.941	37.454	37.967	38.481	38.994	39.507	40.020	40.533
80	41.046	41.559	42.072	42.585	43.098	43.611	44.124	44.637	45.151	45.664
90	46.177	46.690	47.203	47.716	48.229	48.742	49.255	49.768	50.281	50.794

1 Metre = 3.078444 Paris Feet.

Metres. Tens.	Metres. Units.									
	0.	**1.**	**2.**	**3.**	**4.**	**5.**	**6.**	**7.**	**8.**	**9.**
	Fr. Feet.	Fr. Feet.	Fr. Feet.	Fr. Feet.	Fr. Feet.	Fr. Feet.	Fr. Feet.	Fr. Feet.	Fr. Feet.	Fr. Feet.
0	0.00	3.08	6.16	9.24	12.31	15.39	18.47	21.55	24.63	27.71
10	30.78	33.86	36.94	40.02	43.10	46.18	49.26	52.33	55.41	58.49
20	61.57	64.65	67.73	70.80	73.88	76.96	80.04	83.12	86.20	89.27
30	92.35	95.43	98.51	101.59	104.67	107.75	110.82	113.90	116.98	120.06
40	123.14	126.22	129.29	132.37	135.45	138.53	141.61	144.69	147.77	150.84
50	153.92	157.00	160.08	163.16	166.24	169.31	172.39	175.47	178.55	181.63
60	184.71	187.79	190.86	193.94	197.02	200.10	203.18	206.26	209.33	212.41
70	215.49	218.57	221.65	224.73	227.80	230.88	233.96	237.04	240.12	243.20
80	246.28	249.35	252.43	255.51	258.59	261.67	264.75	267.82	270.90	273.98
90	277.06	280.14	283.22	286.30	289.37	292.45	295.53	298.61	301.69	304.77
100	307.84	310.92	314.00	317.08	320.16	323.24	326.32	329.39	332.47	335.55
110	338.63	341.71	344.79	347.86	350.94	354.02	357.10	360.18	363.26	366.33
120	369.41	372.49	375.57	378.65	381.73	384.81	387.88	390.96	394.04	397.12
130	400.20	403.28	406.35	409.43	412.51	415.59	418.67	421.75	424.83	427.90
140	430.98	434.06	437.14	440.22	443.30	446.37	449.45	452.53	455.61	458.69
150	461.77	464.85	467.92	471.00	474.08	477.16	480.24	483.32	486.39	489.47
160	492.55	495.63	498.71	501.79	504.86	507.94	511.02	514.10	517.18	520.26
170	523.34	526.41	529.49	532.57	535.65	538.73	541.81	544.88	547.96	551.04
180	554.12	557.20	560.28	563.36	566.43	569.51	572.59	575.67	578.75	581.83
190	584.90	587.98	591.06	594.14	597.22	600.30	603.38	606.45	609.53	612.61
200	615.69	618.77	621.85	624.92	628.00	631.08	634.16	637.24	640.32	643.39
210	646.47	649.55	652.63	655.71	658.79	661.87	664.94	668.02	671.10	674.18
220	677.26	680.34	683.41	686.49	689.57	692.65	695.73	698.81	701.89	704.96
230	708.04	711.12	714.20	717.28	720.36	723.43	726.51	729.59	732.67	735.75
240	738.83	741.90	744.98	748.06	751.14	754.22	757.30	760.38	763.45	766.53
250	769.61	772.69	775.77	778.85	781.92	785.00	788.08	791.16	794.24	797.32
260	800.40	803.47	806.55	809.63	812.71	815.79	818.87	821.94	825.02	828.10
270	831.18	834.26	837.34	840.42	843.49	846.57	849.65	852.73	855.81	858.89
280	861.96	865.04	868.12	871.20	874.28	877.36	880.43	883.51	886.59	889.67
290	892.75	895.83	898.91	901.98	905.06	908.14	911.22	914.30	917.38	920.45
300	923.53	926.61	929.69	932.77	935.85	938.93	942.00	945.08	948.16	951.24
310	954.32	957.40	960.47	963.55	966.63	969.71	972.79	975.87	978.95	982.02
320	985.10	988.18	991.26	994.34	997.42	1000.49	1003.57	1006.65	1009.73	1012.81
330	1015.89	1018.96	1022.04	1025.12	1028.20	1031.28	1034.36	1037.44	1040.51	1043.59
340	1046.67	1049.75	1052.83	1055.91	1058.98	1062.06	1065.14	1068.22	1071.30	1074.38
350	1077.46	1080.53	1083.61	1086.69	1089.77	1092.85	1095.93	1099.00	1102.08	1105.16
360	1108.24	1111.32	1114.40	1117.48	1120.55	1123.63	1126.71	1129.79	1132.87	1135.95
370	1139.02	1142.10	1145.18	1148.26	1151.34	1154.42	1157.49	1160.57	1163.65	1166.73
380	1169.81	1172.89	1175.97	1179.04	1182.12	1185.20	1188.28	1191.36	1194.44	1197.51
390	1200.59	1203.67	1206.75	1209.83	1212.91	1215.99	1219.06	1222.14	1225.22	1228.30
	0.	**1.**	**2.**	**3.**	**4.**	**5.**	**6.**	**7.**	**8.**	**9.**

1 Metre = 3.078444 Paris Feet.

Metres. Tens.	Metres. Units									
	0.	**1.**	**2.**	**3.**	**4.**	**5.**	**6.**	**7.**	**8.**	**9.**
	Fr. Feet.	Fr. Feet.	Fr. Feet.	Fr. Feet.	Fr. Feet.	Fr. Feet.	Fr. Feet.	Fr. Feet.	Fr Feet.	Fr. Feet.
400	1231.38	1234.46	1237.53	1240.61	1243.69	1246.77	1249.85	1252.93	1256.01	1259.08
410	1262.16	1265.24	1268.32	1271.40	1274.48	1277.55	1280.63	1283.71	1286.79	1289.87
420	1292.95	1296.02	1299.10	1302.18	1305.26	1308.34	1311.42	1314.50	1317.57	1320.65
430	1323.73	1326.81	1329.89	1332.97	1336.04	1339.12	1342.20	1345.28	1348.36	1351.44
440	1354.52	1357.59	1360.67	1363.75	1366.83	1369.91	1372.99	1376.06	1379.14	1382.22
450	1385.30	1388.38	1391.46	1394.54	1397.61	1400.69	1403.77	1406.85	1409.93	1413.01
460	1416.08	1419.16	1422.24	1425.32	1428.40	1431.48	1434.55	1437.63	1440.71	1443.79
470	1446.87	1449.95	1453.03	1456.10	1459.18	1462.26	1465.34	1468.42	1471.50	1474.57
480	1477.65	1480.73	1483.81	1486.89	1489.97	1493.05	1496.12	1499.20	1502.28	1505.36
490	1508.44	1511.52	1514.59	1517.67	1520.75	1523.83	1526.91	1529.99	1533.07	1536.14
500	1539.22	1542.30	1545.38	1548.46	1551.54	1554.61	1557.69	1560.77	1563.85	1566.93
510	1570.01	1573.08	1576.16	1579.24	1582.32	1585.40	1588.48	1591.56	1594.63	1597.71
520	1600.79	1603.87	1606.95	1610.03	1613.10	1616.18	1619.26	1622.34	1625.42	1628.50
530	1631.58	1634.65	1637.73	1640.81	1643.89	1646.97	1650.05	1653.12	1656.20	1659.28
540	1662.36	1665.44	1668.52	1671.60	1674.67	1677.75	1680.83	1683.91	1686.99	1690.07
550	1693.14	1696.22	1699.30	1702.38	1705.46	1708.54	1711.61	1714.69	1717.77	1720.85
560	1723.93	1727.01	1730.09	1733.16	1736.24	1739.32	1742.40	1745.48	1748.56	1751.63
570	1754.71	1757.79	1760.87	1763.95	1767.03	1770.11	1773.18	1776.26	1779.34	1782.42
580	1785.50	1788.58	1791.65	1794.73	1797.81	1800.89	1803.97	1807.05	1810.13	1813.20
590	1816.28	1819.36	1822.44	1825.52	1828.60	1831.67	1834.75	1837.83	1840.91	1843.99
600	1847.07	1850.14	1853.22	1856.30	1859.38	1862.46	1865.54	1868.62	1871.69	1874.77
610	1877.85	1880.93	1884.01	1887.09	1890.16	1893.24	1896.32	1899.40	1902.48	1905.56
620	1908.64	1911.71	1914.79	1917.87	1920.95	1924.03	1927.11	1930.18	1933.26	1936.34
630	1939.42	1942.50	1945.58	1948.66	1951.73	1954.81	1957.89	1960.97	1964.05	1967.13
640	1970.20	1973.28	1976.36	1979.44	1982.52	1985.60	1988.67	1991.75	1994.83	1997.91
650	2000.99	2004.07	2007.15	2010.22	2013.30	2016.38	2019.46	2022.54	2025.62	2028.69
660	2031.77	2034.85	2037.93	2041.01	2044.09	2047.17	2050.24	2053.32	2056.40	2059.48
670	2062.56	2065.64	2068.71	2071.79	2074.87	2077.95	2081.03	2084.11	2087.19	2090.26
680	2093.34	2096.42	2099.50	2102.58	2105.66	2108.73	2111.81	2114.89	2117.97	2121.05
690	2124.13	2127.20	2130.28	2133.36	2136.44	2139.52	2142.60	2145.68	2148.75	2151.83
700	2154.91	2157.99	2161.07	2164.15	2167.22	2170.30	2173.38	2176.46	2179.54	2182.62
710	2185.70	2188.77	2191.85	2194.93	2198.01	2201.09	2204.17	2207.24	2210.32	2213.40
720	2216.48	2219.56	2222.64	2225.72	2228.79	2231.87	2234.95	2238.03	2241.11	2244.19
730	2247.26	2250.34	2253.42	2256.50	2259.58	2262.66	2265.73	2268.81	2271.89	2274.97
740	2278.05	2281.13	2284.21	2287.28	2290.36	2293.44	2296.52	2299.60	2302.68	2305.75
750	2308.83	2311.91	2314.99	2318.07	2321.15	2324.23	2327.30	2330.38	2333.46	2336.54
760	2339.62	2342.70	2345.77	2348.85	2351.93	2355.01	2358.09	2361.17	2364.24	2367.32
770	2370.40	2373.48	2376.56	2379.64	2382.72	2385.79	2388.87	2391.95	2395.03	2398.11
780	2401.19	2404.26	2407.34	2410.42	2413.50	2416.58	2419.66	2422.74	2425.81	2428.89
790	2431.97	2435.05	2438.13	2441.21	2444.28	2447.36	2450.44	2453.52	2456.60	2459.68
	0.	**1.**	**2.**	**3.**	**4.**	**5.**	**6.**	**7.**	**8.**	**9.**

1 Metre = 3.078444 Paris Feet.

Metres. Tens.	Metres. Units. 0.	1.	2.	3.	4.	5.	6.	7.	8.	9.
	Fr. Feet.	Fr. Feet.	Fr. Feet.	Fr. Feet.	Fr. Feet.	Fr. Feet.	Fr. Feet.	Fr. Feet.	Fr. Feet.	Fr. Feet.
800	2462.76	2465.83	2468.91	2471.99	2475.07	2478.15	2481.23	2484.30	2487.38	2490.46
810	2493.54	2496.62	2499.70	2502.77	2505.85	2508.93	2512.01	2515.09	2518.17	2521.25
820	2524.32	2527.40	2530.48	2533.56	2536.64	2539.72	2542.79	2545.87	2548.95	2552.03
830	2555.11	2558.19	2561.27	2564.34	2567.42	2570.50	2573.58	2576.66	2579.74	2582.81
840	2585.89	2588.97	2592.05	2595.13	2598.21	2601.29	2604.36	2607.44	2610.52	2613.60
850	2616.68	2619.76	2622.83	2625.91	2628.99	2632.07	2635.15	2638.23	2641.30	2644.38
860	2647.46	2650.54	2653.62	2656.70	2659.78	2662.85	2665.93	2669.01	2672.09	2675.17
870	2678.25	2681.32	2684.40	2687.48	2690.56	2693.64	2696.72	2699.80	2702.87	2705.95
880	2709.03	2712.11	2715.19	2718.27	2721.34	2724.42	2727.50	2730.58	2733.66	2736.74
890	2739.82	2742.89	2745.97	2749.05	2752.13	2755.21	2758.29	2761.36	2764.44	2767.52
900	2770.60	2773.68	2776.76	2779.83	2782.91	2785.99	2789.07	2792.15	2795.23	2798.31
910	2801.38	2804.46	2807.54	2810.62	2813.70	2816.78	2819.85	2822.93	2826.01	2829.09
920	2832.17	2835.75	2838.33	2841.40	2844.48	2847.56	2850.64	2853.72	2856.80	2859.87
930	2862.95	2866.03	2869.11	2872.19	2875.27	2878.35	2881.42	2884.50	2887.58	2890.66
940	2893.74	2896.82	2899.89	2902.97	2906.05	2909.13	2912.21	2915.29	2918.36	2921.44
950	2924.52	2927.60	2930.68	2933.76	2936.84	2939.91	2942.99	2946.07	2949.15	2952.23
960	2955.31	2958.38	2961.46	2964.54	2967.62	2970.70	2973.78	2976.86	2979.93	2983.01
970	2986.09	2989.17	2992.25	2995.33	2998.40	3001.48	3004.56	3007.64	3010.72	3013.80
980	3016.88	3019.95	3023.03	3026.11	3029.19	3032.27	3035.35	3038.42	3041.50	3044.58
990	3047.66	3050.74	3053.82	3056.89	3059.97	3063.05	3066.13	3069.21	3072.29	3075.37

Metres.	French Feet	Metres.	French Feet.	Metres.	French Feet.	Metres.	French Feet.
1000	3078.44	5000	15392.22	9000	27706.00	13000	40019.78
2000	6156.89	6000	18470.67	10000	30784.44	14000	43098.22
3000	9235.33	7000	21549.11	11000	33862.89	15000	46176.67
4000	12313.78	8000	24627.56	12000	36941.33	16000	49255.11

Metres.	Decimetres. 0.	1.	2.	3.	4.	5.	6.	7.	8.	9.
	Fr. Feet	Fr. Feet.	Fr. Feet.	Fr. Feet.	Fr. Feet.	Fr. Feet	Fr. Feet.	Fr. Feet.	Fr. Feet.	Fr. Feet.
0	0.0000	0.3078	0.6157	0.9235	1.2314	1.5392	1.8471	2.1549	2.4628	2.7706
1	3.0784	3.3863	3.6941	4.0020	4.3098	4.6177	4.9255	5.2334	5.5412	5.8490
2	6.1569	6.4647	6.7726	7.0804	7.3883	7.6961	8.0040	8.3118	8.6196	8.9275
3	9.2353	9.5432	9.8510	10.1589	10.4667	10.7746	11.0824	11.3902	11.6981	12.0059
4	12.3138	12.6216	12.9295	13.2373	13.5452	13.8530	14.1608	14.4687	14.7765	15.0844
5	15.3922	15.7001	16.0079	16.3158	16.6236	16.9314	17.2393	17.5471	17.8550	18.1628
6	18.4707	18.7785	19.0864	19.3942	19.7020	20.0099	20.3177	20.6256	20.9334	21.2413
7	21.5491	21.8570	22.1648	22.4726	22.7805	23.0883	23.3962	23.7040	24.0119	24.3197
8	24.6276	24.9354	25.2432	25.5511	25.8589	26.1668	26.4746	26.7825	27.0903	27.3982
9	27.7060	28.0138	28.3217	28.6295	28.9374	29.2452	29.5531	29.8609	30.1688	30.4766

1 Metre = 3.28089917 English Feet.

Metres.	Metres. (Units.)									
	0.	1.	2.	3.	4.	5.	6.	7.	8.	9.
	Eng.Feet.	Eng.Feet.	Eng.Feet.	Eng.Feet.	Eng.Feet.	Eng.Feet.	Eng.Feet.	Eng.Feet.	Eng.Feet.	Eng.Feet.
0	0.0	3.28	6.56	9.84	13.12	16.40	19.69	22.97	26.25	29.53
10	32.81	36.09	39.37	42.65	45.93	49.21	52.49	55.78	59.06	62.34
20	65.62	68.90	72.18	75.46	78.74	82.02	85.30	88.58	91.87	95.15
30	98.43	101.71	104.99	108.27	111.55	114.83	118.11	121.39	124.67	127.96
40	131.24	134.52	137.80	141.08	144.36	147.64	150.92	154.20	157.48	160.76
50	164.04	167.33	170.61	173.89	177.17	180.45	183.73	187.01	190.29	193.57
60	196.85	200.13	203.42	206.70	209.98	213.26	216.54	219.82	223.10	226.38
70	229.66	232.94	236.22	239.51	242.79	246.07	249.35	252.63	255.91	259.19
80	262.47	265.75	269.03	272.31	275.60	278.88	282.16	285.44	288.72	292.00
90	295.28	298.56	301.84	305.12	308.40	311.69	314.97	318.25	321.53	324.81
100	328.09	331.37	334.65	337.93	341.21	344.49	347.78	351.06	354.34	357.62
110	360.90	364.18	367.46	370.74	374.02	377.30	380.58	383.87	387.15	390.43
120	393.71	396.99	400.27	403.55	406.83	410.11	413.39	416.67	419.96	423.24
130	426.52	429.80	433.08	436.36	439.64	442.92	446.20	449.48	452.78	456.04
140	459.33	462.61	465.89	469.17	472.45	475.73	479.01	482.29	485.57	488.85
150	492.13	495.42	498.70	501.98	505.26	508.54	511.82	515.10	518.38	521.66
160	524.94	528.22	531.51	534.79	538.07	541.35	544.63	547.91	551.19	554.47
170	557.75	561.03	564.31	567.60	570.88	574.16	577.44	580.72	584.00	587.28
180	590.56	593.84	597.12	600.40	603.69	606.97	610.25	613.53	616.81	620.09
190	623.37	626.65	629.93	633.21	636.49	639.78	643.06	646.34	649.62	652.90
200	656.18	659.46	662.74	666.02	669.30	672.58	675.87	679.15	682.43	685.71
210	688.99	692.27	695.55	698.83	702.11	705.39	708.67	711.96	715.24	718.52
220	721.80	725.08	728.36	731.64	734.92	738.20	741.48	744.76	748.05	751.33
230	754.61	757.89	761.17	764.45	767.73	771.01	774.29	777.57	780.85	784.13
240	787.42	790.70	793.98	797.26	800.54	803.82	807.10	810.38	813.66	816.94
250	820.22	823.51	826.79	830.07	833.35	836.63	839.91	843.19	846.47	849.75
260	853.03	856.31	859.60	862.88	866.16	869.44	872.72	876.00	879.28	882.56
270	885.84	889.12	892.40	895.69	898.97	902.25	905.53	908.81	912.09	915.37
280	918.65	921.93	925.21	928.49	931.78	935.06	938.34	941.62	944.90	948.18
290	951.46	954.74	958.02	961.30	964.58	967.87	971.15	974.43	977.71	980.99
300	984.27	987.55	990.83	994.11	997.39	1000.67	1003.96	1007.24	1010.52	1013.80
310	1017.08	1020.36	1023.64	1026.92	1030.20	1033.48	1036.76	1040.05	1043.33	1046.61
320	1049.89	1053.17	1056.45	1059.73	1063.01	1066.29	1069.57	1072.85	1076.13	1079.42
330	1082.70	1085.98	1089.26	1092.54	1095.82	1099.10	1102.38	1105.66	1108.94	1112.22
340	1115.51	1118.79	1122.07	1125.35	1128.63	1131.91	1135.19	1138.47	1141.75	1145.03
350	1148.31	1151.60	1154.88	1158.16	1161.44	1164.72	1168.00	1171.28	1174.56	1177.84
360	1181.12	1184.40	1187.69	1190.97	1194.25	1197.53	1200.81	1204.09	1207.37	1210.65
370	1213.93	1217.21	1220.49	1223.78	1227.06	1230.34	1233.62	1236.90	1240.18	1243.46
380	1246.74	1250.02	1253.30	1256.58	1259.87	1263.15	1266.43	1269.71	1272.99	1276.27
390	1279.55	1282.83	1286.11	1289.39	1292.67	1295.96	1299.24	1302.52	1305.80	1309.08
	0.	1.	2.	3.	4.	5.	6.	7.	8.	9.

400 to 799.

Metres.	Metres. (Units.)									
	0.	**1.**	**2.**	**3.**	**4.**	**5.**	**6.**	**7.**	**8.**	**9.**
	Eng.Feet.	Eng.Feet.	Eng.Feet.	Eng.Feet.	Eng.Feet.	Eng.Feet.	Eng.Feet.	Eng.Feet.	Eng.Feet.	Eng.Feet.
400	1312.36	1315.64	1318.92	1322.20	1325.48	1328.76	1332.05	1335.33	1338.61	1341.89
410	1345.17	1348.45	1351.73	1355.01	1358.29	1361.57	1364.85	1368.13	1371.42	1374.70
420	1377.98	1381.26	1384.54	1387.82	1391.10	1394.38	1397.66	1400.94	1404.22	1407.51
430	1410.79	1414.07	1417.35	1420.63	1423.91	1427.19	1430.47	1433.75	1437.03	1440.31
440	1443.60	1446.88	1450.16	1453.44	1456.72	1460.00	1463.28	1466.56	1469.84	1473.12
450	1476.40	1479.69	1482.97	1486.25	1489.53	1492.81	1496.09	1499.37	1502.65	1505.93
460	1509.21	1512.49	1515.78	1519.06	1522.34	1525.62	1528.90	1532.18	1535.46	1538.74
470	1542.02	1545.30	1548.58	1551.87	1555.15	1558.43	1561.71	1564.99	1568.27	1571.55
480	1574.83	1578.11	1581.39	1584.67	1587.96	1591.23	1594.52	1597.80	1601.08	1604.36
490	1607.64	1610.92	1614.20	1617.48	1620.76	1624.05	1627.33	1630.61	1633.89	1637.17
500	1640.45	1643.73	1647.01	1650.29	1653.57	1656.85	1660.13	1663.42	1666.70	1669.98
510	1673.26	1676.54	1679.82	1683.10	1686.38	1689.66	1692.94	1696.22	1699.51	1702.79
520	1706.07	1709.35	1712.63	1715.91	1719.19	1722.47	1725.75	1729.03	1732.31	1735.60
530	1738.88	1742.16	1745.44	1748.72	1752.00	1755.28	1758.56	1761.84	1765.12	1768.40
540	1771.69	1774.97	1778.25	1781.53	1784.81	1788.09	1791.37	1794.65	1797.93	1801.21
550	1804.49	1807.78	1811.06	1814.34	1817.62	1820.90	1824.18	1827.46	1830.74	1834.02
560	1837.30	1840.58	1843.87	1847.15	1850.43	1853.71	1856.99	1860.27	1863.55	1866.83
570	1870.11	1873.39	1876.67	1879.96	1883.24	1886.52	1889.80	1893.08	1896.36	1899.64
580	1902.92	1906.20	1909.48	1912.76	1916.05	1919.33	1922.61	1925.89	1929.17	1932.45
590	1935.73	1939.01	1942.29	1945.57	1948.85	1952.13	1955.42	1958.70	1961.98	1965.26
600	1968.54	1971.82	1975.10	1978.38	1981.66	1984.94	1988.22	1991.51	1994.79	1998.07
610	2001.35	2004.63	2007.91	2011.19	2014.47	2017.75	2021.03	2024.31	2027.60	2030.88
620	2034.16	2037.44	2040.72	2044.00	2047.28	2050.56	2053.84	2057.12	2060.40	2063.69
630	2066.97	2070.25	2073.53	2076.81	2080.09	2083.37	2086.65	2089.93	2093.21	2096.49
640	2099.78	2103.06	2106.34	2109.62	2112.90	2116.18	2119.46	2122.74	2126.02	2129.30
650	2132.58	2135.87	2139.15	2142.43	2145.71	2148.99	2152.27	2155.55	2158.83	2162.11
660	2165.39	2168.67	2171.96	2175.24	2178.52	2181.80	2185.08	2188.36	2191.64	2194.92
670	2198.20	2201.48	2204.76	2208.05	2211.33	2214.61	2217.89	2221.17	2224.45	2227.73
680	2231.01	2234.29	2237.57	2240.85	2244.13	2247.42	2250.70	2253.98	2257.26	2260.54
690	2263.82	2267.10	2270.38	2273.66	2276.94	2280.22	2283.51	2286.79	2290.07	2293.35
700	2296.63	2299.91	2303.19	2306.47	2309.75	2313.03	2316.31	2319.60	2322.88	2326.16
710	2329.44	2332.72	2336.00	2339.28	2342.56	2345.84	2349.12	2352.40	2355.69	2358.97
720	2362.25	2365.53	2368.81	2372.09	2375.37	2378.65	2381.93	2385.21	2388.49	2391.78
730	2395.06	2398.34	2401.62	2404.90	2408.18	2411.46	2414.74	2418.02	2421.30	2424.58
740	2427.87	2431.15	2434.43	2437.71	2440.99	2444.27	2447.55	2450.83	2454.11	2457.39
750	2460.67	2463.96	2467.24	2470.52	2473.80	2477.08	2480.36	2483.64	2486.92	2490.20
760	2493.48	2496.76	2500.05	2503.33	2506.61	2509.89	2513.17	2516.45	2519.73	2523.01
770	2526.29	2529.57	2532.85	2536.14	2539.42	2542.70	2545.98	2549.26	2552.54	2555.82
780	2559.10	2562.38	2565.66	2568.94	2572.22	2575.51	2578.79	2582.07	2585.35	2588.63
790	2591.91	2595.19	2598.47	2601.75	2605.03	2608.31	2611.60	2614.88	2618.16	2621.44
	0.	**1.**	**2.**	**3.**	**4.**	**5.**	**6.**	**7.**	**8.**	**9.**

800 to 1199.

Metres.	Metres. (Units.)									
	0.	**1.**	**2.**	**3.**	**4.**	**5.**	**6.**	**7.**	**8.**	**9.**
	Eng.Feet.	Eng.Feet.	Eng.Feet.	Eng.Feet.	Eng.Feet.	Eng.Feet.	Eng.Feet.	Eng.Feet.	Eng.Feet.	Eng.Feet.
800	2624.72	2628.00	2631.28	2634.56	2637.84	2641.12	2644.40	2647.69	2650.97	2654.25
810	2657.53	2660.81	2664.09	2667.37	2670.65	2673.93	2677.21	2680.49	2683.78	2687.06
820	2690.34	2693.62	2696.90	2700.18	2703.46	2706.74	2710.02	2713.30	2716.58	2719.87
830	2723.15	2726.43	2729.71	2732.99	2736.27	2739.55	2742.83	2746.11	2749.39	2752.67
840	2755.96	2759.24	2762.52	2765.80	2769.08	2772.36	2775.64	2778.92	2782.20	2785.48
850	2788.76	2792.05	2795.33	2798.61	2801.89	2805.17	2808.45	2811.73	2815.01	2818.29
860	2821.57	2824.85	2828.14	2831.42	2834.70	2837.98	2841.26	2844.54	2847.82	2851.10
870	2854.38	2857.66	2860.94	2864.22	2867.51	2870.79	2874.07	2877.55	2880.63	2883.91
880	2887.19	2890.47	2893.75	2897.03	2900.31	2903.60	2906.88	2910.16	2913.44	2916.72
890	2920.00	2923.28	2926.56	2929.84	2933.12	2936.40	2939.69	2942.97	2946.25	2949.53
900	2952.81	2956.09	2959.37	2962.65	2965.93	2969.21	2972.49	2975.78	2979.06	2982.34
910	2985.62	2988.90	2992.18	2995.46	2998.74	3002.02	3005.30	3008.58	3011.87	3015.15
920	3018.43	3021.71	3024.99	3028.27	3031.55	3034.83	3038.11	3041.39	3044.67	3047.96
930	3051.24	3054.52	3057.80	3061.08	3064.36	3069.64	3070.92	3074.20	3077.48	3080.76
940	3084.05	3087.33	3090.61	3093.89	3097.17	3100.45	3103.73	3107.01	3110.29	3113.57
950	3116.85	3120.14	3123.42	3126.70	3129.98	3133.26	3136.54	3139.82	3143.10	3146.38
960	3149.66	3152.94	3156.22	3159.51	3162.79	3166.07	3169.35	3172.63	3175.91	3179.19
970	3182.47	3185.75	3189.03	3192.31	3195.60	3198.88	3202.16	3205.44	3208.72	3212.00
980	3215.28	3218.56	3221.84	3225.12	3228.40	3231.69	3234.97	3238.25	3241.53	3244.81
990	3248.09	3251.37	3254.65	3257.93	3261.21	3264.49	3267.78	3271.06	3274.34	3277.62
1000	3280.90	3284.18	3287.46	3290.74	3294.02	3297.30	3300.58	3303.87	3307.15	3310.43
1010	3313.71	3316.99	3320.27	3323.55	3326.83	3330.11	3333.39	3336.67	3339.96	3343.24
1020	3346.52	3349.80	3353.08	3356.36	3359.64	3362.92	3366.20	3369.48	3372.76	3376.05
1030	3379.33	3382.61	3385.89	3389.17	3392.45	3395.73	3399.01	3402.29	3405.57	3408.85
1040	3412.14	3415.42	3418.70	3421.98	3425.26	3428.54	3431.82	3435.10	3438.38	3441.66
1050	3444.94	3448.22	3451.51	3454.79	3458.07	3461.35	3464.63	3467.91	3471.19	3474.47
1060	3477.75	3481.03	3484.31	3487.60	3490.88	3494.16	3497.44	3500.72	3504.00	3507.28
1070	3510.56	3513.84	3517.12	3520.40	3523.69	3526.97	3530.25	3533.53	3536.81	3540.09
1080	3543.37	3546.65	3549.93	3553.21	3556.49	3559.78	3563.06	3566.34	3569.62	3572.90
1090	3576.18	3579.46	3582.74	3586.02	3589.30	3592.58	3595.87	3599.15	3602.43	3605.71
1100	3608.99	3612.27	3615.55	3618.83	3622.11	3625.39	3628.67	3631.96	3635.24	3638.52
1110	3641.80	3645.08	3648.36	3651.64	3654.92	3658.20	3661.48	3664.76	3668.05	3671.33
1120	3674.61	3677.89	3681.17	3684.45	3687.73	3691.01	3694.29	3697.57	3700.85	3704.14
1130	3707.42	3710.70	3713.98	3717.26	3720.54	3723.82	3727.10	3730.38	3733.66	3736.94
1140	3740.22	3743.51	3746.79	3750.07	3753.35	3756.63	3759.91	3763.19	3766.47	3769.75
1150	3773.03	3776.31	3779.60	3782.88	3786.16	3789.44	3792.72	3796.00	3799.28	3802.56
1160	3805.84	3809.12	3812.40	3815.69	3818.97	3822.25	3825.53	3828.81	3832.09	3835.37
1170	3838.65	3841.93	3845.21	3848.49	3851.78	3855.06	3858.34	3861.62	3864.90	3868.18
1180	3871.46	3874.74	3878.02	3881.30	3884.58	3887.87	3891.15	3894.43	3897.71	3900.99
1190	3904.27	3907.55	3910.83	3914.11	3917.39	3920.67	3923.96	3927.24	3930.52	3933.80
	0.	**1.**	**2.**	**3.**	**4.**	**5.**	**6.**	**7.**	**8.**	**9.**

1200 to 1599.

Metres.	Metres. (Units.)									
	0.	1.	2.	3.	4.	5.	6.	7.	8.	9.
	Eng.Feet.	Eng.Feet.	Eng.Feet.	Eng.Feet.	Eng.Feet.	Eng.Feet.	Eng.Feet.	Eng.Feet.	Eng.Feet.	Eng.Feet.
1200	3937.08	3940.36	3943.64	3946.92	3950.20	3953.48	3956.76	3960.05	3963.33	3966.61
1210	3969.89	3973.17	3976.45	3979.73	3983.01	3986.29	3989.57	3992.85	3996.14	3999.42
1220	4002.70	4005.98	4009.26	4012.54	4015.82	4019.10	4022.38	4025.66	4028.94	4032.23
1230	4035.51	4038.79	4042.07	4045.35	4048.63	4051.91	4055.19	4058.47	4061.75	4065.03
1240	4068.31	4071.60	4074.88	4078.16	4081.44	4084.72	4088.00	4091.28	4094.56	4097.84
1250	4101.12	4104.40	4107.69	4110.97	4114.25	4117.53	4120.81	4124.09	4127.37	4130.65
1260	4133.93	4137.21	4140.49	4143.78	4147.06	4150.34	4153.62	4156.90	4160.18	4163.46
1270	4166.74	4170.02	4173.30	4176.58	4179.87	4183.15	4186.43	4189.71	4192.99	4196.27
1280	4199.55	4202.83	4206.11	4209.39	4212.67	4215.96	4219.24	4222.52	4225.80	4229.08
1290	4232.36	4235.64	4238.92	4242.20	4245.48	4248.76	4252.05	4255.33	4258.61	4261.89
1300	4265.17	4268.45	4271.73	4275.01	4278.29	4281.57	4284.85	4288.14	4291.42	4294.70
1310	4297.98	4301.26	4304.54	4307.82	4311.10	4314.38	4317.66	4320.94	4324.23	4327.51
1320	4330.79	4334.07	4337.35	4340.63	4343.91	4347.19	4350.47	4353.75	4357.03	4360.31
1330	4363.60	4366.88	4370.16	4373.44	4376.72	4380.00	4383.28	4386.56	4389.84	4393.12
1340	4396.40	4399.69	4402.97	4406.25	4409.53	4412.81	4416.09	4419.37	4422.65	4425.93
1350	4429.21	4432.49	4435.78	4439.06	4442.34	4445.62	4448.90	4452.18	4455.46	4458.74
1360	4462.02	4465.30	4468.58	4471.87	4475.15	4478.43	4481.71	4484.99	4488.27	4491.55
1370	4494.83	4498.11	4501.39	4504.67	4507.96	4511.24	4514.52	4517.80	4521.08	4524.36
1380	4527.64	4530.92	4534.20	4537.48	4540.76	4544.05	4547.33	4550.61	4553.89	4557.17
1390	4560.45	4563.73	4567.01	4570.29	4573.57	4576.85	4580.14	4583.42	4586.70	4589.98
1400	4593.26	4596.54	4599.82	4603.10	4606.38	4609.66	4612.94	4616.23	4619.51	4622.79
1410	4626.07	4629.35	4632.63	4635.91	4639.19	4642.47	4645.75	4649.03	4652.31	4655.60
1420	4658.88	4662.16	4665.44	4668.72	4672.00	4675.28	4678.56	4681.84	4685.12	4688.40
1430	4691.69	4694.97	4698.25	4701.53	4704.81	4708.09	4711.37	4714.65	4717.93	4721.21
1440	4724.49	4727.78	4731.06	4734.34	4737.62	4740.90	4744.18	4747.46	4750.74	4754.02
1450	4757.30	4760.58	4763.87	4767.15	4770.43	4773.71	4776.99	4780.27	4783.55	4786.83
1460	4790.11	4793.39	4796.67	4799.96	4803.24	4806.52	4809.80	4813.08	4816.36	4819.64
1470	4822.92	4826.20	4829.48	4832.76	4836.05	4839.33	4842.61	4845.89	4849.17	4852.45
1480	4855.73	4859.01	4862.29	4865.57	4868.85	4872.14	4875.42	4878.70	4881.98	4885.26
1490	4888.54	4891.82	4895.10	4898.38	4901.66	4904.94	4908.23	4911.51	4914.79	4918.07
1500	4921.35	4924.63	4927.91	4931.19	4934.47	4937.75	4941.03	4944.31	4947.60	4950.88
1510	4954.16	4957.44	4960.72	4964.00	4967.28	4970.56	4973.84	4977.12	4980.40	4983.69
1520	4986.97	4990.25	4993.53	4996.81	5000.09	5003.37	5006.65	5009.93	5013.21	5016.49
1530	5019.78	5023.06	5026.34	5029.62	5032.90	5036.18	5039.46	5042.74	5046.02	5049.30
1540	5052.58	5055.87	5059.15	5062.43	5065.71	5068.99	5072.27	5075.55	5078.83	5082.11
1550	5085.39	5088.67	5091.96	5095.24	5098.52	5101.80	5105.08	5108.36	5111.64	5114.92
1560	5118.20	5121.48	5124.76	5128.05	5131.33	5134.61	5137.89	5141.17	5144.45	5147.73
1570	5151.01	5154.29	5157.57	5160.85	5164.14	5167.42	5170.70	5173.98	5177.26	5180.54
1580	5183.82	5187.10	5190.38	5193.66	5196.94	5200.23	5203.51	5206.79	5210.07	5213.35
1590	5216.63	5219.91	5223.19	5226.47	5229.75	5233.03	5236.32	5239.60	5242.88	5246.16
	0.	1.	2.	3.	4.	5.	6.	7.	8.	9.

1600 to **2000.**

Metres.	Metres. (Units.)									
	0.	**1.**	**2.**	**3.**	**4.**	**5.**	**6.**	**7.**	**8.**	**9.**
	Eng.Feet.	Eng.Feet.	Eng.Feet.	Eng.Feet.	Eng.Feet.	Eng.Feet.	Eng.Feet.	Eng.Feet.	Eng.Feet.	Eng.Feet.
1600	5249.44	5252.72	5256.00	5259.28	5262.56	5265.84	5269.12	5272.40	5275.69	5278.97
1610	5282.25	5285.53	5288.81	5292.09	5295.37	5298.65	5301.93	5305.21	5308.49	5311.78
1620	5315.06	5318.34	5321.62	5324.90	5328.18	5331.46	5334.74	5338.02	5341.30	5344.58
1630	5347.87	5351.15	5354.43	5357.71	5360.99	5364.27	5367.55	5370.83	5374.11	5377.39
1640	5380.67	5383.96	5387.24	5390.52	5393.80	5397.08	5400.36	5403.64	5406.92	5410.20
1650	5413.48	5416.76	5420.05	5423.33	5426.61	5429.89	5433.17	5436.45	5439.73	5443.01
1660	5446.29	5449.57	5452.85	5456.14	5459.42	5462.70	5465.98	5469.26	5472.54	5475.82
1670	5479.10	5482.38	5485.66	5488.94	5492.23	5495.51	5498.79	5502.07	5505.35	5508.63
1680	5511.91	5515.19	5518.47	5521.75	5525.03	5528.32	5531.60	5534.88	5538.16	5541.44
1690	5544.72	5548.00	5551.28	5554.56	5557.84	5561.12	5564.40	5567.69	5570.97	5574.25
1700	5577.53	5580.81	5584.09	5587.37	5590.65	5593.93	5597.21	5600.49	5603.78	5607.06
1710	5610.34	5613.62	5616.90	5620.18	5623.46	5626.74	5630.02	5633.30	5636.58	5639.87
1720	5643.15	5646.43	5649.71	5652.99	5656.27	5659.55	5662.83	5666.11	5669.39	5672.67
1730	5675.96	5679.24	5682.52	5685.80	5689.08	5692.36	5695.64	5698.92	5702.20	5705.48
1740	5708.76	5712.05	5715.33	5718.61	5721.89	5725.17	5728.45	5731.73	5735.01	5738.29
1750	5741.57	5744.85	5748.14	5751.42	5754.70	5757.98	5761.26	5764.54	5767.82	5771.10
1760	5774.38	5777.66	5780.94	5784.23	5787.51	5790.79	5794.07	5797.35	5800.63	5803.91
1770	5807.19	5810.47	5813.75	5817.03	5820.32	5823.60	5826.88	5830.16	5833.44	5836.72
1780	5840.00	5843.28	5846.56	5849.84	5853.12	5856.40	5859.69	5862.97	5866.25	5869.53
1790	5872.81	5876.09	5879.37	5882.65	5885.93	5889.21	5892.49	5895.78	5899.06	5902.34
1800	5905.62	5908.90	5912.18	5915.46	5918.74	5922.02	5925.30	5928.58	5931.87	5935.15
1810	5938.43	5941.71	5944.99	5948.27	5951.55	5954.83	5958.11	5961.39	5964.67	5967.96
1820	5971.24	5974.52	5977.80	5981.08	5984.36	5987.64	5990.92	5994.20	5997.48	6000.76
1830	6004.05	6007.33	6010.61	6013.89	6017.17	6020.45	6023.73	6027.01	6030.29	6033.57
1840	6036.85	6040.14	6043.42	6046.70	6049.98	6053.26	6056.54	6059.82	6063.10	6066.38
1850	6069.66	6072.94	6076.23	6079.51	6082.79	6086.07	6089.35	6092.63	6095.91	6099.19
1860	6102.47	6105.75	6109.03	6112.32	6115.60	6118.88	6122.16	6125.44	6128.72	6132.00
1870	6135.28	6138.56	6141.84	6145.12	6148.40	6151.69	6154.97	6158.25	6161.53	6164.81
1880	6168.09	6171.37	6174.65	6177.93	6181.21	6184.49	6187.78	6191.06	6194.34	6197.62
1890	6200.90	6204.18	6207.46	6210.74	6214.02	6217.30	6220.58	6223.87	6227.15	6230.43
1900	6233.71	6236.99	6240.27	6243.55	6246.83	6250.11	6253.39	6256.67	6259.96	6263.24
1910	6266.52	6269.80	6273.08	6276.36	6279.64	6282.92	6286.20	6289.48	6292.76	6296.05
1920	6299.33	6302.61	6305.89	6309.17	6312.45	6315.73	6329.01	6322.29	6325.57	6328.85
1930	6332.14	6335.42	6338.70	6341.98	6345.26	6348.54	6351.82	6355.10	6358.38	6361.66
1940	6364.94	6368.23	6371.51	6374.79	6378.07	6381.35	6384.63	6387.91	6391.19	6394.47
1950	6397.75	6401.03	6404.32	6407.60	6410.88	6414.16	6417.44	6420.72	6424.00	6427.28
1960	6430.56	6433.84	6437.12	6440.41	6443.69	6446.97	6450.25	6453.53	6456.81	6460.09
1970	6463.37	6466.65	6469.93	6473.21	6476.49	6479.78	6483.06	6486.34	6489.62	6492.90
1980	6496.18	6499.46	6502.74	6506.02	6509.30	6512.58	6515.87	6519.15	6522.43	6525.71
1990	6528.99	6532.27	6535.55	6538.83	6542.11	6545.39	6548.67	6551.96	6555.24	6558.52
2000	6561.80	6565.08	6568.36	6571.64	6574.92	6578.20	6581.48	6584.76	6588.05	6591.33
	0.	**1.**	**2.**	**3.**	**4.**	**5.**	**6.**	**7.**	**8.**	**9.**

2000 to 2399.

Metres.	Metres. (Units)									
	0.	**1.**	**2.**	**3.**	**4.**	**5.**	**6.**	**7.**	**8.**	**9.**
	Eng.Feet.	Eng.Feet.	Eng.Feet.	Eng.Feet.	Eng.Feet.	Eng.Feet.	Eng.Feet.	Eng.Feet.	Eng.Feet.	Eng.Feet.
2000	6561.80	6565.08	6568.36	6571.64	6574.92	6578.20	6581.48	6584.76	6588.05	6591.33
2010	6594.61	6597.89	6601.17	6604.45	6607.73	6611.01	6614.29	6617.57	6620.85	6624.14
2020	6627.42	6630.70	6633.98	6637.26	6640.54	6643.82	6647.10	6650.38	6653.66	6656.94
2030	6660.23	6663.51	6666.79	6670.07	6673.35	6676.63	6679.91	6683.19	6686.47	6689.75
2040	6693.03	6696.32	6699.60	6702.88	6706.16	6709.44	6712.72	6716.00	6719.28	6722.56
2050	6725.84	6729.12	6732.41	6735.69	6738.97	6742.25	6745.53	6748.81	6752.09	6755.37
2060	6758.65	6761.93	6765.21	6768.49	6771.78	6775.06	6778.34	6781.62	6784.90	6788.18
2070	6791.46	6794.74	6798.02	6801.30	6804.58	6807.87	6811.15	6814.43	6817.71	6820.99
2080	6824.27	6827.55	6830.83	6834.11	6837.39	6840.67	6843.96	6847.24	6850.52	6853.80
2090	6857.08	6860.36	6863.64	6866.92	6870.20	6873.48	6876.76	6880.05	6883.33	6886.61
2100	6889.89	6893.17	6896.45	6899.73	6903.01	6906.29	6909.57	6912.85	6916.14	6919.42
2110	6922.70	6925.98	6929.26	6932.54	6935.82	6939.10	6942.38	6945.66	6948.94	6952.23
2120	6955.51	6958.79	6962.07	6965.35	6968.63	6971.91	6975.19	6978.47	6981.75	6985.03
2130	6988.32	6991.60	6994.88	6998.16	7001.44	7004.72	7008.00	7011.28	7014.56	7017.84
2140	7021.12	7024.41	7027.69	7030.97	7034.25	7037.53	7040.81	7044.09	7047.37	7050.65
2150	7053.93	7057.21	7060.49	7063.78	7067.06	7070.34	7073.62	7076.90	7080.18	7083.46
2160	7086.74	7090.02	7093.30	7096.58	7099.87	7103.15	7106.43	7109.71	7112.99	7116.27
2170	7119.55	7122.83	7125.11	7129.39	7132.67	7135.96	7139.24	7142.52	7145.80	7149.08
2180	7152.36	7155.64	7158.92	7162.20	7165.48	7168.76	7172.05	7175.33	7178.61	7181.89
2190	7185.17	7188.45	7191.73	7195.01	7198.29	7201.57	7204.85	7208.14	7211.42	7214.70
2200	7217.98	7221.26	7224.54	7227.82	7231.10	7234.38	7237.66	7240.94	7244.23	7247.51
2210	7250.79	7254.07	7257.35	7260.63	7263.91	7267.19	7270.47	7273.75	7277.03	7280.32
2220	7283.60	7286.88	7290.16	7293.44	7296.72	7300.00	7303.28	7306.56	7309.84	7313.12
2230	7316.41	7319.69	7322.97	7326.25	7329.53	7332.81	7336.09	7339.37	7342.65	7345.93
2240	7349.21	7352.49	7355.78	7359.06	7362.34	7365.62	7368.90	7372.18	7375.46	7378.74
2250	7382.02	7385.30	7388.58	7391.87	7395.15	7398.43	7401.71	7404.99	7408.27	7411.55
2260	7414.83	7418.11	7421.39	7424.67	7427.96	7431.24	7434.52	7437.80	7441.08	7444.36
2270	7447.64	7450.92	7454.20	7457.48	7460.76	7464.05	7467.33	7470.61	7473.89	7477.17
2280	7480.45	7483.73	7487.01	7490.29	7493.57	7496.85	7500.14	7503.42	7506.70	7509.98
2290	7513.26	7516.54	7519.82	7523.10	7526.38	7529.66	7532.94	7536.23	7539.51	7542.79
2300	7546.07	7549.35	7552.64	7555.91	7559.19	7562.47	7565.75	7569.03	7572.32	7575.60
2310	7578.88	7582.16	7585.44	7588.72	7592.00	7595.28	7598.56	7601.84	7605.12	7608.41
2320	7611.69	7614.97	7618.25	7621.53	7624.81	7628.09	7631.37	7634.65	7637.93	7641.21
2330	7644.50	7647.78	7651.06	7654.34	7657.62	7660.90	7664.18	7667.46	7670.74	7674.02
2340	7677.30	7680.58	7683.87	7687.15	7690.43	7693.71	7696.99	7700.27	7703.55	7706.83
2350	7710.11	7713.39	7716.67	7719.96	7723.24	7726.52	7729.80	7733.08	7736.36	7739.64
2360	7742.92	7746.20	7749.48	7752.76	7756.05	7759.33	7762.61	7765.89	7769.17	7772.45
2370	7775.73	7779.01	7782.29	7785.57	7788.85	7792.14	7795.42	7798.70	7801.98	7805.26
2380	7808.54	7811.82	7815.10	7818.38	7821.66	7824.94	7828.23	7831.51	7834.79	7838.07
2390	7841.35	7844.63	7847.91	7851.19	7854.47	7857.75	7861.03	7864.32	7867.60	7870.88
	0.	**1.**	**2.**	**3.**	**4.**	**5.**	**6.**	**7.**	**8.**	**9.**

2400 to 2799.

Metres.	Metres. (Units.)									
	0.	**1.**	**2.**	**3.**	**4.**	**5.**	**6.**	**7.**	**8.**	**9.**
	Eng.Feet.	Eng.Feet.	Eng.Feet.	Eng.Feet.	Eng.Feet.	Eng.Feet.	Eng.Feet.	Eng.Feet.	Eng.Feet.	Eng.Feet.
2400	7874.16	7877.44	7880.72	7884.00	7887.28	7890.56	7893.84	7897.12	7900.41	7903.69
2410	7906.97	7910.25	7913.53	7916.81	7920.09	7923.37	7926.65	7929.93	7933.21	7936.50
2420	7939.78	7943.06	7946.34	7949.62	7952.90	7956.18	7959.46	7962.74	7966.02	7969.30
2430	7972.59	7975.87	7979.15	7982.43	7985.71	7988.99	7992.27	7995.55	7998.83	8002.11
2440	8005.39	8008.67	8011.96	8015.24	8018.52	8021.80	8025.08	8028.36	8031.64	8034.92
2450	8038.20	8041.48	8044.76	8048.05	8051.33	8054.61	8057.89	8061.17	8064.45	8067.73
2460	8071.01	8074.29	8077.57	8080.85	8084.14	8087.42	8090.70	8093.98	8097.26	8100.54
2470	8103.82	8107.10	8110.38	8113.66	8116.94	8120.22	8123.51	8126.79	8130.07	8133.35
2480	8136.63	8139.91	8143.19	8146.47	8149.75	8153.03	8156.32	8159.60	8162.88	8166.16
2490	8169.44	8172.72	8176.00	8179.28	8182.56	8185.84	8189.12	8192.41	8195.69	8198.97
2500	8202.25	8205.53	8208.81	8212.09	8215.37	8218.65	8221.93	8225.21	8228.50	8231.78
2510	8235.06	8238.34	8241.62	8244.90	8248.18	8251.46	8254.74	8258.02	8261.30	8264.59
2520	8267.87	8271.15	8274.43	8277.71	8280.99	8284.27	8287.55	8290.83	8294.11	8297.39
2530	8300.67	8303.96	8307.24	8310.52	8313.80	8317.08	8320.36	8323.64	8326.92	8330.20
2540	8333.48	8336.76	8340.05	8343.33	8346.61	8349.89	8353.17	8356.45	8359.73	8363.01
2550	8366.29	8369.37	8372.85	8376.14	8379.42	8382.70	8385.98	8389.26	8392.54	8395.82
2560	8399.10	8402.38	8405.66	8408.94	8412.23	8415.51	8418.79	8422.07	8425.35	8428.63
2570	8431.91	8435.19	8438.47	8441.75	8445.03	8448.32	8451.60	8454.88	8458.16	8461.44
2580	8464.72	8468.00	8471.28	8474.56	8477.84	8481.12	8484.41	8487.69	8490.97	8494.25
2590	8497.53	8500.81	8504.09	8507.37	8510.65	8513.93	8517.21	8520.50	8523.78	8527.06
2600	8530.34	8533.62	8536.90	8540.18	8543.46	8546.74	8550.02	8553.30	8556.58	8559.87
2610	8563.15	8566.43	8569.71	8572.99	8576.27	8579.55	8582.83	8586.11	8589.39	8592.67
2620	8595.96	8599.24	8602.52	8605.80	8609.08	8612.36	8615.64	8618.92	8622.20	8625.48
2630	8628.76	8632.05	8635.33	8638.61	8641.89	8645.17	8648.45	8651.73	8655.01	8658.29
2640	8661.57	8664.85	8668.14	8671.42	8674.70	8677.98	8681.26	8684.54	8687.82	8691.10
2650	8694.38	8697.66	8700.94	8704.23	8707.51	8710.79	8714.07	8717.35	8720.63	8723.91
2660	8727.19	8730.47	8733.75	8737.03	8740.32	8743.60	8746.88	8750.16	8753.44	8756.72
2670	8760.00	8763.28	8766.56	8769.84	8773.12	8776.41	8779.69	8782.97	8786.25	8789.53
2680	8792.81	8796.09	8799.37	8802.65	8805.93	8809.21	8812.50	8815.78	8819.06	8822.34
2690	8825.62	8828.90	8832.18	8835.46	8838.74	8842.02	8845.30	8848.59	8851.87	8855.15
2700	8858.43	8861.71	8864.99	8868.27	8871.55	8874.83	8878.11	8881.39	8884.67	8887.96
2710	8891.24	8894.52	8897.80	8901.08	8904.36	8907.64	8910.92	8914.20	8917.48	8920.76
2720	8926.05	8927.33	8930.61	8933.89	8937.17	8940.45	8943.73	8947.01	8950.29	8953.57
2730	8956.85	8960.14	8963.42	8966.70	8969.98	8973.26	8976.54	8979.82	8983.10	8986.38
2740	8989.66	8992.94	8996.23	8999.51	9002.79	9006.07	9009.35	9012.63	9015.91	9019.19
2750	9022.47	9025.75	9029.03	9032.32	9035.60	9038.88	9042.16	9045.44	9048.72	9052.00
2760	9055.28	9058.56	9061.84	9065.12	9068.41	9071.69	9074.97	9078.25	9081.53	9084.81
2770	9088.09	9091.37	9094.65	9097.93	9101.21	9104.50	9107.78	9111.06	9114.34	9117.62
2780	9120.90	9124.18	9127.46	9130.74	9134.02	9137.30	9140.59	9143.87	9147.15	9150.43
2790	9153.71	9156.99	9160.27	9163.55	9166.83	9170.11	9173.39	9176.68	9179.96	9183.24
	0.	**1.**	**2.**	**3.**	**4.**	**5.**	**6.**	**7.**	**8.**	**9.**

2800 to 3000.

Metres.	Metres. (Units.)									
	0.	1.	2.	3.	4.	5.	6.	7.	8.	9.
	Eng.Feet.	Eng.Feet.	Eng.Feet.	Eng.Feet.	Eng.Feet.	Eng.Feet.	Eng.Feet.	Eng.Feet.	Eng.Feet.	Eng.Feet.
2800	9186.52	9189.80	9193.08	9196.36	9199.64	9202.92	9206.20	9209.48	9212.76	9216.05
2810	9219.33	9222.61	9225.89	9229.17	9232.45	9235.73	9239.01	9242.29	9245.57	9248.85
2820	9252.14	9255.42	9258.70	9261.98	9265.26	9268.54	9271.82	9275.10	9278.38	9281.66
2830	9284.94	9288.23	9291.51	9294.79	9298.07	9301.35	9304.64	9307.91	9311.19	9314.47
2840	9317.75	9321.03	9324.32	9327.60	9330.88	9334.16	9337.44	9340.72	9344.00	9347.28
2850	9350.56	9353.84	9357.12	9360.41	9363.69	9366.97	9370.25	9373.53	9376.81	9380.09
2860	9383.37	9386.65	9389.93	9393.21	9396.50	9399.78	9403.06	9406.34	9409.62	9412.90
2870	9416.18	9419.46	9422.74	9426.02	9429.30	9432.59	9435.87	9439.15	9442.43	9445.71
2880	9448.99	9452.27	9455.55	9458.83	9462.11	9465.39	9468.68	9471.96	9475.24	9478.52
2890	9481.80	9485.08	9488.36	9491.64	9494.92	9498.20	9501.48	9504.76	9508.05	9511.33
2900	9514.61	9517.89	9521.17	9524.45	9527.73	9531.01	9534.29	9537.57	9540.85	9544.14
2910	9547.42	9550.70	9553.98	9557.26	9560.54	9563.82	9567.10	9570.38	9573.66	9576.94
2920	9580.23	9583.51	9586.79	9590.07	9593.35	9596.63	9599.91	9603.19	9606.47	9609.75
2930	9613.03	9616.32	9619.60	9622.88	9626.16	9629.44	9632.72	9636.00	9639.28	9642.56
2940	9645.84	9649.12	9652.41	9655.69	9658.97	9662.25	9665.53	9668.81	9672.09	9675.37
2950	9678.62	9681.93	9685.21	9688.50	9691.78	9695.06	9698.34	9701.62	9704.90	9708.18
2960	9711.46	9714.74	9718.02	9721.30	9724.59	9727.87	9731.15	9734.43	9737.71	9740.99
2970	9744.27	9747.55	9750.83	9754.11	9757.39	9760.68	9763.96	9767.24	9770.52	9773.80
2980	9777.08	9780.36	9783.64	9786.92	9790.20	9793.48	9796.76	9800.05	9803.33	9806.61
2990	9809.89	9813.17	9816.45	9819.73	9823.01	9826.29	9829.57	9832.85	9836.14	9839.42
3000	9842.70	9845.98	9849.26	9852.54	9855.82	9859.10	9862.38	9865.66	9868.94	9872.23

Proportional Parts.

Metres.	Decimetres.									
	0.	1.	2.	3.	4.	5.	6.	7.	8.	9.
	Eng.Feet.	Eng.Feet.	Eng.Feet.	Eng.Feet.	Eng.Feet.	Eng.Feet	Eng.Feet	Eng.Feet	Eng.Feet.	Eng.Feet.
0	0.0000	0.3281	0.6562	0.9843	1.3124	1.6404	1.9685	2.2966	2.6247	2.9528
1	3.2809	3.6090	3.9371	4.2652	4.5933	4.9213	5.2494	5.5775	5.9056	6.2337
2	6.5618	6.8899	7.2180	7.5461	7.8742	8.2022	8.5303	8.8584	9.1865	9.5146
3	9.8427	10.1708	10.4989	10.8270	11.1551	11.4831	11.8112	12.1393	12.4674	12.7955
4	13.1236	13.4517	13.7798	14.1079	14.4360	14.7640	15.0921	15.4202	15.7483	16.0764
5	16.4045	16.7326	17.0607	17.3888	17.7169	18.0449	18.3730	18.7011	19.0292	19.3573
6	19.6854	20.0135	20.3416	20.6697	20.9978	21.3258	21.6539	21.9820	22.3101	22.6382
7	22.9663	23.2944	23.6225	23.9506	24.2787	24.6067	24.9348	25.2629	25.5910	25.9191
8	26.2472	26.5753	26.9034	27.2315	27.5596	27.8876	28.2157	28.5438	28.8719	29.2000
9	29.5281	29.8562	30.1843	30.5124	30.8405	31.1685	31.4966	31.8247	31.1528	32.4809
	0.	1.	2.	3.	4.	5.	6.	7.	8.	9.

VIII. CONVERSION OF METRES INTO AMERICAN FEET AND DECIMALS.

1 Metre = 3.28070878 American Feet.

Metres. Thousands.	Hundreds.									
	0.	**100.**	**200.**	**300.**	**400.**	**500.**	**600.**	**700.**	**800.**	**900.**
	Am. Feet	Am Feet	Am. Feet	Am. Feet	Am. Feet	Am. Feet	Am. Feet	Am. Feet	Am. Feet	Am. Feet
0	0.0	328.1	656.1	984.2	1312.3	1640.4	1968.4	2296.5	2624.6	2952.6
1000	3280.7	3608.8	3936.9	4264.9	4593.0	4921.1	5249.1	5577.2	5905.3	6233.3
2000	6561.4	6889.5	7217.6	7545.6	7873.7	8201.8	8529.8	8857.9	9186.0	9514.1
3000	9842.1	10170.2	10498.3	10826.3	11154.4	11482.5	11810.6	12138.6	12466.7	12794.8
4000	13122.8	13450.9	13779.0	14107.0	14435.1	14763.2	15091.3	15419.3	15747.4	16075.5
5000	16403.5	16731.6	17059.7	17387.8	17715.8	18043.9	18372.0	18700.0	19028.1	19356.2
6000	19684.3	20012.3	20340.4	20668.5	20996.5	21324.6	21652.7	21980.7	22308.8	22636.9
7000	22965.0	23293.0	23621.1	23949.2	24277.2	24605.3	24933.4	25261.5	25589.5	25917.6
8000	26245.7	26573.7	26901.8	27229.9	27558.0	27886.0	28214.1	28542.2	28870.2	29198.3
9000	29526.4	29854.4	30182.5	30510.6	30838.7	31166.7	31494.8	31822.9	32150.9	32479.0

Tens.	Units.									
	0.	**1.**	**2.**	**3.**	**4.**	**5.**	**6.**	**7.**	**8.**	**9.**
	Am. Feet	Am Feet.	Am. Feet	Am. Feet.	Am. Feet	Am Feet.	Am. Feet	Am. Feet.	Am Feet.	Am. Feet.
0	0.000	3.281	6.561	9.842	13.123	16.404	19.684	22.965	26.246	29.526
10	32.807	36.088	39.369	42.649	45.930	49.211	52.491	55.772	59.053	62.333
20	65.614	68.895	72.176	75.456	78.737	82.018	85.298	88.579	91.860	95.141
30	98.421	191.702	104.983	108.263	111.544	114.825	118.106	121.386	124.667	127.948
40	131.228	134.509	137.790	141.070	144.351	147.632	150.913	154.193	157.474	160.755
50	164.035	167.316	170.597	173.878	177.158	180.439	183.720	187.000	190.281	193.562
60	196.843	200.123	203.404	206.685	209.965	213.246	216.527	219.807	223.088	226.369
70	229.650	232.930	236.211	239.492	242.772	246.053	249.334	252.615	255.895	259.176
80	262.457	265.737	269.018	272.299	275.580	278.860	282.141	285.422	288.702	291.983
90	295.264	298.544	301.825	305.106	308.387	311.667	314.948	318.229	321.509	324.790

IX. CONVERSION OF METRES INTO RHINE OR PRUSSIAN FEET AND DECIMALS.

1 Metre = 3 1861995 Rhine Feet.

Metres. Thousands	Hundreds.									
	0.	**100.**	**200.**	**300.**	**400.**	**500.**	**600.**	**700.**	**800.**	**900.**
	Rhine Ft.	Rhine Ft	Rhine Ft	Rhine Ft	Rhine Ft.	Rhine Ft	Rhine Ft.	Rhine Ft.	Rhine Ft	Rhine Ft.
0	0.0	318.6	637.2	955.9	1274.5	1593.1	1911.7	2230.3	2549.0	2867.6
1000	3186.2	3504.8	3823.4	4142.1	4460.7	4779.3	5097.9	5416.5	5735.2	6053.8
2000	6372.4	6691.0	7009.6	7328.3	7646.9	7965.5	8284.1	8602.7	8921.4	9240.0
3000	9558.6	9877.2	10195.8	10514.5	10833.1	11151.7	11470.3	11788.9	12107.6	12426.2
4000	12744.8	13063.4	13382.0	13700.7	14019.3	14337.9	14656.5	14975.1	15293.8	15612.4
5000	15931.0	16249.6	16568.2	16886.9	17205.5	17524.1	17842.7	18161.3	18480.0	18798.6
6000	19117.2	19435.8	19754.4	20073.1	20391.7	20710.3	21028.9	21347.5	21666.2	21984.8
7000	22303.4	22622.0	22940.6	23259.3	23577.9	23896.5	24215.1	24533.7	24852.4	25171.0
8000	25489.6	25808.2	26126.8	26445.5	26764.1	27082.7	27401.3	27719.9	28038.6	28357.2
9000	28675.8	28994.4	29313.0	29631.7	29950.3	30268.9	30587.5	30906.1	31224.8	31543.4

TO CONVERT

PARIS OR FRENCH FEET

INTO DIFFERENT MEASURES OF LENGTH.

X. CONVERSION OF PARIS OR FRENCH FEET INTO TOISES.

1 French Foot = 0.1666666 Toise.

French Feet. Thousands.	Hundreds. 0.	100.	200.	300.	400.	500.	600.	700.	800.	900.
	Toises.	Toises.	Toises.	Toises.	Toises.	Toises	Toises.	Toises.	Toises.	Toises.
0	0.00	16.67	33.33	50.00	66.67	83.33	100.00	116.67	133.33	150.00
1000	166.67	183.33	200.00	216.67	233.33	250.00	266.67	283.33	300.00	316.67
2000	333.33	350.00	366.67	383.33	400.00	416.67	433.33	450.00	466.67	483.33
3000	500.00	516.67	533.33	550.00	566.67	583.33	600.00	616.67	633.33	650.00
4000	666.67	683.33	700.00	716.67	733.33	750.00	766.67	783.33	800.00	816.67
5000	833.33	850.00	866.67	883.33	900.00	916.67	933.33	950.00	966.67	983.33
6000	1000.00	1016.67	1033.33	1050.00	1066.67	1083.33	1100.00	1116.67	1133.33	1150.00
7000	1166.67	1183.33	1200.00	1216.67	1233.33	1250.00	1266.67	1283.33	1300.00	1316.67
8000	1333.33	1350.00	1366.67	1383.33	1400.00	1416.67	1433.33	1450.00	1466.67	1483.33
9000	1500.00	1516.67	1533.33	1550.00	1566.67	1583.33	1600.00	1616.67	1633.33	1650.00
10000	1666.67	1683.33	1700.00	1716.67	1733.33	1750.00	1766.67	1783.33	1800.00	1816.67
11000	1833.33	1850.00	1866.67	1883.33	1900.00	1916.67	1933.33	1950.00	1966.67	1983.33
12000	2000.00	2016.67	2033.33	2050.00	2066.67	2083.33	2100.00	2116.67	2133.33	2150.00
13000	2166.67	2183.33	2200.00	2216.67	2233.33	2250.00	2266.67	2283.33	2300.00	2316.67
14000	2333.33	2350.00	2366.67	2383.33	2400.00	2416.67	2433.33	2450.00	2466.67	2483.33
15000	2500.00	2516.67	2533.33	2550.00	2566.67	2583.33	2600.00	2616.67	2633.33	2650.00
16000	2666.67	2683.33	2700.00	2716.67	2733.33	2750.00	2766.67	2783.33	2800.00	2816.67
17000	2833.33	2850.00	2866.67	2883.33	2900.00	2916.67	2933.33	2950.00	2966.67	2983.33
18000	3000.00	3016.67	3033.33	3050.00	3066.67	3083.33	3100.00	3116.67	3133.33	3150.00
19000	3166.67	3183.33	3200.00	3216.67	3233.33	3250.00	3266.67	3283.33	3300.00	3316.67
20000	3333.33	3350.00	3366.67	3383.33	3400.00	3416.67	3433.33	3450.00	3466.67	3483.33
21000	3500.00	3516.67	3533.33	3550.00	3566.67	3583.33	3600.00	3616.67	3633.33	3650.00
22000	3666.67	3683.33	3700.00	3716.67	3733.33	3750.00	3766.67	3783.33	3800.00	3816.67
23000	3833.33	3850.00	3866.67	3883.33	3900.00	3916.67	3933.33	3950.00	3966.67	3983.33
24000	4000.00	4016.67	4033.33	4050.00	4066.67	4083.33	4100.00	4116.67	4133.33	4150.00
25000	4166.67	4183.33	4200.00	4216.67	4233.33	4250.00	4266.67	4283.33	4300.00	4316.67
26000	4333.33	4350.00	4366.67	4383.33	4400.00	4416.67	4433.33	4450.00	4466.67	4483.33

1 Paris Foot = 0.32483943 Metres.

French Feet. Thousands.	Hundreds. 0.	100.	200.	300.	400.	500.	600.	700.	800.	900.
	Metres.	Metres.	Metres.	Metres.	Metres.	Metres.	Metres.	Metres.	Metres.	Metres.
0	000.00	32.48	64.97	97.45	129.94	162.42	194.90	227.39	259.87	292.36
1000	324.84	357.32	389.81	422.29	454.78	487.26	519.74	552.23	584.71	617.19
2000	649.68	682.16	714.65	747.13	779.61	812.10	844.58	877.07	909.55	942.03
3000	974.52	1007.00	1039.49	1071.97	1104.45	1136.94	1169.42	1201.91	1234.39	1266.87
4000	1299.36	1331.84	1364.33	1396.81	1429.29	1461.78	1494.26	1526.75	1559.23	1591.71
5000	1624.20	1656.68	1689.16	1721.65	1754.13	1786.62	1819.10	1851.58	1884.07	1916.55
6000	1949.04	1981.52	2014.00	2046.49	2078.97	2111.46	2143.94	2176.42	2208.91	2241.39
7000	2273.88	2306.36	2338.84	2371.33	2403.81	2436.30	2468.78	2501.26	2533.75	2566.23
8000	2598.72	2631.20	2663.68	2696.17	2728.65	2761.14	2793.62	2826.10	2858.59	2891.07
9000	2923.55	2956.04	2988.52	3021.01	3053.49	3085.97	3118.46	3150.94	3183.43	3215.91
10000	3248.39	3280.88	3313.36	3345.85	3378.33	3410.81	3443.30	3475.78	3508.27	3540.75
11000	3573.23	3605.72	3638.20	3670.69	3703.17	3735.65	3768.14	3800.62	3833.11	3865.59
12000	3898.07	3930.56	3963.04	3995.52	4028.01	4060.49	4092.98	4125.46	4157.94	4190.43
13000	4222.91	4255.40	4287.88	4320.36	4352.85	4385.33	4417.82	4450.30	4482.78	4515.27
14000	4547.75	4580.24	4612.72	4645.10	4677.59	4710.07	4742.56	4775.04	4807.52	4840.01
15000	4872.59	4905.08	4937.56	4970.04	5002.53	5035.01	5067.49	5099.98	5132.46	5164.95
16000	5197.43	5229.91	5262.40	5294.88	5327.37	5359.85	5392.33	5424.82	5457.30	5489.79
17000	5522.27	5554.75	5587.24	5619.72	5652.21	5684.69	5717.17	5749.66	5782.14	5814.63
18000	5847.11	5879.59	5912.08	5944.56	5977.05	6009.53	6042.01	6074.50	6106.98	6139.47
19000	6171.95	6204.43	6236.92	6269.40	6301.88	6334.37	6366.85	6399.34	6431.82	6464.30
20000	6496.79	6529.27	6561.76	6594.24	6626.72	6659.21	6691.69	6724.18	6756.66	6789.14
21000	6821.63	6854.11	6886.60	6919.08	6951.56	6984.05	7016.53	7049.02	7081.50	7113.98
22000	7146.47	7178.95	7211.44	7243.92	7276.40	7308.89	7341.37	7373.86	7406.34	7438.82
23000	7471.31	7503.79	7536.27	7568.76	7601.24	7633.73	7666.21	7698.69	7731.18	7763.66
24000	7796.15	7828.63	7861.11	7893.60	7926.08	7958.57	7991.05	8023.53	8056.02	8088.50
25000	8120.99	8153.47	8185.95	8218.44	8250.92	8283.41	8315.89	8448.37	8380.86	8413.34
26000	8445.83	8478.31	8510.79	8543.28	8575.76	8608.24	8640.73	8673.21	8705.70	8738.18
27000	8770.66	8803.15	8835.63	8868.12	8900.60	8933.08	8965.57	8998.05	9030.54	9063.02

Tens.	Units. 0.	1.	2.	3.	4.	5.	6.	7.	8.	9.
	Metres.	Metres.	Metres.	Metres.	Metres.	Metres.	Metres.	Metres.	Metres.	Metres.
0	0.0000	0.3248	0.6497	0.9745	1.2994	1.6242	1.9490	2.2739	2.5987	2.9236
10	3.2484	3.5732	3.8981	4.2229	4.5478	4.8726	5.1974	5.5223	5.8471	6.1719
20	6.4968	6.8216	7.1465	7.4713	7.7961	8.1210	8.4458	8.7707	9.0955	9.4203
30	9.7452	10.0700	10.3949	10.7197	11.0445	11.3694	11.6942	12.0191	12.3439	12.6687
40	12.9936	13.3184	13.6433	13.9681	14.2929	14.6178	14.9426	15.2675	15.5923	15.9171
50	16.2420	16.5668	16.8916	17.2165	17.5413	17.8662	18.1910	18.5158	18.8407	19.1655
60	19.4904	19.8152	20.1400	20.4649	20.7897	21.1146	21.4394	21.7642	22.0891	22.4139
70	22.7388	23.0636	23.3884	23.7133	24.0381	24.3630	24.6878	25.0126	25.3375	25.6623
80	25.9872	26.3120	26.6368	26.9617	27.2865	27.6114	27.9362	28.2610	28.5859	28.9107
90	29.2355	29.5604	29.8852	30.2101	30.5349	30.8597	31.1846	31.5094	31.8343	32.1591

1 French Foot = 1.065765424 English Feet.

French Feet. Thousands.	Hundreds.									
	0.	**100.**	**200.**	**300.**	**400.**	**500.**	**600.**	**700.**	**800.**	**900.**
	Eng. feet	Eng. feet.	Eng feet.	Eng feet.	Eng. feet.	Eng. feet.	Eng. feet.	Eng. feet.	Eng. feet.	Eng. feet.
0	0.0	106.6	213.2	319.7	426.3	532.9	639.5	746.0	852.6	959.2
1000	1065.8	1172.3	1278.9	1385.5	1492.1	1598.6	1705.2	1811.8	1918.4	2025.0
2000	2131.5	2238.1	2344.7	2451.3	2557.8	2664.4	2771.0	2877.6	2984.1	3090.7
3000	3197.3	3303.9	3410.4	3517.0	3623.6	3730.2	3836.8	3943.3	4049.9	4156.5
4000	4263.1	4369.6	4476.2	4582.8	4689.4	4795.9	4902.5	5009.1	5115.7	5222.3
5000	5328.8	5435.4	5542.0	5648.6	5755.1	5861.7	5968.3	6074.9	6181.4	6288.0
6000	6394.6	6501.2	6607.7	6714.3	6820.9	6927.5	7034.1	7140.6	7247.2	7353.8
7000	7460.4	7566.9	7673.5	7780.1	7886.7	7993.2	8099.8	8206.4	8313.0	8419.5
8000	8526.1	8632.7	8739.3	8845.9	8952.4	9059.0	9165.6	9272.2	9378.7	9485.3
9000	9591.9	9698.5	9805.0	9911.6	10018.2	10124.8	10231.3	10337.9	10444.5	10551.1
10000	10657.7	10764.2	10870.8	10977.4	11084.0	11190.5	11297.1	11403.7	11510.3	11616.8
11000	11723.4	11830.0	11936.6	12043.1	12149.7	12256.3	12362.9	12469.5	12576.0	12682.6
12000	12789.2	12895.8	13002.3	13108.9	13215.5	13322.1	13428.6	13535.2	13641.8	13748.4
13000	13855.0	13961.5	14068.1	14174.7	14281.3	14387.8	14494.4	14601.0	14707.6	14814.1
14000	14920.7	15027.3	15133.9	15240.4	15347.0	15453.6	15560.2	15666.8	15773.3	15879.9
15000	15986.5	16093.1	16199.6	16306.2	16412.8	16519.4	16625.9	16732.5	16839.1	16945.7
16000	17052.2	17158.8	17265.4	17372.0	17478.6	17585.1	17691.7	17798.3	17904.9	18011.4
17000	18118.0	18224.6	18331.2	18437.7	18544.3	18650.9	18757.5	18864.0	18970.6	19077.2
18000	19183.8	19290.4	19396.9	19503.5	19610.1	19716.7	19823.2	19929.8	20036.4	20143.0
19000	20249.5	20356.1	20462.7	20569.3	20675.8	20782.4	20889.0	20995.6	21102.2	21208.7
20000	21315.3	21421.9	21528.5	21635.0	21741.6	21848.2	22054.8	22161.3	22167.9	22274.5
21000	22381.1	22487.7	22594.2	22700.8	22807.4	22914.0	23020.5	23127.1	23233.7	23340.3
22000	23446.8	23553.4	23660.0	23766.6	23873.1	23979.7	24086.3	24192.9	24299.5	24406.0
23000	24512.6	24619.2	24725.8	24832.3	24938.9	25045.5	25152.1	25258.6	25365.2	25471.8
24000	25578.4	25684.9	25791.5	25898.1	26004.7	26111.3	26217.8	26324.4	26431.0	26537.6
25000	26644.1	26750.7	26857.3	26963.9	27070.4	27177.0	27283.6	27390.2	27496.7	27603.3
26000	27709.9	27816.5	27923.1	28029.6	28136.2	28242.8	28349.4	28455.9	28562.5	28669.1
27000	28775.7	28882.2	28988.8	29095.4	29202.0	29308.5	29415.1	29521.7	29628.3	29734.9

Tens.	Units.									
	0.	**1.**	**2.**	**3.**	**4.**	**5.**	**6.**	**7.**	**8.**	**9.**
	Eng. feet.	Eng. feet.	Eng. feet.	Eng. feet.	Eng. feet.	Eng. feet.	Eng. feet.	Eng. feet.	Eng. feet.	Eng. feet
0	0.000	1.066	2.132	3.197	4.263	5.329	6.395	7.460	8.526	9.592
10	10.658	11.723	12.789	13.855	14.921	15.986	17.052	18.118	19.184	20.250
20	21.315	22.381	23.447	24.513	25.578	26.644	27.710	28.776	29.841	30.907
30	31.973	33.039	34.104	35.170	36.236	37.302	38.368	39.433	40.499	41.565
40	42.631	43.696	44.762	45.828	46.894	47.959	49.025	50.091	51.157	52.223
50	53.288	54.354	55.420	56.486	57.551	58.617	59.683	60.749	61.814	62.880
60	63.946	65.012	66.077	67.143	68.209	69.275	70.341	71.406	72.472	73.538
70	74.604	75.669	76.735	77.801	78.867	79.932	80.998	82.064	83.130	84.195
80	85.261	86.327	87.393	88.459	89.524	90.590	91.656	92.722	93.787	94.853
90	95.919	96.985	98.050	99.116	100.182	101.248	102.313	103.379	104.445	105.511

1 French Foot = 1.06570358 American Foot.

French Feet. Thousands.	Hundreds.									
	0.	**100.**	**200.**	**300.**	**400.**	**500.**	**600.**	**700.**	**800.**	**900.**
	Am. Feet	Am. Feet	Am. Feet.	Am. Feet.	Am Feet	Am. Feet	Am. Feet	Am. Feet.	Am. Feet	Am Feet
0	0.0	106.6	213.1	319.7	426.3	532.9	639.4	746.0	852.6	959.1
1000	1065.7	1172.3	1278.8	1385.4	1492.0	1598.6	1705.1	1811.7	1918.3	2024.8
2000	2131.4	2238.0	2344.5	2451.1	2557.7	2664.3	2770.8	2877.4	2984.0	3090.5
3000	3197.1	3303.7	3410.3	3516.8	3623.4	3730.0	3836.5	3943.1	4049.7	4156.2
4000	4262.8	4369.4	4476.0	4582.5	4689.1	4795.7	4902.2	5008.8	5115.4	5221.9
5000	5328.5	5435.1	5541.7	5648.2	5754.8	5861.4	5967.9	6074.5	6181.1	6287.7
6000	6394.2	6500.8	6607.4	6713.9	6820.5	6927.1	7033.6	7140.2	7246.8	7353.4
7000	7459.9	7566.5	7673.1	7779.6	7886.2	7992.8	8099.3	8205.9	8312.5	8419.1
8000	8525.6	8632.2	8738.8	8845.3	8951.9	9058.5	9165.1	9271.6	9378.2	9484.8
9000	9591.3	9697.9	9804.5	9911.0	10017.6	10124.2	10230.8	10337.3	10443.9	10550.5
10000	10657.0	10763.6	10870.2	10976.7	11083.3	11189.9	11296.5	11403.0	11509.6	11616.2
11000	11722.7	11829.3	11935.9	12042.5	12149.0	12255.6	12362.2	12468.7	12575.3	12681.9
12000	12788.4	12895.0	13001.6	13108.2	13214.7	13321.3	13427.9	13534.4	13641.0	13747.6
13000	13854.1	13960.7	14067.3	14173.9	14280.4	14387.0	14493.6	14600.1	14706.7	14813.3
14000	14919.9	15026.4	15133.0	15239.6	15346.1	15452.7	15559.3	15665.8	15772.4	15879.0
15000	15985.6	16092.1	16198.7	16305.3	16411.8	16518.4	16625.0	16731.5	16838.1	16944.7
16000	17051.3	17157.8	17264.4	17371.0	17477.5	17584.1	17690.7	17797.2	17903.8	18010.4
17000	18117.0	18223.5	18330.1	18436.7	18543.2	18649.8	18756.4	18863.0	18969.5	19076.1
18000	19182.7	19289.2	19395.8	19502.4	19608.9	19715.5	19822.1	19928.7	20035.2	20141.8
19000	20248.4	20354.9	20461.5	20568.1	20674.6	20781.2	20887.8	20994.4	21100.9	21207.5
20000	21314.1	21420.6	21527.2	21633.8	21740.4	21846.9	21953.5	22060.1	22166.6	22273.2
21000	22379.8	22486.3	22592.9	22699.5	22806.1	22912.6	23019.2	23125.8	23232.3	23338.9
22000	23445.5	23552.0	23658.6	23765.2	23871.8	23978.3	24084.9	24191.5	24298.0	24404.6
23000	24511.2	24617.8	24724.3	24830.9	24937.5	25044.0	25150.6	25257.2	25363.7	25470.3
24000	25576.9	25683.5	25790.0	25896.6	26003.2	26109.7	26216.3	26322.9	16429.4	26536.0
25000	26642.6	26749.2	26855.7	26962.3	27068.9	27175.4	27282.0	27388.6	27495.2	27601.7
26000	27708.3	27814.9	27921.4	28028.0	28134.6	28241.1	28347.7	28454.3	28560.9	28667.4
27000	28774.0	28880.6	28987.1	29093.7	29200.3	29306.8	29413.4	29520.0	29626.6	29733.1
28000	29839.7	29946.3	30052.8	30159.4	30266.0	30372.6	30479.1	30585.7	30692.3	30799.8

Tens.	Units.									
	0.	**1.**	**2.**	**3.**	**4.**	**5.**	**6.**	**7.**	**8.**	**9.**
	Am. Feet.	Am. Feet.	Am Feet.	Am. Feet.	Am Feet	Am Feet.	Am. Feet.	Am Feet.	Am Feet	Am. Feet.
0	0.00	1.07	2.13	3.20	4.26	5.33	6.39	7.46	8.53	9.59
10	10.66	11.72	12.79	13.85	14.92	15.99	17.05	18.12	19.18	20.25
20	21.31	22.38	23.45	24.51	25.58	26.64	27.71	28.77	29.84	30.91
30	31.97	33.04	34.10	35.17	36.23	37.30	38.37	39.43	40.50	41.56
40	42.63	43.69	44.76	45.83	46.89	47.96	49.02	50.09	51.15	52.22
50	53.29	54.35	55.42	56.48	57.55	58.61	59.68	60.75	61.81	62.88
60	63.94	65.01	66.07	67.14	68.21	69.27	70.34	71.40	72.47	73.53
70	74.60	75.66	76.73	77.80	78.86	79.93	80.99	82.06	83.12	84.19
80	85.26	86.32	87.39	88.45	89.52	90.58	91.65	92.72	93.78	94.85
90	95.91	96.98	98.04	99.11	100.18	101.24	102.31	103.37	104.44	105.50

TO CONVERT

ENGLISH YARDS AND FEET

INTO DIFFERENT MEASURES OF LENGTH.

XIV. CONVERSION OF ENGLISH YARDS INTO FRENCH TOISES.

1 English Yard = 0.4691464 Toise.

English Yards.	Hundreds.									
Thousands.	**0.**	**100.**	**200.**	**300.**	**400.**	**500.**	**600.**	**700.**	**800.**	**900.**
	Toises.	Toises.	Toises	Toises.	Toises.	Toises.	Toises	Toises.	Toises.	Toises.
0	0.00	46.91	93.83	140.74	187.66	234.57	281.49	328.40	375.32	422.23
1000	469.15	516.06	562.98	609.89	656.80	703.72	750.63	797.55	844.46	891.38
2000	938.29	985.21	1032.12	1079.04	1125.95	1172.87	1219.78	1266.70	1313.61	1360.52
3000	1407.44	1454.35	1501.27	1548.18	1595.10	1642.01	1688.93	1735.84	1782.76	1829.67
4000	1876.59	1923.50	1970.41	2017.33	2064.24	2111.16	2158.07	2204.99	2251.90	2298.82
5000	2345.73	2392.65	2439.56	2486.48	2533.39	2580.31	2627.22	2674.13	2721.05	2767.96
6000	2814.88	2861.79	2908.71	2955.62	3002.54	3049.45	3096.37	3143.28	3190.20	3237.11
7000	3284.02	3330.94	3377.85	3424.77	3471.68	3518.60	3565.51	3612.43	3659.34	3706.26
8000	3753.17	3800.09	3847.00	3893.92	3940.83	3987.74	4034.66	4081.57	4128.49	4175.40
9000	4222.32	4269.23	4316.15	4363.06	4409.98	4456.89	4503.81	4550.72	4597.63	4644.55

XV. CONVERSION OF ENGLISH YARDS INTO METRES.

1 English Yard = 0.91438347 Metre.

English Yards.	Hundreds.									
Thousands.	**0.**	**100.**	**200.**	**300.**	**400.**	**500.**	**600.**	**700.**	**800.**	**900.**
	Metres.	Metres.	Metres.	Metres.	Metres.	Metres.	Metres.	Metres.	Metres.	Metres.
0	0.00	91.44	182.88	274.32	365.75	457.19	548.63	640.07	731.51	822.95
1000	914.38	1005.82	1097.26	1188.70	1280.14	1371.58	1463.01	1554.45	1645.89	1737.33
2000	1828.77	1920.21	2011.64	2103.08	2194.52	2285.96	2377.40	2468.84	2560.27	2651.71
3000	2743.15	2834.59	2926.03	3017.47	3108.90	3200.34	3291.78	3383.22	3474.66	3566.10
4000	3657.53	3748.97	3840.41	3931.85	4023.29	4114.73	4206.16	4297.60	4389.04	4480.48
5000	4571.92	4663.36	4754.79	4846.23	4937.67	5029.11	5120.55	5211.99	5303.42	5394.86
6000	5486.30	5577.74	5669.18	5760.62	5852.05	5943.49	6034.93	6126.37	6217.81	6309.25
7000	6400.68	6492.12	6583.56	6675.00	6766.44	6857.88	6949.31	7040.75	7132.19	7223.63
8000	7315.07	7406.51	7497.94	7589.38	7680.82	7772.26	7863.70	7955.14	8046.57	8138.01
9000	8229.45	8320.89	8412.33	8503.77	8595.20	8686.64	8778.08	8869.52	8960.96	9052.40

1 English Foot = 0.30479449 Metre.

English Feet. Thousands.	Hundreds.									
	0.	**100.**	**200.**	**300.**	**400.**	**500.**	**600.**	**700.**	**800.**	**900.**
	Metres.	Metres.	Metres.	Metres.	Metres.	Metres.	Metres.	Metres.	Metres.	Metres.
0	000.000	30.4794	60.9589	91.4383	121.918	152.397	182.877	213.356	243.836	274.315
1000	304.794	335.274	365.763	396.233	426.712	457.192	487.671	518.151	548.630	579.110
2000	609.589	640.068	670.548	701.027	731.507	761.986	792.466	822.945	853.425	883.904
3000	914.383	944.863	975.342	1005.82	1036.30	1066.78	1097.26	1127.74	1158.22	1188.70
4000	1219.18	1249.66	1280.14	1310.62	1341.10	1371.58	1402.05	1432.53	1463.01	1493.49
5000	1523.97	1554.45	1584.93	1615.41	1645.89	1676.37	1706.85	1737.33	1767.81	1798.29
6000	1828.77	1859.25	1889.73	1920.21	1950.68	1981.16	2011.64	2042.12	2072.60	2103.08
7000	2133.56	2164.04	2194.52	2225.00	2255.48	2285.96	2316.44	2346.92	2377.40	2407.88
8000	2438.36	2468.84	2499.31	2529.79	2560.27	2590.75	2621.23	2651.71	2682.19	2712.67
9000	2743.15	2773.63	2804.11	2834.59	2865.07	2895.55	2926.03	2956.51	2986.99	3017.47
10000	3047.94	3078.42	3108.90	3139.38	3169.86	3200.34	3230.82	3261.30	3291.78	3322.26
11000	3352.74	3383.22	3413.70	3444.18	3474.66	3505.14	3535.62	3566.10	3596.57	3627.05
12000	3657.53	3688.01	3718.49	3748.97	3779.45	3809.93	3840.41	3870.89	3901.37	3931.85
13000	3962.33	3992.81	4023.29	4053.77	4084.25	4114.73	4145.21	4175.68	4206.16	4236.64
14000	4267.12	4297.60	4228.08	4358.56	4389.04	4419.52	4450.00	4480.48	4510.96	4541.44
15000	4571.92	4602.40	4632.88	4663.36	4693.84	4724.31	4754.79	4785.27	4815.75	4846.23
16000	4876.71	4907.19	4937.67	4968.15	4998.63	5029.11	5059.59	5090.07	5120.55	5151.03
17000	5181.51	5211.99	5242.47	5272.94	5303.42	5333.90	5364.38	5394.86	5425.34	5455.82
18000	5486.30	5516.78	5547.26	5577.74	5608.22	5638.70	5669.18	5699.66	5730.14	5760.62
19000	5791.10	5821.57	5852.05	5882.53	5913.01	5943.49	5973.97	6004.45	6034.93	6065.41
20000	6095.89	6126.37	6156.85	6187.33	6217.81	6248.29	6278.77	6309.25	6339.73	6370.20
21000	6400.68	6431.16	6461.64	6492.12	6522.60	6553.08	6583.56	6614.04	6644.52	6675.00
22000	6705.48	6735.96	6766.44	6796.92	6827.40	6857.88	6888.36	6918.83	6949.31	6979.79
23000	7010.27	7040.75	7071.23	7101.71	7132.19	7162.67	7193.15	7223.63	7254.11	7284.59
24000	7315.07	7345.55	7376.03	7406.51	7436.99	7467.47	7497.94	7528.42	7558.90	7589.38
25000	7619.86	7650.34	7680.82	7711.30	7741.78	7772.26	7802.74	7833.22	7863.70	7894.18
26000	7924.66	7955.14	7985.62	8016.10	8046.57	8077.05	8107.53	8138.01	8168.49	8198.97
27000	8229.45	8259.93	8290.41	8320.89	8351.37	8381.85	8412.33	8442.81	8473.29	8503.77
28000	8534.25	8564.73	8595.20	8625.68	8656.16	8686.64	8717.12	8747.60	8778.08	8808.56

Tens.	Units.									
	0.	**1.**	**2.**	**3.**	**4.**	**5.**	**6.**	**7.**	**8.**	**9.**
	Metres.	Metres.	Metres.	Metres.	Metres.	Metres.	Metres.	Metres.	Metres.	Metres.
0	0.00000	0.30479	0.60959	0.91438	1.21918	1.52397	1.82877	2.13356	2.43836	2.74315
10	3.04794	3.35274	3.65753	3.96233	4.26712	4.57192	4.87671	5.18151	5.48630	5.79110
20	6.09589	6.40068	6.70548	7.01027	7.31507	7.61986	7.92466	8.22945	8.53425	8.83904
30	9.14383	9.44863	9.75342	10.0582	10.3630	10.6678	10.9726	11.2774	11.5822	11.8870
40	12.1918	12.4966	12.8014	13.1062	13.4110	13.7158	14.0205	14.3253	14.6301	14.9349
50	15.2397	15.5445	15.8493	16.1541	16.4589	16.7637	17.0685	17.3733	17.6781	17.9829
60	18.2877	18.5925	18.8973	19.2021	19.5068	19.8116	20.1164	20.4212	20.7260	21.0308
70	21.3356	21.6404	21.9452	22.2500	22.5548	22.8596	23.1644	23.4692	23.7740	24.0788
80	24.3836	24.6884	24.9931	25.2979	25.6027	25.9075	26.2123	26.5171	26.8219	27.1267
90	27.4315	27.7363	28.0411	28.3459	28.6507	28.9555	29.2603	29.5651	29.8699	30.1747

XVII. CONVERSION OF ENGLISH FEET INTO FRENCH OR PARIS FEET AND DECIMALS.

1 English Foot = 0.93829277 Paris Foot.

English Feet. Thousands.	Hundreds.									
	0.	**100.**	**200.**	**300.**	**400.**	**500.**	**600.**	**700.**	**800.**	**900.**
	Par. Feet.	Par. Feet.	Par. Feet.	Par. Feet.	Par. Feet.	Par. Feet.	Par. Feet.	Par. Feet.	Par. Feet	Par. Feet.
0	000.0	93.8	187.7	281.5	375.3	469.1	563.0	656.8	750.6	844.5
1000	938.3	1032.1	1126.0	1219.8	1313.6	1407.4	1501.3	1595.1	1688.9	1782.8
2000	1876.6	1970.4	2064.2	2158.1	2251.9	2345.7	2439.6	2533.4	2627.2	2721.0
3000	2814.9	2908.7	3002.5	3096.4	3190.2	3284.0	3377.9	3471.7	3565.5	3659.3
4000	3753.2	3847.0	3940.8	4034.7	4128.5	4222.3	4316.1	4410.0	4503.8	4597.6
5000	4691.5	4785.3	4879.1	4973.0	5066.8	5160.6	5254.4	5348.3	5442.1	5535.9
6000	5629.8	5723.6	5817.4	5911,2	6005.1	6098.9	6192.7	6286.6	6380.4	6474.2
7000	6568.0	6661.9	6755.7	6849.5	6943.4	7037.2	7131.0	7224.9	7318.7	7412.5
8000	7506.3	7600.2	7694.0	7787.8	7881.7	7975.5	8069.3	8163.1	8257.0	8350.8
9000	8444.6	8538.5	8632.3	8726.1	8820.0	8913.8	9007.6	9101.4	9195.3	9289.1
10000	9382.9	9476.8	9570.6	9664.4	9758.2	9852.1	9945.9	10039.7	10133.6	10227.4
11000	10321.2	10415.0	10508.9	10602.7	10696.5	10790.4	10884.2	10978.0	11071.9	11165.7
12000	11259.5	11353.3	11447.2	11541.0	11634.8	11728.7	11822.5	11916.3	12010.1	12104.0
13000	12197.8	12291.6	12385.5	12479.3	12573.1	12667.0	12760.8	12854.6	12948.4	13042.3
14000	13136.1	13229.9	13323.8	13417.6	13511.4	13605.2	13699.1	13792.9	13886.7	13980.6
15000	14074.4	14168.2	14262.0	14355.9	14449.7	14543.5	14637.4	14731.2	14825.0	14918.9
16000	15012.7	15106.5	15200.3	15294.2	15388.0	15481.8	15575.7	15669.5	15763.3	15857.1
17000	15951.0	16044.8	16138.6	16232.5	16326.3	16420.1	16514.0	16607.8	16701.6	16795.4
18000	16889.3	16983.1	17076.9	17170.8	17264.6	17358.4	17452.2	17546.1	17639.9	17733.7
19000	17827.6	17921.4	18015.2	18109.0	18202.9	18296.7	18390.5	18484.4	18578.2	18672.0
20000	18765.9	18859.7	18953.5	19047.3	19141.2	19235.0	19328.8	19422.7	19516.5	19610.3
21000	19704.1	19798.0	19891.8	19985.6	20079.5	20173.3	20267.1	20361.0	20454.8	20548.6
22000	20642.4	20736.3	20830.1	20923.9	21017.8	21111.6	21205.4	21299.2	21393.1	21486.9
23000	21580.7	21674.6	21768.4	21862.2	21956.0	22049.9	22143.7	22237.5	22331.4	22425.2
24000	22519.0	22612.9	22706.7	22800.5	22894.3	22988.2	23082.0	23175.8	23269.7	23363.5
25000	23457.3	23551.1	23645.0	23738.8	23832.6	23926.5	24020.3	24114.1	24208.0	24301.8
26000	24395.6	24489.4	24583.3	24677.1	24770.9	24864.8	24958.6	25052.4	25146.2	25240.1
27000	25333.9	25427.7	25521.6	25615.4	25709.2	25803.1	25896.9	25990.7	26084.5	26178.4
28000	26272.2	26366.0	26459.9	26553.7	26647.5	26741.3	26835.2	26929.0	27022.8	27116.7

Tens.	Units.									
	0.	**1.**	**2.**	**3.**	**4.**	**5.**	**6.**	**7.**	**8.**	**9.**
	Par. Feet.	Par. Feet.	Par. Feet.	Par. Feet.	Par. Feet.	Par. Feet.	Par. Feet.	Par. Feet.	Par. Feet.	Par. Feet.
0	0.00	0.94	1.88	2.81	3.75	4.69	5.63	6.57	7.51	8.44
10	9.38	10.32	11.26	12.20	13.14	14.07	15.01	15.95	16.89	17.83
20	18.77	19.70	20.64	21.58	22.52	23.46	24.40	25.33	26.27	27.21
30	28.15	29.09	30.03	30.96	31.90	32.84	33.78	34.72	35.66	36.59
40	37.53	38.47	39.41	40.35	41.28	42.22	43.16	44.10	45.04	45.98
50	46.91	47.85	48.79	49.73	50.67	51.61	52.54	53.48	54.42	55.36
60	56.30	57.24	58.17	59.11	60.05	60.99	61.93	62.87	63.80	64.74
70	65.68	66.62	67.56	68.50	69.43	70.37	71.31	72.25	73.19	74.13
80	75.06	76.00	76.94	77.88	78.82	79.75	80.69	81.63	82.57	83.51
90	84.45	85.38	86.32	87.26	88.20	89.14	90.08	91.01	91.95	92.89

XVIII. CONVERSION OF ENGLISH FEET INTO AMERICAN FEET AND DECIMALS.

1 English Foot = 0.99994197 American Foot.

Eng. Feet. Thousands	Hundreds. 0.	100.	200.	300.	400.	500.	600.	700.	800.	900.
	Am Feet	Am. Feet.	Am Feet.	Am Feet	Am. Feet.	Am Feet	Am Feet	Am. Feet.	Am Feet.	Am Feet.
0	0.00	99.99	199.99	299.98	399.98	499.97	599.97	699.96	799.95	899.95
1000	999.94	1099.94	1199.93	1299.92	1399.92	1499.91	1599.91	1699.90	1799.90	1899.89
2000	1999.88	2099.88	2199.87	2299.87	2399.86	2499.85	2599.85	2699.84	2799.84	2899.83
3000	2999.83	3099.82	3199.81	3299.81	3399.80	3499.80	3599.79	3699.79	3799.78	3899.77
4000	3999.77	4099.76	4199.76	4299.75	4399.74	4499.74	4599.73	4699.73	4799.72	4899.72
5000	4999.71	5099.70	5199.70	5299.69	5399.69	5499.68	5599.68	5699.67	5799.66	5899.66
6000	5999.65	6099.65	6199.64	6299.63	6399.63	6499.62	6599.62	6699.61	6799.61	6899.60
7000	6999.59	7099.59	7199.58	7299.58	7399.57	7499.56	7599.56	7699.55	7799.55	7899.54
8000	7999.54	8099.53	8199.52	8299.52	8399.51	8499.51	8599.50	8699.50	8799.49	8899.48
9000	8999.48	9099.47	9199.47	9299.46	9399.45	9499.45	9599.44	9699.44	9799.43	9899.43
10000	9999.42	10099.4	10199.4	10299.4	10399.4	10499.4	10599.4	10699.4	10799.4	10899.4
11000	10999.4	11099.4	11199.4	11299.3	11399.3	11499.3	11599.3	11699.3	11799.3	11899.3
12000	11999.3	12099.3	12199.3	12299.3	12399.3	12499.3	12599.3	12699.3	12799.3	12899.2
13000	12999.2	13099.2	13199.2	13299.2	13399.2	13499.2	13599.2	13699.2	13799.2	13899.2
14000	13999.2	14099.2	14199.2	14299.2	14399.2	14499.2	14599.2	14699.1	14799.1	14899.1
15000	14999.1	15099.1	15199.1	15299.1	15399.1	15499.1	15599.1	15699.1	15799.1	15899.1

The following Table of Differences between English and American Feet, for every hundred feet, will make it easy to convert English into American Feet, or American into English Feet, by adding to, or subtracting from, the number of feet to be converted, which is contained in the first column, the numbers found in the other columns.

XIX. DIFFERENCES BETWEEN ENGLISH AND AMERICAN FEET.

To obtain English Feet *add*. To obtain American Feet *subtract*.

10000 American Feet = 10000.5803 English Feet.

Number of Feet. Thousands.	Hundreds. 0.	200.	400.	600.	800.	Number of Feet. Thousands.	Hundreds. 0.	200.	400.	600.	800.
	Diff feet	Diff. feet	Diff feet	Diff. feet.	Diff feet		Diff. feet.	Diff. feet	Diff. feet	Diff feet.	Diff. feet.
0	±0.000	±0.012	±0.023	±0.035	±0.046	15000	±0.870	±0.882	±0.894	±0.905	±0.917
1000	0.058	0.070	0.082	0.093	0.105	16000	0.928	0.940	0.952	0.963	0.975
2000	0.116	0.128	0.139	0.151	0.162	17000	0.987	0.998	1.010	1.021	1.033
3000	0.174	0.186	0.197	0.209	0.221	18000	1.045	1.056	1.068	1.079	1.091
4000	0.232	0.244	0.255	0.267	0.279	19000	1.103	1.114	1.126	1.137	1.149
5000	0.290	0.302	0.313	0.325	0.337	20000	1.161	1.172	1.184	1.195	1.207
6000	0.348	0.360	0.371	0.383	0.395	21000	1.219	1.230	1.242	1.253	1.265
7000	0.406	0.418	0.429	0.441	0.453	22000	1.277	1.288	1.300	1.311	1.323
8000	0.464	0.476	0.487	0.499	0.511	23000	1.335	1.346	1.358	1.370	1.381
9000	0.522	0.534	0.546	0.557	0.569	24000	1.393	1.404	1.416	1.428	1.439
10000	0.580	0.592	0.604	0.615	0.627	25000	1.451	1.462	1.474	1.486	1.497
11000	0.638	0.650	0.662	0.673	0.685	26000	1.509	1.520	1.532	1.544	1.555
12000	0.696	0.708	0.720	0.731	0.743	27000	1.567	1.578	1.590	1.602	1.613
13000	0.754	0.766	0.778	0.789	0.801	28000	1.625	1.636	1.648	1.660	1.671
14000	0.812	0.824	0.836	0.847	0.859	29000	1.683	1.694	1.606	1.718	1.729

TO CONVERT

AMERICAN YARDS AND FEET

INTO DIFFERENT MEASURES OF LENGTH

XX. CONVERSION OF AMERICAN YARDS INTO FRENCH TOISES.

1 American Yard = 0.4691736 Toise.

American Yards. Thousands.	Hundreds. 0.	100.	200.	300.	400.	500.	600.	700.	800.	900.
	Toises.	Toises.	Toises.	Toises.	Toises.	Toises.	Toises.	Toises.	Toises.	Toises.
0	0.00	46.92	93.83	140.75	187.67	234.59	281.50	328.42	375.34	422.26
1000	469.17	516.09	563.01	609.93	656.84	703.76	750.68	797.60	844.51	891.43
2000	938.35	985.26	1032.18	1079.10	1126.02	1172.93	1219.85	1266.77	1313.69	1360.60
3000	1407.52	1454.44	1501.36	1548.27	1595.19	1642.11	1689.02	1735.94	1782.86	1829.78
4000	1876.69	1923.61	1970.53	2017.45	2064.36	2111.28	2158.20	2205.12	2252.03	2298.95
5000	2345.87	2392.79	2439.70	2486.62	2533.54	2580.45	2627.37	2674.29	2721.21	2768.12
6000	2815.04	2861.96	2908.88	2955.79	3002.71	3049.63	3096.55	3143.46	3190.38	3237.30
7000	3284.22	3331.13	3378.05	3424.97	3471.88	3518.80	3565.72	3612.64	3659.55	3706.47
8000	3753.39	3800.31	3847.22	3894.14	3941.06	3987.98	4034.89	4081.81	4128.73	4175.65
9000	4222.56	4269.48	4316.40	4363.31	4410.23	4457.15	4504.07	4550.98	4597.90	4644.82

XXI. CONVERSION OF AMERICAN YARDS INTO METRES.

1 American Yard = 0.91443654 Metre.

American Yards. Thousands.	Hundreds. 0.	100.	200.	300.	400.	500.	600.	700.	800.	900.
	Metres.	Metres.	Metres	Metres.	Metres.	Metres.	Metres.	Metres.	Metres.	Metres.
0	0.00	91.44	182.89	274.33	365.77	457.22	548.66	640.11	731.55	822.99
1000	914.44	1005.88	1097.32	1188.77	1280.21	1371.65	1463.10	1554.54	1645.99	1737.43
2000	1828.87	1920.32	2011.76	2103.20	2194.65	2286.09	2377.54	2468.98	2560.42	2651.87
3000	2743.31	2834.75	2926.20	3017.64	3109.08	3200.53	3291.97	3383.42	3474.86	3566.30
4000	3657.75	3749.19	3840.63	3932.08	4023.52	4114.96	4206.41	4297.85	4389.30	4480.74
5000	4572.18	4663.63	4755.07	4846.51	4937.96	5029.40	5120.84	5212.29	5303.73	5395.18
6000	5486.62	5578.06	5669.51	5760.95	5852.39	5943.84	6035.28	6126.72	6218.17	6309.61
7000	6401.06	6492.50	6583.94	6675.39	6766.83	6858.27	6949.72	7041.16	7132.61	7224.05
8000	7315.49	7406.94	7498.38	7589.82	7681.27	7772.71	7864.15	7955.60	8047.04	8138.49
9000	8229.93	8321.37	8412.82	8504.26	8595.70	8687.15	8778.59	8870.03	8961.48	9052.92

1 American Foot = 0.30481218 Metre.

Amer. Feet. Thousands.	Hundreds.									
	0.	**100.**	**200.**	**300.**	**400.**	**500.**	**600.**	**700.**	**800.**	**900.**
	Metres.	Metres.	Metres.	Metres.	Metres.	Metres.	Metres.	Metres.	Metres.	Metres.
0	0.00	30.48	60.96	91.44	121.92	152.41	182.89	213.37	243.85	274.33
1000	304.81	335.29	365.77	396.26	426.74	457.22	487.70	518.18	548.66	579.14
2000	609.62	640.11	670.59	701.07	731.55	762.03	792.51	822.99	853.47	883.96
3000	914.44	944.92	975.40	1005.88	1036.36	1066.84	1097.32	1127.81	1158.29	1188.77
4000	1219.25	1249.73	1280.21	1310.69	1341.17	1371.65	1402.14	1432.62	1463.10	1493.58
5000	1524.06	1554.54	1585.02	1615.50	1645.99	1676.47	1706.95	1737.43	1767.91	1798.39
6000	1828.87	1859.35	1889.84	1910.32	1940.80	1971.28	2001.76	2032.24	2062.72	2093.20
7000	2123.69	2154.17	2184.65	2225.13	2255.61	2286.09	2316.57	2347.05	2377.54	2408.02
8000	2438.50	2468.98	2499.46	2529.94	2560.42	2590.90	2621.38	2651.87	2682.35	2712.83
9000	2743.31	2773.79	2804.27	2834.75	2865.23	2895.72	2926.20	2956.68	2987.16	3017.64
10000	3048.12	3078.60	3109.08	3139.57	3170.05	3200.53	3231.01	3261.49	3291.97	3322.45
11000	3352.93	3383.42	3413.90	3444.38	3474.86	3505.34	3535.82	3566.30	3596.78	3627.26
12000	3657.75	3688.23	3718.71	3749.19	3779.67	3810.15	3840.63	3871.11	3901.60	3932.08
13000	3962.56	3993.04	4023.52	4054.00	4084.48	4114.96	4145.45	4175.93	4206.41	4236.89
14000	4267.37	4297.85	4328.33	4358.81	4389.30	4419.78	4450.26	4480.74	4511.22	4541.70
15000	4572.18	4602.66	4633.15	4663.63	4694.11	4724.59	4755.07	4785.55	4816.03	4846.51
16000	4876.99	4907.48	4937.96	4968.44	4998.92	5029.40	5059.88	5090.36	5120.84	5151.33
17000	5181.81	5212.29	5242.77	5273.25	5303.73	5334.21	5364.69	5395.18	5425.66	5456.14
18000	5486.62	5517.10	5547.58	5578.06	5608.54	5639.03	5669.51	5699.99	5730.47	5760.95
19000	5791.43	5821.91	5852.39	5882.88	5913.36	5943.84	5974.32	6004.80	6035.28	6065.76
20000	6096.24	6126.72	6157.21	6187.69	6218.17	6248.65	6279.13	6309.61	6340.09	6370.57
21000	6401.06	6431.54	6462.02	6492.50	6522.98	6553.46	6583.94	6614.42	6644.91	6675.39
22000	6705.87	6736.35	6766.83	6797.31	6827.79	6858.27	6888.76	6919.24	6949.72	6980.20
23000	7010.68	7041.16	7071.64	7102.12	7132.61	7163.09	7193.57	7224.05	7254.53	7285.01
24000	7315.49	7345.97	7376.45	7406.94	7437.42	7467.90	7498.38	7528.86	7559.34	7589.82
25000	7620.30	7650.79	7681.27	7711.75	7742.23	7772.71	7803.19	7833.67	7864.15	7894.64
26000	7925.12	7955.60	7986.08	8016.56	8047.04	8077.52	8108.00	8138.49	8168.97	8199.45
27000	8229.93	8260.41	8290.89	8321.37	8351.85	8382.33	8412.82	8443.30	8473.78	8504.26
28000	8534.74	8565.22	8595.70	8626.18	8656.67	8687.15	8717.63	8748.11	8778.59	8809.07

Tens.	Units.									
	0.	**1.**	**2.**	**3.**	**4.**	**5.**	**6.**	**7.**	**8.**	**9.**
	Metres.	Metres.	Metres.	Metres.	Metres.	Metres.	Metres.	Metres.	Metres.	Metres.
0	0.0000	0.3048	0.6096	0.9144	1.2192	1.5241	1.8289	2.1337	2.4385	2.7433
10	3.0481	3.3529	3.6577	3.9626	4.2674	4.5722	4.8770	5.1818	5.4866	5.7914
20	6.0962	6.4011	6.7059	7.0107	7.3155	7.6203	7.9251	8.2299	8.5347	8.8396
30	9.1444	9.4492	9.7540	10.0588	10.3636	10.6684	10.9732	11.2781	11.5829	11.8877
40	12.1925	12.4973	12.8021	13.1069	13.4117	13.7165	14.0214	14.3262	14.6310	14.9358
50	15.2406	15.5454	15.8502	16.1550	16.4599	16.7647	17.0695	17.3743	17.6791	17.9839
60	18.2887	18.5935	18.8984	19.1032	19.4080	19.7128	20.0176	20.3224	20.6272	20.9320
70	21.2369	21.5417	21.8465	22.2513	22.5561	22.8609	23.1657	23.4705	23.7754	24.0802
80	24.3850	24.6898	24.9946	25.2994	25.6042	25.9090	26.2138	26.5187	26.8235	27.1283
90	27.4331	27.7379	28.0427	28.3475	28.6523	28.9572	29.2620	29.5668	29.8716	30.1764

1 American Foot = 0.93834723 Paris Foot.

Amer. Feet. Thousands.	Hundreds. 0.	100.	200.	300.	400.	500.	600.	700.	800.	900.
	Par. Feet.	Par. Feet.	Par. Feet.	Par. Feet.	Par. Feet.	Par. Feet.	Par. Feet.	Par. Feet.	Par. Feet.	Par. Feet.
0	0.0	93.8	187.7	281.5	375.3	469.2	563.0	656.8	750.7	844.5
1000	938.3	1032.2	1126.0	1219.9	1313.7	1407.5	1501.4	1595.2	1689.0	1782.9
2000	1876.7	1970.5	2064.4	2158.2	2252.0	2345.9	2439.7	2533.5	2627.4	2721.2
3000	2815.0	2908.9	3002.7	3096.5	3190.4	3284.2	3378.1	3471.9	3565.7	3659.6
4000	3753.4	3847.2	3941.1	4034.9	4128.7	4222.6	4316.4	4410.2	4504.1	4597.9
5000	4691.7	4785.6	4879.4	4973.2	5067.1	5160.9	5254.7	5348.6	5442.4	5536 2
6000	5630.1	5723.9	5817.8	5911.6	6005.4	6099.3	6193.1	6286.9	6380.8	6474.6
7000	6568.4	6662.3	6756.1	6849.9	6943.8	7037.6	7131.4	7225.3	7319.1	7412.9
8000	7506.8	7600.6	7694.4	7788.3	7882.1	7976.0	8069.8	8163.6	8257.5	8351.3
9000	8445.1	8539.0	8632.8	8726.6	8820.5	8914.3	9008.1	9102.0	9195.8	9289.6
10000	9383.5	9477.3	9571.1	9665.0	9758.8	9852.6	9946.5	10040.3	10134.2	10228.0
11000	10321.8	10415.7	10509.5	10603.3	10697.2	10791.0	10884.8	10978.7	11072.5	11166.3
12000	11260.2	11354.0	11447.8	11541.7	11635.5	11729.3	11823.2	11917.0	12010.8	12104.7
13000	12198.5	12292.3	12386.2	12480.0	12573.9	12667.7	12761.5	12855.4	12949.2	13043.0
14000	13136.9	13230.7	13324.5	13418.4	13512.2	13606.0	13699.9	13793.7	13887.5	13981.4
15000	14075.2	14169.0	14262.9	14356.7	14450.5	14544.4	14638.2	14732.1	14825.9	14919.7
16000	15013.6	15107.4	15201.2	15295.1	15388.9	15482.7	15576.6	15670.4	15764.2	15858.1
17000	15951.9	16045.7	16139.6	16233.4	16327.2	16421.1	16514.9	16608.7	16702.6	16796.4
18000	16890.3	16984.1	17077.9	17171.8	17265.6	17359.4	17453.3	17547.1	17640.9	17734.8
19000	17828.6	17922.4	18016.3	18110.1	18203.9	18297.8	18391.6	18485.4	18579.3	18673.1
20000	18766.9	18860.8	18954.6	19048.4	19142.3	19236.1	19330.0	19423.8	19517.6	19611.5
21000	19705.3	19799.1	19893.0	19986.8	20080.6	20174.5	20268.3	20362.1	20456.0	20549.8
22000	20643.6	20737.5	20831.3	20925.1	21019.0	21112.8	21206.6	21300.5	21394.3	21488.2
23000	21582.0	21675.8	21769.7	21863.5	21957.3	22051.2	22145.0	22238.8	22332.7	22426.5
24000	22520.3	22614.2	22708.0	22801.8	22895.7	22989.5	23083.3	23177.2	23271.0	23364.8
25000	23458.7	23552.5	23646.4	23740.2	23834.0	23927.9	24021.7	24115.5	24209.4	24303.2
26000	24397.0	24490.9	24584.7	24678.5	24772.4	24866.2	24960.0	25053.9	25147.7	25241.5
27000	25335.4	25429.2	25523.0	25616.9	25710.7	25804.5	25898.4	25992.2	26086.1	26179.9
28000	26273.7	26367.6	26461.4	26555.2	26649.1	26742.9	26836.7	26930.6	27024.4	27118.2

Tens.	Units. 0.	1.	2.	3.	4.	5.	6.	7.	8.	9.
	Par Feet.	Par Feet.	Par. Feet.	Par Feet	Par. Feet.	Par Feet.	Par. Feet.	Par Feet	Par Feet	Par Feet
0	0.00	0.94	1.88	2.82	3.75	4.69	5.63	6.57	7.51	8.45
10	9.38	10.32	11.26	12.20	13.14	14.08	15.01	15.95	16.89	17.83
20	18.77	19.71	20.64	21.58	22.52	23.46	24.40	25.34	26.27	27.21
30	28.15	29.09	30.03	30.97	31.90	32.84	33.78	34.72	35.66	36.60
40	37.53	38.47	39.41	40.35	41.29	42.23	43.16	44.10	45.04	45.98
50	46.92	47.86	48.79	49.73	50.67	51.61	52.55	53.49	54.42	55.36
60	56.30	57.24	58.18	59.12	60.05	60.99	61.93	62.87	63.81	64.75
70	65.68	66.62	67.56	68.50	69.44	70.38	71.31	72.25	73.19	74.13
80	75.07	76.01	76.94	77.88	78.82	79.76	80.70	81.64	82.57	83.51
90	84.45	85.39	86.33	87.27	88.20	89.14	90.08	91.02	91.96	92.90

TO CONVERT

KLAFTER AND FEET OF VIENNA

INTO DIFFERENT MEASURES OF LENGTH.

1 KLAFTER OF VIENNA = 6 FEET OF VIENNA = 840.76134 FRENCH OR PARIS LINES.

From this value are derived the equations used in computing the following tables.

XXIV. CONVERSION OF KLAFTER OF VIENNA INTO FRENCH TOISES.

1 Klafter = 0.9731034 Toise.

Klafter of Vienna. Tens.	Units. 0.	1.	2.	3.	4.	5.	6.	7.	8.	9.
	Toises.	Toises.	Toises.	Toises.	Toises.	Toises.	Toises.	Toises.	Toises.	Toises.
0	0.000	0.973	1.946	2.919	3.892	4.866	5.839	6.812	7.785	8.758
10	9.731	10.704	11.677	12.650	13.623	14.597	15.570	16.543	17.516	18.489
20	19.462	20.435	21.408	22.381	23.354	24.328	25.301	26.274	27.247	28.220
30	29.193	30.166	31.139	32.112	33.086	34.059	35.032	36.005	36.978	37.951
40	38.924	39.897	40.870	41.843	42.817	43.790	44.763	45.736	46.709	47.682
50	48.655	49.628	50.601	51.574	52.548	53.521	54.494	55.467	56.440	57.413
60	58.386	59.359	60.332	61.306	62.279	63.252	64.225	65.198	66.171	67.144
70	68.117	69.090	70.063	71.037	72.010	72.983	73.956	74.929	75.902	76.875
80	77.848	78.821	79.794	80.768	81.741	82.714	83.687	84.660	85.633	86.606
90	87.579	88.552	89.526	90.499	91.472	92.445	93.418	94.391	95.364	96.337

Klafter. Thousands.	Hundreds. 0.	100.	200.	300.	400.	500.	600.	700.	800.	900.
	Toises.	Toises.	Toises.	Toises.	Toises.	Toises.	Toises.	Toises.	Toises.	Toises.
0	0.00	97.31	194.62	291.93	389.24	486.55	583.86	681.17	778.48	875.79
1000	973.10	1070.41	1167.72	1265.03	1362.34	1459.65	1556.97	1654.28	1751.59	1848.90
2000	1946.21	2043.52	2140.83	2238.14	2335.45	2432.76	2530.07	2627.38	2724.69	2822.00
3000	2919.31	3016.62	3113.93	3211.24	3308.55	3405.86	3503.17	3600.48	3797.79	3795.10
4000	3892.41	3989.72	4087.03	4184.34	4281.65	4378.97	4476.28	4573.59	4670.90	4768.21
5000	4865.52	4962.83	5060.14	5157.45	5254.76	5852.07	5449.38	5546.69	5644.00	5741.31

XXV. CONVERSION OF KLAFTER OF VIENNA INTO METRES.

1 Klafter = 1.8966142 Metre.

Klafter of Vienna Tens.	Klafter. Units.									
	0.	**1.**	**2.**	**3.**	**4.**	**5.**	**6.**	**7.**	**8.**	**9.**
	Metres.	Metres.	Metres.	Metres.	Metres.	Metres.	Metres.	Metres.	Metres.	Metres.
0	0.000	1.897	3.793	5.690	7.586	9.483	11.380	13.276	15.173	17.070
10	18.966	20.863	22.759	24.656	26.553	28.449	30.346	32.242	34.139	36.036
20	37.932	39.829	41.726	43.622	45.519	47.415	49.312	51.209	53.105	55.002
30	56.898	58.795	60.692	62.588	64.485	66.381	68.278	70.175	72.071	73.968
40	75.865	77.761	79.658	81.554	83.451	85.348	87.244	89.141	91.037	92.934
50	94.831	96.727	98.624	100.521	102.417	104.314	106.210	108.107	110.004	111.900
60	113.797	115.693	117.590	119.487	121.383	123.280	125.177	127.073	128.970	130.866
70	132.763	134.660	136.556	138.453	140.349	142.246	144.143	146.039	147.936	149.833
80	151.729	153.626	155.522	157.419	159.316	161.212	163.109	165.005	166.902	168.799
90	170.695	172.592	174.489	176.385	178.282	180.178	182.075	183.972	185.868	187.765

Klafter. Thousands.	Hundreds.									
	0.	**100.**	**200.**	**300.**	**400.**	**500.**	**600.**	**700.**	**800.**	**900.**
	Metres.	Metres.	Metres.	Metres.	Metres.	Metres.	Metres.	Metres	Metres.	Metres.
0	0.00	189.66	379.32	568.98	758.65	948.31	1137.97	1327.63	1517.29	1706.95
1000	1896.61	2086.28	2275.94	2465.60	2655.26	2844.92	3034.58	3224.24	3413.91	3603.57
2000	3793.23	3982.89	4172.55	4362.21	4551.87	4741.54	4931.20	5120.86	5310.52	5500.18
3000	5689.84	5879.50	6069.17	6258.83	6448.49	6638.15	6827.81	7017.47	7207.13	7396.80
4000	7586.46	7776.12	7965.78	8155.44	8345.10	8534.76	8724.43	8914.09	9103.75	9293.41

XXVI. CONVERSION OF KLAFTER OF VIENNA INTO PARIS OR FRENCH FEET.

1 Klafter = 5.8386204 Paris or French Feet.

Klafter of Vienna. Tens.	Klafter. Units.									
	0.	**1.**	**2.**	**3.**	**4.**	**5.**	**6.**	**7.**	**8.**	**9.**
	Par.Feet.	Par.Feet.	Par.Feet.	Par.Feet.	Par.Feet.	Par.Feet.	Par.Feet.	Par.Feet.	Par.Feet.	Par.Feet.
0	0.00	5.84	11.68	17.52	23.35	29.19	35.03	40.87	46.71	52.55
10	58.39	64.22	70.06	75.90	81.74	87.58	93.42	99.26	105.10	110.93
20	116.77	122.61	128.45	134.29	140.13	145.97	151.80	157.64	163.48	169.32
30	175.16	181.00	186.84	192.67	198.51	204.35	210.19	216.03	221.87	227.71
40	233.54	239.38	245.22	251.06	256.90	262.74	268.58	274.42	280.25	286.09
50	291.93	297.77	303.61	309.45	315.29	321.12	326.96	332.80	338.64	344.48
60	350.32	356.16	361.99	367.83	373.67	379.51	385.35	391.19	397.03	402.86
70	408.70	414.54	420.38	426.22	432.06	437.90	443.74	449.57	455.41	461.25
80	467.09	472.93	478.77	484.61	490.44	496.28	502.12	507.96	513.80	519.64
90	525.48	531.31	537.15	542.99	548.83	554.67	560.51	566.35	572.18	578.02

Klafter. Thousands	Hundreds.									
	0.	**100.**	**200.**	**300.**	**400.**	**500.**	**600.**	**700.**	**800.**	**900.**
	Par.Feet.	Par.Feet.	Par Feet.	Par.Feet	Par.Feet.	Par Feet.	Par.Feet.	Par Feet.	Par.Feet	Par.Feet.
0	0.00	583.86	1167.72	1751.59	2335.45	2919.31	3503.17	4087.03	4670.90	5254.76
1000	5838.62	6422.48	7006.34	7590.21	8174.07	8757.93	9341.79	9925.65	10509.5	11093.4
2000	11677.2	12261.1	12845.0	13428.8	14012.7	14596.6	15180.4	15764.3	16348.1	16932.0
3000	17515.9	18099.7	18683.6	19267.4	19851.3	20435.2	21019.0	21602.9	22186.8	22770.6
4000	23354.5	23938.3	24522.2	25106.1	25689.9	26273.8	26857.7	27441.5	28025.4	28609.2

XXVII. CONVERSION OF KLAFTER OF VIENNA INTO ENGLISH FEET.

1 Klafter = 6 22260 English Feet.

Klafter of Vienna Tens.	Klafter. Units. 0.	1.	2.	3.	4.	5.	6.	7.	8.	9.
	Eng feet.	Eng feet.	Eng. feet.	Eng. feet.	Eng. feet.	Eng. feet	Eng feet.	Eng. feet.	Eng. feet.	Eng feet.
0	0.00	6.22	12.45	18.67	24.89	31.11	37.34	43.56	49.78	56 00
10	62.23	68.45	74.67	80.89	87.12	93.34	99.56	105.78	112.01	118.23
20	124.45	130.67	136.90	143.12	149.34	155.56	161.79	168.01	174.23	180.46
30	186.68	192.90	199.12	205.35	211.57	217.79	224.01	230.24	236.46	242.68
40	248.90	255.13	261.35	267.57	273.79	280.02	286.24	292.46	298.68	304.91
50	311.13	317.35	323.58	329.80	336.02	342.24	348.47	354.69	360.91	367.13
60	373.36	379.58	385.80	392.02	398.35	404.47	410.69	416.91	423.14	429.36
70	435.58	441.80	448.03	454.25	460.47	466.69	472.92	479.14	485.36	491.59
80	497.81	504.03	510.25	516.48	522.70	528.92	535.14	541.37	547.59	553.81
90	560.03	566.26	572.48	578.70	584.92	591.15	597.37	603.59	609.81	616.04

Klafter. Thousands	Hundreds. 0.	100.	200.	300.	400.	500.	600.	700.	800.	900.
	Eng. feet.	Eng. feet.	Eng. feet.	Eng. feet.	Eng feet.	Eng. feet.	Eng. feet.	Eng. feet.	Eng. feet	Eng feet
0	0.00	622.26	1244.52	1866.78	2489.04	3111.30	3733.56	4355.82	4978.08	5600.34
1000	6222.60	6844.86	7467.12	8089.38	8711.64	9333.90	9956.16	10578.4	11200.7	11822.9
2000	12445.2	13067.5	13689.7	14312.0	14934.2	15556.5	16178.8	16801.0	17423.3	18045.5
3000	18667.8	19290.1	19912.3	20534.6	21156.8	21779.1	22401.4	23023.6	23645.9	24268.1
4000	24890.4	25512.7	26134.9	26757.2	27379.4	28001.7	28624.0	29246.2	39868.5	30490.7

XXVIII. CONVERSION OF FEET OF VIENNA INTO METRES.

1 Foot of Vienna = 0.3161024 Metre.

Feet of Vienna. Thousands.	Hundreds. 0.	100.	200.	300.	400.	500.	600.	700.	800.	900.
	Metres.	Metres.	Metres.	Metres.	Metres.	Metres.	Metres.	Metres.	Metres.	Metres.
0	0.00	31.61	63.22	94.83	126.44	158.05	189.66	221.27	252.88	284.49
1000	316.10	347.71	379.32	410.93	442.54	474.15	505.76	537.37	568.98	600.59
2000	632.20	663.82	695.43	727.04	758.65	790.26	821.87	853.48	885.09	916.70
3000	948.31	979.92	1011.53	1043.14	1074.75	1106.36	1137.97	1169.58	1201.19	1232.80
4000	1264.41	1296.02	1327.63	1359.24	1390.85	1422.46	1454.07	1485.68	1517.29	1548.90
5000	1580.51	1612.12	1643.73	1675.34	1706.95	1738.56	1770.17	1801.78	1833.39	1865.00
6000	1896.61	1928.22	1959.83	1991.45	2023.06	2054.67	2086.28	2117.89	2149.50	2181.11
7000	2212.72	2244.33	2275.94	2307.55	2339.16	2370.77	2402.38	2433.99	2465.60	2497.21
8000	2528.82	2560.43	2592.04	2623.65	2655.26	2686.87	2718.48	2750.09	2781.70	2813.31
9000	2844.92	2876.53	2908.14	2939.75	2971.36	3002.97	3034.58	3066.19	3097.80	3129.41
10000	3161.02	3192.63	3224.24	3255.85	3287.46	3319.08	3350.69	3382.30	3413.91	3445.52
11000	3477.13	3508.74	3540.35	3571.96	3603.57	3635.18	3666.79	3698.40	3730.01	3761.62
12000	3793.23	3824.84	3856.45	3888.06	3919.67	3951.28	3982.89	4014.50	4046.11	4077.72
13000	4109.33	4140.94	4172.55	4204.16	4235.77	4267.38	4298.99	4330.60	4362.21	4393.82
14000	4425.43	4457.04	4488.65	4520.26	4551.87	4583.48	4615.10	4646.71	4678.32	4709.93
15000	4741.54	4773.15	4804.76	4836.37	4867.98	4899.59	4931.20	4962.81	4994.42	5026.03

XXIX. CONVERSION OF FEET OF VIENNA INTO PARIS OR FRENCH FEET AND DECIMALS.

1 Foot of Vienna = 0.9731034 Paris Foot.

Feet of Vienna.	Hundreds.									
Thousands.	**0.**	**100.**	**200.**	**300.**	**400.**	**500.**	**600.**	**700.**	**800.**	**900.**
	Par. Feet.	Par. Feet.	Par. Feet.	Par. Feet.	Par. Feet.	Par. Feet.	Par. Feet.	Par. Feet.	Par. Feet.	Par. Feet.
0	0.0	97.3	194.6	291.9	389.2	486.6	583.9	681.2	778.5	875.8
1000	973.1	1070.4	1167.7	1265.0	1362.3	1459.7	1557.0	1654.3	1751.6	1848.9
2000	1946.2	2043.5	2140.8	2238.1	2335.4	2432.8	2530.1	2627.4	2724.7	2822.0
3000	2919.3	3016.6	3113.9	3211.2	3308.6	3405.9	3503.2	3600.5	3697.8	3795.1
4000	3892.4	3889.7	4087.0	4184.3	4281.7	4379.0	4476.3	4573.6	4670.9	4768.2
5000	4865.5	4962.8	5060.1	5157.4	5254.8	5352.1	5449.4	5546.7	5644.0	5741.3
6000	5838.6	5935.9	6033.2	6130.6	6227.9	6325.2	6422.5	6519.8	6617.1	6714.4
7000	6811.7	6909.0	7006.3	7103.7	7201.0	7298.3	7395.6	7492.9	7590.2	7687.5
8000	7784.8	7882.1	7979.4	8076.8	8174.1	8271.4	8368.7	8466.0	8563.3	8660.6
9000	8757.9	8855.2	8952.6	9049.9	9147.2	9244.5	9341.8	9439.1	9536.4	9633.7
10000	9731.0	9828.3	9925.7	10023.0	10120.3	10217.6	10314.9	10412.2	10509.5	10606.8
11000	10704.1	10801.4	10898.8	10996.1	11093.4	11190.7	11288.0	11385.3	11482.6	11579.9
12000	11677.2	11774.6	11871.9	11969.2	12066.5	12163.8	12261.1	12358.4	12455.7	12553.0
13000	12650.3	12747.7	12845.0	12942.3	13039.6	13136.9	13234.2	13331.5	13438.8	13526.1
14000	13623.4	13720.8	13818.1	13915.4	14012.7	14110.0	14207.3	14304.6	14401.9	14499.2
15000	14596.6	14693.9	14791.2	14888.5	14985.8	15083.1	15180.4	15277.7	15375.0	15472.3
16000	15569.7	15667.0	15764.3	15861.6	15958.9	16056.2	16153.5	16250.8	16348.1	16445.4

XXX. CONVERSION OF FEET OF VIENNA INTO ENGLISH FEET AND DECIMALS.

1 Foot of Vienna = 1.03710 English Foot.

Feet of Vienna.	Hundreds.									
Thousands.	**0.**	**100.**	**200.**	**300.**	**400.**	**500.**	**600.**	**700.**	**800.**	**900.**
	Eng. feet.	Eng. feet.	Eng. feet.	Eng. feet.	Eng. feet.	Eng. feet.	Eng. feet.	Eng. feet.	Eng. feet.	Eng. feet.
0	0.0	103.7	207.4	311.1	414.8	518.5	622.3	726.0	829.7	933.4
1000	1037.1	1140.8	1244.5	1348.2	1451.9	1555.6	1659.4	1763.1	1866.8	1970.5
2000	2074.2	2177.9	2281.6	2385.3	2489.0	2592.7	2696.5	2800.2	2903.9	3007.6
3000	3111.3	3215.0	3318.7	3422.4	3526.1	3629.8	3733.6	3837.3	3941.0	4044.7
4000	4148.4	4252.1	4355.8	4459.5	4563.2	4666.9	4770.7	4874.4	4978.1	5081.8
5000	5185.5	5289.2	5392.9	5496.6	5600.3	5704.0	5807.8	5911.5	6015.2	6118.9
6000	6222.6	6326.3	6430.0	6533.7	6637.4	6741.1	6844.9	6948.6	7052.3	7156.0
7000	7259.7	7363.4	7467.1	7570.8	7674.5	7778.2	7882.0	7985.7	8089.4	8193.1
8000	8296.8	8400.5	8504.2	8607.9	8711.6	8815.3	8919.1	9022.8	9126.5	9230.2
9000	9333.9	9437.6	9541.3	9645.0	9748.7	9852.4	9956.2	10059.9	10163.6	10267.3
10000	10371.0	10474.7	10578.4	10682.1	10785.8	10889.5	10993.3	11097.0	11200.7	11304.4
11000	11408.1	11511.8	11615.5	11719.2	11822.9	11926.6	12030.4	12134.1	12297.8	12341.5
12000	12445.2	12548.9	12652.6	12756.3	12860.0	12963.8	13067.5	13171.2	13274.9	13378.6
13000	13482.3	13586.0	13689.7	13793.4	13897.1	14000.9	14104.6	14208.3	14312.0	14415.7
14000	14519.4	14623.1	14726.8	14830.5	14934.2	15038.0	15141.7	15245.4	15349.1	15452.8
15000	15556.5	15660.2	15763.9	15867.6	15971.3	16075.0	16178.8	16282.5	16386.2	16489.9
16000	16593.6	16697.3	16801.0	16904.7	17008.4	17112.1	17215.9	17319.6	17423.3	17527.0

TO CONVERT

RHINE OR PRUSSIAN FEET

INTO DIFFERENT MEASURES OF LENGTH.

The Rhine Foot is used in Physical Geography, though not so extensively as the French or Paris Foot, in the northwestern part of Germany, Denmark, and Holland. Its legal value in the Prussian system of weights and measures is 139.13 French or Paris Lines, from which are derived the equations used in computing the following tables.

XXXI. CONVERSION OF RHINE OR PRUSSIAN FEET INTO FRENCH TOISES.

1 Rhine Foot = 0.1610301 Toise.

Rhine Feet. Thousands.	Hundreds.									
	0.	**100.**	**200.**	**300.**	**400.**	**500.**	**600.**	**700.**	**800.**	**900.**
	Toises.	Toises.	Toises.	Toises.	Toises.	Toises.	Toises.	Toises.	Toises.	Toises.
0	0.00	16.10	32.21	48.31	64.41	80.52	96.62	112.72	128.82	144.93
1000	161.03	177.13	193.24	209.34	225.44	241.55	257.65	273.75	289.85	305.96
2000	322.06	338.16	354.27	370.37	386.47	402.58	418.68	434.78	450.88	466.99
3000	483.09	499.19	515.30	531.40	547.50	563.61	579.71	595.81	611.91	628.02
4000	634.12	650.22	666.33	692.43	608.53	724.64	740.74	756.84	772.94	789.05
5000	805.15	821.25	837.36	853.46	869.56	885.67	901.77	917.87	933.97	950.08
6000	966.18	982.28	998.39	1014.49	1030.59	1046.70	1062.80	1078.90	1095.00	1111.11
7000	1127.21	1143.31	1159.42	1175.52	1191.62	1207.73	1223.83	1239.93	1256.03	1272.14
8000	1288.24	1304.34	1320.45	1336.55	1352.65	1368.76	1384.86	1400.96	1417.06	1433.17
9000	1449.27	1465.37	1481.48	1497.58	1513.68	1529.79	1545.89	1561.99	1578.09	1594.20

XXXII. CONVERSION OF RHINE OR PRUSSIAN FEET INTO METRES.

1 Rhine Foot = 0.31385354 Metre.

Rhine Feet. Thousands.	Rhine Feet. Hundreds. 0.	100.	200.	300.	400.	500.	600.	700.	800.	900.
	Metres.	Metres.	Metres.	Metres.	Metres.	Metres.	Metres.	Metres.	Metres.	Metres.
0	0.00	31.39	62.77	94.16	125.54	156.93	188.31	219.70	251.08	282.47
1000	313.85	345.24	376.62	408.01	439.39	470.78	502.17	533.55	564.94	596.32
2000	627.71	659.09	690.48	721.86	753.25	784.63	816.02	847.40	878.79	910.18
3000	941.56	972.95	1004.33	1035.72	1067.10	1098.49	1129.87	1161.26	1192.64	1224.03
4000	1255.41	1286.80	1318.18	1349.57	1380.96	1412.34	1443.73	1475.11	1506.50	1537.88
5000	1569.27	1600.65	1632.04	1663.42	1694.81	1726.19	1757.58	1788.97	1820.35	1851.74
6000	1883.12	1914.51	1945.89	1977.28	2008.66	2040.05	2071.43	2102.82	2134.20	2165.59
7000	2196.97	2228.36	2259.75	2291.13	2322.52	2353.90	2385.29	2416.67	2448.06	2479.44
8000	2510.83	2542.21	2573.60	2604.98	2636.37	2667.76	2699.14	2730.53	2761.91	2793.30
9000	2824.68	2856.07	2887.45	2918.84	2950.22	2981.61	3012.99	3044.38	3075.76	3107.15

XXXIII. OF RHINE OR PRUSSIAN FEET INTO FRENCH FEET AND DECIMALS.

1 Rhine Foot = 0.96618056 French Foot.

Rhine Feet. Thousands.	Rhine Feet. Hundreds. 0.	100.	200.	300.	400.	500.	600.	700.	800.	900.
	Fr. Feet.	Fr Feet.	Fr. Feet.	Fr. Feet.	Fr. Feet.	Fr. Feet.	Fr. Feet.	Fr. Feet.	Fr. Feet.	Fr. Feet.
0	0.00	96.62	193.24	289.85	386.47	483.09	579.71	676.33	772.94	869.56
1000	966.18	1062.80	1159.42	1256.03	1352.65	1449.27	1545.89	1642.51	1739.13	1835.74
2000	1932.36	2028.98	2125.60	2222.22	2318.83	2415.45	2512.07	2608.69	2705.31	2801.92
3000	2898.54	2995.16	3091.78	3188.40	3285.01	3381.63	3478.25	3574.87	3671.49	3768.10
4000	3864.72	3961.34	4057.96	4154.58	4251.19	4347.81	4444.43	4541.05	4637.67	4734.28
5000	4830.90	4927.52	5024.14	5120.76	5217.38	5313.99	5410.61	5507.23	5603.85	5700.47
6000	5797.08	5893.70	5990.32	6086.94	6183.56	6280.17	6376.79	6473.41	6570.03	6666.65
7000	6763.26	6859.88	6956.50	7053.12	7149.74	7246.35	7342.97	7439.59	7536.21	7632.83
8000	7729.44	7826.06	7922.68	8019.30	8115.92	8212.53	8309.15	8405.77	8502.39	8599.01
9000	8695.63	8792.24	8888.86	8985.48	9082.10	9178.72	9275.33	9371.95	9468.57	9565.19

XXXIV. OF RHINE OR PRUSSIAN FEET INTO ENGLISH FEET AND DECIMALS.

1 Rhine Foot = 1.0297218 English Foot.

Rhine Feet. Thousands.	Rhine Feet. Hundreds. 0.	100.	200.	300.	400.	500.	600.	700.	800.	900.
	Eng. feet.	Eng. feet.	Eng. feet.	Eng. feet.	Eng. feet.	Eng. feet.	Eng. feet.	Eng. feet.	Eng. feet.	Eng. feet.
0	0.00	102.97	205.94	308.92	411.89	514.86	617.83	720.81	823.78	926.75
1000	1029.72	1132.69	1235.67	1338.64	1441.61	1544.58	1647.55	1750.53	1853.50	1956.47
2000	2059.44	2162.42	2265.39	2368.36	2471.33	2574.30	2677.28	2780.25	2883.22	2986.19
3000	3089.17	3192.14	3295.11	3398.08	3501.05	3604.03	3707.00	3809.97	3912.94	4015.92
4000	4118.89	4221.86	4324.83	4427.80	4530.78	4633.75	4736.72	4839.69	4942.66	5045.64
5000	5148.61	5251.58	5354.55	5457.53	5560.50	5663.47	5766.44	5869.41	5972.39	6075.36
6000	6178.33	6281.30	6384.28	6487.25	6590.22	6693.19	6796.16	6899.14	7002.11	7105.08
7000	7208.05	7311.02	7414.00	7516.97	7619.94	7722.91	7825.89	7928.86	8031.83	8134.80
8000	8237.77	8340.75	8443.72	8546.69	8649.66	8752.64	8855.61	8958.58	9061.55	9164.52
9000	9267.50	9370.47	9473.44	9576.41	9679.38	9782.36	9885.33	9988.30	10091.3	10194.2

D

TO CONVERT

SPANISH OR MEXICAN VARAS AND FEET

INTO DIFFERENT MEASURES OF LENGTH.

XXXV. CONVERSION OF SPANISH OR MEXICAN VARAS INTO METRES.

1 Vara = 0.847965 Metre.

Varas. Thousands.	Hundreds. 0.	100.	200.	300.	400.	500.	600.	700.	800.	900.
	Metres.	Metres.	Metres.	Metres.	Metres.	Metres.	Metres.	Metres.	Metres.	Metres.
0	0.00	84.80	169.59	254.39	339.19	423.98	508.78	593.58	678.37	763.17
1000	847.96	932.76	1017.56	1102.35	1187.15	1271.95	1356.74	1441.54	1526.34	1611.13
2000	1695.93	1780.73	1865.52	1950.32	2035.12	2119.91	2204.71	2289.51	2374.30	2459.10
3000	2543.89	2628.69	2713.49	2798.28	2883.08	2967.88	3052.67	3137.47	3222.27	3307.06
4000	3391.86	3476.66	3561.45	3646.25	3731.05	3815.84	3900.64	3985.44	4070.23	4155.03
5000	4239.82	4324.62	4409.42	4494.21	4579.01	4663.81	4748.60	4833.40	4918.20	5002.99
6000	5087.79	5172.59	5257.38	5342.18	5426.98	5511.77	5596.57	5681.37	5766.16	5850.96
7000	5935.75	6020.55	6105.35	6190.14	6274.94	6359.74	6444.53	6529.33	6614.13	6698.92
8000	6783.72	6868.52	6953.31	7038.11	7122.91	7207.70	7292.50	7377.30	7462.09	7546.89
9000	7631.68	7716.48	7801.28	7886.07	7970.87	8055.67	8140.46	8225.26	8310.06	8394.85

XXXVI. OF SPANISH OR MEXICAN VARAS INTO ENGLISH FEET AND DECIMALS.

1 Vara = 2.78209 English Feet.

Varas. Thousands.	Hundreds. 0.	100.	200.	300.	400.	500.	600.	700.	800.	900.
	Eng. feet.	Eng. feet.	Eng. feet.	Eng. feet.	Eng. feet.	Eng. feet.	Eng. feet.	Eng. feet.	Eng. feet.	Eng. feet.
0	0.0	278.2	556.4	834.6	1112.8	1391.0	1669.3	1947.5	2225.7	2503.9
1000	2782.1	3060.3	3338.5	3616.7	3894.9	4173.1	4451.3	4729.6	5007.8	5286.0
2000	5564.2	5842.4	6120.6	6398.8	6677.0	6955.2	7233.4	7511.6	7789.9	8068.1
3000	8346.3	8624.5	8902.7	9180.9	9459.1	9737.3	10015.5	10293.7	10571.9	10850.2
4000	11128.4	11406.6	11684.8	11963.0	12241.2	12519.4	12797.6	13075.8	13354.0	13632.2
5000	13910.4	14188.7	14466.9	14745.1	15023.3	15301.5	15579.7	15857.9	16136.1	16414.3
6000	16692.5	16970.7	17249.0	17527.2	17805.4	18083.6	18361.8	18640.0	18918.2	19196.4
7000	19474.6	19752.8	20031.0	20309.3	20587.5	20865.7	21143.9	21422.1	21700.3	21978.5
8000	22256.7	22534.9	22813.1	23091.3	23369.6	23647.8	23926.0	24204.2	24482.4	24760.6
9000	25038.8	25317.0	25595.2	25873.4	26151.6	26429.9	26708.1	26986.3	27164.5	27442.7

XXXVII. CONVERSION OF CASTILIAN FEET INTO METRES.

1 Castilian Foot = 0.282655 Metre.

Castilian Feet. Thousands.	Hundreds. 0.	100.	200.	300.	400.	500.	600.	700.	800.	900.
	Metres.	Metres.	Metres.	Metres.	Metres.	Metres.	Metres.	Metres.	Metres.	Metres.
0	0.00	28.27	56.53	84.80	113.06	141.33	169.59	197.86	226.12	254.39
1000	282.65	310.92	339.19	367.45	395.72	423.98	452.25	480.51	508.78	537.04
2000	565.31	593.58	621.84	650.11	678.37	706.64	734.90	763.17	791.43	819.70
3000	847.96	876.23	904.50	932.76	961.03	989.29	1017.56	1045.82	1074.09	1102.35
4000	1130.62	1158.89	1187.15	1215.42	1243.68	1271.95	1300.21	1328.48	1356.74	1385.01
5000	1413.27	1441.54	1469.81	1498.07	1526.34	1554.60	1582.87	1611.13	1639.40	1667.66
6000	1695.93	1724.20	1752.46	1780.73	1808.99	1837.26	1865.52	1893.79	1922.05	1950.32
7000	1978.58	2006.85	2035.12	2063.38	2091.65	2119.91	2148.18	2176.44	2204.71	2232.97
8000	2261.24	2289.51	2317.77	2346.04	2374.30	2402.57	2430.83	2459.10	2487.36	2515.63
9000	2543.89	2572.16	2600.43	2628.69	2656.96	2685.22	2713.49	2741.75	2770.02	2798.28

XXXVIII. CONVERSION OF CASTILIAN FEET INTO PARIS OR FRENCH FEET.

1 Castilian Foot = 0.8701382 Paris Foot.

Castilian Feet. Thousands.	Hundreds. 0.	100.	200.	300.	400.	500.	600.	700.	800.	900.
	Par. Feet.	Par. Feet.	Par. Feet.	Par. Feet.	Par. Feet.	Par. Feet.	Par. Feet.	Par. Feet.	Par. Feet.	Par. Feet.
0	0.00	87.01	174.03	261.04	348.06	435.07	522.08	609.10	696.11	783.12
1000	870.14	957.15	1044.17	1131.18	1218.19	1305.21	1392.22	1479.23	1566.25	1653.26
2000	1740.28	1827.29	1914.30	2001.32	2088.33	2175.35	2262.36	2349.37	2436.39	2523.40
3000	2610.41	2697.43	2784.44	2871.46	2958.47	3045.48	3132.50	3219.51	3306.52	3393.54
4000	3480.55	3567.57	3654.58	3741.59	3828.61	3915.62	4002.64	4089.65	4176.66	4263.68
5000	4350.69	4437.70	4524.72	4611.73	4698.75	4785.76	4872.77	4959.79	5046.80	5133.82
6000	5220.83	5307.84	5394.86	5481.87	5568.88	5655.90	5742.91	5829.93	5916.94	6003.95
7000	6090.97	6177.98	6265.00	6352.01	6439.02	6526.04	6613.05	6700.06	6787.08	6874.09
8000	6961.11	7048.12	7135.13	7222.15	7309.16	7396.17	7483.19	7570.20	7657.22	7744.23
9000	7831.24	7918.26	8005.27	8092.29	8179.30	8266.31	8353.33	8440.34	8527.35	8614.37

XXXIX. CONVERSION OF CASTILIAN FEET INTO AMERICAN FEET.

1 Castilian Foot = 0.9273093 American Foot.

Castilian Feet. Thousands.	Hundreds. 0.	100.	200.	300.	400.	500.	600.	700.	800.	900.
	Am. Feet.	Am. Feet.	Am. Feet.	Am. Feet.	Am. Feet.	Am. Feet.	Am. Feet.	Am. Feet.	Am. Feet.	Am. Feet.
0	0.00	92.73	185.46	278.19	370.92	463.65	556.39	649.12	741.85	834.58
1000	927.31	1020.04	1112.77	1205.50	1298.23	1390.96	1483.69	1576.43	1669.16	1761.89
2000	1854.62	1947.35	2040.08	2132.81	2225.54	2318.27	2411.00	2503.74	2596.47	2689.20
3000	2781.93	2874.66	2967.39	3060.12	3152.85	3245.58	3338.31	3431.04	3523.78	3616.51
4000	3709.24	3801.97	3894.70	3987.43	4080.16	4172.89	4265.62	4358.35	4451.08	4543.82
5000	4636.55	4729.28	4822.01	4914.74	5007.47	5100.20	5192.93	5285.66	5378.39	5471.12
6000	5563.86	5656.59	5749.32	5842.05	5934.78	6027.51	6120.24	6212.97	6305.70	6398.43
7000	6491.17	6583.90	6676.63	6769.36	6862.09	6954.82	7047.55	7140.28	7233.01	7325.74
8000	7418.47	7511.21	7603.94	7696.67	7789.40	7882.13	7974.86	8067.59	8160.32	8253.05
9000	8345.78	8438.51	8531.25	8623.98	8716.71	8809.44	8902.17	8994.90	9087.63	9180.36

TO CONVERT

FRACTIONAL PARTS OF A TOISE AND OF A FOOT

INTO EACH OTHER.

XL. CONVERSION OF INCHES INTO DUODECIMAL LINES.

1 Inch = 12 Lines.

Inches. Tens.	Inches. Units.									
	0.	**1.**	**2.**	**3.**	**4.**	**5.**	**6.**	**7.**	**8.**	**9.**
	Lines.	Lines.	Lines.	Lines.	Lines.	Lines.	Lines.	Lines.	Lines.	Lines.
0	0	12	24	36	48	60	72	84	96	108
10	120	132	144	156	168	180	192	204	216	228
20	240	252	264	276	288	300	312	324	336	348
30	360	372	384	396	408	420	432	444	456	468
40	480	492	504	516	528	540	552	564	576	588
50	600	612	624	636	648	660	672	684	696	708
60	720	732	744	756	768	780	792	804	816	828
70	840	852	864	876	888	900	912	924	936	948
80	960	972	984	996	1008	1020	1032	1044	1056	1068
90	1080	1092	1104	1116	1128	1140	1152	1164	1176	1188
100	1200	1212	1224	1236	1248	1260	1272	1284	1296	1308

XLI. CONVERSION OF DECIMALS OF A TOISE INTO FEET AND INCHES.

1 Toise = 6 Feet = 72 Inches = 864 Lines.

Toises. Tens.	Hundredths of a Toise.									
	0.	**1.**	**2.**	**3.**	**4.**	**5.**	**6.**	**7.**	**8.**	**9.**
	ft. in. lin.	ft. in. lin.	ft. in. lin.	ft. in. lin.	ft. in lin.	ft. in. lin.	ft. in. lin.	ft. in. lin.	ft. in. lin.	ft. in. lin.
0.0	0.0.0,00	0. 0. 8,64	0. 1. 5,28	0. 2. 1,92	0. 2.10,56	0. 3.7,20	0. 4. 3,84	0.5. 0,48	0. 5. 9,12	0. 6. 5,76
0.1	0.7.2,40	0. 7.11,04	0. 8. 7,68	0. 9. 4,32	0.10. 0,96	0.10.9,60	0.11. 6,24	1.0. 2,88	1. 0.11,52	1. 1. 8,16
0.2	1.2.4,80	1. 3. 1,44	1. 3.10,08	1. 4. 6,72	1. 5. 3,36	1. 6.0,00	1. 6. 8,64	1.7. 5,28	1. 8. 1,92	1. 8.10,56
0.3	1.9.7,20	1.10. 3,84	1.11. 0,48	1.11. 9,12	2. 0. 5,76	2. 1.2,40	2. 1.11,04	2.2. 7,68	2. 3. 4,32	2. 4. 0,96
0.4	2.4.9,60	2. 5. 6,24	2. 6. 2,88	2. 6.11,52	2. 7. 8,16	2. 8.4,80	2. 9. 1,44	2.9.10,08	2.10. 6,72	2.11. 3,36
0.5	3.0.0,00	3. 0. 8,64	3. 1. 5,28	3. 2. 1,92	3. 2.10,56	3. 3.7,20	3. 4. 3,84	3.5. 0,48	3. 5. 9,12	3. 6. 5,76
0.6	3.7.2,40	3. 7.11,04	3. 8. 7,68	3. 9. 4,32	3.10. 0,96	3.10.9,60	3.11. 6,24	4.0. 2,88	4. 0.11,52	4. 1. 8,16
0.7	4.2.4,80	4. 3. 1,44	4. 3.10,08	4. 4. 6,72	4. 5. 3,36	4. 6.0,00	4. 6. 8,64	4.7. 5,28	4. 8. 1,92	4. 8.10,56
0.8	4.9.7,20	4.10. 3,84	4.11. 0,48	4.11. 9,12	5. 0. 5,76	5. 1.2,40	5. 1.11,04	5.2. 7,68	5. 3. 4,32	5. 4. 0,96
0.9	5.4.9,60	5. 5. 6,24	5. 6. 2,88	5. 6.11,52	5. 7. 8,16	5. 8.4,80	5. 9. 1,44	5.9.10,08	5.10. 6,72	5.11. 3,36

XLII. CONVERSION OF DECIMALS OF A FOOT INTO INCHES AND DECIMALS.

Feet. Tens.	Hundredths of a Foot. 0.	1.	2.	3.	4.	5.	6.	7.	8.	9.
	Inches.	Inches.	Inches.	Inches.	Inches.	Inches.	Inches.	Inches.	Inches.	Inches.
0.0	0.00	0.12	0.24	0.36	0.48	0.60	0.72	0.84	0.96	1.08
0.1	1.20	1.32	1.44	1.56	1.68	1.80	1.92	2.04	2.16	2.28
0.2	2.40	2.52	2.64	2.76	2.88	3.00	3.12	3.24	3.36	3.48
0.3	3.60	3.72	3.84	3.96	4.08	4.20	4.32	4.44	4.56	4.68
0.4	4.80	4.92	5.04	5.16	5.28	5.40	5.52	5.64	5.76	5.88
0.5	6.00	6.12	6.24	6.36	6.48	6.60	6.72	6.84	6.96	7.08
0.6	7.20	7.32	7.44	7.56	7.68	7.80	7.92	8.04	8.16	8.28
0.7	8.40	8.52	8.64	8.76	8.88	9.00	9.12	9.24	9.36	9.48
0.8	9.60	9.72	9.84	9.96	10.08	10.20	10.32	10.44	10.56	10.68
0.9	10.80	10.92	11.04	11.16	11.28	11.40	11.52	11.64	11.76	11.88

XLIII. CONVERSION OF DECIMALS OF A FOOT INTO INCHES AND DUODECIMAL LINES.

Feet. Tens.	Hundredths of a Foot. 0.	1.	2.	3.	4.	5.	6.	7.	8.	9.
	In. Line.	In. Line.	In. Line.	In. Line.	In. Line.	In. Line.	In. Line.	In. Line.	In. Line.	In. Line.
0.0	0.0,00	0. 1,44	0. 2,88	0. 4,32	0. 5,76	0.7,20	0. 8,64	0.10,08	0.11,52	1. 0,96
0.1	1.2,40	1. 3,84	1. 5,28	1. 6,72	1. 8,16	1.9,60	1.11,04	2. 0,48	2. 1,92	2. 3,36
0.2	2.4,80	2. 6,24	2. 7,68	2. 9,12	2.10,56	3.0,00	3. 1,44	3. 2,88	3. 4,32	3. 5,76
0.3	3.7,20	3. 8,64	3.10,08	3.11,52	4. 0,96	4.2,40	4. 3,84	4. 5,28	4. 6,72	4. 8,16
0.4	4 9,60	4.11,04	5. 0,48	5. 1,92	5. 3,36	5.4,80	5. 6,24	5. 7,68	5. 9,12	5.10,56
0.5	6.0,00	6. 1,44	6. 2,88	6. 4,32	6. 5,76	6.7,20	6. 8,64	6.10,08	6.11,52	7. 0,96
0.6	7.2,40	7. 3,84	7. 5,28	7. 6,72	7. 8,16	7.9,60	7.11,04	8. 0,48	8. 1,92	8. 3,36
0.7	8.4,80	8. 6,24	8. 7,68	8. 9,12	8.10,56	9.0,00	9. 1,44	9. 2,88	9. 4,32	9. 5,76
0.8	9.7,20	9. 8,64	9.10,08	9.11,52	10. 0,96	10.2,40	10. 3,84	10. 5,28	10. 6,72	10. 8,16
0.9	10.9,60	10.11,04	11. 0,48	11. 1,92	11. 3,36	11.4,80	11. 6,24	11. 7,68	11. 9,12	11.10,56

XLIV. CONVERSION OF INCHES AND DUODECIMAL LINES INTO DECIMALS OF A FOOT.

1 Inch = 0.08333 of a Foot. 1 Line = 0.006944 of a Foot.

Inches.	Lines. 0.	1.	2.	3.	4.	5.	6.	7.	8.	9.	10.	11.
	Foot.	Foot.	Foot.	Foot.	Foot.	Foot.	Foot.	Foot.	Foot.	Foot.	Foot.	Foot.
0	0.0000	0.0069	0.0139	0.0208	0.0278	0.0347	0.0417	0.0486	0.0556	0.0625	0.0694	0.0764
1	0.0833	0.0903	0.0972	0.1042	0.1111	0.1181	0.1250	0.1319	0.1389	0.1458	0.1528	0.1597
2	0.1667	0.1736	0.1806	0.1875	0.1944	0.2014	0.2083	0.2153	0.2222	0.2292	0.2361	0.2431
3	0.2500	0.2569	0.2639	0.2708	0.2778	0.2847	0.2917	0.2986	0.3056	0.3125	0.3194	0.3264
4	0.3333	0.3403	0.3472	0.3542	0.3611	0.3681	0.3750	0.3819	0.3889	0.3958	0.4028	0.4097
5	0.4167	0.4236	0.4306	0.4375	0.4444	0.4514	0.4583	0.4653	0.4722	0.4792	0.4861	0.4931
6	0.5000	0.5069	0.5139	0.5208	0.5278	0.5347	0.5417	0.5486	0.5556	0.5625	0.5694	0.5764
7	0.5833	0.5903	0.5972	0.6042	0.6111	0.6181	0.6250	0.6319	0.6389	0.6458	0.6528	0.6597
8	0.6667	0.6736	0.6806	0.6875	0.6944	0.7014	0.7083	0.7153	0.7222	0.7292	0.7361	0.7431
9	0.7500	0.7569	0.7639	0.7708	0.7778	0.7847	0.7917	0.7986	0.8056	0.8125	0.8194	0.8264
10	0.8333	0.8403	0.8472	0.8542	0.8611	0.8681	0.8750	0.8819	0.8889	0.8958	0.9028	0.9097
11	0.9167	0.9236	0.9306	0.9375	0.9444	0.9514	0.9583	0.9653	0.9722	0.9792	0.9861	0.9931

METEOROLOGICAL TABLES.

V.

METEOROLOGICAL CORRECTIONS,

OR

TABLES

FOR CORRECTING SERIES OF OBSERVATIONS FOR THE PERIODIC AND NON-PERIODIC VARIATIONS.

CONTENTS.

[The figures refer to the folio at the bottom of the page.—The letters near them mean, D. = calculated by Dove; Gl. = Glaisher; G. = Guyot; L. = Lefroy. For the letters before the latitudes, see page 12.]

Temperature.

Hourly Corrections for Periodic Variations.

NORTH AMERICA.

SOUTH AMERICA.

ASIA.

EUROPE.

AFRICA AND AUSTRALIA.

Monthly Corrections for Non-periodic Variations.

Force of Vapor and Relative Humidity.

Hourly Corrections for Periodic Variations.

METEOROLOGICAL CORRECTIONS.

One of the prominent objects of a prolonged series of meteorological observations is to determine the mean condition of the atmosphere, during a given interval of time, such as a day, a month, or a year, as to its temperature, moisture, and barometric pressure. In order to furnish the true means of these elements, free from the periodic changes which depend upon the daily course of the sun and upon the seasons, the observations ought to be made at equal intervals of time, and be so often repeated as actually to represent the sum of the variations which took place during the stated time. It is generally admitted that observations taken at every one of the twenty-four hours of the day give means which do not sensibly differ from the means which would be obtained from a still larger number of observations during the same time; so that means derived from hourly observations may be considered as the true daily, monthly, and annual means of the year in which the observations were taken.

However, as the means of a given month, or year, will generally be found somewhat to differ from those of another year, at the same place, from causes which are not of a periodic nature, it is obvious that the absolute means can only be derived from the means of a series of years, in which the differences arising from these non-periodic variations may be considered as sufficiently balancing each other.

Hourly observations can be expected only from a very few stations, favored with peculiar arrangements for the purpose. By far the larger number of observers must necessarily confine themselves to three or four observations a day. The means, therefore, deduced from such a set of observations, generally differ from the true means which would be given by hourly observations, by a quantity which varies with the hours selected for the observations. If that quantity, however, is known by having been previously determined for every hour, or set of hours, by a long series of hourly observations taken at some station in a similar climatic situation, it is evident that, whatever be the hours at which observations are taken, the means derived from them can always be reduced to the true means by correcting them for that difference.

The following tables furnish such corrections, both for periodic and non-periodic variations of temperature, and for stations situated in various latitudes. They give the quantities which must be added to, or subtracted from, the hourly means, in order to obtain the true means of the day, of the month, and of the year.

Two tables of the same description, for moisture, which may be considered as specimens of the kind, close the set.

Two other tables, for correcting the mean barometric pressures, are found at the end of the Hypsometrical Tables, pp. 92, 93.

CORRECTIONS FOR TEMPERATURE.

HOURLY CORRECTIONS FOR PERIODIC VARIATIONS,

OR

TABLES

FOR REDUCING THE MEANS OF THE OBSERVATIONS TAKEN AT ANY HOUR OF THE DAY TO THE TRUE MEAN TEMPERATURE OF THE DAY, OF THE MONTH, AND OF THE YEAR.

HOURLY CORRECTIONS FOR PERIODIC VARIATIONS,

OR

CORRECTIONS TO BE APPLIED TO THE MEANS OF THE HOURS OF OBSERVATION, OR SETS OF HOURS, IN ORDER TO OBTAIN THE TRUE MEAN TEMPERATURES OF THE RESPECTIVE DAYS, MONTHS, AND OF THE YEAR.

THE following set contains all the tables for correcting the means of observations on atmospheric temperature for the effect of diurnal variation which have been published by Dove, together with a few others of the same description. Dove's tables are found in two papers, published in the *Memoirs of the Royal Academy of Berlin* for 1846 and for 1856, and in the first *Report on the Observations of the Meteorological Institute of Prussia*, Berlin, 1851.

In the first paper are twenty-nine tables, in Reaumur's scale, nine of which have been republished, in Fahrenheit's scale, in the *Proceedings of the British Association* for **1847**, and will also be found below. In that series the corrections have been formed by finding first the differences between the hourly and the true means, and then computing the observations by Bessel's formula, in order to eliminate the accidental irregularities due to the shortness of the period during which the observations were taken. Calling x the horary angle reckoned from noon, Bessel's formula is

$$t x = u + u' \sin (x + U') + u'' \sin (2 x + U'') + u''' \sin (3 x + U''').$$

The stations at which hourly observations were made are Trevandrum, Madras, Bombay, Salzuflen, Prague, St. Petersburg, Catharinenburg, Barnaul, Nertchinsk, Matoschkin-Schar, Strait of Kara, and Boothia Felix. Bi-hourly observations were taken at Brussels, Greenwich, and Toronto; in all others the night observations are wanting, and were obtained by interpolation. Moreover, in several stations the number of observations was small, at Madras even only thirty-six days. The tables of that series may be readily distinguished from those belonging to the same stations in the second, by their containing the corrections for several sets of hours, which are not found in the tables of the other.

In Dove's second series, and in all other tables, the corrections given are simply the differences, with reverse signs, between the hourly and the true means, excepting, however, the stations of Toronto, in which the corrections were computed, by Bessel's formula, by Colonel Sabine; of Prague, by Jelineck; of Salzburg, and those of Geneva and St. Bernard, by Plantamour.

The observations from which these tables are derived were made hourly at Hobarton during 8 years; at the Cape of Good Hope, for 5¼ years; St. Helena, 5 years; Madras, 5 years; Bombay, 4 years; Calcutta, 1½ years; Toronto, 6 years; Philadelphia, 3 years; Makerstoun, 3 years; Utrecht, 1¾ years; Prague, 10½ years; Munich, 7 years; Salzburg, 6 years; St. Petersburg, 10 years; Catherinenburg, 6 years; Barnaul, 5 years; Tiflis, 4 years; Nertchinsk, 6 years; Peking, 4 years; Sitka, 5 years. In the following stations the observations were bi-hourly: — Washington, for 1½ years; Greenwich, 7 years; Dublin, 4 years; Brussels, 9 years; Geneva and St. Bernard, 4 years; Schwerin, 3 years.

The observations made in England, and in her colonies, are found in the various government publications. Those of the Russian stations are taken from the *Annuaire Météorologique et Magnétique des Ingénieurs des Mines*, and in the *Annales de l'Observatoire Physique Central de Russie.* The observations made at Prague, Munich, Geneva, with those at St. Bernard, Makerstoun, Greenwich, Brussels, and Washington, were published by their respective Observatories; those of Utrecht, by Buys-Ballot; of Dublin, by Lloyd, in his *Notes on the Meteorology of Ireland;* those of Schwerin were communicated in manuscript by Dippe; the observations at Melville Island are published in No. 42 of the Parliamentary papers for 1854; and those at Bossekop, by Martins and Bravais, in the *Voyage de la Commission Scientifique du Nord.*

The tables of this second series being mostly deduced from longer series of observations than those in the first, when the same station is found in both, the table in the second is generally to be preferred.

Glaisher's table for Greenwich has been taken from the *Greenwich Observations.* Captain Lefroy kindly furnished the tables for Toronto and Lake Athabasca. To him the author is also indebted for the observations made at Montreal by Mr. McCord, from which Table X. was computed. Table III., for Philadelphia, was deduced by the writer from the observations made at Girard College under the direction of Prof. A. D. Bache.

In order to facilitate the selection of the tables, they are marked in the table of contents with capitals, which have the following signification:—

A and B mean that the tables have been derived from hourly and bi-hourly observations, and have been computed by Bessel's formula; C, that the tables contain values obtained by interpolation.

A′, B′, and C′ indicate the tables based respectively on hourly and bi-hourly or partly interpolated observations, which give simply the differences between the hourly and the true means.

The figures added to the letters indicate the number of years during which the observations used in forming the table were carried on. The stations are arranged, in each continent, in the order of their latitude.

Use of the Tables.

In order to reduce meteorological means obtained from any set of hours to the true means, the table best suited to the purpose must first be selected. The diurnal variation changing with the seasons, the latitude, the altitude, and the distance from the sea-shore, the station which comes nearest, in all these respects, to the station the observations of which are to be corrected, must be adopted.

Suppose the thermometer has been observed at Baltimore, during the month of January, at 7 A. M., 1 P. M., and 7 P. M., and the monthly means of these hours to be respectively 27°, 35°, and 31° Fahrenheit. We take Table III., Philadelphia, it being the nearest in latitude and climatic situation. We find the correction for the hours 7, 1, and 7, and we have

	Observed Means.	Corrections.		True Means.
For 7 A. M.	27°	+ 3°.63	=	30°.63
For 1 P. M.	35°	— 3°.87	=	31°.13
For 7 P. M.	31°	— 1°.13	=	29°.87
Sums,	93°	— 1°.37	=	91°.63
Means,	31°	— 0°.46	=	30°.54 True Mean for January.

It is obvious that the corrections can be applied, either separately to each hour, as is done above, or collectively, in taking the mean of the three hourly corrections and applying it to the mean of the three observations, as in the last line, which is the more convenient method. Therefore, in order to find the correction for any set of hours, it suffices to take the mean of the corrections given in the table for the hours composing the set. The true daily means can be found in the same way, and the true yearly means can be derived from the corrected monthly means, or by applying the corrections given in the last column.

HOURLY CORRECTIONS

FOR

PERIODIC VARIATIONS.

NORTH AMERICA.—SOUTH AMERICA.

I.

North America. — Washington. *Lat.* 38° 54′ N. *Long.* 77° 3′ W. *Greenw.*

Corrections to be applied to the Means of the Hours of Observation to obtain the true Mean Temperatures of the respective Days, Months, and of the Year. — Dove.

Degrees of Reaumur.

Hour.	Jan.	Feb.	March.	April.	May.	June.	July.	Aug.	Sept.	Oct.	Nov.	Dec.	Year.
A.M. 0 12′	1.15	1.26	1.60	1.95	2.33	2.87	2.94	2.31	2.39	1.73	0.85	0.96	1.86
2 12′	1.28	1.86	2.14	2.40	3.15	3.21	3.25	3.07	2.75	2.27	1.34	1.12	2.32
4 12′	1.45	2.18	2.67	2.75	3.56	3.64	3.83	3.49	3.15	2.89	1.92	1.54	2.76
6 12′	1.88	2.32	2.76	2.59	2.20	2.23	2.12	2.81	3.02	3.19	2.18	1.81	2.43
8 12′	1.48	1.76	1.68	1.05	0.32	−0.16	0.09	0.28	1.04	1.69	1.88	1.68	1.07
10 12′	−0.18	−0.58	−0.88	−0.76	−1.24	−1.82	−1.32	−1.81	−1.31	−1.25	−0.17	−0.15	−0.96
P.M. 0 12′	−1.47	−2.05	−2.36	−2.39	−2.64	−2.69	−2.55	−2.97	−2.92	−2.89	−1.90	−1.57	−2.37
2 12′	−2.60	−3.15	−3.35	−3.41	−3.57	−3.84	−3.49	−3.83	−3.74	−3.64	−2.44	−2.50	−3.30
4 12′	−2.32	−3.05	−3.20	−3.51	−3.66	−4.29	−4.16	−3.59	−3.65	−3.29	−2.08	−2.19	−3.25
6 12′	−0.76	−1.25	−1.73	−2.18	−2.44	−1.60	−2.24	−1.74	−1.88	−1.84	−1.59	−1.01	−1.69
8 12′	−0.23	0.02	−0.05	0.06	0.27	0.44	−0.21	−0.26	−0.23	0.18	−0.22	−0.26	−0.04
10 12′	0.33	0.69	0.76	1.42	1.67	2.04	1.26	1.79	1.41	0.98	0.23	0.43	1.08
Means.	1.32	1.52	6.26	9.02	12.64	18.34	19.29	17.78	16.04	7.47	5.20	1.63	

II.

N. America.—Philadelphia. *Lat.* 39° 58′ N. *Long.* 75° 11′ W. *Gr.*—Dove.

Degrees of Reaumur.

Hour.	Jan.	Feb.	March.	April.	May.	June.	July.	Aug.	Sept.	Oct.	Nov.	Dec.	Year.
Midn.	0.64	1.27	1.33	1.81	2.06	2.34	2.10	1.94	2.12	1.70	1.31	0.62	1.60
1	0.94	1.48	1.61	2.20	2.32	2.63	2.45	2.19	2.04	1.87	1.22	0.81	1.81
2	1.00	1.67	1.85	2.58	2.64	2.86	2.69	2.41	2.22	2.18	1.43	0.98	2.04
3	1.13	1.95	2.00	2.76	2.96	3.20	2.88	2.44	2.43	2.36	1.50	1.12	2.23
4	1.24	2.05	2.08	2.97	3.27	3.40	3.04	2.74	2.56	2.58	1.74	1.28	2.41
5	1.36	2.13	2.50	3.06	3.32	3.28	3.11	2.89	2.68	2.78	1.83	1.38	2.53
6	1.50	2.24	2.44	2.84	2.63	2.54	2.56	2.64	2.65	2.95	1.89	1.44	2.36
7	1.60	2.28	2.24	2.15	1.68	1.45	1.53	1.84	1.92	2.40	1.88	1.36	1.86
8	1.40	1.46	1.26	1.17	0.65	0.40	0.54	0.67	0.78	1.08	1.21	1.14	0.98
9	0.78	0.57	0.35	0.23	−0.39	−0.52	−0.36	−0.20	−0.18	−0.15	0.26	0.52	0.08
10	0.02	−0.39	−0.46	−0.71	−1.06	−1.23	−1.00	−1.05	−1.08	−1.17	−0.56	−0.22	−0.74
11	−0.68	−1.20	−1.38	−1.54	−1.74	−1.93	−1.74	−1.84	−1.90	−1.96	−1.27	−0.92	−1.50
Noon.	−1.21	−1.77	−1.97	−2.16	−2.24	−2.51	−2.26	−2.34	−2.45	−2.61	−1.77	−1.28	−2.05
1	−1.73	−2.36	−2.45	−2.86	−2.71	−3.06	−2.66	−2.67	−2.88	−3.14	−2.26	−1.63	−2.53
2	−2.04	−2.66	−2.74	−3.29	−3.11	−3.32	−2.97	−3.01	−3.22	−3.45	−2.52	−1.84	−2.85
3	−2.10	−2.82	−3.07	−3.42	−3.36	−3.40	−3.15	−3.11	−3.26	−3.45	−2.48	−1.85	−2.96
4	−1.98	−2.69	−2.99	−3.44	−3.46	−3.44	−3.06	−2.98	−3.17	−3.33	−2.24	−1.63	−2.87
5	−1.30	−2.18	−2.52	−3.14	−3.26	−3.05	−2.94	−2.70	−2.77	−2.46	−1.46	−1.10	−2.41
6	−0.91	−1.37	−1.60	−2.49	−2.46	−2.47	−2.30	−2.03	−1.77	−1.33	−0.82	−0.64	−1.68
7	−0.51	−0.80	−0.88	−1.23	−1.28	−1.38	−1.44	−1.02	−0.76	−0.52	−0.33	−0.31	−0.87
8	−0.20	−0.21	−0.20	−0.29	−0.06	0.06	0.03	0.01	0.28	0.18	−0.14	−0.04	−0.05
9	0.07	0.11	0.90	0.35	0.65	0.82	0.57	0.60	0.81	0.65	0.29	0.09	0.49
10	0.33	0.48	0.77	0.93	1.24	1.37	1.08	1.09	1.33	1.24	0.45	0.27	0.88
11	0.56	0.75	0.96	1.44	1.74	1.91	1.55	1.44	1.64	1.63	0.79	0.40	1.23
Mean.	0.30	1.12	5.18	8.75	12.18	16.22	18.19	17.52	14.66	8.72	3.67	0.58	

The numbers without sign must be added; those with the sign — must be subtracted.

III.

North America. — Philadelphia. *Lat.* 39° 58′ N. *Long.* 75° 11′ W. *Greenw.*

Corrections to be applied to the Means of the Hours of Observation to obtain the true Mean Temperatures of the respective Days, Months, and of the Year. — Guyot.

Degrees of Fahrenheit.

Hour.	Jan.	Feb.	March.	April.	May.	June.	July.	Aug.	Sept.	Oct.	Nov.	Dec.	Year.
Midnight.	1.47	2.90	2.90	4.13	4.68	5.28	4.70	4.37	4.47	3.80	2.70	1.40	3.57
1	2.13	3.37	3.63	4.88	5.25	5.93	5.57	4.93	4.60	4.17	2.73	1.83	4.08
2	2.20	3.57	4.17	5.88	5.95	6.45	6.10	5.43	5.00	4.87	3.20	2.20	4.59
3	2.57	4.43	4.50	6.28	6.68	7.23	6.53	5.50	5.47	5.27	3.37	2.53	5.03
4	2.80	4.67	4.70	6.75	7.38	7.68	6.90	6.17	5.77	5.77	3.90	2.87	5.45
5	3.07	4.83	5.63	6.95	7.48	7.40	7.03	6.50	6.03	6.23	4.10	3.10	5.70
6	3.40	5.10	5.50	6.45	5.93	5.73	5.80	5.93	5.97	6.60	4.23	3.23	5.32
7	3.63	5.17	5.03	4.90	3.80	3.28	3.50	4.13	4.33	5.37	4.20	3.07	4.20
8	3.17	3.33	2.80	2.50	1.48	0.90	1.27	1.50	1.93	2.40	2.70	2.57	2.16
9	1.77	1.33	0.80	0.58	−0.85	−1.15	−0.77	−0.43	−0.40	−0.37	0.57	1.17	0.19
10	0.07	−0.83	−1.03	−1.53	−2.38	−2.75	−2.20	−2.37	−2.43	−2.67	−1.27	−0.50	−1.66
11	−1.40	−2.63	−3.10	−3.40	−3.90	−4.33	−3.87	−4.13	−4.27	−4.43	−2.87	−2.07	−3.62
Noon.	−2.70	−3.93	−4.43	−4.72	−5.03	−5.63	−5.03	−5.27	−5.50	−5.90	−4.00	−2.87	−4.58
1	−3.87	−5.27	−5.50	−6.38	−6.08	−6.88	−5.93	−6.00	−6.47	−7.10	−5.10	−3.67	−5.69
2	−4.57	−5.97	−6.17	−7.12	−6.98	−7.45	−6.63	−6.83	−7.20	−7.80	−5.67	−4.13	−6.40
3	−4.70	−6.30	−6.90	−7.63	−7.55	−7.63	−7.03	−7.00	−7.33	−7.80	−5.60	−4.17	−6.64
4	−4.43	−6.00	−6.73	−7.65	−7.78	−7.73	−6.83	−6.70	−7.13	−7.53	−5.07	−3.67	−6.44
5	−2.90	−4.87	−5.67	−7.00	−7.33	−6.85	−6.57	−6.07	−6.23	−5.57	−3.30	−2.47	−5.40
6	−2.03	−3.03	−3.60	−5.55	−5.53	−5.55	−5.13	−4.57	−3.97	−3.03	−1.87	−1.43	−3.77
7	−1.13	−1.77	−1.97	−2.70	−2.88	−3.10	−3.20	−2.30	−1.70	−1.20	−0.77	−0.70	−1.95
8	−0.43	−0.43	−0.43	−0.60	−0.13	0.15	0.08	0.03	0.63	0.37	0.15	−0.10	−0.11
9	0.17	0.30	0.73	0.85	1.48	1.85	1.33	1.37	1.83	1.43	0.63	0.20	1.01
10	0.77	1.13	1.73	2.15	2.80	3.10	2.47	2.47	3.00	2.77	1.00	0.60	2.00
11	1.27	1.73	2.17	3.30	3.93	4.30	3.53	3.23	3.70	3.63	1.77	0.90	2.78
6, 6	0.69	1.04	0.95	0.45	0.20	0.09	0.34	0.68	1.00	1.79	1.18	0.90	0.78
7, 7	1.25	1.70	1.53	1.10	0.46	0.09	0.15	0.92	1.32	2.09	1.72	1.19	1.13
8, 8	1.37	1.45	1.18	0.85	0.68	0.53	0.67	0.77	1.01	1.38	1.35	1.24	1.04
9, 9	0.97	0.82	0.76	0.72	0.32	0.35	0.28	0.47	0.72	0.53	0.60	0.69	0.66
10, 10	0.42	0.15	0.35	0.31	0.21	0.18	0.14	0.05	0.29	0.05	−0.13	0.05	0.17
7, 2, 9	−0.22	−0.17	−0.15	−0.53	−0.57	−0.77	−0.61	−0.44	−0.35	−0.33	−0.28	−0.29	−0.39
6, 2, 8	−0.53	−0.43	−0.37	−0.42	−0.39	−0.52	−0.37	−0.29	−0.20	−0.28	−0.43	−0.67	−0.41
6, 2, 10	−0.13	0.09	0.53	0.74	0.58	0.46	0.55	0.52	0.59	0.52	−0.15	−0.10	0.44
6, 2, 6	−1.07	−0.72	−1.42	−2.07	−2.19	−2.42	−1.43	−1.82	−1.73	−1.41	−1.10	−0.78	−1.44
7, 2	−0.47	−0.40	−0.57	−1.11	−1.59	−2.09	−1.57	−1.35	−1.44	−1.22	−0.74	−0.53	−1.09
8, 2	−0.70	−1.32	−1.68	−2.31	−2.75	−3.28	−2.68	−2.67	−2.90	−2.70	−1.49	−0.78	−2.10
8, 1	−0.35	−0.97	−1.35	−1.94	−2.30	−2.99	−2.33	−2.25	−2.53	−2.35	−1.20	−0.55	−1.76
7, 1	−0.12	−0.05	−0.24	−0.74	−1.14	−1.80	−1.22	−0.94	−1.07	−0.87	−0.45	−0.30	−0.75
9, 12, 3, 9	−1.37	−2.15	−2.45	−2.73	−2.99	−3.14	−2.88	−2.83	−2.85	−3.16	−2.39	−1.42	−2.53

N. America. — Frankfort Arsenal. *Lat.* 39° 57′ N. *Long.* 75° 8′ W. *Greenw.*

Corrections to be applied to the Means of the Hours of Observation to obtain the true Mean Temperatures of the respective Days, Months, and of the Year. — Dove.

Degrees of Reaumur.

Hours.	Jan.	Feb.	March.	April.	May.	June.	July.	Aug.	Sept.	Oct.	Nov.	Dec.	Mean.
Morn. 1	1.34	1.46	1.75	1.87	2.60	3.41	3.07	2.69	2.63	2.40	1.18	1.34	2.15
2	1.51	1.73	2.13	2.33	3.05	3.73	3.51	3.04	3.05	2.67	1.27	1.50	2.46
3	1.82	1.98	2.56	2.88	3.43	3.92	3.83	3.32	3.49	2.94	1.41	1.66	2.77
4	2.13	2.23	2.90	3.29	3.57	3.84	3.84	3.36	3.73	3.13	1.51	1.80	2.94
5	2.31	2.46	2.95	3.31	3.32	3.36	3.40	2.99	3.54	3.12	1.73	1.87	2.86
6	2.25	2.35	2.62	2.83	2.65	2.46	2.52	2.21	2.84	2.82	1.38	1.80	2.39
7	1.88	2.01	1.91	1.94	1.66	1.26	1.34	1.15	1.71	2.19	1.06	1.52	1.64
8	1.22	1.33	0.94	0.85	0.57	−0.03	0.08	0.01	0.36	1.26	0.58	0.97	0.68
9	0.34	0.30	−0.07	−0.20	−0.45	−1.20	−1.06	−1.00	−0.96	0.12	−0.02	0.18	−0.34
10	−0.62	−0.72	−1.00	−1.05	−1.29	−2.11	−1.96	−1.78	−2.06	−1.13	−0.70	−0.76	−1.27
11	−1.54	−1.77	−1.76	−1.69	−1.97	−2.74	−2.64	−2.34	−2.89	−2.33	−1.12	−1.70	−2.04
Noon. . .	−2.30	−2.60	−2.32	−2.22	−2.35	−3.17	−3.16	−2.78	−3.47	−3.35	−1.96	−2.45	−2.68
1	−2.85	−3.01	−2.74	−2.72	−3.07	−3.51	−3.58	−3.16	−3.86	−4.05	−2.38	−2.87	−3.15
2	−3.02	−3.18	−3.01	−3.19	−3.52	−3.77	−3.87	−3.48	−4.07	−4.36	−2.54	−2.89	−3.41
3	−2.92	−2.93	−3.10	−3.53	−3.78	−3.89	−3.94	−3.61	−4.02	−4.22	−2.40	−2.54	−3.41
4	−2.53	−2.44	−2.95	−3.55	−3.70	−3.75	−3.67	−3.42	−3.63	−3.66	−1.96	−1.94	−3.10
5	−1.90	−1.87	−2.50	−3.11	−3.20	−3.23	−3.00	−2.81	−2.84	−2.75	−1.52	−1.23	−2.50
6	−1.14	−1.11	−1.78	−2.23	−2.31	−2.33	−2.00	−1.83	−1.72	−1.65	−0.56	−0.55	−1.60
7	−0.37	−0.46	−0.92	−1.09	−1.19	−1.16	−0.83	−0.67	−0.48	−0.54	0.14	0.01	−0.63
8	0.29	0.12	−0.06	0.02	−0.10	0.07	0.28	0.43	0.66	0.43	0.69	0.42	0.27
9	0.76	0.66	0.61	0.85	0.80	1.17	1.17	1.29	1.49	1.17	1.02	0.71	0.98
10	1.02	0.93	1.05	1.32	1.43	2.02	1.79	1.84	1.96	1.66	1.15	0.90	1.42
11	1.13	1.18	1.31	1.50	1.85	2.61	2.24	2.15	2.18	1.96	0.91	1.06	1.67
Midn. . .	1.19	1.36	1.48	1.62	2.01	3.04	2.63	2.40	2.35	2.18	1.15	1.20	1.88
6. 6	0.56	0.62	0.42	0.30	0.17	0.07	0.26	0.19	0.56	0.58	0.41	0.62	0.40
7. 7	0.76	0.78	0.50	0.42	0.24	0.05	0.26	0.24	0.62	0.83	0.60	0.76	0.51
8. 8	0.76	0.72	0.44	0.43	0.24	0.02	0.18	0.22	0.51	0.85	0.63	0.70	0.48
9. 9	0.55	0.48	0.27	0.33	0.18	−0.02	0.06	0.14	0.26	0.64	0.50	0.44	0.32
10.10	0.20	0.11	0.03	0.13	0.07	−0.05	−0.08	0.03	−0.05	0.26	0.23	0.07	0.08
7. 2. 9	−0.13	−0.17	−0.16	−0.13	−0.35	−0.45	−0.45	−0.35	−0.29	−0.33	−0.15	−0.22	−0.27
6. 2. 8	−0.16	−0.24	−0.15	−0.11	−0.32	−0.41	−0.36	−0.28	−0.19	−0.37	−0.16	−0.22	−0.25
6. 2.10	0.08	0.03	0.22	0.32	0.19	0.24	0.15	0.19	0.24	0.04	0.00	−0.06	0.14
6. 2. 6	−0.64	−0.65	−0.72	−0.86	−1.06	−1.21	−1.12	−1.03	−0.98	−1.06	−0.57	−0.55	−0.87
7. 2	−0.57	−0.59	−0.55	−0.63	−0.93	−1.26	−1.27	−1.17	−1.18	−1.09	−0.74	−0.69	−0.89
8. 2	−0.90	−0.93	−1.04	−1.17	−1.48	−1.90	−1.90	−1.74	−1.86	−1.55	−0.98	−0.96	−1.37
8. 1	−0.82	−0.84	−0.90	−0.94	−1.25	−1.77	−1.75	−1.58	−1.75	−1.40	−0.90	−0.95	−1.24
7. 1	−0.49	−0.50	−0.42	−0.39	−0.71	−1.13	−1.12	−1.10	−1.08	−0.93	−0.66	−0.68	−0.76
9.12.3.9	−1.03	−1.14	−1.22	−1.28	−1.45	−1.77	−1.75	−1.53	−1.74	−1.57	−0.84	−1.03	−1.36
7. 2.2(9)	0.10	0.04	−0.03	0.11	−0.07	−0.04	−0.05	0.06	0.16	0.04	0.14	0.01	0.04
Dail. ext.	−0.36	−0.36	−0.08	−0.12	−0.11	0.02	−0.05	−0.13	−0.17	−0.62	−0.41	−0.51	−0.24

The numbers without sign must be added; those with the sign — must be subtracted.

N. America. — Frankfort Arsenal. *Lat.* 39° 57′ N. *Long.* 75° 8′ W. *Greenw.*

Corrections to be applied to the Means of the Hours of Observation to obtain the true Mean Temperatures of the respective Days, Months, and of the Year. — Dove.

Degrees of Fahrenheit.

Hours.	Jan.	Feb.	March.	April.	May.	June.	July.	Aug.	Sept.	Oct.	Nov.	Dec.	Mean.
Morn. 1	3.02	3.29	3.94	4.21	5.85	7.67	6.91	6.05	5.92	5.40	2.66	3.02	4.84
2	3.40	3.89	4.79	5.24	6.86	8.39	7.90	6.84	6.86	6.01	2.86	3.38	5.54
3	4.10	4.46	5.76	6.48	7.72	8.82	8.62	7.47	7.85	6.62	3.17	3.74	6.23
4	4.79	5.02	6.53	7.40	8.03	8.64	8.64	7.56	8.39	7.04	3.40	4.05	6.62
5	5.20	5.54	6.64	7.45	7.74	7.56	7.65	6.73	7.97	7.02	3.89	4.21	6.44
6	5.06	5.29	5.90	6.37	5.96	5.54	5.67	4.97	6.39	6.35	3.11	4.05	5.38
7	4.23	4.52	4.30	4.37	3.74	2.84	3.02	2.59	3.85	4.93	2.39	3.42	3.69
8	2.75	2.99	2.12	1.91	1.28	−0.07	0.18	0.02	0.81	2.84	1.31	2.18	1.53
9	0.77	0.68	−0.16	−0.45	−1.01	−2.70	−2.39	−2.25	−2.16	0.27	−0.05	0.41	−0.77
10	−1.40	−1.62	−2.25	−2.36	−2.90	−4.75	−4.41	−4.01	−4.64	−2.54	−1.58	−1.71	−2.86
11	−3.47	−3.98	−3.96	−3.80	−4.43	−6.17	−5.94	−5.27	−6.50	−5.24	−2.52	−3.83	−4.59
Noon. . .	−5.18	−5.85	−5.22	−5.00	−5.29	−7.13	−7.11	−6.26	−7.81	−7.54	−4.41	−5.51	−6.03
1	−6.41	−6.77	−6.17	−6.12	−6.91	−7.90	−8.06	−7.11	−8.69	−9.11	−5.36	−6.46	−7.09
2	−6.80	−7.16	−6.77	−7.18	−7.92	−8.48	−8.71	−7.83	−9.16	−9.81	−5.72	−6.58	−7.67
3	−6.57	−6.59	−6.98	−7.94	−8.51	−8.75	−8.87	−8.12	−9.05	−9.50	−5.40	−5.72	−7.67
4	−5.69	−5.49	−6.64	−7.99	−8.33	−8.44	−8.26	−7.70	−8.17	−8.24	−4.41	−4.37	−6.98
5	−4.28	−4.21	−5.63	−7.00	−7.20	−7.27	−6.75	−6.32	−6.39	−6.19	−3.42	−2.77	−5.63
6	−2.57	−2.50	−4.01	−5.02	−5.20	−5.24	−4.50	−4.12	−3.87	−3.71	−1.26	−1.24	−3.60
7	−0.83	−1.04	−2.07	−2.45	−2.68	−2.61	−1.87	−1.51	−1.08	−1.22	0.32	0.02	−1.42
8	0.65	0.27	−0.14	0.05	−0.23	0.16	0.63	0.97	1.49	0.97	1.55	0.95	0.61
9	1.71	1.48	1.37	1.91	1.80	2.63	2.63	2.90	3.35	2.63	2.30	1.60	2.21
10	2.30	2.09	2.36	1.97	3.22	4.55	4.03	4.14	4.41	3.74	2.59	2.03	3.20
11	2.54	2.66	2.95	3.38	4.16	5.87	5.04	4.84	4.91	4.41	2.05	2.39	3.76
Midn. . .	2.68	3.06	3.33	3.65	4.52	6.84	5.92	5.40	5.29	4.91	2.59	2.10	4.23
6. 6	1.26	1.40	0.95	0.68	0.38	0.16	0.59	0.43	1.26	1.31	0.92	1.40	0.90
7. 7	1.71	1.76	1.13	0.95	0.54	0.11	0.59	0.54	1.40	1.87	1.35	1.71	1.15
8. 8	1.71	1.62	0.99	0.97	0.54	0.05	0.41	0.50	1.15	1.91	1.42	1.58	1.08
9. 9	1.24	1.08	0.61	0.74	0.41	−0.05	0.14	0.32	0.59	1.44	1.13	0.99	0.72
10.10	0.45	0.25	0.07	0.29	0.16	−0.11	−0.18	0.07	−0.11	0.59	0.52	0.16	0.18
7. 2. 9	−0.29	−0.38	−0.36	−0.29	−0.79	−1.01	−1.01	−0.79	−0.65	−0.74	−0.34	−0.50	−0.61
6. 2. 8	−0.36	−0.54	−0.39	−0.25	−0.72	−0.92	−0.81	−0.63	−0.43	−0.83	−0.36	−0.50	−0.56
6. 2.10	0.18	0.07	0.50	0.72	0.43	0.54	0.34	0.43	0.54	0.09	0.00	−0.14	0.32
6. 2. 6	−1.44	−1.46	−1.62	−1.94	−2.39	−2.72	−2.52	−2.32	−2.21	−2.39	−1.28	−1.24	−1.96
7. 2	−1.28	−1.33	−1.24	−1.42	−2.09	−2.84	−2.86	−2.63	−2.66	−2.45	−1.67	−1.55	−2.00
8. 2	−2.03	−2.09	−2.34	−2.63	−3.33	−4.28	−4.28	−3.92	−4.19	−3.49	−2.21	−2.16	−3.08
8. 1	−1.85	−1.89	−2.03	−2.12	−2.81	−3.98	−3.94	−3.56	−3.94	−3.75	−2.03	−2.14	−2.79
7. 2	−1.10	−1.13	−0.95	−0.88	−1.60	−2.54	−2.52	−2.27	−2.43	−2.09	−1.49	−1.53	−1.71
9.12.3.9	−2.32	−2.57	−2.75	−2.88	−3.26	−3.98	−3.94	−3.44	−3.92	−3.53	−1.89	−2.32	−3.06
7. 2.2(9)	0.23	0.09	0.07	0.25	−0.16	−0.09	−0.11	0.14	0.36	0.09	0.32	0.02	0.09
Dail. ext.	−0.81	−0.81	−0.18	−0.27	−0.25	0.04	−0.11	−0.29	−0.38	−1.39	−0.92	−1.15	−0.54

The numbers without sign must be added; those with the sign — must be subtracted.

E

N. America. — Toronto. *Lat.* 43° 39′ 35″ N. *Long.* 79° 21′ 30″ W. *Greenw.*

Corrections to be applied to the Means of the Hours of Observation to obtain the true Mean Temperatures of the respective Days, Months, and of the Year. — Dove.

Degrees of Fahrenheit.

Hours.	Jan.	Feb.	March.	April.	May.	June.	July.	Aug.	Sept.	Oct.	Nov.	Dec.	Mean.
Morn. 1	1.87	0.92	3.04	4,43	5.90	5.94	6.30	5.06	5.74	4.16	1.91	1.04	3.87
2	2.16	1.33	3.56	5.11	6.64	6.62	7.13	5.68	6.68	4.68	2.14	1.13	4.41
3	2.39	1.91	4.19	5.76	7.36	7.29	8.01	6.82	7.63	5.04	2.39	1.40	5.02
4	2.68	2.66	4.75	6.17	7.65	7.56	8.44	7.61	8.19	5.20	2.61	1.78	5.45
5	3.02	3.40	4.95	5.94	7.07	6.98	7.88	7.49	7.94	5.02	2.68	2.16	5.38
6	3.29	3.92	4.61	4.97	5.49	5.38	6.14	6.14	6.71	4.48	2.52	2.39	4.68
7	3.26	3.98	3.65	3.38	3.17	3.04	3.49	3.67	4.52	3.44	2.05	2.27	3.33
8	2.72	3.40	2.12	1.42	0.68	0.43	0.52	0.68	1.78	1.91	1.15	1.71	1.55
9	1.58	2.33	0.29	−0.50	−1.51	−1.85	−2.12	−2.09	−1.06	−0.05	−0.07	0.79	−0.36
10	0.00	0.61	−1.60	−2.07	−3.08	−3.47	−4.01	−4.14	−3.62	−2.25	−1.46	−0.34	−2.12
11	−1.71	−1.15	−3.26	−3.26	−4.14	−4.46	−5.15	−5.33	−5.72	−4.39	−2.79	−1.44	−3.58
Noon. . .	−3.11	−2.66	−4.55	−4.19	−5.00	−5.18	−5.90	−5.96	−7.25	−6.12	−3.78	−2.30	−4.66
1	−3.89	−3.67	−5.36	−5.00	−5.99	−5.94	−6.59	−6.50	−8.33	−7.11	−4.28	−2.77	−5 45
2	−3.98	−4.07	−5.72	−5.76	−7.16	−6.89	−7.47	−7.11	−8.89	−7.25	−4.14	−2.86	−5.94
3	−3.53	−3.92	−5.60	−6.35	−8.15	−7.74	−8.28	−7.70	−8.87	−6.53	−3.51	−2.66	−6.08
4	−2.84	−3.38	−5.02	−6.48	−8.51	−8.08	−8.55	−7.81	−8.12	−5.18	−2.52	−2.23	−5.72
5	−2.14	−2.63	−4.03	−5.94	−7.76	−7.43	−7.83	−6.95	−6.59	−3.53	−1.44	−1.71	−4.84
6	−1.62	−1.89	−2.75	−4.66	−5.83	−5.65	−5.94	−5.00	−4.43	−1.91	−0.45	−1.13	−3.44
7	−1.24	−1.24	−1.31	−2.81	−3.08	−3.04	−3.17	−2.25	−1.94	−0.50	0.32	−0.54	−1.73
8	−0.88	−0.68	0.05	−0.77	−0.16	−0.18	−0.18	0.65	0.43	0.65	0.86	0.02	−0.02
9	−0.43	−0.25	1.15	1.06	2.30	2.30	2.39	2.97	2.30	1.53	1.17	0.47	1.42
10	0.16	0.11	1.89	2.41	3.94	3.98	4.14	4.32	3.58	2.25	1.37	0.81	2.41
11	0.83	0.38	2.34	3.26	4.82	4.93	5.11	4.77	4.37	2.90	1.53	0.97	3.02
Midn. . .	1.42	0.63	2.66	3.85	5.33	5.45	5.64	4.84	5.00	3.56	1.71	1.01	3.42
6. 6	0.83	1.01	0.95	0.16	−0.18	0.14	0.11	0.56	1.13	1.28	1.04	0.63	0.61
7. 7	1.01	1.27	1.17	0.29	−0.05	0.00	0.16	0.72	1.28	1.49	1.19	0.86	0.81
8. 8	0.92	1.37	1.08	0.34	0.27	0.14	0.16	0.68	1.10	1.28	1.01	0.86	0.77
9. 9	0.59	0.99	0.72	0.29	0.41	0.23	0.14	0.45	0.63	0.74	0.56	0.63	0.54
10.10	0.07	0.36	0.14	0.16	0.43	0.27	0.07	0.09	−0.02	0.00	−0.05	0.23	0.14
7. 2. 9	−0.38	−0.11	−0.32	−0.45	−0.56	−0.52	−0.54	−0.16	−0.70	−0.77	−0.32	−0.05	−0.41
6. 2. 8	−0.52	−0.27	−0.36	−0.52	−0.61	−0.56	−0.50	−0.11	−0.59	−0.70	−0.25	−0.16	−0.43
6. 2.10	−0.18	−0.02	0.27	0.54	0.77	0.83	0.95	1.13	0.47	−0.18	−0.09	0.11	0.38
6. 2. 6	−0.77	−0.68	−1.28	−1.82	−2.50	−2.39	−2.43	−1.98	−2.21	−1.55	−0.70	−0.54	−1.58
7. 2	−0.36	−0.05	−1.04	−1.19	−2.00	−1.94	−2.00	−1.73	−2.18	−1.91	−1.06	−0.29	−1.31
8. 2	−0.63	−0.34	−1.80	−2.18	−3.24	−3.24	−3.49	−3.22	−3.56	−2.68	−1.51	−0.59	−2.21
8. 1	−0.59	−0.14	−1.62	−1.80	−2.66	−2.77	−3.04	−2.93	−3.29	−2.61	−1.58	−0.54	−1.96
7. 1	−0.32	0.16	−0.86	−0.81	−1.42	−1.46	−1.55	−1.42	−1.91	−1.85	−1.13	−0.25	−1.06
9.12.3.9	−1.37	−1.15	−2.18	−2.50	−3.08	−3.13	−3.49	−3.20	−3.71	−2.79	−1.55	−0.92	−2.43
7. 2.2(9)	−0.41	−0.16	0.07	−0.07	0.16	0.18	0.20	0.63	0.07	−0.18	0.07	0.09	0.05

The numbers without sign must be added; those with the sign — must be subtracted.

N. America. — Toronto. *Lat.* 43° 39′ 35″ N. *Long.* 79° 21′ 30″ W. *Greenw.*

Corrections to be applied to the Means of the Hours of Observation to obtain the true Mean Temperatures of the respective Days, Months, and of the Year. — Dove.

Degrees of Reaumur.

Hours.	Jan.	Feb.	March.	April.	May.	June.	July.	Aug.	Sept.	Oct.	Nov.	Dec.	Mean.
Morn. 1	0.83	0.41	1.35	1.97	2.62	2.64	2.80	2.25	2.55	1.85	0.85	0.46	1.72
2	0.96	0.59	1.58	2.27	2.95	2.94	3.17	2.57	2.97	2.08	0.95	0.50	1.96
3	1.06	0.85	1.86	2.56	3.27	3.24	3.56	3.03	3.39	2.24	1.06	0.62	2.23
4	1.19	1.18	2.11	2.74	3.40	3.36	3.75	3.38	3.64	2.31	1.16	0.79	2.42
5	1.34	1.51	2.20	2.64	3.14	3.10	3.50	3.33	3.53	2.23	1.19	0.96	2.39
6	1.46	1.74	2.05	2.21	2.44	2.39	2.73	2.73	2.98	1.99	1.12	1.06	2.08
7	1.45	1.77	1.62	1.50	1.41	1.35	1.55	1.63	2.01	1.53	0.91	1.01	1.48
8	1.21	1.51	0.94	0.63	0.30	0.19	0.23	0.30	0.79	0.85	0.51	0.76	0.69
9	0.70	0.99	0.13	−0.22	−0.67	−0.82	−0.94	−0.93	−0.47	−0.02	−0.03	0.85	−0.16
10	−0.00	0.27	−0.71	−0.92	−1.37	−1.54	−1.78	−1.84	−1.61	−1.00	−0.65	−0.15	−0.94
11	−0.76	−0.51	−1.45	−1.45	−1.84	−1.98	−2.29	−2.37	−2.54	−1.95	−1.24	−0.64	−1.59
Noon. . .	−1.38	−1.18	−2.02	−1.86	−2.22	−2.30	−2.62	−2.65	−3.22	−2.72	−1.68	−1.02	−2.07
1	−1.73	−1.63	−2.38	−2.22	−2.66	−2.64	−2.93	−2.89	−3.70	−3.16	−1.90	−1.23	−2.42
2	−1.77	−1.81	−2.54	−2.56	−3.18	−3.06	−3.32	−3.16	−3.95	−3.22	−1.84	−1.27	−2.64
3	−1.57	−1.74	−2.49	−2.82	−3.62	−3.44	−3.68	−3.42	−3.94	−2.90	−1.56	−1.18	−2.70
4	−1.26	−1.50	−2.23	−2.88	−3.78	−3.59	−3.80	−3.47	−3.61	−2.30	−1.12	−0.99	−2.54
5	−0.95	−1.17	−1.79	−2.64	−3.45	−3.30	−3.48	−3.09	−2.93	−1.57	−0.64	−0.76	−2.15
6	−0.72	−0.84	−1.22	−2.07	−2.59	−2.51	−2.64	−2.22	−1.97	−0.85	−0.20	−0.50	−1.53
7	−0.55	−0.55	−0.58	−1.25	−1.37	−1.35	−1.41	−1.00	−0.86	−0.22	0.14	−0.24	−0.77
8	−0.39	−0.30	0.02	−0.34	−0.07	−0.08	−0.08	0.29	0.19	0.29	0.38	0.01	−0.01
9	−0.19	−0.11	0.51	0.47	1.02	1.02	1.06	1.32	1.02	0.68	0.52	0.21	0.63
10	0.07	0.05	0.84	1.07	1.75	1.77	1.84	1.92	1.59	1.00	0.61	0.36	1.07
11	0.37	0.17	1.04	1.45	2.14	2.19	2.27	2.12	1.94	1.29	0.68	0.43	1.34
Midn. . .	0.63	0.28	1.18	1.71	2.37	2.42	2.53	2.15	2.22	1.58	0.76	0.45	1.52
6. 6	0.37	0.45	0.42	0.07	−0.08	−0.06	0.05	0.25	0.50	0.57	0.46	0.28	0.27
7. 7	0.45	0.61	0.52	0.13	0.02	0.00	0.07	0.32	0.57	0.66	0.53	0.38	0.36
8. 8	0.41	0.61	0.48	0.15	0.12	0.06	0.07	0.30	0.49	0.57	0.45	0.38	0.34
9. 9	0.26	0.44	0.32	0.13	0.18	0.10	0.06	0.20	0.28	0.33	0.25	0.28	0.24
10.10	0.03	0.16	0.06	0.07	0.19	0.12	0.03	0.04	−0.01	0.00	−0.02	0.10	0.06
7. 2. 9	−0.17	−0.05	−0.14	−0.20	−0.25	−0.23	−0.24	−0.07	−0.31	−0.34	−0.14	−0.02	−0.18
6. 2. 8	−0.23	−0.12	−0.16	−0.23	−0.27	−0.25	−0.22	−0.05	−0.26	−0.31	−0.11	−0.07	−0.19
6. 2.10	−0.08	−0.01	0.12	0.24	0.34	0.37	0.42	0.50	0.21	−0.08	−0.04	0.05	0.17
6. 2. 6	−0.34	−0.30	−0.57	−0.81	−1.11	−1.06	−1.08	−0.88	−0.98	−0.69	−0.31	−0.24	−0.70
7. 2	−0.16	−0.02	−0.46	−0.53	−0.89	−0.86	−0.89	−0.77	−0.97	−0.85	−0.47	−0.13	−0.58
8. 2	−0.28	−0.15	−0.80	−0.97	−1.44	−1.44	−1.55	−1.43	−1.58	−1.19	−0.67	−0.26	−0.98
8. 1	−0.26	−0.06	−0.72	−0.80	−1.18	−1.23	−1.35	−1.30	−1.46	−1.16	−0.70	−0.24	−0.87
7. 1	−0.14	0.07	−0.38	−0.36	−0.63	−0.65	−0.69	−0.63	−0.85	−0.82	−0.50	−0.11	−0.47
9.12.3.9	−0.61	−0.51	−0.97	−1.11	−1.37	−1.39	−1.55	−1.42	−1.65	−1.24	−0.69	−0.41	−1.08
7. 2.2(9)	−0.18	−0.07	0.03	−0.03	0.07	0.08	0.09	0.28	0.03	−0.08	0.03	0.04	0.02
Dail. ext.	−0.16	−0.02	−0.17	−0.07	−0.19	−0.12	−0.03	−0.05	−0.16	−0.46	−0.36	−0.11	−0.14

The numbers without sign must be added; those with the sign — must be subtracted.

VIII.

NORTH AMERICA. — TORONTO. *Lat.* 43° 40′ N. *Long.* 79° 21′ W. *Greenw.*

Corrections to be applied to the Means of the Hours of Observation to obtain the true Mean Temperatures of the respective Days, Months, and of the Year. — LEFROY.

Degrees of Fahrenheit.

Hour.	Jan.	Feb.	March.	April.	May.	June.	July.	Aug.	Sept.	Oct.	Nov.	Dec	Year.
Midnight.	1.47	1.73	2.63	3.22	5.02	5.15	6.37	5.33	5.96	3.22	1.80	0.90	3.57
1	1.95	2.09	3.11	3.79	5.93	6.00	7.13	6.06	4.57	3.80	2.10	1.50	4.00
2	2.05	2.46	3.47	4.48	6.77	6.70	7.68	6.69	5.17	4.13	2.36	1.85	4.48
3	2.20	2.82	3.76	5.08	7.45	7.50	8.41	7.29	5.59	4.31	2.66	1.96	4.92
4	2.28	3.20	4.07	5.38	7.93	8.06	9.03	7.63	6.18	4.64	2.85	2.04	5.27
5	2.46	3.62	4.35	5.75	7.83	7.88	9.02	7.89	6.77	4.77	2.76	2.07	5.43
6	1.83	4.23	4.75	5.48	5.40	5.21	5.92	6.57	6.17	4.71	2.52	2.39	4.60
7	1.94	4.34	3.93	3.22	2.43	2.41	2.38	3.28	3.68	3.94	2.52	2.55	3.05
8	1.66	3.29	1.89	1.09	0.06	0.10	−0.31	0.21	1.02	1.66	1.53	2.12	1.25
9	0.63	1.02	−0.25	−1.01	−2.11	−1.82	−2.39	−2.26	−1.52	−1.01	0.01	0.92	−0.82
10	−0.59	−0.95	−1.91	−2.45	−3.81	−3.49	−3.98	−4.18	−3.47	−2.93	−1.41	−0.53	−2.47
11	−1.70	−2.44	−3.14	−3.85	−4.92	−4.77	−5.49	−5.57	−4.85	−4.33	−2.44	−1.72	−3.77
Noon.	−2.48	−3.56	−4.15	−4.86	−5.87	−5.88	−6.72	−6.39	−5.95	−5.36	−3.34	−2.52	−4.76
1	−2.92	−4.49	−4.79	−5.72	−6.83	−6.59	−7.58	−7.11	−6.58	−5.76	−3.74	−3.06	−5.43
2	−3.20	−4.88	−5.31	−6.14	−7.13	−7.03	−8.26	−7.62	−6.96	−6.04	−3.82	−3.31	−5.81
3	−3.16	−4.90	−5.15	−6.16	−7.20	−7.37	−8.34	−7.98	−7.01	−5.85	−3.64	−3.13	−5.82
4	−2.63	−4.47	−4.65	−5.81	−7.17	−7.60	−8.25	−7.79	−6.75	−5.17	−2.83	−2.47	−5.47
5	−1.68	−3.30	−3.92	−5.12	−6.80	−7.18	−7.93	−7.20	−5.78	−3.40	−1.58	−1.49	−4.61
6	−0.90	−1.87	−2.35	−3.42	−5.05	−5.73	−6.57	−5.39	−3.16	−1.37	−0.76	−0.82	−3.12
7	−0.40	−0.98	−0.91	−0.94	−2.19	−2.99	−3.28	−1.64	−0.43	−0.25	−0.15	−0.47	−1.22
8	−0.12	−0.13	0.03	0.66	0.43	0.33	0.68	1.23	0.81	0.48	0.19	−0.12	0.38
9	0.07	0.52	1.00	1.78	2.31	2.44	2.99	2.70	1.90	1.25	0.44	0.18	1.46
10	0.44	1.06	1.63	2.59	3.29	3.80	4.24	3.73	2.94	1.97	0.78	0.47	2.24
11	0.77	1.60	2.01	3.07	4.20	4.76	5.21	4.54	3.61	2.68	1.13	0.59	2.85
6, 6	0.46	1.18	1.20	1.03	0.17	−0.26	−0.32	0.59	1.50	1.67	1.38	0.78	0.74
7, 7	0.77	1.67	1.51	1.14	0.12	−0.29	−0.45	0.82	1.62	1.84	1.18	1.04	0.91
8, 8	0.77	1.58	0.96	0.87	0.24	0.21	0.18	0.72	0.91	1.45	0.98	1.15	0.82
9, 9	0.35	0.77	0.37	0.38	0.10	0.31	0.30	0.22	0.19	0.10	0.22	0.55	0.32
10, 10	−0.07	0.05	−0.14	−0.07	−0.26	0.25	0.13	−0.22	−0.26	−0.48	−0.31	−0.03	−0.11
6, 2, 10	−0.31	0.14	0.36	0.64	0.52	0.66	0.63	0.89	0.72	0.21	−0.17	−0.15	0.34
7, 2, 9	−0.40	−0.01	−0.09	−0.38	−0.80	−0.73	−0.96	−0.55	−0.46	−0.28	−0.29	−0.19	−0.43
9, 12, 3, 9	−1.23	−1.73	−2.01	−2.56	−3.22	−3.16	−3.61	−3.48	−3.14	−2.74	−1.63	−1.14	−2.48
Mean.	25.82	23.70	29.79	41.99	52.92	60.67	66.39	65.86	57.55	44.14	36.18	27.40	44.37

The numbers without sign must be added; those with the sign — must be subtracted.

IX.

NORTH AMERICA. — TORONTO. *Lat.* 43° 40′ N. *Long.* 79° 21′ W. *Gr.*

Corrections to be applied to the Means of the Hours of Observation to obtain the true Mean Temperatures of the respective Days, Months, and of the Year. — DOVE.

Degrees of Reaumur.

Hour.	Jan.	Feb.	March.	April.	May.	June.	July.	Aug.	Sept.	Oct.	Nov.	Dec.	Year.
Midn.	0.68	0.81	1.10	1.45	2.24	2.36	2.91	2.43	1.76	1.44	0.81	0.40	1.53
1	0.88	0.98	1.31	0.78	2.62	2.67	3.29	2.72	2.03	1.71	0.94	0.66	1.80
2	0.92	1.13	1.48	2.08	2.99	2.98	3.54	3.02	2.29	1.85	1.06	0.83	2.01
3	0.99	1.32	1.61	2.17	3.31	3.32	3.86	3.32	2.49	1.92	1.20	0.88	2.20
4	1.03	1.45	1.78	2.36	3.52	3.58	4.14	3.48	2.76	2.06	1.28	0.90	2.36
5	1.11	1.61	2.01	2.52	3.49	3.49	4.16	3.57	3.04	2.13	1.23	0.91	2.44
6	0.79	1.86	2.13	2.47	2.40	2.32	2.74	2.92	2.74	2.04	1.11	1.09	2.05
7	0.83	1.92	1.75	1.45	1.08	1.07	1.11	1.60	1.60	1.70	1.11	1.16	1.36
8	0.73	1.47	0.87	0.45	0.09	0.03	−0.05	0.15	0.38	0.70	0.64	0.97	0.56
9	0.30	0.44	−0.10	−0.43	−0.94	−0.81	−1.03	−0.96	−0.69	−0.49	−0.04	0.45	−0.36
10	−0.25	−0.45	−0.87	−1.11	−1.69	−1.55	−1.78	−1.84	−1.57	−1.35	−0.68	−0.20	−1.11
11	−0.77	−1.16	−1.41	−1.72	−2.20	−2.12	−2.47	−2.48	−2.20	−1.96	−1.13	−0.75	−1.70
Noon.	−1.12	−1.69	−1.87	−2.18	−2.62	−2.61	−3.05	−3.04	−2.64	−2.36	−1.48	−1.11	−2.15
1	−1.34	−2.07	−2.16	−2.60	−3.03	−2.93	−3.46	−3.25	−2.90	−2.55	−1.66	−1.42	−2.45
2	−1.46	−2.25	−2.41	−2.76	−3.18	−3.12	−3.84	−3.51	−3.08	−2.70	−1.69	−1.49	−2.62
3	−1.44	−2.24	−2.32	−2.80	−3.21	−3.29	−3.92	−3.66	−3.09	−2.60	−1.62	−1.38	−2.63
4	−1.21	−2.00	−2.11	−2.62	−3.19	−3.40	−3.93	−3.60	−3.00	−2.28	−1.22	−1.09	−2.47
5	−0.77	−1.47	−1.78	−2.30	−3.02	−3.13	−3.72	−3.35	−2.57	−1.50	−0.68	−0.67	−2.08
6	−0.40	−0.82	−1.03	−1.50	−2.24	−2.55	−3.08	−2.51	−1.38	−0.59	−0.32	−0.36	−1.40
7	−0.17	−0.38	−0.38	−0.37	−0.96	−1.33	−1.54	−0.74	−0.18	−0.10	−0.06	−0.21	−0.53
8	−0.03	0.00	0.05	0.33	0.24	0.13	0.33	0.56	0.39	0.23	0.08	−0.04	0.19
9	0.06	0.28	0.50	0.81	1.02	1.09	1.38	1.26	0.85	0.57	0.20	0.07	0.67
10	0.23	0.53	0.79	1.16	1.45	1.69	1.93	1.72	1.32	0.90	0.36	0.20	1.02
11	0.37	0.76	1.08	1.38	1.86	2.12	2.45	2.07	1.60	1.20	0.52	0.25	1.31
Mean.	−2.97	−3.88	−0.98	4.72	9.29	12.75	15.11	15.00	11.37	5.42	1.88	−2.03	

X.

NORTH AMERICA. — MONTREAL. *Lat.* 45° 30′ N. *Long.* 73° 22′ E. *Gr.*

Degrees of Fahrenheit.

Hour.	Aug.	Sept.	Oct.	Nov.	Dec.	Jan.	Feb.	March.	April.	May.	June.	July.	Year.
Midn.	4.00	3.89	2.83	1.36	1.68	1.10	1.28	1.31	2.52	4.55	5.25	4.39	2.85
2	5.39	4.34	4.01	1.59	1.00	2.36	2.69	2.88	4.37	6.95	7.42	7.17	4.20
4	6.34	5.60	4.84	1.81	1.38	2.88	3.36	5.56	7.09	6.95	7.18	7.57	4.96
6	5.99	4.59	4.83	1.36	1.32	3.54	3.90	5.22	5.56	6.61	5.55	5.46	4.50
8	2.79	2.19	2.52	0.78	0.92	3.10	3.22	3.30	3.44	3.06	0.88	0.60	2.24
10	−1.74	−1.48	−0.99	−0.41	0.21	−0.21	−0.81	−0.03	−0.79	−0.97	−1.75	−2.85	−0.93
Noon.	−5.63	−5.43	−4.22	−1.87	−1.22	−2.82	−3.50	−4.23	−5.01	−7.10	−5.17	−5.46	−4.30
2	−7.93	−6.60	−6.96	−2.37	−2.54	−4.07	−5.43	−6.49	−5.99	−8.76	−7.72	−7.36	−6.02
4	−7.72	−6.70	−5.62	−2.52	−3.22	−3.88	−3.60	−5.96	−5.79	−8.35	−7.00	−7.51	−5.65
6	−5.63	−2.80	−2.79	−1.04	−1.30	−1.77	−1.50	−3.43	−3.88	−3.87	−5.02	−5.40	−3.20
8	−0.70	0.10	−0.25	0.03	0.02	−0.90	−0.59	−1.23	−0.81	−1.61	−1.10	−0.67	−0.65
10	1.99	2.39	1.42	1.18	0.89	0.17	0.22	−0.30	0.64	−1.87	2.47	2.64	1.30
Mean.	66.40	57.70	48.31	30.39	23.42	8.10	20.84	27.31	42.27	56.61	64.38	70.39	43.01

The numbers without sign must be added; those with the sign — must be subtracted.

X.

North America. — Montreal, *Continued.*

Corrections to be applied to the Means of the Hours of Observation to obtain the true Mean Temperatures of the respective Days, Months, and of the Year.

Degrees of Fahrenheit.

Hour.	Aug.	Sept.	Oct.	Nov.	Dec.	Jan.	Feb.	March.	April.	May.	June.	July.	Year.
A.M. 1	5.03	4.92	2.53	1.16	0.88	1.43	1.61	4.38	3.12	4.85	4.55	5.07	3.30
3	5.99	5.20	3.61	1.58	1.79	1.30	2.72	5.18	5.14	6.51	5.10	6.80	4.25
5	6.44	5.43	4.45	2.08	2.21	1.87	3.95	6.84	6.54	6.56	6.30	7.76	5.05
7	2.10	3.47	3.61	2.01	2.08	1.98	5.22	7.07	3.84	3.56	4.72	3.04	3.56
9	−0.58	0.73	0.77	0.63	1.14	1.16	3.99	2.96	0.71	0.50	−0.02	0.22	1.02
11	−3.61	−2.20	−2.73	−1.35	−0.49	−1.08	−0.17	−2.51	−2.48	−2.79	−3.42	−3.21	−2.17
P.M. 1	−6.61	−5.12	−5.41	−3.47	−2.38	−1.49	−4.80	−7.41	−4.93	−5.78	−5.97	−6.08	−4.95
3	−7.34	−6.65	−5.80	−3.22	−2.78	−2.36	−6.08	−9.03	−6.33	−6.46	−6.93	−8.01	−5.91
5	−5.47	−5.83	−3.15	−1.19	−1.44	−0.63	−4.12	−6.48	−5.63	−6.62	−6.18	−6.53	−4.43
7	−1.45	−0.62	−1.00	−0.44	−0.70	−0.60	−1.23	−2.40	−2.93	−3.50	−3.17	−2.88	−1.74
9	1.58	1.32	0.32	0.13	−0.71	−0.66	−0.96	−0.75	0.44	0.61	1.58	1.17	0.34
11	3.10	3.02	2.47	1.48	0.22	0.61	0.24	1.78	2.06	2.52	3.55	3.39	2.02
Mean.	69.69	57.53	44.70	32.76	15.91	18.96	14.52	22.50	34.47	51.33	65.08	67.42	41.24

XI.

North America. — Sitka. *Lat.* 57° 3′ N. *Long.* 135° 18′ W. *Gr.* — Dove.

Degrees of Reaumur.

Hour.	Jan.	Feb.	March.	April.	May.	June.	July.	Aug.	Sept.	Oct.	Nov.	Dec.	Year.
Midn.	0.33	0.58	0.97	1.51	1.80	1.81	1.68	1.34	1.07	1.19	0.41	0.28	1.08
1	0.34	0.66	1.09	1.68	2.04	2.06	1.88	1.53	1.18	1.11	0.46	0.33	1.20
2	0.35	0.72	1.17	1.81	2.20	2.25	2.04	1.66	1.33	1.18	0.49	0.33	1.29
3	0.51	0.78	1.36	1.89	2.43	2.49	2.16	1.77	1.24	0.64	0.48	0.18	1.33
4	0.45	0.86	1.47	2.02	2.55	2.57	2.20	1.82	1.29	0.68	0.49	0.18	1.38
5	0.45	0.83	1.57	2.07	2.39	2.47	2.95	1.89	1.33	0.70	0.49	0.14	1.52
6	0.45	0.84	1.56	1.89	1.76	1.77	1.67	1.62	1.33	0.78	0.46	0.18	1.26
7	0.52	0.82	1.37	1.13	0.96	1.08	0.96	1.09	1.05	0.58	0.40	0.17	0.85
8	0.48	0.76	0.75	0.31	0.00	0.26	0.26	0.40	0.47	0.53	0.33	0.12	0.39
9	0.39	0.49	−0.08	−0.63	−0.82	−0.52	−0.58	−0.26	−0.17	0.12	0.23	0.10	−0.15
10	0.16	−0.03	−0.69	−1.12	−1.35	−1.28	−1.27	−0.95	−0.73	−0.28	0.00	−0.11	−0.64
11	−0.19	−0.60	−1.29	−1.68	−1.75	−1.70	−1.97	−1.57	−1.28	−0.75	−0.35	−0.11	−1.11
Noon.	−0.57	−1.05	−1.71	−2.13	−2.17	−2.11	−2.11	−2.04	−1.65	−1.14	−0.72	−0.32	−1.48
1	−0.83	−1.36	−1.74	−2.33	−2.35	−2.35	−2.25	−2.33	−1.56	−1.38	−0.84	−0.46	−1.65
2	−0.95	−1.44	−1.99	−2.28	−2.40	−2.42	−2.31	−2.16	−1.86	−1.42	−1.00	−0.50	−1.73
3	−0.95	−1.47	−1.94	−2.10	−2.28	−2.31	−2.13	−2.00	−1.72	−1.37	−0.94	−0.44	−1.64
4	−0.78	−1.20	−1.67	−1.91	−2.04	−2.09	−1.94	−1.76	−1.56	−1.13	−0.75	−0.32	−1.43
5	−0.50	−0.85	−1.17	−1.63	−1.73	−1.76	−1.65	−1.43	−1.24	−0.88	−0.45	−0.20	−1.12
6	−0.25	−0.45	−0.82	−1.13	−1.37	−1.48	−1.26	−1.02	−0.64	−0.50	−0.21	−0.10	−0.77
7	−0.15	−0.10	−0.29	−0.48	−0.76	−1.00	−0.81	−0.49	−0.28	−0.16	−0.04	−0.03	−0.38
8	−0.01	0.11	0.13	0.15	−0.23	−0.41	−0.22	0.12	0.19	0.06	0.07	0.01	0.00
9	0.15	0.30	0.44	0.70	0.48	0.27	0.33	0.66	0.52	0.21	0.22	0.12	0.37
10	0.23	0.37	0.64	1.07	1.02	0.97	0.99	0.96	0.76	0.30	0.29	0.19	0.65
11	0.31	0.48	0.84	1.28	1.57	1.46	1.38	1.19	0.90	0.95	0.43	0.22	0.93
Mean.	−1.39	−1.07	0.55	3.51	6.21	9.10	10.24	10.28	7.96	5.26	2.52	1.73	

The numbers without sign must be added; those with the sign — must be subtracted.

XII.

Arctic America. — Boothia Felix. *Lat.* 69° 59′ N. *Long.* 92° 1′ W. *Greenw.*

Corrections to be applied to the Means of the Hours of Observation to obtain the true Mean Temperatures of the respective Days, Months, and of the Year. — Dove.

Degrees of Reaumur.

Hours.	Jan.	Feb.	March.	April.	May.	June.	July.	Aug.	Sept.	Oct.	Nov.	Dec.	Mean.
Morn. 1	0.08	0.42	1.61	2.17	2.64	2.38	1.78	1.34	0.56	0.30	0.02	0.12	1.12
2	0.10	0.28	1.85	2.25	2.75	2.55	1.78	1.30	0.62	0.32	0.18	0.13	1.15
3	0.11	0.25	2.10	2.30	2.61	2.45	1.65	1.17	0.66	0.33	0.29	0.10	1.12
4	0.11	0.21	2.30	2.26	2.23	2.05	1.35	1.02	0.66	0.34	0.31	0.06	1.02
5	0.10	0.22	2.38	2.02	1.76	1.39	0.99	0.86	0.56	0.32	0.24	0.02	0.87
6	0.10	0.26	2.23	1.53	1.02	0.65	0.61	0.70	0.46	0.27	0.13	−0.04	0.64
7	0.09	0.29	1.77	0.81	0.35	−0.04	0.26	0.50	0.27	0.17	0.02	−0.07	0.37
8	0.08	0.22	0.98	−0.06	−0.32	−0.58	−0.03	0.24	0.05	0.01	0.01	−0.10	0.04
9	0.06	0.05	−0.06	−0.98	−0.95	−0.99	−0.37	−0.10	−0.12	−0.20	−0.04	−0.10	−0.32
10	0.02	−0.26	−1.22	−1.81	−1.54	−1.33	−0.70	−0.49	−0.43	−0.41	−0.14	−0.10	−0.70
11	−0.02	−0.58	−2.28	−2.45	−2.06	−1.66	−1.05	−0.86	−0.65	−0.59	−0.26	−0.11	−1.05
Noon. . .	−0.05	−0.87	−3.05	−2.86	−2.46	−2.02	−1.43	−1.16	−0.82	−0.69	−0.32	−0.12	−1.32
1	−0.11	−1.02	−3.38	−3.03	−2.66	−2.33	−1.70	−1.34	−0.93	−0.68	−0.30	−0.14	−1.47
2	−0.14	−0.98	−3.26	−2.96	−2.65	−2.48	−1.86	−1.38	−0.94	−0.57	−0.19	−0.13	−1.46
3	−0.15	−0.78	−2.78	−2.67	−2.40	−2.38	−1.78	−1.32	−0.93	−0.38	−0.04	−0.10	−1.31
4	−0.14	−0.46	−2.06	−2.18	−1.98	−1.98	−1.56	−1.18	−0.68	−0.18	0.06	−0.05	−1.03
5	−0.11	−0.14	−1.29	−1.50	−1.45	−1.36	−1.18	−1.01	−0.44	0.01	0.24	0.01	−0.69
6	−0.09	0.13	−0.57	−0.74	−0.88	−0.66	−0.78	−0.78	−0.17	0.14	0.31	0.07	−0.34
7	−0.06	0.32	0.01	0.06	−0.34	−0.01	−0.34	−0.50	0.08	0.22	0.36	0.10	−0.01
8	−0.05	0.43	0.44	0.78	0.20	0.51	0.07	−0.16	0.26	0.25	0.38	0.11	0.27
9	−0.03	0.50	0.76	1.35	0.74	0.92	0.50	0.24	0.38	0.26	0.38	0.10	0.51
10	−0.02	0.51	0.99	1.74	1.28	1.26	0.90	0.66	0.44	0.26	0.35	0.10	0.71
11	0.02	0.52	1.19	1.95	1.82	1.63	1.20	1.01	0.48	0.26	0.28	0.09	0.87
Midn. . .	0.05	0.49	1.38	2.08	2.30	2.04	1.59	1.25	0.51	0.28	0.15	0.12	1.02
6. 6	0.01	0.20	0.83	0.40	0.07	−0.01	−0.09	−0.04	0.15	0.21	0.09	0.02	0.15
7. 7	0.02	0.31	0.89	0.44	0.01	−0.03	−0.04	−0.00	0.18	0.20	0.17	0.02	0.18
8. 8	0.02	0.33	0.71	0.36	−0.06	−0.04	0.02	0.04	0.16	0.13	0.20	0.01	0.16
9. 9	0.02	0.28	0.35	0.19	−0.11	−0.04	0.07	0.07	0.13	0.03	0.17	−0.00	0.10
10.10	−0.00	0.13	−0.12	−0.04	−0.13	−0.04	0.10	0.09	0.01	−0.08	0.11	−0.00	0.00
7. 2. 9	−0.03	−0.06	−0.24	−0.27	−0.52	−0.53	−0.37	−0.21	−0.10	−0.05	0.06	−0.03	−0.20
6. 2. 8	−0.03	−0.10	−0.20	−0.22	−0.48	−0.44	−0.39	−0.28	−0.07	−0.02	0.02	−0.02	−0.19
6. 2.10	−0.02	−0.07	−0.01	0.10	−0.12	−0.19	−0.12	−0.01	−0.01	−0.01	0.01	−0.02	−0.04
6. 2. 6	−0.04	−0.20	−0.53	−0.72	−0.84	−0.83	−0.68	−0.49	−0.22	−0.05	−0.00	−0.03	−0.39
7. 2	−0.03	−0.35	−0.75	−1.08	−1.15	−1.26	−0.80	−0.44	−0.34	−0.20	−0.11	−0.10	−0.55
8. 2	−0.03	−0.38	−1.14	−1.51	−1.49	−1.53	−0.95	−0.57	−0.45	−0.28	−0.09	−0.12	−0.71
8. 1	−0.02	−0.40	−1.20	−1.55	−1.49	−1.46	−0.87	−0.55	−0.44	−0.34	−0.15	−0.12	−0.72
7. 1	−0.01	−0.37	−0.81	−1.11	−1.16	−1.19	−0.72	−0.42	−0.33	−0.26	−0.16	−0.11	−0.55
9.12.3.9	−0.04	−0.28	−1.28	−1.29	−1.27	−1.12	−0.77	−0.59	−0.37	−0.25	−0.01	−0.06	−0.61
7. 2.2(9)	−0.03	0.08	0.01	0.14	−0.21	−0.17	−0.15	−0.10	0.02	0.03	0.14	−0.00	−0.02
Dail. ext.	−0.02	−0.25	−0.50	−0.37	0.05	0.04	−0.04	−0.02	−0.14	−0.18	0.03	−0.01	−0.16

The numbers without sign must be added; those with the sign — must be subtracted.

XIII.

N. America. — Lake Athabasca. *Lat.* 59° N. *Long.* 111° W. *Greenw.*

Corrections to be applied to the Means of the Hours of Observation to obtain the true Mean Temperatures of the respective Days, Months, and of the Year. — Lefroy.

The corrections for April and May are derived from observations made at Fort Simpson, Lat. 62° N.

Degrees of Fahrenheit.

Hour.	April.	May.	October.	November.	December.	January.	February.
daily ext.	1.58	1.71	0.33	0.25	−0.17	0.77	1.19
6, 6	1.15	0.51	1.07	0.59	0.27	0.84	1.19
7, 7	1.50	0.16	0.76	0.54	0.30	0.58	1.31
8, 8	1.72	0.18	0.69	0.55	0.62	0.95	1.27
9, 9	0.54	0.30	0.37	0.32	0.84	0.80	0.78
10, 10	−0.43	−0.08	−0.32	−0.06	0.34	0.12	0.31
11, 11	−1.68	−1.20	−0.57	−0.37	0.10	−0.62	−0.23
6, 2, 10	0.47	0.46	−0.31	−0.21	−0.22	−0.17	−0.05
7, 3, 11	0.46	0.59	−0.40	−0.16	0.17	0.06	−0.26
Mean.	32.48	44.56	21.44	9.76	0.40	−23.00	4.79

XIV.

Arctic America. — Melville Island. *Lat.* 74° 47′ N. *Long.* 110° 48′ W. *Gr.* — Dove.

Degrees of Reaumur.

Hour.	January.	February.	March.	October.	Hour.	November.	December.
A.M. 1	0.12	0.10	1.04	0.04	A.M. 2	−0.12	−0.09
3	0.18	0.05	1.22	0.12	4	−0.02	−0.06
5	0.07	0.25	0.90	0.24	6	0.00	0.11
7	0.11	0.29	0.57	0.20	8	−0.22	0.07
9	−0.13	−0.24	0.29	−0.15	10	−0.38	0.11
11	−0.35	−0.43	−1.33	−0.46	12	−0.41	0.24
P.M. 1	−0.22	−0.65	−1.72	−0.43	P.M. 2	−0.27	0.14
3	−0.25	−0.52	−1.00	0.22	4	0.16	0.00
5	0.04	0.04	−0.43	−0.24	6	0.27	−0.12
7	0.04	0.24	0.06	−0.10	8	0.38	−0.26
9	0.11	0.35	0.33	0.11	10	0.36	−0.12
11	0.40	0.49	0.66	0.43	12	0.25	0.00
Mean.	−29.75	−27.58	−22.73	−14.32	Mean.	−18.65	−25.75

XV.

Spitzbergen. — Hecla Cove. *Lat.* 79° 55′ N. *Long.* 16° 49′ E. *Gr.* — Dove.

Degrees of Reaumur.

Hour.	June.	July.	August.	Hour.	June.	July.	August.
A.M. 1	0.63	0.62	0.42	P.M. 1	−0.67	−0.67	−0.63
3	0.43	0.84	0.54	3	−0.58	−0.42	−0.58
5	0.26	0.51	0.53	5	−0.27	−0.44	−0.32
7	−0.12	−0.02	0.25	7	0.26	−0.17	−0.06
9	−0.29	−0.09	−0.09	9	0.21	0.06	0.14
11	−0.47	−0.49	−0.45	11	0.61	0.26	0.24
				Mean.	1.71	3.63	2.84

The numbers without sign must be added; those with the sign — must be subtracted.

S. America. — Rio Janeiro. *Lat.* 22° 54′ S. *Long.* 43° 16′ W. *Greenw.*

Corrections to be applied to the Means of the Hours of Observation to obtain the true Mean Temperatures of the respective Days, Months, and of the Year. — Dove.

Degrees of Fahrenheit.

Hours.	Jan.	Feb.	March.	April.	May.	June.	July.	Aug.	Sept.	Oct.	Nov.	Dec.	Mean.
Morn. 1	0.74	1.51	1.80	0.90	1.13	0.56	1.85	1.31	1.04	0.97	1.76	1.31	1.24
2	1.64	2.41	2.48	1.64	2.12	1.53	2.75	2.00	1.69	1.64	2.32	2.05	2.03
3	2.50	3.11	3.02	2.32	2.93	2.43	3.47	2.66	2.27	2.21	2.75	2.66	2.70
4	3.08	3.90	3.24	2.79	3.38	3.04	3.87	3.04	2.59	2.50	2.93	2.99	3.06
5	3.22	3.29	3.15	2.90	3.40	3.29	3.83	3.08	2.66	2.52	2 79	2.99	3.08
6	2.93	2.84	2.75	2.75	3.06	3.20	3.47	2.79	2.41	2.27	2.32	2.68	2.79
7	2.30	2.21	2.14	2.30	2.48	2.84	2.70	2.25	2.00	1.82	1.67	2.12	2.23
8	1.49	1.49	1.40	1.71	1.85	2.39	1.96	1.60	1.46	1.28	0.90	1.40	1.58
9	0.68	0.72	0.59	1.04	1.15	1.82	1.15	0.90	0.86	0.68	0.14	0.59	0.86
10	−0.07	−0.05	−0.23	0.32	0.50	1.13	0.32	0.23	0.18	0.05	−0.56	−0.23	−0.14
11	−0.77	−0.86	−1.01	−0.45	−0.23	0.32	−0.50	−0.50	−0.54	−0.59	−1.22	−1.04	−0.61
Noon. . .	−1.40	−1.64	−1.71	−1.22	−0.99	−0.65	−1.31	−1.19	−1.26	−1.22	−1.80	−1.82	−1.35
1	−2.00	−2.30	−2.30	−1.94	−1.71	−1.67	−2.16	−1.91	−1.89	−1.78	−2.32	−2.43	−2.03
2	−2.41	−2.75	−2.66	−2.41	−2.30	−2.48	−2.88	−2.48	−2.34	−2.16	−2.66	−2.81	−2.52
3	−2.59	−2.88	−2.84	−2.66	−2.66	−2.99	−3.40	−2.84	−2.50	−2.27	−2.79	−2.86	−2.77
4	−2.45	−2.70	−2.77	−2.57	−2.75	−3.04	−3.60	−2.93	−2.36	−2.12	−2.66	−2.59	−2.70
5	−2.05	−2.30	−2.50	−2.21	−2.54	−2.75	−3.47	−2.68	−2.00	−1.78	−2.25	−2.09	−2.39
6	−1.51	−1.82	−2.12	−1.76	−2.21	−2.23	−3.04	−2.23	−1.55	−1.37	−1.67	−1.49	−1.91
7	−1.04	−1.40	−1.67	−1.28	−1.89	−1.76	−2.39	−1.67	−1.13	−1.04	−1.08	−0.99	−1.44
8	−0.72	−1.13	−1.22	−0.95	−1.67	−1.42	−1.85	−1.13	−0.83	−0.77	−0.59	−0.61	−1.08
9	−0.59	−0.92	−0.77	−0.72	−1.44	−1.26	−1.22	−0.70	−0.61	−0.61	−0.14	−0.38	−0.79
0	−0.56	−0.63	−0.25	−0.52	−1.13	−1.13	−0.59	−0.32	−0.41	−0.45	0.23	−0.16	−0.50
11	−0.41	−0.14	0.36	−0.25	−0.63	−0.86	0.09	0.09	−0.09	−0.16	0.65	0.14	0.09
Midn. . .	0.00	0.59	1.06	0.23	0.14	−0.29	0.92	0.61	0.38	0.32	1.15	0.65	0.47
6. 6	0.72	0.52	0.32	0.50	0.43	0.50	0.30	0.29	0.43	0.45	0.34	0.61	0.45
7. 7	0.63	0.41	0.25	0.52	0.29	0.54	0.16	0.29	0.45	0.41	0.29	0.56	0.41
8. 8	0.38	0.18	0.09	0.38	0.09	0.50	0.07	0.25	0.32	0.27	0.16	0.41	0.25
9. 9	0.05	−0.11	−0.09	−0.16	−0.16	0.29	−0.05	0.11	0.14	0.05	0.00	0.11	0.05
10.10	−0.32	−0.34	−0.25	−0.11	−0.32	0.00	−0.14	−0.05	−0.11	−0.20	−0.18	−0.20	−0.18
7. 2. 9	−0.23	−0.50	−0.43	−0.27	−0.43	−0.29	−0.47	−0.32	−0.32	−0.32	−0.38	−0.36	−0.36
6. 2. 8	−0.07	−0.34	−0.38	−0.20	−0.29	−0.23	−0.43	−0.27	−0.25	−0.23	−0.32	−0.25	−0.27
6. 2.10	−0.02	−0.18	−0.05	−0.07	−0.11	−0.14	0.00	0.00	−0.11	−0.11	−0.05	−0.09	−0.07
6. 2. 6	−0.34	−0.59	−0.68	−0.47	−0.47	−0.50	−0.81	−0.63	−0.50	−0.43	−0.68	−0.54	−0.56
7. 2	−0.07	−0.27	−0.27	−0.07	0.09	0.18	−0.09	−0.11	−0.18	−0.18	−0.50	−0.36	−0.16
8. 2	−0.47	−0.63	−0.63	−0.36	−0.23	−0.05	−0.47	−0.45	−0.45	−0.45	−0.88	−0.72	−0.47
8. 1	−0.27	−0.41	−0.45	−0.11	0.07	0.36	−0.11	−0.16	−0.23	−0.25	−0.72	−0.52	−0.23
7. 1	0.16	−0.05	−0.09	0.18	0.38	0.59	−0.27	0.18	0.07	0.02	−0.34	−0.16	−0.11
9.12.3.9	−0.97	−1.19	−1.19	−0.90	−0.99	−0.77	−1.19	−0.97	−0.88	−0.86	−1.15	−1.13	−1.01
7. 2.2(9)	−0.32	−0.61	−0.52	−0.38	−0.68	−0.54	−0.65	−0.41	−0.38	−0.38	−0.32	−0.36	−0.47
Dail. ext.	0.32	0.27	0.20	0.14	0.34	0.14	0.14	0.09	0.09	0.14	0.07	0.07	0.16

The numbers without sign must be added; those with the sign — must be subtracted.

XVII.

S. America. — Rio Janeiro. *Lat.* 22° 54′ S. *Long.* 43° 16′ W. *Greenw.*

Corrections to be applied to the Means of the Hours of Observation to obtain the true Mean Temperatures of the respective Days, Months, and of the Year. — Dove.

Degrees of Reaumur.

Hours.	Jan.	Feb.	March.	April.	May.	June.	July.	Aug.	Sept.	Oct.	Nov.	Dec.	Mean.
Morn. 1	0.33	0.67	0.80	0.40	0.50	0.25	0.82	0.58	0.46	0.43	0.78	0.58	0.55
2	0.73	1.07	1.10	0.73	0.94	0.68	1.22	0.89	0.75	0.73	1.03	0.91	0.90
3	1.11	1.38	1.34	1.03	1.30	1.08	1.54	1.18	1.01	0.98	1.22	1.18	1.20
4	1.37	1.51	1.44	1.24	1.50	1.35	1.72	1.35	1.15	1.11	1.30	1.33	1.36
5	1.43	1.46	1.40	1.29	1.51	1.46	1.70	1.37	1.18	1.12	1.24	1.33	1.37
6	1.30	1.26	1.22	1.22	1.36	1.42	1.54	1.24	1.07	1.01	1.03	1.19	1.24
7	1.02	0.98	0.95	1.02	1.10	1.26	1.20	1.00	0.89	0.81	0.74	0.94	0.99
8	0.66	0.66	0.62	0.76	0.82	1.06	0.87	0.71	0.65	0.57	0.40	0.62	0.70
9	0.30	0.32	0.26	0.46	0.51	0.81	0.51	0.40	0.38	0.30	0.06	0.26	0.38
10	–0.03	–0.02	–0.10	0.14	0.22	0.50	0.14	0.10	0.08	0.02	–0.25	–0.10	0.06
11	–0.34	–0.38	–0.45	–0.20	–0.10	0.14	–0.22	–0.22	–0.24	–0.26	–0.54	–0.46	–0.27
Noon. . .	–0.62	–0.73	–0.76	–0.54	–0.44	–0.29	–0.58	–0.53	–0.56	–0.54	–0.80	–0.81	–0.60
1	–0.89	–1.02	–1.02	–0.86	–0.76	–0.74	–0.96	–0.85	–0.84	–0.79	–1.03	–1.08	–0.90
2	–1.07	–1.22	–1.18	–1.07	–1.02	–1.10	–1.28	–1.10	–1.04	–0.96	–1.18	–1.25	–1.12
3	–1.15	–1.28	–1.26	–1.18	–1.18	–1.33	–1.51	–1.26	–1.11	–1.01	–1.24	–1.27	–1.23
4	–1.09	–1.20	–1.23	–1.14	–1.22	–1.35	–1.60	–1.30	–1.05	–0.94	–1.18	–1.15	–1.20
5	–0.91	–1.02	–1.11	–0.98	–1.13	–1.22	–1.54	–1.19	–0.89	–0.79	–1.00	–0.93	–1.06
6	–0.67	–0.81	–0.94	–0.78	–0.98	–0.99	–1.35	–0.99	–0.69	–0.61	–0.74	–0.66	–0.85
7	–0.46	–0.62	–0.74	–0.57	–0.84	–0.78	–1.06	–0.74	–0.50	–0.46	–0.48	–0.44	–0.64
8	–0.32	–0.50	–0.54	–0.42	–0.74	–0.63	–0.82	–0.50	–0.37	–0.34	–0.26	–0.27	–0.48
9	–0.26	–0.41	–0.34	–0.32	–0.64	–0.56	–0.54	–0.31	–0.27	–0.27	–0.06	–0.17	–0.35
10	–0.25	–0.28	–0.11	–0.23	–0.50	–0.50	–0.26	–0.14	–0.18	–0.20	0.10	–0.07	–0.22
11	–0.18	–0.06	0.16	–0.11	–0.28	–0.38	0.04	–0.04	–0.04	–0.07	0.29	0.06	–0.04
Midn. . .	0.00	0.26	0.47	0.10	0.06	–0.13	0.41	0.27	0.17	0.14	0.51	0.29	0.21
6. 6	0.32	0.23	0.14	0.22	0.19	0.22	0.10	0.13	0.19	0.20	0.15	0.27	0.20
7. 7	0.28	0.18	0.11	0.23	0.13	0.24	0.07	0.13	0.20	0.18	0.13	0.25	0.18
8. 8	0.17	0.08	0.04	0.17	0.04	0.22	0.03	0.11	0.14	0.12	0.07	0.18	0.11
9. 9	0.02	–0.05	–0.04	0.07	–0.07	0.13	–0.02	0.05	0.06	0.02	–0.00	0.05	0.02
10.10	–0.14	–0.15	–0.11	–0.05	–0.14	–0.00	–0.06	–0.02	–0.05	–0.09	–0.08	–0.09	–0.08
7. 2. 9	–0.10	–0.22	–0.19	–0.12	–0.19	–0.13	–0.21	–0.14	–0.14	–0.14	–0.17	–0.16	–0.16
6. 2. 8	–0.03	–0.15	–0.17	–0.09	–0.13	–0.10	–0.19	–0.12	–0.11	–0.10	–0.14	–0.11	–0.12
6. 2.10	–0.01	–0.08	–0.02	–0.03	–0.05	–0.06	–0.00	–0.00	–0.05	–0.05	–0.02	–0.04	–0.03
6. 2. 6	–0.15	–0.26	–0.30	–0.21	–0.21	–0.22	–0.36	–0.28	–0.22	–0.19	–0.30	–0.24	–0.25
7. 2	–0.03	–0.12	–0.12	–0.03	0.04	0.08	–0.04	–0.05	–0.08	–0.08	–0.22	–0.16	–0.07
8. 2	–0.21	–0.28	–0.28	–0.16	–0.10	–0.02	–0.21	–0.20	–0.20	–0.20	–0.39	–0.32	–0.21
8. 1	–0.12	–0.18	–0.20	–0.05	0.03	0.16	–0.05	–0.07	–0.10	–0.11	–0.32	–0.23	–0.10
7. 1	0.07	–0.02	–0.04	0.08	0.17	0.26	0.12	0.08	0.03	0.01	–0.15	–0.07	0.05
9.12.3.9	–0.43	–0.53	–0.53	–0.40	–0.44	–0.34	–0.53	–0.43	–0.39	–0.38	–0.51	–0.50	–0.45
7. 2.2(9)	–0.14	–0.27	–0.23	–0.17	–0.30	–0.24	–0.29	–0.18	–0.17	–0.17	–0.14	–0.16	–0.21
Dail. ext.	0.14	0.12	0.09	0.06	0.15	0.06	0.06	0.04	0.04	0.06	0.03	0.03	0.07

The numbers without sign must be added; those with the sign — must be subtracted.

HOURLY CORRECTIONS

FOR

PERIODIC VARIATIONS.

ASIA.

XVIII.

INDIA. — TREVANDRUM. *Lat.* 8° 31′ N. *Long.* 74° 50′ E. *Greenw.*

Corrections to be applied to the Means of the Hours of Observation to obtain the true Mean Temperatures of the respective Days, Months, and of the Year. — DOVE.

Degrees of Fahrenheit.

Hours.	Jan.	Feb.	March.	April.	May.	June.	July.	Aug.	Sept.	Oct.	Nov.	Dec.	Mean.
Morn. 1	4.41	4.03	3.80	3.85	3.26	2.66	2.41	2.88	2.99	3.06	3.33	4.25	3.42
2	5.13	4.95	4.64	4.46	3.80	3.02	2.75	3.24	3.44	3.44	3.83	4.86	3.96
3	6.03	6.12	5.67	5.15	4.39	3.47	3.17	3.74	3.98	3.92	4.46	5.67	4.66
4	6.95	7.31	6.64	5.74	4.82	3.80	3.58	4.21	4.48	4.34	5.04	6.50	5.29
5	7.56	8.15	7.13	5.81	4.82	3.83	3.76	4.41	4.61	4.46	5.22	6.93	5.56
6	7.34	8.01	6.73	5.11	4.14	3.35	3.49	4.07	4.14	4.01	4.73	6.57	5.15
7	6.01	6.59	5.20	3.53	2.81	2.34	2.68	3.06	3.02	2.88	3.40	5.11	3.89
8	3.56	3.92	2.66	1.22	0.95	0.90	1.35	1.49	1.26	1.13	1.40	2.70	1.87
9	0.41	0.50	−0.47	−1.42	−1.13	−0.74	−0.27	−0.45	−0.81	−0.99	−0.92	−0.29	−0.54
10	−2.84	−2.97	−3.53	−3.89	−3.04	−2.30	−1.91	−2.41	−2.86	−3.06	−3.11	−3.24	−2.93
11	−5.51	−5.85	−5.94	−5.76	−4.48	−3.53	−3.33	−4.05	−4.50	−4.73	−4.75	−5.58	−4.84
Noon. . .	−7.25	−7.58	−7.36	−6.82	−5.33	−4.34	−4.32	−5.18	−5.54	−5.72	−5.67	−7.00	−6.01
1	−7.92	−8.17	−7.72	−7.04	−5.60	−4.68	−4.79	−5.69	−5.87	−5.94	−5.90	−7.49	−6.41
2	−7.76	−7.83	−7.22	−6.59	−5.38	−4.61	−4.77	−5.60	−5.60	−5.54	−5.60	−7.25	−6.14
3	−7.09	−6.98	−6.26	−5.65	−4.79	−4.19	−4.30	−5.04	−4.86	−4.66	−4.95	−6.57	−5.45
4	−6.17	−5.99	−5.06	−4.46	−3.94	−3.47	−3.51	−4.10	−3.80	−3.53	−4.12	−5.67	−4.48
5	−5.13	−4.88	−3.83	−3.11	−2.88	−2.52	−2.52	−2.90	−2.59	−2.32	−3.15	−4.61	−8.38
6	−3.92	−3.74	−2.57	−1.71	−1.69	−1.42	−1.40	−1.58	−1.31	−1.10	−2.03	−3.35	−2.16
7	−2.50	−2.45	−1.31	−0.34	−0.50	−0.32	−0.29	−0.27	−0.11	0.00	−0.81	−1.89	−0.90
8	−0.92	−1.04	−0.07	0.92	0.63	0.70	0.68	0.90	0.92	0.97	0.38	−0.32	0.32
9	0.68	0.38	1.06	1.91	1.53	1.46	1.40	1.76	1.69	1.71	1.42	1.19	1.35
10	2.05	1.64	1.96	2.61	2.16	1.96	1.85	2.30	2.18	2.25	2.21	2.43	2.14
11	3.08	2.57	2.63	3.06	2.57	2.23	2.09	2.54	2.48	2.57	2.68	3.26	2.66
Midn. . .	3.83	3.31	3.17	3.42	2.88	2.41	2.23	2.68	2.70	2.81	2.99	3.80	3.02
6. 6	1.71	2.14	2.09	1.71	1.24	0.97	1.04	1.24	1.42	1.46	1.35	1.60	1.51
7. 7	1.76	2.07	1.96	1.60	1.17	1.01	1.19	1.40	1.44	1.44	1.28	1.62	1.49
8. 8	1.33	1.44	1.31	1.06	0.79	0.79	1.01	1.19	1.08	1.06	0.88	1.19	1.10
9. 9	0.54	0.43	0.29	0.25	0.20	0.36	0.56	0.65	0.43	0.36	0.25	0.45	0.41
10.10	−0.41	−0.65	−0.79	−0.63	−0.45	−0.18	−0.02	−0.07	−0.34	−0.41	−0.45	−0.40	−0.41
7. 2. 9	−0.36	−0.29	−0.32	−0.38	−0.34	−0.27	−0.23	−0.27	−0.29	−0.32	−0.27	−0.32	−0.32
6. 2. 8	−0.45	−0.29	−0.18	−0.18	−0.20	−0.18	−0.20	−0.20	−0.18	−0.18	−0.16	−0.34	−0.23
6. 2.10	0.54	0.61	0.50	0.38	0.32	0.23	0.18	0.25	0.25	0.25	0.45	0.59	0.38
6. 2. 6	−1.44	−1.19	−1.01	−1.06	−0.97	−0.90	−0.90	−1.04	−0.92	−0.88	−0.97	−1.35	−1.06
7. 2	−0.88	−0.63	−1.01	−1.53	−1.28	−1.15	−1.06	−1.28	−1.31	−1.33	−1.10	−1.08	−1.13
8. 2	−2.12	−1.96	−2.30	−2.70	−2.23	−1.87	−1.71	−2.07	−2.18	−2.21	−2.12	−2.27	−2.14
8. 1	−2.18	−2.14	−2.54	−2.93	−2.34	−1.89	−1.73	−2.12	−2.32	−2.41	−2.25	−2.41	−2.27
7. 1	−0.97	−0.79	−1.26	−1.76	−1.40	−1.17	−1.06	−1.33	−1.44	−1.53	−1.26	−1.19	−1.26
9.12.3.9	−3.31	−3.42	−3.26	−2.99	−2.43	−1.96	−1.87	−2.23	−2.39	−2.41	−2.54	−3.17	−2.66
7. 2.2(9)	−0.11	−0.11	0.02	0.20	0.14	0.16	0.18	0.25	0.20	0.20	0.16	0.07	0.11

The numbers without sign must be added; those with the sign — must be subtracted.

India. — Trevandrum. *Lat.* 8° 31′ N. *Long.* 74° 50′ E. *Greenw.*

Corrections to be applied to the Means of the Hours of Observation to obtain the true Mean Temperatures of the respective Days, Months, and of the Year. — Dove.

Degrees of Reaumur.

Hours.	Jan.	Feb.	March.	April.	May.	June.	July.	Aug.	Sept.	Oct.	Nov.	Dec.	Mean.
Morn. 1	1.96	1.79	1.69	1.71	1.45	1.18	1.07	1.28	1.33	1.36	1.48	1.89	1.52
2	2.28	2.20	2.06	1.98	1.69	1.34	1.22	1.44	1.53	1.53	1.70	2.16	1.76
3	2.68	2.72	2.52	2.29	1.95	1.54	1.41	1.66	1.77	1.74	1.98	2.52	2.07
4	3.09	3.25	2.95	2.55	2.14	1.69	1.59	1.87	1.90	1.93	2.24	2.89	2.35
5	3.36	3.62	3.17	2.58	2.14	1.70	1.67	1.96	2.05	1.98	2.32	3.08	2.47
6	3.26	3.56	2.99	2.27	1.85	1.49	1.55	1.81	1.84	1.78	2.10	2.92	2.29
7	2.67	2.93	2.31	1.57	1.25	1.04	1.19	1.36	1.34	1.28	1.51	2.27	1.73
8	1.58	1.74	1.18	0.54	0.42	0.40	0.60	0.66	0.56	0.50	0.62	1.20	0.83
9	0.18	0.22	−0.21	−0.63	−0.50	−0.33	−0.12	−0.20	−0.36	−0.44	−0.41	−0.13	−0.24
10	−1.26	−1.32	−1.57	−1.73	−1.35	−1.02	−0.85	−1.07	−1.27	−1.36	−1.38	−1.44	−1.30
11	−2.45	−2.60	−2.64	−2.56	−1.99	−1.57	−1.48	−1.80	−2.00	−2.10	−2.11	−2.48	−2.15
Noon. . .	−3.22	−3.37	−3.27	−3.03	−2.37	−1.93	−1.92	−2.30	−2.46	−2.54	−2.52	−3.11	−2.67
1	−3.52	−3.63	−3.43	−3.13	−2.49	−2.08	−2.13	−2.53	−2.61	−2.64	−2.62	−3.33	−2.85
2	−3.45	−3.48	−3.21	−2.93	−2.39	−2.05	−2.12	−2.49	−2.49	−2.46	−2.49	−3.22	−2.73
3	−3.15	−3.10	−2.78	−2.51	−2.13	−1.86	−1.91	−2.24	−2.16	−2.07	−2.20	−2.92	−2.42
4	−2.74	−2.66	−2.25	−1.98	−1.75	−1.54	−1.56	−1.82	−1.69	−1.57	−1.83	−2.52	−1.99
5	−2.28	−2.17	−1.70	−1.38	−1.28	−1.12	−1.12	−1.29	−1.15	−1.03	−1.40	−2.05	−1.50
6	−1.74	−1.66	−1.14	−0.76	−0.75	−0.63	−0.62	−0.70	−0.58	−0.49	−0.90	−1.49	−0.96
7	−1.11	−1.09	−0.58	−0.15	−0.22	−0.14	−0.13	−0.12	−0.05	0.00	−0.36	−0.84	−0.40
8	−0.41	−0.46	−0.03	0.41	0.28	0.31	0.30	0.40	0.41	0.43	0.17	−0.14	0.14
9	0.30	0.17	0.47	0.85	0.68	0.65	0.62	0.78	0.75	0.76	0.63	0.53	0.60
10	0.91	0.73	0.87	1.16	0.96	0.87	0.82	1.02	0.97	1.00	0.98	1.08	0.95
11	1.37	1.14	1.17	1.36	1.14	0.99	0.93	1.13	1.10	1.14	1.19	1.45	1.18
Midn. . .	1.70	1.47	1.41	1.52	1.28	1.07	0.99	1.19	1.20	1.25	1.33	1.69	1.34
6. 6	0.76	0.95	0.93	0.76	0.55	0.43	0.46	0.55	0.63	0.65	0.60	0.71	0.67
7. 7	0.78	0.92	0.87	0.71	0.52	0.45	0.53	0.62	0.64	0.64	0.57	0.72	0.66
8. 8	0.59	0.64	0.58	0.47	0.35	0.35	0.45	0.53	0.48	0.47	0.39	0.53	0.49
9. 9	0.24	0.19	0.13	0.11	0.09	0.16	0.25	0.29	0.19	0.16	0.11	0.20	0.18
10.10	−0.18	−0.29	−0.35	−0.28	−0.20	−0.08	−0.01	−0.03	−0.15	−0.18	−0.20	−0.18	−0.18
7. 2. 9	−0.16	−0.13	−0.14	−0.17	−0.15	−0.12	−0.10	−0.12	−0.13	−0.14	−0.12	−0.14	−0.14
6. 2. 8	−0.20	−0.13	−0.08	−0.08	−0.09	−0.08	−0.09	−0.09	−0.08	−0.08	−0.07	−0.15	−0.10
6. 2.10	0.24	0.27	0.22	0.17	0.14	0.10	0.08	0.11	0.11	0.11	0.20	0.26	0.17
6. 2. 6	−0.64	−0.53	−0.45	−0.47	−0.43	−0.40	−0.40	−0.46	−0.41	−0.39	−0.43	−0.60	−0.47
7. 2	−0.39	−0.28	−0.45	−0.68	−0.57	−0.51	−0.47	−0.57	−0.58	−0.59	−0.49	−0.48	−0.50
8. 2	−0.94	−0.87	−1.02	−1.20	−0.99	−0.83	−0.76	−0.92	−0.97	−0.98	−0.94	−1.01	−0.95
8. 1	−0.97	−0.95	−1.13	−1.30	−1.04	−0.84	−0.77	−0.94	−1.03	−1.07	−1.00	−1.07	−1.01
7. 1	−0.43	−0.35	−0.56	−0.78	−0.62	−0.52	−0.47	−0.59	−0.64	−0.68	−0.56	−0.53	−0.56
9.12.3.9	−1.47	−1.52	−1.45	−1.33	−1.08	−0.87	−0.83	−0.99	−1.06	−1.07	−1.13	−1.41	−1.18
7. 2.2(9)	−0.05	−0.05	0.01	0.09	0.06	0.07	0.08	0.11	0.09	0.09	0.07	0.03	0.05
Dail. ext.	−0.08	−0.01	−0.13	−0.28	−0.18	−0.19	−0.23	−0.29	−0.28	−0.33	−0.15	−0.13	−0.19

The numbers without sign must be added; those with the sign — must be subtracted.

INDIA. — MADRAS. *Lat.* 13° 4′ N. *Long.* 80° 19′ E. *Greenw.*

Corrections to be applied to the Means of the Hours of Observation to obtain the true Mean Temperatures of the respective Days, Months, and of the Year. — DOVE.

Degrees of Fahrenheit.

Hour.	Jan.	Feb.	March.	April.	May.	June.	July.	Aug.	Sept.	Oct.	Nov.	Dec.	Year.
Midnight.	2.05	2.54	2.25	3.65	2.74	3.03	2.90	2.86	2.34	1.84	2.05	1.89	2.50
1	2.54	3.26	2.90	3.08	3.31	3.50	3.10	3.01	2.70	2.27	2.54	2.25	2.87
2	2.96	3.95	3.60	3.57	3.72	3.86	3.55	3.39	3.10	2.79	3.03	2.63	3.35
3	3.33	4.52	4.25	4.07	4.07	4.27	3.93	3.69	3.55	3.12	3.50	2.96	3.77
4	3.62	5.06	4.79	4.40	4.45	4.68	4.31	3.98	3.95	3.46	3.91	3.19	4.15
5	3.81	5.49	5.24	4.45	4.68	4.95	4.66	4.34	4.23	3.71	4.23	3.60	4.45
6	4.05	5.64	5.11	3.78	3.86	4.21	4.31	4.07	3.82	3.28	4.05	3.73	4.16
7	2.43	3.33	2.54	1.78	2.07	2.51	2.92	2.79	2.43	1.80	2.00	2.38	2.41
8	-0.04	0.29	0.16	-0.18	-0.11	0.38	1.06	0.99	0.72	0.13	-0.56	0.00	0.23
9	-2.02	-1.93	-1.89	-2.41	-2.43	-1.73	-0.76	-0.90	-1.12	-1.26	-2.49	-1.73	-1.72
10	-3.26	-3.60	-3.67	-4.14	-4.68	-3.67	-2.67	-2.74	-2.96	-2.34	-3.53	-3.05	-3.36
11	-4.02	-4.81	-4.81	-4.83	-5.75	-5.02	-4.25	-4.16	-4.54	-3.17	-4.09	-3.62	-4.42
Noon.	-4.43	-5.06	-5.35	-5.66	-5.87	-5.85	-5.51	-5.28	-5.04	-3.76	-4.31	-3.93	-5.01
1	-4.40	-5.35	-5.42	-5.53	-5.64	-6.05	-6.07	-5.75	-5.04	-3.73	-4.25	-3.86	-5.09
2	-4.14	-5.30	-4.99	-4.95	-4.99	-5.69	-6.02	-5.40	-4.66	-3.55	-3.73	-3.60	-4.75
3	-3.46	-4.85	-4.27	-4.07	-4.00	-4.61	-4.92	-4.59	-3.73	-3.03	-3.05	-2.88	-3.95
4	-2.41	-3.64	-3.10	-2.65	-2.45	-3.57	-3.73	-3.44	-2.56	-2.38	-1.98	-2.04	-2.83
5	-1.19	-2.27	-1.66	-1.03	-1.01	-1.91	-2.18	-1.84	-1.44	-1.26	-0.88	-1.01	-1.47
6	-0.38	-1.10	-0.52	0.20	0.11	-0.58	-0.81	-0.70	-0.52	-0.63	-0.25	-0.38	-0.46
7	0.09	-0.36	0.17	0.83	0.76	0.36	0.16	0.13	0.07	-0.18	0.09	0.00	0.18
8	0.54	0.27	0.58	0.99	1.19	0.97	0.83	0.74	0.47	0.16	0.47	0.34	0.63
9	0.94	0.81	0.97	1.57	1.57	1.42	1.35	1.17	0.99	0.49	0.74	0.67	1.06
10	1.39	1.33	1.39	1.89	1.96	2.11	1.87	1.64	1.39	0.90	1.08	1.03	1.50
11	1.84	1.87	1.84	2.25	2.34	2.41	2.29	2.14	1.89	1.28	1.46	1.44	1.92
6, 6	1.83	2.27	2.29	1.99	1.98	1.81	1.75	1.65	1.65	1.32	1.90	1.67	1.84
7, 7	1.26	1.48	1.35	1.30	1.41	1.43	1.54	1.46	1.25	0.81	1.04	1.19	1.29
8, 8	0.25	0.28	0.37	0.40	0.54	0.67	0.94	0.81	0.59	0.14	-0.04	0.17	0.43
9, 9	-0.54	-0.56	-0.46	-0.42	-0.43	-0.15	0.29	0.13	-0.06	-0.38	-0.87	-0.53	-0.33
10, 10	-0.93	-1.13	-1.14	-1.12	-1.36	-0.78	-0.40	-0.55	-0.78	-0.70	-1.22	-1.01	-0.93
7, 1	-0.98	-1.01	-1.44	-1.87	-1.78	-1.77	-1.57	-1.48	-1.30	-0.86	-1.12	-0.74	-1.33
7, 2, 9	-0.26	-0.39	-0.49	-0.53	-0.45	-0.59	-0.58	-0.48	-0.41	-0.42	-0.33	-0.18	-0.43
6, 2, 10	0.43	0.56	0.50	0.24	0.28	0.21	0.05	0.10	0.18	0.21	0.47	0.39	0.30
Mean.	76.77	78.25	82.24	85.73	87.10	87.01	86.22	84.51	83.50	81.18	78.53	76.75	

The numbers without sign must be added; those with the sign — must be subtracted.

XXI.

INDIA. — MADRAS. *Lat.* 13° 4′ N. *Long.* 80° 19′ E. *Greenw.*

Corrections to be applied to the Means of the Hours of Observation to obtain the true Mean Temperatures of the respective Days, Months, and of the Year. — DOVE.

Degrees of Reaumur.

Hours.	Jan.	Feb.	March.	April.	May.	June.	July.	Aug.	Sept.	Oct.	Nov.	Dec.	Mean.
Morn. 1	1.41	1.22	1.32	1.06	1.26	1.15	0.93	0.83	1.26	1.18	1.04	1.38	1.17
2	1.79	1.64	1.42	1.36	1.59	1.42	1.09	1.40	1.52	1.46	1.32	1.50	1.46
3	2.14	2.10	1.50	1.76	1.94	1.70	1.26	1.66	1.67	1.70	1.70	1.68	1.73
4	2.38	2.42	1.58	2.10	2.17	1.90	1.42	1.66	1.70	1.88	1.90	1.93	1.92
5	2.42	2.43	1.61	2.20	2.18	1.95	1.42	1.45	1.62	1.88	2.02	2.17	1.95
6	2.22	2.05	1.48	1.91	1.86	1.77	1.33	1.10	1.39	1.64	1.81	2.25	1.73
7	1.76	1.30	1.14	1.24	1.19	1.30	1.12	0.75	1.02	1.14	1.27	2.00	1.27
8	1.05	0.36	0.54	0.30	0.27	0.70	0.78	0.46	0.47	0.40	0.50	1.32	0.60
9	0.15	−0.59	−0.23	−0.71	−0.75	−0.06	0.35	0.16	−0.23	−0.46	−0.35	0.27	−0.20
10	−0.82	−1.38	−1.04	−1.56	−1.67	−0.82	−0.21	−0.18	−1.02	−1.26	−1.10	−0.94	−1.00
11	−1.74	−1.94	−1.70	−2.12	−2.31	−1.46	−0.86	−0.62	−1.77	−1.83	−1.75	−2.20	−1.69
Noon. . .	−2.48	−2.23	−2.06	−2.36	−2.58	−1.94	−1.52	−1.12	−2.29	−2.18	−2.12	−2.76	−2.14
1	−2.90	−2.34	−2.10	−2.34	−2.48	−2.20	−2.13	−1.57	−2.47	−2.17	−2.25	−2.98	−2.33
2	−2.97	−2.30	−1.88	−2.14	−2.13	−2.24	−2.47	−1.82	−2.27	−1.91	−2.18	−2.76	−2.26
3	−2.68	−2.12	−1.52	−1.84	−1.62	−2.07	−2.48	−1.77	−1.77	−1.50	−1.98	−2.25	−1.97
4	−2.14	−1.81	−1.14	−1.46	−1.11	−1.74	−2.12	−1.43	−1.12	−1.08	−1.61	−1.65	−1.53
5	−1.47	−1.34	−0.83	−1.00	−0.65	−1.28	−1.44	−0.94	−0.50	−0.70	−1.10	−1.13	−1.03
6	−0.81	−0.78	−0.58	−0.48	−0.27	−0.78	−0.65	−0.46	−0.06	−0.38	−0.58	−0.72	−0.55
7	−0.26	−0.18	−0.35	0.04	0.02	−0.30	0.08	−0.14	0.18	−0.14	−0.14	−0.39	−0.13
8	0.13	0.30	−0.08	0.49	0.26	0.12	0.62	−0.04	0.27	0.06	0.36	−0.06	0.20
9	0.38	0.62	0.42	0.71	0.45	0.42	0.86	−0.06	0.33	0.26	0.64	0.30	0.44
10	0.58	0.77	0.60	0.90	0.61	0.63	0.91	−0.06	0.44	0.46	0.81	0.66	0.61
11	0.79	0.84	0.91	0.91	0.78	0.79	0.87	0.11	0.66	0.67	0.83	0.99	0.76
Midn. . .	1.06	0.96	1.16	0.92	0.98	0.94	0.84	0.47	0.95	0.91	0.89	1.22	0.94
6. 6	0.71	0.64	0.45	0.72	0.80	0.50	0.34	0.32	0.67	0.63	0.62	0.77	0.60
7. 7	0.75	0.56	0.40	0.64	0.61	0.50	0.60	0.31	0.60	0.50	0.57	0.81	0.57
8. 8	0.59	0.33	0.23	0.40	0.27	0.41	0.70	0.21	0.37	0.23	0.43	0.63	0.40
9. 9	0.27	0.02	0.10	−0.00	−0.15	0.18	0.61	0.05	0.05	−0.10	0.15	0.29	0.12
10.10	−0.12	−0.31	−0.22	−0.33	−0.53	−0.10	−0.35	−0.12	−0.29	−0.40	−0.15	−0.14	−0.20
7. 2. 9	−0.28	−0.13	−0.11	−0.06	−0.16	−0.17	−0.16	−0.38	−0.31	−0.17	−0.09	−0.15	−0.18
6. 2. 8	−0.21	0.02	−0.16	0.09	−0.00	−0.12	−0.17	−0.25	−0.20	−0.07	−0.01	−0.19	−0.11
6. 2.10	−0.06	0.17	0.07	0.22	0.11	0.05	−0.08	−0.26	−0.15	0.06	0.15	0.05	0.03
6. 2. 6	−0.52	−0.34	−0.33	−0.24	−0.18	−0.42	−0.60	−0.39	−0.31	−0.22	−0.32	−0.41	−0.36
7. 2	−0.61	−0.50	−0.37	−0.45	−0.47	−0.47	−0.68	−0.54	−0.63	−0.32	−0.46	−0.38	−0.50
8. 2	−0.96	−0.97	−0.67	−0.92	−0.93	−0.77	−0.85	−0.68	−0.90	−0.76	−0.84	−0.72	−0.83
8. 1	−0.93	−0.99	−0.78	−1.02	−1.11	−0.75	−0.68	−0.56	−1.00	−0.89	−0.88	−0.83	−0.87
7. 1	−0.57	−0.52	−0.48	−0.55	−0.65	−0.45	−0.51	−0.41	−0.73	−0.52	−0.49	−0.49	−0.53
9.12.3.9	−1.16	−1.08	−0.85	−1.05	−1.13	−0.01	−0.70	−0.70	−0.99	−0.97	−0.95	−1.11	−0.97
7. 2.2(9)	−0.11	0.06	0.03	0.13	−0.01	−0.03	0.09	−0.30	−0.15	−0.06	0.09	−0.04	−0.03
Dail. ext.	−0.28	0.05	−0.25	−0.08	−0.20	−0.15	−0.53	−0.08	−0.39	−0.15	−0.12	−0.37	−0.19

The numbers without sign must be added; those with the sign — must be subtracted.

XXII.

India. — Bombay. *Lat.* 18° 56′ N. *Long.* 72° 54′ E. *Greenw.*

Corrections to be applied to the Means of the Hours of Observation to obtain the true Mean Temperatures of the respective Days, Months, and of the Year. — Dove.

Degrees of Fahrenheit.

Hours.	Jan.	Feb.	March.	April.	May.	June.	July.	Aug.	Sept.	Oct.	Nov.	Dec.	Mean.
Morn. 1	1.49	1.40	0.99	1.13	1.42	1.15	0.79	0.97	0.86	1.49	2.03	1.55	1.26
2	1.80	1.69	1.33	1.51	1.78	1.40	0.88	1.13	0.97	1.87	2.18	1.87	1.53
3	2.27	2.21	1.91	2.05	2.14	1.69	0.90	1.24	1.24	2.32	2.45	2.41	1.91
4	2.86	2.84	2.59	2.48	2.32	1.91	0.90	1.31	1.53	2.75	2.81	3.11	2.27
5	3.47	3.40	3.04	2.61	2.23	1.96	0.86	1.31	1.71	2.95	3.11	3.78	2.54
4	3.83	3.62	3.06	2.34	1.80	1.80	0.79	1.24	1.67	2.79	3.15	4.16	2.52
7	3.69	3.33	2.54	1.67	1.15	1.42	0.65	1.04	1.22	2.21	2.79	4.01	2.14
8	2.97	2.48	1.58	0.77	0.36	0.88	0.38	0.74	0.79	1.28	1.91	3.24	1.44
9	1.69	1.22	0.38	−0.14	−0.41	0.23	0.00	0.32	0.09	0.16	0.63	1.87	0.50
10	0.07	−0.23	−0.77	−0.90	−1.06	−0.43	−0.52	−0.20	−0.65	−0.95	−0.83	0.16	−0.52
11	−1.55	−1.55	−1.67	−1.49	−1.55	−6.08	−6.08	−0.79	−1.28	−1.91	−2.21	−1.60	−1.49
Noon. . .	−2.86	−2.61	−2.30	−1.91	−1.94	−1.64	−1.55	−1.35	−1.80	−2.59	−3.29	−3.08	−2.25
1	−3.69	−3.29	−2.66	−2.25	−2.21	−2.12	−1.82	−1.78	−2.12	−2.99	−3.92	−4.10	−2.75
2	−3.98	−3.60	−2.84	−2.50	−2.34	−2.41	−1.78	−2.00	−2.25	−3.13	−4.07	−4.59	−2.95
3	−3.85	−3.65	−2.86	−2.61	−2.32	−2.45	−1.44	−1.98	−2.16	−2.99	−3.85	−4.55	−2.90
4	−3.42	−3.42	−2.72	−2.50	−2.09	−2.25	−0.92	−1.69	−1.87	−2.66	−3.33	−4.12	−2.59
5	−2.84	−2.95	−2.34	−2.07	−1.64	−1.78	−0.38	−1.24	−1.37	−2.14	−2.61	−3.38	−2.07
6	−2.18	−2.27	−1.71	−1.37	−1.04	−1.15	0.09	−0.72	−0.74	−1.46	−1.78	−2.45	−1.40
7	−1.49	−1.44	−0.88	−0.54	−0.38	−0.47	0.38	−0.23	0.05	−0.72	−0.88	−1.46	−0.68
8	−0.79	−0.56	−0.07	0.23	0.18	0.14	0.50	0.16	0.47	−0.02	0.00	−0.52	−0.02
9	−0.11	0.23	0.56	0.72	0.59	0.54	0.54	1.43	0.86	0.52	0.77	0.29	0.50
10	0.47	0.81	0.90	0.92	0.83	0.79	0.54	0.59	0.99	0.88	1.35	0.86	0.83
11	0.92	1.10	0.97	0.92	0.99	0.96	0.61	0.72	0.97	1.08	1.71	1.19	1.01
Midn. . .	1.24	1.26	0.92	0.95	1.15	0.99	0.70	0.83	0.88	1.26	1.91	1.37	1.13
6. 6	0.81	0.68	0.68	0.50	0.38	0.34	0.43	0.25	0.45	0.68	0.70	0.86	0.56
7. 7	1.10	0.95	0.83	0.56	0.38	0.47	0.52	0.41	0.63	0.74	0.95	1.28	0.74
8. 8	1.08	0.97	0.77	0.50	0.27	0.50	0.45	0.45	0.63	0.63	0.95	1.35	0.72
9. 9	0.79	0.72	0.47	0.29	0.09	0.38	0.27	0.36	0.47	0.34	0.70	1.08	0.50
10.10	0.27	0.29	0.07	0.00	−0.11	0.18	0.02	0.20	0.18	−0.05	0.25	0.52	0.16
7. 2. 9	−0.14	−0.02	0.09	−0.05	−0.20	−0.16	−0.20	−0.18	−0.07	−0.14	−0.18	−0.09	−0.11
6. 2. 8	−0.32	−0.18	0.05	0.02	−0.11	−0.16	−0.16	−0.20	−0.05	−0.11	−0.32	−0.32	−0.16
6. 2.10	0.11	0.27	0.38	0.25	0.09	0.07	−0.16	−0.07	0.14	0.18	0.14	0.14	0.14
6. 2. 6	−0.79	−0.74	−0.50	−0.52	−0.52	−0.59	−0.29	−0.50	−0.45	−0.61	−0.90	−0.97	−0.61
7. 2	−0.16	−0.14	−0.16	−0.43	−0.61	−0.50	−0.56	−0.50	−0.52	−0.47	−0.65	−0.29	−0.41
8. 2	−0.52	−0.56	−0.63	−0.88	−0.99	−0.77	−0.70	−0.63	−0.74	−0.92	−1.08	−0.68	−0.77
8. 1	−0.36	−0.41	−0.54	−0.79	−0.92	−0.63	−0.72	−0.52	−0.68	−0.86	−1.01	−0.43	−0.65
7. 1	0.00	0.02	−0.07	−0 29	−0.54	−0.36	−0.59	−0.38	−0.45	−0.41	−0.56	−0.05	−0.32
9.12.3.9	−1.28	−1.22	−1.06	−0.99	−1.01	−0.83	−0.61	−0.65	−0.77	−1.24	−1.44	−1.37	−1.04
7. 2.2(9)	−0.14	0.05	0.20	0.16	0.00	0.02	−0.02	−0.02	0.18	0.02	0.07	0.00	0.05

The numbers without sign must be added; those with the sign — must be subtracted.

XXIII.

INDIA. — BOMBAY. *Lat.* 18° 56′ N. *Long.* 72° 54′ E. *Greenw.*

Corrections to be applied to the Means of the Hours of Observation to obtain the true Mean Temperatures of the respective Days, Months, and of the Year. — DOVE.

Degrees of Reaumur.

Hours.	Jan.	Feb.	March.	April.	May.	June.	July.	Aug.	Sept.	Oct.	Nov.	Dec.	Mean.
Morn. 1	0.66	0.62	0.44	0.50	0.63	0.51	0.35	0.43	0.38	0.66	0.90	0.69	0.56
2	0.80	0.75	0.59	0.67	0.79	0.62	0.39	0.50	0.43	0.83	0.97	0.83	0.68
3	1.01	0.98	0.85	0.91	0.95	0.75	0.40	0.55	0.55	1.03	1.09	1.07	0.85
4	1.27	1.26	1.15	1.10	1.03	0.85	0.40	0.58	0.68	1.22	1.25	1.38	1.01
5	1.54	1.51	1.35	1.16	0.99	0.87	0.38	0.58	0.76	1.31	1.38	1.68	1.13
6	1.70	1.61	1.36	1.04	0.80	0.80	0.35	0.55	0.74	1.24	1.40	1.85	1.12
7	1.64	1.48	1.13	0.74	0.51	0.63	0.29	0.46	0.54	0.98	1.24	1.78	0.95
8	1.32	1.10	0.70	0.34	0.16	0.39	0.17	0.33	0.35	0.57	0.85	1.44	0.64
9	0.75	0.54	0.17	-0.06	-0.18	0.10	-0.	0.14	0.04	0.07	0.28	0.83	0.22
10	0.03	-0.10	-0.34	-0.40	-0.47	-0.19	-0.23	-0.09	-0.29	-0.42	-0.37	0.07	-0.23
11	-0.69	-0.69	-0.74	-0.66	-0.69	-0.48	-0.48	-0.35	-0.57	-0.85	-0.98	-0.71	-0 66
Noon. . .	-1.27	-1.16	-1.02	-0.85	-0.86	-0.73	-0.69	-0.60	-0.80	-1.15	-1.46	-1.37	-1.00
1	-1.64	-1.46	-1.18	-1.00	-0.98	-0.94	-0.81	-0.79	-0.94	-1 33	-1.74	-1.82	-1.22
2	-1.77	-1.60	-1.26	-1.11	-1.04	-1.07	-0.79	-0.89	-1.00	-1 39	-1.81	-2.04	-1.31
3	-1.71	-1.62	-1.27	-1.16	-1.03	-1.09	-0.64	-0.88	-0.96	-1.33	-1.71	-2.02	-1.29
4	-1.52	-1.52	-1.21	-1.11	-0.93	-1.00	-0.41	-0.75	-0.83	-1.18	-1.48	-1.83	-1.15
5	-1.26	-1.31	-1.04	-0.92	-0.73	-0.79	-0.17	-0.55	-0.61	-0.95	-1.16	-1.50	-0.92
6	-0.97	-1.01	-0.76	-0.61	-0.46	-0.51	0.04	-0.32	-0.33	-0.65	-0.79	-1.09	-0.62
7	-0.66	-0.64	-0.39	-0.24	-0.17	-0.21	0.17	-0.10	0.02	-0.32	-0.39	-0.65	-0.30
8	-0.35	-0.25	0.03	0.10	0.08	0.06	0.22	0.07	0.21	-0.01	0.	-0.23	-0 01
9	-0.05	0.10	0.25	0.32	0.26	0.24	0.24	0.19	0.38	0.23	0.34	0.13	0.22
10	0.21	0.36	0.40	0.41	0.37	0.35	0.24	0.26	0.44	0.39	0 60	0.38	0.37
11	0.41	0.49	0.43	0.41	0.44	0.40	0.27	0.32	0.43	0.48	0.76	0.53	0.45
Midn. . .	0.55	0.56	0.41	0.42	0.51	0.44	0.31	0.37	0.39	0.56	0.85	0.61	0.50
6. 6	0.36	0.30	0.30	0.22	0.17	0.15	0.19	0.11	0.20	0.30	0.31	0 38	0.25
7. 7	0.49	0.42	0.37	0.25	0.17	0.21	0.23	0.18	0.28	0.33	0.42	0 57	0.33
8. 8	0.48	0.43	0.34	0.22	0.12	0.22	0.20	0.20	0.28	0.28	0 42	0.60	0.32
9. 9	0.35	0.32	0.21	0.13	0.04	0.17	0.12	0.16	0.21	0.15	0.31	0.48	0.22
10.10	0.12	0.13	0.03	0.00	-0.05	0.08	0.01	0.09	0.08	-0.02	0.11	0.23	0.07
7. 2. 9	-0.06	-0.01	0.04	-0.02	-0.09	-0.07	-0.09	-0.08	-0.03	-0.06	-0.08	-0.04	-0.05
6. 2. 8	0.14	-0.08	0.02	0.01	-0.05	-0.07	-0.07	-0.09	-0.02	-0.05	-0.14	-0.14	-0.07
6. 2.10	0.05	0.12	0.17	0.11	0.04	0.03	-0.07	-0.03	0.06	0.08	0.06	0.06	0.06
6. 2. 6	-0.35	-0.33	-0.22	-0.23	-0.23	-0.26	-0.13	-0.22	-0.20	-0.27	-0.40	-0.43	-0.27
7. 2	-0.07	-0.06	-0.07	-0.19	-0.27	-0.22	-0.25	-0.22	-0.23	-0.21	-0.29	-0.13	-0.18
8. 2	-0.23	-0.25	-0.28	-0.39	-0.44	-0.34	-0.31	-0.28	-0.33	-0.41	-0.48	-0.30	-0.34
8. 1	-0.16	-0.18	-0.24	-0.33	-0.41	-0.28	-0.32	-0 23	-0.30	-0.38	-0.45	-0.19	-0.29
7. 1	0.00	0.01	-0.03	-0.13	-0.24	-0.16	-0.26	-0.17	-0.20	-0.18	-0.25	-0.02	-0.14
9.12.2.9	-0.57	-0.54	-0.47	-0.44	-0.45	-0.37	-0.27	-0.29	-0.34	-0.55	-0.64	-0.61	-0.46
7. 2.2(9)	-0.06	0.02	0.09	0.07	0.00	0.01	-0.01	-0.01	0.08	0.01	0.03	0 00	0.02
Dail. ext.	-0.04	-0.01	0.05	0.01	0.00	-0.11	-0.21	-0.16	-0.12	-0.04	-0.21	-0.10	-0.09

The numbers without sign must be added; those with the sign — must be subtracted.

XXIV.

India. — Madras. *Lat.* 13° 4′ N. *Long.* 80° 19′ E. *Greenw.*

Corrections to be applied to the Means of the Hours of Observation to obtain the true Mean Temperatures of the respective Days, Months, and of the Year. — Dove.

Degrees of Reaumur.

Hour.	Jan.	Feb.	March.	April.	May.	June.	July.	Aug.	Sept.	Oct.	Nov.	Dec.	Year.
Midn.	0.91	1.13	1.00	1.62	1.22	1.35	1.19	1.27	1.04	0.82	0.91	0.84	1.11
1	1.13	1.45	1.29	1.37	1.47	1.56	1.38	1.34	1.20	1.01	1.13	1.00	1.28
2	1.32	1.76	1.60	1.59	1.65	1.72	1.58	1.51	1.38	1.24	1.35	1.17	1.49
3	1.48	2.01	1.88	1.81	1.81	1.90	1.75	1.64	1.58	1.39	1.56	1.32	1.68
4	1.61	2.25	2.13	1.96	1.98	2.08	1.92	1.77	1.76	1.54	1.74	1.42	1.85
5	1.74	2.44	2.33	1.98	2.08	2.20	2.07	1.93	1.88	1.65	1.88	1.60	1.98
6	1.80	2.51	2.27	1.68	1.72	1.87	1.92	1.81	1.70	1.46	1.80	1.66	1.85
7	1.08	1.48	1.13	0.79	0.92	1.12	1.30	1.24	1.08	0.80	1.89	1.06	1.07
8	−0.02	0.13	0.07	−0.08	−0.05	0.17	0.47	0.44	0.32	0.06	−0.25	0.00	0.10
9	−0.90	−0.86	−0.84	−1.07	−1.08	−0.77	−0.34	−0.40	−0.50	−0.56	−1.11	−0.77	−0.77
10	−1.45	−1.60	−1.63	−1.84	−2.08	−1.63	−1.19	−1.22	−1.32	−1.04	−1.57	−1.36	−1.49
11	−1.79	−2.14	−2.14	−2.15	−2.56	−2.23	−1.89	−1.85	−2.02	−1.41	−1.82	−1.61	−1.47
Noon.	−1.97	−2.25	−2.38	−2.52	−2.61	−2.60	−2.45	−2.35	−2.24	−1.67	−1.92	−1.75	−2.23
1	−1.96	−2.38	−2.41	−2.46	−2.51	−2.69	−2.70	−2.56	−2.24	−1.66	−1.89	−1.72	−2.26
2	−1.84	−2.36	−2.22	−2.20	−2.22	−2.53	−2.67	−2.40	−2.07	−1.58	−1.66	−1.60	−2.11
3	−1.54	−2.16	−1.90	−1.81	−1.78	−2.05	−2.19	−2.04	−1.66	−1.35	−1.36	−1.28	−1.76
4	−1.07	−1.62	−1.38	−1.18	−1.09	−1.59	−1.66	−1.53	−1.14	−1.06	−0.88	−0.91	−1.26
5	−0.53	−1.01	−0.74	−0.46	−0.45	−0.85	−0.97	−0.82	−0.64	−0.56	−0.39	−0.45	−0.66
6	−0.17	−0.49	−0.23	0.09	0.05	−0.26	−0.36	−0.31	−0.23	−0.28	−0.11	−0.17	−0.21
7	0.04	−0.16	0.07	0.37	0.34	0.16	0.07	0.06	0.03	−0.08	0.04	0.00	0.08
8	0.24	0.12	0.26	0.44	0.53	0.43	0.37	0.33	0.21	0.07	0.21	0.15	0.28
9	0.42	0.36	0.43	0.70	0.70	0.63	0.60	0.52	0.44	0.22	0.33	0.30	0.47
10	0.62	0.59	0.62	0.84	0.87	0.94	0.83	0.73	0.62	0.40	0.48	0.46	0.67
11	0.82	0.83	0.82	1.00	1.04	1.07	1.02	0.95	0.84	0.57	0.65	0.64	0.85
Mean.	19.90	20.56	22.33	23.88	24.49	24.45	24.10	23.34	22.89	21.86	20.68	19.89	

XXV.

India. — Bombay. *Lat.* 18° 56′ N. *Long.* 72° 54′ E. *Greenw.* — Dove.

Degrees of Reaumur.

Hour.	Jan.	Feb.	March.	April.	May.	June.	July.	Aug.	Sept.	Oct.	Nov.	Dec.	Year.
Midn.	1.76	1.68	1.43	1.40	1.30	0.80	0.57	0.59	0.92	1.36	1.74	1.93	1.29
1	1.91	1.88	1.65	1.54	1.40	0.89	0.65	0.64	0.98	1.52	1.80	2.00	1.40
2	2.04	2.04	1.80	1.75	1.54	0.88	0.63	1.16	1.09	1.62	1.97	2.18	1.56
3	2.18	2.22	1.90	1.92	1.69	0.94	0.65	0.81	1.18	1.74	2.11	2.28	1.63
4	2.39	2.44	2.26	2.02	1.81	1.04	0.76	0.82	1.25	1.89	2.23	2.41	1.78
5	2.65	2.68	2.42	2.26	1.92	1.09	0.83	0.90	1.25	1.96	2.40	2.62	1.92
6	2.88	2.88	2.60	2.20	1.65	1.03	0.84	0.84	1.21	2.00	2.55	2.66	1.94
7	2.53	2.37	1.61	0.76	0.44	0.60	0.55	0.51	0.61	1.02	1.47	2.08	1.21
8	0.72	0.48	−1.04	−0.62	−0.51	−0.01	0.02	0.08	−0.20	−0.31	−0.12	0.20	−0.11
9	−1.04	−1.05	−1.49	−1.53	−1.30	−0.46	−0.46	−0.45	−0.84	−1.53	−1.40	−1.00	−1.05
10	−2.40	−2.29	−2.28	−2.00	−1.73	−0.79	−0.74	−0.76	−1.32	−2.17	−2.38	−2.14	−1.75
11	−3.08	−2.98	−2.54	−2.20	−2.08	−1.18	−1.07	−1.12	−1.51	−2.38	−3.18	−2.94	−2.19

The numbers without sign must be added; those with the sign — must be subtracted.

XXV.

India. — Bombay, *Continued.*

Corrections to be applied to the Means of the Hours of Observation to obtain the true Mean Temperatures of the respective Days, Months, and of the Year. — Dove.

Degrees of Reaumur.

Hour.	Jan.	Feb.	March.	April.	May.	June.	July.	Aug.	Sept.	Oct.	Nov.	Dec.	Year.
Noon.	−3.40	−3.29	−2.52	−2.44	−2.32	−1.40	−1.09	−1.34	−1.72	−2.39	−3.26	−3.32	−2.37
1	−3.02	−3.12	−2.67	−2.53	−2.28	−1.50	−1.12	−1.35	−1.77	−2.22	−2.96	−3.35	−2.32
2	−2.78	−2.89	−2.56	−2.32	−2.14	−1.52	−0.97	−1.35	−1.55	−2.09	−2.55	−2.97	−2.14
3	−2.38	−2.54	−2.25	−2.05	−1.85	−1.31	−0.85	−1.09	−1.37	−1.79	−2.22	−2.59	−1.86
4	−1.96	−2.07	−1.72	−1.49	−1.36	−0.89	−0.63	−0.76	−0.95	−1.38	−1.55	−2.03	−1.40
5	−1.30	−1.41	−1.08	−0.96	−0.83	−0.49	−0.36	−0.34	−0.36	−0.61	−0.67	−1.09	−0.79
6	−0.64	−0.44	−0.16	0.00	0.09	−0.02	0.03	0.13	0.14	0.01	−0.14	−0.52	−0.13
7	−0.28	−0.07	0.19	0.43	0.63	0.22	0.21	0.26	0.28	0.30	0.09	−0.23	0.17
8	0.00	0.23	0.48	0.66	0.87	0.39	0.28	0.34	0.44	0.53	0.36	0.10	0.39
9	0.58	0.63	0.80	0.83	0.92	0.44	0.36	0.41	0.58	0.76	0.85	0.75	0.66
10	1.16	1.15	1.04	1.09	0.95	0.52	0.41	0.52	0.78	0.96	1.32	1.35	0.94
11	1.47	1.48	1.20	1.24	1.17	0.71	0.48	0.56	0.89	1.18	1.58	1.65	1.13
Mean.	18.38	19.30	21.00	22.50	23.43	22.35	21.67	21.45	21.42	22.08	21.28	19.54	

XXVI.

India. — Calcutta. *Lat.* 22° 33′ 5″ N. *Long.* 88° 19′ 2″ E. *Greenw.* — Dove.

Degrees of Reaumur.

Hour.	Jan.	Feb.	March.	April.	May.	June.	July.	Aug.	Sept.	Oct.	Nov.	Dec.	Year.
Midn.	1.86	1.69	2.06	1.60	1.90	1.12	0.69	0.69	0.71	1.00	1.24	1.51	1.34
1	2.24	2.00	2.37	1.96	2.06	1.12	0.80	0.78	0.76	1.17	1.47	1.77	1.54
2	2.53	2.22	2.62	2.18	2.21	1.16	0.91	0.85	0.84	1.26	1.69	2.00	1.71
3	2.80	2.44	2.84	2.27	2.32	1.29	1.02	0.92	0.93	1.26	1.82	2.31	1.85
4	3.06	2.71	3.08	2.40	2.41	1.29	1.11	0.96	1.04	1.46	2.00	2.40	1.99
5	3.33	2.89	3.28	2.47	2.50	1.34	1.24	1.07	1.16	1.53	2.22	2.66	2.14
6	3.53	3.11	3.42	2.53	2.41	1.34	1.24	1.12	1.16	1.62	2.36	2.80	2.22
7	3.71	3.24	3.42	2.22	1.90	1.03	0.96	0.89	0.93	0.86	2.31	2.93	2.03
8	2.73	2.20	1.97	1.18	0.81	0.45	0.42	0.32	0.27	0.31	0.93	1.68	1.11
9	0.91	0.71	0.46	0.11	−0.34	−0.13	−0.16	−0.22	−0.24	−0.47	−0.13	0.35	0.07
10	−0.78	−0.62	−0.98	−0.44	−1.39	−0.66	−0.69	−0.33	−0.73	−0.58	−1.02	−0.76	−0.75
11	−2.09	−1.64	−2.14	−1.82	−2.14	−1.15	−1.13	−1.08	−1.16	−1.60	−1.91	−1.87	−1.64
Noon.	−3.31	−2.62	−3.16	−2.67	−2.76	−1.60	−1.51	−1.51	−1.40	−1.94	−2.44	−2.80	−2.31
1	−4.14	−3.28	−3.87	−3.09	−3.12	−1.68	−1.58	−1.55	−1.44	−2.05	−2.80	−3.29	−2.66
2	−4.52	−3.64	−4.25	−3.47	−3.32	−1.73	−1.29	−1.80	−1.63	−2.12	−3.07	−3.69	−2.88
3	−4.65	−3.87	−4.40	−3.62	−3.43	−1.92	−1.24	−1.20	−1.27	−1.83	−2.98	−3.69	−2.84
4	−3.78	−3.69	−4.23	−3.40	−3.10	−1.53	−0.96	−0.95	−0.91	−1.49	−2.18	−2.76	−2.41
5	−3.07	−3.13	−3.36	−2.73	−2.43	−1.20	−0.64	−0.68	−0.56	−0.92	−1.60	−2.18	−1.88
6	−1.87	−1.91	−1.96	−1.42	−1.23	−0.57	−0.31	−0.31	−0.16	−0.25	−0.76	−1.34	−1.01
7	−0.96	−0.93	−0.78	−0.31	−0.14	−0.11	−0.07	−0.09	0.04	0.13	−0.22	−0.63	−0.34
8	−0.20	−0.22	0.00	0.40	0.68	0.20	0.09	0.25	0.22	0.42	0.27	−0.05	0.17
9	0.42	0.38	0.73	0.89	1.08	0.49	0.22	0.45	0.33	0.60	0.62	0.44	0.55
10	0.95	0.80	1.22	1.20	1.46	0.63	0.36	0.56	0.47	0.75	1.07	0.93	0.87
11	1.37	1.20	1.66	1.54	1.64	0.74	0.49	0.65	0.60	0.88	1.16	1.20	1.09
Mean.	15.49	17.57	21.19	22.51	24.01	23.29	22.68	22.86	22.42	21.73	18.88	16.36	

The numbers without sign must be added ; those with the sign — must be subtracted.

XXVII.

Asia. — Tiflis. *Lat.* 41° 41′ N. *Long.* 45° 17′ E. *Greenw.*

Corrections to be applied to the Means of the Hours of Observation to obtain the true Mean Temperatures of the respective Days, Months, and of the Year. — Dove.

Degrees of Reaumur.

Hour.	Jan.	Feb.	March.	April.	May.	June.	July.	Aug.	Sept.	Oct.	Nov.	Dec.	Year.
Midn.	0.87	1.01	1.54	1.81	1.95	2.38	2.43	2.22	1.60	1.38	0.99	0.80	1.58
1	1.02	1.15	1.80	2.10	2.28	2.67	2.79	2.52	1.81	1.64	1.16	0.94	1.82
2	1.17	1.33	2.02	2.40	2.58	2.94	3.13	2.82	2.08	1.88	1.37	1.04	2.06
3	1.32	1.47	2.23	2.64	2.84	3.22	3.49	3.13	2.29	2.11	1.59	1.14	2.28
4	1.46	1.57	2.39	2.94	3.14	3.43	3.73	3.44	2.59	2.39	1.73	1.25	2.51
5	1.60	1.69	2.58	3.12	3.09	3.09	3.55	3.59	2.74	2.62	1.85	1.35	2.57
6	1.76	1.75	2.63	2.89	2.39	2.35	2.77	3.06	2.63	2.77	1.99	1.40	2.37
7	1.87	1.75	2.14	2.19	1.53	1.28	1.50	2.16	1.99	2.38	1.85	1.42	1.84
8	1.40	1.23	1.23	0.99	0.53	0.35	0.70	1.05	1.07	1.52	1.44	1.19	1.06
9	0.05	0.50	0.16	−0.22	−0.51	−0.65	−0.32	−0.21	−0.03	0.30	0.54	0.49	0.01
10	−0.41	−0.46	−0.94	−1.20	−1.41	−1.66	−1.35	−1.32	−1.15	−0.47	−0.46	−0.19	−0.92
11	−1.17	−1.33	−1.85	−2.06	−2.19	−2.40	−2.27	−2.20	−2.01	−1.77	−1.31	−1.11	−1.81
Noon.	−1.91	−1.94	−2.64	−2.77	−2.89	−2.42	−2.99	−2.89	−2.67	−2.53	−2.07	−1.76	−2.46
1	−2.37	−2.45	−3.12	−3.29	−3.21	−3.42	−3.53	−3.60	−3.17	−3.07	−2.50	−2.21	−3.00
2	−2.59	−2.65	−3.25	−3.37	−3.34	−3.50	−3.68	−3.85	−3.41	−3.56	−2.81	−2.38	−3.20
3	−2.33	−2.58	−3.21	−3.41	−3.25	−3.51	−3.82	−3.98	−3.37	−3.41	−2.55	−2.08	−3.12
4	−1.78	−2.07	−2.78	−3.20	−2.97	−3.39	−3.82	−3.72	−2.95	−2.81	−1.87	−1.43	−2.73
5	−0.99	−1.24	−2.08	−2.46	−2.65	−2.86	−3.47	−3.20	−1.53	−1.85	−1.27	−0.90	−2.04
6	−0.57	−0.60	−1.11	−1.56	−1.47	−1.81	−2.36	−2.01	−1.18	−1.17	−0.73	−0.49	−1.26
7	−0.17	−0.19	−0.48	−0.69	−0.45	−0.63	−0.86	−0.85	−0.46	−0.50	−0.35	−0.13	−0.48
8	0.15	0.19	0.12	−0.02	0.26	0.23	0.13	−0.02	0.18	0.11	−0.02	0.19	0.12
9	0.33	0.44	0.51	0.64	0.83	0.92	0.87	0.72	0.61	0.50	0.24	0.36	0.58
10	0.55	0.65	0.91	1.05	1.28	1.51	1.44	1.33	1.00	0.81	0.48	0.53	0.96
11	0.69	0.89	1.25	1.45	1.63	1.95	1.96	1.80	1.32	1.10	0.76	0.68	1.29
Mean.	−0.20	3.00	5.64	9.99	13.54	16.10	19.01	19.43	15.03	11.40	5.07	2.45	

XXVIII.

China. — Peking. *Lat.* 39° 54′ N. *Long.* 116° 26′ E. *Greenw.* — Dove.

Degrees of Reaumur.

Hour.	Jan.	Feb.	March.	April.	May.	June	July.	Aug.	Sept.	Oct.	Nov.	Dec.	Year
Midn.	1.16	1.70	1.83	1.75	2.19	2.24	1.61	1.49	1.69	1.64	1.19	1.25	1.64
1	1.47	2.07	2.19	2.26	2.76	2.73	1.89	1.80	2.04	2.05	1.47	1.39	2.01
2	1.66	2.35	2.78	2.67	3.20	3.12	2.23	2.04	2.32	2.37	1.68	1.65	2.34
3	1.93	2.55	2.93	3.18	3.72	3.47	2.50	2.31	2.55	2.62	1.88	1.83	2.62
4	2.13	2.81	3.27	3.57	4.13	3.82	2.74	2.54	2.97	2.92	2.01	2.46	2.95
5	2.41	2.94	3.57	3.89	4.30	3.88	2.78	2.71	3.10	3.19	2.20	2.10	3.09
6	2.58	3.15	3.65	3.81	3.37	2.86	2.10	2.46	2.96	3.43	2.32	2.18	2.91
7	2.63	3.21	3.19	2.91	2.30	1.95	1.34	1.65	2.10	2.98	2.30	2.29	2.40
8	2.23	2.37	1.84	1.65	1.19	1.07	0.52	0.76	0.87	1.68	1.39	1.73	1.44
9	0.77	0.70	0.49	0.34	0.00	0.03	−0.12	−0.20	−0.24	0.15	0.19	0.31	0.20
10	−0.57	−0.65	−0.81	−0.79	−1.20	−1.06	−0.97	−1.09	−1.36	−1.05	−0.84	−0.97	−0.95
11	−1.35	−1.90	−1.93	−2.03	−1.24	−2.17	−1.71	−1.67	−2.17	−2.18	−1.74	−1.96	−1.84

The numbers without sign must be added; those with the sign — must be subtracted.

XXVIII.

China. — Peking, *Continued.*

Corrections to be applied to the Means of the Hours of Observation to obtain the true Mean Temperatures of the respective Days, Months, and of the Year. — Dove.

Degrees of Reaumur.

Hour.	Jan.	Feb.	March.	April.	May.	June.	July.	Aug.	Sept.	Oct.	Nov.	Dec.	Year.
Noon.	−2.83	−2.80	−2.95	−2.92	−3.05	−2.92	−2.24	−2.02	−2.77	−3.03	−2.39	−2.64	−2.71
1	−3.01	−3.54	−3.54	−3.59	−3.74	−3.55	−2.65	−2.64	−3.10	−3.65	−2.87	−3.18	−3.25
2	−3.37	−3.84	−4.03	−3.98	−4.08	−3.97	−2.88	−2.90	−3.38	−3.96	−3.07	−3.41	−3.57
3	−3.40	−3.94	−4.12	−4.06	−4.24	−4.00	−2.85	−2.94	−3.44	−3.97	−2.88	−2.74	−3.55
4	−2.88	−3.65	−3.92	−3.86	−4.03	−3.74	−2.74	−2.79	−3.06	−2.43	−2.23	−2.50	−3.15
5	−1.79	−2.83	−3.21	−3.24	−3.65	−3.31	−2.36	−2.20	−2.34	−2.34	−1.18	−1.34	−2.48
6	−0.97	−1.79	−2.20	−2.34	−3.04	−2.44	−1.76	−1.45	−1.18	−1.12	−0.59	−0.64	−1.63
7	−0.48	−0.15	−1.05	−1.13	−1.18	−1.21	−0.72	−0.45	−0.50	−0.54	−0.48	−0.26	−0.68
8	−0.02	−0.27	−0.30	−0.33	−0.19	−0.11	0.12	0.08	0.09	−0.02	0.01	0.18	−0.06
9	0.30	0.26	0.26	0.24	0.59	0.59	0.63	0.51	0.57	0.42	0.30	0.54	0.43
10	0.57	0.73	0.83	0.84	1.15	1.14	1.04	0.83	0.97	0.86	0.59	0.77	0.86
11	0.90	1.20	1.30	1.28	1.67	1.65	1.35	1.18	1.32	1.00	0.81	1.01	1.22
Mean.	−3.57	−2.04	3.42	9.66	15.83	19.61	21.27	19.30	15.68	9.61	1.79	−2.44	

XXIX.

Siberia. — Nertchinsk. *Lat.* 51° 18′ N. *Long.* 117° 20′ E. *Gr.* — Dove.

Degrees of Reaumur.

Hour.	Jan.	Feb.	March.	April.	May.	June.	July.	Aug.	Sept.	Oct.	Nov.	Dec.	Year.
Midn.	0.78	1.38	1.92	2.53	3.10	3.13	2.63	2.51	2.12	1.66	0.96	0.75	1.96
1	1.06	1.61	2.25	2.95	3.71	3.55	3.00	2.87	2.58	1.98	1.22	0.94	2.31
2	1.24	1.84	2.65	3.36	4.20	3.98	3.34	3.25	2.93	2.27	1.42	1.16	2.64
3	1.45	2.15	3.02	3.75	4.78	4.32	3.64	3.57	3.28	2.57	1.70	1.33	2.96
4	1.70	2.40	3.38	4.09	5.04	4.29	3.86	3.79	3.62	2.80	1.91	1.45	3.19
5	1.93	2.72	3.70	4.15	3.97	3.27	3.17	3.68	3.97	3.00	2.06	1.63	3.10
6	2.08	2.94	3.89	2.96	2.31	2.03	1.99	2.61	3.63	3.16	2.15	1.76	2.63
7	2.26	3.00	2.88	1.43	0.82	0.74	1.01	1.31	2.07	2.46	2.35	1.95	1.86
8	2.20	1.82	1.36	0.19	−0.53	−0.45	−1.28	0.11	0.66	0.84	1.61	1.98	0.71
9	0.56	−0.20	−0.12	−1.32	−1.77	−1.59	−1.25	−1.08	−0.72	−0.69	−0.03	0.62	−0.63
10	−0.96	−1.27	−1.71	−2.35	−2.73	−2.52	−2.13	−2.10	−1.99	−1.82	−1.17	−0.89	−1.80
11	−1.90	−2.34	−2.61	−3.08	−3.34	−3.17	−2.79	−2.91	−2.94	−2.78	−2.12	−1.85	−2.65
Noon.	−2.70	−3.16	−3.43	−3.70	−3.82	−3.62	−3.28	−3.49	−3.71	−3.41	−2.84	−2.58	−3.31
1	−3.06	−3.75	−3.96	−4.01	−4.08	−3.80	−3.58	−3.76	−4.09	−3.75	−3.09	−2.85	−3.65
2	−3.00	−3.80	−4.23	−4.08	−4.10	−3.73	−3.66	−3.92	−4.20	−3.66	−2.97	−2.52	−3.66
3	−2.50	−3.47	−4.03	−3.84	−3.99	−3.59	−3.48	−3.79	−3.86	−3.26	−2.27	−1.87	−3.33
4	−1.54	−2.73	−3.53	−3.48	−3.55	−3.24	−3.02	−3.21	−3.34	−2.43	−1.34	−0.96	−2.70
5	−0.71	−1.61	−2.75	−2.85	−3.02	−3.73	−2.38	−2.56	−2.48	−1.42	−0.87	−0.43	−1.98
6	−0.28	−0.63	−1.71	−1.97	−2.27	−2.06	−1.73	−1.68	−1.22	−0.50	−0.10	−0.17	−1.20
7	0.02	0.01	−0.34	−0.34	−0.93	−0.93	−0.82	−0.66	−0.49	−0.24	−0.17	−0.70	−0.47
8	0.13	0.39	0.24	0.61	0.27	0.97	0.37	0.41	0.34	0.30	0.06	0.08	0.29
9	0.27	0.63	0.66	1.19	1.34	1.32	1.24	1.30	0.89	0.64	0.34	0.22	0.84
10	0.43	0.86	1.06	1.72	1.92	2.02	1.78	1.70	1.30	1.01	0.54	0.43	1.23
11	0.57	1.16	1.47	2.17	2.63	2.63	2.29	2.14	1.71	1.31	0.75	0.56	1.62
Mean.	−21.94	−17.84	−8.35	0.04	7.51	1.78	13.91	11.91	6.55	−1.80	−13.44	−21.36	

The numbers without sign must be added; those with the sign — must be subtracted.

XXX.

Siberia. — Nertchinsk. *Lat.* 51° 18′ N. *Long.* 119° 21′ E. *Greenw.*

Corrections to be applied to the Means of the Hours of Observation to obtain the true Mean Temperatures of the respective Days, Months, and of the Year. — Dove.

Degrees of Reaumur.

Hours.	Jan.	Feb.	March.	April.	May.	June.	July.	Aug.	Sept.	Oct.	Nov.	Dec.	Mean.
Morn. 1	0.91	1.42	2.07	2.69	4.07	4.29	3.07	3.00	2.16	2.31	0.76	0.66	2.28
2	1.00	1.68	2.57	3.29	4.69	4.71	3.46	3.48	2.96	2.79	0.96	0.74	2.69
3	1.15	2.08	3.16	3.78	5.08	4.90	3.75	3.89	3.27	3.26	1.26	0.84	3.04
4	1.42	2.52	3.63	3.97	4.98	4.70	3.76	4.04	3.81	3.61	1.66	1.07	3.26
5	1.78	2.84	3.73	3.69	4.24	3.96	3.37	3.72	3.94	3.66	2.06	1.41	3.20
6	2.07	2.80	3.28	2.88	2.86	2.67	2.54	2.89	3.15	3.30	2.30	1.75	2.71
7	2.06	2.28	2.31	1.63	1.07	0.99	1.37	1.62	2.38	2.47	2.18	1.87	1.85
8	1.60	1.28	0.99	0.16	0.78	−0.79	0.06	0.15	0.87	1.24	1.58	1.59	0.66
9	0.65	−0.05	−0.41	−1.26	−2.33	−2.34	−1.19	−1.25	−0.70	−0.23	0.55	0.87	−0.64
10	−0.59	−1.43	−1.67	−2.42	−3.40	−3.41	−1.98	−2.38	−1.74	−1.70	−0.69	−0.17	−1.80
11	−1.79	−2.58	−2.64	−3.22	−3.98	−3.97	−2.92	−3.15	−2.99	−2.96	−1.84	−1.23	−2.77
Noon. . .	−2.61	−3.29	−3.25	−3.64	−4.19	−4.12	−3.38	−3.61	−3.49	−3.84	−2.60	−2.01	−3.34
1	−2.87	−3.49	−3.61	−3.76	−4.22	−4.05	−3.64	−3.83	−3.69	−4.25	−2.81	−2.30	−3.54
2	−2.56	−3.27	−3.74	−3.65	−4.18	−3.92	−3.72	−3.88	−4.00	−4.20	−2.50	−2.08	−3.48
3	−1.89	−2.76	−3.65	−3.33	−4.03	−3.77	−3.62	−3.75	−3.54	−3.77	−1.87	−1.54	−3.13
4	−1.14	−2.12	−3.31	−2.84	−3.69	−3.54	−3.29	−3.40	−3.24	−3.08	−1.17	−0.92	−2.65
5	−0.56	−1.45	−2.65	−2.17	−3.04	−3.07	−2.68	−2.76	−2.68	−2.24	−0.61	−0.47	−2.03
6	−0.23	−0.81	−1.78	−1.39	−2.08	−2.30	−1.82	−1.86	−1.54	−1.36	−0.27	−0.25	−1.31
7	−0.11	−0.21	−0.77	−0.56	−0.92	−1.23	−0.81	−0.80	−0.86	−0.54	−0.12	−0.23	−0.60
8	−0.04	0.31	0.18	0.20	0.26	0.00	0.20	0.24	0.17	0.17	−0.25	−0.24	0.12
9	0.09	0.74	0.90	0.82	1.29	1.21	1.06	1.11	0.97	0.74	0.05	−0.17	0.73
10	0.31	1.02	1.34	1.29	2.11	2.25	1.51	1.74	1.17	1.18	0.20	0.02	1.18
11	0.57	1.19	1.56	1.71	2.78	3.09	2.23	2.19	1.73	1.54	0.39	0.28	1.61
Midn. . .	0.78	1.29	1.76	2.15	3.41	3.75	2.65	2.57	1.88	1.90	0.58	0.52	1.94
6. 6	0.92	1.00	0.75	0.75	0.39	0.19	0.36	0.52	0.80	0.97	1.01	0.75	0.70
7. 7	0.98	1.04	0.77	0.53	0.07	−0.12	0.28	0.41	0.76	0.97	1.03	0.82	0.63
8. 8	0.78	0.80	0.58	0.18	−0.26	−0.39	0.13	0.20	0.52	0.71	0.77	0.67	0.39
9. 9	0.37	0.34	0.24	−0.22	−0.52	−0.56	−0.06	−0.07	0.13	0.26	0.30	0.35	0.05
10.10	−0.14	−0.20	−0.16	−0.57	−0.65	−0.58	−0.24	−0.32	−0.29	−0.26	−0.25	−0.07	−0.31
7. 2. 9	−0.14	−0.08	−0.18	−0.40	−0.61	−0.57	−0.43	−0.38	−0.22	−0.33	−0.09	−0.13	−0.30
6. 2. 8	−0.18	−0.05	−0.09	−0.19	−0.35	−0.42	−0.33	−0.25	−0.23	−0.24	−0.08	−0.19	−0.22
6. 2.10	−0.06	0.18	0.29	0.17	0.26	0.33	0.11	0.25	0.11	0.09	0.00	−0.01	0.14
6. 2. 6	−0.24	−0.43	−0.75	−0.72	−1.13	−1.18	−1.00	−0.95	−0.80	−0.75	−0.16	−0.19	−0.69
7. 2	−0.41	−0.61	−0.65	−0.07	−1.58	−1.53	−1.14	−1.11	−0.66	−0.89	−0.32	−0.22	−0.85
8. 2	−0.94	−1.34	−1.60	−1.85	−2.32	−2.26	−1.78	−1.88	−1.69	−1.78	−0.97	−0.71	−1.59
8. 1	−0.08	0.12	0.09	−0.10	−0.13	−0.13	−0.06	−0.01	0.08	−0.06	−0.06	−0.14	−0.04
7. 1	−0.25	−0.50	−0.72	−1.01	−1.56	−1.47	−1.18	−1.13	−0.81	−0.87	−0.16	−0.11	−0.81
9.12.3.9	−0.48	−1.00	−1.38	−1.75	−2.48	−2.36	−1.83	−1.87	−1.57	−1.48	−0.46	−0.25	−1.41
7. 2.2(9)	−0.64	−1.11	−1.31	−1.80	−2.50	−2.42	−1.79	−1.84	−1.41	−1.51	−0.62	−0.36	−1.44
Dail. ext.	−0.40	−0.33	−0.01	−0.11	0.43	0.39	0.02	0.08	−0.03	−0.30	−0.26	−0.22	−0.14

The numbers without sign must be added; those with the sign — must be subtracted.

XXXI.

Siberia. — Barnaul. *Lat.* 53° 20′ N. *Long.* 83° 27′ E. *Greenw.*

Corrections to be applied to the Means of the Hours of Observation to obtain the true Mean Temperatures of the respective Days, Months, and of the Year. — Dove.

Degrees of Fahrenheit.

Hours.	Jan.	Feb.	March.	April.	May.	June.	July.	Aug.	Sept.	Oct.	Nov.	Dec.	Mean.
Morn. 1	2.54	1.85	4.70	5.49	8.82	7.83	8.37	7.11	5.45	3.06	2.48	1.82	4 95
2	2.81	2.14	5.47	6.30	10.19	8.87	9.77	8.35	6.50	3.78	2.97	2.00	5.76
3	2.70	2.48	6.28	7.07	10.96	9.59	10.69	9.52	7.65	4.52	3.35	2.07	6.41
4	2.39	2.81	7.02	7.45	10.76	9.14	10.67	10.15	8.48	5.15	3.71	2.18	6.66
5	2.07	3.13	7.43	7.09	9.32	7.58	9.50	9.77	8.60	5.47	4.01	2.45	6.37
6	1.96	3.33	9.38	5.87	6.68	5.45	7.18	8.12	5.58	5.29	4.16	2.79	5.65
7	2.00	3.20	5.90	3.87	3.38	2.50	4.05	5.36	2.70	4.46	3.96	2.99	3.94
8	1.98	2.59	3.71	1.37	−0.11	−0.18	0.70	1.96		2.97	3.15	2.70	1.96
9	1.53	1.37	0.86	−1.28	−3.02	−2.48	−2.32	−1.44	−0.56	0.99	1.64	1.73	−0.25
10	0.45	−0.36	−2.18	−3.74	−5.06	−4.61	−4.68	−4.32	−3.67	−1.22	−0.41	0.11	−2.48
11	−1.22	−2.30	−4.91	−5.78	−6.35	−5.99	−6.35	−6.48	−6.21	−3.31	−2.61	−1.76	−4.43
Noon. . .	−3.08	−4.03	−6.89	−7.34	−7.20	−7.31	−7.52	−7.97	−7.99	−5.00	−4.48	−3.42	−6.01
1	−4.59	−5.13	−7.97	−8.35	−8.03	−8.39	−8.42	−8.96	−8.96	−6.05	−5.58	−4.39	−7.07
2	−5.27	−5.38	−8.21	−8.71	−8.78	−8.78	−9.16	−9.63	−9.23	−6.39	−5.72	−4.48	−7.47
3	−4.93	−4.77	−7.76	−8.39	−9.41	−8.91	−9.56	−9.88	−8.82	−6.05	−5.02	−3.78	−7.27
4	−3.78	−3.56	−6.84	−7.34	−9.50	−8.01	−9.36	−9.50	−7.81	−5.22	−3.85	−2.68	−6.46
5	−2.25	−2.14	−5.65	−5.58	−8.66	−6.32	−8.35	−8.28	−6.26	−4.05	−2.57	−1.60	−5.15
6	−0.90	−0.83	−6.46	−3.35	−6.82	−4.39	−6.48	−6.19	−4.25	−2.75	−1.55	−0.83	−3.74
7	0.02	0.09	−2.61	−1.04	−4.16	−1.94	−4.01	−3.51	−2.07	−1.49	−0.86	−0.43	−1.82
8	0.47	0.63	−0.97	1.04	−1.31	0.11	−1.31	−0.68	0.02	−0.36	−0.41	−0.23	−0.25
9	0.70	0.92	0.63	2.61	1.46	1.80	1.24	1.80	1.76	0.54	0.00	0.00	1.13
10	0.95	1.10	2.00	3.62	3.78	3.49	3.38	3.67	2.99	1.28	0.52	0.38	2.27
11	1.42	1.28	3.13	4.25	5.69	4.75	5.20	4.97	3.85	1.87	1.15	0.92	3.22
Midn. . .	2.03	1.55	3.98	4.82	7.36	6.26	6.82	6.03	4.59	2.45	1.85	1.44	4.10
6. 6	0.54	1.24	1.46	1.26	−0.07	0.54	0.34	0.97	1.69	1.28	1.31	0 99	0.97
7. 7	1.01	1.64	1.64	1.42	−0.41	0.27	0.02	0.92	1.76	1.49	1.55	1 28	1.06
8. 8	1.24	1.62	1.37	1 22	−0.72	−0 05	−0.29	0.65	1.35	1.31	1.37	1.24	0.86
9. 9	1.10	1.15	0.74	0.68	−0.79	−0.34	−0.54	0.18	0.59	0.77	0.83	0.86	0.43
10.10	0.70	0.38	−0.09	−0.07	−0.63	−0.56	−0.65	−0.34	−0.34	0.05	0.07	0.25	−0.11
7. 2. 9	−0.86	−0.43	−0.56	−0.74	−1.31	−1.49	−1.28	−0.83	−0.63	−0.47	−0.59	−0.50	−0.81
6. 2. 8	−0.95	−0.47	0.07	−0.61	−1.13	−1.08	−1.10	−0.72	−0.52	−0.50	−0 65	−0.63	−0.70
6. 2.10	−0.79	−0.32	1.06	0.27	0.56	0 05	0.47	0.72	0.47	0.07	−0.34	−0.43	0.16
6. 2. 6	−1.10	−0.97	−1.76	−2.07	−2.97	−2.57	−2.81	−2.57	−1.94	−1.28	−1.04	0 83	−1.85
7. 2	−1.64	−1.09	−1.16	−2.42	−2.70	−3.14	−2.56	−2.14	−1.83	−0 97	−0.88	−0.75	−1.77
8. 2	−1.65	−1.40	−2 25	−3.67	−4.45	−4.48	−4.23	−3.84	−3.27	−1.71	−1.29	−0.89	−2.76
8. 1	−1.31	−1.27	−2.13	−3.49	−4.07	−4.29	−3.86	−3.50	−3.13	−1.54	−1.22	−0.85	−2.56
7. 1	−1.30	−0.97	−1.04	−2.24	−2 33	−2.95	−2.19	−1.80	−1.69	−0.80	−0.81	−0.70	−1.57
9.12.3.9	−1.45	−1.62	−3.29	−3.60	−4.55	−4.23	−4.55	−4.37	−3.92	−2.39	−1.96	−1.37	−3.11
7. 2.2(9)	−0.47	−0.09	−0.27	0.09	−0.63	−0.68	−0.65	−0.18	−0.05	−0.23	−0.45	−0.38	−0.34
Dail. ext.	−1.24	−1.04	0.59	−0.63	0.74	0.34	0.56	0.14	−0 32	−0.47	−0 79	−0.74	−0.41

The numbers without sign must be added; those with the sign — must be subtracted.

E

XXXII.

Siberia. — Barnaul. *Lat.* 53° 20′ N. *Long.* 83° 27′ E. *Greenw.*

Corrections to be applied to the Means of the Hours of Observation to obtain the true Mean Temperatures of the respective Days, Months, and of the Year. — Dove.

Degrees of Reaumur.

Hours.	Jan.	Feb.	March.	April.	May.	June.	July.	Aug.	Sept.	Oct.	Nov.	Dec.	Mean.
Morn. 1	1.13	0.82	2.09	2.44	3.92	3.48	3.72	3.16	2.42	1.36	1.10	0.81	2.20
2	1.25	0.95	2.43	2.80	4.53	3.94	4.34	3.71	2.89	1.68	1.32	0.89	2.56
3	1.20	1.10	2.79	3.14	4.87	4.26	4.73	4.23	3.40	2.01	1.49	0.92	2.85
4	1.06	1.25	3.12	3.31	4.78	4.06	4.74	4.51	3.77	2.29	1.65	0.97	2.96
5	0.92	1.39	3.30	3.15	4.14	3.37	4.22	4.34	3.82	2.43	1.78	1.09	2.83
6	0.87	1.48	4.17	2.61	2.97	2.42	3.19	3.61	3.40	2.35	1.85	1.24	2.51
7	0.89	1.42	2.62	1.72	1.50	1.11	1.80	2.38	2.48	1.98	1.76	1.33	1.75
8	0.88	1.15	1.65	0.61	−0.05	−0.08	0.31	0.87	1.20	1.32	1.40	1.20	0.87
9	0.68	0.61	0.38	−0.57	−1.34	−1.10	−1.03	−0.64	−0.25	0.44	0.73	0.77	−0.11
10	0.20	−0.16	−0.97	−1.66	−2.25	−2.05	−2.08	−1.92	−1.63	−0.54	−0.18	0.05	−1.10
11	−0.54	−1.02	−2.18	−2.57	−2.82	−2.66	−2.82	−2.88	−2.76	−1.47	−1.16	−0.78	−1.97
Noon. . .	−1.37	−1.79	−3.06	−3.26	−3.20	−3.25	−3.34	−3.54	−3.55	−2.22	−1.99	−1.52	−2.67
1	−2.04	−2.28	−3.54	−3.71	−3.57	−3.73	−3.74	−3.98	−3.98	−2.69	−2.48	−1.95	−3.14
2	−2.34	−2.39	−3.65	−3.87	−3.90	−3.90	−4.07	−4.28	−4.10	−2.84	−2.54	−1.99	−3.32
3	−2.19	−2.12	−3.45	−3.73	−4.18	−3.96	−4.25	−4.39	−3.92	−2.69	−2.23	−1.68	−3.23
4	−1.68	−1.58	−3.04	−3.26	−4.22	−3.56	−4.16	−4.22	−3.47	−2.32	−1.71	−1.19	−2.87
5	−1.00	−0.95	−2.51	−2.48	−3.85	−2.81	−3.71	−3.68	−2.78	−1.80	−1.14	−0.71	−2.29
6	−0.40	−0.37	−2.87	−1.49	−3.03	−1.95	−2.88	−2.75	−1.89	−1.22	−0.69	−0.37	−1.66
7	0.01	0.04	−1.16	−0.46	−1.85	−0.86	−1.78	−1.56	−0.92	−0.66	−0.38	−0.19	−0.81
8	0.21	0.28	−0.43	0.46	−0.58	0.05	−0.58	−0.30	0.01	−0.16	−0.18	−0.10	−0.11
9	0.31	0.41	0.28	1.16	0.65	0.80	0.55	0.80	0.78	0.24	0.00	0.00	0.50
10	0.42	0.49	0.89	1.61	1.68	1.55	1.50	1.63	1.33	0.57	0.23	0.17	1.01
11	0.63	0.57	1.39	1.89	2.53	2.11	2.31	2.21	1.71	0.83	0.51	0.41	1.43
Midn. . .	0.90	0.69	1.77	2.14	3.27	2.78	3.03	2.68	2.04	1.09	0.82	0.64	1.82
6. 6	0.24	0.55	0.65	0.56	−0.03	0.24	0.15	0.43	0.75	0.57	0.58	0.44	0.43
7. 7	0.45	0.73	0.73	0.63	−0.18	0.12	0.01	0.41	0.78	0.66	0.69	0.57	0.47
8. 8	0.55	0.72	0.61	0.54	−0.32	−0.02	−0.13	0.29	0.60	0.58	0.61	0.55	0.38
9. 9	0.49	0.51	0.33	0.30	−0.35	−0.15	−0.24	0.08	0.26	0.34	0.37	0.38	0.19
10.10	0.31	0.17	−0.04	−0.03	−0.28	−0.25	−0.29	−0.15	−0.15	0.02	0.03	0.11	−0.05
7. 2. 9	−0.38	−0.19	−0.25	−0.33	−0.58	−0.66	−0.57	−0.37	−0.28	−0.21	−0.26	−0.22	−0.36
6. 2. 8	−0.42	−0.21	0.03	−0.27	−0.50	−0.48	−0.49	−0.32	−0.23	−0.22	−0.29	−0.28	−0.31
6. 2.10	−0.35	−0.14	0.47	0.12	0.25	0.02	0.21	0.32	0.21	0.03	−0.15	−0.19	0.07
6. 2. 6	−0.62	−0.43	−0.78	−0.92	−1.32	−1.14	−1.25	−1.14	−0.86	−0.57	−0.46	−0.37	−0.82
7. 2	−0.73	−0.49	−0.52	−1.80	−1.20	−1.40	−1.14	−0.95	−0.81	−0.43	−0.39	−0.33	−0.79
8. 2	−0.73	−0.62	−1.00	−1.63	−1.98	−1.99	−1.88	−1.71	−1.45	−0.76	−0.57	−0.40	−1.23
8. 1	−0.58	−0.57	−0.95	−1.55	−1.81	−1.91	−1.72	−1.56	−1.39	−0.69	−0.54	−0.38	−1.14
7. 1	−0.58	−0.43	−0.46	−1.00	−1.04	−1.31	−0.97	−0.80	−0.75	−0.36	−0.36	−0.31	−0.70
9.12.3.9	−0.64	−0.72	−1.46	−1.60	−2.02	−1.88	−2.02	−1.94	−1.74	−1.06	−0.87	−0.61	−1.38
7. 2.2(9)	−0.21	−0.04	−0.12	0.04	−0.28	−0.30	−0.29	−0.08	−0.02	−0.10	−0.20	−0.17	−0.15
Dail. ext.	−0.55	−0.46	0.26	−0.28	0.33	0.15	0.25	0.06	−0.14	−0.21	−0.35	−0.33	−0.18

The numbers without sign must be added; those with the sign — must be subtracted.

XXXIII.

SIBERIA. — BARNAUL. *Lat.* 53° 20′ N. *Long.* 83° 27′ E. *Greenw.*

Corrections to be applied to the Means of the Hours of Observation to obtain the true Mean Temperatures of the respective Days, Months, and of the Year. — DOVE.

Degrees of Reaumur.

Hour.	Jan.	Feb.	March.	April.	May.	June.	July.	Aug.	Sept.	Oct.	Nov.	Dec.	Year.
Midn.	0.99	1.98	2.43	2.65	3.70	3.75	3.48	3.10	2.80	1.99	1.06	0.77	2.39
1	1.15	2.21	2.77	3.03	4.11	4.30	4.07	3.50	3.20	2.24	1.22	0.86	2.72
2	1.26	2.36	3.13	3.24	4.47	4.83	4.49	3.90	3.63	2.50	1.39	0.95	3.00
3	1.41	2.47	3.34	3.49	4.72	4.95	4.77	4.29	3.92	2.69	1.46	1.01	3.21
4	1.56	2.56	3.61	3.59	4.20	4.41	4.40	4.23	4.11	2.89	1.51	1.07	3.18
5	1.55	2.68	3.70	2.78	2.85	3.12	3.34	3.60	3.90	2.91	1.57	1.10	2.76
6	1.61	2.69	2.90	1.58	1.44	1.75	1.88	2.29	3.06	2.68	1.59	1.09	2.05
7	1.53	2.30	1.63	0.46	0.28	0.49	0.50	0.85	1.54	1.84	1.50	1.18	1.17
8	0.94	1.15	0.13	−0.69	−0.80	−0.65	−0.54	−0.51	−0.08	0.87	0.95	0.93	0.14
9	0.27	−0.47	−1.35	−1.80	−1.94	−1.78	−1.81	−1.79	−1.62	−0.73	−0.03	0.11	−1.08
10	−0.79	−1.90	−2.36	−2.68	−2.71	−2.75	−2.70	−2.80	−2.84	−1.96	−1.12	−0.83	−2.12
11	−1.69	−2.95	−3.31	−3.27	−3.39	−3.39	−3.44	−3.41	−3.75	−2.81	−1.93	−1.62	−2.91
Noon.	−2.35	−3.89	−3.78	−3.66	−3.73	−3.98	−3.90	−3.81	−4.19	−3.48	−2.42	−2.04	−3.44
1	−2.61	−4.25	−4.11	−3.68	−4.04	−4.19	−4.09	−4.11	−4.41	−3.72	−2.57	−2.12	−3.66
2	−2.39	−4.23	−4.07	−3.65	−4.13	−4.34	−4.21	−4.10	−4.34	−3.64	−2.39	−1.70	−3.60
3	−1.88	−3.62	−3.69	−3.39	−4.09	−4.19	−3.89	−3.91	−4.11	−3.17	−1.66	−1.09	−3.22
4	−1.19	−2.30	−2.67	−2.62	−3.51	−3.57	−3.65	−3.68	−3.21	−2.53	−1.05	−0.76	−2.56
5	−0.81	−1.30	−1.69	−1.82	−3.09	−3.04	−3.07	−2.78	−2.29	−1.49	−0.71	−0.53	−1.89
6	−0.41	−0.56	−0.84	−0.62	−1.92	−2.19	−2.09	−1.54	−1.05	−0.72	−0.33	−0.28	−1.05
7	−0.20	0.09	0.35	0.27	−0.46	−0.84	−0.69	−0.20	−0.17	−0.08	−0.03	−0.02	−0.17
8	0.12	0.69	0.39	0.99	0.77	0.51	0.52	0.67	0.60	0.31	0.23	0.19	0.50
9	0.32	1.08	0.88	1.50	1.64	1.48	1.42	1.46	1.26	0.82	0.42	0.39	1.06
10	0.73	1.47	1.46	2.02	2.42	2.31	2.22	2.04	1.85	1.29	0.58	0.58	1.58
11	0.78	1.76	1.92	2.35	3.11	3.05	2.88	2.58	2.36	1.68	0.83	0.75	2.00
Mean.	−14.71	−13.47	−5.47	1.77	7.78	13.62	14.98	12.76	7.53	1.58	−8.36	−13.07	4.94

The numbers without sign must be added; those with the sign — must be subtracted.

HOURLY CORRECTIONS

FOR

PERIODIC VARIATIONS.

EUROPE.

XXXIV.

Italy. — Rome. *Lat.* 41° 54′ N. *Long.* 12° 25′ E. *Greenw.*

Corrections to be applied to the Means of the Hours of Observation to obtain the true Mean Temperatures of the respective Days, Months, and of the Year. — Dove.

Degrees of Reaumur.

Hours.	Jan.	Feb.	March.	April.	May.	June.	July.	Aug.	Sept.	Oct.	Nov.	Dec.	Mean.
Morn. 1	0.90	1.08	1.22	1.55	1.88	2.44	2.17	2.20	1.63	1.50	1.15	0.93	1.55
2	0.99	1.26	1.50	1.84	2.10	2.59	2.41	2.49	1.91	1.75	1.29	1.02	1.76
3	1.14	1.58	1.96	2.31	2.56	3.02	2.99	3.00	2.38	2.12	1.53	1.19	2.15
4	1.36	1.99	2.46	2.80	3.06	3.51	3.68	3.54	2.91	2.58	1.87	1.43	2.60
5	1.60	2.36	2.80	3.07	3.30	3.71	4.06	3.79	3.25	2.96	2.22	1.70	2.90
6	1.77	2.52	2.76	2.92	3.04	3.36	3.81	3.53	3.17	3.10	2.42	1.87	2.86
7	1.74	2.33	2.24	2.25	2.19	2.38	2.82	2.62	2.58	2.82	2.33	1.83	2.34
8	1.40	1.73	1.29	1.15	0.93	0.98	1.27	1.22	1.51	2.05	1.82	1.47	1.40
9	0.72	0.78	0.10	−0.15	−0.47	−0.51	−0.44	−0.35	0.15	0.86	0.93	0.78	0.20
10	−0.24	−0.38	−1.08	−1.39	−1.68	−1.75	−1.89	−1.78	−1.23	−0.58	−0.22	−0.15	−1.03
11	−1.27	−1.54	−2.06	−2.36	−2.53	−2.59	−2.87	−2.84	−2.41	−2.00	−1.41	−1.14	−2.09
Noon. . .	−2.15	−2.49	−2.71	−2.98	−3.01	−3.08	−3.38	−3.49	−3.24	−3.14	−2.39	−1.99	−2.84
1	−2.69	−3.07	−3.02	−3.27	−3.23	−3.40	−3.61	−3.81	−3.70	−3.82	−3.00	−2.52	−3.26
2	−2.78	−3.25	−3.04	−3.28	−3.31	−3.70	−3.76	−3.92	−3.80	−3.99	−3.16	−2.66	−3.39
3	−2.44	−3.03	−2.84	−3.10	−3.31	−3.97	−3.89	−3.87	−3.59	−3.69	−2.93	−2.44	−3.26
4	−1.83	−2.51	−2.45	−2.72	−3.14	−4.05	−3.88	−3.62	−3.11	−3.04	−2.41	−1.95	−2.89
5	−1.11	−1.81	−1.89	−2.15	−2.70	−3.70	−3.53	−3.05	−2.38	−2.21	−1.76	−1.35	−2.30
6	−0.45	−1.05	−1.20	−1.39	−1.91	−2.79	−2.67	−2.18	−1.48	−1.32	−1.09	−0.75	−1.52
7	0.05	−0.34	−0.44	−0.53	−0.84	−1.42	−1.38	−1.01	−0.51	−0.50	−0.48	−0.24	−0.64
8	0.39	0.25	0.26	0.30	0.29	0.13	0.08	0.21	0.38	0.19	0.05	0.17	0.23
9	0.59	0.67	0.78	0.94	1.22	1.46	1.33	1.22	1.05	0.71	0.46	0.46	0.91
10	0.71	0.90	1.07	1.31	1.76	2.29	2.10	1.86	1.43	1.05	0.76	0.66	1.33
11	0.78	0.99	1.15	1.44	1.93	2.57	2.33	2.11	1.54	1.24	0.95	0.79	1.49
Midn. . .	0.84	1.02	1.15	1.46	1.88	2.51	2.24	2.14	1.55	1.36	1.06	0.86	1.51
6. 6	0.66	0.74	0.78	0.76	0.57	0.28	0.57	0.68	0.85	0.89	0.67	0.56	0.67
7. 7	0.90	1.00	0.90	0.86	0.68	0.48	0.72	0.80	1.03	1.16	0.92	0.80	0.85
8. 8	0.89	0.99	0.77	0.72	0.61	0.55	0.67	0.71	0.95	1.12	0.94	0.82	0.81
9. 9	0.65	0.72	0.44	0.40	0.37	0.48	0.45	0.43	0.60	0.78	0.70	0.62	0.55
10.10	0.24	0.26	−0.01	−0.04	0.04	0.27	0.10	0.04	0.10	0.23	0.27	0.26	0.15
7. 2. 9	−0.15	−0.08	−0.01	−0.03	0.03	0.05	0.13	−0.03	−0.06	−0.15	−0.12	−0.12	−0.05
6. 2. 8	−0.21	−0.16	−0.01	−0.02	0.01	−0.07	0.04	−0.06	−0.08	−0.23	−0.23	−0.21	−0.10
6. 2.10	−0.10	0.06	0.26	0.32	0.50	0.65	0.72	0.49	0.27	0.05	0.01	−0.04	0.27
6. 2. 6	−0.49	−0.59	−0.49	−0.58	−0.73	−1.04	−0.87	−0.86	−0.70	−0.74	−0.61	−0.51	−0.68
7. 2	−0.52	−0.46	−0.40	−0.52	−0.56	−0.66	−0.47	−0.65	−0.61	−0.59	−0.42	−0.42	−0.52
8. 2	−0.69	−0.76	−0.88	−1.07	−1.19	−1.36	−1.25	−1.35	−1.15	−0.97	−0.67	−0.60	−1.00
8. 1	−0.65	−0.67	−0.87	−1.06	−1.15	−1.21	−1.17	−1.30	−1.10	−0.89	−0.59	−0.53	−0.93
7. 1	−0.48	−0.37	−0.39	−0.51	−0.52	−0.51	−0.40	−0.60	−0.56	−0.50	−0.34	−0.35	−0.46
9.12.3.9	−0.82	−1.02	−1.17	−1.32	−1.39	−1.53	−1.60	−1.62	−1.41	−1.32	−0.98	−0.80	−1.25
7. 2.2(9)	0.04	0.11	0.19	0.21	0.33	0.40	0.43	0.29	0.22	0.06	0.02	0.02	0.19
Dail.ext.	−0.51	−0.37	−0.12	−0.11	−0.01	−0.17	0.09	−0.07	−0.28	−0.45	−0.37	−0.40	−0.25

The numbers without sign must be added; those with the sign — must be subtracted.

XXXV.

Italy. — Padua. *Lat.* 45° 24′ N. *Long.* 11° 52′ E. *Greenw.*

Corrections to be applied to the Means of the Hours of Observation to obtain the true Mean Temperatures of the respective Days, Months, and of the Year. — Dove.

Degrees of Reaumur.

Hours.	Jan.	Feb.	March.	April.	May.	June.	July.	Aug.	Sept.	Oct.	Nov.	Dec.	Mean.
Morn. 1	0.58	0.57	0.89	1.23	2.43	2.21	2.86	2.27	1.59	0.86	1.04	0.83	1.45
2	0.58	0.81	1.20	1.49	2.70	2.40	3.20	2.70	1.85	1.03	1.16	0.96	1.67
3	0.76	0.97	1.42	1.66	3.00	2.68	3.53	3.05	2.10	1.20	1.26	0.98	1.88
4	0.79	1.13	1.68	1.97	3.14	2.71	3.78	3.44	2.34	1.39	1.35	1.05	2.06
5	1.06	1.31	1.89	2.26	2.97	2.39	3.34	3.44	2.66	1.58	1.42	1.12	2.12
6	1.13	1.46	2.06	2.22	1.96	1.22	2.07	2.93	2.54	1.54	1.49	1.16	1.82
7	1.25	1.58	1.86	1.82	0.66	0.08	0.56	1.82	1.78	1.37	1.58	1.23	1.30
8	1.07	1.42	0.66	1.03	−0.23	−0.65	−0.25	0.58	0.79	0.81	0.97	1.00	0.60
9	0.70	0.82	0.61	0.18	−1.07	−1.24	−1.63	−1.65	−0.58	0.18	0.02	0.33	−0.28
10	0.10	−0.08	−0.83	−0.42	−1.70	−1.66	−2.29	−1.90	−1.03	−0.51	−0.81	−0.26	−0.95
11	−0.58	−0.62	−0.87	−0.85	−2.30	−2.23	−2.77	−2.38	−1.56	−0.99	−1.51	−1.05	−1.48
Noon. . .	−0.98	−1.24	−1.32	−1.27	−2.74	−2.52	−3.16	−2.97	−2.14	−1.41	−2.02	−1.50	−1.94
1	−1.38	−1.45	−1.54	−1.68	−2.88	−2.61	−3.53	−3.34	−2.54	−1.74	−2.42	−1.90	−2.25
2	−1.51	−1.62	−1.74	−1.92	−2.94	−2.62	−3.74	−3.73	−2.84	−2.01	−2.55	−2.06	−2.44
3	−1.45	−1.65	−1.90	−2.14	−2.94	−2.59	−3.54	−3.81	−2.87	−2.04	−2.22	−1.68	−2.40
4	−1.18	−1.34	−1.71	−2.10	−2.67	−2.20	−2.82	−3.23	−2.38	−1.94	−1.53	−1.14	−2.02
5	−0.87	−0.98	−1.39	−1.98	−2.08	−1.60	−2.44	−2.49	−1.60	−1.05	−0.73	−0.74	−1.50
6	−0.59	−0.79	−1.02	−1.51	−1.20	−1.00	−1.41	−1.34	−0.83	−0.54	−0.15	−0.33	−0.89
7	−0.32	−0.62	−0.73	−1.12	−0.26	−0.12	−0.46	−0.32	−0.18	−0.14	0.12	−0.15	−0.36
8	−0.07	−0.42	−0.43	−0.47	−0.14	0.38	1.01	0.50	−0.10	0.05	0.33	0.04	0.06
9	0.05	−0.14	−0.10	−0.11	1.11	1.38	1.54	1.01	0.23	0.26	0.49	0.26	0.50
10	0.18	0.09	0.24	0.27	1.44	1.72	1.67	1.36	0.58	0.52	0.72	0.46	0.77
11	0.29	0.31	0.48	0.60	1.75	1.86	2.14	1.78	0.84	0.68	0.86	0.59	1.02
Midn. . .	0.37	0.49	0.72	0.85	2.02	2.10	2.43	2.23	1.36	0.78	0.94	0.70	1.25
6. 6	0.27	0.34	0.52	0.36	0.38	0.11	0.33	0.80	0.86	0.50	0.67	0.42	0.46
7. 7	0.47	0.48	0.57	0.35	0.20	−0.02	0.05	0.75	0.80	0.62	0.85	0.54	0.47
8. 8	0.50	0.50	0.12	0.28	−0.19	−0.14	0.38	0.54	0.35	0.43	0.65	0.52	0.33
9. 9	0.38	0.34	0.26	0.04	0.02	0.07	−0.05	−0.32	−0.18	0.22	0.26	0.30	0.11
10.10	0.14	0.01	−0.30	−0.08	−0.13	0.03	−0.31	−0.27	−0.23	0.01	−0.05	0.10	−0.09
7. 2. 9	−0.07	−0.06	0.01	−0.07	−0.39	−0.39	−0.55	−0.30	−0.28	−0.13	−0.16	−0.19	−0.21
6. 2. 8	−0.15	−0.19	−0.04	−0.06	−0.37	−0.34	−0.22	−0.10	−0.13	−0.14	−0.24	−0.29	−0.19
6. 2.10	−0.07	−0.02	0.19	0.19	0.15	0.11	−0.00	0.19	0.09	0.02	−0.11	−0.15	0.05
6. 2. 6	−0.32	−0.32	−0.23	−0.40	−0.73	−0.80	−1.03	−0.71	−0.38	−0.34	−0.40	−0.41	−0.51
7. 2	−0.13	−0.02	0.06	−0.05	−1.14	−1.27	−1.59	−0.96	−0.53	−0.32	−0.49	−0.42	−0.57
8. 2	−0.22	−0.10	−0.54	−0.45	−1.59	−1.64	−2.00	−1.58	−1.03	−0.60	−0.79	−0.53	−0.92
8. 1	−0.16	−0.02	−0.44	−0.33	−1.56	−1.63	−1.89	−1.38	−0.88	−0.47	−0.73	−0.45	−0.83
7. 1	−0.07	0.07	0.16	0.07	−1.11	−1.27	−1.49	−0.76	−0.38	−0.19	−0.42	−0.34	−0.48
9.12.3.9	−0.42	−0.55	−0.68	−0.84	−1.41	−1.24	−1.70	−1.86	−1.34	−0.75	−0.93	−0.65	−1.03
7. 2.2(9)	−0.04	−0.08	−0.02	−0.08	−0.02	0.06	−0.03	0.03	−0.15	−0.03	−0.00	−0.08	−0.04
Dail. ext.	−0.13	−0.04	0.08	0.06	0.10	0.05	0.02	−0.19	−0.11	−0.23	−0.49	−0.42	−0.16

The numbers without sign must be added; those with the sign — must be subtracted.

XXXVI.

Switzerland. — Geneva. *Lat.* 46° 12′ N. *Long.* 6° 9′ E. *Greenw.*

Corrections to be applied to the Means of the Hours of Observation to obtain the true Mean Temperatures of the respective Days, Months, and of the Year. — Dove.

Degrees of Reaumur.

Hour.	Jan.	Feb.	March.	April.	May.	June.	July.	Aug.	Sept.	Oct.	Nov.	Dec.	Year.
Midn.	0.50	0.68	1.38	1.68	2.16	2.77	2.54	2.38	1.86	1.44	0.80	0.48	1.56
1	0.62	0.83	1.88	2.14	2.72	3.32	3.19	3.08	2.41	1.71	0.97	0.54	1.95
2	0.74	1.01	2.34	2.53	3.16	3.68	3.70	3.68	2.93	1.95	1.14	0.61	2.29
3	0.83	1.22	2.70	2.76	3.40	3.74	3.89	4.03	3.34	2.14	1.30	0.70	2.50
4	0.92	1.46	2.89	2.78	3.34	3.50	3.80	4.00	3.49	2.22	1.43	0.81	2.55
5	0.98	1.66	2.83	2.54	2.93	2.88	3.26	3.52	3.30	2.14	1.51	0.91	2.37
6	1.02	1.75	2.49	2.03	2.22	2.03	2.39	2.65	2.72	1.85	1.48	0.97	1.97
7	0.97	1.66	1.90	1.33	1.28	1.05	1.38	1.54	1.84	1.34	1.26	0.92	1.37
8	0.78	1.33	1.09	0.50	0.27	0.08	0.26	0.37	0.78	0.65	0.84	0.70	0.64
9	0.46	0.74	0.17	−0.34	−0.69	−0.82	−0.71	−0.70	−0.30	−0.15	0.23	0.34	−0.16
10	−0.02	−0.01	−0.77	−1.10	−1.51	−1.57	−1.53	−1.58	−1.26	−0.98	−0.47	−0.16	−0.91
11	−0.57	−0.80	−1.61	−1.75	−2.17	−2.18	−2.24	−2.29	−2.06	−1.70	−1.14	−0.67	−1.60
Noon.	−1.06	−1.49	−2.26	−2.23	−2.66	−2.70	−2.74	−2.85	−2.66	−2.22	−1.66	−1.10	−2.14
1	−1.40	−1.98	−2.70	−2.55	−2.98	−3.10	−3.18	−3.29	−3.08	−2.53	−1.94	−1.37	−2.51
2	−1.50	−2.18	−2.87	−2.67	−3.12	−3.35	−3.48	−3.58	−3.29	−2.58	−1.94	−1.41	−2.66
3	−1.41	−2.10	−2.81	−2.61	−3.07	−3.42	−3.51	−3.65	−3.28	−2.41	−1.74	−1.26	−2.61
4	−1.14	−1.82	−2.54	−2.37	−2.80	−3.25	−3.37	−3.43	−3.04	−2.06	−1.38	−0.97	−2.35
5	−0.79	−1.37	−2.10	−1.97	−2.32	−2.78	−2.90	−2.92	−2.57	−1.59	−0.99	−0.64	−1.91
6	−0.46	−0.94	−1.59	−1.46	−1.70	−2.11	−2.22	−2.18	−1.91	−1.06	−0.62	−0.32	−1.38
7	−0.20	−0.51	−1.06	−0.90	−1.00	−1.29	−1.40	−1.31	−1.16	−0.53	−0.30	−0.07	−0.81
8	−0.01	−0.14	−0.54	−0.34	−0.29	−0.42	−0.49	−0.46	−0.42	−0.02	−0.03	0.11	−0.26
9	0.12	0.14	0.05	0.20	0.38	0.47	0.34	0.32	0.26	0.42	0.20	0.24	0.26
10	0.25	0.37	0.42	0.70	0.91	1.30	1.10	1.02	0.83	0.82	0.42	0.34	0.71
11	0.37	0.54	0.90	1.20	1.51	2.07	1.87	1.70	1.35	1.15	0.62	0.41	1.14
Mean	−0.53	1.24	3.41	6.77	10.37	13.31	14.30	13.58	11.46	7.48	3.76	0.58	

XXXVII.

Switzerland. — Geneva. *Lat.* 46° 12′ N. *Long.* 6° 9′ E. *Gr.* — Dove.

Degrees of Reaumur.

Hour.	Jan.	Feb.	March.	April.	May.	June.	July.	Aug.	Sept.	Oct.	Nov.	Dec.	Year.
Midn.	0.45	0.69	1.26	1.44	1.54	1.98	2.12	1.63	1.44	0.94	0.50	0.59	1.21
2	0.70	0.96	2.21	2.62	2.60	3.20	3.18	2.83	2.72	1.46	0.73	0.66	1.99
4	1.01	1.33	2.91	3.36	3.11	3.55	3.82	3.51	3.26	1.90	1.02	0.80	2.46
6	1.19	1.49	2.70	2.87	2.26	2.38	2.47	2.82	2.79	1.74	1.13	0.97	2.07
8	1.22	1.22	1.42	0.74	0.27	0.13	0.22	0.49	0.72	0.94	0.90	0.95	0.77
10	−0.02	−0.25	−0.68	−1.70	−1.30	−1.34	−1.25	−1.01	−1.10	−0.73	−0.26	−0.14	−0.73
Noon.	−0.13	−1.30	−1.97	−2.14	−2.42	−2.54	−2.50	−2.34	−2.38	−1.86	−1.18	−1.22	−1.91
2	−1.69	−1.70	−2.82	−2.94	−2.97	−3.09	−3.11	−3.17	−3.03	−2.35	−1.55	−1.46	−2.49
4	−1.30	−1.61	−2.70	−2.94	−2.46	−2.87	−2.89	−3.04	−2.86	−1.53	−1.19	−1.05	−2.20
6	−0.54	−0.90	−1.79	−2.06	−1.40	−1.89	−2.24	−2.04	−1.74	−0.88	−0.45	−0.43	−1.36
8	−0.09	−0.21	−0.89	−0.70	−0.10	−0.25	−0.58	−0.38	−0.38	−0.08	0.03	0.10	−0.29
10	0.20	0.28	0.34	0.40	0.86	0.78	0.78	0.69	0.57	0.47	0.29	0.18	0.49
Mean	1.20	0.47	2.28	6.81	9.48	12.82	14.43	13.74	10.66	7.73	3.30	0.12	

The numbers without sign must be added; those with the sign — must be subtracted.

XXXVIII.

SWITZERLAND. — ST. BERNARD. *Lat.* 45° 52′ N. *Long.* 9° 22′ E. *Gr.*

Corrections to be applied to the Means of the Hours of Observation to obtain the true Mean Temperatures of the respective Days, Months, and of the Year. — DOVE.

Degrees of Reaumur.

Hour.	Jan.	Feb.	March.	April.	May.	June.	July.	Aug.	Sept.	Oct.	Nov.	Dec.	Year.
Midn.	0.48	0.81	1.34	1.96	2.10	1.72	1.62	1.30	0.76	1.02	0.59	0.31	1.17
1	0.63	0.91	1.58	2.22	2.45	1.99	1.93	1.53	0.97	1.17	0.66	0.33	1.36
2	0.81	1.09	1.82	2.40	2.73	2.15	2.14	1.82	1.17	1.30	0.78	0.40	1.55
3	0.99	1.26	1.98	2.46	2.81	2.24	2.24	1.94	1.34	1.36	0.89	0.50	1.67
4	1.08	1.38	2.02	2.34	2.67	2.14	2.17	1.91	1.41	1.34	0.98	0.52	1.66
5	1.08	1.34	1.84	2.00	2.28	1.88	1.90	1.70	1.35	1.19	0.98	0.66	1.52
6	0.91	1.14	1.42	1.45	1.72	1.42	1.44	1.34	1.14	0.92	0.86	0.62	1.20
7	0.60	0.74	0.79	0.70	0.81	0.81	0.82	0.76	0.77	0.83	0.61	0.50	0.73
8	0.17	0.18	0.00	−0.16	−0.08	0.09	0.10	0.12	0.29	0.06	0.26	0.26	0.11
9	−0.31	−0.48	−0.85	−1.06	−1.10	−0.66	−0.66	−0.53	−0.26	−0.46	−0.22	−0.06	−0.55
10	−0.78	−1.13	−1.63	−1.86	−1.94	−1.36	−1.34	−1.13	−0.78	−0.94	−0.68	−0.41	−1.16
11	−1.14	−1.66	−2.23	−2.50	−2.58	−1.95	−1.90	−1.60	−1.22	−1.33	−1.09	−0.71	−1.66
Noon.	−1.34	−1.98	−2.58	−2.87	−2.96	−2.34	−2.26	−1.90	−1.51	−1.58	−1.36	−0.94	−1.97
1	−1.38	−2.04	−2.62	−2.98	−3.06	−2.51	−2.40	−2.02	−1.62	−1.66	−1.47	−1.03	−2.07
2	−1.24	−1.86	−2.38	−2.78	−2.89	−2.44	−2.33	−1.94	−1.56	−1.59	−1.39	−0.99	−1.95
3	−0.98	−1.47	−1.92	−2.36	−2.51	−2.21	−2.08	−1.74	−1.35	−1.38	−1.16	−0.82	−1.66
4	−0.65	−0.97	−1.34	−1.79	−1.98	−1.80	−1.70	−1.42	−1.05	−1.07	−0.83	−0.57	−1.26
5	−0.32	−0.43	−0.73	−1.17	−1.40	−1.32	−1.26	−1.06	−0.70	−0.72	−0.46	−0.27	−0.82
6	−0.05	0.04	−0.19	−0.54	−0.81	−0.80	−0.80	−0.70	−0.38	−0.36	−0.10	0.00	−0.39
7	0.14	0.39	0.25	0.04	−0.25	−0.28	−0.34	−0.34	−0.11	−0.03	0.19	0.21	−0.01
8	0.25	0.60	0.56	0.54	0.27	0.20	0.09	0.00	0.10	0.24	0.38	0.34	0.30
9	0.30	0.69	0.78	0.96	0.76	0.63	0.50	0.32	0.27	0.47	0.49	0.38	0.55
10	0.34	0.72	0.96	1.33	1.22	1.02	0.89	0.64	0.42	0.67	0.53	0.38	0.76
11	0.38	0.74	1.14	1.66	1.68	1.40	1.26	0.97	0.58	0.85	0.55	0.33	0.96
Mean.	−8.26	−6.62	−5.72	−2.97	0.74	3.55	4.82	4.32	2.40	−0.91	−3.95	−5.86	

XXXIX.

SWITZERLAND. — ST. BERNARD. *Lat.* 45° 52′ N. *Long.* 9° 22′ E. *Gr.* — DOVE.

Degrees of Reaumur.

Hour.	Jan.	Feb.	March.	April.	May.	June.	July.	Aug.	Sept.	Oct.	Nov.	Dec.	Year.
Midn.	0.34	0.55	0.75	1.19	1.26	1.39	1.02	1.08	0.81	0.66	0.33	0.28	0.80
2	0.52	0.78	1.14	1.64	1.75	1.88	1.62	1.53	1.16	0.94	0.42	0.27	1.14
4	0.82	1.06	1.50	1.84	1.91	1.98	1.82	1.71	1.34	1.17	0.65	0.42	1.35
6	0.65	0.86	1.20	1.50	1.53	1.46	1.46	1.27	0.98	0.88	0.50	0.32	1.05
8	0.48	0.26	0.14	−0.08	−0.25	0.01	0.22	0.16	0.08	0.28	0.27	0.15	0.14
10	−0.35	−0.91	−1.06	−1.26	−1.39	−1.18	−1.11	−0.94	−0.86	−0.68	−0.54	−0.23	−0.88
Noon.	−1.40	−1.66	−1.74	−2.11	−2.15	−1.92	−1.81	−1.77	−1.58	−1.45	−1.26	−0.91	−1.65
2	−1.37	−1.55	−1.89	−2.12	−2.12	−2.23	−2.01	−1.97	−1.54	−1.52	−1.23	−1.22	−1.73
4	−0.42	−0.71	−1.14	−1.55	−1.47	−1.65	−1.49	−1.30	−0.88	−0.86	−0.37	−0.02	−0.99
6	0.09	0.17	0.09	−0.26	−0.35	−0.71	−0.57	−0.46	−0.26	−0.07	0.08	0.22	−0.17
8	0.25	0.44	0.49	0.49	0.50	0.35	0.30	0.26	0.26	0.22	0.70	0.30	0.38
10	0.37	0.55	0.55	0.71	0.76	0.64	0.56	0.43	0.46	0.43	0.40	0.40	0.52
Mean.	−6.08	−8.83	−6.66	−3.01	−0.42	2.71	4.82	4.70	2.07	−0.36	−5.46	−6.18	

The numbers without sign must be added; those with the sign — must be subtracted.

AUSTRIA. — KREMSMÜNSTER. *Lat.* 48° 3′ N. *Long.* 14° 7′ E. *Greenw.*

Corrections to be applied to the Means of the Hours of Observation to obtain the true Mean Temperatures of the respective Days, Months, and of the Year. — DOVE.

Degrees of Reaumur.

Hours.	Jan.	Feb.	March.	April.	May.	June.	July.	Aug.	Sept.	Oct.	Nov.	Dec.	Mean.
Morn. 1	0.58	0.90	1.05	1.14	2.30	2.77	1.86	1.94	1.52	1.26	0.61	0.40	1.36
2	0.66	1.03	1.30	1.36	2.66	3.08	2.16	2.26	1.94	1.58	0.72	0.42	1.60
3	0.71	1.07	1.57	1.63	2.84	3.14	2.35	2.50	2.32	1.82	0.78	0.42	1.76
4	0.78	1.12	1.80	1.88	2.78	2.90	2.34	2.54	2.58	1.97	0.83	0.42	1.83
5	0.84	1.19	1.90	1.99	2.44	2.32	2.08	2.30	2.60	1.98	0.88	0.46	1.75
6	0.88	1.24	1.82	1.88	1.86	1.54	1.54	1.80	2.34	1.91	0.93	0.54	1.52
7	0.84	1.26	1.50	1.41	1.11	0.68	0.94	1.11	1.81	1.63	0.92	0.59	1.15
8	0.67	1.07	0.96	0.87	0.31	−0.15	0.23	0.35	1.09	1.21	0.80	0.56	0.66
9	0.35	0.67	0.30	0.14	−0.45	−0.86	−0.42	−0.37	0.28	0.62	0.51	0.38	0.10
10	−0.10	0.01	−0.41	−0.58	−1.10	−1.42	−0.95	−0.98	−0.52	−0.13	0.06	0.05	−0.56
11	−0.58	−0.72	−1.06	−1.20	−1.65	−1.84	−1.39	−1.47	−1.23	−0.92	−0.47	−0.38	−1.08
Noon. . .	−0.98	−1.37	−1.56	−1.65	−2.09	−2.17	−1.75	−1.86	−1.81	−1.68	−0.97	−0.78	−1.56
1	−1.22	−1.78	−1.89	−1.93	−2.42	−2.42	−2.05	−2.21	−2.28	−2.25	−1.30	−1.03	−1.90
2	−1.26	−1.90	−2.02	−2.06	−2.62	−2.58	−2.26	−2.38	−2.56	−2.53	−1.40	−1.09	−2.05
3	−1.12	−1.69	−1.99	−2.04	−2.67	−2.62	−2.33	−2.46	−2.65	−2.49	−1.28	−0.94	−2.02
4	−0.86	−1.32	−1.79	−1.89	−2.51	−2.49	−2.22	−2.34	−2.52	−2.17	−1.01	−0.66	−1.98
5	−0.59	−0.92	−1.48	−1.60	−2.15	−2.16	−1.88	−2.00	−2.18	−1.69	−0.68	−0.35	−1.47
6	−0.35	−0.57	−1.08	−1.18	−1.62	−1.66	−1.38	−1.49	−1.66	−1.14	−0.41	−0.11	−1.05
7	−0.18	−0.36	−0.65	−0.68	−0.98	−1.03	−0.76	−0.86	−1.05	−0.66	−0.22	0.02	−0.62
8	−0.04	−0.19	−0.23	−0.17	−0.34	−0.35	−0.15	−0.24	−0.46	−0.26	−0.11	0.09	−0.20
9	0.07	−0.02	0.13	0.28	0.28	0.34	0.38	0.30	0.05	0.06	−0.02	0.12	0.16
10	0.20	0.18	0.42	0.61	0.84	1.02	0.82	0.76	0.46	0.34	0.11	0.18	0.49
11	0.34	0.46	0.63	0.82	1.36	1.68	1.19	1.15	0.80	0.63	0.27	0.25	0.80
Midn. . .	0.47	0.70	0.83	0.97	1.85	2.27	1.52	1.53	1.14	0.94	0.46	0.34	1.08
6. 6	0.27	0.34	0.37	0.35	0.12	−0.06	0.08	0.16	0.34	0.39	0.26	0.22	0.24
7. 7	0.33	0.45	0.43	0.37	0.07	−0.18	0.09	0.13	0.38	0.48	0.35	0.29	0.27
8. 8	0.32	0.44	0.37	0.35	−0.02	−0.10	0.04	0.06	0.32	0.48	0.35	0.24	0.24
9. 9	0.21	0.33	0.22	0.21	−0.09	−0.26	−0.02	−0.04	0.17	0.34	0.25	0.25	0.13
10.10	0.05	0.10	0.01	0.02	−0.13	−0.20	−0.07	−0.11	−0.03	0.11	0.09	0.12	0.00
7. 2. 9	−0.12	−0.22	−0.13	−0.12	−0.41	−0.52	−0.31	−0.32	−0.23	−0.28	−0.17	−0.16	−0.25
6. 2. 8	−0.14	−0.28	−0.14	−0.12	−0.37	−0.46	−0.29	−0.27	−0.23	−0.29	−0.19	−0.15	−0.24
6. 2.10	−0.06	−0.16	0.07	0.14	0.03	−0.01	0.03	0.06	0.08	−0.09	−0.12	−0.12	−0.01
6. 2. 6	−0.24	−0.41	−0.43	−0.45	−0.79	−0.90	−0.70	−0.69	−0.63	−0.94	−0.36	−0.15	−0.56
7. 2	−0.21	−0.32	−0.26	−0.33	−0.76	−0.95	−0.66	−0.63	−0.38	−0.45	−0.24	−0.25	−0.45
8. 2	−0.30	−0.42	−0.53	−0.60	−1.16	−1.22	−1.02	−1.02	−0.74	−0.66	−0.30	−0.27	−0.69
8. 1	−0.28	−0.36	−0.47	−0.53	−1.06	−1.14	−0.91	−0.93	−0.60	−0.52	−0.25	−0.24	−0.61
7. 1	−0.19	−0.26	−0.20	−0.26	−0.66	−0.87	−0.56	−0.55	−0.24	−0.31	−0.19	−0.22	−0.38
9.12.3.9	−0.42	−0.60	−0.78	−0.82	−1.23	−1.33	−1.03	−1.10	−1.03	−0.87	−0.44	−0.31	−0.83
7. 2.2(9)	−0.07	−0.17	−0.07	−0.02	−0.24	−0.31	−0.15	−0.17	−0.14	−0.19	−0.13	−0.07	−0.14
Dail. ext.	−0.19	−0.32	−0.06	−0.04	0.09	0.36	0.01	0.04	−0.03	−0.28	−0.24	−0.25	−0.08

The numbers without sign must be added; those with the sign — must be subtracted.

XLI.

Austria. — Salzburg. *Lat.* 47° 48′ N. *Long.* 13° 1′ E. *Greenw.*

Corrections to be applied to the Means of the Hours of Observation to obtain the true Mean Temperatures of the respective Days, Months, and of the Year. — Dove.

Degrees of Reaumur.

Hour.	Jan.	Feb.	March.	April.	May.	June.	July.	Aug.	Sept.	Oct.	Nov.	Dec.	Year.
Midn.	0.54	0.70	1.06	1.31	2.03	2.07	1.87	1.57	1.21	1.02	0.48	0.42	1.19
1	0.59	0.79	1.29	1.58	2.37	2.27	2.13	1.81	1.45	1.15	0.65	0.50	1.38
2	0.72	0.97	0.51	1.79	2.64	2.56	2.36	2.05	1.61	1.27	0.81	0.59	1.49
3	0.82	1.08	1.75	2.04	2.90	2.73	2.64	2.24	1.87	1.41	0.88	0.70	1.75
4	0.96	1.09	1.89	2.21	3.10	2.82	2.62	2.23	2.04	1.52	0.91	0.69	1.84
5	1.03	1.28	2.01	2.37	3.10	2.75	2.59	2.24	2.14	1.72	1.03	0.81	1.92
6	1.06	1.34	2.14	2.28	2.76	2.45	2.31	2.26	2.18	1.77	1.03	0.87	1.87
7	1.09	1.36	2.06	1.86	1.89	1.53	1.61	1.74	1.94	1.74	1.06	0.94	1.57
8	1.12	1.24	1.58	1.06	0.84	0.63	0.67	0.89	1.15	1.26	1.07	1.00	1.04
9	0.91	0.75	0.76	0.14	−0.10	−0.25	0.20	0.04	0.33	0.48	0.64	0.74	0.39
10	0.38	0.04	−0.06	−0.67	−0.92	−1.10	−0.97	−0.76	−0.53	−0.35	0.06	0.21	−0.39
11	−0.26	−0.62	−0.96	−1.39	−1.80	−1.87	−1.63	−1.40	−1.25	−1.17	−0.62	−0.35	−1.11
Noon.	−0.90	−1.19	−1.75	−1.99	−2.36	−2.90	−2.14	−2.13	−2.00	−1.84	−1.25	−0.93	−1.78
1	−1.47	−1.68	−2.26	−2.48	−2.82	−2.84	−2.59	−2.59	−1.48	−2.39	−1.68	−1.47	−2.15
2	−1.70	−1.96	−2.55	−2.74	−3.08	−3.03	−2.77	−2.73	−2.71	−2.55	−1.85	−1.64	−2.44
3	−1.68	−2.04	−2.61	−2.74	−3.21	−3.04	−2.90	−2.75	−2.67	−2.51	−1.75	−1.55	−2.45
4	−1.40	−1.80	−2.55	−2.60	−3.27	−3.00	−2.90	−2.85	−2.56	−2.21	−1.37	−1.19	−2.31
5	−1.00	−1.46	−2.26	−2.10	−2.97	−2.64	−2.64	−2.46	−2.09	−1.63	−0.85	−0.72	−1.90
6	−0.60	−0.76	−1.51	−1.52	−2.27	−2.10	−2.05	−1.78	−1.31	−0.83	−0.35	−0.42	−1.29
7	−0.31	−0.27	−0.76	−0.75	−1.43	−1.21	−1.24	−0.85	−0.48	−0.29	−0.10	−0.15	−0.65
8	−0.25	−0.02	−0.16	−0.07	−0.43	−0.13	−0.24	0.06	0.15	0.16	0.11	0.04	−0.06
9	−0.04	0.20	0.17	0.51	0.48	0.71	0.67	0.70	0.50	0.48	0.24	0.17	0.40
10	0.12	0.43	0.46	0.81	1.03	1.41	1.22	1.09	0.78	0.76	0.34	0.33	0.73
11	0.28	0.53	0.76	1.08	1.50	1.70	1.56	1.38	0.76	1.03	0.52	0.41	0.96
Mean.	−2.71	1.14	2.49	6.90	10.42	13.22	13.93	13.66	10.30	7.37	1.52	1.63	

XLII.

Germany. — Munich. *Lat.* 48° 9′ N. *Long.* 11° 37′ E. *Greenw.* — Dove.

Degrees of Reaumur.

Hour.	Jan.	Feb.	March.	April.	May.	June.	July.	Aug.	Sept.	Oct.	Nov.	Dec.	Year.
Midn.	0.71	0.92	1.54	2.27	2.58	2.49	2.84	2.37	2.17	1.53	0.91	0.46	1.73
1	0.90	1.04	1.83	2.37	3.02	3.06	3.27	2.64	2.33	1.59	0.87	0.58	1.96
2	0.97	1.18	2.04	2.62	3.30	3.39	3.56	2.94	2.61	1.67	0.94	0.67	2.16
3	1.04	1.30	2.16	2.89	3.61	3.66	3.80	3.19	2.81	1.78	1.00	0.77	2.33
4	1.03	1.33	2.25	3.12	3.85	3.82	4.05	3.41	2.98	1.91	1.04	0.85	2.47
5	1.07	1.43	2.37	3.29	3.69	3.25	3.71	3.50	3.16	2.01	1.12	0.92	2.46
6	1.14	1.52	2.56	2.93	2.61	2.11	2.41	2.79	3.08	2.14	1.13	0.99	2.12
7	1.17	1.55	2.17	1.80	1.21	0.77	0.93	1.48	2.22	1.84	1.13	0.97	1.44
8	1.10	1.14	1.14	0.36	−0.07	−0.35	−0.28	0.18	0.59	0.99	0.75	0.88	0.54
9	0.46	0.36	−0.11	−0.79	−1.00	−1.21	−1.25	−1.05	−0.74	−0.24	0.06	0.41	−0.42
10	−0.72	−0.61	−1.18	−1.80	−1.99	−1.96	−2.12	−1.88	−1.70	−1.34	−0.79	−0.42	−1.38
11	−1.06	−1.46	−2.04	−2.39	−2.59	−2.69	−2.66	−2.58	−2.61	−2.19	−1.49	−0.97	−2.06

The numbers without sign must be added ; those with the sign — must be subtracted.

XLII.

Germany. — Munich, *Continued.*

Corrections to be applied to the Means of the Hours of Observation to obtain the true Mean Temperatures of the respective Days, Months, and of the Year. — Dove.

Degrees of Reaumur.

Hour.	Jan.	Feb.	March.	April.	May.	June.	July.	Aug.	Sept.	Oct.	Nov.	Dec.	Year.
Noon	–1.70	–1.93	–2.67	–2.99	–3.28	–2.98	–3.14	–3.09	–3.18	–2.69	–1.94	–1.02	–2.55
1	–2.08	–2.31	–3.01	–3.27	–3.59	–3.41	–3.48	–3.55	–3.58	–3.08	–2.23	–1.83	–2.95
2	–2.15	–2.40	–3.24	–3.60	–3.77	–3.79	–3.75	–3.72	–3.74	–3.15	–2.05	–1.85	–3.10
3	–1.83	–2.15	–3.17	–3.45	–3.77	–3.54	–3.83	–3.58	–3.56	–2.87	–1.75	–1.43	–2.91
4	–1.08	–1.67	–2.64	–3.18	–3.41	–3.34	–3.49	–3.30	–3.24	–2.27	–1.02	–0.76	–2.45
5	–0.46	–0.95	–1.98	–2.51	–2.87	–2.80	–3.07	–2.76	–2.56	–1.27	–0.43	–0.34	–1.83
6	–0.16	–0.37	–0.94	–1.53	–2.05	–1.94	–2.32	–1.81	–1.29	–0.44	–0.12	–0.13	–1.09
7	0.04	–0.07	–0.20	–0.36	–0.74	–0.84	–2.99	–0.47	–0.30	0.08	0.20	0.06	–0.47
8	0.23	0.22	0.28	0.40	0.41	0.61	0.40	0.55	0.37	0.56	0.44	0.14	0.38
9	0.39	0.45	0.55	0.91	1.13	1.35	1.20	1.15	0.93	0.88	0.57	0.23	0.81
10	0.49	0.59	1.02	1.31	1.65	1.86	1.87	1.60	1.40	1.14	0.74	0.33	1.17
11	0.61	0.77	1.33	1.69	2.18	2.28	2.41	2.06	1.80	1.34	0.85	0.40	1.48
Mean.	–2.15	–0.12	0.75	5.57	9.29	12.74	13.65	12.93	9.45	6.28	1.55	–1.28	

XLIII.

Bohemia. — Prague. *Lat.* 50° 5′ N. *Long.* 14° 25′ E. *Greenw.* — Dove.

Degrees of Reaumur.

Hour.	Jan.	Feb.	March.	April.	May.	June.	July.	Aug	Sept.	Oct.	Nov.	Dec.	Year.
Midn	0.30	0.52	1.03	1.47	1.70	1.68	1.72	1.17	1.23	0.84	0.36	0.25	1.02
1	0.40	0.60	1.14	1.68	1.97	1.97	2.05	1.78	1.49	1.02	0.45	0.32	1.24
2	0.50	0.71	1.29	1.95	2.25	2.23	2.34	2.10	1.72	1.19	0.54	0.39	1.43
3	0.55	0.83	1.44	2.17	2.46	2.47	2.60	2.38	1.96	1.31	0.61	0.50	1.61
4	0.65	0.89	1.60	2.39	2.75	2.71	2.91	2.63	2.19	1.49	0.70	0.56	1.79
5	0.71	0.99	1.72	2.64	2.96	2.86	3.07	2.88	2.43	1.65	0.77	0.65	1.94
6	0.77	1.00	1.81	2.75	2.96	2.71	2.92	2.93	2.61	1.73	0.82	0.72	1.98
7	0.68	0.99	1.53	2.32	2.11	1.88	2.13	2.34	2.29	1.65	0.79	0.73	1.62
8	0.73	0.88	1.28	1.29	0.98	0.82	1.02	1.30	1.62	1.29	0.66	0.70	1.05
9	0.62	0.57	0.63	0.32	0.06	–0.14	0.17	0.21	0.60	0.70	0.41	0.54	0.39
10	0.26	0.15	–0.11	–0.53	–0.91	–0.93	–0.95	–0.77	–0.51	–0.10	–0.12	0.17	–0.36
11	–0.16	–0.45	–0.77	–1.51	–1.60	–1.58	–1.62	–1.50	–1.46	–0.86	–0.46	–0.22	–1.02
Noon.	–0.60	–0.92	–1.37	–2.09	–2.16	–2.08	–2.16	–2.18	–2.02	–1.53	–0.86	–0.65	–1.55
1	–0.93	–1.27	–1.83	–2.48	–2.56	–2.48	–2.59	–2.61	–2.56	–2.01	–1.13	–0.95	–1.95
2	–1.10	–1.50	–2.20	–2.74	–2.80	–2.73	–2.83	–2.89	–2.84	–2.31	–1.25	–1.07	–2.19
3	–1.11	–1.51	–2.29	–2.88	–2.90	–2.79	–2.93	–3.01	–2.96	–2.32	–1.28	–0.99	–2.25
4	–0.93	–1.35	–2.20	–2.76	–2.82	–2.71	–2.92	–2.85	–2.78	–2.10	–0.87	–0.79	–2.09
5	–0.68	–0.97	–1.83	–2.46	–2.53	–2.56	–2.83	–2.66	–2.35	–1.58	–0.62	–0.55	–1.80
6	–0.44	–0.61	–1.26	–1.91	–2.17	–2.10	–2.36	–2.11	–1.64	–1.01	–0.36	–0.37	–1.36
7	–0.31	–0.32	–0.70	–1.12	–1.49	–1.37	–1.59	–1.23	–0.87	–0.54	–0.19	–0.21	–0.83
8	–0.23	–0.06	–0.24	–0.33	–0.51	–0.39	–0.58	–0.34	–0.24	–0.10	0.01	–0.19	–0.27
9	0.01	0.12	0.09	0.20	0.27	0.30	0.22	0.20	0.27	0.23	0.16	0.06	0.18
10	0.10	0.26	0.40	0.72	0.80	0.91	0.90	0.81	0.74	0.51	0.29	0.16	0.55
11	0.19	0.39	0.66	1.12	1.24	1.28	1.32	1.20	1.08	0.85	0.43	0.25	0.83
Mean.	–1.69	0.64	2.20	7.27	11.27	14.47	15.66	15.01	11.52	7.94	3.02	–0.12	

The numbers without sign must be added; those with the sign — must be subtracted.

Bohemia. — Prague. *Lat.* 50° 5′ N. *Long.* 14° 24′ E. *Greenw.*

Corrections to be applied to the Means of the Hours of Observation to obtain the true Mean Temperatures of the respective Days, Months, and of the Year. — Dove.

Degrees of Reaumur.

Hours.	Jan.	Feb.	March.	April.	May.	June.	July.	Aug.	Sept.	Oct.	Nov.	Dec.	Mean.
Morn. 1	0.45	0.76	0.86	1.73	1.47	1.90	1.93	1.59	1.46	1.06	0.73	0.45	1.20
2	0.52	0.88	1.05	2.06	1.77	2.22	2.24	1.85	1.69	1.18	0.79	0.52	1.40
3	0.54	0.98	1.24	2.45	2.08	2.62	2.36	2.04	1.85	1.23	0.82	0.54	1.56
4	0.53	1.06	1.42	2.82	2.31	3.02	2.27	2.10	1.95	1.24	0.78	0.55	1.67
5	0.50	1.14	1.55	3.02	2.35	3.22	2.01	2.01	1.97	1.22	0.78	0.60	1.70
6	0.49	1.15	1.60	2.92	2.12	3.03	1.62	1.76	1.90	1.19	0.80	0.70	1.61
7	0.47	1.09	1.51	2.43	1.62	2.40	1.16	1.36	1.69	1.10	0.77	0.80	1.37
8	0.42	0.91	1.24	1.59	0.92	1.40	0.66	0.83	1.28	0.90	0.69	0.82	0.97
9	0.29	0.55	0.77	0.53	0.15	0.24	0.10	0.19	0.64	0.51	0.42	0.67	0.42
10	0.08	−1.01	0.16	−0.56	−0.57	−0.85	−0.52	−0.51	−0.20	−0.07	−0.02	0.31	−0.31
11	−0.21	−1.19	−0.52	−1.52	−1.16	−1.68	−1.19	−1.23	−1.14	−0.78	−0.55	−1.18	−0.95
Noon. . .	−0.52	−1.10	−1.16	−2.25	−1.60	−2.23	−1.84	−1.86	−2.00	−1.47	−1.10	−0.70	−1.49
1	−0.76	−1.51	−1.63	−2.74	−1.91	−2.55	−2.37	−2.33	−2.63	−1.99	−1.47	−1.08	−1.91
2	−0.88	−1.70	−1.89	−3.00	−2.14	−2.76	−2.66	−2.57	−2.89	−2.21	−1.58	−1.23	−2.13
3	−0.85	−1.64	−1.92	−3.08	−2.26	−2.92	−2.65	−2.53	−2.76	−2.08	−1.44	−1.13	−2.11
4	−0.71	−1.39	−1.75	−2.97	−2.26	−2.98	−2.36	−2.23	−2.31	−1.68	−1.08	−0.87	−1.88
5	−0.51	−1.05	−1.45	−2.65	−2.08	−2.86	−1.86	−1.75	−1.70	−1.14	−0.67	−0.56	−1.52
6	−0.31	−0.66	−1.10	−2.13	−1.71	−2.45	−1.28	−1.18	−1.07	−0.60	−0.31	−0.31	−1.09
7	−0.16	−0.34	−0.73	−1.42	−1.17	−1.75	−0.73	−0.62	−0.52	−0.17	−0.04	−0.17	−0.65
8	−0.06	−0.09	−0.40	−0.64	−0.56	−0.85	−0.24	−0.12	−0.08	0.13	0.10	−0.11	−0.24
9	0.02	0.11	0.10	0.11	0.03	0.06	0.19	0.30	0.26	0.34	0.20	0.07	0.12
10	0.11	1.35	0.18	0.71	0.52	0.81	0.61	0.65	0.57	0.51	0.32	0.01	0.53
11	0.22	1.10	0.42	1.15	0.89	1.32	1.05	0.97	0.87	0.70	0.44	0.14	0.77
Midn. . .	0.34	0.61	0.65	1.46	1.18	1.64	1.51	1.28	1.17	0.89	0.61	0.31	0.97
6. 6	0.09	0.24	0.25	0.40	0.21	0.29	0.17	0.29	0.42	0.29	0.25	0.19	0.26
7. 7	0.15	0.38	0.39	0.50	0.22	0.33	0.22	0.37	0.59	0.47	0.37	0.32	0.36
8. 8	0.18	0.41	0.42	0.47	0.18	0.27	0.21	0.36	0.60	0.51	0.39	0.35	0.36
9. 9	0.16	0.33	0.34	0.32	0.90	0.15	0.15	0.25	0.45	0.42	0.31	0.30	0.27
10.10	0.09	0.17	0.17	0.08	−0.03	−0.02	0.04	0.07	0.18	0.22	0.15	0.16	0.11
7. 2. 9	−0.13	−0.17	−0.16	−0.15	−0.16	−0.10	−0.44	−0.30	−0.31	−0.26	−0.20	−0.17	−0.21
6. 2. 8	−0.15	−0.21	−0.23	−0.24	−0.19	−0.19	−0.43	−0.31	−0.36	−0.30	−0.23	−0.21	−0.25
6. 2.10	−0.09	0.27	−0.04	0.21	0.17	0.36	−0.14	−0.05	−0.14	−0.17	−0.15	−0.17	0.01
6. 2. 6	−0.23	−0.40	−0.46	−0.74	−0.58	−0.73	−0.77	−0.66	−0.69	−0.54	−0.36	−0.28	−0.54
7. 2	−0.21	−0.31	−0.19	−0.29	−0.26	−0.18	−0.75	−0.61	−0.60	−0.56	−0.41	−0.22	−0.38
8. 2	−0.23	−0.40	−0.33	−0.71	−0.61	−0.68	−1.00	−0.87	−0.81	−0.66	−0.45	−0.21	−0.58
8. 1	−0.17	−0.30	−0.20	−0.58	−0.50	−0.58	−0.86	−0.75	−0.68	−0.55	−0.39	−0.13	−0.97
7. 1	−0.15	−0.21	−0.06	−0.16	−0.15	−0.08	−0.61	−0.49	−0.47	−0.45	−0.35	−0.14	−0.28
9.12.3.9	−0.27	−0.52	−0.60	−1.17	−0.92	−1.21	−1.05	−0.98	−0.97	−0.68	−0.48	−0.31	−0.76
7. 2.2(9)	−0.09	−0.10	−0.15	−0.09	−0.12	−0.06	−0.28	−0.15	−0.17	−0.11	−0.10	−0.14	−0.13
Dail. ext.	−0.17	−0.18	−0.16	−0.03	0.05	0.12	−0.15	−0.24	−0.46	−0.49	−0.38	−0.21	−0.22

The numbers without sign must be added: those with the sign — must be subtracted.

XLV.

England. — Plymouth. *Lat.* 50° 22′ N. *Long.* 4° 7′ W. *Greenw.*

Corrections to be applied to the Means of the Hours of Observation to obtain the true Mean Temperatures of the respective Days, Months, and of the Year. — Dove.

Degrees of Fahrenheit.

Hours.	Jan.	Feb.	March.	April.	May.	June.	July.	Aug.	Sept.	Oct.	Nov.	Dec.	Mean.
Morn. 1	0.86	1.46	2.32	4.01	5.13	4.34	4.75	4.16	3.24	2.66	1.58	0.95	2.95
2	0.90	1.67	2.63	4.43	5.94	4.82	5.38	4.79	3.60	2.79	1.69	0.86	3.29
3	0.99	1.87	3 02	4.91	6.62	5.13	5.69	5.45	4.03	3.02	1.80	0.74	3.60
4	1.15	2.12	3.31	5.13	6.75	5.00	5.58	5.76	4.34	3.31	1.96	0.81	3.76
5	1.37	2.36	3.40	4.91	6.03	4.57	4.82	5.42	4.25	3.51	2.09	1.04	3.65
6	1.53	2.48	3.08	3.98	4.37	2.79	3.35	4.21	3.62	3.38	2.14	1.31	3.02
7	1.46	2.30	2.25	2.39	2.00	0.95	1.92	2.25	2.32	2.66	1.89	1.40	1.94
8	1.10	1.67	0.97	0.29	−0.54	−1.01	−0.65	−0.11	0.50	1.26	1.24	1.13	0.50
9	0.36	0.59	−0.63	−1.94	−2.88	−2.70	−2.57	−2.39	−1.53	−0.65	0.16	0.41	−1.15
10	−0.61	−0.83	−2.23	−3.94	−4.57	−3.87	−4.03	−4.21	−3.44	−2.70	−1.17	−0.61	−2.68
11	−1.58	−2.25	−3.56	−5.40	−5.63	−4.64	−5.02	−5.36	−4.93	−4.41	−2.43	−1.67	−3.92
	−2.32	−3.33	−4.43	−6.17	−6.12	−4.93	−5.58	−5.87	−5.74	−5.40	−3.29	−2.43	−4.64
Noon. 1	−2.63	−3.85	−4.70	−6.37	−6.37	−5.02	−5.81	−5.96	−5.92	−5.51	−3.56	−2.70	−4.86
2	−2.50	−3.69	−4.43	−5.99	−6.37	−4.91	−5.76	−5.72	−5.49	−4.84	−3.22	−2.45	−4.61
3	−1.96	−3.02	−3.74	−5.22	−6.12	−4.64	−5.40	−5.27	−4.64	−3.71	−2.45	−1.85	−4.01
4	−1.26	−2.07	−2.81	−4.14	−5.47	−4.03	−4.64	−4.52	−3.49	−2.45	−1.55	−1.10	−3.13
5	−0.59	−1.10	−1.76	−2.86	−4.32	−3.04	−3.47	−3.44	−2.18	−1.33	−0.77	−0.41	−2.12
6	−0.07	−0.38	−0.74	−1.42	−2.68	−1.73	−2.00	−2.23	−0.88	−0.45	−0.23	0.05	−1.06
7	0.29	0.09	0.14	0.00	−0.81	−0.27	−0.38	−0.47	0.36	0.23	0.07	0.34	−0.05
8	0.50	0.36	0.86	1.26	0.99	0.74	1.10	1.06	1.37	0.79	0.25	0.52	0.81
9	0.63	0.56	1.35	2.25	2.36	2.21	2.27	2.23	2.12	1.33	0.47	0.72	1.55
10	0.72	0.77	1.69	2.93	3.29	2.93	3.11	2.97	2.59	1.85	0.77	0.88	2.05
11	0.79	0.99	1.89	3.35	3.89	3.44	3.67	3.40	2.84	2.23	1.08	1.01	2.39
Midn. . .	0.83	1.26	2.07	3.67	4.46	3.87	4.21	3.71	2.99	2.48	1.37	1.04	2.66
6. 6	0.74	1.06	1.17	1.28	0.86	0.54	0.68	0.99	1.37	1.46	0.97	0.68	0.99
7. 7	0.88	1 19	1.19	1.19	0.61	0.34	0.52	0.90	1.35	1.44	0.99	0.88	0.95
8. 8	0 81	1.01	0.92	0.79	0.23	0.14	0.23	0.47	0.95	1.04	0.74	0.83	0.65
9. 9	0.50	0.59	0.36	0.16	−0.27	−0.25	−0.16	−0.09	0.29	0.34	0.32	0.56	0.20
10.10	0.07	−0.05	−0.27	−0.52	−0.65	−0.47	−0.47	−0.63	−0.43	−0.43	−0.20	0.14	−0.32
7. 2. 9	−0.14	−0.27	−0.27	−0.45	−0.68	−0.59	−0.70	−0.41	−0.36	−0.29	−0.29	−0.11	−0.38
6. 2. 8	−0.16	−0.29	−0.16	−0.25	−0.34	−0.45	−0.43	−0.16	−0.16	−0.23	−0.27	−0.20	−0.25
6. 2.10	−0.09	−0.16	0.11	0.32	0.43	0.27	0.23	0.50	0.25	0.14	−0.11	−0.09	0.16
6. 2. 6	−0.34	−0.54	−0.70	−1.15	−1.55	−1.28	−1.46	−1.24	−0.92	−0.63	−0.43	−0.36	−0.88
7. 2	−0.52	−0.70	−1.10	−1.80	−2.18	−1.98	−2.18	−1.73	−1.60	−1.10	−0.68	−0.54	−1.35
8. 2	−0.70	−1.01	−1.73	−2.86	−3.47	−2.97	−3.22	−2.93	−2.50	−1.80	−0.99	−0.68	−2.07
8. 1	−0.77	−1.10	−1.87	−3.04	−3.47	−3.02	−3.24	−3.04	−2.72	−2.14	−1.17	−0.79	−2.21
7. 1	−0.59	−0.79	−1.24	−2.00	−2.18	−2.05	−2.21	−1.87	−1.80	−1.44	−0.83	−0.65	−1.46
9.12.3.9	−0.83	−1.31	−1.87	−2.77	−4.20	−2.52	−2.81	−2.84	−2.45	−2.12	−1.28	−0.79	−2.07
7. 2.2(9)	0.07	−0.07	0.14	0.23	0.09	0.11	0.05	0.25	0.27	0.11	−0.09	0.09	0.11

The numbers without sign must be added; those with the sign — must be subtracted.

XLVI.

England. — Plymouth. *Lat.* 50° 22′ N. *Long.* 4° 7′ W. *Greenw.*

Corrections to be applied to the Means of the Hours of Observation to obtain the true Mean Temperatures of the respective Days, Months, and of the Year. — Dove.

Degrees of Reaumur.

Hours.	Jan.	Feb.	March.	April.	May.	June.	July.	Aug.	Sept.	Oct.	Nov.	Dec.	Mean.
Morn. 1	0.38	0.65	1.03	1.78	2.28	1.93	2.11	1.85	1.44	1.18	0.70	0.42	1.31
2	0.40	0.74	1.17	1.97	2.64	2.14	2.39	2.13	1.60	1.24	0.75	0.38	1.46
3	0.44	0.83	1.34	2.18	2.94	2.28	2.53	2.42	1.79	1.34	0.80	0.33	1.60
4	0.51	0.94	1.47	2.28	3.00	2.22	2.48	2.56	1.93	1.47	0.87	0.36	1.67
5	0.61	1.05	1.51	2.18	2.68	2.03	2.14	2.41	1.89	1.56	0.93	0.46	1.62
6	0.68	1.10	1.37	1.77	1.94	1.24	1.49	1.87	1.61	1.50	0.95	0.58	1.34
7	0.65	1.02	1.00	1.06	0.89	0.42	0.63	1.00	1.03	1.18	0.84	0.62	0.86
8	0.49	0.74	0.43	0.13	−0.24	−0.45	−0.29	−0.05	0.22	0.56	0.55	0.50	0.22
9	0.16	0.26	−0.28	−0.86	−1.28	−1.20	−1.14	−1.06	−0.68	−0.29	0.07	0.18	−0.51
10	−0.27	−0.37	−0.99	−1.75	−2.03	−1.72	−1.79	−1.87	−1.53	−1.20	−0.52	−0.27	−1.19
11	−0.70	−1.00	−1.58	−2.40	−2.50	−2.06	−2.23	−2.38	−2.19	−1.96	−1.08	−0.74	−1.74
Noon. . .	−1.03	−1.48	−1.97	−2.74	−2.72	−2.19	−2.48	−2.61	−2.55	−2.40	−1.46	−1.08	−2.06
1	−1.17	−1.71	−2.09	−2.83	−2.83	−2.23	−2.58	−2.65	−2.63	−2.45	−1.58	−1.20	−2.16
2	−1.11	−1.64	−1.97	−2.66	−2.83	−2.18	−2.56	−2.54	−2.44	−2.15	−1.43	−1.09	−2.05
3	−0.87	−1.34	−1.66	−2.32	−2.72	−2.06	−2.40	−2.34	−2.06	−1.65	−1.09	−0.82	−1.78
4	−0.56	−0.92	−1.25	−1.84	−2.43	−1.79	−2.06	−2.01	−1.55	−1.09	−0.69	−0.49	−1.39
5	−0.26	−0.49	−0.78	−1.27	−1.92	−1.35	−1.54	−1.53	−0.97	−0.59	−0.34	−0.18	−0.94
6	−0.03	−0.17	−0.33	−0.63	−1.19	−0.77	−0.89	−0.99	−0.39	−0.20	−0.10	0.02	−0.47
7	0.13	0.04	0.06	−0.00	−0.36	−0.12	−0.17	−0.21	0.16	0.10	0.03	0.15	−0.02
8	0.22	0.16	0.38	0.56	0.44	0.33	0.49	0.47	0.61	0.35	0.11	0.23	0.36
9	0.28	0.25	0.60	1.00	1.05	0.98	1.01	0.99	0.94	0.59	0.21	0.32	0.69
10	0.32	0.34	0.75	1.30	1.46	1.30	1.38	1.32	1.15	0.82	0.34	0.39	0.91
11	0.35	0.44	0.84	1.49	1.73	1.53	1.63	1.51	1.26	0.99	0.48	0.45	1.06
Midn. . .	0.37	0.56	0.92	1.63	1.98	1.72	1.87	1.65	1.33	1.10	0.61	0.46	1.18
6. 6	0.33	0.47	0.52	0.57	0.38	0.24	0.30	0.44	0.61	0.65	0.43	0.30	0.44
7. 7	0.39	0.53	0.53	0.53	0.27	0.15	0.23	0.40	0.60	0.64	0.44	0.39	0.42
8. 8	0.36	0.45	0.41	0.35	0.10	−0.06	0.10	0.21	0.42	0.46	0.33	0.37	0.29
9. 9	0.22	0.26	0.16	0.07	−0.12	−0.11	−0.07	−0.04	0.13	0.15	0.14	0.25	0.09
10.10	0.03	−0.02	−0.12	−0.23	−0.29	−0.21	−0.21	−0.28	−0.19	−0.19	−0.09	0.06	−0.14
7. 2. 9	−0.06	−0.12	−0.12	−0.20	−0.30	−0.26	−0.31	−0.18	−0.16	−0.13	−0.13	−0.05	−0.17
6. 2. 8	−0.07	−0.13	−0.07	−0.11	−0.15	−0.20	−0.19	−0.07	−0.07	−0.10	−0.12	−0.09	−0.11
6. 2.10	−0.04	−0.07	0.05	0.14	0.19	0.12	0.10	0.22	0.11	0.06	−0.05	−0.04	0.07
6. 2. 6	−0.15	−0.24	−0.31	−0.51	−0.69	−0.57	−0.65	−0.55	−0.41	−0.28	−0.19	−0.16	−0.39
7. 2	−0.23	−0.31	−0.42	−0.80	−0.97	−0.88	−0.97	−0.77	−0.71	−0.49	−0.30	−0.24	−0.60
8. 2	−0.31	−0.45	−0.77	−1.27	−1.54	−1.32	−1.43	−1.30	−1.11	−0.80	−0.44	−0.30	−0.92
8. 1	−0.34	−0.49	−0.83	−1.35	−1.54	−1.34	−1.44	−1.35	−1.21	−0.95	−0.52	−0.35	−0.98
7. 1	−0.26	−0.35	−0.55	−0.89	−0.97	−0.91	−0.98	−0.83	−0.80	−0.64	−0.37	−0.29	−0.65
9.12.3.9	−0.37	−0.58	−0.83	−1.23	−1.42	−1.12	−1.25	−1.26	−1.09	−0.94	−0.57	−0.35	−0.92
7. 2.2(9)	0.03	−0.03	0.06	0.10	0.04	0.05	0.02	0.11	0.12	0.05	−0.04	0.04	0.05
Dail. ext.	−0.25	−0.31	−0.29	−0.28	0.09	0.03	−0.03	−0.05	−0.35	−0.45	−0.32	−0.29	−0.25

The numbers without sign must be added; those with the sign — must be subtracted.

E

XLVII.

Belgium. — Brussels. *Lat.* 50° 51′ N. *Long.* 4° 22′ E. *Greenw.*

Corrections to be applied to the Means of the Hours of Observation to obtain the true Mean Temperatures of the respective Days, Months, and of the Year. — Dove.

Degrees of Reaumur.

Hours.	Jan.	Feb.	March.	April.	May.	June.	July.	Aug.	Sept.	Oct.	Nov.	Dec.	Mean.
Morn. 1	0.58	0.67	1.19	2.23	2.57	2.83	2.34	2.49	1.71	0.85	0.49	0.73	1.56
2	0.60	0.73	1.36	2.59	2.89	3.12	2.57	2.84	2.00	0.99	0.49	0.39	1.71
3	0.60	0.79	1.54	2.99	3.17	3.18	2.74	3.20	2.33	1.15	0.54	0.08	1.86
4	0.60	0.86	1.70	3.29	3.28	3.14	2.74	3.42	2.57	1.31	0.65	0.02	1.97
5	0.62	0.92	1.79	3.29	3.06	2.71	2.47	3.32	2.58	1.40	0.77	0.25	1.93
6	0.64	0.97	1.74	2.86	2.45	2.00	1.88	2.82	2.28	1.35	0.85	0.65	1.71
7	0.61	0.93	1.50	2.01	1.52	1.10	1.06	1.94	1.67	1.11	0.81	0.97	1.27
8	0.46	0.75	1.03	0.86	0.44	0.16	0.15	0.82	0.82	0.68	0.58	0.97	0.64
9	0.18	0.39	0.39	−0.35	−0.59	−0.61	−0.69	−0.34	−0.14	0.08	0.19	0.56	−0.08
10	−0.22	−0.13	−0.36	−1.42	−1.43	−1.35	−1.33	−1.37	−1.06	−0.60	−0.31	−0.13	−0.81
11	−0.65	−0.71	−1.11	−2.23	−2.06	−1.86	−1.77	−2.19	−1.86	−1.23	−0.80	−0.84	−1.44
Noon. . .	−1.01	−1.23	−1.72	−2.77	−2.52	−2.27	−2.06	−2.81	−2.48	−1.71	−1.16	−1.29	−1.92
1	−1.20	−1.57	−2.13	−3.11	−2.89	−2.65	−2.29	−3.27	−2.88	−1.96	−1.32	−1.33	−2.22
2	−1.19	−1.65	−2.29	−3.29	−3.21	−2.97	−2.51	−3.58	−3.05	−1.95	−1.27	−1.03	−2.33
3	−0.99	−1.49	−2.21	−3.33	−3.40	−3.25	−2.69	−3.69	−2.98	−1.71	−1.05	−0.59	−2.28
4	−0.70	−1.14	−1.93	−3.18	−3.36	−3.16	−2.71	−3.53	−2.63	−1.31	−0.75	−0.26	−2.06
5	−0.39	−0.72	−1.51	−2.76	−2.97	−2.83	−2.47	−3.02	−2.05	−0.84	−0.45	−0.16	−1.68
6	−0.15	−0.33	−1.03	−2.05	−2.21	−2.17	−1.91	−2.19	−1.30	−0.39	−0.18	−0.25	−1.18
7	0.02	−0.03	−0.55	−1.13	−1.20	−1.28	−1.11	−1.15	−0.49	−0.01	0.03	−0.37	−0.61
8	0.12	0.17	−0.10	−0.16	−0.12	−0.31	−0.20	−0.09	0.23	0.28	0.19	−0.33	−0.03
9	0.21	0.31	0.28	0.69	0.82	0.68	0.64	0.82	0.78	0.48	0.32	0.05	0.50
10	0.31	0.41	0.59	1.31	1.51	1.37	1.31	1.48	1.13	0.60	0.41	0.37	0.90
11	0.42	0.50	0.83	1.70	1.96	1.97	1.77	1.89	1.33	0.68	0.47	0.75	1.19
Midn. . .	0.52	0.59	1.02	1.96	2.28	2.44	2.08	2.19	1.49	0.75	0.49	0.89	1.39
6. 6	0.25	0.32	0.35	0.41	0.12	−0.09	−0.01	0.31	0.49	0.48	0.33	0.20	0.26
7. 7	0.31	0.45	0.47	0.44	0.16	−0.09	−0.02	0.39	0.59	0.55	0.42	0.30	0.33
8. 8	0.29	0.46	0.47	0.35	0.16	−0.07	−0.03	0.37	0.53	0.48	0.39	0.32	0.31
9. 9	0.20	0.35	0.34	0.17	0.12	0.04	−0.02	0.24	0.32	0.28	0.25	0.25	0.21
10.10	0.05	0.14	0.11	0.05	0.04	0.01	−0.01	0.05	0.03	0.00	0.05	0.12	0.05
7. 2. 9	−0.12	−0.14	−0.17	−0.20	−0.29	−0.40	−0.27	−0.27	−0.20	−0.12	−0.05	−0.04	−0.19
6. 2. 8	−0.14	−0.17	−0.22	−0.20	−0.29	−0.43	−0.28	−0.28	−0.18	−0.11	−0.08	−0.24	−0.22
6. 2.10	−0.08	−0.09	0.01	0.29	0.25	0.13	0.23	0.24	0.12	0.00	−0.00	−0.00	0.09
6. 2. 6	−0.23	−0.34	−0.53	−0.83	−0.99	−1.05	−0.85	−0.98	−0.69	−0.33	−0.20	−0.21	−0.60
7. 2	−0.29	−0.36	−0.40	−0.64	−0.85	−0.94	−0.73	−0.82	−0.69	−0.42	−0.23	−0.03	−0.53
8. 2	−0.37	−0.45	−0.63	−1.22	−1.39	−1.41	−1.18	−1.38	−1.12	−0.64	−0.35	−0.03	−0.85
8. 1	−0.37	−0.41	−0.55	−1.13	−1.23	−1.25	−1.07	−1.23	−1.03	−0.64	−0.37	−0.18	−0.79
7. 1	−0.30	−0.32	−0.32	−0.55	−0.69	−0.78	−0.62	−0.67	−0.61	−0.43	−0.26	−0.18	−0.48
9.12.2.9	−0.40	−0.51	−0.82	−1.44	−1.42	−1.36	−1.20	−1.51	−1.21	−0.72	0.43	−0.34	−0.95
7. 2.2(9)	−0.04	−0.03	−0.06	0.03	−0.01	−0.13	−0.04	−0.00	0.05	0.03	0.05	−0.04	−0.02
Dail. ext.	−0.28	−0.34	−0.25	−0.02	−0.06	−0.04	0.02	−0.14	−0.24	−0.28	−0.24	−0.18	−0.18

The numbers without sign must be added; those with the sign — must be subtracted.

XLVIII.

Belgium. — Brussels. *Lat.* 50° 51′ N. *Long.* 4° 22′ E. *Greenw.*

Corrections to be applied to the Means of the Hours of Observation to obtain the true Mean Temperatures of the respective Days, Months, and of the Year. — Dove.

Degrees of Reaumur.

Hour.	Jan.	Feb.	March.	April.	May.	June.	July.	Aug.	Sept.	Oct.	Nov.	Dec.	Year.
Midn.	0.30	0.60	1.09	1.72	2.27	2.46	2.20	1.88	1.52	0.92	0.51	0.30	1.31
2	0.56	0.82	1.39	2.19	3.00	2.82	2.77	2.44	2.03	1.20	0.77	0.47	1.70
4	0.64	0.97	1.66	2.64	3.32	3.53	3.14	2.76	2.38	1.44	0.83	0.62	1.99
6	0.66	1.03	1.83	2.43	2.44	2.27	2.30	2.44	2.47	1.56	0.93	0.63	1.75
8	0.67	0.84	1.02	0.76	0.49	0.41	0.32	0.68	1.03	0.96	0.79	0.63	0.72
9	0.36	0.33	0.21	−0.38	0.61	−0.61	−0.63	−0.39	−0.14	0.07	0.21	0.34	0.00
10	0.07	−0.09	−0.54	−1.18	−1.43	−1.32	−1.36	−1.26	−1.19	−0.78	−0.36	−0.08	−0.79
Noon.	−0.92	−1.27	−1.78	−2.42	−2.61	−2.47	−2.35	−2.47	−2.46	−1.87	−1.27	−0.83	−1.89
2	−1.15	−1.65	−2.30	−2.95	−3.22	−3.21	−2.92	−3.08	−3.04	−2.17	−1.42	−1.04	−2.35
4	−0.72	−1.19	−2.04	−2.63	−3.15	−3.18	−2.90	−2.93	−2.70	−1.61	−0.90	−0.65	−2.05
6	−0.21	−0.49	−0.94	−1.71	−2.44	−2.57	−2.38	−1.87	−1.21	−0.37	−0.28	−0.18	−1.22
8	−0.08	−0.05	−0.00	0.13	0.05	−0.16	−0.15	0.17	0.21	0.23	0.07	−0.03	0.03
9	0.13	0.17	0.31	0.63	0.76	0.80	0.79	0.76	0.64	0.43	0.24	0.07	0.48
10	0.20	0.30	0.58	1.04	1.25	1.45	1.39	1.27	1.01	0.54	0.38	0.14	0.80
Mean.	0.52	2.45	3.56	7.27	10.37	13.10	13.69	13.58	11.22	7.69	4.72	1.89	

XLIX.

Germany. — Schwerin. *Lat.* 53° 36′ N. *Long.* 11° 30′ E. *Gr.* — Dove.

Degrees of Reaumur.

Hour.	Jan.	Feb.	March.	April.	May.	June.	July.	Aug.	Sept.	Oct.	Nov.	Dec.	Year.
Midn.	0.05	0.49	0.92	1.66	1.97	2.10	2.12	1.92	1.70	0.87	0.21	0.16	1.18
2	0.08	0.69	1.20	2.17	2.44	2.69	2.72	2.41	2.19	1.14	0.24	0.34	1.53
4	0.27	0.83	1.43	2.53	2.96	2.97	2.96	2.62	2.54	1.51	0.42	0.48	1.79
6	0.35	0.86	1.62	2.67	2.07	1.80	1.94	2.13	2.70	1.67	0.62	0.48	1.55
8	0.59	1.19	1.24	0.98	0.56	0.25	0.12	0.32	0.95	1.21	0.70	0.63	0.73
10	0.17	0.18	−0.11	−0.97	−1.15	−1.20	−1.26	−1.17	−1.12	−0.34	0.01	0.13	−0.57
Noon.	−0.42	−0.97	−1.32	−2.34	−2.47	−2.36	−2.20	−2.29	−2.42	−1.80	−0.77	−0.43	−1.65
2	−0.61	−0.72	−2.21	−3.50	−3.38	−3.23	−3.26	−3.45	−3.58	−2.54	−0.91	−0.68	−2.42
4	−0.43	−1.22	−2.13	−2.86	−2.70	−2.62	−2.76	−2.76	−3.03	−1.85	−0.62	−0.62	−1.97
6	−0.02	−0.42	−0.95	−1.54	−1.62	−1.71	−1.70	−1.37	−1.32	−0.55	−0.23	−0.27	−0.98
8	−0.07	−0.07	−0.11	0.13	0.11	−0.02	0.08	0.34	0.26	0.16	0.02	−0.14	0.06
10	0.06	0.21	0.45	1.01	1.15	1.28	1.29	1.30	1.19	0.57	0.24	−0.02	0.73
Mean.	−1.05	−2.00	1.18	5.26	8.45	12.19	13.50	13.02	10.42	7.48	1.42	−1.38	

The numbers without sign must be added; those with the sign — must be subtracted.

PRUSSIA. — MÜHLHAUSEN. *Lat.* 51° 13′ N. *Long.* 10° 27′ E. *Greenw.*

Corrections to be applied to the Means of the Hours of Observation to obtain the true Mean Temperatures of the respective Days, Months, and of the Year. — DOVE.

Degrees of Reaumur.

Hours.	Jan.	Feb.	March.	April.	May.	June.	July.	Aug.	Sept.	Oct.	Nov.	Dec.	Mean.
Morn. 1	0.71	1.28	1.10	1.84	2.40	3.56	2.91	2.49	1.95	1.39	0.47	0.58	1.72
2	0.75	1.30	1.28	2.19	2.80	3.97	3.30	2.80	2.20	1.65	0.53	0.59	1.95
3	0.77	1.33	1.46	2.40	3.06	4.16	3.50	3.06	3.29	1.85	0.60	0.60	2.17
4	0.82	1.40	1.60	2.74	3.06	3.98	3.42	3.14	2.70	1.99	0.66	0.62	2.18
5	0.86	1.47	1.62	2.61	2.67	3.40	3.00	2.98	2.73	2.05	0.68	0.66	2.06
6	0.91	1.50	1.46	2.25	2.06	2.49	2.22	2.51	2.46	1.93	0.63	0.67	1.76
7	0.86	1.36	1.11	1.41	1.15	1.32	1.20	1.73	1.03	1.50	0.46	0.59	1.14
8	0.62	0.98	0.55	0.58	0.16	0.11	0.09	0.86	0.87	0.84	0.16	0.46	0.52
9	0.21	0.33	−0.02	−0.38	−0.75	−1.02	−0.97	−0.36	−0.26	−0.03	−0.22	0.03	−0.29
10	−0.38	−0.50	−0.70	−1.16	−1.50	−1.98	−1.82	−1.38	−1.40	−0.99	−0.62	−0.54	−1.08
11	−0.93	−1.35	−1.30	−1.97	−2.06	−2.77	−2.46	−2.24	−2.42	−1.88	−0.92	−0.77	−1.76
Noon. . .	−1.38	−2.02	−1.76	−2.42	−2.44	−3.39	−2.94	−2.89	−3.14	−2.53	−1.09	−1.06	−2.26
1	−1.58	−2.38	−2.02	−2.80	−2.71	−3.86	−3.26	−3.29	−3.52	−2.82	−1.08	−1.15	−2.54
2	−1.52	−2.38	−2.07	−2.94	−2.87	−4.14	−3.42	−3.46	−3.54	−2.99	−0.89	−1.10	−2.61
3	−1.24	−2.07	−1.90	−2.85	−2.89	−4.13	−3.36	−3.39	−3.23	−2.48	−0.66	−0.81	−2.42
4	−0.84	−1.56	−1.58	−2.39	−2.69	−3.78	−3.06	−3.07	−2.65	−1.89	−0.39	−0.50	−2.03
5	−0.44	−1.02	−1.11	−1.95	−2.19	−3.06	−2.52	−2.51	−1.89	−1.21	−0.14	−0.23	−1.52
6	−0.20	−0.54	−0.62	−1.20	−1.59	−2.10	−1.76	−1.76	−1.06	−0.58	0.02	−0.02	−0.95
7	−0.04	−0.17	−0.18	−0.47	−0.83	−1.02	−0.85	−0.90	−0.24	−0.03	0.06	0.12	−0.38
8	0.18	0.13	0.16	0.09	−0.08	0.05	0.03	−0.05	0.50	0.38	0.22	0.26	0.16
9	0.27	0.41	0.45	0.53	0.58	1.01	0.81	0.71	0.99	0.70	0.26	0.32	0.59
10	0.37	0.66	0.64	0.89	1.10	1.76	1.46	1.24	1.35	0.91	0.34	0.40	0.93
11	0.53	0.89	0.78	1.14	1.56	2.42	2.01	1.78	1.58	1.10	0.38	0.47	1.22
Midn. . .	0.64	1.08	0.94	1.58	1.98	3.05	3.29	2.16	1.75	1.26	0.42	0.54	1.56
6. 6	0.36	0.48	0.42	0.53	0.24	0.20	0.23	0.38	0.70	0.68	0.33	0.33	0.41
7. 7	0.41	0.60	0.47	0.47	0.16	0.15	0.18	0.42	0.40	0.74	0.26	0.36	0.38
8. 8	0.40	0.56	0.36	0.34	0.04	0.08	0.06	0.41	0.69	0.61	0.19	0.36	0.34
9. 9	0.24	0.37	0.22	0.08	−0.09	−0.01	−0.08	0.18	0.37	0.34	0.02	0.18	0.15
10.10	−0.01	0.08	−0.03	−0.14	−0.20	−0.11	−0.18	−0.07	−0.03	−0.04	−0.14	−0.07	−0.08
7. 2. 9	−0.13	−0.20	−0.17	−0.23	−0.38	−0.60	−0.47	−0.34	−0.51	−0.26	−0.06	−0.06	−0.29
6. 2. 8	−0.14	−0.25	−0.15	−0.20	−0.30	−0.53	−0.39	−0.33	−0.19	−0.23	−0.01	−0.06	−0.23
6. 2.10	−0.08	−0.07	0.01	0.07	0.10	0.04	0.09	0.10	0.09	−0.05	0.03	−0.01	0.03
6. 2. 6	−0.27	−0.47	−0.41	−0.63	−0.80	−1.25	−0.99	−0.90	−0.71	−0.55	−0.08	−0.15	−0.60
7. 2	−0.33	−0.51	−0.48	−0.77	−0.86	−1.41	−1.11	−0.87	−1.26	−0.75	−0.22	−0.26	−0.74
8. 2	−0.45	−0.70	−0.76	−1.18	−1.36	−2.02	−1.67	−1.30	−1.34	−1.08	−0.37	−0.32	−1.05
8. 1	−0.48	−0.70	−0.74	−1.11	−1.28	−1.88	−1.59	−1.22	−1.33	−0.99	−0.46	−0.35	−1.01
7. 1	−0.36	−0.51	−0.46	−0.70	−0.78	−1.27	−1.03	−0.78	−1.25	−0.66	−0.31	−0.28	−0.70
9.12.3.9	−0.54	−0.84	−0.81	−1.28	−1.38	−1.88	−1.62	−1.48	−1.41	−1.09	−0.43	−0.38	−1.10
7. 2.2(9)	−0.03	−0.05	−0.02	−0.12	−0.14	−0.20	−0.15	−0.08	−0.13	−0.02	0.02	0.03	−0.07
Dail. ext.	−0.34	−0.44	−0.23	−0.10	0.09	0.01	0.04	−0.16	−0.13	−0.47	−0.21	−0.24	−0.22

The numbers without sign must be added; those with the sign — must be subtracted.

LI.

HOLLAND. — UTRECHT. *Lat.* 52° 5′ N. *Long.* 5° 8′ E. *Greenw.*

Corrections to be applied to the Means of the Hours of Observation to obtain the true Mean Temperatures of the respective Days, Months, and of the Year. — DOVE.

Degrees of Reaumur.

Hour.	Jan.	Feb.	March.	April.	May.	June.	July.	Aug.	Sept.	Oct.	Nov.	Dec.	Year.
Midn.	0.36	0.62	1.13	1.71	2.56	2.74	2.64	1.87	1.91	1.07	0.76	0.11	1.44
1	0.37	0.74	1.18	1.87	2.86	3.29	2.67	1.91	2.10	1.11	0.70	0.19	1.58
2	0.46	0.82	1.24	2.00	3.00	3.21	2.82	2.02	2.21	1.18	0.78	0.32	1.67
3	0.51	0.87	1.27	2.10	3.02	3.25	2.97	2.07	2.34	1.25	0.82	0.42	1.74
4	0.57	0.90	1.31	2.16	2.70	2.84	2.76	2.06	2.45	1.31	0.82	0.44	1.69
5	0.61	0.97	1.26	1.92	1.80	1.82	1.86	1.80	2.42	1.42	0.90	0.50	1.44
6	0.66	0.98	1.02	1.30	0.67	0.44	0.33	1.05	1.87	1.22	0.91	0.46	0.91
7	0.64	0.84	0.62	0.37	−0.38	−0.70	−0.77	0.04	0.72	0.39	0.78	0.38	0.24
8	0.50	0.56	−0.01	−0.40	−1.17	−1.50	−1.28	−0.68	−0.39	0.12	0.29	0.31	−0.30
9	0.13	−0.07	−0.53	−1.20	−1.68	−2.02	−1.69	−1.33	−1.12	−0.50	−0.22	0.14	−0.84
10	−0.26	−0.49	−1.05	−1.71	−2.06	−2.42	−2.02	−1.65	−1.79	−1.12	−0.71	−0.14	−1.29
11	−0.62	−0.97	−1.50	−2.16	−2.46	−2.78	−2.27	−1.87	−2.34	−1.68	−1.15	−0.33	−1.68
Noon.	−0.85	−1.34	−1.77	−2.41	−2.78	−2.94	−2.53	−2.16	−2.83	−1.98	−1.49	−0.62	−1.97
1	−0.98	−1.58	−1.88	−2.42	−2.94	−3.00	−2.61	−2.40	−3.07	−2.11	−1.62	−0.75	−2.11
2	−1.02	−1.54	−1.82	−2.42	−2.88	−2.94	−2.60	−2.30	−2.99	−1.99	−1.43	−0.66	−2.05
3	−0.81	−1.21	−1.54	−2.24	−2.58	−2.64	−1.58	−2.13	−2.68	−1.64	−1.08	−0.47	−1.72
4	−0.60	−0.89	−1.25	−1.82	−2.06	−2.20	−2.00	−1.79	−2.06	−1.10	−0.70	−0.23	−1.39
5	−0.35	−0.48	−0.75	−1.23	−1.42	−1.53	−1.62	−1.30	−1.34	−0.52	−0.42	−0.17	−0.93
6	−0.19	−0.21	−0.24	−0.47	−0.76	−0.74	−0.76	−0.61	−0.52	−0.11	−0.18	−0.10	−0.41
7	−0.05	−0.03	0.14	0.20	0.07	0.17	0.02	0.14	0.10	0.22	−0.02	−0.03	0.06
8	0.05	0.12	0.48	0.72	0.85	1.01	0.82	0.86	0.62	0.53	0.18	0.02	0.52
9	0.22	0.23	0.74	1.13	1.51	1.77	1.50	1.24	1.17	0.84	0.40	0.06	0.90
10	0.36	0.40	0.94	1.41	1.92	2.25	1.96	1.52	1.51	1.01	0.58	0.04	1.16
11	0.36	0.67	1.02	1.58	2.16	2.53	2.17	1.70	1.76	1.14	1.06	0.02	1.35
Mean.	−2.83	4.18	3.20	7.14	10.55	12.95	13.75	12.90	10.87	6.88	4.65	0.76	

LII.

ENGLAND. — GREENWICH. *Lat.* 51° 28′ 38″ N. *Long.* 0° 0′. — DOVE.

Degrees of Reaumur.

Hour.	Jan.	Feb.	March.	April.	May.	June.	July.	Aug.	Sept.	Oct.	Nov.	Dec.	Year.
A.M. 1	0.44	0.75	1.44	2.32	2.72	3.24	2.73	2.49	2.05	1.34	0.67	0.47	1.72
3	0.62	0.94	1.66	2.66	3.04	3.70	3.11	2.82	2.40	1.42	0.80	0.56	1.98
5	0.75	1.06	1.92	2.84	2.84	3.25	2.91	2.89	2.58	1.54	0.87	0.56	2.00
7	0.86	1.08	1.60	1.31	0.75	0.80	0.88	1.22	1.65	1.26	0.88	0.60	1.07
9	0.41	0.24	−0.22	−0.82	−1.30	−1.52	−1.14	−1.14	−0.76	−0.30	0.11	0.24	−0.50
11	−0.74	−1.03	−1.90	−2.48	−2.60	−2.91	−2.67	−2.64	−2.57	−1.88	−1.06	−0.73	−1.93
P.M. 1	−1.25	−1.73	−2.62	−3.31	−3.36	−3.75	−3.17	−3.40	−3.28	−2.40	−1.64	−1.20	−2.59
3	−1.10	−1.59	−2.43	−3.08	−3.02	−3.60	−3.09	−3.20	−2.94	−2.04	−1.26	−0.85	−2.35
5	−0.36	−0.63	−1.33	−2.04	−2.05	−2.51	−2.24	−2.11	−1.65	−0.73	−0.38	−0.24	−1.37
7	0.03	0.05	0.09	−0.16	−0.29	−0.58	−0.50	−0.11	0.04	0.11	0.09	0.00	−0.10
9	0.10	0.32	0.71	0.99	1.20	1.40	1.13	1.22	0.89	0.63	0.40	0.21	0.77
11	0.23	0.54	1.11	1.77	2.06	2.52	2.08	1.96	1.60	1.07	0.53	0.37	1.33
Mean.	2.48	2.53	4.53	6.71	9.62	12.47	13.08	12.98	11.12	7.71	5.47	3.09	

The numbers without sign must be added; those with the sign — must be subtracted.

ENGLAND. — GREENWICH. *Lat.* 51° 29′ N. *Long.* 0° 0′.

Corrections to be applied to the Means of the Hours of Observation to obtain the true Mean Temperatures of the respective Days, Months, and of the Year. — DOVE.

Degrees of Reaumur.

Hours.	Jan.	Feb.	March.	April.	May.	June.	July.	Aug.	Sept.	Oct.	Nov.	Dec.	Mean.
Morn. 1	0.38	0.68	1.29	2.21	2.72	3.13	2.61	2.61	1.89	1.28	0.60	0.40	1.65
2	0.63	0.82	1.44	2.31	2.85	3.30	2.71	2.68	2.06	1.45	0.75	0.52	1.79
3	0.83	0.95	1.62	2.44	2.91	3.41	2.74	2.78	2.22	1.56	0.88	0.59	1.91
4	0.93	1.02	1.82	2.54	2.85	3.40	2.71	2.86	2.34	1.60	0.95	0.62	1.97
5	0.93	1.03	1.95	2.46	2.60	3.14	2.53	2.81	2.35	1.56	0.95	0.62	1.91
6	0.84	0.97	1.93	2.17	2.08	2.52	2.11	2.48	2.15	1.42	0.89	0.60	1.68
7	0.71	0.84	1.66	1.56	1.25	1.53	1.38	1.77	1.67	1.15	0.75	0.57	1.24
8	0.53	0.61	1.11	0.66	0.20	0.28	0.40	0.72	0.88	0.71	0.52	0.48	0.59
9	0.30	0.26	0.30	−0.37	−0.92	−1.02	−0.71	−0.55	−0.13	0.09	0.19	0.28	−0.19
10	−0.01	−0.20	−0.66	−1.43	−1.94	−2.12	−1.73	−1.78	−1.23	−0.66	−0.26	−0.04	−1.01
11	−0.39	−0.75	−1.60	−2.30	−2.70	−2.89	−2.51	−2.79	−2.22	−1.43	−0.77	−0.46	−1.73
Noon. . .	−0.79	−1.27	−2.35	−2.87	−3.13	−3.28	−2.94	−3.43	−2.94	−2.07	−1.25	−0.87	−2.27
1	−1.12	−1.66	−2.79	−3.17	−3.26	−3.39	−3.04	−3.69	−3.28	−2.45	−1.59	−1.17	−2.55
2	−1.28	−1.81	−2.85	−3.14	−3.16	−3.34	−2.91	−3.63	−3.23	−2.48	−1.69	−1.25	−2.56
3	−1.21	−1.67	−2.57	−2.92	−2.90	−3.21	−2.67	−3.34	−2.86	−2.17	−1.51	−1.10	−2.34
4	−0.95	−1.29	−2.05	−2.54	−2.54	−3.01	−2.38	−2.89	−2.28	−1.63	−1.10	−0.76	−1.95
5	−0.58	−0.78	−1.40	−1.97	−2.06	−2.67	−2.30	−2.30	−1.60	−1.01	−0.59	−0.36	−1.45
6	−0.22	−0.26	−0.75	−1.34	−1.45	−2.10	−1.57	−1.56	−0.91	−0.43	−0.10	−0.01	−0.89
7	0.03	0.14	−0.17	−0.60	−0.71	−1.26	−0.96	−0.69	−0.27	0.02	0.24	0.20	−0.34
8	0.11	0.37	0.30	0.17	0.11	−0.24	−0.19	0.24	0.29	0.32	0.41	0.26	0.18
9	0.08	0.46	0.65	0.84	0.92	0.81	0.64	1.11	0.77	0.52	0.44	0.23	0.62
10	0.03	0.48	0.89	1.42	1.62	1.74	1.41	1.81	1.17	0.69	0.41	0.19	0.99
11	0.04	0.49	1.05	1.81	2.16	2.42	2.01	2.27	1.47	0.87	0.40	0.20	1.27
Midn. . .	0.16	0.56	1.17	2.03	2.51	2.86	2.40	2.51	1.70	1.08	0.46	0.28	1.48
6. 6	0.31	0.36	0.59	0.42	0.31	0.21	0.27	0.46	0.62	0.50	0.39	0.30	0.40
7. 7	0.37	0.49	0.75	0.48	0.27	0.13	0.21	0.54	0.70	0.59	0.50	0.38	0.45
8. 8	0.32	0.49	0.71	0.42	0.16	0.02	0.10	0.48	0.59	0.52	0.47	0.37	0.39
9. 9	0.19	0.36	0.48	0.24	0.00	−0.10	−0.04	0.28	0.32	0.31	0.31	0.25	0.22
10.10	0.01	0.14	0.12	−0.00	−0.16	−0.19	−0.16	0.01	−0.03	0.02	0.08	0.07	−0.01
7. 2. 9	−0.16	−0.17	−0.18	−0.25	−0.33	−0.33	−0.30	−0.25	−0.26	−0.27	−0.17	−0.15	−0.24
6. 2. 8	−0.11	−0.16	−0.21	−0.27	−0.32	−0.35	−0.33	−0.30	−0.26	−0.25	−0.13	−0.13	−0.24
6. 2.10	−0.14	−0.12	−0.01	0.15	0.18	0.31	−0.20	0.22	0.03	−0.12	−0.13	−0.15	0.04
6. 2. 6	−0.22	−0.37	−0.56	−0.77	−0.84	−0.97	−0.79	−0.90	−0.66	−0.50	−0.30	−0.22	−0.59
7. 2	−0.29	−0.49	−0.60	−0.79	−0.96	−0.91	−0.77	−0.93	−0.78	−0.67	−0.47	−0.34	−0.67
8. 2	−0.38	−0.60	−0.87	−1.24	−1.48	−1.53	−1.26	−1.46	−1.18	−0.89	−0.59	−0.39	−0.99
8. 1	−0.30	−0.53	−0.84	−1.26	−1.53	−1.56	−1.31	−1.49	−1.20	−0.87	−0.54	−0.35	−0.98
7. 1	−0.21	−0.41	−0.57	−0.81	−1.01	−0.93	−0.83	−0.96	−0.81	−0.65	−0.42	−0.30	−0.66
9.12.3.9	−0.41	−0.56	−0.99	−1.33	−1.51	−1.68	−1.42	−1.55	−1.29	−0.91	−0.53	−0.37	−1.05
7. 2.2(9)	−0.10	−0.01	0.03	0.03	−0.02	−0.05	−0.06	0.09	−0.01	−0.07	−0.02	−0.06	−0.02
Dail. ext.	−0.18	−0.39	−0.45	−0.32	−0.18	0.01	−0.15	−0.42	−0.47	−0.44	−0.37	−0.32	−0.30

The numbers without sign must be added; those with the sign — must be subtracted.

LIV.

ENGLAND. — GREENWICH. *Lat.* 51° 29′ N. *Long.* 0° 0′.

Corrections to be applied to the Means of the Hours of Observation to obtain the true Mean Temperatures of the respective Days, Months, and of the Year. — GLAISHER.

Degrees of Fahrenheit.

Hours.	Jan.	Feb.	March.	April.	May.	June.	July.	Aug.	Sept.	Oct.	Nov.	Dec.	Mean.
Midn. . .	1.0	1.6	2.9	4.8	5.4	6.2	5.0	5.1	4.0	2.9	1.7	0.9	3.5
1	0.9	1.8	3.0	5.2	6.0	7.1	5.5	5.5	4.5	3.0	1.8	1.0	3.8
2	1.2	2.0	3.3	5.7	6.4	8.0	6.0	6.0	5.5	3.4	2.0	1.2	4.2
3	1.3	2.1	3.6	6.2	6.7	8.7	6.4	6.3	6.4	3.6	2.0	1.3	4.5
4	1.6	2.3	3.9	6.6	6.7	9.3	6.6	6.5	6.6	3.8	2.1	1.4	4.8
5	1.8	2.2	4.0	6.7	6.3	8.8	6.2	6.5	6.2	3.8	2.0	1.4	4.7
6	1.9	2.3	3.9	6.0	4.8	6.4	4.5	5.5	5.3	3.5	1.9	1.4	3.9
7	1.9	2.1	3.6	4.3	2.6	3.0	2.5	3.3	4.0	2.8	1.7	1.5	2.8
8	1.5	1.6	2.5	2.0	0.5	0.0	0.0	0.9	2.1	1.6	1.0	1.3	1.2
9	1.0	0.7	0.2	−0.9	−2.0	−2.5	−2.0	−1.6	−0.4	0.0	0.4	0.9	−0.5
10	0.2	−0.5	−1.9	−3.2	−4.0	−4.5	−4.0	−3.5	−3.0	−2.0	−0.6	0.0	−2.2
11	−1.3	−2.1	−3.5	−5.3	−5.5	−5.8	−5.4	−5.4	−5.0	−3.8	−2.0	−1.3	−3.9
Noon. . .	−2.3	−3.2	−5.0	−6.8	−6.7	−7.3	−6.4	−6.5	−6.4	−5.1	−3.1	−2.1	−5.1
1	−2.9	−3.9	−5.8	−7.9	−7.5	−8.1	−6.7	−7.5	−7.1	−5.5	−3.5	−2.4	−5.7
2	−3.0	−3.9	−5.8	−8.2	−7.7	−8.6	−6.7	−7.7	−7.1	−4.9	−3.6	−2.3	−5.8
3	−2.5	−3.6	−5.5	−7.7	−7.3	−8.4	−6.5	−7.0	−6.6	−3.7	−3.0	−1.9	−5.3
4	−1.9	−2.8	−4.5	−6.7	−6.1	−7.4	−5.8	−5.5	−5.5	−2.8	−2.1	−1.3	−4.4
5	−1.1	−1.6	−3.3	−5.4	−4.8	−6.1	−4.9	−3.6	−4.2	−1.7	−1.2	−0.8	−3.2
6	−0.6	−0.6	−1.8	−3.5	−3.0	−4.5	−3.5	−2.0	−2.5	−0.8	−0.4	−0.4	−2.0
7	−0.3	0.3	−0.4	−1.1	−1.0	−2.4	−1.5	−0.5	−0.6	0.0	0.1	−0.1	−0.6
8	0.1	0.6	0.9	0.7	0.9	0.0	0.3	1.0	1.0	0.7	0.6	0.2	0.6
9	0.4	1.0	1.7	2.0	2.3	1.8	1.9	2.4	1.8	1.3	1.0	0.4	1.5
10	0.6	1.3	2.3	3.2	3.5	3.6	3.3	3.3	2.7	1.9	1.3	0.5	2.3
11	0.7	1.5	2.6	4.1	4.5	5.0	4.2	4.3	3.4	2.4	1.5	0.8	2.9
6. 6	0.6	0.9	1.0	1.2	0.9	0.9	0.5	1.7	1.4	1.3	0.8	0.5	0.9
7. 7	0.8	1.2	1.6	1.6	0.8	0.3	0.5	1.4	1.7	1.4	0.9	0.7	1.1
8. 8	0.8	1.1	1.7	1.3	0.7	0.0	0.1	0.9	1.5	1.1	0.8	0.8	0.9
9. 9	0.7	0.8	0.9	0.5	0.1	−0.3	−0.0	0.4	0.7	0.6	0.7	0.6	0.5
10.10	0.4	0.4	0.2	0.0	−0.2	−0.4	−0.4	−0.1	−0.1	−0.0	0.4	0.2	0.0
7. 2. 9	−0.2	−0.3	−0.2	−0.6	−0.9	−1.2	−0.8	−0.7	−0.4	−0.2	−0.3	−0.1	−0.5
6. 2. 8	−0.3	−0.3	−0.3	−0.5	−0.7	−0.7	−0.6	−0.4	−0.3	−0.2	−0.4	−0.2	−0.4
6. 2.10	−0.2	−0.1	0.1	0.3	0.2	0.5	0.4	0.3	0.3	0.2	−0.1	−0.1	0.1
6. 2. 6	−0.6	−0.7	−1.2	−1.9	−1.9	−2.2	−1.9	−1.4	−1.4	−0.7	−0.7	−0.4	−1.3
7. 2	−0.5	−0.9	−1.1	−1.9	−2.5	−2.8	−2.1	−2.2	−1.5	−1.0	−0.9	−0.4	−1.5
8. 2	−0.7	−1.1	−1.6	−3.1	−3.6	−4.3	−3.3	−3.4	−2.5	−1.7	−1.3	−0.5	−2.3
8. 1	−0.7	−1.1	−1.6	−2.9	−3.5	−4.0	−3.4	−3.3	−2.5	−1.9	−1.3	−0.5	−2.2
7. 1	−0.5	−0.9	−1.1	−1.8	−2.4	−2.6	−2.1	−2.1	−1.5	−1.4	−0.9	−0.4	−1.5
9.12.3.9	−0.8	−1.3	−2.1	−3.3	−3.4	−4.1	−3.2	−3.2	−2.9	−1.9	−1.2	−0.7	−2.4

The numbers without sign must be added; those with the sign — must be subtracted.

Prussia. — Halle. *Lat.* 51° 30′ N. *Long.* 11° 57′ E. *Greenw.*

Corrections to be applied to the Means of the Hours of Observation to obtain the true Mean Temperatures of the respective Days, Months, and of the Year. — Dove.

Degrees of Reaumur.

Hours.	Jan.	Feb.	March.	April.	May.	June.	July.	Aug.	Sept.	Oct.	Nov.	Dec.	Mean.
Morn. 1	0.53	1.00	1.36	2.52	3.98	3.91	3.72	3.32	2.70	2.01	0.95	0.46	2.21
2	0.56	1.14	1.58	2.86	4.10	3.94	3,82	3.57	2.99	2.22	0.97	0.48	2.35
3	0.60	1.26	1.74	3.00	3.78	3.62	3.56	3.56	3.12	2.37	1.01	0.50	2.34
4	0.66	1.34	1.82	2.94	3.10	2.95	2.97	3.27	3.02	2.41	1.03	0.54	2.17
5	0.72	1.36	1.72	2.62	2.18	2.09	2.14	2.64	2.62	2.25	1.00	0.55	1.82
6	0.72	1.30	1.42	1.98	1.30	1.18	1.24	1.90	1.97	1.90	0.92	0.58	1.37
7	0.65	1.10	0.94	1.07	0.32	0.25	0.23	0.84	0.98	1.32	0.74	0.55	0.75
8	0.36	0.53	0.20	0.03	−0.56	−0.58	−0.57	−0.20	0.12	0.33	0.30	0.28	0.02
9	0.05	−0.08	−0.66	−0.98	−1.34	−1.34	−1.30	−1.20	−1.14	−0.71	−0.31	−0.09	−0.76
10	−0.45	−0.76	−1.18	−1.86	−2.09	−2.01	−1.99	−2.10	−2.03	−1.66	−0.87	−0.54	−1.46
11	−0.82	−1.29	−1.73	−2.58	−2.66	−2.68	−2.65	−2.90	−2.72	−2.44	−1.35	−0.90	−2.06
Noon. . .	−1.09	−1.77	−2.06	−3.08	−3.14	−3.07	−3.16	−3.35	−3.11	−2.86	−1.66	−1.08	−2.45
1	−1.17	−2.02	−2.22	−3.32	−3.33	−3.35	−3.46	−3.53	−3.30	−3.01	−1.73	−1.09	−2.63
2	−1.06	−1.86	−2.10	−3.26	−3.37	−3.46	−3.54	−3.57	−3.27	−2.76	−1.52	−0.94	−2.56
3	−0.86	−1.49	−1.86	−2.90	−3.13	−3.23	−3.29	−3.30	−2.98	−2.32	−1.14	−0.74	−2.27
4	−0.53	−1.01	−1.42	−2.39	−2.74	−2.74	−2.76	−2.84	−2.50	−1.81	−0.75	−0.42	−1.83
5	−0.30	−0.59	−0.91	−1.78	−2.24	−2.22	−2.16	−1.97	−1.83	−1.20	−0.40	−0.20	−1.32
6	−0.13	−0.29	−0.52	−0.96	−1.58	−1.50	−1.39	−1.38	−1.12	−0.69	−0.14	−0.03	−0.81
7	−0.00	−0.09	−0.06	−0.34	−0.86	−0.73	−0.55	−0.59	−0.38	−0.21	0.04	0.09	−0.31
8	0.11	0.13	0.26	0.32	−0.10	0.07	0.26	0.15	0.29	0.25	0.21	0.22	−0.18
9	0.21	0.30	0.59	0.88	0.68	0.90	1.09	0.90	0.87	0.68	0.39	0.34	0.65
10	0.31	0.46	0.79	1.33	1.64	1.81	1.87	1.61	1.42	1.12	0.59	0.37	1.11
11	0.41	0.65	0.98	1.78	2.61	2.69	2.64	2.30	1.90	1.47	0.76	0.40	1.55
Midn. . .	0.48	0.83	1.16	2.17	3.43	3.42	3.29	2.86	2.33	1.77	0.89	0.43	1.92
6. 6	0.21	0.39	0.41	0.42	−0.03	−0.07	−0.01	0.34	0.40	0.53	0.30	0.18	0.26
7. 7	0.30	0.51	0.45	0.51	−0.14	−0.16	−0.08	0.26	0.43	0.61	0.39	0.28	0.28
8. 8	0.33	0.51	0.44	0.37	−0.27	−0.24	−0.16	0.13	0.30	0.56	0.39	0.32	0.22
9. 9	0.24	0.33	0.23	0.18	−0.33	−0.26	−0.16	−0.03	0.21	0.29	0.26	0.25	0.10
10.10	0.13	0.11	−0.04	−0.05	−0.33	−0.22	−0.11	−0.15	−0.14	−0.02	0.04	0.13	−0.05
7. 2. 9	−0.11	−0.20	−0.18	−0.34	−0.71	−0.70	−0.65	−0.49	−0.35	−0.29	−0.20	−0.10	−0.36
6. 2. 8	−0.15	−0.25	−0.19	−0.35	−0.67	−0.66	−0.62	−0.49	−0.35	−0.32	−0.23	−0.15	−0.37
6. 2.10	−0.08	−0.12	0.03	0.06	−0.16	−0.12	−0.08	−0.00	0.06	−0.03	−0.11	−0.07	−0.05
6. 2. 6	−0.25	−0.42	−0.47	−0.83	−1.13	−1.16	−1.16	−0.95	−0.84	−0.65	−0.38	−0.25	−0.71
7. 2	−0.23	−0.36	−0.40	−0.67	−1.02	−1.09	−1.11	−0.82	−0.67	−0.56	−0.41	−0.26	−0.63
8. 2	−0.26	−0.46	−0.64	−1.13	−1.51	−1.55	−1.62	−1.35	−1.16	−0.85	−0.50	−0.27	−0.94
8. 1	−0.22	−0.34	−0.56	−1.01	−1.41	−1.41	−1.47	−1.26	−1.07	−0.77	−0.46	−0.27	−0.85
7. 1	−0.19	−0.24	−0.32	−0.55	−0.92	−0.95	−0.96	−0.73	−0.57	−0.48	−0.37	−0.25	−0.54
9.12.3.9	−0.35	−0.62	−0.84	−1.37	−1.67	−1.66	−1.63	−1.63	−1.40	−1.16	−0.59	−0.34	−1.10
7. 2.2(9)	−0.06	−0.12	−0.07	−0.18	−0.56	−0.51	−0.43	−0.33	−0.19	−0.15	−0.10	−0.02	−0.23
Dail. ext.	−0.23	−0.33	−0.20	−0.16	0.37	0.24	0.14	0.00	−0.09	−0.30	−0.35	−0.26	−0.14

The numbers without sign must be added; those with the sign — must be subtracted.

LVI.

HANOVER. — GÖTTINGEN. *Lat.* 51° 32′ N. *Long.* 9° 56′ E. *Greenw.*

Corrections to be applied to the Means of the Hours of Observation to obtain the true Mean Temperatures of the respective Days, Months, and of the Year. — DOVE.

Degrees of Reaumur.

Hours.	Jan.	Feb.	March.	April.	May.	June.	July.	Aug.	Sept.	Oct.	Nov.	Dec.	Mean.
Morn. 1	0.90	1.13	1.58	2.24	3.31	3.43	3.56	3.35	2.31	1.58	0.69	0.60	2.06
2	0.92	1.14	1.77	2.49	3.70	3.71	3.82	3.70	2.68	1.75	0.74	0.59	2.25
3	0.94	1.16	2.01	2.79	3.93	3.73	3.92	3.92	3.23	1.94	0.82	0.58	2.41
4	0.99	1.20	2.22	3.04	3.91	3.57	3.79	3.89	3.63	2.10	0.92	0.58	2.49
5	1.15	1.26	2.29	3.08	3.55	3.10	3.36	3.52	3.62	2.15	1.00	0.62	2.39
6	1.12	1.20	2.10	2.73	2.62	2.22	2.59	2.79	3.50	1.99	1.08	0.66	2.05
7	1.13	1.14	1.77	2.24	1.78	1.21	1.40	1.69	2.62	1.58	0.94	0.65	1.51
8	1.12	0.80	1.02	0.89	0.75	0.49	0.48	0.56	1.36	1.08	0.53	0.54	0.80
9	0.50	–0.08	–0.14	–0.16	–0.47	–0.55	–0.65	–0.68	–0.22	–0.21	0.10	0.30	–0.19
10	–0.37	–0.88	–1.09	–1.32	–1.53	–1.60	–2.22	–1.84	–1.45	–0.82	–0.42	–0.02	–1.13
11	–1.26	–1.78	–1.87	–2.30	–2.59	–2.53	–2.74	–2.83	–2.45	–1.74	–0.99	–0.74	–1.99
Noon. . .	–1.83	–2.17	–2.43	–2.98	–3.30	–3.19	–3.48	–3.52	–3.37	–2.50	–1.46	–1.12	–2.61
1	–2.02	–2.32	–2.81	–3.37	–3.82	–3.72	–3.78	–3.82	–3.80	–2.89	–1.58	–1.42	–2.95
2	–2.03	–2.23	–3.05	–3.56	–3.98	–4.03	–4.09	–4.15	–4.00	–2.98	–1.60	–1.28	–3.08
3	–1.74	–1.98	–2.88	–3.48	–3 95	–3.91	–4.00	–4.03	–4.03	–2.84	–1.32	–1.02	–2.93
4	–1.23	–1.35	–2.48	–3.24	–3.67	–3.65	–3.82	–3.71	–3.62	–2.40	–0.90	–0.66	–2.56
5	–0.79	–0.59	–1.79	–2.64	–3.13	–3.09	–3.18	–3.15	–2.94	–1.74	–0.54	–0.36	–2.00
6	–0.33	–0.04	–1.06	–1.86	–2.40	–2.20	–2.40	–2.32	–1.97	–0.94	–0.23	–0.14	–1.32
7	–0.05	0.31	–0.26	–0.80	–1.44	–1.16	–1.30	–1.09	–0.87	–0.30	0.01	0.06	–0.57
8	0.24	0.58	0.34	0.04	0.22	–0.15	0.03	0.13	0.05	0.24	0.17	0.20	0.14
9	0.40	0.82	0.78	0.77	0.88	0.79	1.09	1.05	0.78	0.71	0.30	0.30	0.72
10	0.57	0.94	1.05	1.30	1.59	1.73	1.87	1.62	1.28	1.02	0.42	0.40	1.15
11	0.71	1.01	1.30	1.75	2.29	2.69	2.62	2.26	1.71	1.35	0.56	0.44	1.56
Midn. . .	0.88	1.07	1.54	2.11	2.52	3.01	3.18	2.93	2.00	1.44	0.62	0.56	1.22
6. 6	0.40	0.58	0.52	0.44	0.11	0.01	0.10	0.24	0.77	0.53	0.43	0.26	0.37
7. 7	0.54	0.73	0.76	0.72	0.17	0.03	0.05	0.30	0.88	0.64	0.48	0.36	0.47
8. 8	0.68	0.69	0.68	0.47	0.27	0.17	0.26	0.35	0.71	0.66	0.35	0.37	0.47
9. 9	0.45	0.37	0.32	0.31	0.21	0.12	0.22	0.19	0.28	0.25	0.20	0.30	0.27
10.10	0.10	0.03	–0.02	–0.01	0.03	0.07	–0.19	–0.11	–0.09	0.10	–0.00	0.19	0.01
7. 2. 9	–0.17	–0.09	–0.17	–0.18	–0.44	–0.68	–0.53	–0.47	–0.20	–0.23	–0.12	–0.11	–0.28
6. 2. 8	–0.22	–0.15	–0.20	–0.26	–0.53	–0.65	–0.49	–0.41	–0.15	–0.25	–0.12	–0.14	–0.30
6. 2.10	–0.11	–0.03	0.03	0.16	0.08	–0.03	0.12	0.09	0.26	0.01	–0.03	–0.07	0.04
6. 2. 6	–0.41	–0.36	–0.67	–0.90	–1.25	–1.34	–1.30	–1.23	–0.82	–0.64	–0.25	–0.25	–0.79
7. 2	–0.45	–0.55	–0.64	–0.66	–1.10	–1.41	–1.35	–1.23	–0.69	–0.70	–0.33	–0.32	–0.79
8. 2	–0.46	–0.72	–1.02	–1.34	–1.62	–1.77	–1.81	–1.80	–1.32	–0.95	–0.54	–0.37	–1.14
8. 1	–0.45	–0.76	–0.90	–1.24	–1.54	–1.62	–1.65	–1.63	–1.22	–0.91	–0.53	–0.44	–1.07
7. 1	–0.45	–0.59	–0.52	–0.57	–1.02	–1.26	–1.19	–1.07	–0.59	–0.66	–0.32	–0.39	–0.72
9.12.3.9	–0.67	–0.85	–1.17	–1.46	–1.71	–1.72	–1.76	–1.80	–1.71	–1.21	–0.60	–0.39	–1.25
7. 2.2(9)	–0 03	0.14	0.07	0.06	–0.11	–0.31	–0.13	–0.09	0.05	0.01	–0.02	–0.01	–0.03
Dail. ext.	–0.44	–0.53	–0.38	–0.24	–0.03	–0.15	–0.09	–0.12	–0.20	–0.42	–0.26	–0.38	–0.30

The numbers without sign must be added; those with the sign — must be subtracted.

E

LVII.

Prussia. — Berlin. *Lat.* 52° 30′ N. *Long.* 13° 24′ E. *Greenw.*

Corrections to be applied to the Means of the Hours of Observation to obtain the true Mean Temperatures of the respective Days, Months, and of the Year. — Dove.

Degrees of Fahrenheit.

Hour.	Jan.	Feb.	March.	April.	May.	June.	July.	Aug.	Sept.	Oct.	Nov.	Dec.	Year.
Midnight.	0.34	0.59	0.90	1.78	2.21	2.15	1.78	1.52	1.50	0.95	0.44	0.34	1.21
1	0.43	0.78	1.13	2.22	3.23	2.80	2.52	2.53	1.99	1.48	0.59	0.37	1.67
2	0.49	0.97	1.38	2.56	3.83	3.38	3.06	3.05	2.41	1.95	0.68	0.43	2.02
3	0.54	1.09	1.64	2.41	4.00	3.46	3.28	3.15	2.76	2.31	0.73	0.46	2.15
4	0.58	1.25	1.85	3.03	3.77	3.18	3.12	3.40	2.94	2.45	0.79	0.52	2.24
5	0.65	1.37	1.97	3.05	3.16	2.59	2.67	3.16	2.89	2.32	0.84	0.56	2.10
6	0.73	1.39	1.92	2.69	3.23	1.73	1.92	2.57	2.56	1.52	0.84	0.71	1.73
7	0.75	1.18	1.62	2.01	1.43	0.94	1.18	1.83	2.03	1.15	0.65	0.63	1.21
8	0.62	0.89	1.14	0.94	0.42	0.41	0.44	0.75	1.03	0.62	0.56	0.60	0.70
9	0.41	0.49	0.44	−0.17	−0.65	−0.40	−0.35	−0.36	−0.09	−0.04	0.35	0.38	0.00
10	0.19	−0.09	−0.25	−1.08	−1.47	−1.14	−1.15	−1.27	−0.81	−0.81	−0.09	0.05	−0.66
11	−0.30	−0.66	−1.02	−1.78	−2.20	−1.72	−1.78	−2.07	−1.90	−1.46	−0.55	−0.36	−1.32
Noon.	−0.55	−1.16	−1.44	−2.37	−2.65	−2.17	−2.26	−2.64	−2.49	−1.87	−0.95	−0.69	−1.77
1	−0.93	−1.48	−1.97	−2.85	−3.00	−2.25	−2.51	−3.06	−2.96	−2.14	−1.26	−0.86	−2.11
2	−1.07	−1.73	−2.20	−2.98	−3.25	−2.72	−2.75	−3.17	−3.16	−2.23	−1.27	−0.96	−2.29
3	−1.03	−1.67	−2.28	−3.27	−3.34	−2.84	−2.82	−3.25	−3.19	−2.12	−1.20	−0.95	−2.32
4	−0.78	−1.46	−2.07	−3.11	−3.11	−2.72	−2.73	−3.14	−3.16	−1.75	−0.81	−0.73	−2.13
5	−0.56	−1.07	−1.70	−2.57	−2.69	−2.60	−2.40	−2.72	−2.34	−1.53	−0.49	−0.48	−1.76
6	−0.38	−0.74	−1.12	−2.05	−2.35	−2.36	−2.16	−2.21	−1.58	−0.87	−0.30	−0.25	−1.36
7	−0.22	−0.46	−0.64	−1.13	−1.55	−1.75	−1.44	−1.22	−0.82	−0.63	−0.11	−0.15	−0.84
8	−0.12	−0.20	−0.27	−0.34	−0.67	−0.57	−0.39	−0.30	−0.16	−0.30	0.05	−0.02	−0.27
9	0.06	0.03	0.05	0.40	0.13	0.31	0.38	0.84	0.36	−0.08	0.10	0.06	0.22
10	0.14	0.21	0.39	0.93	0.82	0.96	1.01	1.06	0.98	0.26	0.20	0.17	0.59
11	0.24	0.38	0.65	1.37	1.56	1.58	1.44	1.57	1.20	0.56	0.36	0.25	0.93
6, 6	0.17	0.32	0.40	0.32	−0.06	−0.31	−0.12	0.19	0.49	0.32	0.27	0.23	0.10
7, 7	0.27	0.36	0.49	0.44	−0.06	−0.40	−0.13	0.30	0.60	0.26	0.27	0.24	0.22
8, 8	0.25	0.35	0.44	0.30	−0.08	−0.03	0.03	0.23	0.44	0.16	0.26	0.29	0.13
9, 9	0.24	0.26	0.30	0.12	0.21	−0.05	0.02	0.24	−0.23	−0.06	0.23	0.22	0.12
10, 10	0.17	0.06	0.07	−0.08	−0.33	−0.09	−0.07	−0.11	−0.09	−0.28	0.06	0.11	0.05
7, 1	−0.09	−0.15	−0.17	−0.42	−0.79	−0.65	−0.63	−0.61	−0.46	−0.49	−0.31	−0.12	−0.42
7, 2, 9	−0.09	−0.17	−0.18	−0.19	−0.56	−0.49	−0.36	−0.17	−0.26	−0.39	−0.17	−0.09	−0.26
6, 2, 10	−0.07	−0.04	0.04	0.21	−0.07	−0.01	0.06	0.15	0.13	−0.15	−0.08	−0.03	−0.07
Daily ext.	−0.16	−0.17	−0.16	−0.11	0.33	0.31	0.23	0.08	−0.13	0.11	−0.22	−0.13	0.00

The numbers without sign must be added; those with the sign — must be subtracted.

LVIII.

GERMANY. — SALZUFLEN. *Lat.* 52° 5′ N. *Long.* 8° 40′ E. *Greenw.*

Corrections to be applied to the Means of the Hours of Observation to obtain the true Mean Temperatures of the respective Days, Months, and of the Year. — DOVE.

Degrees of Reaumur.

Hours.	Jan.	Feb.	March.	April.	May.	June.	July.	Aug.	Sept.	Oct.	Nov.	Dec.	Mean.
Morn. 1	0.00	1.10	1.05	2.11	2.41	2.57	2.05	1.71	2.12	1.24	0.90	0.31	1.46
2	0.55	1.22	1.20	2.44	2.93	2.85	2.27	2.01	2.44	1.55	1.26	0.48	1.77
3	0.60	1.27	1.34	2.64	3.29	2.98	2.39	2.23	2.74	1.82	1.53	0.65	1.96
4	0.62	1.26	1.38	2.62	3.37	2.86	2.32	2.26	2.87	1.98	1.64	0.78	2.00
5	0.72	1.18	1.29	2.35	3.08	2.47	1.99	2.00	2.71	1.97	1.58	0.83	1.85
6	0.62	1.01	1.06	1.80	2.41	1.83	1.42	1.48	2.18	1.75	1.37	0.79	1.48
7	0.51	0.75	0.70	1.05	1.45	1.02	0.70	0.79	1.34	1.34	1.04	0.64	0.94
8	0.31	0.41	0.25	0.20	0.38	0.15	−0.06	0.08	0.30	0.75	0.62	0.38	0.31
9	0.08	−0.03	−0.22	−0.63	−0.59	−0.67	−0.74	−0.54	−0.65	0.09	0.14	0.06	−0.31
10	−0.33	−0.53	−0.68	−1.36	−1.42	−1.38	−1.15	−1.02	−1.47	−0.63	−0.35	−0.24	−0.88
11	−0.74	−1.02	−1.06	−1.93	−1.96	−1.94	−1.62	−1.38	−2.08	−1.27	−0.02	−0.48	−1.29
Noon. . .	−0.91	−1.42	−1.39	−2.32	−2.31	−2.39	−1.90	−1.72	−2.41	−1.78	−1.18	−0.62	−1.70
1	−1.01	−1.68	−1.59	−2.54	−2.53	−2.72	−2.13	−2.03	−2.75	−2.09	−1.48	−0.64	−1.93
2	−0.94	−1.74	−1.65	−2.60	−2.66	−2.91	−2.30	−2.30	−2.90	−2.18	−1.56	−0.58	−2.03
3	−0.79	−1.58	−1.56	−2.49	−2.72	−2.92	−2.36	−2.42	−2.90	−2.06	−1.46	−0.50	−1.98
4	−0.50	−1.29	−1.33	−2.21	−2.65	−2.71	−2.24	−2.30	−2.70	−1.76	−1.22	−0.41	−1.78
5	−0.20	−0.90	−0.98	−1.77	−2.39	−2.26	−1.89	−1.87	−2.25	−1.34	−0.92	−0.35	−1.43
6	−0.10	−0.51	−0.56	−1.22	−1.94	−1.62	−1.32	−1.22	−1.55	−0.90	−0.65	−0.32	−0.99
7	0.01	−0.17	−0.15	−0.62	−1.34	−0.86	−0.62	0.47	−0.71	−0.47	−0.44	−0.30	−0.51
8	0.08	0.11	0.19	−0.04	−0.65	−0.09	0.09	0.21	0.12	−0.11	−0.31	−0.27	−0.06
9	0.14	0.34	0.45	0.48	0.04	0.62	0.70	0.71	0.81	0.18	−0.20	−0.21	0.34
10	0.21	0.55	0.63	0.94	0.68	1.22	1.26	1.03	1.30	0.42	−0.06	−0.12	0.67
11	0.22	0.74	0.77	1.34	1.27	1.74	1.52	1.25	1.61	0.66	0.18	0.01	0.94
Midn. . .	0.40	0.93	0.90	1.74	1.84	2.18	1.80	1.45	1.86	0.94	0.50	0.15	1.22
6. 6	0.26	0.25	0.25	0.29	0.24	0.11	0.05	0.13	0.32	0.43	0.36	0.24	0.24
7. 7	0.26	0.29	0.28	0.22	0.06	0.08	0.04	0.16	0.32	0.44	0.30	0.17	0.22
8. 8	0.20	0.26	0.22	0.08	−0.14	0.03	0.02	0.15	0.21	0.32	0.16	0.06	0.13
9. 9	0.11	0.16	0.12	−0.08	−0.28	−0.03	−0.02	0.09	0.08	0.14	−0.03	−0.08	0.02
10.10	−0.06	0.01	−0.03	−0.21	−0.37	−0.08	0.06	0.01	−0.09	−0.11	−0.21	−0.18	−0.11
7. 2. 9	−0.10	−0.22	−0.17	−0.36	−0.39	−0.42	−0 30	−0.27	−0.25	−0.22	−0.24	−0.05	−0.25
6. 2. 8	−0.08	−0.21	−0.13	−0.28	−0.30	−0.39	−0.26	−0.20	−0.20	−0.18	−0.17	−0.02	−0 20
6. 2.10	−0.04	−0.06	0.01	0.05	0.14	0.05	0.13	0.07	0.19	−0.00	−0.08	0.03	0.04
6. 2. 6	−0.14	−0.41	−0.38	−0.67	−0.73	−0.90	−0.73	−0.68	−0.76	−0.44	−0.28	−0.04	−0.51
7. 2	−0.22	−0.50	−0.48	−0.78	−0.61	−0.95	−0.80	−0.76	−0.78	−0.42	−0.26	0.03	−0.54
8. 2	−0.32	−0.67	−0.70	−1.20	−1.14	−1.38	−1.18	−1.11	−1.30	−0.72	−0.47	−0.10	−0.86
8. 1	−0.35	−0.64	−0.67	−1.17	−1.08	−1.29	−1.10	−0.98	−1.23	−0.67	−0.43	−0.13	−0.81
7. 1	−0.25	−0.47	−0.45	−0.75	−0.54	−0.85	−0.72	−0.62	−0.71	−0.38	−0.22	−0.00	−0.50
9.12.3.9	−0.37	−0.67	−0.68	−1.24	−1.40	−1.34	−1.08	−0.99	−1.29	−0.89	−0.68	−0.32	−0.91
7. 2.2(9)	−0.04	−0.08	−0.01	−0.15	−0.28	−0.16	−0.05	−0.02	0.02	−0.12	−0.23	−0.09	−0.10
Dail. ext.	−0.15	−0.24	−0.14	0.02	0.33	0.03	0.02	−0.08	−0.02	−0.10	0.04	0.10	−0.02

The numbers without sign must be added; those with the sign — must be subtracted.

LIX.

Prussia. — Stettin. *Lat.* 53° 25′ N. *Long.* 14° 34′ E. *Greenw.*

Corrections to be applied to the Means of the Hours of Observation to obtain the true Mean Temperatures of the respective Days, Months, and of the Year. — Dove.

Degrees of Reaumur.

Hour.	Jan.	Feb.	March.	April.	May.	June.	July.	Aug.	Sept.	Oct.	Nov.	Dec.	Year.
Midnight.	0.26	0.54	0.98	1.66	2.21	2.21	1.83	1.93	1.53	0.88	0.50	0.39	1.24
1	0.38	0.59	1.17	1.91	2.66	2.46	2.25	2.24	1.61	1.01	0.44	0.46	1.43
2	0.43	0.70	1.30	2.15	3.03	2.84	2.62	2.54	1.87	1.13	0.47	0.50	1 63
3	0.49	0.88	1.41	2.39	3.39	3.10	2.95	2.83	2.11	1.24	0.51	0.56	1.99
4	0.53	0.89	1.51	2.60	3.58	3.08	3.07	3.08	2.33	1.33	0.55	0.61	1.92
5	0.57	0.97	1.63	2.67	3.45	2.78	2.85	3.10	2.46	1.40	0.58	0.64	1.92
6	0.55	0.94	1.62	2.40	2.78	2.12	2.21	2.78	2.45	1.42	0.60	0.56	1.70
7	0.46	0.83	1.37	1.70	1.63	1.17	1.31	2.02	1.98	1.25	0.52	0.46	1.23
8	0.36	0.66	0.90	0.66	0.33	0.20	0.35	0.96	1.11	0.79	0.43	0.38	0.59
9	0.22	0.36	0.23	–0.42	–0.88	–0.72	–0.53	–0.26	–0.05	0.16	0.13	0.23	–0.13
10	–0.04	–0.02	–0.44	–1.36	–1.87	–1.54	–1.33	–1.40	–1.11	–0.55	–0.22	–0.03	–0 83
11	–0.36	–0.53	–1.06	–2.07	–2.62	–2.18	–1.96	–2.23	–1.96	–1.23	–0.60	–0.35	–1.43
Noon.	–0.63	–0.93	–1.59	–2.50	–3.09	–2.59	–2.46	–2.93	–2.58	–1.68	–0.90	–0.64	–1.88
1	–0.81	–1.26	–1.92	–2.80	–3.36	–2.90	–2.81	–3.38	–2.88	–1.98	–1.06	–0.86	–2.17
2	–0.90	–1.39	–2.08	–2.94	–3.50	–2.99	–2.99	–3.50	–2.99	–2.06	–1.06	–0.91	–2.28
3	–0.78	–1.34	–2.06	–2.84	–3.35	–2.90	–2.80	–3.38	–2.82	–1.88	–0.94	–0.86	–2.16
4	–0.63	–1.15	–1.84	–2.54	–2.99	–2.99	–2.60	–3.03	–2.44	–1.43	–0.68	–0.70	–1.92
5	–0.41	–0.83	–1.43	–2.02	–2.46	–2.46	–2.15	–2.40	–1.85	–0.99	–0.39	–0.48	–1.49
6	–0.25	–0.46	–0.90	–1.32	–1.74	–1.74	–1.62	–1.68	–1.14	–0.46	–0.19	–0.30	–0.98
7	–0.11	–0.23	–0.40	–0.55	–0.89	–0.89	–0.93	–0.78	–0.52	–0.10	–0.00	–0.18	–0.46
8	0.01	–0.04	–0.02	0.10	–0.14	–0.14	–0.17	0.02	0.06	0.17	0.18	–0.06	–0.00
9	0.08	0.16	0.32	0.68	0.73	0.73	0.48	0.74	0.60	0.39	0.30	0.07	0.31
10	0.20	0.30	0.61	1.10	1.30	1.30	1.03	1.20	1.00	0.58	0.43	0.22	0.77
11	0.25	0.42	0.79	1.42	1.76	1.76	1.47	1.60	1.31	0.74	0.50	0.32	1.03
6, 6	0.15	0.24	0.36	0.54	0.52	0.19	0.29	0.55	0.65	0.48	0.21	0.13	0.36
7, 7	0.17	0.30	0.48	0.57	0.37	0.14	0.19	0.62	0.73	0.57	0.26	0.14	0.38
8, 8	0.19	0.31	0.44	0.38	0.10	0.03	0.09	0.49	0.59	0.48	0.31	0.16	0.30
9, 9	0.15	0.26	0.28	0.13	–0.08	0.01	–0.03	0.24	0.28	0.28	0.22	0.15	0.16
10, 10	0.08	0.14	0.09	–0.13	–0.29	–0.12	–0.15	–0.10	–0.06	0.02	0.11	0.10	–0.03
7, 1	–0.17	–0.21	–0.67	–0.55	–0.86	–0.86	–0.75	–0.68	–0.45	–0.36	–0.27	–0.20	–0.50
7, 2, 9	–0.13	–0.13	–0.13	–0.19	–0.38	–0.36	–0.40	–0.25	–0.14	–0.14	–0.08	–0.14	–0.21
6, 2, 10	–0.05	–0.05	0.38	0.19	0.19	0.14	0.08	0.16	0.15	–0.02	–0.01	–0.05	0.09
Daily ext.	–0.16	–0.21	–0.23	–0.14	0.04	0.06	0.04	–0.20	–0.27	–0.32	–0.23	–0.15	–0.15

The numbers without sign must be added; those with the sign — must be subtracted.

Sleswick. — Apenrade. *Lat.* 55° 3′ N. *Long.* 9° 25′ E. *Greenw.*

Corrections to be applied to the Means of the Hours of Observation to obtain the true Mean Temperatures of the respective Days, Months, and of the Year. — Dove.

Degrees of Reaumur.

Hours.	Jan.	Feb.	March.	April.	May.	June.	July.	Aug.	Sept.	Oct.	Nov.	Dec.	Mean.
Morn. 1	0.26	0.69	0.98	1.73	3.18	3.82	2.50	2.61	2.16	1.06	0.54	0.31	1.65
2	0.31	0.78	1.14	1.83	3.17	3.90	2.38	2.66	2.29	1.19	0.59	0.35	1.72
3	0.38	0.79	1.26	1.98	3.02	3.82	2.13	2.66	2.54	1.30	0.64	0.37	1.74
4	0.42	0.75	1.34	2.10	2.71	3.50	1.78	2.64	2.62	1.37	0.66	0.38	1.69
5	0.44	0.69	1.31	2.02	2.22	2.89	1.35	2.18	2.43	1.36	0.69	0.40	1.50
6	0.50	0.62	1.18	1.63	1.54	1.94	0.86	1.56	2.02	1.25	0.69	0.40	1.18
7	0.47	0.54	0.90	1.15	0.70	0.83	0.30	0.77	1.18	0.97	0.61	0.37	0.73
8	0.39	0.38	0.50	0.41	−0.23	−0.34	−0.29	−0.18	0.18	0.52	0.42	0.27	0.17
9	0.23	0.10	−0.02	−0.42	−1.14	−1.38	−0.87	−1.10	−0.83	−0.10	0.10	0.10	−0.44
10	−0.06	−0.32	−0.66	−1.22	−1.90	−2.16	−1.40	−1.98	−1.71	−0.79	−0.30	−0.15	−1.00
11	−0.36	−0.78	−1.15	−1.90	−2.49	−2.66	−1.80	−2.42	−2.38	−1.38	−0.68	−0.43	−1.41
Noon. . .	−0.62	−1.19	−1.62	−2.42	−2.86	−2.98	−2.09	−2.74	−2.79	−1.94	−0.98	−0.66	−1.91
1	−0.78	−1.40	−1.90	−2.75	−3.08	−3.24	−2.23	−2.89	−3.03	−2.15	−1.10	−0.78	−2.11
2	−0.69	−1.34	−1.96	−2.89	−3.16	−3.49	−2.27	−2.90	−3.08	−2.07	−1.02	−0.75	−2.14
3	−0.61	−1.06	−1.78	−2.79	−3.10	−3.68	−2.21	−2.78	−2.93	−1.74	−0.82	−0.59	−2.01
4	−0.38	−0.64	−1.41	−2.43	−2.86	−3.62	−2.02	−2.39	−2.54	−1.23	−0.59	−0.38	−1.71
5	−0.16	−0.23	−0.92	−1.80	−2.40	−3.34	−1.70	−2.02	−1.93	−0.71	−0.29	−0.15	−1.30
6	−0.03	0.05	−0.42	−0.99	−1.70	−2.57	−1.18	−1.23	−1.13	−0.25	−0.12	0.02	−0.80
7	0.01	0.18	0.02	−0.12	−0.79	−1.42	−0.57	−0.47	−0.26	0.10	0.02	0.10	−0.27
8	0.03	0.18	0.33	0.66	0.22	−0.07	0.18	0.40	0.56	0.34	0.03	0.14	0.25
9	0.01	0.17	0.54	1.25	1.22	1.25	0.97	1.21	1.21	0.51	0.09	0.15	0.71
10	0.02	0.22	0.66	1.57	2.05	2.33	1.63	1.72	1.61	0.65	0.18	0.18	1.07
11	0.07	0.33	0.76	1.69	2.66	3.10	2.14	2.25	1.83	0.85	0.30	0.21	1.35
Midn. . .	0.15	0.52	0.86	1.70	3.02	3.57	2.43	1.68	1.97	0.92	0.42	0.26	1.46
6. 6	0.24	0.34	0.38	0.32	−0.08	−0.32	−0.16	0.17	0.45	0.50	0.29	0.21	0.19
7. 7	0.24	0.36	0.46	0.52	−0.05	−0.30	−0.14	0.15	0.46	0.54	0.30	0.24	0.23
8. 8	0.21	0.28	0.42	0.54	−0.01	−0.21	−0.06	0.11	0.37	0.43	0.23	0.21	0.21
9. 9	0.11	0.14	0.26	0.42	0.04	−0.07	0.05	0.06	0.19	0.21	0.10	0.13	0.14
10.10	−0.02	−0.05	−0.00	0.18	0.08	0.09	0.12	−0.13	−0.05	−0.07	−0.06	0.02	0.01
7. 2. 9	−0.08	−0.21	−0.17	−0.16	−0.41	−0.47	−0.33	−0.31	−0.23	−0.20	−0.11	−0.08	−0.23
6. 2. 8	−0.05	−0.18	−0.15	−0.20	−0.47	−0.54	−0.41	−0.31	−0.17	−0.16	−0.10	−0.07	−0.23
6. 2.10	−0.06	−0.17	−0.04	0.10	0.14	0.26	0.07	0.13	0.18	−0.06	−0.05	−0.06	0.04
6. 2. 6	−0.07	−0.22	−0.40	−0.75	−1.11	−1.37	−0.86	−0.86	−0.73	−0.36	−0.15	−0.11	−0.58
7. 2	−0.11	−0.40	−0.53	−0.87	−1.23	−1.33	−0.99	−1.07	−0.95	−0.55	−0.21	−0.19	−0.70
8. 2	−0.15	−0.48	−0.73	−1.24	−1.70	−1.92	−1.28	−1.54	−1.45	−0.78	−0.30	−0.24	−0.98
8. 1	−0.20	−0.51	−0.70	−1.17	−1.66	−1.79	−1.26	−1.54	−1.43	−0.82	−0.34	−0.26	−0.97
7. 1	−0.16	−0.43	−0.50	−0.80	−1.19	−1.21	−0.97	−1.06	−0.93	−0.59	−0.25	−0.21	−0.69
9.12.3.9	−0.25	−0.50	−0.72	−1.10	−1.47	−1.70	−1.05	−1.35	−1.34	−0.82	−0.40	−0.25	−0.91
7. 2.2(9)	−0.06	−0.12	0.01	0.19	0.01	−0.04	−0.01	0.07	0.13	−0.02	−0.06	−0.02	0.01
Dailv ext.	−0.14	−0.31	−0.31	−0.40	0.01	0.11	0.12	−0.12	−0.23	−0.39	−0.21	−0.19	−0.20

The numbers without sign must be added; those with the sign — must be subtracted.

Scotland. — Leith. *Lat.* 55° 59′ N. *Long.* 3° 10′ E. *Greenw.*

Corrections to be applied to the Means of the Hours of Observation to obtain the true Mean Temperatures of the respective Days, Months, and of the Year. — Dove.

Degrees of Fahrenheit.

Hours.	Jan.	Feb.	March.	April.	May.	June.	July.	Aug.	Sept.	Oct.	Nov.	Dec.	Mean.
Morn. 1	0.38	0.86	1.76	3.02	3.04	3.29	4.10	2.95	2.54	1.10	1.26	0.72	2.09
2	0.61	0.77	1.98	3.92	3.47	3.62	4.28	3.20	2.77	1.19	1.53	0.65	2.33
3	0.68	0.77	2.41	4.57	3.96	3.74	4.66	3.49	3.29	1.31	1.40	0.61	2.57
4	0.95	0.95	2.59	5.31	4.41	3.98	5.11	3.71	3.65	1.33	1.46	0.70	2.84
5	1.06	1.17	2.75	5.49	4.28	3.94	4.59	3.65	3.78	1.62	1.37	0.77	2.87
6	1.06	1.31	2.79	5.36	3.51	3.04	3.56	3.26	3.51	2.03	1.28	0.59	2.61
7	0.97	1.24	2.48	3.47	2.66	2.25	2.39	2.25	2.75	1.62	1.06	0.68	1.98
8	0.88	1.26	1.80	2.18	1.40	1.10	1.15	1.08	1.46	0.97	1.04	0.54	1.24
9	0.61	0.77	0.81	−0.27	0.11	−0.18	−0.23	−0.50	−0.14	0.32	0.56	0.32	0.18
10	0.16	−0.07	0.18	−2.00	−1.06	−1.31	−1.37	−1.26	−1.10	−0.83	−0.34	−0.02	−0.75
11	−0.34	−0.97	−1.22	−3.02	−2.00	−2.30	−2.25	−2.03	−2.21	−1.71	−1.33	−0.86	−1.69
Noon. . .	−1.04	−1.69	−2.61	−3.92	−2.75	−2.79	−3.58	−2.99	−3.13	−2.36	−1.96	−1.33	−2.51
1	−1.42	−2.25	−2.97	−4.37	−3.35	−3.15	−3.67	−3.44	−3.92	−2.79	−2.30	−1.51	−2.93
2	−1.58	−2.23	−3.29	−4.73	−3.78	−3.83	−4.07	−3.65	−4.28	−2.84	−2.57	−1.55	−3.20
3	−1.60	−2.27	−3.38	−5.09	−3.85	−4.37	−4.37	−3.65	−4.16	−2.57	−2.63	−1.13	−3.26
4	−1.19	−1.73	−3.33	−4.79	−4.19	−3.94	−4.46	−3.87	−3.56	−1.96	−1.69	−0.83	−2.96
5	−0.68	−0.95	−2.84	−4.25	−4.03	−3.71	−4.57	−3.76	−3.56	−1.31	−1.04	−0.50	−2.60
6	−0.45	−0.47	−2.14	−3.83	−3.51	−3.29	−4.41	−3.47	−2.30	−0.59	−0.68	−0.27	−2.12
7	−0.09	−0.09	−1.17	−2.45	−2.61	−2.52	−3.58	−1.69	−0.97	0.05	−0.25	0.18	−1.27
8	0.14	0.32	−0.45	−0.81	−1.17	−0.79	−1.31	−0.41	−0.16	0.59	0.05	0.29	−0.31
9	0.23	0.61	0.25	0.38	0.32	0.50	0.43	0.59	0.59	0.72	0.32	0.36	0.44
10	0.18	0.88	0.77	1.08	0.86	1.89	1.71	1.58	1.24	1.15	0.79	0.41	1.04
11	0.32	0.99	1.31	2.18	1.69	2.16	2.52	2.23	1.67	1.60	1.19	0.54	1.53
Midn. . .	0.38	1.01	1.44	2.68	2.32	2.68	3.44	2.77	2.27	1.49	1.42	0.59	1.87
6. 6	0.32	0.43	0.34	0.77	0.00	−0.14	−0.43	−0.11	0.61	0.72	0.32	0.16	0.25
7. 7	0.45	0.59	0.65	0.52	0.02	−0.14	−0.61	0.29	0.90	0.83	0.41	0.43	0.36
8. 8	0.52	0.79	0.68	0.70	0.11	0.16	−0.09	0.34	0.65	0.79	0.54	0.43	0.47
9. 9	0.43	0.70	0.54	0.07	0.23	0.16	0.11	0.05	0.23	0.52	0.45	0.34	0.32
10.10	0.18	0.41	0.47	−0.47	−0.11	0.29	0.18	0.16	0.07	0.16	0.23	0.20	0.15
7. 2. 9	−0.14	−0.14	−0.18	−0.29	−0.27	−0.36	−0.43	−0.27	−0.32	−0.16	−0.41	−0.18	−0.26
6. 2. 8	−0.14	−0.20	−0.32	−0.07	−0.47	−0.52	−0.61	−0.27	−0.32	−0.07	−0.41	−0.23	−0.30
6. 2.10	−0.11	−0.02	0.09	0.56	0.20	0.36	0.41	0.41	0.16	0.11	−0.16	−0.18	0.15
6. 2. 6	−0.32	−0.47	−0.88	−1.06	−1.26	−1.35	−1.64	−1.28	−1.01	−0.47	−0.65	−0.41	−0.90
7. 2	−0.32	−0.50	−0.41	−0.63	−0.56	−0.79	−0.86	−0.70	−0.77	−0.61	−0.77	−0.45	−0.61
8. 2	−0.36	−0.50	−0.74	−1.28	−1.19	−1.37	−1.46	−1.28	−1.42	−0.95	−0.77	−0.52	−0.99
8. 1	−0.27	−0.50	−0.59	−1.10	−0.99	−1.04	−1.26	−1.19	−1.24	−0.92	−0.63	−0.50	−0.85
7. 1	−0.23	−0.52	−0.25	−0.45	−0.36	−0.45	−0.65	−0.61	−0.59	−0.59	−0.63	−0.41	−0.47
9.12.3.9	−0.45	−0.65	−1.24	−2.23	−1.55	−1.71	−1.94	−1.64	−1.71	−0.97	−0.92	−0.45	−1.29
7. 2.2(9)	−0.05	0.07	−0.09	−0.14	−0.14	−0.16	−0.20	−0.07	−0.09	0.07	−0.23	−0.05	−0.09
Dail. ext.	−0.27	−0.49	−0.29	0.20	−0.11	−0.20	0.27	−0.09	−0.25	−0.40	−0.56	−0.40	−0.20

The numbers without sign must be added; those with the sign — must be subtracted.

Scotland. — Leith. *Lat.* 55° 59′ N. *Long.* 3° 10′ E. *Greenw.*

Corrections to be applied to the Means of the Hours of Observation to obtain the true Mean Temperatures of the respective Days, Months, and of the Year. — Dove.

Degrees of Reaumur.

Hours.	Jan.	Feb.	March.	April.	May.	June.	July.	Aug.	Sept.	Oct.	Nov.	Dec.	Mean.
Morn. 1	0.17	0.38	0.78	1.34	1.35	1.46	1.82	1.31	1.13	0.49	0.56	0.32	0.93
2	0.27	0.34	0.88	1.74	1.54	1.61	1.90	1.42	1.23	0.53	0.68	0.29	1.04
3	0.30	0.34	1.07	2.03	1.76	1.66	2.07	1.55	1.46	0.58	0.62	0.27	1.14
4	0.42	0.42	1.15	2.36	1.96	1.77	2.27	1.65	1.62	0.59	0.65	0.31	1.26
5	0.47	0.52	1.22	2.44	1.90	1.75	2.04	1.62	1.68	0.72	0.61	0.34	1.28
6	0.47	0.58	1.24	2.38	1.56	1.35	1.58	1.45	1.56	0.90	0.57	0.26	1.16
7	0.43	0.55	1.10	1.54	1.18	1.00	1.06	1.00	1.22	0.72	0.47	0.30	0.88
8	0.39	0.56	0.80	0.97	0.62	0.49	0.51	0.48	0.65	0.43	0.46	0.24	0.55
9	0.27	0.34	0.36	−0.12	0.05	−0.08	−0.10	−0.22	−0.06	0.14	0.25	0.14	0.08
10	0.07	−0.03	0.08	−0.89	−0.47	−0.58	−0.61	−0.56	−0.49	−0.37	−0.15	−0.01	−0.33
11	−0.15	−0.43	−0.54	−1.34	−0.89	−1.02	−1.00	−0.90	−0.98	−0.76	−0.59	−0.38	−0.75
Noon. . .	−0.46	−0.75	−1.16	−1.74	−1.22	−1.24	−1.59	−1.33	−1.39	−1.05	−0.87	−0.59	−1.12
1	−0.63	−1.00	−1.32	−1.94	−1.49	−1.40	−1.63	−1.53	−1.74	−1.24	−1.02	−0.67	−1.30
2	−0.70	−0.99	−1.46	−2.10	−1.68	−1.70	−1.81	−1.62	−1.90	−1.26	−1.14	−0.69	−1.42
3	−0.71	−1.01	−1.50	−2.26	−1.71	−1.94	−1.94	−1.62	−1.85	−1.14	−1.17	−0.50	−1.45
4	−0.53	−0.77	−1.48	−2.13	−1.86	−1.75	−1.98	−1.72	−1.58	−0.87	−0.75	−0.37	−1.32
5	−0.30	−0.42	−1.26	−1.89	−1.79	−1.65	−2.03	−1.67	−1.58	−0.58	−0.46	−0.22	−1.15
6	−0.20	−0.21	−0.95	−1.70	−1.56	−1.46	−1.96	−1.54	−1.02	−0.26	−0.30	−0.12	−0.94
7	−0.04	−0.04	−0.52	−1.09	−1.16	−1.12	−1.59	−0.75	−0.43	0.02	−0.11	0.08	−0.56
8	0.06	0.14	−0.20	−0.36	−0.52	−0.35	−0.58	−0.18	−0.07	0.26	0.02	0.13	−0.14
9	0.10	0.27	0.11	0.17	0.14	0.22	0.19	0.26	0.26	0.32	0.14	0.16	0.20
10	0.08	0.39	0.34	0.48	0.38	0.84	0.76	0.70	0.55	0.51	0.35	0.18	0.46
11	0.14	0.44	0.58	0.97	0.75	0.96	1.12	0.99	0.74	0.71	0.53	0.24	0.68
Midn. . .	0.17	0.45	0.64	1.19	1.03	1.19	1.53	1.23	1.01	0.66	0.63	0.26	0.83
6. 6	0.14	0.19	0.15	0.34	0.00	−0.06	−0.19	−0.05	0.27	0.32	0.14	0.07	0.11
7. 7	0.20	0.26	0.29	0.23	0.01	−0.06	−0.27	0.13	0.40	0.37	0.18	0.19	0.16
8. 8	0.23	0.35	0.30	0.31	0.05	0.07	−0.04	0.15	0.29	0.35	0.24	0.19	0.21
9. 9	0.19	0.31	0.24	0.03	0.10	0.07	0.05	0.02	0.10	0.23	0.20	0.15	0.14
10.10	0.08	0.18	0.21	−0.21	−0.05	0.13	0.08	0.07	0.03	0.07	0.10	0.09	0.06
7. 2. 9	−0.06	−0.06	−0.08	−0.13	−0.12	−0.16	−0.19	−0.12	−0.14	−0.07	−0.18	−0.08	−0.12
6. 2. 8	−0.06	−0.09	−0.14	−0.03	−0.21	−0.23	−0.27	−0.12	−0.14	−0.03	−0.18	−0.10	−0.13
6. 2.10	−0.05	−0.01	0.04	0.25	0.09	0.16	0.18	0.18	0.07	0.05	−0.07	−0.08	0.07
6. 2. 6	−0.14	−0.21	−0.39	−0.47	−0.56	−0.60	−0.73	−0.57	−0.45	−0.21	−0.29	−0.18	−0.40
7. 2	−0.14	−0.22	−0.18	−0.28	−0.25	−0.35	−0.38	−0.31	−0.34	−0.27	−0.34	−0.20	−0.27
8. 2	−0.16	−0.22	−0.33	−0.57	−0.53	−0.61	−0.65	−0.57	−0.63	−0.42	−0.34	−0.23	−0.44
8. 1	−0.12	−0.22	−0.26	−0.49	−0.44	−0.46	−0.56	−0.53	−0.55	−0.41	−0.28	−0.22	−0.38
7. 1	−0.10	−0.23	−0.11	−0.20	−0.16	−0.20	−0.29	−0.27	−0.26	−0.26	−0.28	−0.18	−0.21
9.12.3.9	−0.20	−0.29	−0.55	−0.99	−0.69	−0.76	−0.86	−0.73	−0.76	−0.43	−0.41	−0.20	−0.57
7. 2.2(9)	−0.02	0.03	−0.04	−0.06	−0.06	−0.07	−0.09	−0.03	−0.04	0.03	−0.10	−0.02	−0.04
Dail. ext.	−0.12	−0.22	−0.13	0.09	0.05	−0.09	0.12	−0.04	−0.11	−0.18	−0.25	−0.18	−0.09

The numbers without sign must be added; those with the sign — must be subtracted.

LXIII.

SCOTLAND. — MAKERSTOWN. *Lat.* 55° 36′ N. *Long.* 2° 31′ W. *Gr.*

Corrections to be applied to the Means of the Hours of Observation to obtain the true Mean Temperatures of the respective Days, Months, and of the Year. — DOVE.

Degrees of Reaumur.

Hour.	Jan.	Feb.	March.	April.	May.	June.	July.	Aug.	Sept.	Oct.	Nov.	Dec.	Year.
Midn.	0.67	0.88	1.24	2.30	2.00	2.25	2.10	1.98	1.95	0.88	0.46	0.24	1.41
1	0.76	0.92	1.37	2.52	2.04	2.43	2.44	2.24	2.15	0.88	0.46	0.16	1.53
2	0.78	1.08	1.37	2.70	2.33	2.54	2.57	2.38	2.26	1.06	0.60	0.18	1.65
3	0.76	1.06	1.48	2.79	2.55	2.65	2.79	2.56	2.35	1.57	0.60	0.29	1.79
4	0.67	1.01	1.66	2.96	2.51	2.43	2.70	2.56	2.48	1.20	0.68	0.40	1.77
5	0.78	0.92	1.77	2.88	2.06	1.96	2.21	2.44	2.46	1.40	0.60	0.44	1.66
6	0.60	0.85	1.73	2.25	1.31	1.12	1.35	1.78	2.22	1.31	0.66	0.51	1.31
7	0.51	0.99	1.26	1.43	0.48	0.32	0.46	0.91	1.24	1.26	0.66	0.44	0.83
8	0.53	0.79	0.46	0.36	−0.25	−0.51	−0.39	−0.09	0.00	0.62	0.66	0.40	0.22
9	0.33	0.08	−0.38	−0.79	−0.94	−1.11	−0.96	−1.02	−1.00	−0.16	0.08	0.22	−0.47
10	−0.22	−0.72	−1.12	−1.86	−1.52	−1.68	−1.59	−1.78	−1.92	−0.96	−0.47	−0.20	−1.17
11	−0.84	−1.21	−1.67	−2.55	−2.09	−2.26	−2.14	−2.33	−2.45	−1.63	−0.94	−0.62	−1.73
Noon.	−1.36	−1.61	−2.09	−3.06	−2.34	−2.48	−2.45	−2.73	−2.67	−2.03	−1.34	−0.93	−2.09
1	−1.71	−2.03	−2.27	−3.44	−2.69	−2.75	−2.48	−2.87	−3.03	−2.25	−1.56	−1.13	−2.35
2	−1.67	−2.05	−2.36	−3.57	−2.65	−2.57	−2.52	−2.93	−3.12	−2.20	−1.47	−0.96	−2.34
3	−1.29	−1.68	−2.32	−3.52	−2.65	−2.28	−2.54	−2.73	−2.85	−1.83	−0.96	−0.60	−2.10
4	−0.71	−1.30	−1.80	−3.05	−2.27	−1.95	−2.28	−2.47	−2.29	−1.23	−0.45	−0.16	−1.66
5	−0.13	−0.50	−1.20	−2.30	−1.76	−1.64	−1.81	−1.78	−1.49	−0.49	−0.07	−0.11	−1.11
6	0.18	−0.08	−0.40	−1.39	−0.98	−0.95	−1.34	−1.07	−0.60	−0.09	0.13	0.18	−0.53
7	0.29	0.15	0.08	−0.19	−0.18	−0.40	−0.59	−0.18	0.06	0.17	0.17	0.18	−0.04
8	0.31	0.37	0.46	0.52	0.62	0.36	0.35	0.56	0.46	0.40	0.28	0.18	0.41
9	0.29	0.52	0.73	1.21	1.15	1.00	0.95	1.09	0.95	0.64	0.37	0.24	0.76
10	0.27	0.64	0.95	1.74	1.46	1.56	1.48	1.58	1.33	0.73	0.46	0.31	1.04
11	0.22	0.79	1.06	2.08	1.77	1.94	1.70	1.89	1.51	0.73	0.40	0.36	1.20
Mean.	1.53	0.35	2.06	5.96	6.86	10.25	10.12	10.00	8.51	6.64	4.60	1.16	

LXIV.

IRELAND. — DUBLIN. *Lat.* 53° 23′ N. *Long.* 6° 20′ W. *Gr.* — DOVE.

Degrees of Reaumur.

Hour.	Jan.	Feb.	March.	April.	May.	June.	July.	Aug.	Sept.	Oct.	Nov.	Dec.	Year.
A.M. 1	0.58	0.53	1.56	2.18	2.53	2.76	2.18	2.22	1.64	1.16	0.53	0.36	1.52
3	0.80	0.71	1.64	2.40	2.89	3.11	2.53	2.40	1.87	1.42	0.67	0.49	1.74
5	0.93	0.98	1.64	2.49	2.31	2.18	2.18	2.53	1.87	1.73	0.76	0.58	1.68
7	0.84	0.93	1.38	0.58	−0.22	−0.89	−0.36	0.40	1.07	1.56	0.80	0.53	0.56
9	0.36	0.18	−0.31	−1.11	−1.24	−1.38	−1.10	−1.16	−0.76	−0.09	0.27	0.36	−0.50
11	−0.98	−0.07	−1.82	−2.40	−2.18	−2.09	−2.04	−2.27	−2.13	−1.91	−0.98	−0.71	−1.71
P.M. 1	−1.60	−1.78	−2.67	−2.93	−2.62	−2.40	−2.27	−2.62	−2.67	−2.44	−1.56	−1.16	−2.23
3	−1.33	−1.47	−2.44	−2.84	−2.71	−2.31	−2.27	−2.49	−2.22	−2.04	−1.11	−0.67	−1.99
5	−0.44	0.44	−1.29	−1.82	−1.82	−1.87	−1.64	−1.73	−1.29	−0.84	−0.27	−0.18	−1.14
7	0.09	0.18	0.18	0.04	−0.27	−0.44	−0.27	−0.09	0.27	0.04	0.04	0.09	−0.01
9	0.22	0.31	0.76	1.20	1.29	1.24	1.20	1.16	0.93	0.58	0.36	0.18	0.79
11	0.36	0.40	1.07	1.73	1.96	2.04	1.87	1.64	1.42	0.84	0.44	0.22	1.17
Mean.	4.09	4.75	5.10	6.66	9.51	11.86	12.48	12.31	10.79	7.73	5.99	4.88	

The numbers without sign must be added; those with the sign — must be subtracted.

LXV.

Russia. — Catharinenburg. *Lat.* 56° 50′ N. *Long.* 60° 34′ E. *Greenw.*

Corrections to be applied to the Means of the Hours of Observation to obtain the true Mean Temperatures of the respective Days, Months, and of the Year. — Dove.

Degrees of Reaumur.

Hours.	Jan.	Feb.	March.	April.	May.	June.	July.	Aug.	Sept.	Oct.	Nov.	Dec.	Mean.
Morn. 1	0.59	0.91	1.84	1.97	3.09	3.69	3.51	2.49	1.99	0.68	0.47	0.65	1.82
2	0.58	0.89	2.09	2.41	3.52	4.15	3.76	2.93	2.27	0.84	0.42	0.67	2.04
3	0.53	0.87	2.42	2.87	3.80	4.35	3.96	3.42	2.60	1.04	0.36	0.64	2.24
4	0.48	0.89	2.80	3.21	3.82	4.17	4.01	3.78	2.89	1.23	0.35	0.61	2.35
5	0.58	0.95	3.11	3.23	3.45	3.54	3.78	3.79	2.98	1.36	0.43	0.63	2.32
6	0.54	1.00	3.15	2.83	2.67	2.49	3.18	3.30	2.74	1.36	0.55	0.72	2.04
7	0.60	0.94	2.76	1.99	1.57	1.18	2.21	2.29	2.11	1.17	0.64	0.81	1.52
8	0.56	0.71	1.90	0.84	0.31	0.17	0.98	0.94	1.16	0.80	0.60	0.80	0.79
9	0.37	0.27	0.65	−0.41	−0.88	−1.35	−0.34	−0.48	0.05	0.28	0.37	0.61	−0.07
10	−0.01	−0.33	−0.75	−1.52	−1.85	−2.23	−1.61	−1.70	−1.03	−0.32	−0.02	0.21	−0.93
11	−0.60	−0.97	−2.03	−2.34	−2.53	−2.79	−2.72	−2.55	−1.93	−0.89	−0.49	−0.34	−1.68
Noon. . .	−0.98	−1.47	−3.00	−2.83	−2.98	−3.13	−3.64	−3.03	−2.58	−1.34	−0.89	−0.90	−2.23
1	−1.30	−1.75	−3.52	−3.04	−3.25	−3.35	−4.33	−3.25	−2.98	−1.62	−1.12	−1.32	−2.57
2	−1.37	−1.77	−3.62	−3.03	−3.41	−3.50	−4.78	−3.34	−3.16	−1.69	−1.13	−1.50	−2.69
3	−1.19	−1.55	−3.39	−2.88	−3.46	−3.56	−4.90	−3.36	−3.17	−1.58	−0.95	−1.40	−2.62
4	−0.84	−1.19	−2.96	−2.60	−3.33	−3.46	−4.62	−3.27	−2.98	−1.31	−0.66	−1.10	−2.36
5	−0.34	−0.79	−2.40	−2.18	−2.95	−3.09	−3.90	−2.98	−2.57	−0.96	−0.37	−0.73	−1.94
6	−0.11	−0.42	−1.77	−1.61	−2.29	−2.43	−2.77	−2.39	−1.93	−0.58	−0.14	−0.39	−1.40
7	0.11	−0.10	−1.08	−0.92	−1.41	−1.52	−1.39	−1.53	−1.12	−0.23	0.01	−0.14	−0.78
8	0.22	0.17	0.36	−0.22	−0.42	−0.48	0.03	−0.53	−0.26	0.06	0.12	0.03	0.14
9	0.30	0.42	0.32	0.42	0.53	0.56	1.28	0.43	0.52	0.26	0.22	0.15	0.45
10	0.37	0.63	0.90	0.91	1.35	1.51	2.22	1.20	1.13	0.40	0.33	0.28	0.95
11	0.36	0.80	1.32	1.29	2.03	2.35	2.84	1.74	1.52	0.48	0.42	0.43	1.30
Midn. . .	0.55	0.89	1.62	1.61	2.59	3.07	3.23	2.12	1.77	0.56	0.48	0.57	1.59
6. 6	0.21	0.27	0.69	0.61	0.19	0.03	0.20	0.45	0.40	0.39	0.21	0.17	0.32
7. 7	0.35	0.42	0.84	0.53	0.08	−0.17	0.41	0.38	0.49	0.47	0.33	0.33	0.37
8. 8	0.39	0.44	0.77	0.31	−0.05	−0.33	0.51	0.20	0.45	0.43	0.36	0.41	0.32
9. 9	0.33	0.34	0.49	0.01	−0.17	−0.39	0.47	−0.03	0.29	0.27	0.29	0.38	0.19
10.10	0.18	0.15	0.08	−0.30	−0.25	−0.36	0.31	−0.25	0.05	0.04	0.15	0.25	0.00
7. 2. 9	−0.16	−0.14	−0.18	−0.21	−0.44	−0.59	−0.43	−0.21	−0.18	−0.09	−0.09	−0.18	−0.20
6. 2. 8	−0.20	−0.20	−0.28	−0.14	−0.39	−0.50	−0.52	−0.19	−0.23	−0.09	−0.15	−0.25	−0.26
6. 2.10	−0.15	−0.05	0.14	0.24	0.20	0.17	0.21	0.39	0.24	0.02	0.08	−0.17	0.10
6. 2. 6	−0.31	−0.40	−0.75	−0.60	−0.01	−1.15	−1.46	−0.81	−0.78	−0.30	−0.24	−0.39	−0.68
7. 2	−0.39	−0.42	−0.43	−0.52	−0.92	−1.16	−1.29	−0.53	−0.53	−0.26	−0.25	−0.35	−0.59
8. 2	−0.41	−0.54	−0.86	−1.10	−1.55	−1.84	−1.90	−1.20	−1.00	−0.45	−0.27	−0.35	−0.96
8. 1	−0.37	−0.52	−0.81	−1.10	−1.47	−1.76	−1.68	−1.16	−0.91	−0.41	−0.26	−0.26	−0.89
7. 1	−0.35	−0.41	−0.38	−0.53	−0.84	−1.09	−1.06	−0.48	−0.44	−0.23	−0.24	−0.26	−0.53
9.12.3.9	−0.38	−0.58	−1.36	−1.43	−1.70	−1.87	−1.90	−1.61	−1.30	−0.60	−0.31	−0.39	−1.12
7. 2.2(9)	−0.04	−0.00	−0.06	−0.05	−0.20	−0.30	−0.00	−0.05	1.49	0.00	−0.01	−0.10	0.06
Dail. ext.	−0.39	−0.39	−0.24	0.10	0.18	0.40	−0.45	0.22	0.17	−0.17	−0.25	−0.35	−0.17

The numbers without sign must be added; those with the sign — must be subtracted.

LXVI.

RUSSIA. — CATHARINENBURG. *Lat.* 56° 50′ N. *Long.* 60° 34′ E. *Greenw.*

Corrections to be applied to the Means of the Hours of Observation to obtain the true Mean Temperatures of the respective Days, Months, and of the Year. — DOVE.

Degrees of Reaumur.

Hour.	Jan.	Feb.	March.	April.	May.	June.	July.	Aug.	Sept.	Oct.	Nov.	Dec.	Year.
Midn.	0.42	1.07	1.70	2.12	2.64	3.06	2.93	2.16	1.96	0.89	0.47	0.47	1.66
1	0.52	1.19	2.00	2.40	3.11	3.51	3.41	2.49	2.31	1.08	0.51	0.50	1.92
2	0.52	1.25	2.23	2.82	3.49	3.90	3.86	2.76	2.58	0.99	0.54	0.52	2.12
3	0.55	1.41	2.53	3.05	3.73	4.15	4.11	3.03	2.83	1.47	0.58	0.54	2.33
4	0.63	1.52	2.75	3.26	3.74	3.92	4.28	3.22	3.06	1.61	0.68	0.58	2.44
5	0.68	1.67	2.85	3.24	3.27	3.35	3.66	3.14	3.22	1.67	0.71	0.61	2.34
6	0.73	1.76	3.06	2.24	2.27	1.99	2.47	2.45	3.04	1.69	0.82	0.64	1.93
7	0.81	1.76	2.59	1.61	0.89	0.61	1.02	1.37	2.27	1.53	0.85	0.65	1.33
8	0.88	1.51	1.46	0.34	−0.24	−0.53	−0.28	0.18	0.85	0.91	0.77	0.58	0.54
9	0.67	0.73	−0.06	−0.81	−1.09	−1.46	−1.45	−0.97	−0.57	−0.03	0.33	0.39	−0.36
10	0.13	−0.45	−1.45	−1.99	−1.94	−2.23	−2.35	−1.72	−1.68	−0.78	−0.22	−0.08	−1.23
11	−0.57	−1.44	−2.39	−2.62	−2.72	−2.93	−3.10	−2.54	−2.50	−1.46	−0.72	−0.71	−1.98
Noon.	−1.04	−2.13	−2.95	−3.09	−3.19	−3.38	−3.58	−2.99	−3.09	−1.73	−1.03	−1.19	−2.45
1	−1.39	−2.58	−3.27	−3.22	−3.28	−3.48	−3.57	−3.04	−3.32	−1.99	−1.25	−1.45	−2.65
2	−1.50	−2.74	−3.38	−3.26	−3.41	−3.59	−3.55	−3.02	−3.36	−2.02	−1.23	−1.39	−2.70
3	−1.28	−2.37	−3.18	−2.86	−3.14	−3.37	−3.40	−3.03	−3.48	−2.23	−1.11	−1.00	−2.54
4	−0.85	−1.97	−2.82	−2.65	−2.99	−3.05	−3.15	−2.83	−3.18	−1.61	−0.79	−0.61	−2.21
5	−0.50	−1.28	−2.20	−2.14	−2.60	−2.49	−2.67	−2.37	−2.48	−0.95	−0.47	−0.33	−1.71
6	−0.22	−0.74	−1.37	−1.46	−1.98	−1.98	−2.14	−1.66	−1.56	−0.56	−0.26	−0.11	−1.17
7	0.00	−0.25	−0.67	−0.59	−0.95	−1.17	−1.29	−0.79	−0.65	−0.22	−0.07	0.02	−0.55
8	0.10	0.08	−0.12	0.13	−0.04	−0.12	−0.16	0.11	0.07	0.06	0.06	0.11	0.02
9	0.17	0.40	0.44	0.65	0.85	0.96	0.83	0.84	0.67	0.36	0.16	0.26	0.55
10	0.24	0.65	0.94	1.13	1.53	1.88	1.67	1.39	1.25	0.53	0.27	0.39	0.99
11	0.34	0.86	1.34	1.58	2.13	2.51	2.36	1.81	1.65	0.74	0.40	0.56	1.36
Mean.	−10.76	−9.50	−5.83	0.47	6.31	12.08	14.53	10.61	6.32	1.41	−6.11	−11.68	

LXVII.

RUSSIA. — ST. PETERSBURG. *Lat.* 59° 56′ N. *Long.* 30° 18′ E. *Gr.* — DOVE.

Degrees of Reaumur.

Hour.	Jan.	Feb.	March.	April.	May.	June.	July.	Aug.	Sept.	Oct.	Nov.	Dec.	Year.
Midn.	0.14	0.38	0.73	1.44	2.08	1.99	1.77	1.68	1.17	0.52	0.15	0.17	1.02
1	0.21	0.44	0.99	1.68	2.43	2.29	2.05	2.02	1.38	0.60	0.17	0.21	1.21
2	0.25	0.46	1.22	1.91	2.70	2.56	2.24	2.24	1.58	0.65	0.15	0.27	1.35
3	0.30	0.52	1.38	2.11	2.91	2.73	2.43	2.48	1.75	0.73	0.25	0.34	1.49
4	0.38	0.63	1.56	2.24	2.86	2.44	2.32	2.59	1.87	0.78	0.30	0.36	1.53
5	0.43	0.72	1.71	2.28	2.38	1.97	1.92	2.40	1.96	0.84	0.34	0.34	1.44
6	0.45	0.76	1.75	1.95	1.72	1.33	1.33	1.96	1.90	0.90	0.37	0.30	1.23
7	0.41	0.78	1.57	1.32	0.93	0.63	0.64	1.19	1.47	0.82	0.37	0.29	0.87
8	0.42	0.60	1.07	0.65	0.14	−0.04	0.05	0.42	0.81	0.57	0.32	0.25	0.44
9	0.35	0.40	0.40	−0.05	−0.59	−0.69	−0.56	−0.40	0.00	0.20	0.17	0.17	−0.05
10	0.13	−0.05	−0.19	−0.78	−1.30	−1.21	−1.12	−1.07	−0.71	−0.22	0.00	0.04	−0.54
11	−0.20	−0.48	−0.86	−1.42	−1.92	−1.71	−1.58	−1.64	−1.27	−0.61	−0.20	−0.14	−1.00

The numbers without sign must be added; those with the sign — must be subtracted.

LXVII.

Russia. — St. Petersburg, *Continued.*

Corrections to be applied to the Means of the Hours of Observation to obtain the true Mean Temperatures of the respective Days, Months, and of the Year. — Dove.

Degrees of Reaumur.

Hour.	Jan.	Feb.	March.	April.	May.	June.	July.	Aug.	Sept.	Oct.	Nov.	Dec.	Year.
Noon.	–0.38	–0.90	–1.31	–1.93	–2.30	–1.99	–1.89	–2.10	–1.72	–0.94	–0.37	–0.30	–1.34
1	–0.63	–0.97	–1.62	–2.10	–2.41	–2.17	–2.03	–2.47	–2.26	–1.75	–0.64	–0.48	–1.63
2	–0.66	–1.04	–1.88	–2.36	–2.65	–2.32	–2.15	–2.60	–2.34	–1.29	–0.63	–0.58	–1.71
3	–0.55	–0.99	–1.94	–2.49	–2.90	–2.45	–2.29	–2.64	–2.31	–1.06	–0.46	–0.40	–1.71
4	–0.33	–0.83	–1.92	–2.65	–2.92	–2.60	–2.41	–2.80	–2.27	–0.86	–0.20	–0.31	–1.68
5	–0.25	–0.45	–1.53	–2.31	–2.48	–2.23	–2.06	–2.45	–1.76	–0.50	–0.16	–0.22	–1.37
6	–0.19	–0.26	–1.02	–1.43	–1.65	–1.41	–1.30	–1.41	–0.95	–0.25	–0.11	–0.14	–0.84
7	–0.18	–0.16	–0.55	–0.61	–0.74	–0.71	–0.63	–0.62	–0.35	–0.09	–0.05	–0.10	–0.40
8	–0.14	–0.03	–0.25	–0.03	0.06	–0.03	0.02	0.09	0.07	0.07	0.01	0.08	–0.01
9	–0.11	0.08	0.03	0.47	0.79	0.67	0.64	0.65	0.40	0.18	0.08	0.03	0.33
10	–0.03	0.17	0.24	0.84	1.22	1.25	1.18	1.05	0.66	0.33	0.08	0.02	0.58
11	0.06	0.30	0.50	1.17	1.76	1.65	1.45	1.40	0.91	0.45	0.11	0.11	0.82
Mean.	–7.41	–6.73	–3.56	1.10	7.01	11.33	13.39	13.58	8.43	3.61	–0.80	–3.75	

LXVIII.

Russia. — Helsingfors. *Lat.* 60° 10′ N. *Long.* 24° 57′ E. *Gr.* — Dove.

Degrees of Reaumur.

Hour.	Jan.	Feb.	March.	April.	May.	June.	July.	Aug.	Sept.	Oct.	Nov.	Dec.	Year.
Midn.	0.06	0.47	1.28	1.61	1.61	2.01	1.65	1.36	0.83	0.37	0.18	0.20	0.97
1	0.13	0.49	1.48	1.87	1.94	2.44	1.90	1.68	1.03	0.45	0.15	0.21	1.15
2	0.16	0.52	1.64	2.07	2.21	2.84	2.17	1.98	1.21	0.55	0.18	0.18	1.31
3	0.23	0.67	1.84	2.21	2.58	3.04	2.45	2.23	1.35	0.65	0.23	0.15	1.47
4	0.35	0.64	1.91	2.37	2.68	2.77	2.42	2.49	1.48	0.62	0.28	0.23	1.52
5	0.38	0.77	1.98	2.34	2.28	2.21	2.05	2.41	1.63	0.67	0.33	0.10	1.43
6	0.38	0.92	2.01	1.74	1.31	1.31	1.33	1.81	1.63	0.75	0.33	0.03	1.13
7	0.41	0.99	1.78	1.14	0.58	0.51	0.55	1.11	1.28	0.73	0.36	0.01	0.79
8	0.43	0.99	1.04	0.17	–0.19	–0.36	–0.10	0.26	0.58	0.57	0.35	0.00	0.31
9	0.38	0.55	0.04	–0.73	–0.86	–0.83	–0.73	–0.56	–0.09	0.33	0.25	0.06	–0.18
10	0.08	–0.20	–0.89	–1.49	–1.39	–1.29	–1.23	–1.12	–0.65	–0.15	0.13	–0.07	–0.69
11	–0.19	–0.93	–1.19	–1.93	–1.76	–1.83	–1.65	–1.59	–1.05	–0.47	–0.19	–0.32	–1.09
Noon.	–0.72	–1.25	–2.36	–2.26	–1.82	–1.76	–1.80	–2.02	–1.67	–0.90	–0.59	–0.42	–1.46
1	–0.79	–1.50	–2.62	–2.46	–2.12	–2.06	–2.13	–2.26	–1.82	–1.08	–0.70	–0.45	–1.67
2	–0.74	–1.60	–2.62	–2.56	–2.19	–2.36	–2.28	–2.31	–1.85	–1.10	–0.64	–0.42	–1.72
3	–0.49	–1.33	–2.46	–2.37	–2.16	–2.49	–2.13	–2.17	–1.75	–0.95	–0.50	–0.22	–1.58
4	–0.24	–0.90	–2.12	–1.89	–1.82	–2.16	–1.75	–1.84	–1.52	–0.77	–0.29	–0.02	–1.28
5	–0.12	–0.43	–1.56	–1.59	–1.49	–1.89	–1.48	–1.64	–1.20	–0.43	–0.17	0.03	–1.00
6	–0.04	–0.21	–0.79	–1.09	–1.09	–1.53	–1.15	–1.19	–0.72	–0.25	–0.09	–0.02	–0.68
7	0.03	0.07	–0.29	–0.49	–0.86	–0.96	–0.68	–0.64	–0.27	–0.13	–0.04	0.01	–0.35
8	0.08	0.20	0.01	0.14	–0.16	–0.36	–0.10	–0.14	0.05	–0.03	0.00	0.11	–0.02
9	0.10	0.25	0.44	0.64	0.44	0.37	0.55	0.28	0.23	0.05	0.06	0.13	0.29
10	0.08	0.35	0.74	1.04	0.94	1.04	1.02	0.71	0.43	0.13	0.10	0.18	0.56
11	0.01	0.42	1.01	1.37	1.34	1.54	1.37	1.06	0.63	0.27	0.18	0.15	0.78
Mean.	–5.02	–7.43	–3.89	–0.06	5.11	10.84	12.75	14.11	9.23	4.55	1.13	–3.42	

The numbers without sign must be added; those with the sign — must be subtracted.

Russia. — Petersburg. *Lat.* 59° 56′ N. *Long.* 30° 18′ E. *Greenw.*

Corrections to be applied to the Means of the Hours of Observation to obtain the true Mean Temperatures of the respective Days, Months, and of the Year. — Dove.

Degrees of Reaumur.

Hours.	Jan.	Feb.	March.	April.	May.	June.	July.	Aug.	Sept.	Oct.	Nov.	Dec.	Mean.
Morn. 1	0.20	0.38	0.92	1.52	2.59	2.40	1.98	2.08	1.39	0.72	0.14	0.17	1.21
2	0.23	0.37	1.10	1.75	2.84	2.69	2.26	2.43	1.67	0.77	0.13	0.27	1.38
3	0.22	0.39	1.30	2.01	3.03	2.90	2.49	2.79	1.97	0.82	0.14	0.33	1.53
4	0.21	0.43	1.49	2.19	3.05	2.91	2.57	3.01	2.20	0.88	0.16	0.35	1.62
5	0.26	0.50	1.59	2.17	2.79	2.60	2.37	2.92	2.25	0.95	0.20	0.35	1.58
6	0.37	0.57	1.56	1.88	2.20	1.98	1.88	2.46	2.06	0.98	0.23	0.34	1.38
7	0.51	0.56	1.36	1.35	1.27	1.13	1.15	1.70	1.62	0.92	0.23	0.33	1.01
8	0.59	0.46	0.99	0.68	0.41	0.24	0.34	0.79	1.01	0.72	0.16	0.31	0.56
9	0.53	0.23	0.47	−0.02	−0.47	−0.53	−0.40	−0.10	0.31	0.36	0.03	0.27	0.06
10	0.38	−0.09	−0.13	−0.65	−1.16	−1.09	−0.97	−0.86	−0.42	−0.09	−0.16	0.18	−0.43
11	0.01	−0.43	−0.74	−1.18	−1.68	−1.49	−1.37	−1.47	−1.12	−0.58	−0.35	0.03	−0.86
Noon. . .	−0.34	−0.73	−1.28	−1.62	−2.09	−1.83	−1.67	−2.01	−1.75	−0.99	−0.49	−0.15	−1.25
1	−0.59	−0.92	−1.68	−2.01	−2.50	−2.20	−1.98	−2.53	−2.29	−1.27	−0.54	−0.32	−1.57
2	−0.68	−0.95	−1.89	−2.33	−2.91	−2.62	−2.31	−3.01	−2.67	−1.36	−0.49	−0.44	−1.81
3	−0.61	−0.86	−1.92	−2.52	−3.25	−2.98	−2.58	−3.35	−2.81	−1.30	−0.35	−0.48	−1.92
4	−0.45	−0.67	−1.75	−2.50	−3.36	−3.12	−2.68	−3.39	−2.65	−1.12	−0.18	−0.44	−1.86
5	−0.27	−0.44	−1.44	−2.10	−3.11	−2.89	−2.46	−3.02	−2.19	−0.88	−0.02	−0.36	−1.61
6	−0.15	−0.22	−1.04	−1.01	−2.44	−2.26	−1.94	−2.26	−1.50	−0.62	0.10	−0.26	−1.18
7	−0.12	−0.02	−0.60	−0.86	−1.37	−1.33	−1.15	−1.25	−0.72	−0.37	0.17	−0.19	−0.65
8	−0.13	0.13	−0.20	−0.10	−0.34	−0.31	−0.29	−0.20	−0.01	−0.12	0.19	−0.14	−0.13
9	−0.14	0.24	0.14	0.54	0.69	0.61	0.49	0.66	0.53	0.11	0.19	−0.12	0.33
10	−0.09	0.32	0.40	0.96	1.47	1.30	1.07	1.24	0.87	0.33	0.18	−0.09	0.66
11	0.02	0.37	0.59	1.20	2.00	1.77	1.45	1.58	1.05	0.50	0.17	−0.02	0.89
Midn. . .	0.12	0.38	0.75	1.35	2.33	2.11	1.73	1.81	1.20	0.63	0.16	0.07	1.05
6. 6	0.11	0.18	0.26	0.14	−0.12	−0.14	−0.03	0.10	0.28	0.18	0.17	0.04	0.10
7. 7	0.20	0.27	0.38	0.25	−0.05	−0.10	−0.00	0.23	0.45	0.28	0.20	0.07	0.18
8. 8	0.23	0.29	0.40	0.29	0.04	−0.04	0.03	0.29	0.50	0.30	0.18	0.09	0.22
9. 9	0.20	0.24	0.31	0.26	0.11	0.04	0.04	0.28	0.42	0.24	0.11	0.08	0.19
10.10	0.12	0.12	0.13	0.15	0.16	0.11	0.05	0.19	0.22	0.12	0.01	0.05	0.12
7. 2. 9	−0.10	−0.05	−0.13	−0.15	−0.32	−0.29	−0.22	−0.22	−0.17	−0.11	−0.02	−0.08	−0.16
6. 2. 8	−0.15	−0.08	−0.18	−0.18	−0.35	−0.32	−0.24	−0.25	−0.21	−0.17	−0.02	−0.08	−0.19
6. 2.10	−0.13	−0.02	0.02	0.17	0.25	0.22	0.21	0.23	0.09	−0.02	−0.03	−0.06	0.08
6. 2. 6	−0.15	−0.20	−0.46	−0.69	−1.05	−0.97	−0.79	−0.94	−0.70	−0.33	−0.05	−0.12	−0.54
7. 2	−0.09	−0.20	−0.27	−0.49	−0.82	−0.75	−0.58	−0.66	−0.53	−0.22	−0.13	−0.06	−0.40
8. 2	−0.05	−0.25	−0.45	−0.83	−1.25	−1.19	−0.99	−1.11	−0.83	−0.32	−0.17	−0.07	−0.63
8. 1	0.00	−0.23	−0.35	−0.67	−1.05	−0.98	−0.82	−0.87	−0.64	−0.28	−0.19	−0.01	−0.51
7. 1	−0.04	−0.18	−0.16	−0.33	−0.62	−0.54	−0.42	−0.42	−0.34	−0.18	−0.16	0.01	−0.28
9.12.3.9	−0.14	−0.28	−0.65	−0.91	−1.28	−1.18	−1.04	−1.20	−0.93	−0.46	−0.16	−0.12	−0.70
7. 2.2(9)	−0.11	0.02	−0.06	0.03	−0.07	−0.07	−0.05	0.00	0.00	−0.06	0.03	−0.09	−0.04
Dail. ext.	−0.05	−0.19	−0.17	−0.17	−0.16	−0.11	−0.06	−0.19	−0.28	−0.19	−0.16	−0.07	−0.15

The numbers without sign must be added; those with the sign — must be subtracted.

LXX.

Russia. — Helsingfors. *Lat.* 60° 10′ N. *Long.* 24° 57′ E. *Greenw.*

Corrections to be applied to the Means of the Hours of Observation to obtain the true Mean Temperatures of the respective Days, Months, and of the Year. — Dove.

Degrees of Reaumur.

Hours.	Jan.	Feb.	March.	April.	May.	June.	July.	Aug.	Sept.	Oct.	Nov.	Dec.	Mean.
Morn. 1	0.47	0.85	1.40	2.10	2.49	3.37	3.16	2.58	1.60	1.06	0.64	0.34	1.67
2	0.79	1.25	1.86	3.18	2.82	3.78	3.48	2.96	2.09	1.45	0.99	0.68	2.11
3	0.99	1.55	2.28	2.79	2.89	3.74	3.45	3.11	2.48	1.70	1.22	0.91	2.26
4	1.13	1.71	2.52	2.77	2.62	3.22	3.02	2.92	2.61	1.74	1.26	0.97	2.21
5	1.06	1.66	2.49	2.41	2.06	2.32	2.25	2.39	2.40	1.51	1.09	0.84	1.87
6	0.86	1.43	2.16	1.76	1.30	1.24	1.23	1.59	1.84	1.10	0.76	0.59	1.32
7	0.58	1.07	1.57	0.92	0.49	0.20	0.17	0.64	1.06	0.59	0.38	0.31	0.67
8	0.28	0.60	0.79	0.05	−0.26	−0.65	−0.78	−0.28	0.21	0.08	0.02	0.07	0.01
9	0.01	0.10	−0.05	−0.74	−0.87	−1.26	−1.51	−1.07	−0.58	−0.38	−0.27	−0.10	−0.56
10	−0.25	−0.42	−0.87	−1.35	−1.34	−1.65	−2.02	−1.68	−1.23	−0.77	−0.48	−0.22	−1.02
11	−0.48	−0.91	−1.56	−1.80	−1.70	−1.93	−2.35	−2.12	−1.71	−1.07	−0.64	−0.32	−1.38
Noon. . .	−0.70	−1.29	−2.06	−2.10	−1.98	−2.16	−2.54	−2.43	−2.04	−1.30	−0.76	−0.43	−1.65
1	−0.86	−1.54	−2.36	−2.30	−2.19	−2.36	−2.65	−2.61	−2.23	−1.42	−0.85	−0.54	−1.83
2	−0.92	−1.60	−2.45	−2.37	−2.32	−2.51	−2.66	−2.66	−2.30	−1.43	−0.88	−0.61	−1.89
3	−0.84	−1.47	−2.32	−2.31	−2.31	−2.55	−2.55	−2.55	−2.20	−1.30	−0.82	−0.60	−1.82
4	−0.73	−1.20	−2.01	−2.10	−2.11	−2.42	−2.27	−2.26	−1.92	−1.05	−0.68	−0.49	−1.60
5	−0.52	−0.87	−1.56	−1.73	−1.77	−2.13	−1.85	−1.80	−1.48	−0.74	−0.48	−0.33	−1.27
6	−0.32	−0.57	−1.07	−1.25	−1.30	−1.71	−1.30	−1.24	−0.95	−0.44	−0.28	−0.18	−0.88
7	−0.19	−0.38	−0.60	−0.72	−0.78	−1.20	−0.68	−0.62	−0.42	−0.22	−0.16	−0.11	−0.51
8	−0.15	−0.25	−0.20	−0.21	−0.24	−0.61	−0.04	−0.03	−0.00	−0.10	−0.12	−0.12	−0.17
9	−0.16	−0.18	0.10	0.26	0.29	0.07	0.61	0.52	0.31	−0.03	−0.12	−0.20	0.12
10	−0.16	−0.08	0.36	0.69	0.82	0.87	1.27	1.03	0.54	0.08	−0.10	−0.25	0.42
11	−0.06	0.12	0.63	1.13	1.40	1.75	1.95	1.54	0.79	0.29	0.02	−0.19	0.78
Midn. . .	0.16	0.44	0.96	1.60	1.97	2.63	2.61	2.08	1.14	0.63	0.28	0.02	1.21
6. 6	0.27	0.43	0.55	0.26	−0.00	−0.24	−0.04	0.18	0.45	0.33	0.24	0.21	0.22
7. 7	0.20	0.35	0.49	0.10	−0.15	−0.50	−0.26	0.01	0.32	0.19	0.11	0.10	0.08
8. 8	0.07	0.18	0.30	−0.08	−0.25	−0.63	−0.41	−0.16	0.11	−0.01	−0.05	−0.03	−0.08
9. 9	−0.08	−0.04	0.03	−0.24	−0.29	−0.60	−0.45	−0.28	−0.14	−0.21	−0.20	−0.15	−0.22
10.10	−0.21	−0.25	−0.26	−0.33	−0.26	−0.39	−0.38	−0.33	−0.35	−0.35	−0.29	−0.24	−0.30
7. 2. 9	−0.17	−0.24	−0.26	−0.40	−0.51	−0.75	−0.63	−0.50	−0.31	−0.29	−0.21	−0.17	−0.37
6. 2. 8	−0.07	−0.14	−0.16	−0.27	−0.42	−0.63	−0.49	−0.37	−0.15	−0.14	−0.08	−0.05	−0.25
6. 2.10	−0.07	−0.08	0.02	0.03	−0.07	−0.13	−0.05	−0.01	0.03	−0.08	−0.07	−0.09	−0.05
6. 2. 6	−0.13	−0.25	−0.45	−0.62	−0.77	−0.99	−0.91	−0.77	−0.47	−0.26	−0.13	−0.07	−0.49
7. 2	−0.17	−0.27	−0.44	−0.73	−0.92	−1.16	−1.25	−1.01	−0.62	−0.42	−0.25	−0.15	−0.62
8. 2	−0.32	−0.50	−0.83	−1.16	−1.29	−1.58	−1.72	−1.47	−1.05	−0.68	−0.43	−0.27	−0.94
8. 1	−0.29	−0.47	−0.79	−1.13	−1.23	−1.51	−1.72	−1.45	−1.01	−0.67	−0.42	−0.24	−0.91
7. 1	−0.14	−0.24	−0.40	−0.69	−0.85	−1.08	−1.24	−0.99	−0.59	−0.42	−0.24	−0.12	−0.58
9.12.3.9	−0.42	−0.71	−1.08	−1.22	−1.22	−1.48	−1.50	−1.38	−1.13	−0.75	−0.49	−0.33	−0.98
7. 2.2(9)	−0.17	−0.22	−0.17	−0.23	−0.31	−0.54	−0.32	−0.25	−0.16	−0.23	−0.19	−0.18	−0.25
Dail. ext.	0.11	0.06	0.04	0.41	0.29	0.62	0.41	0.23	0.16	0.16	0.19	0.18	0.19

The numbers without sign must be added; those with the sign — must be subtracted.

Norway. — Christiania. *Lat.* 59° 55′ N. *Long.* 10° 43′ E. *Greenw.*

Corrections to be applied to the Means of the Hours of Observation to obtain the true Mean Temperatures of the respective Days, Months, and of the Year. — Dove.

Degrees of Reaumur.

Hours.	Jan.	Feb.	March.	April.	May.	June.	July.	Aug.	Sept.	Oct.	Nov.	Dec.	Mean.
Morn. 1	0.16	0.89	1.07	1.56	2.55	2.58	2.21	2.04	1.64	0.74	0.52	0.22	1.35
2	0.21	0.94	1.30	1.88	2.85	3.15	2.53	2.23	1.88	0.82	0.50	0.21	1.54
3	0.27	1.17	1.51	2.03	3.23	3.23	2.64	2.41	2.03	0.94	0.49	0.28	1.69
4	0.32	1.49	1.67	2.12	3.21	3.05	2.62	2.60	2.07	1.06	0.55	0.30	1.84
5	0.38	1.60	1.82	2.23	2.55	2.39	2.09	2.44	2.14	1.16	0.51	0.22	1.63
6	0.47	1.54	1.69	1.81	1.63	1.31	1.37	1.98	2.10	1.16	0.60	0.11	1.31
7	0.51	1.67	1.71	1.28	0.71	0.43	0.58	1.00	1.50	1.13	0.46	0.19	0.93
8	0.54	1.42	1.29	0.56	0.07	−0.32	−0.22	0.10	0.62	0.75	0.38	0.15	0.44
9	0.48	1.11	0.36	−0.06	−0.52	−0.86	−0.78	−0.59	0.01	0.15	0.17	0.16	−0.03
10	0.24	0.27	−0.35	−0.67	−1.19	−1.57	−1.26	−1.23	−0.78	−0.48	−0.23	0.11	−0.59
11	−0.17	−0.69	−0.96	−1.38	−1.66	−2.05	−1.74	−1.67	−1.44	−1.00	−0.76	−0.20	−1.14
Noon. . .	−0.67	−1.32	−1.48	−1.80	−2.17	−2.29	−2.02	−2.11	−2.02	−1.30	−1.06	−0.40	−1.55
1	−0.87	−1.90	−1.74	−2.22	−2.46	−2.50	−2.21	−2.35	−2.41	−1.59	−1.15	−0.42	−1.82
2	−1.04	−2.22	−1.95	−2.32	−2.46	−2.40	−2.20	−2.50	−2.54	−1.67	−1.15	−0.35	−1.90
3	−0.91	−2.29	−2.16	−2.26	−2.54	−2.47	−2.21	−2.50	−2.50	−1.58	−0.88	−0.23	−1.88
4	−0.62	−2.00	−1.99	−2.11	−2.53	−2.29	−2.00	−2.32	−2.35	−1.33	−0.55	−0.12	−1.68
5	−0.35	−1.42	−1.58	−1.80	−2.20	−2.14	−1.87	−1.97	−1.80	−0.90	−0.23	−0.06	−1.36
6	−0.12	−1.10	−1.10	−1.27	−1.82	−1.70	−1.48	−1.48	−1.21	−0.52	−0.02	−0.03	−0.99
7	−0.01	−0.60	−0.65	−0.70	−1.35	−0.98	−0.89	−0.78	−0.57	−0.24	0.11	−0.10	−0.58
8	0.12	−0.32	−0.20	−0.14	−0.44	−0.31	−0.30	−0.10	0.02	0.18	0.23	−0.13	−0.12
9	0.16	0.09	0.09	0.36	0.24	0.44	0.45	0.55	0.36	0.36	0.27	−0.05	0.28
10	0.27	0.34	0.36	0.70	0.93	1.20	1.06	1.08	0.81	0.58	0.33	−0.04	0.63
11	0.31	0.52	0.53	0.99	1.46	1.76	1.63	1.41	1.06	0.75	0.43	0.10	0.91
Midn. . .	0.33	0.86	0.77	1.20	1.90	2.31	2.00	1.75	1.38	0.95	0.48	0.09	1.17
6. 6	0.18	0.22	0.30	0.27	−0.10	−0.20	−0.06	0.25	0.45	0.32	0.29	0.04	0.16
7. 7	0.25	0.54	0.53	0.29	−0.32	−0.28	−0.16	0.11	0.47	0.45	0.29	0.05	0.18
8. 8	0.33	0.55	0.55	0.21	−0.19	−0.32	−0.26	0.00	0.32	0.47	0.31	0.01	0.16
9. 9	0.32	0.60	0.23	0.15	−0.14	−0.21	−0.17	−0.02	0.19	0.26	0.22	0.06	0.12
10.10	0.26	0.31	0.01	0.05	−0.13	−0.18	−0.10	−0.08	0.02	0.05	0.05	0.04	0.02
7. 2. 9	−0.12	−0.15	−0.05	−0.23	−0.50	−0.51	−0.39	−0.32	−0.23	−0.06	−0.14	−0.07	−0.23
6. 2. 8	−0.15	−0.33	−0.15	−0.22	−0.42	−0.47	−0.38	−0.21	−0.14	−0.11	−0.11	−0.12	−0.23
6. 2.10	−0.10	−0.11	0.03	0.06	0.03	0.04	0.08	0.19	0.12	0.02	−0.07	−0.09	0.02
6. 2. 6	−0.23	−0.59	−0.45	−0.59	−0.76	−0.67	−0.70	−0.67	−0.55	−0.34	−0.19	−0.11	−0.49
7. 2	−0.27	−0.28	−0.12	−0.52	−0.88	−0.99	−0.81	−0.75	−0.52	−0.27	−0.35	−0.08	−0.49
8. 2	−0.25	−0.40	−0.33	−0.88	−1.20	−1.04	−0.99	−1.20	−0.96	−0.46	−0.39	−0.10	−0.68
8. 1	−0.17	−0.24	−0.23	−0.83	−1.20	−1.09	−1.00	−1.13	−0.90	−0.42	−0.39	−0.14	−0.64
7. 1	−0.18	−0.12	−0.02	−0.47	−0.88	−1.04	−0.82	−0.68	−0.46	−0.23	−0.35	−0.12	−0.50
9.12.3.9	−0.56	−0.60	−0.80	−0.94	−1.25	−1.29	−1.14	−1.16	−1.04	−0.59	−0.38	−0.13	−0.82
7. 2.2(9)	−0.05	−0.09	−0.02	−0.08	−0.32	−0.27	−0.18	−0.10	−0.08	0.05	−0.04	−0.07	−0.11
Dail. ext.	−0.25	−0.31	−0.17	−0.05	0.35	0.39	0.22	0.05	−0.20	−0.26	−0.28	−0.06	−0.05

The numbers without sign must be added; those with the sign — must be subtracted.

NORWAY. — DRONTHEIM. *Lat.* 63° 26′ N. *Long.* 10° 25′ E. *Greenw.*

Corrections to be applied to the Means of the Hours of Observation to obtain the true Mean Temperatures of the respective Days, Months, and of the Year. — DOVE.

Degrees of Reaumur.

Hours.	Jan.	Feb.	March.	April.	May.	June.	July.	Aug.	Sept.	Oct.	Nov.	Dec.	Mean.
Morn. 1	0.29	0.41	0.77	1.94	2.63	2.64	2.53	2.51	1.37	0.89	0.27	0.33	1.38
2	0.25	0.50	0.95	2.09	2.97	2.76	2.75	2.68	1.48	0.91	0.31	0.31	1.50
3	0.22	0.64	1.11	2.19	3.13	2.82	2.77	2.91	1.59	0.97	0.23	0.42	1.58
4	0.20	0.71	1.27	2.32	3.03	2.82	2.65	2.77	1.55	1.07	0.28	0.34	1.58
5	0.13	0.75	1.37	2.05	2.76	2.52	2.35	2.58	1.59	0.86	0.30	0.42	1.47
6	0.11	0.82	1.42	1.67	2.30	1.96	1.86	2.13	1.49	0.71	0.14	0.43	1.25
7	0.04	0.58	1.35	1.36	1.68	1.39	1.17	1.58	1.07	0.42	0.00	0.36	0.92
8	0.08	0.23	1.17	0.94	0.83	0.61	0.40	1.02	0.57	0.06	-0.02	0.36	0.52
9	0.00	-0.08	0.41	-0.02	-0.28	-0.03	-0.14	0.22	-0.07	-0.29	-0.14	0.19	-0.02
10	-0.09	-0.48	-0.13	-0.85	-1.29	-0.92	-1.30	-1.22	-0.89	-0.59	-0.16	0.02	-0.65
11	-0.16	-0.78	-0.65	-1.90	-2.09	-2.01	-1.95	-2.63	-1.34	-0.88	-0.33	-0.12	-1.24
Noon. . .	-0.59	-1.08	-1.35	-2.57	-2.81	-2.43	-2.77	-3.21	-2.05	-1.20	-0.38	-0.42	-1.75
1	-0.80	-1.22	-1.70	-2.66	-3.28	-3.25	-3.20	-3.39	-2.12	-1.14	-0.44	-0.42	-1.97
2	-0.68	-1.15	-1.70	-2.46	-3.27	-3.32	-3.07	-3.36	-2.28	-1.09	-0.42	-0.47	-1.94
3	-0.48	-0.80	-1.54	-2.22	-3.25	-3.05	-3.06	-3.21	-1.85	-1.07	-0.28	-0.37	-1.76
4	-0.36	-0.56	-1.37	-1.83	-2.90	-2.78	-2.41	-2.81	-1.43	-0.86	-0.16	-0.29	-1.48
5	-0.29	-0.36	-1.07	-1.30	-2.20	-2.45	-2.02	-2.23	-1.09	-0.50	-0.06	-0.22	-1.15
6	-0.17	-0.11	-0.75	-0.90	-1.70	-1.84	-1.15	-1.27	-0.79	-0.51	0.08	-0.23	-0.78
7	0.09	-0.04	-0.54	-0.57	-1.03	-1.00	-0.61	-0.68	-0.32	-0.28	0.09	-0.30	-0.43
8	0.27	0.17	-0.27	-0.20	-0.37	0.04	0.01	0.11	0.03	-0.02	0.17	-0.19	-0.02
9	0.45	0.37	0.00	0.16	0.50	0.41	0.66	0.51	0.43	0.22	0.05	-0.11	0.30
10	0.52	0.53	0.23	0.61	1.10	1.08	1.17	1.18	0.75	0.55	0.13	-0.06	0.65
11	0.47	0.50	0.43	0.90	1.61	1.63	1.48	1.67	1.02	0.74	0.11	0.02	0.88
Midn. . .	0.45	0.49	0.63	1.27	1.92	2.07	1.88	2.13	1.28	1.14	0.19	0.02	1.12
6. 6	-0.03	0.36	0.34	0.39	0.30	0.06	0.36	0.43	0.35	0.10	0.11	0.10	0.24
7. 7	0.07	0.27	0.41	0.40	0.33	0.20	0.28	0.45	0.38	0.07	0.05	0.03	0.24
8. 8	0.18	0.20	0.45	0.37	0.23	0.33	0.21	0.57	0.30	0.02	0.08	0.09	0.25
9. 9	0.23	0.15	0.21	0.07	0.11	0.19	0.26	0.37	0.18	-0.04	-0.05	0.04	0.14
10.10	0.22	0.03	0.05	-0.12	-0.10	0.08	-0.07	-0.02	-0.07	-0.02	-0.02	-0.02	0.00
7. 2. 9	-0.06	-0.07	-0.12	-0.31	-0.36	-0.51	-0.41	-0.42	-0.26	-0.15	-0.12	-0.07	-0.24
6. 2. 8	-0.10	-0.05	-0.18	-0.33	-0.45	-0.44	-0.40	-0.37	-0.25	-0.13	-0.04	-0.08	-0.23
6. 2.10	-0.02	0.07	-0.02	-0.06	0.04	-0.09	-0.01	-0.02	-0.01	0.06	-0.05	-0.03	-0.01
6. 2. 6	-0.25	-0.62	-0.39	-0.56	-0.89	-1.06	-0.79	-0.83	-0.53	-0.30	-0.07	-0.09	-0.53
7. 2	-0.32	-0.29	-0.18	-0.55	-0.80	-0.97	-0.95	-0.89	-0.61	-0.34	-0.21	-0.06	-0.51
8. 2	-0.30	0.46	-0.27	-0.76	-1.22	-1.36	-1.34	-1.17	-0.86	-0.52	-0.22	-0.06	-0.63
8. 1	-0.36	-0.50	-0.27	-0.86	-1.23	-1.32	-1.40	-1.19	-0.78	-0.54	-0.23	-0.03	-0.73
7. 1	-0.38	-0.32	-0.18	-0.65	-0.80	-0.93	-1.02	-0.91	-0.53	-0.36	-0.22	-0.03	-0.53
9.12.3.9	-0.16	-0.40	-0.62	-1.16	-1.46	-1.28	-1.33	-1.42	-0.89	-0.59	-0.19	-0.18	-0.87
7. 2.2(9)	0.07	0.04	-0.09	-0.19	-0.15	-0.28	-0.15	-0.19	-0.09	-0.16	-0.02	-0.12	-0.11
Dail. ext.	-0.14	-0.20	-0.14	-0.17	-0.08	-0.25	-0.22	-0.24	-0.35	-0.07	-0.07	-0.02	-0.16

The numbers without sign must be added; those with the sign — must be subtracted.

Strait of Kara. *Lat.* 70° 37′ N. *Long.* 57° 47′ E. *Greenw.*

Corrections to be applied to the Means of the Hours of Observation to obtain the true Mean Temperatures of the respective Days, Months, and of the Year. — Dove.

Degrees of Reaumur.

Hours.	Jan.	Feb.	March.	April.	May.	June.	July.	Aug.	Sept.	Oct.	Nov.	Dec.	Mean.
Morn. 1	0.27	0.38	1.66	2.53	2.26	1.86	1.37	0.62	0.33	0.00	0.08	0.55	0.99
2	0.24	0.38	1.78	2.67	2.22	1.68	1.24	0.58	0.40	0.02	0.14	0.42	0.98
3	0.22	0.40	1.86	2.66	2.06	1.41	1.03	0.53	0.49	0.02	0.14	0.26	0.92
4	0.23	0.42	1.88	2.44	1.82	1.12	0.79	0.47	0.58	0.06	0.15	0.11	0.84
5	0.25	0.42	1.80	1.98	1.48	0.82	0.54	0.38	0.61	0.17	0.22	−0.00	0.72
6	0.27	0.33	1.55	1.30	1.01	0.49	0.25	0.26	0.58	0.29	0.36	−0.15	0.55
7	0.29	0.16	1.10	0.52	0.40	0.10	−0.05	0.10	0.42	0.35	0.52	−0.29	0.30
8	0.30	0.08	0.42	−0.27	−0.30	−0.33	−0.35	−0.07	0.27	0.32	0.64	−0.42	0.01
9	0.26	0.30	−0.43	−0.98	−1.01	−0.78	−0.66	−0.23	0.01	0.18	0.66	−0.54	−0.32
10	0.18	−0.50	−1.32	−1.58	−1.63	−1.19	−0.85	−0.36	−0.28	0.02	0.55	−0.61	−0.63
11	0.04	−0.64	−2.07	−2.13	−2.06	−1.48	−0.98	−0.46	−0.54	−0.25	0.33	−0.62	−0.91
Noon. . .	−0.12	−0.70	−2.56	−2.41	−2.27	−1.62	−1.04	−0.55	−0.72	−0.37	0.18	−0.54	−1.07
1	−0.31	−0.70	−2.70	−2.67	−2.26	−1.62	−1.03	−0.63	−0.81	−0.43	−0.13	−0.44	−1.14
2	−0.49	−0.64	−2.52	−2.81	−2.11	−1.54	−1.00	−0.71	−0.78	−0.36	−0.25	−0.31	−1.13
3	−0.60	−0.53	−2.10	−2.75	−1.88	−1.40	−0.95	−0.76	−0.66	−0.23	−0.30	−0.21	−1.03
4	−0.63	−0.38	−1.54	−2.46	−1.61	−1.25	−0.90	−0.69	−0.49	−0.10	−0.32	−0.11	−0.87
5	−0.58	−0.21	−0.98	−1.91	−1.30	−1.05	−0.78	−0.59	−0.30	0.02	−0.35	−0.04	−0.67
6	−0.46	−0.02	−0.47	−1.18	−0.90	−0.76	−0.59	−0.38	−0.13	0.07	−0.41	0.06	−0.43
7	−0.26	0.14	−0.04	−0.37	−0.40	−0.35	−0.29	−0.09	0.06	0.08	−0.48	0.18	−0.15
8	−0.06	0.32	0.34	0.42	0.20	0.18	0.11	0.22	0.11	0.07	−0.52	0.33	0.14
9	0.11	0.42	0.67	1.08	0.83	0.78	0.54	0.46	0.17	0.06	−0.49	0.48	0.43
10	0.22	0.46	0.98	1.59	1.42	1.31	0.94	0.62	0.20	0.06	−0.38	0.61	0.67
11	0.28	0.44	1.25	1.98	1.88	1.71	1.23	0.68	0.23	0.06	−0.20	0.66	0.85
Midn. . .	0.29	0.40	1.48	2.29	2.16	1.90	1.38	0.66	0.27	0.01	−0.03	0.64	0.95
6. 6	0.10	0.16	0.54	0.06	0.06	−0.14	−0.17	−0.06	0.23	0.18	−0.03	−0.05	0.06
7. 7	0.02	0.15	0.53	0.08	−0.00	−0.13	−0.17	0.01	0.24	0.22	0.02	−0.06	0.08
8. 8	0.12	0.12	0.38	0.08	−0.05	−0.08	−0.12	0.08	0.19	0.20	0.06	−0.05	0.08
9. 9	0.19	0.06	0.12	0.05	−0.09	−0.00	−0.06	0.12	0.09	0.12	0.09	−0.03	0.05
10.10	0.20	−0.02	−0.17	0.01	−0.11	0.06	0.05	0.13	−0.04	0.04	0.09	−0.00	0.02
7. 2. 9	−0.03	−0.02	−0.25	−0.40	−0.29	−0.22	−0.17	−0.05	−0.06	0.02	−0.07	−0.04	−0.13
6. 2. 8	−0.09	−0.00	−0.21	−0.36	−0.30	−0.29	−0.21	−0.08	−0.03	−0.00	−0.14	−0.04	−0.15
6. 2.10	−0.00	0.05	−0.00	0.03	0.11	0.09	0.06	0.06	−0.00	−0.00	−0.09	0.05	0.03
6. 2. 6	−0.23	−0.11	−0.48	−0.90	−0.67	−0.60	−0.45	−0.28	−0.11	−0.00	−0.10	−0.13	0.34
7. 2	−0.10	−0.24	−0.71	−1.15	−0.86	−0.72	−0.53	−0.31	−0.18	−0.01	0.14	−0.30	−0.41
8. 2	−0.10	−0.36	−1.05	−1.54	−1.21	−0.94	−0.68	−0.39	−0.26	−0.02	0.20	−0.37	−0.56
8. 1	−0.01	−0.39	−1.14	−1.47	−1.28	−0.98	−0.69	−0.35	−0.27	−0.06	0.26	−0.43	−0.57
7. 1	−0.01	−0.27	−0.80	−1.08	−0.93	−0.76	−0.54	−0.27	−0.20	−0.04	0.20	−0.37	−0.42
9.12.3.9	−0.09	−0.28	−1.11	−1.27	−1.08	−0.76	−0.53	−0.27	−0.30	−0.09	−0.01	−0.20	−0.50
7. 2.2(9)	0.01	0.09	−0.02	−0.03	−0.01	0.03	0.01	0.08	−0.01	0.03	−0.18	0.09	0.01
Dail. ext.	−0.17	−0.12	−0.41	−0.07	−0.01	0.14	0.17	−0.04	−0.10	−0.04	0.07	0.02	−0.08

The numbers without sign must be added; those with the sign — must be subtracted.

LXXIV.

Novaia Zemlia. — Matoschkin Schar. *Lat.* 73° —′ N. *Long.* 57° 20′ E. *Gr.*

Corrections to be applied to the Means of the Hours of Observation to obtain the true Mean Temperatures of the respective Days, Months, and of the Year. — Dove.

Degrees of Reaumur.

Hours.	Jan.	Feb.	March.	April.	May.	June.	July.	Aug.	Sept.	Oct.	Nov.	Dec.	Mean.
Morn. 1	–0.22	0.16	0.46	1.63	2.42	1.70	1.18	0.73	1.08	–0.49	–0.14	–0.11	0.70
2	–0.30	0.09	0.70	1.34	2.28	1.54	1.20	0.79	0.88	–0.47	–0.14	0.05	0.66
3	–0.31	0.01	0.91	1.15	1.89	1.26	1.11	0.80	0.62	–0.22	–0.10	0.17	0.61
4	–0.26	–0.06	1.02	1.09	1.41	0.93	0.94	0.72	0.46	0.02	–0.00	0.26	0.54
5	–0.14	–0.09	0.99	0.81	0.85	0.61	0.73	0.55	0.46	0.20	0.10	0.34	0.45
6	–0.03	–0.09	0.86	0.63	0.26	0.30	0.47	0.30	0.56	0.26	0.20	0.41	0.34
7	0.06	–0.07	0.62	0.09	–0.38	–0.02	0.18	0.01	0.58	0.18	0.26	0.45	0.16
8	0.10	–0.05	0.34	–0.50	–1.03	–0.38	–0.13	–0.30	0.38	0.06	0.26	0.46	–0.07
9	0.10	–0.05	0.02	–1.14	–1.65	–0.78	–0.46	–0.58	–0.00	–0.06	0.24	0.43	–0.33
10	0.07	–0.06	–0.28	–1.78	–2.17	–1.16	–0.75	–0.79	–0.71	–0.19	0.18	0.37	–0.61
11	0.05	–0.10	–0.58	–2.02	–2.53	–1.45	–0.97	–0.91	–1.24	–0.14	0.15	0.28	–0.79
Noon. . .	0.05	–0.13	–0.78	–2.09	–2.67	–1.58	–1.08	–0.93	–1.46	–0.12	0.11	0.18	–0.88
1	0.06	–0.14	–0.93	–1.93	–2.58	–1.52	–1.06	–0.85	–1.32	–0.10	0.08	0.10	–0.85
2	0.09	–0.14	–0.96	–1.62	–2.28	–1.32	–0.96	–0.70	–0.89	–0.09	0.02	–0.02	–0.74
3	0.10	–0.11	–0.88	–1.26	–1.83	–1.05	–0.81	–0.52	–0.40	–0.07	–0.04	–0.11	–0.58
4	0.10	–0.07	–0.71	–0.80	–1.30	–0.78	–0.66	–0.32	–0.07	–0.02	–0.10	–0.20	–0.41
5	0.10	–0.03	–0.50	–0.54	–0.72	–0.57	–0.54	–0.14	–0.02	0.10	–0.18	–0.26	–0.28
6	0.10	0.02	–0.30	–0.26	–0.14	–0.38	–0.43	–0.00	–0.17	0.26	–0.20	–0.36	–0.16
7	0.10	0.06	–0.16	0.30	0.46	–0.16	–0.30	0.12	–0.35	0.40	–0.18	–0.43	–0.01
8	0.12	0.10	–0.09	0.70	1.04	0.15	–0.11	0.21	–0.36	0.46	–0.14	–0.48	0.13
9	0.12	0.15	–0.06	1.24	1.59	0.56	0.14	0.30	–0.12	0.36	–0.10	–0.49	0.31
10	0.08	0.19	–0.02	1.50	2.06	1.02	0.46	0.39	0.33	0.18	–0.08	–0.44	0.47
11	–0.00	0.21	0.09	1.75	2.40	1.42	0.78	0.50	0.79	–0.15	–0.08	–0.34	0.61
Midn. . .	–0.11	0.20	0.23	1.72	2.55	1.66	1.03	0.62	1.06	–0.39	–0.11	–0.22	0.69
6. 6	0.04	0.04	0.28	0.19	0.06	–0.04	0.02	0.15	0.20	0.26	0.00	0.03	0.10
7. 7	0.08	0.01	0.23	0.20	0.04	–0.09	–0.06	0.07	0.12	0.29	0.04	0.01	0.08
8. 8	0.11	0.03	0.13	0.10	0.01	–0.12	–0.12	–0.05	0.01	0.26	0.06	–0.01	0.03
9. 9	0.11	0.05	–0.02	0.05	–0.03	–0.11	–0.16	–0.14	–0.06	0.15	0.07	–0.03	–0.01
10.10	0.08	0.07	–0.15	–0.14	–0.06	–0.07	–0.15	–0.20	–0.19	–0.01	0.05	–0.04	–0.07
7. 2. 9	0.09	–0.02	–0.13	–0.10	–0.36	–0.26	–0.21	–0.13	–0.14	0.15	0.06	–0.02	–0.09
6. 2. 8	0.06	–0.04	–0.06	–0.10	–0.33	–0.29	–0.20	–0.06	–0.23	0.21	0.03	–0.03	–0.09
6. 2.10	0.05	–0.01	–0.04	0.17	0.01	–0.00	–0.01	–0.00	–0.00	0.12	0.05	–0.02	0.03
6. 2. 6	0.05	–0.07	–0.13	–0.42	–0.72	–0.47	–0.31	–0.13	–0.17	0.14	0.01	0.01	–0.18
7. 2	0.08	–0.11	–0.17	–0.77	–1.33	–0.67	–0.39	–0.35	–0.16	0.05	0.14	0.22	–0.29
8. 2	0.10	–0.10	–0.31	–1.06	–1.66	–0.85	–0.55	–0.50	–0.26	–0.02	0.14	0.22	–0.40
8. 1	0.08	–0.10	–0.30	–1.22	–1.81	–0.95	–0.60	–0.58	–0.47	–0.02	0.17	0.28	–0.46
7. 1	0.06	–0.11	–0.16	–0.92	–1.48	–0.77	–0.44	–0.42	–0.37	0.04	0.17	0.28	–0.34
9.12.3.9	0.09	–0.04	–0.43	–0.81	–1.14	–0.71	–0.55	–0.43	–0.50	0.03	0.05	–0.00	–0.37
7. 2.2(9)	0.10	0.02	–0.12	0.24	0.13	–0.06	–0.13	–0.02	–0.14	0.20	0.02	–0.14	0.01
Dail. ext.	–0.10	0.04	0.03	–0.17	–0.06	0.06	0.06	–0.07	–0.19	–0.02	0.03	–0.02	–0.09

The numbers without sign must be added; those with the sign — must be subtracted.

LXXV.

Norway. — Bossekop. *Lat.* 69° 58′ N. *Long.* 22° E. *Greenw.*

Corrections to be applied to the Means of the Hours of Observation to obtain the true Mean Temperatures of the respective Days, Months, and of the Year. — Dove.

Degrees of Reaumur.

Hour.	Jan.	Feb.	March.	April.	Sept.	Oct.	Nov.	Dec.	80 Days without Sun.
A.M. 2	−0.26	0.36	1.37	. . .	1.20	0.66	0.04	0.35	0.04
4	−0.11	0.30	1.78	. . .	1.01	0.53	−0.03	0.42	0.10
6	0.00	0.50	1.90	. . .	1.22	0.73	0.04	0.28	0.08
8	0.09	0.26	1.18	0.36	0.62	0.41	0.07	0.10	0.02
10	−0.13	−0.19	−1.09	−0.85	−1.01	−0.29	−0.15	−0.14	−0.19
Noon.	0.18	−0.79	−2.39	−1.29	−1.66	−1.05	−0.13	−0.09	−0.03
2	0.20	−1.02	−2.85	−1.22	−1.69	−1.02	−0.09	−0.34	−0.10
4	0.30	−0.11	−2.38	−0.82	−1.54	−0.50	0.09	−0.38	0.06
6	0.18	0.06	−0.57	−0.10	−0.27	−0.17	0.18	−0.23	0.09
8	0.12	0.16	0.46	0.70	0.39	0.09	0.14	−0.26	0.02
10	−0.34	0.21	1.19	1.44	0.79	0.13	−0.03	0.14	−0.10
12	−0.27	0.22	1.39	1.83	0.89	0.49	−0.13	0.17	−0.10
Mean.	−7.67	−6.39	−7.55	−0.77	5.91	−1.62	−6.55	−5.66	−7.66

LXXV′.

Norway. — Bossekop. *Lat.* 69° 58′ N. *Long.* 22° E. *Greenw.*

Centigrade Degrees.

Hour.	Jan.	Feb.	March.	April.	Sept.	Oct.	Nov.	Dec.	80 Days without Sun.
A.M. 2	−0.32	0.45	1.71	. . .	1.50	0.82	0.05	0.44	0.05
4	−0.14	0.37	2.22	. . .	1.26	0.66	−0.04	0.52	0.12
6	0.00	0.62	2.37	. . .	1.52	0.91	0.05	0.35	0.10
8	0.11	0.32	1.47	0.45	0.77	0.51	0.09	0.12	0.02
10	−0.16	−0.24	−1.36	−1.06	−1.26	−0.36	−0.19	−0.17	−0.24
Noon.	0.22	−0.99	−2.98	−1.62	−2.07	−1.31	−0.16	−0.11	−0.04
2	0.25	−1.27	−3.56	−1.52	−2.11	−1.27	−0.11	−0.42	−0.12
4	0.37	−0.14	−2.97	−1.02	−1.92	−0.62	0.11	−0.47	0.07
6	0.22	0.07	−0.71	−0.12	−0.34	−0.21	0.22	−0.29	0.11
8	0.15	0.20	0.57	0.87	0.49	0.11	0.17	−0.32	0.02
10	−0.42	0.26	1.48	1.80	0.99	0.16	−0.04	0.17	−0.12
12	−0.34	0.27	1.73	2.29	1.11	0.61	−0.16	0.21	−0.12
Mean.	−9.59	−7.99	−9.44	−0.96	7.39	−2.02	−8.19	−7.07	−9.57

HOURLY CORRECTIONS

FOR

PERIODIC VARIATIONS.

AFRICA.—AUSTRALIA.

LXXVI.

Africa. — St. Helena. *Lat.* 15° 55′ S. *Long.* 5° 43′ W. *Greenw.*

Corrections to be applied to the Means of the Hours of Observation to obtain the true Mean Temperatures of the respective Days, Months, and of the Year. — Dove.

Degrees of Reaumur.

Hour.	Jan.	Feb.	March.	April.	May.	June.	July.	Aug.	Sept.	Oct.	Nov.	Dec.	Year.
Midn.	0.76	0.70	0.63	0.58	0.52	0.43	0.48	0.43	0.52	0.62	0.71	0.73	0.59
1	0.85	0.76	0.71	0.66	0.61	0.48	0.53	0.48	0.56	0.71	0.78	0.81	0.66
2	0.93	0.84	0.77	0.70	0.66	0.54	0.56	0.53	0.62	0.78	0.86	0.90	0.72
3	1.03	0.92	0.86	0.76	0.73	0.59	0.62	0.63	0.69	0.86	0.95	0.98	0.80
4	1.06	1.00	0.92	0.81	0.80	0.65	0.66	0.66	0.76	0.91	0.99	1.02	0.85
5	1.11	1.04	0.93	0.86	0.83	0.67	0.69	0.73	0.79	0.94	1.02	1.08	0.89
6	1.15	1.07	0.98	0.93	0.83	0.68	0.72	0.74	0.83	0.99	1.07	1.09	0.92
7	1.16	1.08	0.97	0.94	0.89	0.71	0.75	0.79	0.81	0.96	1.03	1.06	0.93
8	0.95	0.99	0.78	0.85	0.88	0.69	0.72	0.72	0.72	0.77	0.80	0.98	0.82
9	0.53	0.63	0.52	0.49	0.46	0.42	0.41	0.43	0.42	0.38	0.40	0.48	0.46
10	−0.05	0.06	−0.07	−0.04	−0.08	−0.04	−0.04	−0.02	−0.05	−0.17	−0.16	−0.09	−0.06
11	−0.62	−0.55	−0.49	−0.51	−0.47	−0.40	−0.40	−0.40	−0.55	−0.66	−0.67	−0.56	−0.52
Noon.	−1.14	−1.06	−0.95	−1.00	−0.96	−0.73	−0.76	−0.80	−0.92	−1.11	−1.12	−1.08	−0.97
1	−1.64	−1.46	−1.28	−1.31	−1.20	−1.04	−1.06	−1.12	−1.25	−1.45	−1.60	−1.52	−1.33
2	−1.81	−1.67	−1.48	−1.46	−1.32	−1.20	−1.26	−1.25	−1.42	−1.67	−1.80	−1.80	−1.51
3	−1.76	−1.78	−1.62	−1.50	−1.35	−1.18	−1.24	−1.31	−1.38	−1.64	−1.84	−1.82	−1.54
4	−1.69	−1.66	−1.54	−1.35	−1.24	−1.03	−1.12	−1.13	−1.20	−1.37	−1.64	−1.76	−1.39
5	−1.48	−1.38	−1.27	−1.06	−0.94	−0.78	−0.84	−0.86	−0.91	−0.99	−1.24	−1.38	−1.09
6	−0.92	−0.91	−0.83	−0.61	−0.47	−0.40	−0.44	−0.42	−0.43	−0.48	−0.66	−0.82	−0.62
7	−0.27	−0.33	−0.28	−0.11	−0.23	−0.03	−0.07	−0.03	0.01	0.02	−0.04	−0.18	−0.13
8	0.26	0.21	0.18	0.20	−0.12	0.17	0.13	0.15	0.23	0.29	0.32	0.30	0.19
9	0.47	0.44	0.34	0.34	0.14	0.26	0.23	0.25	0.32	0.26	0.48	0.48	0.33
10	0.60	0.55	0.48	0.44	0.41	0.32	0.33	0.32	0.38	0.49	0.56	0.58	0.46
11	0.69	0.64	0.55	0.51	0.45	0.39	0.38	0.38	0.46	0.55	0.64	0.67	0.53
Mean.	14.21	15.04	15.22	14.93	13.80	12.48	11.55	11.19	11.14	11.66	12.37	13.23	

LXXVII.

Africa.—Cape of Good Hope. *Lat.* 33° 56′ S. *Long.* 19° 39′ E. *Gr.*—Dove.

Degrees of Reaumur.

Hour	Jan.	Feb.	March.	April.	May.	June.	July.	Aug.	Sept.	Oct.	Nov.	Dec.	Year.
Midn.	1.69	1.50	1.51	1.37	1.00	0.88	1.04	0.85	1.07	1.45	1.62	1.85	1.32
1	2.80	1.64	1.64	1.49	1.07	1.01	1.20	1.03	1.25	1.62	1.79	2.01	1.55
2	1.89	1.74	1.81	1.61	1.14	1.09	1.33	1.14	1.39	1.72	1.98	2.16	1.58
3	2.01	1.92	1.92	1.70	1.24	1.16	1.43	1.23	1.54	1.82	2.12	2.30	1.70
4	2.10	2.00	2.05	1.88	1.34	1.30	1.53	1.37	1.63	1.92	2.21	2.42	1.81
5	1.96	2.13	2.13	1.93	1.46	1.42	1.59	1.53	1.59	1.93	1.92	2.01	1.80
6	1.06	1.53	1.97	1.98	1.59	1.48	1.73	1.55	1.62	1.26	0.85	0.86	1.46
7	0.15	0.70	1.21	1.39	1.41	1.47	1.57	1.22	0.81	0.39	−0.02	−0.20	0.84
8	−0.53	−0.01	0.16	0.36	0.53	0.86	0.77	0.64	−0.06	−0.46	−0.67	−0.81	0.06
9	−1.10	−0.80	−0.76	−0.68	−0.39	−0.12	−0.24	−0.42	−0.82	−1.24	−1.25	−1.36	−0.77
10	−1.72	−1.65	−1.66	−1.48	−1.10	−0.90	−1.09	−1.08	−1.41	−1.82	−1.80	−1.90	−1.47
11	−2.23	−2.31	−2.37	−2.10	−1.64	−1.46	−1.72	−1.63	−1.85	−2.25	−2.24	−2.25	−2.00

The numbers without sign must be added; those with the sign — must be subtracted.

LXXVII.

Africa. — Cape of Good Hope, *Continued.*

Corrections to be applied to the Means of the Hours of Observation to obtain the true Mean Temperatures of the respective Days, Months, and of the Year. — Dove.

Degrees of Reaumur.

Hour.	Jan.	Feb.	March.	April.	May.	June.	July.	Aug.	Sept.	Oct.	Nov.	Dec.	Year.
Noon.	−2.48	−2.72	−2.66	−2.56	−2.09	−1.92	−2.11	−1.88	−2.15	−2.45	−2.46	−2.52	−2.33
1	−2.54	−2.74	−2.95	−2.81	−2.20	−2.07	−2.33	−2.04	−2.23	−2.55	−2.48	−2.61	−2.46
2	−2.42	−2.54	−2.86	−2.79	−2.14	−2.06	−2.33	−1.97	−2.18	−2.44	−2.30	−2.44	−2.37
3	−2.16	−2.20	−2.51	−2.42	−1.84	−1.86	−2.13	−1.77	−1.82	−2.08	−2.01	−2.16	−2.08
4	−1.75	−1.70	−1.78	−1.75	−1.28	−1.28	−1.49	−1.32	−1.28	−1.52	−1.66	−1.90	−1.56
5	−1.21	−1.09	−1.03	−0.71	−0.61	−0.64	−0.76	−0.57	−0.56	−0.71	−1.05	−1.28	−0.85
6	−0.16	−0.13	−0.10	−0.03	−0.21	−0.29	−0.33	−0.17	0.00	0.20	−0.01	−0.15	−0.12
7	0.65	0.54	0.35	0.22	0.09	−0.05	−0.03	0.12	0.30	0.57	0.60	0.63	0.33
8	0.95	0.79	0.61	0.48	0.36	0.19	0.26	0.32	0.51	0.86	0.92	0.96	0.60
9	1.14	1.00	0.92	0.73	0.54	0.40	0.48	0.46	0.69	1.09	1.10	1.20	0.81
10	1.30	1.14	1.14	1.00	0.78	0.61	0.69	0.65	0.97	1.26	1.31	1.46	1.03
11	1.55	1.32	1.29	1.22	0.95	0.81	0.91	0.76	1.02	1.44	1.48	1.67	1.20
Mean.	15.81	15.96	15.00	13.61	11.38	9.84	9.96	10.06	11.01	12.43	13.54	14.82	

LXXVIII.

Australia. — Hobarton. *Lat.* 42° 53′ S. *Long.* 147° 21′ E. *Gr.* — Dove.

Degrees of Reaumur.

Hour.	Jan.	Feb.	March.	April.	May.	June.	July.	Aug.	Sept.	Oct.	Nov.	Dec.	Year.
Midn.	2.34	1.95	1.78	1.31	0.88	0.66	0.72	1.10	1.51	1.99	2.44	2.45	1.59
1	2.59	2.17	1.99	1.41	1.03	0.76	0.86	1.36	1.71	2.19	2.67	2.76	1.79
2	2.89	2.32	2.19	1.62	1.11	0.88	1.01	1.43	1.93	2.45	2.77	2.95	1.96
3	3.09	2.53	2.39	1.75	1.23	0.97	1.16	1.58	2.06	2.68	2.98	3.24	2.14
4	3.20	2.68	2.49	1.85	1.31	1.15	1.28	1.69	2.20	2.80	3.11	3.38	2.26
5	3.33	2.82	2.54	1.99	1.44	1.15	1.40	1.82	2.32	2.85	2.99	3.13	2.31
6	2.62	2.59	2.64	2.11	1.55	1.29	1.50	1.91	2.34	2.60	2.24	2.24	2.14
7	1.48	1.75	2.10	2.00	1.60	1.37	1.50	1.90	1.84	1.61	1.16	1.03	1.61
8	0.27	0.68	1.08	1.30	1.27	1.26	1.31	1.32	0.93	0.41	0.01	−0.24	0.80
9	−0.88	−0.56	−0.17	0.24	0.45	0.60	0.60	0.44	−0.21	−0.70	−1.13	−1.27	−0.22
10	−1.92	−1.61	−1.28	−0.85	−0.46	−0.18	−0.21	−0.52	−1.21	−1.68	−2.10	−2.16	−1.18
11	−2.75	−2.34	−2.24	−1.78	−1.29	−0.96	−1.01	−1.53	−2.09	−2.54	−2.89	−2.85	−2.02
Noon.	−3.51	−3.22	−3.03	−2.58	−2.00	−1.67	−1.67	−2.28	−2.70	−3.10	−3.43	−3.36	−2.71
1	−3.82	−3.52	−3.48	−2.95	−2.42	−2.08	−2.17	−2.73	−3.14	−3.48	−3.72	−3.67	−3.10
2	−3.91	−3.54	−3.63	−3.11	−2.53	−2.22	−2.38	−2.91	−3.25	−3.48	−3.67	−3.56	−3.18
3	−3.60	−3.36	−3.43	−2.87	−2.32	−2.02	−2.23	−2.71	−3.10	−3.32	−3.33	−3.45	−2.98
4	−3.20	−2.94	−2.92	−2.23	−1.69	−1.43	−1.73	−2.20	−2.53	−3.04	−3.12	−3.12	−2.51
5	−2.57	−2.22	−2.02	−1.35	−0.92	−0.73	−1.01	−1.37	−1.59	−2.02	−2.30	−2.56	−1.72
6	−1.38	−1.04	−0 84	−0.56	−0.36	−0.25	−0.48	−0.64	−0.65	−0.80	−1.01	−1.38	−1.78
7	−0.13	−0.20	−0.04	−0.05	0.01	0.00	0.12	−0.13	0.01	0.05	0.20	−0.09	−0.02
8	0.82	0.68	0.45	0.32	0.27	0.24	0.14	0.21	0.46	0.55	0.90	0.89	0.49
9	1.31	1.13	0.82	0.57	0.42	0.24	0.34	0.57	0.79	1.00	1.41	1.51	0.84
10	1.71	1.47	1.19	0.84	0.62	0.40	0.50	0.79	1.08	1.34	1.75	1.91	1.13
11	2.05	1.77	1.47	1.06	0.77	0.54	0.64	0.93	1.31	1.63	2.05	2.25	1.37
Mean.	13.38	13.96	11.96	9.41	7.69	5.93	5.21	6.24	7.97	9.39	11.38	12.95	

The numbers without sign must be added; those with the sign — must be subtracted.

CORRECTIONS FOR TEMPERATURE.

MONTHLY AND YEARLY

CORRECTIONS FOR NON-PERIODIC VARIATIONS,

OR

TABLES

FOR REDUCING THE MONTHLY AND YEARLY MEANS OF SINGLE YEARS TO THE MEANS DERIVED FROM A SERIES OF YEARS.

TABLES

FOR REDUCING THE MONTHLY AND YEARLY MEANS OF SINGLE YEARS TO THE MEANS DERIVED FROM A SERIES OF YEARS.

OBSERVATION shows that the monthly and annual mean temperature of a place somewhat varies from year to year. No law, however, has been as yet discovered as to the course of these oscillations. It follows that the means derived from observations carried on during a single year are but approximations to the true means. These last must be obtained from observations made for a series of years, during which these irregular variations become insensible by compensating each other; and it is obvious that their accuracy increases with the number of years which compose the series.

Professor Dove, having proved by his researches that these abnormal temperatures above and below the average of a whole month, or of a year, are apt to be felt simultaneously on extensive tracts of country, concluded that the means of a single year could be made available for obtaining the true means of the place, by being corrected for the non-periodic variations by means of normal stations in *the same meteorological region*, in which those elements had been more accurately determined by the observations of a long series of years. Comparing, namely, the means of a given year with the means derived from the whole series, we find a difference in + or —, which, applied, with *reverse* signs, to the means of the same year in the neighboring station to be corrected, will reduce, with a good degree of probability, the means of that particular year to the means which would have been obtained from a long series of years similar to that of the normal station.

The following tables, LXXIX. to XCVII., have been selçcted from those given by Dove in his five papers on the non-periodic variations of the atmospheric temperature, to be found in the *Memoirs of the Academy of Sciences of Berlin* for the years 1838, 1839, 1842, 1848, and 1853, to which we must refer for further details. They furnish normal stations for various latitudes; the columns contain the corrections for every month, viz. the differences, with *reverse* signs, between the monthly means in the year indicated in the first and last columns, and the means derived from the whole series, which are contained in the line at the bottom.

LXXIX.

Region of the Monsoons. — Madras.

For Reducing the Monthly and Yearly Means of Single Years to the Means derived from Series of Years.

Degrees of Reaumur.

Year.	Jan.	Feb.	March.	April.	May.	June.	July.	Aug.	Sept.	Oct.	Nov.	Dec.	Year.
1796	0.00	0.24	0.00	0.36	–0.10	–1.48	–1.16	–1.15	–0.31	–0.28	–0.47	–0.51	1796
1797	. .	. .	0.66	0.53	0.39	0.56	0.09	0.85	–0.09	–0.33	0.16	–0.02	1797
1798	–0.13	1.12	0.40	. .	. .	0.39	0.53	–0.31	0.27	0.56	–0.16	0.20	1798
1799	–0.13	–0.08	0.62	0.36	0.26	–0.06	–1.20	0.00	–0.36	0.38	–1.44	0.25	1799
1800	0.40	0.41	0.57	1.20	–0.23	–0.50	–1.02	–0.40	–0.58	0.20	0.47	–0.60	1800
1801	0.44	0.01	1.77	. .	. .	–0.59	–0.67	0.63	–0.49	–0.02	–0.20	0.25	1801
1802	0.44	0.86	1.77	1.02	–0.36	0.65	0.58	–0.04	1.60	0.43	–0.02	–0.28	1802
1803	0.22	0.24	0.80	0.53	–0.32	0.08	0.18	0.80	0.80	0.38	0.33	0.65	1803
1804	1.64	1.48	0.75	1.38	0.70	0.70	1.24	0.00	0.58	0.38	0.91	0.29	1804
1805	0.27	0.41	0.66	–0.36	0.61	0.52	–0.76	–0.22	–0.27	–0.33	0.69	0.65	1805
1806	0.00	–0.39	–0.09	0.09	–0.41	–1.61	0.00	–0.13	1.07	0.47	0.96	0.12	1806
1807	0.22	–1.54	–3.20	–5.47	–1.79	0.48	1.20	–0.17	–0.09	–0.64	–0.20	0.78	1807
1813	0.80	0.37	0.13	0.96	1.12	–0.32	0.44	–0.22	–0.18	0.25	–0.38	–1.04	1813
1814	–0.36	–0.39	–0.58	0.04	–2.99	1.10	1.38	0.29	–0.22	0.07	–0.20	–0.37	1814
1815	–0.98	0.32	–0.67	2.00	1.55	–1.39	–0.98	0.27	0.31	–0.73	–0.91	–0.82	1815
1816	–1.09	–1.76	–1.56	–0.93	0.44	0.39	–0.44	–0.71	–0.67	–0.20	0.33	–0.51	1816
1817	–0.58	–0.70	–0.67	–0.62	0.12	–0.19	0.67	0.29	–0.71	–0.55	–0.96	0.52	1817
1818	0.22	0.32	–0.80	–0.04	1.41	0.65	–1.33	–2.00	–0.18	–0.55	–0.56	–0.37	1818
1819	–1.78	–1.28	–0.76	–0.13	0.48	0.88	0.44	0.98	–0.31	0.03	0.78	0.16	1819
1820	–0.67	–0.30	–0.85	0.58	–1.16	–0.32	0.18	0.23	–0.09	0.47	0.69	0.47	1820
1821	1.02	0.64	1.06	–1.51	0.26	0.08	0.58	0.94	–0.04	–0.02	0.20	0.20	1821
Means.	19.19	20.07	21.30	22.41	24.41	24.96	23.84	23.43	23.03	22.16	20.74	19.48	Means.
1822	–0.36	0.37	0.41	–0.28	0.07	–0.95	–0.76	0.72	–0.37	–0.70	–0.35	–0.19	1822
1823	0.31	0.37	–0.21	0.30	0.15	0.29	0.22	0.17	–0.60	0.72	0.27	0.97	1823
1824	0.71	0.59	0.27	0.52	–0.02	0.60	1.55	0.88	1.36	–0.93	0.14	0.26	1824
1825	–0.09	0.37	–0.21	0.12	0.24	–0.29	0.04	–0.36	0.03	0.32	0.59	–0.59	1825
1826	0.80	0.24	0.45	0.92	0.78	–1.17	0.04	–0.36	0.25	0.81	0.36	0.30	1826
1827	–0.09	–0.29	–0.17	0.17	–1.27	–0.46	–0.01	–0.09	–0.15	–0.13	0.54	0.08	1827
1828	1.07	0.51	–0.57	–0.59	–0.42	0.34	–0.23	0.04	–0.60	–0.17	0.81	0.21	1828
1829	0.09	–0.69	–0.35	0.08	–0.11	0.16	–0.89	–0.01	0.16	0.54	0.23	0.12	1829
1830	–0.27	–0.74	0.01	–0.32	–2.73	–0.15	–0.36	–0.23	0.25	1.12	0.68	0.53	1830
1831	0.31	1.49	1.66	0.48	1.89	1.36	0.04	0.67	0.70	0.41	0.41	0.53	1831
1832	–0.49	–0.29	1.26	1.73	2.51	2.65	1.64	2.40	0.34	–0.25	0.46	. .	1832
1833	0.36	0.91	–0.19	0.97	0.83	0.83	1.33	0.40	0.16	0.41	0.19	1.06	1833
1834	0.18	0.60	0.55	–0.58	1.31	0.12	–0.98	–0.18	–0.15	–0.03	0.01	–0.01	1834
1835	–0.66	–0.73	–0.57	–1.07	–0.24	–0.86	–0.67	–0.45	–0.46	–0.74	–0.48	–0.94	1835
1836	–0.75	–0.73	–1.41	–0.72	0.60	0.12	–0.58	–1.29	–0.24	0.15	–0.92	–1.03	1836
1837	–0.31	–0.02	0.06	–0.63	–1.17	–0.41	–0.40	–0.05	0.03	–0.34	–0.17	–0.85	1837
1838	–1.24	–0.69	–0.30	0.04	–0.33	–0.24	0.75	–0.05	0.65	–0.12	–0.57	–0.41	1838
1839	0.36	–0.11	–0.12	–0.45	–0.15	–0.41	–0.49	–0.93	–0.68	0.68	–0.83	0.71	1839
1840	–0.13	–0.42	–0.70	0.17	0.29	0.25	–0.45	0.27	–0.77	0.19	–1.14	–0.32	1840
1841	0.05	–0.16	–0.17	–0.58	–1.17	–0.81	0.35	–0.71	0.74	–1.28	–0.34	–0.27	1841
1842	–0.09	–0.51	–0.08	–0.49	0.47	0.07	0.00	0.09	–0.86	–0.30	–0.30	–0.23	1842
1843	0.23	–0.02	–0.03	0.22	–1.53	–1.04	–0.22	0.44	0.16	–0.52	0.23	–0.32	1843
Means.	20.53	21.31	22.92	24.27	25.62	25.35	24.31	23.73	23.70	22.92	21.32	20.67	Means.

The numbers without sign must be subtracted; those with the sign — must be added.

E

LXXX.

Sicily. — Palermo.

For Reducing the Monthly and Yearly Means of Single Years to the Means derived from Series of Years.

Degrees of Reaumur.

Year.	Jan.	Feb.	March.	April.	May.	June.	July.	Aug.	Sept.	Oct.	Nov.	Dec.	Year.
	°	°	°	°	°	°	°	°	°	°	°	°	
1791	. .	. .	. .	. .	−0.44	−0.32	−0.52	0.95	−0.67	−0.36	1.73	0.00	1791
1792	1.18	0.51	0.09	0.12	−0.48	1.12	−1.01	−1.65	−0.63	0.22	−0.83	−0.96	1792
1793	−1.68	−0.38	−0.33	−1.63	−2.04	−1.83	−1.28	−1.14	0.86	−0.54	−0.25	0.44	1793
1794	−0.04	−0.69	−0.51	0.59	−0.79	−1.92	−0.81	−0.48	−0.14	−0.69	−0.29	−0.47	1794
1795	−1.62	1.27	0.78	−0.10	0.12	−0.59	−1.12	−0.34	−0.72	. .	−0.23	0.18	1795
1796	0.78	0.58	−0.84	−1.56	−0.19	−0.59	−0.39	−0.01	0.40	1.11	0.00	0.98	1796
1797	−0.24	−0.29	−1.15	−0.19	−0.24	−0.70	−0.56	0.39	0.15	0.13	−0.12	0.16	1797
1798	0.03	0.20	0.78	−0.90	−0.99	−0.45	0.72	−0.41	0.00	−1.00	1.97	0.31	1798
1799	−1.75	1.38	0.52	0.64	−0.35	0.08	0.37	0.75	0.48	1.40	−0.32	0.40	1799
1800	2.27	2.96	0.69	2.46	0.63	−0.14	0.26	−0.41	−0.58	0.02	−0.18	0.09	1800
1801	−0.11	0.76	1.45	0.24	−0.10	−0.16	1.26	−0.56	−0.07	1.04	1.04	1.64	1801
1802	0.09	−0.16	0.47	−1.01	−0.30	2.50	0.17	0.72	0.42	0.77	1.51	1.40	1802
1803	1.67	−1.69	. .	2.08	−1.08	0.66	0.04	0.52	0.31	−0.65	1.42	0.42	1803
1804	4.63	−0.82	0.16	0.21	0.14	1.30	1.12	0.12	−0.14	0.31	1.22	1.40	1804
1805	0.80	0.69	−0.68	−1.59	−1.59	1.21	−0.65	−0.34	−1.52	0.06	−1.85	−1.02	1805
1806	−1.15	0.64	−0.04	−0.50	0.41	0.10	−0.14	−0.85	−1.16	−0.43	−0.14	0.40	1806
1807	−1.06	0.16	0.34	−1.21	0.74	0.90	1.37	0.92	2.80	1.26	1.95	−0.07	1807
1808	−0.24	−1.22	−0.86	−1.36	−0.48	−0.43	0.88	0.04	2.42	−1.92	−0.29	−2.31	1808
1809	0.87	−0.31	0.23	−0.50	−0.48	0.86	1.46	−0.23	−0.67	−1.67	−1.36	−0.98	1809
1810	0.01	−0.27	2.49	0.28	0.50	−0.63	−0.54	−0.19	−0.29	−0.67	0.06	−0.91	1810
1811	−0.15	0.69	−0.91	0.24	0.43	1.46	0.97	0.26	0.04	0.95	0.00	−0.76	1811
1812	−1.51	0.40	0.00	−0.39	−0.61	0.15	−1.32	−0.21	−0.69	−0.16	0.35	−0.18	1812
1813	−1.51	−1.02	−0.80	−0.52	0.79	0.32	−0.92	−1.25	−1.00	1.31	0.04	−1.18	1813
1814	0.54	−3.04	−0.88	0.04	−1.46	−0.59	−0.96	−0.56	−2.03	−0.49	−0.52	−0.42	1814
1815	−0.46	0.07	0.29	0.90	0.61	−0.63	−1.12	−2.01	−0.78	0.22	0.08	−0.78	1815
1816	−0.40	−0.31	−0.71	−0.54	0.05	−1.94	−0.65	−0.48	−0.80	−1.09	−0.63	−1.24	1816
1817	−0.11	−0.09	−0.15	. .	. .	0.32	−0.39	0.46	−0.34	0.11	−0.47	−0.02	1817
1818	−0.66	0.87	. .	1.21	0.19	−1.10	−0.25	−0.45	0.24	−0.78	0.33	0.62	1818
1819	−1.02	0.18	0.72	0.97	−0.12	−0.21	−0.28	−0.34	−0.32	0.82	1.11	0.82	1819
1820	1.89	−0.11	−0.97	0.37	2.03	0.68	0.48	. .	. .	. .	−0.65	0.29	1820
1821	1.92	−0.76	0.49	0.50	0.85	−0.74	−0.30	−0.21	0.51	−0.74	−0.72	0.69	1821
1822	−1.28	−1.11	−0.53	. .	0.68	2.97	1.48	1.46	1.88	1.51	0.06	0.18	1822
1823	0.52	1.78	−0.80	0.28	0.99	0.30	−0.36	0.35	−0.34	−0.76	−1.63	−0.53	1823
1824	−0.91	0.42	−1.04	−1.01	1.25	−0.25	−0.70	1.86	0.13	1.51	0.64	0.51	1824
1825	−1.04	−1.02	−0.17	0.12	0.30	−0.45	−0.10	0.46	0.55	−1.00	−0.05	1.67	1825
1826	−0.88	0.56	−0.29	−0.59	−1.08	−0.74	0.39	0.52	1.35	0.46	−0.87	−0.24	1826
1827	0.07	0.83	0.82	−0.51	0.18	−1.30	0.80	1.33	−0.73	0.50	−1.76	−0.04	1827
1828	−0.16	0.20	0.23	0.29	1.99	1.28	2.48	1.10	0.74	−0.34	0.06	−0.37	1828
1829	0.79	−1.90	1.12	2.49	−0.09	−0.47	0.16	−0.12	0.41	−0.38	−0.35	−0.16	1829
Means.	8.35	8.27	9.40	11.52	14.35	17.12	19.25	19.48	17.60	14.78	11.69	9.44	Means.

The numbers without sign must be subtracted; those with the sign — must be added.

LXXXI.

North Italy. — Milan.

For Reducing the Monthly and Yearly Means of Single Years to the Means derived from Series of Years.

Degrees of Reaumur.

Year.	Jan.	Feb.	March.	April.	May.	June.	July.	Aug.	Sept.	Oct.	Nov.	Dec.	Year.
	°	°	°	°	°	°	°	°	°	°	°	°	
1763	−1.32	1.58	−0.60	0.27	−2.28	−0.79	0.68	1.11	−1.11	−1.69	−0.56	1.02	1763
1764	1.68	1.98	−0.70	−0.63	1.32	0.91	−0.12	−1.29	−1.21	−1.19	−0.26	1.52	1764
1765	3.88	−0.92	0.60	0.47	−1.08	0.11	−2.62	−1.69	−0.11	0.11	0.24	−0.98	1765
1766	−3.42	−1.52	−0.40	0.57	0.02	1.31	−1.32	−0.19	−1.21	−0.49	2.14	−0.68	1766
1767	−4.22	0.38	0.10	−0.93	−1.08	−1.19	0.78	−0.69	. .	. .	. .	−0.88	1767
1768	−0.82	−1.22	−1.50	0.37	−0.58	−2.19	0.68	0.51	. .	. .	0.64	−0.78	1768
1769	1.88	−0.42	−0.50	−1.63	−0.48	1.01	−0.52	1.51	. .	−2.19	1.24	0.62	1769
1770	−0.52	0.98	−0.60	−0.33	−0.58	0.81	−0.72	0.01	1.99	0.51	1.04	−0.68	1770
1771	1.78	−0.52	−0.60	−1.43	1.02	−0.19	0.68	1.51	0.49	−0.69	−1.06	2.32	1771
1772	1.58	2.48	2.50	0.57	−0.58	1.61	1.38	0.41	0.29	2.01	1.94	2.02	1772
1773	1.58	−0.42	−0.80	−0.03	−0.48	. .	−1.72	−1.29	0.69	1.61	0.34	1.82	1773
1774	0.48	0.08	0.70	0.77	−0.28	0.51	−0.12	1.31	−0.31	−1.09	−0.96	−2.68	1774
1775	0.38	2.08	1.60	0.47	−0.58	0.71	0.78	−0.09	−0.31	−1.79	−0.16	−0.88	1775
1776	−0.32	−0.02	1.30	0.97	−1.28	0.11	0.48	0.41	−0.71	0.11	−0.36	−1.18	1776
1777	−1.52	−1.42	1.30	−0.23	−1.08	−0.79	−1.22	0.51	0.19	0.41	1.24	−1.98	1777
1778	0.38	0.08	−1.90	1.47	0.62	−0.29	0.98	0.81	−0.81	−0.09	0.64	1.72	1778
1779	−3.52	1.98	0.00	1.07	1.72	−1.39	0.18	−0.19	1.59	1.81	−0.16	1.82	1779
1780	−0.62	−1.92	2.70	−0.43	1.72	1.51	0.78	0.11	−0.51	1.81	−0.16	−1.08	1780
1781	−0.12	0.38	1.90	1.47	0.22	0.01	1.78	0.41	0.39	−0.89	0.04	1.42	1781
1782	2.18	−2.42	−0.70	−1.03	−1.08	1.21	2.08	0.91	−0.31	−1.79	−2.46	−0.58	1782
1783	0.98	1.18	−0.60	0.97	0.42	−0.99	1.08	−0.29	−0.31	1.51	0.24	−1.88	1783
1784	0.48	−2.02	0.50	−2.03	2.62	2.11	1.38	0.61	1.49	−1.49	−0.46	−1.18	1784
1785	0.58	−1.12	−3.80	−1.23	0.72	1.21	0.68	0.61	2.69	0.41	0.74	2.02	1785
1786	0.18	0.68	−0.90	0.87	0.72	0.81	−0.52	−0.89	1.09	−1.89	−0.36	−0.48	1786
1787	−0.32	0.08	0.90	−0.03	−1.98	1.71	−0.02	1.61	0.09	0.81	0.84	1.72	1787
1788	2.78	1.08	2.30	1.37	−0.18	1.51	2.78	−0.39	0.99	0.21	−0.86	−2.88	1788
1789	−1.72	0.98	−1.70	1.37	2.22	−0.79	0.28	0.11	0.29	0.31	−1.26	−2.38	1789
1790	−0.12	1.48	−0.20	−1.73	1.62	0.71	−0.72	1.21	0.19	2.21	1.24	0.02	1790
1791	2.48	1.08	1.20	1.87	−0.18	−0.49	0.58	1.51	0.09	−0.29	−0.46	1.92	1791
1792	0.98	−0.12	1.30	1.87	−0.18	0.21	0.08	0.11	−0.41	0.71	0.54	−0.08	1792
1793	−1.22	−0.02	0.40	−1.43	−0.38	0.01	1.78	−0.29	2.49	1.31	1.44	2.22	1793
1794	2.28	3.08	2.00	2.37	−0.08	0.81	1.78	0.21	−1.11	−0.49	1.84	−0.38	1794
1795	−3.72	−3.12	−0.20	1.37	1.52	−0.79	−1.42	0.91	0.49	1.71	−0.16	1.52	1795
1796	2.48	1.18	−1.70	−0.13	−0.28	−0.29	−0.12	0.71	1.39	0.41	1.24	−1.38	1796
1797	0.78	0.18	−1.40	0.67	1.22	−1.59	1.18	2.51	1.09	−0.59	0.94	1.32	1797
1798	1.78	2.08	0.20	0.27	0.72	−0.09	0.48	0.51	0.29	−0.39	−0.86	−2.08	1798
1799	−3.22	0.88	0.60	−1.23	−0.98	−1.49	−0.62	0.41	1.39	0.51	−0.96	−1.18	1799
1800	1.78	4.58	−1.10	2.67	1.32	−1.59	0.38	−0.09	0.49	0.01	1.24	−0.08	1800
1801	1.38	1.08	1.50	0.77	0.32	−0.39	−0.62	−0.79	0.49	0.61	0.04	0.02	1801
1802	0.18	1.18	0.70	0.87	−0.08	1.71	0.28	2.21	1.09	2.81	1.04	1.52	1802
1803	2.38	−3.82	0.30	1.47	−0.88	1.11	0.78	1.11	−0.91	−0.49	0.54	0.22	1803

The numbers without sign must be subtracted; those with the sign — must be added.

LXXXI.

North Italy. — Milan (*continued*).

For Reducing the Monthly and Yearly Means of Single Years to the Means derived from Series of Years.

Degrees of Reaumur.

Year.	Jan.	Feb.	March.	April.	May.	June.	July.	Aug.	Sept.	Oct.	Nov.	Dec.	Year.
	°	°	°	°	°	°	°	°	°	°	°	°	
1804	3.98	−1.32	−0.60	−0.03	1.32	2.11	. .	−0.39	0.49	0.71	−0.36	−0.18	1804
1805	−0.12	−0.02	−0.10	−2.03	−0.78	0.21	−0.42	−0.29	0.79	−1.19	−2.36	−1.58	1805
1806	0.18	1.68	0.10	−1.53	0.32	1.01	−0.52	−1.19	−0.41	−0.19	1.34	1.92	1806
1807	0.58	0.28	−2.40	−1.33	1.32	0.21	1.18	1.71	0.19	1.71	1.34	−0.08	1807
1808	−1.02	−1.62	−3.80	−1.23	1.62	−0.49	1.98	−0.69	0.39	−2.39	0.24	−2.08	1808
1809	0.48	1.98	−1.40	−2.63	1.02	0.51	−0.52	0.21	−0.51	−0.19	−0.96	0.22	1809
1810	0.08	−0.72	1.90	−0.23	0.22	−1.49	−2.12	−0.79	0.59	1.11	0.34	1.82	1810
1811	−0.72	1.48	1.70	1.47	1.82	−0.29	1.18	−0.49	0.39	2.21	2.24	−0.38	1811
1812	−3.32	−0.32	−0.40	−1.73	0.92	1.01	−0.82	−0.59	−1.41	0.11	−2.96	−2.38	1812
1813	−0.12	1.08	0.50	1.07	1.52	−0.99	−2.12	−1.09	−1.11	0.21	−0.56	1.32	1813
1814	−0.12	−4.42	−1.40	0.77	−1.98	−1.09	−0.12	−1.09	−1.91	−0.69	1.04	1.82	1814
1815	−2.02	−0.32	1.90	0.77	1.22	−0.49	−1.22	−1.39	0.29	0.81	−1.26	−1.78	1815
1816	−0.52	−2.92	−1.20	−0.93	−0.58	−1.49	−2.22	−3.69	−0.51	0.41	−1.46	−1.88	1816
1817	−2.52	2.08	0.30	−2.83	−1.28	0.21	−3.52	−0.59	0.89	−1.89	0.44	−0.18	1817
1818	0.48	3.34	0.70	0.37	−3.80	0.26	0.53	−0.79	−0.23	0.48	1.13	−0.39	1818
1819	0.00	0.73	1.48	1.35	−0.02	−0.53	0.32	−0.50	0.48	0.46	0.93	0.30	1819
1820	−0.79	0.58	−0.56	1.60	1.03	−0.48	−0.48	1.76	−0.09	−0.30	−0.72	−0.03	1820
1821	0.80	−0.18	−0.52	0.59	0.10	−2.20	−1.46	0.43	1.01	−0.29	0.78	0.35	1821
1822	1.81	1.28	2.10	0.99	1.05	3.31	0.53	0.26	0.78	0.66	1.38	−0.48	1822
1823	−1.92	−0.25	−0.37	−0.55	0.93	−0.78	−0.60	0.53	1.18	0.11	−1.37	0.01	1823
1824	1.01	1.49	−0.40	−0.85	−0.16	−1.57	1.33	0.90	0.71	0.23	1.25	2.07	1824
1825	1.39	0.62	−2.38	1.21	−0.17	0.32	0.05	0.53	0.86	−0.81	0.82	3.92	1825
1826	−2.18	0.44	0.76	−0.72	−1.23	−0.09	0.18	1.55	0.71	1.48	−0.56	1.16	1826
1827	0.36	−1.72	1.12	0.73	0.46	−1.26	1.20	−0.60	−1.06	1.37	−1.46	0.25	1827
1828	1.38	−0.36	1.49	0.64	0.45	1.27	1.38	0.19	0.47	0.38	−0.81	0.60	1828
1829	−0.04	−2.79	0.05	0.05	−0.03	0.23	0.22	−1.15	−0.89	−0.40	−1.68	−1.90	1829
1830	−3.72	−3.45	1.66	2.66	0.66	−0.33	1.71	1.02	−0.79	−0.81	0.99	0.60	1830
1831	0.38	−0.51	0.73	0.19	−1.12	−0.56	0.12	−1.05	−1.08	1.77	0.24	1.34	1831
1832	0.41	0.52	−0.21	−0.68	−2.07	−1.27	0.03	0.49	−1.15	−0.47	−0.41	−1.71	1832
1833	−0.47	1.18	−0.59	−1.23	2.00	0.37	−2.28	−2.77	−3.20	−1.36	0.14	1.37	1833
1834	0.17	−0.80	−0.19	−1.97	0.53	−0.31	−0.23	−1.16	0.41	−0.79	−0.23	−0.94	1834
1835	1.03	0.76	−0.44	−0.88	−1.01	−1.47	−2.59	−2.35	−2.01	−2.24	−3.27	−2.69	1835
1836	−2.51	−2.08	−0.05	−1.07	−3.41	−0.49	−0.97	−1.09	−2.48	−0.65	−2.16	−0.02	1836
1837	−0.83	−4.75	−3.12	−2.41	−3.58	0.64	−1.29	0.37	−2.40	−1.93	−1.76	−0.36	1837
1838	−2.16	−2.39	−0.72	−2.74	−0.98	−0.75	−0.78	−1.55	−1.58	−1.74	0.08	−0.80	1838
Means.	0.52	2.82	6.40	10.03	14.08	17.09	18.92	18.39	15.31	10.79	5.76	2.08	Means.

The numbers without sign must be subtracted; those with the sign — must be added.

LXXXII.

Switzerland. — Geneva.

For Reducing the Monthly and Yearly Means of Single Years to the Means derived from Series of Years.

Degrees of Reaumur.

Year.	Jan.	Feb.	March.	April.	May.	June.	July.	Aug.	Sept.	Oct.	Nov.	Dec.	Year.
	°	°	°	°	°	°	°	°	°	°	°	°	
1768	-0.86	-0.01	-2.38	-0.14	-0.61	-1.71	-0.92	-1.25	-1.42	0.99	0.77	0.08	1768
1769	0.92	0.16	-1.05	0.13	-0.71	-1.57	-1.11	-1.43	-0.89	-3.00	1.62	0.61	1769
1770	-1.25	-1.30	-1.72	-2.40	-1.20	-1.34	-2.97	-1.24	0.49	-0.81	0.35	0.42	1770
1771	0.53	-0.67	0.17	-2.68	0.34	-1.84	0.07	-1.45	-0.57	-0.17	-1.80	1.66	1771
1772	0.61	2.57	1.76	-0.41	-2.23	0.35	-0.72	-0.47	0.52	1.34	1.29	1.16	1772
1773	1.47	-1.84	-1.18	-1.16	-1.64	-0.87	-2.03	-1.70	-0.12	-0.35	-0.12	1.01	1773
1774	1.22	0.91	2.38	0.73	-0.94	-0.47	-1.27	0.42	-1.08	-1.45	-1.18	-2.03	1774
1775	0.89	1.89	0.99	-1.60	-1.90	0.54	-0.84	-0.82	-0.02	-0.24	0.33	-0.68	1775
1776	-1.78	1.92	1.85	0.18	-1.96	0.09	0.20	0.28	-1.50	0.26	-0.26	-0.09	1776
1777	-0.41	-0.76	2.46	-1.23	-1.51	-0.38	-1.12	0.45	-0.41	1.27	0.21	-1.72	1777
1778	0.03	-0.93	0.86	0.78	-0.09	-0.76	1.59	0.68	-1.85	0.32	0.76	1.85	1778
1779	-3.43	-0.28	-0.14	1.70	0.97	-1.22	-0.61	-0.45	0.48	1.77	0.57	2.70	1779
1780	-1.48	-1.63	2.35	-0.86	0.97	1.14	0.95	1.16	0.14	0.51	-1.02	-1.25	1780
1781	0.96	1.09	0.37	2.15	1.78	0.28	-1.17	0.30	0.76	-0.47	1.69	2.97	1781
1782	2.22	-3.74	-0.53	-0.95	-1.76	0.18	-1.10	-0.72	-0.97	-1.06	-1.83	-2.04	1782
1783	2.01	1.68	-0.27	-0.71	-0.06	-1.13	1.75	-0.94	0.17	0.93	0.81	1.03	1783
1784	-1.06	-2.03	-0.43	-2.51	1.73	1.69	0.84	-1.62	1.40	-1.87	-0.76	-3.38	1784
1785	0.58	-3.26	-6.75	-5.48	-0.19	0.30	-0.33	-1.75	0.94	-0.40	0.11	0.42	1785
1786	0.41	0.08	-1.62	0.69	-0.25	1.92	-0.99	-1.22	-0.59	-1.71	-0.69	0.19	1786
1787	-1.99	-1.15	1.76	-0.30	-1.98	0.79	-0.70	0.21	-0.27	0.41	0.72	2.83	1787
1788	1.01	2.06	2.19	1.04	1.13	1.04	1.61	-0.45	0.71	-0.83	-2.15	-4.48	1788
1789	-1.17	1.12	-1.97	1.19	1.71	-1.25	-0.80	-0.19	-0.57	-0.59	-1.59	-0.17	1789
1790	0.36	0.75	0.99	-0.73	1.52	0.94	-1.10	0.63	-0.84	1.95	1.13	0.78	1790
1791	2.40	0.04	-0.02	2.86	0.61	1.04	0.98	2.30	0.98	0.72	-1.37	1.30	1791
1792	1.22	-0.28	2.11	1.81	-0.12	1.14	1.03	0.83	-0.09	1.37	0.86	0.45	1792
1793	-0.52	1.05	1.77	0.08	-0.05	0.20	3.12	2.49	-0.12	1.24	0.61	1.19	1793
1794	0.14	2.21	1.91	3.26	0.76	1.10	2.11	0.39	-0.74	-0.28	0.86	-1.75	1794
1795	-4.85	0.37	0.26	1.76	1.32	1.34	-0.73	1.34	1.54	1.92	-0.96	1.11	1795
1796	1.25	0.72	-2.15	-0.06	0.60	0.60	0.37	0.80	1.61	0.41	-0.14	-1.92	1796
1797	0.11	-1.41	-1.08	1.49	2.14	-1.28	2.21	1.28	0.71	-0.08	0.71	1.61	1797
1798	0.53	-1.17	-1.02	0.83	1.00	1.29	0.45	0.81	0.48	-0.29	0.44	-0.96	1798
1799	-1.57	1.71	-0.16	-1.73	-1.70	-1.16	-0.13	0.67	0.21	-0.40	-0.54	-2.59	1799
1800	1.64	0.06	-1.66	2.43	2.40	-0.83	1.48	0.82	0.96	-1.55	0.63	-0.27	1800
Means.	-0.43	0.75	3.08	7.19	11.21	14.03	15.44	14.85	11.49	7.32	3.34	0.57	Means.
1796	2.27	0.07	-2.14	-0.25	-0.91	-0.64	-1.10	0.16	0.70	0.08	-0.68	-1.70	1796
1797	0.45	-0.85	-0.66	0.97	0.67	-2.03	1.27	0.71	-0.48	-0.26	0.47	1.58	1797
1798	0.68	-0.25	-0.40	0.96	-0.22	0.32	-0.64	-0.14	0.12	-0.08	-0.76	-1.36	1798
1799	-1.44	1.93	-0.26	-1.60	-1.50	-0.49	-0.46	0.33	0.19	-0.26	-1.24	-3.30	1799
1800	2.06	0.03	-1.53	2.88	1.66	-0.97	1.62	0.70	0.41	-1.16	0.67	-0.32	1800
1801	1.81	0.13	1.43	0.74	0.43	-0.26	0.42	0.15	0.90	0.84	0.67	0.95	1801
1802	-3.98	-0.38	0.94	1.18	0.53	1.66	-0.12	2.68	1.72	2.51	0.63	0.58	1802
1803	-0.26	-2.58	0.24	2.05	-1.42	0.89	2.20	2.25	-0.79	-0.57	1.04	1.88	1803
1804	4.58	-1.58	-0.19	0.30	1.50	2.02	0.04	0.47	0.59	0.22	1.36	-0.59	1804

The numbers without sign must be subtracted; those with the sign — must be added

LXXXII.

Switzerland. — Geneva (*continued*).

For Reducing the Monthly and Yearly Means of Single Years to the Means derived from Series of Years.

Degrees of Reaumur.

Year.	Jan.	Feb.	March.	April.	May.	June.	July.	Aug.	Sept.	Oct.	Nov.	Dec.	Year.
	°	°	°	°	°	°	°	°	°	°	°	°	
1805	-0.41	-0.23	-0.41	-1.35	-1.22	-0.49	-0.41	-0.73	0.18	-1.45	-2.19	-1.64	1805
1806	3.23	1.83	0.12	-1.80	1.33	1.66	0.08	-0.39	0.11	1.10	1.34	2.42	1806
1807	-1.10	0.24	-2.65	-1.47	1.42	0.43	2.66	3.03	-0.58	1.42	0.39	-2.48	1807
1808	-0.49	-3.14	-2.58	-1.87	1.14	-1.33	0.70	0.59	-0.14	-2.40	-0.28	-2.99	1808
1809	2.23	1.95	0.19	-3.68	-0.06	0.12	-0.43	-0.32	-1.00	-1.05	-1.76	0.70	1809
1810	-3.14	-3.34	3.08	-0.28	0.29	-0.45	-1.01	-0.70	1.37	1.26	1.05	1.19	1810
1811	-2.22	1.98	1.46	1.34	1.23	1.82	1.53	-0.14	0.70	2.21	0.71	-0.85	1811
1812	-3.92	1.40	-0.02	-1.54	0.27	0.02	-0.37	-0.69	-0.43	0.13	-1.80	-2.74	1812
1813	-1.74	1.51	-0.69	0.53	0.54	-0.84	-2.10	-1.02	-1.14	0.78	-0.49	0.32	1813
1814	-1.32	-3.92	-1.44	0.96	-1.74	-0.26	0.37	-0.66	-1.74	-0.87	0.95	2.34	1814
1815	-2.24	1.43	2.17	1.06	0.82	0.08	0.20	-0.59	0.54	1.43	-1.57	-0.39	1815
1816	-0.13	-1.33	2.54	-0.48	-0.54	-1.41	-2.40	-2.14	-0.47	0.59	-1.05	-0.02	1816
1817	2.50	2.38	0.29	-2.11	-1.34	1.35	-0.25	-0.71	2.16	-1.58	0.85	-0.45	1817
1818	0.54	0.69	0.13	-0.08	-1.26	0.66	1.41	-0.41	-0.89	-0.29	1.60	-0.26	1818
1819	1.86	0.98	0.82	1.00	-0.21	-0.19	0.07	-0.34	0.42	0.07	-0.40	0.95	1819
1820	0.10	0.54	-1.24	2.07	0.39	-0.59	-0.65	0.84	-1.93	-0.31	-2.16	0.02	1820
1821	1.98	-1.31	0.94	0.71	-1.19	-1.54	-1.17	0.62	0.26	0.27	2.34	3.36	1821
1822	0.20	1.27	3.06	0.47	1.32	3.85	0.27	-0.85	-0.07	0.69	1.60	-2.32	1822
1823	-1.17	1.46	-0.29	-0.42	0.17	-1.62	-1.54	-1.04	-0.42	-2.10	-1.97	1.04	1823
1824	-0.78	-0.30	-1.84	-2.05	-1.50	-2.05	0.17	-1.49	-1.23	-1.58	0.03	1.30	1824
1825	-0.07	-0.55	-1.09	1.69	-0.63	0.26	-0.40	-0.11	0.88	0.30	0.54	2.76	1825
Means.	-0.42	1.87	4.70	8.79	13.45	15.81	17.67	17.66	14.70	9.73	5.23	1.27	Means.
1826	-3.23	1.12	1.47	0.34	-1.04	-0.06	0.90	2.57	1.22	0.95	-1.19	0.03	1826
1827	1.49	-2.15	1.02	1.29	0.90	-0.07	1.95	0.66	0.24	0.99	-2.02	2.56	1827
1828	2.82	1.06	0.70	0.81	1.22	0.89	0.59	-0.80	0.86	0.96	0.62	0.94	1828
1829	-0.85	-0.63	0.10	0.25	-0.06	-0.83	0.15	-0.89	-0.71	-1.52	-1.28	-3.87	1829
1830	-4.14	-1.74	1.20	2.70	0.57	-0.49	0.53	-0.01	-0.94	-0.86	0.46	-0.90	1830
1831	-1.10	0.46	1.67	1.54	0.53	-0.11	-0.02	-0.02	-0.29	2.16	0.58	0.90	1831
1832	0.10	0.36	-0.25	0.45	-0.40	-0.83	0.81	2.29	-0.39	0.07	-0.02	0.62	1832
1833	-0.06	3.36	-0.50	-0.68	2.67	1.23	-1.29	-1.17	-0.17	0.59	0.39	3.28	1833
1834	5.06	1.47	0.35	-0.70	2.28	1.53	1.94	1.07	2.74	0.68	0.71	-1.18	1834
1835	1.15	1.40	-0.44	-0.06	0.56	0.15	1.69	0.40	0.22	-1.34	-2.27	-2.66	1835
1836	0.48	-0.04	1.82	-0.95	-2.14	0.17	0.57	0.28	-0.62	0.14	-0.02	0.50	1836
1837	0.37	0.52	-2.94	-1.89	-2.18	1.21	-0.58	1.41	-1.16	-0.39	-1.06	-0.46	1837
1838	-3.64	-0.91	0.25	-1.75	-0.11	-0.71	-0.56	-1.23	-0.58	-0.61	1.18	-0.57	1838
1839	0.55	-0.07	-0.42	-1.55	-0.97	1.14	0.24	-1.73	-0.58	1.11	1.43	2.81	1839
1840	2.60	0.02	-3.22	0.71	-0.10	-0.37	-2.32	-0.01	-0.56	-1.74	1.43	-3.14	1840
1841	0.45	-0.25	0.77	-0.69	1.82	-1.71	-1.98	-1.37	0.09	0.90	0.26	0.89	1841
1842	-5.18	-2.84	0.56	-0.58	0.02	1.00	-0.19	0.78	-1.08	-2.18	-1.03	-0.71	1842
1843	1.50	2.22	-0.34	0.27	-1.60	-2.56	-2.35	-0.73	0.60	-0.24	0.25	-0.83	1843
1844	..	..	..	..	..	..	..	..	..	..	..	..	1844
1845	1.70	-3.33	-1.77	0.49	-1.98	0.47	0.06	-1.53	1.10	0.40	1.54	1.74	1845
Means.	-0.72	0.98	4.16	7.03	10.77	13.61	14.96	14.58	11.84	7.98	3.98	1.30	Means.

The numbers without sign must be subtracted; those with the sign — must be added

LXXXIII.

South Germany. — Vienna.

For Reducing the Monthly and Yearly Means of Single Years to the Means derived from Series of Years.

Degrees of Reaumur.

Year.	Jan.	Feb.	March.	April.	May.	June.	July.	Aug.	Sept.	Oct.	Nov.	Dec.	Year.
	°	°	°	°	°	°	°	°	°	°	°	°	
1775	−1.43	1.86	1.21	−2.35	−2.77	1.32	−0.41	1.29	0.84	0.26	0.29	−1.09	1775
1776	−4.30	0.57	0.70	−1.11	−2.30	−0.42	−0.24	0.25	−1.42	−1.53	−1.32	−2.19	1776
1777	−1.79	−1.24	0.32	−2.93	−0.22	−0.10	−1.17	0.57	−1.38	−0.53	0.35	−1.00	1777
1778	1.92	−1.04	0.18	1.89	0.04	−0.43	1.18	0.95	−0.89	−0.54	0.87	3.61	1778
1779	−1.75	3.15	2.27	3.05	1.24	−1.32	−1.35	−0.07	0.65	1.00	0.43	3.01	1779
1780	−1.68	3.04	2.73	−1.38	−0.18	−0.92	−0.70	−0.48	−1.08	0.51	0.19	−1.99	1780
1781	−0.87	0.05	0.77	0.86	0.25	1.44	−0.06	2.31	1.40	−0.45	1.84	0.34	1781
1782	2.72	−2.63	0.60	−0.06	0.54	1.82	2.74	0.35	0.86	−0.76	−1.50	0.62	1782
1783	3.59	4.12	−0.08	0.65	1.81	1.94	1.66	1.81	2.12	1.59	0.58	−2.56	1783
1784	−3.51	−1.87	−0.42	−1.36	1.69	0.86	0.47	0.49	1.98	−2.56	0.70	0.03	1784
1785	−0.73	−0.93	−5.63	−3.04	−0.67	−1.47	−0.83	−0.86	2.11	−0.55	0.41	0.17	1785
1786	0.52	0.16	−0.04	1.84	−1.12	0.25	−1.54	−1.85	−0.92	−2.11	−2.12	0.60	1786
1787	−0.39	1.47	0.65	−1.46	−2.11	1.11	−0.40	0.35	−0.78	1.10	0.93	2.82	1787
1788	2.22	0.17	0.81	0.05	−0.36	1.18	2.28	−1.72	1.00	−0.29	−1.39	−6.79	1788
1789	−0.49	2.00	−2.43	1.19	2.15	−0.49	0.40	−0.60	0.37	0.77	0.73	0.21	1789
1790	0.86	2.87	0.31	−1.11	1.20	1.56	−1.10	0.31	−0.83	−0.76	−0.43	2.09	1790
1791	4.29	1.01	1.63	1.33	−0.44	−0.33	−0.37	0.67	−0.84	−0.40	−0.46	0.89	1791
1792	0.56	−1.24	0.47	0.38	−0.96	0.62	0.38	0.26	−0.93	−1.11	−0.24	0.56	1792
1793	−1.55	1.27	−1.00	−2.40	−1.23	−1.08	1.81	1.86	−0.07	1.13	0.64	1.99	1793
1794	2.24	2.99	1.95	3.74	1.35	1.55	2.92	−0.75	−1.38	−0.19	0.33	−0.95	1794
1795	−4.94	−1.29	0.23	1.81	−0.05	1.44	−1.95	0.31	−0.17	2.75	−1.00	2.28	1795
1796	5.23	1.32	−2.73	−1.52	0.48	−0.04	0.14	0.58	1.96	0.84	−0.14	−1.48	1796
1797	1.58	1.02	−0.71	2.10	2.94	0.68	1.95	2.17	2.01	1.23	0.54	1.11	1797
1798	1.96	2.83	1.40	0.65	0.26	0.84	0.14	1.29	1.62	−0.47	−0.68	−3.68	1798
1799	−5.34	−2.08	−0.83	−0.43	−0.45	−1.16	−0.58	1.00	−0.50	0.45	0.58	−2.94	1799
1800	0.74	−0.19	−3.31	5.57	1.90	−1.45	−0.44	1.49	0.27	−0.40	1.57	0.10	1800
1801	1.85	−0.21	2.47	0.80	1.83	−0.85	−1.18	−1.32	1.37	1.94	1.71	0.99	1801
1802	−0.43	−1.34	0.89	0.73	−1.14	1.33	1.02	1.65	0.38	2.10	1.84	1.40	1802
1803	−2.68	−3.46	−0.50	2.49	−1.59	−0.75	0.23	0.08	−2.12	−0.45	1.24	0.27	1803
1804	3.42	−0.59	−2.44	0.05	0.29	−0.10	0.25	−0.51	0.80	0.48	−2.47	−2.40	1804
1805	−0.48	−1.18	−1.28	−2.16	−1.85	−0.79	−1.26	−1.61	−0.04	−2.89	−2.19	0.24	1805
1806	4.04	2.12	1.07	−2.07	1.84	−0.02	−0.16	−0.62	0.56	−0.80	1.60	3.48	1806
1807	1.08	1.96	−1.54	−1.18	1.23	−0.34	1.25	4.74	0.17	1.37	1.96	0.46	1807
1808	1.20	−0.51	−4.99	−1.20	1.42	0.15	1.30	1.80	1.13	−0.97	−0.32	−3.68	1808
1809	−0.08	1.54	−1.13	−2.51	0.89	0.27	0.23	0.79	0.11	−1.31	−0.75	1.67	1809
1810	−0.71	−0.03	2.03	−0.74	0.50	−1.65	0.82	0.15	2.26	−0.18	−0.09	2.01	1810
1811	−3.58	−0.91	2.08	0.75	3.12	4.62	2.56	0.99	0.42	3.63	1.20	0.19	1811
1812	−2.13	0.53	0.67	−2.67	0.65	0.35	−0.87	−0.52	−1.32	2.04	−0.84	−3.96	1812
1813	−1.84	2.07	−0.76	1.56	0.36	−1.82	−1.34	−1.80	−1.34	−0.37	−0.24	0.68	1813
1814	−0.34	−4.37	−0.55	1.54	−2.19	−1.76	0.66	−0.21	−2.45	−0.73	0.32	2.19	1814
1815	−1.03	2.39	2.06	0.10	0.52	0.28	−1.51	−1.29	−1.20	0.06	−1.07	−2.87	1815
1816	1.84	−0.80	−0.19	0.09	−0.95	−0.73	−1.58	−1.39	−0.95	−0.73	−0.39	−1.45	1816
1817	3.24	3.78	0.51	−4.08	0.53	2.18	−0.08	−0.25	0.56	−2.29	1.09	0.16	1817
1818	2.77	0.78	1.84	2.01	−0.11	0.55	0.13	−0.71	0.41	0.84	0.60	−1.31	1818
1819	1.22	2.04	1.94	1.17	−0.75	1.01	0.66	−0.35	0.71	−0.12	0.51	−1.21	1819

The numbers without sign must be subtracted; those with the sign — must be added.

LXXXIII.

South Germany. — Vienna (*continued*).

For Reducing the Monthly and Yearly Means of Single Years to the Means derived from Series of Years.

Degrees of Reaumur.

Year.	Jan.	Feb.	March.	April.	May.	June.	July.	Aug.	Sept.	Oct.	Nov.	Dec.	Year.
	°	°	°	°	°	°	°	°	°	°	°	°	
1820	−2.47	0.36	−0.86	1.78	1.97	−1.18	−0.96	2.36	−0.71	0.16	−0.36	−1.49	1820
1821	2.22	−1.56	−0.72	1.57	−0.81	−3.08	−1.83	−0.76	0.51	−0.12	1.93	2.90	1821
1822	2.85	1.63	3.44	1.05	1.21	1.50	1.16	−0.27	0.06	2.12	0.44	−0.27	1822
1823	−4.55	0.68	0.80	−0.29	0.42	−0.68	−1.35	0.15	0.36	1.13	0.29	1.35	1823
1824	1.77	2.31	0.09	−0.72	−0.74	−0.60	−0.22	−0.53	1.36	0.60	1.56	4.00	1824
1825	3.15	0.50	−1.59	1.02	−0.14	−0.31	−0.72	−0.47	−0.62	−1.71	1.74	3.11	1825
1826	−3.65	−2.12	0.91	−0.12	−2.42	−0.38	1.34	2.06	0.69	0.89	−0.32	1.78	1826
1827	0.69	−2.92	1.61	1.65	1.33	1.19	1.67	−1.06	−0.57	0.82	−3.48	0.83	1827
1828	0.19	−2.22	0.88	1.30	−0.16	0.21	0.63	−1.49	−0.70	−0.82	0.48	1.57	1828
1829	−1.66	−3.79	−1.87	−0.23	−2.26	−2.69	−0.32	−2.62	−0.31	−2.12	−3.62	−6.11	1829
1830	−5.31	−3.23	−0.44	0.94	−0.39	0.33	0.02	−0.04	−1.81	−1.68	0.76	1.13	1830
1831	−1.42	0.26	0.43	2.23	−0.90	−1.86	0.33	−1.01	−1.96	2.02	−0.16	−0.04	1831
1832	0.55	0.61	0.04	−0.16	−1.90	−1.46	−1.29	0.32	−0.86	0.04	−1.57	−1.36	1832
1833	−3.35	2.33	0.24	−1.40	2.57	1.20	−2.26	−2.80	−1.22	−0.55	0.23	4.03	1833
1834	4.67	0.32	−0.29	−1.17	2.24	1.65	2.61	1.26	2.85	−0.08	−0.89	1.25	1834
1835	1.71	1.46	0.46	−1.10	0.27	−0.07	0.92	0.19	0.09	−0.76	−3.77	−1.39	1835
1836	−0.08	0.29	3.84	0.00	−2.95	0.30	−0.48	−0.78	−0.89	0.91	−1.00	2.44	1836
1837	0.20	−2.39	−1.96	−1.18	−2.57	−1.38	−2.96	0.84	−2.22	−0.82	−0.74	−0.95	1837
1838	−5.10	−4.14	−0.50	−2.44	−0.76	−0.74	−1.39	−2.29	−0.03	−1.75	−0.65	−0.84	1838
1839	1.12	0.73	−2.31	−3.85	−2.04	1.06	0.36	−2.23	0.23	1.05	1.55	0.70	1839
1840	1.03	−0.88	−3.76	−0.55	−1.59	−1.05	−1.56	−1.94	−0.11	−2.03	2.09	−7.72	1840
1841	0.33	−3.24	0.65	0.93	2.19	−1.02	0.55	−1.10	0.24	2.04	0.28	2.27	1841
Means.	−1.22	0.63	3.85	8.66	13.31	15.72	17.14	16.77	13.25	8.51	3.67	0.39	Means.

LXXXIV. South Germany. — Ratisbon.

Year.	Jan.	Feb.	March.	April.	May.	June.	July.	Aug.	Sept.	Oct.	Nov.	Dec.	Year.
1773	3.00	−0.28	−0.04	−0.28	0.25	0.34	−1.23	−0.60	0.47	1.20	1.06	2.35	1773
1774	1.63	0.85	2.17	1.97	−0.10	−0.17	−1.11	0.16	−1.29	−0.63	−2.98	−2.32	1774
1775	0.67	2.87	1.13	−2.41	−3.42	−0.51	−1.91	. .	−0.73	−2.19	−0.14	−0.64	1775
1776	−3.04	1.19	. .	. .	. .	. .	. .	. .	. .	. .	. .	. .	1776
1777	−1.47	−0.68	2.37	−1.29	−0.16	0.28	−1.02	1.24	0.01	1.07	1.31	−1.17	1777
1778	1.88	0.21	0.89	1.98	1.76	0.81	3.20	2.38	−1.33	−0.36	1.36	3.06	1778
1779	−2.51	1.43	2.27	2.89	1.88	−0.34	−0.38	0.95	1.40	2.18	1.47	3.74	1779
1780	−0.83	−1.52	2.87	−0.92	0.87	1.30	0.64	1.65	1.32	1.25	0.25	−0.75	1780
1781	. .	. .	1.52	1.88	0.82	1.52	0.48	2.36	2.45	−1.03	0.53	0.56	1781
1782	2.46	−2.93	3.32	3.15	3.74	1.92	2.02	−0.28	0.80	−1.67	−2.72	−0.32	1782
1783	3.48	2.22	−0.95	0.26	0.93	0.92	1.73	0.45	0.13	1.00	−0.34	−2.38	1783
1784	−4.07	−3.45	−1.69	−2.76	1.67	0.60	0.23	0.34	2.21	−2.46	0.36	−1.21	1784
1785	−1.20	−2.85	−6.49	−4.37	−1.08	−0.83	−1.42	−1.91	2.05	−0.77	0.28	0.10	1785
1786	0.66	0.04	−2.05	1.32	−1.54	1.28	−2.31	−1.85	−1.30	−1.93	−2.68	−0.18	1786
1787	−1.03	4.29	0.75	−1.53	−2.51	0.92	−1.31	0.36	0.07	1.78	0.89	2.06	1787
1788	1.86	−0.61	−0.30	−0.59	−0.35	0.88	1.66	−1.53	1.44	−0.20	−2.14	−8.30	1788
1789	−1.93	1.41	−2.90	0.64	1.62	−1.30	−0.29	−0.28	−0.53	0.27	0.25	0.64	1789
1790	1.99	1.73	0.91	−1.21	1.20	1.42	−1.49	−0.08	−0.87	−0.37	−0.26	0.89	1790
1791	3.24	0.14	1.00	1.81	−0.76	−0.35	−0.35	1.14	−0.15	0.50	−2.43	0.84	1791

The numbers without sign must be subtracted; those with the sign — must be added.

LXXXIV.

South Germany. — Ratisbon (*continued*).

For Reducing the Monthly and Yearly Means of Single Years to the Means derived from Series of Years.

Degrees of Reaumur.

Year.	Jan.	Feb.	March.	April.	May.	June.	July.	Aug.	Sept.	Oct.	Nov.	Dec.	Year.
	°	°	°	°	°	°	°	°	°	°	°	°	
1792	−0.57	−0.21	1.41	1.17	−1.12	0.87	0.66	0.88	−1.01	0.05	0.17	0.89	1792
1793	−1.17	1.26	0.53	−1.81	−1.23	−0.76	1.56	1.09	−0.47	1.87	0.95	1.26	1793
1794	2.30	3.01	3.05	3.04	1.19	1.69	2.85	−0.80	−1.00	1.24	0.67	−0.78	1794
1795	−5.05	−0.89	−0.10	1.96	−0.82	1.39	−2.22	0.29	1.08	3.11	−0.98	2.26	1795
1796	4.26	1.59	−1.63	−0.77	−0.25	0.05	0.47	0.93	2.24	0.28	−0.38	−2.04	1796
1797	1.46	1.52	−0.17	2.09	2.60	−0.72	2.14	1.73	0.75	0.06	1.00	1.59	1797
1798	1.88	1.94	0.18	1.02	0.66	1.65	0.50	1.26	1.08	−0.73	−0.68	−2.69	1798
1799	−5.61	0.14	−0.29	−1.87	−1.37	−0.83	−0.91	−2.86	−0.60	−0.23	−0.35	−3.81	1799
1800	1.15	−0.63	−2.62	4.66	1.80	−1.42	0.32	1.53	0.43	−0.69	1.30	0.63	1800
1801	2.72	0.10	1.82	0.76	2.45	−0.75	−0.30	0.13	1.15	1.60	1.45	0.81	1801
1802	−3.20	−0.90	0.29	0.62	0.16	1.66	−0.04	2.80	0.73	2.40	0.63	0.71	1802
1803	−1.24	−2.05	−0.06	2.70	−1.78	−0.31	1.70	1.31	−0.96	−0.33	0.29	0.93	1803
1804	3.84	−0.86	−1.18	−0.49	1.17	0.88	0.29	−0.14	1.27	1.09	−0.71	−1.68	1804
1805	−1.41	−1.00	−0.34	−1.27	−1.76	−0.88	−0.88	−1.60	0.74	−2.03	−1.81	−0.21	1805
1806	4.22	2.45	0.40	−2.24	2.47	0.16	−0.49	0.15	0.86	0.04	1.94	3.58	1806
1807	1.19	1.18	−1.17	−1.32	1.24	0.46	2.87	4.63	−0.94	1.58	1.03	1.54	1807
1808	1.08	−0.73	−2.79	−1.93	2.02	−0.45	1.61	1.19	0.33	−1.97	−0.23	−5.46	1808
1809	0.33	2.19	−0.40	−2.92	0.71	−0.25	0.02	0.23	−0.31	−0.76	−0.86	0.93	1809
1810	−1.72	−2.39	0.86	−0.63	−0.05	−1.00	−0.41	0.17	2.72	0.52	0.04	1.89	1810
1811	−2.93	−0.16	2.09	1.48	2.23	2.85	1.75	0.24	0.43	2.24	1.43	−0.25	1811
1812	−1.33	1.05	0.28	−2.87	0.13	−1.15	−2.18	−1.44	−1.39	0.60	−1.99	−4.72	1812
1813	−3.03	0.99	−1.15	0.46	−0.60	−1.86	−1.73	−2.10	−1.47	−0.50	−0.75	−0.33	1813
1814	−1.37	−4.71	−2.93	0.49	−2.79	−2.39	−0.12	−1.12	−2.45	−1.50	0.65	1.77	1814
1815	−1.30	1.05	1.18	−0.37	−0.46	−0.74	−2.23	−2.07	−1.35	−0.70	−1.37	−2.26	1815
1816	1.36	−1.83	−1.23	−0.93	−2.69	−2.21	−2.42	−2.56	−2.04	−0.93	−1.49	−0.75	1816
1817	2.51	2.42	−1.14	−5.01	−1.93	0.61	−1.79	−1.89	0.56	−3.22	0.63	−0.70	1817
1818	2.08	0.29	−0.16	0.27	−1.72	−0.02	−0.48	−2.27	−1.09	−0.71	0.41	−2.08	1818
1819	1.49	0.60	0.64	−0.09	−0.76	0.15	−0.05	−0.35	−0.28	−0.78	−0.99	−1.34	1819
1820	−2.43	−0.35	−2.26	0.38	−0.47	−2.89	−1.66	0.93	−2.28	−1.22	−1.63	−1.66	1820
1821	1.17	−3.06	−1.51	0.99	−2.48	−3.01	−2.77	−1.18	−0.06	−0.99	1.51	2.53	1821
1822	2.21	0.63	1.92	0.26	0.53	2.43	0.49	−0.87	−0.56	0.73	0.48	−2.29	1822
1823	−4.17	0.86	0.11	−1.72	0.20	−0.97	−1.05	0.43	0.38	0.02	−0.61	1.05	1823
1824	0.92	0.38	−1.02	−2.10	−1.74	−1.07	−0.14	−0.51	0.97	−0.26	1.44	3.93	1824
1825	2.80	0.39	−1.05	2.12	0.93	0.56	0.51	0.32	1.31	0.21	3.09	4.15	1825
1826	−3.57	−0.34	1.39	0.17	−1.04	1.05	1.99	3.61	1.61	1.38	−0.28	0.92	1826
1827	0.09	−4.95	1.00	1.31	1.20	1.05	2.06	−0.57	0.93	1.35	−1.30	2.93	1827
1828	2.12	0.27	0.51	0.31	−0.77	0.39	0.85	−2.47	−1.95	−0.20	0.61	2.28	1828
1829	−0.85	−3.13	−1.38	0.20	−1.14	−1.11	−0.20	−2.37	−1.35	−1.41	−3.79	−5.79	1829
1830	−5.98	−3.61	1.17	0.77	−0.12	−0.94	0.50	−1.28	−1.05	−0.88	1.14	−0.58	1830
1831	−2.09	−0.82	0.70	3.60	−0.48	−1.36	−0.09	−0.12	−1.57	2.40	2.27	0.26	1831
1832	0.77	1.28	0.09	0.21	−2.45	−0.87	−1.18	0.59	−0.98	0.16	−0.71	0.25	1832
1833	−3.05	3.32	0.21	−1.45	2.29	1.06	−1.70	−3.06	−2.01	−0.90	2.78	3.95	1833
1834	5.52	−0.43	−0.25	−1.69	0.99	0.44	4.89	2.48	1.21	0.50	0.74	1.36	1834
Means.	−2.42	−0.09	3.09	7.55	11.94	13.72	14.88	14.62	11.69	7.11	2.22	−0.71	Means.

The numbers without sign must be subtracted; those with the sign — must be added.

E

LXXXV.

South Germany. — Stuttgard.

For Reducing the Monthly and Yearly Means of Single Years to the Means derived from Series of Years.

Degrees of Reaumur.

Year.	Jan.	Feb.	March.	April.	May.	June.	July.	Aug.	Sept.	Oct.	Nov.	Dec.	Year.
	°	°	°	°	°	°	°	°	°	°	°	°	
1792	0.64	−1.23	1.78	1.36	−1.12	−0.30	1.04	1.70	−0.70	1.30	−0.73	0.44	1792
1793	−1.41	1.64	0.37	−1.36	−1.24	−0.31	2.52	1.84	−0.84	1.71	0.66	2.10	1793
1794	2.02	3.72	2.76	3.26	0.41	1.30	2.75	0.12	−1.34	0.32	0.85	−1.72	1794
1795	−4.88	0.36	0.65	2.81	0.64	1.43	−1.01	1.33	1.93	3.69	−0.49	3.76	1795
1796	6.17	1.90	−2.51	−0.67	−0.32	0.10	−0.36	0.23	2.34	0.13	−0.87	−2.18	1796
1797	2.46	0.08	−0.27	1.88	1.16	−1.30	2.86	1.01	1.04	0.36	1.82	3.02	1797
1798	0.65	1.44	0.69	1.16	0.66	1.21	0.14	0.66	1.36	0.53	0.33	−2.05	1798
1799	−3.46	1.77	−0.78	−1.30	−0.89	−0.75	−0.98	0.32	−0.04	−0.15	0.42	−4.70	1799
1800	3.03	−0.92	−2.04	4.56	2.03	−1.81	0.00	0.79	0.84	−0.37	1.61	−0.18	1800
1801	3.95	0.97	1.98	0.24	0.94	−0.54	0.91	1.42	2.32	2.89	1.30	1.46	1801
1802	−2.55	−0.02	0.80	2.13	0.23	1.53	−0.24	2.22	0.62	2.08	0.81	1.41	1802
1803	−0.81	−1.90	−1.31	1.49	−2.28	0.05	1.21	1.28	−1.78	−0.90	0.46	1.36	1803
1804	4.61	−0.98	−1.05	−0.22	0.78	0.92	−0.35	−0.66	2.88	0.74	0.56	−1.56	1804
1805	−1.03	−0.28	−0.60	−1.38	−2.16	−1.35	−1.28	−1.44	0.38	−2.73	−2.47	0.06	1805
1806	−2.78	2.77	1.10	−1.99	1.37	−0.27	−0.62	−0.59	−0.27	0.03	1.67	4.85	1806
1807	0.76	1.58	−2.43	−1.14	−1.02	−0.21	2.15	3.23	−0.74	1.71	1.37	−0.94	1807
1808	1.95	−1.23	−3.55	−1.35	1.96	−0.98	0.54	0.77	−0.34	−1.49	−0.17	−4.02	1808
1809	1.56	3.64	0.69	−2.58	0.84	−0.65	−0.36	0.16	0.30	−1.05	−1.65	2.12	1809
1810	−1.56	−2.45	2.11	−4.12	−0.25	−0.95	−0.26	−0.48	2.08	0.09	1.44	0.97	1810
1811	−3.01	0.49	2.31	0.97	1.41	1.41	0.75	−0.38	−0.04	3.02	1.28	−0.04	1811
1812	−2.36	1.26	−0.23	−3.17	0.64	0.37	−1.85	−1.17	−0.49	0.01	−2.28	−4.81	1812
1813	−2.25	0.28	−0.42	0.53	0.13	−1.45	−1.96	−3.00	−1.63	−0.34	−1.15	−0.70	1813
1814	−1.96	−3.96	−3.67	1.09	−2.14	−1.52	0.26	−0.76	−1.56	−1.34	0.69	2.13	1814
1815	−1.92	1.26	2.15	0.39	0.71	−0.42	−1.95	−1.44	−0.48	0.03	−2.30	−1.44	1815
1816	0.69	−2.37	−0.60	−0.68	−2.22	−2.63	−2.45	−2.34	−0.85	−0.17	−2.45	−0.82	1816
1817	3.31	1.47	−0.71	−3.71	−1.78	0.92	−1.54	−0.97	1.08	−3.25	1.01	−0.49	1817
1818	2.67	0.40	0.41	1.33	−0.82	1.06	0.15	−0.91	−0.70	−0.91	1.62	−2.04	1818
1819	−0.61	1.54	0.89	1.21	0.29	0.36	1.02	0.36	−0.92	−0.02	−0.95	1.16	1819
1820	−1.64	−0.06	−2.16	1.31	−0.05	−1.91	−1.37	1.10	−1.84	−1.11	−2.81	−0.66	1820
1821	2.13	−2.72	0.19	1.52	−1.98	−2.26	−1.97	0.15	0.77	−0.62	2.42	3.25	1821
1822	2.11	1.58	2.61	0.51	1.43	2.90	0.08	−0.85	−0.48	1.33	1.82	−3.09	1822
1823	−2.76	1.25	−0.05	−0.77	0.92	−1.42	−1.19	0.25	−0.38	−1.03	−1.37	1.70	1823
1824	0.79	0.79	−0.89	−1.61	−1.05	−0.98	0.30	−0.39	0.68	0.44	2.52	3.81	1824
1825	1.92	−0.37	−1.36	1.85	0.27	0.26	0.19	0.02	−0.61	−0.64	1.27	2.74	1825
1826	−4.81	1.06	1.16	0.19	−0.93	0.54	1.70	1.78	1.51	1.43	−1.07	0.54	1826
1827	−0.49	−5.36	1.47	1.22	1.60	0.23	1.10	−0.45	0.08	0.67	−2.41	2.98	1827
1828	3.10	−0.35	0.61	0.54	0.43	0.97	1.02	−1.10	0.07	−0.61	−0.17	1.19	1828
1829	−2.45	−3.10	−0.58	0.46	−0.36	−0.61	0.45	−1.13	−1.50	−1.66	−2.88	−5.91	1829

The numbers without sign must be subtracted; those with the sign — must be added.

LXXXV.

South Germany.—Stuttgard (*continued*).

For Reducing the Monthly and Yearly Means of Single Years to the Means derived from Series of Years.

Degrees of Reaumur.

Year.	Jan.	Feb.	March.	April.	May.	June.	July.	Aug.	Sept.	Oct.	Nov.	Dec.	Year.
	°	°	°	°	°	°	°	°	°	°	°	°	
1830	–6.40	–3.47	1.62	2.06	0.90	–0.38	1.05	0.00	–1.43	–0.89	0.90	–0.74	1830
1831	–0.73	1.25	1.68	1.44	–0.11	–0.09	1.22	–0.10	–1.15	2.92	0.15	1.26	1831
1832	0.10	–0.76	–0.64	0.13	–0.93	–0.59	–0.22	0.75	–1.05	–0.78	–1.41	0.05	1832
1833	–2.56	2.99	–0.99	–1.31	3.38	2.06	–1.46	–2.81	–1.35	–1.03	–0.13	3.18	1833
1834	5.05	0.14	–0.60	–1.87	2.16	1.33	2.74	0.90	1.73	–0.06	0.12	–0.27	1834
1835	1.53	1.20	–0.15	–0.90	–0.53	0.28	1.62	–0.21	0.60	–1.20	–3.22	–2.85	1835
1836	0.45	–1.27	3.18	–0.90	–2.14	0.64	0.22	0.46	–1.21	0.47	–0.01	1.04	1836
1837	0.90	0.24	–2.68	–2.84	–2.23	1.16	–1.19	1.17	–1.94	–0.60	–0.40	0.04	1837
1838	–4.43	–2.08	0.21	–2.32	–0.45	–0.03	–0.36	–0.90	0.64	–0.36	1.03	–1.34	1838
1839	0.78	–0.03	–1.31	–2.71	–1.07	2.30	0.58	–1.55	0.64	0.84	1.07	1.95	1839
1840	1.82	0.15	–2.90	1.36	0.29	0.16	–1.42	–0.23	–0.26	–2.39	1.10	–5.61	1840
1841	0.89	–1.98	2.09	0.53	3.42	–1.55	–1.83	–0.69	1.65	1.24	1.22	2.85	1841
1842	–1.50	–1.05	1.36	–0.47	1.35	1.74	0.35	2.64	0.07	–2.47	–1.82	–0.20	1842
1843	2.07	1.54	0.15	0.60	–1.06	–1.48	–0.68	0.20	–0.15	0.02	0.67	0.23	1843
1844	0.31	–0.91	–0.30	1.42	–1.01	1.39	–1.99	–2.17	0.56	0.39	0.91	–3.18	1844
Means.	–0.80	1.64	3.97	7.80	11.87	14.03	15.48	15.02	12.05	8.05	4.11	1.25	Means.

LXXXVI. South Germany—Carlsruhe.

Year.	Jan.	Feb.	March.	April.	May.	June.	July.	Aug.	Sept.	Oct.	Nov.	Dec.	Year.
1779	–3.98	1.18	1.26	2.19	1.14	–0.64	0.80	1.72	2.40	2.75	1.71	2.94	1779
1780	–2.23	–2.20	3.27	–1.17	0.52	0.41	0.27	1.39	0.18	0.83	0.00	–1.32	1780
1781	0.45	1.82	0.99	2.26	0.87	1.63	0.63	1.20	1.11	–0.94	–0.38	1.09	1781
1782	3.13	–3.95	–0.67	–1.10	–1.44	0.93	1.00	–1.62	–1.69	–2.08	–3.90	–1.07	1782
1783	3.33	1.35	–1.60	0.05	–0.14	0.47	1.69	–0.38	–0.71	–0.35	–1.04	–3.26	1783
1784	–4.85	–3.17	–1.67	–2.72	0.42	–0.13	–0.48	–1.97	–0.63	–3.71	–0.71	–2.14	1784
1785	–0.27	–3.06	–5.49	–3.26	–1.28	–0.44	–0.46	. .	. .	. .	–0.63	–0.60	1785
1786	. .	. .	. .	0.76	–1.35	1.20	–1.77	. .	. .	–1.97	–3.13	–0.18	1786
1788	. .	. .	. .	. .	. .	. .	. .	. .	. .	. .	. .	–8.65	1788
1789	–0.91	1.74	–3.15	. .	2.19	–1.49	0.34	–0.14	–1.17	. .	. .	0.91	1789
1798	. .	. .	. .	. .	. .	. .	0.20	0.39	0.59	0.55	0.14	–1.90	1798
1799	–3.09	0.92	–0.59	–1.15	–1.43	–0.70	–0.60	0.20	0.01	0.12	–0.12	–4.22	1799
1800	2.53	–1.60	–2.25	3.40	1.46	–2.10	–0.26	1.20	0.89	–0.43	1.15	0.21	1800
1801	3.13	0.68	1.95	0.32	0.98	–0.98	–0.15	–0.49	0.76	0.98	1.17	2.21	1801
1802	–2.69	0.64	0.83	1.08	–0.60	1.37	–0.97	2.33	0.23	1.38	–0.38	0.53	1802
1803	–1.27	–2.92	–1.30	1.11	–2.75	–0.71	0.63	0.54	–2.37	–0.90	0.52	1.99	1803
1804	4.53	–1.30	–1.12	–0.47	0.94	0.96	–0.56	–0.70	0.10	0.90	0.08	–2.05	1804
1805	–1.49	–0.61	–0.70	–0.89	–1.66	–0.76	–1.21	–1.17	0.16	–2.18	–2.92	–0.39	1805
1806	4.11	1.89	0.58	–2.23	1.41	–0.09	0.03	–0.35	–0.36	–0.67	1.62	4.71	1806

The numbers without sign must be subtracted; those with the sign — must be added.

LXXXVI.

South Germany. — Carlsruhe (*continued*).

For Reducing the Monthly and Yearly Means of Single Years to the Means derived from Series of Years.

Degrees of Reaumur.

Year.	Jan.	Feb.	March.	April.	May.	June.	July.	Aug.	Sept.	Oct.	Nov.	Dec.	Year.
	°	°	°	°	°	°	°	°	°	°	°	°	
1807	0.02	1.11	−2.83	−1.34	1.19	−0.28	2.34	3.15	−1.57	1.30	1.21	−0.53	1807
1808	1.38	−1.25	−3.54	−1.63	2.40	−0.41	1.94	0.96	−0.54	−1.28	−0.24	−3.75	1808
1809	1.24	3.25	0.52	−3.16	0.41	−1.22	−0.61	−0.33	−0.80	−1.36	−1.90	1.60	1809
1810	−3.19	−2.78	1.26	−0.17	−0.60	−0.62	−0.65	−0.46	1.58	−0.05	0.84	1.69	1810
1811	−2.40	[illegible].17	2.79	1.65	2.32	1.55	0.78	−0.29	0.53	2.92	1.21	0.48	1811
1812	−2.09	1.47	−0.16	−2.96	0.83	−0.50	−1.53	−0.27	−0.24	1.33	−1.42	−3.80	1812
1813	−0.84	2.15	0.57	1.50	0.12	−1.01	−1.75	−1.78	−1.15	0.27	−0.11	−0.89	1813
1814	−1.51	−3.24	−1.56	1.82	−1.68	−1.66	0.07	−1.12	−1.07	−0.68	0.82	2.67	1814
1815	−2.35	2.31	2.67	0.75	1.10	−0.57	−1.75	−1.03	0.09	0.62	−1.99	−1.02	1815
1816	1.38	−2.00	−0.27	0.27	−2.23	−2.48	−2.61	−2.16	−0.89	−0.33	−2.14	0.17	1816
1817	3.56	2.16	−0.36	−3.14	−1.63	0.87	−1.47	−1.40	1.60	−2.82	1.78	0.13	1817
1818	2.91	1.12	0.59	1.53	−1.44	1.00	0.34	−1.01	−0.50	−0.63	1.49	−2.11	1818
1819	1.85	1.30	0.75	1.42	0.49	0.15	0.42	0.64	0.49	−0.15	−0.75	0.31	1819
1820	−1.09	0.55	−1.35	2.19	0.22	−2.16	−0.96	0.66	−1.20	−0.61	−1.80	0.00	1820
1821	2.31	−1.59	0.73	1.78	−1.87	−2.01	−2.03	0.14	0.17	−0.68	2.72	3.52	1821
1822	2.52	2.96	4.04	1.75	2.11	3.77	0.56	−0.14	0.46	1.19	2.66	−1.31	1822
1823	−2.23	2.20	1.05	−0.09	1.23	−1.02	−1.14	0.87	0.24	−0.27	−0.11	2.95	1823
1824	1.39	1.98	−0.10	−0.89	−1.13	−0.65	0.32	−0.22	1.04	0.66	2.68	4.09	1824
1825	1.92	0.28	−0.76	1.43	−0.15	−0.41	0.85	0.49	1.15	0.15	1.51	3.05	1825
1826	−3.48	1.35	1.13	0.20	−1.25	1.06	2.12	2.86	1.75	1.94	−0.21	0.93	1826
1827	−0.55	−5.10	1.19	1.50	1.25	1.01	2.06	0.00	1.15	1.34	−2.01	2.85	1827
1828	3.18	0.41	1.17	0.82	0.74	1.19	0.91	−1.22	0.48	0.28	−0.33	1.85	1828
1829	−2.12	−2.33	−0.05	0.72	−0.04	0.21	0.50	−1.17	−0.88	−0.60	−1.88	−4.97	1829
1830	−5.83	−2.98	2.14	2.21	0.81	−0.22	0.86	−0.13	−1.01	−0.17	1.31	−0.32	1830
1831	−0.98	0.96	1.68	1.83	−0.50	−0.61	0.38	0.40	−0.81	3.26	0.30	1.64	1831
1832	0.10	−0.27	0.23	0.96	−0.88	−0.49	−0.01	1.02	−0.59	0.18	−0.68	0.95	1832
1833	−2.63	3.41	−0.71	−0.78	2.94	1.45	−1.24	−2.23	−1.08	−0.17	0.51	4.43	1833
1834	5.74	0.29	0.76	−1.12	1.87	1.12	2.76	1.35	1.82	0.53	0.79	0.29	1834
1835	1.77	1.74	0.11	−0.90	−0.68	0.13	1.46	−0.17	−0.03	−0.83	−2.92	−2.23	1835
1836	0.43	−0.85	3.27	−0.66	−1.99	0.47	0.21	0.47	1.67	0.82	0.66	1.56	1836
1837	1.30	0.80	−1.86	−2.33	−2.05	1.26	−0.86	1.24	−1.66	0.28	0.46	0.72	1837
1838	−4.35	−2.13	0.21	−2.36	−0.33	−0.21	−0.36	−1.20	0.44	−0.02	1.10	−0.52	1838
1839	0.88	0.67	−0.73	−2.24	−0.54	2.28	0.16	−0.47	0.09	1.20	1.49	2.20	1839
1840	1.37	−0.69	−2.82	1.28	−0.51	0.23	−1.39	0.34	−0.22	−2.04	1.81	−5.32	1840
Means.	−0.17	1.95	4.39	8.31	12.40	14.43	15.80	15.41	12.60	8.30	4.16	1.35	Means.

The numbers without sign must be subtracted; those with the sign — must be added.

E

LXXXVII.

North Germany. — Berlin.

For Reducing the Monthly and Yearly Means of Single Years to the Means derived from Series of Years.

Degrees of Reaumur.

Year.	Jan.	Feb.	March.	April.	May.	June.	July.	Aug.	Sept.	Oct.	Nov.	Dec.	Year.
	°	°	°	°	°	°	°	°	°	°	°	°	
1719	2.44	0.21	1.50	0.69	1.45	2.38	3.13	1.86	0.08	0.66	2.09	−1.02	1719
1720	2.27	0.40	−0.14	0.70	1.34	0.94	2.01	0.31	0.10	1.62	−0.03	1.47	1720
1721	2.38	−1.80	−1.53	2.23	−0.91	1.21	−0.67	−0.17	0.54	0.40	1.69	0.07	1721
1728	1.50	−2.28	2.39	0.65	1.24	0.26	−0.38	−1.36	−0.10	0.66	−0.58	−1.51	1728
1729	−3.18	−1.46	−3.57	−2.11	. .	. .	. .	. .	. .	. .	. .	. .	1729
1730	1.64	0.20	0.29	0.70	0.00	0.12	−0.62	−0.03	−0.69	−2.55	1.99	−0.48	1730
1731	−2.00	−1.78	−0.67	−1.67	−1.33	−0.89	−1.44	−0.62	−0.25	1.85	0.67	0.26	1731
1732	−1.50	1.34	1.05	1.34	0.29	−1.54	−1.95	−0.98	−0.84	1.14	−0.78	−3.99	1732
1733	2.69	2.54	0.86	1.59	−1.77	−2.71	−0.38	−0.97	−2.02	−0.53	0.21	2.46	1733
1734	0.40	2.51	1.86	0.55	−0.54	−1.26	−0.62	−0.93	−0.54	0.65	−2.85	−1.03	1734
1735	1.79	0.30	1.81	1.49	−0.87	−0.33	−1.38	−0.84	0.91	−1.01	−1.07	−0.17	1735
1736	−0.08	−0.92	−0.73	0.85	−0.88	−0.87	−0.24	0.64	−0.98	0.23	−0.09	1.18	1736
1737	1.83	0.55	1.57	−1.36	0.77	0.11	−0.77	−1.65	−0.10	−0.39	−0.83	−0.05	1737
1738	−0.55	0.55	1.11	1.54	−0.08	−0.42	−0.79	−0.38	−0.05	0.88	−2.21	0.90	1738
1739	−0.17	2.06	1.11	−1.65	0.64	−0.96	0.99	−1.23	0.91	−2.62	−5.35	−0.01	1739
1740	−6.61	−6.54	−3.28	−3.45	−3.49	−1.70	−0.96	−0.62	1.62	−3.12	−2.35	−0.18	1740
1741	−0.93	1.88	−0.71	−1.38	−1.90	−1.59	0.17	−0.54	−0.20	1.22	1.77	−0.16	1741
1742	−1.23	1.08	−0.99	−2.16	−1.83	−0.72	−0.66	−1.26	−1.78	0.19	0.70	−3.22	1742
1743	1.32	0.99	−0.53	−1.94	0.28	1.05	−1.46	0.32	−0.50	−1.44	2.77	0.84	1743
1744	−1.98	−2.42	−0.09	2.33	0.10	−1.47	0.25	−0.60	0.94	2.10	1.25	−0.39	1644
1745	−1.92	−1.26	−0.10	0.20	0.73	1.01	0.01	0.17	0.10	1.15	2.17	−2.36	1745
1746	0.12	0.03	−1.88	−0.39	0.43	−0.72	1.41	−0.43	0.44	−1.06	−0.53	1.89	1746
1747	−0.17	3.49	−2.09	0.70	−0.67	2.34	−0.33	0.18	1.43	0.43	0.21	1.04	1747
1748	−1.17	−1.70	−2.29	0.22	1.53	2.11	0.56	2.85	−0.14	0.00	1.79	3.19	1748
1749	2.28	0.47	−1.52	−0.14	1.58	0.21	0.39	1.64	0.33	0.05	−0.63	1.28	1749
1750	1.19	3.22	3.87	1.26	0.30	1.06	1.97	1.56	0.26	−0.55	. .	−0.06	1750
1751	−0.45	−1.70	2.79	−0.86	3.59	2.39	1.78	3.12	0.42	−0.04	. .	. .	1751
Means.	−0.19	0.69	2.65	6.51	10.63	12.82	14.02	13.14	11.06	6.53	3.15	1.24	Means.
1755	−4.56	−6.47	. .	0.54	. .	. .	. .	−0.25	. .	. .	. .	2.14	1755
1756	4.13	2.63	1.85	1.77	0.37	2.55	1.50	−0.35	1.61	1.62	−0.38	−1.43	1756
1757	1.17	2.37	1.71	. .	−0.39	1.47	3.25	0.22	−1.70	−2.88	1.21	−1.25	1757
1758	−2.57	−0.17	0.13	−0.21	1.08	0.18	−0.86	0.55	−1.11	−0.97	0.16	0.38	1758
1759	3.26	1.79	1.18	−0.01	−1.45	0.87	1.15	0.60	−0.45	1.09	−2.21	−3.85	1759
1760	−0.56	−1.48	−0.81	0.34	0.33	0.57	−0.29	0.03	0.87	0.98	0.12	2.05	1760
1761	0.97	1.65	2.51	−0.01	1.55	1.95	−0.62	1.88	2.30	−1.02	−0.12	−3.08	1761
1762	2.11	−0.01	−1.88	1.88	0.42	0.27	−0.19	−1.45	0.23	−1.34	−0.32	−1.82	1762
1763	−2.25	3.02	−0.40	−0.55	−0.34	0.17	0.92	1.32	−0.86	−0.87	−0.25	2.67	1763
1764	2.91	2.88	−0.10	−0.30	1.71	−1.94	1.43	−0.60	−1.70	−0.63	−1.32	−1.54	1764
1765	1.64	−2.90	1.70	0.78	−2.50	−0.88	−1.92	1.12	−1.16	1.20	0.15	0.03	1765
1766	−0.10	−0.12	1.01	2.07	1.17	0.30	−0.36	−0.25	0.73	−0.43	0.48	−0.26	1766
1767	−5.54	1.74	0.01	−1.58	−1.03	−1.65	−0.23	0.88	0.42	0.95	2.03	−1.75	1767

The numbers without sign must be subtracted; those with the sign — must be added.

North Germany. — Berlin (*continued*).

For Reducing the Monthly and Yearly Means of Single Years to the Means derived from Series of Years.

Degrees of Reaumur.

Year.	Jan.	Feb.	March.	April.	May.	June.	July.	Aug.	Sept.	Oct.	Nov.	Dec.	Year.
	°	°	°	°	°	°	°	°	°	°	°	°	
1768	−3.52	−0.98	−1.28	−0.11	−0.68	−0.06	0.28	−0.08	−1.03	−0.48	0.54	0.47	1768
1769	1.22	−0.74	0.75	0.14	−1.01	−1.01	−0.71	−1.07	0.58	−2.26	0.26	0.84	1769
1770	−0.20	−0.21	−3.16	−1.09	−0.11	−1.20	−0.35	−0.08	0.62	0.81	0.16	1.92	1770
1771	−1.24	−3.28	−3.40	−3.27	2.04	−0.21	−0.82	−2.12	−0.46	0.76	−1.47	0.95	1771
1772	0.66	1.20	0.86	−0.86	−2.50	−0.10	−1.40	−0.41	0.60	1.62	1.89	1.38	1772
1773	2.50	−1.00	−0.61	0.49	1.37	−1.22	−0.85	0.06	0.57	1.89	−0.92	2.21	1773
Means.	−0.13	1.64	3.87	7.71	11.94	15.23	16.18	15.34	12.12	7.73	4.38	1.85	Means.
1774	1.50	2.26	2.29	1.56	−0.05	0.69	−1.56	−1.92	−1.61	0.71	−3.70	−0.75	1774
1775	0.95	3.20	2.53	−0.65	−0.72	3.26	1.88	1.61	2.00	1.23	−0.84	2.16	1775
1776	−5.55	2.42	2.10	−0.13	−2.11	1.19	1.21	0.32	0.12	−0.47	0.70	0.54	1776
1777	0.04	−1.67	0.67	−1.12	0.52	0.04	−0.60	−0.01	−1.71	0.23	2.23	0.75	1777
1778	−0.58	−1.72	1.09	1.98	0.67	0.30	1.02	0.66	−0.67	−1.69	1.44	3.84	1778
1779	0.33	3.82	2.99	2.39	0.61	−0.30	0.74	1.71	1.59	1.95	0.90	2.26	1779
1780	−1.06	−2.02	3.37	−1.27	0.72	0.24	0.45	0.99	−0.03	1.46	−0.34	−0.70	1780
1781	−0.44	0.53	2.05	1.85	1.19	1.97	2.02	2.56	1.60	−0.39	0.80	0.01	1781
1782	3.15	−2.86	−0.39	−0.87	0.33	1.78	1.52	0.21	1.75	−0.30	−1.13	0.78	1782
1783	3.19	3.67	−0.58	0.86	1.38	2.71	1.45	0.71	0.36	0.34	0.50	−1.51	1783
1784	−3.97	−3.54	−1.68	−2.30	0.58	0.20	−0.75	−1.35	0.02	−2.21	1.29	−0.94	1784
1785	0.47	−3.28	−5.74	−2.54	−1.48	−0.84	−0.70	−1.12	0.61	−0.34	1.09	−1.42	1785
1786	1.81	−0.93	−2.32	1.60	−1.25	0.54	−1.71	−1.26	−1.86	−1.97	−3.64	−0.16	1786
1787	−0.29	1.38	2.05	−1.31	−0.77	0.99	−0.65	−0.59	−0.17	1.32	0.69	2.07	1787
1788	2.46	−1.26	−1.47	0.10	0.45	1.64	1.64	−1.21	1.20	−0.35	−0.79	−8.64	1788
1789	−1.93	1.46	−4.45	0.01	1.85	0.14	0.11	0.36	1.85	0.64	0.89	3.55	1789
1790	3.05	2.82	2.19	−1.67	1.70	0.58	−1.13	−0.54	−0.48	−0.44	−0.30	1.92	1790
1791	3.91	1.52	1.47	1.74	−1.16	0.19	0.78	1.08	−0.78	0.22	−0.89	1.35	1791
1792	0.53	−1.89	0.80	1.45	−0.81	0.83	1.59	0.46	−0.98	−0.30	−0.01	1.14	1792
1793	−0.70	2.14	0.61	−0.68	−0.58	−1.34	1.68	0.22	−0.83	1.99	0.99	2.05	1793
1794	1.18	2.56	3.66	3.12	0.18	1.77	2.79	−0.59	−1.62	0.37	1.53	−2.14	1794
1795	−5.23	−0.36	−0.84	2.88	−1.78	2.10	−0.92	−0.37	1.27	3.36	0.10	3.14	1795
1796	6.51	0.68	−1.70	−0.34	−0.46	0.38	0.48	1.33	1.74	0.07	−0.60	−1.82	1796
1797	1.60	1.89	0.66	1.09	1.41	−0.23	1.55	1.26	2.02	0.55	−0.80	1.81	1797
1798	1.79	1.57	−0.07	1.29	0.76	1.20	0.38	0.92	1.24	−0.17	−0.45	−3.54	1798
1799	−2.97	−4.47	−1.65	−2.12	−2.27	−1.53	−1.05	−0.32	−0.65	−0.70	0.48	−4.41	1799
1800	−1.12	−3.61	−4.09	4.43	2.33	−3.06	−1.99	0.22	0.67	−0.41	1.47	0.00	1800
1801	1.88	−1.02	1.84	0.05	3.00	−1.37	−0.61	−0.68	1.01	1.40	0.93	0.84	1801
1802	−1.00	0.50	1.65	0.45	−2.37	−1.01	−1.54	1.54	−0.08	3.04	0.78	1.81	1802
1803	−5.33	2.02	−0.16	2.84	−1.36	−1.46	2.03	1.80	−1.82	−0.45	0.68	−0.39	1803
1804	1.51	−1.48	−3.11	−1.06	1.04	−0.54	0.10	−0.73	1.17	−0.02	−2.40	−3.92	1804
1805	−3.90	−1.94	−0.48	−1.58	1.36	−1.53	−1.18	−1.83	0.55	−3.53	−2.58	1.24	1805
1806	3.02	0.94	0.19	−2.82	0.99	−2.26	−1.35	−0.98	0.41	−0.12	1.47	4.14	1806
1807	1.62	0.18	−1.97	−1.43	−0.42	−1.50	0.42	3.72	−2.15	0.08	1.11	1.53	1807

The numbers without sign must be subtracted; those with the sign — must be added.

North Germany. — Berlin (*continued*).

For Reducing the Monthly and Yearly Means of Single Years to the Means derived from Series of Years.

Degrees of Reaumur.

Year.	Jan.	Feb.	March.	April.	May.	June.	July.	Aug.	Sept.	Oct.	Nov.	Dec.	Year.
	°	°	°	°	°	°	°	°	°	°	°	°	
1808	0.83	−1.07	−3.39	−2.80	0.80	−0.42	1.19	0.69	−0.54	−1.56	−1.10	−4.40	1808
1809	−3.31	1.64	−1.09	−3.34	0.99	−0.89	−0.48	0.36	0.29	−0.99	0.02	2.23	1809
1810	−0.99	−1.66	0.40	−1.41	−1.88	−1.93	−0.05	−0.47	1.16	−1.33	0.09	1.22	1810
1811	−2.93	−0.72	2.01	−0.15	3.07	2.67	0.94	−0.59	−0.72	2.21	0.35	1.50	1811
1812	−1.14	−0.27	−1.05	−3.98	−1.20	−0.68	−2.37	−0.78	−1.81	1.14	−1.57	−5.52	1812
1813	−1.20	2.38	0.24	1.00	−0.73	−1.23	−1.27	−2.07	−0.82	−1.30	0.05	1.02	1813
1814	−2.12	−5.52	−2.78	1.00	−2.92	−1.99	1.02	−1.34	−2.23	−1.21	0.55	1.26	1814
1815	−2.81	1.14	1.56	−0.45	−0.15	0.61	−2.98	−1.57	−1.95	0.42	−0.69	−1.37	1815
1816	0.95	−2.27	−0.68	−0.21	−2.68	−1.54	−1.32	−2.59	−1.64	−1.23	−1.96	−0.39	1816
1817	2.58	1.79	−0.19	−3.86	−0.49	1.04	−1.57	−0.55	1.43	−2.57	2.37	−0.14	1817
1818	2.54	0.19	1.56	0.53	0.22	0.95	0.72	−1.41	0.14	−0.58	−0.60	−0.89	1818
1819	2.51	1.57	1.59	0.85	1.00	2.28	1.42	1.60	0.81	−0.41	−0.66	−2.61	1819
1820	−3.08	0.34	−0.02	1.52	0.91	−2.38	−2.08	1.23	−0.75	0.99	−1.57	−1.88	1820
1821	1.52	−1.05	0.14	3.28	−0.48	−2.17	−1.51	−0.78	0.91	1.33	3.27	3.44	1821
Means.	−1.59	0.30	2.28	6.89	11.36	13.73	15.16	15.00	11.83	7.16	2.61	−0.32	Means.
1822	3.39	3.67	3.22	1.55	0.59	0.58	0.77	−0.19	−1.24	1.36	1.58	−3.18	1822
1823	−7.56	−0.25	0.41	−1.32	−0.26	−0.78	−1.76	1.03	−0.34	0.66	1.01	1.12	1823
1824	3.67	2.45	0.29	−0.52	−1.04	−0.75	−0.56	−0.58	1.27	0.56	1.96	2.69	1824
1825	3.92	0.92	−2.26	0.86	−0.15	−1.10	−0.47	0.05	0.54	−0.12	1.30	2.03	1825
1826	−3.44	1.98	1.15	−0.19	−0.24	1.20	3.03	3.00	0.35	0.71	−0.33	0.49	1826
1827	0.25	−4.90	1.25	2.29	1.98	1.33	0.80	−0.04	1.09	0.83	−2.24	1.16	1827
1828	−0.26	−0.55	0.67	1.22	0.33	0.30	1.17	−0.71	−0.15	−0.28	0.17	0.47	1828
1829	−2.87	−2.67	−1.23	0.41	−0.29	0.12	0.41	−0.56	−0.16	−1.62	−2.54	−8.25	1829
1830	−4.21	−2.70	1.09	1.53	0.30	0.07	0.35	−0.26	−0.57	−0.69	1.47	−1.79	1830
1831	−1.81	0.75	0.40	2.21	−0.94	−1.34	0.36	0.20	−1.22	1.77	−0.54	0.11	1831
1832	0.76	1.12	0.42	0.32	−1.43	−0.33	−2.40	0.22	−1.22	−0.35	−0.63	−0.24	1832
1833	−0.86	3.16	−0.18	−1.82	3.46	1.33	−0.45	−3.12	−0.48	−0.93	0.14	2.48	1833
1834	4.73	1.31	1.00	−0.68	1.82	1.23	3.65	2.34	0.74	−0.28	0.56	0.36	1834
1835	2.81	2.37	0.57	−0.91	−0.86	0.13	0.21	−0.59	1.22	−0.97	−2.71	−1.77	1835
1836	1.37	1.11	3.42	0.07	−2.55	0.20	−1.08	−1.49	−1.06	1.00	−1.10	0.26	1836
1837	1.91	0.38	−1.98	−1.68	−1.42	−0.69	−1.11	1.20	−0.92	0.37	0.72	−0.87	1837
1838	−6.30	−3.63	0.42	−1.42	−0.24	0.35	−0.22	−1.78	1.27	−0.89	−1.14	−0.33	1838
1839	0.79	1.50	−1.98	−2.54	0.58	0.95	0.77	−0.44	1.10	0.15	1.10	−1.49	1839
1840	−0.09	0.65	0.23	−0.07	−0.03	0.16	0.27	−0.07	−0.05	0.09	−0.05	−0.33	1840
1841	−0.01	−4.03	0.91	1.01	2.51	−0.88	−1.10	−0.01	0.58	1.29	0.75	1.62	1841
1842	−1.34	0.39	0.93	−1.52	0.75	−0.54	−0.84	3.13	0.42	−1.55	−2.82	0.71	1842
1843	2.40	2.45	−1.09	0.44	−2.01	−1.00	−0.41	1.17	−0.64	−0.66	1.42	1.96	1843
1844	1.00	−0.96	−1.50	0.48	0.56	−1.00	−2.35	−1.60	0.36	−0.24	0.56	2.41	1844
1845	1.65	−4.55	−6.24	0.28	−1.48	0.49	0.90	−0.94	−0.93	−0.18	1.26	0.33	1845
Means.	−1.90	−0.15	2.74	6.88	10.92	13.94	15.04	14.43	11.75	7.97	3.25	1.32	Means.

The numbers without sign must be subtracted; those with the sign — must be added.

LXXXVIII.

Denmark. — Copenhagen.

For Reducing the Monthly and Yearly Means of Single Years to the Means derived from Series of Years.

Degrees of Reaumur.

Year.	Jan.	Feb.	March.	April.	May.	June.	July.	Aug.	Sept.	Oct.	Nov.	Dec.	Year.
	°	°	°	°	°	°	°	°	°	°	°	°	
1767	-3.89	0.34	0.52	-1.49	-1.83	-1.99	-1.40	-0.47	0.22	-0.75	1.58	-0.29	1767
1768	-0.67	-0.14	-1.01	0.11	-0.82	-0.50	-0.18	-0.59	-1.37	-0.49	0.57	1.66	1768
1769	1.74	0.79	1.44	0.30	-0.70	-0.50	-0.42	-1.21	-0.17	-1.92	0.03	0.37	1769
1770	0.19	1.64	-2.57	-0.77	-0.32	-0.88	0.21	0.36	1.11	1.38	-0.48	0.69	1770
1771	-1.20	-2.18	-3.96	-3.14	0.27	1.50	-0.33	-2.04	-0.90	0.03	-1.11	1.10	1771
1772	-0.88	-1.71	-2.53	-1.72	-1.94	-0.51	-0.56	-0.59	0.83	1.49	2.39	1.33	1772
1773	1.78	-0.46	0.35	0.33	0.83	-0.54	0.50	0.81	0.45	1.73	0.81	0.94	1773
1774	-2.37	0.34	0.87	0.83	-0.09	0.37	0.04	-0.65	-1.07	-0.31	-5.39	-2.55	1774
1775	-0.51	1.79	1.72	0.21	-0.09	2.13	1.39	1.72	2.60	0.78	-1.97	0.71	1775
1776	-5.22	1.18	1.49	0.73	-0.87	1.61	2.30	1.35	0.59	0.67	0.77	0.69	1776
1782	2.38	-0.61	-0.99	-0.62	-0.63	3.43	0.12	0.32	0.99	-1.09	-1.42	0.07	1782
1783	0.81	2.57	-0.38	2.01	1.97	2.36	3.05	1.56	1.67	1.91	-0.06	-0.87	1783
1784	-2.02	-0.59	-2.41	-1.51	0.24	0.07	-0.31	-0.21	0.18	-0.78	1.16	-0.76	1784
1785	0.53	-2.27	-2.96	-1.04	-1.52	0.78	-0.33	-0.46	0.10	-0.05	1.55	-0.20	1785
1786	0.13	0.06	-2.69	0.83	-1.08	1.48	-0.35	-0.41	-0.74	-1.21	-2.91	-0.04	1786
1787	0.94	2.21	2.09	-0.20	0.07	0.01	0.06	-0.31	0.58	1.81	-0.40	0.26	1787
1788	2.02	0.63	-1.14	0.95	1.00	1.28	-0.93	0.38	1.71	-0.31	-0.19	-6.92	1788
1798	1.15	2.27	1.31	2.48	2.71	2.06	2.00	2.15	1.09	1.01	0.01	-2.29	1798
1799	-0.71	-4.50	-1.94	-1.59	-2.12	-0.44	-0.18	-0.43	0.21	0.56	1.27	-2.55	1799
1800	-0.96	-2.07	-3.57	2.60	1.77	-1.69	-0.89	0.42	0.21	1.19	1.78	1.20	1800
1801	1.28	0.75	2.82	1.44	2.93	-0.10	1.30	0.58	0.69	2.17	1.97	0.46	1801
1802	-0.56	1.04	1.90	. .	-1.78	-2.26	-3.12	-0.56	-0.87	0.98	0.45	0.32	1802
1803	-3.02	-1.58	-0.39	1.86	-1.69	-2.02	-0.21	-0.14	-1.76	-0.90	-0.31	-1.36	1803
1804	2.01	-1.47	-1.82	-0.58	0.25	-0.57	-0.30	0.12	1.23	0.77	-1.74	-2.85	1804
1805	-1.79	-2.02	0.26	-1.03	-2.14	-3.46	-1.48	-1.03	0.77	-2.53	-0.56	0.77	1805
1806	1.90	1.64	-0.49	-1.59	0.03	-2.28	-1.79	-0.08	1.32	0.35	1.27	2.54	1806
1807	1.75	1.46	-0.55	-0.56	-0.37	-1.60	-0.17	2.54	-2.22	0.02	0.19	0.77	1807
1808	1.04	-0.77	-1.30	-1.40	0.19	0.02	1.26	1.34	1.10	-0.14	-0.85	-2.42	1808
1809	-2.64	0.30	-0.42	-2.52	0.60	-0.91	-0.89	0.47	0.30	-0.44	-0.23	1.65	1809
1810	0.60	-0.28	0.05	-1.19	-2.69	-1.01	-0.07	-0.29	0.51	-0.79	-0.22	0.10	1810
1811	-0.65	0.23	2.46	-0.71	1.75	0.96	2.07	-0.32	-0.26	1.28	1.12	1.07	1811
1812	0.40	1.21	-1.55	-2.62	-1.63	-0.97	-2.38	-0.61	-1.67	1.36	-1.14	-3.56	1812
1813	0.23	2.66	1.50	0.55	-1.01	-1.00	0.44	-0.89	-0.53	-2.01	0.20	0.97	1813
1814	-3.81	-4.01	-2.15	0.28	-2.99	-1.97	0.13	-0.87	-1.05	-0.58	1.22	0.85	1814
1815	-0.67	1.47	1.82	0.30	-0.26	-1.26	-1.95	-0.81	-1.11	0.59	0.20	-0.66	1815
1816	0.72	-1.56	-0.05	-0.49	-2.69	-1.87	-0.49	-1.86	-0.89	-0.72	-0.95	-0.31	1816

The numbers without sign must be subtracted; those with the sign — must be added.

LXXXVIII.

Denmark. — Copenhagen (*continued*).

For Reducing the Monthly and Yearly Means of Single Years to the Means derived from Series of Years.

Degrees of Reaumur.

Year.	Jan.	Feb.	March.	April.	May.	June.	July.	Aug.	Sept.	Oct.	Nov.	Dec.	Year.
	°	°	°	°	°	°	°	°	°	°	°	°	
1817	2.79	2.98	1.13	−1.10	−0.01	−1.04	−1.53	−1.38	0.62	−2.24	1.41	−1.71	1817
1818	1.99	1.73	2.40	−1.05	−0.05	0.97	1.29	−0.24	0.69	0.87	1.48	0.20	1818
1819	3.46	2.30	2.39	1.56	1.25	1.69	1.58	3.28	1.46	−0.79	−1.03	−1.26	1819
1820	−1.67	0.51	0.52	1.55	0.25	−1.16	−0.36	−0.33	−0.60	−0.64	−0.56	−0.87	1820
1821	0.36	0.16	0.24	2.24	−0.43	−1.77	−1.81	−0.86	0.67	1.62	1.68	2.47	1821
1822	2.56	3.82	3.64	2.28	1.59	0.87	0.24	−0.17	−0.64	1.52	2.63	0.56	1822
1823	−2.60	−0.08	0.70	−0.04	0.51	0.15	−0.94	0.41	0.47	1.02	1.88	1.74	1823
1824	3.65	2.36	0.97	0.91	0.14	1.22	−0.61	−0.48	1.62	0.09	1.20	2.18	1824
1826	..	..	..	..	4.30	5.91	7.76	6.63	..	..	..	2.04	1826
1827	0.16	−2.30	0.59	2.14	1.44	1.93	0.09	−0.29	1.48	1.16	−1.20	2.30	1827
1828	−0.07	0.43	1.37	0.58	1.31	1.34	1.36	0.26	0.41	0.46	0.61	0.50	1828
1829	−1.14	−3.06	−0.95	−1.00	1.84	1.50	−0.23	−1.01	0.03	−1.43	−2.91	−3.60	1829
1830	−2.26	−2.85	1.39	0.69	−0.18	−0.86	0.30	−0.81	−0.95	0.15	1.75	−0.22	1830
1831	−1.60	0.61	−0.16	1.87	0.31	0.85	2.52	1.55	−0.59	2.71	−0.65	1.91	1831
1832	1.52	1.73	1.55	1.84	−0.23	1.29	−0.94	−0.06	−0.98	0.60	−0.47	0.58	1832
1833	0.05	1.50	−0.45	−0.72	2.32	0.72	0.79	−2.27	0.08	0.63	0.77	1.32	1833
1834	2.26	1.71	2.23	0.90	1.98	0.72	3.60	3.26	0.11	−0.05	0.22	0.59	1834
1835	1.87	2.16	1.66	−0.02	−0.92	1.17	1.03	−0.57	0.09	−0.85	−1.44	−0.88	1835
1836	0.29	0.63	2.71	0.14	−0.17	0.24	−0.89	−1.86	−1.62	−0.48	−1.34	0.09	1836
1837	0.17	0.54	−1.08	−1.50	−1.10	0.05	−0.21	0.60	−0.80	−0.06	−0.91	−0.75	1837
1838	−2.83	−4.85	−0.56	−2.63	−0.97	−0.70	−0.09	−2.25	−0.44	−1.82	−2.01	−0.25	1838
1839	−0.17	−0.38	−2.06	−2.80	0.49	0.13	0.23	−1.24	−0.40	0.11	−0.19	−2.12	1839
1840	−0.63	−0.39	−0.64	0.35	−2.64	−2.17	−3.25	−1.79	−1.95	−3.77	−0.51	−2.65	1840
1841	−1.14	−2.52	0.97	0.62	2.21	−1.37	−2.56	−0.97	−0.71	−0.38	−0.36	2.37	1841
1842	−0.26	1.43	2.05	0.61	1.78	−0.11	−0.99	2.73	0.31	−0.88	−1.57	2.36	1842
1843	1.82	0.79	−0.33	0.46	−0.96	−0.25	−0.67	1.03	−0.20	−1.23	0.86	2.99	1843
1844	0.07	−2.48	−1.50	0.74	1.49	−1.12	−2.17	−1.42	−0.62	−0.29	0.46	−1.43	1844
1845	1.24	−4.16	−4.45	0.54	−1.01	0.20	0.22	−0.86	−1.26	−1.04	1.28	0.59	1845
Means.	−1.16	−0.80	0.55	4.45	8.98	12.45	13.81	13.50	10.86	7.05	3.12	0.68	Means.

The numbers without sign must be subtracted; those with the sign — must be added.

LXXXIX.

France. — Paris.

For Reducing the Monthly and Yearly Means of Single Years to the Means derived from Series of Years.

Degrees of Reaumur.

Year.	Jan.	Feb.	March.	April.	May.	June.	July.	Aug.	Sept.	Oct.	Nov.	Dec.	Year.
	°	°	°	°	°	°	°	°	°	°	°	°	
1806	3.35	1.38	0.28	−1.54	2.07	0.77	0.64	−0.38	0.53	−0.26	1.69	4.00	1806
1807	0.34	1.39	−2.74	−0.63	1.28	−0.52	1.94	2.34	−2.08	1.15	−0.74	−1.75	1807
1808	0.42	−1.42	−2.19	−2.23	2.55	−0.30	2.14	0.66	−0.78	−1.74	0.58	−1.87	1808
1809	2.95	2.91	0.42	−2.72	0.54	−1.38	−1.08	−0.36	−0.81	−1.09	−1.54	1.04	1809
1810	−2.90	−1.11	1.16	−0.42	−0.62	−0.06	−0.74	−0.70	1.75	0.25	0.80	1.30	1810
1811	−1.83	2.31	1.90	1.58	2.14	0.25	0.44	−0.66	0.95	2.55	1.38	0.72	1811
1812	−0.32	1.63	−0.82	−1.92	0.88	−0.77	−0.96	−0.46	−0.17	0.51	−1.95	−3.71	1812
1813	−1.18	1.33	−0.23	0.71	0.48	−1.26	−1.12	−1.42	−1.38	0.29	−0.63	−0.47	1813
1814	−1.70	−3.37	−2.30	1.30	−1.67	−1.17	0.46	−0.91	−0.26	−1.22	−0.51	2.02	1814
1815	−1.98	2.39	2.29	0.36	0.18	−0.89	−0.93	−0.54	−0.11	0.77	−2.70	−1.34	1815
1816	0.54	−1.69	−0.71	0.10	−1.40	−1.83	−2.53	−2.37	−1.26	0.29	−2.24	0.07	1816
1817	2.48	2.22	−0.20	−2.02	−1.70	0.61	−1.34	−1.66	0.99	−3.16	1.80	−1.12	1817
1818	1.94	−0.21	−0.15	1.20	−0.65	1.75	1.14	−0.18	0.05	0.38	1.98	−1.23	1818
1819	2.43	0.95	0.16	1.31	0.02	−0.85	0.30	0.78	0.58	−0.12	−1.66	−0.30	1819
1820	−2.02	−0.98	−1.42	1.20	−0.30	−1.37	−0.35	0.11	−1.19	−0.93	−1.30	−0.22	1820
1821	1.02	−2.58	0.54	1.34	−1.95	−2.05	−1.39	1.20	0.85	−0.14	2.70	3.10	1821
1822	1.96	1.52	2.62	1.01	1.72	3.26	0.09	0.42	0.18	1.72	1.82	−3.42	1822
1823	−1.79	0.88	−0.14	−0.62	0.50	−1.69	−1.23	0.46	0.00	−0.58	−0.84	1.58	1823
1824	0.61	0.68	−1.00	−0.54	−1.52	−0.61	−0.02	−0.17	0.89	0.54	2.30	2.74	1824
1825	1.23	0.06	−0.94	1.54	−0.22	−0.05	1.24	0.70	1.77	0.75	0.40	2.18	1825
1826	−2.77	1.73	0.56	0.27	−1.48	1.35	1.59	2.10	1.11	1.70	−1.08	1.72	1826
1827	−1.63	−4.14	1.14	1.14	0.18	−0.09	0.85	−0.43	0.46	1.52	−0.77	2.58	1827
1828	3.28	0.80	0.29	0.50	0.46	0.34	0.34	−0.74	0.74	−0.30	0.51	0.89	1828
1829	−3.16	−0.97	−0.75	−0.08	0.32	0.05	−0.10	−1.30	−1.53	−1.01	−1.64	−5.70	1829
1830	−3.42	−2.59	2.54	1.68	0.11	−0.82	0.16	−1.23	−1.50	−0.44	0.83	−0.82	1830
1831	0.13	1.53	1.85	1.30	−0.20	−0.12	0.86	0.12	−0.35	2.83	−0.10	1.50	1831
1832	−0.36	−0.59	−0.93	0.65	−1.05	0.22	0.68	1.87	−0.10	0.06	−0.10	0.53	1832
1833	−1.73	2.34	−1.82	−0.38	2.54	1.06	−0.24	−1.65	−1.53	0.57	−0.61	3.46	1833
1834	4.34	−0.42	0.67	−0.70	1.59	0.70	1.25	0.69	1.24	0.29	−0.05	−0.02	1834
1835	1.35	1.69	−0.14	−0.38	−0.55	0.18	1.92	1.42	0.36	−0.92	−1.10	−2.84	1835
1836	0.55	−1.03	1.62	−1.02	−1.67	1.06	0.56	0.30	−1.24	−0.04	0.66	0.36	1836
1837	0.39	0.97	−3.26	−3.34	−2.79	1.14	−0.32	1.26	−0.84	0.04	−0.62	0.60	1837
1838	−5.21	−5.03	0.26	−2.52	−0.23	−0.68	−0.32	−0.42	−0.12	−0.04	0.74	−1.48	1838
1839	0.75	0.73	−0.62	−1.70	−0.71	1.62	−0.04	−0.86	0.00	−0.56	1.10	1.60	1839
1840	1.23	−0.47	−2.58	2.26	0.49	1.02	−1.08	0.98	−0.64	−1.40	0.99	−4.76	1840
1841	0.47	−1.35	1.94	0.42	2.25	−1.26	−1.68	−0.50	2.28	0.12	0.02	1.48	1841
1842	−2.65	0.33	1.30	0.26	0.05	2.66	0.52	3.18	−0.12	−2.28	−1.10	0.36	1842
1843	2.07	−0.39	1.06	0.50	−0.31	−0.86	−0.48	0.70	0.96	0.12	0.54	0.60	1843
1844	0.83	−1.31	0.18	2.22	−1.35	0.54	−1.12	−2.34	0.24	−0.36	0.26	−3.40	1844
Means.	1.53	3.35	5.33	7.90	11.59	13.66	14.96	14.82	12.52	9.00	5.41	2.92	Means.

The numbers without sign must be subtracted; those with the sign — must be added.

XC.

Holland. — Zwanenburg.

For Reducing the Monthly and Yearly Means of Single Years to the Means derived from Series of Years.

Degrees of Reaumur.

Year.	Jan.	Feb.	March.	April.	May.	June.	July.	Aug.	Sept.	Oct.	Nov.	Dec.	Year.
	°	°	°	°	°	°	°	°	°	°	°	°	
1743	0.60	1.40	−0.15	−2.69	−0.40	0.59	−1.15	0.05	−0.22	−2.27	1.83	−0.23	1743
1744	−0.91	−2.36	−0.74	−0.80	−0.89	−0.26	−1.02	−1.23	−0.71	0.39	0.66	0.21	1744
1745	0.15	−1.64	−0.70	−0.43	−0.04	−0.69	−0.92	−1.20	0.02	−0.27	−0.50	−2.16	1745
1746	−0.82	−1.70	−2.19	−1.20	1.36	−0.62	0.04	−1.28	−0.65	−2.09	−2.80	1.02	1746
1747	−0.47	2.16	−2.29	−0.18	−0.52	0.92	−0.65	−0.21	0.34	−0.49	1.62	1.60	1747
1748	−0.24	−2.63	−4.14	−2.12	−0.31	1.45	0.08	0.39	−0.03	0.28	1.68	3.46	1748
1749	2.68	0.11	−1.09	−0.52	1.11	−2.30	−0.10	0.23	−0.11	−0.35	−0.45	1.65	1749
1750	−0.34	2.60	2.88	−0.06	0.14	−0.10	0.97	−0.45	0.75	−1.25	−1.63	−0.31	1750
1751	1.09	−2.29	1.33	−0.60	−1.21	−0.10	−0.78	−0.52	−1.19	−0.48	−1.31	0.33	1751
1752	1.71	−0.56	0.72	−0.63	−1.10	0.95	−0.48	−0.09	0.39	0.07	0.90	1.37	1752
1753	−1.80	−0.11	1.34	0.01	−0.30	1.19	−0.34	−1.00	0.30	0.59	−0.88	0.67	1753
1754	0.64	−1.14	−2.23	−1.40	0.41	−0.49	−1.33	−0.16	−0.44	0.61	0.05	−0.36	1754
1755	−1.98	−3.19	−1.24	1.72	−1.37	1.89	−0.31	−1.33	−1.12	−0.08	−0.03	1.22	1755
1756	3.20	1.32	0.38	−1.57	−1.53	0.97	0.80	−0.50	0.74	−0.31	−1.13	−2.60	1756
1757	−2.22	−0.59	0.00	1.00	−1.01	−0.11	2.37	0.36	−0.21	−1.09	1.43	−0.09	1757
1758	−1.28	0.37	0.41	−0.39	1.95	0.29	−1.41	0.99	−0.17	0.21	0.05	0.36	1758
1759	2.86	2.13	1.49	0.86	−0.58	0.99	1.66	0.71	−0.07	1.05	−1.54	−2.68	1759
1760	−1.64	−0.69	0.15	0.77	−0.22	1.31	−0.15	−0.40	1.14	0.28	1.08	2.67	1760
1761	1.78	1.90	2.37	0.47	0.92	0.86	−0.61	1.16	0.67	−1.75	0.34	−1.59	1761
1762	2.10	0.09	−1.25	2.37	0.93	0.67	0.30	−1.31	−0.04	−1.98	−1.37	−2.02	1762
1763	−4.88	0.79	−0.34	−0.24	−1.04	0.28	−0.08	0.22	−0.56	−0.99	0.56	1.52	1763
1764	3.37	2.52	0.17	0.52	1.71	0.02	1.43	−0.32	−1.14	−0.74	−0.45	−1.01	1764
1765	2.24	−2.13	2.30	1.62	0.27	1.22	−0.84	0.85	−0.05	1.24	0.08	−0.82	1765
1766	−0.22	−0.78	0.72	1.67	0.37	0.35	0.20	0.45	0.49	0.32	0.46	−0.68	1766
1767	−3.34	2.34	1.08	−0.63	−1.36	−0.94	−0.80	0.36	0.98	0.71	2.15	−1.33	1767
1768	−1.94	0.93	−0.07	−0.09	−0.02	0.54	0.65	0.33	−1.27	−0.37	0.70	0.72	1768
1769	1.19	0.09	0.85	0.99	−0.21	−0.53	0.53	−0.06	0.48	−1.71	0.58	1.43	1769
1770	1.45	0.92	−1.12	−1.04	−0.15	−0.34	0.02	1.20	1.59	0.19	0.06	2.01	1770
1771	−0.50	−1.44	−2.33	−2.59	1.72	0.26	−0.29	−1.01	0.04	0.89	0.69	1.68	1771
1772	0.11	0.21	0.68	−0.50	−1.11	1.19	0.57	0.36	0.83	2.68	2.36	1.16	1772
1773	3.38	−0.57	1.36	0.81	0.35	0.31	−0.16	1.17	0.66	1.79	1.51	1.76	1773
1774	0.58	1.62	2.18	1.30	0.08	0.96	1.12	0.51	−0.30	1.23	−1.84	−0.45	1774
1775	1.31	3.40	2.12	1.04	−0.12	2.19	0.78	0.88	1.84	1.25	−1.53	1.65	1775
1776	−4.40	1.20	1.99	1.45	−0.85	1.11	1.56	0.47	−0.01	1.31	0.46	0.05	1776
1777	−0.23	−1.57	1.14	−0.56	0.15	−0.19	−0.07	0.88	0.60	0.73	1.97	−0.60	1777
1778	−1.26	−1.70	−0.55	0.36	0.71	0.43	1.43	0.54	−1.58	−2.02	1.08	2.90	1778
1779	−0.28	2.55	1.79	1.21	0.61	−0.77	0.60	1.51	1.27	1.61	0.19	0.53	1779
1780	−1.54	−0.56	2.68	−0.78	1.07	−0.51	−0.25	2.04	1.08	1.03	−0.07	−1.09	1780

The numbers without sign must be subtracted; those with the sign — must be added.

XC.

Holland. — Zwanenburg (*continued*).

For Reducing the Monthly and Yearly Means of Single Years to the Means derived from Series of Years.

Degrees of Reaumur.

Year.	Jan.	Feb.	March.	April.	May.	June.	July.	Aug.	Sept.	Oct.	Nov.	Dec.	Year.
	°	°	°	°	°	°	°	°	°	°	°	°	
1781	−0.97	1.18	1.18	1.23	0.37	2.47	1.04	1.56	0.91	0.75	0.36	−0.39	1781
1782	2.88	−1.88	−0.56	−1.11	−1.09	0.77	0.34	−0.54	0.50	−0.93	−2.43	−0.89	1782
1783	2.39	2.13	−1.31	1.24	−0.05	0.92	2.75	0.93	0.44	0.73	0.48	−2.74	1783
1784	−3.26	−3.01	−2.04	−2.16	1.23	0.15	−0.37	−0.80	0.94	−2.30	0.80	−1.60	1784
1785	−0.06	−2.34	−3.32	−1.54	−0.96	−0.46	−0.01	−0.59	1.14	0.40	0.41	−1.70	1785
1786	0.35	−0.08	−3.19	0.44	−0.59	0.72	−1.80	−0.75	−1.55	−1.49	−3.59	−0.23	1786
1787	−0.23	1.24	1.82	−0.90	−1.11	−0.11	−0.82	−0.56	..	..	..	..	1787
1788	2.20	−0.42	−1.15	0.24	0.58	1.05	0.87	−0.66	0.30	0.33	−0.73	−6.23	1788
1789	−2.66	0.98	−3.65	−1.64	0.56	−0.65	−0.58	−0.07	−0.40	−1.13	−1.10	1.84	1789
1790	2.20	2.51	1.53	−2.00	0.89	−0.72	−1.76	−1.25	−1.73	−0.86	−1.71	0.89	1790
1791	2.74	1.29	1.23	1.34	−1.21	−1.25	−1.20	−0.14	−0.74	−0.60	−0.79	−0.53	1791
1792	1.06	−0.38	0.03	1.70	−1.11	−0.93	−0.07	0.27	−1.53	−1.13	−0.14	1.05	1792
1793	0.52	1.59	0.03	−1.40	−1.61	−1.70	0.67	−0.65	−1.68	0.98	−0.17	1.60	1793
1794	−0.21	2.09	2.58	2.59	−0.76	−0.43	1.52	−0.87	−1.14	−0.54	0.41	−2.08	1794
1795	−4.52	−1.53	−0.92	0.85	−1.88	−0.18	−2.29	−0.08	1.51	2.39	0.37	2.87	1795
1796	4.72	1.76	−0.99	1.00	−0.63	−0.50	−0.91	0.02	0.64	−0.80	−0.46	−2.07	1796
1797	0.84	0.52	−0.18	0.81	0.52	−1.18	1.38	0.01	−0.78	−0.60	0.32	1.59	1797
1798	1.45	1.73	0.31	1.22	0.11	0.77	−0.05	0.36	0.19	0.68	−0.17	−3.49	1798
1799	−2.11	−2.00	−1.77	−2.19	−1.68	−1.83	−1.47	−1.08	−0.72	−0.63	0.59	−3.54	1799
1800	−0.65	−1.76	−1.97	2.08	1.85	−2.10	−1.32	0.04	0.50	0.02	1.12	−0.46	1800
1801	1.97	−0.59	1.61	0.26	0.68	−1.43	−0.76	0.32	0.45	1.16	0.53	0.47	1801
1802	−0.75	0.24	0.56	0.55	−1.10	−0.28	−1.69	1.08	0.03	1.15	0.54	1.19	1802
1803	−3.04	−2.29	0.00	2.06	−1.55	−0.92	1.43	0.75	−1.11	0.06	0.29	0.43	1803
1804	3.30	0.13	−0.92	−0.84	1.35	0.26	0.03	−0.20	1.57	0.62	−1.79	−2.84	1804
1805	−1.22	−0.36	−0.07	−0.56	−2.16	−1.97	−1.18	0.05	1.47	−2.00	−1.69	0.94	1805
1806	3.14	1.58	0.25	−1.95	1.79	−0.52	0.13	0.67	1.41	0.23	2.52	4.12	1806
1807	2.36	1.74	−1.32	−0.37	1.09	−0.17	1.64	2.53	−1.40	1.63	−0.15	0.84	1807
1808	1.19	0.07	−1.71	−2.02	2.07	−0.46	2.62	1.64	0.24	−1.35	−0.05	−1.50	1808
1809	..	..	..	−2.53	1.30	−1.03	−0.47	0.09	−0.27	−1.32	−0.99	0.68	1809
1810	−1.94	−1.39	−0.36	−0.41	−1.76	−0.96	0.05	−0.07	0.99	−0.63	−0.03	1.06	1810
1811	−2.75	0.55	1.41	1.16	2.75	1.53	0.47	−0.30	−0.49	2.40	1.80	1.05	1811
1812	0.81	1.20	−1.21	−2.48	0.16	−0.68	−1.28	−0.56	−0.62	0.46	−2.11	−4.00	1812
1813	−0.84	1.53	0.16	0.04	0.85	−0.32	−0.12	−0.91	−0.75	−1.32	−0.76	−1.31	1813
1814	−3.33	−4.20	−2.89	1.27	−2.01	−1.86	0.44	−0.66	−0.72	−1.44	−0.17	0.17	1814
1815	−2.69	0.96	2.23	0.59	0.59	−0.08	−1.63	−0.77	−0.54	0.07	−0.97	−1.90	1815
1816	0.52	−1.64	−0.78	−0.28	−1.48	−2.28	−1.31	−1.85	−1.14	−0.12	−2.06	−0.45	1816
1817	2.36	2.31	0.19	−2.12	−1.38	0.84	−0.83	−1.30	0.69	−3.16	1.83	−0.67	1817
1818	1.96	−0.40	0.40	−0.21	−0.56	1.69	0.99	−0.64	−0.36	−0.34	0.74	−1.22	1818

The numbers without sign must be subtracted; those with the sign — must be added.

XC.

Holland. — Zwanenburg (*continued*).

For Reducing the Monthly and Yearly Means of Single Years to the Means derived from Series of Years.

Degrees of Reaumur.

Year.	Jan.	Feb.	March.	April.	May.	June.	July.	Aug.	Sept.	Oct.	Nov.	Dec.	Year.
	°	°	°	°	°	°	°	°	°	°	°	°	
1819	1.47	1.04	0.68	0.84	0.91	0.50	0.56	1.12	0.55	−0.79	−1.18	−2.18	1819
1820	−2.89	−1.49	−1.21	0.72	0.12	−1.71	−1.06	−0.11	−0.93	−0.81	−1.84	−1.59	1820
1821	−0.67	−1.32	−0.16	1.66	−1.24	−1.91	−1.81	−0.14	0.72	0.20	1.72	2.08	1821
1822	2.64	1.93	2.26	0.40	1.53	1.65	0.24	−0.19	−0.88	0.74	1.99	−2.95	1822
1823	−6.29	−0.94	0.11	−1.19	0.44	−1.88	−0.89	0.12	−0.37	−0.66	0.90	1.65	1823
1824	2.30	0.20	−0.22	−0.78	−0.47	−0.40	−0.19	0.00	1.03	0.36	1.52	2.59	1824
1825	2.63	0.60	−1.42	0.43	0.12	0.00	−0.04	−0.36	1.20	1.04	1.03	1.70	1825
1826	−2.57	0.97	0.87	0.17	−0.59	1.52	2.12	2.01	0.30	1.95	0.16	1.99	1826
1827	−0.65	−3.83	0.58	0.93	0.40	−0.24	0.14	−0.55	−0.14	0.88	−0.91	2.79	1827
1828	0.75	−0.75	1.05	0.43	0.49	0.70	0.79	−0.64	0.43	0.24	−0.18	1.96	1828
1829	−3.35	−2.47	−1.43	−0.45	0.10	−0.37	−0.42	−1.35	−1.52	−0.43	−1.61	−5.77	1829
1830	−2.70	−4.01	0.50	0.75	0.13	−1.45	0.59	−1.17	−1.45	0.34	1.00	−1.80	1830
1831	−1.07	0.04	1.24	1.61	−0.10	−0.09	0.90	0.66	−0.14	3.16	0.66	1.72	1831
1832	−0.77	−1.34	−0.43	0.55	−1.49	−0.07	−1.74	−0.12	−0.64	0.48	−1.37	0.72	1832
1833	−2.12	1.33	−1.62	−0.68	2.22	0.92	−0.48	−2.08	−0.99	0.11	0.44	3.07	1833
1834	4.21	0.40	1.15	−0.87	1.31	0.87	1.80	1.00	0.86	0.68	−0.31	1.42	1834
1835	1.21	1.81	0.47	−0.76	−1.09	0.92	0.47	0.07	−0.22	−0.77	−1.44	−0.44	1835
Means.	0.99	3.14	3.86	6.80	10.12	12.45	13.97	14.13	12.30	8.61	4.84	2.16	Means.

XCI. England. — London.

Degrees of Reaumur.

Year.	Jan.	Feb.	March.	April.	May.	June.	July.	Aug.	Sept.	Oct.	Nov.	Dec.	Year.
1794	−0.96	2.72	1.23	1.64	−0.99	−0.43	1.83	−0.38	−1.35	−0.61	0.36	−1.10	1794
1795	−5.04	−2.08	−1.26	−0.23	−0.46	−1.98	−0.04	0.11	1.76	1.61	−0.88	2.46	1795
1796	4.42	0.50	−1.00	1.10	−1.26	−1.00	−1.28	−0.51	1.23	−1.45	−0.97	−3.76	1796
1797	−0.01	−1.44	−1.51	−0.45	−0.70	−1.56	0.62	−0.82	−0.97	−1.34	−0.44	0.93	1797
1798	−3.44	−0.28	−0.12	1.41	0.44	1.31	−0.10	0.88	−0.11	0.09	−1.24	−2.39	1798
1799	−1.00	−1.05	−1.74	−1.94	−1.39	−1.34	−0.79	−1.40	−1.19	−1.02	0.13	−2.79	1799
1800	0.59	−2.04	−1.70	1.14	0.66	−1.37	0.66	1.23	0.42	−0.86	−0.15	−0.24	1800
1801	1.64	−0.08	1.26	−0.35	−0.10	−0.09	−0.48	0.76	0.88	0.33	−1.08	−1.37	1801
1802	−1.21	0.11	−0.04	1.14	−1.50	−0.66	−2.20	1.74	0.49	0.23	−0.89	−0.56	1802
1803	−0.92	−1.03	0.51	0.88	−1.12	−0.89	0.97	0.41	−1.77	−0.40	−0.31	0.98	1803
1804	3.39	−0.73	0.00	−0.95	1.80	1.07	−0.57	−0.20	1.16	0.66	0.68	−1.52	1804
1805	−0.52	0.04	0.34	−0.20	−1.38	−1.49	−0.89	0.60	1.15	−1.06	−1.17	0.08	1805
1806	2.27	1.27	−0.23	−1.21	1.00	0.64	−0.06	0.38	0.16	0.54	2.11	3.64	1806
1807	0.64	0.54	−1.80	−0.14	1.05	−0.34	1.07	1.36	−1.61	1.44	−1.60	−1.19	1807
1808	0.64	−1.01	−1.80	−1.43	1.99	0.02	1.87	0.82	−0.55	−1.76	0.58	−1.32	1808
1809	−0.11	2.36	0.65	−2.05	1.23	−0.38	−0.75	−1.09	−0.24	−0.08	−1.33	0.72	1809

The numbers without sign must be subtracted; those with the sign — must be added

England. — London (*continued*).

For Reducing the Monthly and Yearly Means of Single Years to the Means derived from Series of Years.

Degrees of Reaumur.

Year.	Jan.	Feb.	March.	April.	May.	June.	July.	Aug.	Sept.	Oct.	Nov.	Dec.	Year.
	°	°	°	°	°	°	°	°	°	°	°	°	
1810	−0.47	0.01	0.38	0.12	−1.44	0.20	−0.44	−0.16	1.32	0.95	0.32	−0.03	1810
1811	−1.09	0.85	1.54	1.64	2.03	0.51	0.36	−0.51	0.83	2.50	1.29	−0.16	1811
1812	0.42	1.43	−0.68	−1.56	−0.19	−1.09	−1.24	−1.89	−0.64	−1.10	−0.75	0.90	1812
1813	−0.51	1.34	0.87	−0.81	0.12	−0.96	−0.97	−1.00	−0.99	−1.05	−0.84	−1.01	1813
1814	−3.80	−2.21	−2.55	1.06	−1.66	−2.03	−0.04	−0.91	−0.72	−1.10	−0.75	0.90	1814
1815	−1.49	1.34	1.94	0.44	1.19	0.24	−0.53	−0.07	2.48	0.55	−1.42	−0.83	1815
1816	0.64	−0.70	−0.64	−0.50	−0.99	−1.27	−2.35	−1.18	0.96	0.28	−1.24	−0.48	1816
1817	1.84	2.05	0.25	−0.63	−1.75	0.77	−1.46	−2.60	−0.81	−1.76	2.14	−0.70	1817
1818	1.67	−1.32	0.03	0.04	−0.06	2.24	2.40	1.98	2.30	2.06	3.20	−0.08	1818
1819	2.29	0.85	1.36	1.37	0.88	−0.69	0.36	1.58	0.70	3.08	−0.75	−0.74	1819
1820	−1.44	−0.66	0.25	1.68	−0.01	−0.74	−0.71	−1.18	−0.99	−0.96	−0.22	0.59	1820
1821	1.04	−0.97	0.87	2.08	−1.26	−1.80	−1.55	0.47	1.28	0.32	2.32	2.32	1821
1822	2.16	2.19	2.78	0.48	1.45	1.57	0.36	0.29	−0.24	1.04	2.36	−1.14	1822
1823	−1.40	0.19	0.16	−0.10	2.16	0.33	0.14	0.78	0.39	−0.56	0.54	0.55	1823
1824	0.78	2.41	−0.73	−0.94	−1.48	−1.40	0.00	−0.29	0.48	−0.03	1.38	1.08	1824
1825	1.31	−0.21	−1.17	1.28	0.08	−0.03	1.47	0.38	1.63	0.32	−0.84	0.59	1825
1826	−1.49	1.61	. .	1.46	1.16	1.97	1.69	1.67	0.30	1.28	−1.11	1.19	1826
1827	−0.96	−3.19	0.74	0.39	−0.08	−0.40	0.74	−0.73	0.21	0.84	−0.28	1.99	1827
1828	1.73	0.54	1.00	0.28	0.70	0.88	0.36	−0.62	0.52	−0.16	0.65	2.37	1828
1829	−1.76	−0.24	−1.08	−0.85	0.50	0.35	−0.48	−1.22	−1.41	−1.16	−1.60	−3.14	1829
1830	−2.31	−2.17	1.98	1.15	−1.39	−1.09	0.65	−1.09	−1.37	0.32	0.63	−2.12	1830
1831	−0.73	1.01	1.16	1.21	−0.21	0.55	1.49	1.29	−0.04	2.39	−0.08	1.21	1831
1832	0.13	−0.86	−0.42	0.35	−0.70	0.57	−0.20	0.18	−0.06	0.52	0.47	1.08	1832
1833	−0.64	1.45	−1.68	−0.10	2.72	0.66	−0.13	−1.31	−1.41	0.24	0.16	2.21	1833
1834	3.73	0.48	1.16	−0.48	1.59	1.20	1.29	0.76	0.70	0.10	0.45	0.35	1834
1835	0.82	0.81	−0.22	0.30	−0.12	0.71	0.87	1.09	0.21	−0.90	0.05	−1.76	1835
1836	0.80	−0.99	0.94	−1.12	−1.28	0.48	0.18	−1.11	−1.50	−1.14	−0.55	0.28	1836
1837	0.73	0.74	−2.22	−2.79	−2.01	0.04	0.05	−0.16	−0.75	0.21	−0.57	1.17	1837
1838	−2.93	−2.57	0.18	−1.50	−0.88	0.02	−0.31	−0.42	−0.92	0.10	−0.68	−0.03	1838
1839	0.73	0.14	−1.08	−2.48	−1.24	0.66	−0.35	−0.73	−1.06	−0.52	0.67	−0.21	1839
1840	1.27	−0.50	−1.97	−0.01	0.14	1.02	−0.77	0.73	−1.10	−1.32	0.60	−2.41	1840
1841	−0.38	−1.41	2.58	0.61	2.08	2.17	−1.02	−0.02	0.16	−0.01	0.40	1.06	1841
1842	−1.02	0.81	1.47	−0.43	0.59	2.84	0.18	2.11	0.19	−1.70	0.36	2.50	1842
Means.	2.38	3.81	5.00	7.30	10.46	12.92	14.26	14.07	12.06	8.88	5.51	3.81	Means.

The numbers without sign must be subtracted; those with the sign — must be added.

XCII.

Scotland. — Kinfauns Castle.

For Reducing the Monthly and Yearly Means of Single Years to the Means derived from Series of Years.

Degrees of Reaumur.

Year.	Jan.	Feb.	March.	April.	May.	June.	July.	Aug.	Sept.	Oct.	Nov.	Dec.	Year.
	°	°	°	°	°	°	°	°	°	°	°	°	
1814	-4.71	-1.63	-1.73	0.56	-2.46	-2.36	-0.60	-1.12	-0.34	-1.16	-1.61	-1.46	1814
1815	-1.69	1.13	0.16	-0.03	0.95	0.24	-0.12	0.23	0.27	0.74	-1.68	-2.47	1815
1816	-0.24	-1.60	-1.49	-1.58	-0.74	-0.60	-0.96	-0.53	-0.99	-0.24	-1.12	-1.84	1816
1817	1.60	1.29	-0.44	0.33	-1.12	0.80	-0.43	-1.11	0.18	-2.22	1.78	-1.38	1817
1818	0.51	-1.03	-1.41	-1.59	1.03	1.43	0.85	-0.22	-0.15	2.34	2.54	0.25	1818
1819	0.85	-0.88	0.67	-0.20	-0.36	-0.85	0.07	2.00	0.30	-0.32	-2.35	-2.60	1819
1820	-2.43	0.95	0.33	1.10	0.20	-0.12	0.39	-0.26	-0.36	-1.20	0.15	0.36	1820
1821	0.55	0.97	0.26	1.12	-1.09	-0.45	-0.01	0.84	1.44	0.83	0.38	0.73	1821
1822	1.85	1.28	1.08	0.79	0.97	2.04	0.50	0.26	-0.81	0.48	1.38	-0.61	1822
1823	-0.91	-1.69	-0.16	-0.60	0.63	-1.01	-0.92	-0.85	-0.15	-0.56	2.02	-0.04	1823
1824	2.64	1.20	-0.56	0.39	0.18	0.26	0.43	0.03	0.24	-2.16	-0.16	0.35	1824
1825	1.94	0.84	0.45	0.82	-0.09	0.31	1.59	1.53	1.85	1.79	-0.32	0.80	1825
1827	0.68	-0.77	0.02	0.73	0.51	0.38	0.16	0.37	1.48	2.48	-0.99	2.23	1827
1828	2.50	1.44	1.63	0.69	1.20	1.23	0.93	1.03	1.23	1.10	2.05	2.73	1828
1829	-0.38	0.96	0.42	-0.48	0.87	1.00	-0.12	-0.44	-1.02	0.34	-0.19	0.02	1829
1830	0.40	-0.22	2.07	0.87	0.60	-0.63	0.50	-1.13	0.11	1.33	0.92	-0.89	1830
1832	1.91	1.27	0.92	1.22	-0.19	0.50	0.24	0.93	1.35	1.53	-0.56	0.40	1832
1833	-1.40	0.51	-0.41	0.32	2.79	0.59	0.67	-0.98	-0.24	0.53	0.12	0.57	1833
1834	2.23	0.97	1.05	0.51	1.01	0.53	0.93	0.34	0.28	0.49	0.14	0.57	1834
1835	-0.27	0.72	-0.08	0.23	-0.58	0.20	-0.17	1.09	-0.10	-1.10	-0.31	-0.34	1835
1836	0.59	-0.67	-0.70	-0.81	0.10	-0.54	-1.16	-1.09	-1.67	-0.86	-0.94	-0.05	1836
1837	-0.07	0.20	-2.26	-2.35	-1.70	-0.05	0.52	-1.13	-1.32	0.23	-1.18	1.74	1837
1838	-2.58	-4.61	-0.83	-1.44	-1.75	-1.03	-0.04	-0.24	-0.53	-0.55	2.73	0.48	1838
1839	-0.90	-0.79	-1.56	-1.24	-1.18	-0.45	-0.34	-0.79	-0.64	-0.17	0.11	-0.35	1839
1840	0.65	-0.26	-0.07	1.00	-0.72	-0.40	-1.30	0.21	-1.29	-0.63	-0.17	-0.58	1840
1841	-2.19	-0.09	2.25	-0.28	0.51	-1.07	-0.83	-0.20	0.51	-1.52	-1.94	-0.49	1841
1842	-1.17	0.49	0.35	-0.07	0.48	0.02	-0.83	1.24	0.32	-1.52	-0.81	1.81	1842
Means.	1.77	2.74	3.87	5.71	8.13	10.58	11.76	11.28	9.52	6.72	4.35	2.96	Means.

XCIII. Finland. — Torneå.

Year.	Jan.	Feb.	March.	April.	May.	June.	July.	Aug.	Sept.	Oct.	Nov.	Dec.	Year.
1801	. .	. .	. .	. .	. .	. .	. .	. .	. .	. .	-0.01	-1.67	1801
1802	-0.57	-0.17	-0.15	0.10	-2.88	-0.66	-2.03	-1.60	-1.60	1.30	-2.10	-4.06	1802
1803	-3.50	-0.90	-0.13	1.57	1.69	-0.44	-0.58	0.93	-0.90	1.18	0.71	-3.67	1803
1804	-2.50	-4.82	-2.34	1.99	1.50	-0.97	0.78	-0.70	-0.21	1.19	1.46	-4.01	1804
1805	3.36	-2.94	-1.15	-0.79	-1.56	-2.90	-1.03	0.62	-1.34	-4.62	-2.83	-2.98	1805
1806	2.91	1.91	-0.03	2.02	1.00	-1.18	-1.90	2.00	1.20	0.13	-0.97	0.74	1806
1807	-3.40	1.94	-1.25	-2.57	-1.93	-0.61	0.34	0.89	-1.41	-2.30	-0.20	-0.92	1807
1808	1.80	-1.50	0.19	-2.31	1.14	2.65	0.58	-0.11	-0.51	3.53	2.24	-3.74	1808
1809	-7.19	-3.99	-2.74	-3.78	-1.91	0.62	-0.50	1.16	-0.34	-0.25	-1.67	8.07	1809
1810	-2.18	-2.36	-2.41	-2.45	-6.45	-0.68	-2.13	-0.68	-1.34	-1.23	-4.13	-2.20	1810
1811	2.98	-2.74	3.64	-2.04	-0.69	0.42	-0.91	-2.66	-1.05	-1.90	-0.10	-2.06	1811
1812	1.18	1.85	-3.37	-1.39	0.55	-2.94	-2.53	-1.20	-2.85	-0.78	-4.18	-1.15	1812
1813	1.32	1.15	1.70	1.88	-0.71	-1.58	1.87	0.08	1.88	-2.89	3.65	-1.43	1813

The numbers without sign must be subtracted; those with the sign — must be added.

XCIII.

Finland. — Torneå (*continued*).

For Reducing the Monthly and Yearly Means of Single Years to the Means derived from Series of Years.

Degrees of Reaumur.

Year.	Jan.	Feb.	March.	April.	May.	June.	July.	Aug.	Sept.	Oct.	Nov.	Dec.	Year.
	°	°	°	°	°	°	°	°	°	°	°	°	
1814	−7.01	2.71	−1.85	0.92	−0.59	2.44	4.65	4.46	2.60	0.44	−0.15	−4.50	1814
1815	1.22	3.16	0.66	5.27	3.22	5.58	4.70	5.03	4.02	3.38	4.30	4.82	1815
1816	2.27	−8.23	−4.25	0.50	−3.05	−0.12	0.18	−0.41	1.97	0.16	1.17	2.29	1816
1817	3.54	−2.13	−2.78	0.19	2.42	−1.14	0.65	−1.34	−0.36	−1.14	−0.07	−2.85	1817
1818	3.46	−3.34	−1.07	−2.61	−3.48	−0.92	2.98	−2.55	0.09	1.08	2.89	5.83	1818
1819	4.47	−0.15	−0.50	−2.07	0.23	1.46	2.90	2.22	1.04	−4.58	−3.62	−2.15	1819
1820	−5.74	−0.22	−0.63	−1.32	−0.73	1.62	0.13	−0.17	0.18	−2.17	−1.94	−2.67	1820
1821	−2.18	1.12	0.50	0.83	1.24	−3.70	−2.44	−1.32	−0.58	3.58	−1.52	−4.13	1821
1822	0.13	6.44	5.68	4.22	1.67	−1.39	−0.89	1.75	−0.14	0.47	−2.05	4.46	1822
1823	−4.01	−1.08	4.15	0.66	0.87	−0.43	−0.09	−0.73	−0.86	2.06	−1.38	1.26	1823
1824	0.71	4.20	1.75	−0.22	−0.40	0.29	−0.89	−0.73	1.25	−2.18	−1.01	−0.96	1824
1825	3.99	1.42	1.83	1.78	−0.29	−0.43	−1.53	−0.17	6.34	2.14	2.35	3.20	1825
1826	1.99	4.70	4.99	0.50	2.65	1.56	2.28	1.70	−0.70	2.67	3.23	3.74	1826
1827	0.03	0.00	0.59	−2.13	2.39	1.79	−2.00	−1.64	1.21	−1.53	−0.56	5.68	1827
1828	−0.50	−0.84	−1.77	−0.66	2.84	0.18	−1.73	−0.73	−2.86	1.18	0.50	1.69	1828
1829	1.26	−4.27	−2.69	−2.53	1.26	−0.31	0.30	−1.82	0.38	−1.78	−0.53	2.86	1829
1830	0.99	0.80	2.08	−0.54	−1.10	−0.66	−0.89	−1.73	−0.88	−0.03	3.44	−1.22	1830
1831	−3.98	−0.07	−2.31	2.01	0.98	1.98	0.81	0.79	−0.54	0.01	2.99	1.69	1831
1832	5.26	8.25	3.64	2.92	0.10	0.51	−1.11	−1.22	−3.67	2.86	. .	. .	1832
Means.	−12.55	−10.76	−7.19	−1.62	4.01	10.59	13.05	10.81	6.22	0.26	−6.27	−10.32	Means.

XCIV. North America. — Albany, N. Y.

Degrees of Reaumur.

Year.	Jan.	Feb.	March.	April.	May.	June.	July.	Aug.	Sept.	Oct.	Nov.	Dec.	Year.
1826	1.92	2.44	1.65	−1.02	3.23	1.07	0.72	1.09	1.57	1.46	0.81	0.35	1826
1827	−2.91	1.07	1.15	1.62	−0.02	0.05	0.55	0.08	0.43	1.14	−1.72	0.77	1827
1828	2.80	4.52	2.10	−0.88	0.76	2.66	−0.41	1.33	0.35	−0.31	0.76	3.17	1828
1829	−0.21	−2.27	−0.87	0.12	2.09	0.03	−1.54	−0.42	−1.93	0.92	0.50	3.63	1829
1830	0.28	−0.11	1.41	3.64	−0.21	−0.92	0.81	0.27	0.19	1.42	3.83	4.71	1830
1831	−1.30	−1.03	2.77	1.89	1.07	2.11	0.32	1.01	1.00	1.52	0.63	−4.94	1831
1832	0.18	−0.87	0.16	−1.29	−1.35	0.19	−0.34	−0.31	0.53	0.67	1.15	0.76	1832
1833	2.34	−1.34	−1.15	1.75	1.55	−2.35	−1.06	−1.47	−0.55	−0.55	−0.61	0.18	1833
1834	−1.18	3.73	0.67	0.68	−0.05	−1.12	1.59	−0.03	0.27	−1.31	−0.36	−1.13	1834
1835	−1.06	−1.50	−0.98	−1.59	−0.57	−0.34	−0.43	−0.90	−2.14	1.45	0.31	−3.06	1835
1836	−0.35	−3.89	−3.48	−2.27	−0.95	−1.30	0.20	−2.39	−0.39	−3.06	−0.62	−0.92	1836
1837	−3.40	−0.72	−1.94	−2.02	−1.23	0.07	−0.95	−0.98	−0.60	−0.89	0.33	−0.49	1837
1838	3.34	−4.01	0.97	−3.07	−1.26	1.78	0.31	0.27	0.36	−0.68	−1.47	−2.11	1838
1839	−0.25	1.62	0.14	0.79	−0.79	−1.79	0.15	−0.14	0.41	0.99	−0.94	−0.19	1839
1840	−3.32	3.14	0.60	1.32	0.96	−0.14	0.94	0.81	−0.91	0.28	0.28	−1.26	1840
1841	1.95	−0.72	−1.19	−2.58	−1.13	1.90	0.	1.23	0.88	−1.72	−0.49	0.86	1841
1842	2.03	3.15	2.06	0.62	−1.96	−0.98	0.28	0.13	−1.09	−0.12	−1.00	−1.69	1842
1843	2.65	−3.06	−4.26	−0.62	−0.62	−0.64	−0.55	0.64	0.85	−1.24	−1.11	0.93	1843
1844	−3.74	−0.15	0.27	2.97	0.47	−0.29	−0.60	−0.19	0.72	0.02	−0.20	0.47	1844
Means.	−3.58	−3.08	1.28	7.04	12.33	16.02	17.80	16.86	13.06	7.64	2.70	−1.65	Means.

The numbers without sign must be subtracted; those with the sign — must be added.

XCV.

North America. — Salem, Mass.

For Reducing the Monthly and Yearly Means of Single Years to the Means derived from Series of Years.

Degrees of Reaumur.

Year.	Jan.	Feb.	March.	April.	May.	June.	July.	Aug.	Sept.	Oct.	Nov.	Dec.	Year.
	°	°	°	°	°	°	°	°	°	°	°	°	
1787	0.40	−1.37	0.24	−0.24	−0.61	−0.84	−1.53	−0.28	−1.13	−1.00	0.58	0.07	1787
1788	−1.38	−2.15	−0.32	−0.47	−0.28	−1.39	0.14	−0.17	0.87	−1.00	2.03	−1.60	1788
1789	0.17	−2.81	−0.65	−0.47	−1.94	0.61	−0.31	0.05	−0.47	−2.56	0.47	1.18	1789
1790	1.17	−1.04	−1.32	−1.47	−0.50	−0.50	−0.75	−1.50	−1.02	−0.56	−0.97	−2.82	1790
1791	0.17	−1.48	0.90	0.64	1.50	1.16	−0.08	0.16	−0.69	−2.23	−0.42	0.07	1791
1792	−2.94	−0.37	1.79	0.87	1.61	−0.84	−0.64	−0.28	−1.80	0.77	0.92	−1.15	1792
1793	1.03	0.70	1.42	1.51	2.55	2.07	0.59	0.75	0.37	−0.09	0.07	−0.10	1793
1794	0.95	−0.25	1.91	1.19	1.16	0.11	0.52	0.58	0.75	−1.26	−0.16	4.35	1794
1795	0.20	−0.50	0.54	0.21	0.39	0.12	−0.31	1.85	1.04	1.24	0.36	1.51	1795
1796	1.18	0.12	−0.37	1.17	−0.11	0.40	0.39	0.80	−0.06	−0.55	−1.26	−3.02	1796
1797	−1.15	2.24	0.55	−0.26	−1.25	0.41	1.40	−0.45	−0.64	−0.83	−1.72	−2.52	1797
1798	0.68	−0.89	0.54	0.76	1.44	0.60	0.46	2.29	0.83	0.81	−1.57	−3.03	1798
1799	0.28	0.08	0.31	0.51	0.63	0.58	0.45	0.99	0.27	−0.16	−0.53	−0.53	1799
1800	0.31	0.24	−0.31	1.92	−0.12	1.22	1.15	0.11	0.04	0.43	−0.93	1.63	1800
1801	0.40	0.46	1.51	0.21	1.69	0.08	0.35	0.49	1.41	0.96	0.17	0.30	1801
1802	3.79	−0.16	0.76	0.31	−1.34	0.13	0.13	0.88	1.19	1.87	1.23	1.19	1802
1803	1.12	2.15	0.67	0.38	−0.81	0.53	−0.08	1.09	−0.24	0.96	−0.71	1.99	1803
1804	−0.48	0.08	−0.48	−0.98	1.55	0.20	−0.25	−0.44	0.28	−1.05	0.16	−1.76	1804
1805	−1.46	1.02	1.92	1.45	0.91	0.11	1.40	0.82	1.23	−0.82	0.13	3.24	1805
1806*	0.48	1.60	−1.83	−2.28	−0.44	−0.19	−1.12	−0.77	−0.52	−0.04	0.15	−0.06	1806
1807	−1.05	−1.13	−1.30	−0.31	−0.80	−0.62	0.05	0.00	−1.08	0.22	−0.65	2.45	1807
1808	0.13	1.41	1.55	0.37	−0.74	0.04	−0.15	−0.86	0.54	−0.72	0.69	0.72	1808
1809	−1.15	−1.73	−1.36	0.31	−0.24	−0.42	−1.90	−0.76	−0.95	3.00	−2.19	2.04	1809
1810	0.11	0.95	−0.68	0.70	0.84	0.04	−0.93	−0.39	0.46	−0.12	−0.24	−0.34	1810
1811	0.30	0.14	1.69	−0.01	0.65	0.43	0.16	0.14	0.58	1.74	0.67	−0.34	1811
1812	−1.51	−1.16	−2.68	−1.05	−3.22	−2.04	−2.13	−1.64	−2.07	−0.30	−0.90	−0.73	1812
1813	−1.09	−0.34	−2.55	0.08	−1.46	−0.95	−1.17	0.44	1.02	−0.62	0.83	−0.70	1813
1814	−0.73	0.80	−0.51	1.08	0.76	−1.58	−0.30	−0.94	−0.57	−0.07	0.39	−1.78	1814
1815	−0.93	−1.98	0.28	−1.47	−1.49	−0.16	1.12	−1.82	−0.50	−0.69	1.07	−0.45	1815
1816	−0.16	0.07	−2.14	−0.44	−1.36	−2.36	−2.49	−1.31	−1.77	0.17	1.79	0.31	1816
1817	−0.71	−3.48	−1.43	−0.73	−0.44	−1.65	−0.52	−0.76	0.18	−0.70	0.78	0.68	1817
1818	−0.51	−3.56	0.14	−2.31	−0.42	1.17	0.85	−0.01	−0.84	0.61	1.92	−1.94	1818
1819	2.45	4.91	−2.30	−1.06	−0.23	1.33	0.64	0.59	1.63	0.64	1.26	−0.43	1819
1820	−1.51	1.00	−0.22	−0.07	−0.23	0.51	1.95	0.26	1.52	−0.17	−0.98	−2.49	1820
1821	−2.75	1.50	−0.80	−0.97	−0.37	0.36	−1.08	0.83	−0.11	−0.05	0.42	−1.31	1821
1822	−1.60	−0.50	1.64	−0.87	1.77	0.09	0.44	0.06	1.84	0.75	0.96	0.12	1822
1823	0.37	−1.99	−0.99	0.20	−1.19	−0.42	−0.19	0.35	−1.63	−0.58	−1.72	0.52	1823
1824	2.28	0.47	−0.11	0.62	−0.84	−0.59	−0.14	−1.08	0.12	0.21	−0.61	1.43	1824
1825	1.30	1.27	2.16	1.49	0.69	1.74	2.36	−0.12	−1.05	0.70	−0.14	0.62	1825
1826	0.96	1.11	0.10	−1.05	2.95	0.04	1.56	−0.13	0.78	0.23	0.19	0.55	1826
1827	−1.49	0.52	0.64	1.56	−0.03	−0.60	−0.35	−0.82	−0.28	1.13	−2.74	0.01	1827
1828	2.42	4.05	1.10	−0.97	−0.68	1.06	0.36	0.96	0.37	−0.19	1.17	2.04	1828
Means.	−2.84	−1.85	1.54	6.36	11.05	15.61	17.97	17.17	13.80	8.56	3.53	−0.63	Means.

The numbers without sign must be subtracted; those with the sign — must be added.

XCVI.

ICELAND. — REIKIAVIK.

For Reducing the Monthly and Yearly Means of Single Years to the Means derived from Series of Years.

Degrees of Reaumur.

Year.	Jan.	Feb.	March.	April.	May.	June.	July.	Aug.	Sept.	Oct.	Nov.	Dec.	Year.
	°	°	°	°	°	°	°	°	°	°	°	°	
1823	1.80	−0.56	0.40	2.09	−0.60	0.06	2.44	1.76	0.84	−1.50	0.18	−0.86	1823
1824	−0.32	0.61	−0.05	2.16	2.95	4.63	3.12	1.53	−0.73	−2.37	−3.64	−3.99	1824
1825	−1.07	−0.40	3.04	0.98	0.50	0.33	1.70	0.66	2.34	1.68	−0.81	−0.92	1825
1826	−0.19	2.84	2.15	−0.79	1.58	−1.10	−0.75	−0.18	1.24	1.12	0.36	1.17	1826
1827	−0.72	1.93	−3.80	−0.86	0.67	0.86	0.14	1.73	0.64	2.29	2.26	0.88	1827
1828	1.98	2.48	1.54	1.29	2.37	0.53	3.15	3.98	3.07	3.26	0.94	2.77	1828
1829	1.02	−0.09	0.20	0.56	0.79	0.26	1.21	2.21	−0.20	−1.16	0.03	1.86	1829
1830	1.89	−0.58	−1.22	−0.72	2.44	0.52	−0.80	0.68	0.85	2.09	−0.35	−2.60	1830
1831	0.28	−0.95	2.58	1.39	−1.76	1.44	−1.89	−1.85	−0.37	0.95	−0.76	1.45	1831
1832	0.71	−0.48	−1.77	0.17	−2.20	−1.87	−2.80	−2.94	−2.59	−0.42	1.22	−0.29	1832
1833	1.41	−0.13	1.93	−0.21	−0.57	−0.40	−1.96	−2.14	−1.22	−0.79	0.31	−1.64	1833
1834	−0.43	0.10	0.73	0.14	−1.35	−1.99	−1.81	−2.41	−1.44	−1.13	0.22	2.76	1834
1835	−4.08	−1.92	−1.55	−1.32	−2.35	−1.97	−1.62	−0.38	−0.64	−2.41	1.58	1.30	1835
1836	−1.86	−3.24	−2.00	−3.01	−0.37	−0.94	−0.59	−2.68	−1.80	−1.67	−1.52	−1.95	1836
1837	−0.42	0.43	−2.23	−1.91	−2.07	−0.32	0.40	..	..	..	..	..	1837
Means.	−1.00	−1.60	−1.07	1.84	5.54	8.67	10.78	9.27	6.42	2.19	−0.60	−1.15	Means.

XCVII. GREENLAND. — GODTHAAB.

Degrees of Reaumur.

Year.	Jan.	Feb.	March.	April.	May.	June.	July.	Aug.	Sept.	Oct.	Nov.	Dec.	Year.
1796	..	..	..	..	..	..	..	..	..	−2.52	1.51	2.19	1796
1797	0.91	−2.08	−0.73	−1.96	1.14	0.27	1.40	1.31	0.77	1.02	2.22	0.87	1797
1798	−1.30	0.53	3.98	0.08	0.37	−0.39	0.39	0.07	−0.37	−0.67	0.83	−0.08	1798
1799	−0.40	3.08	−1.87	0.47	0.37	−0.71	−0.47	−0.72	0.62	−0.43	−0.91	4.72	1799
1800	2.75	0.22	2.32	−0.68	1.52	1.05	0.35	0.88	−0.42	0.48	0.05	0.07	1800
1801	−0.86	2.63	0.00	−1.00	−2.86	−1.61	0.89	0.92	−0.39	0.19	0.22	1.94	1801
1802	1.85	−2.99	−3.76	−2.68	−0.44	..	..	..	..	..	..	..	1802
1816	..	..	..	..	..	..	0.09	−0.98	−0.12	−0.15	−0.01	−6.91	1816
1817	−1.55	−2.46	−4.17	0.37	−1.32	−0.79	−1.63	−0.28	−0.41	−1.65	−0.52	−1.73	1817
1818	−5.58	−5.13	−4.00	2.56	−0.90	−0.84	0.52	0.15	−0.71	−1.97	−1.82	−0.42	1818
1819	−2.74	0.94	−0.35	0.98	−0.91	−0.97	−3.78	−2.29	−2.30	1.78	1.38	3.15	1819
1820	4.16	0.14	0.35	−2.15	0.97	0.66	−0.96	−1.57	−0.72	−0.06	1.60	1.19	1820
1821	0.04	0.42	1.30	1.00	−0.07	0.68	..	..	..	..	..	..	1821
1841	..	..	..	..	..	..	..	..	0.45	0.14	−0.27	0.23	1841
1842	1.13	−1.15	−1.12	1.56	2.03	0.37	0.89	0.34	1.39	1.95	−0.37	−1.37	1842
1843	0.11	4.74	4.65	2.18	1.18	1.16	1.52	0.72	1.57	1.66	−2.89	−3.93	1843
1844	−0.13	0.40	−0.51	−3.10	−1.29	0.79	0.78	1.39	0.66	0.19	−1.08	0.01	1844
1845	1.54	0.76	3.98	2.34	0.24	0.32	..	..	..	..	..	..	1845
Means.	−8.72	−8.64	−7.29	−4.44	0.07	3.15	4.41	3.93	1.62	−0.96	−4.47	−6.45	Means.

The numbers without sign must be subtracted; those with the sign — must be added.

CORRECTIONS

FOR

FORCE OF VAPOR AND RELATIVE HUMIDITY.

HOURLY CORRECTIONS FOR PERIODIC VARIATIONS,

OR

TABLES

FOR REDUCING THE MEANS OF THE OBSERVATIONS TAKEN AT ANY HOUR OF THE DAY TO THE TRUE MEAN FORCE OF VAPOR AND RELATIVE HUMIDITY OF THE DAY, OF THE MONTH, AND OF THE YEAR.

ENGLAND.—GREENWICH. *Lat.* 51° 29′ N.; *Long.* 0° 0′.

Corrections to be applied to the Means of the Hours of Observation, or Sets of Hours, to obtain the true Mean *Force of Vapor* for the respective Months. (GLAISHER.)

English Inches.

Hours.	Jan.	Feb.	March.	April.	May.	June.	July.	Aug.	Sept.	Oct.	Nov.	Dec.	Mean.
	Inch.	Inch.	Inch.	Inch.	Inch.	Inch.	Inch.	Inch.	Inch.	Inch.	Inch.	Inch.	Inch.
Midn. . .	.006	.006	.008	.017	.026	.031	.028	.025	.024	.018	.010	.009	.017
1	.011	.008	.010	.021	.028	.037	.031	.031	.030	.020	.012	.010	.021
2	.015	.010	.011	.024	.031	.043	.036	.035	.035	.021	.015	.010	.024
3	.015	.011	.013	.027	.032	.048	.038	.039	.037	.023	.017	.011	.026
4	.015	.013	.015	.029	.031	.047	.037	.040	.040	.025	.019	.011	.027
5	.015	.014	.016	.029	.027	.037	.031	.038	.040	.023	.021	.011	.025
6	.014	.015	.016	.025	.019	.022	.019	.029	.033	.021	.021	.010	.020
7	.013	.014	.014	.016	.007	.008	.007	.014	.022	.018	.018	.009	.013
8	.010	.010	.010	.005	-.005	-.004	-.004	.000	.010	.011	.012	.007	.005
9	.007	.006	.005	.005	-.016	-.015	-.014	-.012	-.005	.005	.005	.005	-.002
10	.002	.000	-.003	-.013	-.024	-.027	-.019	-.021	-.019	-.005	-.004	.001	-.010
11	-.004	-.005	-.007	-.020	-.028	-.036	-.025	-.027	-.027	-.009	-.010	-.004	-.017
Noon. . .	-.007	-.009	-.012	-.026	-.030	-.042	-.029	-.030	-.030	-.015	-.017	-.007	-.021
1	-.008	-.013	-.013	-.027	-.030	-.045	-.033	-.032	-.030	-.018	-.019	-.008	-.023
2	-.007	-.015	-.013	-.027	-.028	-.043	-.034	-.034	-.029	-.017	-.020	-.008	-.023
3	-.007	-.012	-.012	-.025	-.026	-.039	-.033	-.031	-.027	-.014	-.016	-.008	-.021
4	-.007	-.010	-.010	-.020	-.021	-.035	-.028	-.027	-.021	-.009	-.010	-.007	-.017
5	-.004	-.006	-.006	-.014	-.015	-.025	-.021	-.020	-.017	-.006	-.005	-.005	-.012
6	-.002	-.004	-.002	-.006	-.010	-.017	-.016	-.015	-.010	-.004	.000	-.003	-.007
7	-.001	-.001	.002	.001	-.004	-.007	-.007	-.006	-.003	.003	.004	-.001	-.002
8	.000	.001	.004	.005	.005	.005	.004	.004	.004	.005	.006	.001	.004
9	.000	.003	.005	.007	.013	.015	.010	.010	.008	.008	.008	.004	.007
10	.001	.004	.007	.010	.017	.023	.017	.015	.013	.011	.009	.005	.011
11	.002	.005	.008	.014	.022	.029	.024	.020	.018	.014	.010	.006	.014
6. 6	.006	.005	.007	.009	.005	.003	.001	.007	.012	.008	.010	.004	.006
7. 7	.006	.006	.008	.009	.001	.000	.000	.004	.009	.011	.011	.004	.005
8. 8	.005	.005	.007	.005	.000	.000	.000	.002	.007	.008	.009	.004	.005
9. 9	.003	.004	.005	.006	-.002	.000	-.002	-.001	.002	.006	.007	.004	.003
10.10	.001	.002	.002	-.002	-.003	-.002	-.001	-.003	-.003	.003	.002	.003	.000
7. 2. 9	.002	.001	.002	-.001	-.003	-.007	-.006	-.003	.000	.003	.002	.002	-.001
6. 2. 8	.002	.000	.002	.001	-.001	-.005	-.004	-.000	.003	.003	.002	.001	.000
6. 2.10	.003	.001	.003	.003	.002	.001	.001	.003	.006	.005	.003	.002	.003
6. 2. 6	.002	-.001	.000	-.003	-.006	-.013	-.010	-.007	-.002	.000	.000	-.000	-.003
7. 2	.003	-.000	.000	-.005	-.011	-.017	-.014	-.010	-.003	.000	-.001	.000	-.005
8. 2	.001	-.002	-.001	-.011	-.017	-.023	-.019	-.017	-.009	-.003	-.004	-.000	-.009
8. 1	.001	-.001	-.001	-.011	-.017	-.025	-.018	-.016	-.010	-.004	-.004	-.000	-.009
7. 1	.002	.000	.000	-.005	-.012	-.018	-.013	-.009	-.004	-.000	-.000	.000	-.005
9.12.3.9	-.002	-.003	-.003	-.010	-.015	-.020	-.016	-.016	-.013	-.004	-.005	-.001	-.009

The numbers without sign must be added; those with the sign — must be subtracted.

England.—Greenwich. *Lat.* 51° 29′ N.; *Long.* 0° 0′.

Corrections to be applied to the Means of the Hours of Observation, or Sets of Hours, to obtain the true Mean *Humidity* for the respective Months. (Glaisher.)

Thousandths.

Hours.	Jan.	Feb.	March.	April.	May.	June.	July.	Aug.	Sept.	Oct.	Nov.	Dec.	Mean.
Midn. . .	-.013	-.021	-.063	-.095	-.087	-.105	-.091	-.096	-.080	-.053	-.018	-.011	-.061
1	.002	-.021	-.065	-.106	-.100	-.114	-.095	-.104	-.080	-.059	-.009	-.012	-.064
2	.004	-.026	-.066	-.116	-.108	-.125	-.107	-.113	-.085	-.066	-.011	-.017	-.069
3	-.003	-.033	-.067	-.123	-.113	-.132	-.116	-.117	-.091	-.070	-.020	-.019	-.075
4	-.013	-.036	-.068	-.126	-.114	-.138	-.120	-.123	-.097	-.075	-.030	-.024	-.080
5	-.019	-.035	-.066	-.125	-.106	-.139	-.120	-.123	-.098	-.077	-.030	-.024	-.080
6	-.021	-.034	-.063	-.112	-.085	-.107	-.097	-.107	-.097	-.071	-.033	-.026	-.071
7	-.020	-.030	-.055	-.080	-.059	-.065	-.055	-.061	-.080	-.058	-.031	-.025	-.052
8	-.020	-.020	-.035	-.065	-.024	-.015	-.005	-.020	-.047	-.037	-.021	-.018	-.027
9	-.017	-.007	-.003	-.034	.018	.035	.041	.030	.000	-.009	-.008	-.007	.003
10	-.004	.009	.031	-.015	.051	.078	.080	.070	.042	.025	.008	.008	.032
11	.011	.028	.060	.022	.083	.100	.104	.102	.082	.060	.027	.022	.058
Noon. . .	.031	.045	.084	.070	.110	.123	.114	.127	.115	.088	.040	.033	.082
1	.054	.058	.100	.132	.126	.137	.119	.142	.131	.109	.050	.046	.100
2	.059	.065	.106	.151	.125	.135	.123	.145	.132	.113	.054	.048	.105
3	.048	.065	.104	.147	.118	.123	.121	.138	.126	.108	.047	.036	.098
4	.036	.053	.087	.128	.108	.113	.111	.120	.103	.089	.032	.024	.084
5	.021	.032	.063	.110	.091	.099	.095	.100	.071	.055	.018	.013	.064
6	.007	.009	.038	.088	.074	.078	.062	.071	.044	.030	.005	.004	.042
7	-.005	-.010	.010	.059	.052	.049	.025	.036	.009	.007	-.005	-.003	.019
8	-.014	-.023	-.010	.020	.022	.010	-.015	.000	-.015	-.011	-.012	-.005	-.004
9	-.016	-.029	-.032	-.030	-.018	-.025	-.040	-.038	-.040	-.025	-.017	-.007	-.026
10	-.019	-.030	-.048	-.058	-.050	-.060	-.068	-.067	-.058	-.039	-.020	-.008	-.044
11	-.018	-.036	-.060	-.080	-.075	-.085	-.080	-.085	-.071	-.048	-.020	-.009	-.055
6. 6	-.007	-.012	-.012	-.012	-.005	-.015	-.017	-.018	-.027	-.020	-.014	-.011	-.015
7. 7	-.012	-.020	-.023	-.010	-.004	-.008	-.015	-.012	-.035	-.026	-.018	-.014	-.017
8. 8	-.017	-.021	-.023	-.022	-.001	-.003	-.010	-.010	-.031	-.024	-.016	-.011	-.016
9. 9	-.016	-.018	-.018	-.032	.000	.005	.000	-.004	-.020	-.017	-.012	-.007	-.012
10.10	-.011	-.010	-.009	-.037	.000	.009	.006	.001	-.008	-.007	-.006	.000	-.006
7. 2. 9	.008	.002	.006	.014	.016	.015	.009	.015	.004	.010	.002	.005	.009
6. 2. 8	.008	.003	.011	.019	.021	.013	.004	.013	.016	.010	.003	.006	.010
6. 2.10	-.006	.000	-.002	-.006	-.003	-.010	-.014	-.009	-.008	.001	.000	.005	-.003
6. 2. 6	.015	.013	.027	.042	.038	.035	.029	.036	.026	.024	.009	.009	.025
7. 2	.019	.017	.026	.036	.033	.035	.034	.042	.026	.027	.012	.011	.026
8. 2	.019	.022	.036	.043	.050	.060	.059	.062	.042	.038	.016	.015	.039
8. 1	.017	.019	.032	.034	.051	.061	.057	.061	.042	.036	.014	.014	.037
7. 1	.017	.014	.023	.026	.033	.036	.032	.041	.025	.026	.009	.010	.024
9.12.3.9	.011	-.018	.038	.038	.032	.064	.059	.064	.050	.040	.016	.014	.037

The numbers without sign must be added; those with the sign — must be subtracted.

METEOROLOGICAL TABLES.

VI.

MISCELLANEOUS TABLES,

USEFUL IN

TERRESTRIAL PHYSICS AND METEOROLOGY.

CONTENTS.

I.

POSITIONS OF THE PRINCIPAL OBSERVATORIES.

[From the American Nautical Almanac.]

(North Latitudes and West Longitudes are considered as positive.)

Place.	Latitude.	Longitude from Washington in Time.	Longitude from Washington in Arc.	Longitude from Greenwich in Arc.
	° ′ ″	h. m. s.	° ′ ″	° ′ ″
Åbo,	+60 26 56.8	− 6 37 20.0	260 40 0.6	337 42 48.6
Altona,	53 32 45.3	5 47 57.4	273 0 39.8	350 3 27.8
Athens,	37 58 20	6 43 6.4	259 13 24.2	336 16 12.2
Berlin,	52 30 16.7	6 1 46.1	269 33 28.1	346 36 16.1
Bilk,	51 12 25	5 35 16.1	276 10 58.1	353 13 46.1
Bonn,	50 43 45.0	5 36 35.7	275 51 5.1	352 53 53.1
Breslau,	51 6 56.0	6 16 21.2	265 54 42.0	342 57 30.0
Brussels,	50 51 10.7	5 25 38.8	278 35 18.0	355 38 6.0
Cambridge (Eng.),	52 12 51.8	5 8 34.7	282 51 18.9	359 54 6.9
Cambridge (Mass.),	+42 22 48.6	0 23 41.5	354 4 36.9	71 7 24.9
Cape of Good Hope,	−33 56 3	6 22 7.2	264 28 12.3	341 31 0.3
Christiania,	+59 54 43.7	− 5 51 6.0	272 13 30.6	349 16 18.6
Cincinnati,	39 5 54	+ 0 29 46.9	7 26 42.8	84 29 30.8
Copenhagen,	55 40 53.0	− 5 58 30.5	270 22 22.5	347 25 10.5
Cracow,	50 3 50.0	6 28 2.4	262 59 23.4	340 2 11.4
Dorpat,	58 22 47.1	6 55 5.8	256 13 33.6	333 16 21.6
Dublin,	53 23 13	4 42 49.2	289 17 42.0	6 20 30.0
Durham,	54 46 6.4	5 1 53.2	284 31 42.0	1 34 30.0
Edinburgh,	55 57 23.2	4 55 28.2	286 7 57.0	3 10 45.0
Florence,	43 46 40.8	5 53 12.9	271 41 47.1	348 44 35.1
Geneva,	46 11 58.8	− 5 32 48.9	276 47 46.8	353 50 34.8
Georgetown,	38 54 26.1	+ 0 0 6.2	0 1 33.0	77 4 21.0
Göttingen,	51 31 47.9	− 5 47 57.3	273 0 40.5	350 3 28.5
Gotha,	50 56 5.2	5 51 6.9	272 13 17.1	349 16 5.1
Greenwich,	51 28 38.2	5 8 11.2	282 57 12.0	0 0 0
Hamburg,	53 33 7	− 5 48 4.8	272 58 48.6	350 1 36.6
Hudson,	41 14 42.6	+ 0 17 32.1	4 23 0.9	81 25 48.9
Kasan,	55 47 23.1	− 8 24 43.1	233 49 13.1	310 52 1.1
Königsberg,	54 42 50.4	6 30 11.6	262 27 6.6	339 29 54.6
Kremsmünster,	48 3 23.8	6 4 44.6	268 48 50.7	345 51 38.7
Leipsic,	51 20 20.7	5 57 39.7	270 35 4.5	347 37 52.5
Leyden,	52 9 28.2	5 26 8.6	278 27 50.6	355 30 38.6
Liverpool,	53 24 47.7	4 56 11.1	285 57 13.7	3 0 1.7
London,	51 31 29.8	5 7 34.1	283 6 28.5	0 9 16.5
Madras,	+13 4 9.2	−10 29 8.2	202 42 57.0	279 45 45.0

Place.	Latitude.	Longitude from Washington in Time.	Longitude from Washington in Arc.	Longitude from Greenwich in Arc.
	° ′ ″	h. m. s.	° ′ ″	° ′ ″
Mannheim,	+49 29 12.9	−5 42 2.7	274 29 19.5	351 32 7.5
Markree,	54 10 31.7	4 34 22.8	291 24 18.0	8 27 6.0
Marseilles,	43 17 49	5 29 40.2	277 34 57.2	354 37 45.2
Milan,	45 28 0.7	5 44 57.8	273 45 32.4	350 48 20.4
Modena,	44 38 52.8	5 51 55.2	272 1 12.5	349 4 0.5
Moscow,	55 45 19.8	7 38 28.5	245 22 52.7	322 25 40.7
Munich,	48 8 45	5 54 37.6	271 20 35.4	348 23 23.4
Naples,	40 51 46.6	6 5 12.1	268 41 58.1	345 44 46.1
Olmütz,	49 35 40.0	6 17 11.3	265 42 10.5	342 44 58.5
Oxford,	51 45 36.0	5 3 8.6	284 12 51.0	1 15 39.0
Padua,	45 24 2.5	5 55 40.2	271 4 56.6	348 7 44.6
Palermo,	+38 6 44	−6 1 36.7	269 35 50.1	346 38 38.1
Paramatta,	−33 48 49.8	+8 47 42.6	131 55 38.3	208 58 26.3
Paris,	+48 50 13.2	−5 17 32.7	280 36 50.1	357 39 38.1
St. Petersburg,	59 56 29.7	7 9 24.7	252 38 49.8	329 41 37.8
Philadelphia,	39 57 7.5	0 7 33.6	358 6 35.4	75 9 23.4
Prague,	50 5 18.5	6 5 53.2	268 31 42.6	345 34 30.6
Pulkowa,	59 46 18.7	7 9 29.9	252 37 31.9	329 40 19.9
Rome,	41 53 54	5 58 5.9	270 28 31.5	347 31 19.5
San Fernando,	+36 27 45	4 43 22.1	289 9 29.1	6 12 17.1
Santiago,	−33 26 24.8	0 25 52.3	353 31 55.5	70 34 43.5
Senftenberg,	+50 5 10.1	6 14 1.1	266 29 43.1	343 32 31.1
Vienna,	48 12 35.5	6 13 43.7	266 34 4.1	343 36 52.1
Washington,	38 53 39.3	0 0 0	0 0 0	77 2 48.0
Wilna,	+54 40 59.1	−6 49 23.0	257 39 15.5	334 42 3.5

II. TO CONVERT PARTS OF THE EQUATOR IN ARC INTO SIDEREAL TIME, OR TO CONVERT TERRESTRIAL LONGITUDE IN ARC INTO TIME.

Degrees.											
Arc.	Time.	Arc.	Time.	Arc.	Time.	Arc.	Time.	Arc.	Time.	Arc.	Time.
°	h. m.	°	h. m.	°	h. m.	°	h. m.	°	h. m.	°	h. m.
1	0 4	41	2 44	81	5 24	121	8 4	161	10 44	201	13 24
2	0 8	42	2 48	82	5 28	122	8 8	162	10 48	202	13 28
3	0 12	43	2 52	83	5 32	123	8 12	163	10 52	203	13 32
4	0 16	44	2 56	84	5 36	124	8 16	164	10 56	204	13 36
5	0 20	45	3 0	85	5 40	125	8 20	165	11 0	205	13 40
6	0 24	46	3 4	86	5 44	126	8 24	166	11 4	206	13 44
7	0 28	47	3 8	87	5 48	127	8 28	167	11 8	207	13 48
8	0 32	48	3 12	88	5 52	128	8 32	168	11 12	208	13 52
9	0 36	49	3 16	89	5 56	129	8 36	169	11 16	209	13 56
10	0 40	50	3 20	90	6 0	130	8 40	170	11 20	210	14 0
11	0 44	51	3 24	91	6 4	131	8 44	171	11 24	211	14 4
12	0 48	52	3 28	92	6 8	132	8 48	172	11 28	212	14 8
13	0 52	53	3 32	93	6 12	133	8 52	173	11 32	213	14 12
14	0 56	54	3 36	94	6 16	134	8 56	174	11 36	214	14 16
15	1 0	55	3 40	95	6 20	135	9 0	175	11 40	215	14 20
16	1 4	56	3 44	96	6 24	136	9 4	176	11 44	216	14 24
17	1 8	57	3 48	97	6 28	137	9 8	177	11 48	217	14 28
18	1 12	58	3 52	98	6 32	138	9 12	178	11 52	218	14 32
19	1 16	59	3 56	99	6 36	139	9 16	179	11 56	219	14 36
20	1 20	60	4 0	100	6 40	140	9 20	180	12 0	220	14 40
21	1 24	61	4 4	101	6 44	141	9 24	181	12 4	221	14 44
22	1 28	62	4 8	102	6 48	142	9 28	182	12 8	222	14 48
23	1 32	63	4 12	103	6 52	143	9 32	183	12 12	223	14 52
24	1 36	64	4 16	104	6 56	144	9 36	184	12 16	224	14 56
25	1 40	65	4 20	105	7 0	145	9 40	185	12 20	225	15 0
26	1 44	66	4 24	106	7 4	146	9 44	186	12 24	226	15 4
27	1 48	67	4 28	107	7 8	147	9 48	187	12 28	227	15 8
28	1 52	68	4 32	108	7 12	148	9 52	188	12 32	228	15 12
29	1 56	69	4 36	109	7 16	149	9 56	189	12 36	229	15 16
30	2 0	70	4 40	110	7 20	150	10 0	190	12 40	230	15 20
31	2 4	71	4 44	111	7 24	151	10 4	191	12 44	231	15 24
32	2 8	72	4 48	112	7 28	152	10 8	192	12 48	232	15 28
33	2 12	73	4 52	113	7 32	153	10 12	193	12 52	233	15 32
34	2 16	74	4 56	114	7 36	154	10 16	194	12 56	234	15 36
35	2 20	75	5 0	115	7 40	155	10 20	195	13 0	235	15 40
36	2 24	76	5 4	116	7 44	156	10 24	196	13 4	236	15 44
37	2 28	77	5 8	117	7 48	157	10 28	197	13 8	237	15 48
38	2 32	78	5 12	118	7 52	158	10 32	198	13 12	238	15 52
39	2 36	79	5 16	119	7 56	159	10 36	199	13 16	239	15 56
40	2 40	80	5 20	120	8 0	160	10 40	200	13 20	240	16 0

TO CONVERT PARTS OF THE EQUATOR IN ARC INTO SIDEREAL TIME, OR TO CONVERT TERRESTRIAL LONGITUDE IN ARC INTO TIME.

Arc.	Time.	Arc.	Time.	Arc.	Time.	Arc.	Time.	Arc.	Time.	Arc.	Time.
					DEGREES.						
°	h. m.	°	h. m.	°	h. m.	°	h. m.	°	h. m.	°	h. m.
241	16 4	261	17 24	281	18 44	301	20 4	321	21 24	341	22 44
242	16 8	262	17 28	282	18 48	302	20 8	322	21 28	342	22 48
243	16 12	263	17 32	283	18 52	303	20 12	323	21 32	343	22 52
244	16 16	264	17 36	284	18 56	304	20 16	324	21 36	344	22 56
245	16 20	265	17 40	285	19 0	305	20 20	325	21 40	345	23 0
246	16 24	266	17 44	286	19 4	306	20 24	326	21 44	346	23 4
247	16 28	267	17 48	287	19 8	307	20 28	327	21 48	347	23 8
248	16 32	268	17 52	288	19 12	308	20 32	328	21 52	348	23 12
249	16 36	269	17 56	289	19 16	309	20 36	329	21 56	349	23 16
250	16 40	270	18 0	290	19 20	310	20 40	330	22 0	350	23 20
251	16 44	271	18 4	291	19 24	311	20 44	331	22 4	351	23 24
252	16 48	272	18 8	292	19 28	312	20 48	332	22 8	352	23 28
253	16 52	273	18 12	293	19 32	313	20 52	333	22 12	353	23 32
254	16 56	274	18 16	294	19 36	314	20 56	334	22 16	354	23 36
255	17 0	275	18 20	295	19 40	315	21 0	335	22 20	355	23 40
256	17 4	276	18 24	296	19 44	316	21 4	336	22 24	356	23 44
257	17 8	277	18 28	297	19 48	317	21 8	337	22 28	357	23 48
258	17 12	278	18 32	298	19 52	318	21 12	338	22 32	358	23 52
259	17 16	279	18 36	299	19 56	319	21 16	339	22 36	359	23 56
260	17 20	280	18 40	300	20 0	320	21 20	340	22 40	360	24 0
					MINUTES.						
′	m. s.	′	m. s.	′	m. s.	′	m. s.	′	m. s.	′	m. s.
1	0 4	11	0 44	21	1 24	31	2 4	41	2 44	51	3 24
2	0 8	12	0 48	22	1 28	32	2 8	42	2 48	52	3 28
3	0 12	13	0 52	23	1 32	33	2 12	43	2 52	53	3 32
4	0 16	14	0 56	24	1 36	34	2 16	44	2 56	54	3 36
5	0 20	15	1 0	25	1 40	35	2 20	45	3 0	55	3 40
6	0 24	16	1 4	26	1 44	36	2 24	46	3 4	56	3 44
7	0 28	17	1 8	27	1 48	37	2 28	47	3 8	57	3 48
8	0 32	18	1 12	28	1 52	38	2 32	48	3 12	58	3 52
9	0 36	19	1 16	29	1 56	39	2 36	49	3 16	59	3 56
10	0 40	20	1 20	30	2 0	40	2 40	50	3 20	60	4 0
					SECONDS.						
″	s.	″	s.	″	s.	″	s.	″	s.	″	s.
1	0.067	11	0.733	21	1.400	31	2.067	41	2.733	51	3.400
2	0.133	12	0.800	22	1.467	32	2.133	42	2.800	52	3.467
3	0.200	13	0.867	23	1.533	33	2.200	43	2.867	53	3.533
4	0.267	14	0.933	24	1.600	34	2.267	44	2.933	54	3.600
5	0.333	15	1.000	25	1.667	35	2.333	45	3.000	55	3.667
6	0.400	16	1.067	26	1.733	36	2.400	46	3.067	56	3.733
7	0.467	17	1.133	27	1.800	37	2.467	47	3.133	57	3.800
8	0.533	18	1.200	28	1.867	38	2.533	48	3.200	58	3.867
9	0.600	19	1.267	29	1.933	39	2.600	49	3.267	59	3.933
10	0.667	20	1.333	30	2.000	40	2.667	50	3.333	60	4.000

III. TO CONVERT SIDEREAL TIME INTO PARTS OF THE EQUATOR IN ARC, OR TO CONVERT TIME INTO TERRESTRIAL LONGITUDE IN ARC.

Hours.

Time.	Arc.	Time.	Arc	Time.	Arc.	Time.	Arc.	Time.	Arc.	Time.	Arc.
h.	°	h.	°	h.	°	h.	°	h.	°	h.	°
1	15	5	75	9	135	13	195	17	255	21	315
2	30	6	90	10	150	14	210	18	270	22	330
3	45	7	105	11	165	15	225	19	285	23	345
4	60	8	120	12	180	16	240	20	300	24	360

Minutes.

m.	° ′	m.	° ′	m.	° ′	m.	° ′	m.	° ′	m.	° ′
1	0 15	11	2 45	21	5 15	31	7 45	41	10 15	51	12 45
2	0 30	12	3 0	22	5 30	32	8 0	42	10 30	52	13 0
3	0 45	13	3 15	23	5 45	33	8 15	43	10 45	53	13 15
4	1 0	14	3 30	24	6 0	34	8 30	44	11 0	54	13 30
5	1 15	15	3 45	25	6 15	35	8 45	45	11 15	55	13 45
6	1 30	16	4 0	26	6 30	36	9 0	46	11 30	56	14 0
7	1 45	17	4 15	27	6 45	37	9 15	47	11 45	57	14 15
8	2 0	18	4 30	28	7 0	38	9 30	48	12 0	58	14 30
9	2 15	19	4 45	29	7 15	39	9 45	49	12 15	59	14 45
10	2 30	20	5 0	30	7 30	40	10 0	50	12 30	60	15 0

Seconds.

s.	′ ″	s.	′ ″	s.	′ ″	s.	′ ″	s.	′ ″	s.	′ ″
1	0 15	11	2 45	21	5 15	31	7 45	41	10 15	51	12 45
2	0 30	12	3 0	22	5 30	32	8 0	42	10 30	52	13 0
3	0 45	13	3 15	23	5 45	33	8 15	43	10 45	53	13 15
4	1 0	14	3 30	24	6 0	34	8 30	44	11 0	54	13 30
5	1 15	15	3 45	25	6 15	35	8 45	45	11 15	55	13 45
6	1 30	16	4 0	26	6 30	36	9 0	46	11 30	56	14 0
7	1 45	17	4 15	27	6 45	37	9 15	47	11 45	57	14 15
8	2 0	18	4 30	28	7 0	38	9 30	48	12 0	58	14 30
9	2 15	19	4 45	29	7 15	39	9 45	49	12 15	59	14 45
10	2 30	20	5 0	30	7 30	40	10 0	50	12 30	60	15 0

Tenths of Seconds.

s.	″	s.	″	s.	″	s.	″	s.	″	s.	″
0.01	0.15	0.18	2.70	0.35	5.25	0.52	7.80	0.69	10.35	0.86	12.90
0.02	0.30	0.19	2.85	0.36	5.40	0.53	7.95	0.70	10.50	0.87	13.05
0.03	0.45	0.20	3.00	0.37	5.55	0.54	8.10	0.71	10.65	0.88	13.20
0.04	0.60	0.21	3.15	0.38	5.70	0.55	8.25	0.72	10.80	0.89	13.35
0.05	0.75	0.22	3.30	0.39	5.85	0.56	8.40	0.73	10.95	0.90	13.50
0.06	0.90	0.23	3.45	0.40	6.00	0.57	8.55	0.74	11.10	0.91	13.65
0.07	1.05	0.24	3.60	0.41	6.15	0.58	8.70	0.75	11.25	0.92	13.80
0.08	1.20	0.25	3.75	0.42	6.30	0.59	8.85	0.76	11.40	0.93	13.95
0.09	1.35	0.26	3.90	0.43	6.45	0.60	9.00	0.77	11.55	0.94	14.10
0.10	1.50	0.27	4.05	0.44	6.60	0.61	9.15	0.78	11.70	0.95	[illegible].25
0.11	1.65	0.28	4.20	0.45	6.75	0.62	9.30	0.79	11.85	0.96	[illegible]4.40
0.12	1.80	0.29	4.35	0.46	6.90	0.63	9.45	0.80	12.00	0.97	[illegible]4.55
0.13	1.95	0.30	4.50	0.47	7.05	0.64	9.60	0.81	12.15	0.98	14.70
0.14	2.10	0.31	4.65	0.48	7.20	0.65	9.75	0.82	12.30	0.99	14.85
0.15	2.25	0.32	4.80	0.49	7.35	0.66	9.90	0.83	12.45	1.00	15.00
0.16	2.40	0.33	4.95	0.50	7.50	0.67	10.05	0.84	12.60		
0.17	2.55	0.34	5.10	0.51	7.65	0.68	10.20	0.85	12.75		

TABLE VIII.

Correction to be added to English Barometers for Capillary Action.

Diameter of Tube.	Correction for	
	Unboiled Tubes.	Boiled Tubes.
Inch.	Inch.	Inch.
0.60	0.004	0.002
0.50	0.007	0.003
0.45	0.010	0.005
0.40	0.014	0.007
0.35	0.020	0.010
0.30	0.028	0.014
0.25	0.040	0.020
0.20	0.060	0.029
0.15	0.088	0.044
0.10	0.142	0.070

TABLE IX.

GIVING THE CORRECTION

TO BE APPLIED TO

BAROMETERS WITH BRASS SCALES,

EXTENDING FROM

THE CISTERN TO THE TOP OF THE MERCURIAL COLUMN,

TO REDUCE THE

OBSERVATION TO THIRTY-TWO DEGREES FAHRENHEIT.

This Table is that adopted by the Committee of Physics and Meteorology of the Royal Society of London. It gives immediately the correction for each degree Fahrenheit, and for each half of an inch from 20 up to 31 inches.

Examples of Calculation.

Barometer: observed height - - - - - 30.231
Attached Thermometer - - - - - - +82.0

See, in the last page, the column of 30 inches; go down as far as the horizontal line corresponding with 82° in the first vertical column, which contains the temperature, you will find there the correction — .143. We have thus:

Barometer: observed height - - - - - 30.231
Substractive correction for 82.0° Fahr. - - - —0.143

Barometer at 32° Fahr. - - - - 30.088

Barometer: observed height - - - - - 29.743
Attached Thermometer 25° Fahr.
The column of 29.5 inches opposite to 25° Fahr. gives an *additive* correction of - - - - - - +0.009

Barometer at - - - - - - 29.752

It will be easy to apply also the correction for fractions of degree Fahr. Ex. gr.

Barometer: observed height - - - - - 28.358
Attached Thermometer 71.3.
In the column of 28.5 inches, we find that the difference between the correction for 71° and that for 72° is .003; dividing this difference proportionally to the fraction, we have for 71.03 a correction of .108+.001, or - - - - - - - - - - —0.109

And Barometer at 32° Fahr. - - - - - 28.239

TABLE IX.

Temp. F.	Inches.								Temp. F.
	20	20.5	21	21.5	22	22.5	23	23.5	
	+	+	+	+	+	+	+	+	
0°	.051	.053	.054	.055	.056	.058	.059	.060	0°
1	.049	.051	.052	.053	.054	.056	.057	.058	1
2	.048	.049	.050	.051	.052	.054	.055	.056	2
3	.046	.047	.048	.049	.050	.052	.053	.054	3
4	.044	.045	.046	.047	.048	.050	.051	.052	4
5	.042	.043	.044	.045	.046	.048	.049	.050	5
6	.040	.042	.042	.044	.044	.046	.047	.048	6
7	.039	.040	.041	.042	.042	.044	.044	.046	7
8	.037	.038	.039	.040	.041	.041	.042	.043	8
9	.035	.036	.037	.038	.039	.039	.040	.041	9
10	.033	.034	.035	.036	.037	.037	.038	.039	10
11	.031	.032	.033	.034	.035	.035	.036	.037	11
12	.030	.030	.031	.032	.033	.033	.034	.035	12
13	.028	.029	.029	.030	.031	.031	.032	.033	13
14	.026	.027	.027	.028	.029	.029	.030	.031	14
15	.024	.025	.026	.026	.027	.027	.028	.029	15
16	.022	.023	.024	.024	.025	.025	.026	.026	16
17	.021	.021	.022	.022	.023	.023	.024	.024	17
18	.019	.019	.020	.020	.021	.021	.022	.022	18
19	.017	.018	.018	.018	.019	.019	.020	.020	19
20	.015	.016	.016	.016	.017	.017	.018	.018	20
21	.014	.014	.014	.015	.015	.015	.015	.016	21
22	.012	.012	.012	.013	.013	.013	.013	.014	22
23	.010	.010	.010	.011	.011	.011	.011	.012	23
24	.008	.008	.009	.009	.009	.009	.009	.010	24
25	.006	.007	.007	.007	.007	.007	.007	.007	25
26	.005	.005	.005	.005	.005	.005	.005	.005	26
27	.003	.003	.003	.003	.003	.003	.003	.003	27
28	.001	.001	.001	.001	.001	.001	.001	.001	28
	—	—	—	—	—	—	—	—	
29	.001	.001	.001	.001	.001	.001	.001	.001	29
30	.003	.003	.003	.003	.003	.003	.003	.003	30
31	.005	.005	.005	.005	.005	.005	.005	.005	31
32	.006	.006	.007	.007	.007	.007	.007	.007	32
33	.008	.008	.008	.009	.009	.009	.009	.010	33
34	.010	.010	.010	.011	.011	.011	.011	.012	34
35	.012	.012	.012	.013	.013	.013	.013	.014	35
36	.013	.014	.014	.014	.015	.015	.016	.016	36
37	.015	.016	.016	.016	.017	.017	.018	.018	37
38	.017	.017	.018	.018	.019	.019	.020	.020	38
39	.019	.019	.020	.020	.021	.021	.022	.022	39
40	.021	.021	.022	.022	.023	.023	.024	.024	40
41	.022	.023	.024	.024	.025	.025	.026	.026	41
42	.024	.025	.025	.026	.027	.027	.028	.028	42
43	.026	.027	.027	.028	.029	.029	.030	.031	43
44	.028	.029	.029	.030	.031	.031	.032	.033	44
45	.030	.030	.031	.032	.033	.033	.034	.035	45
46	.031	.032	.033	.034	.035	.035	.036	.037	46
47	.033	.034	.035	.036	.036	.037	.038	.039	47
48	.035	.036	.037	.038	.038	.039	.040	.041	48
49	.037	.038	.039	.040	.040	.041	.042	.043	49
50	.038	.039	.040	.041	.042	.043	.044	.045	50

TABLE IX—Continued.

Temp. F.	Inches.								Temp. F.
	20	20.5	21	21.5	22	22.5.	23	23.5	
	—	—	—	—	—	—	—	—	
51°	.040	.041	.042	.043	.044	.045	.046	.047	51°
52	.042	.043	.044	.045	.046	.047	.048	.049	52
53	.044	.045	.046	.047	.048	.049	.050	.052	53
54	.046	.047	.048	.049	.050	.051	.052	.054	54
55	.047	.049	.050	.051	.052	.053	.055	.056	55
56	.049	.050	.052	.053	.054	.055	.057	.058	56
57	.051	.052	.054	.055	.056	.057	.059	.060	57
58	.053	.054	.055	.057	.058	.059	.061	.062	58
59	.055	.056	.057	.059	.060	.061	.063	.064	59
60	.056	.058	.059	.061	.062	.063	.065	.066	60
61	.058	.060	.061	.062	.064	.065	.067	.068	61
62	.060	.061	.063	.064	.066	.067	.069	.070	62
63	.062	.063	.065	.066	.068	.069	.071	.072	63
64	.063	.065	.067	.068	.070	.071	.073	.075	64
65	.065	.067	.068	.070	.072	.073	.075	.077	65
66	.067	.069	.070	.972	.074	.075	.077	.079	66
67	.069	.071	.072	.074	.076	.077	.079	.081	67
68	.071	.072	.074	.076	.078	.079	.081	.083	68
69	.072	.074	.076	.078	.080	.081	.083	.085	69
70	.074	.076	.078	.080	.082	.083	.085	.087	70
71	.076	.078	.080	.082	.083	.085	.087	.089	71
72	.078	.080	.082	.084	.085	.087	.089	.091	72
73	.079	.081	.083	.085	.087	.089	.091	.093	73
74	.081	.083	.085	.087	.089	.091	.093	.095	74
75	.083	.085	.087	.089	.091	.093	.095	.098	75
76	.085	.087	.089	.091	.093	.095	.097	.100	76
77	.087	.089	.091	.093	.095	.097	.100	.102	77
78	.088	.091	.093	.095	.097	.099	.102	.104	78
79	.090	.092	.095	.097	.099	.101	.104	.106	79
80	.092	.094	.096	.099	.101	.103	.106	.108	80
81	.094	.096	.098	.101	.103	.105	.108	.110	81
82	.095	.098	.100	.103	.105	.107	.110	.112	82
83	.097	.100	.102	.104	.107	.109	.112	.114	83
84	.099	.101	.104	.106	.109	.111	.114	.116	84
85	.101	.103	.106	.108	.111	.113	.116	.118	85
86	.103	.105	.108	.110	.113	.115	.118	.120	86
87	.104	.107	.109	.112	.115	.117	.120	.123	87
88	.106	.109	.111	.114	.117	.119	.122	.125	88
89	.108	.111	.113	.116	.119	.121	.124	.127	89
90	.110	.112	.115	.118	.121	.123	.126	.129	90
91	.111	.114	.117	.120	.122	.125	.128	.131	91
92	.113	.116	.119	.122	.124	.127	.130	.133	92
93	.115	.118	.121	.124	.126	.129	.132	.135	93
94	.117	.120	.122	.125	.128	.131	.134	.137	94
95	.118	.121	.124	.127	.130	.133	.136	.139	95
96	.120	.123	.126	.129	.132	.135	.138	.141	96
97	.122	.125	.128	.131	.134	.137	.140	.143	97
98	.124	.127	.130	.133	.136	.139	.142	.145	98
99	.125	.129	.132	.135	.138	.141	.144	.147	99
100	.127	.130	.134	.137	.140	.143	.146	.150	100

TABLE IX—CONTINUED.

Temp. F.	Inches.								Temp. F.
	24	24.5	25	25.5	26	26.5	27	27.5	
°	+	+	+	+	+	+	+	+	°
0	.061	.063	.064	.065	.067	.068	.069	.071	0
1	.059	.061	.062	.063	.064	.065	.067	.068	1
2	.057	.058	.060	.061	.062	.063	.064	.066	2
3	.055	.056	.057	.059	.060	.061	.062	.063	3
4	.053	.054	.055	.056	.057	.058	.059	.061	4
5	.051	.052	.053	.054	.055	.056	.057	.058	5
6	.049	.050	.051	.052	.053	.054	.055	.056	6
7	.046	.047	.048	.049	.050	.051	.052	.053	7
8	.044	.045	.046	.047	.048	.049	.050	.051	8
9	.042	.043	.044	.045	.046	.046	.047	.048	9
10	.040	.041	.042	.042	.043	.044	.045	.046	10
11	.038	.039	.039	.040	.041	.042	.042	.043	11
12	.036	.036	.037	.038	.039	.039	.040	.041	12
13	.033	.034	.035	.036	.036	.037	.038	.038	13
14	.031	.032	.033	.033	.034	.035	.035	.036	14
15	.029	.030	.030	.031	.032	.032	.033	.033	15
16	.027	.028	.028	.029	.029	.030	.030	.031	16
17	.025	.025	.026	.026	.027	.027	.028	.028	17
18	.023	.023	.024	.024	.025	.025	.925	.026	18
19	.021	.021	.021	.022	.022	.023	.023	.024	19
20	.018	.019	.019	.020	.020	.020	.021	.021	20
21	.016	.017	.017	.017	.018	.018	.018	.019	21
22	.014	.014	.015	.015	.015	.016	.016	.016	22
23	.012	.012	.012	.013	.013	.013	.013	.014	23
24	.010	.010	.010	.010	.011	.011	.011	.011	24
25	.008	.008	.008	.008	.008	.008	.009	.009	25
26	.005	.006	.006	.006	.006	.006	.006	.006	26
27	.003	.003	.003	.003	.004	.004	.004	.004	27
28	.001	.001	.001	.001	.001	.001	.001	.001	28
	—	—	—	—	—	—	—	—	
29	.001	.001	.001	.001	.001	.001	.001	.001	29
30	.003	.003	.003	.004	.004	.004	.004	.004	30
31	.005	.006	.006	.006	.006	.006	.006	.006	31
32	.008	.008	.008	.008	.008	.008	.008	.009	32
33	.010	.010	.010	.010	.011	.011	.011	.011	33
34	.012	.012	.012	.013	.013	.013	.013	.014	34
35	.014	.014	.015	.015	.015	.015	.016	.016	35
36	.016	.017	.017	.017	.017	.018	.018	.019	36
37	.018	.019	.019	.019	.020	.020	.021	.021	37
38	.020	.021	.021	.022	.022	.023	.023	.023	38
39	.023	.023	.024	.024	.024	.025	.025	.026	39
40	.025	.025	.026	.026	.027	.027	.028	.028	40
41	.027	.027	.028	.029	.029	.030	.030	.031	41
42	.029	.030	.030	.031	.031	.032	.033	.033	42
43	.031	.032	.032	.033	.034	.034	.035	.036	43
44	.033	.034	.035	.035	.036	.037	.037	.038	44
45	.035	.036	.037	.038	.038	.039	.040	.041	45
46	.038	.038	.039	.040	.041	.042	.042	.043	46
47	.040	.041	.041	.042	.043	.044	.045	.046	47
48	.042	.043	.044	.045	.045	.046	.047	.048	48
49	.044	.045	.046	.047	.048	.049	.050	.005	49
50	.046	.047	.048	.049	.050	.051	.052	.053	50

TABLE IX—CONTINUED.

Temp. F.	Inches.								Temp. F.
	24	24.5	25	25.5	26	26.5	27	27.5	
°	—	—	—	—	—	—	—	—	°
51	.048	.049	.050	.051	.052	.053	.054	.055	51
52	.050	.052	.053	.054	.055	.056	.057	.058	52
53	.053	.054	.055	.056	.057	.058	.059	.060	53
54	.055	.056	.057	.058	.059	.060	.062	.063	54
55	.057	.058	.059	.060	.062	.063	.064	.065	55
56	.059	.060	.061	.063	.064	.065	.066	.068	56
57	.061	.062	.064	.065	.066	.068	.069	.070	57
58	.063	.065	.066	.067	.069	.070	.071	.073	58
59	.065	.067	.068	.070	.071	.072	.074	.075	59
60	.068	.069	.070	.072	.073	.075	.076	.077	60
61	.070	.071	.073	.074	.075	.077	.078	.080	61
62	.072	.073	.075	.076	.078	.079	.081	.082	62
63	.074	.076	.077	.079	.080	.082	.083	.085	63
64	.076	.078	.079	.081	.082	.084	.086	.087	64
65	.078	.080	.082	.083	.085	.086	.088	.090	65
66	.080	.082	.084	.085	.087	.089	.090	.092	66
67	.083	.084	.086	.088	.089	.091	.093	.095	67
68	.085	.086	.088	.090	.092	.094	.095	.097	68
69	.087	.089	.090	.092	.094	.096	.098	.100	69
70	.089	.091	.093	.095	.096	.098	.100	.102	70
71	.091	.093	.095	.097	.099	.101	.102	.104	71
72	.093	.095	.097	.099	.101	.103	.105	.107	72
73	.095	.097	.099	.101	.103	.105	.107	.109	73
74	.097	.099	.102	.104	.106	.108	.110	.112	74
75	.100	.102	.104	.106	.108	.110	.112	.114	75
76	.102	.104	.106	.108	.110	.112	.114	.117	76
77	.104	.106	.108	.110	.112	.115	.117	.119	77
78	.106	.108	.110	.113	.115	.117	.119	.122	78
79	.108	.110	.113	.115	.117	.119	.122	.124	79
80	.110	.113	.115	.117	.119	.122	.124	.126	80
81	.112	.115	.117	.119	.122	.124	.126	.129	81
82	.114	.117	.119	.122	.124	.126	.129	.131	82
83	.117	.119	.121	.124	.126	.129	.131	.134	83
84	.119	.121	.124	.126	.129	.131	.134	.136	84
85	.121	.123	.126	.128	.131	.133	.136	.139	85
86	.123	.126	.128	.131	.133	.136	.138	.141	86
87	.125	.128	.130	.133	.136	.138	.141	.143	87
88	.127	.130	.133	.135	.138	.141	.143	.146	88
89	.129	.132	.135	.137	.140	.143	.146	.148	89
90	.131	.134	.137	.140	.142	.145	.148	.151	90
91	.134	.136	.139	.142	.145	.148	.150	.153	91
92	.136	.139	.141	.144	.147	.150	.153	.156	92
93	.138	.141	.144	.147	.149	.152	.155	.158	93
94	.140	.143	.146	.149	.152	.155	.157	.161	94
95	.142	.145	.148	.151	.154	.157	.160	.163	95
96	.144	.147	.150	.153	.156	.159	.162	.165	96
97	.146	.149	.152	.156	.159	.162	.165	.168	97
98	.148	.152	.155	.158	.161	.164	.167	.170	98
99	.151	.154	.157	.160	.163	.166	.169	.173	99
100	.153	.156	.159	.162	.165	.169	.172	.175	100

TABLE IX—Continued.

Temp. F.	Inches.							Temp. F.
	28	28.5	29	29.5	30	30.5	31	
°	+	+	+	+	+	+	+	°
0	.072	.073	.074	.076	.077	.078	.080	0
1	.069	.071	.072	.073	.074	.076	.077	1
2	.067	.068	.069	.070	.072	.073	.074	2
3	.064	.065	.067	.068	.069	.070	.071	3
4	.062	.063	.064	.065	.066	.067	.068	4
5	.059	.060	.061	.062	.063	.065	.066	5
6	.057	.058	.059	.060	.061	.062	.063	6
7	.054	.055	.056	.057	.058	.059	.060	7
8	.052	.053	.054	.054	.055	.056	.057	8
9	.049	.050	.051	.052	.053	.054	.054	9
10	.047	.047	.048	.049	.050	.051	.052	10
11	.044	.045	.046	.046	.047	.048	.049	11
12	.042	.042	.043	.044	.045	.045	.046	12
13	.039	.040	.040	.041	.042	.043	.043	13
14	.037	.037	.038	.038	.039	.040	.040	14
15	.034	.035	.035	.036	.036	.037	.038	15
16	.032	.032	.033	.033	.034	.034	.035	16
17	.029	.030	.030	.031	.031	.032	.032	17
18	.026	.027	.027	.028	.028	.029	.029	18
19	.024	.024	.025	.025	.026	.026	.027	19
20	.021	.022	.022	.023	.023	.023	.024	20
21	.019	.019	.020	.020	.020	.021	.021	21
22	.016	.017	.017	.017	.018	.018	.018	22
23	.014	.014	.014	.015	.015	.015	.015	23
24	.011	.012	.012	.012	.012	.012	.013	24
25	.009	.009	.009	.009	.009	.010	.010	25
26	006	.006	.007	.007	.007	.007	.007	26
27	04	.004	.004	.004	.004	.004	.004	27
28	1	.001	.001	.001	.001	.001	.001	28
		—	—	—	—	—	—	
29	.001	.001	.001	.001	.001	.001	.001	29
30	.004	.064	.004	.004	.004	.004	.004	30
31	.006	.006	.007	.007	.007	.007	.007	31
32	.009	.009	.009	.009	.009	.010	.010	32
33	.011	.012	.012	.012	.012	.012	.012	33
34	.014	.014	.014	.015	.015	.015	.015	34
35	.016	.017	.017	.017	.018	.018	.018	35
36	.019	.019	.020	.020	.020	.021	.021	36
37	.021	.022	.022	.022	.023	.023	.024	37
38	.024	.024	.025	.025	.026	.026	.026	38
39	.026	.027	.027	.028	.028	.029	.029	39
40	.029	.029	.030	.030	.031	.031	.032	40
41	.031	.032	.033	.033	.034	.034	.035	41
42	.034	.034	.035	.036	.036	.037	.037	42
43	.036	.037	.038	.038	.039	.040	.040	43
44	.030	.040	.040	.041	.042	.042	.043	44
45	.041	.042	.043	.044	.044	.045	.046	45
46	.044	.045	.045	.046	.047	.048	.049	46
47	.046	.047	.048	.049	.050	.051	.051	47
48	.049	.050	.051	.052	.052	.053	.054	48
49	.051	.052	.053	.054	.055	.056	.057	49
50	.054	.055	.056	.057	.058	.059	.060	50

TABLE IX—CONTINUED.

Temp. F.	Inches.							Temp. F.
	28	28.5	29	29.5	30	30.5	31	
°	—	—	—	—	—	—	—	°
51	.056	.057	.058	.059	.060	.061	.062	51
52	.059	.060	.061	.062	.063	.064	.065	52
53	.061	.063	.064	.065	.066	.067	.068	53
54	.064	.065	.066	.067	.068	.070	.071	54
55	.066	.068	.069	.070	.071	.072	.073	55
56	.069	.070	.071	.073	.074	.075	.076	56
57	.071	.073	.074	.075	.076	.078	.079	57
58	.074	.075	.077	.078	.079	.081	.082	58
59	.076	.078	.079	.080	.082	.083	.085	59
60	.079	.080	.082	.083	.085	.086	.087	60
61	.081	.083	.084	.086	.087	.089	.090	61
62	.084	.085	.087	.088	.090	.091	.093	62
63	.086	.088	.089	.091	.093	.094	.096	63
64	.089	.090	.092	.094	.095	.097	.098	64
65	.091	.093	.095	.096	.098	.100	.101	65
66	.094	.096	.097	.099	.101	.102	.104	66
67	.096	.098	.100	.102	.103	.105	.107	67
68	.099	.101	.102	.104	.106	.108	.109	68
69	.101	.103	.105	.107	.109	.110	.112	69
70	.104	.106	.108	.109	.111	.113	.115	70
71	.106	.108	.110	.112	.114	.116	.118	71
72	.109	.111	.113	.115	.117	.119	.120	72
73	.111	.113	.115	.117	.119	.121	.123	73
74	.114	.116	.118	.120	.122	.124	.126	74
75	.116	.118	.120	.122	.125	.127	.129	75
76	.119	.121	.123	.125	.127	.129	.131	76
77	.121	.123	.126	.128	.130	.132	.134	77
78	.124	.126	.128	.130	.133	.135	.137	78
79	.126	.128	.131	.133	.135	.127	.140	79
80	.129	.131	.133	.136	.138	.140	.143	80
81	.131	.134	.136	.138	.141	.143	.145	81
82	.134	.136	.138	.141	.143	.146	.148	82
83	.136	.139	.141	.143	.146	.148	.151	83
84	.139	.141	.144	.146	.149	.151	.154	84
85	.141	.144	.146	.149	.151	.154	.156	85
86	.144	.146	.149	.151	.154	.156	.159	86
87	.146	.149	.151	.154	.157	.159	.162	87
88	.149	.151	.154	.157	.159	.162	.165	88
89	.151	.154	.156	.159	.162	.165	.167	89
90	.153	.156	.159	.162	.164	.167	.170	90
91	.156	.159	.162	.165	.167	.170	.173	91
92	.158	.161	.164	.167	.170	.172	.175	92
93	.161	.164	.167	.170	.172	.175	.178	93
94	.163	.166	.169	.172	.175	.177	.180	94
95	.166	.169	.172	.175	.178	.180	.183	95
96	.168	.171	.174	.178	.181	.183	.186	96
97	.171	.174	.177	.180	.183	.186	.189	97
98	.173	.176	.179	.183	.186	.188	.191	98
99	.176	.179	.182	.185	.188	.191	.194	99
100	.178	.181	.184	.188	.191	.194	.197	100

NOMENCLATURE OF CLOUDS, PRIMARY FORMS.

PLATE. I

Stratus or Fallcloud.

Cirrus or Curlcloud.

Cumulus or Stackencloud.

Nimbus or Raincloud.

NOMENCLATURE OF CLOUDS, SECONDARY FORMS.

PLATE. II.

Cirrus or *Curlcloud.*

Cirrocumulus or *Sondercloud*

Cirrostratus or *Wanecloud* } *various forms*

Cumulostratus or *Twaincloud.*

LITH. OF SARONY, NEW YORK.

IV. FOR CONVERTING SIDEREAL TIME INTO MEAN SOLAR TIME, AND MEAN TIME INTO SIDEREAL TIME.

HOURS.			MINUTES.						SECONDS.			
Hours	Mean Time.	Sidereal Time.	Minutes.	Mean Time.	Sidereal Time.	Minutes.	Mean Time.	Sidereal Time.	Seconds.	Mean or Sidereal Time.	Seconds.	Mean or Sidereal Time.
	m s.	m. s.		s.	s.		s.	s.		s.		s.
1	0 9.83	0 9.86	1	0.16	0.16	31	5.08	5.09	1	0.00	31	0.09
2	0 19.66	0 19.71	2	0.33	0.33	32	5.24	5.26	2	0.01	32	0.09
3	0 29.49	0 29.57	3	0.49	0.49	33	5.41	5.42	3	0.01	33	0.09
4	0 39.32	0 39.43	4	0.66	0.66	34	5.57	5.59	4	0.01	34	0.09
5	0 49.15	0 49.28	5	0.82	0.82	35	5.75	5.75	5	0.01	35	0.10
6	0 58.98	0 59.14	6	0.98	0.99	36	5.90	5.91	6	0.02	36	0.10
7	1 8.81	1 9.00	7	1.15	1.15	37	6.06	6.08	7	0.02	37	0.10
8	1 18.64	1 18.85	8	1.31	1.31	38	6.23	6.24	8	0.02	38	0.10
9	1 28.47	1 28.71	9	1.47	1.48	39	6.39	6.41	9	0.03	39	0.11
10	1 38.30	1 38.57	10	1.64	1.64	40	6.55	6.57	10	0.03	40	0.11
11	1 48.13	1 48.42	11	1.80	1.81	41	6.72	6.74	11	0.03	41	0.11
12	1 57.96	1 58.28	12	1.97	1.97	42	6.88	6.90	12	0.03	42	0.12
13	2 7.78	2 8.13	13	2 13	2.14	43	7.05	7.06	13	0.04	43	0.12
14	2 17.61	2 17.99	14	2.29	2.30	44	7.21	7.23	14	0.04	44	0.12
15	2 27.44	2 27.85	15	2.46	2.46	45	7.37	7.39	15	0.04	45	0.12
16	2 37.27	2 37.70	16	2.62	2.63	46	7.54	7.56	16	0.04	46	0.13
17	2 47.10	2 47.56	17	2.79	2.79	47	7.70	7.72	17	0.05	47	0.13
18	2 56.93	2 57.42	18	2.95	2.96	48	7.86	7.89	18	0.05	48	0.13
19	3 6.76	3 7.27	19	3.11	3.12	49	8.03	8.05	19	0.05	49	0.13
20	3 16.59	3 17.13	20	3.28	3.29	50	8.19	8.21	20	0.06	50	0.14
21	3 26.42	3 26.99	21	3.44	3.45	51	8.36	8.38	21	0.06	51	0.14
22	3 36.25	3 36.84	22	3.60	3.61	52	8.52	8.54	22	0.06	52	0.14
23	3 46.08	3 46.70	23	3.77	3.79	53	8.68	8.71	23	0.06	53	0.15
24	3 55.91	3 56.56	24	3.93	3.94	54	8.85	8.87	24	0.07	54	0.15
25	[illegible] 5.74	4 6.41	25	4.10	4.11	55	9.01	9.04	25	0.07	55	0.15
26	4 15.57	4 16.27	26	4.26	4.27	56	9.17	9.20	26	0.07	56	0.15
27	4 25.40	4 26.13	27	4.42	4.43	57	9.34	9.36	27	0.07	57	0.16
28	4 35.23	4 35.98	28	4.59	4.60	58	9.50	9.53	28	0.08	58	0.16
29	4 45.06	4 45.84	29	4.75	4.76	59	9.67	9.69	29	0.08	59	0.16
30	4 54.89	4 55.69	30	4.92	4.93	60	9.83	9.86	30	0.08	60	0.16

V.

CORRECTION OF THE TIME OBTAINED BY OBSERVATION OF THE SUN, IN ORDER TO HAVE THE TRUE TIME OF THE CLOCK.

Day of Month.	Jan.	Feb.	Mar.	Apr.	Apr.	May.	June.	June.	July.	Aug.	Sept.	Oct.	Nov.	Dec.	Dec.	Day of Month.
	Add.	Add.	Add.	Add.	Subt.	Subt.	Subt.	Add.	Add.	Add.	Subt.	Subt.	Subt.	Subt.	Add.	
	Min.	Min.	Min.	Min.	Min.	Min.	Min.	Min.	Min.	Min.	Min.	Min.	Min.	Min.	Min.	
1	4	14	13	4	..	3	3	..	3	6	0	10	16	11	..	1
2	4	14	12	4	..	3	2	..	4	6	0	11	16	10	..	2
3	5	14	12	3	..	3	2	..	4	6	1	11	16	10	..	3
4	5	14	12	3	..	3	2	..	4	6	1	11	16	10	..	4
5	6	14	12	3	..	4	2	..	4	6	1	12	16	9	..	5
6	6	14	12	2	..	4	2	..	4	6	2	12	16	9	..	6
7	7	14	11	2	..	4	2	..	4	5	2	12	16	8	..	7
8	7	15	11	2	..	4	1	..	5	5	2	12	16	8	..	8
9	8	15	11	2	..	4	1	..	5	5	3	13	16	7	..	9
10	8	15	11	1	..	4	1	..	5	5	3	13	16	7	..	10
11	9	15	10	1	..	4	1	..	5	5	3	13	16	6	..	11
12	9	15	10	1	..	4	1	..	5	5	4	13	16	6	..	12
13	9	15	10	1	..	4	0	..	5	5	4	14	16	5	..	13
14	10	14	9	0	..	4	0	..	5	4	5	14	15	5	..	14
15	10	14	9	0	..	4	0	..	6	4	5	14	15	4	..	15
16	10	14	9	0	..	4	0	..	6	4	5	14	15	4	..	16
17	11	14	9	0	..	4	0	..	6	4	6	15	15	3	..	17
18	11	14	8	..	1	4	..	1	6	4	6	15	15	3	..	18
19	11	14	8	..	1	4	..	1	6	3	6	15	14	2	..	19
20	11	14	8	..	1	4	..	1	6	3	7	15	14	2	..	20
21	12	14	7	..	1	4	..	1	6	3	7	15	14	1	..	21
22	12	14	7	..	2	4	..	2	6	3	7	15	14	1	..	22
23	12	14	7	..	2	4	..	2	6	2	8	16	13	0	..	23
24	12	13	6	..	2	3	..	2	6	2	8	16	13	0	..	24
25	13	13	6	..	2	3	..	2	6	2	8	16	13	0	..	25
26	13	13	6	..	2	3	..	2	6	2	9	16	12	..	1	26
27	13	13	5	..	2	3	..	3	6	1	9	16	12	..	1	27
28	13	13	5	..	3	3	..	3	6	1	9	16	12	..	2	28
29	14	13	5	..	3	3	..	3	6	1	10	16	11	..	2	29
30	14	..	4	..	3	3	..	3	6	0	10	16	11	..	3	30
31	14	..	4	..	..	3	..	..	6	0	..	16	..	..	3	31

VI.

TABLE FOR COMPUTING TERRESTRIAL SURFACES.

THE tables under No. VI. were published by Delcros in the *Annuaire Météorologique de la France pour* 1850, p. 65 *et seq.*

The formula from which they have been computed reads as follows: —

$$S = \frac{a\,b\,\pi}{90} \left\{ \begin{array}{l} \sin \tfrac{1}{2}\phi \cos (L + \tfrac{1}{2}\phi) \\ -\tfrac{1}{3}\left[2 \cdot \left(\frac{a-b}{a+b}\right) + \left(\frac{a-b}{a+b}\right)^2\right] \sin (\phi + \tfrac{1}{2}\phi) \cos [3\,L + (\phi + \tfrac{1}{2}\phi)] \\ +\tfrac{1}{5}\left[3 \cdot \left(\frac{a-b}{a+b}\right)^2 + \left(\frac{a-b}{a+b}\right)^3\right] \sin (2\,\phi + \tfrac{1}{2}\phi) \cos [5\,L + (2\,\phi + \tfrac{1}{2}\phi)] \\ -\text{etc.}; \end{array} \right.$$

in which $a = \frac{1}{2}$ great axis of the globe; $b = \frac{1}{2}$ small axis; $L =$ the latitude of the lower limit of a quadrilateral surface; $L' =$ the latitude of the upper limit of the same; $\phi = L' - L$; $S =$ the area of a quadrilateral surface of one degree in longitude; $\pi =$ the ratio of the circumference to the diameter.

Substituting the numerical values, the quarter of the meridian being = 10,000,724 legal metres; the $\frac{1}{2}$ great axis, or a, = 6,376,989 metres; the $\frac{1}{2}$ small axis, or b, = 6,356,323 metres; the ratio of the axis $\frac{1}{308.64}$; and making $\phi = 1°$ nonagesimal, the formula becomes,

$$S = \left\{ \begin{array}{l} 224.996360 \cos (\;\;L + 0°\ 30') \\ -0.730851 \cos (3\,L + 1°\ 30') \\ +0.001784 \cos (5\,L + 2°\ 30') \\ -0.000004 \cos (7\,L + 3°\ 30') \\ +\text{etc.} \end{array} \right.$$

The first three terms of the formula give the results with sufficient accuracy.

In order to avoid too large a number of figures, the results are given in square miles, the linear base of which is a mile equal to $\frac{1}{15}$ of the mean degree of the meridian. That mile is thus $= \left(\frac{10000724}{90 \times 15}\right) = 7407.942$ metres. In order to convert the results into new geographical miles, of which 60 = 1°, multiply by 16, log = 1.2041200; into common French leagues, 25 = 1°, multiply by 2.777778, log = 0.4436975; into nautical leagues, 20 = 1°, multiply by 1.777778, log = 0.2498775; into English statute miles, 69.163 = 1°, by 21.711034, log = 1.3366868.

USE OF THE TABLES.

Table I., which gives the number of square miles contained in the quadrilateral surfaces of one degree in latitude and longitude, successively from the equator to the pole, will be more frequently used. Table II. has been computed for maps on a smaller scale; and Tables III. and IV. for maps of very small scale, covering large areas, in which surfaces of one degree could not be estimated with sufficient accuracy. If the scale is large enough to have the minutes traced on, then Table V. is to be used.

For computing a surface by Table I., which may serve as an example for all the others, find first the lowest parallel circle which crosses, on the map, the surface to be estimated; suppose it is 40° lat. N., and the zone within 40° and 41° lat. N. contains four integral degrees of longitude, that is, four surfaces of one degree each way; [illegible] in the first column of the table, on the line beginning with latitude 40°, and in the vertical column headed 4, take the value of these four surfaces, viz. 685.88. Then take likewise the value of the number of surfaces between 41° and 42° lat. N., and so on. The fractional parts left outside of the integral degrees are best estimated, with the compass, in decimals, the values of which can be found in the columns of the multiples, by properly moving the decimal point to the left. Having taken them in that way, and summing them up with all the integral surfaces, we obtain the total surface required.

TABLE I. QUADRILATERAL SURFACES OF 1 DEGREE IN LATITUDE AND IN LONGITUDE ON THE TERRESTRIAL ELLIPSOID.

Limiting LATITUDES.		Multiples of these Quadrilateral Surfaces from 1 to 9.								
Inf.	Sup.	1.	2.	3.	4.	5.	6.	7.	8.	9.
0	1	224.259	448.52	672.78	897.04	1121.29	1345.55	1569.81	1794.07	2018.33
1	2	224.192	448.38	672.58	896.77	1120.96	1345.15	1569.35	1793.54	2017.73
2	3	224.059	448.12	672.18	896.24	1120.30	1344.36	1568.42	1792.47	2016.53
3	4	223.860	447.72	671.58	895.44	1119.30	1343.16	1567.02	1790.88	2014.74
4	5	223.594	447.19	670.78	894.37	1117.97	1341.56	1565.16	1788.75	2012.34
5	6	223.261	446.52	669.78	893.05	1116.31	1339.57	1562.83	1786.09	2009.35
6	7	222.863	445.73	668.59	891.45	1114.31	1337.18	1560.04	1782.90	2005.76
7	8	222.398	444.80	667.19	889.59	1111.99	1334.39	1556.78	1779.18	2001.58
8	9	221.867	443.73	665.60	887.47	1109.33	1331.20	1553.07	1774.93	1996.80
9	10	221.270	442.54	663.81	885.08	1106.35	1327.62	1548.89	1770.16	1991.43
10	11	220.607	441.21	661.82	882.43	1103.03	1323.64	1544.25	1764.85	1985.46
11	12	219.878	439.76	659.63	879.51	1099.39	1319.27	1539.15	1759.02	1978.90
12	13	219.084	433.17	657.25	876.34	1095.42	1314.50	1533.59	1752.67	1971.76
13	14	218.225	436.45	654.67	872.90	1091.12	1309.35	1527.57	1745.80	1964.02
14	15	217.300	434.60	651.90	869.20	1086.50	1303.80	1521.10	1738.40	1955.70
15	16	216.311	432.62	648.93	865.24	1081.55	1297.86	1514.17	1730.48	1946.80
16	17	215.257	430.51	645.77	861.03	1076.28	1291.54	1506.80	1722.05	1937.31
17	18	214.138	428.28	642.41	856.55	1070.69	1284.83	1498.97	1713.10	1927.24
18	19	212.955	425.91	638.87	851.82	1064.78	1277.73	1490.69	1703.64	1916.60
19	20	211.709	423.42	636.13	846.84	1058.54	1270.25	1481.96	1693.67	1905.38
20	21	210.399	420.80	631.20	841.59	1051.99	1262.39	1472.79	1683.19	1893.59
21	22	209.025	418.05	627.08	836.10	1045.13	1254.15	1463.18	1672.20	1881.23
22	23	207.589	415.18	622.77	830.36	1037.95	1245.54	1453.12	1660.71	1868.30
23	24	206.090	412.18	618.27	824.36	1030.45	1236.54	1442.63	1648.72	1854.81
24	25	204.529	409.06	613.59	818.12	1022.65	1227.18	1431.71	1636.24	1840.76
25	26	202.907	405.81	608.72	811.63	1014.53	1217.44	1420.35	1623.25	1826.16
26	27	201.223	402.45	603.67	804.89	1006.11	1207.34	1408.56	1609.78	1811.00
27	28	199.477	398.95	598.43	797.91	997.39	1196.86	1396.34	1595.82	1795.30
28	29	197.672	395.34	593.02	790.69	988.36	1186.03	1383.70	1581.38	1779.05
29	30	195.806	391.61	587.42	783.23	979.03	1174.84	1370.64	1566.45	1762.26
30	31	193.881	387.76	581.64	775.52	969.40	1163.29	1357.17	1551.05	1744.93
31	32	191.897	383.79	575.69	767.59	959.48	1151.38	1343.28	1535.17	1727.07
32	33	189.854	379.71	569.56	759.41	949.27	1139.12	1328.98	1518.83	1708.68
33	34	187.753	375.51	563.26	750.01	938.76	1126.52	1314.27	1502.02	1689.77
34	35	185.594	371.19	556.78	742.38	927.97	1113.57	1299.16	1484.75	1670.35
35	36	183.379	366.76	550.14	733.52	916.89	1100.27	1283.65	1467.03	1650.41
36	37	181.107	362.21	543.32	724.43	905.53	1086.64	1267.75	1448.86	1629.96
37	38	178.780	357.56	536.34	715.12	893.90	1072.68	1251.46	1430.24	1609.02
38	39	176.397	352.79	529.19	705.59	881.98	1058.38	1234.78	1411.18	1587.57
39	40	173.960	347.92	521.88	695.84	869.80	1043.76	1217.72	1391.68	1565.64
40	41	171.469	342.94	514.41	685.88	857.34	1028.81	1200.28	1371.75	1543.22
41	42	168.925	337.85	506.77	675.70	844.62	1013.55	1182.47	1351.40	1520.22
42	43	166.328	332.66	498.98	665.31	831.64	997.97	1164.30	1330.62	1496.95
43	44	163.680	327.36	491.04	654.72	818.40	982.08	1145.76	1309.44	1473.12
44	45	160.980	321.96	482.94	643.92	804.90	965.88	1126.86	1287.84	1448.82

TABLE I. (*Continued.*) QUADRILATERAL SURFACES OF 1 DEGREE IN LATITUDE AND IN LONGITUDE ON THE TERRESTRIAL ELLIPSOID.

Limiting LATITUDES.		Multiples of these Quadrilateral Surfaces from 1 to 9.								
Inf.	Sup.	1.	2.	3.	4.	5.	6.	7.	8.	9.
45	46	158.231	316.46	474.69	632.92	791.15	949.39	1107.62	1265.85	1424.08
46	47	155.432	310.86	466.30	621.73	777.16	932.59	1088.02	1243.46	1398.89
47	48	152.584	305.17	457.75	610.34	762.92	915.51	1068.09	1220.67	1373.26
48	49	149.689	299.38	449.07	598.75	748.44	899.13	1047.82	1197.51	1347.20
49	50	146.746	293.49	440.24	586.98	733.73	880.48	1027.22	1173.97	1320.71
50	51	143.757	287.51	431.27	575.03	718.78	862.54	1006.30	1150.06	1293.81
51	52	140.723	281.45	422.17	562.89	703.61	844.34	985.06	1125.78	1266.51
52	53	137.644	275.29	412.93	550.58	688.22	825.86	963.51	1101.15	1238.80
53	54	134.522	269.04	403.57	538.09	672.61	807.13	941.65	1076.17	1210.70
54	55	131.357	262.71	394.07	525.43	656.78	788.14	919.50	1050.86	1182.21
55	56	128.150	256.30	384.45	512.60	640.75	768.90	897.05	1025.20	1153.35
56	57	124.903	249.81	374.71	499.61	624.51	749.42	874.32	999.22	1124.13
57	58	121.616	243.23	364.85	486.46	608.08	729.69	851.31	972.92	1094.54
58	59	118.289	236.58	354.87	473.16	591.45	709.74	828.03	946.32	1064.61
59	60	114.926	229.85	344.78	459.70	574.63	689.55	804.48	919.41	1034.33
60	61	111.525	223.05	334.58	446.10	557.63	669.15	780.68	892.20	1003.73
61	62	108.089	216.18	324.27	432.35	540.44	648.53	756.62	864.71	972.80
62	63	104.618	209.24	313.85	418.47	523.09	627.71	732.32	836.94	941.56
63	64	101.113	202.23	303.34	404.45	505.56	606.68	707.79	808.90	910.02
64	65	97.575	195.15	292.73	390.30	487.88	585.45	683.03	780.60	878.18
65	66	94.007	188.01	282.02	376.03	470.03	564.04	658.05	752.05	846.06
66	67	90.408	180.82	271.22	361.63	452.04	542.45	632.85	723.26	813.67
67	68	86.779	173.56	260.34	347.12	433.90	520.68	607.46	694.23	781.01
68	69	83.123	166.25	249.37	332.49	415.61	498.74	581.86	664.98	748.11
69	70	79.439	158.88	238.32	317.76	397.20	476.64	556.08	635.52	714.95
70	71	75.730	151.46	227.19	302.92	378.65	454.38	530.11	605.84	681.57
71	72	71.996	143.99	215.99	287.99	359.98	431.98	503.98	575.97	647.97
72	73	68.239	136.48	204.72	272.96	341.20	409.44	477.68	545.91	614.15
73	74	64.460	128.92	193.38	257.84	322.30	386.76	451.22	515.68	580.14
74	75	60.659	121.32	181.98	242.64	303.30	363.96	424.62	485.28	545.94
75	76	56.839	113.68	170.52	227.36	284.20	341.04	397.88	454.72	511.55
76	77	53.001	106.00	159.00	212.00	265.00	318.00	371.00	424.00	477.01
77	78	49.145	98.29	147.43	196.58	245.72	294.87	344.01	393.16	442.30
78	79	45.272	90.54	135.82	181.09	226.36	271.63	316.91	362.18	407.45
79	80	41.386	82.77	124.16	165.54	206.93	248.31	289.70	331.08	372.47
80	81	37.485	74.97	112.46	149.94	187.43	224.91	262.40	299.88	337.37
81	82	33.572	67.14	100.72	134.29	167.86	201.43	235.01	268.58	302.15
82	83	29.649	59.30	88.95	118.59	148.24	177.89	207.54	237.19	266.84
83	84	25.715	51.43	77.15	102.86	128.58	154.29	180.01	205.72	231.44
84	85	21.773	43.55	65.32	87.09	108.87	130.64	152.41	174.19	195.96
85	86	17.824	35.65	53.47	71.30	89.12	106.95	124.77	142.59	160.42
86	87	13.869	27.74	41.61	55.48	69.35	83.22	97.09	110.96	124.82
87	88	9.910	19.82	29.73	39.64	49.55	59.46	69.37	79.28	89.19
88	89	5.947	11.89	17.84	23.79	29.74	35.68	41.63	47.58	53.53
89	90	1.983	3.97	5.95	7.93	9.91	11.90	13.88	15.86	17.84

TABLE II. Quadrilateral Surfaces of 2 Degrees in Latitude and in Longitude on the Terrestrial Ellipsoid.

Limiting LATITUDES.		Multiples of these Quadrilateral Surfaces from 1 to 9.								
Inf.	Sup.	1.	2.	3.	4.	5.	6.	7.	8.	9.
0	2	896.903	1793.81	2690.71	3587.61	4484.51	5381.42	6278.32	7175.22	8072.13
2	4	895.838	1791.68	2687.51	3583.35	4479.19	5375.03	6270.87	7166.71	8062.54
4	6	893.710	1787.42	2681.13	3574.84	4468.55	5362.26	6255.97	7149.68	8043.39
6	8	890.520	1781.04	2671.56	3562.08	4452.60	5343.12	6233.64	7124.16	8014.68
8	10	886.272	1772.54	2658.82	3545.09	4431.36	5317.63	6203.91	7090.18	7976.45
10	12	880.969	1761.94	2642.91	3523.88	4404.85	5285.82	6166.79	7047.76	7928.72
12	14	874.617	1749.23	2623.85	3498.47	4373.09	5247.70	6122.32	6996.94	7871.55
14	16	867.221	1734.44	2601.66	3468.88	4336.11	5203.33	6070.55	6937.77	7804.99
16	18	858.789	1717.58	2576.37	3435.16	4293.95	5152.74	6011.52	6870.31	7729.10
18	20	849.328	1698.66	2547.98	3397.31	4246.64	5095.97	5945.30	6794.63	7643.95
20	22	838.848	1677.70	2516.54	3355.39	4194.24	5033.09	5871.94	6710.78	7549.63
22	24	827.359	1654.72	2482.08	3309.44	4136.80	4964.16	5791.51	6618.87	7446.23
24	26	814.872	1629.74	2444.62	3259.49	4074.36	4889.23	5704.11	6518.98	7333.85
26	28	811.400	1602.80	2404.20	3205.60	4007.00	4808.40	5609.80	6411.20	7212.60
28	30	786.956	1573.91	2360.87	3147.83	3934.78	4721.74	5508.69	6295.65	7082.61
30	32	771.555	1543.11	2314.67	3086.22	3857.78	4629.33	5400.89	6172.44	6944.00
32	34	755.213	1510.43	2265.64	3020.85	3776.06	4531.28	5286.49	6041.70	6796.92
34	36	737.946	1475.89	2213.84	2951.78	3689.73	4427.68	5165.62	5903.57	6641.51
36	38	719.773	1439.55	2159.32	2879.09	3598.87	4318.64	5038.41	5758.19	6477.96
38	40	700.713	1401.43	2102.14	2802.85	3503.57	4204.28	4904.99	5605.71	6306.42
40	42	680.787	1361.57	2042.36	2723.15	3403.93	4084.72	4765.51	5446.29	6127.08
42	44	660.016	1320.03	1980.05	2640.06	3300.08	3960.09	4620.11	5280.13	5940.14
44	46	638.423	1276.85	1915.27	2553.69	3192.11	3830.54	4468.96	5107.38	5745.81
46	48	616.032	1232.06	1848.10	2464.13	3080.16	3696.19	4312.23	4928.26	5544.29
48	50	592.869	1185.74	1778.61	2371.48	2964.34	3557.21	4150.08	4742.95	5335.82
50	52	568.960	1137.92	1706.88	2275.84	2844.80	3413.76	3982.72	4551.68	5120.64
52	54	544.332	1088.66	1632.99	2177.33	2721.66	3265.99	3810.32	4354.65	4898.99
54	56	519.014	1038.03	1557.04	2076.06	2595.07	3114.09	3633.10	4152.11	4671.13
56	58	493.037	986.07	1479.11	1972.15	2465.18	2958.22	3451.26	3944.29	4437.33
58	60	466.430	932.86	1399.29	1865.72	2332.15	2798.58	3265.01	3731.44	4197.87
60	62	439.228	878.46	1317.68	1756.91	2196.14	2635.37	3074.59	3513.82	3953.05
62	64	411.461	822.92	1234.38	1645.84	2057.30	2468.76	2880.23	3291.69	3703.15
64	66	383.164	766.33	1149.49	1532.66	1915.82	2298.99	2682.15	3065.32	3448.48
66	68	354.374	708.75	1063.12	1417.50	1771.87	2126.24	2480.62	2834.99	3189.36
68	70	325.124	650.25	975.37	1300.50	1625.62	1950.75	2275.87	2601.00	2926.12
70	72	295.453	590.91	886.36	1181.81	1477.27	1772.72	2068.17	2363.63	2659.08
72	74	265.398	530.80	796.20	1061.59	1326.99	1592.39	1857.79	2123.19	2388.59
74	76	234.998	469.99	704.99	939.99	1174.99	1409.99	1644.98	1879.98	2114.98
76	78	204.290	408.58	612.87	817.16	1021.45	1225.74	1430.03	1634.32	1838.61
78	80	173.316	346.63	519.95	693.26	866.58	1039.90	1213.21	1386.53	1559.85
80	82	142.115	284.23	426.34	568.46	710.57	852.69	994.80	1136.92	1279.03
82	84	110.728	221.46	332.18	442.91	553.64	664.37	775.09	885.82	996.55
84	86	79.195	168.39	237.59	316.78	395.98	475.17	554.37	633.56	712.76
86	88	47.559	95.12	142.68	190.24	237.79	285.35	332.91	380.47	428.03
88	90	15.860	31.72	47.58	63.44	79.30	95.16	111.02	126.88	142.74

TABLE III. Quadrilateral Surfaces of 5 Degrees in Latitude and in Longitude on the Terrestrial Ellipsoid.

Limiting LATITUDES.		Multiples of these Quadrilateral Surfaces from 1 to 9.								
Inf.	Sup.	1.	2.	3.	4.	5.	6.	7.	8.	9.
0	5	5599.821	11199.64	16799.46	22399.29	27999.11	33598.93	39198.75	44798.57	50398.39
5	10	5558.288	11116.58	16674.87	22233.15	27791.44	33349.73	38908.02	44466.31	50024.60
10	15	5475.466	10950.93	16426.40	21901.87	27377.33	32852.80	38328.27	43803.73	49279.20
15	20	5351.846	10703.69	16055.54	21407.39	26759.23	32111.08	37462.93	42814.77	48166.62
20	25	5188.165	10376.33	15564.49	20752.66	25940.82	31128.99	36317.15	41505.32	46693.48
25	30	4985.425	9970.85	14956.27	19941.70	24927.12	29912.55	34897.97	39883.40	44868.82
30	35	4744.891	9485.78	14234.67	18979.57	23724.46	28469.35	33214.24	37959.13	42704.02
35	40	4468.110	8936.22	13404.33	17872.44	22340.55	26808.66	31276.77	35744.88	40212.99
40	45	4156.909	8313.82	12470.73	16627.64	20784.54	24941.45	29098.36	33255.27	37412.18
45	50	3813.408	7626.82	11440.22	15253.63	19067.04	22880.45	26693.86	30507.26	34320.67
50	55	3440.013	6880.03	10320.04	13760.05	17200.06	20640.08	24080.09	27520.10	30960.12
55	60	3039.419	6078.84	9118.26	12157.68	15197.09	18236.51	21275.93	24315.35	27354.77
60	65	2614.598	5229.20	7843.80	10458.39	13072.99	15687.59	18302.19	20916.79	23531.39
65	70	2168.779	4337.56	6506.34	8675.12	10843.89	13012.67	15181.45	17350.23	19519.01
70	75	1705.427	3410.85	5116.28	6821.71	8527.13	10232.56	11937.99	13643.42	15348.84
75	80	1228.213	2456.43	3684.64	4912.85	6141.07	7369.28	8597.49	9825.71	11053.92
80	85	740.973	1481.95	2222.92	2963.89	3704.86	4445.84	5186.81	5927.78	6668.76
85	90	247.668	495.34	743.00	990.67	1238.34	1486.01	1733.68	1981.34	2229.01

TABLE IV. Quadrilateral Surfaces of 10 Degrees in Latitude and in Longitude on the Terrestrial Ellipsoid.

Limiting LATITUDES.		Multiples of these Quadrilateral Surfaces from 1 to 9.								
Inf.	Sup.	1.	2.	3.	4.	5.	6.	7.	8.	9.
0	10	22316.220	44632.44	66948.66	89264.88	111581.10	133897.32	156213.54	178529.76	200845.98
10	20	21654.626	43309.25	64963.88	86618.50	108273.13	129927.76	151582.38	173237.01	194891.63
20	30	20347.180	40694.36	61041.54	81388.72	101735.90	122083.08	142430.26	162777.44	183124.62
30	40	18426.004	36852.01	55278.01	73704.02	92130.02	110556.02	128982.03	147408.03	165834.04
40	50	15940.634	31881.27	47821.90	63762.54	79703.17	95643.80	111584.44	127525.07	143465.71
50	60	12958.864	25917.73	38876.59	51835.46	64794.32	77753.18	90712.05	103670.91	116629.78
60	70	9566.755	19133.51	28700.26	38267.02	47833.77	57400.53	66967.28	76534.04	86100.79
70	80	5867.281	11734.56	17601.84	23469.12	29336.40	35203.69	41070.97	46938.25	52805.53
80	90	1977.282	3954.56	5931.85	7909.13	9886.41	11863.69	13840.97	15818.26	17795.54

TABLE V. Mean Quadrilateral Surfaces of 1, 10, 20, and 30 Minutes in Latitude and in Longitude deduced from each Quadrilateral of 1 Degree in Table I.

Limiting LATITUDES.		Mean Surfaces measuring in Latitude and in Longitude.				Limiting LATITUDES.		Mean Surfaces measuring in Latitude and in Longitude.			
Inf.	Sup.	1′.	10′.	20′.	30′.	Inf.	Sup.	1′.	10′.	20′.	30′.
0	1	0.0623	6.229	24.918	56.065	45	46	0.0440	4.395	17.581	39.558
1	2	0.0623	6.228	24.910	56.048	46	47	0.0432	4.318	17.270	38.858
2	3	0.0622	6.224	24.895	56.015	47	48	0.0424	4.238	16.954	38.146
3	4	0.0622	6.218	24.873	55.965	48	49	0.0416	4.158	16.632	37.422
4	5	0.0621	6.211	24.844	55.898	49	50	0.0408	4.076	16.305	36.686
5	6	0.0620	6.202	24.807	55.815	50	51	0.0399	3.993	15.973	35.939
6	7	0.0619	6.191	24.763	55.716	51	52	0.0391	3.909	15.636	35.181
7	8	0.0618	6.178	24.711	55.599	52	53	0.0382	3.823	15.294	34.411
8	9	0.0616	6.163	24.652	55.467	53	54	0.0374	3.737	14.947	33.630
9	10	0.0615	6.146	24.586	55.317	54	55	0.0365	3.649	14.595	32.839
10	11	0.0613	6.128	24.512	55.152	55	56	0.0356	3.560	14.239	32.038
11	12	0.0611	6.108	24.431	54.970	56	57	0.0347	3.470	13.878	31.226
12	13	0.0609	6.086	24.343	54.771	57	58	0.0338	3.378	13.513	30.404
13	14	0.0606	6.062	24.247	54.556	58	59	0.0329	3.286	13.143	29.572
14	15	0.0604	6.036	24.144	54.325	59	60	0.0319	3.192	12.770	28.731
15	16	0.0601	6.009	24.035	54.078	60	61	0.0310	3.098	12.392	27.881
16	17	0.0598	5.979	23.917	53.814	61	62	0.0300	3.002	12.010	27.022
17	18	0.0595	5.948	23.793	53.534	62	63	0.0291	2.906	11.624	26.154
18	19	0.0592	5.915	23.662	53.239	63	64	0.0281	2.809	11.235	25.278
19	20	0.0588	5.881	23.523	52.927	64	65	0.0271	2.710	10.842	24.394
20	21	0.0584	5.844	23.378	52.600	65	66	0.0261	2.611	10.445	23.502
21	22	0.0581	5.806	23.225	52.256	66	67	0.0251	2.511	10.045	22.602
22	23	0.0577	5.766	23.065	51.897	67	68	0.0241	2.411	9.642	21.695
23	24	0.0572	5.725	22.899	51.523	68	69	0.0231	2.309	9.236	20.781
24	25	0.0568	5.681	22.725	51.132	69	70	0.0221	2.207	8.827	19.860
25	26	0.0564	5.636	22.545	50.727	70	71	0.0210	2.104	8.414	18.933
26	27	0.0559	5.590	22.358	50.306	71	72	0.0200	2.000	8.000	17.999
27	28	0.0554	5.541	22.164	49.869	72	73	0.0190	1.896	7.582	17.060
28	29	0.0549	5.491	21.964	49.418	73	74	0.0179	1.791	7.162	16.115
29	30	0.0544	5.439	21.756	48.952	74	75	0.0168	1.685	6.740	15.165
30	31	0.0539	5.386	21.542	48.470	75	76	0.0158	1.579	6.315	14.210
31	32	0.0533	5.330	21.322	47.974	76	77	0.0147	1.472	5.889	13.250
32	33	0.0527	5.274	21.095	47.463	77	78	0.0137	1.365	5.461	12.286
33	34	0.0522	5.215	20.861	46.938	78	79	0.0126	1.258	5.030	11.318
34	35	0.0516	5.155	20.622	46.399	79	80	0.0115	1.150	4.598	10.346
35	36	0.0509	5.094	20.375	45.845	80	81	0.0104	1.041	4.165	9.371
36	37	0.0503	5.031	20.123	45.277	81	82	0.0093	0.933	3.730	8.393
37	38	0.0497	4.966	19.864	44.695	82	83	0.0082	0.824	3.294	7.412
38	39	0.0490	4.900	19.600	44.099	83	84	0.0071	0.714	2.857	6.429
39	40	0.0483	4.832	19.329	43.490	84	85	0.0060	0.605	2.419	5.443
40	41	0.0476	4.763	19.052	42.867	[illegible]	86	0.0049	0.495	1.980	4.456
41	42	0.0469	4.692	18.769	42.231	86	87	0.0039	0.385	1.541	3.467
42	43	0.0462	4.620	18.481	41.582	87	88	0.0028	0.275	1.101	2.477
43	44	0.0455	4.547	18.187	40.920	88	89	0.0017	0.165	0.661	1.487
44	45	0.0447	4.472	17.887	40.245	89	90	0.0006	0.055	0.220	0.496

www.ingramcontent.com/pod-product-compliance
Lightning Source LLC
LaVergne TN
LVHW021055110826
845150LV00001B/73

* 9 7 8 1 4 2 5 5 6 7 1 2 5 *